ALGEBRA AND TRIGONOMETRY

Quadratic equation: $\qquad ax^2 + bx + c = 0 \qquad\qquad x = \dfrac{-b + \sqrt{b^2 - 4ac}}{2a}$

Right triangle:

Pythagorean theorem: $\quad c^2 = a^2 + b^2$

$$\sin \theta = \frac{a}{c}$$

$$\cos \theta = \frac{b}{c}$$

$$\tan \theta = \frac{a}{b}$$

$$\sin(\alpha \pm \beta) = \sin \alpha \cos \beta \pm \cos \alpha \sin \beta$$

$$\cos(\alpha \pm \beta) = \cos \alpha \cos \beta \mp \sin \alpha \sin \beta$$

$$\cos \alpha + \cos \beta = 2 \cos\left(\frac{\alpha + \beta}{2}\right) \cos\left(\frac{\alpha - \beta}{2}\right)$$

COMMONLY USED PREFIXES FOR POWERS OF 10

Power	Prefix	Abbreviation	Example
10^{-18}	atto-	a	
10^{-15}	femto-	f	
10^{-12}	pico-	p	picofarad (pF)
10^{-9}	nano-	n	nanometer (nm)
10^{-6}	micro-	μ	micrometer (μm)
10^{-3}	milli-	m	millimeter (mm)
10^{-2}	centi-	c	centimeter (cm)
10^{-1}	deci-	d	decibel (dB)

Power	Prefix	Abbreviation	Example
10^1	deka-	da	(rarely used in physics)
10^2	hecto-	h	(rarely used in physics)
10^3	kilo-	k	kilogram (kg)
10^6	mega-	M	megawatt (MW)
10^9	giga-	G	gigajoule (GJ)
10^{12}	tera-	T	
10^{15}	peta-	P	
10^{18}	exa-	E	

CONVERSION FACTORS OCCASIONALLY NEEDED

Length

1 inch = 2.54 cm
1 mile = 1.653 km
1 light-year = 9.461×10^{15} m
1 AU = 1.496×10^{11} m

Time

1 d = 8.6400×10^4 s
1 y = 3.156×10^7 s

Speed

1 km/h = 0.2778 m/s

Volume

1 m³ = 1000 liters

Energy

1 eV = 1.602×10^{-19} J

Pressure

1 atm = 1.013×10^5 Pa

UNIVERSITY PHYSICS

UNIVERSITY PHYSICS

RONALD LANE REESE
Washington and Lee University

Brooks/Cole Publishing Company

I(T)P® An International Thomson Publishing Company

Pacific Grove • Albany • Belmont • Boston • Cincinnati
Johannesburg • London • Madrid • Melbourne • Mexico City
New York • Scottsdale • Singapore • Tokyo • Toronto

Sponsoring Editors: *Keith Dodson and Beth Wilbur*
Editorial Assistant: *Nancy Conti*
Production Coordinator: *Tessa McGlasson Avila*
Production Management: *Electronic Publishing Services Inc., NYC*
Marketing: *Steve Catalano*
Interior Design: *Electronic Publishing Services Inc., NYC*

Cover Design: *Roy R. Neuhaus*
Cover Photo: *Ronald Lane Reese*
Interior Illustration: *Electronic Publishing Services Inc., NYC*
Photo Researcher: *Electronic Publishing Services Inc., NYC*
Typesetting: *Electronic Publishing Services Inc., NYC*
Printing and Binding: *Von Hoffman Printing Company*

For more information, contact:

BROOKS/COLE PUBLISHING COMPANY
511 Forest Lodge Road
Pacific Grove, CA 93950
USA

International Thomson Publishing Europe
Berkshire House 168-173
High Holborn
London WC1V 7AA
England

Thomas Nelson Australia
102 Dodds Street
South Melbourne, 3205
Victoria, Australia

Nelson Canada
1120 Birchmount Road
Scarborough, Ontario
Canada M1K 5G4

International Thomson Editores
Seneca 53
Col. Polanco
11560 México, D. F., México

International Thomson Publishing GmbH
Königswinterer Strasse 418
53227 Bonn
Germany

International Thomson Publishing Asia
60 Albert Street
#15-01 Albert Complex
Singapore 189969

International Thomson Publishing Japan
Hirakawacho Kyowa Building, 3F
2-2-1 Hirakawacho
Chiyoda-ku, Tokyo 102
Japan

Printed in the United States of America.

10 9 8 7 6 5 4 3 2 1

Library of Congress Cataloging-in-Publication Data

The Library of Congress has cataloged the combined volume as follows:

Reese, Ronald Lane
    University physics / Ronald Lane Reese.
        p.   cm.
    Includes index.
    ISBN 0-534-24655-9
    1. Physics.   I. Title.
QC21.5.R435   1998
530—dc21   98-41666
                              CIP

Magna opera Domini:
exquisita in omnes voluntates ejus

[*Psalmi* CXI, v 2]*

The past:
In loving memory of my mother
Edith Lemberg Reese
(1906–1984)

and mother-in-law
Bertha Marie Carlson
(1907–1981)

and in honor of my father
Harold Augustus Reese Sr.
(1906–)

The present:
With grateful thanks for the love
and devotion of my wife
Edith Joanne Carlson Reese

The future:
For the priceless blessings
of a wonderful son and daughter
Daniel Austin Reese
Anna-Loren Reese

* The Hexaplar Psalter, Samuel Bagster and Sons, London, 1843.

PREFACE

GOALS

In recent years much active discussion and debate has revolved around just what body of knowledge and skills science and engineering students should take from a university physics course. An obvious related issue is how best to achieve the desired learning goals. As the primary instructional resource for the student outside the classroom and the professor's office, the textbook naturally has been a focus of these discussions. Over the last 30 years or so a "standard model" of a university physics textbook has evolved to the point of extensive refinement. Several generations of future scientists and engineers have been introduced to the powerful ideas of physics through these texts, and certainly we should acknowledge the many strengths of these books while considering *how we can improve on them as learning tools for today's physics students.*

When writing this text, I decided from the onset that the text should follow the twin educational commandments of Alfred North Whitehead: "Do not teach too many subjects, and what you teach, teach thoroughly." I believe that Whitehead's statement, at a basic level, reflects two of the most common themes emerging from what has come to be known as the physics reform movement. Thus, while this text was not written as a reform textbook, it nonetheless embraces the spirit of many reform goals, such as better integration of modern physics topics, a stronger emphasis on conceptual understanding, and an attention to different learning styles. Most importantly, however, this book is written for students, to allow them not only to learn the tools that physics provides but also to see why they work and the beauty of the ideas that underlie them.

TEXT OVERVIEW

A Focused Perspective

One of the great triumphs of physics is the amount of understanding that comes from a relatively small investment of fundamental ideas and principles. Students, however, often see the course as a random assortment of 25 to 30 topics deemed worthy of chapter status. Unifying concepts, such as conservation laws and field theory, can be lost amidst the mountainous amount of material. Students frequently fail to see just how little must be known to describe as much of nature as possible. Thus, a central goal of this text is to help students develop a thorough *understanding* of the *principles* of the basic areas of physics: kinematics, dynamics, waves, thermodynamics, electromagnetism, optics, relativity, and modern physics. It is better to build technical knowledge upon a firm foundation of fundamental principles than on a large collection of mere formulas.

Since most of us do not innately discern simplifying patterns and connections when faced with the seemingly complex, we become good and experienced students of physics through steady practice. This is a fundamental pedagogical issue, and one that this text addresses clearly through focusing on many of the difficulties encountered by students when studying physics, problems mentioned by Arnold Arons in *A Guide to Introductory Physics Teaching*, and by others in the educational literature of physics. Thus, the book

- continually integrates the most significant material from previous chapters into new material, in keeping with Arons's admonition to "spiral back" frequently, for greater insight and retention.
- provides an accurate conceptual understanding of fundamental physical principles by placing great emphasis on these principles and how they arose.
- recognizes and points out the limits of applicability of the theories and equations of physics. It can be just as important, after all, to know what doesn't apply as what does.
- stresses connections between topics by incorporating many aspects of contemporary physics into a mix of traditional topics. This goal is carried through in all aspects of the text—exposition, examples, questions, problems, and investigative projects.

A Thorough Development

Some recent texts have jockeyed to outdo each other by reducing the number of overall pages. While brevity is often a laudatory goal, it can sometimes also work to defeat other, more important purposes. For true conceptual understanding to take place, a "fewer pages is more" approach can make the physics learning experience similar to trying to extrapolate the beauty and subtleties of a Shakespearean drama by reading a summary of the plot line. This text, while no longer than many other university physics texts, has been written with the primary philosophy that students need a text that lays a careful, detailed groundwork for strong conceptual understanding and the development of mature problem-solving skills. For example, much research has recently been done on the different learning styles that students apply when first studying new material, but for a text to try to implement pedagogical structure to these different learning styles and goals (such as multiple problem-solving approaches or collaborative learning techniques), it is inevitable that the lesser goal of brevity must be sacrificed. In a similar vein, students often complain that the examples in the text do not prepare them well for the more challenging homework problems, where more than one idea may be addressed. Page length can be kept down by focusing on just the most straightforward examples, but students also need to see how the principles can be applied to more involved scenarios. I have placed special emphasis on thoroughly preparing students for the homework sets through strong emphasis (and reemphasis) on problem-solving techniques, by frequent references to and explanations of common misunderstandings, and by providing a set of examples that address both single-concept problem solving and the application of fundamental principles to longer, multiconcept problems. The ability to question whether results are reasonable has been fostered throughout these examples.

The text contains an ample selection of sections from which individual instructors can design a course compatible with their academic institution and student audience. Numerous sections, typically at the end of chapters, are listed as optional (designated by a *) and may be omitted by instructors preferring a leaner course. Others may want to choose their own path, including some of the optional sections while omitting others.

Features

STYLE

Physics is a great story, and in this text I have attempted to tell that story in as lively, clear, and precise a manner as possible. Students sometimes fail to see how the topics connect to each other or to the world outside the classroom. Thus, I have placed great emphasis on introducing each new topic by describing how it relates to experiences and phenomena with which the student is already familiar or to topics previously discussed. There is also no reason that reading a physics text shouldn't be fun (or at least not a chore). My philosophy is that occasional lapses into whimsy are a small price to pay if the result is that students stay more engaged with the reading. Finally, by filling in the details that are sometimes left unstated, this text should help students better bridge the gaps where misconceptions can arise.

STRATEGIC EXAMPLES AND OTHER EXAMPLES

A strong emphasis has been placed on beginning almost all Examples with a few, fundamental principles and equations, rather than specialized equations of secondary importance. Strategic Examples address the application of fundamental principles to longer problems; they are discussed in great detail, which students find particularly helpful in developing their own problem-solving abilities. Moreover, many of the end-of-chapter problems mirror the methodological details of the Strategic Examples.

A unique feature of this text is that many of the Examples are solved in more than one way. All too often students suffer from the perception that they must be doing a problem incorrectly because a fellow student or even the professor has set it up differently. By working selected problems using different choices of signs, coordinate axes, or even overall approach, these Examples

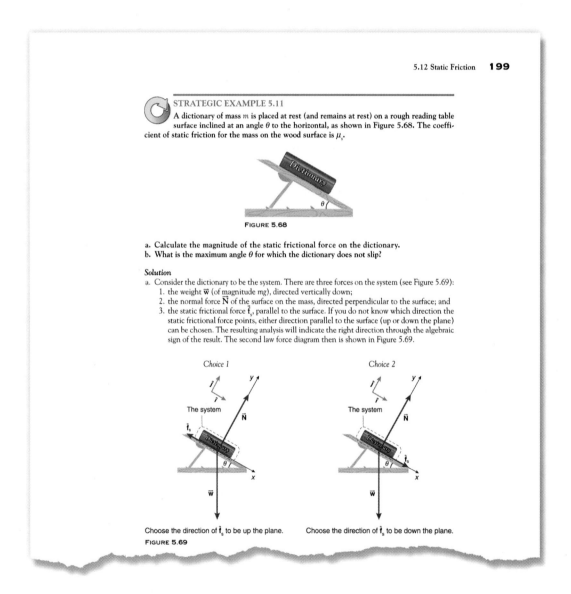

STRATEGIC EXAMPLE 5.11

A dictionary of mass m is placed at rest (and remains at rest) on a rough reading table surface inclined at an angle θ to the horizontal, as shown in Figure 5.68. The coefficient of static friction for the mass on the wood surface is μ_s.

FIGURE 5.68

a. Calculate the magnitude of the static frictional force on the dictionary.
b. What is the maximum angle θ for which the dictionary does not slip?

Solution

a. Consider the dictionary to be the system. There are three forces on the system (see Figure 5.69):
   1. the weight $\vec{w}$ (of magnitude mg), directed vertically down;
   2. the normal force $\vec{N}$ of the surface on the mass, directed perpendicular to the surface; and
   3. the static frictional force $\vec{f}_s$, parallel to the surface. If you do not know which direction the static frictional force points, either direction parallel to the surface (up or down the plane) can be chosen. The resulting analysis will indicate the right direction through the algebraic sign of the result. The second law force diagram then is shown in Figure 5.69.

Choice 1   Choice 2

The system   The system

Choose the direction of $\vec{f}_s$ to be up the plane.   Choose the direction of $\vec{f}_s$ to be down the plane.

FIGURE 5.69

also help students develop intuition on how certain choices can simplify later calculations compared with other alternatives.

CONCEPTUAL NOTES

Throughout the text, key points of each section have been high-lighted with shading. Importantly, these Conceptual Notes are not just the most important equations—they focus on the principal ideas and concepts (and sometimes equations) that a student should take from each section. My students have found the Conceptual Notes very useful as a reviewing tool for tests and quizzes.

PROBLEM-SOLVING TACTICS

In addition to useful problem-solving hints, the Problem-Solving Tactics also provide warnings to students about common errors and how to avoid them. Often these important tips of the trade

are also integrated into the text discussion. For example, specific Problem-Solving Tactics are often cross-referenced in some of the Examples. At the end of each chapter a summary of that chapter's Problem-Solving Tactics is included, with a page reference to the related text discussion for each tactic.

QUESTIONS

A common student lament is that "I understand the material; I just can't do the problems." The questions within and at the end of each chapter test a student's understanding of concepts before the student is asked to apply these concepts to more complex or quantitative situations in the problems. Some of these questions entail a short qualitative explanation, whereas others may require a short back-of-the-envelope calculation or even a quick and dirty experiment to determine the approximate magnitude of a quantity.

PROBLEMS

The development of successful problem-solving techniques is an essential goal for all introductory physics students. To help students hone these abilities, many of the problems involve a multistep approach in which students are guided through the problem with specific questions that enable them first to find all the pieces and then to put them together. Additionally, many of the problems mimic the approach of the Strategic Examples so that there is a strong correlation between the presentation of the material and the problems that a student is expected to be able to solve.

Since the real world is awash with information, the problems occasionally include irrelevant or superfluous information. This teaches a student to discriminate between what is needed and what is not. It also may be necessary to consult an appropriate table in the chapter or on the front and back inside covers to find numerical values of standard constants or parameters. In both the examples and problems, attention is paid to the consistent use of significant figures.

Three levels of difficulty are provided, with unbulleted, bulleted (•), and double-bulleted (⁑) problems representing straightforward, moderate, and more difficult problems, respectively. Those problems with red numbers include answers at the back of the book and are solved in the Student's Solutions Manual. I have personally solved all of the problems, and the problem sets have been additionally fine-tuned by actual classroom usage over the course of several years.

INVESTIGATIVE PROJECTS

These projects are highly amenable to collaborative group work and are of the following types:

- *Expanded Horizons*—These projects are well suited to journal club research, discussion, and supplementary reading.
- *Lab and Field Work*—Doing physics is an important part of studying physics. In these projects, students are asked to design and carry out experiments either with other students or the professor.
- *Communicating Physics*—A key developmental goal for any student is the ability to write about and discuss technical topics. Practice on these written and oral communication skills is provided by these project topics, which are ideal for writing-intensive assignments, public speaking, and community service opportunities.

The projects are interesting to read, even if never performed or assigned, since they indicate the breadth and depth of applications of chapter material. As such, they can help stimulate inquiry, class discussion, and faculty–student interaction. Most of the projects are provided with references that serve as a guide (and entryway) to the appropriate literature.

SUMMARIES

Each chapter concludes with an extensive summary that, when combined with the Conceptual Notes and the Summary of Problem-Solving Tactics, provides an ideal in-text study guide for the student.

QUOTATIONS

I have used these frequently throughout the text to cast the subject matter in a different light, be it serious or whimsical. Great writing (communication!) from the past is central to a real understanding of any discipline, even physics.

MATHEMATICAL LEVEL

This text assumes a familiarity with calculus comparable to what a student would obtain from a high school calculus course (with or without advanced placement credit). Of course, additional calculus is useful when taken concurrently with this course. No prior knowledge of physics is presumed.

PRECISION

Effective, unambiguous communication in physics requires clear and consistent use of the technical vocabulary and a solid understanding of the meaning of the technical notation. This tenet has informed the presentation throughout the book.

Chapter Contents

CHAPTER 1 PRELUDES

An overview of physics is presented along with an introduction to measurement standards of the SI unit system, distinguishing them from common units of convenience. The various meanings of the equal sign are discussed as well as estimation and order of magnitude calculations. The distinction is made between precision and accuracy. The notion of significant figures is discussed in the context of common mathematical operations such as multiplication (and division) and addition (and subtraction). Having made these points, the text does not ignore their use and makes consistent use of significant figures throughout its examples and problems so that students realize their importance even outside a laboratory context.

CHAPTER 2 A MATHEMATICAL TOOLBOX: AN INTRODUCTION TO VECTOR ANALYSIS

The proper and consistent use of vectors is very important to success in physics. This chapter and the rest of the book distinguish clearly among a vector, its magnitude, and its components with respect to a chosen coordinate system. Vector addition and subtraction are designated by boldface + and − to distinguish the operations clearly from their scalar counterparts, a source of much student confusion in problem solving.

CHAPTER 3 KINEMATICS I: RECTILINEAR MOTION

The notion of a particle is addressed. A one-dimensional vector approach is used so its extension to two- and three-dimensional motion in Chapter 4 is seamless and painless. The choices that a student must make in establishing a coordinate system for a problem and the consequences of that choice in tailoring the (few) fundamental kinematic equations to a problem are stressed throughout. Consistent use of vectors and vector terminology takes the mystery out of the choice for the signs associated with the various terms in the equations of kinematics.

CHAPTER 4 KINEMATICS II: MOTION IN TWO AND THREE DIMENSIONS

The vector approach of Chapter 2 easily allows extension to motion in two and three dimensions. Relative velocity addition is examined. Uniform and nonuniform circular motion are approached by introducing both the angular velocity and angular acceleration vectors so students are ready for more advanced work with these vectors in upper-division mechanics or dynamics courses in physics and engineering.

CHAPTER 5 NEWTON'S LAWS OF MOTION

An overview of fundamental particles and forces is presented. The concept of force and its measurement are introduced stressing how important it is to define clearly the system under

486 Chapter 10 *Spin and Orbital Motion*

INVESTIGATIVE PROJECTS

A. Expanded Horizons

1. Investigate the dynamics associated with throwing a bola, a traditional South American hunting weapon.
   D. L. Mathieson, "Wrap up rotational motion by throwing a bola," *The Physics Teacher*, 30, #3, pages 180–181 (March 1992).

2. Yo-yos are fascinating examples of spin and orbital motion. Investigate the physics of the yo-yo.
   William Boudreau, "Cheap and simple yo-yos," *The Physics Teacher*, 28, #2, page 92 (February 1990).
   Wolfgang Bürger, "The yo-yo: a toy flywheel," *American Scientist*, 72, #2, pages 137–142 (March–April 1984).
   Edward Zuckerman, "Quest for the perfect yo-yo," *Science Digest*, 93, #7, pages 54–55, 60 (July 1985).

3. Investigate the use of an ultracentrifuge to determine molecular masses of complex molecules.
   I. W. Richardson and Ejler B. Neergaard, *Physics for Biology and Medicine* (Wiley-Interscience, New York, 1972), pages 158–160.

4. Investigate the dynamics associated with boomerangs.
   Vernon Barger and Martin Olsson, *Classical Mechanics: A Modern Perspective* (2nd edition, McGraw-Hill, New York, 1995), pages 195–202.
   Allen L. King, "Project boomerang," *American Journal of Physics*, 43, #9, pages 770–773 (September 1975).
   Henk Vos, "Straight boomerang of balsa wood and its physics," *American Journal of Physics*, 53, #6, pages 524–527 (June 1985).
   Michael Hanson, "The flight of the boomerang," *The Physics Teacher*, 28, #3, pages 142–147 (March 1990).
   Jacques Thomas, "Why boomerangs boomerang," *New Scientist*, 99, #1376, pages 838–843 (

vestigate the physi
H. D. N

9. Measurements of the moment of inertia are common in physics laboratories. On the other hand, it is quite another story to *measure* the moment of inertia of the Moon or a planet such as Venus or Mars; they are not easily tinkered with in the lab! Measurements of such moments of inertia enable astronomers and geologists to model the interiors of the moons and planets. How might the moments of inertia of moons and planets be measured?
   William B. Hubbard, *Planetary Interiors* (Van Nostrand Reinhold, New York, 1984).
   Ralph Snyder, "Two-density model of the Earth," *American Journal of Physics*, 54, #6, pages 511–513 (June 1986).

10. Investigate the physics of the rotating, orbiting space colony stations proposed by Gerald K. O'Neill.
    Gerald K. O'Neill, "The colonization of space," *Physics Today*, 27, #9, pages 32–40 (September 1974).

11. High angular speed flywheels are used as energy storage devices in hybrid race cars. Investigate the state of this technology.
    William B. Scott, "Satellite control concepts bolster civil, defense systems," *Aviation Week and Space Technology*, 142, pages 43, 46 (6 March 1995).

B. Lab and Field Work

12. Many science museums have a dramatic pendulum in their entrance foyers called a Foucault pendulum. The slow rotation of the plane of oscillation of the pendulum is experimental proof that the Earth rotates. Investigate the dynamics of a Foucault pendulum to explain why this is the case. Imagine such a

---

Investigative Projects **487**

C. Communicating Physics

16. Take two quarters (or any two identical coins). Orient the two coins so they are both heads up with both profiles of George Washington upright as indicated in Figure I.16a. Roll one of the coins, say the one on the left, around *half* the circumference of the other coin as indicated. It is amazing to discover that the profile of the rolling coin is *not* upside down but right

FIGURE I.16

side up after this maneuver. Explain why. This phenomenon is related to another curious aspect of rolling motion. Suppose a wheel (see Figure I.16b) rolls through one complete turn, carrying it from point A to point B. The distance between A and B is, therefore, equal to the circumference of the wheel. Notice, however, that point C, along the radius from the center of the wheel *also* executes one revolution as the wheel moves from A to B. Since the distance AB is equal to the distance CD, "we are confronted with the evident absurdity that the circumference of the small circle is equal to the circumference of the large circle." Resolve this paradox.
   See James R. Newman, *The World of Mathematics* (Simon & Schuster, New York, 1956), volume 3, pages 1937–1939; the quotation is from this work, page 1938.

17. Call or visit a tire retailer and determine the manufacturer of a dynamic tire balancing machine. Communicate with engineers at the manufacturer to learn the physics associated with dynamic balancing of tires. Write a summary of your findings appropriate for an audience of your peers.
    Richard C. Smith, "Static vs spin balancing of automobile wheels," *American Journal of Physics*, 40, #1, pages 199–201 (January 1972).

18. The tippe top is a toy that has fascinated even great physicists like Niels Bohr and Wolfgang Pauli, perhaps even you. Demonstrate and explain its peculiar dynamics.
    Richard J. Cohen, "The tippe top revisited," *American Journal of Physics*, 45, #1, pages 12–17 (January 1977).
    Ivars Peterson, "Topsy-turvy top," *Science News*, 146, page 108 (13 August 1994).

---

consideration. The significance and importance of all three of Newton's laws of motion are addressed. Both second law and third law force diagrams are discussed. The intricacies of special forces such as weight, tension, and static and kinetic friction are explored. Inertial and noninertial reference frames are contrasted.

CHAPTER 6 THE GRAVITATIONAL FORCE AND THE GRAVITATIONAL FIELD

Newton's law of universal gravitation is *not* presented as a *fait accompli* as if inscribed in stone; rather the process by which Newton deduced the law is explored. The gravitational shell theorems are discussed and applied (and proved in an appendix). Kepler's laws of planetary motion are discussed, along with a convenient simplification to the third law commonly used in astronomy (using customized units: years and astronomical units). Newton's form for Kepler's third law is derived. The concept of the gravitational field is introduced so that parallels with it may be exploited later when studying electricity and magnetism. Gauss's law for the gravitational field is proved, so that further parallels with electromagnetism can be made. Many of the problems consider contemporary astronomical applications.

CHAPTER 7 HOOKE'S FORCE LAW AND SIMPLE HARMONIC OSCILLATION

Hooke's law for springlike forces is introduced. A horizontal and a vertical spring (with the additional gravitational force) are compared. Simple harmonic oscillation and its relationship to uniform circular motion are discussed. The simple pendulum is introduced as well as the oscillatory gravitational motion through a uniform sphere. Damped simple harmonic oscillation and forced oscillation with resonance are explored.

CHAPTER 8 WORK, ENERGY, AND THE CWE THEOREM

Students typically think of work and energy as the same thing. The similarities and distinctions between work and energy are explored as well as the concept of power. The classical work–energy theorem (called the CWE theorem) and its limitations caused by the neglect of thermal effects are thoroughly examined to set the stage for a more general and encompassing conservation of energy theorem when we consider thermodynamics in Chapter 13. The importance of the choice made for the zero of a potential energy function is emphasized. The connection between the local form for the gravitational potential energy (mgy, with $\hat{\jmath}$ up) and the more general form ($-GMm/r^2$) is discussed. Applications to astrophysical problems such as the escape speed and black holes are explored. The concept of energy diagrams also is introduced at this early juncture to set the stage for their later use in modern physics.

CHAPTER 9 IMPULSE, MOMENTUM, AND COLLISIONS

The general principles are stressed rather than a plethora of specialized equations for collisions. The contemporary idea of force transmission by particle exchange is explored by means of a classical example for repulsive forces. The center of mass is introduced and the dynamics of a system of particles is explored.

CHAPTER 10 SPIN AND ORBITAL MOTION

The similarities and distinctions between spin and orbital motion are explored. The rotational dynamics of rigid bodies with at least one symmetry axis through the center of mass is examined, emphasizing the parallels to analogous equations in linear dynamics. The shape of the spinning Earth and the precession of tops and of the spinning Earth also are explored. Rolling motion and a model of a wheel also are examined to explain the difficult but common observation that less force is needed to roll rather than to drag a massive system.

CHAPTER 11 SOLIDS AND FLUIDS

The mechanical properties of solids and fluids are investigated. The variation of pressure with depth or height in a liquid is examined, leading to an equation giving students great freedom to approach a problem with many different coordinate choices. Archimedes' principle and the stability of floating systems are explored (why things such as submarines, ships, or poles will float in one orientation but are unstable in another orientation). Bernoulli's principle is derived from the CWE theorem. Capillary action, nonideal fluids, and viscous flow also are discussed.

CHAPTER 12 WAVES

General waves and their wavefunction, waveform, and oscillatory behavior are discussed. Nonsinusoidal periodic waves are discussed before sinusoidal periodic waves so the distinction between the oscillatory behavior of the wave at a fixed place can be clearly distinguished from its waveform at a fixed time. The classical wave equation is derived so that it can be contrasted with the Schrödinger equation in Chapter 27. Waves on strings are introduced. A unique section explores the nature of a sound wave and the relationship between the particle position and the pressure or density wave. The measurement of sound intensity and sound level is discussed so that students become aware of common sound levels that can damage their hearing. The acoustic Doppler effect treats motion of the source and/or observer as well as the effect of a wind. Superpositions of waves to form standing waves are applied to strings and both open and closed pipes. The superposition leading to wave beats is explored as well as the distinction between phase and group speeds. A simplified introduction to Fourier analysis leads to wave uncertainty relations that appear later in Chapter 27 as the Heisenberg uncertainty principle.

CHAPTER 13 TEMPERATURE, HEAT TRANSFER, AND THE FIRST LAW OF THERMODYNAMICS

The definition of a simple thermodynamic system is presented. The intuitive yet difficult concepts of temperature and heat transfer are introduced. Thermal effects in solids, liquids, and gases as well as mechanisms of heat transfer are examined. A general statement of energy conservation is developed that specializes to the CWE theorem of mechanics and to the first law of thermodynamics. Various thermodynamic processes for gases are explored.

CHAPTER 14 KINETIC THEORY

The kinetic theory of an ideal gas is presented as well as its limitations. The notion of degrees of freedom and the effect of quantum mechanics on the effective number of degrees of freedom are discussed. Adiabatic processes for ideal gases also are presented.

CHAPTER 15 THE SECOND LAW OF THERMODYNAMICS

The need for this great unifying principle is discussed in the context of why some things happen and others do not. Thermodynamic models for engines and refrigerators are presented and related to the second law. The nonintuitive concept of entropy is carefully developed as well as a classical model that explores the Boltzmann statistical interpretation of the meaning of entropy.

CHAPTER 16 ELECTRIC CHARGES, ELECTRIC FORCES, AND THE ELECTRIC FIELD

The chapter begins with an exploration of how electrical effects were distinguished from magnetic effects. The question of just what is meant by the term electric charge is confronted by a careful exploration of the experiments that led to the discovery of the two types of charge property and why Franklin's subsequent naming of them (as positive and negative charges) was particularly useful and convenient. Charge quantification is distinguished from charge quantization. The concept of the electric field is developed by exploring the similarities and differences between electricity and gravitation. Gauss's law for the electric field is developed from the parallel law for the gravitational field.

CHAPTER 17 ELECTRIC POTENTIAL ENERGY AND THE ELECTRIC POTENTIAL

The often confused and subtle distinction between these two concepts is thoroughly explored. The electron-volt energy unit is introduced and its convenience illustrated. Lightning rods also are discussed

CHAPTER 18 CIRCUIT ELEMENTS, INDEPENDENT VOLTAGE SOURCES, AND CAPACITORS

CHAPTER 19 ELECTRIC CURRENT, RESISTANCE, AND DC CIRCUIT ANALYSIS

The terminology and methodology used for circuit analysis conforms to the standard conventions used in electrical engineering so the transition between physics and electronics can be made easily. The text clearly explains why positive charge flowing one way is equivalent to negative charge flowing in the opposite direction, a point of much mystification to students.

CHAPTER 20 MAGNETIC FORCES AND THE MAGNETIC FIELD

The need for a magnetic field is introduced by contrasting it with the electric field and its effects on electric charge. North and south magnetic poles are defined clearly rather than assumed to be obvious or innate. Applications include a velocity selector, mass spectrometer, and the Hall effect for determining the sign of the charge carriers of a current. Magnetic forces on currents lead to the torque on a current loop and the electric motor. The source of a magnetic field is introduced by exploring the parallels with both gravitational and electric fields. Gauss's law for the magnetic field, Ampere's law, the concept of a displacement current, and the Ampere–Maxwell law are discussed. The magnetic field of the Earth and how its reversals were discovered (via sea-floor sediments) also are explored to connect with another exciting discipline of the sciences that students see as remote from physics. *Nothing* is remote from physics!

CHAPTER 21 FARADAY'S LAW OF ELECTROMAGNETIC INDUCTION

The technological importance of Faraday's law is presented, leading to the development of an ac generator. The Maxwell equations are celebrated. The Maxwell equations in a vacuum are examined, leading to self-sustaining electromagnetic waves and the identification of such waves with light. Inductors and ideal transformers as standard circuit elements are explored using the standard engineering conventions.

CHAPTER 22 SINUSOIDAL AC CIRCUIT ANALYSIS

The typical approach to ac circuits in physics makes them seem impossibly complicated to students. In contrast, a brief introduction to complex variables permits the treatment of sinusoidal ac circuits via an extension of dc circuit analysis techniques, as is standard practice in electrical engineering. The use of current, potential difference, and voltage source phasors and the concept of impedance mean that sinusoidal ac circuit analysis then is reduced to the algebra of complex numbers.

CHAPTER 23 GEOMETRIC OPTICS

The simple Cartesian sign convention is used for mirrors, single surface refraction, and lenses, rather than a host of different, complex, and difficult to memorize mirror and lens conventions. Applications include the vertebrate eye, cameras, microscopes, and telescopes.

CHAPTER 24 PHYSICAL OPTICS

Interference via wavefront division (single, double, and multiple slit experiments as well as diffraction gratings) and amplitude division (thin-film interference) all are explored. Polarization and optical activity are discussed.

CHAPTER 25 THE SPECIAL THEORY OF RELATIVITY

Classical Galilean relativity is reviewed as well as the need for change. With the two postulates of special relativity, time dilation and length contraction are explored and used to derive the Lorentz transformation equations. The apparent relativistic paradox that *each* reference frame measures clocks in the *other* reference frame to run slow and lengths parallel to the motion to be shorter is confronted directly and resolved with a specific example. The existence of superluminal jets in astrophysics is found to be an optical illusion. The relativistic Doppler effect is explored, leading to the startling realization that for a source approaching with a nonzero impact parameter, the transition from a blue to a red shift occurs *before* the source is transverse to the line of sight. Questions of energy, momentum, the CWE theorem, and the relationship among mass, energy, and particles all are explored. The reason that the speed of light is an unreachable speed limit for material particles is discussed. The so-called mass–energy equivalence is clearly and properly addressed. Space–time diagrams are introduced and used to show why travel into the past (an idea with much student interest in view of contemporary culture) is forbidden in special relativity. The electromagnetic implications of relativity also are examined. The general theory of relativity and its classical tests are discussed using a qualitative approach.

CHAPTER 26 AN APERITIF: MODERN PHYSICS

The fortuitous discoveries of the electron, x-rays, and radioactivity are explored. The nuclear model of the atom is developed from the viewpoint that it was quite a radical proposal by Rutherford, rather than being simply obvious. The photoelectric effect and Compton scattering are used to justify the existence of the photon. The Bohr model and its limitations are explored. The biological effects of radiation and dosage units are discussed. The de Broglie hypothesis is introduced and questions raised about the meaning of a particle-wave.

CHAPTER 27 AN INTRODUCTION TO QUANTUM MECHANICS

The Heisenberg uncertainty principle is explored as well as the famous double slit experiment. The meaning of the wavefunction is assessed. Heuristic arguments lead to the Schrödinger equation.

A COMPLETE ANCILLARY PACKAGE

The following comprehensive teaching and learning package accompanies this book.

For the Student

MEDIA RESOURCES

Brooks/Cole Physics Resource Center is Brooks/Cole's website for physics, which contains a homepage for *University Physics*. All information is arranged according to the text's table of contents. Students can access flash cards for all glossary terms, supplementary practice and conceptual problems, practice quizzes for every chapter, and hyperlinks that relate to each chapter's contents.

InfoTrac® College Edition is an online library available FREE with each copy of each volume of *University Physics*. (Due to license restrictions, *InfoTrac College Edition* is only available to college students in North America upon the purchase of a new book.) It gives students access to full-length articles—not simply abstracts—from more than 700 scholarly and popular periodicals, updated daily and dating back as much as four years. Student subscribers receive a personalized account ID that gives them four months of unlimited Internet access—at any hour of the day—to readings from *Discover, Science World,* and *American Health* magazines.

OTHER STUDY AIDS

Student Solutions Manual in two volumes by Ronald Lane Reese, Mark D. Semon, and Robin B. S. Brooks includes answers and solutions to every other odd-numbered end-of-chapter problem.

FOR THE INSTRUCTOR

Complete Solutions Manual in two volumes, by Ronald Lane Reese, Mark D. Semon, and Robin B. S. Brooks, contains answers and solutions to all end-of-chapter problems in the text.

ASSESSMENT TOOLS AND MATERIALS

Test Items for University Physics, by Frank Steckel, includes a copy of the test questions provided electronically in *Thomson World Class Learning™ Testing Tools,* Review Exercise Worksheets, and answers to the test item questions. The notation of the test items carefully follows that of the main text.

Thomson World Class Learning™ Testing Tools is a fully integrated suite of test creation, delivery, and classroom management tools. This invaluable set of tools includes World Class Test, Test Online, and World Class Manager software. World Class Test allows instructors to create dynamic questions that regenerate the values of variables and calculations between multiple versions of the same test. Tests, practice tests, and quizzes created in World Class Test can be delivered via paper, diskette or local hard drive, LAN (Local Area Network), or the Internet. All testing results can then be integrated into a complete classroom management tool with scoring, gradebook, and reporting capabilities.

With World Class Test, instructors can create a test from an existing bank of objective questions including multiple-choice, true/false, and matching questions or instructors can also easily edit existing questions and add their own questions and graphics. The online system can automatically score *objective* questions. *Subjective* essay and fill-in-the-blank questions that the instructor evaluates can also be added. Results can be scored, merged with final test results, and entered automatically into the gradebook.

Using *World Class Course,* you can quickly and easily create and update a web page specifically for a course or class. Post your own course information, office hours, lesson information, assignments, sample tests, and links to rich web content.

PRESENTATION TOOLS AND ONLINE RESOURCES

Transparencies in full color include more than 200 illustrations from the text, enlarged for use in the classroom and lecture halls.

CNN Physics Video, produced by Turner Learning, can stimulate and engage your students by launching a lecture, sparking a discussion, or demonstrating an application. Each physics-related segment from recent CNN broadcasts clearly demonstrates the relevancy of physics to everyday life.

With *Brooks/Cole's PhysicsLink,* a cross-platform CD-ROM, creating lectures has never been easier. Using multi-tiered indexing, search capabilities, and a comprehensive resource bank that includes glossary, graphs, tables, illustrations, photographs, and animations, instructors can conduct a quick search to incorporate these materials into presentations and tests. And, any *PhysicsLink* file can be posted to the web for easy student reference.

WebAssignOnline homework, a versatile, web-based homework delivery system, saves time grading and recording homework assignments and provides students with individual practice and instant feedback on their work. It delivers, collects, grades, and records customized homework assignments over the Internet. Assignments can be customized so each student can receive a unique question to solve. Access to *WebAssign* is secured by passwords and each student has access only to his or her record. *WebAssign* ©1998–99 by North Carolina State University.

ACKNOWLEDGMENTS

A wise man will hear, and will increase learning;
and a man of understanding shall attain unto wise counsels.

Proverbs 1:5

My interest in physics was sparked long ago by two gifted mentors at Middlebury College: Benjamin F. Wissler, whose wonderfully friendly Cheshire-cat-like grin I still can see today, and Chung-Ying Chih, whose appreciation of elegance was always apparent in his approach to physics. Their excitement for the subject was contagious and their rigor and demands upon their students legendary. Subsequent mentors included Herman Z. Cummins (then at The Johns Hopkins University, now at CUNY) and my colleagues at Washington and Lee University and elsewhere, gentlemen and gentlewomen whose gifts for teaching and research continue to be admirable role models, worthy of emulation.

The unrequited help of many persons involved with this project gives me great faith in the benevolence of humanity. My colleagues in the sciences, mathematics, and the libraries at Washington and Lee University withstood incessant questions about all manner of subjects. The conviviality, camaraderie, and good humor of the Department of Physics and Engineering are especially appreciated. All have fostered an academic environment where teaching and scholarship are emphasized in an atmosphere of mutual respect, dignity, and honor.

The reviewers of this manuscript are listed alphabetically below. Their thorough and insightful reading and frank and honest critiques were invaluable to the creation of this book.

Royal G. Albridge, Vanderbilt University
C. David Andereck, Ohio State University
Gordon Aubrecht, Ohio State University
Rene Bellwied, Wayne State University
Van Bluemel, Worcester Polytechnic Institute
Neal M. Cason, University of Notre Dame
Kenneth C. Clark, University of Washington
Richard M. Heinz, University of Indiana
Daniel G. Montague, Willamette University
Richard Muirhead, University of Washington
Richard Ditteon, Rose-Hulman Institute of Technology
Charles Scheer, University of Texas, Austin
Mark Semon, Bates College
William S. Smith, Boise State University
Karl Trappe, University of Texas, Austin
Ronald E. Zammit, California Polytechnic
    State University

Two reviewers deserve special accolades. Their dedication went well beyond the call.

Professor Kenneth C. Clark meticulously read and reread, critiqued and recritiqued *every* draft of the manuscript from its humblest beginnings, making innumerable suggestions for clean and clever ways to elucidate many phenomena. Special thanks also go to Professor Mark Semon, whose attention to detail in his reviews was similar: insightful, thoughtful, and brimming with constructive suggestions.

At Brooks/Cole, the dedication and diligence of the entire staff and the pre-production and production teams warrant praise. Their expertise exemplify the work of true professionals. Among them are Physics Editor Beth Wilbur, Senior Developmental Editor Keith Dodson, Production Editor Tessa McGlasson Avila, Assistant Editor Melissa Henderson, Editorial Assistants Georgia Jurickovich and Nancy Conti, and Cover Designer Roy Neuhaus. Also my sincere thanks to past editors Harvey Pantzis and Lisa Moller, Developmental Editors Maxine Effenson Chuck and Casey FitzSimons, Marketing Manager Steve Catalano, and Cartoonist Tom Wentzel.

The entire staff at Electronic Publishing Services deserves kudos for a wonderful job creating the book. They were a real pleasure to work with: Senior Production Editor Rob Anglin, Photo Researcher Francis Hogan, Copyeditor and Accuracy Checker Andrew Schwartz, Art Editor Michael Gutch, Creative Supervisor Stephanie McWilliams, Illustrator Matthew McAdams, their fine coterie of artists, Page Layout Specialists Linda Harms and Brent Burket, Operations Manager Patty O'Connell, Proofreader Cheryl Smith Robbins, and Indexer Lee Gable.

To my students, both recent and more venerable, I owe a *huge* and special debt. The many students here at Washington and Lee (as well as some at Bates College) who endured a photocopied manuscript and critiqued the early drafts deserve kudos and heartfelt thanks. Students Jennifer Strawbridge and William Kanner greatly assisted with the page proofs. I am humbly and profoundly grateful to all of you for your good humor, patience, enthusiasm, and encouragement. God bless you all.

During the decade this manuscript has been in preparation, the personal friendship of many people was a constant blessing. I particularly want to thank Charlotte and Chuck Gilmore, Susan and Doug Blevins, Karen Swan, Ned Wisnefski, Fran and Dick Hodges, Pastor Mark Graham and all other ministers at St. John Lutheran Church (Roanoke, Virginia), Bethany Arnold, Ed Reed, Lynn and Fred Genheimer, and the Chanthavongsa family. Finally, I treasure the warmth and love of my extended family: foremost, my dear and venerable father, as well as Harold Augustus Reese Jr., Christine Reese McCulloch, Elaine Hildebrand, Tom McCulloch, Robert Richards, Betty Fake, Liz Tisdale, Jackie and Mel Lockwood, Kristen Lockwood, Kim and Guy Boros, and Tory and George Bolten.

Clearly, any errors remaining within the text are my responsibility. I would be very grateful to readers who bring errors of any kind to my attention [reeser@wlu.edu]. I truly welcome all your comments, critiques, and suggestions.

Shalom aleichem

PREFACE TO STUDENTS

The supreme task of the physicist is to arrive at those universal elementary laws from which the cosmos can be built up by pure deduction. There is no logical path to these laws; only intuition, resting upon sympathetic understanding of experience, can reach them.

Albert Einstein (1879–1955)*

What a wonderful and exciting privilege it is to study and to teach physics! Physics is the bedrock of all the sciences and technology. Whether it be chemistry, geology, biology, medicine, engineering, or astronomy, our descriptions of nature involve understanding how particles move and interact individually and collectively. This understanding is the fundamental domain of physics. So, if you want to become a scientist, engineer, doctor, or a natural philosopher, or simply to understand nature at its most fundamental level as an intelligent citizen-scientist, you need to begin with a voyage through the foundations of physics.

The mission of this text is

- to present the subject in a logical, clear, and comprehensible style;
- to stress how *little* needs to be known in order to understand as much as possible;
- to recognize that many aspects of physics, while quantifiable, remain fundamentally abstract and mysterious—we do not have the answers to many profound questions; and
- to lighten the stress over technical gobbledygook with occasional humor.

I hope this book conveys the excitement of a fascinating search for a fundamental understanding of nature. The discipline of physics is, after all, the observation, explanation, and integration

---

*"Principles of Research," *Ideas and Opinions by Albert Einstein*, edited by Carl Seelig (Crown Publishers, New York, 1954), page 226.

into a conceptual whole of as much as we can see, perceive, and infer during our all-too-short intellectual rendezvous with this amazing universe.

The beauty and coherence of physics may be obvious to professors, but may be less clear to you, our students, since physics likely seems less romantic than the beauty of the starlit sky. Understanding the celestial dances of the firmament, though, really is physics. So is understanding why the night sky is dark, why clouds are white, why the sky is blue, why bubbles appear in a bottle of beer or a glass of champagne, and what drives the wondrous biochemical processes that make life itself possible and gives us the opportunity to wonder at the beauty, not only of each other, but also of the natural world surrounding us.

We have an exciting time of study ahead. From a practical viewpoint, careful and thoughtful multiple readings of the text material are encouraged. Your first reading might omit the example problems in order to gain a conceptual overview of the material sufficient to address the Questions; a second reading then can include sufficient examples (particularly the Strategic Examples) for you to gain the necessary familiarity to approach the Problems.

In the conceptual development, examples, and problems, great emphasis is placed on the choices you have to make and the consequences of those choices. This is designed to gradually build confidence in your ability to tackle new and different situations. It will also teach you to be alert for the unexpected, or unanticipated result. Most significant discoveries in science begin with simply noticing something unexpected or peculiar. "What's that?" has led to many a significant "Aha!" in the history of science and technology.

ABOUT THE AUTHOR

Professor Reese teaches physics and astronomy at Washington and Lee University in Lexington, Virginia. He received his undergraduate degree in physics from Middlebury College and his Ph.D. in physics from The Johns Hopkins University. He has been teaching introductory physics, astronomy, and various advanced physics courses for almost 30 years. He also has performed consulting research on the interaction of visible and microwave electromagnetic radiation with matter at the Naval Research Laboratory in Washington, D.C., and has been a Visiting Fellow at University College, Oxford, during several sabbaticals.

Brief Contents

CONTENTS

VOLUME I

CHAPTER 3
Kinematics I   73
RECTILINEAR MOTION

CHAPTER 4
Kinematics II   117
MOTION IN TWO AND THREE DIMENSIONS

CHAPTER 5

Newton's Laws of Motion   169

CHAPTER 6

**The Gravitational Force
and the Gravitational Field   231**

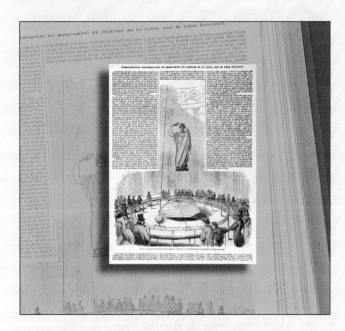

CHAPTER 7
Hooke's Force Law and Simple Harmonic Oscillation   281

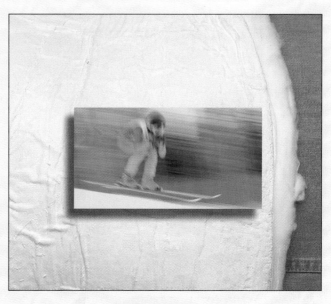

CHAPTER 8
Work, Energy, and the CWE Theorem   319

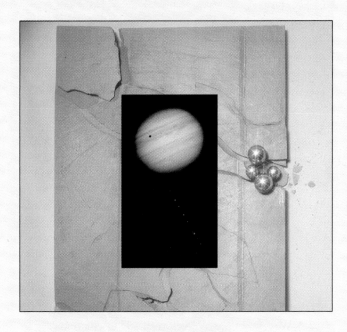

CHAPTER 9
Impulse, Momentum, and Collisions   381

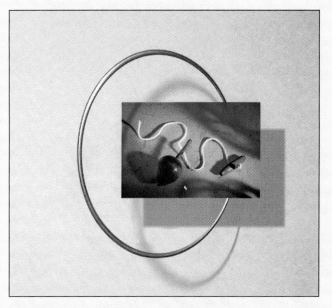

CHAPTER 10
Spin and Orbital Motion   425

CHAPTER 11
Solids and Fluids 489

CHAPTER 12
Waves 531

CHAPTER 13
**Temperature, Heat Transfer, and the First
Law of Thermodynamics   587**

CHAPTER 14
Kinetic Theory   639

CHAPTER 15
The Second Law of Thermodynamics   667

APPENDIX A
Proofs of the Gravitational Shell Theorems   A.1

VOLUME II

CHAPTER 16
Electric Charges, Electrical Forces, and the Electric Field   705

CHAPTER 17
Electric Potential Energy and the Electric Potential 767

CHAPTER 18
Circuit Elements, Independent Voltage Sources, and Capacitors 805

CHAPTER 19
Electric Current, Resistance,
and dc Circuit Analysis   835

CHAPTER 20
Magnetic Forces and
the Magnetic Field   895

CHAPTER 21
Faraday's Law of
Electromagnetic Induction 953

CHAPTER 22
Sinusoidal ac Circuit Analysis 1005

CHAPTER 23
Geometric Optics   1041

CHAPTER 24
Physical Optics   1103

CHAPTER 27
An Introduction to
Quantum Mechanics   1249

EPILOGUE   1268

APPENDIX A
Proofs of the Gravitational Shell
Theorems   A.1

CHAPTER 1
PRELUDES

But thou shalt have a perfect and just weight, a perfect and just measure shalt thou have.

Deuteronomy 25:15

Art, religion, and the sciences are lenses through which we perceive the universe and attempt to come to terms with its existence—and our own. The word **science** stems from the Latin root *scientia*, meaning the state of knowing, or *knowledge*. In this context, art is the science of the way we see and depict beauty. Religion is the science of the mystical and supernatural. What we now mean by the word science, though, is the knowledge of the natural and physical world. One of the most incomprehensible things about the physical universe is that it *is* comprehensible, an observation made by Albert Einstein, among others.

1.1 NATURE AND MATHEMATICS: PHYSICS AS NATURAL PHILOSOPHY

Et harum scientarum porta et clavis est Mathematica.
[Of these sciences, the gate and key is mathematics.]
Roger Bacon (c. 1214–1294)*

The word **physics** comes from the Latin *physica* and the Greek φυσικά, meaning the knowledge or science of natural things. The study of physics is a tribute to the human intellect and its use of knowledge and reason to invent mathematical models of reality. These models mimic the actual happenings of nature, from the perceptible regimes of motion, forces, waves, and light to the less perceptible regimes of molecules, atoms, nuclei, elementary particles, the duality of particle-waves, and at the other extreme, the creation and evolution of the universe itself.

Physics is one of the liberal arts, combining elements of reason, philosophy, mathematics, language, and rhetoric. Historically, physics was called **natural philosophy**; we gave up much nuance when this phrase was abandoned during the 19th century. The term natural philosophy reflects the creative and dynamic interplay that exists in physics among experiment, theory, logic, insight, inspiration, symmetry, beauty, and language. It is in this spirit that we, as students and professors alike, study nature, constantly observing, changing, and modeling the world around us with all the real and theoretical tools that we, as an intelligent species, can invent. All of us are natural philosophers; we *all* are physicists.

Physics is timeless. The universe evolves and changes. Likewise, our *descriptions* of physics also change and improve as we develop better ways of observing and studying the universe with keener insight, more sophisticated equipment, more realistic approximations, and more encompassing theories. The underlying mechanisms of nature on the *macroscopic* level are immutable and exist with or without human scrutiny. On the other hand, at the *atomic* and *subatomic* levels, just the opposite may be true: the role of the observer and of measurement are inextricably interwoven into the quantum mechanical behavior of such systems, which is quite confounding!

Developing Theories

Physics and mathematics are the twin pillars on which we build modern science and its offspring, technology. While mathematics is essential to studying physics and engineering, mathematics

Opus Majus, Part 4, Chapter 1, translated by Robert Belle Burke (University of Pennsylvania Press, Philadelphia, 1928), Volume I, page 116.

itself is neither. In fact, we use the language of mathematics to describe and predict the behavior of the natural world, with nature itself as both guide and arbiter of our observations and physical theories. Most theories arise from and are nourished by observations and experiments[†]; all theories depend on experiments for confirmation or refutation. The goal of theory is to integrate as many observations of nature as possible under one conceptual umbrella—*to discover how little must be known to explain as much as possible*.

Most physical theories are *approximations* or models that accurately describe natural phenomena under certain limiting conditions and idealized circumstances. For example, classical dynamics is valid when the speeds involved are much less than the speed of light. Relativistic dynamics is appropriate for speeds approaching that of light and reduces to classical dynamics for small speeds. Physicists construct models whose primary purpose is to consistently account for and predict natural phenomena. A guiding principle of physics is expressed colloquially as the **KIS principle**: *Keep It Simple*. As long as it is consistent with observations in nature, simplicity is preferred over complexity. Observation is a necessary but not a sufficient condition for a good or a unique theory.

Theorists are like turtles: they must stick their necks out (i.e., make predictions) to get anywhere. We all do this; there is no gain without risk, no progress without change. Sometimes we get more physics out of a theory than we originally put into it; surprises are part of the game.

The characteristics of a good physical theory are shown in Figure 1.1.

Physics and Equations

If you feel that understanding physics amounts to memorizing myriad equations and formulas, and then searching for a "magic bullet" among them, you have missed the forest for the trees. Physics is only incidentally a process of learning what formula to use to solve a problem. After all, computers can solve equations easily, but the results are useless if the equations are the wrong

[†]It can be argued that Einstein's general theory of relativity arose more from aesthetic considerations and symmetry arguments than from experiments.

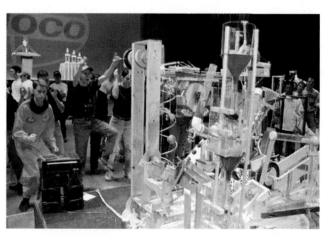

If a solution is too complex, it may not be the answer. A theory should account for observations with minimal complexity.

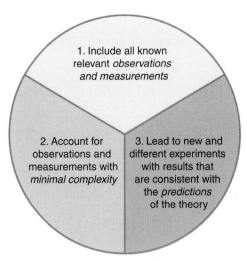

FIGURE 1.1 Characteristics of a good physical theory.

ones for the problem or are applied beyond their limitations. Computers can solve the wrong equations perfectly correctly. We must devise the theoretical equations to begin with and then tell the computers which equations to solve and under what conditions, and that is where physicists, scientists, and engineers come into play. "Doing physics" is the reasoning that leads from observations and experiments to a mathematical model or explanation of them. It is the reasoning process and the explanation that constitute the real physics. Mathematical results can be checked against experiments, but mathematical equations can sometimes camouflage what really is happening.

Language and Physics

Physics uses language as well as equations. It is through the written word that we communicate ideas and elucidate the natural world. If words were not necessary, this text would simply be a list of equations and formulas; surely such a mere list is not sufficient to understand nature.

For better or worse, physicists are like lawyers. Both use common words in special ways, with narrow and definite meanings that typically differ from their colloquial usage. For example, words such as velocity, force, momentum, work, and power have distinct meanings in physics, and these may vary from your intuitive understanding of them.

> *"When I use a word," Humpty Dumpty said in a rather scornful tone, "it means just what I choose it to mean—neither more nor less."*
> Lewis Carroll (1832–1898)*

When *writing* physics, one must be as precise as when crafting a legal document. When *reading* physics or the law, you likewise must be careful, deliberate—and in a good mood too.

QUESTION 1

Mathematicians usually do not like to be called scientists. In what ways is mathematics similar to yet different from science or physics in particular?

Through the Looking Glass (Random House, New York, 1946), Chapter 6, page 94.

1.2 CONTEMPORARY PHYSICS: CLASSICAL AND MODERN

The history of physics encompasses two broad time periods in which **classical physics** and **modern physics** developed. Classical physics developed between 1600 and 1900. It embraces the general areas of physics known as mechanics, thermodynamics, and electromagnetism, culminating in relativity. Modern physics began developing between about 1890 and 1930, when it was realized that classical physics could not account for the newly discovered behavior of nature at the atomic and molecular level. Modern physics includes relativity[†] as well as quantum mechanics and most of the subsequent new physics discovered and developed during the 20th century.

Classical and modern physics together constitute **contemporary physics**. We use both classical and modern physics to account for physical phenomena of current interest. You should not think of classical physics as dead physics; we use classical physics today just as much as a hundred years ago. The flight of a baseball, the motion of a pendulum, the sound of a beautiful instrument, or a trip to the Moon are described using classical physics. On the other hand, we need modern physics to explain the structure of atoms and molecules, the process of light emission and absorption, and the origin of the universe itself.

The Emergence of Classical Physics

Classical mechanics has its roots in the work of Johannes Kepler, who, between 1600 and 1619, became the first person to describe quantitatively and accurately the elliptic paths of the planets around the Sun (see Figure 1.2).

The problem of the shape of the planetary orbits was not easy to solve, because all historical observations of the planets were made from the Earth, itself a moving planet. Although a crude *heliocentric* (Sun-centered) *model* of the solar system using circular orbits was proposed by the Greek astronomer Aristarchus in the third century B.C. and resurrected by Copernicus in 1542

[†]It can be argued that relativity is "the end" of classical physics, as well as "the beginning" of modern physics.

(not to scale)

FIGURE 1.2 The path of a planet around the Sun is elliptical, with the Sun at one focus of the ellipse.

(see Figure 1.3), the planetary positions predicted by Copernicus were *not* as accurate as those found using Ptolemy's complicated *geocentric* (Earth-centered) *model* of the second century A.D. (see Figure 1.4).

The pure Copernican model was inaccurate because Copernicus refused to abandon the flawed concept of early Greek thought that the natural path of a moving "perfect" celestial object, such as a planet, had to be in the shape of the "perfect" geometrical figure—a circle—with the planet moving at constant speed. Ironically, using the criteria for a good physical theory outlined in the previous section, a scientist living at the time of Copernicus would have been forced to reject the Copernican theory with its circular planetary orbits, since they did *not* account for the *observed* motions of the planets as accurately as the existing, but more complicated, geocentric theory of Ptolemy.

The change from the accurate but complex Ptolemaic geocentric model of the solar system to the accurate and simple heliocentric model of Kepler illustrates the characteristics of a good theory. With Kepler's work, accuracy *and* simplicity blossomed simultaneously from the heliocentric hypothesis. Kepler's model and mathematical methods began the development of physics as we know it. His work was a watershed of quantitative thought and a new worldview.

Observation of nature and experimental results typically precede the general theories that account for them. The extensive observations of the planets by Tycho Brahe just prior to 1600 provided Kepler with the data he needed to model the solar system accurately and precisely.*

---

*We distinguish between precision and accuracy in Section 1.8.

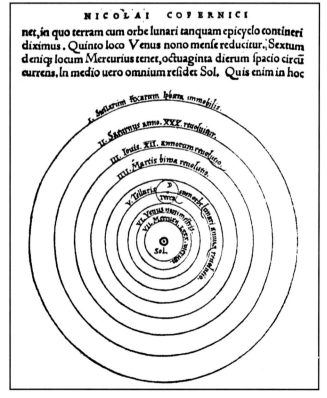

FIGURE 1.3 The Copernican heliocentric model of the solar system used circular orbits.

FIGURE 1.4 The earlier Ptolemaic geocentric model of the solar system also used circles.

In most instances, natural phenomena exist long before we observe them critically. For example, in the 17th century, Newton did not *discover* gravity; he *described* it quantitatively. Gravity was here from the beginning. Likewise, atoms existed long before we ever were able to describe them.

Kinematics and Dynamics

Kinematics, a quantitative description of motion, was invented by Galileo Galilei in the early 1600s after much observation and experimentation. We say *invented* rather than *discovered*, since physical theories do not lie lurking in nature, waiting to be discovered like an unknown planet or plant. Rather, *physical theories are inventions of the human intellect* that describe and account for observations of nature. Despite this distinction between the words invent and discover, the words will henceforth be used interchangeably in this text.

Galileo's kinematics enabled Isaac Newton to invent several *laws of motion* (a theory that explained the causes of motion, called *dynamics*) as well as a *law of universal gravitation* encompassing the ideas of both Galileo and Kepler. Newton's great synthesis was published in 1686 as the *Philosophiae Naturalis Principia Mathematica* (*Mathematical Principles of Natural Philosophy*), more commonly called simply the *Principia*. It was one of the most influential scientific books ever published, establishing classical kinematics and dynamics as the first comprehensive physical theory and illustrating the characteristics of a good theory shown in Figure 1.1. With its great predictive power, Newton's theory revolutionized our view of the natural world.

Electricity and Magnetism

Fundamental experiments in *electricity* and *magnetism* were performed by many investigators during the 18th and 19th centuries. Benjamin Franklin invented the names positive and negative for the property of matter we call electric charge. Charles Coulomb deduced the form of the electric force law between pointlike charges at rest. Christian Oersted studied magnetism. The experimental genius Michael Faraday discovered the *law of electromagnetic induction*, which made possible the electronic age. All this work culminated in 1864 with a seminal treatise by James Clerk Maxwell that integrated the heretofore separate theories of electrical and magnetic phenomena into a single *theory of electromagnetism*. Unexpectedly, his theory predicted the existence of electromagnetic waves, found to be identical to light. Electromagnetic theory was the second great physical theory of natural philosophy.

Thermodynamics

Thermodynamics arose during the 19th century from a need to understand the parameters associated with designing better steam engines during the Industrial Revolution. The fundamental research of Sadi Carnot into engine efficiency, of James Prescott Joule into the dissipation of mechanical energy and heat transfer, of Rudolf Clausius into entropy, of Ludwig Boltzmann into kinetic theory and the statistical meaning of entropy, and the culminating theoretical work of J. Willard Gibbs established the physical *laws of thermodynamics*.

Modern Physics

The dawn of the 20th century saw physicists heady with successes in mechanics, electromagnetism, and thermodynamics, almost smug with the successes of the seemingly all-encompassing theories of Newton, Maxwell, Boltzmann, and Gibbs. Some even predicted the end of physics was nigh. There seemed to be little left to discover.

Then, in 1905, Albert Einstein realized that Maxwell's electromagnetic theory implied that the Newtonian synthesis of motion and classical dynamics, the first triumph of classical physics and an unimpeachable theory for over two centuries, was only an *approximation* of a more general theory of space, time, and motion. This incredible discovery, now known as *relativity*, is a splendid example of getting more physics out of our equations than originally was put into them.

Several other unexpected and seemingly unrelated discoveries between 1895 and 1910 collectively shook the foundations of physics and our view of reality. Among these discoveries were the following:

a. The accidental discovery of a new penetrating radiation, called *x-rays*, in 1895 implied that there was more to learn about the emission of electromagnetic radiation and its interaction with matter. In 1901, Wilhelm Conrad Röntgen was awarded the first Nobel prize in physics for the discovery.*
b. The discovery of *radioactivity*, the spontaneous emission of particles by certain atoms, in 1896 clearly indicated that atoms, the heretofore "indivisible" elements of nature, had a divisible structure. Radioactivity also showed that microscopic processes were statistical in nature, similar to the statistical formulation of thermodynamics by Ludwig Boltzmann and J. Willard Gibbs. In 1903 Antoine Henri Becquerel was

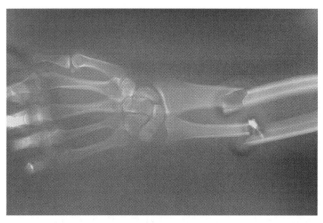
Medical applications of x-rays are legion.

awarded the Nobel prize in physics for the discovery of radioactivity. Very few individuals have won multiple Nobel prizes, but Marie Curie received *two* awards for her studies of radioactivity: one in physics in 1903 (along with her husband) and one in chemistry in 1911.[†‡]
c. The discovery of the negatively charged *electron* particle in 1896 indicated that atoms are composite structures made from more fundamental subunits, some of which must have negative and positive charge, since atoms themselves are

---

[†]John Bardeen won two awards in physics (in 1956 and again in 1972), and Linus Pauling won an award in chemistry (1954) and one in peace (1962).

[‡]One of Marie Curie's two daughters, Irène Joliet-Curie, also won a Nobel prize in chemistry, in 1935. Like mother, like daughter.

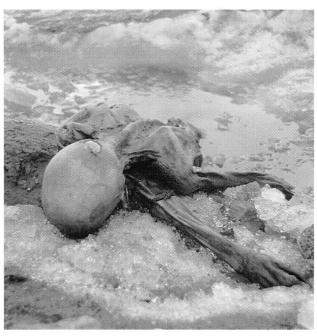
Radioactive carbon is used by archaeologists to date many historical remains such as this prehistoric hunter, discovered in a melting glacier in the Alps.

---

*The Nobel prizes are universally recognized as the most prestigious prizes in the fields of physics, chemistry, physiology and medicine, peace, and literature. The lucrative prizes were established by the will of Alfred Nobel (1833–1896), the inventor of dynamite. A Nobel prize in economics was established later.

electrically neutral. In 1906, J. J. Thomson received the Nobel prize for this discovery.

d. The shocking discovery of the *quantum of action* in 1901 resolved a troubling problem of an unexpectedly large amount of energy, known picturesquely as the "ultraviolet catastrophe," in the classical explanation of the electromagnetic radiation emitted by hot objects. The explanation introduced a hitherto unknown and mysterious fundamental physical constant into physics. The quantum of action now is called *Planck's constant, h*. In 1918, Max Planck became a Nobel laureate.

e. A new theory in 1905 of the *photoelectric effect*, the emission of electrons by metals when illuminated by appropriate light, led to a new and radically different view of light as a *particle-wave* called a *photon*. It was principally for this theoretical work that Einstein received the Nobel prize in 1921.

f. Between 1905 and 1912, the discovery of the *nucleus* of the atom, the invention of the atomic model of hydrogen, and the explanation of hydrogen's light emissions using the ideas of *quantization of angular momentum and energy* signaled that a fundamentally new type of physics was needed to explain the affairs of nature on the molecular, atomic, and nuclear scales. Ernest Rutherford received the Nobel award in chemistry in 1908 for the discovery of the nucleus, and Niels Bohr received the prize in physics in 1922 for his model of the hydrogen atom.*

The new physics that integrated all these and many other phenomena under one conceptual umbrella was *quantum mechanics*, developed during the 1920s by such giants of 20th-century physics as Erwin Schrödinger, Wolfgang Pauli, Werner Heisenberg, P. A. M. Dirac, Niels Bohr, and Enrico Fermi, all of whom eventually became Nobel laureates in physics.

Early in the 20th century, an idea emerged that gave many disparate areas of physics a unified leitmotif. You likely know that the three certainties of existence are death, taxes,

---

*One of Bohr's sons, Aage Bohr, also won the Nobel prize in physics (in 1975). Like father, like son.

Medical imaging technologies are based on many discoveries in modern physics.

and change. When an isolated physical system changes in some way, certain properties of the physical system such as its total energy, charge, momentum, or angular momentum may be left unchanged. When such quantities are unchanged, we say they are *conserved* and there exists a *conservation law* for that property. The discovery of conservation laws for energy, charge, momentum, angular momentum, and other quantities was neither obvious nor easy. About 1920, an unsung heroine, Amalie Noether, realized that each conservation law is a manifestation of an abstract mathematical symmetry associated with the physical system. The conservation laws and their associated symmetry principles are common to both classical and modern physics and are unifying threads in the tapestry of physics.

Thus the era that saw the development of modern physics coincides with the first half of the 20th century. In the latter half of the 20th century, physicists also have made progress toward a theory to unify the *four fundamental forces of nature:* gravitation, electromagnetism, and two other forces, called the strong and weak nuclear forces, whose existence became apparent only as our knowledge of nuclear physics progressed. Although great strides have been made toward such a "theory of everything," it still is only a partially realized goal. The search awaits your participation.

1.3 STANDARDS FOR MEASUREMENT

There shall be one measure of wine throughout Our kingdom, and one of ale, and one measure of corn, to wit, the London quarter. . . . As with measures so shall it also be with weights.
  Paragraph 35 of the *Magna Carta* (A.D. 1215)

The Congress shall have the Power To . . . fix the Standard of Weights and Measures
  The Constitution of the United States of America, Article I, Section 8 (1789)

These quotations from some of our most cherished documents reflect the importance of **measurement standards**. These standards are significant not only to science but also to society—no one likes to be shortchanged at the grocery store or the petrol station.

Experiments are the key to the development of physical theories. Observations of experiments, of course, involve measurements of various physical quantities in terms of how many standard units of the quantity are present. It is quite remarkable to realize that centuries of the scientific enterprise indicate that *all* physical quantities ultimately can be described in terms of a surprisingly small number of primary physical parameters, including mass, length, time, electric current, and temperature. Physical concepts, such as area, volume, density, force, acceleration, pressure, and energy, that are expressed in terms of these primary parameters are known as **derived quantities**.

The primary parameters are not trivial to define or explain. Nonetheless, it is apparent that we need a set of standards for the measurement of them. The most widely used standards for measurement in physics are those of *Le Système International d'Unités* (the International System of Units, when translated from the French), known more simply as the **SI system**. The system is a

TABLE 1.1 The Primary Parameters of Mass, Length, and Time in the SI Unit System

Parameter	SI unit	Abbreviation
Mass	kilogram	kg
Length	meter	m
Time	second	s

refinement of the **metric system** that originated in France during the late 18th century amid the radical furor of the French Revolution.* The word *metric* comes from the Greek word μέτρον (*metron*), meaning measure.

The SI system is the standard system of units used in this text. The SI units used to quantify the primary parameters of mass, length, and time are indicated in Table 1.1.

The SI units for temperature and electric current (from which we get the unit for electric charge) will be defined later in the text, as the need arises. For now, the three parameters mass, length, and time are sufficient for our purposes.

Common prefixes associated with various SI units are indicated in Table 1.2.

Definitions are like belts. The shorter they are, the more elastic they need to be.

Stephen Toulmin (1922–)[†]

Mass

Newton was the first person to distinguish between mass and its weight; prior to Newton, the two terms had been considered synonymous.

Mass and weight are different physical concepts.

Mass is a measure of the resistance of an object to changes in its motion.[‡] It is still uncertain just what gives rise to this property of matter. The mass of an object has the same measure *anywhere* in

---

*The original standards of the metric system were supplanted as improvements in technology allowed more reproducible standards; the process is ongoing.
[†]*Foresight and Understanding* (Greenwood Press, Westport, Connecticut, 1981), Chapter 2, page 18.
[‡]An operational definition of mass is presented in Chapter 5.

TABLE 1.2 Common Prefixes Attached to SI Units

Factor	Prefix	Symbol
10^{12}	tera-	T
10^{9}	giga-	G
10^{6}	mega-	M
10^{3}	kilo-	k
10^{1}	deka-	dk**
10^{-1}	deci-	d
10^{-2}	centi-	c
10^{-3}	milli-	m
10^{-6}	micro-	μ
10^{-9}	nano-	n
10^{-12}	pico-	p
10^{-15}	femto-	f

**Rarely used in physics.

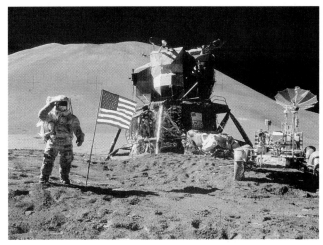

FIGURE 1.5 Your mass is independent of location.

the universe; your mass is the same whether you are on the Earth, the Moon, or the asteroid Hamburga, as indicated in Figure 1.5.[§]

On the other hand, **weight** is the measure of the gravitational force on a mass due to the simultaneous presence of *another* mass, be it the Earth, the Moon, the Sun, or an asteroid. A given mass has *different* weights on the surface of the Earth, far above or below the surface of the Earth, on the Moon, or near an asteroid (see Figure 1.6).

Weight depends on location in space; mass does not.

The standard SI unit for the measurement of mass is the **kilogram** (kg). See Figures 1.7 and 1.8. The kilogram is the mass of a single, special, custom-fabricated cylinder made of an alloy of platinum (90%) and iridium (10%).

[§]The asteroids are a collection of small, planet-like objects, most of which orbit the Sun between the orbits of Mars and Jupiter. More than 5000 asteroids are known. Most are numbered and named. Hamburga is number 449. An amusing astronomical parlor game is to create phrases from the asteroid names. Asteroids 904, 1603, 673, 991, and 449 spell out the phrase "Rockefellia Neva Edda McDonalda Hamburga."

FIGURE 1.6 Your weight on the Earth is different from your weight on the Moon.

FIGURE 1.7 One kilogram of brass and one kilogram of cushion have the same mass.

The kilogram is the only SI unit still defined by a manufactured artifact rather than by a natural phenomenon or property. *The* cylinder defining the kilogram resides in a special environmental chamber at the International Bureau of Weights and Measures in Sèvres, France, near Paris.

Replicas of the standard kilogram are kept at the National Institute of Standards and Technology near Washington, D.C. (see Figure 1.8), but the standard that in fact defines the kilogram is the single cylinder at Sèvres. The height of the cylinder is equal to its diameter (both approximately 3.8948 centimeters at 0 degrees celsius) with slightly beveled edges. The platinum–iridium alloy was chosen because of its stability under variation in temperature and its resistance to chemical changes, such as oxidation (the chemical binding of a substance with oxygen).

FIGURE 1.8 A replica of the standard kilogram.

The size of the cylinder was chosen so that its mass was close to an older, obsolete definition of the kilogram, based on the amount of pure liquid water in a cube 10 cm on a side kept at special environmental conditions. Table 1.3 surveys a range of masses in kilograms.

Related to the mass standard is another SI unit known as the **mole** (mol). Twelve eggs are defined to be one dozen eggs, regardless of their mass, and 500 sheets of paper are defined to be a ream of paper, regardless of their size. A mole likewise is a fixed count of particles, atoms, or molecules, regardless of the particles being counted.

> One **mole** of any kind of particle, element, or molecule is Avogadro's number of the particles, atoms, or molecules. **Avogadro's number** is the number of atoms in exactly 12 grams of $_6^{12}C$, an isotope of carbon.

The subscript in the designation $_6^{12}C$ is the **atomic number** of the element: the number of protons in the nucleus of the atom. The number of protons in the nucleus identifies the particular chemical element; each carbon atom always has 6 protons in its nucleus but can have different numbers of neutrons. The superscript is the **atomic mass number** of the element and represents the sum of the number of protons and neutrons in the nucleus of the atom. The atomic mass number identifies a specific **isotope** of an element. The various isotopes of an element have the same number of protons in the nucleus but different numbers of neutrons. The common isotopes of carbon are $_6^{12}C$, with 6 protons and 6 neutrons in each nucleus, and $_6^{14}C$, with 6 protons and 8 neutrons in the nucleus of each atom.

Experiments indicate that Avogadro's number N_A is approximately

$$N_A = 6.022\ 137 \times 10^{23} \text{ particles/mole}$$

One mole of $_6^{12}C$ has a mass of exactly 12 grams and contains Avogadro's number of atoms. One mole of gold (Au) has Avogadro's number of gold atoms; one mole of water (H_2O) has Avogadro's number of water molecules. The mass in grams of

TABLE 1.3 A Range of Masses

The entire universe (very approximate)	$\sim 10^{53}$ kg
The Milky Way Galaxy	$\approx 3 \times 10^{41}$ kg
Sun	1.99×10^{30} kg
Earth	5.98×10^{24} kg
Moon of the Earth	7.36×10^{22} kg
Aircraft carrier	$\sim 3 \times 10^{8}$ kg
Boeing 747 jet (empty)	$\approx 1.6 \times 10^{5}$ kg
Physics student	≈ 60 kg
Grapefruit	~ 1 kg
Computer diskette	2.0×10^{-2} kg
Credit card	5×10^{-3} kg
Sand grain	$\sim 10^{-7}$ kg
Single living cell	$\sim 10^{-12}$ kg
HIV virus	$\sim 10^{-19}$ kg
Uranium atom	3.95×10^{-25} kg
Proton	1.67×10^{-27} kg
Electron	9.11×10^{-31} kg

one mole of a substance is numerically equal to the atomic mass number of the isotope or the **molecular mass** of the molecule.* The mass of one mole of any substance is called its **molar mass**. For example, one mole of gold $^{197}_{79}$Au has a mass of 197 g (= 0.197 kg) because the atomic mass number of this gold isotope is 197. Thus the molar mass of $^{197}_{79}$Au is 197 g and contains Avogadro's number of gold atoms. Water is composed of two hydrogen atoms, $^{1}_{1}$H, and one oxygen atom, $^{16}_{8}$O. The molecular mass is the sum of the atomic mass numbers of all the atoms in the molecule, here 1 + 1 + 16 = 18. Therefore, one mole of water has a mass of 18 g (= 1.8×10^{-2} kg) (see Figure 1.9). A mole of water contains Avogadro's number of water molecules.

QUESTION 2

From the standpoint of physics, what is wrong with these labels?

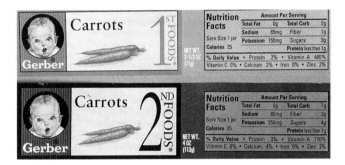

EXAMPLE 1.1

How many moles and how many gold atoms are in a gold brick of $^{197}_{79}$Au with a mass of 1.00 kg?

Solution

The mass of a mole of $^{197}_{79}$Au is 197 g = 0.197 kg. Therefore the 1.00 kg of gold has

$$\frac{1.00 \text{ kg}}{0.197 \text{ kg/mol}} = 5.08 \text{ mol}$$

Each mole has Avogadro's number of gold atoms, and so the number of gold atoms in the brick is

$$(5.08 \text{ mol})(6.02 \times 10^{23} \text{ atoms/mol}) = 3.06 \times 10^{24} \text{ atoms}$$

Time

For what is time? Who is able easily and briefly to explain that? Who is able so much as in thought to comprehend it, so as to express himself concerning it? And yet what in our usual discourse do we more familiarly and knowingly make mention of than time? And surely we understand it well enough, when we speak of it; we understand it also, when in speaking with another we hear it named. What is time then? If nobody asks me, I know; but if I were desirous to explain it to one that should ask me, plainly I know not.

St. Augustine (A.D. 354–430)[†]

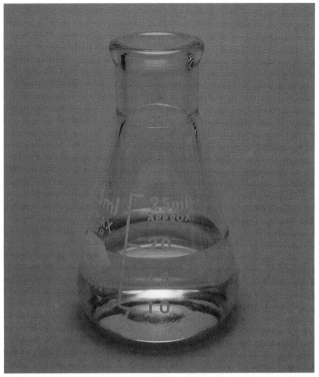

FIGURE 1.9 One mole of water has a mass of 18 g.

It has been said, perhaps in jest, that time is nature's way of preventing everything from happening at once.

The standard SI unit for the measurement of time is the **second** (s). Originally, a second was defined to be $1/(24 \times 60 \times 60)$ = 1/86 400 of the average rotational period of the Earth with respect to the Sun (called a mean solar day). However, the rotational motion of the Earth is slightly variable and gradually, yet systematically, is slowing down.[‡] Therefore, the motion of the Earth is an unacceptable basis for a precise and invariant definition of a second. The development of atomic clocks led to the present determination of the second.

A second is the time interval required for a specific kind of light emitted from the cesium atom $^{133}_{55}$Cs to undergo 9 192 931 770 oscillations. Do not memorize this number; it can be looked up if necessary.

While this definition may seem peculiar, it is based on the natural cesium atom, and provides a standard that is reproducible anywhere in the universe with suitable—if complex—experimental equipment.

The rotation of the Earth gradually gets out of step with the current definition of the second. Hence, an occasional *leap*

---

*The molar mass of an element listed in the periodic table of the elements accounts for the relative abundances of the various isotopes of the element (and other factors) and thus indicates molar masses to fractions of a gram per mole.

[†] *St. Augustine's Confessions*, translated by William Watts (Harvard University Press, Cambridge, Massachusetts, 1974), Book XI, #XIV, pages 237, 239.
[‡] The slowing down of the rotational motion of the Earth is caused by the tides; we discuss this briefly in Chapter 10, Problem 53.

second, similar to the extra day in February during leap years, is inserted into the calendar every few years, usually on the last day of June or December, so that noon according to the atomic clocks corresponds with noon astronomically.

Some time intervals of interest, expressed in seconds, are noted in Table 1.4.

Length

The SI standard unit for length is the **meter** (m). Originally, the meter was defined to be exactly one ten-millionth (1/10 000 000 $\equiv 10^{-7}$) of the distance between the north geographic pole of the Earth and the equator, measured along the curved path of the terrestrial meridian (a line of constant longitude) through Paris (specifically, the Paris Observatory). (See Figure 1.10.)

This definition proved to be unacceptable for two reasons: (1) the distance between the pole and equator is difficult to measure, and (2) the north and south geographic poles of the Earth wander slightly and erratically over the surface of the Earth, again increasing the difficulty of a precise measurement of the distance between the pole and equator.* The distance between two marks scratched on a straight platinum–iridium bar (made from the same alloy casting as the standard kilogram) then was used to define the meter, but this definition also was abandoned because of the difficulty of measuring accurately and precisely the distance between the positions of the centers of the two scratches on the bar. To avoid these difficulties, the meter now is determined using the defined speed of light and the definition of the second.

---

*The axis about which the Earth rotates defines the positions of the north and south geographic poles. (Of course, there is no real pole or axis sticking out of the Earth at the poles!) The area over which the position of the poles wanders is about the size of a tennis court. The reasons for the polar wanderings are quite complex but are related to seasonal changes in the distribution of mass on the surface of the Earth (relocations of snow and ice as they melt to liquid water) and complicated but unknown internal motions of materials deep within the Earth.

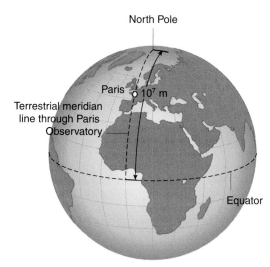

FIGURE 1.10 The old definition of the meter involved the distance between the north geographic pole and the equator.

A meter is the distance through which light travels in a vacuum during a time interval of exactly 1/299 792 458 second. You can look up this number if needed; do not memorize it.

The definition of the meter implies that the speed of light in vacuum is *exactly* 299 792 458 m/s.

A survey of various distances in meters is indicated in Table 1.5.

QUESTION 3

Describe the characteristics of a universe in which one of the primary physical parameters—mass, length, and time—did not exist.

TABLE 1.4 Time Intervals

Age of the universe	$\approx 6 \times 10^{17}$ s
Age of the solar system	$\approx 1.6 \times 10^{17}$ s
Age of a centenarian	3.2×10^9 s
Age of an 18-year-old physics student	5.7×10^8 s
One day	86 400 s
One 50-minute lecture	3.0×10^3 s
Interval between heartbeats	≈ 1 s
One vibration corresponding to the musical note middle C	3.83×10^{-3} s
Time for light to travel one meter	3.33×10^{-9} s
Time for light to travel the approximate diameter of an atom	$\sim 10^{-18}$ s
Time for light to travel the approximate diameter of a proton (10^{-23} s is called a jiffy)	$\sim 10^{-23}$ s
Lifetime of the most unstable particles	$\sim 10^{-23}$ s
Planck time (the earliest time after the creation of the universe at which the laws of physics, as we currently know them, can be applied to the universe)	$\sim 10^{-43}$ s

TABLE 1.5 Lengths

Most distant object in the visible universe	$\sim 10^{26}$ m
Distance to the Andromeda Galaxy (a neighbor of the Milky Way Galaxy)	$\approx 2 \times 10^{22}$ m
Diameter of Milky Way Galaxy	$\sim 10^{21}$ m
Distance from Sun to nearest star	$\approx 4 \times 10^{16}$ m
Radius of the solar system	$\approx 6 \times 10^{12}$ m
Average distance between the Earth and the Sun	1.496×10^{11} m
Radius of the Sun	6.96×10^8 m
Average distance between the Earth and Moon	3.84×10^8 m
Average radius of the Earth	6.37×10^6 m
Deepest ocean depth	$\approx 10^4$ m
Height of an average human	≈ 2 m
Diameter of a pencil lead	$\approx 10^{-3}$ m
Thickness of a sheet of paper	$\sim 10^{-4}$ m
Size of the bacterium *Escherichia coli* (E. coli)	$\approx 10^{-6}$ m
Diameter of an atom	$\sim 10^{-10}$ m
Diameter of the nucleus	$\sim 10^{-15}$ m

QUESTION 4

How can we be certain that other primary physical parameters do or do not exist? Upon what factors should such a judgment be made?

1.4 UNITS OF CONVENIENCE AND UNIT CONVERSIONS

Occasionally, it is convenient to express physical quantities in **units of convenience** rather than in SI units. The common units such as the minute (min), hour (h), day (d), year (y), centimeter (cm), kilometer* (km), and gram (g) are units used for convenience, but they are not SI units. It is just easier to say you have a mass of one gram (1 g) rather than ten-to-the-minus-three kilograms (10^{-3} kg). Likewise, it is more convenient to say a trip lasts three hours (3.00 h) rather than 1.08×10^4 s. Common units of convenience are summarized in Table 1.6.

In this text we will write the units associated with each physical quantity in standard Roman type; for example, kilogram is represented as kg, meter as m, and second as s. In this way you can appropriately track the units through a calculation. Do not confuse *unit* abbreviations with algebraic *variables* representing various physical quantities that are denoted in italics. For example, m is used to represent the mass of an object, whereas m is the SI unit known as the meter. In later chapters we will see that V is the SI unit called the volt, whereas V is the algebraic symbol for both a volume and the electric potential. To avoid confusion between units and variables that use the same letter, you might choose to designate all units with parentheses [e.g., (m)], but that decision is up to you and your professor.

Unit conversions involve changing the measure of a quantity from one system of units to another, say from units of convenience to SI units or vice versa. You will need to learn to do unit conversions easily and with confidence, because this is an important skill in any science. Unit conversions are similar to expressing a collection of nickels, dimes, and quarters (units of convenience) in terms of dollars (the monetary standard).

When expressing unit conversions in words, the equations

$$1000 \text{ m} = 1 \text{ km}$$
$$1000 \text{ g} = 1 \text{ kg}$$
$$100 \text{ cm} = 1 \text{ m}$$

are expressed as

> one thousand meters *per* kilometer,
> one thousand grams *per* kilogram, and
> one hundred centimeters *per* meter.

The meaning of the word *per* is simply "in one," and so the expressions really mean, respectively,

> one thousand meters *in one* kilometer,
> one thousand grams *in one* kilogram, and
> one hundred centimeters *in one* meter.

The word *per* also is used with certain derived units. For example:

> The **density** of an object is the mass contained in one unit of volume.

The *measure* of the density depends on the specific units employed for the measure of mass and volume. The density can be expressed in grams per cubic centimeter (g/cm^3) or, in the SI unit system, the mass in kilograms contained in a volume of one cubic meter of the material, kilograms per cubic meter (kg/m^3). The density of various substances is listed in Table 1.7.

When we say, for example, the density of mercury is 13.6×10^3 kg/m^3, we usually say "thirteen thousand, six hundred kilograms *per* cubic meter." What we mean is there is a mass of 13.6×10^3 kg in a volume of one cubic meter of mercury.

> In this text, we will write the mathematical expressions of physical laws in forms appropriate for the use of SI units, with only rare exceptions.

Notice in the following examples that *units are treated as algebraic quantities* when making conversions and in the equations of physics.

PROBLEM-SOLVING TACTIC

1.1 In solving most problems, be sure to convert units of convenience into standard SI units.

---

*The proper pronunciation of this unit is kil'ə mēt'ər, *not* kə läm'ət ər.

TABLE 1.6 Units of Convenience and SI Units

	Abbreviation	SI units		Abbreviation	SI units		Abbreviation	SI units
Mass			*Length*			*Time*		
milligram	mg	$\equiv 10^{-6}$ kg	millimeter	mm	$\equiv 10^{-3}$ m	minute	min	$\equiv 60$ s
gram	g	$\equiv 10^{-3}$ kg	centimeter	cm	$\equiv 10^{-2}$ m	hour	h	$\equiv 3.6 \times 10^3$ s
kilogram	kg	SI standard unit	meter	m	SI standard unit	second	s	SI standard unit
metric ton	mt	$\equiv 10^3$ kg	kilometer	km	$\equiv 10^3$ m	day*	d	$\equiv 86\,400$ s
						year†	y	$\approx 3.16 \times 10^7$ s

*Occasionally, on those days when a leap second is introduced to keep the atomic definition of the second in step with the rotation of the Earth, the day has 86 401 s.

†The number is approximate, because the time it takes the Earth to orbit the Sun is approximately 365.256 37 d, called one sidereal year.

TABLE 1.7 Density of Various Substances

Substance	Density (kg/m³)
Air	1.3
Aluminum	2.70×10^3
Brass	8.4×10^3
Carbon	
Graphite	2.25×10^3
Diamond	3.5×10^3
Copper	8.9×10^3
Ethyl alcohol	0.79×10^3
Gasoline	0.67×10^3
Gold	19.3×10^3
Invar	8.00×10^3
Iron	7.8×10^3
Lead	11.0×10^3
Mercury	13.6×10^3
Uranium	18.7×10^3
Water	
Liquid	1.00×10^3
Ice	0.917×10^3
Sea water	1.02×10^3
Wood	
Red oak (*Quercus borealis*)	0.66×10^3
Sugar maple (*Acer saccharum*)	0.68×10^3
Eastern white pine (*Pinus strobus*)	0.37×10^3
Redwood (*Sequoia sempervirens*)	0.44×10^3

PROBLEM-SOLVING TACTICS

1.6 Realize that there may be more than one way to solve the problem. Of course, you may not realize this initially, but if a colleague does the same problem independently and comes up with the same answer using a different approach, it is likely you both are right. Such independent solutions are useful for checking calculations and typically provide additional insight.

1.7 Ask yourself if the answer you obtain is reasonable. When making any calculation, even when figuring your income taxes, it is a good idea to question whether the result seems reasonable. This may be difficult initially, but as you practice and study, your feeling for the approximate numerical values of various physical quantities develops.

QUESTION 5

Metric units are used in many manufacturing processes. This Bösendorfer grand piano is 2.25 m in length and 156 cm in width at its keyboard with 92 keys. The piano has a mass of 370 kg and a useful lifetime measured in gigaseconds, if properly cared for. Which of the specifications quoted here is not an SI unit?

PROBLEM-SOLVING TACTICS

1.2 Read and understand the problem. A slow and careful reading of any problem is critical to understand and distinguish between what information is given and what is to be determined. Be sure you understand any technical terms in the problem. A good way to test your understanding of a problem is to try to explain it orally in your own words to a roommate or friend. The process of explaining it to someone else usually indicates whether you know what you are talking about . . . or not! Try it frequently.

1.3 If appropriate, sketch a simple diagram of the actual arrangement of problem features. Label the relevant quantities with letter symbols. This stimulates concentration and planning.

1.4 Assess whether additional information is needed. It may be necessary to look up some information, such as conversion factors for various units, values of appropriate constants, or data from tables. The inside covers of the text have the numerical values of many useful constants in SI units.

1.5 Assess whether any given information is irrelevant. The real world is adrift in a sea of information, not knowledge. Some information presented within a problem may not be needed to answer the questions posed. Sorting relevant from irrelevant information is a critical skill to develop in any field, not only in the sciences.

STRATEGIC EXAMPLE 1.2

The funerary mask of the famous pharaoh Tutankhamen is made of almost pure gold. A reference indicates that the density of the most common gold isotope $^{197}_{79}\text{Au}$ is 19.3 grams per cubic centimeter. Convert the density to appropriate SI units.

Solution

Apply Problem-Solving Tactic 1.2: *Read and understand the problem.* We begin with 19.3 g/cm³ and are asked to find the density in SI units of kg/m³.

Keep in mind Problem-Solving Tactic 1.4: *Assess whether additional information is needed.* Since grams and centimeters are not SI units, you need the conversion factors for grams to kilograms:

$$10^3 \text{ g} = 1 \text{ kg}$$

and for centimeters to meters:

$$10^2 \text{ cm} = 1 \text{ m}$$

Now ponder Problem-Solving Tactic 1.5: *Assess whether any given information is irrelevant.* The information contained in the particular designation of the isotope of gold $^{197}_{79}$Au is not needed to solve the problem. The given density means there are 19.3 grams per cubic centimeter, or 19.3 grams in a volume of one cubic centimeter. Expand the third power of the centimeter unit:

$$19.3 \text{ g/cm}^3 = 19.3 \text{ g/(cm} \cdot \text{cm} \cdot \text{cm)}$$

Now convert each of the units of convenience into the appropriate SI unit:

$$19.3 \text{ g/cm}^3 = [19.3 \text{ g/(cm} \cdot \text{cm} \cdot \text{cm)}](10^{-3} \text{ kg/g})(10^2 \text{ cm/m})$$
$$\times (10^2 \text{ cm/m})(10^2 \text{ cm/m})$$
$$= 19.3 \times 10^3 \text{ kg/m}^3$$

You could also write this in scientific notation as

$$= 1.93 \times 10^4 \text{ kg/m}^3$$

Thus there are 1.93×10^4 kilograms per cubic meter—that is, 1.93×10^4 kilograms in a volume of one cubic meter of gold. The mass of one cubic meter of gold is 1.93×10^4 kilograms.

When finished with a calculation, try to bear in mind Problem-Solving Tactic 1.7: *Ask yourself if the answer is reasonable.* The given density represents the mass of gold in a volume of one cubic centimeter of material. The density in SI units is the mass in a volume of one cubic meter of material. Certainly, one cubic meter of gold has a mass much larger than one cubic centimeter of gold, and so the large numerical value of the density in SI units is reasonable.

EXAMPLE 1.3

Given the atomic mass number of gold (197), how many gold atoms are in a pure gold coin that has a mass 3.11×10^{-2} kg? This is one troy ounce of gold.

Solution
One mole of gold has a mass in grams numerically equal to its atomic mass number. Thus, one mole of gold has a mass of 197 g, or 1.97×10^{-1} kg using the proper SI unit, the kilogram; this is the molar mass of gold. The number of moles of gold in the coin is the mass ratio:

$$\frac{3.11 \times 10^{-2} \text{ kg}}{1.97 \times 10^{-1} \text{ kg/mol}} = 1.58 \times 10^{-1} \text{ mol}$$

A mole has Avogadro's number of particles. Therefore, the number N of gold atoms in the coin is the number of moles times the number of particles (here atoms) per mole (Avogadro's number):

$$N = (1.58 \times 10^{-1} \text{ mol})(6.02 \times 10^{23} \text{ atoms/mol})$$
$$= 9.51 \times 10^{22} \text{ atoms}$$

There are 9.51×10^{22} atoms in the gold coin.

EXAMPLE 1.4

How many gold atoms are in a pure gold nugget that has a volume of 2.10 cubic centimeters? The density of gold is 1.93×10^4 kg/m^3, and its atomic mass number is 197.

Solution
First, convert the 2.10 cubic centimeters to cubic meters:

$$2.10 \text{ cm}^3 = \frac{2.10 \text{ cm} \cdot \text{cm} \cdot \text{cm}}{(10^2 \text{ cm/m})(10^2 \text{ cm/m})(10^2 \text{ cm/m})}$$
$$= 2.10 \times 10^{-6} \text{ m}^3$$

The mass of the nugget is the product of its density times its volume:

$$(1.93 \times 10^4 \text{ kg/m}^3)(2.10 \times 10^{-6} \text{ m}^3) = 4.05 \times 10^{-2} \text{ kg}$$

Since the atomic mass number is 197, the mass of one mole of gold (its molar mass) is 197 g, or

$$(197 \text{ g/mol})(10^{-3} \text{ kg/g}) = 1.97 \times 10^{-1} \text{ kg/mol}$$

The number of moles of gold the nugget represents is the mass of the sample divided by the mass per mole:

$$\frac{4.05 \times 10^{-2} \text{ kg}}{1.97 \times 10^{-1} \text{ kg/mol}} = 2.06 \times 10^{-1} \text{ mol}$$

The number of atoms is the number of moles times the number of atoms per mole (Avogadro's number):

$$(2.06 \times 10^{-1} \text{ mol})(6.02 \times 10^{23} \text{ atoms/mol})$$
$$= 1.24 \times 10^{23} \text{ atoms}$$

Hence there are 1.24×10^{23} atoms in the gold nugget.

1.5 THE MEANING OF THE WORD DIMENSION

The word **dimension** has two different meanings in physics. One meaning is *geometrical*: a dimension is any of the least number of coordinates needed to specify the location of a point in space. Thus we live in a three-dimensional world, because we use three coordinates to locate a point in space. Certain physical concepts have the word dimension associated with them in this context: distances (not necessarily in a straight line) are one-dimensional, areas (not necessarily flat) are two-dimensional, and volumes are three-dimensional. (See Figure 1.11.) In Chapter 3, we will speak of motion in one, two, or three dimensions depending on the number of independent coordinates needed to specify the position of the particle under consideration.

The second meaning of the word dimension is quite different. We say the *dimensions* of the density of a material are *mass per unit volume*. Likewise, the dimensions of speed are those of distance divided by a time interval. The *measure* of density or speed depends on the unit system used. (See Table 1.8.) Density can be expressed in g/cm^3 or in SI units of kg/m^3. The speedometer in your car likely indicates speed in miles per hour (mi/h) or kilometers per hour (km/h). The measure of speed in the SI unit system is meters per second (m/s).

Equations in physics are such that

- both sides must have the same dimensions; and

- both sides must be expressed in the same units.

We will almost always use the SI system of units.

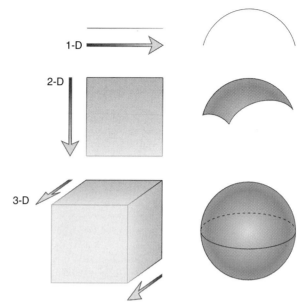

FIGURE 1.11 One, two, and three dimensions.

PROBLEM-SOLVING TACTIC

1.8 Check the dimensions or the units. The process of checking whether the dimensions or the units are the same on both sides of an equation or calculation is one aspect of **dimensional analysis**. If the dimensions or the units on both sides of an equation are *not* the same, you are guaranteed that something is incorrect. This is an extremely useful way of checking your work. Unfortunately, of course, Murphy's law* applies: if they *do* check out on both sides of an equation, there still is no guarantee that the result of your calculation is correct, since the analysis of the problem may have been incorrect. Sigh.

*There are many statements of Murphy's law. Among them: if something can go wrong, it will.

Some units, such as hectares and acres, are inherently two-dimensional.

Some units, such as liters or quarts, are inherently three-dimensional.

TABLE 1.8 Distinction Between Dimensions and Units

Dimensions	SI unit	Units of convenience
Mass	kg	g
Length	m	cm
Time	s	h, d, y
Mass per unit volume	kg/m^3	g/cm^3

Although both sides of every equation in physics must have the same units, we do *not* divide out the units to leave a purely numerical relationship. When we have an algebraic expression such as

$$3x^2 - xy = 5xz$$

there is an x in every term of the equation. Hence, algebraically, we divide by x to obtain

$$3x - y = 5z$$

On the other hand, when we have a physical expression such as

$$3 \text{ m} + 5 \text{ m} = 8 \text{ m}$$

we do *not* divide by the units, here m, on each side of the equation. To do so would leave a purely numerical equation,

$$3 + 5 = 8$$

where the information about the particular units has disappeared. To be meaningful, numerical relationships in physics must indicate the units.

Conversions of units have identical dimensions on each side of the equal sign, but not the same units.

For example, when we have the conversion

$$100 \text{ cm} = 1 \text{ m}$$

both sides of the conversion have the same dimensions (length) but not the same units, since centimeters are on the left-hand side and meters on the right-hand side of the equation. (See Figure 1.12.)

QUESTION 6

Artists such as M. C. Escher (1898–1972) occasionally play interesting tricks with geometrical dimensions. What makes this picture plausible yet impossible?

EXAMPLE 1.5

Your supervisor asks you for the formula for the surface area A of a right-circular cylinder. You and two colleagues suggest the following three versions of the formula, where r is the radius of the cylinder and ℓ is its length:

(1) $A = 2\pi r^2 + 2\pi r\ell$ (your suggestion)
(2) $A = r^2 + r\ell$ (colleague #2)
(3) $A = 2\pi r + 2\pi r\ell$ (colleague #3)

Which of the expressions are dimensionally correct?

Solution
Use Problem-Solving Tactic 1.8: *Check the dimensions and/or units.* Since the equations for the area of various geometric figures are independent of the specific units used, we check the dimensions of the equations. An area has the dimensions of a length squared. Notice that each term on the right-hand side of equations (1) and (2) has the product of two lengths. These two formulas are *both* dimensionally correct.

Notice that a dimensional analysis provides *no* information about the existence or correctness of purely numerical factors such as the 2 and π in all the equations. The first term of the third equation has only a single length, while the second term has the product of two lengths; it is the first term that makes the equation dimensionally inconsistent, because areas have dimensions of length squared. Hence the third equation for the area cannot be correct, and colleague #3 is fired by the supervisor for incompetence. Your suggestion, equation (1), actually is the correct expression for the surface area.

1.6 THE VARIOUS MEANINGS OF THE EQUAL SIGN

There was nothing there now except a single Commandment. It ran:

> ALL ANIMALS ARE EQUAL
> BUT SOME ANIMALS ARE MORE EQUAL
> THAN OTHERS
> George Orwell (1903–1950)*

The equal sign is used in several contexts with distinctly different meanings. It is important to be aware of the subtle distinctions among the ways the equal sign is used in physics and other sciences.

The Equal Sign Is Used in Definitions, Occasionally Symbolized by ≡

In Chapter 3 we will introduce the concept of the average speed, v_{ave}, of a moving particle. When you drive home from campus, your average speed is the total distance d traveled (as

---

Animal Farm (Harcourt, Brace and Co., New York, 1946), page 112.

---

100 cm   =   1 m

FIGURE 1.12 A conversion.

indicated by a trip odometer in your car) divided by the time interval Δt needed for the trip. Here we use the uppercase delta, Δ, to mean *difference in value of*. The quantity Δt thus means difference in the value of *t*, the elapsed time. The average speed is *defined* as

$$v_{ave} \equiv \frac{d}{\Delta t}$$

Rather than writing $d/\Delta t$, we use a new symbol v_{ave}. The special equal sign, ≡ , means that the quantity on the left-hand side is *defined* to be what is on the right-hand side of the symbol; the quantity $d/\Delta t$ is what we mean by the symbol v_{ave}.

The Equal Sign Is Used as an Equivalence in Conversions of Units

When we say that

$$1 \text{ m} = 100 \text{ cm}$$

we do not mean that 1 = 100. A conversion means that both sides are equivalent measures of the same physical concept (here, a length or distance). As we noted in Section 1.5, *the two sides of a conversion have the same dimensions but not the same units*. The *dimensions* on both sides of the conversion 1 m = 100 cm are *lengths*; the *units* for measuring length are different (m on the left, cm on the right). Likewise, when we say that

$$1 \text{ kg} = 1000 \text{ g}$$

we do not mean that 1 = 1000; rather, we mean one kilogram represents an equivalent measure of the same mass as one thousand grams. Both sides of the conversion have the same dimensions (mass), but not the same units.

Consider density. When we say 1 g/cm³ = 10^3 kg/m³, we again have the same dimensions on each side (the mass divided by a length cubed), but different units. The mass of one cubic centimeter of material is not the same mass as that in one cubic meter of the material, but they each have the same density, no matter how we express it.

The Equal Sign Is Used in the Laws of Physics

Looking ahead, in Chapter 5 we will learn that one version of Newton's second law of motion is summarized by the algebraic equation

$$F_{total} = ma$$

where F_{total} is the magnitude of the total force on the mass *m* and *a* is the magnitude of its acceleration. The use of the equal sign in this equation is subtle. It means that if a total force of this magnitude acts on the mass, then *m* experiences an acceleration whose magnitude is calculated using this equation (namely, $a = F_{total}/m$). The expression *ma* is *not* a force, although it has the same dimensions as a force (because both sides of an equation in physics must have the same dimensions). The quantity *ma* is *equal* to the magnitude of the total force, but is *not* itself a force. The magnitude of the total force F_{total} on the mass (the left-hand side of the equation) produces a *response ma* (the right-hand side of the equation). What we mean is the magnitude of the total force F_{total} produces

Since all these masses of brass are made of the same material, they all have the same density.

a movement of *m*, which we find to be an acceleration of magnitude *a*. The two sides of the equation are, to use a clear but historically unfortunate phrase, *separate but equal*. Some of the fundamental laws of physics have this peculiar dichotomy on either side of the equal sign, because they are concerned with the relationships between different physical things.

> This text and more advanced texts typically use the standard equal sign (=) interchangeably for definitions, conversions, and the laws of physics.

Two other symbols are used to indicate, not equality in a strict sense, but *approximate equality*.

1. The symbol ~ means *on the order of*. For example, Table 1.3 indicates that the mass of the universe is on the order of 10^{53} kg, written ~10^{53} kg; it is known perhaps only to within a factor of 10. The symbol ~ indicates that the quantity is not precisely known. Thus we say the **order of magnitude** of the mass of the universe is about 10^{53} kg.
2. The symbol ≈ means *is approximately equal to*. For example, Table 1.3 indicates that the mass of the Milky Way Galaxy is approximately 3×10^{41} kg. This is written as ≈3×10^{41} kg.

Approximately equal, ≈, while not precise, indicates that the quantity is known with more precision than does ~.

1.7 ESTIMATION AND ORDER OF MAGNITUDE

. . . so easy it seemed
Once found, which yet unfound most would have thought
Impossible

John Milton (1608–1674)*

How many bricks were used to construct your science building? How many rice grains are in one kilogram of rice? These are not

---

Paradise Lost, Book VI, lines 499–501.

easy numbers to come by unless one enjoys tedious counting. Nobody really cares about the rice grain problem, but someone had to estimate the number of bricks needed for the building in order to make a cost estimate and place the order for them.

In the sciences and other disciplines, it is often necessary to quantify something that at first glance may seem difficult to quantify. The great physicist Enrico Fermi was a master of the art of estimation; he even estimated the explosive yield of the first atomic bomb test on 16 July 1945 at Alamogordo, New Mexico, by observing the effect of the shockwave from the blast on bits of paper dropped from his hand (at some distance from ground zero!).

The importance of being able to make estimates transcends the discipline of physics and is useful in virtually any endeavor. You may want to estimate how much wallpaper to buy to paper your room. You may be asked to estimate a budget for a development project, or determine the popularity of a politician via polls. No matter what your intended career path, it is important for you to develop the ability to find an approximate value for many quantities.

Why are estimates useful?

1. It may be impossible or impractical to measure the quantity directly. For example, how many domestic cats are there in the United States?
2. The precise value of the quantity may not be needed. For example, what is the exterior wall area of your home so that you can purchase an appropriate amount of paint?
3. An estimate may be needed before a measurement is made. The estimate may determine the technique or instrument needed to perform a particular measurement. Should a meter stick or a long metric tape be used to measure the size of a swimming pool?
4. An estimate can serve as a check on a measurement or calculation. The estimate can be used to see if an instrument is working properly or a complex software package is yielding an appropriate answer to a sophisticated calculation. For instance, if an instrument for measuring the power output of a light source indicates 200 watts for a lighted match, your eyebrows should be raised about the calibration or performance of the instrument—200 watts is brighter than most desk lamps, hardly the case for a lighted match.

Quick estimates attempt to determine whether the value of the quantity to be estimated is on the order of 1, 10, 100, 10^{-1}, 10^{-2}, and so forth, to within perhaps a factor of 2 or 3. Such rough estimates, or order of magnitude calculations, also are colloquially referred to as back-of-the-envelope calculations; simply grab a piece of paper (the back of an envelope) and proceed. More precise estimates are needed for many applications in construction, design, and engineering.

EXAMPLE 1.6

The marketing manager of a pet food company needs an estimate of the number of domestic cats in the United States. Make such an estimate.

Solution
There are approximately three hundred million (3×10^8) people in the United States. If we assume there are an average

of three people per household, this means there are about 1×10^8 family units. Some street smarts based on what is known about the local neighborhood tells us that perhaps one in five families has a cat, excluding multiple cats per family. Therefore, there are approximately

$$\text{number of cats } = \left(\frac{1}{5} \text{ cats/family} \right) (10^8 \text{ families})$$
$$= 2 \times 10^7 \text{ cats}$$

EXAMPLE 1.7

The American Association of Physics Teachers is planning a mailing to every physics professor in the United States. Estimate the number of physics professors teaching at colleges and universities in the U.S.

Solution
There are about three thousand (3×10^3) colleges and universities in the United States. Some of these likely have large physics departments with several dozen professors; some have small departments with perhaps 2 or 3 faculty. Since there are many more smaller colleges than larger colleges, perhaps 7 professors is a suitable average guesstimate for the number of physics faculty members per college or university. Thus the total is about

number of physics professors
$$\approx (7 \text{ faculty/college})(3 \times 10^3 \text{ colleges})$$
$$\approx 2 \times 10^4 \text{ faculty}$$

or twenty thousand physics faculty nationwide.

EXAMPLE 1.8

The human population of the globe is about 6.0×10^9 people. If the entire human population were distributed uniformly over the dry land surface area of the Earth, approximately how far apart would the people be separated from one another?

Solution
Recall Problem-Solving Tactic 1.4: *Assess whether additional information is needed.* First, the surface area of the Earth needs to be calculated. To find the surface area, we need to know the radius of the Earth, which is not given in the statement of the problem, but can be found in the tables on the inside covers of the text. The average radius of the Earth is 6.37×10^6 m. The surface area of a sphere of radius r is $4\pi r^2$, and so the surface area of the Earth is

$$4\pi(6.37 \times 10^6 \text{ m})^2 = 5.10 \times 10^{14} \text{ m}^2$$

We also need to know the proportion of the area of the Earth that is dry land. This proportion is *not* given in the tables on the inside covers, but may be found by consulting an atlas; perhaps, though, you recall from your geography classes that about 3/4 of the surface of the Earth is covered with water, so that the dry land surface area is about 1/4 of the total surface area. Hence the dry land surface area of the Earth is about

$$(0.25)(5.10 \times 10^{14} \text{ m}^2) = 1.3 \times 10^{14} \text{ m}^2$$

The dry land surface area per person is then

$$\frac{1.3 \times 10^{14} \text{ m}^2}{6.0 \times 10^9 \text{ persons}} = 2.2 \times 10^4 \text{ m}^2/\text{person}$$

We can imagine each person at the center of a square of this area; the distance between each person is the same as the length of a side of such a square. The square root of 2.2×10^4 m^2 is about 1.5×10^2 m. The approximate distance between the people is about 1.5×10^2 m, or one hundred and fifty meters.

1.8 THE DISTINCTION BETWEEN PRECISION AND ACCURACY

Some quantities are known exactly, others less so. When we say we have one apple, there is no question that we have precisely one apple (as long as we are not eating it), since simple counting determines the number exactly. Although they can be counted, the number of apples in a cubic meter will vary depending on the size of the apples used to fill the volume. Other numbers, such as special irrational numbers in mathematics, have symbols to represent their precise value. Among them are the famous numbers

- the ratio between the circumference and diameter of a circle:

$$\pi = 3.141\,592\,6536\ldots$$

- the base of natural logarithms:

$$e = 2.718\,281\,8284\ldots$$

and other irrational numbers such as

- the ratio between the diagonal and side of a square:

$$\sqrt{2} = 1.414\,2136\ldots$$

- other irrational square roots, such as

$$\sqrt{3} = 1.732\,0508\ldots$$

Such numbers can be used in decimal form to whatever precision is desired.*

Physical quantities, however, involve measurements and they can *never* be made with infinite precision.

In physics we distinguish between the words **precision** and **accuracy**. To see the distinction, consider a rifle target as in Figure 1.13. Let the bull's-eye in the center of the target represent the actual or true value of the quantity to be measured, the value it really has if we could only measure it. The actual value may in fact be unknown. Individual mea-

surements of the quantity, under identical conditions, are represented schematically by individual shots on the target value. If the shots are closely clustered on the target but not near the bull's-eye (as in Figure 1.13), these results are described by saying the measurements are *precise*, because repetitive measurements yield substantially the same result. They are *not accurate*, however, since the value is not the bull's-eye—the actual value. In this case, there is a **systematic error** in the measurements. It is caused, perhaps, by a misaligned rifle sight, if one is talking about actual rifles and targets. In other cases systematic errors are caused by equipment that is not properly calibrated.

On the other hand, if the individual measurements (shots) are randomly scattered around the target (as in Figure 1.14), the measurements are *not precise* because they are scattered all about the target value (the bull's-eye). But a suitable average of these measurements yields a result that is *accurate*, since the values are centered on the bull's-eye and their average is close to the actual value of the quantity.

The hope of all experimentalists is shown in Figure 1.15. These results are both precise (closely clustered) and accurate (near the actual or true value). Physicists usually can agree on precision but may disagree about accuracy.

> The precision of a measurement is indicated by the number of digits written. The number of digits is called the number of **significant figures**.

For example, to determine the mass of an apple, a measurement is necessary. Thus, if the scale in Figure 1.16 is used to determine the mass, we estimate the mass as 0.2 kg.† There is one significant figure; the initial zero merely helps to locate the decimal point. The mass can be written in scientific notation as 2×10^{-1} kg, although 0.2 kg conveys the same information (one significant figure) in a more familiar form.

If the scale in Figure 1.17 is used, the mass is determined to be 0.25 kg = 2.5×10^{-1} kg; there are two significant figures. The more sensitive scale in Figure 1.18 might result in determining the mass to three significant figures: 0.250 kg = 2.50×10^{-1} kg.

---

†The scale actually measures the magnitude of the force of the Earth on the apple—its weight—which is proportional to its mass, as we will see later (in Chapters 5, 6, and 7). The dial on the scale is *calibrated* to indicate the mass directly.

---

*Of course, some square roots are exact: $\sqrt{49} = 7$.

FIGURE 1.13
Precise but not accurate results.

FIGURE 1.14
Not precise but accurate results.

FIGURE 1.15
Precise and accurate results.

0.2 kg

0.25kg

FIGURE 1.16 One significant figure.

FIGURE 1.17 Two significant figures.

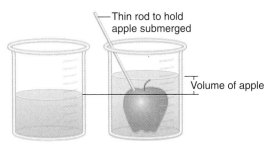

FIGURE 1.19 Measure the volume.

The last zero here *is* significant, since it implies that the last digit is known (it is a zero and not some other number).

Now measure the *volume* of the same apple by dunking it in a graduated cylinder, as in Figure 1.19. Suppose the measurement of its volume yields 356 cm^3 or, converting to cubic meters, 3.56×10^{-4} m^3. Notice that the volume here has three significant figures. To determine the *average density* of the apple, we need to divide its mass by its volume; we will make this calculation in Example 1.9.

> The result of a calculation involving multiplication or division has no more significant digits than the quantity with the *least* number of significant digits.

When data with different numbers of significant figures are added or subtracted, we must be more careful. See Examples 1.10 and 1.11.

> For addition and subtraction, the result must have the same number of significant digits as the term with the smallest number of decimal places in the sum.

In other words, just because a calculator or computer indicates many digits, it does not follow that the result is more precise or has more significant figures than the data entered into the calculator.

> The data, not your calculator display, determine the precision of the result.

> To avoid confusion in this text, we assume that all data quoted have the indicated number of significant figures.

For example, in this text when we write that the mass of an object is 5.00 kg, three significant figures should be understood. If the mass is given as 5.0 kg, then only two significant figures are known.

> Purely numerical factors, such as 2, 4, 10, are written with no decimal point; they can be assumed to be exact.

Although the trigonometric functions such as sine, cosine, and tangent are mathematical in nature, in practice angles are determined from measurements. We will assume that all trigonometric functions have a number of significant figures equal to the number of significant figures in the angle itself.* See Example 1.12.

When making multiplication and division calculations, it is customary to round the answer to the appropriate number of significant figures at the conclusion of the calculation rather than at each step. With addition and subtraction, we must be more careful.

One gauge of the **uncertainty** of a measurement typically is half the place value of the last significant digit quoted. Thus, if you measure the length of your shoe and quote the result as 28 cm, the uncertainty of the measurement is half of one centimeter, the place value of the last significant digit in 28 cm (the 8 is in the ones place). The uncertainty of the measurement thus is ±0.5 cm. Uncertainties are quoted with ± values: 28 cm ±0.5 cm, meaning the actual value of the measurement is somewhere between 27.5 cm and 28.5 cm. If you make a more precise measurement and determine the length of the shoe to be 28.3 cm, the uncertainty of the measurement is half the value of 0.1 cm, or ±0.05 cm, since the last significant figure (3) is in the tenths place.

FIGURE 1.18 Three significant figures.

---

*This is not strictly true because the trigonometric functions are nonlinear functions of the angle. Nonetheless, the assumption is a reasonable first approximation that we will use throughout the text.

PROBLEM-SOLVING TACTICS

1.9 Do not confuse significant figures with decimal places. Use scientific notation as much as possible. The set of different numbers

$$4.56$$
$$0.456 = 4.56 \times 10^{-1}$$
$$0.0456 = 4.56 \times 10^{-2}$$
$$0.00456 = 4.56 \times 10^{-3}$$

all have three significant figures. The zeros are used merely to locate the decimal point and are *not* significant figures. The great advantage of scientific notation is that the number of significant figures is immediately apparent.

1.10 Quote your results using the appropriate number of significant figures. Some calculators can be set to display only a certain number of significant figures. When in this mode, the calculator actually is carrying along in its operations many more digits than are displayed. The display indicates the result automatically rounded to the preset number of significant figures. If you do not have or use this feature of your calculator, you must round any result to the appropriate number of significant figures. All examples and answers to problems in this text are quoted with the number of significant figures appropriate to the data given; you should try to do so as well.

QUESTION 7

Draw a bull's-eye target diagram with shots that are neither precise nor accurate.

QUESTION 8

How would you describe the results of this toddler trying to feed himself in terms of accuracy and precision?

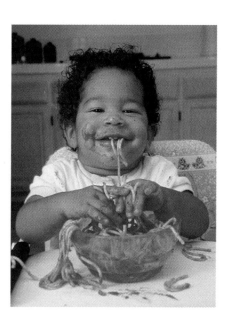

EXAMPLE 1.9

Determine the average density of the apple shown in Figures 1.16–1.18 for the three cases where its mass is known to one, two, and three significant figures and the volume is known (in all cases) to be 3.56×10^{-4} m^3, that is, to three significant figures.

Solution

Remember Problem-Solving Tactic 1.10: *Use the appropriate number of significant figures* to present your results.

1. If the mass is known to one significant figure, then

$$\text{average density} = \frac{\text{mass}}{\text{volume}} = \frac{0.2 \text{ kg}}{3.56 \times 10^{-4} \text{ m}^3}$$

If you run this through your calculator, the result is

$$\text{average density} = 5.617\ 9775 \times 10^2 \text{ kg/m}^3$$

However, since the mass is known to only one significant figure while the volume is known to three, the average density must be quoted with only one significant figure. The other digits displayed on the calculator are meaningless. Hence, to one significant figure, the average density is, after rounding,

$$\text{average density} = 6 \times 10^2 \text{ kg/m}^3$$

One measure of the uncertainty of the result is half the place value of the last significant figure. So here the uncertainty of the result is $\pm 0.5 \times 10^2$ kg/m^3.

2. If the mass is known to two significant figures, the average density is

$$\text{average density} = \frac{0.25 \text{ kg}}{3.56 \times 10^{-4} \text{ m}^3}$$

Your calculator will yield

$$\text{average density} = 7.022\ 4719 \times 10^2 \text{ kg/m}^3$$

Since the mass is known to two significant figures, the average density can be quoted with only two significant figures:

$$\text{average density} = 7.0 \times 10^2 \text{ kg/m}^3$$

The uncertainty of the result is $\pm 0.05 \times 10^2$ kg/m^3, which is half the place value of the last significant figure.

3. If the mass is known to three significant figures, the average density can be quoted with three significant figures, because the volume also has three significant figures:

$$\text{average density} = \frac{0.250 \text{ kg}}{3.56 \times 10^{-4} \text{ m}^3}$$
$$= 7.02 \times 10^2 \text{ kg/m}^3$$

The uncertainty of the result is $\pm 0.005 \times 10^2$ kg/m^3, half the place value of the last significant digit.

EXAMPLE 1.10

Add the distance 2.30×10^{-2} m = 2.30 cm **to** 4×10^{-4} m = 0.04 cm.

Solution

The first distance has *three* significant figures (2, 3, 0). The last zero *is* significant because it implies the digit is known to be a zero. The second distance has *one* significant figure, 4 (the zeros in 0.04 cm merely locate the decimal point). Perform the addition by adding the columns:

$$
\begin{array}{r}
2.30 \text{ cm} \\
+ \ 0.04 \text{ cm} \\
\hline
2.34 \text{ cm} = 2.34 \times 10^{-2} \text{ m}
\end{array}
$$

The result has three significant figures because both numbers were known with the *same precision* (to a tenth of a millimeter).

EXAMPLE 1.11

Add the distances 1.00 m and 6.78×10^{-4} m.

Solution
Each measurement has three significant figures. But in this case the first distance is not known with the same precision as the second one. The addition of the distances is

$$
\begin{array}{r}
1.00 \text{ m} \\
+ \ 0.000678 \text{ m} \\
\hline
1.00 \text{ m}
\end{array}
$$

The result must be quoted as 1.00 m, *not* 1.000678 m, since the latter number (with *seven* significant figures) wrongly implies the distances were known with the same precision.

EXAMPLE 1.12

The angle between two stars is measured to be 49.0°.

a. How many significant figures are there in the measurement?
b. Find the sine, cosine, and tangent of the angle.

Solution
a. The angle has three significant figures. The zero is significant, since the angle was not measured to be either 48.9° or 49.1°.
b. Since the angle has three significant figures, you keep three significant figures in the trigonometric functions of the angle (see Figure 1.20):

$$
\sin 49.0° = \frac{\text{opposite}}{\text{hypotenuse}} = 0.755 \quad (\text{not } 0.7547)
$$

$$
\cos 49.0° = \frac{\text{adjacent}}{\text{hypotenuse}} = 0.656 \quad (\text{not } 0.6561)
$$

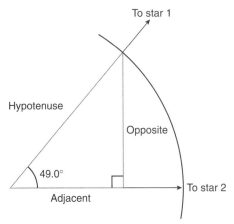

FIGURE 1.20

and

$$
\tan 49.0° = \frac{\text{opposite}}{\text{adjacent}} = 1.15 \quad (\text{not } 1.150)
$$

STRATEGIC EXAMPLE 1.13

A tabletop in a corporate boardroom is measured with a meter stick and found to have a length of 2.36 m and a width of 1.98 m (see Figure 1.21). The thickness is measured with a vernier caliper to be 3.02 cm (see Figure 1.22).

a. What is the area of the tabletop?
b. What is the volume of the tabletop?
c. What is the perimeter of the top?
d. What is the sum of the length and the thickness?

FIGURE 1.21

FIGURE 1.22

Solution
a.
$$
\begin{aligned}
\text{area} &= (\text{length})(\text{width}) \\
&= (2.36 \text{ m})(1.98 \text{ m}) \\
&= 4.67 \text{ m}^2 \quad \textit{not } 4.6728 \text{ m}^2
\end{aligned}
$$

b.
$$
\begin{aligned}
\text{volume} &= (\text{length})(\text{width})(\text{depth}) \\
&= (2.36 \text{ m})(1.98 \text{ m})(0.0302 \text{ m}) \\
&= 0.141 \text{ m}^3 \quad \textit{not } 0.1411 \text{ m}^3
\end{aligned}
$$

c.
$$
\begin{aligned}
\text{perimeter} &= 2(\text{length}) + 2(\text{width}) \\
&= 4.72 \text{ m} + 3.96 \text{ m} \\
&= 8.68 \text{ m}
\end{aligned}
$$

d.
$$
\begin{aligned}
(\text{length}) + (\text{thickness}) &= 2.36 \text{ m} + 0.0302 \text{ m} \\
&= 2.39 \text{ m} \quad \textit{not } 2.3902 \text{ m}
\end{aligned}
$$

CHAPTER SUMMARY

The discipline of physics is divided into two broad areas:

- *classical physics*, which encompasses classical mechanics, electromagnetism, and thermodynamics; and
- *modern physics*, encompassing relativity and quantum mechanics.

Contemporary physics uses both classical and modern physics.

The standard system of units used for measurement in this text is *Le Système International d'Unités*, known as the *SI system*. Four of the fundamental units of this system are shown in Table 1.9.

Other units, such as the centimeter (cm), kilometer (km), gram (g), hour (h), day (d), and year (y), also are employed for convenience; it is important to know how to convert these *units of convenience* into the appropriate SI units.

Mass and *weight* are different physical concepts. Weight depends on location in space; mass does not.

There is a distinction between the word *dimension* (length, mass, time, speed, etc.) and the *units* used to specify the dimension (meter, kilogram, second, meters per second). Each side of an equation in physics must have the same dimensions and the same units. On the other hand, conversions, such as 1 m = 100 cm and 1 d = 86 400 s, have the same dimensions on each side, but not the same units.

The equal sign (=) is used for several purposes in physics:

a. in definitions (instead of the symbol ≡),
b. in conversions of units, and
c. in mathematical expressions representing physical laws.

The symbol (~) means *on the order of*, while the symbol (≈) means *approximately equal to*.

There is a distinction between *precision* and *accuracy* in physical measurements. *Precision* indicates the extent to which repetitive measurements of the same physical parameter, under identical conditions, yield the same value. The number of digits that have meaning in a measurement (*significant figures*) is crucial for correctly conveying the precision to which the quantity is known. *It is important to know and use the appropriate number of significant figures* associated with a measurement and subsequent calculations using such measurements.

Accuracy is the extent to which the measurements yield the true or actual value of the physical quantity; in many instances the actual value is unknown.

The ability to make *estimates* and *order of magnitude* calculations is important in physics—as well as in virtually any endeavor.

TABLE 1.9 Fundamental Units of the SI System

Physical quantity	SI unit	Definition
Mass	kilogram (kg)	One kilogram is the mass of a single, special cylinder of a platinum–iridium alloy.
Quantity	mole (mol)	One mole of a substance is Avogadro's number N_A of particles of the substance; Avogadro's number is 6.022×10^{23} particles/mol.
Time	second (s)	One second is the time interval for a specific kind of light emitted by the $^{133}_{55}$Cs atom to undergo 9 192 931 770 oscillations. Do not memorize this number; one can look it up if needed.
Length or distance	meter (m)	One meter is the distance through which light travels in exactly 1/299 792 458 seconds. Likewise, do not memorize this number; it can be looked up if needed.

SUMMARY OF PROBLEM-SOLVING TACTICS

1.1 **(page 11)** In solving most problems, be sure to convert units of convenience into standard SI units.
1.2 **(page 12)** Read and understand the problem.
1.3 **(page 12)** Sketch a diagram of the problem features.
1.4 **(page 12)** Assess whether additional information is needed.
1.5 **(page 12)** Assess whether any given information is irrelevant.
1.6 **(page 12)** Realize there may be more than one way to solve the problem.
1.7 **(page 12)** Ask yourself if the answer is reasonable.
1.8 **(page 14)** Check the dimensions or the units.
1.9 **(page 20)** Do not confuse significant figures with decimal places. Use scientific notation as much as possible.
1.10 **(page 20)** Quote your results using the appropriate number of significant figures.

QUESTIONS

1. (page 3); 2. (page 9); 3. (page 10); 4. (page 11);
   5. (page 12); 6. (page 15); 7. (page 20); 8. (page 20)

9. Is the difference between physics and engineering the same as that between science and technology?

10. It has been said that science is always wrong. In what sense is this statement both true and false?

11. In what sense is the time signature in a musical score a time standard? (See Figure Q.11.) In what sense is it not?

12. How is the word meter used in a musical context? What is a metronome?

Time signature

FIGURE Q.11

13. Is physics the only science with basic fundamental units of measurement? If not, are such basic units in other areas based on the SI system of units or other basic units? When

answering, consider several disciplines such as chemistry, biology, geology, astronomy, economics, psychology, mathematics, and various areas of engineering.

14. Discover the meaning of the peculiar English unit of length known as the *cubit*. What is the conversion from cubits to meters?

15. One of your elementary school teachers likely said that you cannot add apples to oranges. What, then, is the meaning of the following:

$$5 \text{ apples} + 3 \text{ oranges} = 8 \text{ fruit}$$

16. Astronomers use a continuous day count called the Julian Day number to calculate the number of days between two calendar dates. Each day has a unique Julian Day number associated with it. For example, 1 January A.D. 2000 has Julian Day number 2 451 545. The Julian Day number tells you how many days have elapsed from 1 January 4713 B.C. (which was Julian Day number 0) to the date in question. So, 2 451 545 days elapsed between 1 January 4713 B.C. and 1 January 2000. No kidding. The Julian Day number for any calendar date is found using tables in *The Astronomical Almanac* (published annually by the U.S. Naval Observatory and the Government Printing Office, Washington, D.C.); look in the index under "Julian Day Numbers." Use the *Astronomical Almanac* to determine the Julian Day number of your birth date and for today. A simple subtraction gives your age in days. From this number, calculate your age in seconds to an appropriate number of significant figures. Just for fun, tell your friends the results!

17. The use of occasional leap seconds means that the Julian Day count (Question 16) does not give the exact time interval between two time instants if there have been one or more intervening leap seconds. Explain why. Suggest an alternative continuous time count (analogous to the Julian Day count) that could give the precise time interval between two widely separated time instants.

18. Time is a physical quantity, since it can be measured. Refer to the quotation by St. Augustine in Section 1.3. If someone were to ask you what time really *is*, how would you respond? Where does time come from and where does it go? What is meant by the expression "at a point in time"?

19. The U.S. dollar is not an SI unit. Nonetheless, express the sticker price of your favorite sports car in kilodollars (kilobucks!). What is the cost per kilogram?

20. What is the approximate value of the national debt of the United States in teradollars as of this year?

21. Does the prefix in the term microwave (as in microwave oven) have the same meaning as the prefix micro- in the SI unit system? If not, what does it mean?

22. Compare and contrast the use of monetary standards in economics to the notion of physical standards for measurement in physics.

23. Just to be ornery, the height of horses is measured in a unit called *hands*; one hand is four inches. (See Figure Q.23.) What is the equivalent of a hand in meters?

24. When people say they have "to get the dimensions of the problem before attacking it," in what sense is the word dimension being used? How does it differ from the meanings of the word in physics?

FIGURE Q.23

25. The kilogram is the only basic unit of the SI system defined in terms of a macroscopic artifact, the standard platinum–iridium cylinder at Sèvres, France. The *atomic mass unit* (u) is defined as 1/12 of the mass of a carbon-12 atom: $^{12}_{6}$C. Show that

$$1 \text{ u} = 1.6605 \times 10^{-27} \text{ kg}$$

Why do you think it is inconvenient to use atomic mass units for the masses of common objects such as tomatoes, students, and sports balls?

26. What factors have made the acceptance and common usage of the SI system of units so difficult in the United States?

27. Do you think it likely that ancient societies such as those in Africa, China, Egypt, Greece, Japan, Persia, and Rome had basic standards for measurement? If so, for what physical concepts?

28. Before wooden meter sticks are imprinted with the centimeter and millimeter markings indicating positions along the stick, it is essential that the wood be cured. Explain the consequences of not properly curing the wood beforehand.

29. Long steel measuring tapes were common in surveying years ago. What kind of systematic errors might be introduced into measurements made with such tapes between summer and winter? Surveying now is done with laser ranging systems.

30. The density of water is about 1.0×10^3 kg/m^3, and mercury has a density of 13.6×10^3 kg/m^3. On the same graph, qualitatively sketch how the mass m of a sample of each is a function of the volume V of the sample. Use m as the ordinate and V as the abscissa. Do you expect a linear relation between m and V or one that is curved? Be sure to indicate which line (curve?) corresponds to mercury and which to water. What is the meaning of the slope of each graph?

31. The prefixes kilo- and mega- in the words kilobyte and megabyte, used in connection with computers, do not have the same meaning as in the SI unit system. What do they mean? Why is there a difference in the meanings?

32. What are the drawbacks associated with the antiquated English system of units that make its use for scientific purposes inconvenient?

33. Suppose an extraterrestrial technical civilization were discovered. How could we communicate with them to compare our basic measurement standards for mass, length, and time with the different standards they likely have?

34. Is it possible to develop a complex social and technical civilization without the use of manipulative organs such as fingers? Can bee and ant colonies be considered to be technical civilizations? Give arguments for and against the proposition.

35. On what basis could it ever be said that the "end of physics" had arrived? (We do not mean the end of this course!)

36. How well can you count seconds? With a partner using a stopwatch or an accurate watch that you cannot see or hear, count out what you think is a sixty second interval and see how close you come.

37. How well can you tell mass by hefting? Search for a rock or dig an amount of dirt or sand that you think approximates a kilogram. Check to see how close you are.

38. Separate two pencils on your desk until they are what you think is a meter apart. Use a meter stick to see how close you are.

39. Approximately how many degrees of longitude correspond to a standard time zone? (See Figure Q.39.) When traveling eastward across the boundary between two time zones, do you set your watch ahead or behind? By how much? (There are locations where the time interval involved in crossing a time zone boundary is *half* the typical time interval. Newfoundland, central Australia, India, and Iran have such time zone boundaries.) Spacecraft circle the Earth in about 90 minutes, traveling eastward. Imagine yourself to be an astronaut in such a spacecraft, dutifully setting your watch accordingly as you traverse the time zone boundaries around the globe. After several orbits of the Earth, have you traveled into the past or future as the watch indicates? Explain. What demarcation is used on the globe so that such apparent time travel is not possible?

41. Suppose you and a friend are backpacking in the wilderness. Each of you has a watch, but the times indicated by the watches disagree. Using the two watches, is it possible to tell which watch indicates the correct time? If so, how? If not, why not? If a third person is along and also has a watch, does this change your conclusion? Explain.

42. If the standard kilogram at Sèvres, France, were destroyed by accident, design, warfare, or catastrophe, would this affect the SI system? If so, how? If not, why not?

43. If your watch does not keep accurate time, what units are typically used to express the rate at which it loses or gains time?

44. The arts are considered creative endeavors. Discuss how physics, engineering, and medicine also are creative enterprises. What about mathematics? Is the type of creativity in the arts, sciences, and mathematics the same? In what ways are they the same or different?

45. Time units such as the hours of the day, the days of the week, and the days of the month are counted using *modulo arithmetic*. For example, the hours of the day are counted from 1 to 12 (or 24) and then begin all over again; we say the hours are counted modulo 12 (or modulo 24). The days of the week are counted 1, 2, 3, 4, 5, 6, and 7, and then begin back at 1 again for the next week, and so are counted modulo 7. Cite three or four other common instances of modulo counting not associated with time units.

46. If one takes appropriate account of the number of significant digits at each step of a calculation, will the result differ from that obtained if the number of significant digits is accounted for only at the end of a calculation? Which procedure should be used?

47. When measuring the position of the edge of a book, a student notes what is seen in Figure Q.47. Is the position of the edge correctly given as 0.265 m, 0.266 m, or 0.2655 m? What is the basis of your judgment?

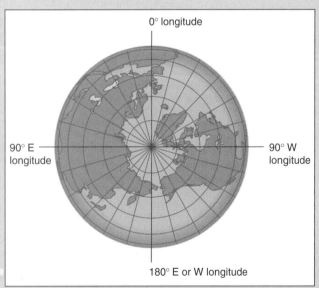

FIGURE Q.39

40. The determination of longitude used to be accomplished through astronomical observation of the position of the Sun in the local sky along with a clock set to keep Greenwich time (England). The meridian through Greenwich is 0° longitude. Describe how longitude can be determined approximately using such observations.

FIGURE Q.47

48. The typical shots hit by several golfers are described in the following way:
    - Golfer A hits shots that are precise, but not accurate.
    - Golfer B hits shots that are accurate, but not precise.
    - Golfer C hits shots that are neither precise nor accurate.
    - Golfer D hits shots that are both precise and accurate.

    Describe the characteristics of the shots made by each golfer.

PROBLEMS

Section 1.3 Standards for Measurement

1. What is the volume of the standard kilogram at Sèvres, France?

2. Determine the density of the platinum and iridium alloy used to fabricate the standard kilogram.

3. (a) With a suitable balance, determine the mass of a penny in kilograms. (b) What is the approximate monetary value of a kilogram of pennies? (c) Using the same scale, is it more precise to determine the mass of a single penny by using 1 or 100 pennies?

4. What time interval does it take light to travel 1.00 m in a vacuum?

5. One can have some fun with some of the SI prefixes. What are the following: 10^{-12} boos; 10^{-9} goats; 10^{-3} tary; 10^{-3} onaires; 10^6 phones; 10^{-6} phones; 10^6 bucks; 10^{-2} pedes; 10^{-6} film; 10^{12} pin; 10^6 lopolis; 10^1 cards; 10^{-6} fish; 10^{-6} waves; 10^{-1} mate; 10^{-6} economics; 2×10^3 mockingbird; 10^{12} bell; 10^{12} fermi; 10^6 lomaniacs; 10^{-3} nery; 10^{-6} scopes; 10^6 bytes

Many of these came from Robert L. Weber, *A Random Walk in Science* (Crane, Russak and Company, New York, 1973), page 61.

6. The hydrogen atom consists of a single proton in the nucleus and an electron. What percentage of the mass of a hydrogen atom is the electron? (Relevant data can be found on the inside covers of the text.)

7. Consider the Earth to be a sphere of radius 6.37×10^3 km. (a) What is the circumference of the Earth in kilometers? (b) What is the distance in meters between the pole and the equator measured along the surface of the Earth? (c) What is the straight-line distance through the Earth between a point on the equator and either pole?

8. Use the information on the inside covers to determine what percentage of the mass of the entire solar system resides in planets.

9. Compute approximate values for the following ratios to see which, if any, are of the same order of magnitude. Table 1.3 on page 8 has appropriate data.

(a) $\dfrac{\text{mass of a proton}}{\text{mass of an electron}}$

(b) $\dfrac{\text{mass of the Sun}}{\text{mass of the Earth}}$

(c) $\dfrac{\text{mass of the Milky Way Galaxy}}{\text{mass of the Sun}}$

(d) $\dfrac{\text{mass of the Universe}}{\text{mass of the Milky Way Galaxy}}$

10. Compute approximate values for the following ratios to see which, if any, are of the same order of magnitude. Table 1.5 on page 10 has appropriate data.

(a) $\dfrac{\text{diameter of an atom}}{\text{diameter of the nucleus}}$

(b) $\dfrac{\text{diameter of the planetary solar system}}{\text{diameter of the Earth}}$

(c) $\dfrac{\text{diameter of the planetary solar system}}{\text{diameter of the Sun}}$

(d) $\dfrac{\text{diameter of the Milky Way Galaxy}}{\text{diameter of the solar system}}$

(e) $\dfrac{\text{diameter of the observable universe}}{\text{diameter of the Milky Way Galaxy}}$

11. Look at Table 1.3 on page 8. (a) Find the length of the edge of a cube of water whose mass is about a factor of 10^{40} smaller than the mass of the universe. The density of water is $1.0 \times 10^3 \text{ kg/m}^3$. (b) Find the length of the edge of a cube of water whose mass is about a factor of 10^{40} greater than the mass of an electron.

12. Look at Table 1.4 on page 10. What time interval in seconds is on the order of a factor of 10^{20} smaller than the age of the universe and on the order of 10^{20} greater than the lifetime of the most unstable particle?

13. Look at Table 1.5 on page 10. (a) What length is about a factor of 10^{20} smaller than the distance to the most distant objects in the universe? (b) What length is a factor of 10^{20} larger than the diameter of a typical nucleus?

‡14. Show that for a right-circular cylinder of fixed volume, its surface area is a minimum when the height of the cylinder is equal to the diameter of the circular cross section. This is the reason the standard kilogram cylinder was constructed with a diameter equal to its height: this minimizes the exposure of the material to the ambient atmosphere.

Sections 1.4 Units of Convenience and Unit Conversions
1.5 The Meaning of the Word Dimension
1.6 The Various Meanings of the Equal Sign

15. The cubical dice common to many games are approximately 1.0 cm on an edge. (See Figure P.15.) If the dice are stacked into a cubical array, what is the length of the edges of the cubical array that contains Avogadro's number of dice?

FIGURE P.15

16. Use a flexible tape measure to determine the circumference and diameter of the circular cross section of a soda can. What is the ratio between the circumference of the circle and its diameter? Does this ratio change if the size of the circle changes?

17. A peanut farmer has a rectangular field of length ℓ and width w and is planning next year's crop. If the length and width of the field are both tripled, (a) by what factor does the amount of fencing needed to enclose the field increase? (b) by what factor does the area under cultivation increase? (See Figure P.17.)

18. A truck tire has a radius three times as big as that of a tire for a compact car. (See Figure P.18.) Each tire rolls through one revolution. By what factor is the distance traveled by the truck tire greater than that traveled by the car tire?

19. A toy manufacturer is planning to make painted, cubical wooden blocks with sides of length ℓ and 3ℓ for a Montessori

FIGURE P.17

FIGURE P.18

preschool program. (See Figure P.19.) (a) What is the ratio of the amount of paint needed for a larger block to that required for a smaller block? (b) What is the ratio of the volume of wood needed for a larger block to that required for a smaller block?

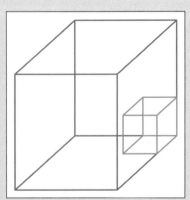

FIGURE P.19

20. Measure the circumference and the mass of a baseball and an inflated basketball. (a) What is the ratio of their radii? (b) What is the ratio of their surface areas? (c) What is the ratio of their volumes? (d) What is the ratio of their masses?

21. While inflating a spherical balloon, you note that the surface area triples with only two "blows." What lungs! (a) By what factor did the radius of the balloon increase? (b) By what factor did its volume increase?

22. Cepheid variable stars alternately increase and decrease their volume over time intervals ranging from days to weeks. If the volume of such a spherical star increases by 50% while in its expanding phase, (a) by what factor does its radius increase? (b) by what factor does its surface area increase?

23. As a community project, your dormitory is designing a square playground with a perimeter of 25.0 meters. The kids want it bigger, and so the perimeter of the square is increased to 36.0 meters. (See Figure P.23.) (a) What is the ratio of the larger perimeter to the smaller perimeter? (b) What is the ratio of the side of the larger square to the side of the smaller square? (c) What is the ratio of the area of the larger square to the area of the smaller square?

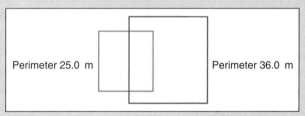

Perimeter 25.0 m   Perimeter 36.0 m

FIGURE P.23

24. (a) What is the ratio of the surface area of the Earth to the surface area of the Moon? Assume both are spherical. (See Figure P.24.) You will find appropriate data on the inside covers of the text. Does your result depend on the units used to measure the area? (b) Calculate the ratio of the volume of the Earth to the volume of the Moon.

FIGURE P.24

25. The speedometer on your slick new powder blue roadster indicates speed in miles per hour. (See Figure P.25.) What conversion factor allows you to change the speed of the car in miles per hour to meters per second (m/s)?

26. The posted speed limit on most rural interstate highways in the United States is 65 miles per hour. (See Figure P.26.) Convert this to kilometers per hour.

27. How many minutes are in 1.00 microcentury? Express the duration of your physics lecture in microcenturies. [This

FIGURE P.25

FIGURE P.28

FIGURE P.30

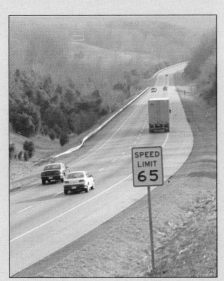

FIGURE P.26

comparison was first made by the Nobel prize–winning physicist Enrico Fermi (1901–1954).]

28. One league is three miles. Jules Verne wrote a book titled *Twenty Thousand Leagues Under the Sea*. How many kilometers is 20 000 leagues? (See Figure P.28.)

29. How many water molecules are there in 1.00 cm³ of water? The density of water is 1.0×10^3 kg/m³, and its molar mass is 18 g.

30. The Himalayas are the highest mountains on the Earth and rise up 8.85×10^3 m above sea level. The deepest ocean trench is the Mariana Trench, the bottom of which lies 1.15×10^4 m below sea level. (See Figure P.30.) The surface roughness of the Earth is measured by the sum of these distances. Express the surface roughness of the Earth as a percentage of the average radius of the Earth (6.37×10^6 m).

The smallness of this percentage means that the Earth is globally quite smooth.

31. How long does it take light to travel from the Sun to the Earth? Express your result in minutes.

32. Determine the number of seconds in one year. Is the result approximate or exact?

33. A *cord* is the standard unit of volume for stacked firewood in the United States. A cord of wood is a pile 4.00 feet high, 4.00 feet wide, and 8.00 feet long. (See Figure P.33.) Determine the volume of a cord in cubic meters.

34. Determine the number of meters in 1.00 mile.

35. If you moved one meter each day of your life, how far have you traveled during your lifetime to date?

36. After checking prices in the local supermarket, determine the cost per kilogram of filet mignon.

•**37.** A gold brick has a mass of 15.0 kg. (See Figure P.37.) The density of gold is 19.3×10^3 kg/m³. (a) What is the volume of the gold brick? (b) If an additional 3.0 kg of gold is added to the brick, what is the *increase* in the volume of the brick?

•**38.** At the end of their "life," stars collapse under gravitational forces and, under certain conditions, become spherical neutron stars or white dwarf stars. (a) A neutron star has a mass typically twice that of the Sun but a radius of only 10 km. Calculate the average density (in SI units) of such a neutron star. (b) A white dwarf star typically has a mass on the order of the mass of the Sun but with a radius on the order of the

FIGURE P.33

FIGURE P.37

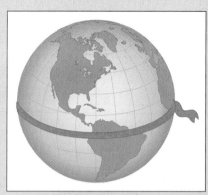

FIGURE P.39

•41. The population of the world is about 6.0 billion. Assume an average person can be modeled as a rectangular parallelepiped with edge sizes 0.200 m, 0.300 m, and 1.80 m. (See Figure P.41.) (a) What is the volume of the entire human population? (b) What are the dimensions of a cube that could accommodate the entire world population? (c) If everyone stood upright and elbow to elbow, what surface area would the entire human population cover? (d) What is the length of the side of a square whose area is equal to that calculated in part (c)?

FIGURE P.41

radius of the Earth. Calculate the average density (in SI units) of a white dwarf with a mass equal to that of the Sun and a radius of 7.0×10^3 km.

•39. Consider the Earth to be a sphere of radius 6370 km. To mark the equator, a long red ribbon is wrapped around the Earth. (See Figure P.39.) Suppose an additional 1.000 km of ribbon is added to the length of the ribbon and the longer ribbon also is shaped into a circle centered on the Earth. What is the height of the new ribbon above the surface of the Earth?
The idea for this problem came from Arnold Arons, *A Guide to Introductory Physics Teaching* (John Wiley, New York, 1990), page 19; the problem originates with Kenneth Clark.

•40. The density of gold is 19.3×10^3 kg/m³, and a mole of gold has a mass of 0.197 kg. (a) Find the volume of a mole of gold. (b) If we somewhat unrealistically model the gold atoms as small cubes, what is the center-to-center distance between the atomic gold cubes?

•42. Assume the radius of the solar system is 50 times the distance between the Sun and the Earth. What is the radius of the solar system expressed in light-hours? A light-hour is the distance light travels during one hour.

•43. The distance between the Earth and Sun is called an astronomical unit (AU). Draw a line to represent the Earth–Sun distance, as in Figure P.43. Imagine a point P along the perpendicular bisector of this line. (a) At what distance along the perpendicular bisector is P if the Earth–Sun distance subtends an angle of 1.00 arc second measured at P? Express your result

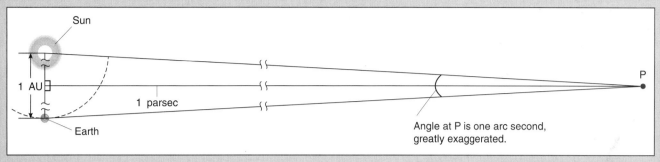

FIGURE P.43

in AU. There are exactly 3600 arc seconds in one degree. (b) In astronomy, the distance calculated in part (a) is defined to be one *parsec* (pc). How many astronomical units in one parsec? How many meters are in 1.00 pc? (c) A light-year (LY) is the distance light travels during one year. How many meters are in 1.00 LY? (d) How many light-years are in 1.00 pc?

•**44.** The brightest star in the night sky is the star α Canis Majoris, also known as Sirius (the Dog Star), and it lies 8.8 light-years away. If we represent the distance between the Earth and the Sun (1 AU) by a line 25 cm long, how far away is Sirius on this scale? (The results of the previous problems may be useful.)

•**45.** Calculate the speed of light in furlongs per fortnight. One furlong is exactly 1/8 of a mile, one mile is exactly 5280 feet, and a fortnight is exactly 2 weeks.

•**46.** A nautical mile is one arc minute of latitude. Given the radius of the Earth, determine the approximate number of meters in a nautical mile. (60 arc minutes = 1°.)

•**47.** A hectare is a metric unit of area equal to exactly 10^4 m². (a) What is the length of the side of a square field with an area of 1.0000 hectare? (b) An acre is an English unit of area equal to exactly 43 560 square feet. What is the length in feet of the side of a square field of area 1.0000 acre? (c) How many acres in 1.00 hectare? (d) How many hectares in 1.00 km²?

•**48.** (a) Use the data shown on the inside covers about the planets of the solar system to calculate the average densities of the planets. (b) The planets typically are grouped into two families: the large *Jovian planets* (Jupiter, Saturn, Uranus, and Neptune) and the *terrestrial planets* (Mercury, Venus, Earth, Mars, and Pluto). What conclusion can be drawn about the average densities of the Jovian planets versus those of the terrestrial planets? (c) How do the average densities of the Jovian planets compare with the average density of the Sun? What might this imply about the nature of the material comprising the Sun and that comprising the Jovian planets?

•**49.** Visit a local bank with a micrometer and a scale; as a courtesy, it may be advisable to phone ahead. With the permission of the bank manager, measure the thickness and the mass of a stack of new, crisp one-dollar bills. Measure the length and width of a single bill. (a) What is the mass of a single one-dollar bill? (b) What is the thickness of a single one-dollar bill? (c) What is the volume of a single one-dollar bill? (d) What is the mass of a stack of 1.00×10^9 one-dollar bills? (e) What is the height of a stack of 1.00×10^9 one-dollar bills? (f) What is the total area that could be covered by 1.00×10^9 one-dollar bills? (g) If 1.00×10^9 one-dollar bills were stored in and filled a cubical box, what is the length of an edge of the box?

•**50.** The national debt is on the order of $\$5.0 \times 10^{12}$. Calculate the height of a stack of one-dollar bills with a monetary value equal to the national debt. Compare your answer with the average distance between the Earth and Moon.

•**51.** You may have constructed a model of the atom when in elementary school. Suppose you model the nucleus with a marble that has a diameter of 1.0 cm. Use the data in Table 1.5 to determine the approximate diameter of the atom if your model is constructed to scale. What does this imply about most models of the atom that you see?

•**52.** You are commissioned by the local science museum to make a scale model of the solar system using a ball of diameter 1.00 m to represent the Sun. (a) Use the data on the inside covers to determine the sizes of the balls that are needed to represent each of the nine planets. (b) If the solar system is constructed to scale, determine the distances of the balls representing each planet from the ball representing the Sun. (c) In this scale model, what is the approximate distance to the star closest to the Sun in space, α Centauri, located 4.2 light-years from the Sun? Express your result in km.

•**53.** The average density of the universe is about 5×10^{-31} g/cm³. (a) Convert this density to SI units of kg/m³. (b) If a 60 kg physics professor is atomized and dispersed uniformly into a spherical volume large enough so the average density of the sphere is equal to the average density of the universe, what is the radius of such a sphere?

•**54.** Leap years occur every four years; coincidentally, leap years also are presidential election years in the United States. The first of January can occur on any day of the week. (a) If you collect calendars, how many different calendars exist? (b) If you collect calendars in *chronological order*, how many years elapse before the collection can be used again in the same order? This time period is called the *solar cycle* of the calendar.

•**55.** If the price of gold is $400 for one troy ounce, what is the price per kilogram? One troy ounce has a mass of 3.11×10^{-2} kg.

Section 1.7 Estimation and Order of Magnitude

56. Estimate the order of magnitude of the mass in kilograms of (a) an adult elephant; (b) a large salmon; (c) a grape; (d) a grain of rice.

57. Estimate the time it takes (a) to swim 1 km; (b) to paddle a kayak around the world; (c) to get to sleep—zzzz . . . .

58. Estimate the distance from your location to (a) the nearest ice cream shop; (b) the sea; (c) the north pole.

59. (a) Estimate the number of heartbeats your heart has performed during your life to date. (b) Estimate the number of breaths you have taken during your life to date.

60. What is the current calendar year? Estimate the number of seconds that have elapsed since the year A.D. 1. Ironically, there is no year 0 in the calendar. The year 1 B.C. is followed immediately by the year A.D. 1. The current dating scheme originated in what we now call the year A.D. 532 by Dionysius Exiguus, although the method was not widely used until the Venerable Bede used it in his *Ecclesiastical History of the English People* several centuries later. Prior to Dionysius and Bede, Western dating was done on the Roman calendar, dating from the approximate founding of Rome by Romulus. The mathematical concept of zero is thought to have originated in India during the 4th century and only reached western Europe via the Arabs during the 12th century. Hence, when Dionysius Exiguus introduced the B.C. and A.D. dating scheme, zero was a number unknown to him. Our bookkeeping system of debits and credits also predated the ideas of zero and negative numbers.

61. Given the mass of the Sun (1.99×10^{30} kg) and the mass of a proton (1.67×10^{-27} kg), express the mass of the Sun in proton masses.

62. Estimate the number of bricks needed to build a wall 100 m in length and 1.0 m in height with bricks on both sides of the wall. (See Figure P.62.)

•63. (a) The Julian calendar (introduced by Julius Caesar in 46 B.C.) has three normal years, each with 365 d, followed by a leap year with 366 d. What is the average number of days per year on this calendar? (b) In 1582, Pope Gregory XIII introduced the Gregorian calendar so that the average duration of the calendar year would be closer to the actual duration of a year, 365.2422 d. He modified the Julian leap year rule in the following way. Century years not divisible by 400 (such as 1700, 1800, and 1900), which normally would be leap years in the Julian scheme since they are integrally divisible by 4 with no remainder, have the leap day omitted and are normal 365 d years. Century years that are divisible by 400 (such as 1600 and 2000) retain the leap year extra day. What is the average duration of the year with the Gregorian calendar reform?

•64. Enrico Fermi, one of the theoretical and experimental geniuses of 20th-century physics, was noted for his ability to make order of magnitude calculations. One of his more famous estimation problems concerned piano tuners. Estimate the number of piano tuners in the major metropolitan area nearest to you. What information source could you consult to see if your answer is about the right order of magnitude?

•65. You are employed as an estimator for a high-speed rail line to be constructed between New York and San Francisco. Such railroads use concrete ties rather than the wood ties of the early days of railroading. (a) Estimate the number of railroad ties needed for a parallel set of dual tracks, one for eastbound traffic, one for westbound traffic. (b) Estimate the volume of concrete needed to construct the ties.

•66. Estimate the number of soft drink containers (bottles and cans) used by the population of the United States during one year.

•67. A new tire of diameter 63.0 cm has a tread depth of 1.00 cm. (See Figure P.67.) The tire is designed to last 80×10^3 km before replacement, when the tread depth will be only 0.20 cm. (a) Approximately how much does the radius of the tire

FIGURE P.62

FIGURE P.67

decrease with each revolution of the tire? (b) How many molecules are worn off the radius of the tire in each revolution of the tire? Assume the average diameter of the molecules making up the tire is about 10^{-10} m.

• **68.** Make and state some reasonable assumptions to estimate the volume of Mt. Everest, the summit of which is at an elevation of 8.85 km above sea level.

• **69.** Make and state some reasonable assumptions and estimate the volume of water in the oceans in cubic kilometers.

• **70.** Estimate the daily vehicular traffic count past the main intersection of your town or city each day.

• **71.** Use a popsicle stick or pencil, held in front of your eye at an appropriate distance, to barely but completely cover the Moon in the sky, as in Figure P.71. Make appropriate measurements and use your knowledge of geometry to measure the angular size of the Moon in degrees. For obvious reasons, *do not do this for the Sun*.

FIGURE P.71

• **72.** Estimate the mass in kilograms of all the people on a fully loaded Boeing 747 jumbo jet.

• **73.** McDonalds claims to have served 100 billion burgers. Estimate the percentage of the population of the United States that visits McDonalds for a hamburger each day. Be sure to specify your assumptions.

• **74.** While lost in a forest, you decide to while away the time while waiting for a rescue team to arrive. You notice that when you look horizontally through the forest, a tree trunk appears in every direction; some trees are nearby, of course, some farther away. The maximum distance that can be seen through the forest is called the *lookout limit*. (See Figure P.74.) Let $\langle A \rangle$ be the average land area occupied by a single tree including its canopy in the forest, and let $\langle w \rangle$ be the average width of a single tree trunk at eye-level height. (a) Show that an estimate of the lookout limit is

$$\frac{\langle A \rangle}{\langle w \rangle}$$

(b) Given the expression for the lookout limit in part (a), show that the number of trees that are within the lookout limit is

$$\frac{\pi \langle A \rangle}{\langle w \rangle^2}$$

(c) Evaluate the lookout limit and the total number of trees that can be seen if $\langle A \rangle = 100$ m^2 and $\langle w \rangle = 25$ cm. This problem is related to the cosmological problem known as the dark night sky paradox. The problem also is called *Olbers's paradox*, after Heinrich Olbers (1758–1840), although Johannes Kepler (1571–1630) may have been the first to state the problem in 1610 in his book *Conversation with the Starry Messenger*. The paradox goes like this: stars, of course, have a finite size; if there are enough stars, no matter what direction you look into space your line of sight will run into a star; thus the night sky should have stars in all directions and be as brilliant as the surface of the sun. So why is it dark at night? The resolution of the problem involves calculating the corresponding lookout limit for the universe: it turns out to be about 10^{23} light years. Since the age of the universe is estimated to be about 10 or 20 billion years, we can at best actually look out only 10 or 20 billion light-years. Then, too, many stars do not live for 20 billion years, so that all the lights are not on at the same time. Hence the night sky is dark.

This problem was inspired by Edward R. Harrison, *Cosmology: The Science of the Universe* (Cambridge University Press, Cambridge, England, 1981), Chapter 12, pages 249–265.

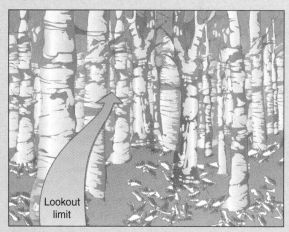

FIGURE P.74

Section 1.8 The Distinction Between Precision and Accuracy

75. The Census Bureau reports that there are 289 689 526 people in the United States as of some date. Then they say the count may be as much as 5 percent in error. Express the population count in scientific notation with an appropriate number of significant figures.

76. (a) What is the sum of the lengths 5.6 m, 0.22 m, 16.1 m, and 0.04 m? (b) What is the product of 3.46 m and 5.891 m? (c) What is π to six significant figures? (d) What is 7.83π? (e) How many significant figures are there in the product $\sqrt{2}\,\pi$? (f) What is $8.42 + \pi$?

77. Determine the sine, cosine, and tangent of each of the following angles with the appropriate number of significant figures:

$$78.0°$$
$$78.02°$$
$$78.024°$$

•**78.** Make a single graph with three curves plotting (i) the value of θ (in radians), (ii) sin θ, and (iii) tan θ, for values of θ (in degrees) from zero to thirty degrees along the abscissa. Note that for small angles, all three curves approach each other. (a) Do the curves ever cross each other over this range of angles? The *small angle approximation* states that for small angles θ, the sine of the angle, the tangent of the angle, and the angle itself (in radians) are approximately equal to each other. That is,

$$\theta \approx \sin\theta \approx \tan\theta \qquad \text{(for small } \theta \text{ in radians)}$$

(b) What is the largest angle θ such that these trigonometric functions are equal to each other to two significant figures? Express the angle in radians and degrees. (c) Why is the cosine of θ not part of the small angle approximation? That is, why is

$$\theta \approx \sin\theta \approx \tan\theta \neq \cos\theta \qquad \text{(for small } \theta \text{ in radians)?}$$

•**79.** Does the small angle approximation (see Problem 78) also imply that the cosecant of θ and the cotangent of θ are approximately equal for small angles θ in radians? In other words, is the following also true for small θ?

$$\frac{1}{\theta} \approx \csc\theta \approx \cot\theta \qquad \text{(for small } \theta \text{ in radians)}$$

On the same grid, plot (i) 1/θ, (ii) csc θ, and (iii) cot θ versus θ (as the abscissa) for angles between about 5° (say 0.09 rad) and 30° (0.52 rad) to confirm your conclusion.

•**80.** Measure the circumference of your neck, and express the result with an appropriate precision and uncertainty.

INVESTIGATIVE PROJECTS

A. Expanded Horizons

1. Investigate the historical origins of the metric and SI systems of units.
   Robert A. Nelson, "Foundations of the international system of units (SI)," *The Physics Teacher*, 19, #9, pages 596–613 (December 1981).
   J. L. Heilbron, "The politics of the meter stick," *American Journal of Physics*, 57, #11, pages 988–992 (November 1989).
   D. F. Bartlett, "Natural units: are they for everybody?" *American Journal of Physics*, 42, #2, pages 148–156 (February 1974).

2. *There is something fascinating about science. One gets such wholesale returns of conjecture out of such a trifling investment of fact.*

   Mark Twain (1835–1910)*

   Notice from Table 1.4 that the ratio between the approximate age of the universe divided by the lifetime of the most unstable particles is on the order of 10^{40}. Notice also from Table 1.5 that the ratio between the largest and smallest distances (the distance to the most distant object divided by the size of a typical nucleus) also is on the order of 10^{40}. Peculiar. Another funny coincidence: from Table 1.3 the mass of the universe divided by the mass of a proton is on the order of 10^{80}, or $(10^{40})^2$. We will see later that the ratio of the magnitude of the electrical force of attraction between an electron and a proton to the magnitude of the gravitational force of their attraction also is about 10^{40}. The approximate numerical coincidence of these diverse and apparently unrelated ratios is remarkable—but unrelated to the dreaded IRS 1040 form. Tables 1.3, 1.4, and 1.5 use the SI system to measure the physical quantities, but the ratios are the *same* whether we use fortnights, furlongs, slugs, or stones to measure the times, lengths, or masses. Some physicists and astronomers speculate that the approximate numerical coincidence of these ratios is more than accidental and may be a manifestation or hint of some yet unknown order in the universe. The reason for the coincidence may be the same as the answer to the following question: why is the universe 20 billion years old? The answer is because it took that long to find that out!

   Such speculation may be whimsical, but it should serve to make you wonder about what, if anything, might be implied by the numerical coincidences. This is the stuff that motivates physicists to search for unifying ideas and explanations. Check this out in the following reference.
   Edward R. Harrison, *Cosmology: The Science of the Universe* (Cambridge University Press, Cambridge, England, 1981), Chapter 17, pages 329–345.

3. Investigate the kinds of clocks used to indicate time with the greatest accuracy and precision and why such clocks are needed and useful.
   Wayne M. Itano and Norman F. Ramsey, "Accurate measurement of time," *Scientific American*, 269, #1, pages 56–65 (July 1993); other references are given in this article.

4. The construction of a calendar is not a trivial exercise. We use a calendar known as the Gregorian calendar, named after Pope Gregory XIII, who introduced it in 1582. Investigate the historical origins of the Gregorian calendar and the motivations for its peculiar leap year rules; see Problem 63.
   Gordon Moyer, "The Gregorian calendar," *Scientific American*, 246, #5, pages 144–152, 178 (May 1982).
   Alexander Philip, *The Calendar: Its History, Structure, and Improvement* (Cambridge University Press, Cambridge, England, 1921).
   Ronald Lane Reese and George Y. Chang, "The date and time of the vernal equinox: a graphical representation of the Gregorian calendar," *American Journal of Physics*, 55, #9, pages 848–849 (September 1987).
   W. C. Elmore, "Perpetual calendar," *American Journal of Physics*, 44, #5, pages 482–483 (May 1976).

5. Adult mammals of different species have a wide range of masses. Consult appropriate sources to determine the average mass and average life expectancy of a variety of mammals. By making an appropriate graph, investigate whether there is any correlation between the average mass of an adult mammal and its average life expectancy.

6. Investigate why clocks are needed to determine the longitude coordinate on the surface of the Earth. Research the work of John Harrison in developing the first accurate marine chronometers for determining longitude at sea, for which he won a state prize of £20 000 sterling in the latter part of the 18th century (but only after a lengthy legal battle).

Life on the Mississippi (J. R. Osgood, Boston, 1883), Chapter XVII.

Eric G. Forbes, "Who discovered longitude at sea?" *Sky and Telescope*, 41, #1, pages 4–6 (January 1971).

Dava Sobel, *Longitude: The True Story of a Lone Genius Who Solved the Greatest Scientific Problem of His Time* (Walker, New York, 1995).

7. The gradual increase in the rotational period of the Earth (the time interval for one complete rotation) was discovered by noting discrepancies between ancient records of the locations where solar eclipses actually were observed and the locations calculated by extrapolating the motions of the Sun and Moon backward in time using a constant rotational period. Investigate the magnitude of the slowdown and its effect on eclipses.

   Robert A. Nelson, "Foundations of the international system of units (SI)," *The Physics Teacher*, 19, #9, pages 596–613 (December 1981).

   Peter Borsche and Jurgen Sündermann, *Tidal Friction and the Earth's Rotation* (Springer-Verlag, New York, 1978).

   F. Richard Stephenson, "Historical eclipses," *Scientific American*, 247, #4, pages 170–183, 194 (October 1982).

8. You no doubt are familiar with the phenomenon of leap years, where an extra day is inserted into the month of February every four years because the orbital period of the Earth about the Sun is *approximately* one-fourth of a day more than an integral number of days (≈ 365.25 d). The extra one-fourth day accumulates to one day every four years. The accumulated day is inserted into February, since it is the shortest month. This process of inserting an extra day is called *intercalation*. The gradual slowdown in the rotation of the Earth (see Investigative Project 7) means that occasional *leap seconds* are introduced to keep time intervals based on atomic standards in step with those based on the astronomical spin and orbital motions of the Earth. Investigate more thoroughly why leap seconds are needed.

   Robert A. Nelson, "Foundations of the international system of units (SI)," *The Physics Teacher*, 19, #9, pages 596–613 (December 1981).

9. Compare and contrast the Gregorian calendar with the religious calendars of the Hebrew and Islamic faiths and those of other religions.

   Frank Parise, *A Book of Calendars* (Facts on File, New York, 1982).

10. Astronomers use a peculiar time scale that is a continuous counting of days that ignores all aspects of the calendar. Each day has a unique *Julian Day number* associated with it that indicates how many days have elapsed since 1 January 4713 B.C. (which was Julian Day number 0). For example, 1 January 1998 is Julian Day 2 450 815, indicating that 2 450 815 days elapsed between 1 January 4713 B.C. and 1 January 1998. The Julian day count is based upon a 7980-year cycle known as the *Julian Period*. Investigate the peculiar 7980-year time interval and the origins of the Julian Period and the Julian Day count.

    Ronald Lane Reese, Edwin D. Craun, and Steven M. Everett, "The origin of the Julian Period: an application of congruences and the Chinese Remainder Theorem," *American Journal of Physics*, 49, #7, pages 658–661 (July 1981).

    Ronald Lane Reese, Edwin D. Craun, and Charles W. Mason, "Twelfth-century origins of the Julian Period," *American Journal of Physics*, 51, #1, page 73 (January 1983).

    Ronald Lane Reese, Edwin D. Craun, and Michael Herrin, "New evidence concerning the origin of the Julian Period," *American Journal of Physics*, 59, #11, page 1043 (November 1991).

11. Archaeologists have speculated about whether prehistoric cultures used a standard of length (a megalithic meter) in their construction activities. Based on his studies of ancient stone circles in England, the archaeoastronomer Alexander Thom has proposed they used a megalithic yard. Report on his investigations and the approximate length of a megalithic yard in meters.

    Alexander Thom, "The megalithic unit of length," *Journal of the Royal Statistical Society* A, 125 Part 2, pages 243–251 (1962).

B. Lab and Field Work

12. You likely have marveled at the number of stars visible to the naked eye on a crisp, clear evening. Using a paper towel tube as an instrument, devise and describe a simple procedure to estimate the number of stars visible to the naked eye. (See Figure I.12.) Make the necessary observations on a clear, moonless evening, in a dark location, well away from local sources of light pollution. Report on your findings. You may be surprised by the result.

FIGURE I.12

13. You can estimate the diameter of a gasoline molecule in the following way. First, determine the density ρ of gasoline by appropriate measurements. Use a micropipette to make a droplet of known volume. Determine the mass m of the droplet of gasoline. Place the droplet of gasoline on a smooth puddle of water, and estimate the surface area A of the slick created by the gasoline on the water. (See Figure I.13.) Assume the layer of gasoline is one molecule thick and that the molecules are spherical in shape. (a) Show that an estimate of the radius r of a gasoline molecule is

$$r \approx \frac{3m}{4\rho A}$$

(b) Use the data secured to estimate the diameter of a gasoline molecule.

FIGURE I.13

14. Visit the office of a local surveyor and investigate the techniques and instruments used in the trade for measuring distances. To what precision can a 100 m distance be measured with such instruments?

15. The variety of automobiles available is quite staggering. Determine the sticker price and mass of, say, twenty different automobiles that span a variety of vehicle types and prices. Make two graphs, plotting the price per kilogram versus the mass and the sticker price of the automobiles. What conclusions can be drawn from your data?

16. With the information in Tables 1.3 and 1.5, plot the logarithm (base 10) of the mass, log m, versus the logarithm (base 10) of the size, log d. Approximate any missing masses or sizes, or find them in an appropriate reference. Do the points on the graph appear randomly over the diagram, or is there a pattern or trend? If there is a trend, can you find an empirical equation that roughly relates log m to log d?

C. Communicating Physics

17. Science is not value neutral. Ethics play a role in science as in politics or any field. Scientists have a collegial duty to each other and to society to be honest with the data and findings they report. Nonetheless, occasional instances of the deliberate falsification of data and results arise (fortunately, few) and are not just of recent vintage. What motivates such conduct, and what are the consequences of such unethical actions by scientists?

18. Science is not without controversy. Various issues highlight honest differences of opinion within the scientific community. Investigate one or more of the following issues, presenting an essay debating the issues involved: (a) the trial of Galileo; (b) the "discovery" of cold fusion; (c) the competition between funds for space research, particle accelerators, and those for social programs; (d) the ethics of radiation experiments on humans without informed consent; (e) whether it is moral to engage in weapons research and/or research for antiballistic missile defense; (f) alleged discrimination against women and minorities in research funding and academic advancement; (g) government versus commercial funding of basic research; (h) spectacular technological *faux pas* such as the flawed mirror of the Hubble space telescope (1990) and the loss of the Mars Observer (1993); (i) patent ownership by employees or the laboratories that employ them. Is discovery an individual or collective endeavor? (j) awarding and sharing credit for simultaneous discoveries; (k) should credit for discovery be assessed on the basis of the actual discovery date or upon the publication date of the results? (l) what research (if any) should be kept secret? (m) is scientific fraud a criminal or civil crime? (n) are computer virus hackers criminals? (o) should scientific results be first publicized to the press and then to the scientific community via research publications, or vice versa? (p) what are the responsibilities and obligations of a scientist to the profession and to the wider community? (q) what are the ethics of the planned obsolescence of products? (r) medical malpractice suits are increasingly common. Are such suits in science and engineering far behind? (s) should science be immune from issues of race and gender?

19. How is science as a creative process similar to yet different from the creative arts?

David Hawkins, "The creativity of science," *Science and the Creative Spirit* (University of Toronto Press, Toronto, 1958), pages 127–165.

20. Write an essay exploring the relationship between the sciences and the humanities.
C. P. Snow, *The Two Cultures and the Scientific Revolution* (Cambridge University Press, New York, 1959 and 1993).
Jacob Bronowski, *Science and Human Values* (Harper-Collins, New York, 1990).

21. Richard Feynman (1918–1988) a Nobel prize–winning physicist, once remarked that "science is the belief in the ignorance of experts." Do you agree or disagree with his opinion? Explain your answer in a well-reasoned essay.
Richard Feynman, "What is science?" *The Physics Teacher*, 7, #6, pages 313–320 (September 1969); quotation is on page 320.

22. When making measurements, explain how one can distinguish between the unavoidable uncertainty associated with every measurement and "human error." Is human error synonymous with a blunder?
John R. Taylor, *An Introduction to Error Analysis: The Study of Uncertainties in Physical Measurements* (2nd edition, University Science Books, Sausalito, California, 1997).

23. How has science been used in literature and literature in science?
F. E. L. Priestley, "Those scattered rays convergent : science and imagination in English literature," *Science and the Creative Spirit*, edited by Harcourt Brown (University of Toronto Press, Toronto, 1958), pages 53–88.
Alan Friedman, "Contemporary American physics fiction," *American Journal of Physics*, 47, #5, pages 392–395 (May 1979).
William H. Davenport, "Resource letter TLA-2: Technology, literature, and the arts," *American Journal of Physics*, 43, #1, pages 4–8 (January 1975).

"IT'S SAD... SHE DISCOVERED A NEW FUNDAMENTAL UNIT, BUT CAN'T FIND ANYTHING TO MEASURE USING IT."

A MATHEMATICAL TOOLBOX
AN INTRODUCTION TO VECTOR ANALYSIS

**For the things of this world cannot be made known
without a knowledge of mathematics**

Roger Bacon (c. 1214–1294)*

While it is one of the languages of physics, mathematics is not what the enterprise of physics is about. Nonetheless, mathematics provides essential analytic tools used to describe our observations of nature and the results of experiments. In our study of motion and its causes, in our investigation of electric and magnetic phenomena, and in many other areas of physics, a working knowledge of vectors is necessary. It is one section of the mathematical toolbox that helps us describe a host of physical phenomena quantitatively.

This chapter inventories the vector portion of the mathematical toolbox. You will find this chapter essential for many succeeding chapters. It lays the foundation for almost every aspect of physics that we will examine. Consequently, this chapter warrants careful study. The time invested here will pay many dividends. When you master the various vector tools and techniques, you can use them with confidence, alacrity, and success in succeeding chapters as we apply them to a wide variety of physical situations. Should you already be familiar with these vector methods, proceed to Chapter 3 after a light skim through the toolbox.

2.1 SCALAR AND VECTOR QUANTITIES

What if angry vectors veer
Round your sleeping head, and form.
                            Robert Penn Warren (1905–1989)[†]

For thine arrows stick fast in me
                                                  Psalm 38:2

Our qualitative and quantitative ways of *describing* physical quantities depend on the phenomenon being considered. Certain physical concepts such as time, mass, and temperature can be defined using one numerical quantity; these are called **scalar quantities**. The particular numerical value of a scalar quantity depends on the specific physical concept represented, the standards used for measurement, and perhaps even the instant a measurement of the quantity is made.

> Physical concepts that require only one numerical quantity for their complete specification are known as scalar quantities.[‡]

Your mass may be 70 kilograms using SI units or, equivalently, 11 stone[§] in another unit system. Your mass typically increases with age and may vary slightly with time, depending on how many pizzas you just consumed. On the other hand, the mass of a fundamental particle, such as an electron or proton, is constant, independent of position and time; such a mass is a constant positive scalar quantity. Mass is intrinsically a positive scalar quantity (or zero); negative masses do not exist in nature.[#]

---

* (Chapter Opener) *Opus Majus*, translated by Robert Belle Burke (University of Pennsylvania Press, Philadelphia, 1928), Volume I, page 128.
[†]"Lullaby," lines 45–48, *Encounter*, May 1957, pages 13–14.
[‡]The word comes from the Latin *scalaris*, which means "resembling a staircase or ladder"; this is a metaphor for the real number line.
[§]One stone is a quaint English unit of mass that is approximately 6.4 kg.
[#]In Chapter 5 we shall see what observations lead us to conclude that mass is never negative.

The temperature outside your window in the dead of winter may be, say, −10 degrees celsius or, equivalently, 263 kelvin or 14 degrees fahrenheit. Chilly. The temperature at other times or locations, such as at Victoria Falls in Zimbabwe, may be different. The temperature at any given location is a positive or negative scalar quantity depending on the particular temperature scale used. We say the temperature is a scalar function of position and time.

Scalar physical quantities abound. The mathematics of such scalar quantities involves the familiar arithmetic, algebra, and calculus that you already know.

Many other physical concepts, such as those dealing with the motion of a system (which we discuss in detail in Chapters 3 and 4) and the pushes and pulls (forces) (Chapter 5) acting on it, have another aspect to them in conjunction with their simple numerical magnitude: *direction*. It makes a big difference *which direction* something is moving or pushed: a soccer ball headed toward your goal is quite different from one headed toward your opponent's goal, as shown in Figure 2.1.

The direction and speed of a ball and the direction and vigor of a push are examples of **vector quantities**.[¶]

> Vector quantities require for their complete specification a positive scalar quantity, called the **magnitude** of the vector, and a direction.

Vector quantities are represented visually by straight arrows, as in Figure 2.2. The arrow has a **head** or **tip** (i.e., the arrow point; the words head and tip can be used interchangeably) and a **tail** point (the opposite endpoint).

The length of the arrow is drawn proportional to the *magnitude*, or positive scalar numerical value, of the physical quantity (such as the number of meters per second a ball is moving). To make

---

[¶]The word *vector* comes from the Latin *vector*, meaning "a carrier." The word arises from the original description of the line from the Sun to a planet, a line that is "carried" around the Sun with the revolutionary motion of the planet.

FIGURE 2.1 Some physical quantities have directional attributes. The direction and speed of a ball have significant consequences in sports.

FIGURE 2.2 Visual representation of a vector: a straight arrow.

the drawing accurately, a scale must be introduced to indicate the relationship between, say, 1 cm of arrow length and the amount of the physical quantity represented by the arrow. The *direction* of the arrow representing the vector indicates the direction associated with the physical quantity. So we visualize a vector quantity as an arrow with a positive magnitude (or length) pointing in some direction. On the other hand, scalar physical quantities such as mass and temperature have no directional attributes and no arrow representations.

We must be careful here. All vectors have a magnitude and a direction, but *not* all concepts for which we can associate a magnitude and a direction are vectors. Later, in Section 2.5, we will examine a concept for which you can conceive a magnitude and an associated direction, but nonetheless is not a vector quantity. Hence, you have to be circumspect about which physical concepts constitute vector quantities. Strictly speaking, physical quantities are vectors if they have a magnitude and direction *and* obey the rules for vector operations, such as addition, that we establish in Section 2.4.

For our purposes in this text, *every physical quantity we consider is either a scalar quantity or a vector quantity* within the context that we will study them.* It is *crucial* that you know whether the physical quantity under consideration is a scalar or a vector quantity, because the arithmetic, algebra, and calculus of vectors are *different* from those of scalars, although the mathematical symbols used to represent the various operations look similar. Since the mathematics of vectors differs from that of scalars, you must learn the rules for the arithmetic, algebra, and calculus of vectors; that is the principal objective of this chapter. Fortunately, the rules for vector operations are not complicated.

An arrow can represent your position with respect to some origin or point of reference such as on a map, as in Figure 2.3. The direction to the top of the map is north while that to the right is east.[†] The length of the arrow representing your position is drawn proportional to the positive scalar distance you are from the origin (say, at the Washington Monument). The arrow direction indicates the direction *from* the origin *to* your location on the map. Such a vector is called a **position vector**, since it locates your position with respect to the chosen origin. We shall

---

*Some physical concepts are neither scalars nor vectors; they have more involved mathematical descriptions and are called *tensors*. The word comes from the Latin *tensor,* meaning "to stretch," reflecting the use of tensor quantities in theories of the elasticity of solids. Certain concepts that we will treat as scalars, such as the index of refraction of an isotropic transparent material and the moment of inertia of a symmetric rigid body, actually are tensors for nonisotropic materials and asymmetric rigid bodies. We do not consider such materials and objects in this text. Scalar and vector quantities actually are special classes of tensor quantities.

[†]The convention for drawing maps with north at the top originated in the second century A.D. with the Alexandrian Greek astronomer Ptolemy. Recall from Chapter 1 that Ptolemy was the person who codified the geocentric theory of the solar system.

FIGURE 2.3 A position vector.

use such position vectors extensively to describe motion in Chapters 3 and 4.

A vector also can represent the direction and speed you are headed in your neat new car. We define the **velocity** as a vector whose magnitude is called the **speed** and whose direction is the direction of motion of the car. In this case the arrow length is drawn proportional to the positive scalar measure of how fast you are moving (your speed), say 20 m/s or 72 km/h. In Figure 2.4, the arrow is drawn with its tail at the center of the car; we consider the car to be represented by a point at this location. If the car is moving at 20 m/s east, we represent this information first by choosing a scale, say 1.0 cm of arrow length to represent 10 m/s. Therefore, the velocity vector, 20 m/s east, is represented by an arrow 2.0 cm long directed east. Note that the scale used to draw the physical size of the car can be different from that used to draw the velocity vector.

The velocity vector of a car moving twice as fast, 40 m/s, but going northeast, then is represented by an arrow twice as long, directed northeast, as in Figure 2.5.

Likewise, if you playfully push your roommate, an arrow is used to represent the direction of the push, with the arrow length

Scale 1.0 cm = 10 m/s for drawing this velocity vector

FIGURE 2.4 The velocity vector 20 m/s directed east.

Scale 1.0 cm = 10 m/s
for drawing this velocity vector

FIGURE 2.5 The velocity vector 40 m/s directed northeast.

drawn proportional to the positive scalar magnitude of the push according to some predetermined standard push. Such an arrow represents a **force vector**.

Such visual representation drawings take considerable time and effort. It is convenient, for many purposes, to talk about the arrows representing vectors without actually drawing them to scale. A vector is named and indicated by a letter symbol in boldface with an arrow symbol over it. A position vector is designated as $\vec{r}$, a velocity vector $\vec{v}$, and a general vector $\vec{A}$. The small arrow symbol over the boldface symbolic letter *always* points to the right when symbolically representing a vector; it has no relationship to the actual direction of the vector in space. The phrase "vector $\vec{A}$" is redundant because the arrow over the boldface letter means it is a vector quantity. Nonetheless, we frequently say "vector $\vec{A}$" for emphasis.

The *positive scalar magnitude* of any vector is represented symbolically in two equivalent ways:

1. By the same letter symbol *without the arrow* and in *lightface italic* type. Thus the magnitude of vector $\vec{r}$ is written as r, the magnitude of vector $\vec{v}$ is written as v, and the magnitude of vector $\vec{A}$ is written A.
2. By surrounding the vector (boldface letter with its arrow) with absolute value signs, such as $|\vec{r}|$, $|\vec{v}|$, or $|\vec{A}|$.

Notationally, then, for *any* vector $\vec{A}$,

$$|\vec{A}| \quad \text{or} \quad A \qquad (2.1)$$

are equivalent ways of saying the same thing: the magnitude of the vector symbolized by $\vec{A}$.

When $\vec{A}$ is written, we mean the *vector*, encompassing both its magnitude and direction. When A or $|\vec{A}|$ is written, only the *magnitude* of the vector is meant with no information about its direction. *These are important notational and semantic distinctions to remember.*

Vector (magnitude *and* direction)	Magnitude of a vector (positive scalar quantity)
$\vec{A}$	$|\vec{A}|$ or A

For example, when we say that the velocity of a car is $\vec{v}$, we mean the velocity vector with its associated magnitude and direction. When we write only v, we mean only the *magnitude* of the velocity vector, a positive scalar quantity known as the speed.

PROBLEM-SOLVING TACTICS

2.1 Scalars are not vectors, and vectors are not scalars. Lurking in your mind as you study should be the question, "Is this physical concept a vector quantity (i.e., with a magnitude and direction) or a scalar quantity?" If you keep this question in mind, you will avoid many of the pitfalls traditionally associated with understanding and applying physics.

2.2 Draw visual representations of vectors carefully. Decide on a scale to represent the vectors; draw them carefully to appropriate lengths with their proper orientations.

2.3 The magnitude of any vector always is a positive scalar quantity (or zero). Remember this!

QUESTION 1 _____

How is the word *vector* used in a biological or medical context? (Consult a dictionary, if necessary.) In what way is its usage in biology and medicine similar to its usage in physics and mathematics? In what way(s) is it different?

2.2 MULTIPLICATION OF A VECTOR BY A SCALAR

If you pull on your car in a certain direction (whether it moves is irrelevant), we represent the pull by a vector $\vec{F}$ whose length is drawn proportional to the magnitude of the pull and directed in the direction of the pull, as in Figure 2.6.*

What happens if three of you, all of equal strength, pull on the car in the same direction? Of course, you get a pull three times what you alone can muster. We represent the total effect by an arrow three times as long as the original arrow of your pull alone, pointing in the same direction—that is, by $3\vec{F}$, as in Figure 2.7. This illustrates the need for the **multiplication of a vector by a scalar**.

---

*In Chapter 5, we will show that such pulls obey the rules of vector arithmetic, and so are vectors. For now, we assume such pulls are legitimate vectors for purposes of illustration.

Visual representation   Symbolic representation

$\vec{F}$

FIGURE 2.6 Vector $\vec{F}$ representing your pull on a car.

Visual representation   Symbolic representation

$3\vec{F}$

FIGURE 2.7 A vector that is three times your pull.

> If any vector $\vec{A}$, not only a push, is multiplied by a scalar α, the result is a vector $|\alpha|$ times as long as $\vec{A}$ itself; the new vector has a magnitude $|\alpha|$ times that of the original vector.

If the scalar multiplier is *positive*, the direction of the new vector is *unchanged*, as in Figure 2.8. However, if the scalar multiplier is *negative*, the direction of the vector is *reversed*, as shown in Figure 2.9.

> The **negative of any vector** $\vec{A}$, $(-1)\vec{A} \equiv -\vec{A}$, is a vector of the same magnitude as $\vec{A}$ but pointing in the opposite direction, as indicated in Figure 2.10.

Thus if $\vec{F}$ represents a push of some magnitude in a specific direction, $-\vec{F}$ is a push of the same magnitude but in the opposite direction. In Chapter 5 we shall find such push pairs (force pairs) are important in studying dynamics; they form the basis for Newton's third law of motion.

Two vectors that point in the same direction are called **parallel vectors** (see Figure 2.11). Two vectors that point in opposite directions are called **antiparallel vectors** (see Figure 2.12). Thus $-\vec{A}$ is antiparallel to $\vec{A}$, whatever the direction of $\vec{A}$.

The scalar zero times any vector is the **zero vector**. The zero vector has zero magnitude and no direction. We write the zero vector simply as a traditional zero, with no arrow over it, as the zero vector has no conceivable direction or imaginable visual representation:

$$0\vec{A} \equiv 0$$

We call the zero vector simply zero.

$\alpha > 0$
$\alpha\vec{A}$
(here α also is greater than 1)
$\vec{A}$

FIGURE 2.8 Multiplication of a vector by a *positive* scalar.

$\alpha\vec{A}$
$\alpha < 0$
(here $|\alpha|$ also is greater than 1)
$\vec{A}$

FIGURE 2.9 Multiplication of a vector by a *negative* scalar.

$\vec{A}$
$-\vec{A}$

FIGURE 2.10 Negative of a vector.

FIGURE 2.11
Parallel vectors.

FIGURE 2.12
Antiparallel vectors.

EXAMPLE 2.1

If $\vec{A}$ is the vector indicated in Figure 2.13, what is the vector $-3.0\vec{A}$?

$\vec{A}$

FIGURE 2.13

Solution
The vector $-3.0\vec{A}$ is a vector with a magnitude 3.0 times that of $\vec{A}$ itself (since magnitudes *always* are positive scalars), but $-3.0\vec{A}$ points in the direction opposite to $\vec{A}$ since the scalar multiplier is negative (see Figure 2.14).

$-3.0\vec{A}$

$\vec{A}$

FIGURE 2.14

2.3 PARALLEL TRANSPORT OF VECTORS

For various mathematical purposes, such as addition or multiplication with other vectors,* you can imagine transporting a vector from one place to another, preserving both its direction in space and its magnitude, as shown in Figure 2.15. This procedure is called **parallel transport** of the vector.

The arrow is merely a visual *representation* of the vector physical quantity. A vector quantity has a meaning independent of the way you describe or draw it and is, in a sense, mathematically footloose.

---

*We consider these topics in the next few sections of the chapter.

FIGURE 2.15 Parallel transport of a vector preserves its length and direction.

2.4 VECTOR ADDITION BY GEOMETRIC METHODS: TAIL-TO-TIP METHOD

If you push on a refrigerator in one direction and a stronger buddy simultaneously pushes on it in another direction, what single push is equivalent to the two pushes at the same time (see Figure 2.16)? We schematically represent the pushes as acting on the center of the refrigerator, with the tails of both vectors located at this point, the same way we represent pulls. In this section we learn how to combine any two (or more) vectors of the same physical type.

Vectors with different magnitudes and directions that represent the same physical concept typically (but not always) are symbolized using the same boldface letter (either upper or lower case), adding subscripts or primes to distinguish them. Thus two different force vectors may be represented by $\vec{\mathbf{F}}_1$ and $\vec{\mathbf{F}}_2$, by $\vec{\mathbf{F}}$ and $\vec{\mathbf{f}}$, or by $\vec{\mathbf{F}}$ and $\vec{\mathbf{F}}'$. Two different velocity vectors may be designated by $\vec{\mathbf{v}}_1$ and $\vec{\mathbf{v}}_2$, or by $\vec{\mathbf{v}}$ and $\vec{\mathbf{v}}'$, two different position vectors by $\vec{\mathbf{r}}_1$ and $\vec{\mathbf{r}}_2$, and so on. You get the idea.

The rules for **vector addition** apply not just to pushes and pulls, but also to any vector quantities of the same physical type. Vectors of the same physical type always are represented in this text visually with arrows of the same color. When added together, two or more vectors representing the same physical concept, such as two or more position vectors or two or more forces, form a new vector called the **vector sum**, which is equivalent to all the others combined. We indicate the operation for forming the vector sum with the boldface addition symbol, **+** , to distinguish it from ordinary scalar addition (+).

Vectors (and scalars for that matter) that represent different physical quantities *cannot* be added together. You cannot add apples to oranges—this rule still applies. Thus do not attempt to add a force vector to a velocity vector yielding some strange interspecies combination. If you do attempt to add the wrong types of vectors together, the dimensions (and units) will be inconsistent, a "red flag" indicating that something is amiss. Such red flags are signals to stop and rethink what has been done; they keep us on the straight, narrow, and legitimate path of vector addition. So, for purposes of vector addition, vectors must be properly of the same physical species—that is, representing the same physical concept.

The simplest case of vector addition is to add a vector to itself: $\vec{\mathbf{F}} + \vec{\mathbf{F}}$. For example, this could represent two persons of equal strength pushing on a refrigerator in the same direction, as in Figure 2.17. The combined effect of this is found to be in the same direction as each push and twice the magnitude of either (see Figure 2.18). That is,

$$\vec{\mathbf{F}} + \vec{\mathbf{F}} = 2\vec{\mathbf{F}}$$

where $2\vec{\mathbf{F}}$ is a new vector, equivalent to the two vectors in the sum, with a magnitude twice that of $\vec{\mathbf{F}}$. This special case is equivalent to the multiplication of the vector $\vec{\mathbf{F}}$ by the scalar 2.

But suppose the two pushes are not of equal magnitude, nor in the same direction, say $\vec{\mathbf{F}}_1$ and $\vec{\mathbf{F}}_2$ in Figure 2.19. What is the effect of the pushes now? That is, what is the vector sum $\vec{\mathbf{F}}_1 + \vec{\mathbf{F}}_2$? The vector sum of the two vectors is defined to be a new vector found using the following geometric procedure, known as the **tail-to-tip method**:

1. Move the vector $\vec{\mathbf{F}}_2$ by parallel transport—that is, preserving its magnitude and direction in space—so that its tail is next to the head or tip of $\vec{\mathbf{F}}_1$ (i.e., *tail-to-tip*), as in Figure 2.20. This is an example of moving vectors using parallel transport for mathematical purposes.
2. Draw a new vector, $\vec{\mathbf{F}}_{\text{total}}$, from the *tail* of the first vector $\vec{\mathbf{F}}_1$ to the *tip* of the second vector $\vec{\mathbf{F}}_2$, thus completing the triangle, as shown in Figure 2.21. The rule for drawing the new vector sum is from the tail of the first vector to the tip of the second vector: again, *tail-to-tip*. The new arrow, $\vec{\mathbf{F}}_{\text{total}}$, is the vector that represents the vector sum of $\vec{\mathbf{F}}_1$ and $\vec{\mathbf{F}}_2$. Algebraically, the process of vector addition is written as

$$\vec{\mathbf{F}}_{\text{total}} = \vec{\mathbf{F}}_1 + \vec{\mathbf{F}}_2 \tag{2.2}$$

Note the use of the boldface plus sign (**+**) to represent vector addition to distinguish it from ordinary scalar addition.

FIGURE 2.17 Two equal vectors.

FIGURE 2.18 The sum of the two vectors in Figure 2.17.

FIGURE 2.16 Two pushes of different magnitude and direction.

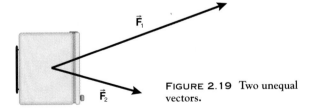

FIGURE 2.19 Two unequal vectors.

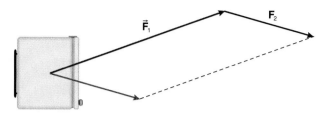

FIGURE 2.20 Parallel transport $\vec{F}_2$ so its tail is next to the head of $\vec{F}_1$.

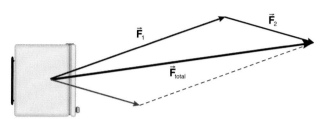

FIGURE 2.21 The vector sum $\vec{F}_{total} = \vec{F}_1 + \vec{F}_2$.

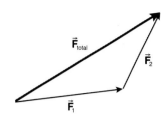

FIGURE 2.22 The magnitude (and direction) of the vector sum depends on the orientations of the vectors to be added.

Vector arrow lengths are drawn proportional to the magnitudes of the vector quantities. Notice in the triangle of Figure 2.21 that the *magnitude* of the vector sum is *not* typically the sum of the magnitudes of the individual vectors included in the sum. That is, in most instances

$$F_{total} \neq F_1 + F_2 \qquad (2.3)$$

where the plus sign here indicates the *scalar addition* of the two scalar magnitudes. The magnitude of the vector sum *is* equal to the sum of the magnitudes *only* when the two vectors to be added are parallel.

The direction of the vector sum also typically is *not* in the direction of either vector in the sum. The magnitude (and direction) of a vector sum depends critically on the relative directions of the two vectors to be added. This is shown explicitly in Figure 2.22 for various orientations of two vectors of fixed magnitudes.

> Vector addition is quite different from the addition of scalar quantities.

The order in which the vectors are added is not important. That is, you obtain the *same* vector whether you take $\vec{F}_1 + \vec{F}_2$ or $\vec{F}_2 + \vec{F}_1$; this is shown in Figure 2.23. In other words, the process of vector addition is commutative:

$$\vec{F}_1 + \vec{F}_2 = \vec{F}_2 + \vec{F}_1 \qquad (2.4)$$

If more than two vectors of the same type are to be added together, say

$$\vec{F}_1 + \vec{F}_2 + \vec{F}_3 + \vec{F}_4$$

they can be added sequentially in pairs in *any* order. Thus, the process of vector addition is associative.

PROBLEM-SOLVING TACTIC

2.5   There is a shortcut for the addition of many vectors. You can add several vectors, such as those in Figure 2.24, more easily by placing the tail of $\vec{F}_2$ next to the tip of $\vec{F}_1$ using parallel transport (see Figure 2.25), the tail of $\vec{F}_3$ next to the tip of $\vec{F}_2$, and the tail of $\vec{F}_4$ next to the tip of $\vec{F}_3$, and so forth (as in Figure 2.25). The vector sum of all the vectors is represented by an arrow from the tail of the first vector $(\vec{F}_1)$ to the tip of the last vector $(\vec{F}_4)$, as shown in Figure 2.26.

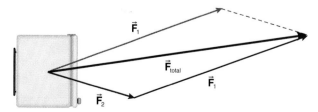

FIGURE 2.23 The order in which you add two vectors does not matter.

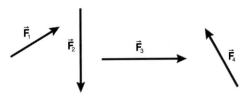

FIGURE 2.24 A collection of vectors to add.

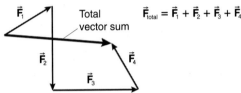

FIGURE 2.25 Parallel transport the vectors tail-to-tip.

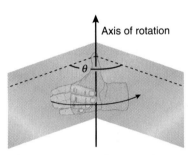

$$\vec{F}_{total} = \vec{F}_1 + \vec{F}_2 + \vec{F}_3 + \vec{F}_4$$

FIGURE 2.26 The vector sum $\vec{F}_{total}$ is drawn from the tail of the first vector to tip of the last.

FIGURE 2.28 A rotation of an object about an axis.

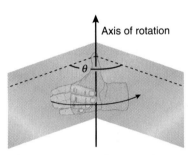

FIGURE 2.29 An arrow representing a rotation.

QUESTION 2

Show how a vector of magnitude 1 can be added to another vector of magnitude 1 to equal a vector of magnitude 1. What geometrical figure do the three vector arrows form? You have learned how one plus one is one (for vectors)!

2.5 DETERMINING WHETHER A QUANTITY IS A VECTOR*

If you love pizzas, you likely think all hot pizzas are tasty; however, that does not mean all tasty things are hot pizzas. Likewise, although all vectors have both magnitude and direction, not all things with a magnitude and direction are vectors.

Simple arrows of any size can indicate direction without implying magnitude, as in Figure 2.27. If an arrow is used to represent a quantity that has a magnitude and an associated direction, the distinction between an arrow that is a vector and one that is not becomes increasingly subtle. Just because some physical concepts *can be* represented by an arrow does *not* ensure that the quantity is a vector.

For example, think of rotating an object about a specified axis, as indicated in Figure 2.28. The angle through which the

FIGURE 2.27 An arrow simply indicating direction is not a vector.

object is rotated is a measure of the magnitude of the rotation: the larger the angle, the larger the magnitude of the rotation. We cannot use the curved arrow in Figure 2.28 to represent the rotation, since vectors are represented visually only by *straight* arrows. A straight arrow will suffice if we direct it along the axis of rotation in the sense of the following *rotational right-hand rule:* Wrap the bunched fingers of your right hand around the axis in the sense the object is rotated about the axis, as shown in Figure 2.29. Then the extended right-hand thumb indicates the direction for the arrow representing the rotation.

We represent the rotation visually by a straight arrow along the axis of rotation whose length is proportional to the angle of rotation and whose direction is determined by the rotational right-hand rule. But even so, such rotation arrows are *not vectors*. Why? We will illustrate using a simple test.

Orient your textbook on a flat table with the binding facing you as in Figure 2.30. The Cartesian coordinate system in Figure 2.30 will assist you in visualizing and performing two rotations in succession:

1. Rotation 1: Rotate the book through an angle of 90° about the y-axis, as shown in Figure 2.31. The back cover of the text now faces you. Call this rotation Rot_1. The arrow representing this rotation is schematically indicated in Figure 2.31; the length of the arrow is proportional to 90°, and the direction of the arrow is toward increasing values of y according to the rotational right-hand rule.

2. Rotation 2: Rotate the book through 90° about the x-axis, as shown in Figure 2.32. The book now lies face up on the table. Call this rotation Rot_2. This rotation is represented by

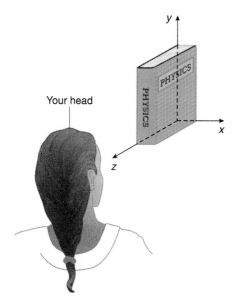

FIGURE 2.30 Your textbook with its binding facing you.

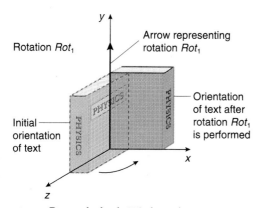

FIGURE 2.31 Rotate the book 90° about the y-axis.

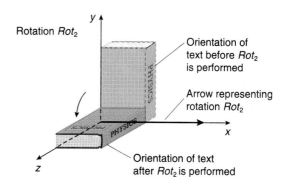

FIGURE 2.32 Rotate the book 90° about the x-axis.

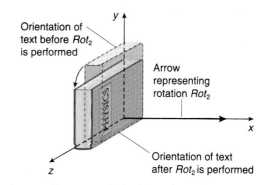

FIGURE 2.33 First rotate the book 90° about the x-axis.

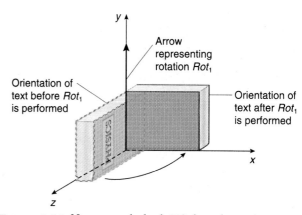

FIGURE 2.34 Next rotate the book 90° about the y-axis.

an arrow the same length as Rot_1 (since the two rotations each were of magnitude 90°) but with the Rot_2 arrow directed toward increasing values of x according to the rotational right-hand rule.

Thus the combination of rotation Rot_1 followed by rotation Rot_2, symbolized by $Rot_1 + Rot_2$, results in the book lying face up on the table as indicated in Figure 2.32. We still must determine whether the + sign here indicates what we mean by the vector addition, +.

We know that adding two *vectors* together in either order does not affect the outcome. Is the same true for these rotations? Let's see. Begin again with the text in the original orientation of Figure 2.30, but *change the order* in which the rotations are performed. That is, do rotation Rot_2 first, followed by Rot_1, symbolized by $Rot_2 + Rot_1$.

After rotation Rot_2 (90° about the x-axis), the text is oriented as shown in Figure 2.33. Then after the next rotation, Rot_1 (90° about the y-axis), the text is oriented as shown in Figure 2.34. The result of this new sequence of the same two rotations orients the textbook on its edge with the front cover

facing away from you. This is *different* from what we obtained with the same rotations performed in the original order!

We could blindly add the rotation arrows using the geometric vector addition rules, but *the result the arrows physically represent is not the same* when we take $Rot_1 + Rot_2$ instead of $Rot_2 + Rot_1$. In other words,

$$Rot_1 + Rot_2 \neq Rot_2 + Rot_1$$

The *order* or sequence of the rotations is critical to the result obtained; the rotations do not commute. The arrows of finite rotation thus do *not* obey the commutative vector addition rule (Equation 2.4). Hence these rotations are *not vector quantities* even though they can be represented visually by an arrow with a magnitude and a direction.

So, in conclusion, just because some physical concept *can* be represented visually by an arrow with a magnitude and a direction does *not* ensure that the physical concept is a vector quantity.

> To be a vector quantity, that quantity must also *obey the rules* we established for, among other things, vector arithmetic.

2.6 VECTOR DIFFERENCE BY GEOMETRIC METHODS

Finding the difference between two vectors is useful in many physical contexts. For example, if you are at the Vietnam Veterans Memorial in Washington, D.C., and are located by a position vector $\vec{r}_1$ with respect to the Washington Monument (as in Figure 2.35) and the White House is located by position vector $\vec{r}_2$ from the same origin, then what is the position vector $\vec{r}_3$ of the White House measured from your location? The three vectors are related by vector addition in the following way:

$$\vec{r}_1 + \vec{r}_3 = \vec{r}_2 \qquad (2.5)$$

FIGURE 2.35 Three position vectors related by a vector sum.

Now transpose $\vec{r}_1$ in Equation 2.5 just as we would if this were ordinary algebra:

$$\vec{r}_3 = \vec{r}_2 - \vec{r}_1 \qquad (2.6)$$

We have to determine what we mean by the boldface minus sign ($-$), symbolizing a **vector difference**. We do this by noting in Figure 2.36 that both the magnitude and direction of vector $\vec{r}_3$ can be found by forming the *vector sum* of $\vec{r}_2$ with $-\vec{r}_1$:

$$\vec{r}_3 = \vec{r}_2 + (-\vec{r}_1) \qquad (2.7)$$

This means that the vector difference is executed geometrically by *adding* the vector $-\vec{r}_1$ to $\vec{r}_2$ using the *vector addition rules* in Section 2.4. That is,

$$\vec{r}_3 = \vec{r}_2 - \vec{r}_1$$
$$\equiv \ \vec{r}_2 + (-\vec{r}_1)$$

> The operation $+ \ (-\vec{r}_1)$ is identical in all respects to the operation implied by $- \ \vec{r}_1$.

We generalize to say that the process of finding the vector difference between *any* two vectors, $\vec{A}_2$ and $\vec{A}_1$ (see Figure 2.37), can be viewed in two steps.

1. To form $\vec{A} \ \equiv \ \vec{A}_2 - \vec{A}_1$, first form the vector $-\vec{A}_1$ by reversing the direction of the arrow of $\vec{A}_1$, preserving its length (magnitude) (see Step 1 in Figure 2.38).
2. Next *add* the vector $-\vec{A}_1$ to the vector $\vec{A}_2$ using the tail-to-tip method of vector addition, as in Step 2 in Figure 2.38.

The magnitude of the vector difference $\vec{A}$ is *not* typically the difference between the magnitudes of the two vectors $\vec{A}_2$ and $\vec{A}_1$. That is, in most instances

$$A \neq A_2 - A_1 \qquad (2.8)$$

Note that Equation 2.8 involves the *magnitudes* of the vectors, and the minus sign is ordinary scalar subtraction. The magnitude of the vector difference is equal to the absolute value of the difference between the individual magnitudes *only* when the two vectors to be subtracted are parallel or antiparallel.

The direction of the vector difference $\vec{A}$ also is *not* typically in the direction of either $\vec{A}_2$ or $\vec{A}_1$. The vector dif-

FIGURE 2.36 Another way to view the three vectors of Figure 2.35.

FIGURE 2.37 Two vectors.

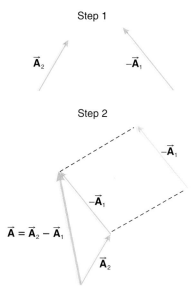

Step 1

Step 2

FIGURE 2.38 Reverse the direction of the vector to be subtracted and then add it to the other vector to form their difference.

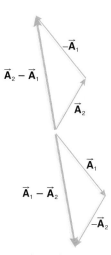

FIGURE 2.39 The vector $\vec{A}_2 - \vec{A}_1$ is the negative of the vector $\vec{A}_1 - \vec{A}_2$.

> **PROBLEM-SOLVING TACTIC**
>
> **2.6** The addition and subtraction of vector quantities is *different* from the addition and subtraction of scalars. This fact is *crucial* to remember!

QUESTION 3

Show how a vector of magnitude 1 can be subtracted from another vector of magnitude 1 to equal a vector of magnitude 1.

2.7 THE SCALAR PRODUCT OF TWO VECTORS

Only vectors of the same physical type can be added to or subtracted from each other. But vectors of *different* types, representing different physical quantities, *can* be multiplied together, as can vectors of the same physical type. There are two distinct ways that two vectors are multiplied together; each way has its own rules, terminology, notation, and usefulness in physics. In this section we examine one of those ways, leaving the other for a subsequent section of the chapter.

We saw in Section 2.2 that a vector multiplied by a scalar α produces a vector with a magnitude $|\alpha|$ times the original vector. Multiplying a vector by a scalar produces another *vector*. Multiplying a vector by a scalar is *different* from what is meant by the **scalar product** in vector analysis. The scalar product is a way to multiply *two vectors* to yield a *scalar* result. The scalar product is useful for describing circular motion (as we will see in Chapter 4) and in many other physical contexts, but perhaps most importantly, in defining the physical concept of the work done by a force (as we shall see in Chapter 8).

The scalar product of any two vectors $\vec{A}$ and $\vec{B}$ is written as $\vec{A} \cdot \vec{B}$, with the product symbolized by the big fat dot •. The scalar product also is commonly referred to as the *dot product* of

---

ference is in the direction of $\vec{A}_2$ only when the two vectors $\vec{A}_2$ and $\vec{A}_1$ are parallel *and* the magnitude of $\vec{A}_2$ is greater than the magnitude of $\vec{A}_1$; if they are parallel and $\vec{A}_1$ has the greater magnitude, then the difference points in the direction of $\vec{A}_1$.

> The vector difference is different from the subtraction of scalar quantities.

Just as in scalar subtraction, the order of the terms in the vector difference affects the result (see Figure 2.39). That is,

$$\vec{A}_2 - \vec{A}_1 \neq \vec{A}_1 - \vec{A}_2$$

Notice from Figure 2.39 that $\vec{A}_2 - \vec{A}_1$ is the negative of $\vec{A}_1 - \vec{A}_2$:

$$\vec{A}_2 - \vec{A}_1 = -(\vec{A}_1 - \vec{A}_2) \qquad (2.9)$$

If we subtract any vector from itself, we obtain the zero vector:

$$\vec{A}_1 - \vec{A}_1 = 0$$

two vectors, indicative of the symbol used to represent it. In this text we shall consistently use the term scalar product for this type of vector multiplication.

> The scalar product of two vectors $\vec{A}$ and $\vec{B}$ is defined as
>
> $$\vec{A} \cdot \vec{B} \equiv AB \cos \theta \qquad (2.10)$$
>
> where A is the magnitude of $\vec{A}$, B is the magnitude of $\vec{B}$, and θ is the angle between the two vectors when they are drawn from a common point (i.e., with their tails together at the common point), as in Figure 2.40.

Parallel transport occasionally must be used (as illustrated in Figure 2.40) so that the two vectors have their tails together.

The angle between two vectors *always* is taken to be the *smaller* angle between the vectors when they are drawn from a common point (angle θ in Figure 2.40, not angle ϕ). With this convention, θ always is less than or equal to 180°.

> The angle between two vectors always is considered a positive quantity and is always less than or equal to 180°.

The magnitudes of the two vectors on the right-hand side of Equation 2.10 are never negative. Since the *smaller* angle between the two vectors always will be somewhere between 0° and 180°, the cosine is either positive (for angles up to 90°) or negative (for angles greater than 90° but less than 180°).

> Thus the scalar product can be either a positive or negative scalar quantity or zero, depending on the angle between the two vectors (see Figure 2.41).

PROBLEM-SOLVING TACTICS

2.7 The result of the scalar product of two vectors is a scalar quantity, not another vector quantity. Scalars and vectors are very different; do not confuse the two.

2.8 To see if two vectors are perpendicular to each other, take their scalar product. If the two vectors are perpendicular to each other, the angle between them is 90°. *Their scalar product is zero, since cos 90° = 0*. This is a wonderful test to see if two (nonzero) vectors are perpendicular.

The scalar product has a distributive property. That is,

$$\vec{A} \cdot (\vec{B}_1 + \vec{B}_2) = \vec{A} \cdot \vec{B}_1 + \vec{A} \cdot \vec{B}_2 \qquad (2.11)$$

This fact is comforting, for we can use our common algebraic knowledge to eliminate the parentheses. Notice, however, that the (+) sign on the *left*-hand side of Equation 2.11 represents the *vector addition* of $\vec{B}_1$ and $\vec{B}_2$, whereas the (+) sign on the *right*-hand side of this equation represents the addition of two *scalars*, $\vec{A} \cdot \vec{B}_1$ and $\vec{A} \cdot \vec{B}_2$.

There is a geometric way to interpret the scalar product. With $\vec{A}$ and $\vec{B}$ drawn with their tails together, if we drop a perpendicular from the tip of $\vec{B}$ to the line containing $\vec{A}$ (see Figure 2.42), the quantity $B \cos \theta$ is defined to be the **projection** or **component** of $\vec{B}$ on the line containing $\vec{A}$ (or on any line parallel to the line containing $\vec{A}$). Figure 2.42 illustrates why the word projection is used in this context. If we imagine sunlight or light from a distant slide projector shining from a direction perpendicular to $\vec{A}$, then the shadow of the vector $\vec{B}$ on the line containing $\vec{A}$ has length equal to the absolute value of the projection (or component) of $\vec{B}$ on the line containing $\vec{A}$.

If the angle θ between the vectors is acute ($\theta < 90°$), the projection is positive; if the angle is obtuse ($\theta > 90°$), the projection is negative since cos θ then is less than zero. The scalar product of the two vectors is the simple scalar multiplication of this projection ($B \cos \theta$) by the magnitude of $\vec{A}$:

$$\vec{A} \cdot \vec{B} = A(B \cos \theta)$$

We could also do the projection the other way around. Drop a perpendicular from the tip of $\vec{A}$ to the line containing $\vec{B}$, as in Figure 2.43. Then the quantity $A \cos \theta$ is the projection or component of $\vec{A}$ on the line containing $\vec{B}$ (or any line parallel to it). The scalar product then is the scalar multiplication of this projection ($A \cos \theta$) by the magnitude of $\vec{B}$:

$$\vec{A} \cdot \vec{B} = (A \cos \theta)B$$

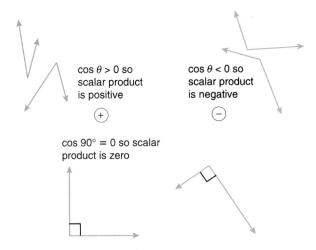

FIGURE 2.41 The scalar product can be positive, negative, or zero depending on the angle between the two vectors. Two vectors that are perpendicular to each other have a scalar product equal to zero.

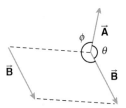

FIGURE 2.40 Parallel transport may be needed to determine the angle θ between two vectors.

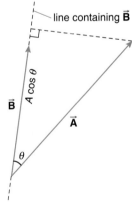

FIGURE 2.42 The projection or component of $\vec{B}$ on the line containing $\vec{A}$ is $B \cos \theta$.

line containing $\vec{B}$

FIGURE 2.43 The projection or component of $\vec{A}$ on the line containing $\vec{B}$ (or any line parallel to it) is $A \cos \theta$.

QUESTION 4

Suppose you have a vector $\vec{A} \neq 0$. If $\vec{A} \cdot \vec{B} = 0$, does this necessarily mean that $\vec{B} = 0$? Explain.

EXAMPLE 2.2

Find the scalar product of the two vectors shown in Figure 2.44.

Solution

Parallel transport one of the vectors until their two tails are at a common point, as in Figure 2.45. The angle between the two vectors is 50.0°. The scalar product of the two vectors is found using Equation 2.10:

$$\vec{A} \cdot \vec{B} = (2.00)(5.00) \cos 50.0°$$
$$= 6.43$$

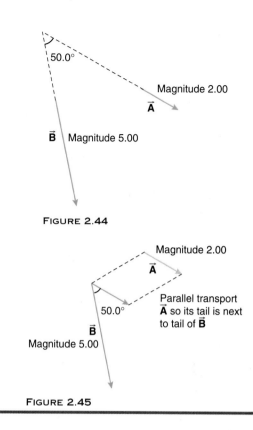

FIGURE 2.44

FIGURE 2.45

2.8 THE CARTESIAN COORDINATE SYSTEM AND THE CARTESIAN UNIT VECTORS

The magnitude of a vector is a positive scalar quantity. The direction of the vector is indicated by the direction of the arrow in its visual representation. One way that you can *specify* the direction of the vector is to choose a coordinate system.

In many situations in physics, we represent the three spatial dimensions in which we live with a **right-handed Cartesian coordinate system**. The three coordinate axes, known as the *x*-, *y*-, and *z*-axes, are perpendicular (i.e., orthogonal) to each other. To draw the three-dimensional coordinate system on a flat surface, at least one axis must be drawn skewed (in perspective) or at an angle, as in Figure 2.46. The axes are labeled so the coordinate values increase as you go from the origin to the *x*, *y*, or *z* label.

Positive values of *x* form what is called the +*x*-axis; negative values of *x* form the −*x*-axis. Similar conventions exist for the *y*- and *z*-axes. The Cartesian coordinate system shown in Figure 2.46 is called *right-handed* for the following reason: if you orient the index finger of your right hand along the direction that *x* increases in value and the middle finger along the direction *y* increases, then your extended thumb will indicate the direction that *z* increases. Be sure to use your right hand for this test; if you try this procedure in Figure 2.46 with your left hand, you will find that your left thumb does *not* point along the direction *z* increases; it will indicate the direction of decreasing values of *z*.

Equivalently, if in Figure 2.47 you bunch the four fingers of your right hand and point them along the direction that *x* increases in value, then turn them toward the direction *y* increases in value, then your extended right thumb will point in the direction that *z* increases in value.

We will always use right-handed coordinate systems.*

PROBLEM-SOLVING TACTIC

2.9 Always use right-handed coordinate systems. To use anything else is making life difficult for yourself and for others who must look over your work. For two-dimensional situations, it is *not* critical to check whether the coordinate system you choose is right-handed since the third, but irrelevant, coordinate dimension can be directed according to whatever choice you make for, say, the x- and y-axes. However, when introducing a three-dimensional Cartesian coordinate system, *always* check to be sure the coordinate system you introduce is right-handed.

Nothing is sacred about the specific spatial directions chosen for the x-, y-, and z-axes *as long as they form a right-handed system*. You need not always choose +x going to the right, +y going up, and +z coming toward you out of the page. A few legitimate and illegitimate coordinate choices are shown in Figure 2.48; you can test each with your right hand to verify that it is legitimate or illegitimate.

We want to be able to express any vector in terms of the directions you choose for the Cartesian coordinate axes. For this purpose, define three fundamental Cartesian vectors $\hat{\imath}$, $\hat{\jmath}$, and $\hat{k}$ as **basis vectors**, since we will see that any vector can be expressed in terms of them. The Cartesian basis vectors $\hat{\imath}$, $\hat{\jmath}$, and $\hat{k}$ also are called **Cartesian unit vectors** because each has a magnitude equal to the pure number 1.[†] The direction of each unit vector is as follows (see Figure 2.49).

> $\hat{\imath}$ indicates the direction that coordinate x increases in value. For $\hat{\imath}$, say "i-hat."
> $\hat{\jmath}$ indicates the direction that coordinate y increases in value. For $\hat{\jmath}$, say "j-hat."
> $\hat{k}$ indicates the direction that coordinate z increases in value. For $\hat{k}$, say "k-hat."

Notice in Figure 2.49 that parallel transport applies to the unit vectors; you need not consider their tails always to be at the origin. The vectors $-\hat{\imath}$, $-\hat{\jmath}$, and $-\hat{k}$ indicate the directions in which x, y, and z, respectively, decrease in value.

Unit vectors carry *no* dimensions and *no* units along with them; they are *dimensionless and unitless* vectors that serve merely to indicate directions. The unit vectors are designated differently from *physical* vectors such as position vectors and velocity

---

*If this seems like a bias in favor of right-handed persons, it is in a sense. However, left-handed persons have a distinct advantage: they can use their right hands to apply the right-hand rules without first setting down a pencil!
[†]In some fields of physics, notably solid state physics and crystallography, basis vectors are used that are neither unit vectors nor perpendicular (orthogonal) vectors. In such fields, it is more convenient to have basis vectors related to the crystal symmetry (see Problem 62).

FIGURE 2.46 A right-handed Cartesian coordinate system.

FIGURE 2.47 Another way to verify the right-handedness of a coordinate system.

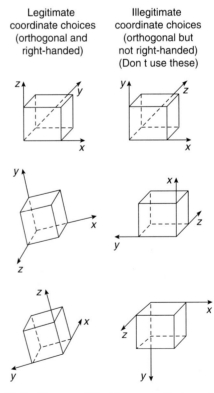

FIGURE 2.48 Legitimate and illegitimate coordinate system choices.

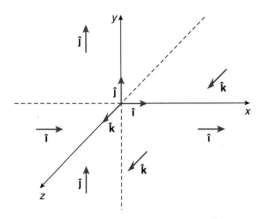

FIGURE 2.49 The Cartesian unit vectors î, ĵ, and k̂.

vectors, which *do* have dimensions and units associated with them. Unit vectors do not have an arrow over them but rather a *hat* or circumflex.

Consider the scalar product of a Cartesian unit vector with itself. For example, the angle between î and î is 0°. Thus the scalar product of î with itself is, by definition (Equation 2.10), the product of the magnitudes of the two vectors times the cosine of the angle between them. Here the magnitude of each vector is 1 and the angle between them is 0°, so that

$$\hat{\imath} \cdot \hat{\imath} = (1)(1) \cos 0°$$
$$= 1$$

Now consider the scalar product of a Cartesian unit vector with a *different* Cartesian unit vector. For example, consider the scalar product of î with ĵ, î · ĵ. The angle between î and ĵ is 90°, since they are perpendicular. Thus their scalar product is

$$\hat{\imath} \cdot \hat{\jmath} = (1)(1) \cos 90°$$
$$= 0$$

All possible scalar products of the Cartesian unit vectors with each other are summarized in Table 2.1.

QUESTION 5

The Cartesian basis vectors î, ĵ, and k̂ are called *unit vectors*, but they have no units associated with them. Explain what is meant by this apparent oxymoron.

2.9 THE CARTESIAN REPRESENTATION OF ANY VECTOR

Since the Cartesian coordinate system is used extensively in physics, we must know how to express any vector using the

Cartesian unit vectors î, ĵ, and k̂ as a basis. In other words, we want to determine how to represent any vector $\vec{A}$ as an appropriate vector sum of the Cartesian unit vectors.

The Case of Two Dimensions

Consider a vector $\vec{A}$ confined to the x–y plane of a Cartesian coordinate system. If necessary, parallel transport the vector so its tail is at the origin, as in Figure 2.50. Take the scalar product of $\vec{A}$ with î for the fun of it; use Equation 2.10:

$$\vec{A} \cdot \hat{\imath} = (A)(1) \cos \theta$$
$$= A \cos \theta$$
$$\equiv A_x \quad (2.12)$$

According to the geometric interpretation of the scalar product, $A \cos \theta$ ($\equiv A_x$) is the projection or **component** of $\vec{A}$ on the line containing the unit vector î (or any line parallel to it); see Figure 2.51. The specific projection or component A_x is known as the Cartesian **x-component** of $\vec{A}$.

Note that A_x is a *scalar* quantity. Directional sense is indicated by î. Thus A_xî is the multiplication of the unit vector î by the scalar A_x. Likewise,

$$\vec{A} \cdot \hat{\jmath} = (A)(1) \cos(90° - \theta)$$
$$= A(\cos 90° \cos \theta + \sin 90° \sin \theta)$$
$$= A \sin \theta$$
$$\equiv A_y \quad (2.13)$$

You can get from the first line of Equation 2.13 to the last line without using the double angle trigonometry formula by noting that in a right triangle, the cosine of one of the acute angles is equal to the sine of the other acute angle, as shown in Figure 2.52.

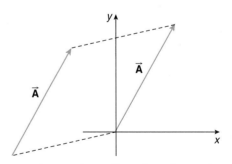

FIGURE 2.50 A vector in two dimensions, the x–y plane.

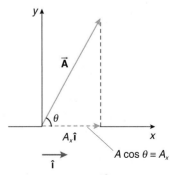

FIGURE 2.51 The x-component of $\vec{A}$.

TABLE 2.1 Scalar Product of the Cartesian Unit Vectors with Each Other

$\hat{\imath} \cdot \hat{\imath} = 1$	$\hat{\imath} \cdot \hat{\jmath} = 0$	$\hat{\imath} \cdot \hat{k} = 0$
$\hat{\jmath} \cdot \hat{\imath} = 0$	$\hat{\jmath} \cdot \hat{\jmath} = 1$	$\hat{\jmath} \cdot \hat{k} = 0$
$\hat{k} \cdot \hat{\imath} = 0$	$\hat{k} \cdot \hat{\jmath} = 0$	$\hat{k} \cdot \hat{k} = 1$

The scalar quantity A_y is the projection or component of $\vec{A}$ on the line containing $\hat{\jmath}$ (or any line parallel to it) and is called the Cartesian **y-component** of $\vec{A}$; see Figure 2.53.

The scalars A_x and A_y are known collectively as the **Cartesian components** of the vector $\vec{A}$ in two dimensions. Notice that since the scalar product is a scalar that is either positive, negative, or zero, *the components are scalar quantities that can be positive, negative, or zero.*

The units and dimensions associated with the Cartesian components of $\vec{A}$ are those associated with vector $\vec{A}$ itself; the units are, say, meters only if $\vec{A}$ represents a directed distance between two points in space.

The *vector $A_x\hat{\imath}$* is the scalar multiplication of the unit vector $\hat{\imath}$ by the scalar A_x; likewise the *vector $A_y\hat{\jmath}$* is the scalar multiplication of the unit vector $\hat{\jmath}$ by the scalar A_y. In two dimensions, the vector $\vec{A}$ is written in terms of its components as the *vector*

sum of the mutually perpendicular vectors $A_x\hat{\imath}$ and $A_y\hat{\jmath}$, as shown in Figure 2.54. An alternative representation of this relationship is shown in Figure 2.55, with all three vectors with their tails at the origin.

PROBLEM-SOLVING TACTIC

2.10 Do not confuse a vector with its Cartesian components. The components are *scalar* quantities. The multiplication of the unit vectors by the Cartesian components leads to a vector sum that represents the vector.

components

$$\vec{A} = A_x\hat{\imath} + A_y\hat{\jmath} \qquad (2.14)$$

vectors

Note several things from Figure 2.54 or 2.55:

a. The Pythagorean theorem means that we have

$$A^2 = A_x^2 + A_y^2 \qquad (2.15)$$

The magnitude of $\vec{A}$ thus is

$$A = (A_x^2 + A_y^2)^{1/2} \qquad \text{(use positive root only)} \qquad (2.16)$$

By convention, since the magnitude of a vector is never negative, *always* take the positive square root when using Equation 2.16.

b. Also note that the tangent of the angle θ that the two-dimensional vector makes with the direction of $\hat{\imath}$ is the ratio of the y- to the x-components:

$$\tan\theta = \frac{\text{opposite side}}{\text{adjacent side}}$$

$$= \frac{A_y}{A_x}$$

so

$$\theta = \tan^{-1}\left(\frac{A_y}{A_x}\right) \qquad (2.17)$$

Hypotenuse

$90° - \theta$

$90° - \theta$

θ

$\cos(90° - \theta) = \sin\theta$

FIGURE 2.52 In any right triangle, the cosine of one acute angle is equal to the sine of the other acute angle.

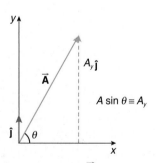

$A_y\hat{\jmath}$

$\vec{A}$

$A\sin\theta \equiv A_y$

$\hat{\imath}$ θ

FIGURE 2.53 The y-component of $\vec{A}$.

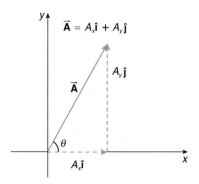

$\vec{A} = A_x\hat{\imath} + A_y\hat{\jmath}$

$\vec{A}$

$A_y\hat{\jmath}$

θ

$A_x\hat{\imath}$

FIGURE 2.54 The two-dimensional vector $\vec{A}$ in Cartesian form.

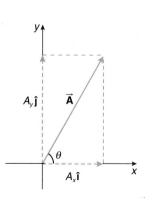

$A_y\hat{\jmath}$ $\vec{A}$

θ

$A_x\hat{\imath}$

FIGURE 2.55 An alternative representation of $\vec{A}$.

2.11 The mathematical sign convention for angles on a two-dimensional Cartesian coordinate grid is used with Equation 2.17. If the angle θ is *positive*, the angle is measured *counterclockwise* from the positive x-axis. If the angle is negative, the angle is measured clockwise from the positive x-axis. Remember, however, that the angle between two vectors always is considered to be positive.

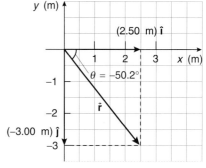

FIGURE 2.56

EXAMPLE 2.3 _____

Given the two-dimensional position vector

$$\vec{r} = (2.50 \text{ m}) \, \hat{i} - (3.00 \text{ m}) \, \hat{j}$$

a. Sketch the vectors $(2.50 \text{ m}) \, \hat{i}$, $(-3.00 \text{ m}) \, \hat{j}$, and $\vec{r}$ on a two-dimensional Cartesian coordinate grid with the tail of each vector at the origin.
b. Find the magnitude of $\vec{r}$.
c. Find the angle that $\vec{r}$ makes with the positive x-direction.
d. If $\vec{r}$ is parallel transported to another location in the sketch, what happens to the answers to parts (b) and (c)?

Solution

a. The sketch of the vectors with their tails at the origin is given in Figure 2.56. By parallel transporting either $(2.50 \text{ m})\hat{i}$ or $(-3.00 \text{ m})\hat{j}$, notice that $\vec{r}$ is the vector sum

$$\vec{r} = (2.50 \text{ m}) \, \hat{i} + (-3.00 \text{ m}) \, \hat{j}$$

b. According to Equation 2.16, the magnitude of $\vec{r}$ is the square root of the sum of the squares of its components. From Section 2.6, $-(3.00 \text{ m}) \, \hat{j}$ is equivalent to $+(-3.00 \text{ m}) \, \hat{j}$. Hence the components of this vector are the scalars

$$r_x \equiv x = 2.50 \text{ m}$$
$$r_y \equiv y = -3.00 \text{ m}$$

The magnitude of the vector then is found from Equation 2.16:

$$r = [(2.50 \text{ m})^2 + (-3.00 \text{ m})^2]^{1/2}$$
$$= 3.91 \text{ m}$$

c. The angle θ that the vector makes with the positive x-direction (the direction of $\hat{i}$) is found from Equation 2.17:

$$\theta = \tan^{-1} \left(\frac{-3.00 \text{ m}}{2.50 \text{ m}} \right)$$
$$= -50.2°$$

or 50.2° measured clockwise from the positive x-axis according to Problem-Solving Tactic 2.11. The angle is indicated in Figure 2.56.

d. If the vector is parallel transported to another location in the sketch of part (a), its magnitude and direction are unchanged. The answers to parts (b) and (c) thus remain the same.

The Case of Three Dimensions

For the three-dimensional case, the three Cartesian components of any vector $\vec{A}$ are the scalar quantities:

$$\vec{A} \cdot \hat{i} \equiv A_x$$
$$\vec{A} \cdot \hat{j} \equiv A_y \qquad (2.18)$$
$$\vec{A} \cdot \hat{k} \equiv A_z$$

They are the three scalar *projections* or *components* of $\vec{A}$ on the lines containing the three Cartesian coordinate axes (or lines parallel to them).

The vector $\vec{A}$ is the *vector sum* of three mutually perpendicular vectors:

components

$$\vec{A} = A_x\hat{i} + A_y\hat{j} + A_z\hat{k} \qquad (2.19)$$

vectors

The vector sum indicated in Equation 2.19 is indicated schematically in Figure 2.57.

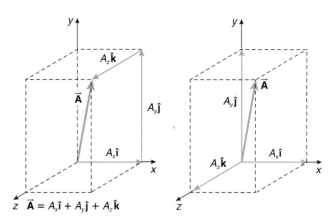

(a) The vector $\vec{A}$ is the vector sum of the vectors $A_x\hat{i}$, $A_y\hat{j}$, and $A_z\hat{k}$.

(b) An equivalent view, parallel transporting $A_y\hat{j}$ and $A_z\hat{k}$ to the axes themselves so that the tails of all the vectors are at a common point.

FIGURE 2.57 Two equivalent views of a vector $\vec{A}$ in three dimensions.

When $\vec{A}$ is described as a sum of a set of mutually perpendicular vectors, we say that $\vec{A}$ has been **resolved** into its *components*. The Cartesian components of $\vec{A}$ are the *scalars* A_x, A_y, and A_z. The vector sum indicated in Equation 2.19 is equivalent to $\vec{A}$ itself.

We can find the magnitude of any vector $\vec{A}$ by taking the scalar product of $\vec{A}$ with itself:

$$\vec{A} \cdot \vec{A} = (A_x \hat{\imath} + A_y \hat{\jmath} + A_z \hat{k}) \cdot (A_x \hat{\imath} + A_y \hat{\jmath} + A_z \hat{k})$$

Because of the distributive property of the scalar product operation, this results in nine scalar products of the unit vectors:

$$\begin{aligned}
\vec{A} \cdot \vec{A} = \; & A_x A_x (\hat{\imath} \cdot \hat{\imath}) + A_x A_y (\hat{\imath} \cdot \hat{\jmath}) + A_x A_z (\hat{\imath} \cdot \hat{k}) \\
& + A_y A_x (\hat{\jmath} \cdot \hat{\imath}) + A_y A_y (\hat{\jmath} \cdot \hat{\jmath}) + A_y A_z (\hat{\jmath} \cdot \hat{k}) \\
& + A_z A_x (\hat{k} \cdot \hat{\imath}) + A_z A_y (\hat{k} \cdot \hat{\jmath}) + A_z A_z (\hat{k} \cdot \hat{k})
\end{aligned}$$

Six of the scalar products of the unit vectors are zero; only three terms survive (refer back to Table 2.1). Something wonderfully simple emerges:

$$\vec{A} \cdot \vec{A} = A_x^2 + A_y^2 + A_z^2$$

But the Pythagorean theorem applied to Figure 2.57 indicates that

$$A^2 = A_x^2 + A_y^2 + A_z^2 \tag{2.20}$$

Thus we have proved that

$$\vec{A} \cdot \vec{A} = A^2 \tag{2.21}$$

After taking the positive square root, we have the magnitude of $\vec{A}$:

$$\boxed{\begin{aligned} A &= (\vec{A} \cdot \vec{A})^{1/2} \\ &= (A_x^2 + A_y^2 + A_z^2)^{1/2} \quad \text{(use positive root only)} \end{aligned}} \tag{2.22}$$

Note from Equation 2.22 that a vector of magnitude zero (the zero vector) must have *all* its components equal to zero.

PROBLEM-SOLVING TACTIC

2.12 The magnitude of a vector $\vec{A}$ that is expressed in terms of its Cartesian components is found by taking the square root of the scalar product of the vector with itself:

$$\begin{aligned} A &= (\vec{A} \cdot \vec{A})^{1/2} \\ &= (A_x^2 + A_y^2 + A_z^2)^{1/2} \quad \text{(use positive root only)} \end{aligned}$$

QUESTION 6

What are the three Cartesian components of the vector $\hat{\imath}$?

EXAMPLE 2.4

a. What are the Cartesian components of the velocity vector

$$\vec{v} = (-2.20 \text{ m/s})\hat{\imath} - (3.30 \text{ m/s})\hat{\jmath} + (6.85 \text{ m/s})\hat{k}$$

b. What is the magnitude of this velocity vector (the speed)?

Solution

a. The Cartesian components are the scalar coefficients of the Cartesian unit vectors. Remember that $- (3.30 \text{ m/s})\hat{\jmath}$ is equivalent to $+ (-3.30 \text{ m/s})\hat{\jmath}$. Hence you have

$$v_x = -2.20 \text{ m/s}$$
$$v_y = -3.30 \text{ m/s}$$
$$v_z = 6.85 \text{ m/s}$$

b. The magnitude of the velocity vector is found by taking the square root of $\vec{v} \cdot \vec{v}$ (which is the same as using Equation 2.22):

$$\begin{aligned} v &= (\vec{v} \cdot \vec{v})^{1/2} \\ &= [(-2.20 \text{ m/s})^2 + (-3.30 \text{ m/s})^2 + (6.85 \text{ m/s})^2]^{1/2} \\ &= 7.92 \text{ m/s} \end{aligned}$$

2.10 MULTIPLICATION OF A VECTOR EXPRESSED IN CARTESIAN FORM BY A SCALAR

The multiplication of a vector $\vec{A}$ by a scalar α changes all the Cartesian components by the same factor:

$$\begin{aligned} \alpha \vec{A} &= \alpha(A_x \hat{\imath} + A_y \hat{\jmath} + A_z \hat{k}) \\ &= \alpha A_x \hat{\imath} + \alpha A_y \hat{\jmath} + \alpha A_z \hat{k} \end{aligned} \tag{2.23}$$

where α is any scalar (positive, negative, or zero).

EXAMPLE 2.5

If $\vec{B} = -2.0\hat{\imath} + 4.0\hat{\jmath} - 3.0\hat{k}$, what is $-2.2\vec{B}$?

Solution

$$\begin{aligned} -2.2\vec{B} &= -2.2(-2.0\hat{\imath} + 4.0\hat{\jmath} - 3.0\hat{k}) \\ &= 4.4\hat{\imath} - 8.8\hat{\jmath} + 6.6\hat{k} \end{aligned}$$

2.11 EXPRESSING VECTOR ADDITION AND SUBTRACTION IN CARTESIAN FORM

As we learned in Sections 2.4 and 2.6, vectors in Cartesian form may be added and subtracted provided the vectors are of the same type, representing the same kind of physical quantity, such as forces added to forces or velocities added to velocities. In particular, when the vectors are in Cartesian form we have

$$\vec{F} + \vec{f} = (F_x \hat{\imath} + F_y \hat{\jmath} + F_z \hat{k}) + (f_x \hat{\imath} + f_y \hat{\jmath} + f_z \hat{k})$$

Collect the various x-, y-, and z-components together:

$$\vec{F} + \vec{f} = (F_x + f_x)\hat{\imath} + (F_y + f_y)\hat{\jmath} + (F_z + f_z)\hat{k} \tag{2.24}$$

Thus the components of the vector sum $\vec{F} + \vec{f}$ are the scalar sums of the respective components of the individual vectors.

The vector difference between $\vec{F}$ and $\vec{f}$ is similar:

$$\begin{aligned} \vec{F} - \vec{f} &= (F_x \hat{\imath} + F_y \hat{\jmath} + F_z \hat{k}) - (f_x \hat{\imath} + f_y \hat{\jmath} + f_z \hat{k}) \\ &= (F_x - f_x)\hat{\imath} + (F_y - f_y)\hat{\jmath} + (F_z - f_z)\hat{k} \end{aligned} \tag{2.25}$$

Thus the components of the vector $\vec{F} - \vec{f}$ are the scalar differences of the respective scalar components of the individual vectors, which should be comforting to you from an algebraic viewpoint.

EXAMPLE 2.6

If
$$\vec{F} = 2\hat{\imath} - 3\hat{\jmath} + 4\hat{k}$$
and
$$\vec{f} = -2\hat{\imath} + 4\hat{\jmath} - 7\hat{k}$$

find (a) $\vec{F} + \vec{f}$ and (b) $\vec{F} - \vec{f}$.

Solution

a. Use Equation 2.24 and add the respective Cartesian components:

$$\vec{F} + \vec{f} = (F_x + f_x)\hat{\imath} + (F_y + f_y)\hat{\jmath} + (F_z + f_z)\hat{k}$$
$$= [2 + (-2)]\hat{\imath} + [-3 + 4]\hat{\jmath} + [4 + (-7)]\hat{k}$$
$$= 0\hat{\imath} + \hat{\jmath} - 3\hat{k}$$
$$= \hat{\jmath} - 3\hat{k}$$

b. Use Equation 2.25 and subtract the respective Cartesian components:

$$\vec{F} - \vec{f} = (F_x - f_x)\hat{\imath} + (F_y - f_y)\hat{\jmath} + (F_z - f_z)\hat{k}$$
$$= [2 - (-2)]\hat{\imath} + [-3 - 4]\hat{\jmath} + [4 - (-7)]\hat{k}$$
$$= 4\hat{\imath} - 7\hat{\jmath} + 11\hat{k}$$

2.12 THE SCALAR PRODUCT OF TWO VECTORS EXPRESSED IN CARTESIAN FORM

When two vectors $\vec{A}$ and $\vec{B}$ are expressed in Cartesian form, the scalar product becomes

$$\vec{A} \cdot \vec{B} = (A_x\hat{\imath} + A_y\hat{\jmath} + A_z\hat{k}) \cdot (B_x\hat{\imath} + B_y\hat{\jmath} + B_z\hat{k})$$

The scalar product simplifies considerably, since six of the nine scalar products of the unit vectors are zero (refer to Table 2.1), leaving only three surviving terms:

$$\boxed{\vec{A} \cdot \vec{B} = A_xB_x + A_yB_y + A_zB_z} \tag{2.26}$$

Notice that the right-hand side of Equation 2.26 is an ordinary *scalar sum*; the result of a scalar product always is a scalar quantity.

The scalar product of two vectors is the (scalar) sum of the products of their respective Cartesian components.

EXAMPLE 2.7

If
$$\vec{A} = 2\hat{\imath} - \hat{\jmath} + 4\hat{k}$$
and
$$\vec{B} = -3\hat{\imath} - 2\hat{\jmath} + 4\hat{k}$$

find $\vec{A} \cdot \vec{B}$.

Solution

Use Equation 2.26 and (scalar) sum the products of the respective Cartesian components:

$$\vec{A} \cdot \vec{B} = A_xB_x + A_yB_y + A_zB_z$$
$$= (2)(-3) + (-1)(-2) + (4)(4)$$
$$= -6 + 2 + 16$$
$$= 12$$

2.13 DETERMINING THE ANGLE BETWEEN TWO VECTORS EXPRESSED IN CARTESIAN FORM

We now have all the tools needed for some interesting geometric calculations. For example, consider two particular vectors $\vec{A}$ and $\vec{B}$ expressed in Cartesian form:

$$\vec{A} = 1.00\hat{\imath} - 2.00\hat{\jmath} + 4.00\hat{k}$$
and
$$\vec{B} = 2.00\hat{\imath} + 1.00\hat{\jmath} - 3.00\hat{k}$$

We want to find the angle θ between them. If we sketch the geometric situation, as in Figure 2.58, with the tails of the two vectors at the origin, the measure of the angle between the two vectors is not at all obvious. It is even difficult to estimate the angle from the three-dimensional sketch.

The angle θ between the two vectors is found by exploiting the scalar product, which involves the angle between them:

$$\vec{A} \cdot \vec{B} = AB \cos\theta$$

1. First, find the magnitude of the vector $\vec{A}$, using Equation 2.22:

$$A = (\vec{A} \cdot \vec{A})^{1/2} = (A_x^2 + A_y^2 + A_z^2)^{1/2}$$
$$= [(1.00)^2 + (-2.00)^2 + (4.00)^2]^{1/2}$$
$$= (1.00 + 4.00 + 16.0)^{1/2}$$
$$= (21.0)^{1/2}$$

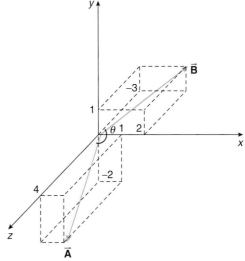

FIGURE 2.58 The vectors $\vec{A} = 1.00\hat{\imath} - 2.00\hat{\jmath} + 4.00\hat{k}$ and $\vec{B} = 2.00\hat{\imath} + 1.00\hat{\jmath} - 3.00\hat{k}$.

2. Then find the magnitude of $\vec{B}$ the same way:

$$B = (\vec{B} \cdot \vec{B})^{1/2} = (B_x^2 + B_y^2 + B_z^2)^{1/2}$$
$$= [(2.00)^2 + (1.00)^2 + (-3.00)^2]^{1/2}$$
$$= (4.00 + 1.00 + 9.00)^{1/2}$$
$$= (14.00)^{1/2}$$

3. Now find the scalar product of $\vec{A}$ and $\vec{B}$ using the Cartesian components of the vectors and Equation 2.26:

$$\vec{A} \cdot \vec{B} = A_x B_x + A_y B_y + A_z B_z$$
$$= (1.00)(2.00) + (-2.00)(1.00) + (4.00)(-3.00)$$
$$= 2.00 - 2.00 - 12.0$$
$$= -12.0$$

4. Now put the pieces together into the definition of the scalar product, Equation 2.10:

$$\vec{A} \cdot \vec{B} = AB \cos \theta$$

The left-hand side of the equation is -12.0 (from the Cartesian evaluation of the scalar product); on the right-hand side we know the magnitude of $\vec{A}$ is $(21.0)^{1/2}$ and $\vec{B}$ is $(14.00)^{1/2}$. Thus

$$-12.0 = (21.0)^{1/2}(14.00)^{1/2} \cos \theta$$

so that

$$\cos \theta = \frac{-12.0}{(21.0)^{1/2}(14.00)^{1/2}}$$
$$= -0.700$$

or

$$\theta = 134°$$

This technique is used to find the angle between *any* two vectors when they are expressed in Cartesian form.

STRATEGIC EXAMPLE 2.8

A carpenter has built a rectangular shipping container of dimensions 1.00 m by 1.00 m by 1.50 m, as shown in Figure 2.59. A triangular piece of wood is to be fitted diagonally across the box. What is the angle θ that the hypotenuse of the triangular piece of wood makes with the base of the container?

Solution

Use the coordinate choice indicated in Figure 2.59. The diagonal of the box (the hypotenuse of the triangle) has the Cartesian vector representation

$$\vec{D} = (1.00 \text{ m})\hat{\imath} + (1.50 \text{ m})\hat{\jmath} - (1.00 \text{ m})\hat{k}$$

The edge of the base of the triangular piece of wood has the Cartesian vector representation

$$\vec{d} = (1.00 \text{ m})\hat{\imath} - (1.00 \text{ m})\hat{k}$$

The angle θ is the angle between these two vectors.

1. The magnitude of $\vec{D}$ is

$$D = [(1.00 \text{ m})^2 + (1.50 \text{ m})^2 + (-1.00 \text{ m})^2]^{1/2}$$
$$= 2.06 \text{ m}$$

2. The magnitude of $\vec{d}$ is

$$d = [(1.00 \text{ m})^2 + (-1.00 \text{ m})^2]^{1/2}$$
$$= 1.41 \text{ m}$$

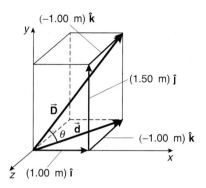

FIGURE 2.59

3. Use Equation 2.26 to find the scalar product $\vec{D} \cdot \vec{d}$:

$$\vec{D} \cdot \vec{d} = (1.00 \text{ m})(1.00 \text{ m}) + (1.50 \text{ m})(0 \text{ m})$$
$$+ (-1.00 \text{ m})(-1.00 \text{ m})$$
$$= 2.00 \text{ m}^2$$

4. Use the definition of the scalar product, Equation 2.10:

$$\vec{D} \cdot \vec{d} = Dd \cos \theta$$

This yields

$$2.00 \text{ m}^2 = (2.06 \text{ m})(1.41 \text{ m})\cos \theta$$

Solve for $\cos \theta$, and you find

$$\cos \theta = 0.689$$

or

$$\theta = 46.4°$$

So the angle between the diagonal and the base is $46.4°$.

2.14 EQUALITY OF TWO VECTORS

Note that if $\vec{F} - \vec{f} = 0$, then $\vec{F} = \vec{f}$. If two vectors are equal to each other, their respective components are equal. Thus the *single* vector equation

$$\vec{F} = \vec{f}$$

really is *three* different scalar equations, one for each of the three Cartesian components:

$$F_x = f_x$$
$$F_y = f_y$$
$$F_z = f_z$$

This is one of the beauties of vector notation; much information can be gleaned from a single vector equation.

QUESTION 7

Your roommate says that two vectors are "equal and opposite." Is this possible? Explain what your friend probably means.

2.15 VECTOR EQUATIONS

It is important to realize that an equation relating various vectors is independent of the particular coordinate system used to represent the vectors. For example, looking ahead to Chapter 5, we say that a vector $\vec{F}_{total}$ is equal to a scalar m times another vector $\vec{a}$:

$$\vec{F}_{total} = m\vec{a}$$

This mathematical relationship is valid independent of the co-ordinate system we choose to represent the various vectors algebraically. Once a coordinate system is chosen, we write *all* the vectors in the equation in terms of the *same* coordinate system.

Nature and natural phenomena on a macroscopic scale do not depend on the choice we scientists make of a coordinate system.*† Nature was "doing its thing" long before we physicists and engineers arrived on the scene with our mathematical toolboxes to describe physical phenomena. We will see that the most general *laws of physics* are equations relating scalars to other scalars (scalar equations) or vectors to other vectors (vector equations).

In view of what we mean by the equality of two vectors, a vector equation such as $\vec{F}_{total} = m\vec{a}$ actually represents *three scalar equations* relating the components of the vectors:

$$F_{x\,total} = ma_x$$
$$F_{y\,total} = ma_y$$
$$F_{z\,total} = ma_z$$

We will see in Chapter 5 that these equations are a statement of Newton's second law of motion.

2.16 THE VECTOR PRODUCT OF TWO VECTORS

The **vector product** of two vectors, as its name implies, creates a new vector from the two that form the product. We will use the vector product extensively to simplify the description of circular motion in Chapter 4, which is why we introduce it into the vector toolbox at this early juncture. The vector product has many other uses in physics as well, particularly in the study of orbital motion and spin motion (Chapter 10) and magnetism (Chapter 20). For now, we play with the vector product more abstractly, recognizing that its utility in physics will be apparent soon.

The vector product of two vectors $\vec{A}$ and $\vec{B}$ is written symbolically as

$$\vec{A} \times \vec{B}$$

where the boldfaced symbol ✕ is used to distinguish the vector product from the scalar product of two vectors (symbolized by •). The vector product commonly is called the *cross product*, reflecting the symbol used to designate this type of vector multiplication. We prefer the term vector product, to emphasize the vector nature of the result.

The new vector that results from the vector product has both a magnitude and a direction. The *magnitude* of the vector product is related to the magnitudes of $\vec{A}$ and $\vec{B}$ and the angle θ between them. As we learned in Section 2.7, the angle θ between the two vectors is the *smaller angle* measured when they are drawn from a common point (i.e., with their tails together), as shown in Figure 2.60. Parallel transport may be needed to get the tails of the two vectors together so θ can be determined. The angle θ is the *same* angle used in the definition of the scalar product.

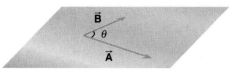

Plane of $\vec{A}$ and $\vec{B}$

FIGURE 2.60 Two vectors and the angle θ between them.

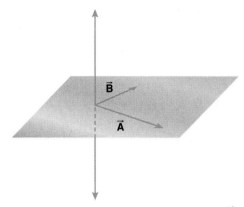

FIGURE 2.61 A perpendicular to the plane formed by $\vec{A}$ and $\vec{B}$ defines two directions.

The *magnitude* of the vector product is defined to be

$$\boxed{|\vec{A} \times \vec{B}| \equiv AB \sin \theta} \qquad (2.27)$$

Notice that the *sine* of the angle is used in the magnitude of the vector product; it was the cosine of the angle that determined the scalar product. The angle between the two vectors always is less than or equal to 180°. The sine is *never negative* over this range of angles. Since the magnitudes of $\vec{A}$ and $\vec{B}$ in the product also are positive scalars, the magnitude of the vector product, given by the right-hand side of Equation 2.27, is a scalar quantity greater than or equal to zero, as is the magnitude of *any* vector.‡

The *direction* of the vector product is defined to be perpendicular to *both* the vectors $\vec{A}$ and $\vec{B}$.

If the two vectors are not coplanar, use the parallel transport property to move either vector, preserving its length and orientation in space, until the tails of the two vectors emerge from a common point. The two vectors then define a plane, as shown in Figure 2.60. The direction of the vector product is perpendicular to this plane. But the perpendicular defines *two* directions (see Figure 2.61) and one must be singled out. The direction universally chosen is that determined by a **vector product right-hand rule**. There are two equivalent ways to envision this rule:

1. Orient the index finger of your right hand along the direction of the first vector of the product and the middle finger along the direction of the second vector, as in Figure 2.62. The

---

*Occasionally, a specific coordinate choice is made by convention.
†In quantum mechanics, nature can be more subtle!

‡On the other hand, the scalar product can be positive (when $\theta < 90°$), negative (when $\theta > 90°$), or zero (when $\theta = 90°$).

extended thumb then indicates the directional sense of the vector product $\vec{A} \times \vec{B}$.

2. Alternatively, take the four bunched fingers of your right hand and imagine swinging them from the direction of the first vector in the vector product to the direction of the second vector, as shown in Figure 2.63. The extended thumb of your right hand then indicates the appropriate direction of the vector product. Always choose the smaller of the two possible angles for swinging your fingers from the first vector to the second vector of the product.

These are the same sorts of right-hand rules that defined the right-handed Cartesian coordinate directions and the relative orientations of the Cartesian unit vectors $\hat{i}$, $\hat{j}$, and $\hat{k}$.

PROBLEM-SOLVING TACTIC

2.13 When taking the vector product of two vectors, remember that the result is a vector perpendicular to both. You must specify the magnitude and direction of the vector that results from the product, or express it in Cartesian form.

There are several geometric interpretations of the *magnitude* of the vector product.

1. The magnitude of the vector product is

$$AB \sin \theta$$

where θ is the angle between the two vectors when their tails are together at a common point. This magnitude is the area of the *parallelogram* whose sides are the two vectors, as shown in Figure 2.64. This area is measured in square meters *only* if

both vectors represent directed lengths expressed in meters; otherwise, the SI units attached to the area are those of vector $\vec{A}$ times those of vector $\vec{B}$.

2. Another way to interpret the magnitude of the vector product is as *twice* the area of the *triangle* formed by connecting the tips of the two vectors, as shown in Figure 2.65.

The two geometric interpretations of the vector product are not, of course, independent of one another since the area of the parallelogram in Figure 2.64 is twice the area of the triangle in Figure 2.65.

Since the angle between any vector $\vec{A}$ and *itself* is 0°, the magnitude of the vector product of $\vec{A}$ with itself is, from Equation 2.27,

$$|\vec{A} \times \vec{A}| = AA \sin 0°$$

Since sin 0° = 0, the magnitude of $\vec{A} \times \vec{A}$ is zero. Since the magnitude is zero, the resulting vector is the zero vector; that is,

$$\vec{A} \times \vec{A} = 0 \qquad \text{for } any \text{ vector } \vec{A} \qquad (2.28)$$

The vector product of any vector with itself is zero.

If the vector product of two vectors is zero and *neither* vector is of zero magnitude, then the sine of the angle between the two vectors must be zero. That is, if

$$\vec{A} \times \vec{B} = 0$$

and $A \neq 0$ and $B \neq 0$, then

$$|\vec{A} \times \vec{B}| = AB \sin \theta = 0$$

implies

$$\sin \theta = 0$$

Thus θ is either 0° or 180°. The two vectors are either parallel (0° angle between them) or antiparallel (180° angle between them).

The order of the terms in a vector product is important.

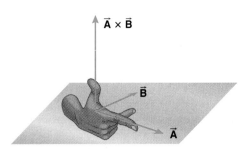

FIGURE 2.62 One way to determine the direction of the vector product $\vec{A} \times \vec{B}$.

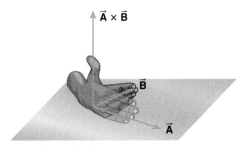

FIGURE 2.63 An equivalent way to find the direction of the vector product $\vec{A} \times \vec{B}$.

Area of parallelogram = $A (B \sin \theta) = |\vec{A} \times \vec{B}|$

FIGURE 2.64 The magnitude of the vector product is the area of the parallelogram formed by the two vectors.

Area of triangle
= 1/2 (base)(height) = 1/2 $A(B \sin \theta)$
= $\dfrac{|\vec{A} \times \vec{B}|}{2}$

FIGURE 2.65 The magnitude of the vector product is twice the area of the triangle formed by connecting the tips of the vectors.

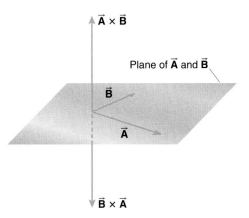

FIGURE 2.66 $\vec{A} \times \vec{B}$ and $\vec{B} \times \vec{A}$ have the same magnitude but point in opposite directions.

The terms of the vector product do *not* commute.

That is,

$$\vec{A} \times \vec{B} \neq \vec{B} \times \vec{A} \qquad (2.29)$$

Although $\vec{A} \times \vec{B}$ and $\vec{B} \times \vec{A}$ have the same magnitude, they point in opposite directions according to the vector product right-hand rule for determining the directions of a vector product. That is,

$$\vec{A} \times \vec{B} = -\vec{B} \times \vec{A} \qquad (2.30)$$

This is illustrated in Figure 2.66.

The order of the terms in a vector product is important, since $\vec{A} \times \vec{B} = -\vec{B} \times \vec{A}$.

The vector product obeys a distributive law as long as the *order* of the terms in the product is preserved. That is,

$$\vec{A} \times (\vec{B}_1 + \vec{B}_2) = (\vec{A} \times \vec{B}_1) + (\vec{A} \times \vec{B}_2)$$

The parentheses are immaterial on the right-hand side of the equation, since the vector products must be performed before the vector addition:

$$\vec{A} \times (\vec{B}_1 + \vec{B}_2) = \vec{A} \times \vec{B}_1 + \vec{A} \times \vec{B}_2 \qquad (2.31)$$

Notice that the **+** sign on both sides of this expression represents *vector* addition.

QUESTION 8

Suppose a vector $\vec{A}$ is not the zero vector. If $\vec{A} \times \vec{B} = 0$, does this necessarily mean that $\vec{B} = 0$? Explain.

EXAMPLE 2.9

For the two vectors indicated in **Figure 2.67**, find the magnitude of their vector product and indicate the direction of the vector product in a sketch.

FIGURE 2.67

Solution

The magnitude of their vector product is found from Equation 2.27:

$$\begin{aligned}
|\vec{A} \times \vec{B}| &= AB \sin \theta \\
&= (2.50)(3.60) \sin 60° \\
&= 7.8
\end{aligned}$$

The direction of the vector product is found using either vector product right-hand rule. Swing the fingers of your right hand from $\vec{A}$ to $\vec{B}$. Your extended right thumb indicates the direction of $\vec{A} \times \vec{B}$, shown in Figure 2.68.

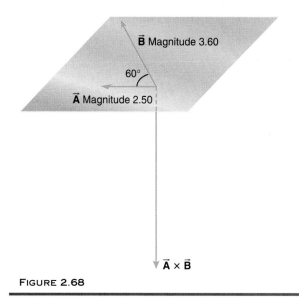

FIGURE 2.68

2.17 THE VECTOR PRODUCT OF TWO VECTORS EXPRESSED IN CARTESIAN FORM

The vector products of the Cartesian unit vectors are useful for the computation of vector products in Cartesian form. To evaluate the vector products of the Cartesian unit vectors, let's consider two specific examples.

1. To find $\hat{\imath} \times \hat{\imath}$, remember that the vector product of *any* vector with itself is zero. Hence $\hat{\imath} \times \hat{\imath} = 0$. Equivalently, since the angle between $\hat{\imath}$ and itself is 0°, the magnitude of $\hat{\imath} \times \hat{\imath}$ is

$$\begin{aligned}
|\hat{\imath} \times \hat{\imath}| &= (1)(1) \sin 0° \\
&= 0
\end{aligned}$$

2. To find $\hat{\imath} \times \hat{\jmath}$, note that the magnitude of the resulting vector is, by definition (Equation 2.27), the product of the magnitudes of the two vectors times the sine of the angle between them. Since both $\hat{\imath}$ and $\hat{\jmath}$ each have magnitude 1, and the angle between them is 90°, the magnitude of the vector product is

$$|\hat{\imath} \times \hat{\jmath}| = (1)(1) \sin 90°$$
$$= 1$$

The unit vectors are dimensionless, as is the sine, and so the magnitude of their vector product also is dimensionless. We use the vector product right-hand rule to find the direction of the vector product $\hat{\imath} \times \hat{\jmath}$; it is perpendicular to both $\hat{\imath}$ and $\hat{\jmath}$. We find that $\hat{\imath} \times \hat{\jmath}$ is directed toward increasing values of z, that is, parallel to $\hat{k}$. Thus

$$\hat{\imath} \times \hat{\jmath} = \hat{k}$$

since $\hat{k}$ is the vector of magnitude 1 that points toward increasing values of z.

The results of all possible vector products of the Cartesian unit vectors with each other are summarized in Table 2.2.

PROBLEM-SOLVING TACTIC

2.14 There is a mnemonic for remembering the vector products of the Cartesian unit vectors. Write the unit vectors $\hat{\imath}$, $\hat{\jmath}$, and $\hat{k}$ twice in succession:

$$\hat{\imath} \quad \hat{\jmath} \quad \hat{k} \quad \hat{\imath} \quad \hat{\jmath} \quad \hat{k}$$

Any time you form the vector product of one unit vector with the one to the right, the result is the next one to the right.

$$\hat{\imath} \times \hat{\jmath} = \hat{k}$$
$$\hat{\jmath} \times \hat{k} = \hat{\imath}$$
$$\hat{k} \times \hat{\imath} = \hat{\jmath}$$

This ordering of the terms is called **cyclic ordering**. If the vector product of one unit vector with another must be read from right to left in the mnemonic (backward or **anticyclic ordering**), the result is the *negative* of the next one to the left in the list. Thus

$$\hat{k} \times \hat{\jmath} = -\hat{\imath}$$
$$\hat{\jmath} \times \hat{\imath} = -\hat{k}$$
$$\hat{\imath} \times \hat{k} = -\hat{\jmath}$$

These relationships can be summarized symbolically by

$$\overset{\rightarrow \; + \; \rightarrow}{\hat{\imath} \quad \hat{\jmath} \quad \hat{k} \quad \hat{\imath} \quad \hat{\jmath} \quad \hat{k}}_{\leftarrow \; - \; \leftarrow}$$

The vector product of two vectors that are expressed in Cartesian form gets complicated:

$$\vec{A} \times \vec{B} = (A_x\hat{\imath} + A_y\hat{\jmath} + A_z\hat{k}) \times (B_x\hat{\imath} + B_y\hat{\jmath} + B_z\hat{k})$$

The distributive property of the vector product is used to evaluate the nine vector products of the Cartesian unit vectors on the right-hand side of the equation. Remember that the *order* of the terms in the vector product is important! After the nine individual vector products of the Cartesian unit vectors are performed, three of which are zero, the result is, after collecting terms,

---

TABLE 2.2 Vector Products of the Cartesian Unit Vectors

$\hat{\imath} \times \hat{\imath} = 0$	$\hat{\imath} \times \hat{\jmath} = \hat{k}$	$\hat{\imath} \times \hat{k} = -\hat{\jmath}$
$\hat{\jmath} \times \hat{\imath} = -\hat{k}$	$\hat{\jmath} \times \hat{\jmath} = 0$	$\hat{\jmath} \times \hat{k} = \hat{\imath}$
$\hat{k} \times \hat{\imath} = \hat{\jmath}$	$\hat{k} \times \hat{\jmath} = -\hat{\imath}$	$\hat{k} \times \hat{k} = 0$

$$\vec{A} \times \vec{B} = (A_yB_z - A_zB_y)\hat{\imath} + (A_zB_x - A_xB_z)\hat{\jmath}$$
$$+ (A_xB_y - A_yB_x)\hat{k} \qquad (2.32)$$

The positive terms have the subscripts x, y, and z in cyclic order ($x\,y\,z\,x\,y\,z$, read left to right), whereas the negative terms have the subscripts in the reverse (anticyclic) order. The x-component of the vector product involves only the y- and z-components of the vectors involved in the product, the y-component of the product involves only the x- and z-components of the vectors in the product, and the z-component of the product involves only the x- and y-components.

If you know how to evaluate a 3×3 determinant, the vector product of two vectors in Cartesian form can be written elegantly as

$$\vec{A} \times \vec{B} = \begin{vmatrix} \hat{\imath} & \hat{\jmath} & \hat{k} \\ A_x & A_y & A_z \\ B_x & B_y & B_z \end{vmatrix} \qquad (2.33)$$

If a determinant is unfamiliar to you, just ignore it.

PROBLEM-SOLVING TACTICS

2.15 Do not memorize Equation 2.32. Evaluate the vector product of two vectors expressed in Cartesian form term-by-term using the mnemonic described in Problem-Solving Tactic 2.14. See Strategic Example 2.10.

2.16 A way to check the calculation of a vector product in Cartesian form is to recall that if two vectors are perpendicular to each other, their scalar product is zero. Since the vector product of two vectors is perpendicular to *both* vectors in the product, the *scalar product* of the vector $\vec{A} \times \vec{B}$ with either $\vec{A}$ or $\vec{B}$ must be zero; that is,

$$(\vec{A} \times \vec{B}) \cdot \vec{A} = 0$$

and

$$(\vec{A} \times \vec{B}) \cdot \vec{B} = 0$$

When calculating the vector product of two vectors, these scalar products serve as a useful check on whether the vector product was taken successfully (see Strategic Example 2.10).

 STRATEGIC EXAMPLE 2.10

If

$$\vec{A} = 2\hat{\imath} + 3\hat{\jmath} - 4\hat{k}$$

and

$$\vec{B} = -3\hat{\imath} + 2\hat{\jmath} - 3\hat{k}$$

(a) find $\vec{A} \times \vec{B}$ and (b) verify that $\vec{A} \times \vec{B}$ is perpendicular to both $\vec{A}$ and $\vec{B}$.

Solution

a. Use Problem-Solving Tactic 2.15; *do not memorize Equation 2.32*, just go ahead and take the vector product (although it is a bit tedious and you must be quite careful to watch out for all the signs):

$$
\begin{aligned}
\vec{A} \times \vec{B} &= (2\hat{i} + 3\hat{j} - 4\hat{k}) \times (-3\hat{i} + 2\hat{j} - 3\hat{k}) \\
&= (2)(-3)(\hat{i} \times \hat{i}) + (2)(2)(\hat{i} \times \hat{j}) + (2)(-3)(\hat{i} \times \hat{k}) \\
&\quad + (3)(-3)(\hat{j} \times \hat{i}) + (3)(2)(\hat{j} \times \hat{j}) + (3)(-3)(\hat{j} \times \hat{k}) \\
&\quad + (-4)(-3)(\hat{k} \times \hat{i}) + (-4)(2)(\hat{k} \times \hat{j}) + (-4)(-3)(\hat{k} \times \hat{k})
\end{aligned}
$$

But $\hat{i} \times \hat{i} = 0, \hat{j} \times \hat{j} = 0$, and $\hat{k} \times \hat{k} = 0$. Use the mnemonic mentioned in Problem-Solving Tactic 2.14 to evaluate the other vector products of the unit vectors; you find

$$
\begin{aligned}
\vec{A} \times \vec{B} &= 4\hat{k} - 6(-\hat{j}) - 9(-\hat{k}) - 9\hat{i} + 12\hat{j} - 8(-\hat{i}) \\
&= -\hat{i} + 18\hat{j} + 13\hat{k}
\end{aligned}
$$

b. Use Problem-Solving Tactic 2.16 to check the calculation of a vector product in Cartesian form. Since $\vec{A} \times \vec{B}$ is perpendicular to $\vec{A}$, the scalar product of $\vec{A} \times \vec{B}$ with $\vec{A}$ should be zero:

$$
\begin{aligned}
(\vec{A} \times \vec{B}) \cdot \vec{A} &= (-\hat{i} + 18\hat{j} + 13\hat{k}) \cdot (2\hat{i} + 3\hat{j} - 4\hat{k}) \\
&= (-1)(2) + (18)(3) + (13)(-4) \\
&= -2 + 54 - 52 \\
&= 0
\end{aligned}
$$

Hence $\vec{A} \times \vec{B}$ is perpendicular to $\vec{A}$. Since $\vec{A} \times \vec{B}$ also is perpendicular to $\vec{B}$, the scalar product of $\vec{A} \times \vec{B}$ with $\vec{B}$ should be zero:

$$
\begin{aligned}
(\vec{A} \times \vec{B}) \cdot \vec{B} &= (-\hat{i} + 18\hat{j} + 13\hat{k}) \cdot (-3\hat{i} + 2\hat{j} - 3\hat{k}) \\
&= (-1)(-3) + (18)(2) + (13)(-3) \\
&= 3 + 36 - 39 \\
&= 0
\end{aligned}
$$

Hence $\vec{A} \times \vec{B}$ is perpendicular to $\vec{B}$.

2.18 VARIATION OF A VECTOR

But many that are first shall be last; and the last shall be first.
Matthew 19:30

Variations in vector quantities are ubiquitous and important in physics and engineering. If a vector quantity varies, its visual and Cartesian representations change. Vectors can change in several ways:

1. The *magnitude* of the vector may change, but its direction is preserved. The change may produce either an increase or decrease in magnitude. For example, in Figure 2.69 the speed of a car may increase or decrease (as indicated by its speedometer), causing a change in the magnitude of the velocity vector (which specifies the speed and direction of motion).

2. The *direction* of the vector changes, but its magnitude is preserved. For example, in Figure 2.70 a car moving at

constant speed around a curve changes its direction of motion, as indicated by a compass attached to the dashboard.

3. The magnitude *and* direction of the vector *both* change. For example, in Figure 2.71 the car may brake or speed up as it executes a curve. The speedometer reading changes, indicating a change in the speed (the magnitude of the velocity), and the compass direction changes (indicating a change in the direction of the velocity vector).

> The change in any vector always is defined to be the vector difference between the new (final) vector and the old (initial) vector.
>
> $$\Delta\vec{A} \equiv \vec{A}_f - \vec{A}_i \qquad (2.34)$$

This definition applies whether the change is in time or space (hence the relevance of the quotation at the beginning of this section). The change in a vector $\vec{A}$ is symbolized by $\Delta\vec{A}$, where the use of the symbolic mathematical Δ indicates "change in" and not multiplication of $\vec{A}$ by some scalar factor Δ. So the expression $\Delta\vec{A}$ should be read as *change in the vector $\vec{A}$* and should be considered a single algebraic vector entity. Notice that by

FIGURE 2.69 A car changing its speed but not its direction is changing its velocity vector.

FIGURE 2.70 A car changing its direction but not its speed is changing its velocity vector.

FIGURE 2.71 A car changing its speed and direction is changing its velocity vector.

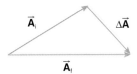

FIGURE 2.72 The change $\Delta\vec{A}$ in a vector $\vec{A}$.

transposing $\vec{A}_i$ in Equation 2.34, the change in the vector added *vectorially* to the initial vector produces the final vector, as shown geometrically in Figure 2.72:

$$\vec{A}_f = \vec{A}_i + \Delta\vec{A} \qquad (2.35)$$

The magnitude and direction of the change in *any* vector quantity are independent of the coordinate system used to describe the change.

This important fact is apparent in Figure 2.72, which makes no reference to any specific choice of a coordinate system.

2.19 SOME ASPECTS OF VECTOR CALCULUS

In our study of motion in the next chapter, we will have frequent need to differentiate vectors that are functions of time. We will be interested in dynamic changes where the independent variable of interest is the time t. Or, for other purposes, the independent variable of interest may be a position coordinate such as the Cartesian coordinates x, y, or z or a radial coordinate r.

Whatever the independent variable, remember that the *same* rules of calculus apply for differentiating functions whether the operation is

$$\frac{d}{dt}, \frac{d}{dx}, \frac{d}{dy}, \frac{d}{dz}, \text{ or } \frac{d}{dr}$$

Differentiation of a vector expressed in Cartesian form proceeds as follows. For example, the time derivative of a vector $\vec{A}$ is

$$\frac{d\vec{A}}{dt} = \frac{d}{dt}(A_x\hat{\imath} + A_y\hat{\jmath} + A_z\hat{k})$$
$$= \frac{d}{dt}(A_x\hat{\imath}) + \frac{d}{dt}(A_y\hat{\jmath}) + \frac{d}{dt}(A_z\hat{k})$$

Each differentiation is performed in turn, according to the standard differentiation rules. The Cartesian unit vectors $\hat{\imath}$, $\hat{\jmath}$, and $\hat{k}$ each have *fixed* magnitude ($= 1$) and fixed direction; they are *constant, unchanging* vectors.* Thus the time derivative of $\vec{A}$ is

$$\frac{d\vec{A}}{dt} = \frac{dA_x}{dt}\hat{\imath} + \frac{dA_y}{dt}\hat{\jmath} + \frac{dA_z}{dt}\hat{k} \qquad (2.36)$$

The derivative of a vector also is a vector, whose components are the derivatives of the respective scalar components of the original vector.

The derivative of *any constant vector* is *zero*. Recall that a constant vector has a constant magnitude *and* a fixed or unchanging direction. A vector that is of constant magnitude but changing direction is *not* a constant vector. A vector of fixed orientation but changing magnitude also is *not* a constant vector.

The distributive property applies to vector differentiation:

$$\frac{d}{dt}(\vec{A}_1 + \vec{A}_2) = \frac{d\vec{A}_1}{dt} + \frac{d\vec{A}_2}{dt} \qquad (2.37)$$

The differentiation of both the scalar and vector products proceeds according to the usual product rule in calculus. For example, differentiating a scalar product proceeds like

$$\frac{d}{dt}(\vec{A} \cdot \vec{B}) = \left(\frac{d\vec{A}}{dt}\right) \cdot \vec{B} + \vec{A} \cdot \left(\frac{d\vec{B}}{dt}\right) \qquad (2.38)$$

Note that the $+$ sign is that for scalar addition.

For the vector product, one must be careful to preserve the order of the terms in the product since they do not commute. Otherwise the rule is the standard product rule:

$$\frac{d}{dt}(\vec{A} \times \vec{B}) = \left(\frac{d\vec{A}}{dt}\right) \times \vec{B} + \vec{A} \times \left(\frac{d\vec{B}}{dt}\right) \qquad (2.39)$$

Note that the $+$ sign here is vector addition.

---

*This assumes that the *directions* of the coordinate axes in space do *not* change with time. A rotating coordinate system does not meet this condition; you will encounter rotating coordinate systems in more advanced courses in mechanics. We will always consider Cartesian coordinate systems with axes maintaining *fixed* directions in space, and so the time derivatives of the Cartesian unit vectors are zero.

Is the following expression legitimate? Explain why or why not.

$$\frac{d}{dt}(\vec{A} \cdot \vec{B}) = \vec{B} \cdot \left(\frac{d\vec{A}}{dt}\right) + \left(\frac{d\vec{B}}{dt}\right) \cdot \vec{A}$$

If the expression is not legitimate, correct it to make it a true statement.

EXAMPLE 2.11

Differentiate the following vector with respect to t:

$$\vec{A} = 5t^2\hat{\imath} + 6t\hat{\jmath} + 8\hat{k}$$

Solution

Differentiate each Cartesian component with respect to t using the standard differentiation rules; you obtain

$$\frac{d\vec{A}}{dt} = 5(2t)\hat{\imath} + 6(1)\hat{\jmath} + 0\hat{k}$$

$$= 10t\hat{\imath} + 6\hat{\jmath}$$

CHAPTER SUMMARY

A *scalar quantity* is completely specified by one numerical quantity that may be positive, negative, or zero.

A *vector quantity* is completely specified by its magnitude and direction in space; it also is specified by its three scalar components in an appropriate coordinate system (such as a Cartesian system).

The *magnitude of a vector* is a *positive scalar* quantity. A vector quantity can be visualized as a straight arrow whose length is drawn proportional to the magnitude of the vector. The magnitude of a vector $\vec{A}$ is symbolized by A or, equivalently, by $|\vec{A}|$.

Multiplying a vector $\vec{A}$ by a scalar quantity α produces a vector $\alpha\vec{A}$ whose magnitude is $|\alpha|$ times that of the original vector. The direction of $\alpha\vec{A}$ is the same as that of $\vec{A}$ if $\alpha > 0$. The direction of $\alpha\vec{A}$ is opposite to that of $\vec{A}$ if $\alpha < 0$.

Two vectors that point in the same direction are called *parallel* vectors. Two vectors that point in opposite directions are called *antiparallel* vectors.

A *unit vector* is a dimensionless vector of magnitude equal to 1. The Cartesian unit vectors $\hat{\imath}$, $\hat{\jmath}$, and $\hat{k}$ are directed toward increasing values of the coordinates x, y, and z, respectively.

A *component of a vector* is the projection of a vector on a specified coordinate axis or line. A component of a vector is a *scalar* quantity. The Cartesian representation of a vector $\vec{A}$ has the form

$$\vec{A} = A_x\hat{\imath} + A_y\hat{\jmath} + A_z\hat{k} \tag{2.19}$$

where A_x, A_y, and A_z are called the Cartesian components of $\vec{A}$. The Cartesian components are scalar quantities that may be positive, negative, or zero.

The rules for vector addition and subtraction are *not* the same as for scalar addition and subtraction. Arithmetic operations with vectors such as vector addition and vector subtraction of two vectors of the same physical type (two forces, two velocities, two accelerations, etc.) can be performed using either

i. geometric techniques via the *tail-to-tip method*; or
ii. analytical techniques when the vectors are expressed in Cartesian form:

$$\vec{F} \pm \vec{f} = (F_x \pm f_x)\hat{\imath} + (F_y \pm f_y)\hat{\jmath} + (F_z \pm f_z)\hat{k} \tag{2.24, 2.25}$$

The scalar product of any two vectors is a scalar quantity symbolized by $\vec{A} \cdot \vec{B}$. The scalar product is defined as

$$\vec{A} \cdot \vec{B} \equiv AB\cos\theta \tag{2.10}$$

where θ is the angle ($\leq 180°$) between the two vectors when they are drawn from a common point. The angle θ between two vectors *always* is considered positive. Because of the cosine term, the scalar product is positive if the angle θ between the two vectors is acute ($\theta < 90°$), negative if obtuse ($\theta > 90°$), and zero if $\theta = 90°$.

If the vectors are expressed in Cartesian form, the scalar product is the scalar sum of the products of their respective components:

$$\vec{A} \cdot \vec{B} = A_xB_x + A_yB_y + A_zB_z \tag{2.26}$$

The magnitude of a vector is the square root of the sum of the squares of its Cartesian components:

$$A = (A_x^2 + A_y^2 + A_z^2)^{1/2} \tag{2.22}$$

If two vectors are perpendicular to each other, their scalar product is zero.

The vector product of any two vectors is a new vector symbolized by $\vec{A} \times \vec{B}$. The vector product is a vector of magnitude

$$|\vec{A} \times \vec{B}| \equiv AB\sin\theta \tag{2.27}$$

whose direction is determined using the vector product right-hand rule. The vector product of any vector with itself is zero:

$$\vec{A} \times \vec{A} = 0 \tag{2.28}$$

The vector product of two parallel or two antiparallel vectors is zero.

The vector product of two vectors in Cartesian form is

$$\vec{A} \times \vec{B} = (A_yB_z - A_zB_y)\hat{\imath} + (A_zB_x - A_xB_z)\hat{\jmath} + (A_xB_y - A_yB_x)\hat{k} \tag{2.32}$$

but is more easily calculated and remembered using the mnemonic

$$\overset{\rightarrow \; + \; \rightarrow}{\hat{\imath} \quad \hat{\jmath} \quad \hat{k} \quad \hat{\imath} \quad \hat{\jmath} \quad \hat{k}}_{\leftarrow \; - \; \leftarrow}$$

for the Cartesian unit vectors.

A vector may vary in space and/or time in three ways:

• its magnitude may vary;
• its direction may vary; or
• both its magnitude and direction may vary.

The change in a vector quantity, $\Delta\vec{A}$, is defined as the vector difference

$$\Delta\vec{A} \equiv \vec{A}_f - \vec{A}_i \tag{2.34}$$

Vectors that vary in space or time or both, as well as scalar products and vector products, can be differentiated using rules similar to the calculus of scalar functions.

With a thorough grounding in vector analysis, you now have a well-stocked toolbox for successful achievement in physics.

SUMMARY OF PROBLEM-SOLVING TACTICS

2.1 **(page 38)**   Scalars are not vectors, and vectors are not scalars.

2.2 **(page 38)**   Draw visual representations of vectors carefully.

2.3 **(page 38)**   The magnitude of any vector quantity always is a positive scalar quantity (or zero).

2.4 **(page 40)**   Parallel transport vectors carefully.

2.5 **(page 41)**   There is a shortcut for the addition of many vectors.

2.6 **(page 45)**   The addition and subtraction of vector quantities is different from the addition and subtraction of scalars.

2.7 **(page 46)**   The result of the scalar product of two vectors is a scalar quantity, not another vector quantity.

2.8 **(page 46)**   To see if two vectors are perpendicular to each other, take their scalar product and see if it is equal to zero.

2.9 **(page 48)**   Always use right-handed coordinate systems.

2.10 **(page 50)**   Do not confuse a vector with its Cartesian components (which are scalars).

2.11 **(page 51)**   The mathematical sign convention for angles on a two-dimensional Cartesian coordinate grid is used for

Equation 2.17. But remember that the angle between two vectors always is considered positive.

2.12 **(page 52)**   The magnitude of any vector is found by taking the square root of the scalar product of the vector with itself.

2.13 **(page 56)**   When taking the vector product of two vectors, remember that the result is a vector perpendicular to both.

2.14 **(page 58)**   There is an easy way to remember the vector products of the Cartesian unit vectors:

$$\overset{\rightarrow\ +\ \rightarrow}{\hat{\imath}\ \ \hat{\jmath}\ \ \hat{k}\ \ \hat{\imath}\ \ \hat{\jmath}\ \ \hat{k}}_{\leftarrow\ -\ \leftarrow}$$

2.15 **(page 58)**   Do not memorize Equation 2.32; evaluate vector products in Cartesian form, term by term.

2.16 **(page 58)**   A way to check the calculation of a vector product is to verify that

$$(\vec{A} \times \vec{B}) \cdot \vec{A} = 0$$
$$(\vec{A} \times \vec{B}) \cdot \vec{B} = 0$$

QUESTIONS

1. (page 38); 2. (page 42); 3. (page 45); 4. (page 47); 5. (page 49); 6. (page 52); 7. (page 54); 8. (page 57); 9. (page 61)

10. Explain the distinction between the operation implied by **+** and the operation implied by +. Likewise, distinguish between the operation implied by **−** and the operation implied by −.

11. Show how a vector of magnitude 2 can be added to a vector of magnitude 1 to yield a vector of magnitude 2. What geometrical figure do the three vector arrows form?

12. What superficially has a magnitude and a direction (other than the rotations discussed in Section 2.5), but is not a true vector because it violates one of the rules of vector arithmetic?

13. Explain why a vector equation has a triple meaning.

14. Distinguish between the meaning of zero as a scalar and the zero vector.

15. Two vectors $\vec{F}_1$ and $\vec{F}_2$ have magnitudes such that $F_1 \neq F_2$. Is it possible for

$$\vec{F}_1 + \vec{F}_2 = 0?$$

16. If two vectors sum to zero:

$$\vec{F} + \vec{f} = 0$$

what can be said (if anything) about the magnitudes and directions of the two vectors?

17. Suppose vector $\vec{F}$ is the sum of two vectors $\vec{F}_1$ and $\vec{F}_2$:

$$\vec{F} = \vec{F}_1 + \vec{F}_2$$

Under what special circumstances does this also mean

$F = F_1 + F_2$? Why is it that F usually is *not* equal to $F_1 + F_2$?

18. Suppose vector $\vec{E}$ is the difference between two vectors:

$$\vec{E} = \vec{E}_1 - \vec{E}_2$$

Under what circumstances does this *also* mean $E = E_1 - E_2$? Under what circumstances is $E = E_2 - E_1$? Why is it that E usually is *not* equal to $E_1 - E_2$?

19. Show that if three vectors add to zero, that is, if

$$\vec{F}_1 + \vec{F}_2 + \vec{F}_3 = 0$$

then if the vectors are parallel transported to form their sum, the three vectors all lie in the same plane.

20. If $\vec{F}_1 + \vec{F}_2 + \vec{F}_3 = 0$, and $F_1 = F_2$, what is the maximum value of F_3? What is the minimum value of F_3?

21. Distinguish clearly between the multiplication of a vector by a scalar and the scalar product of two vectors.

22. Suppose you have a vector $\vec{A} \neq 0$. If $\vec{A} \cdot \vec{B} = 0$ *and* $\vec{A} \times \vec{B} = 0$, does this necessarily mean that $\vec{B} = 0$? Explain.

23. For any two arbitrary vectors $\vec{A}$ and $\vec{B}$, explain why the following two expressions are (a) both scalars and (b) both equal to zero:

$$\vec{A} \cdot (\vec{A} \times \vec{B})$$
$$\vec{B} \cdot (\vec{A} \times \vec{B})$$

No detailed calculations need be made.

24. Explain why the magnitude of a vector is always greater than or equal to the absolute value of its smallest Cartesian component.

25. What are the Cartesian components of the vector $\hat{\imath} - \hat{\jmath}$?

26. Is the following expression legitimate?

$$\vec{A} \times (\vec{F} + \vec{f}) = \vec{A} \times \hat{f} + \vec{A} \times \vec{F}$$

Explain why or why not.

27. Why is the following expression meaningless?

$$\vec{A} \times (\vec{B} \cdot \vec{C})$$

28. If the magnitude of $\vec{F}$ is 6 and that of $\vec{f}$ is 2, what restrictions (if any) can be placed on each of the following expressions: (a) $|\vec{F} + \vec{f}|$; (b) $|\vec{F} - \vec{f}|$; (c) $F + f$; (d) $F - f$; (e) $\vec{F} \cdot \vec{f}$; (f) $|\vec{F} \times \vec{f}|$.

29. Is it possible for a vector of zero magnitude to have a nonzero component? Explain.

30. If $\vec{A}$ is a vector and β is a scalar, explain why each of the following expressions is meaningless: (a) $\vec{A} \pm \beta$; (b) $\vec{A} \cdot \beta$; (c) $\vec{A} \times \beta$.

31. The Cartesian unit vectors $\hat{\imath}$, $\hat{\jmath}$, and $\hat{k}$ form a right-handed coordinate system. Do the vectors $-\hat{\imath}$, $-\hat{\jmath}$, and $-\hat{k}$ form a right-handed system?

32. A student is asked to take two given vectors and (a) form their sum; (b) find their components; (c) form their scalar product; and (d) form their vector product. For which of these vector operations is it absolutely necessary to specify a coordinate system?

33. In the context of our study of vectors, explain why the following product of two vectors has no meaning:

$$\vec{A}\vec{B}$$

34. One of the following expressions is correct and the other is incorrect. Explain.

$$\frac{d}{dt}(\vec{A} \times \vec{B}) = \vec{A} \times \left(\frac{d\vec{B}}{dt}\right) + \left(\frac{d\vec{A}}{dt}\right) \times \vec{B}$$

$$\frac{d}{dt}(\vec{A} \times \vec{B}) = \vec{B} \times \left(\frac{d\vec{A}}{dt}\right) + \vec{A} \times \left(\frac{d\vec{B}}{dt}\right)$$

By using appropriate minus signs, correct the incorrect expression.

35. Time intervals Δt have a magnitude

$$\Delta t = t_f - t_i$$

and a direction we can define as from the initial instant t_i to the final instant t_f. Why is a time interval not a vector?

36. Temperature can go up or down, yet it is not a vector. Explain why.

37. For two given vectors $\vec{A}$ and $\vec{B}$, show that there is more than one vector $\vec{C}$ that satisfies

$$A = \vec{B} \cdot \vec{C}$$

Also show that there is more than one vector $\vec{D}$ that satisfies

$$\vec{A} = \vec{B} \times \vec{D}$$

Explain why these results imply that we cannot define the vector division symbolized by $\vec{A}/\vec{B}$.

38. The previous question indicates why vector division is not well defined. On the other hand, for a given vector $\vec{A}$, explain why the following expressions are legitimate: (a) $1/A$; (b) $\vec{A}/4$.

PROBLEMS

Sections 2.1 Scalar and Vector Quantities
2.2 Multiplication of a Vector by a Scalar
2.3 Parallel Transport of Vectors
2.4 Vector Addition by Geometric Methods: Tail-to-Tip Method
2.5 Determining Whether a Quantity Is a Vector*
2.6 Vector Difference by Geometric Methods

1. For the two vectors $\vec{F}_1$ and $\vec{F}_2$ indicated in Figure P.1, show geometrically that the arrow representing $\vec{F}_1 + \vec{F}_2$ has the same length and orientation as the arrow representing $\vec{F}_2 + \vec{F}_1$. Thus the order in which two vectors are added is immaterial.

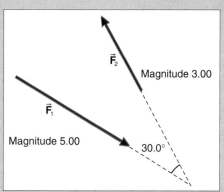

FIGURE P.1

2. For the two vectors indicated in Figure P.2, use geometric methods to find (a) $\vec{E}_1 + \vec{E}_2$; (b) $\vec{E}_1 - \vec{E}_2$; (c) $\vec{E}_2 - \vec{E}_1$. Also determine (d) $E_1 + E_2$; (e) $E_1 - E_2$; (f) $E_2 - E_1$; (g) $|\vec{E}_1 + \vec{E}_2|$; (h) $|\vec{E}_1 - \vec{E}_2|$; (i) $|\vec{E}_2 - \vec{E}_1|$.

FIGURE P.2

3. For the vectors shown in Figure P.3, find (a) $\vec{F} + \vec{f}$; (b) $F + f$; (c) $|\vec{F} + \vec{f}|$; (d) $\vec{F} - \vec{f}$; (e) $F - f$; (f) $|\vec{F} - \vec{f}|$.

4. Draw the vectors shown in Figure P.4 to full scale. Then use the diagram to find (a) $F + f$; (b) $|\vec{F} + \vec{f}|$; (c) $F - f$; (d) $|\vec{F} - \vec{f}|$.

FIGURE P.3

FIGURE P.4

• 5. On a level field, you begin at a wishing well and stroll east an unknown number of meters. Turning exactly northwest, you stroll again until you are 100 m due north of the wishing well. Sketch the situation. Determine the magnitudes of two vectors, one pointing east, one pointing exactly northwest that, when added, form a vector of magnitude 100 m pointing north.

• 6. For no good reason, you take 5.0 paces south, 3.0 paces west, and 6.0 paces exactly northeast. All the paces are of equal length. (a) Carefully sketch the situation. (b) At the end of the stroll, how many paces are you located south of the starting point? (c) How many paces are you east of the starting point? (d) How many paces are you from where you began? (e) Which of the answers to (b), (c), and (d), if any, depend on the *order* of the sequence of steps? (f) At the end of the small journey, in what direction are you relative to the starting point and the direction east? Specify an angle.

• 7. Let the vectors $\vec{r}_1$, $\vec{r}_2$, $\vec{r}_3$, $\vec{r}_4$, $\vec{r}_5$, and $\vec{r}_6$ be the position vectors from the center to the vertices of the regular hexagon as shown in Figure P.7. (a) Express each side (ab, bc, etc.) of the hexagon in terms of these vectors. (b) Show that

$$\vec{r}_1 + \vec{r}_2 + \vec{r}_3 + \vec{r}_4 + \vec{r}_5 + \vec{r}_6 = 0 \text{ m}$$

FIGURE P.7

Section 2.7 The Scalar Product of Two Vectors

8. Show that the expression

$$\frac{\vec{A} \cdot \vec{B}}{B}$$

represents the component of vector $\vec{A}$ on the line containing vector $\vec{B}$.

• 9. Suppose vector $\vec{R}$ has the same *magnitude* as vector $\vec{r}$ but they are *not* parallel or antiparallel to each other. Using an appropriate scalar product, show that the vector *sum* of the two vectors, $\vec{R} + \vec{r}$, is perpendicular to the vector *difference* of the two vectors, $\vec{R} - \vec{r}$. Does it matter which way the difference is taken? That is, will $\vec{R} + \vec{r}$ also be perpendicular to $\vec{r} - \vec{R}$?

Section 2.8 The Cartesian Coordinate System and the Cartesian Unit Vectors

10. Which of the coordinate systems in Figure P.10 are right-handed Cartesian coordinate systems and which are not?

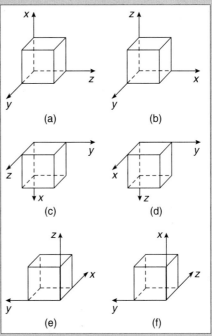

FIGURE P.10

Sections 2.9 The Cartesian Representation of Any Vector
2.10 Multiplication of a Vector Expressed in Cartesian Form by a Scalar
2.11 Expressing Vector Addition and Subtraction in Cartesian Form
2.12 The Scalar Product of Two Vectors Expressed in Cartesian Form
2.13 Determining the Angle Between Two Vectors Expressed in Cartesian Form

11. (a) What is the magnitude of the vector $\hat{i} + \hat{j} + \hat{k}$? (b) What is the magnitude of the vector $\hat{i} - \hat{j} + \hat{k}$? (c) Find a vector with only two Cartesian components that has the same magnitude. Is there only one solution?

12. A wizard creates the vector $\vec{A} = 2.0\hat{i} + 4.0\hat{j} - 3.0\hat{k}$. (a) Sketch a three-dimensional Cartesian coordinate system and indicate

the orientation of the vector, placing its tail at the origin. (b) What are the Cartesian components of the vector? (c) What is the magnitude of the vector? (d) Relabel the x-axis as the y-axis, the y-axis as the z-axis, and the z-axis as the x-axis. Is the new coordinate system right-handed? What is the new Cartesian form for the vector? What is the magnitude of the new Cartesian representation of the vector?

13. Use the definition of the scalar product to verify that the Cartesian unit vectors have the scalar products with each other that are given in Table 2.1.

14. If
$$\vec{A} = 2.0\hat{i} - 4.0\hat{j} + 6.0\hat{k}$$
and
$$\vec{B} = -1.0\hat{i} + 3.0\hat{j} + 3.0\hat{k}$$
(a) Find $\vec{A} \cdot \vec{B}$. (b) Find the magnitude of $\vec{A}$. (c) Find the magnitude of $\vec{B}$.

•15. An origin is on the ground and the x-axis is directed east, the y-axis is directed north, and the z-axis is directed vertically upward. The position of a turkey (*Meleagris gallopavo*) is determined to be
$$\vec{r}_1 = (3.0 \text{ m})\hat{i} + (4.0 \text{ m})\hat{j} + (5.0 \text{ m})\hat{k}$$
A blue-footed booby (*Sula nebouxii*) is located at
$$\vec{r}_2 = (-4.0 \text{ m})\hat{i} - (2.0 \text{ m})\hat{j} + (3.0 \text{ m})\hat{k}$$
(a) What is the difference in their heights above the ground? (b) How far is the turkey from the origin? (c) How far apart are the two birds?

•16. Imagine three position vectors:
$$\vec{r}_1 = (1.00 \text{ m})\hat{i} + (1.00 \text{ m})\hat{j} + (1.00 \text{ m})\hat{k}$$
$$\vec{r}_2 = (2.00 \text{ m})\hat{i} - (1.00 \text{ m})\hat{j} + (3.00 \text{ m})\hat{k}$$
$$\vec{r}_3 = (1.00 \text{ m})\hat{i} - (2.00 \text{ m})\hat{j} - (2.00 \text{ m})\hat{k}$$

with their tails at the origin. The heads of the vectors locate the heads of three partridges (*Perdix perdix*) in a pear tree, P_1, P_2, and P_3, respectively. Schematically sketch the situation. (a) What are the coordinates of the points P_1, P_2, and P_3? (b) Find the vector $\vec{\ell}_1$ that points *from* P_1 to P_2. (c) Find the vector $\vec{\ell}_2$ that points *from* P_2 to P_3. (d) Find the vector $\vec{\ell}_3$ that points *from* P_3 to P_1. (e) Show explicitly that $\vec{\ell}_1 + \vec{\ell}_2 + \vec{\ell}_3 = 0$ m.

•17. Find a vector that points in the same direction as the vector
$$\hat{i} + \hat{j} + \hat{k}$$
but is of magnitude 1. This problem illustrates how to construct a unit vector pointing along any arbitrary direction.

•18. Find a unit vector that points in the same direction as
$$\hat{i} + \hat{j} + 2\hat{k}$$

•19. Show that a unit vector in the same direction as vector $\vec{A}$, where $\vec{A}$ is an unspecified vector, is $\vec{A}/A$.

•20. Consider the two vectors:
$$\vec{L} = 1\hat{i} + 2\hat{j} + 3\hat{k}$$
and
$$\vec{\ell} = 4\hat{i} + 5\hat{j} + 6\hat{k}$$

Find the value of the scalar α such that the vector
$$\vec{L} - \alpha\vec{\ell}$$
is *perpendicular* to $\vec{L}$.

•21. Let vector $\vec{r} = \vec{r}_1 + \vec{r}_2$. Sketch this relationship geometrically. Evaluate the expression
$$r^2 = (\vec{r}_1 + \vec{r}_2) \cdot (\vec{r}_1 + \vec{r}_2)$$
You may recognize the result as the law of cosines from trigonometry.

•22. Use the general Cartesian representation of the vectors to show that the scalar product of $\vec{A}$ with the vector sum of $\vec{F}$ and $\vec{f}$, that is,
$$\vec{A} \cdot (\vec{F} + \vec{f})$$
is the same as the scalar sum of $\vec{A} \cdot \vec{F}$ and $\vec{A} \cdot \vec{f}$.

•23. Find the angle between the two vectors
$$\vec{A} = 3.00\hat{i} + 4.00\hat{j} + 5.00\hat{k}$$
and
$$\vec{B} = -1.00\hat{i} - 1.00\hat{j} - 1.00\hat{k}$$

•24. Find the angle between the vectors
$$\vec{A} = 1.0\hat{i} + 2.0\hat{j} - 1.0\hat{k}$$
and
$$\vec{B} = -1.0\hat{i} + 3.0\hat{j} + 5.0\hat{k}$$

•25. (a) Find the angle that the vector $\vec{A} = 3.00\hat{i} + 4.00\hat{j} + 5.00\hat{k}$ makes with the direction $\hat{i}$. (b) What angle does vector $\vec{A}$ make with $-\hat{i}$?

•26. Find the angle θ between the body diagonal of a cube and one of its edges (see Figure P.26) using the vector techniques of the scalar product.

FIGURE P.26

•27. Find the angle ϕ between the body diagonal of a cube and a face diagonal of the cube using vector techniques. See Figure P.27.

FIGURE P.27

•28. The principal body diagonal of a rectangular parallelopiped connects one corner with the opposite corner through the parallelopiped as shown in Figure P.28. Find the angle between the principal body diagonal of the rectangular parallelopiped and the side of length 8.0 cm.

•29. Consider the two vectors

$$\vec{A} = (\cos \alpha)\hat{i} + (\sin \alpha)\hat{j}$$

and

$$\vec{B} = (\cos \beta)\hat{i} + (\sin \beta)\hat{j}$$

where α and β are the respective angles that each vector makes with the +x-axis. (a) Show that each of these vectors has a magnitude equal to 1. (b) Make a schematic sketch of the vectors in the x–y plane. Make use of these vectors and the scalar product to derive the trigonometric identity

$$\cos(\alpha - \beta) = \cos \alpha \cos \beta + \sin \alpha \sin \beta$$

•30. Once upon a time, there was a vivid, vivacious vector $\vec{v}$ that had the Cartesian form

$$\vec{v} = 3.0\hat{i} + 5.0\hat{j} - 4.0\hat{k}$$

and lived happily ever after. (a) What is the z-component of $\vec{v}$? (b) Calculate the magnitude of $\vec{v}$. (c) What angle does $\vec{v}$ make with the +y-axis? (d) Formulate a unit vector that points in the same direction as $\vec{v}$.

•31. Let the angle that a vector $\vec{B}$ makes with the +x-axis be α; the angle that $\vec{B}$ makes with the +y-axis be β; and the angle that $\vec{B}$ makes with the +z-axis be γ, as shown in Figure P.31. The *cosines* of these angles are known as the *direction cosines*. Show that

$$\cos^2\alpha + \cos^2\beta + \cos^2\gamma = 1$$

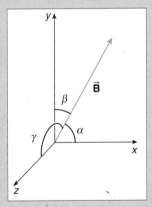

FIGURE P.31

‡32. Three vectors of magnitudes 5.00, 6.00, and 7.00 vector sum to zero. (a) Sketch the situation. (b) Find the angle between

•   the vector of magnitude 5.00 and the vector of magnitude 6.00;
•   the vector of magnitude 6.00 and the vector of magnitude 7.00;
•   the vector of magnitude 7.00 and the vector of magnitude 5.00.

Remember that the angle between two vectors is the angle between them when they are drawn from a common point, with their tails tied together.

‡33. Two vectors of magnitude 12 and 16 have a vector sum of magnitude 20. (a) Find the angle the two vectors of magnitudes 12 and 16 make with each other. (b) Find the angle that each makes with their vector sum.

‡34. You can use the scalar product to determine the shortest distance between two locations on the Earth measured along the surface. Figure P.34 represents a section of the northern hemisphere of the Earth. A point such as A on the surface of the Earth is located (in spherical coordinates) by the three coordinates (R, θ, ϕ) where R is the radius of the Earth (here assumed to be spherical with $R = 6.37 \times 10^3$ km). (a) Show that with the indicated Cartesian coordinate system, the Cartesian position vector of point A is

$$\vec{R}_A = R[(\sin \theta \cos \phi)\hat{i} + (\sin \theta \sin \phi)\hat{j} + (\cos \theta)\hat{k}]$$

where the angle θ is the complement of the latitude L of point A:

$$\theta = 90° - L$$

and ϕ is the longitude (with the convention that longitudes *east* are positive angles and longitudes *west* are negative angles. (b) Write the Cartesian position vector of New York (latitude 40.7° N, longitude 74.0° W). (c) Write the Cartesian position vector of Copenhagen, Denmark (latitude 55.7° N, longitude 12.5° E). (d) Use the scalar product technique to find the angle β between these two position vectors. (e) If β is measured in radians, show that the shortest distance between the cities measured on the surface of the Earth is $R\beta$. (f) Evaluate the distance between New York and Copenhagen in kilometers.

This problem was inspired by an article by B. Cameron Reed, "Dot products and great-circle distances," *The Physics Teacher*, 26, #5, pages 280–281 (May 1988).

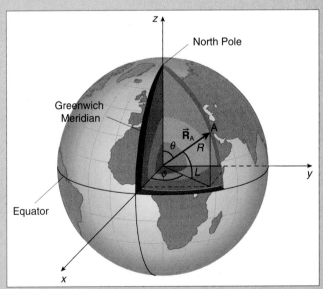

FIGURE P.34

‡35. Let the vectors $\vec{r}_1$, $\vec{r}_2$, $\vec{r}_3$, $\vec{r}_4$, and $\vec{r}_5$ be the position vectors from the center to the vertices of the Pentagon in Washington, D.C., as shown in Figure P.35. (a) Express each side (ab, bc,

etc.) of the Pentagon in terms of these vectors. (b) Show that

$$\vec{r}_1 + \vec{r}_2 + \vec{r}_3 + \vec{r}_4 + \vec{r}_5 = 0 \text{ m}$$

FIGURE P.35

Sections 2.14 Equality of Two Vectors
2.15 Vector Equations

36. If $\vec{F} = 3.00\vec{a}$ and $F_x = 4.00$, $F_y = -6.00$, and $F_z = 9.00$, find $\vec{a}$.

37. If

$$3.00\hat{i} + 4.00\hat{j} - 6.00\hat{k}$$
$$= (2.00 - \alpha)\hat{i} - (6.00 + \beta)\hat{j} + (3.00 - \gamma)\hat{k}$$

find the values of α, β, and γ.

Sections 2.16 The Vector Product of Two Vectors
2.17 The Vector Product of Two Vectors Expressed in Cartesian Form

38. Evaluate the following expressions: (a) $(\hat{i} + \hat{j}) \cdot (\hat{i} + \hat{k})$; (b) $(\hat{i} + \hat{j}) \times (\hat{i} + \hat{k})$.

39. Show that the magnitude of the vector defined by

$$\vec{A} \times (\vec{A} \times \vec{B})$$

is $A^2B \sin \theta$, where θ is the angle between $\vec{A}$ and $\vec{B}$.

40. Find the magnitude of the vector $(\hat{i} + \hat{j}) \times \hat{k}$.

41. Find a vector perpendicular to $\hat{i} + \hat{j}$. Is your solution unique?

42. Show that

$$\vec{A} \times \vec{B} = (A_x\hat{i} + A_y\hat{j} + A_z\hat{k}) \times (B_x\hat{i} + B_y\hat{j} + B_z\hat{k})$$

reduces to

$\vec{A} \times \vec{B}$

$$= (A_yB_z - A_zB_y)\hat{i} + (A_zB_x - A_xB_z)\hat{j} + (A_xB_y - A_yB_x)\hat{k}$$

43. Given the two vectors

$$\vec{A} = -\hat{i} + 2\hat{j} + 3\hat{k}$$

and

$$\vec{B} = 2\hat{i} + \hat{j} - \hat{k}$$

(a) Find $\vec{A} \times \vec{B}$. (b) Verify that $\vec{A} \times \vec{B}$ is indeed perpendicular to *both* $\vec{A}$ and $\vec{B}$ by showing

$$(\vec{A} \times \vec{B}) \cdot \vec{A} = 0$$

and

$$(\vec{A} \times \vec{B}) \cdot \vec{B} = 0$$

(c) Find the vector product $\vec{B} \times \vec{A}$. Think! You do *not* have to reevaluate the whole expression; the answer is almost right before your eyes if you evaluated $\vec{A} \times \vec{B}$ correctly.

44. Deep inside an ancient physics text, you discover two vectors:

$$\vec{A} = 2\hat{i} - \hat{j} + 4\hat{k}$$

and

$$\vec{B} = 5\hat{i} + 2\hat{j} - 2\hat{k}$$

Not content with these hoary relics, you are asked to find a new vector that is *perpendicular to both* $\vec{A}$ and $\vec{B}$. Is the solution unique?

45. You are given a vector

$$\vec{A} = -3.20\hat{i} + 2.40\hat{j}$$

and a vector $\vec{B}$ in the x–y plane of magnitude 5.20. Vector $\vec{B}$ makes an angle of 120° with the positive x-axis. (a) Find $\vec{A} \cdot \vec{B}$. (b) Find $\vec{A} \times \vec{B}$.

46. Find a vector that is perpendicular to both the vector

$$\hat{i} + \hat{j} + \hat{k}$$

and the vector

$$3\hat{i} - \hat{j} + 2\hat{k}$$

Are there other possible solutions?

•**47.** Find a vector that is perpendicular to the vector

$$3\hat{i} - 2\hat{j} + 4\hat{k}$$

Is the solution unique? Explain your answer with the assistance of a sketch.

•**48.** Create a vector of magnitude unity perpendicular to both $2.0\hat{i} + 1.5\hat{j}$ and $2.0\hat{j} + 3.2\hat{k}$. Is there only one solution?

•**49.** The vectors

$$\vec{A} = \hat{i} + 2\hat{j} - \hat{k}$$

and

$$\vec{C} = 3\hat{j} + 6\hat{k}$$

were created to appease the academic gods. (a) Make an offering by finding $\vec{A} \cdot \vec{C}$. (b) Find a vector $\vec{B}$ so that

$$\vec{A} \times \vec{B} = \vec{C}$$

Is the solution for $\vec{B}$ unique? (c) Using the result of part (b), what is $\vec{B} \cdot \vec{C}$?

•**50.** Show that

$$(\vec{A} \cdot \vec{B})^2 + |\vec{A} \times \vec{B}|^2 = A^2B^2$$

•**51.** For the three vectors indicated in Figure P.51, find (a) $\vec{B}_1 + \vec{B}_2 + \vec{B}_3$; (b) $\vec{B}_1 + \vec{B}_2 - 2\vec{B}_3$; (c) $\vec{B}_1 \cdot \vec{B}_2$; (d) $\vec{B}_1 \times \vec{B}_2$; (e) $\vec{B}_3 \times \vec{B}_1$; (f) $\vec{B}_3 \cdot \vec{B}_1$.

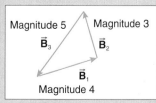

FIGURE P.51

•52. By evaluating the indicated vector and scalar products, show that

$$[\vec{A} \cdot (\vec{B} \times \hat{i})]\hat{i} + [\vec{A} \cdot (\vec{B} \times \hat{j})]\hat{j} + [\vec{A} \cdot (\vec{B} \times \hat{k})]\hat{k}$$
$$= (A_y B_z - A_z B_y)\hat{i} + (A_z B_x - A_x B_z)\hat{j} + (A_x B_y - A_y B_x)\hat{k}$$

Thus, in view of Equation 2.32,

$$[\vec{A} \cdot (\vec{B} \times \hat{i})]\hat{i} + [\vec{A} \cdot (\vec{B} \times \hat{j})]\hat{j} + [\vec{A} \cdot (\vec{B} \times \hat{k})]\hat{k} = \vec{A} \times \vec{B}$$

•53. Use the Cartesian representation of each vector to show that $\vec{A} \times (\vec{B}_1 + \vec{B}_2)$ is equivalent (i.e., identical) to $\vec{A} \times \vec{B}_1 + \vec{A} \times \vec{B}_2$, so that the vector product obeys a distributive law.

$$\vec{A} \times (\vec{B}_1 + \vec{B}_2) = \vec{A} \times \vec{B}_1 + \vec{A} \times \vec{B}_2$$

•54. Three points on a large jungle gym have Cartesian coordinates (x, y, z) given by

$$P_1: \ 1.00 \text{ m}, \ 2.00 \text{ m}, \ 3.00 \text{ m}$$
$$P_2: -2.00 \text{ m}, \ 1.00 \text{ m}, \ 4.00 \text{ m}$$
$$P_3: \ 2.00 \text{ m}, -3.00 \text{ m}, \ 5.00 \text{ m}$$

(a) Let $\vec{\ell}_1$ be the vector from P_1 to P_2. Find the Cartesian representation of $\vec{\ell}_1$. (b) Let $\vec{\ell}_2$ be the vector from P_1 to P_3. Find the Cartesian representation of $\vec{\ell}_2$. (c) Relate the area of the triangle formed by the three points to the *magnitude* of the vector product of $\vec{\ell}_1$ and $\vec{\ell}_2$, and find the numerical value of the area.

•55. While cleaning out the lab, you discover two soiled vectors named

$$\vec{p}_1 = 2.1\hat{i} - 5.3\hat{j} + 3.4\hat{k}$$

and

$$\vec{p}_2 = 3.6\hat{i} + 2.8\hat{j} - 0.9\hat{k}$$

Find (a) $\vec{p}_1 + \vec{p}_2$; (b) $\vec{p}_1 - \vec{p}_2$; (c) the magnitude of $\vec{p}_1$; (d) the magnitude of $\vec{p}_2$; (e) the scalar product of $\vec{p}_1$ and $\vec{p}_2$; (f) the vector product $\vec{p}_1 \times \vec{p}_2$; (g) the angle between $\vec{p}_1$ and $\vec{p}_2$.

•56. Three vectors are expressed in Cartesian form:

$$\vec{A} = 4\hat{j}$$
$$\vec{B} = \hat{i} - 3\hat{j} + 2\hat{k}$$
$$\vec{C} = 2\hat{i} - \hat{j} + 3\hat{k}$$

(a) For the pure thrill of it, evaluate

$$\vec{A} \times (\vec{B} \times \vec{C})$$

and show that it is equivalent to

$$(\vec{A} \cdot \vec{C})\vec{B} - (\vec{A} \cdot \vec{B})\vec{C}$$

This equivalence is true for *any* three vectors, not just the ones given. We state this without proof, but you should be able to verify the identity of the two expressions for three general vectors expressed in Cartesian form. (b) What is $\vec{A} \cdot [\vec{A} \times (\vec{B} \times \vec{C})]$? What does this result mean?

•57. Consider the vectors

$$\vec{A}_1 = 4.0\hat{j}$$
$$\vec{A}_2 = 1.0\hat{i} - 3.0\hat{j} + 2.0\hat{k}$$
$$\vec{A}_3 = 2.0\hat{i} - 1.0\hat{j} + 3.0\hat{k}$$

(a) Find $\vec{A}_1 + \vec{A}_2 - 2\vec{A}_3$. (b) Find $\vec{A}_1 \cdot (\vec{A}_2 \times \vec{A}_3)$. (c) For no apparent reason, calculate the angle between $\vec{A}_2$ and $\vec{A}_3$.

‡58. Consider the following vector relationship:

$$\vec{F} = q(\vec{v} \times \vec{B})$$

where q is a scalar. In our study of magnetism (Chapter 20), we will see that this vector relationship gives the force $\vec{F}$ on a charge q moving at a velocity $\vec{v}$ in a magnetic field $\vec{B}$. Suppose you conduct some experiments and find the following:

If the vector $\vec{v} = \hat{i}$, then $\vec{F}/q = -4\hat{j} - 2\hat{k}$.

If the vector $\vec{v} = \hat{j}$, then $\vec{F}/q = 4\hat{i} - 3\hat{k}$.

If the vector $\vec{v} = \hat{k}$, then $\vec{F}/q = 2\hat{i} + 3\hat{j}$.

Use this information to find the vector $\vec{B}$.

‡59. This problem illustrates how vector notation can be used to write some of the laws of physics. In geometric optics, if a ray of light (1) is incident on a boundary between two transparent media as in Figure P.59, some of the light is reflected, producing a reflected ray (2), and some of the light is transmitted, producing a refracted ray (3). Among other things, the law of reflection states that

$$\theta_1 = \theta_2$$

and the law of refraction states that

$$n_1 \sin \theta_1 = n_2 \sin \theta_3$$

where n_1 and n_2 are two pure numbers characteristic of the respective media (they are known as the indices of refraction). Let $\hat{r}_1$, $\hat{r}_2$, and $\hat{r}_3$ be dimensionless *unit vectors* pointing in the direction of the respective rays as indicated in Figure P.59. Furthermore, let $\hat{r}$ be a dimensionless *unit vector* perpendicular to the interface as shown. (a) Use these unit vectors to write the law of reflection as a scalar product. (b) Use these unit vectors, the indices of refraction, and the concept of a vector product to write a vector equation whose magnitude yields the law of refraction.

This problem was inspired by a problem in Mary L. Boas, *Mathematical Methods in the Physical Sciences* (2nd edition, Wiley, New York, 1983), page 243.

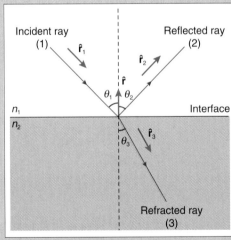

FIGURE P.59

‡60. The story in Figure P.60A appears in George Gamow's book *One Two Three . . . Infinity* (Bantam, New York, 1961), Chapter II.

There was a young and adventurous man who found among his great-grandfather's papers a piece of parchment that revealed the location of a hidden treasure. The instructions read:

"Sail to _____ North latitude and _____ West longitude[1] where thou wilt find a deserted island. There lieth a large meadow, not pent, on the north shore of the island where standeth a lonely oak and a lonely pine.[2] There thou wilt see also an old gallows on which we once were wont to hang traitors. Start thou from the gallows and walk to the oak counting thy steps. At the oak thou must turn *right* by a right angle and take the same number of steps. Put here a spike in the ground. Now must thou return to the gallows and walk to the pine counting thy steps. At the pine thou must turn *left* by a right angle and see that thou takest the same number of steps, and put another spike into the ground. Dig halfway between the spikes; the treasure is there."

The instructions were quite clear and explicit, so our young man chartered a ship and sailed to the South Seas. He found the island, the field, the oak and the pine, but to his great sorrow the gallows were gone. Too long a time had passed since the document had been written; rain and sun and wind had disintegrated the wood and returned it to the soil, leaving no trace even of the place where it once had stood.

Our adventurous young man fell into despair, then in an angry frenzy began to dig at random all over the field. But all his efforts were in vain: the island was too big! So he sailed back with empty hands. And the treasure is probably still there.

[1]The actual figures of longitude and latitude were given in the document but are omitted in this text, in order not to give away the secret.

[2]The names of the trees are also changed for the same reason as above. Obviously there would be other varieties of trees on a tropical treasure island.

FIGURE P.60A

The location of the treasure can be found by using vector analysis in the following way. Construct a Cartesian coordinate system as indicated in Figure P.60B with an origin halfway between the oak and pine trees with the x- and y-axes directed as indicated. The position vectors of the pine and oak trees then are

$$\vec{r}_{pine} = \ell\hat{i}$$
$$\vec{r}_{oak} = -\ell\hat{i}$$

Let the unknown location of the gallows be represented by the vector $\vec{R}$:

$$\vec{R} = R_x\hat{i} + R_y\hat{j}$$

(a) Show that the vector $\vec{L}_1$ *from* the gallows *to* the oak tree is given by

$$\vec{L}_1 = \vec{r}_{oak} - \vec{R}$$
$$= -(\ell + R_x)\hat{i} - R_y\hat{j}$$

Follow the map directions and use the concept of the vector product to show that the location of spike 1 is

$$\vec{R}_1 = \vec{L}_1 \times \hat{k}$$
$$= -R_y\hat{i} + (\ell + R_x)\hat{j}$$

with the tail of $\vec{R}_1$ located at the oak. (b) Show that the vector $\vec{L}_2$ *from* the gallows *to* the pine tree is given by

$$\vec{L}_2 = \vec{r}_{pine} - \vec{R}$$
$$= (\ell - R_x)\hat{i} - R_y\hat{j}$$

Follow the map directions and use the concept of the vector product to show that the location of spike 2 is

$$\vec{R}_2 = \hat{k} \times \vec{L}_2$$
$$= R_y\hat{i} + (\ell - R_x)\hat{j}$$

with the tail of $\vec{R}_2$ located at the pine. (c) Show that the position vector $\vec{r}$ of the treasure is

$$\vec{r} = \ell\hat{j}$$

or along the y-axis at a distance from the origin equal to half the distance between the trees.

This problem is adapted from R. Ramirez-Bon, "An interesting problem solved by vectors," *The Physics Teacher*, 28, #9, pages 594–595 (December 1990).

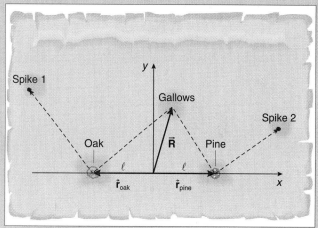

FIGURE P.60B

61. Three noncoplanar vectors with their tails at a common point can define a parallelopiped as shown in Figure P.61. (a) Show that $|\vec{B} \times \vec{C}|$ is the area of a parallelogram whose sides have lengths B and C. (b) Show that the expression $\vec{A} \cdot (\vec{B} \times \vec{C})$ represents the volume of the parallelopiped whose sides are formed from the vectors as in Figure P.61.

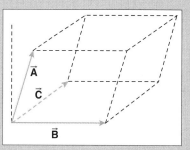

FIGURE P.61

62. Crystals are of interest to geologists, crystallographers, physicists, engineers, and jewelry makers. It is customary to locate the atoms defining the elementary or primitive structural arrangement of a crystal by means of what are called *primitive lattice vectors*. For example, many elements, such as sodium, chromium, and zirconium, have a crystalline structure known as the body-centered cubic lattice. For this structure, the primitive lattice vectors are

$$\vec{a} = \frac{\ell}{2}(\hat{\imath} + \hat{\jmath} - \hat{k})$$

$$\vec{b} = \frac{\ell}{2}(-\hat{\imath} + \hat{\jmath} + \hat{k})$$

$$\vec{c} = \frac{\ell}{2}(\hat{\imath} - \hat{\jmath} + \hat{k})$$

where ℓ is the side of the elementary cubic structure and $\hat{\imath}$, $\hat{\jmath}$, and $\hat{k}$ are the usual Cartesian unit vectors, here directed parallel to the edges of the elementary cubic structure. The locations of all the atoms in the crystal can be represented by a linear combination of these primitive lattice vectors with integral coefficients. The result of Problem 61 indicates that the volume of the primitive cell of the lattice, whose sides are the vectors $\vec{a}$, $\vec{b}$, and $\vec{c}$, is

$$\vec{a} \cdot (\vec{b} \times \vec{c})$$

This is a real, physical volume since the lattice vectors represent physical lengths. (a) Use the primitive lattice vectors defined for the body-centered cubic lattice to show that the volume of the primitive cell is

$$\vec{a} \cdot (\vec{b} \times \vec{c}) = \frac{\ell^3}{2}$$

Another useful set of vectors associated with the crystal are known as the *reciprocal lattice vectors*. They are defined in the following way:

$$\vec{a}' \equiv 2\pi \frac{\vec{b} \times \vec{c}}{\vec{a} \cdot (\vec{b} \times \vec{c})}$$

$$\vec{b}' \equiv 2\pi \frac{\vec{c} \times \vec{a}}{\vec{a} \cdot (\vec{b} \times \vec{c})}$$

$$\vec{c}' \equiv 2\pi \frac{\vec{a} \times \vec{b}}{\vec{a} \cdot (\vec{b} \times \vec{c})}$$

(For some purposes the factor of 2π is omitted from the definition, so that one needs to be careful about *what* definition is used for the reciprocal lattice vectors.) (b) Find the reciprocal lattice vectors for the body-centered cubic lattice. (c) Find the volume of the primitive cell formed by the reciprocal lattice vectors. That is, find

$$\vec{a}' \cdot (\vec{b}' \times \vec{c}')$$

(The result is not a real physical volume, since the dimensions of the reciprocal lattice vectors are not lengths but *inverse* lengths.) The primitive cell formed by the reciprocal lattice vectors is known in the trade as the *first Brillouin zone*.

Sections 2.18 Variation of a Vector
2.19 Some Aspects of Vector Calculus

63. Differentiate the vector $\vec{A}(t) = 3t^2\hat{\imath} + 2t\hat{\jmath} + 8\hat{k}$ with respect to t.

•64. Differentiate the vector

$$\vec{A}(t) = [3\sin(5t)]\hat{\imath} + [4\cos(2t)]\hat{\jmath} + 3e^{-4t}\hat{k}$$

with respect to t.

•65. Differentiate the vector

$$\vec{A}(t) = 5\sqrt{t}\,\hat{\imath} + 6t^{3/2}\hat{\jmath}$$

with respect to t.

•66. Differentiate the vector

$$\vec{r}(t) = [A\cos(\omega t)]\hat{\imath} + [A\sin(\omega t)]\hat{\jmath}$$

with respect to t, where ω is a constant.

•67. Use the Cartesian representation of each vector to show explicitly that

$$\frac{d}{dt}(\vec{F} + \vec{f})$$

the time derivative of the sum of $\vec{F}$ and $\vec{f}$, is the same as the vector sum of

$$\frac{d\vec{F}}{dt}$$

and

$$\frac{d\vec{f}}{dt}$$

In other words, first add $\vec{F}$ and $\vec{f}$ as vectors and then take the derivative of their sum. Then show that this result is identical to that found by taking the vector sum of the time derivative of $\vec{F}$ and the time derivative of $\vec{f}$.

•68. Use the Cartesian representation of the vectors to show that the time derivative of the scalar product of $\vec{A}$ and $\vec{B}$,

$$\frac{d}{dt}(\vec{A} \cdot \vec{B})$$

is the same as the scalar sum of

$$\frac{d\vec{A}}{dt} \cdot \vec{B}$$

and

$$\vec{A} \cdot \frac{d\vec{B}}{dt}$$

•69. Use the Cartesian representation of the vectors to show that the time derivative of the vector product of $\vec{A}$ and $\vec{B}$,

$$\frac{d}{dt}(\vec{A} \times \vec{B})$$

is the same as the vector sum of

$$\frac{d\vec{A}}{dt} \times \vec{B}$$

and

$$\vec{A} \times \frac{d\vec{B}}{dt}$$

Thus the derivative of a vector product obeys the usual product rule in calculus as long as we preserve the order of the terms.

INVESTIGATIVE PROJECTS

A. Expanded Horizons

1. Investigate the similarities between the addition and subtraction of two-dimensional vectors and the algebraic addition and subtraction of complex numbers.

    Ruel V. Churchill, *Complex Variables and Applications* (2nd edition, McGraw-Hill, New York, 1960), pages 4–5.

    W. E. Baylis, J. Huschilt, and Jiansu Wei, "Why *i*?" *American Journal of Physics*, 60, #9, pages 788–797 (September 1992).

2. Vector notation as we know it was invented only about a century ago. Should you be interested in researching its historical roots, a good place to begin is the following:

    A. P. Wills, *Vector Analysis with an Introduction to Tensor Analysis* (Dover, New York, 1958), Chapter 1.

    Michael J. Crowe, *A History of Vector Analysis* (Dover, New York, 1994).

3. Investigate the use of vectors in crystallography. In particular, discover the form for the basis vectors for various crystal symmetries and the corresponding reciprocal lattice vectors (see Problem 62).

    Charles Kittel, *Introduction to Solid State Physics* (7th edition, John Wiley, New York, 1996), Chapters 1 and 2.

    George R. Mitchell, "The reciprocal lattice—a demonstration," *American Journal of Physics*, 46, #5, pages 574–575 (May 1978).

4. Investigate the mathematical field of *Clifford algebra* and how it relates to the vector algebra of this chapter.

    Anthony Garrett, "An advertisement for Clifford algebra," *Physics World*, 5, #9, pages 13–14 (September 1992). This short article references tutorial articles about Clifford algebra.

    David Hestenes, *Clifford Algebra to Geometric Calculus: A Unified Language for Mathematics and Physics* (Kluwer Academic Press, Dordrecht, Netherlands 1987).

5. Investigate the analytical or algebraic extension of vector ideas in three dimensions to tensors using matrix techniques.

    Robert B. Leighton, *Principles of Modern Physics* (McGraw-Hill, New York, 1959), pages 15–22.

6. Study the general area of Cartesian tensors and the Einstein "dummy index" (repeated index) notational convention as a way of representing summations in vector and tensor analysis.

    Mary L. Boas, *Mathematical Methods in the Physical Sciences* (2nd edition, John Wiley, New York, 1983), pages 437–441.

    Robert B. Leighton, *Principles of Modern Physics* (McGraw-Hill, New York, 1959), pages 20–21.

7. Investigate the distinction between (a) scalars and *pseudoscalars* and (b) vectors and *pseudovectors* (also known as *axial vectors*).

    Llewelyn G. Chambers, *A Course in Vector Analysis* (Chapman and Hall, London, 1969), pages 58–60.

    Walter Hauser, "Vector products and pseudovectors," *American Journal of Physics*, 54, #2, pages 168–172 (February 1986).

8. Investigate the ideas of vector analysis in coordinate systems with axes that are *not* perpendicular to each other (as is common in crystallography). In particular, investigate the distinction between *covariant* and *contravariant* components of a vector.

    Mary L. Boas, *Mathematical Methods in the Physical Sciences* (2nd edition, John Wiley, New York, 1983), pages 447–449.

    George Arfken, *Mathematical Methods for Physicists* (Academic Press, New York, 1970), pages 184–187.

B. Communicating Physics

9. A local newspaper editor calls you as a technical expert and asks you to write a short but interesting feature for the science page that distinguishes between scalars and vectors for the edification of the nontechnical general public. Write such an article, keeping within the space limitations set by the editor: 300 words. Suggest an interesting title for your article.

CHAPTER 3

KINEMATICS I
RECTILINEAR MOTION

**Examine the beauty of mobility in a body—Numbers are turned
in time. Go within to the art whence these numbers proceed.
Look there for time and for place.**

St. Augustine (A.D. 354–430) *

If nothing moved, we might as well be rocks. Motion implies change, and it is change that makes life—and physics—viable and interesting. We move, soccer balls move (hopefully toward the right goal), the Earth moves, electrons move, and light *really* moves. The whole universe is a bewildering array of incessant motion. For this reason we begin our study of physics with a description of motion, called **kinematics**.[†]

Our study of motion begins with that of a single **particle**, which is the basic building block for the study of the motion of more complex systems. A particle is anything whose geometric size is ignored; it is represented by a point in space. Protons, electrons, atoms, molecules, soccer balls, cars, boats, *you*, the Earth, Moon, Sun, and stars all are particles if we represent them as pointlike objects whose geometric size is ignored.

The study of motion has roots deep in antiquity. Observations of the night sky with the moving Moon and wandering planets were some of the first inspirations for studying motion. The study was frustrating. Developing a consistent theory of motion was difficult because, as we mentioned in Chapter 1, physical theories do not exist in nature just waiting to be discovered by some lucky scientist, like a prospector sifting for precious gold nuggets. Physical theories are *creations* of the human intellect; they must be *invented* rather than discovered. In this sense, physics is like the arts and literature: it is a creative process. It builds on the best of the past to create new visions of nature.

The poetry of motion!

Kenneth Grahame (1859–1932)[‡]

When Johannes Kepler (1571–1630) first was able to decipher accurately and precisely the paths and positions of the wandering planets, including the Earth, during the early 1600s,[§]

---

[*](Chapter Opener) *De Libero Arbitrio*, translated by Francis E. Tourscher (The Peter Reilly Company, Philadelphia, 1937), Book III, Chapter 16, page 205.

[†]The word comes from the Greek word κίνημα, meaning motion; the word *cinema* has the same Greek root.

[‡]*The Wind in the Willows* (Charles Scribner's, New York, 1908), Chapter II, "The Open Road," page 41.

[§]Kepler's three empirical laws of planetary motion will be examined in Chapter 6.

it seemed possible that the motions of the universe around us, even motions here on the Earth, were within the grasp of the human intellect. Once the Earth was seen by Nicholas Copernicus (1473–1543) and Johannes Kepler to be just another planet of the Sun, there was no longer any reason to think that mathematics could be applied only to the celestial realm and not to physical phenomena in our own backyard.

It was Galileo Galilei (1564–1642) who revolutionized the description of the motion of particles. The work of Galileo was a great turning point in the development of quantitative science, for it was one of the opening salvos of a scientific revolution that continues to this day. Galileo broke radically with the 2000-year-old—but incorrect—Aristotelian ideas that held that while mathematics (with the purity of geometry and numbers) could be applied to the "perfect," godlike, celestial realm of the stars (astronomy), it could not be applied to the mundane, changing phenomena of nature on the Earth.

Galileo's genius lay not only in realizing that mathematics was the key to the sciences, but also in bridging the intellectual chasm between observations (experiments) and mathematical models. He realized that *approximations* were needed to limit complexity. The natural world brims with effects that hinder generalizations, friction and the finite size of material objects being the foremost culprits impeding the development of an accurate, consistent theory of motion. Galileo was among the first to design experiments to test his ideas, experiments that enabled him to probe or isolate one factor from another in the description of motion. Galileo invented the description of motion using inspiration, experiment, and extrapolations from experiments in the real world to idealizations that approximated reality, such as situations without friction.

In this chapter and the next, we examine a quantitative theory of kinematics whose roots lie with Galileo. The underlying theory of the *causes of motion*, called **dynamics**, was developed after Galileo by Isaac Newton (1642–1727) and published in 1686 as the *Principia*. The Newtonian theory of dynamics will be examined in Chapter 5.

Johannes Kepler deciphered the structure of the solar system.

Galileo Galilei made many significant contributions to kinematics.

FIGURE 3.1 Examples of rectilinear motion.

3.1 RECTILINEAR MOTION

I saw before me a vast multitude of small Straight Lines . . . interspersed with other Beings still smaller . . . all moving to and fro in one and the same Straight Line.

Edwin Abbott Abbott (1838–1926)*

Roll a marble across the floor, drop a rock from a high bridge into a river below, toss a ball straight up and watch it ascend and descend, or shoot an electron down the straight beam tube of the Stanford Linear Accelerator (SLAC). All are examples of **rectilinear motion**, the motion of a particle along a straight line. Other examples of rectilinear motion are shown in Figure 3.1.

The generalization to two- and three-dimensional kinematics involves repetitions of the one-dimensional case two or three times, as we will see in Chapter 4. Although it is possible to study rectilinear motion without using vectors, the generalization to two and three dimensions is easier if we use vectors from the beginning, even for the one-dimensional situation. A vector approach also makes the algebraic signs of the numerical values of various terms in the few significant equations of kinematics a matter of the choice we make for the directions of the coordinate axes.

QUESTION 1

Is the universe as a whole a particle? Is it the "mother of all particles"? Is the universe *the* fundamental particle? Explain your response.

3.2 POSITION AND CHANGES IN POSITION

To meet a friend arriving by plane, you need to know two things: *when* and *where* it arrives. Incomplete information such as knowing the time of arrival but not the gate number, or vice versa, is frustrating and inconvenient. To describe the motion of a particle, we also begin by specifying where it is located at a specified time. To say that it is "here" at "such-and-such" a time defines an **event**. An event in physics is no more complicated than the information on an airport arrival marquee or an invitation to an exciting party: where and when need to be specified.

Imagine yourself in your car moving along a straight dragstrip, as in Figure 3.2. Let the car be represented as a particle, say, by a point at the center of the car. Let the *x*-axis of a Cartesian coordinate system be the one-dimensional straight path along which the car is moving, with $\hat{\mathbf{i}}$ directed to the right.

The origin may be chosen to be at any convenient point on the line, perhaps at the starting line of the dragstrip. The **position** of the car on the *x*-axis is indicated by the *coordinate x*

Flatland (Princeton University Press, Princeton, New Jersey, 1991), Part II, Section 13, pages 53–54.

of the particle (the center of the car) at any instant *t*. We use subscripts (or primes) to distinguish specific *x*-coordinates and specific instants. Imagine the particle to be at the coordinate x_i at the instant t_i, as in Figure 3.2; these need not be the origin or when the clock reads zero. Rather, x_i and t_i represent where the car is when we initialize, or begin our observations. The location of the car at this instant is specified by a one-dimensional position vector $\vec{\mathbf{r}}_i$ that locates the coordinate of the particle with respect to the (arbitrarily chosen) origin of the coordinate system as shown in Figure 3.3.

Specifying $\vec{\mathbf{r}}_i$ and t_i defines an initial event: where the car is with respect to the origin of the coordinate grid and when it is there. It is important to realize that the moving particle spends zero time at this position. The position vector is an *instantaneous* position vector that locates the particle at an instant in time.

> **Time instants** have no time duration and do not correspond to **time intervals**, even of very short duration.

Later, at the instant t_f, the car is found at the coordinate x_f; its new (instantaneous) position vector is $\vec{\mathbf{r}}_f = x_f\hat{\mathbf{i}}$, as shown in Figure 3.4. Specifying $\vec{\mathbf{r}}_f$ and t_f defines another event—another where and when.

FIGURE 3.2 A car moving along a straight dragstrip.

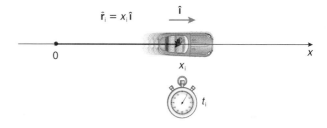

FIGURE 3.3 The initial position vector of the car is $\vec{\mathbf{r}}_i = x_i\hat{\mathbf{i}}$.

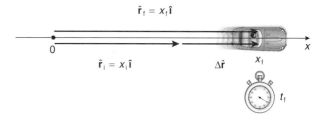

FIGURE 3.4 A subsequent position vector of the car is $\vec{\mathbf{r}}_f = x_f\hat{\mathbf{i}}$.

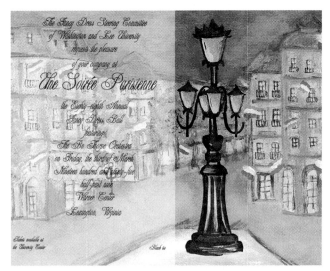

An event specifies where and when like this invitation to a Fancy Dress Ball.

A **time interval** Δt involves the subtraction of two different clock readings (time instants). The time interval between two events associated with the motion is

$$\Delta t \equiv t_f - t_i \qquad (3.1)$$

and is a *positive* scalar quantity since $t_f > t_i$. The change in the position vector $\Delta \vec{r}$ is the vector difference between the new (final) position vector and the old (initial) position vector:

$$\Delta \vec{r} \equiv \vec{r}_f - \vec{r}_i \qquad (3.2)$$

For rectilinear motion, this becomes

$$
\begin{aligned}
\Delta \vec{r} &= x_f \hat{\imath} - x_i \hat{\imath} \\
&= (x_f - x_i)\hat{\imath} \\
&= \Delta x\, \hat{\imath} \qquad (3.3)
\end{aligned}
$$

By transposing $\vec{r}_i$ in Equation 3.2, the three vectors are related by the vector sum:

$$\vec{r}_f = \vec{r}_i + \Delta \vec{r}$$

which is shown in Figure 3.4.

The magnitude of the change in the position vector is *not* necessarily equal to the total distance d traveled by the particle during the time interval Δt between the two events.

The change in the position vector is concerned *only* with the initial and final positions of the particle, not with the details of what happened to the particle during the time interval. Think of the total distance traveled as the reading of a trip odometer initially set to zero at the initial position of the particle. On the other hand, the magnitude of the change in the position vector is concerned only with where the particle ended up relative to where it began.

The distance traveled differs from the magnitude of the change in the position vector for rectilinear motion if the particle changes direction or moves back and forth along the axis while moving from $\vec{r}_i$ to $\vec{r}_f$.

Indeed, for a round trip, the initial and final position vectors of the particle are the same, and $|\Delta \vec{r}| = 0$ m, but the total distance d traveled by the car is not zero. Likewise, in another context, a beeline from your dorm to a pizza parlor and back during an interval Δt means $|\Delta \vec{r}| = 0$ m, but a nonzero trip distance d during the interval. Also, if you toss a marble vertically into the air and catch it at the position it was released, then $|\Delta \vec{r}| = 0$ m while $d \neq 0$ m for the interval Δt.

As we noted in Section 2.18, the magnitude and direction of the change in the position vector $\Delta \vec{r}$, indeed, of *any* vector, are independent of the coordinate system chosen to describe the change. However, the signs of the x-components of $\vec{r}$ and of $\Delta \vec{r}$ depend on the coordinate values of x_f and x_i; these values depend on the choice you make for the direction of $\hat{\imath}$ and the location you choose for the origin.

Since the magnitude and direction of the change in the position vector are independent of the coordinate choice, you can make things easier for yourself if you follow Problem-Solving Tactic 3.1.

PROBLEM-SOLVING TACTIC

3.1 For rectilinear motion, always choose a Cartesian coordinate axis to be coincident with the line followed by the particle.

For simplicity, we will always use the x-axis for rectilinear motion. The Cartesian origin can be chosen at any convenient position along the line of motion; it need not correspond to where the particle is when it begins its motion or to its position when we begin our observations. The direction corresponding to increasing values of the coordinate, that is, the direction of the unit vector $\hat{\imath}$, is ours to choose; it need not correspond to the initial direction of travel of the particle. With such coordinate conventions and choices, though, the description of rectilinear motion will be simpler than with other coordinate choices.

Notice that we have choices to make for

• the location of the origin and

• the direction for $\hat{\imath}$.

Such choices must be specified when solving problems.

The easiest way to specify such choices is with a sketch, as in Figure 3.5. Place 0 where you choose the origin to be located. Place $\hat{\imath}$ somewhere in the sketch or label x at the end of the co-ordinate line, or both. Bear in mind:

a. The position vector $\vec{r}$ locates the particle at an instant t;

b. The magnitude and direction of the *change* in the position vector $\Delta \vec{r}$ that occurs during a time interval Δt are independent of the specific coordinate system we choose; and

FIGURE 3.5 Specify the location of the origin and the direction of $\hat{\imath}$ and/or the x-axis.

c. The total distance d traveled by the particle during the interval Δt is a positive scalar quantity that, even for rectilinear motion, may differ from the magnitude of the change in the position vector if there is back-and-forth motion on the one-dimensional line during the interval Δt.

The position vector of a moving particle is a function of time, written as $\vec{r}(t)$. For one-dimensional (i.e., rectilinear) motion with the x-axis coincident with the line of motion of the particle,

$$\vec{r}(t) = x(t)\hat{\imath} \tag{3.4}$$

The scalar function $x(t)$ describes the time dependence of the coordinate x of the particle. At any specific instant, say t_1, the instantaneous position vector of the particle is $\vec{r}(t_1)$; we find this vector function by evaluating the scalar function $x(t)$ at the particular instant t_1:

$$\vec{r}(t_1) = x(t_1)\hat{\imath}$$

A graph of the x-component of the position vector of the particle as a function of time is called an x versus t graph. For your car moving along a dragstrip, an x versus t graph might appear as in Figure 3.6. The line on the graph representing $x(t)$ is called a **world-line**. An event is a point on the world-line.

PROBLEM-SOLVING TACTIC

3.2 Read and construct graphs carefully. Notice what is plotted versus what or, if you are making the graph, be sure to label what each axis represents. When we say to "plot x versus t," plot x vertically (as the ordinate) and t horizontally (as the abscissa). Pay attention to the units used along each axis of a graph or, if you are making the graph, indicate the appropriate units for the scale you use along each axis.

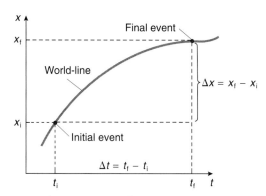

FIGURE 3.6 An x versus t graph.

QUESTION 2

Explain why the world-line of a particle on the x versus t graph in Figure Q.2 is meaningless.

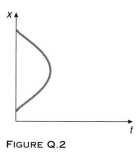

FIGURE Q.2

EXAMPLE 3.1

Your pet beagle is running happily along a wharf and you note her position vector at two instants, as in Figure 3.7.

a. Describe the two events and indicate them on an x versus t graph.
b. Specify the initial position vector.
c. Specify the final position vector.
d. Specify the time interval, the change in the position vector $\Delta\vec{r}$, and its magnitude and direction. Sketch $\Delta\vec{r}$.
e. What, if anything, can be said about the total distance traveled by the beagle-particle during the time interval between the two instants?

Solution

a. The dog initially had the x-coordinate 4.00 m when the stopwatch reading was 3.00 s. Later, when the stopwatch read 5.00 s, the x-coordinate of the beagle was -2.00 m. The two events are represented by points on the x versus t graph shown in Figure 3.8.
b. The initial position vector is $\vec{r}_i = (4.00 \text{ m})\hat{\imath}$.
c. The final position vector is $\vec{r}_f = (-2.00 \text{ m})\hat{\imath}$.

FIGURE 3.7 The position vectors of a beagle-particle at two instants indicated by a digital stopwatch.

FIGURE 3.8

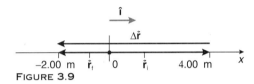

FIGURE 3.9

d. The time interval between the events is

$$\Delta t = t_f - t_i$$
$$= 5.00 \text{ s} - 3.00 \text{ s}$$
$$= 2.00 \text{ s}$$

The change in the position vector is

$$\Delta \vec{r} = \vec{r}_f - \vec{r}_i$$
$$= (-2.00 \text{ m})\hat{\imath} - (4.00 \text{ m})\hat{\imath}$$
$$= (-6.00 \text{ m})\hat{\imath}$$

The magnitude of this vector is 6.00 m, and it is directed to the left in Figure 3.9.

e. You know the positions of the beagle at two instants of time but are given no information about the detailed motion of the dog during the time interval between the two instants. If the dog proceeded from the initial position to the final position without any back-and-forth motion, then the total distance traveled is 6.00 m; if any back-and-forth motion or departures from the straight path occurred during the interval, the total distance traveled is greater than 6.00 m.

EXAMPLE 3.2

The position vector of a fleet-footed African cheetah (*Acinonyx jubatus*) is given by the following function of time:

$$\vec{r}(t) = x(t)\hat{\imath}$$
$$= [4.00 \text{ m} - (2.00 \text{ m/s})t - (3.00 \text{ m/s}^2)t^2]\hat{\imath}$$

where the time t is expressed in seconds on a stopwatch.

a. What is the position vector of the cheetah-particle at the instant the stopwatch commences, when $t = 0$ s?
b. What is the position vector of the cheetah when the stopwatch indicates $t = 3.00$ s?
c. What is the change in the position vector during the three second interval between these time instants?

d. Make a schematic sketch of the two position vectors and the change in the position vector.
e. Make an accurate sketch of the world-line of the cheetah on an x versus t graph between the two events described in parts (a) and (b).

Solution

a. To find the position vector when $t = 0$ s, substitute this instant into the expression for $\vec{r}(t)$:

$$\vec{r}(0 \text{ s}) = [4.00 \text{ m} - (2.00 \text{ m/s})(0 \text{ s})$$
$$- (3.00 \text{ m/s}^2)(0 \text{ s})^2]\hat{\imath}$$
$$= (4.00 \text{ m})\hat{\imath}$$

This is a vector of magnitude 4.00 m, directed in the same sense as $\hat{\imath}$, and it locates the position of the particle when $t = 0$ s.

b. Likewise, to find the position vector when $t = 3.00$ s, substitute the instant $t = 3.00$ s into the expression for $\vec{r}(t)$:

$$\vec{r}(3.00 \text{ s}) = [4.00 \text{ m} - (2.00 \text{ m/s})(3.00 \text{ s})$$
$$- (3.00 \text{ m/s}^2)(3.00 \text{ s})^2]\hat{\imath}$$
$$= (4.00 \text{ m} - 6.00 \text{ m} - 27.0 \text{ m})\hat{\imath}$$
$$= (-29.0 \text{ m})\hat{\imath}$$

This is a vector of magnitude 29.0 m, directed in the same sense as $-\hat{\imath}$, and represents the new position of the cheetah when $t = 3.00$ s.

c. The change in the position vector is

$$\Delta \vec{r} = \vec{r}_f - \vec{r}_i$$
$$= (-29.0 \text{ m})\hat{\imath} - (4.0 \text{ m})\hat{\imath}$$
$$= (-33.0 \text{ m})\hat{\imath}$$

This means the change in the position vector is a vector of magnitude 33.0 m, parallel to $-\hat{\imath}$ (to the left in Figure 3.10).

d. The three vectors are indicated schematically in Figure 3.10.

e. A graph of $x(t)$ versus t is constructed by substituting various values of t into the given equation and determining the corresponding values of x. This is summarized in the following table:

t (s)	x (m)
0	4.00
1.00	−1.00
2.00	−12.0
3.00	−29.0

FIGURE 3.10

FIGURE 3.11

FIGURE 3.12

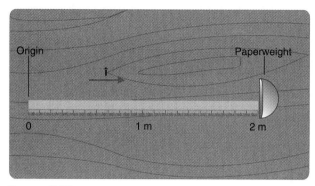

FIGURE 3.13

The curve is plotted in Figure 3.11. The two events described in parts (a) and (b) are indicated as the initial and final points on the world-line.

EXAMPLE 3.3

Lay a 2-meter stick on your desk and roll a marble slowly alongside it as illustrated in Figure 3.12. The marble is found at the 25 cm mark when a clock reads 1 h 15 min 20 s. The marble rebounds from a paperweight at the end of the 2-meter stick and is discovered at the 125 cm mark when the clock reads 1 h 15 min 25 s. In appropriate SI units, find the following:

a. the position vector of the marble at each of the two instants relative to the zero centimeter mark of the scale. Let the stick represent the x-axis;
b. the change in the position vector $\Delta \vec{r}$; and
c. the total distance traveled during the time interval between the two events.
d. Locate the two events on an appropriate x versus t graph. Sketch an appropriate world-line connecting the two events, describing the salient features of the world-line.

Solution

a. Use Problem-Solving Tactic 3.1 and *choose a coordinate axis for rectilinear motion to be coincident with the line followed by the particle*. Choose the origin of a Cartesian coordinate x to be coincident with the zero of the meter stick, since the problem seeks to find the position vectors with respect to this point (see Figure 3.13). You also must choose a direction for $\hat{\imath}$, that is, for increasing values of x. Here it is convenient to choose $\hat{\imath}$ to the

right, so that the meter stick positions correspond to increasing x-coordinates.

The position vector $\vec{r}_i$ at the instant t_i (clock reading 1 h 15 min 20 s) is

$$\vec{r}_i = (0.25 \text{ m})\hat{\imath}$$

using proper SI units. This is a vector of magnitude 0.25 m, directed in the same direction as $\hat{\imath}$. The position vector $\vec{r}_f$ at the instant t_f (clock reading 1 h 15 min 25 s) is

$$\vec{r}_f = (1.25 \text{ m})\hat{\imath}$$

This is a vector of magnitude 1.25 m, directed in the same direction as $\hat{\imath}$.
b. According to Equation 3.2, the change in the position vector is

$$\Delta \vec{r} \equiv \vec{r}_f - \vec{r}_i$$
$$= (1.25 \text{ m})\hat{\imath} - (0.25 \text{ m})\hat{\imath}$$
$$= (1.00 \text{ m})\hat{\imath}$$

The change in the position vector is a vector quantity. Here the magnitude of the change in the position vector is 1.00 m and its direction is to the right, parallel to the direction chosen for $\hat{\imath}$.
c. The total distance d traveled during the 5-second interval is the total distance from its initial position to the paperweight plus the distance from the paperweight to the final position. This is

$$d = 1.75 \text{ m} + 0.75 \text{ m}$$
$$= 2.50 \text{ m}$$

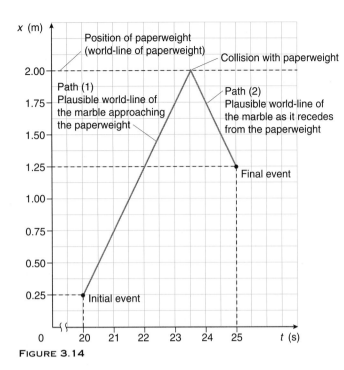

FIGURE 3.14

The total distance traveled always is a positive scalar quantity or zero; it is never negative. Notice that the total distance traveled (2.50 m) is not the same as the magnitude of the change in the position vector (1.00 m), since the particle executed back-and-forth motion along the line of its motion during the interval.

d. An *x* versus *t* graph is a graph of the world-line *x*(*t*), with events represented as points on the world-line. Figure 3.14 shows the two events. Since the paperweight is located at the coordinate *x* = 2.00 m and the marble rebounds from the paperweight, a plausible world-line is shown. On path (1) of the world-line, the marble is approaching the paperweight; on path (2), the marble is receding from the paperweight, retracing its path. The sharp point on the world-line represents the collision event of the marble with the paperweight; here it is assumed the collision is instantaneous, taking zero time.

3.3 AVERAGE SPEED AND AVERAGE VELOCITY

Tempus est mensura motus rerum mobilium

[*Time is the measure of movement*]

Aristotle (384–322 B.C.)*

If your car is moving fast, it travels the length of a dragstrip during a short time interval Δ*t*; conversely, if the car is moving slowly, it takes a longer time interval to traverse the length of the strip.

Les Auctoritates Aristotlelis, un florilège médiéval étude historique et édition critique, editor Jacqueline Hamesse (Louvain, Paris, 1974), page 151; from Aristotle's *Physics.*

The **average speed** of a particle during a time interval Δ*t* is defined as the total distance *d* traveled by the particle divided by the time interval Δ*t* it takes to traverse the distance:

$$\text{average speed} = \frac{|d|}{|\Delta t|} = \frac{\text{total distance traveled}}{\text{time interval}} \quad (3.5)$$

The average speed is a positive scalar quantity, not a vector quantity.

Since both the total distance (the trip odometer reading) and the time interval are positive scalars, the absolute value signs in Equation 3.5 are superfluous; they are included only for emphasis. The dimensions of the average speed are distance divided by time; the SI units are m/s.

There is no reason that you could not define a measure of the motion to be the inverse of the ratio of Equation 3.5, $|\Delta t|/|d|$. You might call this ratio the average "sluggishness" of a particle. The average sluggishness is a measure of the time interval needed to traverse a given distance (say, 30 seconds per kilometer for your car), rather than the average speed, which is a measure of the distance traversed during a given time interval (120 kilometers per hour for your car). A large value of the average sluggishness means it takes a particle a long time interval to traverse a small distance. The contrast in usefulness between the average sluggishness and the average speed illustrates why we say that the theory of motion was *invented*, not discovered.

The **average velocity** $\vec{v}_{ave}$ of a particle during the time interval Δ*t* is defined as the ratio of the change in the position vector of the particle to the time interval during which the change occurred:

$$\vec{v}_{ave} = \frac{\Delta \vec{r}}{\Delta t} \quad (3.6)$$

The average velocity is a vector quantity.

The average velocity is a vector because it is a scalar multiple (1/Δ*t*) of the change in the position vector. The dimensions of the average velocity vector are the same as those of the average speed, distance divided by time—in SI units, m/s. However, do not confuse the average speed (which involves the total distance traveled, *d*) with the average velocity (which involves the change in the position vector, Δ$\vec{r}$); they are quite different concepts.

The magnitude of the average velocity is *not* necessarily equal to the average speed during the same time interval.

The inequality arises because the total distance *d* traveled by the particle during the time interval may be different from the magnitude of the change in the position vector that occurs during the same time interval. The quantity |Δ$\vec{r}$| always is less than or equal to *d*. For rectilinear motion, the inequality arises if the particle reverses direction or undergoes back-and-forth motion during the interval Δ*t*.

The position vector for one-dimensional motion has only a single component, because we always choose the coordinate axis to be coincident with the line followed by the particle.

The x-component of the average velocity, $v_{x\ ave}$, during an interval Δt, its only component for rectilinear motion,

$$\vec{\mathbf{v}}_{ave} = v_{x\ ave}\hat{\mathbf{i}}$$

is found from Equations 3.6 and 3.3 to be

$$v_{x\ ave} = \frac{x_f - x_i}{\Delta t}$$
$$= \frac{\Delta x}{\Delta t}$$

On an x versus t graph, such as Figure 3.15, the average velocity component $v_{x\ ave}$ is the slope of the straight line connecting the initial and final events. The average velocity component $v_{x\ ave}$ is found using *only* the initial and final events on the x versus t graph and *does not involve the detailed shape of the world-line connecting the events.*

The average velocity component $v_{x\ ave}$ is positive if $x_f > x_i$; then the average velocity is in the same sense as the direction chosen for $\hat{\mathbf{i}}$. If $x_f < x_i$, the average velocity component is negative, meaning the average velocity vector is in the same sense as $-\hat{\mathbf{i}}$ in the chosen coordinate system.

PROBLEM-SOLVING TACTIC

3.3 The sign of $v_{x\ ave}$ depends on the choice you make for the direction of $\hat{\mathbf{i}}$ and the values of x_f and x_i. Different coordinate choices change the way you *describe* the motion but not the underlying physics. The magnitude and direction in space of the *vector* $\vec{\mathbf{v}}_{ave}$ do not depend on the coordinate choice.

QUESTION 3

If your average car speed is 100 km/h, what is the corresponding measure of the average sluggishness of the car?

QUESTION 4

Your dog runs in a straight line east chasing a squirrel. Indicate a coordinate choice so that $v_{x\ ave}$ of your dog is positive. Indicate another coordinate choice so that $v_{x\ ave}$ is negative.

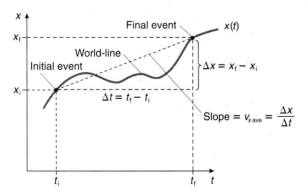

FIGURE 3.15 The average velocity component $v_{x\ ave}$ is the slope of the straight line connecting the initial and final events on an x versus t graph.

EXAMPLE 3.4

a. Find the average speed of the marble in Example 3.3 during the 5 s interval.
b. Find the average velocity of the marble of Example 3.3 during the 5 s interval.

Solution

a. The average speed is found by dividing the total distance the particle travels ($d = 2.50$ m) between the two events by the time interval between the events ($\Delta t = 5$ s):

$$\text{average speed} = \frac{2.50\ \text{m}}{5\ \text{s}}$$
$$= 0.5\ \text{m/s}$$

b. The average velocity is the ratio of the change in the position vector to the time interval during which the change occurred:

$$\vec{\mathbf{v}}_{ave} = \frac{\Delta\vec{\mathbf{r}}}{\Delta t}$$
$$= \frac{(1.00\ \text{m})\hat{\mathbf{i}}}{5\ \text{s}}$$
$$= (0.2\ \text{m/s})\hat{\mathbf{i}}$$

Notice that the average velocity is a vector, since the change in the position vector is a vector. The *magnitude* of the average velocity vector here is 0.2 m/s; this is *not the same as the average speed* [found to be 0.5 m/s in part (a)].

EXAMPLE 3.5

The Iditarod dogsled race from Anchorage to Nome, Alaska, covers a distance of 1864 km. The winning time is about 11 days.

a. Find the average speed of the winner in SI units.
b. Qualitatively compare the average speed during the interval of the entire race with the likely average speed during the actual time during which the team was on the course and not resting or sleeping.

Solution

a. The average speed is the total distance traveled divided by the time interval during which the distance was traversed:

$$\text{average speed} = \frac{1864\ \text{km}}{11\ \text{days}}$$
$$= \frac{(1864\ \text{km})(1000\ \text{m/km})}{(11\ \text{days})(86\ 400\ \text{s/day})}$$
$$= 2.0\ \text{m/s}$$

This means that the winning team could complete the race by traveling at 2.0 m/s throughout the entire 11-day time interval.

b. Because the team had to stop, rest, and eat along the route, the time interval actually spent running the race was shorter than 11 days (perhaps about 7 days). Hence the average speed during the many intervals the team actually was on the race course was greater than the average speed for the duration of the entire trip.

STRATEGIC EXAMPLE 3.6

You drive your car 2.00 km down a dragstrip, then 2.50 km in the opposite direction, completing the excursion in 180 s.

a. Choose an appropriate coordinate system to describe the motion.
b. Determine the initial and final position vectors and the change in the position vector during the time interval.
c. Find the average velocity.
d. Find the average speed.
e. Explain why the average speed of the car is not equal to the magnitude of the average velocity.

Solution

a. Choose a coordinate system with the *x*-axis coincident with the dragstrip and an origin at the initial position of the car (Figure 3.16).

FIGURE 3.16

b. The initial position vector for the car is

$$\vec{r}_i = x_i \hat{\imath}$$

| $\vec{r}_i = (0\ \text{m})\hat{\imath}$ | $\vec{r}_i = (0\ \text{m})\hat{\imath}$ |

The final position vector of the car is

$$\vec{r}_f = x_f \hat{\imath}$$

| $\vec{r}_f = (-500\ \text{m})\hat{\imath}$ | $\vec{r}_f = (500\ \text{m})\hat{\imath}$ |

The change in the position vector is

$$\Delta\vec{r} = \vec{r}_f - \vec{r}_i$$

| $\Delta\vec{r} = (-500\ \text{m})\hat{\imath} - (0\ \text{m})\hat{\imath}$ $= (-500\ \text{m})\hat{\imath}$ | $\Delta\vec{r} = (500\ \text{m})\hat{\imath} - (0\ \text{m})\hat{\imath}$ $= (500\ \text{m})\hat{\imath}$ |

Notice that the magnitude of the change in the position vector is 500 m and is directed to the left for *both* coordinate choices. This illustrates how the magnitude and direction in space of the *change* in a vector are the same regardless of the coordinate system you choose.

c. The average velocity is the change in the position vector divided by the time interval during which the change occurred:

$$\vec{v}_{ave} = \frac{\Delta\vec{r}}{\Delta t}$$

| $\vec{v}_{ave} = \dfrac{(-500\ \text{m})\hat{\imath}}{180\ \text{s}}$ $= (-2.78\ \text{m/s})\hat{\imath}$ | $\vec{v}_{ave} = \dfrac{(500\ \text{m})\hat{\imath}}{180\ \text{s}}$ $= (2.78\ \text{m/s})\hat{\imath}$ |

Note that these results are consistent with Problem-Solving Tactic 3.3: *the sign of* $v_{x\,ave}$ *depends on the choice you make for the direction of* $\hat{\imath}$ *and the values of* x_f *and* x_i. However, the average velocity here has a magnitude of 2.78 m/s and is directed to the left, independent of whatever choice you made for the coordinate system.

d. The average speed is the total distance traveled divided by the time interval:

$$\text{average speed} = \frac{4.50 \times 10^3 \text{ m}}{180 \text{ s}}$$

$$= 25.0 \text{ m/s}$$

e. The magnitude of the average velocity, here 2.78 m/s, is not equal to the average speed, here 25.0 m/s, because the total distance traveled is not equal to the magnitude of the change in the position vector: the car went back and forth along the dragstrip during the interval Δt.

3.4 INSTANTANEOUS SPEED AND INSTANTANEOUS VELOCITY

In many instances we want to know more than averages; we want details about the dynamic motion of the particle on an instant-by-instant basis. Such considerations are the determining factors in "awarding" speeding tickets to motorists. Instantaneous information is gleaned by imagining the particle to be equipped with a speedometer like that in your car. In your car, a glance at the numerical reading of the speedometer at an *instant* t indicates the **instantaneous speed** of the car at that instant. The instantaneous speed of the particle is more commonly referred to as simply the **speed** v.

The number indicated by the speedometer indicates how far the car *would* travel during the next *one* unit interval of time *if* it continued moving *at the same speedometer reading* for that one unit interval of time. For example, if the speedometer in your car indicates the speed is 100 km/h, the car will move 100 km during the entire next 1 hour interval if the speed remains *constant* (on cruise control) during this interval. Traffic jams, traffic lights, or the unnerving sight of the flashing lights of the state police in the rearview mirror may intervene to prevent you from actually keeping the car at the same speedometer reading for the next one hour. But, *at the instant* the speedometer reading is noted, the speed of the car is 100 km/h. Converting the speedometer reading to meters per second, 100 km/h = 27.8 m/s, you see the car will move 27.8 meters during the next 1 second *if* the car continues uninterrupted at the same speed during the next one second interval.

Your car may have a compass that provides information about the instantaneous directional heading of the car.

The **instantaneous velocity** $\vec{v}$ is a vector quantity whose magnitude v is equal to the instantaneous speed and whose instantaneous direction is in the same sense that the particle is moving at that instant.

The instantaneous velocity more commonly is called simply the **velocity** of the particle at the instant t. The velocity vector always is tangent to the trajectory or path followed by the particle, although this point is moot for rectilinear motion.

Speed and velocity are *not* the same physical concepts, even though they are both commonly measured with the same SI units, m/s. *Velocity is a vector quantity whose positive scalar magnitude is the speed.* Be sure you know and remember the distinction between the two terms.

Another way to think of the instantaneous velocity uses the notion of limits from calculus. Let $\vec{r}(t) = x(t)\hat{\imath}$ be the function of time that represents the one-dimensional position vector of the particle at any instant of time. Choose a particular instant $t_i = t$, so that

$$\vec{r}_i = \vec{r}(t)$$
$$= x(t)\hat{\imath}$$

A short time interval Δt later, at the instant $t_f = t + \Delta t$, the position vector is

$$\vec{r}_f = \vec{r}(t + \Delta t)$$
$$= x(t + \Delta t)\hat{\imath}$$

The *change in the position vector* during the time interval Δt then is

$$\Delta\vec{r} = \vec{r}(t + \Delta t) - \vec{r}(t)$$
$$= [x(t + \Delta t) - x(t)]\hat{\imath}$$

The average velocity $\vec{v}_{ave}$ of the particle during the time interval Δt is, using Equation 3.6,

$$\vec{v}_{ave} = \frac{\Delta\vec{r}}{\Delta t}$$

$$= \frac{[x(t + \Delta t) - x(t)]\hat{\imath}}{\Delta t}$$

Take the limit of this expression as the time interval Δt approaches zero; the average velocity thus is taken over shorter and shorter time interval deviations from the instant t:

$$\lim_{\Delta t \to 0 \text{ s}} \frac{\vec{r}(t + \Delta t) - \vec{r}(t)}{\Delta t} = \lim_{\Delta t \to 0 \text{ s}} \frac{[x(t + \Delta t) - x(t)]\hat{\imath}}{\Delta t}$$

This process is shown graphically in the x versus t graph of Figure 3.17.

You may recognize this limit as one way of defining the derivative in calculus. This limiting process defines the instantaneous velocity $\vec{v}(t)$ of the particle at the instant t. In other words,

$$\vec{v}(t) \equiv \lim_{\Delta t \to 0 \text{ s}} \vec{v}_{ave} = \frac{d\vec{r}}{dt}$$

$$= \frac{dx}{dt}\hat{\imath} \qquad \text{(for rectilinear motion)} \qquad (3.7)$$

As Δ*t* approaches zero, the slope of the line
between *x*(*t*) and *x*(*t*₁) approaches the slope of
the tangent to that graph at *t*, which is *v*ₓ(*t*).

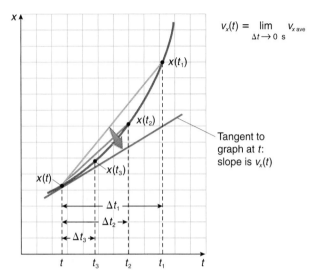

$$v_x(t) = \lim_{\Delta t \to 0 \text{ s}} v_{x \text{ ave}}$$

Tangent to
graph at *t*:
slope is *v*ₓ(*t*)

FIGURE 3.17 The limiting process on an *x* versus *t* graph.

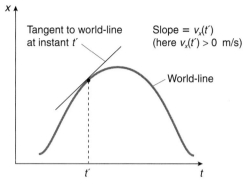

Slope = *v*ₓ(*t'*)
(here *v*ₓ(*t'*) > 0 m/s)

Tangent to world-line
at instant *t'*

World-line

FIGURE 3.18 The instantaneous velocity component at an instant
t' is equal to the slope of the tangent to the world-line at that
instant. Note that the slope of the graph may change during a time
interval, indicating that the velocity component is positive at some
instants, negative at others, or even zero.

Your car

A friend's car

FIGURE 3.19

For rectilinear motion, the instantaneous velocity vector
$\vec{v}(t) = v_x(t)\hat{i}$ has a single component $v_x(t)$, where, from Equation 3.7,

$$v_x(t) = \frac{dx(t)}{dt} \tag{3.8}$$

Since instants have no duration, the particle has the instantaneous velocity $v(t)$ for zero seconds.

In calculus you learned that the derivative of a function is
the *slope* of a graph of the function. Exploiting this geometric
interpretation of the derivative from calculus, Equation 3.8
means that the instantaneous velocity component $v_x(t')$ at any
instant t' is the slope of the tangent to the world-line of the
particle at that instant t'. See Figure 3.18. Note the distinction
in Figure 3.17 between the average velocity *during* a time
interval and the instantaneous velocity *at* an instant.

PROBLEM-SOLVING TACTIC

**3.4 The choice you make for the coordinate system
influences the sign of the velocity component *v*ₓ.** As you
are now aware, the choices you make for a coordinate
system have consequences in the signs associated with the
components of vectors used to describe the motion or
the physics. This will be the case throughout our study of
physics, not only in kinematics.

EXAMPLE 3.7 ━━━━━━━

**You and a friend both leave campus by car and head home in
opposite directions along a straight stretch of inclined roadway,**
as shown in Figure 3.19. A short interval after getting under
way, at time instant 2 h 26 min 36.0 s, both of you glance at
the speedometers in your cars. Yours indicates 60 km/h while
your friend's indicates 90 km/h.

a. Introduce an appropriate Cartesian coordinate system and
   write the instantaneous velocities of each car in terms of the
   coordinate system you choose. Use proper SI units.
b. Indicate on a sketch what is meant by the position vectors
   of the two cars. Is there sufficient information to determine
   the position vectors at the same instant?

Solution

a. The speedometers are read *at* a particular *instant* in time, the
   clock time 2 h 26 min 36.0 s, and so the speeds quoted are
   instantaneous speeds. Convert the speeds into proper SI
   units of m/s:

$$90 \text{ km/h} = \frac{(90 \text{ km/h})(10^3 \text{ m/km})}{(3.600 \times 10^3 \text{ s/h})}$$

Notice that the factors of 10^3 in the numerator and denominator divide out, and so

$$90 \text{ km/h} = \frac{90}{3.600} \text{ m/s}$$
$$= 25 \text{ m/s}$$

In a similar way,

$$60 \text{ km/h} = 17 \text{ m/s}$$

FIGURE 3.20

Since the road is straight, choose a coordinate system with the direction of increasing x (the +x-axis) in the direction your car is moving, with the origin at some convenient pothole in the highway, as in Figure 3.20. Choose the x-axis to be along the road even though the road is not horizontal, because we want the axis to be along the line followed by the cars. Then the instantaneous velocities of the cars are

$$\text{your velocity:}\quad \vec{v}_{you} = (17 \text{ m/s})\hat{i}$$
$$\text{your friend's velocity:}\quad \vec{v}_{friend} = (25 \text{ m/s})(-\hat{i})$$
$$= (-25 \text{ m/s})\hat{i}$$

Draw the velocity vectors with their tails at the centers of the cars since you represent the cars as particles, imagining each car as a point.

b. The position vectors of the two cars are indicated in Figure 3.20. Since no information is given about where the two cars are located along the axis at the given instant, there is insufficient information to determine the Cartesian form for the two position vectors.

STRATEGIC EXAMPLE 3.8

The one-dimensional position vector of a sports car is given by

$$\vec{r}(t) = x(t)\hat{i}$$
$$= [4.00 \text{ m} - (2.00 \text{ m/s})t + (3.00 \text{ m/s}^2)t^2]\hat{i}$$

a. Make an x versus t graph including at least the interval from $t = 0$ s to $t = 3.00$ s.
b. Find the velocity $\vec{v}(t)$ as a function of time.
c. Specify the Cartesian component $v_x(t)$ of the velocity at any instant t.
d. Make a graph of $v_x(t)$ versus t.
e. Find the velocity at the instant when $t = 2.00$ s.
f. Determine at what time the velocity is zero.
g. Describe the motion of the particle.

Solution

a. An x versus t graph depicts $x(t)$ versus t. The graph is parabolic in shape since x is a quadratic function of t. See Figure 3.21.

The slope of the tangent to the curve at any instant t is the velocity component $v_x(t)$ at that instant. Notice that as t increases from zero, the slope of the world-line first is negative, then is zero at the instant represented by the lowest point on the graph, and then is positive. This indicates that

the velocity component v_x of the car first is negative, then zero, and then positive.

b. To find the instantaneous velocity as a function of time, differentiate the position vector $\vec{r}(t)$ with respect to t:

$$\vec{v}(t) = \frac{d\vec{r}(t)}{dt} = \frac{dx(t)}{dt}\hat{i}$$
$$= \frac{d}{dt}[4.00 \text{ m} - (2.00 \text{ m/s})t + (3.00 \text{ m/s}^2)t^2]\hat{i}$$
$$= [0 - 2.00 \text{ m/s} + (3.00 \text{ m/s}^2)2t]\hat{i}$$
$$= [-2.00 \text{ m/s} + (6.00 \text{ m/s}^2)t]\hat{i}$$

c. The x-component of the velocity $v_x(t)$ is the scalar function

$$v_x(t) = -2.00 \text{ m/s} + (6.00 \text{ m/s}^2)t$$

d. Notice that $v_x(t)$ is a linear function of t, since t appears only to the first power. The graph of $v_x(t)$, therefore, is a straight line with slope 6.00 m/s^2 and intercept -2.00 m/s. See Figure 3.22.
e. The velocity at the instant when $t = 2.00$ s is found by substituting this value for t into $\vec{v}(t)$:

$$\vec{v}(2.00 \text{ s}) = [-2.00 \text{ m/s} + (6.00 \text{ m/s}^2)(2.00 \text{ s})]\hat{i}$$
$$= (10.0 \text{ m/s})\hat{i}$$

The particle is moving with a speed of 10.0 m/s in the direction parallel to $\hat{i}$ (toward increasing values of x) at the instant when $t = 2.00$ s. The slope of the world-line is 10.0 m/s at the instant when $t = 2.00$ s (see Figure 3.21).

f. The instantaneous velocity is zero at those times when its component $v_x(t)$ is zero:

$$v_x(t) = 0 \text{ m/s} = -2.00 \text{ m/s} + (6.00 \text{ m/s}^2)t$$

and so

$$t = \frac{2.00 \text{ m/s}}{6.00 \text{ m/s}^2}$$
$$= 0.333 \text{ s}$$

So the particle is instantaneously at rest when $t = 0.333$ s. The slope of the tangent to the world-line (Figure 3.21) is zero at this instant, implying the velocity component is zero when $t = 0.333$ s.

g. Notice from the graph in Figure 3.22 that for times $t < 0.333$ s, the x-component v_x of the velocity vector is negative. This means that particle is moving in the same sense as $-\hat{i}$ for these times, or toward decreasing values of the coordinate x. For times $t > 0.333$ s, the x-component of the velocity is positive, which means the particle is moving in the same sense as $\hat{i}$, that is, in the direction of increasing values of x. Thus the particle reversed its direction at the instant $t = 0.333$ s. The car was, perhaps, initially backing up, then reversed direction when $t = 0.333$ s. The location of the particle at this instant is found from the position vector by substituting $t = 0.333$ s:

$$\vec{r}(t) = x(t)\hat{i}$$
$$= [4.00 \text{ m} - (2.00 \text{ m/s})t + (3.00 \text{ m/s}^2)t^2]\hat{i}$$

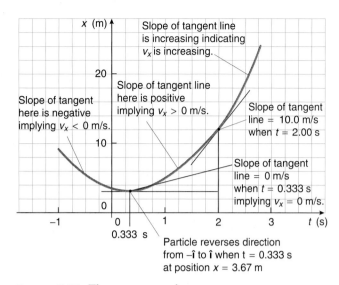

FIGURE 3.21 The x versus t graph.

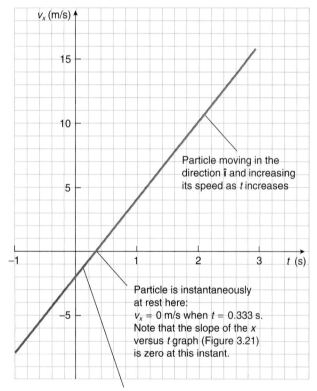

FIGURE 3.22 The v_x versus t graph.

Hence

$$\vec{r}(0.333 \text{ s}) = (4.00 \text{ m} - 0.666 \text{ m} + 0.333 \text{ m})\hat{\imath}$$
$$= (3.67 \text{ m})\hat{\imath}$$

The particle is on the $+x$-axis 3.67 m from the origin when $t = 0.333$ s. This corresponds to the minimum on the x versus t graph shown in Figure 3.21. The velocity of the particle reverses direction from $-\hat{\imath}$ to $\hat{\imath}$ at the coordinate $x = 3.67$ m, where its velocity is instantaneously zero. Recall that *instants are not intervals of time*, not even very short ones.

3.5 AVERAGE ACCELERATION

You likely have felt the dramatic changes in the velocity of a jet when it is moving along a runway prior to takeoff or just after touchdown when landing. Likewise, when you toss a ball vertically upward, the velocity of the ball changes both in magnitude *and* direction over the ensuing seconds before impact. These examples illustrate that there is more to motion than just position and velocity, even for one-dimensional motion. *Changes in velocity are important too.* Any change in the velocity of a particle with time indicates the presence of an **acceleration**.*

Acceleration, though, is more than just the change in velocity. Certainly changing the velocity of a sports car from 0 km/h east to 120 km/h east during a 5 s interval on a dragstrip (see Figure 3.23) is quite different from executing the same change in velocity during a leisurely 90 s interval in a jalopy. Hence the time interval during which the change in velocity is accomplished also is relevant. We want to quantify this distinction.

Let the velocity of a particle change from $\vec{v}_i$ at an instant t_i to $\vec{v}_f$ at a later instant t_f.

> One way to quantify the effect of a change in the velocity during the interval is to form the ratio of the change in the velocity $\Delta\vec{v} \equiv \vec{v}_f - \vec{v}_i$ to the interval $\Delta t = t_f - t_i$. This ratio is called the **average acceleration** $\vec{a}_{ave}$ during the interval Δt:
>
> $$\vec{a}_{ave} \equiv \frac{\Delta\vec{v}}{\Delta t} \qquad (3.9)$$

If a given change in velocity occurs during a short time interval, the average acceleration is larger than if it occurs during a longer time interval. The average acceleration is a vector quantity, since the change in velocity is a vector quantity.

---

*We will never use the word deceleration, meaning a decrease in the magnitude of the velocity during a time interval. *All* changes in velocity manifest accelerations.

FIGURE 3.23 A car changing its velocity on a dragstrip experiences an acceleration.

The units associated with the average acceleration in the SI unit system are those of the change in the velocity, m/s, divided by those for the time interval, s—that is, m/s^2.

For rectilinear motion, the velocity has a single component, $v_x(t)$, and so the change in the velocity also has a single component. This in turn implies that the average acceleration has a single component, $a_{x\,ave}$:

$$\vec{a}_{ave} = a_{x\,ave}\hat{i}$$

If a graph is made of $v_x(t)$ versus t, as in Figure 3.24, the average acceleration component during the interval Δt between t_i and t_f is the slope of the straight line connecting the two points with coordinates $(t_i, v_x(t_i))$ and $(t_f, v_x(t_f))$.

For notational simplicity, we write

$$v_x(t_i) \equiv v_{x\,i}$$

and

$$v_x(t_f) \equiv v_{x\,f}$$

FIGURE 3.24 The average acceleration is the slope of the straight line connecting the initial and final points on a v_x versus t graph.

so that

$$a_{x\,ave} \equiv \frac{v_{x\,f} - v_{x\,i}}{\Delta t} \qquad (3.10)$$

PROBLEM-SOLVING TACTIC

3.5 The sign of the average acceleration component $a_{x\,ave}$ depends on the choice you make for the direction of the unit vector $\hat{i}$ and the values of $v_{x\,f}$ and $v_{x\,i}$.

STRATEGIC EXAMPLE 3.9

A car moving up a long, straight hill has the speed 30.0 m/s at the instant 15.00 s. Later, at the instant 24.00 s, the speed of the car is observed to be 6.00 m/s in the opposite direction. See Figure 3.25.

a. Indicate an appropriate coordinate system in a sketch.
b. Find the average acceleration of the car during the time interval between the two instants.
c. Find the magnitude of the average acceleration of the car during the time interval.

FIGURE 3.25

Solution

a. Model the car as a particle. Choose the *x*-axis to be the line along which the particle is moving. As shown in Figure 3.26, there are two possible choices for the direction of $\hat{\imath}$:

Choice 1	*Choice 2*
Choose $\hat{\imath}$ to be up the hill. Choose the origin to be anywhere along the slope.	Choose $\hat{\imath}$ to be down the hill. Choose the origin to be anywhere along the slope.

FIGURE 3.26

b. The average acceleration is the change in the velocity vector divided by the time interval during which the change occurs. The change in the velocity vector is

$$\Delta \vec{v} = \vec{v}_f - \vec{v}_i$$

$\Delta \vec{v} = (-6.00 \text{ m/s})\hat{\imath} - (30.0 \text{ m/s})\hat{\imath}$ $= (-36.0 \text{ m/s})\hat{\imath}$	$\Delta \vec{v} = (6.00 \text{ m/s})\hat{\imath} - (-30.0 \text{ m/s})\hat{\imath}$ $= (36.0 \text{ m/s})\hat{\imath}$

The time interval between the two instants is $\Delta t = 9.00$ s. The average acceleration during the time interval is

$$\vec{a}_{\text{ave}} \equiv \frac{\Delta \vec{v}}{\Delta t}$$

$= \dfrac{(-36.0 \text{ m/s})\hat{\imath}}{9.00 \text{ s}}$ $= (-4.00 \text{ m/s}^2)\hat{\imath}$	$= \dfrac{(36.0 \text{ m/s})\hat{\imath}}{9.00 \text{ s}}$ $= (4.00 \text{ m/s}^2)\hat{\imath}$
This vector is directed in the same sense as $-\hat{\imath}$, that is, toward the left.	This vector is directed in the same sense as $\hat{\imath}$, that is, toward the left.

This illustrates Problem-Solving Tactic 3.4: *the sign of the average acceleration component* $a_{x \text{ ave}}$ *depends on the choice you make for the direction of the unit vector* $\hat{\imath}$. Note that the average acceleration has the same magnitude and direction in space with either coordinate choice.

c. Since there is only one component, the *magnitude* of the average acceleration is

$$a_{\text{ave}} = (a_{x \text{ ave}}^2)^{1/2}$$

$= [(-4.00 \text{ m/s}^2)^2]^{1/2}$ $= 4.00 \text{ m/s}^2$	$= [(4.00 \text{ m/s}^2)^2]^{1/2}$ $= 4.00 \text{ m/s}^2$

The magnitude of *any* vector is a scalar quantity greater than or equal to zero.

3.6 INSTANTANEOUS ACCELERATION

The thrill of riding in a sports car as it accelerates quickly or the horror of having to brake your car severely to avoid an accident is familiar to most of us. In the previous section we learned how to calculate the average acceleration. In this section we will discuss what is meant by an instantaneous acceleration.

In Section 3.4 we found it useful to compute the average velocity over shorter and shorter time intervals. In the limit as $\Delta t \to 0$ s, the average velocity became the instantaneous velocity at the instant t. We shall also find it convenient to use a similar limiting process with the average acceleration during a time interval Δt to find the **instantaneous acceleration**, commonly called simply the **acceleration**.

Let $\vec{\mathbf{v}}(t) = v_x(t)\hat{\mathbf{i}}$ be the vector that describes the instantaneous velocity of a particle as a function of time in one-dimensional motion. The scalar function $v_x(t)$ is the x-component of the velocity vector, its only component for rectilinear motion. The instantaneous velocity at any particular time instant t is found by substituting the particular value of t into $\vec{\mathbf{v}}(t) = v_x(t)\hat{\mathbf{i}}$. After a time interval Δt, the instant is $t + \Delta t$ and the instantaneous velocity is

$$\vec{\mathbf{v}}(t + \Delta t) = v_x(t + \Delta t)\hat{\mathbf{i}}$$

The *change* in the velocity vector during this interval is found from

$$\begin{aligned}\Delta\vec{\mathbf{v}} &= \vec{\mathbf{v}}_f - \vec{\mathbf{v}}_i \\ &= \vec{\mathbf{v}}(t + \Delta t) - \vec{\mathbf{v}}(t) \\ &= [v_x(t + \Delta t) - v_x(t)]\hat{\mathbf{i}}\end{aligned}$$

Use Equation 3.9 to find the average acceleration during the time interval Δt:

$$\begin{aligned}\vec{\mathbf{a}}_{ave} &= \frac{\Delta\vec{\mathbf{v}}}{\Delta t} \\ &= \frac{[v_x(t + \Delta t) - v_x(t)]\hat{\mathbf{i}}}{\Delta t}\end{aligned}$$

Now calculate the average acceleration over shorter and shorter intervals Δt from the instant t itself. This process is illustrated graphically in Figure 3.27.

Take the limit of the average acceleration as the time interval Δt approaches 0 s; we obtain the instantaneous acceleration at the instant t, $\vec{\mathbf{a}}(t)$:

$$\begin{aligned}\vec{\mathbf{a}}(t) &\equiv \lim_{\Delta t \to 0 \text{ s}} \frac{\Delta\vec{\mathbf{v}}}{\Delta t} \\ &= \lim_{\Delta t \to 0 \text{ s}} \frac{[v_x(t + \Delta t) - v_x(t)]\hat{\mathbf{i}}}{\Delta t}\end{aligned}$$

As Δt approaches zero, the slope of the line between $v(t)$ and $v(t_1)$ approaches the slope of the tangent to the graph at t, which is $a_x(t)$.

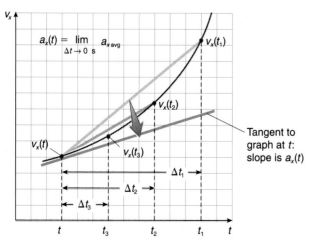

FIGURE 3.27 Graphical depiction of the limiting process.

This limit is the time derivative of the function $v_x(t)\hat{\mathbf{i}}$, that is, the time derivative of the instantaneous velocity vector.

> The instantaneous acceleration $\vec{\mathbf{a}}(t)$ of the particle is the time rate of change of the velocity vector:
>
> $$\vec{\mathbf{a}}(t) \equiv \frac{d\vec{\mathbf{v}}(t)}{dt} \qquad (3.11)$$
>
> The acceleration is a vector quantity.

Just as with the average acceleration, the SI units associated with the instantaneous acceleration are those of the velocity, m/s, divided by the time, s—or m/s².

For rectilinear motion, $\vec{\mathbf{v}}(t)$ has only one component $v_x(t)$, so that

$$\vec{\mathbf{a}}(t) = \frac{dv_x(t)}{dt}\hat{\mathbf{i}} \qquad (3.12)$$

But from Equation 3.8 we saw that

$$v_x(t) = \frac{dx(t)}{dt}$$

Make this substitution for $v_x(t)$ into Equation 3.12. You find that the instantaneous acceleration is the second derivative of the instantaneous position vector with respect to time:

$$\vec{\mathbf{a}}(t) = \frac{dv_x(t)}{dt}\hat{\mathbf{i}} = \frac{d^2x(t)}{dt^2}\hat{\mathbf{i}} = \frac{d^2\vec{\mathbf{r}}(t)}{dt^2} \qquad (3.13)$$

The acceleration vector for rectilinear motion has one component $a_x(t)$:

$$\vec{\mathbf{a}}(t) = a_x(t)\hat{\mathbf{i}}$$

where from Equation 3.13 we see that

$$a_x(t) = \frac{dv_x(t)}{dt} = \frac{d^2x(t)}{dt^2} \qquad (3.14)$$

Thus the x-component of the acceleration, $a_x(t)$, is the first time derivative of the x-component of the velocity, $v_x(t)$, or the second time derivative of the x-component of the position vector, $x(t)$.

The x-component of the acceleration, a_x, at any instant is the *slope* of the tangent line to the graph of v_x versus t at the instant in question, as indicated in Figure 3.28.

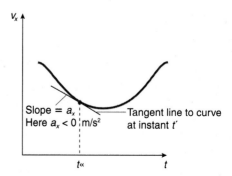

FIGURE 3.28 The acceleration component is the slope of the graph of the velocity component versus time.

FIGURE 3.29 A plumb bob.

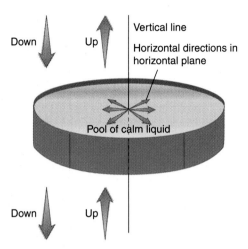

FIGURE 3.30 Horizontal and vertical directions.

To detect and measure the instantaneous acceleration, we need an **accelerometer**, a device analogous to a speedometer used to measure speed (the magnitude of the velocity). An accelerometer detects accelerations and, when suitably calibrated, indicates their magnitude. On the surface of the Earth, a crude but effective accelerometer consists of a simple mass suspended by a string, a device called a **plumb bob**, shown in Figure 3.29. This device detects accelerations directed *parallel* to the surface of the Earth— that is, in a **horizontal direction**. A horizontal direction is defined to be any direction parallel to the smooth surface of a liquid at rest on the Earth's surface (see Figure 3.30). Any plane parallel to the liquid surface is called a horizontal plane. A line perpendicular to a smooth liquid surface defines a **vertical direction**. The vertical directions up and down are defined using such a line.

Horizontal and vertical directions are purely local directions because of the curvature of the Earth itself. The vertically suspended plumb bob cannot be used to detect the existence of accelerations in the vertical direction.*

For example, if the plumb bob is attached to a particle either at rest or moving at constant velocity in a horizontal direction, the plumb bob hangs vertically. Try it yourself by tying a small mass to a piece of string and hanging it from the rearview mirror of your car. When at rest or traveling at constant speed in a fixed horizontal direction, the plumb bob hangs vertically (see Figure 3.31). In both situations, there is no acceleration in the horizontal plane since the velocity vector is not changing

---

*Other devices, such as a mass on a simple vertical spring, are used as accelerometers to detect accelerations in a vertical direction.

with time. On the other hand, when getting up to speed (see Figure 3.32) or stopping (Figure 3.33) when a traffic light changes, the plumb bob no longer hangs vertically.

If the velocity vector changes in magnitude, in direction, or in both magnitude and direction in a horizontal direction, the plumb bob no longer hangs in the vertical direction. If a snapshot of the plumb bob at an instant indicates it is not hanging in the vertical direction, there is a nonzero acceleration in the horizontal direction at that instant. Such effects also are apparent when observing a plumb bob in an airplane during the takeoff roll or braking along a horizontal runway.

The deviation of the plumb bob from the vertical is a measure of the magnitude of the acceleration; the direction of the deviation of the plumb bob from the vertical direction is *opposite* to the direction of the acceleration.† If the plumb bob

---

† In Chapter 5, we will see why the plumb bob is deflected in a direction opposite to that of the acceleration as well as how to calibrate the accelerometer so the acceleration in m/s² can be found from the angle of deviation of the plumb bob from the vertical.

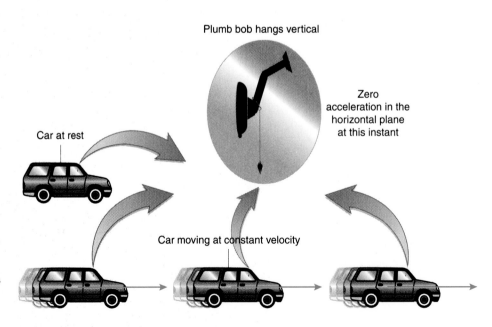

FIGURE 3.31 If a plumb bob hangs vertical, there is zero acceleration in the horizontal direction.

Plumb bob not vertical

Direction of deviation of plumb bob from vertical is opposite to the direction of the acceleration of the particle in the horizontal plane.

$\vec{a}$

FIGURE 3.32 Car speeding up in rectilinear motion.

Plumb bob not vertical

Direction of deviation of plumb bob from vertical is opposite to the direction of the acceleration of the particle in the horizontal plane.

$\vec{a}$

FIGURE 3.33 Car slowing down in rectilinear motion.

is vertical at the instant you observe it, the instantaneous acceleration is zero in any direction in the horizontal plane.

$\hat{i}$

50.0 km/h = 13.9 m/s when t = 2.00 s

0 km/h when t = 7.00 s

Red traffic light

2.00 s

7.00 s

FIGURE 3.34

STRATEGIC EXAMPLE 3.10

A traffic light turns red when a stopwatch indicates t = 2.00 s, and a car initially traveling at a constant velocity in a straight line toward the light at speed 50.0 km/h takes 5.00 s to come to a stop. Let î be in the direction the car is initially moving, as in Figure 3.34. Figure 3.35 is a graph modeling the x-component of the velocity of the car as a function of time.

a. What is the acceleration of the car at each instant between the times t = 0 s and t = 2.00 s on the stopwatch?
b. Estimate the instantaneous acceleration of the car at the instant t = 3.00 s on the stopwatch.
c. What is happening to the x-component of the acceleration of the car during the 5.00 s interval it takes the car to stop? Is the magnitude of the instantaneous acceleration of the car constant during the 5.00 s interval? Explain your answers.
d. Notice in Figure 3.35 that *two* tangent lines can be drawn to the curve at the instant t = 2.00 s. One of the tangent lines has zero slope, the other a negative slope. Does this mean that the particle has two different accelerations at that instant? Explain your answer.

Solution
a. On a plot of $v_x(t)$ versus t, the x-component of the acceleration is the slope of a line drawn tangent to the curve at the instant in question. The slope of the tangent lines at each

instant between t = 0 s and t = 2.00 s is zero. At each instant during this time interval, the instantaneous acceleration of the car is zero; the car has a constant x-component of the velocity.

b. Since you are to estimate the acceleration when t = 3.00 s, draw a tangent to the curve at this instant. See Figure 3.36. The tangent line has a negative slope that is approximately

$$a_x \approx \frac{-8.0 \text{ m/s}}{2.00 \text{ s}}$$

$$\approx -4.0 \text{ m/s}^2$$

c. Figure 3.37 shows a series of tangent lines over the 5.00 s interval between 2.00 s and 7.00 s. Since the slopes of these lines change, the x-component of the acceleration of the car is not constant; it is becoming less negative. The magnitude of the acceleration (a positive scalar quantity, since the magnitude of a vector always is a positive scalar or zero) decreases over the interval.

d. The particle cannot have two different accelerations at the same instant. The graph *models* the behavior of the particle. Near the instant t = 2.00 s, a more realistic graph of the x-component of the velocity is shown in the blowup in

$v_x(t)$ (m/s)

50.0 km/h = 13.9 m/s

15
14
13
12
11
10
9
8
7
6
5
4
3
2
1

0   1   2   3   4   5   6   7   8 t (s)

FIGURE 3.35

FIGURE 3.36

FIGURE 3.38

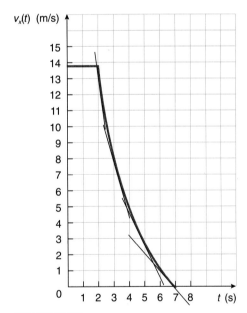

FIGURE 3.37

Figure 3.38. There is no discontinuity in the slope when $t = 2.00$ s; rather, there is a rounding of the cusp so that the slope changes continuously (and rapidly) rather than discontinuously.

3.7 RECTILINEAR MOTION WITH A CONSTANT ACCELERATION

The motion of an airplane accelerating along a runway for takeoff or after landing, the motion of a car braking for a red light or speeding up for a green light, and the **free-fall** motion of vertically moving objects near the surface of the Earth (when we

ignore the effects of air resistance) are examples of the many kinematic situations that commonly are modeled to a first approximation using a constant acceleration.

A constant acceleration means that *neither* the magnitude *nor* the direction of the acceleration vector changes with time or location.

We continue to choose the x-axis to be coincident with the line followed by the particle in rectilinear motion, whether the motion be horizontal (as when a car accelerates or brakes on a level highway), vertical (as when a ball is tossed upward or a rock is dropped off a bridge), or in another direction (as a car accelerating up or down a straight hill). As indicated in Figure 3.39, you have the freedom to choose:

1. The location of the origin of the coordinate system. The origin need not be at the position of the particle when we begin observations.
2. The direction taken for $\hat{\imath}$.

Write the *constant* acceleration vector as

$$\vec{a} = a_x \hat{\imath} \qquad (3.15)$$

fixed in magnitude and direction in both space and time; that is, the acceleration component a_x is *a constant scalar quantity*, independent of both x and t. The numerical value of the acceleration component, a_x, may be positive or negative depending on the choice you make for the direction of $\hat{\imath}$ in a particular problem. We express the acceleration component a_x in SI units of m/s².

In many applications, the *initial* time t_i when we begin our observations is chosen to be *zero* ($t_i = 0$ s) and corresponds to starting a stopwatch for a problem.

The particle is not necessarily at rest when $t = 0$ s.

One choice	Another choice

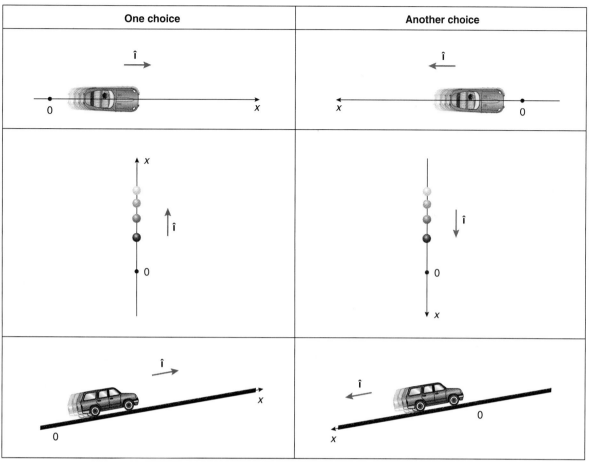

FIGURE 3.39 The x-axis is taken as the line of motion and the origin is chosen at any convenient point. The direction of $\hat{\imath}$ can be taken either way along the line of motion; these decisions are yours to make for each problem.

The instant $t = 0$ s is when we begin our observations of the particle. Likewise, we usually let the time instant t_f be a general time instant t and omit the subscript. With this convention, it is important to realize that t now has a *dual meaning*:

1. t means a specific time *instant* on a stopwatch—thus $x(t)$ and $v_x(t)$ are the position and velocity components at the instant t on the stopwatch; and
2. Since we chose the initial time instant to be the stopwatch reading 0 s, the value of t also is numerically equal to the *time interval* between the zero clock reading and the instant t:

$$\Delta t = t_f - t_i$$
$$= t - 0 \text{ s}$$
$$= t$$

As before, the x-component of the velocity vector at instant t is written $v_x(t)$, while the corresponding x-component of the position vector is $x(t)$.

We also introduce several notational abbreviations:

a. The quantity v_{x0} is the x-component of the velocity of the particle when we begin our observations when $t = 0$ s:

$$v_x(0 \text{ s}) \equiv v_{x0}$$

b. The quantity x_0 is the x-component of the position vector of the particle when we begin observations when $t = 0$ s:

$$x(0 \text{ s}) \equiv x_0$$

PROBLEM-SOLVING TACTIC

3.6 Know the meaning of the notation. To understand the meaning of technical notation is central to applying physics to a problem. Rather than just looking at the notation, envision its meaning in a short phrase or sentence. Say it aloud if possible (without disturbing others); if you can verbalize the meaning of a technical symbol, you probably know what it means. For example, when you see x_0, think "the x-coordinate of the particle when $t = 0$ s." When you see v_{x0}, think "the x-component of the velocity when $t = 0$ s." Study the kinematic notation just introduced; it will be used extensively.

Since the acceleration component a_x is constant, Equation 3.14,

$$\frac{dv_x(t)}{dt} = a_x$$

indicates that $v_x(t)$ must vary linearly with t:

$$v_x(t) = v_{x0} + a_x t \qquad (3.16)$$

Notice that when $t = 0$ s in Equation 3.16, the velocity component is v_{x0}, consistent with the new notation we have introduced. Differentiating Equation 3.16 with respect to t yields Equation 3.14.

A schematic graph of Equation 3.16 is linear (see Figure 3.40). The vertical intercept is the velocity component v_{x0} when $t = 0$ s. The constant slope of the linear graph is the constant acceleration component a_x. In Figure 3.40, $a_x > 0$ m/s². If a_x were negative, the line would have a negative slope and $v_x(t)$ would decrease as t increased.

Since the velocity component $v_x(t)$ varies linearly with t, the x-component of the *average velocity* during the interval $\Delta t = t - 0$ s $= t$ is*

$$v_{x \text{ ave}} = \frac{v_{x0} + v(t)}{2} \qquad (3.17)$$

Adapted to our new notation, Equation 3.6 for the average velocity, written in terms of its (single) component, is

$$v_{x \text{ ave}} = \frac{x(t) - x_0}{t - 0 \text{ s}} \qquad (3.18)$$

Equate Equations 3.17 and 3.18 for the average velocity component:

$$\frac{v_{x0} + v(t)}{2} = \frac{x(t) - x_0}{t}$$

Substituting for $v(t)$ using Equation 3.16, we find

$$\frac{v_{x0} + v_{x0} + a_x t}{2} = \frac{x(t) - x_0}{t}$$

Solve for $x(t)$:

$$x(t) = x_0 + v_{x0}t + \frac{1}{2}a_x t^2 \qquad (3.19)$$

---

*If the acceleration is not constant, Equation 3.17 cannot be used to find the average velocity component.

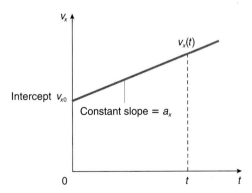

FIGURE 3.40 For a constant acceleration, a graph of v_x versus t is linear with intercept v_{x0} and slope a_x.

Therefore, one-dimensional motion with a constant acceleration is described by two fundamental equations that determine the components of the velocity, and position vectors:

One-dimensional motion with a constant acceleration

$$a_x = \text{constant} \qquad \text{where } \vec{a} = a_x\hat{\imath}$$

$$v_x(t) = v_{x0} + a_x t \qquad \text{where } \vec{v}(t) = v_x(t)\hat{\imath} \qquad (3.16)$$

$$x(t) = x_0 + v_{x0}t + \frac{1}{2}a_x t^2 \qquad \text{where } \vec{r}(t) = x(t)\hat{\imath} \qquad (3.19)$$

Remember:

x_0 is the x-coordinate of the particle when $t = 0$ s; it also is the x-component of the position vector of the particle when $t = 0$ s.

v_{x0} is the x-component of the velocity vector of the particle when $t = 0$ s

PROBLEM-SOLVING TACTICS

3.7 Know the limitations associated with using certain equations. Equations 3.16 and 3.19 are the important ones for motion with a *constant* acceleration (both in magnitude and direction). However, before using them, be sure the problem involves an acceleration that is constant, or can be modeled with a constant acceleration to a first approximation. If the acceleration is not constant, then these equations are useless and, if used, produce predictions that may not correspond with observations.

3.8 Avoid needless memorization of equations. Every problem involving one-dimensional motion with a constant acceleration can be solved using only Equations 3.16 and 3.19. No other equations are needed. You should begin any problem having a constant acceleration by considering only Equations 3.16 and 3.19.

3.9 Tailor the general equations to the specific choice you make for the coordinate system. This tailoring process will become apparent after studying several examples that follow. The flexibility involved in choosing the location and orientation of the Cartesian coordinate system is considered in Strategic Example 3.12. If you understand how to tailor the one-dimensional motion equations at constant acceleration to the specific choice of a Cartesian coordinate system for the problem at hand, no problem will stump you. Everything that can be asked about such motion is extracted from Equations 3.16 and 3.19. Remember, there is no sacred choice for the coordinate system. This allows you the freedom to tackle the situation however you want. That is part of the beauty of physics; it allows independent observers to choose how to describe the situation. The critical skill to develop, then, is the ability to determine and set up appropriate and correct equations.

The experiment performed in Example 3.11 (following) shows that near the surface of the Earth,[†] any object in vertical free-fall (ignoring air resistance effects) continually experiences a constant acceleration *directed vertically down*, regardless of whether the particle is moving vertically up or vertically down. We will see

---

[†]"Near" means any vertical distances traversed by the particle are negligible compared with the radius of the Earth (6370 km).

in Chapter 6 how this constant acceleration arises from the gravitational force of the Earth on the particle. The letter g designates the *magnitude* of this gravitational acceleration near the surface of the Earth; g is called the magnitude of the **local acceleration due to gravity**. In this text, we take the value of g to be*

$$g = 9.81 \text{ m/s}^2$$

PROBLEM-SOLVING TACTIC

3.10 Since g is the magnitude of the local gravitational acceleration, g always is a *positive* scalar quantity in this text.

If you choose $\hat{\imath}$ to be vertically up (see Figure 3.41), then

$$a_x = -g$$

If you choose $\hat{\imath}$ to be vertically down (see Figure 3.42), then

$$a_x = +g$$

QUESTION 5

Toss a ball straight up, and you will see it rise awhile and fall awhile. Is the acceleration vector always in the same direction, or is it directed one way for the flight up and another way for the flight down?

QUESTION 6

Some people might say that an acceleration component is negative and others might say, equally correctly, that it is positive. Explain the difference in viewpoints.

EXAMPLE 3.11

A student drops an egg from rest at an initial height of 2.00 m above the floor. An electronic timer indicates that the egg begins its impact with the floor 0.6385 s after release. Determine the magnitude of the acceleration of the egg during its fall, assuming it is constant. This is a reasonable assumption.

Solution

Use Problem-Solving Tactic 3.1, which states that *for rectilinear motion, always choose a Cartesian coordinate axis to be coincident with*

---

*Since the Earth is only approximately spherical, the value of g depends slightly on your latitude on the surface of the Earth, varying from 9.78 m/s² at the equator to 9.83 m/s² at the poles. For simplicity, we ignore this slight variation of g in this text.

$$\vec{a} = a_x\hat{\imath} = -g\hat{\imath}$$

FIGURE 3.41 Negative gravitational acceleration component.

$$\vec{a} = a_x\hat{\imath} = g\hat{\imath}$$

FIGURE 3.42 Positive gravitational acceleration component.

the line followed by the particle. Choose the x-axis to be coincident with the vertical path followed by the egg. Remember, there is nothing sacred about the orientation you choose for the x-axis in space; it can be horizontal, vertical, or whatever. You simply need to specify what orientation you choose. You can choose the origin to be anywhere along the vertical path. Suppose you choose the origin to be at the point of release of the bottom of the egg (or you can consider the egg as a particle of negligible geometric size). The choice for the direction of $\hat{\imath}$ can be either up or down; choose, say, the upward direction for $\hat{\imath}$, as in Figure 3.43.

Also use Problem-Solving Tactic 3.9 and *tailor the general equations to the specific choice made for the coordinate system.* The egg is released when $t = 0$ s with the bottom of the egg at the initial position $x_0 = 0$ m. The egg was released at rest, so that the initial x-component of the velocity is $v_{x0} = 0$ m/s. Hence Equation 3.19 for the x-component of the position vector at any instant t becomes

$$
\begin{aligned}
x(t) &= x_0 + v_{x0}t + \frac{1}{2}a_xt^2 \\
&= 0 \text{ m} + (0 \text{ m/s})\,t + \frac{1}{2}a_xt^2 \\
&= \frac{1}{2}a_xt^2
\end{aligned}
$$

At the instant $t = 0.6385$ s, the x-coordinate of the bottom of the egg is -2.00 m; this is the x-component of the position vector at that instant. Hence

$$-2.00 \text{ m} = \frac{1}{2}a_x(0.6385 \text{ s})^2$$

Solving for a_x, you obtain

$$a_x = -9.81 \text{ m/s}^2$$

The acceleration is

$$
\begin{aligned}
\vec{a} &= a_x\hat{\imath} \\
&= (-9.81 \text{ m/s}^2)\hat{\imath}
\end{aligned}
$$

The acceleration is directed in the same sense as $-\hat{\imath}$. With the choice you made for the direction of $\hat{\imath}$ (up), the acceleration is directed vertically downward. The egg falls with an acceleration whose magnitude is 9.81 m/s².

FIGURE 3.43 The coordinate system choice.

STRATEGIC EXAMPLE 3.12

A geologist at the top of a 100-meter-deep crevasse cannot resist the temptation to hurl a gneiss rock down to the bottom. The rock has an initial downward speed of 10.0 m/s as it leaves the geologist's hand when $t = 0$ s. Find

a. the time interval during which the particle is in flight, called the time of flight; and
b. the speed of the particle at the instant just before impact.

Solution

The Cartesian coordinate system used to solve this problem can be chosen with the direction for increasing values of x directed either up or down. The origin also can be chosen to be anywhere along the vertical axis. To convince you of this freedom, we will solve the problem two ways using different choices for the coordinate system. Obviously, the time of flight and speed (a positive scalar) the instant just before impact are indifferent about the choice you make, and so the *same* answers will be obtained whatever choice is made for the coordinate system. However, the *description* of the motion is different in each case, since the description is tailored to the specific choice made for the coordinate system.

Choose a Cartesian coordinate system in either of two different ways as shown in Figures 3.44 and 3.45: The acceleration vector is directed downward and is of magnitude $g = 9.81$ m/s^2.

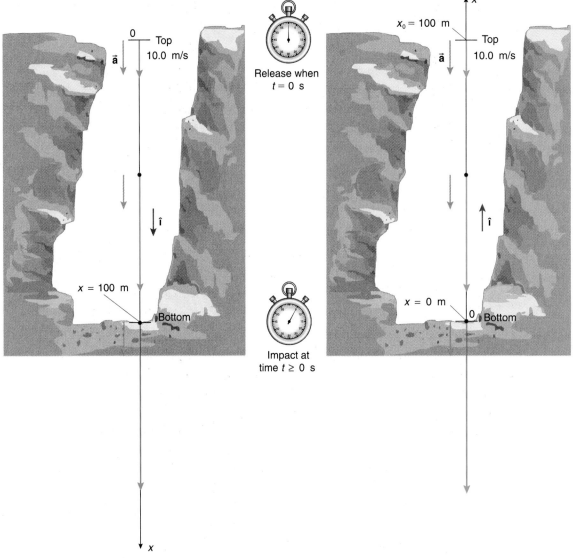

FIGURE 3.44 The origin is at the point of release and the direction of increasing values of x, $\hat{\imath}$, is vertically down.

FIGURE 3.45 The origin is at the bottom of the crevasse and $\hat{\imath}$ is vertically up.

Choice 1	Choice 2
The acceleration is toward increasing values of x, in the same direction as $\hat{\imath}$. Hence the acceleration vector is	The acceleration is toward decreasing values of x, in the same direction as $-\hat{\imath}$. Hence the acceleration is
$$\vec{a} = a_x\hat{\imath}$$ $$= (+9.81 \text{ m/s}^2)\hat{\imath}$$	$$\vec{a} = a_x\hat{\imath}$$ $$= (-9.81 \text{ m/s}^2)\hat{\imath}$$
The x-component of the acceleration is	The x-component of the acceleration is
$$a_x = +9.81 \text{ m/s}^2$$	$$a_x = -9.81 \text{ m/s}^2$$

Because the rock is not dropped from rest, but hurled down the crevasse, there is a nonzero initial velocity

directed toward increasing values of x, in the same direction as $\hat{\imath}$:	directed toward decreasing values of x, in the same direction as $-\hat{\imath}$:
$$v_{x0}\hat{\imath} = (10.0 \text{ m/s})\hat{\imath}$$	$$v_{x0}\hat{\imath} = (-10.0 \text{ m/s})\hat{\imath}$$

so the x-component of the initial velocity, v_{x0}, is

$$v_{x0} = +10.0 \text{ m/s}$$	$$v_{x0} = -10.0 \text{ m/s}$$

The origin of the coordinate system was chosen to be

at the point where the rock was released, so the initial x-component of the position vector, x_0, is zero:	at the bottom of the crevasse, so the initial x-component of the position vector, x_0, is 100 m:
$$x_0 = 0 \text{ m}$$	$$x_0 = 100 \text{ m}$$

The instant $t = 0$ s is when the rock is released. We made this choice to start the clock for any problem when we developed the two equations for one-dimensional motion with a constant acceleration, Equations 3.16 and 3.19.

At the instant t immediately before impact,

the rock is located at the coordinate	the rock is located at the coordinate
$$x(t) = +100 \text{ m}$$	$$x(t) = 0 \text{ m}$$

Notice that t plays a dual role: as an *interval* of time (from the $t = 0$ s initial clock reading, $\Delta t = t - 0$ s $= t$) and as an *instant* when the particle has the position vector component $x(t)$ and velocity component $v_x(t)$.

The equation for the x-component of the position vector, Equation 3.19,

$$x(t) = x_0 + v_{x0}t + \frac{1}{2}a_x t^2$$

becomes (at the instant just before impact)

$$100 \text{ m} = 0 \text{ m} + (10.0 \text{ m/s})\,t$$ $$+ \frac{1}{2}(9.81 \text{ m/s}^2)\,t^2$$	$$0 \text{ m} = 100 \text{ m} + (-10.0 \text{ m/s})\,t$$ $$+ \frac{1}{2}(-9.81 \text{ m/s})\,t^2$$

Notice that we use the *same* fundamental equations for both coordinate choices, but the values of the various substitutions are *different* because the coordinate systems differ.

With some arranging of terms, we have

$$\frac{9.81 \text{ m/s}^2}{2}t^2 + (10.0 \text{ m/s})\,t - 100 \text{ m} = 0 \text{ m}$$	$$\frac{9.81 \text{ m/s}^2}{2}t^2 + (10.0 \text{ m/s})\,t - 100 \text{ m} = 0 \text{ m}$$

which is quadratic in t. Notice that the equation for t is the same for *both* coordinate choices, illustrating once again that the physics is independent of the way you impose a coordinate system on a problem. The equation for t is solved using the quadratic formula. The two roots are

$$t = 3.61 \text{ s} \quad \text{and} \quad t = -5.65 \text{ s}$$

Since the impact time obviously is after the release (which was when $t = 0$ s), the positive solution is the one desired*:

$$t = 3.61 \text{ s}$$

The component of the velocity vector at the instant just before impact is found from Equation 3.16:

$$v_x = v_{x0} + a_x t$$

Substituting for v_{x0}, a_x, and t, you obtain

$v_x = 10.0 \text{ m/s} + (9.81 \text{ m/s}^2)(3.61 \text{ s})$ $= 45.4 \text{ m/s}$	$v_x = -10.0 \text{ m/s} + (-9.81 \text{ m/s}^2)(3.61 \text{ s})$ $= -45.4 \text{ m/s}$

The velocity vector at the instant of impact is $\vec{v} = v_x\hat{i}$:

$\vec{v} = (45.4 \text{ m/s})\hat{i}$ and is in the same sense as $\hat{i}$, vertically downward.	$\vec{v} = (-45.4 \text{ m/s})\hat{i}$ and is in the same sense as $-\hat{i}$, vertically downward.

The speed is the *magnitude* of the velocity vector, or 45.4 m/s. Notice that the velocity has the same magnitude and direction in space for *both* coordinate choices, but the way you *describe* the downward velocity vector depends on your choice of a Cartesian coordinate system.

EXAMPLE 3.13

How far has the rock in Example 3.12 traveled when its speed is 30.0 m/s?

Solution

Since there is no back-and-forth motion of the rock as it falls and the problem is one-dimensional, the magnitude of the change in the position vector is equivalent to the distance the rock travels. Specifically, you want the *change* in its position vector between the instant of release ($t = 0$ s) and the instant its speed is 30.0 m/s.

Choose the point of release as the origin of the coordinate system with the increasing values of x in the upward direction as shown in Figure 3.46. The coordinate choice differs from either of the choices used in Example 3.12. We illustrate this new choice to show the flexibility you have when making coordinate choices.

The initial position vector of the rock when $t = 0$ s then is $(0 \text{ m})\hat{i}$, so that $x_0 = 0$ m. Since $\hat{i}$ is up, the constant downward directed acceleration

$$\vec{a} = a_x\hat{i}$$

is

$$\vec{a} = (-9.81 \text{ m/s}^2)\hat{i}$$

and so the x-component of the acceleration is

$$a_x = -9.81 \text{ m/s}^2$$

From Equation 3.16, the velocity component $v_x(t)$ is

$$v_x(t) = v_{x0} + a_x t$$

The initial velocity component v_{x0} is -10.0 m/s, because the rock was hurled in the direction of decreasing values of x, in

the same direction as $-\hat{i}$. This problem is concerned with what is happening when the *speed* is 30.0 m/s. Since the rock is moving downward (toward decreasing values of x with your choice of the coordinate system), you must find the instant t when the velocity component $v_x(t)$ is -30.0 m/s. Use this information in Equation 3.16 to find

$$-30.0 \text{ m/s} = -10.0 \text{ m/s} + (-9.81 \text{ m/s}^2)t$$

Initial position vector is $(0 \text{ m}) \hat{i}$, so $x_0 = 0$ m.

Release when $t = 0$ s

10.0 m/s($-\hat{i}$)

Instant when speed is 30 m/s

30.0 m/s($-\hat{i}$)

$x = -100$ m

Impact

FIGURE 3.46

---

*The negative root corresponds to when the particle was at $x = 100$ m (Choice 1) or $x = 0$ m (Choice 2) if it had been previously launched vertically upward from the bottom of the crevasse so that it would have a downward speed of 10 m/s at the top of the crevasse when $t = 0$ s.

Solving for t, you find

$$t = 2.04 \text{ s}$$

The position of the rock at any time is found from the equation for the x-component of the position vector, Equation 3.19:

$$x = x_0 + v_{x0}t + \frac{1}{2}a_x t^2$$

Since the rock was released from the origin, the initial position vector component x_0 is 0 m. Recall that $v_{x0} = -10.0$ m/s. Make these substitutions in Equation 3.19:

$$x = 0 \text{ m} + (-10.0 \text{ m/s})(2.04 \text{ s}) + \frac{1}{2}(-9.81 \text{ m/s}^2)(2.04 \text{ s})^2$$

or

$$x = -20.4 \text{ m} - 20.4 \text{ m}$$
$$= -40.8 \text{ m}$$

Thus the position vector of the rock when its *speed* is 30.0 m/s is

$$\vec{r} = x\hat{\imath}$$
$$= (-40.8 \text{ m})\hat{\imath}$$

The change in the position vector of the rock is then

$$\Delta\vec{r} = \vec{r}_f - \vec{r}_i$$
$$= (-40.8 \text{ m})\hat{\imath} - (0 \text{ m})\hat{\imath}$$
$$= (-40.8 \text{ m})\hat{\imath}$$

Hence the magnitude of the change in the position vector of the rock is 40.8 m when its *speed* is 30.0 m/s. Since the rock did not move back and forth along its path, 40.8 m is the distance through which the rock fell while achieving the speed of 30.0 m/s.

EXAMPLE 3.14

You are driving downhill at 110 km/h along a straight section of rural road when a fawn of a white-tailed deer (*Odocoileus virginianus*) springs in front of your car. Fortunately, you manage to stop the car 3.50 s after your foot first hits the brake pedal, and without hitting Bambi. The roadway makes an angle of 5.00° with the horizontal. Assume the acceleration of the car is constant during the 3.50 s time interval.

a. **What is the acceleration?**
b. **What is the average acceleration during the 3.50 s interval?**
c. **Through what distance did the car move during the 3.50 s interval?**

Solution
a. The 110 km/h initial speed is equivalent to

$$110 \text{ km/h} = (110 \text{ km/h})\frac{1000 \text{ m/km}}{3600 \text{ s/h}}$$
$$= 30.6 \text{ m/s}$$

Remember that for one-dimensional motion, you choose the x-axis to be coincident with the path followed by the particle, whatever its orientation. Set up a coordinate system with the

origin at the initial position of the car when the braking began and $\hat{\imath}$ in the direction of motion of the car down the hill (see Figure 3.47). The initial velocity component v_{x0} of the car when $t = 0$ s is

$$v_{x0} = +30.6 \text{ m/s}$$

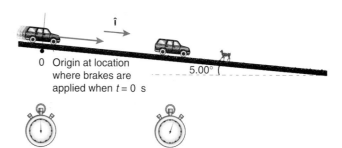

$t = 0$ s $\qquad$ $t = 3.50$ s

FIGURE 3.47

The velocity component of the car at time instant t is given by Equation 3.16:

$$v_x(t) = v_{x0} + a_x t$$

When $t = 3.50$ s, the velocity component v_x is zero, so that Equation 3.16 becomes

$$0 \text{ m/s} = 30.6 \text{ m/s} + a_x(3.50 \text{ s})$$

Solving for a_x, you find

$$a_x = -8.74 \text{ m/s}^2$$

Hence the constant acceleration is

$$\vec{a} = a_x\hat{\imath}$$
$$= (-8.74 \text{ m/s}^2)\hat{\imath}$$

The acceleration vector is directed opposite to the direction the car is moving during the time interval.

b. The average acceleration is given by Equation 3.9:

$$\vec{a}_{ave} = \frac{\Delta\vec{v}}{\Delta t}$$
$$= \frac{\vec{v}_f - \vec{v}_i}{\Delta t}$$
$$= \frac{(0 \text{ m/s})\hat{\imath} - (30.6 \text{ m/s})\hat{\imath}}{3.50 \text{ s}}$$
$$= (-8.74 \text{ m/s}^2)\hat{\imath}$$

Notice that the average acceleration during the 3.50 s interval is the same as the constant instantaneous acceleration at any instant during the interval. If the acceleration is constant, the average acceleration over any interval is equal to the constant acceleration during the interval.

c. Since there is no back-and-forth motion of the car during the time interval, the distance the car travels is equal to the magnitude of the change in its position vector. Note that the initial x-component of the position vector of the car, x_0, is 0 m, because the origin was chosen at this point. Use

$v_{x0} = 30.6$ m/s and $a_x = -8.74$ m/s^2 from part (a); then Equation 3.19 becomes

$$x = 0 \text{ m} + (30.6 \text{ m/s})\, t + \frac{1}{2}(-8.74 \text{ m/s}^2)\, t^2$$

The position x of the car when it stops is found by substituting $t = 3.50$ s:

$$x = (30.6 \text{ m/s})(3.50 \text{ s}) + \frac{1}{2}(-8.74 \text{ m/s}^2)(3.50 \text{ s})^2$$

$$= 107 \text{ m} - 53.5 \text{ m} = 53 \text{ m}$$

The change in the position vector of the car is

$$\Delta \vec{\mathbf{r}} = \Delta x\, \hat{\imath}$$
$$= (53 \text{ m})\hat{\imath} - (0 \text{ m})\hat{\imath}$$
$$= (53 \text{ m})\hat{\imath}$$

The distance traveled by the car is the magnitude of this vector, 53 m, since the car did not move back and forth along the roadway during the 3.50 s interval it took to stop.

3.8 GEOMETRIC INTERPRETATIONS*

We found in Section 3.4 that the x-component of the velocity, $v_x(t)$, is related to the x-component of the position vector by

$$v_x(t) = \frac{dx(t)}{dt}$$

Hence the value of $v_x(t)$ at any instant t is the slope of a graph of $x(t)$ versus time at the instant t. Likewise, we saw in Section 3.6 that the x-component of the acceleration $a_x(t)$ is found from the velocity component via Equation 3.14:

$$a_x(t) = \frac{dv_x(t)}{dt}$$

Hence the value of a_x at any instant t is the slope of a graph of v_x versus time at the instant t. Thus a knowledge of the position vector component $x(t)$ enables us to find the velocity component $v_x(t)$ and then the acceleration component $a_x(t)$ by differentiation with respect to time.

We could equally well reverse the process and proceed from a knowledge of the acceleration component to the velocity component and then to the position vector component via integration. In other words, if the acceleration component $a_x(t)$ is known, then we integrate Equation 3.14 over time to find the velocity component $v_x(t)$:

$$\int_{v_x(t_i)}^{v_x(t_f)} dv_x = \int_{t_i}^{t_f} a_x(t)\, dt$$

where $v_x(t_i)$ is the x-component of the velocity at time t_i and $v_x(t_f)$ is the velocity component at time t_f. After integrating

the left-hand side according to the rules of calculus:

$$v_x(t_f) - v_x(t_i) = \int_{t_i}^{t_f} a_x(t)\, dt \qquad (3.20)$$

As we know from calculus, an integral is the geometric area under the curve of the function being integrated (the integrand). In other words, if we make a graph of the function $a_x(t)$ versus t, as shown in Figure 3.48, the *change* in the value of $v_x(t)$—that is, $\Delta v_x \equiv v_x(t_f) - v_x(t_i)$—is the area under the curve representing $a_x(t)$ between the times t_i and t_f.

Areas under the curve can be positive or negative. From calculus, if the curve is above the horizontal axis (see Figure 3.49), and we are integrating toward increasing values of the abscissa (here t), then the area between the curve and the axis is positive. If the curve dips below the axis, the area between the curve and the axis is negative.

Likewise, from Equation 3.8, the change in the position vector component is the time integral of the velocity component $v_x(t)$:

$$\int_{x(t_i)}^{x(t_f)} dx = \int_{t_i}^{t_f} v_x(t)\, dt$$

where $x(t_i)$ is the x-component of the position vector at time t_i and $x(t_f)$ is the x-component of the position vector at time t_f. After performing the integration on the left-hand side, we have

$$x(t_f) - x(t_i) = \int_{t_i}^{t_f} v_x(t)\, dt \qquad (3.21)$$

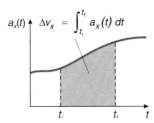

FIGURE 3.48 The change in the velocity component is the area under the curve of a_x versus t.

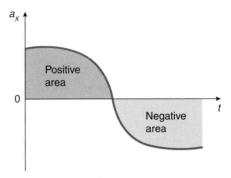

FIGURE 3.49 Areas under a curve can be positive or negative.

On a graph of $v_x(t)$ versus t, as in Figure 3.50, the *change* in the value of the x-component of the position vector—that is, $\Delta x = x(t_f) - x(t_i)$—is the area under the curve representing $v_x(t)$ between the times t_i and t_f.

If we let $t_i = 0$ s and $t_f = t$, then Equations 3.20 and 3.21 become

$$v_x(t) - v_{x0} = \int_{0\,s}^{t} a_x(t)\, dt \qquad (3.22)$$

and

$$x(t) - x_0 = \int_{0\,s}^{t} v_x(t)\, dt \qquad (3.23)$$

Such equations can be applied to situations in which the acceleration is not a constant vector.

If the acceleration is *constant*, we can use Equations 3.22 and 3.23 to get Equations 3.16 and 3.19 in a way different from that used in Section 3.7. According to Equation 3.22, to find the *change* in the x-component of the velocity, you integrate the constant acceleration component a_x:

$$v_x(t) - v_{x0} = \int_{0\,s}^{t} a_x\, dt = a_x \int_{0\,s}^{t} dt$$

where v_{x0} is the x-component of the velocity when $t = 0$ s, and $v_x(t)$ is the velocity component at instant t. After integrating, we have

$$v_x(t) - v_{x0} = a_x[t - 0\ \text{s}] \qquad (3.24)$$
$$\Delta v_x = a_x t$$

Geometrically, in Figure 3.51 Δv_x is the

$$\text{rectangular area} = (\text{width})(\text{length})$$
$$= a_x t$$

Note that if $a_x < 0$ m/s^2, the product (width)(length) $= a_x t$ yields a negative area under the curve (between the curve and the axis).

Transposing v_{x0} in Equation 3.24, we obtain Equation 3.16 once again:

$$v_x(t) = v_{x0} + a_x t$$

To find the *change* in the x-component of the position vector, use Equation 3.23 and integrate the x-component of the velocity vector between 0 s and t:

$$x(t) - x_0 = \int_{0\,s}^{t} v_x(t)\, dt = \int_{0\,s}^{t} (v_{x0} + a_x t)\, dt$$

where both v_{x0} and a_x are constants. After integrating and transposing the x_0 term, we have Equation 3.19 once again:

$$x(t) = x_0 + v_{x0}t + \frac{1}{2}a_x t^2$$

The geometric interpretation of this equation on a v_x versus t graph is shown in Figure 3.52.

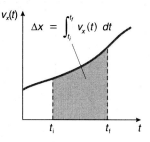

FIGURE 3.50 The change in the value of the position vector component is the area under the curve of v_x versus t.

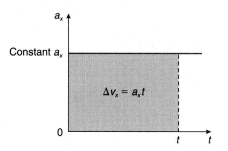

FIGURE 3.51 For a constant acceleration, the change in the velocity component is the rectangular area under the a_x versus t graph.

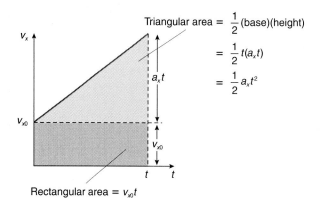

FIGURE 3.52 Geometric interpretation of Equation 3.19 on a v_x versus t graph. The change in the position vector component Δx is the area under the v_x versus t graph.

CHAPTER SUMMARY

Kinematics is the description of motion. *Dynamics* is the study of the causes of motion. *Rectilinear motion* is one-dimensional motion along a straight line.

A *particle* is anything whose geometrical size is ignored.

The location of a particle with respect to the origin of an arbitrarily chosen Cartesian coordinate system is specified by an *instantaneous position vector* $\vec{r}(t)$. For rectilinear motion, $\vec{r}(t) = x(t)\hat{\imath}$.

The *average speed* during an interval Δt is the total distance travelled divided by Δt. The *average velocity* of a particle during a time interval Δt is defined to be the ratio of the *change* in its position vector $\Delta \vec{r}$ during the time interval to the time interval itself:

$$\vec{v}_{\text{ave}} \equiv \frac{\Delta \vec{r}}{\Delta t} \qquad (3.6)$$

The *instantaneous velocity* of a particle, more commonly called the *velocity* $\vec{v}$, is the limit of its average velocity as the time interval Δt approaches zero; this is equivalent to the time derivative of the instantaneous position vector

$$\vec{v}(t) = \lim_{\Delta t \to 0\ \text{s}} \frac{\Delta \vec{r}}{\Delta t} = \frac{d\vec{r}(t)}{dt}$$

$$= \frac{dx(t)}{dt}\,\hat{\imath} \quad \text{(for rectilinear motion)} \qquad (3.7)$$

The magnitude of the velocity is called the *speed*. The instantaneous velocity component at any time instant is the slope of the graph of the corresponding position vector component versus time at that instant. That is, v_x at instant t' is the slope of the graph of $x(t)$ versus t at the particular instant t'.

The *average acceleration* of a particle during a time interval Δt is the ratio of the change in its velocity during the time interval to the time interval itself:

$$\vec{a}_{\text{ave}} \equiv \frac{\Delta \vec{v}}{\Delta t} \qquad (3.9)$$

The *instantaneous acceleration*, commonly called the *acceleration* $\vec{a}$, is the limit of the average acceleration as the time interval Δt approaches zero; this is equivalent to the time derivative of the velocity, or the second time derivative of the position vector:

$$\vec{a}(t) = \lim_{\Delta t \to 0\ \text{s}} \frac{\Delta \vec{v}}{\Delta t} = \frac{d\vec{v}(t)}{dt} = \frac{d^2\vec{r}(t)}{dt^2} \qquad (3.13)$$

$$= \frac{dv_x(t)}{dt}\,\hat{\imath} = \frac{d^2x(t)}{dt^2}\,\hat{\imath} \quad \text{(for rectilinear motion)}$$

The instantaneous acceleration component at any instant is the slope of the graph of the corresponding velocity component versus time at that instant. That is, a_x at instant t' is the slope of the graph of $v_x(t)$ versus t at the particular instant t'.

If the motion is with a *constant* acceleration and the initial time instant t_i is set equal to zero, then the velocity and position vector components of the particle at any instant t are

$$v_x(t) = v_{x0} + a_x t \qquad a_x \text{ constant } (3.16)$$

$$x(t) = x_0 + v_{x0}t + \frac{1}{2}a_x t^2 \qquad a_x \text{ constant } (3.19)$$

where a_x is the constant x-component of the acceleration, v_{x0} is the x-component of the velocity when $t = 0$ s, and x_0 is the x-component of the position vector of the particle when $t = 0$ s. The signs and numerical values of a_x, v_{x0}, and x_0 depend on the particular choices you make for the location of the origin for the x-axis of a Cartesian coordinate system, the direction chosen for $\hat{\imath}$, and the specifics of the problem at hand.

After particles are thrown vertically (up or down) or dropped vertically near the surface of the Earth, they experience *free-fall* and (neglecting air resistance) a *constant* acceleration, called the *local acceleration due to gravity*, of magnitude $g = 9.81$ m/s² directed vertically downward. The sign of the vertical acceleration component a_x depends on the choice you make for the direction of $\hat{\imath}$:

$$a_x = +g \qquad \text{if } \hat{\imath} \text{ is chosen vertically down}$$

$$a_x = -g \qquad \text{if } \hat{\imath} \text{ is chosen vertically up}$$

The free-fall acceleration due to gravity *always* is directed vertically down, whether the particle is rising, falling, or at the instant, between rising and falling, when it is at rest.

If a graph is made of $a_x(t)$ versus t, then the *change* in the velocity component between times t_i and t_f is the area under the graph of $a_x(t)$ between those times. That is,

$$v_x(t_f) - v_x(t_i) = \int_{t_i}^{t_f} a_x(t)\, dt \qquad (3.20)$$

If a graph is made of $v_x(t)$ versus t, then the *change* in the position vector component between times t_i and t_f is the area under the graph of $v_x(t)$ between those times. That is,

$$x(t_f) - x(t_i) = \int_{t_i}^{t_f} v_x(t)\, dt \qquad (3.21)$$

SUMMARY OF PROBLEM-SOLVING TACTICS

3.1 **(page 76)** Always choose a coordinate axis for rectilinear motion that is coincident with the line followed by the particle.

3.2 **(page 77)** Read and construct graphs carefully.

3.3 **(page 81)** The sign of $v_{x\,\text{ave}}$ depends on the choice you make for the direction of $\hat{\imath}$.

3.4 **(page 84)** The choice you make for the coordinate system influences the sign of the velocity component v_x.

3.5 **(page 87)** The sign of the average acceleration component $a_{x\,\text{ave}}$ depends on the direction you choose for $\hat{\imath}$.

3.6 **(page 93)** Know the meaning of the notation.

3.7 **(page 94)** Know the limitations associated with using certain equations.

3.8 **(page 94)** Avoid needless memorizing of equations.

3.9 **(page 94)** Tailor the general equations to the specific choice you make for the coordinate system.

3.10 **(page 95)** Since $g = 9.81$ m/s² is the magnitude of the local acceleration due to gravity, g always is a positive scalar quantity.

QUESTIONS

1. (page 75); 2. (page 77); 3. (page 81); 4. (page 81); 5. (page 95); 6. (page 95).

7. For rectilinear motion, if the change in the position vector $\Delta\vec{r}$ is in the same direction as $\hat{i}$, is it possible to conclude that the particle was moving toward increasing values of x during the *entire* time interval between t_i and t_f? Explain.

8. For rectilinear motion, if the change in the position vector $\Delta\vec{r}$ is in the same direction as $-\hat{i}$, is it possible to conclude that the particle was moving toward decreasing values of x during the *entire* time interval between t_i and t_f? Explain.

9. How would you sketch a situation such that the initial and final position vectors of a particle have the same magnitude, that is, $|\vec{r}_i| = |\vec{r}_f|$, but with $\Delta\vec{r} \neq 0$ m, so that $|\Delta\vec{r}| \neq 0$ m?

10. Explain why the magnitude of the change in the position vector $\Delta\vec{r}$ never can be greater than the total distance traveled. What implications does this result have for the relationship between the average speed and the magnitude of the average velocity?

11. For rectilinear motion, must all particles have *continuous* $x(t)$, $v_x(t)$, and $a_x(t)$ graphs?

12. Sketch a graph that has as the ordinate the x-coordinate of a particle and as abscissa the clock time t. Indicate schematically but clearly on the graph what is meant by (a) an event; (b) a time interval Δt; (c) no change in position during the time interval; (d) a positive change in the position vector component during the interval, where *both* the initial and final values of the position vector component are positive; (e) a positive change in the position vector component during the interval, where *both* the initial and final values of the position vector component are negative; (f) a positive change in the position component during the interval, where the initial position vector component is negative and the final position vector component is positive.

13. If the world-lines of two particles are drawn on the same graph and the two world-lines intersect, what does this mean physically? Can such intersections of world-lines happen to physical objects of finite size as opposed to their representations as particles?

14. A car is traveling on a straight road up a hill of constant slope. (a) If the needle of a common circularly shaped speedometer in the car maintains a fixed orientation pointing toward the numerical value 100 km/h on the scale, what is the car doing physically? Is its acceleration necessarily zero? (b) If the speedometer needle is moving clockwise, what is the car doing physically? (c) If the speedometer needle is moving counterclockwise, what is the car doing physically?

15. What data would you need to estimate the average speed at which a cornstalk grows? Make such an estimate. If you are more familiar with lawn grass, estimate the speed at which it grows.

16. The position coordinate x of a particle as a function of time is indicated in the graphs of Figure Q.16. Describe the motion of the particle in words. Try to mimic the motion of the particle with your hand.
Diagrams from Arnold B. Arons, *A Guide to Introductory Physics Teaching* (John Wiley, New York, 1990), pages 43–44.

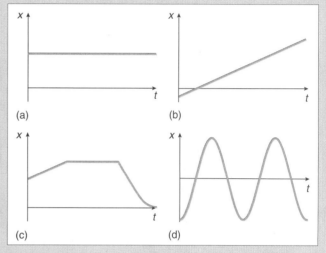

FIGURE Q.16

17. Sketch schematic x versus t graphs for each of the following rectilinear motion situations: (a) Your car is parked for ten minutes. (b) A father pushes a baby stroller across the street. (c) You walk in a straight line up to a postal window, purchase a stamp, and walk away along the same line. (d) A tethered dog runs back and forth along the length of a 15 m wire.

18. Sketch a graph that has the velocity component v_x of a particle with respect to a Cartesian origin as the ordinate and the time t as abscissa. Indicate schematically but clearly on the graph what is meant by (a) a time interval Δt; (b) no change in the velocity component during the time interval; (c) a positive change in the velocity component during the interval, where *both* the initial and final values of the velocity components are positive; (d) a positive change in the velocity component during the interval, where *both* the initial and final values of the velocity components are negative; (e) a positive change in the velocity component during the interval, where the initial velocity component is negative and the final velocity component is positive; (f) a negative change in the velocity component during the interval, where *both* the initial and final values are positive; (g) a negative change in the velocity component during the interval, where *both* the initial and final values of the velocity components are negative; (h) a negative change in the velocity component during the interval, where the initial velocity component is positive and the final velocity component is negative. (i) Is it possible to have a negative change in the velocity component during the interval, with the initial velocity component negative and the final velocity component positive? Explain.

19. In Section 3.3, we called the ratio

$$\frac{\Delta t}{d}$$

a measure of the average sluggishness of a particle, where d is the total distance traveled by a particle during the interval Δt. If this ratio is small, what is the particle doing physically? If the ratio is large, what is the particle doing physically? Use the data provided in Example 3.3 to find the marble's measure of this ratio in SI units.
Inspired by Arnold B. Arons, *A Guide to Introductory Physics Teaching* (John Wiley, New York, 1990), page 23.

20. Examine the ratio

$$\frac{\Delta t}{\Delta \vec{\mathbf{r}}}$$

In view of our study of vectors in Chapter 2, why is this ratio mathematically meaningless in contrast to the ratio examined in the previous question? Does the ratio

$$\frac{\Delta t}{|\Delta \vec{\mathbf{r}}|}$$

make mathematical sense? Why or why not? What does this ratio measure? How does this ratio differ from the ratio defining the average sluggishness? Why does the ratio $\Delta t/|\Delta \vec{\mathbf{r}}|$ provide less information than the ratio $\Delta \vec{\mathbf{r}}/\Delta t$ that was used to define the average velocity?

21. Consider the ratio

$$\frac{\Delta t}{|\Delta \vec{\mathbf{v}}|}$$

where $|\Delta \vec{\mathbf{v}}|$ is the magnitude of the change in the velocity of a particle during a time interval Δt. (a) Under what physical circumstances is this ratio large? small? (b) Invent an appropriate term to describe what this quantity represents physically. (c) Why does this ratio contain less information than the ratio

$$\frac{\Delta \vec{\mathbf{v}}}{\Delta t}$$

used to define the average acceleration?

22. Use the plumb bob accelerometer mentioned in Section 3.6 under the following circumstances. Describe the movement of the plumb bob with respect to the local vertical direction as viewed by a observer at rest with respect to the plumb bob. (a) A traffic light turns green and a car accelerates along a level, straight highway. (b) An elevator begins moving upward from rest to a speed of 2.0 m/s. (c) An automobile moves with a constant speed along a straight, level stretch of interstate highway.

23. Figure Q.23 is a graph of the x-component of the position vector of a particle as a function of time. (a) Does the particle have a nonzero velocity? If so, at what time instants? (b) Is the x-component of the velocity of the particle constant? Explain why or why not. (c) What is the velocity component v_x of the

particle at the instant $t = 0$ s? (d) What is the velocity component v_x of the particle at the instant $t = 3.00$ s? (e) At what instant is the position vector component x equal to zero? What is the velocity component v_x of the particle at this instant? (f) What is the x-component of the acceleration of the particle?

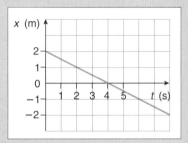

FIGURE Q.23

24. Figure Q.24 is a graph of the x-component of the velocity of a particle as a function of time. (a) Does the particle have a nonzero x-component of the acceleration? (b) Is the x-component of the acceleration of the particle *constant*? Explain why or why not. (c) What is the velocity component v_x of the particle at the instant $t = 0$ s? (d) What is the acceleration component a_x of the particle at the instant $t = 0$ s? (e) At what instant is the velocity component v_x zero? (f) What is the acceleration component a_x of the particle at the instant $t = 2.00$ s?

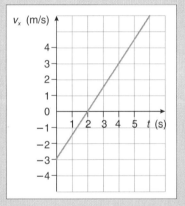

FIGURE Q.24

25. Toss an eraser vertically upward. At the instant when the eraser reaches its greatest elevation: (a) is its velocity equal to zero? Explain. (b) is its acceleration equal to zero? Explain.

26. Sketch a series of graphs of the velocity component v_x of a particle versus time that schematically indicate: (a) a situation where the particle has a zero velocity component at some instant but a nonzero *positive* acceleration component. Indicate this instant on the sketch; (b) a situation where the particle has a zero velocity component at some instant but a nonzero *negative* acceleration component. Indicate this instant on the sketch; (c) a situation where the particle has a velocity component equal to zero and an acceleration component equal to zero at some instant.

27. A car moves left to right along a level, straight highway. Photographs of the car are taken, separated by equal intervals of time Δt. Which of the series of photographs in Figure Q.27, if any, depicts the car (a) moving at constant velocity? (b) moving with a nonzero acceleration? (c) with a positive *x*-component of the velocity if **î** is to the right? (d) with a positive *x*-component of the velocity if **î** is to the left? (e) with a negative *x*-component of the velocity if **î** is to the right? (f) with a negative *x*-component of the velocity if **î** is to the left? (g) with a positive *x*-component of the acceleration if **î** is to the right? (h) with a positive *x*-component of the acceleration if **î** is to the left? (i) with a negative *x*-component of the acceleration if **î** is to the right? (j) with a negative *x*-component of the acceleration if **î** is to the left?

(a)

(b)

(c)

FIGURE Q.27

28. The velocity components v_x of two particles are plotted schematically on the same graph in Figure Q.28. (a) At what times, if any, do the objects have the same velocity component? Indicate such times with the designation t_A. (b) At what times, if any, do the particles have the same acceleration components a_x? Indicate such times with the designation t_B. (c) With just the information given, is it possible to tell whether the two particles simultaneously ever have the same position vector component? Explain your answer.

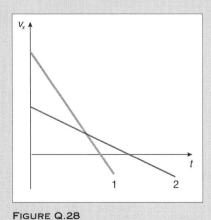

FIGURE Q.28

29. If a particle has the position vector $\vec{r}_i$ at time t_i and at time $t_f > t_i$ the particle has the *same* position vector, does this necessarily imply that the particle did not move during the interval Δt between t_i and t_f? Explain your answer.

30. Someone drops an egg from rest off a cliff while someone at the bottom of the same cliff throws a rock vertically upward. What can be said (if anything) about the accelerations of the two objects during their flights?

31. A particle is found to have an average velocity equal to zero during some interval Δt. What can be said (if anything) about the position vectors of the particle at the beginning and end of the time interval? What can be said (if anything) about changes in the instantaneous position vector *during* the time interval? Explain your answers.

32. A hammer is hurled vertically down from the roof while another hammer is simultaneously hurled vertically upward from the same initial location with the *same* initial speed. (a) If **î** is chosen to be directed vertically *upward*, roughly sketch a graph of the *x*-component of the velocity v_x of each hammer as a function of time. (No numbers need be placed on the graph.) (b) Make a similar graph if **î** is chosen to be vertically *downward*. (No numbers need be placed on the graph.)

33. A particle undergoes rectilinear motion. The velocity component v_x of the particle as a function of time is indicated in the graphs shown in Figure Q.33. Describe the motion of the particle in words and try to mimic the motion of the particle with your hand.
Inspired by Arnold B. Arons, *A Guide to Introductory Physics Teaching* (John Wiley, New York, 1990), page 44.

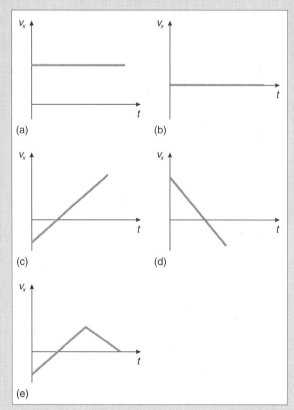

(a)

(b)

(c)

(d)

(e)

FIGURE Q.33

34. Figure Q.34 is a graph of the rectilinear positions of two box turtles (*Terrapene carolina*) as a function of time during a race. (a) Indicate on the sketch the times that the two turtles have

the same position. Label such times t_A. (b) Indicate on the sketch the times that the two turtles have the same velocity. Label such times t_B. (c) If either turtle is (instantaneously) at rest during the time interval shown, indicate which one and indicate the times this occurs on the sketch by t_C. (d) Is either turtle undergoing zero acceleration at some instant? If so, which one? Also indicate the corresponding instants. (e) Is either turtle undergoing accelerated motion? If so, which one? Is it possible to tell from the information given whether this acceleration is constant or not?
Inspired by Arnold Arons, *A Guide to Introductory Physics Teaching* (John Wiley, New York, 1990), page 46.

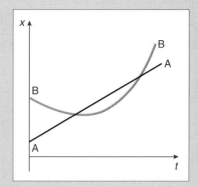

FIGURE Q.34

PROBLEMS

Ignore the effects of air resistance in all of the problems in this chapter.

Sections 3.1 Rectilinear Motion
  3.2 Position and Changes in Position
  3.3 Average Speed and Average Velocity

1. You throw a stick for your dog Sirius to fetch. Sirius runs from right next to you straight out to the stick, lying 25 m away, and brings it directly back to you like a good dog. The trip takes the dog 12 s and you reward Sirius with a pat on the head and a tasty dog biscuit. (a) What is the change in the position vector of the dog? (b) What is the total distance traveled by the dog? (c) What is the average speed of the dog? (d) What is the average velocity of the dog?

2. A pilot flies a Boeing 737 from Seattle, Washington, to Portland, Oregon, taking 60 min. See Figure P.2. After a 45 min layover in Portland, she flies on to San Francisco, south of Portland, with a flight time of 2.0 h. After a 1.0 h layover in San Francisco, the plane flies nonstop back to Seattle in 2.5 h. (a) Set up a coordinate system to analyze the motion, indicating the locations of all three cities. Assume all three cities lie on the same straight line. (b) After determining the distance between the cities from Figure P.2 and the accompanying scale, specify the position vectors of each city with respect to your chosen origin. (c) Determine the average speed in km/h of each flight. (d) Determine the average speed in km/h during the entire time interval from Seattle back to Seattle. (e) Determine the average velocity in km/h of each flight. (f) Determine the average velocity in km/h during the entire time interval from Seattle back to Seattle.

3. The Andromeda and Milky Way galaxies are approximately 2×10^6 light-years apart. (One light-year is the distance traveled by light during 1 year.) How long does it take for a cosmic ray proton, traveling at a constant speed equal to 99.99% the speed of light, to travel between these galaxies?

4. Many years ago, the Apollo lunar astronauts left a collection of corner cube reflectors on the lunar surface that reflect incident light back in the same direction from which it came. See Figure P.4. By measuring the time it takes a light pulse to travel from an Earth-based laser to the corner cube array on the Moon and back, we can measure the distance between the laser and the corner cube with great precision (with an uncertainty of only about a centimeter). Such measurements have

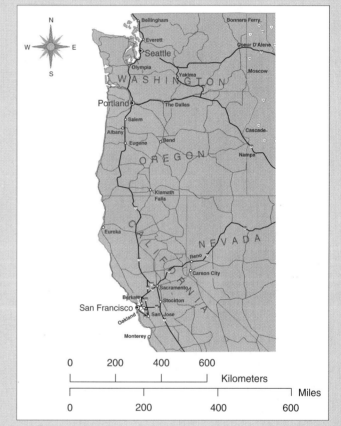

FIGURE P.2

confirmed that the Moon is slowly but systematically moving away from the Earth at a rate of several centimeters per year. (a) Approximately how long does it take light to make the round trip if the distance to the Moon is approximately 3.80×10^5 km? The speed of light is approximately 3.00×10^8 m/s. (b) For a round trip, what is the change in the position vector of the light? (c) What is the average velocity of the light for the round trip?

5. A salesperson hawking kitchen widgets drives about 4.0×10^4 km each year. Assuming an average speed of 60 km/h while in the car, how many weeks a year does the salesperson spend traveling in the car?

FIGURE P.4

FIGURE P.11

6. The slow geologic motion of the continental plates occurs at speeds on the order of one-tenth the rate at which your fingernails grow. Estimate the time in centuries it takes for the continental plates to move 1 meter.

• 7. A base runner is 3.0 m from first base, awaiting the next pitch to his teammate. The pitcher glances over his shoulder and fires the ball to first base at a speed of 120 km/h. Assume the ball travels in a straight line. The distance from the pitcher to the first baseman is 19 m. What is the minimum average speed needed for the base runner to successfully make it back to first base before the ball?

• 8. Your flight leaves in 65 min and you are 40 km from the airport— and fuming. You need 5.0 min to park the car and run to the gate. Unfortunately, heavy traffic on the way keeps your average speed to 25 km/h for the first 20 min of the drive. What must be your average speed for the rest of the trip to make the plane? Is this a minimum or a maximum average speed to make the plane?

• 9. When you see a deer run unexpectedly in front of your car, your reaction time (the time interval it takes you to step on the brake after first seeing danger) is about 0.30 s and your body height is 2.0 m. If your reaction time is caused by the time it takes electrical nerve impulses to be sent from your brain to your feet, what is the minimum speed of such signals? The actual speed of such nerve impulses is on the order of 10^2 m/s. About how long do they actually take to propagate from your head to your feet? What other factors might account for the relatively large reaction time?

• 10. Your average running speed is 8.0 m/s and your father's is 7.0 m/s. You both are located 75 m from the family minivan and begin running toward it at the same time. What is your separation distance when you win the race to the car?

**Sections 3.4 Instantaneous Speed and Instantaneous Velocity
3.5 Average Acceleration
3.6 Instantaneous Acceleration
3.7 Rectilinear Motion with a Constant Acceleration**

11. A reckless, unroped mountain climber (see Figure P.11) falls off a vertical cliff at an elevation of 50 m from the rocks below. How long does the climber have to contemplate meeting his maker? What is the impact speed? As an old saying goes, it's not the fall that hurts, it's the sudden stop (that is, the large acceleration during the short interval of the stop).

12. Calculate the magnitude of the acceleration of a sports car that can drag race from rest to 100 km/h in 5.00 s. Assume the acceleration is constant, although typically this is not a good assumption for automobiles. Compute the ratio of the magnitude of this acceleration to the magnitude of the local acceleration due to gravity (g = 9.81 m/s^2). Make a graph of the speed of the car versus time.

13. A large supertanker, initially moving at 30.0 km/h, is brought to a halt in 15.00 min after a lookout spies an icefield ahead. (a) Set up an appropriate coordinate system and determine the acceleration. (b) Through what distance did the tanker move while stopping?

14. A rubber chicken at a circus is tossed vertically upward with an initial speed of 10.0 m/s from an initial elevation of 2.00 m above the ground. (a) Sketch the situation. (b) Choose a coordinate system clearly indicating the origin and the direction chosen for an appropriate unit vector. (c) Tailor the one-dimensional, constant acceleration equations (Equations 3.16 and 3.19) to the particular choice you made for the coordinate system. (d) Find the time for the chicken to reach its maximum height. (e) What is the maximum height of the bird?

15. Solve the problem of Example 3.12 with the origin of the coordinate system midway down the crevasse and with $\hat{i}$ vertically upward (see Figure P.15). (a) For this coordinate choice, what are x_0, v_{x0}, and a_x? (b) What is the x-coordinate of the rock when it reaches the bottom of the crevasse? (c) Solve for the time it takes the rock to fall, the velocity (a vector), and the speed (a scalar) at the instant just before impact. The results for the time and speed should be the same as those obtained in Example 3.12.

16. The high-speed train in France known as the TGV (*Train à Grande Vitesse*, or Train of Great Speed) accelerates from rest to 200 km/h in 120 s along a straight section of track (see Figure P.16). (a) What is the magnitude of its acceleration, assuming it is constant? (b) Through what distance did it move while accelerating to this speed?

FIGURE P.15

FIGURE P.16

FIGURE P.19

FIGURE P.20

17. When $t = 0$ s, a flower pot on a window sill drops from rest to the sidewalk some 30.0 m below and shatters into smithereens, luckily missing all pedestrians. (a) How long was the pot in flight? (b) What is its maximum speed? (c) How fast is the pot moving when it is at a height of 10.0 m above the ground? (d) How far above the ground is the flower pot when $t = 1.00$ s?

18. After a grueling history test, your flight on U.S. Snarelines is ready for takeoff from Jack Daniels International Airport on its way to Tahiti. Sun and fun! The interval of the takeoff is 40.0 s and the plane uses 1.70 km of the runway. You find yourself with an uncontrollable urge to make a calculation. Find (a) the magnitude of the acceleration (assuming it is constant) in m/s^2 and (b) the takeoff speed in both m/s and km/h.

19. A Martian drops your shoe into the Olympus Mons volcanic caldera, (see Figure P. 19) and it falls 30.0 m in 4.00 s. (a) Find the magnitude of the (constant) acceleration due to gravity on the surface of the planet. (b) By forming a suitable ratio, compare this result with the magnitude of the local acceleration due to gravity on the surface of the Earth (9.81 m/s^2).

20. A thief is cruising along a straight stretch of level highway in your stolen sports car at 120 km/h. See Figure P.20. A flashing light in the rearview mirror triggers a constant acceleration to 180 km/h in 4.00 s. Go get 'im! (a) Set up an appropriate coordinate system to analyze this problem. (b) Determine the appropriate component of the acceleration consistent with your coordinate choice. (c) Determine the distance over which the car moved while accelerating from 120 to 180 km/h.

21. The x-component of the position vector of a particle is shown in the graph in Figure P.21 as a function of time. (a) Estimate the velocity component v_x at the instant 3.0 s. (b) Is the velocity

component zero at any time? If so, estimate the time. If not, explain why not. (c) Is the particle always moving in the same direction along the x-axis? If so, explain what leads you to this conclusion. If not, determine the positions at which the particle changes direction.

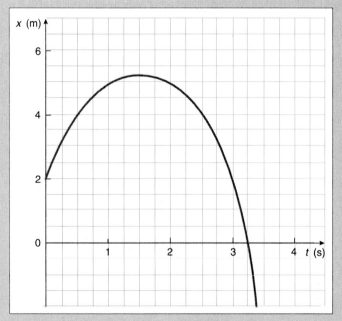

FIGURE P.21

22. The *x*-component of the velocity of a particle is graphed in Figure P.22 as a function of time. (a) Is the acceleration of the particle constant? Explain your answer. (b) Estimate the acceleration of the particle at the instant 1.0 s. (c) At what instants (if any) is the particle at rest? If such instants exist, is the acceleration of the particle also zero at these instants? Explain your answer.

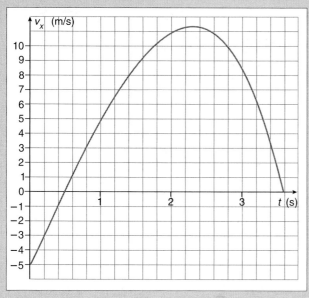

FIGURE P.22

23. When a radio wave impinges on the antenna of your car, electrons in the antenna move back and forth along the antenna with a velocity component v_x as shown schematically in Figure P.23. Roughly sketch the same graph and indicate the time instants when (a) the velocity component is zero; (b) the acceleration component a_x is zero; (c) the acceleration has its maximum magnitude.

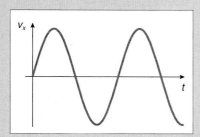

FIGURE P.23

24. A bullfrog (*Rana catesbeiana*) zaps a nearby tasty mosquito (*Culex pungens*) for dinner. The tongue of the frog flicks out 8.0 cm and is back with the mosquito inside the frog's mouth in 0.10 s total time. (a) What is the average velocity of the tongue during this interval? (b) What is the average speed of the tongue during the interval?

25. X-rays are created by slamming high-speed electrons into materials such as tungsten. The rapid acceleration of the electrons produces the x-rays. If an electron initially moving at a speed 3×10^6 m/s is brought to rest in 1×10^{-18} s, what is the magnitude of the average acceleration of the electron?

•26. A rock is thrown vertically upward at a speed v_0. It barely misses the edge of a cliff (see Figure P.26) and strikes the fluid in a foul-smelling river at a speed $2v_0$. How high is the cliff above the surface of the water? Express your answer in terms of v_0 and g.

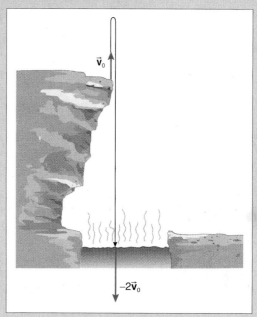

FIGURE P.26

•27. A ladybug (*Anatis quindecimpunctatum*) is observed crawling along a meter stick. The position of the ladybug at various times is recorded by an astute entomologist, yielding the following data:

Time (s)	Position (cm)
0.0	5.4
2.0	7.6
4.0	9.2
5.0	10.0
7.0	9.2
10.0	8.6
12.0	8.2
14.0	7.8

(a) Graph the position versus time. For (b)–(e), quote your results in appropriate SI units. Take the *x*-axis to be the meter stick with its origin at 0 m and with î in the direction the scale increases. Use the data or the graph to calculate the requested information. (b) What is the average velocity of the ladybug during the interval between 10.0 s and 14.0 s? (c) What is the velocity of the ladybug when $t = 12.0$ s? (d) At about what times is the velocity of the ladybug approximately 0 m/s? (e) What is the average acceleration of the ladybug during the interval from 2.0 to 12.0 s?

•28. A water balloon is thrown straight up and reaches its maximum height after 2.00 s. Sketch the situation and introduce an appropriate coordinate system to answer the following questions. (a) What is the acceleration of the balloon after it leaves the hand and is rising? (b) What is the acceleration of the balloon as it is falling? (c) What is the acceleration of the balloon at the instant it is motionless at its maximum height? (d) What was the initial velocity component of the balloon? (e) How high did the balloon rise above the point of its release?

•29. You are tooling along a straight, level road in a new sports car at a leisurely 30.0 m/s. Suddenly you realize that there is a barrier blocking a bridge ahead. You hit the brakes when $t = 0$ s with the car only 160 m from the barrier across the road. When $t = 5.0$ s, the car is still moving toward the bridge barrier but at only 15.0 m/s. You break into a sweat. Assume the acceleration of the car is constant. (a) At what time does the car come to rest? (b) Will the car collide with the bridge barrier? If not, how far is the car from the bridge barrier when the car stops?

•30. You are trailing the leader in a long cross-country ski race by only 3.0 m with both of you traveling at 20 km/h with a flat, straight 1000 m distance between the leader and the finish line. The leader is unable to increase her speed, but you feel a new surge of adrenalin and manage to win by a whisker. Assuming a constant acceleration, what was the magnitude of your acceleration during the last part of the race?

•31. You accidentally drop a peanut from a tower when $t = 0$ s. Oops. One second (1.00 s) later you decide that another peanut deserves the same treatment, and it too falls from the tower with zero initial speed. (a) Sketch on a single graph the speeds of the peanuts as functions of time. (b) While the peanuts are in flight, do they stay the same *distance* apart? Explain why or why not. (c) What is the distance between the peanuts when $t = 3.00$ s?

•32. Charles Walters* suggests slowing his motorcycle in an unusual way. He is traveling across a straight, level stretch of interstate in Kansas. Charles travels at 60.0 km/h for 1 minute, then quickly brakes briefly to achieve a speed of 59.0 km/h for the next minute, braking briefly to a speed of 58.0 km/h for the next minute, and so on until zero speed is achieved. (a) What total distance is covered in this process? (b) Introduce an appropriate coordinate system and determine the average acceleration of the motorcycle during this process.

•33. The x-component of the velocity of a garbage barge is indicated in the simplified graph shown in Figure P.33. (a) What is the average acceleration of the barge during the time interval between 0 min and 1.0 min on the clock? What is the instantaneous acceleration of the barge when $t = 0.5$ min? (b) What is the average acceleration of the barge during the time interval between 1.0 and 2.0 min on the clock? What is the instantaneous acceleration of the barge when $t = 1.5$ min? (c) What is the average acceleration of the barge during the time interval between 2.0 and 3.0 min on the clock? What is the instantaneous acceleration at the instant $t = 2.5$ min? (d) Explain what is happening to the barge in descriptive English. (e) Sketch a graph of the position x of the barge as a function of time consistent with and as complete as the information provided allows. Assume the barge initially is at the Cartesian origin.

•34. You throw your class ring vertically upward in a room with a high cathedral ceiling. It leaves your hand 1.2 m above the floor and reaches a maximum height of 3.0 m before falling to the floor. (a) How long a time interval did the ring rise? (b) What was the speed of the ring when it left your hand? (c) What is the duration of the entire time of flight? (d) What is the speed of the ring the instant before impact?

---

*Mr. Walters, a freshman, suggested this problem to the author.

FIGURE P.33

•35. A cheetah (*Acinonyx jubatus*) is chasing an impala (*Aepyceros melampus*) for lunch. (See Figure P.35.) The cheetah is found at the positions along the +x-axis at 100, 200, 280, 340, and 380 m when a stopwatch reads $t = 20.0, 30.0, 40.0, 50.0,$ and 60.0 s, respectively. (a) Calculate the x-component of the average velocity during each 10.0 s interval. (b) Make a histogram plot of the x-component of the average velocity versus t. (c) Is it reasonable to conclude that the instantaneous acceleration is constant during the interval from $t = 20.0$ s to 60.0 s? (d) If you assume the instantaneous acceleration is constant, find a_x. (e) Assume that a_x also is constant for times less than 20.0 s and greater than 60.0 s. Find v_{x0} and x_0. (f) Find $x(t)$. (g) At what time will the exhausted cheetah come to rest if the data trend continues? (h) Where does the cheetah come to rest?

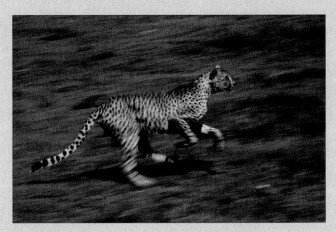

FIGURE P.35

•36. Figure P.36 is a graph of the instantaneous velocity component v_x of a chicken fleeing the axe of a farmer. The chicken is at $x = 0.0$ m when $t = 0.0$ s. (a) When is the chicken at rest? (b) What is the acceleration of the chicken when $t = 3.0$ s? (c) Make a graph of the acceleration component a_x as a function of time over the same time interval as in Figure P.36. (d) Where is the chicken when $t = 5.0$ s?

•37. The x-component of the velocity of an African kudu (*Tragelaphus imberbis*) is graphed in Figure P.37 as a function of time. (a) When is the animal at rest? (b) When is its acceleration equal to zero (if it ever is)? (c) What is its acceleration when $t = 4.5$ s? (d) How far does the animal travel during the first 6.0 s?

•38. A baseball is projected vertically upward with a speed equal to 29.4 m/s. Choose the direction of increasing values of x to be vertically upward. A stopwatch is started the instant the ball is released. (a) What is the velocity component v_x of the ball at each of the following instants: $t = 0.00$ s; $t = 1.00$ s; $t = 2.00$ s;

FIGURE P.36

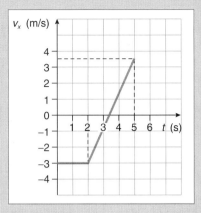

FIGURE P.37

Assume 2 significant figures

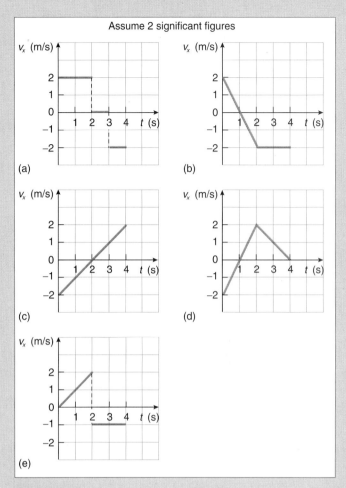

(a)   (b)

(c)   (d)

(e)

FIGURE P.40

$t = 3.00$ s; $t = 4.00$ s? (b) Make a graph of v_x versus time from $t = 0.00$ s to 4.00 s. (c) What is the acceleration component a_x of the ball at each of the instants above?

•**39.** A rifle fires a bullet with a muzzle speed of 350 m/s directly into a fixed block of hardwood. The bullet is brought to a stop embedded 6.0 cm into the wood. Assume the acceleration of the bullet is constant. Calculate the magnitude of the acceleration and the time it takes the block to stop the bullet.

•**40.** The graphs in Figure P.40 depict the velocity component v_x of a rat (*Rattus rattus*) in a one-dimensional maze as a function of time. Your task is to make graphs of the corresponding (i) acceleration component a_x versus time and (ii) the x-component of the position vector versus time. In all cases assume $x = 0$ m when $t = 0$ s.

•**41.** The draw on Robin Hood's archery bow is 0.70 m. (See Figure P.41.) (a) Calculate the magnitude of the acceleration of an arrow that leaves the bow with a speed of 30 m/s; assume the arrow experiences a constant acceleration over the 0.70 m distance before it leaves the bow. (b) If the arrow is shot vertically upward and leaves the bow at a distance 1.80 m from the ground, find the maximum height to which the arrow rises, and the total time of flight until the arrow hits the ground (assume the archer moves out of the way!).

•**42.** *If it were possible to maintain a uniform acceleration with a magnitude equal to 9.81 m/s², how long would it take to accelerate an object from rest to a speed approximately equal to the speed of light (3.00 × 10⁸ m/s)? Express your result in years. Ignore any effects caused by the theory of relativity.*

•**43.** A partner in the law firm of Filch, Robb & Steele notices an ambulance go by at a constant speed of 50 m/s along a level, straight roadway, headed for an accident. As the ambulance

30 m/s

0.70 m

1.80 m

FIGURE P.41

passes, the lawyer sets out in hot pursuit of a new client, with zero initial speed and stopwatch initially set to zero, maintaining a constant acceleration of magnitude 1.0 m/s². (a) Set up an appropriate coordinate system to analyze the situation. (b) Set up an equation that describes the position of the ambulance as a function of time on the stopwatch. (c) Set up an equation that describes the position of the lawyer as a function of time on the stopwatch. (d) How much time elapses before the lawyer catches the ambulance? (e) What distance did the ambulance and lawyer travel during this time? (f) Draw a graph showing the x-component of the position vectors of the ambulance and lawyer as a function of time. Indicate the event that represents the lawyer catching the ambulance.

•44. Have your roommate hold a meter stick by one end as indicated in Figure P.44. Initially, the top of your hand is at, say, the 10.0 cm mark, ready to catch the stick when it is dropped. Without warning, your compatriot drops the stick. Without reaching downward, you catch it as quickly as possible and find it has fallen a distance d. Show that a measure of your reaction time $t_{reaction}$, the time between when you see the stick begin to fall and the time you catch it, is (in seconds)

$$t_{reaction} = (0.452 \text{ s/m}^{1/2})\sqrt{d}$$

(a)   (b)

FIGURE P.44

•45. Alas, having dallied too long at Attitude Adjustment Hour (AAH!), two commuters arrive at a commuter rail platform just in time to see their train start from rest with an acceleration of magnitude 0.250 m/s². They are 30.0 m from the end of the last car, straight along the track, and both run at the furious (and constant) speed of 4.00 m/s to catch the train. A stopwatch monitors the time it takes them to catch the train. (a) Introduce an appropriate coordinate system. (b) At what *two* times will the commuters and the end of the rear car have the same position coordinate? (c) At each of the times calculated in part (b), describe what is passing what.

•46. For one-dimensional motion with a constant acceleration, show that the fundamental equations

$$v_x = v_{x0} + a_x t \qquad (3.16)$$

$$x = x_0 + v_{x0}t + \frac{1}{2}a_x t^2 \qquad (3.19)$$

imply that

$$v_x^2 = v_{x0}^2 + 2a_x(x - x_0)$$

•47. You toss a quarter into the air, releasing it at a distance of 2.0 m above the ground and simultaneously starting a stopwatch. The quarter is in the air for 3.0 s before striking the ground. (a) Sketch the situation. (b) Introduce an appropriate coordinate system, clearly indicating the origin and the direction

of an appropriate unit vector. (c) Tailor Equations 3.16 and 3.19 to your coordinate choice. (d) What was the initial velocity component of the quarter? (e) How high above the ground did the quarter rise? (f) At what speed did the quarter strike the ground?

•48. A dazed physics student is found to have the following positions at the indicated times:

t (s)	x (m)
0.00	0.00
1.00	1.00
2.00	3.00

(a) Is it possible for these data to be consistent with a *constant* acceleration? Explain the reasoning you use to come to your conclusion. (b) If a constant acceleration is possible, find the constant acceleration component a_x and v_{x0}.

•49. The Standard Linear Accelerator (SLAC) at Stanford University in California is 3.0 km long (see Figure P.49). Electrons are accelerated along the length of the straight beam tube. Assume the acceleration of the electrons is constant along the length of the accelerator. If the electrons begin at rest and emerge with a speed of 6.0×10^6 m/s, find (a) the time it takes each electron to traverse the tube and (b) the magnitude of the acceleration. The local acceleration due to gravity may be ignored; why is this a reasonable assumption?

FIGURE P.49

•50. You are asked to move a large basket of eggs initially at rest at the origin to a position at rest 16 m away along the x-axis. The eggs are speeded up with an acceleration of magnitude 1.0 m/s² for the first half of the trip and slowed down with an acceleration of equal magnitude for the latter half of the trip. (a) How long does the trip take? (b) Make a graph of the velocity component v_x of the eggs as a function of time. (c) What is the average velocity for the first half of the trip? for the second half?

•51. When a stopwatch is started, a tennis ball is tossed vertically up with an initial speed of 20.0 m/s. When the ball reaches its maximum height, a second ball is tossed straight up with the same initial speed of 20.0 m/s so that they collide in midair. Both balls are released at an elevation 1.50 m above the ground. (a) Choose $\hat{\imath}$ to be directed vertically up, with the origin at the ground. Sketch on the same graph the v_x components of the velocities of the two balls as functions of time. (b) Show that the speed with which the two balls approach each other before impact is constant and equal to 20.0 m/s. (c) At what elevation will the two tennis balls collide?

•52. A pitcher throws a searing 150 km/h fastball to the catcher, located 19.0 m away. Assume the ball follows a straight line from the pitcher to the catcher. (Ignore any vertical motion of the ball on the way to the plate.) (a) How long does it take the ball to reach the catcher? (b) As the ball leaves the pitcher's hand, a runner on first base attempts to steal second base, covering the 27.4 m distance in 4.00 s. After catching the ball from the pitcher, the catcher hurls the ball, after a 0.8 s delay, at 130 km/h to second base, 38.8 m away from him. (Again assume the ball follows a straight path.) Assuming the second baseman catches the ball and makes the tag, is the runner out or safe?

•53. A ball is tossed vertically upward with a speed of 25.0 m/s. (a) Indicate an appropriate coordinate system for analyzing its motion. (b) What is its maximum altitude above the point from which it was thrown? (c) How long does it take to return to the point of release? (d) Make a graph of the acceleration component a_x of the particle versus time, indicating appropriate numerical values along the axes. (e) Make a graph of the velocity component v_x of the particle versus time, indicating appropriate numerical values along the axes. (f) Make a graph of the position vector component x as a function of time, indicating appropriate numerical values on the axes.

•54. If speed is measured in body lengths per second, it may come as no surprise that the fastest insect in the world is the common American cockroach (*Periplaneta americana*). It can achieve speeds of about 50 body lengths per second. Assume a body length of 4.0 cm—jumbo size! (a) What is its fastest speed in m/s? (b) If the roach achieves this speed in 0.10 s, starting from rest, what is the magnitude of its acceleration assuming it is constant? Compare the magnitude of this acceleration with g. (c) If a 2.0 m tall human could run with a speed of 50 body lengths per second, what would be the approximate time for a 100 m sprint? The world record for a 100 m sprint is about 9.85 s.

•55. A basketball is dropped from rest 3.0 m above the court. After bouncing from the floor, the ball reaches a height of 1.5 m. (a) What is the speed of the ball the instant before hitting the floor? (b) What is the speed of the ball just after leaving the floor? (c) The ratio of speed after the bounce to the speed before the bounce is called the *coefficient of restitution* ε of the ball. Find ε for the basketball. (d) If the ball is in contact with the floor for 0.025 s, what is the magnitude of the average acceleration of the ball during this time interval?

•56. A traffic light turns green, and you zip off from rest in your car along a straight highway with an acceleration of magnitude 1.50 m/s² for 10.0 s, maintain a constant velocity for the next 30.0 s, and then slow to a stop with an acceleration of magnitude 2.00 m/s². How far did you travel?

•57. A pine cone drops from the mouth of an eastern gray squirrel (*Sciurus carolinensis*) while it is scampering at a speed 3.0 m/s up the vertical trunk of a tall pine tree. You notice that it takes the cone 1.5 s to reach the ground. How high was the squirrel when it dropped the cone while running?

•58. A naval gun projects an artillery shell from rest to a speed of 2.00 km/s over the 4.00 m length of a gun barrel. Assume the acceleration of the shell is constant over the length of the barrel. Determine the magnitude of the acceleration of the shell and the time interval the shell is in the barrel after firing.

•59. A kid drops from a tree limb 3.0 m above a swimming hole. (a) With what speed does the kid hit the water? (b) If the kid is brought to rest 1.5 m below the surface of the water, what is the magnitude of the acceleration of the child after contact with the water? Assume the acceleration is constant.

•60. You are standing on the high bridge over the New River gorge in southern West Virginia. The rail of the bridge is 269 m above the surface of the river below. Your buddy drops a rock from the bridge rail. You throw a rock down 1.00 s later from the same location and both rocks strike the water simultaneously. (a) How long is the first rock in flight? (b) With what speed did you throw the second rock?

•61. The worst nightmare of a skydiver is a parachute failure. There are then two options: scream all the way down or enjoy the view for a few all-too-fleeting moments. On rare occasions, people actually survive such falls, not typically by the intercession of divine providence so much as by the intercession of deep snow, steep slopes, thick brush, or the like. Assume the human body can withstand, without fatal consequences, brief accelerations with magnitudes of about 75 times the local acceleration due to gravity (75g). If a skydiver hits at 50 m/s, over what minimum distance is the skydiver brought to a stop under the action of such a large acceleration?

•62. A runner accelerates from the starting blocks to a speed of 10.0 m/s over the first 6.0 m of track in a 100 m sprint. (a) During what time interval does she cover the first 6.0 m if her acceleration is constant? (b) What is the magnitude of her acceleration during this time?

‡63. Hang a mass on a vertical spring; imagine an origin at the place where the mass is at rest with $\hat{\imath}$ directed *down* as shown in Figure P.63. Now set the mass into oscillation in the vertical direction by lowering the mass a distance A and letting it go. The subsequent motion is called simple harmonic oscillation and will be investigated in some detail in Chapter 7. We shall see that the position vector of a particle executing such one-dimensional oscillatory motion is given by the expression

$$\vec{r}(t) = [A\cos(\omega t)]\hat{\imath}$$

where A is expressed in meters, and ω is expressed in radians per second; both are constants. (a) Find the velocity vector and the acceleration vector as functions of time. (b) Is the acceleration constant? (c) What are the *greatest magnitudes* of the velocity and acceleration vectors? (d) What is the earliest (nonnegative) time that the position vector attains maximum magnitude? When the position vector has its greatest magnitude, what is the magnitude of the velocity vector? What is the magnitude of the acceleration vector at the same time? (e) At what time ($t \geq 0$ s) does the position vector first attain a magnitude of 0 m? At this time, what are the magnitudes of the velocity and acceleration vectors?

FIGURE P.63

‡64. When a ball bounces from the floor, the ratio of the speed at the instant just after the bounce to the speed at the instant just before the bounce is called the coefficient of restitution ε of the ball. (a) Show that if a ball, initially at rest, is dropped from a height h_0 and rebounds to a height h_1, the coefficient of restitution is

$$\varepsilon = \left(\frac{h_1}{h_0}\right)^{1/2}$$

(b) The height of rebound is difficult to measure precisely. An alternative way of determining the coefficient of restitution is to drop a ball from height h_0 and measure the total time T from its release to the instant of its *third* impact with the floor. Show that the coefficient of restitution is related to T by

$$T = \sqrt{\frac{2h_0}{g}}\,(1 + 2\varepsilon + 2\varepsilon^2)$$

This problem was inspired by an experiment "Measuring gravity with a superball" by Mark St. John, *Thinking Like a Physicist* (Group in Science and Mathematics, University of California, Berkeley, 1978).

‡65. It is always fun to drop rocks down into a well. An enterprising physics student realizes that it is possible to determine the distance to the water surface of the well with a stopwatch. From a vantage point at the top of the well, the student drops a rock (with zero initial speed) into a deep wishing well and discovers that the total time from release to reception of the sound from the ensuing splash is 3.00 s. Allowing for the finite speed of sound in air, 343 m/s at 20 °C, calculate the distance to the water surface.

‡66. A stone is projected vertically upward at speed v_0 from the top of a cliff of height h. The stone just misses the edge of the cliff on the way down. A second stone is dropped from rest from the top of the cliff. How long should you wait before releasing the second stone so that they both arrive at the bottom of the cliff simultaneously?

‡67. Make and state some reasonable assumptions to estimate the magnitude of the acceleration experienced by the occupants of a car traveling at 120 km/h if it collides with a massive bridge abutment. Explain how air bags are used to decrease the magnitudes of such accelerations.

Section 3.8 Geometric Interpretations*

•68. A particle initially at the origin and moving with an initial velocity $(2.00 \text{ m/s})\hat{\imath}$ experiences an acceleration given by

$$\vec{a} = (3.00 \text{ m/s}^2)\hat{\imath}$$

(a) Make a graph of a_x versus t over the range of times from $t = 0$ s to 5.00 s. What is the area under the curve of a_x? What is the significance of the area under the graph from $t = 0$ s to 5.00 s? (b) Find the change in the velocity component over the same time interval and make an accurate graph of v_x versus t. What is the area under the curve between 0 s and 5.00 s? What is the significance of the area under the graph of the velocity component between $t = 0$ s and 5.00 s? (c) Find the change in the position vector component over the same time interval and make a graph of x versus t. What is the slope of this graph when $t = 0$ s?

•69. The x-component of the velocity of a particle in free-fall motion near the surface of the Earth is shown in Figure P.69. (a) Was $\hat{\imath}$ chosen to be vertically up or down? (b) Was the particle initially moving up or down? (c) At what instant was the particle at its greatest elevation? (d) What is the area under the curve between 0 s and 5.0 s? (e) What is the change in the x-component of the position vector of the particle during the time interval between 0 s and 5.0 s?

FIGURE P.69

‡70. A particle initially traveling with a velocity $v_{x0}\hat{\imath}$ is subjected (beginning when $t = 0$ s) to an acceleration

$$\vec{a} = a_0 e^{-\alpha t}\hat{\imath}$$

where α and a_0 are positive scalar constants, independent of time. (a) What are the SI units of the constant α? of a_0? (b) Find $v_x(t)$. (c) After a long time (i.e., as $t \rightarrow \infty$ s), what is the speed of the particle?

‡71. One model for the motion of a particle moving in a resistive medium suggests that the speed decreases exponentially according to the expression

$$v(t) = v_0 e^{-\beta t}$$

where v_0 is the speed of the particle when $t = 0$ s and β is a positive constant. (a) How long will it take the particle to reach half its initial speed? (b) What distance does the particle traverse during the time interval calculated in part (a)? (c) Through what distance does the particle move before it is brought to rest?

‡72. A theorist in your research group comes up with the following equation for the x-component of the position vector of a particle:

$$x(t) = (20.0 \text{ m/s})te^{-(2.00\text{s}^{-1})t}$$

You are assigned the following tasks: (a) Make a graph of x versus t for the interval from $t = 0$ s to 1.00 s. (b) Determine the (finite) position at which the particle is instantaneously at rest and see if the acceleration of the particle also is zero at this location. (c) Ascertain if the particle is initially moving parallel to $\hat{i}$ or in the opposite direction. (d) Find the acceleration component of the particle when $t = 0$ s.

INVESTIGATIVE PROJECTS

A. Expanded Horizons

1. The Greek philosopher Zeno of Elea (c. fifth century B.C.) posed a paradox now known as *Zeno's paradox*. The Greek hero Achilles starts at point A and attempts to catch a slowly moving tortoise initially at point B. When Achilles reaches point B, the tortoise has moved to point C; when Achilles reaches C, the tortoise is at yet another point D, and so on *ad infinitum*. How does Achilles ever manage to catch the tortoise? Investigate Zeno's paradox and its resolution.

   William I. McLaughlin, "Resolving Zeno's paradoxes," *Scientific American*, 271, #5, pages 84–89 (November 1994).

   James R. Newman, *The World of Mathematics* (Simon & Schuster, New York, 1956), Volume 1, pages 41, 94.

2. Investigate the physiological effects of accelerations on the human body. In particular, determine the approximate magnitude of accelerations that are lethal to humans.

   James G. Parker and Vita R. West (editors), *Bioastronautics Data Book* (2nd edition, NASA, Washington, D.C., 1973).

   Bernard Hoop, "Resource letter PPPP-1: Physical principles of physiological phenomena," *American Journal of Physics*, 55, #3, pages 204–210 (March 1987).

3. Investigate the idea of the time rate of change of the acceleration, known as the *jerk* $\vec{j}$ (no kidding!). That is, the instantaneous jerk at instant t is:

$$\vec{j}(t) \equiv \frac{d\vec{a}(t)}{dt}$$

   where $\vec{a}$ is the acceleration. Does the jerk have any physical manifestations or consequences that warrant its use in kinematics? Is there any value in continuing the process of further differentiation, such as

$$\frac{d\vec{j}}{dt}$$

   On what basis should such a judgment be made?

   Steven H. Schot, "Jerk: the time rate of change of acceleration," *American Journal of Physics*, 46, #11, pages 1090–1094 (November, 1978).

   T. R. Sandin, "The jerk," *The Physics Teacher*, 28, #1, pages 36–40 (January, 1990).

4. Consult appropriate sources such as *The Guiness Book of Records* to determine the approximate fastest speeds for each of the following life forms: human, flying insect, crawling insect (see Problem 54), fish, marine mammal, land mammal, bird. If the fastest speeds are divided by the approximate length or size of the creature, do the resulting speeds per body size have about the same order of magnitude?

B. Lab and Field Work

5. If there is a safe, high place from which to drop a rock so that it falls vertically, design and perform an experiment to determine the magnitude of the local acceleration due to gravity g near the surface of the Earth. What is the precision of your result and what is the uncertainty associated with your measurement? If your physics laboratory has a photogate timer, compare the result you obtain using it with that obtained from measurements of a falling rock.

6. Design and perform an experiment to estimate the magnitude of the acceleration of a sprinter leaving the starting blocks at the beginning of a race. If a safe, public dragstrip is nearby, you also might try to determine the maximum magnitude of the acceleration of your car as its speed changes from 0 km/h to 60 km/h.

7. There likely is a subset of your class that keeps fit by jogging. Using as large a group of such students as possible, time a 100 m sprint by each student. See if there is a relationship between their times and body mass. Is there a difference between any such relationship for males and females?

8. Devise a method for forming water drops (raindrops) of constant size and known mass, perhaps using a graduated micropipette, available in most chemistry and biology departments. Design and perform an experiment to determine if such raindrops reach a limiting (constant) speed when falling through a known distance such as the height of your laboratory. Do the results depend on the mass of the raindrops? You might also see what happens if the water drops fall through a graduated cylinder of cooking or olive oil.

   Howard E. Evans II, "Raindrops keep falling on my head," *The Physics Teacher*, 29, #2, pages 120–121 (February 1991).

9. The kinematics of traffic flow are interesting to examine. Imagine a straight section of interstate with a speed limit of 105 km/h ($\approx$ 65 miles per hour). Assume cars on the road travel at the speed limit if traffic conditions permit. If traffic is light, the average number of cars past any given kilometer post marker during 1 min is equal to the average number of cars per kilometer. Make a qualitative graph of the average number of cars per minute past a kilometer marker as a function of the average number of cars per kilometer. Take into account the average size of a car (say 5 m) as well as the fact that if the traffic becomes too heavy, gridlock ensues. Your state highway department may have data available to support your plot; or secure such data yourself with a well-crafted experiment.

This project was inspired by a letter by Clyde E. Holvenstot to *American Scientist*, *82*, #6, pages 504–505 (November–December 1994).

10. Design and perform an experiment to measure the magnitude of the acceleration of an arrow as it is released from rest in an archery bow. Is the magnitude constant?

    W. C. Marlow, "Bow and arrow dynamics," *American Journal of Physics*, *49*, #4, pages 320–333 (April 1981).

C. Communicating Physics

11. A student observes the *x*-component of the position vector of a particle undergoing rectilinear motion at constant acceleration. The data are graphed as a function of time with the results shown in Figure I.11. Discuss how the student should treat the data point that appears inconsistent with the other data. Can the student justify throwing out the "bad" data point? If so, how? If not, explain why. Should the bad data point simply be ignored, and not discussed?

    John R. Taylor, "Rejection of data: Chauvenet's criterion," Chapter 6 in *An Introduction to Error Analysis: The Study of Uncertainties in Physical Measurements* (2nd edition, University Science Books, Sausalito, California, 1997).

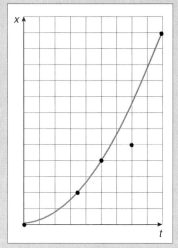

FIGURE I.11

12. The effect of air resistance on free-falling objects causes them not to fall with a constant acceleration of magnitude *g*. Write a research proposal to your physics department for an experiment to measure the effects of air resistance on free-falling objects. Indicate in your proposal (a) how you would attempt to measure the effect; (b) requests for appropriate equipment, indicating why the equipment is needed for the experiment;

(c) what objects will be used in the free-fall; (d) how your experimental results can distinguish between free-fall at constant acceleration and cases where the effects of air resistance are apparent; and (e) an estimate of the time needed to conduct the experiment and prepare a report of your findings. Perhaps if your proposal is judged well, your professor will excuse you from the regular laboratory to perform the experiment and report your findings.

13. Imagine a technical civilization that develops a theory of kinematics based not on speed measured in meters per second, but on the sluggishness alluded to in Section 3.3, measured in seconds per meter. Explain in some detail how such a civilization might formulate a concept analogous to the magnitude of the acceleration. Invent an appropriate word for the concept. How might the directional attributes of these new concepts be treated? Would such concepts obey the rules for vector arithmetic?

RECTILINEAR NONMOTION

CHAPTER 4

KINEMATICS II
MOTION IN TWO AND THREE DIMENSIONS

Such a life, with all vision limited to a Point and all motion to a
Straight Line, seemed to me inexpressibly dreary; . . .

Edwin Abbott Abbott (1838–1926)*

Life is pretty dull confined to one dimension: we would all be particles living and moving in the same rut. Sports fields and amusement park rides would be one-dimensional lines much like the gutter of a bowling alley. Boring. In this chapter we exploit our familiarity with vectors to extend our knowledge of rectilinear motion to the richness of motion in two- and three-dimensional space. We will see that the two- or three-dimensional motions of a particle are natural extensions of the one-dimensional kinematics of Chapter 3. We must learn to describe the kinematics of the multidimensional motion of a particle with dexterity, since such motion frequently is encountered in both our natural and technological environments.

The circular motion of a particle is two-dimensional motion in a plane. We see examples of circular motion in everything from kitchen blenders to merry-go-rounds, from tires and wheels to CD players and gears, from computer disk drives to centrifuges, from pinwheels to windmills, from cyclotron particle accelerators in physics to the rotation of planets, stars, and galaxies in the universe. Perhaps surprisingly, we shall see in this chapter that while the circular motion of a particle occurs in a two-dimensional plane in space, the description of the motion involves only one time-dependent coordinate.

4.1 THE POSITION, VELOCITY, AND ACCELERATION VECTORS IN TWO DIMENSIONS

Of the nature of flatland

Edwin Abbott Abbott[†]

When you toss a crumpled paper from your desk to a wastebasket or shoot a few hoops (see Figure 4.1), the motion of the

---

*(Chapter Opener) *Flatland* (Princeton University Press, Princeton, New Jersey, 1991), Part I, Section 1, page 3.
[†]*Flatland*, Part II, Section 13, page 56.

FIGURE 4.1 A basketball shot is motion in two dimensions.

particle is confined to a two-dimensional vertical plane. When you drive north and then east on a level plane, your motion also is confined to a plane in two dimensions. But since this travel consists of two successive rectilinear parts, it hardly qualifies as fully two dimensional. To the extent that a motion cannot be reduced to some one chosen dimension, we can consider it necessarily two or three dimensional.

When you *come about* while sailing, which means to reverse direction on the sea, or if you turn a level corner in your car, as illustrated in Figure 4.2, the motion is confined to a two-dimensional horizontal plane needing both x- and y-coordinates. Let the plane of the two-dimensional motion be the *x–y* plane of a Cartesian coordinate system. You have the freedom to choose orientations of the two coordinate axes in any convenient directions as long as they are mutually perpendicular, as shown in Figure 4.3.

> **PROBLEM-SOLVING TACTIC**
>
> **4.1 The specific coordinate choice is yours to make, but it needs to be explicitly indicated in a sketch and used consistently throughout a given problem.** You learned this in the previous chapter; it bears repeating, though. For two-dimensional motion, it is *not* necessary to check whether the coordinate system you choose is right-handed, since the third, but irrelevant, direction (z) always can be chosen to make the system right-handed. For example, $\hat{k}$ is directed perpendicularly out of the page in Figures 4.3a and b, but into the page in Figure 4.3c. However, the z-axis is not needed to describe the two-dimensional motion, and so it is superfluous to specify its direction explicitly.

For two-dimensional motion, the two coordinates x and y locate the position of the particle whose motion is to be described. Since the particle is moving, both coordinates are functions of time: $x(t)$ and $y(t)$. The position vector of the particle locates the particle with respect to the origin you choose and is the vector sum:

$$\vec{r}(t) = x(t)\hat{i} + y(t)\hat{j} \tag{4.1}$$

Later, at the instant $t + \Delta t$, the particle has the coordinates $x(t + \Delta t)$ and $y(t + \Delta t)$. The new position vector is

$$\vec{r}(t + \Delta t) = x(t + \Delta t)\hat{i} + y(t + \Delta t)\hat{j} \tag{4.2}$$

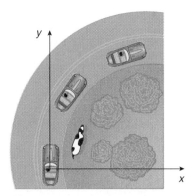

FIGURE 4.2 When you turn a level corner in your car, the motion is in two dimensions.

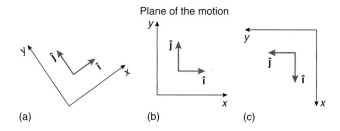

FIGURE 4.3 Some possible coordinate choices for motion in a plane.

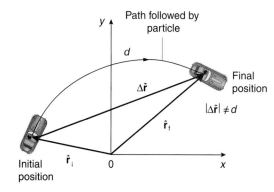

FIGURE 4.4 The total distance traveled, d, typically is not equal to the magnitude of the change in the position vector, $|\Delta \vec{r}|$.

The change in the position vector $\Delta\vec{r}$ of the particle is the vector difference

$$\Delta\vec{r} \equiv \vec{r}_f - \vec{r}_i$$
$$= \vec{r}(t + \Delta t) - \vec{r}(t)$$
$$= [x(t + \Delta t) - x(t)]\hat{\imath} + [y(t + \Delta t) - y(t)]\hat{\jmath}$$

As with any vector, the magnitude and direction of the change in the position vector of a particle are independent of the choice you make for the coordinate system.

Just as in one-dimensional motion, the average speed involves the total distance d traveled by the particle during the interval Δt:

$$\text{average speed} \equiv \frac{d}{\Delta t}$$

The average velocity of the particle during the interval Δt involves the change in the position vector and is, from Equation 3.6,

$$\vec{v}_{\text{ave}} \equiv \frac{\Delta\vec{r}}{\Delta t}$$
$$= \frac{\vec{r}(t + \Delta t) - \vec{r}(t)}{\Delta t} \qquad (4.3)$$

As in one dimension, the magnitude of the change in the position vector typically is not equal to the total distance traveled by the particle during the interval Δt, as shown in Figure 4.4. Consequently, the average speed typically is not equal to the magnitude of the average velocity.*

If we take the limit of average velocity for smaller and smaller time intervals Δt, we approach the instantaneous velocity $\vec{v}(t)$ as $\Delta t \to 0$ s:

$$\vec{v}(t) = \lim_{\Delta t \to 0 \text{ s}} \frac{\vec{r}(t + \Delta t) - \vec{r}(t)}{\Delta t} = \frac{d\vec{r}(t)}{dt} \qquad (4.4)$$

Just as in rectilinear motion, the instantaneous velocity of the particle is the time derivative of the instantaneous position vector. As with rectilinear motion, the magnitude of the velocity

vector is the speed of the particle. The velocity vector always is tangent to the path followed by the particle in space. According to Equation 4.4, the velocity is

$$\vec{v}(t) = \frac{d\vec{r}(t)}{dt}$$
$$= \frac{dx(t)}{dt}\hat{\imath} + \frac{dy(t)}{dt}\hat{\jmath}$$
$$\equiv v_x(t)\hat{\imath} + v_y(t)\hat{\jmath} \qquad (4.5)$$

where

$$v_x(t) \equiv \frac{dx(t)}{dt}$$
$$v_y(t) \equiv \frac{dy(t)}{dt} \qquad (4.6)$$

The velocity components are the time derivatives of the corresponding components of the position vector.

Just as in rectilinear motion, the average acceleration during an interval Δt is related to the change in the velocity by

$$\vec{a}_{\text{ave}} = \frac{\Delta\vec{v}}{\Delta t} \qquad (4.7)$$

The instantaneous acceleration is the limit of the average acceleration as the time interval approaches zero:

$$\vec{a}(t) = \lim_{\Delta t \to 0 \text{ s}} \vec{a}_{\text{ave}} = \lim_{\Delta t \to 0 \text{ s}} \frac{\Delta\vec{v}}{\Delta t}$$

This in turn is the time derivative of the velocity vector:

$$\vec{a}(t) = \frac{d\vec{v}(t)}{dt} = \frac{dv_x(t)}{dt}\hat{\imath} + \frac{dv_y(t)}{dt}\hat{\jmath}$$
$$\equiv a_x(t)\hat{\imath} + a_y(t)\hat{\jmath} \qquad (4.8)$$

where

$$a_x(t) \equiv \frac{dv_x(t)}{dt}$$
$$a_y(t) \equiv \frac{dv_y(t)}{dt} \qquad (4.9)$$

---

*Recall that for rectilinear motion, if the particle did not undergo back-and-forth motion during the interval Δt, the magnitude of the position vector was equal to the total distance traveled by the particle.

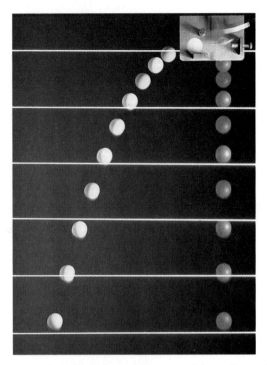

FIGURE 4.5 A marble is dropped from rest while another marble is simultaneously projected horizontally. The vertical motion is independent of the horizontal motion.

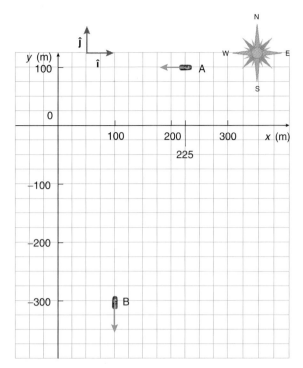

FIGURE 4.6

So the relationships between the position vector, the velocity, and the acceleration are the same as those found for the rectilinear case. The only difference is that we now have two dimensions, which means that the various vectors have two components.

With motion in two or more dimensions, a new question arises. Do the instantaneous values of a_y, v_y, and y depend on the values of a_x, v_x, or x at the same instant or other instants? In other words, are the motions along *perpendicular* coordinate axes *independent* of each other? To answer this question, we turn, as we must, to experiments, the final arbiters for questions in physics. Imagine what happens if a marble is dropped from rest while an identical marble is simultaneously projected horizontally (see Figure 4.5). The vertical motions of the two marbles are identical: they strike the floor simultaneously. The vertical motion is unaffected by the horizontal motion.

Such a kinematic experiment and others like it imply an important fact:

Perpendicular motions are independent of each other.

The implication of this independence means we can analyze the motion along each coordinate axis *separately*, as we will see in the next section.

QUESTION 1

A particle moves at constant speed. Specify the additional circumstance(s) that lets you conclude its acceleration is zero.

EXAMPLE 4.1

Let î be directed east and ĵ be directed north, as in Figure 4.6. When a stopwatch indicates 2.00 s, your car is located at position A and moving west at a speed of 20.0 m/s. Later,

when the stopwatch indicates 17.00 s, the car is located at position B and moving south at 25.0 m/s.

a. Find the two position vectors of the car. Indicate them accurately on a sketch.
b. What is the change in the position vector? What is the magnitude of the change in the position vector? Indicate $\Delta \vec{r}$ on your sketch.
c. Determine the average velocity of the car during the time interval. What is the magnitude of the average velocity of the car?
d. What is the change in the velocity of the car?
e. What is the average acceleration of the car?

Solution

a. In Figure 4.6, the *x*- and *y*-coordinates of the car are initially $x = 225$ m and $y = 100$ m. The initial position vector thus is

$$\vec{r}_i = (225 \text{ m})\hat{\imath} + (100 \text{ m})\hat{\jmath}$$

Later the coordinates are $x = 100$ m and $y = -300$ m, and so the final position vector is

$$\vec{r}_f = (100 \text{ m})\hat{\imath} - (300 \text{ m})\hat{\jmath}$$

These position vectors are indicated in Figure 4.7.

b. The change in the position vector is

$$\Delta \vec{r} = \vec{r}_f - \vec{r}_i$$
$$= [(100 \text{ m})\hat{\imath} - (300 \text{ m})\hat{\jmath}] - [(225 \text{ m})\hat{\imath} + (100 \text{ m})\hat{\jmath}]$$
$$= (-125 \text{ m})\hat{\imath} - (400 \text{ m})\hat{\jmath}$$

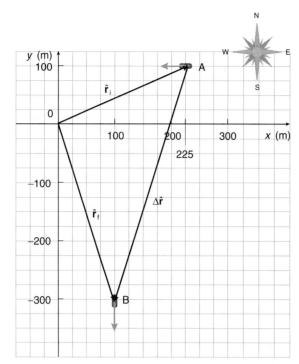

FIGURE 4.7

4.2 TWO-DIMENSIONAL MOTION WITH A CONSTANT ACCELERATION

Throw a rock into a pond or toss a ball to a friend. Shoot a basketball from the foul line or pitch a baseball. Slide a puck across an inclined airtable or watch a pinball move between stations in the maze. All are examples of **projectile motion**. If frictional effects are very small, all such motions have a constant acceleration. (Remember that a constant acceleration vector has a fixed magnitude and direction at all times and locations during the motion.)

PROBLEM-SOLVING TACTIC

4.2 If the acceleration vector is constant, choose one of the coordinate axes to be the line containing the constant acceleration or a line parallel to it. With this choice the acceleration has only a *single* component. The description of the motion then is simpler than otherwise would be the case.

We will choose the y-axis to be the line containing the constant acceleration vector. Thus the acceleration vector is

$$\vec{a} = a_y \hat{j}$$

where the single component a_y is a constant scalar, and $a_x = 0 \text{ m/s}^2$.

PROBLEM-SOLVING TACTICS

4.3 The sign of the acceleration component a_y depends on the direction you choose for $\hat{j}$ along the line containing the acceleration vector. See Figure 4.8.

4.4 The origin may be taken at any convenient location along the line chosen for the y-axis.

Since motions along perpendicular directions are independent of each other, the equations we developed in the last chapter for studying rectilinear motion with a constant acceleration (Equations 3.16 and 3.19) can be used for *each* of the perpendicular coordinate directions with appropriate changes made for labeling the different coordinate axes.

If a particle has an acceleration component a_x equal to zero, the x-component of its velocity is, from Equation 3.16,

$$v_x(t) = v_{x0} + a_x t$$
$$= v_{x0} + (0 \text{ m/s}^2)t$$
$$= v_{x0} \quad \text{(i.e., is constant)} \quad (4.10)$$

This vector also is indicated in Figure 4.7. The magnitude of this vector is

$$|\Delta \vec{r}| = [(-125 \text{ m})^2 + (-400 \text{ m})^2]^{1/2}$$
$$= 419 \text{ m}$$

c. The average velocity is found from Equation 4.3:

$$\vec{v}_{ave} = \frac{\Delta \vec{r}}{\Delta t}$$
$$= \frac{(-125 \text{ m}) \hat{i} - (400 \text{ m}) \hat{j}}{17.00 \text{ s} - 2.00 \text{ s}}$$
$$= (-8.33 \text{ m/s}) \hat{i} - (26.7 \text{ m/s}) \hat{j}$$

The magnitude of the average velocity vector is

$$v_{ave} = [(-8.33 \text{ m/s})^2 + (-26.7 \text{ m/s})^2]^{1/2}$$
$$= 28.0 \text{ m/s}$$

d. The change in the velocity of the car is

$$\Delta \vec{v} = \vec{v}_f - \vec{v}_i$$
$$= (-25.0 \text{ m/s})\hat{j} - (-20.0 \text{ m/s})\hat{i}$$
$$= (20.0 \text{ m/s})\hat{i} - (25.0 \text{ m/s})\hat{j}$$

e. The average acceleration of the car is, from Equation 4.7,

$$\vec{a}_{ave} = \frac{\Delta \vec{v}}{\Delta t}$$
$$= \frac{(20.0 \text{ m/s}) \hat{i} - (25.0 \text{ m/s}) \hat{j}}{17.00 \text{ s} - 2.00 \text{ s}}$$
$$= (1.33 \text{ m/s}^2) \hat{i} - (1.67 \text{ m/s}^2) \hat{j}$$

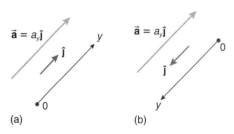

FIGURE 4.8 (a) If you choose $\hat{j}$ to be parallel to $\vec{a}$, then $a_y > 0 \text{ m/s}^2$. (b) If you choose $\hat{j}$ to be antiparallel to $\vec{a}$, then $a_y < 0 \text{ m/s}^2$.

The x-component of the velocity is constant throughout the motion and equal to the initial x-component of the velocity v_{x0}, because $a_x = 0$ m/s².

A graph of the x-component of the velocity versus time (see Figure 4.9) is a horizontal line with zero slope, again since $a_x = 0$ m/s².

Equation 3.19 for the x-coordinate of the particle—with $a_x = 0$ m/s²—indicates that

$$x(t) = x_0 + v_{x0}t \qquad (4.11)$$

where, as always, v_{x0} is the initial x-component of the velocity of the particle when $t = 0$ s and x_0 is the corresponding initial x-component of the position vector. The x-coordinate changes linearly with time, and so the x versus t graph is a straight line with slope v_{x0} and intercept x_0, as shown in Figure 4.10.

QUESTION 2

What would the x versus t graph (with $a_x = 0$ m/s²) look like if $x_0 > 0$ m and $v_{x0} < 0$ m/s?

The equations for $y(t)$ and $v_y(t)$ have the identical form as the rectilinear motion equations for a constant acceleration (Equations 3.16 and 3.19), with merely a notational change to accommodate the different coordinate:

$$v_y(t) = v_{y0} + a_y t \qquad (4.12)$$

$$y(t) = y_0 + v_{y0}t + \frac{1}{2}a_y t^2 \qquad (4.13)$$

where v_{y0} is the initial y-component of the velocity when $t = 0$ s and y_0 is the initial y-component of the position vector. A graph of $v_y(t)$ versus t is a straight *inclined* line since $v_y(t)$ depends linearly on t (see Figure 4.11). The line has intercept v_{y0} and slope a_y.

QUESTION 3

What would the $v_y(t)$ versus t graph look like if $v_{y0} > 0$ m/s and $a_y < 0$ m/s²?

A graph of $y(t)$ versus t is parabolic, since $y(t)$ is quadratic in t (see Figure 4.12). The slope at any instant t_1 is the y-component of the velocity of the particle, $v_y(t_1)$. The intercept is y_0.

We begin our observations when $t = 0$ s. The initial velocity vector has the Cartesian form

$$\vec{v}_0 = v_{x0}\hat{i} + v_{y0}\hat{j} \qquad (4.14)$$

and the initial position vector of the particle (when $t = 0$ s) has the Cartesian form

$$\vec{r}_0 = x_0\hat{i} + y_0\hat{j} \qquad (4.15)$$

PROBLEM-SOLVING TACTIC

4.5 Notice that the equations for $x(t)$ and $v_x(t)$ have the same form as those for $y(t)$ and $v_y(t)$ but with an acceleration component a_x equal to zero. By setting $a_x = 0$ m/s², we see that Equations 3.16 and 3.19 for $x(t)$ and $v_x(t)$ emerge from the general equations for rectilinear motion with a constant acceleration. There is no need to remember two equations for the x-axis motion and two more for the y-axis motion. Just remember Equations 3.16 and 3.19 and apply them to the x- and y-motions for two-dimensional motion.

The equation for the trajectory $y(x)$ of the particle in space (i.e., the path of the particle in space) is found by solving Equation 4.11 for t:

$$t = \frac{x - x_0}{v_{x0}}$$

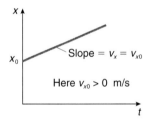

FIGURE 4.10 A graph of x versus t when $a_x = 0$ m/s².

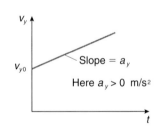

FIGURE 4.11 A graph of $v_y(t)$ versus t.

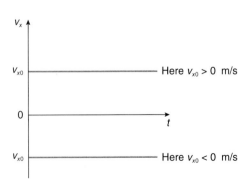

FIGURE 4.9 The x-component of the velocity is constant.

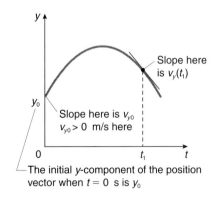

FIGURE 4.12 A graph of $y(t)$ versus t when $a_y < 0$ m/s².

and substituting for t in Equation 4.13. We obtain

$$y(x) = y_0 + v_{y0} \frac{x - x_0}{v_{x0}} + \frac{1}{2} a_y \left(\frac{x - x_0}{v_{x0}} \right)^2$$

The equation for the trajectory $y(x)$ is quadratic in x, indicating that the path in space is *parabolic* in shape.* Two-dimensional motion with a constant acceleration produces parabolic trajectories (see Figure 4.13).

In Figure 4.13, the y-axis is a line containing the acceleration $\vec{a}$, in accordance with Problem-Solving Tactic 4.2. Also notice that the velocity vector at any instant is tangent to the trajectory (path) of the particle in space. The acceleration vector is constant at all points in space (and time). The x-component of the velocity is constant throughout the motion since the acceleration is purely in the y-direction.

Particles thrown or dropped near the surface of the Earth[†] are common cases of projectile motion. While in flight, we say they experience free-fall motion. For simplicity, we neglect the effects of air resistance. The simplification is an approximation to the real flight of such projectiles and means that they experience the constant, vertically downward local acceleration due to gravity, of magnitude $g = 9.81$ m/s².

For such projectiles, invoking Problem-Solving Tactic 4.2, choose the y-axis to be a vertical line, since it is along this line that the constant acceleration is directed. You may choose the specific left or right location of the y-axis, as well as the elevation of the origin for y. The x-axis is horizontal; its elevation is at the location of the origin chosen for y. The x-component of the acceleration is zero: $a_x = 0$ m/s². Problem-Solving Tactic 4.3 indicates that the algebraic sign of the vertical acceleration component a_y depends on the direction chosen for $\hat{j}$ (see Figure 4.14).

The procedure for applying these ideas to the flight of projectiles, neglecting any resistive effects, is best illustrated by careful study of Examples 4.2 and 4.3 following.

---

*Do not confuse the parabolic shape of the trajectory in space, $y(x)$, with the parabolic shape of the $y(t)$ versus t graph: they represent different things!

[†] As we mentioned in Chapter 3, near the surface of the Earth means the total vertical extent of the motion is negligible compared with the radius of the Earth, 6370 km.

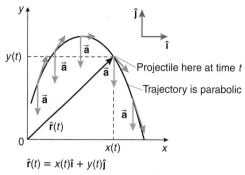

FIGURE 4.13 Two-dimensional motion with a constant acceleration produces a parabolic trajectory. Here $a_y < 0$ m/s².

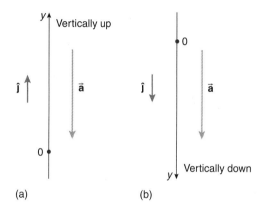

FIGURE 4.14 Free-fall motion. (a) If $\hat{j}$ is up, then a_y is negative: $\vec{a} = a_y\hat{j} = -g\hat{j}$, and so $a_y = -g = -9.81$ m/s². (b) If $\hat{j}$ is down, then a_y is positive: $\vec{a} = a_y\hat{j} = g\hat{j}$, and so $a_y = +g = 9.81$ m/s².

QUESTION 4

A juggler throws a juggling pin across a high room and simultaneously races across the room and barely manages to catch it. A neat trick if you can do it. (a) Which object (the pin or the juggler) had the greater average velocity during the interval or is it the same for both? Explain your answer. (b) Which object had the greater average speed during the interval or is it the same for both? Explain your answer.

STRATEGIC EXAMPLE 4.2

You kick a soccer ball at a speed of 10.0 m/s off a 100 m high cliff. The initial velocity vector is 30.0° above the horizontal direction as the ball leaves your foot, as shown in Figure 4.15.

a. Choose an appropriate coordinate system and write the equations for $v_x(t)$, $x(t)$, $v_y(t)$, and $y(t)$ consistent with your choice.
b. Determine the time of flight.
c. Find the speed of the ball the instant just before impact.
d. Find the horizontal distance of the impact point from the point the ball was kicked.
e. Determine the greatest height of the projectile above the plain on which the ball strikes.

Solution

a. We solve this problem with two different choices for a Cartesian coordinate system to show you the freedom you have in solving problems. Use Problem-Solving Tactic 4.2 and *choose the y-axis*

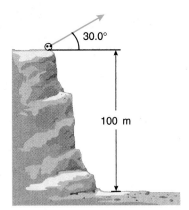

FIGURE 4.15

to be a line containing the constant acceleration vector. Likewise, using Problem-Solving Tactic 4.4, *the origin may be taken at any convenient location along the line chosen for the y-axis.*

Introduce a Cartesian coordinate system as in Figure 4.16.

Choice 1

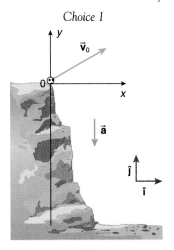

Let the origin be at the point the projectile leaves your foot, with $\hat{\jmath}$ vertically upward, and $\hat{\imath}$ horizontal and to the right.

FIGURE 4.16

Choice 2

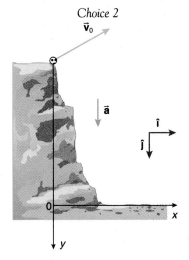

Let the origin be at the bottom of the cliff directly below the point the ball was kicked, with $\hat{\jmath}$ vertically down and $\hat{\imath}$ horizontal and to the right.

Recall from Problem-Solving Tactic 4.3 that *the sign of the acceleration component a_y depends on the direction you choose for $\hat{\jmath}$.* With the coordinate choice determined, the downward acceleration vector is

$$\vec{a} = (-9.81 \text{ m/s}^2)\hat{\jmath} \qquad\qquad \vec{a} = (+9.81 \text{ m/s}^2)\hat{\jmath}$$

The acceleration has a single nonzero component:

$$a_x = 0 \text{ m/s}^2 \qquad\qquad a_x = 0 \text{ m/s}^2$$
$$a_y = -9.81 \text{ m/s}^2 \qquad\qquad a_y = 9.81 \text{ m/s}^2$$

The initial velocity vector $\vec{v}_0$ of the ball when $t = 0$ s is expressed in Cartesian form as

$$\vec{v}_0 = [(10.0 \text{ m/s}) \cos 30°]\hat{\imath} \qquad\qquad \vec{v}_0 = [(10.0 \text{ m/s}) \cos 30°]\hat{\imath}$$
$$+ [(10.0 \text{ m/s}) \sin 30°]\hat{\jmath} \qquad\qquad - [(10.0 \text{ m/s}) \sin 30°]\hat{\jmath}$$

Hence the two components of the initial velocity are

$$v_{x0} = (10.0 \ \text{m/s}) \cos 30.0°$$
$$= 8.66 \ \text{m/s}$$
$$v_{y0} = (10.0 \ \text{m/s}) \sin 30.0°$$
$$= 5.00 \ \text{m/s}$$

$$v_{x0} = (10.0 \ \text{m/s}) \cos 30.0°$$
$$= 8.66 \ \text{m/s}$$
$$v_{y0} = (-10.0 \ \text{m/s}) \sin 30.0°$$
$$= -5.00 \ \text{m/s}$$

The initial position vector of the ball

is zero since it is at the origin when $t = 0$ s.

is $(-100 \ \text{m})\hat{\jmath}$ since it is at the top of the cliff when $t = 0$ s.

Therefore, the components of the initial position vector

are both zero:
$$x_0 = 0 \ \text{m}$$
$$y_0 = 0 \ \text{m}$$

are
$$x_0 = 0 \ \text{m}$$
$$y_0 = -100 \ \text{m}$$

Since the motions along the coordinate axes are independent of each other, you examine each direction separately. Look at the motion along the x-axis, where $a_x = 0 \ \text{m/s}^2$. From Equation 4.10 with

$$v_{x0} = 8.66 \ \text{m/s}$$

you obtain

$$v_x(t) = 8.66 \ \text{m/s}$$

$$v_{x0} = 8.66 \ \text{m/s}$$

you obtain

$$v_x(t) = 8.66 \ \text{m/s}$$

This result is independent of time. The horizontal component of the velocity is constant throughout the entire motion because the acceleration component a_x is zero. The acceleration of the ball is in the vertical direction, y. The x-component of the position vector at any time is found using Equation 4.11, or, after substituting for x_0, v_{x0}, and a_x (= 0 m/s^2):

$$x(t) = 0 + (8.66 \ \text{m/s})t$$
$$= (8.66 \ \text{m/s})t \qquad (1a)$$

$$x(t) = 0 + (8.66 \ \text{m/s})t$$
$$= (8.66 \ \text{m/s})t \qquad (1b)$$

The x-component of the position vector changes linearly with time since the velocity component v_x is constant.

Now examine the y-axis motion. The y-component of the velocity at any time is given by Equation 4.12:

$$v_y(t) = v_{y0} + a_y t$$

Substitute for v_{y0} and a_y from the preceding:

$$v_y(t) = 5.00 \ \text{m/s} + (-9.81 \ \text{m/s}^2)t$$
$$= 5.00 \ \text{m/s} - (9.81 \ \text{m/s}^2)t \qquad (2a)$$

$$v_y(t) = -5.00 \ \text{m/s} + (9.81 \ \text{m/s}^2)t \qquad (2b)$$

The y-component of the position vector is given by Equation 4.13:

$$y(t) = y_0 + v_{y0}t + \frac{1}{2} a_y t^2$$

which, after substituting, becomes

$$y(t) = 0 \ \text{m} + (5.00 \ \text{m/s})\,t$$
$$+ \frac{1}{2}(-9.81 \ \text{m/s}^2)\,t^2$$
$$= (5.00 \ \text{m/s})\,t - \frac{9.81 \ \text{m/s}^2}{2}\,t^2 \qquad (3a)$$

$$y(t) = -100 \ \text{m} - (5.00 \ \text{m/s})\,t$$
$$+ \frac{9.81 \ \text{m/s}^2}{2}\,t^2 \qquad (3b)$$

This completes the tailoring of the kinematic equations for motion with a constant acceleration to the specific choice for the Cartesian coordinate system used to solve the problem. Now you bring in other relevant information.

b. At the impact point, the y-coordinate of the projectile is

$$y = -100 \text{ m} \qquad\qquad y = 0 \text{ m}$$

Substitute this value for y into equation (3) to determine the instant t of impact:

$$-100 \text{ m} = (5.00 \text{ m/s})\, t - \frac{9.81 \text{ m/s}^2}{2}\, t^2 \qquad\qquad 0 \text{ m} = -100 \text{ m} - (5.00 \text{ m/s})\, t$$
$$+ \frac{9.81 \text{ m/s}^2}{2}\, t^2$$

or

$$\frac{9.81 \text{ m/s}^2}{2}\, t^2 - (5.00 \text{ m/s})\, t - 100 \text{ m} \qquad\qquad \frac{9.81 \text{ m/s}^2}{2}\, t^2 - (5.00 \text{ m/s})\, t - 100 \text{ m}$$
$$= 0 \text{ m} \qquad\qquad\qquad\qquad = 0 \text{ m}$$

The equation can be solved for t using the quadratic formula. Notice that you get the *same* equation for t with *both* coordinate choices; this is to be expected since the instant of impact will be the same, regardless of your coordinate choice. The numerical value of t also is equal to the time interval of the flight since, by convention, the clock was set to $t = 0$ s at launch. The two solutions for t are

$$5.05 \text{ s} \qquad \text{and} \qquad -4.03 \text{ s}$$

The positive solution must be chosen because the impact occurs *after* the instant of the kick (which was $t = 0$ s). Thus the time of flight is

$$t = 5.05 \text{ s}$$

The negative root corresponds to the other time the ball has the y-coordinate

$$y = -100 \text{ m} \qquad\qquad y = 0 \text{ m}$$

on the parabolic trajectory (see Figure 4.17). You can discard the negative root for t, because you began your observations when $t = 0$ s and the impact certainly comes after the kick.

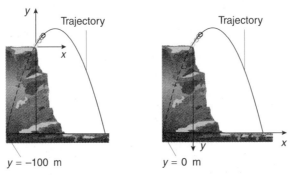

FIGURE 4.17 **The meaning of the negative solution for *t*.**

c. The x-component of the velocity is constant throughout the flight and is

$$v_x = 8.66 \text{ m/s} \qquad\qquad v_x = 8.66 \text{ m/s}$$

The y-component of the velocity at the instant of impact (when $t = 5.05$ s) is

$$v_y = 5.00 \text{ m/s} - (9.81 \text{ m/s}^2)t$$
$$= 5.00 \text{ m/s} - (9.81 \text{ m/s}^2)(5.05 \text{ s})$$
$$= -44.5 \text{ m/s}$$

$$v_y = -5.00 \text{ m/s} + (9.81 \text{ m/s}^2)t$$
$$= -5.00 \text{ m/s} + (9.81 \text{ m/s}^2)(5.05 \text{ s})$$
$$= +44.5 \text{ m/s}$$

So the velocity vector at the instant just before impact is

$$\vec{v} = (8.66 \text{ m/s})\hat{\imath} - (44.5 \text{ m/s})\hat{\jmath}$$

$$\vec{v} = (8.66 \text{ m/s})\hat{\imath} + (44.5 \text{ m/s})\hat{\jmath}$$

The *speed* at the instant just before impact is the magnitude of the velocity vector:

$$v = (v_x^2 + v_y^2)^{1/2}$$

$$v = [(8.66 \text{ m/s})^2 + (-44.5 \text{ m/s})^2]^{1/2}$$
$$= 45.3 \text{ m/s}$$

$$v = [(8.66 \text{ m/s})^2 + (44.5 \text{ m/s})^2]^{1/2}$$
$$= 45.3 \text{ m/s}$$

Notice that the speed at impact is the same with both coordinate choices.

d. The horizontal distance of the impact point from the point the ball was kicked is the x-component of the position vector of the impact point, found by substituting $t = 5.05$ s into equation (1) for the x-component of the position vector:

$$x_{impact} = 0 \text{ m} + (8.66 \text{ m/s})(5.05 \text{ s})$$
$$= 43.7 \text{ m}$$

$$x_{impact} = 0 \text{ m} + (8.66 \text{ m/s})(5.05 \text{ s})$$
$$= 43.7 \text{ m}$$

Since the x-axis origin in both coordinate systems is at the same horizontal position, you expect the x-coordinates of the impact point to be the same for both coordinate choices.

e. To determine the greatest height of the projectile above the plain on which the ball strikes, you need to know the instant the projectile reaches this point in space. The only thing you know about the projectile when it is at its greatest elevation is that its vertical velocity component v_y is zero. Use equation (2) for v_y at any instant t:

$$v_y(t) = 5.00 \text{ m/s} - (9.81 \text{ m/s}^2)t$$

$$v_y(t) = -5.00 \text{ m/s} + (9.81 \text{ m/s}^2)t$$

Solve for t when $v_y(t) = 0$ m/s:

$$0 \text{ m/s} = 5.00 \text{ m/s} - (9.81 \text{ m/s}^2)t$$

$$0 \text{ m/s} = 5.00 \text{ m/s} - (9.81 \text{ m/s}^2)t$$

Solving for t, for both coordinate choices you obtain
$$t = 0.510 \text{ s}$$

Since you are interested in the vertical height at this time, we find the y-component of the position vector when $t = 0.510$ s from equation (3):

$$y(t) = (5.00 \text{ m/s})t - \frac{9.81 \text{ m/s}^2}{2}t^2$$

$$y(t) = -100 \text{ m} - (5.00 \text{ m/s})t + \frac{9.81 \text{ m/s}^2}{2}t^2$$

Substitute $t = 0.510$ s to find the y-coordinate of the position vector at the peak of the trajectory:

$$y_{max} = 1.27 \text{ m}$$

$$y_{max} = -101 \text{ m}$$

Since the origin for y was at the point of release, located 100 m above the plain, the elevation of the projectile above the plain below is 100 m + 1.27 m = 101 m, since you can keep only three significant figures in the addition.

Since the origin for y was at the plain, the elevation of the projectile is the absolute value of this component of the position vector, or 101 m.

You get the same physical results regardless of the coordinate system used, but your description of the motion depends on the choices you make for the coordinate system.

EXAMPLE 4.3

You throw a baseball with an initial speed v_0 at an elevation angle θ across level terrain, as indicated in Figure 4.18. Your roommate catches the ball at the same height above the level ground that you released it. Find expressions for the time of flight t and for the horizontal range R_{horiz} in terms of v_0, θ, and g.

Solution

Use the Cartesian coordinate system given in Figure 4.18. The general equations, 4.10–4.13, for motion at constant acceleration in two dimensions are

$$v_x(t) = v_{x0}$$
$$x(t) = x_0 + v_{x0}t$$
$$v_y(t) = v_{y0} + a_y t$$
$$y(t) = y_0 + v_{y0}t + \frac{1}{2}a_y t^2$$

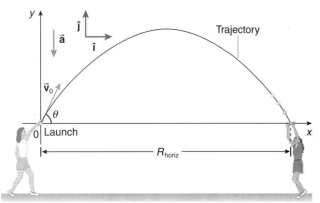

FIGURE 4.18

With the coordinate choice and with the acceleration (of magnitude $g = 9.81$ m/s^2) vertically downward, parallel to $-\hat{\mathbf{j}}$, you make the following identifications:

for the components of the initial position vector when $t = 0$ s:

$$x_0 = 0 \text{ m} \qquad y_0 = 0 \text{ m}$$

for the components of the initial velocity vector when $t = 0$ s:

$$v_{x0} = v_0 \cos\theta \qquad v_{y0} = v_0 \sin\theta$$

and for the constant acceleration component:

$$a_y = -g$$

Equations 4.10–4.13 describing the x- and y-motion then become

$$v_x(t) = v_0 \cos\theta$$
$$x(t) = (v_0 \cos\theta)t$$
$$v_y(t) = (v_0 \sin\theta) - gt$$
$$y(t) = (v_0 \sin\theta)t - \frac{1}{2}gt^2$$

At the instant just before the catch, the x-coordinate of the particle is R_{horiz} and its y-coordinate is 0 m. Hence at the instant t just before the catch the four preceding equations are

$$v_x(t) = v_0 \cos\theta \tag{1}$$
$$R_{\text{horiz}} = (v_0 \cos\theta)t \tag{2}$$

$$v_y(t) = (v_0 \sin\theta) - gt \tag{3}$$
$$0 \text{ m} = (v_0 \sin\theta)t - \frac{1}{2}gt^2 \tag{4}$$

Since $t = 0$ s corresponds to launch, t also is numerically equal to the time of flight in these equations.

Because you are interested in finding the time of flight t and the horizontal range R_{horiz}, you need to solve equations (2) and (4) simultaneously for R_{horiz} and t. To do this, first factor the equation (4) into the form

$$0 \text{ m} = t\left(v_0 \sin\theta - \frac{g}{2}t\right)$$

This equation has two roots that tell you when the y-coordinate of the projectile is zero:

$$t = 0 \text{ s}$$

which corresponds to launch, and

$$t = \frac{2v_0 \sin\theta}{g}$$

which corresponds to the catch when $x = R_{\text{horiz}}$. The latter root is the one desired.

The horizontal range is found by putting the time of flight into equation (2) for the horizontal range:

$$R_{\text{horiz}} = (v_0 \cos\theta)t$$
$$= (v_0 \cos\theta)\frac{2v_0 \sin\theta}{g}$$
$$= \frac{2v_0^2 \sin\theta \cos\theta}{g}$$

The double angle trigonometric identity

$$2\sin\theta\cos\theta = \sin 2\theta$$

enables you to put the expression for the horizontal range into the form

$$R_{\text{horiz}} = \frac{v_0^2 \sin 2\theta}{g} \tag{5}$$

Equation (5) is the expression for the range of a projectile *only when it is launched from and strikes on the same horizontal plane.* Many projectiles hit before or after the impact time for the horizontal range given by equation (5), as shown in Figure 4.19. In

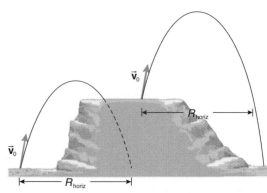

FIGURE 4.19 Projectiles that strike before or after they would on a horizontal range.

such situations, equation (5) for the horizontal range is useless and the horizontal distance covered by the projectile before impact must be determined from the specifics of the given situation.

4.3 MOTION IN THREE DIMENSIONS

Stranger. Exactly: you see you do not even know what Space is. You think it is of Two Dimensions only; but I have come to announce to you a Third . . . .

Edwin Abbott Abbott*

When navigating a difficult rapid with a kayak, you are moving in three dimensions. Likewise, when charged particles emitted by the Sun† encounter the magnetic field in the vicinity of the Earth, their motion resembles a curved spiral and they corkscrew into the atmosphere of the Earth, producing beautiful auroras from their collisions with molecules in the air (see Figure 4.20).

To extend the concepts of the position vector, velocity, and acceleration to motion in three dimensions, we introduce the third Cartesian coordinate axis, z. The position vector then is

$$\vec{r}(t) = x(t)\hat{\imath} + y(t)\hat{\jmath} + z(t)\hat{k} \qquad (4.16)$$

The definitions of the average speed,

$$\text{average speed} = \frac{d}{\Delta t}$$

the average velocity,

$$\vec{v}_{ave} = \frac{\Delta \vec{r}}{\Delta t}$$

and average acceleration,

$$\vec{a}_{ave} = \frac{\Delta \vec{v}}{\Delta t}$$

are the same as in one- or two-dimensional motion. Likewise, the velocity,

$$\vec{v}(t) = \frac{d\vec{r}(t)}{dt}$$

and acceleration,

$$\vec{a}(t) = \frac{d\vec{v}(t)}{dt}$$

Flatland, Part II, Section 16, page 69.
†Such particle flows constitute what is called the *solar wind*.

FIGURE 4.20 An aurora.

also have the same form as in one or two dimensions, showing again the convenience of using vector notation. For three-dimensional motion, each vector has three possible Cartesian components.

4.4 RELATIVE VELOCITY ADDITION AND ACCELERATIONS

Kids find it fun to run up the down escalator or down the up escalator; the stairs are moving one way while they are moving the other way, each with different speeds. Likewise, if you ever have been in a raft, canoe, or kayak on a swiftly flowing stream, you know that your velocity with respect to the banks can differ greatly from your paddling velocity with respect to the water, sometimes with frightening consequences. Pilots in airplanes encounter similar effects; they continually must account for the speed and direction of the wind to determine their headings. Here we see how such effects are neatly accounted for using a vector description of motion.

Coordinate systems (such as the Cartesian coordinate system) and clocks are the "hangers" on which we drape a problem. A coordinate system with a set of synchronized clocks, all ticking at the same rate, is called a **reference frame**. Many choices are possible for a coordinate system. Thus, although the coordinate system is central to setting up the problem, the particular choice for the direction of, say, the Cartesian x-, y-, and z-axes is immaterial to the underlying physics, as we have seen. Different choices for the coordinate system lead to different *descriptions* of the motion, which may be more or less complex than for other coordinate choices, but the underlying *physics* is nonetheless the same.

Until now, the various coordinate system choices we made were all fixed or at rest with respect to each other. Here we relax that condition. Suppose you employ a fixed coordinate system with an origin at 0, perhaps on the ground, while your friends in a hot-air balloon, moving along with the wind, use a coordinate system with an origin at 0′ at the gondola, as in Figure 4.21. An airplane, represented as a particle, is flying at point P. We describe the position of the airplane by means of position vectors either with respect to 0 or with respect to 0′, as illustrated in Figure 4.21. Let $\vec{r}$ be the position vector of the airplane at P with respect to 0,

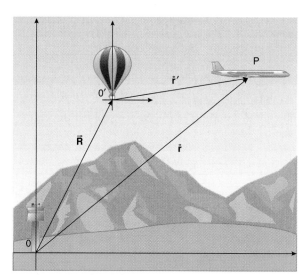

FIGURE 4.21 A plane located with respect to two different coordinate systems.

and let $\vec{\mathbf{r}}'$ be its position vector with respect to $0'$. Vector $\vec{\mathbf{R}}$ locates the gondola origin $0'$ with respect to the origin 0 on the ground.

From Figure 4.21 we can see that the three vectors are related by the vector sum

$$\vec{\mathbf{r}} = \vec{\mathbf{R}} + \vec{\mathbf{r}}' \tag{4.17}$$

If *both* the airplane at point P and the gondola origin $0'$ are moving, then all three position vectors are functions of time. In particular, we can take the time derivative of Equation 4.17 to obtain

$$\frac{d\vec{\mathbf{r}}}{dt} = \frac{d\vec{\mathbf{R}}}{dt} + \frac{d\vec{\mathbf{r}}'}{dt} \tag{4.18}$$

The quantity

$$\frac{d\vec{\mathbf{r}}}{dt}$$

is the velocity of the particle at P *with respect to* 0. Call this velocity $\vec{\mathbf{v}}_{P0}$, to be read as "the velocity of P with respect to 0." The quantity

$$\frac{d\vec{\mathbf{r}}'}{dt}$$

is the velocity of the particle *with respect to* $0'$. Call this velocity $\vec{\mathbf{v}}_{P0'}$, read as "the velocity of P with respect to $0'$." And finally, the quantity

$$\frac{d\vec{\mathbf{R}}}{dt}$$

is the velocity of origin $0'$ *with respect to* 0. Call this velocity $\vec{\mathbf{v}}_{0'0}$, read as "the velocity of $0'$ with respect to 0." With these definitions, Equation 4.18 becomes

$$\vec{\mathbf{v}}_{P0} = \vec{\mathbf{v}}_{0'0} + \vec{\mathbf{v}}_{P0'} \tag{4.19}$$

Because the order of the terms in a vector sum is immaterial, Equation 4.19 can be rearranged slightly into the form

$$\boxed{\vec{\mathbf{v}}_{P0} = \vec{\mathbf{v}}_{P0'} + \vec{\mathbf{v}}_{0'0}} \tag{4.20}$$

Equation 4.20 is called the **relative velocity addition equation**.

There is an easy way to remember the relative velocity addition equation. On the right-hand side of the equation, the subscript in the middle (here $0'$) is imagined to cancel out, leaving on the left-hand side a velocity of the first subscript (P) with respect to the last subscript (0):

$$\vec{\mathbf{v}}_{P0} = \vec{\mathbf{v}}_{P0'} + \vec{\mathbf{v}}_{0'0}$$

As you watch a car approach a bridge on which you are standing, the occupants of the car see you and the bridge approaching them in the opposite direction at the same speed. Hence the velocity of P with respect to 0 is the *negative* of the velocity of 0 with respect to P. That is

$$\vec{\mathbf{v}}_{P0} = -\vec{\mathbf{v}}_{0P}$$

So, if we interchange the two *subscripts*, a minus sign is introduced.

If a coordinate system $0'$ is moving at *constant velocity* with respect to 0, then $\vec{\mathbf{v}}_{0'0}$ is a constant vector. When $\vec{\mathbf{v}}_{0'0}$ is a constant vector, its time derivative is zero:

$$\frac{d\vec{\mathbf{v}}_{0'0}}{dt} = 0 \text{ m/s}^2$$

Then if we differentiate Equation 4.20, we find that

$$\frac{d\vec{\mathbf{v}}_{P0}}{dt} = \frac{d\vec{\mathbf{v}}_{P0'}}{dt} + 0 \text{ m/s}^2$$
$$= \frac{d\vec{\mathbf{v}}_{P0'}}{dt} \tag{4.21}$$

The term on the left-hand side of Equation 4.21 is the acceleration $\vec{\mathbf{a}}_{P0}$ of the particle at P with respect to the coordinate system with origin at 0. The right-hand side of Equation 4.21 is the acceleration $\vec{\mathbf{a}}_{P0'}$ of the particle with respect to the coordinate system with origin at $0'$.

The accelerations of a particle with respect to two coordinate systems that are moving at constant velocity with respect to each other are the same:

$$\vec{\mathbf{a}}_{P0} = \vec{\mathbf{a}}_{P0'} \quad \text{(if $\vec{\mathbf{v}}_{0'0}$ is constant)} \tag{4.22}$$

EXAMPLE 4.4

The water in a scenic river flows at a constant velocity $\vec{\mathbf{v}}_{\text{water ground}}$. A canoeist paddles at velocity $\vec{\mathbf{v}}_{\text{canoeist water}}$ with respect to the water. In what direction θ relative to the current should the canoeist paddle so that the resulting velocity of the canoe with respect to the ground is perpendicular to the river bank? The canoeist will then go straight across the river. See Figure 4.22. In canoeing parlance, the angle θ is called the angle of attack. Discuss the implications if the speed of the current, $v_{\text{water ground}}$, is greater than the paddling speed $v_{\text{canoeist water}}$ of the canoeist.

FIGURE 4.22

FIGURE 4.23

Whitewater ferrying

FIGURE 4.24 The faster the current, the smaller the angle of attack.

Solution
According to the relative velocity addition (Equation 4.20), the velocities are related by

$$\vec{v}_{\text{canoeist ground}} = \vec{v}_{\text{canoeist water}} + \vec{v}_{\text{water ground}}$$

To have the canoeist go straight across the river, the velocity vectors must form the right triangle shown in Figure 4.23. Use the right triangle in Figure 4.23 to find the cosine of the angle of attack in terms of the speeds:

$$\cos \theta = \frac{v_{\text{water ground}}}{v_{\text{canoeist water}}}$$

The faster the current, the smaller the angle θ of attack (see Figure 4.24). This is familiar to any whitewater canoeist (the maneuver is known as a ferry).

  Notice that if $v_{\text{water ground}} > v_{\text{canoeist water}}$, then the preceding equation for the cosine indicates $\cos \theta > 1$, which is mathematically impossible for real angles, because a side of a right triangle cannot exceed the length of the hypotenuse. (Try to take the inverse cosine of a number greater than 1; your calculator will indicate an error.) Physically this means that if $v_{\text{water ground}} > v_{\text{canoeist water}}$, then no θ can be found. The canoeist cannot paddle fast enough for the boat to go straight across the river; in trying to cross it, the paddler then is inevitably swept downstream and perhaps over a waterfall.

STRATEGIC EXAMPLE 4.5

  An airplane is in level flight at a speed of 600 km/h with respect to the surrounding air. A wind is blowing toward the northeast direction at a speed of 100 km/h, as shown in Figure 4.25.

a. **What is the vector relationship among the velocity of the plane with respect to the ground, the velocity of the plane with respect to the air, and the wind velocity?**
b. **In what direction must the pilot aim the plane so that the velocity direction of the plane with respect to the ground is due north?**

Solution
a. Let the velocity of the airplane (P) with respect to the surrounding air (0′) be

$$\vec{v}_{\text{P0'}} \equiv \vec{v}_{\text{plane air}}$$

Let the velocity of the air (0′) with respect to the ground (0) be written as

$$\vec{v}_{\text{0'0}} \equiv \vec{v}_{\text{air ground}}$$

Call the velocity of the plane (P) with respect to the ground (0)

$$\vec{v}_{\text{P0}} \equiv \vec{v}_{\text{plane ground}}$$

Then according to Equation 4.20, the various velocities are related by the vector sum

$$\vec{v}_{\text{plane ground}} = \vec{v}_{\text{plane air}} + \vec{v}_{\text{air ground}}$$

In accordance with the conceptual note following Equation 4.20, notice that on the right-hand side of this equation the subscript in the middle is imagined to cancel out, leaving the velocity of the first subscript with respect to the last subscript on the left-hand side of the equation.

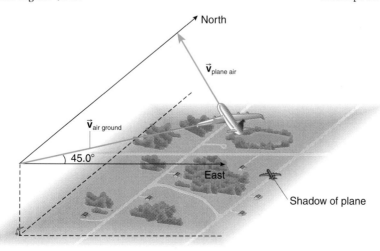

FIGURE 4.25

b. There are two ways to solve the problem.

Method 1

This method is geometric. The three vectors are related according to the sketch in Figure 4.26. Divide the triangle formed by the vectors in Figure 4.26 into two right triangles as shown. The common side of the two right triangles has a magnitude of

$$\frac{100}{\sqrt{2}} \text{ km/h}$$

or 70.7 km/h. You can use the Pythagorean theorem to find the remaining side of the upper right triangle in Figure 4.26, since the hypotenuse (600 km/h) and one side (70.7 km/h) are known; the remaining side is of magnitude 596 km/h. To this add the component of wind vector along the y-axis (70.7 km/h), so that the magnitude of the vector lying along the y-axis is

$$596 \text{ km/h} + 70.7 \text{ km/h} = 667 \text{ km/h}$$

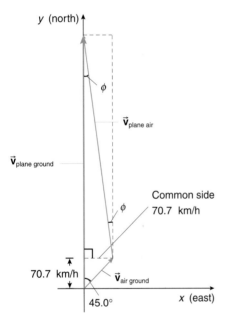

FIGURE 4.26

The angle that the velocity vector $\vec{v}_{\text{plane air}}$ makes with the north direction then is found using trigonometry:

$$\sin \phi = \frac{70.7 \text{ km/h}}{600 \text{ km/h}}$$

$$\phi = 6.77°$$

The pilot should aim about 6.8° west of north.

Method 2

Establish the coordinate system as indicated in Figure 4.27 with the +x-axis directed east and the +y-axis directed north. Consider what you know about the three vectors individually. You know both the magnitude of the wind (100 km/h) and its direction (northeast). Thus you can write $\vec{v}_{\text{air ground}}$ as its magnitude times a unit vector pointing in the northeast

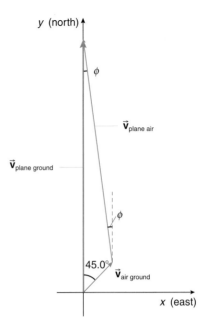

FIGURE 4.27

direction. The vector $\hat{\imath} + \hat{\jmath}$ points in the northeast direction, but is not 1 unit long (its magnitude is $\sqrt{2}$). Thus the vector

$$\frac{\hat{\imath} + \hat{\jmath}}{\sqrt{2}}$$

is a unit vector pointing in the northeast direction. Hence the wind is represented in Cartesian form as

$$\vec{v}_{\text{air ground}} = (100 \text{ km/h}) \frac{\hat{\imath} + \hat{\jmath}}{\sqrt{2}}$$
$$= (70.7 \text{ km/h}) \hat{\imath} + (70.7 \text{ km/h}) \hat{\jmath}$$

Another way to find the vector representing the wind is to realize that the velocity of the wind makes an angle of 45° with both the x- and y-axes, because the wind is directed to the northeast. Since the wind speed is 100 km/h, the vector representing the wind velocity is

$$\vec{v}_{\text{air ground}} = [(100 \text{ km/h}) \cos 45.0°]\hat{\imath}$$
$$+ [(100 \text{ km/h}) \sin 45.0°]\hat{\jmath}$$
$$= (70.7 \text{ km/h})\hat{\imath} + (70.7 \text{ km/h})\hat{\jmath}$$

You want the resulting velocity of the plane with respect to the ground to be in the northern direction (in the same direction as $+\hat{\jmath}$), but you do not know the magnitude of this vector. Call its magnitude v. Thus you write (see Figure 4.27)

$$\vec{v}_{\text{plane ground}} = v\hat{\jmath}$$

You know the magnitude of the velocity of the plane with respect to the air (600 km/h), but not its direction. The vector relationship among the velocities is, from Equation 4.20,

$$v\hat{\jmath} = \vec{v}_{\text{plane air}} + (70.7 \text{ km/h})\hat{\imath} + (70.7 \text{ km/h})\hat{\jmath} \quad (1)$$

Rearrange equation (1) to isolate $\vec{v}_{\text{plane air}}$:

$$\vec{v}_{\text{plane air}} = (-70.7 \text{ km/h})\hat{\imath} + (v - 70.7 \text{ km/h})\hat{\jmath} \quad (2)$$

The square of the magnitude of $\vec{v}_{plane\ air}$ is $(600\ km/h)^2$. Use equation (2) to find the squared magnitude of $\vec{v}_{plane\ air}$:

$$(600\ km/h)^2 = (-70.7\ km/h)^2 + (v - 70.7\ km/h)^2$$

which is quadratic in the unknown v. Use the quadratic formula and take the positive root to obtain a positive result for the speed v; you will find that*

$$v = 666\ km/h$$

This result is the speed of the plane with respect to the ground. You find $\vec{v}_{plane\ air}$ by using this value for v in equation (2) for $\vec{v}_{plane\ air}$:

$$\begin{aligned}\vec{v}_{plane\ air} &= (-70.7\ km/h)\hat{\imath} + (666\ km/h - 70.7\ km/h)\hat{\jmath} \\ &= (-70.7\ km/h)\hat{\imath} + (595\ km/h)\hat{\jmath}\end{aligned}$$

The angle ϕ that this vector makes with the northern direction (see Figure 4.27) is

$$\tan\phi = \frac{70.7\ km/h}{595\ km/h}$$

or

$$\phi = 6.78°$$

The pilot should aim in a direction about 6.8° west of north.

4.5 UNIFORM CIRCULAR MOTION: A FIRST LOOK

Round and round the circle

T. S. Eliot (1888–1965)[†]

Circular motion at constant speed is called **uniform circular motion**. Imagine yourself standing at rest on the outer rim of a constantly spinning merry-go-round. Your orbital path is a circle in the two-dimensional horizontal plane. In physics **spin** describes the motion of an extended object about an axis through itself.[‡] Thus we say the merry-go-round spins about an axis through its center (see Figure 4.28).

The motion of a particle about a point that is not coincident with the particle is called **orbital motion**. An example is you

---

*The discrepancy with Method 1 is due to rounding.
[†]*The Family Reunion*, Part II, Scene III, in T. S. Eliot, *The Complete Poems and Plays 1909–1950* (Harcourt, Brace and World, New York, 1971), page 293.
[‡]We define spin more precisely in Chapter 10.

FIGURE 4.28 A merry-go-round spins about an axis through its center.

(modeled as a particle) traveling in a circular path about the center of the merry-go-round (see Figure 4.29).

Thus, in another context, we say the Earth *spins* on its axis each day, but *orbits* the Sun on a path that is approximately circular (see Figure 4.30).[§] In this example the spin axis and orbital axis are not the same.

We wish to quantify the rate of the spin of the merry-go-round or the rate at which a particle orbits the center of a circular trajectory. Every point of the merry-go-round turns through the same angle $\Delta\theta$ during an interval Δt. Hence the angular rate of the orbital motion of a particle at rest on the merry-go-round equals the angular rate of spin of the merry-go-round. The angular rate of spin or of orbital motion is called the **angular speed**, designated by ω (lower case Greek omega) and is specified in either of the following ways:

1. For convenience, we use the number of complete turns per unit of time. When a particle completes one circular trajectory, we say it has completed one **revolution** of its motion. Hence the angular speed may be specified in revolutions per minute (rev/min), which is not necessarily an integer. You may have a tachometer in your car that indicates rpm, the number of revolutions per minute that the crankshaft within the engine is spinning.
2. For most calculations in the sciences, it is preferable to use the number of radians per unit of time, for example, radians per second (rad/s), since the second is the SI unit of time and radians are the mathematically natural units for specifying angles. The angular speed in radians per second specifies how many radians of angle the particle would sweep out during one second if the particle moved at the same angular speed throughout the one second interval.

[§]We treat the orbital path of the Earth about the Sun approximately as a circle, though it is more accurately an ellipse. In Chapter 6 we analyze elliptical orbital motion.

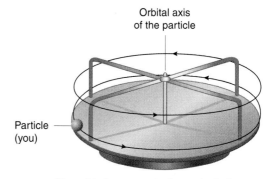

FIGURE 4.29 You orbit the same axis about which the merry-go-round spins.

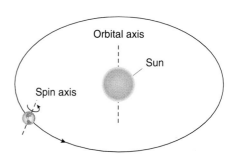

FIGURE 4.30 The Earth spins on its axis and orbits the Sun about a different axis.

Assume the angular speed ω of the merry-go-round is constant. Your velocity vector while riding around the circle is tangent to the circular path and is continually changing its direction, though its magnitude (the speed) is constant. Your speed in circular motion also is known as the **tangential speed** v, since your velocity vector always is tangent to the circular path. But a changing velocity implies a nonzero acceleration. What is the direction of your acceleration? To find out, use the plumb bob accelerometer introduced in Section 3.6. Recall that the deflection of the plumb bob and the acceleration are in *opposite* directions, and the greater the deflection angle, the greater the magnitude of the acceleration. Note that the plumb bob deflection is always away from the center of the merry-go-round (see Figure 4.31), and so the acceleration must be *toward the center* of the circular path.

Such an acceleration toward the center of a circle is called a **centripetal acceleration** $\vec{a}_c$.* We need to find how the magnitude of the centripetal acceleration depends on the tangential speed v and the distance r from the center of the circular path being followed by the particle.

Consider a particle moving in a circle of radius r at constant speed v. Let the particle move a short arc length $v\,\Delta t$ along the circle during a time interval Δt, as shown in Figure 4.32. Viewed from the center of the circle, the angle $\Delta\theta$ subtended by the arc length is, in radians,

$$\Delta\theta = \frac{v\,\Delta t}{r} \qquad (4.23)$$

---

*The word centripetal comes from the Latin *centripetus* and means "center seeking."

FIGURE 4.31 The plumb bob accelerometer indicates that the direction of the acceleration is toward the center of the circular motion.

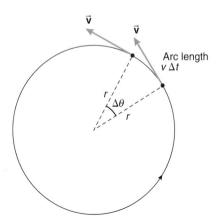

FIGURE 4.32 A particle moves through a short arc length during a short interval Δt.

The angle through which the velocity vector turns during the interval Δt also is $\Delta\theta$, since the velocity always is tangent to the trajectory of the particle. Parallel transport the velocity vector at the beginning of the interval to the velocity vector at the end of the interval to form a velocity vector triangle, as in Figure 4.33.

The magnitude of the change in the velocity vector is the base of the isosceles triangle whose sides are v and whose apex angle is $\Delta\theta$. If the time interval is short, we can use the isosceles triangle formed by the velocity vectors to find another expression for $\Delta\theta$:

$$\Delta\theta \approx \frac{\Delta v}{v} \qquad (4.24)$$

This approximation becomes better as the time interval Δt and the velocity change $\Delta\vec{v}$ become smaller. The magnitude of the average acceleration a_{ave} during the time interval is

$$a_{ave} = \frac{\Delta v}{\Delta t}$$

Using Equation 4.24, we get

$$a_{ave} \approx \frac{v\,\Delta\theta}{\Delta t}$$

Substituting for $\Delta\theta$ using Equation 4.23, we have

$$a_{ave} \approx \frac{v}{\Delta t}\frac{v\,\Delta t}{r}$$

$$\approx \frac{v^2}{r}$$

The direction of the average acceleration is the direction of $\Delta\vec{v}$, which (from Figures 4.32 and 4.33) is directed toward the center of the circle as Δt approaches zero. If the time interval is short, the average acceleration approaches the instantaneous acceleration and is equal to it in the limit as $\Delta t \to 0$ s.

> The centripetal acceleration of the particle is of magnitude v^2/r and is directed toward the center of the circle:
>
> $$a_c = \frac{v^2}{r} \qquad (4.25)$$
>
> The magnitude of the centripetal acceleration varies directly as the square of the speed and inversely with the radius.

QUESTION 5

You move to a larger radius on a merry-go-round and test for the magnitude of the centripetal acceleration with the plumb bob accelerometer. Based on such an experiment, does the magnitude of the centripetal acceleration increase, decrease, or remain the same?

FIGURE 4.33 Parallel transport to form a velocity triangle.

EXAMPLE 4.6

Your car is executing a circular turn of radius 200 m at the constant speed 60.0 km/h. What is the magnitude of the centripetal acceleration of the car in appropriate SI units?

Solution

Convert the speed of the car to SI units:

$$60.0 \text{ km/h} = \frac{(60.0 \text{ km/h}) (10^3 \text{ m/km})}{3600 \text{ s/h}}$$

$$= 16.7 \text{ m/s}$$

The magnitude of the centripetal acceleration is found from Equation 4.25:

$$a_c = \frac{v^2}{r}$$

$$= \frac{(16.7 \text{ m/s})^2}{200 \text{ m}}$$

$$= 1.39 \text{ m/s}^2$$

EXAMPLE 4.7

You are located at the outer rim of a merry-go-round of diameter 20.0 m and are orbiting at 6.00 rev/min.

a. What is your angular speed in rad/s?
b. What is your tangential speed in m/s?
c. What is the magnitude of your centripetal acceleration?

Solution

a. To find the angular speed ω, convert 6.00 rev/min to rad/s:

$$\omega = \frac{(6.00 \text{ rev/min}) (2\pi \text{ rad/rev})}{60 \text{ s/m}}$$

$$= 0.628 \text{ rad/s}$$

b. The radius of the merry-go-round is 10.0 m. Its circumference is

$$2\pi r = 62.8 \text{ m}$$

The angular speed of 6.00 rev/min means you travel a distance of 6.00 times the circumference during one minute. Your speed v is

$$v = \frac{6.00 \ (62.8 \text{ m})}{60 \text{ s}}$$

$$= 6.28 \text{ m/s}$$

c. The magnitude of the centripetal acceleration is found from Equation 4.25:

$$a_c = \frac{v^2}{r}$$

$$= \frac{(6.28 \text{ m/s})^2}{10.0 \text{ m}}$$

$$= 3.94 \text{ m/s}^2$$

EXAMPLE 4.8

Imagine yourself standing at rest at a latitude angle β on the spinning Earth of radius R, as shown in Figure 4.34.

a. What is the radius of your circular orbital motion?

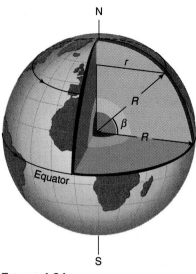

FIGURE 4.34

b. If the Earth rotates on its axis during a time T, what is your orbital speed in terms of R, β, and T?
c. What is your centripetal acceleration?
d. At what latitudes would your centripetal acceleration be zero?

Solution

a. From the geometry of Figure 4.34, the circular path along which you are moving has a radius $r = R \cos \beta$.
b. During the time interval T, you travel the circumference of a circle of radius $R \cos \beta$. Hence your speed v is

$$v = \frac{2\pi (R \cos \beta)}{T}$$

c. The centripetal acceleration is of magnitude

$$a_c = \frac{v^2}{r}$$

$$= \frac{4\pi^2 R \cos \beta}{T^2}$$

d. The centripetal acceleration has zero magnitude when $\cos \beta = 0$, and so $\beta = \pm 90°$. These latitudes correspond to the north and south geographic poles.

4.6 THE ANGULAR VELOCITY VECTOR

The angular speed of a particle in circular motion is the measure of the number of radians per second through which the particle moves. It is convenient to represent the angular speed ω as the magnitude of a vector called the **angular velocity $\vec{\omega}$** of the particle. The angular velocity vector will provide a convenient visual representation of the circular motion of a particle.

In Section 2.5 we saw that an arrow representing the *angle* of a rotation is *not* a vector, since the arrow did not obey the rules for vector arithmetic when rotations were taken about different axes in succession. However, if rotations all are about the same *fixed* axis in space, the angle of rotation *can* be modeled visually as a vector arrow as we attempted in Section 2.5, since then the arrows are confined to the single line in space representing the axis. It may come as a surprise to learn that the angular velocity that we define here *is* a vector quantity for more general rotations in space. Proving this assertion is not easy and is beyond the level and scope of this text.

Since the direction of motion of the particle continually is changing in the plane of its circular motion, no unique direction in this plane is a logical choice for the direction of the angular velocity vector. But at least the axis direction remains the same. Thus we choose a direction perpendicular to the plane of the circle, using a **circular motion right-hand rule**: wrap the four bunched fingers of your right hand around the circle in the sense that the particle is moving (see Figure 4.35). Then the extended thumb of your right hand indicates the appropriate perpendicular direction to take as the direction of the vector $\vec{\omega}$. We draw the angular velocity vector at the center of the circular path of the particle.

> The direction of the angular velocity vector always is determined according to this circular motion right-hand rule. As we have discussed many times now, how you *describe* this direction in space depends on the choice you make for a coordinate system. We will *always* choose the z-axis to be the line containing $\vec{\omega}$.

> **PROBLEM-SOLVING TACTIC**
>
> **4.6** Choose the z-axis to be the line containing $\vec{\omega}$. The magnitude of the angular velocity vector is the angular speed. Hence as a particle begins from rest and speeds up along a circular path (see Figure 4.36), the visual representation of the angular velocity vector $\vec{\omega}(t)$ gets longer, reflecting the increased angular speed of the particle, but the vector maintains the same direction in space.

4.7 THE GEOMETRY AND COORDINATES FOR DESCRIBING CIRCULAR MOTION

In Section 4.5 we learned that circular motion at constant speed is nonetheless accelerated motion. In this section we examine general circular motion, encompassing situations in which the particle is moving in a circle, but not necessarily at constant speed, such as your motion on a merry-go-round as it starts spinning from rest. We will see how to exploit vectors to describe circular motion of all types in an elegant, useful way.

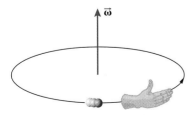

FIGURE 4.35 The circular motion right-hand rule for determining the direction of the angular velocity vector $\vec{\omega}$.

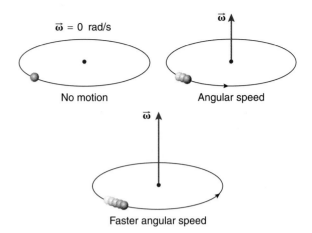

FIGURE 4.36 Increasing angular speed.

For a particle moving in a circle of fixed radius, the choice to make for the origin of an appropriate coordinate system is the center of the circle. You know that the choice of a coordinate system and its origin is arbitrary, at least in principle. However, some coordinate choices make the description of the problem far easier than others; this certainly is the case here in choosing the origin at the center of the circle. Any other choice makes the description of the circular motion of the particle a mathematical quagmire. So we relinquish freedom here, remembering that, all else being equal, simplicity is preferred over complexity.

It is customary to let the plane of the circular motion be the x–y plane of a Cartesian coordinate system; the z-axis of the coordinate system then is *perpendicular* to the plane of the motion and along the line containing the angular velocity vector $\vec{\omega}$. The angular velocity vector $\vec{\omega}(t)$ thus has a single component $\omega_z(t)$:

$$\vec{\omega}(t) = \omega_z(t)\hat{k} \qquad (4.26)$$

> **PROBLEM-SOLVING TACTIC**
>
> **4.7** The direction you choose for $\hat{k}$ determines the sign of ω_z. See Figure 4.37.

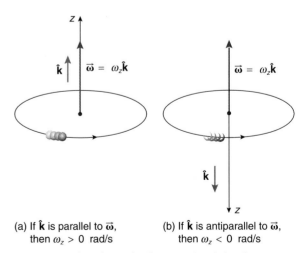

(a) If $\hat{k}$ is parallel to $\vec{\omega}$, then $\omega_z > 0$ rad/s

(b) If $\hat{k}$ is antiparallel to $\vec{\omega}$, then $\omega_z < 0$ rad/s

FIGURE 4.37 The relationship between $\vec{\omega}$ and the choice you make for $\hat{k}$.

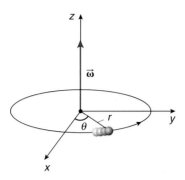

FIGURE 4.38 The angle θ is measured with respect to the x-axis.

The motion of a particle around a circle of fixed radius r is described by specifying an angular coordinate θ with respect to a fixed reference direction, taken to be the x-axis of a Cartesian coordinate system whose origin is at the center of the circle (see Figure 4.38). In this way, r and θ are the polar coordinates of the particle. Since the radius r of the fixed circle is constant, there is only one coordinate that changes with time, $\theta(t)$. Circular motion is *one-variable motion* since the angle θ is the only coordinate, other than the fixed radius of the circle, needed to locate the position of the particle. We will systematically exploit the constancy of r and the one-variable nature of circular motion for the kinematic description of circular motion.

4.8 THE POSITION VECTOR FOR CIRCULAR MOTION

With the x, y, and z coordinate conventions of the previous section, the position vector of the particle is confined to the x–y plane and is a two-dimensional vector, as shown in Figure 4.39:

$$\vec{\mathbf{r}}(t) = x(t)\hat{\mathbf{i}} + y(t)\hat{\mathbf{j}} \qquad (4.27)$$

The fact that the path is circular is not clear from $x(t)$ and $y(t)$. However, using the polar coordinates r and $\theta(t)$ we see that the Cartesian components are

$$x(t) = r \cos \theta(t)$$

and

$$y(t) = r \sin \theta(t)$$

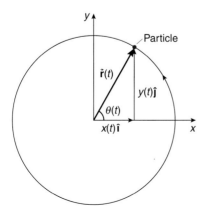

FIGURE 4.39 The position vector is confined to the x–y plane.

Hence the position vector is

$$\vec{\mathbf{r}}(t) = [r \cos \theta(t)]\hat{\mathbf{i}} + [r \sin \theta(t)]\hat{\mathbf{j}} \qquad (4.28)$$

Using Equation 4.28 to find the magnitude of $\vec{\mathbf{r}}(t)$, we get

$$\begin{aligned} |\vec{\mathbf{r}}(t)| &= \{[r \cos \theta(t)]^2 + [r \sin \theta(t)]^2\}^{1/2} \\ &= \{r^2[\cos^2 \theta(t) + \sin^2 \theta(t)]\}^{1/2} \\ &= r \end{aligned}$$

(The last step follows from the trigonometric identity $\cos^2 \theta + \sin^2 \theta = 1$.) Hence the magnitude of the position vector $\vec{\mathbf{r}}(t)$ is the fixed radius r of the circle and does *not* depend on t.

> For circular motion the *magnitude* r of the position vector $\vec{\mathbf{r}}(t)$ does *not* change with time. *Only* the angular coordinate $\theta(t)$ is a function of time.

4.9 THE VELOCITY AND ANGULAR VELOCITY IN CIRCULAR MOTION

We know that the velocity of a particle is the time rate of change (the time derivative) of its position vector. With Equation 4.28 for the position vector, the velocity of the particle is

$$\begin{aligned} \vec{\mathbf{v}}(t) &= \frac{d\vec{\mathbf{r}}(t)}{dt} \\ &= \frac{d}{dt}\left\{[r \cos \theta(t)]\,\hat{\mathbf{i}} + [r \sin \theta(t)]\,\hat{\mathbf{j}}\right\} \end{aligned}$$

Remember that the polar coordinate r is constant for circular motion. Hence the velocity is

$$\begin{aligned} \vec{\mathbf{v}}(t) &= r \frac{d}{dt}[\cos \theta(t)]\,\hat{\mathbf{i}} + r \frac{d}{dt}[\sin \theta(t)]\,\hat{\mathbf{j}} \\ &= r[-\sin \theta(t)]\frac{d\theta(t)}{dt}\,\hat{\mathbf{i}} + r[\cos \theta(t)]\frac{d\theta(t)}{dt}\,\hat{\mathbf{j}} \quad (4.29) \end{aligned}$$

The velocity vector is tangent to the circular trajectory and, because the particle is moving in a circle, $\vec{\mathbf{v}}(t)$ is perpendicular to the position vector $\vec{\mathbf{r}}(t)$ at all times. We can verify this by taking the *scalar product* of $\vec{\mathbf{v}}(t)$ with $\vec{\mathbf{r}}(t)$ using Equations 4.28 and 4.29; after some algebra, we see that the scalar product $\vec{\mathbf{r}}(t) \cdot \vec{\mathbf{v}}(t)$ indeed is zero, independent of t, which means that the two vectors are perpendicular at all times.

The speed of the particle is the magnitude of the velocity vector. The square of the magnitude of the velocity vector is the sum of the squares of its Cartesian components:

$$v^2 = \vec{\mathbf{v}} \cdot \vec{\mathbf{v}} = v_x^2 + v_y^2$$

Using Equation 4.29 for $\vec{\mathbf{v}}(t)$, we have

$$\begin{aligned} [v(t)]^2 &= \left[-r \sin \theta(t)\frac{d\theta(t)}{dt}\right]^2 + \left[r \cos \theta(t)\frac{d\theta(t)}{dt}\right]^2 \\ &= r^2\left[\frac{d\theta(t)}{dt}\right]^2 [\sin^2 \theta(t) + \cos^2 \theta(t)] \\ &= r^2\left[\frac{d\theta(t)}{dt}\right]^2 \end{aligned}$$

The speed $v(t)$ of the particle therefore is

$$v(t) = r \left| \frac{d\theta(t)}{dt} \right| \qquad (4.30)$$

The quantity

$$\left| \frac{d\theta(t)}{dt} \right|$$

is the absolute value of the time rate of change of the angular co-ordinate θ of the particle; this is the angular speed $\omega(t)$, the magnitude of the angular velocity vector. Hence Equation 4.30 can be written

$$v(t) = r\omega(t) \qquad (4.31)$$

Take the magnitude of Equation 4.26:

$$\omega(t) = |\omega_z(t)| $$

Compare this with Equations 4.30 and 4.31; we make the identification

$$\omega_z(t) \equiv \frac{d\theta(t)}{dt} \qquad (4.32)$$

with $\omega_z(t)$ specified in rad/s.

In Figure 4.40, note that the velocity vector $\vec{v}(t)$ is, at any instant t, perpendicular to *both* the angular velocity $\vec{\omega}(t)$ and the position vector $\vec{r}(t)$. This is reminiscent of the vector product of two vectors. Consider the vector product

$$\vec{\omega}(t) \times \vec{r}(t)$$

Since the angle between $\vec{\omega}(t)$ and $\vec{r}(t)$ always is 90°, the magnitude of their vector product is

$$
\begin{aligned}
|\vec{\omega}(t) \times \vec{r}(t)| &= \omega(t)r \sin 90° \\
&= \omega(t)r \\
&= |\omega_z(t)|r \qquad (4.33)
\end{aligned}
$$

Remember that r, the magnitude of $\vec{r}(t)$, is *constant* and has no time dependence for circular motion. Compare Equation 4.33 with Equation 4.31; we conclude that the magnitude of the vector product $\vec{\omega}(t) \times \vec{r}(t)$ is just the speed $v(t)$. Use the vector product right-hand rule and notice that the vector product $\vec{\omega}(t) \times \vec{r}(t)$ is in the *same* direction as $\vec{v}(t)$ at any instant. Since $\vec{\omega}(t) \times \vec{r}(t)$ has the same magnitude as $\vec{v}(t)$ and points in the same direction as $\vec{v}(t)$ at any instant, $\vec{\omega}(t) \times \vec{r}(t)$ must be the same as the vector $\vec{v}(t)$.

For circular motion, the three vectors $\vec{v}(t)$, $\vec{\omega}(t)$, and $\vec{r}(t)$ *always* are related to each other via the vector product

$$\boxed{\vec{v}(t) = \vec{\omega}(t) \times \vec{r}(t)} \qquad (4.34)$$

Since the three vectors are mutually perpendicular, taking magnitudes in Equation 4.34 gives

$$\boxed{v(t) = r\omega(t)} \qquad (4.35)$$

Equations 4.34 and 4.35 are appropriate for

1. circular motion at constant speed v or, equivalently, constant angular velocity $\vec{\omega}$—that is, uniform circular motion; and

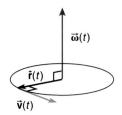

FIGURE 4.40 The velocity always is perpendicular to both the angular velocity and the position vector.

2. circular motion with a changing speed $v(t)$ and changing angular velocity vector $\vec{\omega}(t)$, called **nonuniform circular motion**.

4.10 UNIFORM CIRCULAR MOTION REVISITED

Whether it be a centrifuge whipping around at 50 000 rev/min, the rotation of the Earth once every 24 h, or the planet Venus orbiting the Sun* every 224.7 d, the circular motion of a particle at a constant speed has manifestations everywhere. Although the particle is moving along the circle at a constant speed v, the velocity vector $\vec{v}(t)$ is *not* constant, because the direction of motion of the particle is changing with time.

Since the speed v is constant and the radius of the circle r is fixed, Equation 4.30 implies that

$$\frac{d\theta}{dt}$$

is constant. Since

$$\omega_z = \frac{d\theta}{dt}$$

the z-component (the only component) of the angular velocity vector is constant.

> For uniform circular motion, the angular velocity vector is a constant vector.

We find the acceleration by taking the time derivative of the velocity vector:

$$\vec{a}(t) = \frac{d\vec{v}(t)}{dt}$$

Using Equation 4.29 for $\vec{v}(t)$, we get

$$\vec{a}(t) = \frac{d}{dt}\left\{ \left[-r\frac{d\theta}{dt}\sin\theta(t)\right]\hat{i} + \left[r\frac{d\theta}{dt}\cos\theta(t)\right]\hat{j} \right\}$$

Since the particle is moving in a circle, the radius r is constant. As we know from Section 4.5, if the particle is undergoing *uniform* circular motion,

$$\frac{d\theta}{dt} = \omega_z$$

---

*The shape of the orbit of a planet about the Sun actually is an ellipse, as we will see in Chapter 6. As a first approximation, the planetary orbits typically are considered to be circular, since their eccentricities are small (except for Pluto, Mercury, and Mars); the orbit of Venus is the most nearly circular of all the planets.

also is a constant. It is only $\theta(t)$ itself that varies with time. With both r and

$$\frac{d\theta}{dt}$$

constants, the differentiation affects only the trigonometric functions:

$$\vec{a}(t) = -\left\{r\frac{d\theta}{dt}[\cos\theta(t)]\frac{d\theta}{dt}\right\}\hat{i} + \left\{r\frac{d\theta}{dt}[-\sin\theta(t)]\frac{d\theta}{dt}\right\}\hat{j}$$

$$= -\left(\frac{d\theta}{dt}\right)^2\left\{[r\cos\theta(t)]\hat{i} + [r\sin\theta(t)]\hat{j}\right\}$$

Notice that the term in curly braces is just the position vector $\vec{r}(t)$ of the particle (Equation 4.28). Hence

$$\vec{a}(t) = -\left(\frac{d\theta}{dt}\right)^2\vec{r}(t)$$

Since

$$\frac{d\theta}{dt} = \omega_z$$

the constant angular velocity component, the expression for the acceleration becomes

$$\vec{a}(t) = -\omega_z^2\vec{r}(t) \qquad (4.36)$$

> The minus sign in Equation 4.36 indicates that the acceleration is antiparallel to the position vector $\vec{r}(t)$, just as our plumb bob accelerometer indicated. That is, $\vec{a}(t)$ is directed toward the center of the circle. This is the centripetal acceleration.

> The *magnitude* of the centripetal acceleration is constant and is written in terms of the (constant) angular velocity component ω_z as
>
> $$\boxed{a_c = \omega_z^2 r} \qquad (4.37)$$
>
> or, in terms of the speed v (by substituting for ω_z using $v = |\omega_z|r$ from Equation 4.31),
>
> $$a_c = \left(\frac{v}{r}\right)^2 r$$
>
> $$= \frac{v^2}{r} \qquad (4.38)$$

For a given circle size, the greater the value of the speed v of the particle, the greater the magnitude of the centripetal acceleration. For a given speed v, the centripetal acceleration is larger in magnitude for smaller circles (smaller r).

An equivalent expression for the centripetal acceleration can be found using Equation 4.34 and differentiating the vector product:

$$\vec{a}(t) = \frac{d\vec{v}(t)}{dt}$$

$$= \frac{d}{dt}[\vec{\omega} \times \vec{r}(t)]$$

Remember that the angular velocity $\vec{\omega}$ is a constant vector for *uniform* circular motion. Thus

$$\vec{a}(t) = \vec{\omega} \times \frac{d\vec{r}(t)}{dt}$$

$$= \vec{\omega} \times \vec{v}(t)$$

Use the vector product right-hand rule and the geometry of Figure 4.40 to note that the direction of $\vec{\omega} \times \vec{v}(t)$ is *opposite* to that of $\vec{r}(t)$. Hence the acceleration is directed toward the center of the circle—the centripetal acceleration:

$$\vec{a}_c(t) = \vec{\omega} \times \vec{v}(t) \qquad (4.39)$$

Because we will be using these results for uniform circular motion on many occasions, it is best to keep them fresh.

> Summarizing, the following are equivalent expressions for the centripetal acceleration in circular motion:
>
> $$\vec{a}_c(t) = \vec{\omega} \times \vec{v}(t) = -\omega_z^2\vec{r}(t) \qquad (4.39, 4.36)$$
>
> and for its constant magnitude:
>
> $$a_c = \frac{v^2}{r} = \omega_z^2 r \qquad (4.38, 4.37)$$

PROBLEM-SOLVING TACTIC

4.8 The seemingly different expressions for the centripetal acceleration are all variations of the same equation.

QUESTION 6

Explain carefully why the centripetal acceleration for uniform circular motion has a constant magnitude but is not a constant vector.

STRATEGIC EXAMPLE 4.9

A particle is moving at the constant speed of 15.0 m/s in a circle of radius 4.00 m.

a. Sketch the situation and indicate the direction of the angular velocity vector.
b. Find the angular speed.
c. Find the angular velocity vector $\vec{\omega}$ and its z-component ω_z.
d. Find the magnitude of the centripetal acceleration.

Solution

a. The circular motion right-hand rule means that the direction of the angular velocity vector is as shown in Figure 4.41. (Note that this is one of two possible sketches.)

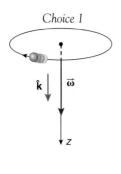

FIGURE 4.41

b. There are two ways to find the angular speed:

Method 1	Method 2
According to Equations 4.30 and 4.32, $$v = r\lvert\omega_z\rvert$$ Since $v = 15.0$ m/s and $r = 4.00$ m, we find that $$\lvert\omega_z\rvert = \frac{15.0 \text{ m/s}}{4.00 \text{ m}}$$ $$= 3.75 \text{ rad/s}$$ since ω_z is expressed in rad/s.	The angular speed indicates how many radians the particle sweeps out during one second at constant speed. Since the speed of the particle is constant and equal to 15.0 m/s, the particle traverses 15.0 m of circumference during one second. So the angular speed in radians per second is $$\omega = \frac{\text{arc length in one second}}{\text{radius}}$$ $$= \frac{15.0 \text{ m/s}}{4.00 \text{ m}}$$ $$= 3.75 \text{ rad/s}$$

c. To write the angular velocity vector, choose the direction of the z-axis to be along the line of the angular velocity (see Figure 4.42).

Choice 1

Choice 2

Take $\hat{\mathbf{k}}$ parallel to $\vec{\omega}$ so the z-component of the angular velocity is *positive*.

Take $\hat{\mathbf{k}}$ antiparallel to $\vec{\omega}$ so the z-component of the angular velocity is *negative*.

FIGURE 4.42

$\omega_z = 3.75$ rad/s $\vec{\omega} = (3.75 \text{ rad/s})\hat{\mathbf{k}}$	$\omega_z = -3.75$ rad/s $\vec{\omega} = (-3.75 \text{ rad/s})\hat{\mathbf{k}}$

d. The magnitude of the centripetal acceleration can be found in two ways:

Method 1	Method 2
$$a_c = \frac{v^2}{r}$$ $$= \frac{(15.0 \text{ m/s})^2}{4.00 \text{ m}}$$ $$= 56.3 \text{ m/s}^2$$	$$a_c = \omega_z^2 r$$ $$= (\pm 3.75 \text{ rad/s})^2(4.00 \text{ m})$$ The $\pm$ sign choice depends on which direction you chose for $\hat{\mathbf{k}}$ in part (c). For either choice: $$a_c = 56.3 \text{ m/s}^2$$

4.11 NONUNIFORM CIRCULAR MOTION AND THE ANGULAR ACCELERATION

It takes some time for a merry-go-round to get up to speed or to come to a stop. Likewise, when a biologist turns on a centrifuge, it may take several minutes for it to get up to speed or to coast to a stop. Similar things happen when you start a washing machine and dryer or monitor the engine rpm on your car as you speed up or slow down in traffic. There are many examples of nonuniform circular motion in the world around us.

If the speed of a particle moving in a fixed circle is not constant, then the velocity vector of the particle is changing in both direction *and* magnitude. For motion confined to a fixed plane in a circle of fixed radius, the *angular velocity vector* changes its magnitude but not its direction.

Here we can exploit our knowledge of vectors to great advantage. The acceleration of the particle is the time rate of change of the velocity vector:

$$\vec{a}(t) = \frac{d\vec{v}(t)}{dt}$$

But from Equation 4.34,

$$\vec{v}(t) = \vec{\omega}(t) \times \vec{r}(t)$$

and so the acceleration is

$$\vec{a}(t) = \frac{d}{dt}[\vec{\omega}(t) \times \vec{r}(t)]$$

Both the position vector and the angular velocity vector change with time in nonuniform circular motion. Thus there are two terms for the time derivative of the vector product. Remember to keep the terms in the vector product in the appropriate order when differentiating:

$$\vec{a}(t) = \frac{d\vec{\omega}(t)}{dt} \times \vec{r}(t) + \vec{\omega}(t) \times \frac{d\vec{r}(t)}{dt} \qquad (4.40)$$

Since

$$\frac{d\vec{r}(t)}{dt} = \vec{v}(t)$$

the second term on the right-hand side of the acceleration equation is

$$\vec{a}_c = \vec{\omega}(t) \times \vec{v}(t) \qquad (4.41)$$

which is the *centripetal acceleration* we encountered in the last section. Since the speed of the particle in the circle is changing with time, the magnitude of the centripetal acceleration is not constant, unlike the case for uniform circular motion. Since $\vec{\omega}(t)$ and $\vec{v}(t)$ are perpendicular, the magnitude of their vector product is

$$a_c(t) = \omega(t)v(t) \qquad (4.42)$$

In view of Equation 4.35, this becomes

$$a_c = \frac{v(t)^2}{r} = |\omega_z(t)|^2 r \qquad (4.43)$$

The centripetal acceleration has the same form as for uniform circular motion, but here its magnitude changes with time.

The first term on the right-hand side of Equation 4.40 is new. Let's see what is involved with this term. The time rate of change of the angular velocity vector is defined to be the **angular acceleration** vector, designated $\vec{\alpha}(t)$. That is,

$$\vec{\alpha}(t) \equiv \frac{d\vec{\omega}(t)}{dt} \qquad (4.44)$$

Since $\vec{\omega}(t)$ is measured in rad/s, the angular acceleration is measured in rad/s^2.

The particle either is speeding up in its circular motion (thus increasing the length of the angular velocity arrow) or slowing down (decreasing the length of the angular velocity vector). The line along which the angular velocity vector lies does not change its orientation, since the motion is in a fixed plane. The angular acceleration vector $\vec{\alpha}(t)$ points in the direction of the *change* in the angular velocity vector $\vec{\omega}(t)$. Thus the angular acceleration vector $\vec{\alpha}(t)$ is directed either *parallel* to $\vec{\omega}(t)$ when the particle is speeding up along the circle, or *antiparallel* to $\vec{\omega}(t)$ when the particle is slowing down along the circle (see Figure 4.43).

Since the angular velocity vector is either parallel or antiparallel to $\hat{k}$ (according to the standard coordinate choice), the angular acceleration vector likewise has only a z-component:

$$\vec{\alpha}(t) = \alpha_z(t)\hat{k}$$

The acceleration term $\vec{\alpha}(t) \times \vec{r}(t)$ in Equation 4.40 has a direction determined using the vector product right-hand rule. Since $\vec{\alpha}(t)$ is perpendicular to the plane of the circle and $\vec{r}(t)$ is in the plane, the direction of the vector product $\vec{\alpha}(t) \times \vec{r}(t)$ always is *tangent* to the circular path and parallel to $\vec{v}(t)$ if the particle is speeding up (see Figure 4.44), or antiparallel to $\vec{v}(t)$ if the particle is slowing down in its circular motion (Figure 4.45).

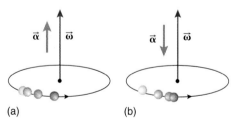

(a)   (b)

FIGURE 4.43 (a) Particle speeding up along the circle means $\vec{\alpha}$ is parallel to $\vec{\omega}$. (b) Particle slowing down along the circle means $\vec{\alpha}$ is antiparallel to $\vec{\omega}$.

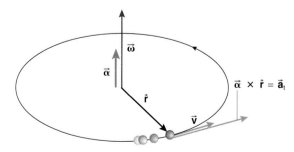

FIGURE 4.44 For a particle speeding up along the circle, $\vec{\alpha}(t) \times \vec{r}(t)$ is parallel to $\vec{v}(t)$.

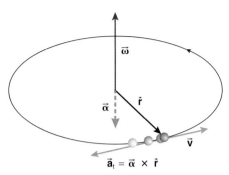

FIGURE 4.45 For a particle slowing down along the circle, $\vec{\alpha}(t) \times \vec{r}(t)$ is antiparallel to $\vec{v}(t)$.

Since $\vec{a}(t) \times \vec{r}(t)$ is tangent to the circle, it is known as the **tangential acceleration** $\vec{a}_t$:

$$\vec{a}_t(t) \equiv \vec{\alpha}(t) \times \vec{r}(t) \qquad (4.45)$$

The tangential acceleration always is perpendicular to the position vector $\vec{r}(t)$, and so the magnitude of the vector product of the two vectors on the right-hand side of Equation 4.45 can be evaluated easily. Thus the magnitude of the tangential acceleration is

$$a_t(t) = \alpha(t) r \sin 90°$$
$$= \alpha(t) r$$
$$= |\alpha_z(t)| r \qquad (4.46)$$

Remember that the magnitude of the position vector is constant, since the particle is moving in a circle of fixed size.

Equation 4.40 for the acceleration is rewritten as

$$\vec{a}(t) = \vec{\alpha}(t) \times \vec{r}(t) + \vec{\omega}(t) \times \vec{v}(t) \qquad (4.47)$$

The total acceleration $\vec{a}(t)$ is the vector sum of the two, mutually perpendicular, tangential and centripetal accelerations:

$$\vec{a}(t) = \vec{a}_t(t) + \vec{a}_c(t) \qquad (4.48)$$

Since both the tangential and centripetal accelerations are in the plane of the circular motion, the total acceleration also is in this plane.

The direction of the total acceleration always is toward the inside of the circle, since the centripetal acceleration *always* is toward the center of the circle (see Figures 4.46 and 4.47).

QUESTION 7

If a particle travels around a circle, can the total acceleration ever be zero at any point along the trajectory? Explain your answer.

EXAMPLE 4.10

A particle moving in a circle of radius 2.00 m has a total acceleration of magnitude $a = 20.0$ m/s² at some instant directed as shown in Figure 4.48.

a. Find the magnitude of the centripetal acceleration at this instant.
b. Find the speed of the particle at this instant.
c. Find the magnitude of the angular velocity of the particle at this instant.

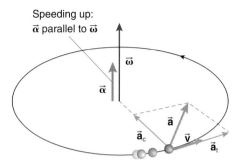

FIGURE 4.46 The bevy of vectors for a particle speeding up in circular motion.

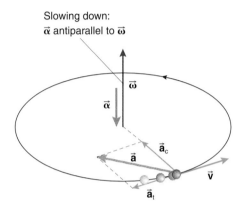

FIGURE 4.47 The bevy of vectors for a particle slowing down in circular motion.

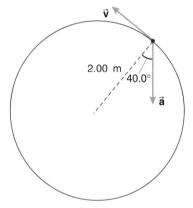

FIGURE 4.48

d. Find the magnitude of the tangential acceleration of the particle at this instant.
e. Find the magnitude of the angular acceleration of the particle at this instant.
f. Sketch the directions of the angular velocity vector and the angular acceleration vector at this instant.

Solution

a. The total acceleration is the vector sum of the mutually perpendicular centripetal and tangential accelerations, as indicated in Figure 4.49. The magnitude of the centripetal

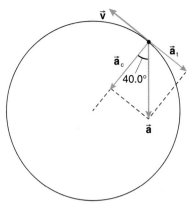

FIGURE 4.49

acceleration is the absolute value of the component of the total acceleration in the radial direction. Hence

$$a_c = a \cos 40.0°$$
$$= (20.0 \ \text{m/s}^2) \cos 40.0°$$
$$= 15.3 \ \text{m/s}^2$$

b. The magnitude of the centripetal acceleration is given by Equation 4.43:

$$a_c = \frac{v^2}{r}$$

so

$$v^2 = ra_c$$

Substitute for r and a_c [found in part (a)]:

$$v^2 = (2.00 \ \text{m})(15.3 \ \text{m/s}^2)$$
$$v = 5.53 \ \text{m/s}$$

c. There are two ways to find the angular speed:

Method 1

The magnitude of the centripetal acceleration is, from Equation 4.37 (realizing that $|\omega_z| = \omega$),

$$a_c = \omega^2 r$$

Substitute for a_c and r:

$$15.3 \ \text{m/s}^2 = \omega^2(2.00 \ \text{m})$$

Solving for ω, you find

$$\omega = 2.77 \ \text{rad/s}$$

Method 2

Use Equation 4.35 realizing that $|\omega_z| = \omega$:

$$v = \omega r$$

Substitute for v and r:

$$5.53 \ \text{m/s} = \omega(2.00 \ \text{m})$$

Solving for ω, you find

$$\omega = 2.77 \ \text{rad/s}$$

d. From Figure 4.49, the tangential acceleration component of the total acceleration is of magnitude

$$a_t = a \sin 40.0°$$
$$= (20.0 \ \text{m/s}^2) \sin 40.0°$$
$$= 12.9 \ \text{m/s}^2$$

e. Equation 4.46 indicates that

$$a_t = \alpha r$$

Substitute for a_t and r:

$$12.9 \ \text{m/s}^2 = \alpha(2.00 \ \text{m})$$

Solve for α:

$$\alpha = 6.45 \ \text{rad/s}^2$$

f. The direction of $\vec{\omega}$ is found using the circular motion right-hand rule. The tangential acceleration here is directed *opposite* to the velocity vector, so that the particle is slowing down in its rotational motion. Thus the angular acceleration vector $\vec{\alpha}$ is directed *opposite* to the direction of $\vec{\omega}$. The orientation of the two vectors is shown in Figure 4.50.

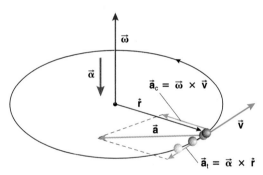

FIGURE 4.50

Notice that the vector product right-hand rule verifies the relative directions of the vectors in the vector relationships

$$\vec{a}_c = \vec{\omega} \times \vec{v}$$

and

$$\vec{a}_t = \vec{\alpha} \times \vec{r}$$

4.12 NONUNIFORM CIRCULAR MOTION WITH A CONSTANT ANGULAR ACCELERATION

Just as in rectilinear motion, many circular motions are modeled with a constant angular acceleration to a first approximation. The kinematics of such circular motion are similar to rectilinear motion, since they *both* are motions with one variable coordinate—in the one case θ, in the other case x. This means we can use the rectilinear motion equations with a constant acceleration as a guide to writing the circular motion equations for a constant angular acceleration:

For rectilinear motion with constant acceleration component a_x:

$$a_x = \frac{dv_x}{dt}$$

we found that

$$v_x(t) = v_{x0} + a_x t$$

Likewise

$$x(t) = x_0 + v_{x0}t + \frac{1}{2}a_x t^2$$

For circular motion with constant angular acceleration component α_z:

$$\alpha_z = \frac{d\omega_z}{dt}$$

Hence, analogously to rectilinear motion,

$$\boxed{\omega_z(t) = \omega_{z0} + \alpha_z t} \tag{4.49}$$

and

$$\boxed{\theta(t) = \theta_0 + \omega_{z0}t + \frac{1}{2}\alpha_z t^2} \tag{4.50}$$

4.9 Notice the formal similarity between the equations for rectilinear motion with a constant acceleration and nonuniform circular motion with a constant angular acceleration. The similarity arises because both motions are functions of a single coordinate. We replace the linear acceleration, velocity, and position with the corresponding angular counterparts:

- the constant acceleration component a_x with the constant *angular* acceleration component α_z;
- the initial velocity component v_{x0} with the initial *angular* velocity component ω_{z0}; and
- the initial position component x_0 with the initial *angular* position coordinate θ_0. For convenience, the angle θ_0, the value of θ when $t = 0$ s, conventionally is set equal to 0 rad.

QUESTION 8

A particle accelerates from rest beginning when $t = 0$ s around a circular track at a constant angular acceleration. (a) What is the ratio

$$\frac{a_c}{a_t}$$

when $t = 0$ s? (b) What does this imply about the initial direction of the *total* acceleration vector of the particle relative to the radial direction? (c) As the particle picks up speed around the circle, how does the direction of the *total* acceleration vector qualitatively change relative to the radial direction?

EXAMPLE 4.11

A centrifuge begins at rest and achieves an angular speed of 5.00×10^4 rev/min after 2.00 min (see Figure 4.51).

a. Sketch the direction of the angular velocity vector.
b. Specify an appropriate direction for the z-axis.
c. Find the z-component (the only component) of the angular acceleration, assuming it is constant.
d. Find the magnitude of the angular acceleration.

Solution

a. The direction of the angular velocity vector is determined from the circular motion right-hand rule (see Figure 4.51); if the spin is drawn in the opposite sense, the angular velocity vector $\vec{\omega}$ also is in the opposite direction.

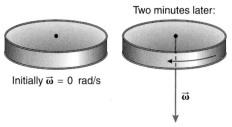

Two minutes later:

Initially $\vec{\omega} = 0$ rad/s

$\vec{\omega}$

FIGURE 4.51

b. The z-axis is taken along the line containing $\vec{\omega}$ (see Figure 4.52).

Choice 1

Choice 2

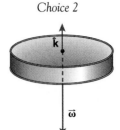

Choose $\hat{k}$ to be in the direction of $\vec{\omega}$.

Choose $\hat{k}$ to be in the opposite direction of $\vec{\omega}$.

FIGURE 4.52

c. The initial angular velocity is zero when $t = 0$ s, so

$$\omega_{z0} = 0 \text{ rad/s} \qquad\qquad \omega_{z0} = 0 \text{ rad/s}$$

The final angular velocity component is

$$\omega_z = \frac{(5.00 \times 10^4 \text{ rev/min})(2\pi \text{ rad/rev})}{60 \text{ s/min}} \qquad \omega_z = \frac{(-5.00 \times 10^4 \text{ rev/min})(2\pi \text{ rad/rev})}{60 \text{ s/min}}$$

$$= 5.24 \times 10^3 \text{ rad/s} \qquad\qquad = -5.24 \times 10^3 \text{ rad/s}$$

Use Equation 4.49,

$$\omega_z = \omega_{z0} + \alpha_z t$$

and make the appropriate substitutions:

$$5.24 \times 10^3 \text{ rad/s} = 0 \text{ rad/s} + \alpha_z(120 \text{ s}) \qquad -5.24 \times 10^3 \text{ rad/s} = 0 \text{ rad/s} + \alpha_z(120 \text{ s})$$

Solving for α_z, you find

$$\alpha_z = 43.7 \text{ rad/s}^2 \qquad\qquad \alpha_z = -43.7 \text{ rad/s}^2$$

d. The magnitude of the angular acceleration is

$$|\alpha_z| = 43.7 \text{ rad/s}^2$$

STRATEGIC EXAMPLE 4.12

Your parents' antique stereo turntable of radius 13.7 cm, initially spinning at 33.0 revolutions per minute, is shut off. The turntable coasts to a stop after 120 s. Assume a constant angular acceleration. Calculate the angular acceleration of the turntable and the number of revolutions through which it spins as it stops.

Solution

Stereo turntables rotate clockwise when viewed from the top. Use the circular motion right-hand rule to determine the direction of the angular velocity vector (see Figure 4.53). You find it points down.

FIGURE 4.53

Choose a direction for $\hat{k}$ (see Figure 4.54):

| *Choice 1* | *Choice 2* |

Choose $\hat{k}$ parallel to $\vec{\omega}$.

Choose $\hat{k}$ antiparallel to $\vec{\omega}$.

FIGURE 4.54

Since the turntable is slowing down, the magnitude of the angular velocity vector is decreasing, and so the angular velocity vector arrow is getting smaller during the interval. The angular acceleration vector points in the direction of the *change* in the angular velocity vector. Thus the angular acceleration vector is pointed opposite to the direction of the angular velocity vector itself, as shown in Figure 4.55.

The angular acceleration $\vec{\alpha}$ points toward decreasing values of z, and so you anticipate that α_z will be negative.

The angular acceleration $\vec{\alpha}$ points toward increasing values of z, and so you anticipate that α_z will be positive.

FIGURE 4.55

Convert the initial angular speed of 33.0 rev/min to radians per second (one revolution of the turntable is equivalent to 2π rad):

$$33.0 \text{ rev/min} = \frac{(33.0 \text{ rev/min})(2\pi \text{ rad/rev})}{60 \text{ s/min}}$$

$$= 3.46 \text{ rad/s}$$

Thus the initial angular velocity component is

| $\omega_{z0} = 3.46$ rad/s | $\omega_{z0} = -3.46$ rad/s |

After 120 s, the final angular velocity is zero. Hence Equation 4.49 for the angular velocity component is, when $t = 120$ s,

| 0 rad/s = 3.46 rad/s + α_z(120 s) | 0 rad/s = −3.46 rad/s + α_z(120 s) |

Solve for the angular acceleration component α_z:

| $\alpha_z = -0.0288$ rad/s^2 | $\alpha_z = 0.0288$ rad/s^2 |

The angular acceleration vector is

$$\vec{\alpha} = \alpha_z \hat{k}$$

$\vec{\alpha} = (-0.0288 \text{ rad/s}^2)\hat{k}$	$\vec{\alpha} = (0.0288 \text{ rad/s}^2)\hat{k}$
and so $\vec{\alpha}$ is in the same direction as $-\hat{k}$.	and so $\vec{\alpha}$ is in the same direction as $\hat{k}$.

In both cases the angular acceleration is *antiparallel* to $\vec{\omega}$, indicating the turntable is slowing down in its circular motion.

The total angle through which the turntable turns as it slows down is found from Equation 4.50 for the angular position:

$$\theta = \theta_0 + \omega_{z0}t + \frac{1}{2}\alpha_z t^2$$

Take the initial angular position of a point on the turntable as zero; hence $\theta_0 = 0$ rad. The angular position θ of the same point on the turntable when $t = 120$ s then is found by substituting for ω_{z0}, α_z, and t:

$\theta = 0$ rad $+ (3.46 \text{ rad/s})(120 \text{ s})$	$\theta = 0$ rad $- (3.46 \text{ rad/s})(120 \text{ s})$
$+ \dfrac{1}{2}(-0.0288 \text{ rad/s}^2)(120 \text{ s})^2$	$+ \dfrac{1}{2}(0.0288 \text{ rad/s}^2)(120 \text{ s})^2$
$= 208$ rad	$= -208$ rad

We indicate x- and y- axes in Figure 4.56, consistent with the choice for $\hat{k}$.

(a)

(b)

FIGURE 4.56

In polar coordinates, the angle θ is positive when measured counterclockwise from the x-axis, when viewed looking down the +z-axis, parallel to $-\hat{k}$.

The positive result for θ indicates that the polar angle θ is measured counterclockwise from the x-axis when sighting along $-\hat{k}$.	The negative result for θ indicates that the polar angle θ is measured clockwise from the x-axis when sighting along $-\hat{k}$.

The absolute value of the angle through which the point on the turntable turns is 208 rad. The number of revolutions is obtained by dividing the angle by 2π rad/rev. The total number of revolutions of the turntable is*

$$\text{number of revolutions} = \frac{\left|+208 \text{ rad}\right|}{2\pi \text{ rad/rev}}$$

$$= 33.1 \text{ rev}$$

---

*Note that the answer is actually exactly 33.0 rev: a record spinning at 33.0 rev/min slowing down to rest in 2 min (120 s) with constant acceleration has average angular speed 16.5 rev/min; since this average angular speed is during 2 min, the record travels 16.5 rev/min × 2 min = 33.0 rev. The discrepancy in the answer is due to rounding.

EXAMPLE 4.13

a. Find the magnitude of the tangential acceleration of a point on the rim of the spinning turntable of Example 4.12 as it slows down. Indicate the direction of the tangential acceleration vector on a sketch that also includes the direction of the velocity vector of the same point on the rim.

b. Determine the magnitude of the centripetal acceleration of a point on the rim the instant the turntable begins to slow down. Sketch the direction of the centripetal acceleration for the same point on the rim as the point considered in part (a).

Solution

a. The magnitude of the tangential acceleration is given by Equation 4.46:

$$a_t = \alpha r = |\alpha_z| r$$

We found the angular acceleration in Example 4.12 to be of magnitude

$$\alpha = |\alpha_z| = 0.0288 \text{ rad/s}^2$$
$$= 2.88 \times 10^{-2} \text{ rad/s}^2$$

The radius of the turntable is 13.7 cm = 0.137 m. Make these substitutions into Equation 4.46:

$$a_t = (2.88 \times 10^{-2} \text{ rad/s}^2)(0.137 \text{ m})$$
$$= 3.95 \times 10^{-3} \text{ m/s}^2$$

There are two ways we can determine the direction of the tangential acceleration for a point on the rim.

Method 1

Since the turntable is slowing down, the tangential acceleration must be directed opposite to the velocity vector of the point on the rim. See Figure 4.57.

FIGURE 4.57

Method 2

Since the turntable is slowing down, $\vec{\alpha}$ must be directed opposite to $\vec{\omega}$. See Figure 4.57. The position vector $\hat{r}$ and velocity of a point on the rim are indicated. Since $\vec{a}_t = \vec{\alpha} \times \hat{r}$ the vector product right-hand rule indicates $\vec{\alpha} \times \hat{r}$ is opposite to $\vec{v}$, as shown in Figure 4.57.

b. The magnitude of the centripetal acceleration of a point on the rim is found using Equation 4.43:

$$a_c = |\omega_z(t)|^2 r$$

The angular speed at the instant the turntable begins to slow down is $|\omega_{z0}|$, which was found in Example 4.12 to be

$$|\omega_{z0}| = 3.46 \text{ rad/s}$$

We know the radius of the turntable is 0.137 m. Make these numerical substitutions:

$$a_c = (3.46 \text{ rad/s})^2 (0.137 \text{ m})$$
$$= 1.64 \text{ m/s}^2$$

The orientation of the centripetal and tangential acceleration vectors for a point on the rim of the turntable is sketched in Figure 4.58. The total acceleration is the vector sum of the tangential and centripetal accelerations.

FIGURE 4.58

STRATEGIC EXAMPLE 4.14

A sports car moves around a horizontal circular curve of radius 200 m, as shown in Figure 4.59. While traversing the curve, the car changes speed at a constant rate from 36.0 km/h (= 10.0 m/s) to 90.0 km/h (= 25.0 m/s) during 20.0 s.

FIGURE 4.59

a. Find the magnitudes of the centripetal acceleration, the tangential acceleration, and the total acceleration of the car (i) the instant when the car has the speed 36.0 km/h, and (ii) the instant the car has the speed 90.0 km/h.

b. Find the angle ϕ between the direction of the total acceleration and the radial direction when the speed of the car is 36.0 km/h, and again when the speed is 90.0 km/h.

c. Find the angle θ through which the car moves during the 20.0 s interval and the length of highway traversed along the curve.

Solution

a. The total acceleration has two mutually perpendicular components at each instant: (1) the centripetal acceleration directed toward the center of the circle; and (2) the tangential acceleration directed tangent to the circle. Both are illustrated in Figure 4.60 for the two given instants. Equation 4.43 indicates the centripetal acceleration has a magnitude equal to

$$a_c = \frac{v^2}{r}$$

where v is the instantaneous speed of the car and r is the radius of the circular turn. At the instant the speed of the car is 36.0 km/h, or 10.0 m/s in SI units, the magnitude of the centripetal acceleration is

$$a_c = \frac{(10.0 \text{ m/s})^2}{200 \text{ m}}$$
$$= 0.500 \text{ m/s}^2$$

At the instant when the car reaches the speed of 90.0 km/h (= 25.0 m/s), the magnitude of the centripetal acceleration also is found by substituting the speed and radius into Equation 4.43:

$$a_c = \frac{(25.0 \text{ m/s})^2}{200 \text{ m}}$$
$$= 3.13 \text{ m/s}^2$$

We see that as the speed of the car increases around the curve, the centripetal acceleration *increases in magnitude*; the direction of the centripetal acceleration vector also changes continuously, but always points toward the center of the circle. Now we find the magnitude of the tangential acceleration.

Method 1

The magnitude of the tangential acceleration is *constant* as the car traverses the curve, since the speed of the car changes at a constant rate from 36.0 km/h to 90.0 km/h. Since the magnitude of the (instantaneous) tangential acceleration is constant, it is equal to the magnitude of the *average* tangential acceleration during the 20.0 s interval. The latter is found by dividing the change in the tangential speed of the car by the time elapsed. Hence the magnitude of the tangential acceleration here is

$$a_t = \frac{25.0 \text{ m/s} - 10.0 \text{ m/s}}{20.0 \text{ s}}$$
$$= 0.750 \text{ m/s}^2$$

and is the same at *every* instant during the 20.0 s time interval. The direction of the tangential acceleration changes continuously as the car traverses the curve, since, as its name implies, this acceleration is directed tangent to the circle.

Method 2

The tangential acceleration also may be found using the vector approach used in describing circular motion, although this way is more involved. Since the car speeds up at a constant rate around the curve, this is a problem of *nonuniform* circular motion at *constant* angular acceleration. From the circular motion right-hand rule, the angular velocity vector of the car as it traverses the curve is directed perpendicularly *into* the page. Let this also be the direction of the +z-axis, as shown in Figure 4.61. When the car has a speed of 36.0 km/h (= 10.0 m/s), the magnitude of the angular velocity vector is found from Equation 4.35:

$$v = r\omega$$

Thus

$$\omega = \frac{v}{r} = \frac{10.0 \text{ m/s}}{200 \text{ m}} = 0.0500 \text{ rad/s}$$

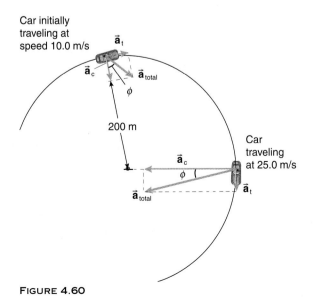

Car initially traveling at speed 10.0 m/s

200 m

Car traveling at 25.0 m/s

FIGURE 4.60

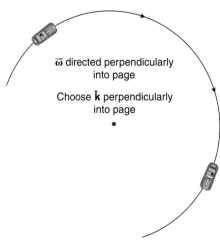

$\vec{\omega}$ directed perpendicularly into page

Choose $\hat{\mathbf{k}}$ perpendicularly into page

FIGURE 4.61

The angular velocity vector $\vec{\omega}$ of the sports car initially (when $t = 0$ s) is

$$\vec{\omega} = \omega_{z0}\hat{k}$$

so its z-component is

$$\omega_{z0} = 0.0500 \ \text{rad/s}$$

When $t = 20.0$ s, the speed of the car is 90.0 km/h ($= 25.0$ m/s) and the angular velocity component has increased to

$$\omega_z = \frac{v}{r} = \frac{25.0 \ \text{m/s}}{200 \ \text{m}} = 0.125 \ \text{rad/s}$$

The angular acceleration component α_z is found from the nonuniform circular motion equation (for constant angular accelerations), Equation 4.49:

$$\omega_z = \omega_{z0} + \alpha_z t$$

Solve for the angular acceleration component by substituting the values for the angular velocity components and the time:

$$0.125 \ \text{rad/s} = 0.0500 \ \text{rad/s} + \alpha_z(20.0 \ \text{s})$$

So

$$\alpha_z = 0.003 \ 75 \ \text{rad/s}^2$$

Since α_z is positive, the angular acceleration vector $\vec{\alpha}$ is directed along the $+z$ axis. Since $\vec{\alpha}$ is in the same direction as $\vec{\omega}$, the car is speeding up in its circular motion. The magnitude of the tangential acceleration is found from Equation 4.46:

$$a_t = \alpha r$$

Substitute for α and r:

$$a_t = (0.003 \ 75 \ \text{rad/s}^2)(200 \ \text{m})$$
$$= 0.750 \ \text{m/s}^2$$

b. The centripetal and tangential accelerations are shown schematically in Figure 4.60 at the beginning and end of the 20.0 s time interval. The total acceleration of the car at each instant is the vector sum of the tangential and centripetal accelerations at each instant. The angle ϕ that the total acceleration vector initially makes with the radial direction is (from Figure 4.60)

$$\tan \phi = \frac{a_t}{a_c}$$

$$= \frac{0.750 \ \text{m/s}^2}{0.500 \ \text{m/s}^2}$$

$$= 1.50$$

or

$$\phi = 56.3°$$

Notice that since the magnitude of the tangential acceleration initially is greater than that of the centripetal acceleration, we would expect the total acceleration to be inclined by more than 45° to the radial direction.

At the final instant, when the speed of car is 90.0 km/h, the angle ϕ that the total acceleration vector makes with the radial direction is (from Figure 4.60)

$$\tan \phi = \frac{a_t}{a_c}$$

$$= \frac{0.750 \ \text{m/s}^2}{3.13 \ \text{m/s}^2}$$

$$= 0.240$$

or

$$\phi = 13.5°$$

The angle now is less than 45°, which indicates that the magnitude of the centripetal acceleration is greater than that of the tangential acceleration at this instant.

c. The angle through which the car moves during the 20.0 s interval is found using the nonuniform circular motion equation for constant angular accelerations, Equation 4.50:

$$\theta = \theta_0 + \omega_{z0}t + \frac{1}{2}\alpha_z t^2$$

where θ_0 is the initial angular position of the car when $t = 0$ s; we set this initial angular position equal to zero: $\theta_0 = 0$ rad. Substituting

$$\omega_{z0} = 0.0500 \ \text{rad/s}$$
$$\alpha_z = 0.003 \ 75 \ \text{rad/s}^2$$

the angular position of the car when $t = 20.0$ s is found to be

$$\theta = 0 \ \text{rad} + (0.050 \ \text{rad/s})(20.0 \ \text{s})$$
$$+ \frac{(0.003 \ 75 \ \text{rad/s}^2)(20.0 \ \text{s})^2}{2}$$
$$= 1.75 \ \text{rad}$$

which is equivalent to 100°. With the definition of an angle in radians,

$$\theta = \frac{\text{arc length}}{\text{radius}} = \frac{s}{r}$$

the arc length s of the curve traversed by the car during the 20.0 s interval is

$$s = r\theta$$
$$= (200 \ \text{m})(1.75 \ \text{rad})$$
$$= 350 \ \text{m}$$

CHAPTER SUMMARY

The general form for the position vector of a moving particle is

$$\vec{r}(t) = x(t)\hat{i} + y(t)\hat{j} + z(t)\hat{k} \qquad (4.16)$$

As in rectilinear motion, the *average velocity* of a moving particle during a time interval Δt is the ratio of the change in the position vector $\Delta \vec{r}$ during the interval to the interval itself:

$$\vec{v}_{ave} \equiv \frac{\Delta \vec{r}}{\Delta t} \qquad (4.3)$$

Likewise, the *velocity* of the particle is the limit of the average velocity as the time interval Δt approaches zero; this is equivalent to the time derivative of the position vector:

$$\vec{v}(t) = \lim_{\Delta t \to 0\ s} \frac{\Delta \vec{r}}{\Delta t} = \frac{d\vec{r}(t)}{dt} \qquad (4.4)$$

The *average acceleration* of the particle during an interval Δt is the ratio of the change in the velocity during the time interval to the interval itself:

$$\vec{a}_{ave} \equiv \frac{\Delta \vec{v}}{\Delta t} \qquad (4.7)$$

The *acceleration* is the limit of the average acceleration as the time interval Δt approaches zero; this is equivalent to the time derivative of the velocity, or the second time derivative of the position vector:

$$\vec{a}(t) = \lim_{\Delta t \to 0\ s} \frac{\Delta \vec{v}}{\Delta t} = \frac{d\vec{v}(t)}{dt} = \frac{d^2\vec{r}(t)}{dt^2} \qquad (4.8)$$

These equations have the same form as those for rectilinear motion, but the vectors now have two or three components depending on whether the particle is moving in two or three dimensions.

Motions along perpendicular directions are independent of each other and can be analyzed separately.

For two-dimensional motion with a constant acceleration, choose the y-axis to be the line containing the acceleration vector. The acceleration then has only a single, constant component a_y, and the x- and y-components of the velocity and position vectors are

$$v_x(t) = v_{x0} \quad \text{and is constant} \qquad (4.10)$$
$$x(t) = x_0 + v_{x0}t \qquad (4.11)$$
$$v_y(t) = v_{y0} + a_y t \qquad (4.12)$$
$$y(t) = y_0 + v_{y0}t + \frac{1}{2}a_y t^2 \qquad (4.13)$$

The signs of the various components depend on the directions you choose for the unit vectors $\hat{i}$ and $\hat{j}$.

If the motion of a particle P is referred to two coordinate systems with origins at 0 and 0′, one of which (0′) is moving, then the velocity $\vec{v}_{P0'}$ of the particle with respect to 0′, and the velocity $\vec{v}_{P0}$ of the particle with respect to 0, are related by the equation

$$\vec{v}_{P0} = \vec{v}_{P0'} + \vec{v}_{0'0} \qquad (4.20)$$

where $\vec{v}_{0'0}$ is the velocity of 0′ with respect to 0.

The *acceleration* of a particle is the *same* with respect to any reference frames moving at *constant velocity* with respect to each other.

The circular motion of a particle in the x–y plane is described by the position vector

$$\vec{r}(t) = [r\cos\theta(t)]\hat{i} + [r\sin\theta(t)]\hat{j} \qquad (4.28)$$

where r, the constant radius of the circle, and $\theta(t)$ are the polar coordinates of the particle. Circular motion is accelerated motion, since the velocity vector changes with time.

The direction of the *angular velocity vector* $\vec{\omega}$ always is chosen according to the *circular motion right-hand rule*. The direction conventionally chosen for the +z-axis is either parallel *or* antiparallel to $\vec{\omega}$; the specific choice is yours to make. Whichever you choose, the angular velocity then has a single component:

$$\vec{\omega} = \omega_z\hat{k}$$

The angular velocity is specified in rad/s. The time rate of change of the polar coordinate $\theta(t)$ is the z-component, the only component, of the angular velocity vector:

$$\omega_z(t) = \frac{d\theta(t)}{dt} \qquad (4.32)$$

The position, velocity, and angular velocity vectors in circular motion are related by the vector product:

$$\vec{v}(t) = \vec{\omega}(t) \times \vec{r}(t) \qquad (4.34)$$

All three vectors are mutually perpendicular at all times.

Uniform circular motion is circular motion at constant speed v. The angular velocity vector $\vec{\omega}$ is constant for such motion. The acceleration is directed toward the center of the circle and is called the *centripetal acceleration*:

$$\vec{a}_c(t) = \vec{\omega} \times \vec{v}(t) = -\omega_z^2\vec{r}(t) \qquad (4.39, 4.36)$$

The magnitude of the centripetal acceleration is constant for uniform circular motion:

$$a_c = \frac{v^2}{r} = \omega_z^2 r \qquad (4.38, 4.37)$$

Nonuniform circular motion is circular motion with a changing speed. The acceleration then has two perpendicular components:

$$\vec{a}(t) = \vec{a}_c(t) + \vec{a}_t(t) \qquad (4.48)$$

The centripetal acceleration is directed toward the center of the circle and is

$$\vec{a}_c(t) \equiv \vec{\omega}(t) \times \vec{v}(t) \qquad (4.41)$$

but is not of constant magnitude, since the speed of the particle is a function of time. The magnitude of the centripetal acceleration is

$$a_c(t) = \frac{v(t)^2}{r} = \omega_z(t)^2 r \qquad (4.43)$$

The *tangential acceleration* is

$$\vec{a}_t(t) \equiv \vec{\alpha}(t) \times \vec{r}(t) \qquad (4.45)$$

where $\vec{\alpha}(t)$ is the *angular acceleration*—that is, the time rate of change of the angular velocity:

$$\vec{\alpha}(t) \equiv \frac{d\vec{\omega}(t)}{dt} \tag{4.44}$$

The tangential acceleration is either parallel to $\vec{v}$ if the particle is speeding up, or antiparallel to $\vec{v}$ if the particle is slowing down.

Both $\vec{\omega}$ and $\vec{\alpha}$ always are perpendicular to the plane of the circular motion. Since we choose the z-axis to be along this line, $\vec{\omega}$ and $\vec{\alpha}$ both have only a z-component. The angular acceleration is specified in rad/s^2. The magnitude of the tangential acceleration is

$$a_t(t) = \alpha(t)r = |\alpha_z(t)|r \tag{4.46}$$

The angular acceleration $\vec{\alpha}(t)$ is

1. parallel to $\vec{\omega}(t)$, when a particle is speeding up in its circular motion; or

2. antiparallel to $\vec{\omega}(t)$, when a particle is slowing down in its circular motion.

For circular motion at *constant angular acceleration* $\vec{\alpha} = \alpha_z\hat{k}$:

a. the z-component (and only component) of the angular velocity at any instant is

$$\omega_z(t) = \omega_{z0} + \alpha_z t \tag{4.49}$$

where ω_{z0} is the z-component of the angular velocity when $t = 0$ s;

b. the angular coordinate of the particle $\theta(t)$ at any instant is

$$\theta(t) = \theta_0 + \omega_{z0}t + \frac{1}{2}\alpha_z t^2 \tag{4.50}$$

where θ_0 is the angular position of the particle when $t = 0$ s; θ_0 usually is taken to be zero (i.e., $\theta_0 = 0$ rad).

The signs of α_z, ω_{z0}, and $\omega_z(t)$ depend on the direction you choose for $\hat{k}$.

SUMMARY OF PROBLEM-SOLVING TACTICS

4.1 (page 118) The specific coordinate choice is yours to make, but it needs to be explicitly indicated in a sketch and used consistently throughout a given problem.

4.2 (page 121) If the acceleration vector is constant, choose one of the coordinate axes to be the line containing the constant acceleration or a line parallel to it; we choose the line containing the constant acceleration to be the y-axis.

4.3 (page 121) The sign of the acceleration component a_y depends on the direction you choose for $\hat{j}$ along the line containing the acceleration vector.

4.4 (page 121) The origin may be taken at any convenient location along the line chosen for the y-axis.

4.5 (page 122) Notice that the equations for $x(t)$ and $v_x(t)$ have the same form as those for $y(t)$ and $v_y(t)$ but with an acceleration component a_x equal to zero.

4.6 (page 136) Choose the z-axis to be the line containing $\vec{\omega}$.

4.7 (page 136) The direction you choose for $\hat{k}$ determines the sign of ω_z.

4.8 (page 139) The seemingly different expressions for the centripetal acceleration are all variations of the same equation.

4.9 (page 144) Notice the formal similarity between the equations for rectilinear motion with a constant acceleration and nonuniform circular motion with a constant angular acceleration.

QUESTIONS

1. (page 120); 2. (page 122); 3. (page 122); 4. (page 123); 5. (page 134); 6. (page 139); 7. (page 142); 8. (page 144)

9. Why is the instantaneous speed the magnitude of the instantaneous velocity, whereas the average speed is *not* the magnitude of the average velocity?

10. Devise an example to show that the average velocity of a particle during a time interval Δt is *not* necessarily equal to

$$\frac{\vec{v}_f + \vec{v}_i}{2}$$

11. (a) Sketch the trajectory of a projectile that has zero speed at the top of its trajectory. (b) Sketch the trajectory of a projectile that has nonzero speed at the top of its trajectory.

12. Describe a situation where the average speed of a particle during a time interval Δt is *not* zero, but its average velocity *is* zero during the same time interval.

13. Is it possible to have the average speed equal to zero but the average velocity nonzero? If so, explain how. If not, explain why not.

14. Describe a physical situation in which (a) the velocity vector of a particle is perpendicular to its position vector; (b) the acceleration vector of the particle is perpendicular to its velocity vector; (c) the acceleration is perpendicular to the position vector; (d) the velocity is in the same direction as the acceleration; (e) the velocity and acceleration are in opposite directions.

15. Three fleas (*Ctenocephalides canis*) have a jumping contest and discover to their amazement that they each jump to the same elevation off the floor (see Figure Q.15) but land in different positions horizontally. Explain your reasoning in answering the following questions about the contest: (a) Which flea had the greatest initial speed? (b) Which flea was in the air for the longest time? (c) Which flea had the greatest magnitude of the acceleration while in flight? (d) Which flea had the horizontal velocity component of greatest absolute value? (e) Which flea had the vertical velocity component of greatest absolute value?

16. At what point along the trajectory of a projectile is its speed a minimum?

FIGURE Q.15

FIGURE Q.21

17. At what point along the trajectory of a projectile is the velocity perpendicular to the acceleration?

18. Can the velocity ever be parallel to the acceleration for some type of projectile motion? Explain your answer.

19. What factors influence the range of an athlete competing in the long jump versus one competing in the high jump? See Figure Q.19.

FIGURE Q.19

20. A vertically falling raindrop strikes loose dirt on a slope. Make a diagram that shows why dirt splashed down the hill travels farther along the slope than dirt splashed up the slope if the initial velocity vectors of the dirt make equal angles with the vertical direction. This process contributes to the slow erosion of mountains.

21. Two identical twins of equal strength are using identical canoes to race across a river of constant width; the current speed is less than the paddling speed of the twins. One twin keeps his canoe aimed directly across the river and is swept downstream as the crossing is made. The other twin ferries across by aiming the canoe so that the actual path followed by the canoe is directly across the river perpendicular to its banks. Who wins the race or is it a tie? Explain your reasoning. See Figure Q.21.

22. Sailors use strips of sailcloth attached to a mast or a stay to deduce wind directions. For a sailboat underway at sea, does the direction of the strip give a true indication of the direction to which the wind is blowing? Explain. See Figure Q.22.

23. While out for a stroll in the rain, why do you tilt the umbrella in the direction you are moving to keep drier?

24. You make one complete orbit riding on a merry-go-round during a time interval Δt. What are your average velocity and average acceleration?

FIGURE Q.22

25. Make a schematic diagram of the Earth with its north and south geographic poles. Indicate on the diagram the direction of the angular velocity vector of a particle at rest on the surface of the Earth.

26. Is the centripetal acceleration of a particle located on the surface of the Earth directed toward the center of the Earth? If so, explain why; if not, explain why not.

27. Where on the surface of the spinning Earth does the centripetal acceleration of a particle at rest on its surface have the greatest magnitude? The least magnitude?

28. Does the angular speed of a particle at rest on the surface of the spinning Earth depend on either the latitude or longitude of the particle? Explain.

29. At what two geographic locations on the surface of the spinning Earth is the centripetal acceleration of a particle at rest on the surface equal to zero?

30. Is the magnitude of the centripetal acceleration of a particle on the surface of the spinning Earth a function of either the latitude or longitude of the particle?

31. As a resident of the spinning Earth, you are carried along with it, orbiting the line connecting the north and south poles. At what two geographic locations on the Earth is your motion spin rather than orbital motion?

32. Do the sweep second, minute, and hour hands of a clock turn at constant angular speed?

33. The angular velocity vector of a particle in circular motion is indicated in Figure Q.33. Indicate the direction the particle is moving around the circle.

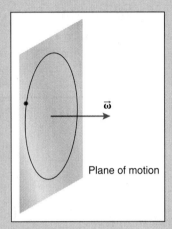

Plane of motion

FIGURE Q.33

34. Is the angular velocity vector representing the daily rotational motion of the Earth in the same direction as the angular velocity vector associated with the orbital motion of the Earth around the Sun?

35. What is the ratio of the angular speed of the spinning Earth to the angular speed associated with the orbital motion of the Earth about the Sun?

36. Is the angular speed of a cassette tape the same throughout its playing time? Explain.

37. For nonuniform circular motion, can the *total* acceleration ever be instantaneously directed as shown in Figure Q.37? Explain your answer.

FIGURE Q.37

38. For circular motion, if the tangential acceleration is zero, the total acceleration is just the centripetal acceleration and the motion is *uniform* circular motion. Is it possible in circular motion to have the centripetal acceleration equal to zero but not the tangential acceleration? Explain.

39. Is it possible to have the angular velocity $\vec{\omega}$ equal to zero at some instant but a nonzero angular acceleration $\vec{\alpha}$? If so, how? If not, why not?

40. Is it possible to have the angular acceleration $\vec{\alpha}$ equal to zero at some instant but not the angular velocity $\vec{\omega}$? If so, how? If not, why not?

41. How would you *qualitatively describe* what is happening physically to an object in circular motion with a *time-dependent* angular acceleration $\vec{\alpha}(t)$? What does such an angular acceleration imply about the appearance of a graph of the angular velocity component versus time?

42. Does a particle at rest on the surface of the Earth experience a *tangential* acceleration because of the rotation of the Earth? If so, in what direction? If not, explain why not.

43. Is the angular speed of the film reel on a motion picture projector constant throughout the interval of the movie? Explain.

PROBLEMS

Neglect any effects caused by air resistance in these problems.

Section 4.1   The Position, Velocity, and Acceleration Vectors in Two Dimensions

1. A gypsy moth caterpillar (*Porthetria dispar*) inches along a crooked branch to a tasty oak leaf, wriggling 15 cm horizontally and then 30 cm along a section of the branch inclined at 30° to the horizontal as shown in Figure P.1. (a) Write the initial and final position vectors of the caterpillar in SI units using the crook in the branch as the origin for a set of horizontal and vertical coordinate axes. (b) If the caterpillar traverses the distance during 1.0 min,

- what is its average speed?
- what is its average velocity?
- what is the magnitude of its average velocity?

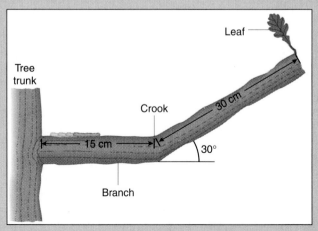

FIGURE P.1

2. A zoologist notices a monkey located at position $(20.0 \text{ m})\hat{\jmath}$ intent on finding a banana. The monkey wanders to a tree situated at $(30.0 \text{ m})\hat{\imath} + (40.0 \text{ m})\hat{\jmath}$ and then proceeds to a nearby vertical vine whose lower end is located at $(40.0 \text{ m})\hat{\imath} + (30.0 \text{ m})\hat{\jmath}$. The x–y plane is the horizontal plane. (a) Sketch the situation on a Cartesian coordinate grid. (b) What is the change in the position vector of the monkey in moving from the tree to the vine? (c) The monkey climbs the vine to a height of 10.0 m and finally spies a banana at the origin. What is the distance between the monkey on the vine and the banana? (d) The monkey leaps from the vine and lands in a swampy pool at $(25.0 \text{ m})\hat{\imath} + (25.0 \text{ m})\hat{\jmath}$. What is the distance between the tree and where the monkey landed in the pool?

3. Superman flies from his well-known telephone booth located at $(5.0 \text{ km})\hat{\imath} + (8.0 \text{ km})\hat{\jmath}$ to the offices of *The Daily Planet* at $(-4.0 \text{ km})\hat{\imath} + (2.0 \text{ km})\hat{\jmath}$ in 2.0 min. (a) Draw the two position vectors on a labeled coordinate grid. (b) What is the change in Superman's position vector in SI units? (c) What is his average velocity?

•4. As a race car driver, you complete one lap at an average speed $v_{1 \text{ ave}}$ and the second lap at the average speed $v_{2 \text{ ave}}$. (a) Show that the average speed v_{ave} over the two laps is

$$v_{\text{ave}} = \frac{2v_{1 \text{ ave}} \, v_{2 \text{ ave}}}{v_{1 \text{ ave}} + v_{2 \text{ ave}}}$$

The average speed v for the two laps is the *harmonic mean* of the average speeds $v_{1 \text{ ave}}$ and $v_{2 \text{ ave}}$ for the two individual laps. The harmonic mean n of any two numbers n_1 and n_2 is defined as the reciprocal of the arithmetic mean of the reciprocals of n_1 and n_2. That is

$$\frac{1}{n} = \frac{\frac{1}{n_1} + \frac{1}{n_2}}{2}$$

This can be rewritten as

$$\frac{2}{n} = \frac{1}{n_1} + \frac{1}{n_2}$$

or

$$n = \frac{2n_1 n_2}{n_1 + n_2}$$

(b) Find the average speed over two laps if the average speed on the first lap is 120 km/h while that on the second lap is 80 km/h. This problem was inspired by an article by P. Glaister, "Natural occurrences of the harmonic mean," *Physics Education*, 27, #4, page 181 (July 1992).

‡ 5. Consider two galaxies traveling radially outward with different speeds as seen from another galaxy at the origin as shown in Figure P.5. The two position vectors

$$\vec{r}_1 = v_1 t \hat{r}_1$$

and

$$\vec{r}_2 = v_2 t \hat{r}_2$$

locate the moving galaxies with respect to the one at the origin 0, where $\hat{r}_1$ and $\hat{r}_2$ are unit vectors in the radially outward directions and v_1 and v_2 are the respective speeds of the galaxies. Note that the speed of each galaxy is proportional to its distance from the origin. The position vectors make an angle θ with respect to each other. Let $\vec{r}$ be the vector from galaxy A to galaxy B. (a) Show that the magnitude of $\vec{r}$ is

$$r = (v_1^2 + v_2^2 - 2v_1 v_2 \cos \theta)^{1/2} t$$

This represents the distance between galaxies A and B. (b) Show that the speed with which galaxies A and B recede from each other (the magnitude of the velocity vector)

$$\vec{v} \equiv \frac{d\vec{r}}{dt}$$

is

$$v = (v_1^2 + v_2^2 - 2v_1 v_2 \cos \theta)^{1/2}$$
$$= \frac{r}{t}$$

Thus observers on galaxy A see the other galaxy B receding from them with a speed that is proportional to the separation of the galaxies. This shows that there is no privileged galaxy at the center of the universe. *Every* galaxy is a center and other galaxies move away from it at speeds proportional to their distances, much like raisins in a loaf of rising raisin bread.

This exercise is the essence of the discovery made by Edwin Hubble in 1929 known as the *expansion of the universe*. A subtlety to the expansion, however, is not apparent in this exercise. The universe is not expanding *into* an empty space that already exists out there, since this would imply an edge to the universe with nothing beyond. The *space* between the galaxies is expanding, carrying the galaxies along with it, much like stretching a rubber sheet with buttons (galaxies) sewn onto it.

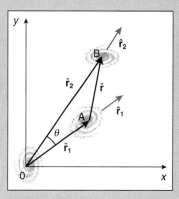

FIGURE P.5

Section 4.2 Two-Dimensional Motion with a Constant Acceleration

•6. An erupting volcano violently projects a boulder from a vent at an angle of 40° to the horizontal direction. The rock lands 6.0 km away as indicated in Figure P.6. (a) Introduce an appropriate coordinate system. (b) Tailor Equations 4.10–4.13 to the given information. (c) Determine the initial speed of the rock and the time of flight.

FIGURE P.6

•7. Slim Pickens hurls a baseball at a speed of 30.0 m/s at an angle of 60.0° to the horizontal as indicated in Figure P.7. The ball leaves Slim's hand at a point 2.50 m above the ground and 20.0 m from the base of a barrier 30.0 m high. (a) Indicate in a sketch an appropriate Cartesian coordinate system for analyzing the problem. (b) Tailor Equations 4.10–4.13 to the given information. (c) Does the ball hit or clear the vertical barrier? (d) Where does the ball first strike? (e) How long is the ball in the air before its first impact?

FIGURE P.7

•8. A rock is tossed off a cliff as indicated in Figure P.8. (a) Introduce a coordinate system to analyze the problem and tailor Equations 4.10–4.13 to the given information. (b) Find the time of flight. (c) Determine the impact speed of the rock.

FIGURE P.8

•9. A soccer ball is kicked off a cliff at a speed of 20.0 m/s as indicated in Figure P.9. (a) In a sketch, indicate a choice for a coordinate system to analyze the problem. (b) Tailor the general Equations 4.10–4.13 to incorporate the specific infor-

mation given about the ball. (c) Determine the time of flight. (d) Determine where the ball hits the ground the first time.

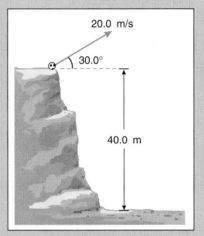

FIGURE P.9

•10. A young basketball player is attempting to make a shot. The ball leaves the hands of the player at an angle of 60° to the horizontal at an elevation of 2.0 m above the floor. The skillful player makes the shot with the ball traveling precisely through the center of the hoop as indicated in Figure P.10. To loud cheers, calculate the speed at which the ball left the hands of the player.

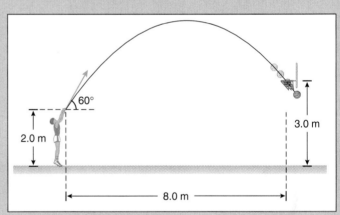

FIGURE P.10

•11. A projectile is fired across a level surface. The initial velocity $\vec{v}_0$ makes an angle θ with the horizontal (see Figure P.11). (a) Show that the maximum height h_{max} of the projectile is

$$h_{max} = \frac{v_0^2 \sin^2 \theta}{2g}$$

(b) The horizontal range of the projectile was found in Example 4.3 to be

$$R_{horiz} = \frac{v_0^2 \sin 2\theta}{2g}$$

What angle θ makes the range of the projectile a maximum? What is the maximum range?

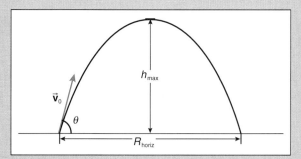

FIGURE P.11

•**12.** A beam of electrons, each with a speed of 5.00×10^6 m/s, is shot horizontally into a region where each electron experiences a vertically upward acceleration of magnitude 5.00×10^{13} m/s². The magnitude of this acceleration is so large that the magnitude of the local acceleration due to gravity may be ignored by comparison. Each electron experiences the large acceleration only while it traverses the region delineated in Figure P.12. (a) Set up an appropriate coordinate system for analyzing the motion of each electron within the region, and tailor Equations 4.10–4.13 to the given information. (b) For how long a time is each electron in the region? (c) What is the vertical deflection of each electron from its initial horizontal path? Such deflections of electron beams are the principle underlying CRT (cathode ray tube) displays such as computer monitors, TVs, and oscilloscopes.

FIGURE P.12

•**13.** A famous soccer player, Hedley Foote, kicks the ball at a speed 15.0 m/s when $t = 0$ s; the initial velocity vector of the ball makes an angle of 60.0° with the horizontal direction. Out of curiosity, calculate the instants when the velocity vector of the ball makes an angle of 45.0° with the horizontal direction.

•**14.** The Chickens are playing the Turkeys in football and the score is tied at $\sqrt{43}$ apiece. Don't ask how. A placekicker is sent out for the Chickens with instructions from Coach Gung-ho to kick a field goal from 45.0 m out. The top of the crossbar on the goalpost is 3.05 m above the level playing field. The moment of truth arrives and the ball leaves the ground at an angle of 30.0° to the horizontal. (a) What is the minimum speed that the ball must have to make the field goal? Assume appropriate aiming. (b) How long does it take the ball to reach the crossbar?

•**15.** A pitcher throws a fastball. The ball is released horizontally from the pitcher's mound when the hand of the pitcher is at an elevation of 1.7 m above the level field. The ball is caught by the catcher 19.5 m away at an elevation of 0.90 m above the ground. See Figure P.15. Calculate the initial speed of the ball as it left the pitcher's hand.

FIGURE P.15

•**16.** A 10-year-old launches a water balloon horizontally with an initial speed of 10.0 m/s from near the top of a tall tree 15.0 m above the ground. (a) What is the time of flight? (b) How far from the base of the tree does the water balloon hit? (c) What is the speed of the water balloon at the instant just before impact?

•**17.** A golfer launches a divot which hits the side of a golf cart as indicated in Figure P.17. (a) Indicate an appropriate coordinate system to analyze the flight of the divot. (b) Determine the initial speed of the divot. (c) Determine the time of flight of the divot.

FIGURE P.17

•**18.** A dirt biker races up a 15.0° incline at a constant speed of 120 km/h. The end of the ramp is 3.00 m off the ground as indicated in Figure P.18. A 4.00 m high obstacle is located 20.0 m from the base of the ramp. (a) Indicate an appropriate coordinate system to attack the problem. (b) Will the daredevil clear the obstacle? If so, where will she land?

FIGURE P.18

•**19.** The Apollo 14 lunar astronaut Alan Shepard was the first golfer on the Moon. He screwed the head of a 6-iron to a geological sampling tool and, using one hand, whacked a golf ball across the lunar surface. The flight time of the ball was 40 s. There is no atmosphere on the Moon, so that the ball did not hook or slice, a golfer's paradise. The local acceleration due to gravity on the surface of the Moon is 0.166 ($\approx 1/6$) that on the surface of the Earth. What was the greatest possible horizontal range of the lunar golf ball?

•20. Two identical pea shooters fire peas with an initial speed of 10.0 m/s. The pea shooters are located a distance d, one above the other as shown in Figure P.20. The lower pea emerges from the shooter at an angle of 20.0° to the horizontal, while the upper pea emerges horizontally. The peas both hit at the same point. (a) Express the horizontal range of the upper pea in terms of the height d. (b) What is the horizontal range of the lower pea? (c) Determine d.

FIGURE P.20

•21. An Englishman named Hew Kennedy has constructed a full-scale reproduction of a medieval siege machine known as a *trebuchet*. A trebuchet originally was used to hurl large masses, such as putrid dead horses, over the walls of medieval castles. However, Kennedy uses his reconstruction to hurl pianos, dead sows, small cars, even flaming gasoline-filled toilets across a pasture on his estate near Shropshire, a feat described by a local paper as "Those Magnificent Men and Their Flaming Latrines." The record hurl for an upright piano is about 140 m; grand pianos average only about 90 m.* (a) Assuming an upright piano is launched at an altitude of 15 m and at an angle of 45° to the horizontal, what is the initial speed of the piano as it leaves the trebuchet? (b) The object to be launched is accelerated for about 1.5 s. Calculate the magnitude of the average acceleration to reach the speed calculated in part (a).

•22. A projectile is to be fired at angle θ to the horizontal. Place a Cartesian origin at the point of departure with the y-axis vertically up and the x-axis horizontal as shown in Figure P.22. The projectile must reach a peak height y_{peak}. Show that the initial speed v_0 of the projectile is

$$v_0 = \left(\frac{2g y_{\text{peak}}}{\sin^2 \theta} \right)^{1/2}$$

*Glynn Mapes, "A Scud it's not, but the trebuchet hurls a mean piano," *The Wall Street Journal*, 30 July 1991 (Volume CCXVIII, #21), pages A1 and A5.

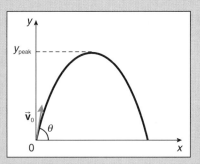

FIGURE P.22

(b) If the peak height of the projectile is to be 10.0 m and $\theta = 60.0°$, show that $v_0 = 16.2$ m/s.

•23. A projectile is fired with an initial speed of 16.2 m/s at 60.0° to the horizontal, as in Figure P.23. Use the coordinate system indicated. (a) Show that the peak height of the projectile is 10.0 m. (b) Show that if you pick any height y between 0 m and 10.0 m, say 5.0 m, there are two times later than 0 s when the projectile achieves this height. (c) Show that there is one time later than 0 s and one time before 0 s when the y-coordinate is less than 0 m, say $y = -5.0$ m. (d) Show that if you attempt to find the times the projectile is at a y-coordinate greater than the peak height of 10.0 m, say 15.0 m, the times are complex numbers. Interpret this result.

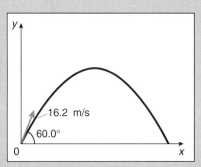

FIGURE P.23

•24. A projectile is launched as shown in Figure P.24 with initial velocity components v_{x0} and v_{y0}. Show that the x-coordinate of the peak of the trajectory, x_{peak}, is

$$x_{\text{peak}} = \frac{v_{x0} \, v_{y0}}{g}$$

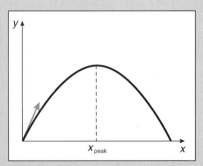

FIGURE P.24

25. A projectile is launched at speed v_0 at an angle θ (with the horizontal) from the bottom of a hill of constant slope β as shown in Figure P.25. Show that the range of the projectile up the slope is

$$\frac{2v_0^2 \cos\theta \, \sin(\theta - \beta)}{g \cos^2 \beta}$$

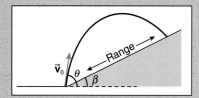

FIGURE P.25

26. The 18th green lies 200 m from a tee as indicated in Figure P.26. Golf pro Sandy Trappe clubs the ball heroically, and it leaves the tee making an angle of 40.0° with the horizontal. Regrettably the ball falls prey to a water hazard 180 m from the tee. (a) What was the speed of the errant ball as it left the tee? (b) If the ball left the tee at the same speed, but at an angle of 45.0° to the horizontal, could the ball have landed on the green? (c) Does the 45.0° launch angle ensure that the ball will travel the maximum horizontal distance? Explain your answer.

FIGURE P.26

27. A flea (*Ctenophalides canis*) jumps from level ground with an initial speed v_0 toward a nearby dog, but fails to reach the dog in one bound. At what angle θ should the flea jump so that its horizontal range is equal to the maximum height of the jump (see Figure P.27)?

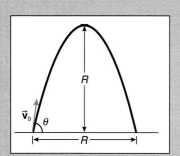

FIGURE P.27

28. Homer Ball, a designated hitter for the Physics Majors, connects with a baseball. The ball is caught 90.0 m from home plate 3.00 s later. The ball is hit and caught at an elevation of 1.00 m above the ground. (a) Find the angle that the ball makes with the horizontal direction as it leaves the bat.

(b) Find the initial speed of the ball. (c) Find the maximum height of the ball above the playing field.

29. A naturalist observes the South African clawed frog (*Xenopus laevis*) leap vertically to a height h. (a) With what speed did the frog leave the ground? Express your answer in terms of h and g. (b) If the frog used the same speed to leap horizontally for maximum range, what distance could it cover? (c) To what height does the frog ascend in making the leap for maximum horizontal range?

30. A football is kicked as pictured in Figure P.30. The initial velocity of the ball makes an angle θ with the horizontal direction. The angle ϕ locates the point of maximum height. Show that

$$\tan\phi = \frac{\tan\theta}{2}$$

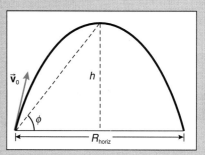

FIGURE P.30

31. A projectile is launched at speed v_0 toward a target located a distance R_{horiz} away over level terrain as indicated in Figure P.31. (a) Show that there are *two* possible launch angles, hence there is a high trajectory and a low trajectory, *except* when the launch angle is 45° for maximum horizontal range, in which case the two launch angles are identical to each other. The two launch angles are symmetrical about the 45° angle. (b) If $R_{\text{horiz}} = 100$ m, and $v_0 = 40.0$ m/s, find the two possible angles for launch. (c) Which launch angle is a quarterback on a football team likely to use?

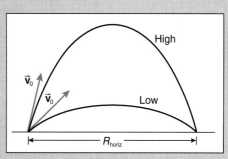

FIGURE P.31

32. (a) A projectile is launched over level terrain as indicated in Figure P.32. Let t_f be the time of flight of the projectile. (a) Show that the maximum height h_{max} to which the projectile rises is

$$h_{\text{max}} = \frac{gt_f^2}{8}$$

Thus if the time of flight of the projectile is known, the

maximum height can be calculated *without* knowing either the initial speed or the initial angle of the launch. (b) This has interesting implications in, say, baseball (if we ignore the effects of air resistance). Suppose a baseball is hit and caught at the same height above the ground, say 1.00 m. If the baseball is in the air for 4.00 s, what is the maximum height of the baseball above the field on its trajectory? Try this out on your friends at the next game you attend.

FIGURE P.32

‡33. One of the classic demonstrations in physics is known as the monkey–blowgun problem. A blowgun fires a small pellet at an object (a toy monkey or a tin can) suspended from the ceiling of a lab. See Figure P.33. The instant the pellet leaves the blowgun, the toy drops from its perch and falls vertically. Show that the pellet invariably hits the toy as long as the vertical distance of the fall is sufficiently large.

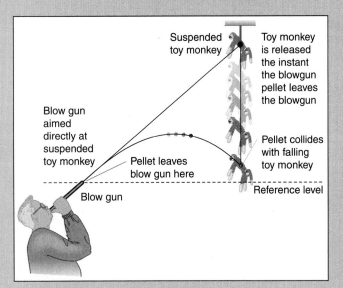

FIGURE P.33

‡34. An artillery mortar accidentally is shot approximately straight up, its initial direction being within 1.0° of the vertical direction. The initial speed of the shell is 400 m/s. Talk about a blunder! The gun crew stands still calculating for 1.00 s and then decides to make tracks in a straight line before the shell lands. (a) Assuming a nonexplosive impact of the shell, how far must they scatter horizontally from the launch point to ensure their safety? (b) With what average speed must they vamoose to get out of potential danger?

‡35. A projectile is launched as indicated in Figure P.35. (a) Show that the y- and x-coordinates of the projectile are related

by the equation

$$y = (\tan \theta)x - \frac{1}{2} \frac{g}{v_0^2 \cos^2 \theta} x^2$$

Since $y(x)$ is quadratic in x, the path of the projectile is a parabola. (b) In analytic geometry, the equation of a parabola passing through the origin, whose directrix is the line $y = c$ and with a focus at coordinates (x_f, y_f), is

$$2(c - y_f)y = (2x_f - x)x$$

Compare this equation with the result of part (a) to show that

$$x_f = \frac{v_0^2}{2g} \sin 2\theta$$

The mathematical definition of a parabola states that any point on the parabola is equidistant from both the directrix and the focus. Since the origin lies on the parabola, apply the definition mentioned in the previous sentence to the origin to show that

$$c^2 = x_f^2 + y_f^2$$

Use this relation in turn to show that

$$y_f = -\frac{v_0^2}{2g} \cos 2\theta$$

Use the equations for x_f and y_f to show that

$$c = \frac{v_0^2}{2g}$$

Recall that c is the distance of the directrix from the origin; notice that this result for the location of the directrix is the *same* as the height to which the projectile would rise if it were launched vertically. (c) Show that the polar coordinates (r_f, θ_f) of the focus are

$$r_f = \frac{v_0^2}{2g}$$

and

$$\theta_f = 2\theta - 90°$$

Since this last result can be rewritten as

$$90° - \theta = \theta - \theta_f$$

the angle that the position vector of the focus makes with the

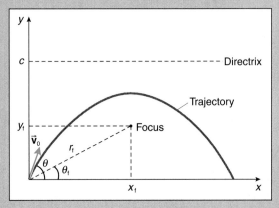

FIGURE P.35

initial angle of flight is equal to the angle between the vertical direction and the initial line of flight. (d) Show that the perpendicular distance of the projectile from the directrix (or the distance of the projectile from the focus) is, at any time,

$$\frac{v^2}{2g}$$

where v is the speed of the projectile.

This problem was inspired by an article by Larry D. Johnson, "The path of a projectile," *The Physics Teacher, 30*, #2, pages 104–105 (February 1992).

‡36. A projectile is fired at initial speed v_0 from an elevation h above a level plain with the initial velocity vector making an angle θ with the horizontal as shown in Figure P.36a. (a) Show that the range R of the projectile satisfies the following equation:

$$0 \text{ m} = h + R \tan \theta - \frac{gR^2}{2v_0^2} \sec^2 \theta$$

(b) By taking the ratio of the velocity components at the impact point, show that the impact angle ϕ satisfies

$$\tan \phi = \frac{(v_0^2 \sin^2 \theta + 2gh)^{1/2}}{v_0 \cos \theta}$$

(c) For a fixed initial speed v_0, show that the angle θ that maximizes the range R is

$$\theta = \tan^{-1} \frac{v_0}{(v_0^2 + 2gh)^{1/2}}$$

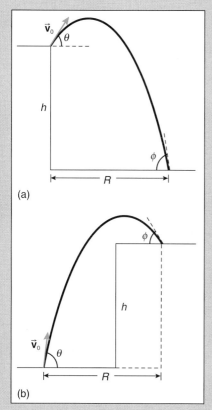

(a)

(b)

FIGURE P.36

(Hint: Find

$$\frac{dR}{d\theta}$$

using the equation in part (a).) Notice the numerator of the fraction in the equation for θ is the initial speed and the denominator is the impact speed of the projectile. (d) Show that the maximum range R_{max} then is

$$R_{max} = \frac{v_0}{g} (v_0^2 + 2gh)^{1/2}$$

(e) Show that when the projectile is launched at the angle for maximum range, the launch angle θ and the impact angle ϕ are related as follows:

$$\tan \phi = \cot \theta$$

This means that the launch and impact angles are complements of each other when the projectile has maximum range. (f) If the projectile is launched to a level plain at height h above the launch point as in Figure P.36b, will the arguments and results of parts (a) through (e) also apply?

This problem was inspired by an article by Carey S. Inouye and Eric W. T. Chong, "Maximum range of a projectile," *The Physics Teacher, 30*, #3, pages 168–169 (March 1992).

Sections 4.3 Motion in Three Dimensions
4.4 Relative Velocity Addition and Accelerations

37. If a swimmer swims at speed 5.00 km/h with respect to the water and a uniform current in a river is flowing at 4.00 km/h, in what direction must the swimmer aim so that the resulting path of the swimmer is directly across a river of width 100 m? How long will the crossing take?

38. A kayaker attempts to cross a channelized river in which the current has a uniform speed of 5.0 km/h across the 50.0 m width of the river. The kayaker is paddling at the frenetic speed of 10.0 km/h with respect to the water and wants to travel straight across the river to avoid a huge waterfall just downstream. (a) What is the resulting speed of the kayak with respect to the ground? (b) At what angle to the stream flow should the kayaker paddle so that the transit is made without moving upstream or downstream? (c) How long does the trip take?

•39. A pilot flies a plane at a speed of 700 km/h with respect to the surrounding air. A wind is blowing to the northeast at a speed of 100 km/h. In what direction must the plane be aimed so that the resulting direction of the plane is due north? What is the ground speed of the aircraft?

•40. Sergeant Pepper is on a bombing run in a helicopter aimed horizontally northward above a road. The speed of the helicopter is 40.0 m/s relative to the air. A wind is blowing to the west at a constant speed of 8.6 m/s relative to the ground. (a) What is the speed of the helicopter relative to the ground? (b) A water bomb is dropped from an altitude of 150 m in the direction of the Dean of Deans but misses its target by a whisker. How long will it take the bomb to hit the ground? (c) Where, relative to the helicopter, will the bomb hit? (d) What is the speed of the bomb the instant before impact?

•41. After an argument with a friend, you stomp off east at 1.50 m/s while he nonchalantly walks north at 1.00 m/s. (a) Introduce an appropriate coordinate system and calculate the velocity at which he recedes from you. (b) Does the answer to (a) depend on where you both are located?

•42. You head directly across a straight river of width 100 m, swimming at a speed of 1.5 m/s relative to the water. You arrive at the opposite bank having been swept downstream a distance of 50 m. (a) How long does the journey take? (b) What is the speed of the current? (c) What is your speed relative to the ground? (d) At what angle with respect to the upstream direction should you have headed to go straight across the stream without slipping downstream?

•43. A pilot of Fly-by-Night Airlines wishes to fly due southeast. A wind is blowing *to* the west at a speed of 120 km/h. The airspeed of the plane is 700 km/h. Determine (a) the speed of the plane with respect to the ground and (b) the angle the plane should be aimed with respect to the direction south.

•44. You are flying an airplane in a strong wind. An air traffic controller reports your velocity to be 700 km/h 30.0° east of north. The wind is reported to be 120 km/h due east. (a) Sketch the situation and introduce an appropriate coordinate system. (b) What is the velocity of the air with respect to the plane? (c) What is the speed of the air with respect to the plane?

•45. As an air traffic controller, you observe Flight 001 of U.S. Snarelines traveling due east at 600 km/h and Flight 002 of Continental Drift Airlines moving exactly northwest at 700 km/h. (a) What is the velocity of Flight 002 as seen by the pilot of Flight 001? (b) What is the speed of Flight 002 as seen by the pilot of Flight 001?

•46. An airplane flies at constant speed v relative to the air on a straight line from Washington, D.C., to Seattle and back, a round-trip distance of 2ℓ. Neglect layover time in Seattle. (a) If there is no wind, what is the round-trip flight time? (b) If there is a headwind of speed v_w going west and a tailwind of the same speed going east, what is the round-trip flight time? (c) Show that the round-trip flight time with a wind is greater than the round-trip flight time with no wind.

‡47. Two identical swimmers have stroke frequencies of one complete stroke per second (a right and left stroke is one complete stroke). They swim at a speed of 3.00 m/s in still water. They start swimming simultaneously from a fixed rock in a uniformly flowing river whose current speed is v. One swimmer goes 100 m downstream and returns to the starting point. The other swimmer moves directly across the stream (along a path perpendicular to the current) to a point 100 m across the river and also returns via the same route. Both swimmers execute perfect turns, taking zero time. The swimmer who went downstream and back returns to the starting point 5.00 s later than the cross-stream swimmer. (a) By how many stroke periods (the time for one right plus left stroke) does the cross-stream swimmer win? (b) Find expressions for the times the swimmers take to complete their courses in terms of the current speed v. (c) Calculate the speed of the current v from the given data.

‡48. Imagine snow falling vertically downward at constant speed v_s. A car is driving horizontally in a straight line through the snow at constant speed v. (a) Show that according to the driver, the snow appears to fall along a direction that makes an

angle θ with the vertical direction where

$$\tan \theta = \frac{v}{v_s}$$

If you tilt an umbrella at this angle while walking at speed v in snow (or rain) falling vertically at speed v_s, you keep driest. (b) This problem also has implications in astronomy. Imagine the snow to be starlight moving at constant speed c toward the solar system (see Figure P.48). The Earth, in its orbital motion, travels at speed v. Observers on the Earth see the light coming from a direction θ that is not in the direction of the true position of the star. The angle is found from

$$\tan \theta = \frac{v}{c}$$

Thus a telescope must be tilted in the direction of the orbital velocity of the Earth by an angle θ to see the star. Over the course of a year, the star executes a small circle of angular radius θ about its true position perpendicular to the plane of the orbit of the Earth. This effect is known as *stellar aberration*. The orbital speed of the Earth is about 30 km/s. Evaluate the angle θ. If the star is in the plane of the orbit, the aberrational path of the star is a line centered on the true position of the star. If the star is between the plane of the orbit and the perpendicular to the orbit, the aberrational path of the star is an ellipse.

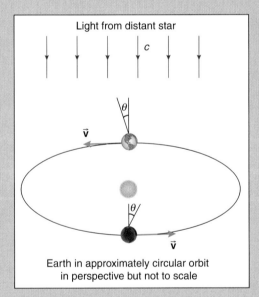

FIGURE P.48

Section 4.5 Uniform Circular Motion: A First Look

49. The tachometer on the dashboard of an automobile indicates the motor is turning at 3000 rev/min. What is the angular speed of the motor in rad/s?

50. Determine the angular speed in rad/s of a particle at rest on the surface of the rotating Earth.

51. A dentist has a new dental drill that spins at 5.00×10^4 rev/min. (a) Calculate the speed of a point on the rim of a drill bit of radius 0.500 mm. (b) Calculate the magnitude of the centripetal acceleration of a point on the rim of the drill bit and compare it with the magnitude of the local acceleration due to gravity, $g = 9.81$ m/s².

52. An object is executing circular motion at a constant speed of 1.00 km/s at an angular speed of 2.00π rad/s. What is the radius of the circular path of the object?

53. A fast sprinter can attain speeds of about 10 m/s. If the sprinter is running around a circular track at this (constant) speed, what radius circle results in a centripetal acceleration with a magnitude equal to $g = 9.81$ m/s²?

54. The orbit of the Earth around the Sun is approximately circular and of radius 1.496×10^8 km. It takes the Earth one year (about 365.25 d) to travel around the Sun. (a) Determine the speed of the Earth in its orbit in km/s. (b) Calculate the angular speed of the Earth in rad/s. (c) What is the magnitude of the centripetal acceleration (in m/s²) of the Earth in its orbit about the Sun?

55. The average distance between the Earth and the Moon is 3.84×10^5 km, about sixty times the radius of the Earth. Assume the orbit of the Moon is circular with the Moon completing one revolution in 27.322 d. (a) Calculate the magnitude of the centripetal acceleration of the Moon in m/s². (b) Calculate the ratio between the magnitudes of the centripetal acceleration of the Moon and the local acceleration due to gravity on the surface of the Earth:

$$\frac{a_c}{g}$$

The numerical value of this ratio is close to $1/3600 = 1/(60)^2$ and had great significance for Isaac Newton when he developed his theory of universal gravitation (as we will see in Chapter 6).

56. In the picturesque (but flawed) Bohr model of the hydrogen atom, the electron is imagined as orbiting the nucleus at a radius of 5.28×10^{-11} m with a constant speed of 2.18×10^6 m/s. (a) Calculate the magnitude of the centripetal acceleration of the electron. (b) How many times greater than the magnitude of our local acceleration due to gravity ($g = 9.81$ m/s²) is the magnitude of this centripetal acceleration? (c) Calculate the number of revolutions per second made by the electron in this orbit.

•57. Neutron stars are rather spectacular. They have radii on the order of only 10 km yet have masses on the order of *twice* the mass of our Sun. Neutron stars also spin quite fast; one such neutron star in the Crab Nebula (in the constellation Taurus) rotates 33 times each second. Calculate the centripetal acceleration of a particle on the equator of such a star.

•58. A physics student is located at latitude 38° North. (a) Calculate the centripetal acceleration of the student due to the rotation of the Earth. (b) At what places on the surface of the Earth is the centripetal acceleration due to the rotation of the Earth a maximum? (c) At what places is it a minimum?

•59. A particle is moving at constant speed 10.0 m/s in circles of various radii. (a) Make a graph of the magnitude of the centripetal acceleration of the particle versus the radius of the circle for radii between $r = 0$ m and 100 m. (b) At what radius is the magnitude of the centripetal acceleration of the particle equal to the magnitude of the local acceleration due to gravity ($g = 9.81$ m/s²)? What is the angular speed ω of the particle with speed 10.0 m/s at this radius?

•60. A particle is moving in a circle of radius 5.00 m. (a) Make a graph of the magnitude of the centripetal acceleration versus the speed of the particle for speeds ranging between 0 m/s and 10.0 m/s. What is the slope of the graph when $v = 0$ m/s? (b) Make a corresponding graph of the magnitude of the centripetal acceleration versus the *angular speed* ω of the particle for the same speeds indicated in part (a). What is the slope of this graph when $\omega = 0$ rad/s?

•61. The TGV (*Train à Grande Vitesse* or Train of Great Speed) high-speed trains between Paris and Lyon achieve speeds exceeding 300 km/h. If the magnitude of the centripetal acceleration of the train is not to exceed 4.0% of the magnitude of the local acceleration due to gravity ($g = 9.81$ m/s²), what is the minimum radius for a circular curve in the track? Explain why this is a *minimum* radius.

•62. The minute hand on Big Ben in London is 4.30 m long. (a) What is the speed of the tip of the minute hand? (b) What is the magnitude of the centripetal acceleration of a point at the tip of the minute hand? (c) Through what distance does the tip of the minute hand move during a year of 365 d?

•63. The Moon orbits the Earth in about 27.3 days, moving in a counterclockwise sense when viewed from above the plane of the orbit (*way* over the north pole of the Earth). Seen from the Earth, the Moon appears projected against a different star background each night. The Earth orbits the Sun in 365.25 days, also moving in a counterclockwise fashion when viewed from above the plane of the orbit. Because of the revolution of the Earth about the Sun, we see the Sun projected against a slightly different star background each day. The Sun and Moon both appear as circles in the sky with an angular size of about 0.5° when seen from the Earth. Assume the Moon and Earth move in circular orbits at constant speed. (a) About how many degrees per day does the Moon move in its orbital motion? (b) About how many degrees per day does the Sun appear to move against the distant star background as seen from the Earth? (c) Assume the Moon and Sun follow the same path when they are viewed against the distant star background (their paths actually are inclined to each other by about 5°). Both move in the same directional sense along this path. About how many days elapse between successive lunar passings of the Sun in the sky? This is the time interval of the phases of the Moon.

‡64. Consider two planets in circular orbits, both planets moving at constant (but different) speeds about the Sun in a counterclockwise sense as indicated in Figure P.64a. The time it takes a planet to complete one circuit of its orbit is called the *sidereal period* of the planet. Let the sidereal period of the *inner* planet, called the inferior planet, be P_{inf} and that of the *outer* planet, called the superior planet, be P_{sup}. Let the planets have the initial configuration as indicated in Figure P.64a with both along a common Sun–planet line. The planets move at different speeds around their different orbits, with the inner planet moving faster. Their motion is analogous to a very unfair race between two mismatched runners on the inside and outside of a circular track with the faster runner on the inside and the slower runner on the outside. (a) Show that the time S it takes for the inner (and faster) planet to gain a full lap on the slower outer planet is

$$\frac{1}{S} = \frac{1}{P_{inf}} - \frac{1}{P_{sup}}$$

The time period S is known as the *synodic period* of the two planets. (b) Let the inner planet be the Earth (P_{inf} = 1 y) and the outer planet be Mars (P_{sup} = 1.88 y). Evaluate S. This is the time interval between successive *oppositions* of Mars (an *opposition* occurs when the planet and the Sun are in opposite directions as seen from the Earth). (c) Let the outer planet now be the Earth and the inner planet be Venus (P_{inf} = 0.62 y). Evaluate S. This is the time interval between successive *inferior conjunctions* of Venus (an *inferior conjunction* is when a planet closer to the Sun than the Earth is between the Sun and Earth). (d) If the planets begin as in Figure P.64b but move in *opposite* directions,* show that the time S when they next align along a common Sun–planet line is

$$\frac{1}{S} = \frac{1}{P_{inf}} + \frac{1}{P_{sup}}$$

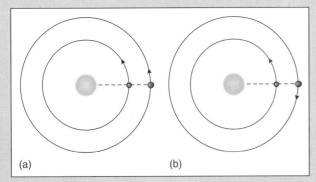

(a)   (b)

FIGURE P.64

Sections 4.6   The Angular Velocity Vector
4.7   The Geometry and Coordinates for Describing Circular Motion
4.8   The Position Vector for Circular Motion
4.9   The Velocity and Angular Velocity in Circular Motion
4.10   Uniform Circular Motion Revisited

65. (a) Calculate the angular speeds (in rad/s) of the second, minute, and hour hands of a clock; assume they rotate smoothly. (b) Do your answers for part (a) depend on the radius of the clock hands? (c) Do the angular velocity vectors of all the hands point in the same direction? Describe the direction of the angular velocity vector relative to the plane of the face of the clock. (d) What is the ratio of the angular speed of the second hand to the angular speed of the minute hand? (e) What is the ratio of the angular speed of the minute hand to the angular speed of the hour hand?

•66. The sweep second hand of a large clock is 10.0 cm long. Choose a coordinate system with its origin at the center of the clock and the y-axis toward the numeral 12 on the clock face and the x-axis toward the numeral 3, as shown in Figure P.66. (a) What is the velocity $\vec{v}$ of the tip of the second hand at the instant it points toward the number 3? (b) What is the change in the position vector of the tip of the second hand during the 15.0 s it takes to sweep from the numeral 12 to the numeral 3?

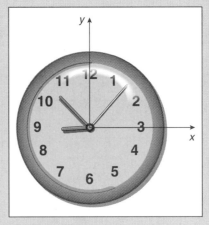

FIGURE P.66

(c) What is the acceleration of the tip of the second hand when it is at the numeral 3?

•67. Use Equation 4.28 for $\vec{r}(t)$:

$$\vec{r}(t) = [r \cos \theta(t)]\hat{\imath} + [r \sin \theta(t)]\hat{\jmath}$$

and Equation 4.26 for $\vec{\omega}$:

$$\vec{\omega}(t) = \omega_z(t)\hat{k}$$

to evaluate the expression

$$\vec{r}(t) \cdot [\vec{\omega}(t) \times \vec{r}(t)]$$

Explain why this result is another way to see that $\vec{v}(t)$ is always perpendicular to $\vec{r}(t)$ for circular motion.

Sections 4.11   Nonuniform Circular Motion and the Angular Acceleration
4.12   Nonuniform Circular Motion with a Constant Angular Acceleration

68. What is the angular acceleration $\vec{\alpha}$ of a particle on the surface of the rotating Earth?

69. An automotive engine shaft revs up from 3.00×10^3 rev/min to 4.50×10^3 rev/min during a 1.20 s interval. (a) Calculate the magnitude of the angular acceleration of the shaft assuming it is constant during the interval. (b) Determine the number of revolutions of the shaft during the interval.

•70. A particle moving clockwise in a circle with a radius of 3.00 m has a total acceleration at some instant of magnitude 15.0 m/s² directed as indicated in Figure P.70. (a) Find the magnitude of the centripetal acceleration at this instant. (b) Find the speed of the particle at this instant. (c) Find the angular speed of the particle at this instant. (d) Determine the magnitude of the tangential acceleration at this instant. (e) Determine the magnitude of the angular acceleration at this instant. (f) Make a sketch indicating the directions of $\vec{\omega}$, $\vec{\alpha}$, $\vec{r}$, and $\vec{v}$. (g) Is the particle speeding up or slowing down in its circular motion?

•71. A round object initially spinning clockwise on a fixed axis with an angular speed of 20.0 rad/s is subjected (when t = 0 s) to an angular acceleration of magnitude 0.500 rad/s² that slows it down in its circular motion. The angular acceleration ceases after 1.00 min. (a) Indicate in a schematic diagram the

---

*All the planets of the solar system move around the Sun in the same sense, not in opposite senses.

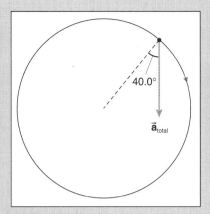

FIGURE P.70

rotational sense of the object, and the directions of the initial angular velocity vector and the angular acceleration vector. (b) Choose an appropriate direction for the +z-axis and indicate this direction on the sketch of part (a). (c) Tailor Equation 4.49,

$$\omega_z = \omega_{z0} + \alpha_z t$$

to the particulars of this problem. (d) Make a graph of ω_z versus t over the range of times from $t = 0$ s to 90.0 s. (e) At what time (if any) is the object instantaneously at rest? (f) *Describe* in words what happens to the object over the ninety second interval of the graph.

•**72.** An ice skater initially is spinning about a vertical axis at 150 rev/min. During 10.0 s, the rotational speed of the skater decreases at a constant rate to 100 rev/min. (a) Choose $\hat{\mathbf{k}}$ to be opposite to the direction of the angular velocity vector. Indicate this choice in an appropriate sketch. (b) What is the initial angular velocity component ω_{z0}? (c) What is the final angular velocity component? (d) Determine the angular acceleration component α_z. (e) Through how many revolutions did the skater turn during the 10.0 s interval?

•**73.** An ultracentrifuge accelerates from rest to 1.00×10^5 rev/min during 7.00 min. (a) Calculate the magnitude of the angular acceleration (assuming it is constant). (b) Through how many revolutions does the rotor turn in reaching this incredible angular speed? (c) The rotor has an effective radius of 8.00 cm. Calculate the magnitude of the centripetal acceleration of a point on the rim of the rotor when spinning at full speed. Compare this with the magnitude of the local acceleration due to gravity. (d) If it takes the rotor 4.00 min to brake to a stop from this angular speed, find the magnitude of the angular acceleration (assuming it is constant) and indicate on a sketch the orientation of the angular acceleration vector relative to the orientation of the angular velocity vector.

•**74.** A ceiling fan with blades 0.50 m long is rotating at 180 rev/min. The fan is shut off and coasts to a stop after 2.5 min. Let $\hat{\mathbf{k}}$ be in the direction of the angular velocity vector. (a) Determine the initial angular velocity vector of the fan in rad/s. (b) Assuming the angular acceleration is constant, determine the angular acceleration vector in rad/s². (c) Through how many revolutions does the fan turn in coming to rest? (d) When the fan was rotating at 180 rev/min, what was the magnitude of the centripetal acceleration of the points on the tips of the blades?

•**75.** A particle moving along a circle of radius 4.00 m has an instantaneous total acceleration of magnitude 40.0 m/s² directed as indicated in Figure P.75. (a) Indicate an appropriate direction for the z-axis. (b) What is the magnitude of the centripetal acceleration? (c) What is the magnitude of the tangential acceleration? (d) What is the angular velocity vector at the instant indicated? (e) What is the angular acceleration vector at the instant indicated? (f) Is the particle speeding up or slowing down in its circular motion? (g) Assume the angular acceleration is constant. Is the tangential acceleration then of constant magnitude? Is the centripetal acceleration of constant magnitude? (h) How long will it take the particle to come to rest?

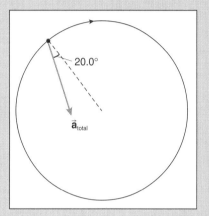

FIGURE P.75

•**76.** A particle moving in a circle of radius 2.00 m begins at rest when time $t = 0$ s and attains a speed of 10.0 m/s at the end of one revolution with a constant angular acceleration. (a) Determine the angular speed after one revolution. (b) Determine the magnitude of the angular acceleration and the time it takes to complete the first revolution. (c) At the instant the speed is 10.0 m/s, find

• the magnitude of the centripetal acceleration,
• the magnitude of the tangential acceleration,
• the magnitude of the *total* acceleration, and
• the angle that the total acceleration vector makes with the radial direction.

•**77.** A particle begins at rest on a circular track and is subjected to a constant angular acceleration of magnitude α beginning when $t = 0$ s. (a) Show that the magnitudes of the tangential and centripetal accelerations of the particle are *equal* when

$$t = \left(\frac{1}{\alpha}\right)^{1/2}$$

independent of the radius of the circular track. (b) What is the angle that the total acceleration vector makes with the radial direction at this time?

•**78.** Show that the ratio of the magnitudes of the tangential and centripetal accelerations in circular motion can be written in the form

$$\frac{a_t}{a_c} = \frac{\alpha}{\omega^2}$$

Notice that the ratio is equal to 1 when the angular speed ω has the value $\sqrt{\alpha}$.

•79. For the car considered in Example 4.14, find the instant during the 20 s interval when the tangential and centripetal accelerations have the same magnitude, so that the total acceleration vector makes an angle of 45° with the radial direction.

•80. A NASA engineer is designing a sophisticated merry-go-round apparatus capable of subjecting astronauts to centripetal accelerations with magnitudes up to five times the magnitude of the local acceleration due to gravity on the surface of the Earth—five gs, using the lingo of the trade. The radius of the machine is to be 20.0 m and the astronauts will be located on the rim of the device. (a) What is the speed of the chamber on the rim when the magnitude of the centripetal acceleration is 5.00g? (b) What is the angular speed of the device when the magnitude of the centripetal acceleration is 5.00g? (c) Determine the number of revolutions per minute executed by the device when the magnitude of the centripetal acceleration at the rim is 5.00g. (d) If the device is to get up to speed during one minute, determine the magnitude of a constant angular acceleration needed to accomplish the task.

•81. A turntable is moving at a constant angular speed of 0.500 rad/s. When $t = 0$ s, two centipedes (*Scutigera forceps*) at rest on opposite sides of the turntable (but not on the turntable) both spy a tasty morsel on the rim of the rotating turntable. See Figure P.81. The morsel is right next to the centipede named Millifoot when $t = 0$ s. Millifoot realizes, of course, that the morsel soon will be carried around by the turntable to the other centipede, named Multiplex, who is salivating eagerly in anticipation of the oncoming meal. Multiplex remains at rest while Millifoot sets off in hot pursuit of the morsel around the circle at a constant angular acceleration and manages to catch the morsel just as it arrives at the position of Multiplex. (a) Set up an appropriate direction for the +z-axis in a sketch. (b) At what time is the morsel next to Multiplex? (c) Determine the angular acceleration component α_z of Millifoot. (d) Make a graph of the angular velocity component of the morsel as a function of time. On the same graph, plot the angular velocity component of Millifoot. Indicate the point on the graph that corresponds to Millifoot catching the morsel.

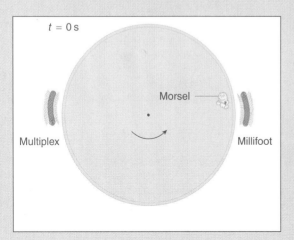

FIGURE P.81

•82. A particle with an initial angular velocity of

$$\omega_{z0}\hat{k} = (20.0 \text{ rad/s})\hat{k}$$

has a constant angular acceleration given by

$$\vec{\alpha} = (-2.00 \text{ rad/s}^2)\hat{k}$$

Consider the interval from $t = 0$ s to 10.0 s in what follows. (a) Make a schematic sketch of the situation indicating the rotational sense of the particle, the angular velocity vector, and the angular acceleration vector. Is the particle speeding up or slowing down in its rotational motion? (b) Make a graph of α_z versus t. (c) Find ω_z as a function of time. (d) What is the physical significance of the area under the α_z versus t curve between $t = 0$ s and 10.0 s? (e) Make a graph of ω_z versus t. (f) What is the physical significance of the area under the graph of ω_z versus t between $t = 0$ s and 10.0 s? (g) Through what angle does the particle turn during the interval?

•83. A car is traveling around a banked, circular curve of radius 150 m on a test track. At the instant when $t = 0$ s, the car is moving north, and its speed is 30.0 m/s but decreasing uniformly, so that after 5.0 s its *angular speed* will be 3/4 that it was when $t = 0$ s. (a) What is the angular speed of the car when $t = 0$ s? (b) What is the angular speed 5.0 s later? (c) Find the magnitude of the centripetal acceleration of the car when $t = 0$ s. (d) Find the magnitude of the centripetal acceleration of the car when $t = 5.00$ s. (e) Find the magnitude of the angular acceleration. (f) Find the magnitude of the tangential acceleration.

‡84. The position vector $\vec{r}(t) = x(t)\hat{i} + y(t)\hat{j}$ of a particle moving in the x–y plane is described by

$$\vec{r}(t) = [A \cos(\omega t)]\hat{i} + [B \sin(\omega t)]\hat{j}$$

where ω is a constant angular speed (in rad/s) and A and B are two distances (in m). (a) Make a sketch to indicate the location of the particle when $t = 0$ s. (b) On the same sketch indicate the location of the particle when $t = \pi/(2\omega)$. Notice that the particle is *not* moving in a circular path if $A \neq B$. (c) Find an expression for the velocity of the particle as a function of time. (d) Is the velocity vector $\vec{v}(t)$ perpendicular to $\vec{r}(t)$ at *all* times? Is the velocity vector perpendicular to $\vec{r}(t)$ at *some* times? If so, what are these times and where is the particle at these times? (e) Find an expression for the acceleration $\vec{a}(t)$ of the particle. (f) Is the acceleration antiparallel to $\vec{r}(t)$ at *all* times? (g) Is the acceleration *perpendicular* to $\vec{v}(t)$ at *all* times? Is the acceleration vector perpendicular to $\vec{v}(t)$ at *some* times? If so, what are these times and where is the particle at these times? (h) Make use of the trigonometric relation

$$\sin^2(\omega t) + \cos^2(\omega t) = 1$$

to find an expression for the orbital path of the particle in terms of the Cartesian coordinates x and y of the position vector, and the constants A and B but *not* the time t. What is the name for the *shape* of this path?

‡85. A particle is moving in a circle at a speed that is *not* constant. Consider Equation 4.28:

$$\vec{r}(t) = [r \cos \theta(t)]\hat{i} + [r \sin \theta(t)]\hat{j}$$

(a) Show that the acceleration vector can be written as

$$\vec{a}(t) = \left\{ [-r\sin\theta(t)]\frac{d^2\theta(t)}{dt^2} - [r\cos\theta(t)]\left[\frac{d\theta(t)}{dt}\right]^2 \right\}\hat{\imath}$$

$$+ \left\{ [r\cos\theta(t)]\frac{d^2\theta(t)}{dt^2} - [r\sin\theta(t)]\left[\frac{d\theta(t)}{dt}\right]^2 \right\}\hat{\jmath}$$

$$\equiv a_x\hat{\imath} + a_y\hat{\jmath}$$

(b) Introduce a new coordinate system with one axis directed radially outward (with unit vector $\hat{r}$) and the other tangent to the circle in the direction of increasing θ (with unit vector $\hat{\theta}$) as indicated in Figure P.85. Show that the component of the acceleration along the radial direction is

$$a_r = a_x\cos\theta + a_y\sin\theta$$

and the component of the acceleration in the tangential direction of increasing θ is

$$a_\theta = -a_x\sin\theta + a_y\cos\theta$$

(c) Let s be the arc length traveled by the particle along the circle, so that

$$s = r\theta$$

Show that the tangential acceleration component is the same as

$$\frac{d^2s}{dt^2}$$

Show that the radial acceleration component is

$$-\frac{\left(\dfrac{ds}{dt}\right)^2}{r}$$

Notice that this is the centripetal acceleration component and is equal to $-v^2/r$.

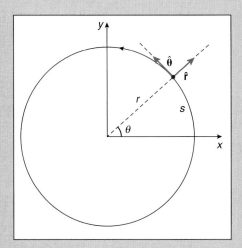

FIGURE P.85

‡86. You likely have noticed that as a cassette tape is played on a boom box or other noise generator, the radius of the tape left on the supply spool decreases slowly with time as the tape is drawn through the machine at a constant speed v_0. Choose $\hat{k}$ to be parallel to $\vec{\omega}$. Since the radius of the tape remaining changes slowly compared with the speed of the tape through the machine, to a good approximation we can write

$$v_0 = r\frac{d\theta}{dt}$$

where

$$\frac{d\theta}{dt} = \omega_z > 0 \text{ rad/s}$$

is the *angular speed* of the tape reel. Notice that the tape speed v_0 is constant, but r changes slowly with time; thus the *angular speed* of the spool also changes slowly with time. Assume that r represents the average radius of each coiled layer (since the tape actually is wound in a *spiral*). Let r_0 be the initial radius of the tape when $t = 0$ s and $\theta_0 = 0$ rad, where θ is the total angle of tape unwound from the spool. Represent the thickness of the tape by b. When the tape has unwound through a total angle θ, the radius remaining is

$$r = r_0 - b\frac{\theta}{2\pi} \tag{1}$$

(a) Show that

$$\frac{dr}{dt} = -\frac{b}{2\pi}\frac{d\theta}{dt}$$

$$= -\frac{b}{2\pi}\frac{v_0}{r}$$

(b) Integrate the result of part (a) to show that

$$\pi(r_0^2 - r^2) = bv_0t$$

What is the geometrical interpretation of the left-hand side of this equation? What is the physical interpretation of the right-hand side of this equation? (c) Solve the result of part (b) for r and substitute this for r in equation (1). Integrate the resulting expression to show that the total angle θ through which the tape has turned at time t is

$$\theta = \frac{2\pi}{b}\left\{ r_0 - \left[r_0^2 - \frac{bv_0t}{\pi} \right]^{1/2} \right\}$$

This problem was inspired by a paper by R. Jordinson, "The angular motion of the supply spool on a tape recorder," *Physics Education, 15,* #5, pages 240–241 (July 1980).

‡87. The position vector of an electron moving in a magnetic field is described by the expression

$$\vec{r}(t) = [r\cos(\omega t)]\hat{\imath} + [r\sin(\omega t)]\hat{\jmath} + v_{z0}t\hat{k}$$

where r, ω, and v_{z0} are positive constants. Make a schematic three-dimensional plot of the trajectory of the electron in space. Describe the motion in a concise sentence. When charged particles emanating from the Sun encounter the magnetic field of the Earth, they move in trajectories similar to this one. When the particles collide with molecules in the atmosphere of the Earth, auroras are produced.

INVESTIGATIVE PROJECTS

A. Expanded Horizons

1. Aspects of kinematics and projectile motion are involved in the sport of juggling. Investigate the physics of this skilled sport.
   Bengt Magnusson and Bruce Tiemann, "The physics of juggling," *The Physics Teacher*, 27, #8, pages 584–588 (November 1989).
   Eric Kincanon, "Juggling and the theorist," *The Physics Teacher*, 28, #4, pages 221–223 (April 1990).

2. Experienced sharpshooters know that if a rifle sight is adjusted for hitting the bull's-eye on a horizontal range, the bullet invariably will hit *above* the bull's-eye when aimed at the bull's-eye when firing at *either* an uphill *or* a downhill target placed the same straight-line distance away. Investigate why this is the case.
   Ole Anton Haugland, "A puzzle in elementary ballistics," *The Physics Teacher*, 21, #4, pages 246–248 (April 1983).
   H. A. Buckmaster, "Ideal ballistic trajectories revisited," *The American Journal of Physics*, 53, #7, pages 638–641 (July 1985).

3. The kinematics of living creatures is interesting. Land-based living things have different numbers of legs for locomotion. Snakes have no legs, humans have two, horses have four, insects have six, arachnids have eight, while centipedes and millipedes have many. How do snakes manage to move? What is the difference between millipedes and centipedes? Do more legs mean a greater speed per body length of the creature? Investigate the work of Robert J. Full on the kinematics of locomotion at the Poly-pedal Laboratory at Stanford University.
   Arthur Herschman, "Animal locomotion as evidence for the universal constancy of muscle tension," *American Journal of Physics*, 42, #9, pages 778–779 (September 1974).

B. Lab and Field Work

4. Most physics laboratories have a device called a ballistic pendulum. Part of the apparatus is a spring gun that can project a metal ball across the room. Devise a method for accurately measuring the range of the ball and use the data to determine the speed of the ball when projected from the spring gun.

5. For high jumps or long jumps, make a few measurements to determine the approximate speed of the athletes as well as the angle they commence flight.
   Herbert H. Lin, "Newtonian mechanics and the human body: some estimates of performance," *American Journal of Physics*, 46, #1, pages 15–18 (January 1978).
   Cliff Frohlich, "Effect of wind and altitude on record performance in foot races, pole vault, and long jump," *American Journal of Physics*, 53, #8, pages 726–730 (August 1985).

6. Devise a method for measuring the angular speed of a fan blade when at the low, medium, and high speed settings. Measure how long it takes the fan to reach these angular speeds from rest, as well as from low to medium, and medium to high settings. Determine the magnitude of the corresponding angular accelerations, assuming each is constant. Do you get the same magnitude angular acceleration for each change of angular speed? Determine the centripetal accelerations of a point on the rim at each of these settings.

7. Secure an automobile or marine tachometer from an automotive or marine graveyard and dissect it to determine how it is able to indicate the number of revolutions per minute.

C. Communicating Physics

8. On the basis of your intuitive ideas about resistive effects and with the knowledge that the density of the atmosphere decreases with altitude above the surface of the Earth, formulate a hypothesis about whether the actual horizontal range of a long-range artillery shell will be greater or less than that predicted by the analysis of projectiles in Example 4.3. Be sure to explain the reasoning that leads to your hypothesis.
   Michael A. Day and Martin H. Walker, "Experimenting with the national guard: field artillery gunnery," *The Physics Teacher*, 31, #3, pages 136–143 (March 1993).

9. Astronomers use the word *rotation* for what a physicist calls spin and the words *revolutionary motion* for what a physicist calls orbital motion. In a cogent paragraph, describe the meaning of the terms in language appropriate for a nontechnical audience such as a junior high school class, illustrating the distinctions with the motion of some everyday object. Use a thesaurus to see if other words convey the same meaning.

See Problem 21.

PHILOSOPHIÆ

NATURALIS

PRINCIPIA

MATHEMATICA.

Autore *JS. NEWTON*, *Trin. Coll. Cantab. Soc.* Matheseos
Professore *Lucasiano*, & Societatis Regalis Sodali.

IMPRIMATUR·
S. PEPYS, *Reg. Soc.* PRÆSES.
Julii 5. 1686.

LONDINI,

Jussu *Societatis Regiæ* ac Typis *Josephi Streater*. Prostat apud
plures Bibliopolas. *Anno* MDCLXXXVII.

CHAPTER 5

NEWTON'S LAWS OF MOTION

Nature, and Nature's Laws lay hid in Night.
God said: *Let Newton be!* and All was *Light*.
*Alexander Pope (1688–1744)**

The description of motion, or kinematics, is half of the picture we wish to paint. Its complement is an understanding of the underlying *causes of changes in motion*: dynamics. What causes an acceleration? For that matter, what causes a constant velocity? You know, from common experience, that if you stop peddling your bicycle on level ground, you soon come to rest. Why? These are profound questions whose answers we address in this chapter.

Historically, dynamics was a sticky problem. When a pitcher accelerates a baseball from rest to its release, an arm is the cause of the acceleration. Likewise, when a car slams into a tree, the immediate cause of the brief, perhaps catastrophic acceleration of the car is the tree. Such causes are descriptive but are not sufficiently general to constitute a general theory for the causes of accelerations. Now think about dropping a pencil from your hand. The pencil falls to the floor. What causes the downward acceleration of the pencil? In this case, when you *release* the pencil, it accelerates, seemingly of its own accord; *nothing obvious* is in physical contact with the pencil during its fall to cause its acceleration![†] Likewise, it is not clear what causes the centripetal acceleration of the Earth in its approximately circular orbit around the Sun, since nothing physically connects the Earth to the Sun. Apparently, whatever causes accelerations can act either with or without direct physical contact.

The preceding examples illustrate why the search for possible causes for changes in motion was so difficult. As with kinematics, our scientific forebears were faced again with *inventing* new, unifying physical concepts to encompass a wide range of seemingly disparate physical phenomena whose only common thread was the kinematical description.

It was Isaac Newton (1642–1727) who, in 1686, first published a consistent theory of the motion of particles, based on the notion of **forces**. His theory is called **Newtonian mechanics**, or classical dynamics. While extraordinarily encompassing, it does have limitations (discussed in Section 5.7), as do all physical theories. Some of these limitations were discovered rather late in the game, during the early 20th century.

---

*(Chapter Opener) *Epitaph Intended for Sir Isaac Newton.*

[†]To say that *gravity* causes the rock to fall does not really provide an answer; this is essentially saying "it falls because it falls."

Isaac Newton also made significant contributions to our understanding of optics, as this stamp commemorates.

In this chapter, we first take a brief, modern look at fundamental particles and the fundamental forces of nature. We then elaborate on the concept of what we mean by a force through its action on a standard mass, the standard kilogram. Subsequently, we examine **Newton's laws of motion**, which form the basis for the study of classical dynamics.

Finally, later in the chapter, we also consider the peculiar aspects of several forces commonly encountered in many practical applications, realizing that these common forces are convenient simplifications of several of the fundamental forces of nature.

5.1 FUNDAMENTAL PARTICLES*

So, Nat'ralists observe, a Flea
Hath smaller Fleas that on him prey,
And these have smaller Fleas to bite 'em,
And so proceed ad infinitum

Jonathan Swift (1667–1745)[‡]

Although it is not obvious, there is an underlying lumpiness or graininess to the masses of everyday macroscopic objects. Even the ancient Greeks thought that there were tiny particles, called atoms,[§] that could not be subdivided into smaller particles. In other words, there are **fundamental particles** that constitute everything from atoms and molecules, to textbooks, soda cans, physics students, fleas, and other systems of interest, use, or annoyance.

Curiously, we now know that some of the fundamental particles in physics have *zero mass*, such as **photons** (the particles of light) and (perhaps) peculiar particles known as **neutrinos** (both of which we will encounter in Chapter 26).[#] Such zero-mass particles *always* travel at the speed of light (although the reasons for this are not apparent at this juncture in our study of physics) and have special dynamical properties that we study when we examine relativity (in Chapter 25).

With the discovery of the electron in 1896, we realized that the elemental atoms of the Greeks had structure. By the early 1930s, the fundamental particles of nature were thought to be electrons, protons (discovered in 1921), neutrons (discovered in 1932), and the massless photons of light (proposed in 1905). However, hosts of new fundamental particles were discovered during the mid-20th century, creating what has been described as an elementary particle zoo because of the variety and number of them (there are hundreds). We group the many particles into three families called **leptons**, **quarks**, and **mediating particles** (see Table 5.1). The familiar electron is a member of the lepton family and is thought to be the particle with the smallest nonzero mass. **Hadrons** are composite particles consisting of various combinations of members of the quark family. For example, the proton and neutron, each composed of three quarks, are hadrons. The proton is composed of two up quarks and one down quark,

---

[‡]*On Poetry: A Rapsody*, lines 337–340. *The Poems of Jonathan Swift*, edited by Harold Williams (Oxford at the Clarendon Press, Oxford, England, 1937), volume II, page 651.

[§]The word comes from the Greek ἄτομος, meaning "indivisible."

[#]It is still not known for certain whether neutrinos have a mass exactly equal to zero. Evidence is increasing for a small, nonzero mass.

TABLE 5.1 Names and Symbolic Representations of What
Are Thought to Be the Fundamental Particles of Nature

Leptons		Quarks		Mediating particles	
e	electron	u	up	γ	photon
v_e	electron neutrino	d	down	g_i	gluons (8 exist)
μ	muon	c	charm	W^-	
v_μ	muon neutrino	s	strange	W^+	vector bosons
τ	tau	t	top	Z^0	
v_τ	tau neutrino	b	bottom		graviton*

* Not yet found experimentally.

or ($u\,u\,d$); the neutron is composed of one up and two down quarks, or ($u\,d\,d$). The quark names are playfully picturesque rather than significant. The photon is an example of a member of the mediating particle family. Other exotic particles in the families need not concern us for now. Frankly, we do not know if this grouping of particles is the whole story or whether there exists yet another underlying arrangement of still more fundamental particles or families.

For each fundamental particle there is a corresponding **antiparticle**. Each particle and antiparticle have identical mass, but other properties differ, such as opposite electric charge. For example, the antielectron, called a **positron**, has the same mass as an ordinary electron, but has a positive charge. An **antiproton** has the same mass as a common proton, but has a negative charge. Some neutral particles are their own antiparticle. Particle and antiparticle pairs are created when there is sufficient energy available to "condense" and form their mass (see Chapter 25). For now, do not be overly concerned with these fundamental particles and their antiparticles for we must first master Newton's laws of motion for common macroscopic particles, such as baseballs and hockey pucks!

5.2 THE FUNDAMENTAL FORCES OF NATURE*

Forces of this kind . . . produce every action in this world.
Roger Bacon (c. 1214–1294)[†]

The search for simplicity and unity in physics is well illustrated by the attempts to enumerate the fundamental forces of nature. The problem is not an easy one.

The **gravitational force** was the first fundamental force to be recognized and described analytically (by Isaac Newton during the 17th century). We consider one aspect of the gravitational force in Section 5.10; Chapter 6 studies the gravitational force in some detail. The gravitational force is an example of a **long-range force**, for its effects are felt over immense distances. Gravitation is the force that dominates our immediate world, the structure of the solar system, and even the dynamics of the universe as a whole. Gravitation truly is a universal force, acting however weakly or strongly on every particle in the universe.

As the nature of electric and magnetic phenomena were investigated during the Renaissance, distinct electric and magnetic

forces were introduced. In the mid-19th century, James Clerk Maxwell integrated both forces into a single theory of electromagnetism. As the 20th century dawned, just two fundamental forces were known to exist in nature: the **electromagnetic force** and the gravitational force. The electromagnetic force, like gravitation, is a long-range force: electrical and magnetic effects can be felt over large distances. The long-range magnetic effects of the Earth are apparent on a simple compass needle, and the long range of electrical forces is spectacularly revealed in thunderstorms. We study the electric force in Chapter 16 and the magnetic force in Chapter 20; their connection is considered in Chapter 25.

The development of the nuclear model of the atom during the early 20th century led to the discovery of two additional fundamental forces, both **short-range forces**. The effects of these new forces are felt only over very small distances, comparable in size to the nuclei of atoms. The newly discovered forces were suggestively (but unimaginatively) named (1) the **strong force** (also known as the **nuclear force**), which manifests itself in interactions within the nucleus of the atom and among quarks; and (2) the **weak force**, which manifests itself in certain types of radioactive decay processes and among the leptons. The strong and weak forces we mention only for completeness; we will not consider them further.

How are short- and long-range forces transmitted between particles? For example, how is the gravitational force of the Earth on the Moon transmitted to the Moon? Likewise, how is the Earth aware of the gravitational presence of the Sun? There is no direct physical contact between these celestial bodies; the space between them is a vacuum, devoid of material particles. The same problem arises with the other fundamental forces. Newton assumed that the gravitational force was felt instantaneously and that changes or rearrangements of the mass of the Earth or Sun were instantaneously apparent at any distance. Such instantaneous response was called **action-at-a-distance**, and this concept troubled physicists for many years because of its implicit assumption that the effects of forces propagate at infinite speeds.

In the 19th century, Michael Faraday invented another useful mechanism for force transmission. He proposed that a massive particle, an electric charge, or a magnet establishes a peculiar condition of space, called a **field**, in the space surrounding itself. *Other* massive particles, charges, or magnets placed *in* the fields feel their effect at the place where these other particles are located in space. Each fundamental force has its own field; thus, we have gravitational fields that act on masses, electric fields that act on electric charges, and magnetic fields that act on moving charges and magnets. We will see the usefulness of the field concept when we study gravitation, electricity, and magnetism in greater detail. Changes are propagated via the field at finite speeds.

Another view, born of the age of quantum mechanics, imagines force transmission via the exchange of surrealistic elementary particles. The exchanged particles, the carriers of the force, are called mediating particles (refer back to Table 5.1) to disguise the difficulty in imagining how nature actually manages to pull this off. We return to this question again in Chapter 9 when we investigate collisions between particles. The mediating particles for each of the four fundamental forces are listed in Table 5.2.

[†]*Opus Majus*, translated by Robert Belle Burke (University of Pennsylvania Press, Philadelphia, 1928), volume I, page 130.

TABLE 5.2 **Fundamental Forces and Their Mediating Particles**

Fundamental force	Mediating particle(s)
1. Gravitational force	graviton (not yet been detected experimentally)
2. Electromagnetic force	photon (the particle of light)
3. Strong force	8 gluons (collectively designated g_i)
4. Weak force	3 vector bosons (called W^+, W^-, and Z^0)

The gravitational and electromagnetic forces are long-range because the mediating particles for these forces have zero mass. The strong and weak forces are short-range because the mediating particles have nonzero mass. In other words, the *greater* the mass of the mediating particle, the *shorter* the range of the force. The connection between the range of the force and the mass of the mediating particle cannot be understood without quantum mechanics, so do not worry if the connection is cryptic to you; it certainly is *not* obvious.

The holy grail of theoretical physics is to show how these four forces naturally arise as special cases of a single all-encompassing theoretical structure or theory. Progress has been made toward such a **Theory of Everything**, but the goal has not yet been achieved. The progress toward unification is depicted schematically in Figure 5.1. The electromagnetic and weak forces were unified into a single, so-called **electroweak force** in the 1960s. This marriage of the forces is similar to, but more mathematically complicated than, the unification of the electric and magnetic forces by Maxwell during the 19th century. The strong force now is well on the way to being integrated into the structure. So far the gravitational force is the tough one, and it has eluded all attempts at integration into a grand scheme. We all can appreciate the common **rule of difficulty**: if you know how to do something, it is easy; if you do *not* know how to do something, it is hard. The unification problem is hard; nobody as yet knows how to do it all. The search continues. Keep your mind open and your imagination honed despite the orthodoxies thrust on you by current physical dogma. Nobel awards are lurking, but are yet unclaimed. Perhaps you will be the one to do it!

But hold on now. We are talking as if the notion of force and its measurement are obvious; they certainly are not. How do we go about defining what is meant by a force and how can it be measured? These topics we now address in Sections 5.3 and 5.4 by returning to the kinematics of a particle and the cause of changes in its motion.

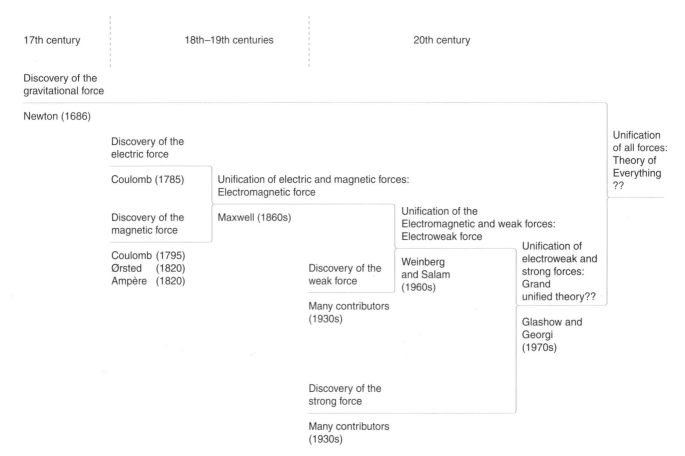

FIGURE 5.1 **The quest for unification of the fundamental forces of physics. (Of course, the forces themselves have existed since very near "the beginning," but we only became aware of them during the indicated time frames.)**

5.3 NEWTON'S FIRST LAW OF MOTION AND A QUALITATIVE CONCEPTION OF FORCE

Recall from Chapter 3 that a particle is any object whose geometric size is ignored; we think of a particle as represented by a point. The particle or particles whose motion is to be studied define a **system** and can be just about anything we want.

PROBLEM-SOLVING TACTIC

5.1 It is essential to specify clearly the system whose motion is to be examined. Think of drawing an imaginary surface around what you want to investigate; what is inside the imaginary surface is the system under consideration.

In this chapter we consider only systems of constant mass.

A system that has an acceleration equal to zero is said to be in a state of **mechanical equilibrium**.

The phrase mechanical equilibrium usually is contracted to simply the word equilibrium in classical mechanics.* The acceleration is the time rate of change of the velocity:

$$\vec{a} \equiv \frac{d\vec{v}}{dt}$$

Hence zero acceleration implies the system has a constant velocity vector, whose magnitude *and* direction do not change with time.

A system with a constant velocity is in equilibrium.

A system that remains at rest has a constant velocity equal to zero, and so is a special case of a system in equilibrium.

Slide a hockey puck across the horizontal surface of your desk, as in Figure 5.2. Since we are interested in the motion of the hockey puck, define it to be the *system*. After release, the puck slows down and eventually comes to a stop. While doing so, it is not in equilibrium since its velocity changes as it slows; the puck experiences a nonzero acceleration. Now slide the puck across a smooth, ice hockey rink. The velocity of the puck still changes, albeit more slowly than before (Figure 5.3). The puck again is not in equilibrium, but its acceleration is less than when it was slid across your desk.

*Later, when we study thermodynamics, we introduce another distinct concept known as *thermodynamic equilibrium*.

Galileo Galilei (1564–1643) first extrapolated such situations to an idealized *gedanken* (thought) experiment on a horizontal surface with *no* **friction**.[†] On such a so-called **frictionless surface**, the motion of the puck continues indefinitely at constant velocity (zero acceleration). The puck then is in equilibrium.

A system in equilibrium defines what we mean by **zero total force** on the system:

system in equilibrium	⇔	zero total force on the system
$\vec{a} = 0$ m/s^2		$\vec{F}_{\text{total}} = 0$

The term *total force* and the term *net force* mean the same thing.

The expressions total force "*on* the system," "*acting on* the system," and "*exerted on* the system" are synonymous and are used interchangeably.

Newton's first law of motion defines a state that we call zero total force on the system: *any system in mechanical equilibrium remains in mechanical equilibrium unless compelled to change that state by a **nonzero** total force acting on the system.*

system *not* in mechanical equilibrium	⇔	nonzero total force on the system
$\vec{a} \neq 0$ m/s^2		$\vec{F}_{\text{total}} \neq 0$

It is difficult to appreciate the revolutionary nature of Newton's first law of motion, since we now have the benefit of three centuries of hindsight. But prior to Galileo and Newton, it was thought that a nonzero total force was needed simply to maintain a constant velocity. Hands-on experience with everything from wine barrels to oxcarts indicated that to keep an object moving at constant velocity on a horizontal surface, a continual force (a push or pull) on the system is needed.[‡] We

[†]We define what we mean by friction more carefully later in the chapter. The word friction stems from the Latin *fricatus*, meaning "rubbing."

[‡]This generalization was made despite the existence of things such as bows and arrows that called the premise into question: the arrow obviously continues to move once it has left the bow.

FIGURE 5.3 Slide a hockey puck across an ice rink.

FIGURE 5.2 Slide a hockey puck across your desk.

now understand that our push or pull is not the only force or the total force in such situations. The only reason we need to continually push to maintain a constant velocity is to cancel the ever-present effects of frictional forces. The first law of motion states that *zero total force* exists when the system has a *constant velocity*; however, a *nonzero* total force is required to *change* the velocity. In other words, a nonzero total force on the system causes its acceleration.

Recall from Section 4.4 that a coordinate system, together with a set of synchronized clocks, all ticking at the same rate, constitute a reference frame. Those reference frames in which Newton's first law is found to be valid are called **inertial reference frames**; frames that are not inertial are called **noninertial reference frames**. These and other reference frames are examined in greater detail later in this chapter (Sections 5.7, 5.8, and 5.16).*

Thus Newton's first law gives a qualitative and quantitative definition of what we mean by zero total force; the law serves to define an inertial reference frame. Next we discuss a way to measure and quantify forces.

5.4 THE CONCEPT OF FORCE AND ITS MEASUREMENT

yea, every force entangles
Itself with strength

William Shakespeare (1564–1616)[†]

The mass of a system is a characteristic scalar property that indicates how sluggishly the system moves in response to forces (think of them as pushes or pulls) of various magnitudes; this property of the system (how slightly it responds to a force) is called **inertia**.[‡] The quantitative measure of inertia in the SI system is the kilogram. The mass of a system is the same whether it is here on the Earth, on the Moon, on Superman's fictional planet Krypton, or near the bright star Arcturus. The concept of mass really was invented to describe how a system responds to forces. Other perfectly legitimate ways of defining this characteristic inertial property of a system are possible, but they represent roads not taken in the history of physics (see Question 1 after reading this section).

Quantifying Force

Take the standard platinum–iridium mass cylinder at Sèvres, France, that defined the kilogram and place it on a horizontal frictionless surface; such a surface can be approximated by using

an airtable in many physics laboratories. The standard mass is the system whose motion we want to examine. Pull horizontally on the standard mass with a spring of negligible mass whose stretch can be measured, as shown in Figure 5.4. The stretch of the spring helps us quantify the concept of force.

A stronger pull by you on the spring produces a greater stretch of the spring. If you vary the strength of the pull (as indicated by the stretch of the spring), you find that different accelerations are imparted to the standard kilogram. You can tell if the strength of the pull is constant by checking to see if the extension of the spring stays the same while the system is pulled. The strength of the pull together with its direction indicates that the pull has the salient characteristics of a vector quantity: a magnitude and a direction. Indeed, experiments show that (a) if you pull in a certain horizontal direction, the mass is accelerated in the *same* horizontal direction as the pull; and (b) if you repeat the experiment using the same mass, with the same strength of the pull, and in the same direction, the *same* acceleration always is produced.

The strength of the pull needed to impart to the standard kilogram an acceleration of magnitude exactly 1 m/s^2 is defined to be a force of magnitude 1 **newton**, abbreviated N. See Figure 5.5.

The stretch of the spring corresponding to an acceleration of magnitude 1 m/s^2 of the standard kilogram can be marked on a scale adjacent to the spring. Thus an extension of the spring by

FIGURE 5.4 Pull on the standard kilogram with a spring.

FIGURE 5.5 A total force of magnitude 1 N produces an acceleration of exactly 1 m/s^2 of the standard kilogram.

---

*Not all reference frames are inertial reference frames. We will see in Section 5.16 that if a reference frame is accelerating (such as a coordinate system attached to an airplane accelerating along a runway), then observations and experiments made with respect to this reference frame do *not* obey Newton's first law. A coordinate system attached to and rotating with a spinning turntable or merry-go-round is another example of a noninertial reference frame.

[†]*Antony and Cleopatra*, Act IV, Scene 14, lines 48–49.

[‡]The word comes from the Latin word *inertia*, meaning "inactivity."

this amount corresponds to a force or pull of magnitude 1 N on *whatever* is attached to the spring.

Now perform another experiment: pull on the spring so the acceleration of the standard kilogram is precisely 2 m/s²; note and mark the extension or stretch of the spring. This extension of the spring defines a force with a magnitude of precisely 2 N (see Figure 5.6). In this way the spring can be calibrated into an instrument capable of indicating the magnitude of a force in newtons. Further experiments are conducted with the calibrated spring on different systems. If the system consists of *two* replicas of the standard kilogram, the system has a mass of 2 kg. If the calibrated spring is used to pull and accelerate the 2 kg system, experiments indicate that

1. a force of magnitude 1 N imparts an acceleration of precisely 0.5 m/s² to the 2 kg mass; and
2. a force of magnitude 2 N imparts an acceleration of magnitude 1 m/s² to the 2 kg mass.

Indeed, the magnitude of the acceleration produced by a given magnitude force is *inversely* proportional to the mass of the system; the direction of the acceleration always is in the direction of the force or pull. For a system of constant mass, a graph of the magnitude of the force in N used in these experiments versus the magnitude of the resulting acceleration in m/s² (see Figure 5.7) is *linear* with zero intercept, and the *constant slope* of the graph is found to be numerically equal to the mass of the system in kg.

The slope of *every* such graph is positive; no system ever has been found with a negative slope for such a graph. Hence mass is always positive.

FIGURE 5.6 If the acceleration of the standard kilogram is exactly 2 m/s², the magnitude of the total force on it is exactly 2 N.

FIGURE 5.7 A graph of the magnitude of the total force on a system versus the magnitude of its acceleration is linear with zero intercept and a slope equal to the mass of the system.

Force Is a Vector Quantity

Further experiments with *two* calibrated springs indicate that if a system on a horizontal frictionless surface is subjected to two pulls of the same magnitude in opposite horizontal directions, no acceleration results (see Figure 5.8). As Newton's first law states, with zero acceleration there must be zero total force on the system. The two applied forces, therefore, sum to zero like vectors, and so the total force on the system is zero.

Likewise, if the two pulls of the same magnitude are applied in the same horizontal direction (see Figure 5.9), the acceleration of the system is twice that with one pull of the same magnitude; the individual forces again add like vectors. An experiment with two forces of different magnitudes and directions (see Figure 5.10) produces an acceleration consistent with a total force that is the vector sum of the two.

These experiments indicate that *forces are vectors* and the acceleration of the system depends on the vector sum of all forces on the system. The vector sum of all the forces on a system is the *total force* $\vec{F}_{total}$ on the system.

The system

FIGURE 5.8 Two forces of equal magnitude and opposite direction produce zero acceleration of the system.

The system

FIGURE 5.9 Two forces of equal magnitude and the same direction produce an acceleration twice that produced by either force acting alone on the system.

The system

FIGURE 5.10 Two forces with different magnitudes and directions produce an acceleration consistent with a total force that is the vector sum of the two.

5.5 NEWTON'S SECOND LAW OF MOTION

. . . it will be found that everything depends on the composition of the forces with which the particles of matter act upon one another; and from these very forces, as a matter of fact, all phenomena of Nature take their origin.

Roger Joseph Boscovich (1711–1787)*

The results of the experiments discussed in Section 5.4 and the graph in Figure 5.7 are summarized by **Newton's second law of motion**, one of the great and unifying laws of physics:

A system of constant mass m, subjected to a totality of individual forces whose vector sum is $\vec{F}_{total}$, experiences an acceleration $\vec{a}$ that is inversely proportional to the mass m of the system and is in the same direction as $\vec{F}_{total}$:

$$\vec{a} = \frac{\vec{F}_{total}}{m}$$

This is usually written as

$$\vec{F}_{total} = m\vec{a} \tag{5.1}$$

In terms of the base units of the SI system, one newton is equivalent to one kilogram meter per second squared ($kg \cdot m/s^2$), since (from Equation 5.1) a force of magnitude 1 N on the standard kilogram produces an acceleration of magnitude $1 \ m/s^2$. That is,

$$1 \text{ N} = (1 \text{ kg})(1 \text{ m/s}^2)$$
$$= 1 \ kg \cdot m/s^2$$

The characteristic scalar property of the system that appears in Newton's second law of motion is called the **inertial mass** of the system or, more colloquially, simply the **mass**.

Keep in mind that mass is an *abstract concept, defined by experiment*. We could just as easily have called the inertial property of matter by other names: "stuff" or "inertial charge"; but we choose to use the word *mass*.[†] The mass of a system relates the response of the system (its acceleration) to the total force *on* the system. Ultimately, we do not know what mass really is, or what gives rise to the property of inertia. Whatever this inherent inertial characteristic of matter really is, we call it mass. We will encounter other such abstract ideas in physics, invented to describe the results of various experiments and observations.[‡]

Newton's second law of motion, Equation 5.1, is a *vector equation*. We saw in Chapter 2 that a vector equation is equivalent to *three scalar equations*, one for each of the components of the vectors (with respect to some coordinate system). In a Cartesian coordinate system, the *x*-component of the left-hand side of the equation is equal to the *x*-component of the right-hand side of the equation, and likewise for the other two

Cartesian directions. That is, Equation 5.1 represents the following three scalar equations for the Cartesian components:

$$\begin{aligned} F_{x \, total} &= ma_x \\ F_{y \, total} &= ma_y \\ F_{z \, total} &= ma_z \end{aligned} \tag{5.2}$$

QUESTION 1

Suppose Newton did things differently and found that a particle has a scalar property χ associated with it such that the scalar multiplication of this characteristic property with the total force on the particle is equal to the acceleration of the particle; that is,

$$\chi \vec{F}_{total} = \vec{a}$$

(a) For a given total force $\vec{F}_{total}$ on the particle, large values of χ produce what kinds of accelerations? (A qualitative answer is OK.) What about small values of χ? (b) Invent a descriptive name that might describe this property χ of the particle. (c) If the magnitude of $\vec{F}_{total}$ is plotted versus the magnitude of $\vec{a}$, what does χ represent on the graph? (d) Suppose two particles, one with this property of value χ_1 and one with value χ_2, are fastened together to make a single system. What is the effective χ value of the combination? (e) What are the SI units associated with this property χ? (f) How is the property χ related to the scalar property known as the mass m of the particle, or is it related at all? (g) Which property, χ or m, is more *convenient* to use, or is this simply a question of which concept (χ or m) we are more familiar with? These questions should convince you that what we call the mass of a system is an inventive theoretical construct of our mind, not something inherently natural, waiting for someone to discover.

Newton's second law of motion, Equation 5.1, is laden with meaning and subtlety that belie its innocuous appearance. Before tackling applications of the law, we emphasize some of its subtle aspects.

1. The left-hand side of the equation is the *vector sum of all forces acting on the system* of mass m. The preposition "on" is key as well. *Only* the vector sum of the forces *on the system*[§] is important for determining the motion of the system. We shall see later that forces *by* the system *on* other things are not involved in the motion of the system itself; they determine the motion of the other things.

2. Since Newton's second law is a vector equation, the direction of the total force vector and the direction of the acceleration vector are the same.

3. If the vector sum of all the forces *on* the system is zero, then the left-hand side of the second law of motion is zero:

$$0 \text{ N} = m\vec{a}$$

Theoria Philosophiae Naturalis (Venice, 1763), Part I, Section 5. [*A Theory of Natural Philosophy* (MIT Press, Cambridge, Massachusetts, 1966), page 20.]

[†]The word is believed to come from an ancient Greek word for "barley-cake."

[‡]Among them is the familiar concept of electric charge. We also do not know what gives rise to the property of matter we call electric charge.

[§]Remember that the phrases "on," "acting on," and "exerted on" a system are synonymous.

which implies that the acceleration vector $\vec{a}$ also is zero, and we get back Newton's first law of motion. You might be tempted to say that the first law is a special case of the second; do not be led into such temptation! As we mentioned in Section 5.2, Newton's first law *defines* what we mean by zero total force and an inertial reference frame, and so the first law has independent significance. As a matter of historical fact, the first law was discovered by Galileo.

4. Having $\vec{F}_{total}$ (the left-hand side of Newton's second law) equal to zero does *not* necessarily mean there are *no* forces on the system; it *does* mean that the vector sum of all forces on the system is zero.* Conversely, if the acceleration $\vec{a}$ of the system is zero, then the total force $\vec{F}_{total}$ on the system is zero and the system is in equilibrium.

5. The two sides of the equation are equal to each other, but the equality is of a peculiar sort. Imagine a wall of separation between the two sides of the equation: on the left are the physical forces acting on the system; on the right is simply the mass times the acceleration of the system, which we can think of as the *response* of the system to the forces on it. Think of the forces involved in $\vec{F}_{total}$ as distinct from the effects they cause. In other words, the total force is *not the same thing* as mass times an acceleration. The product $m\vec{a}$ is not what we mean by force; $m\vec{a}$ is the physical result of the action of forces on the system. The left-hand side represents a kind of action: the forces *on* the system. The right-hand side represents the response of the system to those forces. This wall of separation is a subtle point, but an important one; it will help you to avoid many pitfalls in applications of the second law of motion to physical situations.

PROBLEM-SOLVING TACTIC

5.2 The two sides of Newton's second law represent different things. The *left*-hand side of the law involves the vector sum of the *forces on the system*. The *right*-hand side is the product of the mass (a characteristic scalar property of the system) and the acceleration (a kinematic aspect of the system), which is the *response* of the system to the forces acting on it.

One of the best examples of the distinction between the left- and right-hand sides of the second law of motion arises in uniform circular motion. In this setting, the product of the mass times the magnitude of the centripetal acceleration

$$m\frac{v^2}{r}$$

has the units of a force (kg·m/s^2 or N); but the product is *not* a force. Many texts and, regrettably, many physicists and engineers, call the product of the mass and the centripetal acceleration a "centripetal force." This is, in fact,

The result of an appropriate, short-lived, contact force on a baseball: smiles and a home run!

poor terminology that should be avoided. We will not use the term "centripetal force," since it is simply *not a force*. The left-hand side of the second law, then, involves forces such as the pulls measured by our springs or, as we shall see later, forces caused by tensions in cords or ropes, gravitational forces, pushes or pulls, and so on. We will examine the characteristics of these individual forces later in the chapter. The quantity mv^2/r is simply the product of the mass times the magnitude of the centripetal acceleration; it *is* the magnitude of the *right*-hand side of the second law equation for uniform circular motion.

6. There are two types of forces. For a ball hit by a bat, it is clear that the force of the bat on the ball exists only during the brief time interval during which they are in physical contact with each other; this is an example of a **contact force**, as we mentioned in the introduction to this chapter. Likewise, when you push a baby stroller with your hands, this force on the stroller exists only as long as your hands are in physical contact with it. On the other hand, the force of the Earth on a falling rock is an example of a force where physical contact between the two objects is not required; gravitation is a **noncontact force**. The other fundamental forces of nature (see Section 5.2) also are noncontact forces; contact forces are useful, macroscopic manifestations and simplifications of one or more of the fundamental forces (see Section 5.15).

EXAMPLE 5.1

A total force on a 10.0 kg mass produces an acceleration of magnitude 3.00 m/s^2 in the direction indicated in Figure 5.11. Find the magnitude and direction of the total force on the mass.

$|\vec{a}| = 3.00$ m/s^2

FIGURE 5.11

---

*Remember our experiment in Figure 5.8 with two springs pulling in opposite directions with equal magnitude forces; there are forces on the system, but their vector sum—that is, the total force—is zero.

Solution

Choose a coordinate system with the $+x$-axis parallel to the acceleration as indicated in Figure 5.12. The system is the 10.0 kg mass. The total force acting on the system is parallel to the acceleration. Apply Newton's second law of motion (Equation 5.1) to the system to find the total force:

$$\vec{F}_{total} = m\vec{a}$$
$$= (10.0 \text{ kg})(3.00 \text{ m/s}^2)\hat{\imath}$$
$$= (30.0 \text{ kg·m/s}^2)\hat{\imath}$$
$$= (30.0 \text{ N})\hat{\imath}$$

$|\vec{a}| = 3.00 \text{ m/s}^2$

The system

FIGURE 5.12

STRATEGIC EXAMPLE 5.2

The only two forces acting on a 5.00 kg mass are indicated in Figure 5.13.

FIGURE 5.13

a. Determine the total force $\vec{F}_{total}$ acting on the mass.
b. Find the magnitude of the total force.
c. Find the acceleration of the mass.
d. Determine the magnitude of the acceleration.

Solution

Let the 5.00 kg mass be the system. Introduce the coordinate system indicated in Figure 5.14, with $\hat{\imath}$ parallel to $\vec{F}_1$.

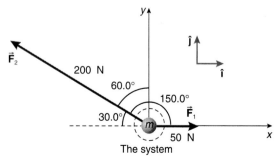

The system

FIGURE 5.14

a. Write each of the forces in Cartesian form. Since $\vec{F}_1$ is parallel to $\hat{\imath}$,

$$\vec{F}_1 = (50 \text{ N})\hat{\imath}$$

The Cartesian representation of $\vec{F}_2$ may be found in two ways.

Method 1

The Cartesian representation may be found from

$$\vec{F}_2 = (\vec{F}_2 \cdot \hat{\imath})\hat{\imath} + (\vec{F}_2 \cdot \hat{\jmath})\hat{\jmath}$$

The angle between $\vec{F}_2$ and $\hat{\imath}$ is 150.0° and the angle between $\vec{F}_2$ and $\hat{\jmath}$ is 60.0°, and so you can evaluate the scalar products:

$$\vec{F}_2 = (200 \cos 150.0°)\hat{\imath} + (200 \cos 60.0°)\hat{\jmath}$$

Method 2

Equivalently, you may find the Cartesian representation more geometrically from Figure 5.14:

$$\vec{F}_2 = (200 \cos 30.0°)(-\hat{\imath}) + (200 \sin 30.0°)\hat{\jmath}$$

Using either method, you find

$$\vec{F}_2 = (-173 \text{ N})\hat{\imath} + (100 \text{ N})\hat{\jmath}$$

The total force on the system is the vector sum of the two forces:

$$\vec{F}_{total} = \vec{F}_1 + \vec{F}_2$$
$$= (50 \text{ N})\hat{\imath} + [(-173 \text{ N})\hat{\imath} + (100 \text{ N})\hat{\jmath}]$$
$$= (-123 \text{ N})\hat{\imath} + (100 \text{ N})\hat{\jmath}$$

b. The magnitude of the total force is the magnitude of this vector:

$$F_{total} = [(-123 \text{ N})^2 + (100 \text{ N})^2]^{1/2}$$
$$= 159 \text{ N}$$

c. Newton's second law of motion enables you to find the acceleration.

Method 1

Use Equation 5.1,

$$\vec{F}_{total} = m\vec{a}$$

and solve for $\vec{a}$:

$$\vec{a} = \frac{(-123 \text{ N})\,\hat{\imath} + (100 \text{ N})\,\hat{\jmath}}{5.00 \text{ kg}}$$
$$= (-24.6 \text{ m/s}^2)\,\hat{\imath} + (20.0 \text{ m/s}^2)\,\hat{\jmath}$$

Method 2

An equivalent way of finding the acceleration is to write the second law for each coordinate direction separately using Equation 5.2. That is, along the x-direction we have a total force component $F_{x\,total} = -123$ N. The x-component of the acceleration, a_x, then is found using Equation 5.2:

$$F_{x\,total} = ma_x$$
$$-123 \text{ N} = (5.00 \text{ kg})a_x$$

Solving for a_x,

$$a_x = -24.6 \text{ m/s}^2$$

Correspondingly, along the y-direction the total force component is $F_{y \text{ total}} = 100$ N. The y-component of the acceleration, a_y, then is found using Equation 5.2:

$$F_{y \text{ total}} = ma_y$$
$$100 \text{ N} = (5.00 \text{ kg})a_y$$

Solving for a_y,

$$a_y = 20.0 \text{ m/s}^2$$

The acceleration vector is

$$\vec{a} = a_x\hat{\imath} + a_y\hat{\jmath}$$
$$= (-24.6 \text{ m/s}^2)\hat{\imath} + (20.0 \text{ m/s}^2)\hat{\jmath}$$

d. The magnitude of the acceleration is the magnitude of this vector:

$$a = [(-24.6 \text{ m/s}^2)^2 + (20.0 \text{ m/s}^2)^2]^{1/2}$$
$$= 31.7 \text{ m/s}^2$$

5.6 NEWTON'S THIRD LAW OF MOTION

Then there's a pair of us!

Emily Dickinson (1830–1886)*

Part of Newton's genius lay in realizing that a complete theory of dynamics needed one additional tenet. An experiment will show why. When the spring pulls on the standard kilogram in Figures 5.4 and 5.5, the standard kilogram pulls on the spring (thus stretching the spring) with a force of equal magnitude but in the opposite direction. Try another experiment: bang your hand on the desk, thus exerting a force *on* the desk. The desk reciprocates and exerts a force *on* your hand, which is the force you feel when you do this exercise. Bang downward *on* the desk harder and the desk responds in kind with a correspondingly increased upward force *on* your hand. Continue banging even harder and eventually the force of the desk on your hand is so great that your hand hurts (ouch!); the force you exert on the desk only "hurts" the desk if you have a black belt in karate.

Newton's third law of motion is an insightful generalization of the common observations just made:

If a system A exerts a force on another system B, then B exerts a force of the same magnitude on A but in the opposite direction. That is,

$$\vec{F}_{A \text{ on } B} = -\vec{F}_{B \text{ on } A} \qquad (5.3)$$

In other words, *forces always occur in pairs*, called **third law force pairs**.

This is nature's way of doing tit-for-tat. The law is colloquially called the action–reaction law, although this phrase conveys little information about the content of the third law and should be banished from the literature.[†]

There are several important aspects of the third law of motion that warrant emphasis:

1. Forces occur in *pairs* (like socks).
2. Each member of a given force pair has the *same magnitude* (the socks of a given pair are the same size).
3. The two members of a given force pair point in *opposite directions* (socks are *not* worn this way!).
4. The paired forces act on *different systems* (socks are worn on different feet).

Once you grasp these aspects of the third law of motion, you will be well on your way to being a master classical mechanic.

PROBLEM-SOLVING TACTICS

5.3 Watch the little words. The words *of*, *on*, and *by* have important ramifications in the Newtonian theory of motion, as we alluded to before. Frequently the preposition *of* is used synonymously with *by*, so that we say the force *of* A on B or, equivalently, the force *by* A on B.

5.4 Remember that the two forces in a third law force pair act on different systems. This is a critical point. If the two members of a third law force pair were to act on the same system, there would never be a nonzero total force on a system and it would always be in equilibrium, which is clearly not the case for all systems.

5.5 Decide what the systems are before you sort out third law force pairs. Consider only the forces acting on a chosen system when applying the second law of motion to it. Only *one* of every third law force pair is involved in applying the *second law of motion* to a particular system. If you are interested in the motion of system A, then the force *of* B on A is relevant; the other member of this third law force pair, the force *of* A on B, is irrelevant to system A, since this force does not act *on* system A.

When you bang your hand on the desk, the two systems are your hand and the desk. According to Newton's third law, the force you exert *on* the desk in the downward direction has a third law counterpart: the upward force of the desk *on* your hand (see Figure 5.15). The two members of this third law force pair have exactly the same magnitude, point in opposite directions, but *act on different systems*.

When you slam a home run, the bat exerts a huge force on the ball, but the ball exerts a force of *identical magnitude* on the bat in the opposite direction. The two forces represent a third law force pair: they are of equal magnitude, are directed in opposite directions, but act on different systems. Likewise, when a little car crashes into a truck, the force of the truck on the little

FIGURE 5.15 A third law force pair when you bang on a desk.

car and the force of the little car on the truck are a Newton's third law force pair; *the two forces have the same magnitude!* The *effect* of the force on each system is another matter.

QUESTION 2

A seer inquires: "According to Newton's third law of motion, forces always occur in pairs that are equal in magnitude and opposite in direction. This means every third law pair of forces vector sums to zero. How, then, can anything ever accelerate?" What is the flaw in this line of reasoning?

5.7 LIMITATIONS TO APPLYING NEWTON'S LAWS OF MOTION

There are several limitations associated with applying Newton's laws of motion.

Reference Frames

Newton's first law defined an inertial reference frame: a reference frame in which a system in equilibrium experiences zero total force. Newton's *second law* only can be used in such inertial reference frames. Let's look at an example of what *is* and what *is not* an inertial reference frame. Place a coin at rest on a level frictionless surface in the laboratory, as in Figure 5.16. The coin remains at rest. Its acceleration is equal to zero and there is zero total force acting on the coin. There *are* forces acting on the coin,* but the vector sum of these forces is zero.

So the laboratory is an inertial reference frame: Newton's first and second laws of motion are obeyed in this reference frame. Any coordinate grid attached to the laboratory serves as an appropriate reference frame for analyzing dynamical problems using Newton's laws.

Now imagine yourself in another laboratory that is spinning around a vertical axis in the first laboratory, much like a merry-go-round. Suppose a horizontal frictionless surface also exists in the spinning laboratory and physicist A. S. Tute in the *nonspinning* lab slowly places a coin (at rest) on the frictionless surface

---

*We shall see later that these forces are the downward gravitational force of the Earth on the coin (the weight of the coin) and the upward force of the surface on the coin.

FIGURE 5.16 There is zero total force acting on a coin at rest on a table.

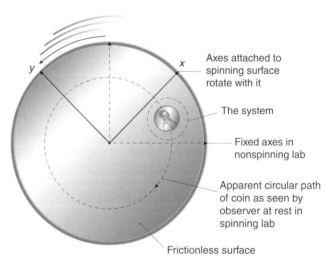

FIGURE 5.17 A coin placed on a frictionless, spinning surface appears to move in a circle with respect to axes attached to the spinning surface.

that is spinning (see Figure 5.17). In the nonspinning lab, the coin remains at rest where it was placed, in accordance with Newton's first and second laws. Yet, lo and behold, the coin is observed by physicist D. Z. Dean in the spinning lab to move in a circle with respect to local coordinate axes attached to the spinning lab!

In the case of the spinning lab, no horizontal force exists on the coin, yet the spinning observer says there indeed is a centripetal acceleration with respect to the coordinate grid in the spinning lab, because the coin is seen to be moving in a circle. Newton's second law of motion is *not* obeyed relative to the reference frame of the spinning lab; hence this is a noninertial reference frame.

> Noninertial reference frames are accelerated reference frames.

> Newton's laws of motion only can be applied in reference frames that are themselves not being accelerated, that is, only in inertial reference frames.

Speed Limits

Newton's second law of motion provides a way of determining the acceleration of a system from the total force acting on it. Newtonian dynamics implicitly assumes that *any* speed, however large, can be attained by the system under the influence of a rectilinear acceleration, given enough time. However, there actually is a speed limit in the universe: the speed of light, always called c (exactly 299 792 458 m/s). The existence of this speed limit implies that Newtonian dynamics must be modified for very large speeds. Such modifications for speeds that are significantly large fractions of the speed of light are a new dynamical theory, called *relativistic dynamics* (which we study in Chapter 25).

> Newtonian mechanics, therefore, is a good *approximation* as long as the speed of the system is much less than the speed of light.

Most common objects of macroscopic size have speeds significantly less than the speed of light, and so Newtonian mechanics is an excellent approximation.

Quantum Mechanics

Experiments originally performed during the early 20th century also indicated that Newtonian mechanics cannot describe or account for many phenomena on the atomic and nuclear scale. This problem instigated the development of a radically new mechanics, called *quantum mechanics*. The relationship between the two types of mechanics was first elucidated by Niels Bohr and others, but the details still are an active area of research today. We briefly consider some of the ideas of quantum mechanics in Chapter 27.

Force Propagation

The third law of motion assumes that forces propagate instantaneously through space and matter at infinite speed during zero time intervals. Change the mass of the Sun, and the third law implies that the effect is felt throughout the universe instantaneously. *This is not strictly true.* If you attach a mass to one end of a long slinky and suddenly pull on the other end, the mass does not accelerate at the instant of the pull, but at sometime later when the force finishes its propagation down the length of the slinky and "arrives" at the mass, as pictured in Figure 5.18.

FIGURE 5.18 Pull on one end of a slinky and a mass at the other end does not respond to the force until a short time interval later.

Likewise, gravitational forces or other action-at-a-distance forces (such as electrical and magnetic forces, discussed later in the text) are not felt everywhere instantaneously. In other words, forces somehow take time to propagate from one place to another. Newton's third law does not account for such propagation delays and, therefore, is not rigorously true over finite distances at all time instants. It is true on a point-by-point basis as the influence or force propagates through space or matter.

> In this text, we assume that the third law of motion is rigorously true; however, modifications must be made at more advanced levels of study to account for the finite propagation speeds.

QUESTION 3

Is the surface of the rotating Earth an inertial reference frame or only *approximately* such a reference frame? Explain.

5.8 INERTIAL REFERENCE FRAMES: DO THEY REALLY EXIST?*

We like to think that the laws of physics (whatever they may be) have the same form in any reference frame. Observers in different reference frames then can approach physical problems using the same mathematical forms for the fundamental laws of nature. Everyone, in any reference frame, can use the same physics book; the descriptions of phenomena may differ depending on the reference frame,[†] but the underlying laws of physics are the same. While a requirement that the laws of physics be the same in *every* reference frame is elegant, aesthetic, and philosophically satisfying, there are untoward mathematical side effects from such a demand. Unfortunately, to make the laws of physics have the same form in *all* reference frames forces us into mathematical quagmires that are inappropriate for an introductory course. Thus we restrict ourselves to a subset of reference frames in which the laws of physics are just a lot easier to write and apply.

As we have seen, there are two distinct types of reference frames:

1. Inertial reference frames are those in which Newton's first law of motion is obeyed. Reference frames either at rest or moving at constant velocity with respect to an inertial reference frame also are inertial frames.
2. Noninertial reference frames are those in which Newton's first law of motion is not obeyed. We mentioned one noninertial reference frame in Section 5.7: a spinning reference frame is an accelerated—and thus noninertial—reference frame. Coordinate systems that are undergoing any acceleration also are noninertial reference frames. Accelerated reference frames (noninertial frames) are considered briefly in Section 5.16; in such accelerated frames Newton's second law of motion must be modified to describe the motion of a system.

[†]For example, the description of the motion of a car depends on whether you are observing the auto from inside the car or outside along the road.

We saw in Section 4.4 that observers in two *different* inertial reference frames measure the *same* acceleration for a given system. Since both inertial reference frames measure the same acceleration for the system, they must agree on the vector sum of the forces $\vec{F}_{total}$ on it. Newton's second law has the same form and describes the motion in *either* inertial reference frame. Since the total force on the system is the same in each inertial frame, Newton's third law of motion also applies within each inertial reference frame.

Do inertial reference frames actually exist? We can only approximately confirm it. The Earth spins on its axis, and so laboratories on the Earth actually are subjected to (small) centripetal accelerations toward the axis of rotation of the Earth. Therefore laboratories on the Earth are, strictly speaking, noninertial reference frames.

To describe precisely the motion of particles moving on or near the surface of the Earth, using coordinate axes attached to the surface, requires slight modifications to Newton's second law. The flavor of such modifications is considered in Section 5.16 for a rectilinear accelerated reference frame.

Likewise, the Earth itself revolves about the Sun, and so the Earth as a whole experiences a (small) centripetal acceleration toward the center of the Sun; thus, the Earth as a whole is a noninertial reference frame. Correspondingly, the Sun is in orbital motion about the center of the Milky Way Galaxy; our Galaxy also has complex motions as part of the so-called Local Group or cluster of galaxies; and so on. Reference frames attached to each of these objects thus are accelerated reference frames and, therefore, are noninertial reference frames in a strict sense. However, the magnitudes of these accelerations are quite small compared with that of the local acceleration due to gravity, $g = 9.81$ m/s^2.

We assume the surface of the Earth is a rigorous inertial reference frame, but realize this is an approximation.

5.9 SECOND LAW AND THIRD LAW FORCE DIAGRAMS

. . . the contemplation falls into unfathomable ecstasies in view of all these decompositions of forces resulting in unity.
Victor Hugo (1802–1885)*

To apply the Newtonian laws of motion to a chosen system, we must *identify all forces acting on the system*. In many situations, the best choice for the system itself is dictated by the problem we want to solve; in other cases, it is less obvious and we have to choose an appropriate system with some thought.

Once the system itself is clearly defined, draw a schematic **second law force diagram**[†] that

1. identifies the system; and
2. illustrates all forces acting on the system with their directions indicated explicitly in the diagram.

Once all forces acting on the system are identified and appropriately directed in a second law force diagram, *choose a coordinate system*. Then write the forces as vectors with components in

---

Les Misérables (1862) (A. L. Burt, New York), Saint Denis, Book III, Chapter 3.
[†]Such second law force diagrams also are known as *free-body diagrams*.

terms of the chosen coordinate system *consistent* with their directions on the force diagram. Appropriate vector sums of all the forces are then made and the motion is described in analytical detail using Newton's second law of motion.

> **PROBLEM-SOLVING TACTICS**
>
> **5.6** Identify the system and indicate the forces acting on it in a sketch (a second law force diagram).
>
> **5.7** If the direction of a particular force is unknown, presume a direction. The analysis will indicate whether the presumed direction is the actual direction or not by the algebraic sign of the answer.

Occasionally, we also draw **third law force diagrams** to indicate schematically the pairs of forces involved in Newton's *third* law. Each such diagram always has only two forces (the third law force pair), which always are the same magnitude, are in opposite directions, and act on two different systems. See Examples 5.3 and 5.4, after reading the next section.

5.10 WEIGHT AND THE NORMAL FORCE OF A SURFACE

But thou hast ordained all things in measure, number, and weight.
Apocrypha, Wisdom of Solomon, 11:20

Remember from Chapter 1 that the mass of a system and its weight are very different. The mass is a scalar measure of the inertia property of the system, and its value is independent of its location in space. On the other hand, the weight $\vec{w}$ is the gravitational force vector on the system and depends on its location in space.

We examine the nature of the gravitational force in some detail in Chapter 6. However, since the gravitational force is ubiquitous on the Earth, we need to develop a preliminary understanding of it at this juncture.

Recall from Chapter 3 that any freely falling object near the surface of the Earth experiences an acceleration of magnitude $g \approx 9.81$ m/s^2, directed vertically downward. This acceleration occurs because the gravitational force of the Earth on the falling object is the total force on it. Apply Newton's second law of motion, Equation 5.1, to the freely falling system. On the *left*-hand side of the equation, the only force acting on the falling mass m (neglecting air resistance) is its weight, so that here $\vec{F}_{total} \equiv \vec{w}$. On the *right*-hand side of the second law, the acceleration is of magnitude g, directed down. Using the *magnitudes* of the vectors, Equation 5.1, $\vec{F}_{total} = m\vec{a}$, becomes

$$w = mg \qquad (5.4)$$

The second law force diagram for the freely falling mass m has the single force $\vec{w}$, indicated in Figure 5.19. Thus the gravitational

The system

FIGURE 5.19 A freely falling system accelerates because the gravitational force is the total force on it.

$\vec{w} = \vec{F}_{total}$

Your weight causes your acceleration when you skydive.

FIGURE 5.20 When you stand on the ground, the total force on you is zero.

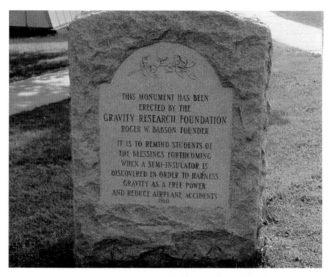

A monument on the campus of Middlebury College, Middlebury, Vermont. A similar monument exists at Colby College, Waterville, Maine.

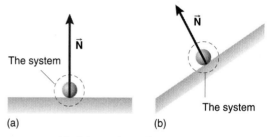

FIGURE 5.21 (a) If the surface is horizontal, then the normal force of the surface on a system is vertical (perpendicular to the surface). (b) If the surface is not horizontal, then the normal force of the surface on a system is *not* vertical, but it still is perpendicular to the surface.

force $\vec{w}$ of the Earth on a system of mass m has a *magnitude w* equal to mg and is directed vertically down.

> The gravitational force of the Earth acts on *every* object on or near the Earth, *whether falling or not*, and is of magnitude mg. Therefore every object near the surface of the Earth has at least one force acting on it, its weight $\vec{w}$.

It is interesting that we do not directly sense our weight. We tend to ignore the most familiar. When you jump off a rock wall or skydive from an airplane (before the parachute opens), you sense no pull in the downward direction as you fall. But the gravitational force of the Earth on you is there nonetheless, whether you are falling or simply standing on the surface of the Earth. You are not aware of your weight because the force never ceases to act on you; we simply take it for granted because it is impossible to eliminate it or insulate ourselves from it.

When you stand at rest on the surface of level ground (see Figure 5.20), you are not accelerating ($\vec{a} = 0$ m/s^2). Hence $\vec{F}_{\text{total}} = 0$ N from Newton's second law of motion.

Call the force that the surface exerts on you $\vec{N}$.* Then since $\vec{a} = 0$ m/s^2, the vector sum of $\vec{N}$ and the ever-present weight $\vec{w}$ must be zero. Equation 5.1, $\vec{F}_{\text{total}} = m\vec{a}$, becomes

$$\vec{N} + \vec{w} = m(0 \text{ m/s}^2)$$
$$= 0 \text{ N}$$

Transposing,

$$\vec{N} = -\vec{w}$$

The minus sign indicates that $\vec{w}$ and $\vec{N}$ are in opposite directions. Thus the surface of the level ground exerts a force on you equal in magnitude to your weight, but directed vertically upward since $\vec{w}$ is directed vertically down. The force you *do* feel is the force $\vec{N}$ of the level ground *on* you.

> The force that a surface exerts perpendicular to the surface is called a **normal force**, designated by $\vec{N}$. The word **normal** means perpendicular to the surface. Normal forces always are directed perpendicular to the surface, whatever its orientation (see Figure 5.21).

*Do not confuse the normal force $\vec{N}$, nor its magnitude N, with the SI unit for force, the newton, which we abbreviate as N. As mentioned in Chapter 1, some people write all units in parentheses to minimize the chances of falling for such ambiguities.

When you jump off the top of a rock wall and briefly experience free-fall, the force you sensed when standing on the rock wall (the normal force of the surface on you) no longer acts on you, since it was a contact force. However, the gravitational force of the Earth on you (your weight) nonetheless still exists, even though you do not sense its pull. It is your weight that causes your acceleration when falling.

While undergoing free-fall, it is colloquially said that you are "weightless," but this terminology is a *misnomer* and dangerous for physics students. The Earth *never ceases* to exert a gravitational force on you in free-fall (or in other circumstances on or near the Earth); since the gravitational force of the Earth on you is your weight, you are *never* weightless.

Imagine yourself standing on a bathroom scale in free-fall (see Figure 5.22). During the free-fall the upward force that the surface of the scale exerts on you is *zero*, because the scale also is in free-fall with you. The scale exerts no upward force on you, and you exert no downward force on the scale. The two forces are a Newton's third law pair of forces, with each force of zero magnitude during free-fall. The scale reads zero during free-fall, and people commonly refer to this observation as weightlessness. The *sensation* of weightlessness, then, is what happens in free-fall. But in such free-fall, you still have weight; you are really normal-force-less, not weightless.

In terms of Newton's third law of motion, you are one system and the Earth is another system. Since the Earth exerts a gravitational force on you called your weight $\vec{w}$, *you* exert a gravitational force of equal magnitude *on the Earth* in the opposite direction, according to Newton's third law of motion, as shown in Figure 5.23. A third law force diagram includes the force of the Earth on you and the force of you on the Earth, since they are a Newton's third law force pair. This pair of forces exists whether

you are in free-fall or not. You and the Earth are parts of another, larger system, a total system.

Likewise, when you stand on level ground, as in Figure 5.24, the ground exerts a normal force on you directed perpendicular to the surface of the ground, in this case vertically upward, and you exert a force of the same magnitude in the opposite direction *on the ground*; these two forces also are a Newton's third law force pair. In this case there also is another third law force pair: the gravitational force of the Earth on you (your weight) and the gravitational force of you on the Earth (this third law pair is identical to Figure 5.23).

QUESTION 4

Does your weight vary depending on whether you stand on a level or inclined surface? Explain your answer carefully.

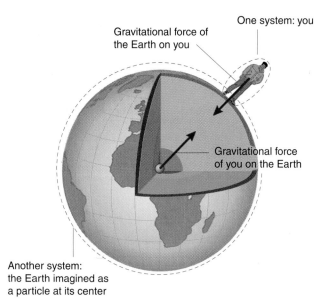

FIGURE 5.23 A third law force diagram of you and the Earth.

FIGURE 5.22 Free-fall with a bathroom scale contacting your feet.

FIGURE 5.24 A third law force pair if you are on the ground.

QUESTION 5

If you stand on a bathroom scale on a level surface and then when it is on an inclined surface, will the reading of the scale be the same in both cases? Why or why not? Explain your answers carefully. Does the scale really measure your weight? If not, what does it really measure?

EXAMPLE 5.3

Your mass is 70.0 kg, and you are standing at rest on level ground.

a. Make a second law force diagram indicating all forces acting on you.
b. What is your weight?
c. What is the normal force of the ground on you?
d. For *each* force on you, make a third law force diagram, identifying the forces in each third law pair.

Solution

a. You are the system and the two forces acting on you are (see Figure 5.25)

FIGURE 5.26

The system is you

FIGURE 5.25

1. the gravitational force of the Earth on you (your weight), directed vertically down, and
2. the normal force of the surface on you, directed perpendicular to the surface, here vertically up.

b. From Equation 5.4, your weight has a magnitude w equal to

$$w = mg$$
$$= (70.0 \text{ kg})(9.81 \text{ m/s}^2)$$
$$= 687 \text{ N}$$

and is directed vertically down.

c. Since you are not accelerating in the vertical direction, $\vec{a} = 0$ m/s^2 and $\vec{F}_{total} = 0$ N according to Equation 5.1. Since here $\vec{F}_{total} = \vec{N} + \vec{w}$, you have $\vec{N} = -\vec{w}$. Hence the normal force of the ground on you is equal in magnitude to your weight (687 N) and is directed vertically upward.

d. One third law pair of forces is the gravitational force of the Earth on you (your weight) and the gravitational force of you on the Earth. These forces are of equal magnitude, point in opposite directions, and act on different systems, as shown in Figure 5.26. The other third law pair of forces is the normal force of the level surface on you and the force of you on the level surface. These forces are of equal magnitude, point in opposite directions, and act on different systems, as shown in Figure 5.27.

FIGURE 5.27

EXAMPLE 5.4

A tired dog named Canis Major, with a weight of magnitude 110 N, comes along and lies down for a nap on a 50.0 kg log lying on level ground, as indicated in Figure 5.28.

a. Draw a second law force diagram schematically indicating all forces acting on the log.
b. Find the magnitudes of these forces.
c. Specify the paired forces that are involved in Newton's third law and draw corresponding third law force diagrams.

FIGURE 5.28

Solution

Let the log be the system.

a. There are three forces acting on the log. The second law force diagram includes

 1. the gravitational force of the Earth on the log (the weight $\vec{\mathbf{w}}$ of the log), of magnitude

$$mg = (50.0 \text{ kg})(9.81 \text{ m/s}^2)$$
$$= 491 \text{ N}$$

 directed vertically downward;

 2. the force of the dog on the log, $\vec{\mathbf{F}}_{dog}$—of magnitude 110 N, directed vertically downward; and

 3. the normal force $\vec{\mathbf{N}}$ of the surface on the log, directed perpendicular to the surface, here vertically upward.

These forces are shown in the second law force diagram of Figure 5.29. Choose $\hat{\mathbf{j}}$ going vertically down, as indicated in Figure 5.29. Newton's second law of motion for the vertical direction states that

$$F_{y \text{ total}}\hat{\mathbf{j}} = ma_y\hat{\mathbf{j}}$$

Since the log is not accelerating, $a_y = 0$ m/s^2, and so

$$(491 \text{ N})\hat{\mathbf{j}} + (110 \text{ N})\hat{\mathbf{j}} - N\hat{\mathbf{j}} = m(0 \text{ m/s}^2)$$

Solving for N, you find

$$N = 601 \text{ N}$$

Notice that here the magnitude of the normal force of the surface on the log is not equal to the magnitude of the weight of the log, even though the surface is horizontal.

FIGURE 5.29

c. The paired forces in Newton's third law are the following:

Pair 1 The gravitational force *of* the Earth *on* the log and the gravitational force *of* the log *on* the Earth, indicated in the third law force diagram of Figure 5.30.

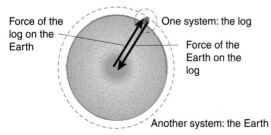

FIGURE 5.30

Pair 2 The force *of* the dog *on* the log and the force *of* the log *on* the dog, shown in Figure 5.31.

FIGURE 5.31

Pair 3 The normal force *of* the surface *on* the log and the force *of* the log *on* the surface, shown in Figure 5.32.

FIGURE 5.32

Notice that the forces in each third law force pair have the same magnitude, are in opposite directions, and act *on* different systems. Only one member of each pair is involved in analyzing the motion of the log.

STRATEGIC EXAMPLE 5.5

A cute pika (*Ochotona princeps*) of mass m slips at rest at the top of an icy frictionless plane inclined at an angle θ to the horizontal direction, as in Figure 5.33.

FIGURE 5.33

a. Make a second law force diagram indicating all forces acting on the pika.
b. What is the magnitude of the normal force on the pika?
c. What is the magnitude of the acceleration of the pika?
d. If the inclination angle increases to 90°, what do you anticipate the magnitude of the acceleration of the pika will be? Is this consistent with part (c) when $\theta = 90$°?

Solution

a. Use Problem-Solving Tactic 5.1 and *specify the system*. Let the system be the pika of mass m. Use Problem-Solving Tactic 5.6

and *identify the forces acting on the system and draw a second law force diagram*. There are two forces acting on the pika (see Figure 5.34):

1. the gravitational force of the Earth on the system (the weight $\vec{w}$ of the pika), of magnitude mg, directed vertically down; and
2. the normal force $\vec{N}$ of the surface on the pika, of magnitude N, directed perpendicular to the inclined surface.

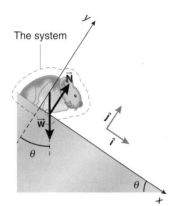

FIGURE 5.34

b. Introduce a coordinate system with axes parallel and perpendicular to the plane, since the motion is parallel to it (see Figure 5.34). The weight $\vec{w}$ makes an angle with a perpendicular to the inclined surface that is equal to the angle θ of the incline. This is shown in Figure 5.34 and makes use of the geometric fact that if two angles θ and ϕ have mutually perpendicular sides (see Figure 5.35), the angles are equal: $\theta = \phi$.

Hence $\vec{w}$ is expressed in Cartesian form as

$$\vec{w} = (mg \sin \theta)\hat{\imath} + (mg \cos \theta)(-\hat{\jmath})$$
$$= (mg \sin \theta)\hat{\imath} - (mg \cos \theta)\hat{\jmath}$$

The normal force is expressed in Cartesian form as

$$\vec{N} = N\hat{\jmath}$$

As the mass slides down the plane, its y-coordinate does not change; thus the velocity *and* acceleration have y-components equal to zero. With the y-component of the acceleration equal

FIGURE 5.35

to zero, apply Newton's second law of motion, Equation 5.2, to the y-direction:

$$F_{y\,\text{total}} = ma_y$$
$$N - mg \cos \theta = m(0 \text{ m/s}^2)$$

Hence

$$N = mg \cos \theta$$

c. Apply Newton's second law of motion, Equation 5.2, to the x-direction:

$$F_{x\,\text{total}} = ma_x$$

Along the x-direction, there is only one force component $(mg \sin \theta)$, and so the second law becomes

$$mg \sin \theta = ma_x$$

Solving for a_x, you obtain

$$a_x = g \sin \theta$$

Since $a_y = 0$ m/s^2, the magnitude of the total acceleration is

$$a = (a_x^2 + a_y^2)^{1/2}$$
$$= [(g \sin \theta)^2 + (0 \text{ m/s}^2)^2]^{1/2}$$
$$= g \sin \theta$$

Therefore the magnitude of the acceleration is $g \sin \theta$.

d. Notice that if the inclination angle is 0°, the magnitude of the acceleration of the pika is zero, since $\sin 0° = 0$. As the inclination angle increases to 90°, the acceleration increases in magnitude. When $\theta = 90°$, the magnitude of the acceleration should equal g, since the pika then is in free-fall. Indeed, when $\theta = 90°$,

$$a_x = g \sin 90° = g$$

A frictionless inclined plane thus provides a way of securing accelerations ranging in magnitude from 0 m/s^2 to 9.81 m/s^2. Galileo exploited this aspect of inclined planes when he began his studies of kinematics. By using small inclination angles, he could make the accelerations small enough for him to use the crude timing devices of the early 17th century.

5.11 TENSIONS IN ROPES, STRINGS, AND CABLES

Mountaineers use ropes for safety and excitement. Fishing would not be very successful without appropriate line. Steel cables are used to support signs and suspension bridges and for ski lifts. The famous cable cars of San Francisco and the gondolas that take you to the top of Table Mountain in Capetown also use cables. Since strings, ropes, and cables have frequent technological applications, it is useful to know how such devices transmit contact forces.

An Ideal String

Real ropes, strings, and cables are modeled, in a first approximation, by an **ideal string**.

An ideal string

a. has negligible mass;
b. does not stretch when pulled;
c. can pull but not push; and
d. pulls at any point only in a direction along the line of the string.

Imagine such an idealized string tied to a wall. A friend at the other end exerts a pull of magnitude *T* newtons horizontally (see Figure 5.36).

Examine the situation in the horizontal direction and consider the string to be the system under study. The string is not accelerating in the horizontal direction; hence Newton's second law of motion implies the total force $\vec{F}_{total}$ on the string in the horizontal direction must be zero. This means that the wall must exert a force on the string, and this force must be equal in magnitude to the force your friend exerts, but must point in the opposite direction; in this way, the two horizontal forces vector sum to zero. A second law force diagram of the forces on the string includes the force your friend exerts on the string and the force of the wall on the string, as shown in Figure 5.37.

Since the acceleration of the string is zero, the two forces are of equal magnitude and opposite direction, because their vector sum must be zero in Newton's second law. Even though the force your friend exerts on the string and the force the wall exerts on the string are equal in magnitude and are opposite in direction, they are *not* a Newton's third law pair of forces since they both act on the same system (the string).

Now *imagine* cutting the string at some point along its length so that it consists of two fictitious pieces. One way to visualize the forces at the cut is to imagine yourself holding the two pieces of the string as they are cut, as indicated in Figure 5.38. Looking at the figure, you must pull to the right on the piece of string on the left, and you must pull to the left on the piece of string on the right. Consider each piece to be a separate system and apply Newton's second law of motion to each. Each piece has zero acceleration, and so the vector sum of the forces in the horizontal direction on *each piece* must be zero. This means that for the piece on the right (see Figure 5.39), *at the point of the cut* there must be a force directed *to the left* on the segment, equal in magnitude to the force your friend exerts on the end of the string. As a result, the total force on this segment is zero in the horizontal direction. Likewise, for the segment on the left (see Figure 5.40), there must be a force *at the cut* that is directed to the right, equal in magnitude to the force of the wall on the string.

Notice that the original pull of magnitude *T* that your friend exerts *to the right, on the right end* of the right segment of the string (Figure 5.39), is equal in magnitude to the force (in Figure 5.40) that the right segment exerts on the left segment *at the cut* and *points in the same direction* as the force your friend exerts on the end of the string. [This is legalese in physics (quite a mouthful of a sentence); read it again slowly and carefully.] Therefore you can think of the force your friend exerts at the *end* of the string as *also* being exerted *at the imaginary cut* on the segment of string on the left side of Figure 5.38. The string, therefore, transmits the pull on its end,

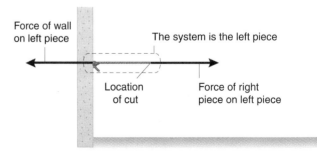

FIGURE 5.38 You hold both ends of the cut string.

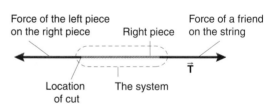

FIGURE 5.39 Second law force diagram for the piece of string on the right.

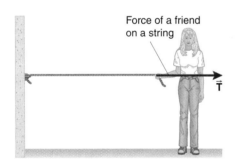

FIGURE 5.36 A friend pulls horizontally on a string tied to a wall.

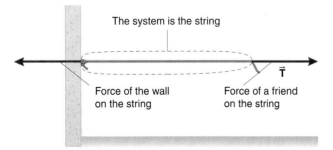

FIGURE 5.37 The horizontal forces on an ideal string tied to a wall and pulled at one end by your friend.

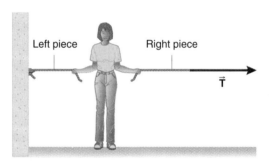

FIGURE 5.40 Second law force diagram for the piece of string on the left.

point by point along its length, since we could imagine the cut to be anywhere along the length of the string.

> The force exerted by any elemental length of the string on its neighbors at any point along the length of a string is called the **tension** *in the string.*

Thus at any point along the string we imagine a cut such that *at the cut*, the tension in the string acts *to the left* on the segment of the string on the right, and *to the right* on the segment of the string on the left, as indicated in Figure 5.41. The forces *at the imagined cut* are of equal magnitude, point in opposite directions, and act on different segments (or systems); thus, they form a third force law pair.

The ideal string transmits a pull on its end, undiminished in magnitude, along its length to the wall. When your friend pulls on the string, the *effect of the string* is as follows:

a. The string exerts a force on your friend in a direction opposite to the force she exerts on the string (see Figure 5.42), since these forces form a third law pair.
b. The string also exerts a force on the wall in a direction opposite to the force exerted by the wall on the string (see Figure 5.43), since these forces form a third law pair.

PROBLEM-SOLVING TACTIC

5.9 An ideal string exerts forces on the things at its ends that are directed toward and along the string; these forces are equal in magnitude. See Figure 5.44.

The forces that ropes, strings, and cables exert on objects at their ends also are called *tensions*, since they arise from the tension forces in such ropes, strings, or cables.

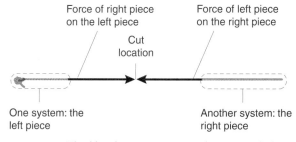

FIGURE 5.41 Third law forces at an imagined cut in an ideal string.

FIGURE 5.42 Third law force diagram at the end of the string pulled by your friend.

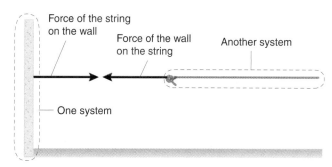

FIGURE 5.43 Third law force diagram at the point the string is attached to the wall.

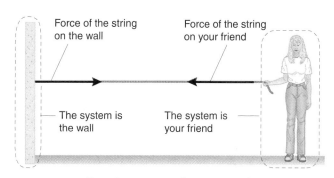

FIGURE 5.44 Force *by* a string *on* things at its ends.

FIGURE 5.45 Forces *on* the string *by* things on its ends.

The *things on the ends of the string* exert forces *on the string*, directed along the string but outward from its ends; these forces are equal in magnitude (see Figure 5.45). The forces in Figure 5.45 are only of interest if you consider the string to be the system, which is not commonly the case (although we do this in the following subsection for a massive string). Typically, the objects on the ends of the string are of concern, not the string itself.

EXAMPLE 5.6

Your dog Sirius does not want to budge from her favorite rug. You pull on her leash with a force of magnitude 150 N as shown in Figure 5.46. Consider the leash to be an ideal string.

a. What is the force of the leash on her?
b. What is the force of the leash on you?

FIGURE 5.46

Solution

An ideal string exerts forces on the objects at its ends that are parallel to and toward the string. Hence the force of the leash on the dog is as shown in Figure 5.47, and the force of the leash on you is as in Figure 5.48.

150 N

The system is the dog

FIGURE 5.47 Force of the leash on the dog.

150 N

The system is you

FIGURE 5.48 Force of the leash on you.

A String with Mass

What is the effect of a string whose mass is not negligible but otherwise exhibits the properties of an ideal string? Imagine a rope of mass m_{rope} attached to a block of mass m_{block} on a frictionless horizontal surface, as in Figure 5.49. Pull on the end of the rope with a horizontal force of magnitude T_1 newtons.

Examine the horizontal forces along the axis of the rope. Imagine cutting the rope at its juncture with the block; call T_2 the magnitude of the tension in the rope at the cut. We want to find the relationship between T_1 and T_2. We examine two different systems: (a) the rope and (b) the block.

a. The forces acting on the rope in the horizontal direction are shown in Figure 5.50. The pull of magnitude T_1 that we exert on the rope to the right and a force of magnitude T_2 on the rope

by the block is directed to the left. Apply Newton's second law of motion to the whole rope as a system. Choose $\hat{\mathbf{i}}$ to be directed as indicated in Figure 5.50. Then Newton's second law becomes

$$F_{x \text{ total on rope}} \hat{\mathbf{i}} = m_{\text{rope}} a_x \hat{\mathbf{i}}$$

or

$$T_1 \hat{\mathbf{i}} + T_2 (-\hat{\mathbf{i}}) = m_{\text{rope}} a_x \hat{\mathbf{i}}$$

Simplifying yields

$$T_1 - T_2 = m_{\text{rope}} a_x \qquad (5.5)$$

b. Now examine the block as a separate system. The only force acting on the block in the horizontal direction is the force of magnitude T_2 exerted on the block by the rope. Since the block is at the end of the rope, the force of the rope on the block is directed toward and along the rope, as shown in Figure 5.51. Applying Newton's second law of motion to the block, you have

$$T_2 \hat{\mathbf{i}} = m_{\text{block}} a_x \hat{\mathbf{i}}$$

Hence

$$T_2 = m_{\text{block}} a_x$$

Solving for a_x,

$$a_x = \frac{T_2}{m_{\text{block}}}$$

If the rope does not stretch (one characteristic of an ideal string), the block and the rope have the same acceleration component. So substitute the expression for a_x into Equation 5.5 for the *rope*, yielding

$$T_1 - T_2 = m_{\text{rope}} \frac{T_2}{m_{\text{block}}}$$

Dividing both sides by T_2 and rearranging the terms slightly,

$$\frac{T_1}{T_2} = 1 + \frac{m_{\text{rope}}}{m_{\text{block}}}$$

The magnitudes of the forces acting on each end of the massive rope are *not* equal. However, if the mass of the rope is small compared with mass of the block, that is, if

$$\frac{m_{\text{rope}}}{m_{\text{block}}} \ll 1$$

the rope approaches the characteristics of an *idealized, massless string* and the magnitude of the tensions at each end of the ideal rope are equal, even if the rope and block are accelerating.

Frictionless tabletop

$\vec{T}_1$

m_{rope}

m_{block}

FIGURE 5.49 Pull on the end of massive string attached to a block.

$\hat{\mathbf{i}}$

The system

m_{rope}

$\vec{T}_2$

$\vec{T}_1$

FIGURE 5.50 The horizontal forces on a massive rope.

The system

$\vec{T}_2$

m_{block}

FIGURE 5.51 Horizontal forces on the block.

FIGURE 5.52 An ideal pulley changes the direction of the tension but not its magnitude.

Ideal, massless ropes, strings, wires, rods, or beams have the same magnitude of tension at all points along their lengths whether the system is in equilibrium or accelerating.

The effects of an ideal, massless, frictionless **pulley** on an ideal rope are the following:

a. The pulley *changes the direction* of the rope and, therefore, the direction *of the tension* force in the rope, as shown in Figure 5.52.
b. The ideal pulley *does not change the magnitude of the tension* in the rope. That is, the tension in the ideal rope is the *same* on both sides of the pulley.
c. An idealized rope does not stretch; therefore, if the rope is moving over the pulley, the speed at which rope comes onto the pulley is equal to the speed at which it leaves.

QUESTION 6

A mass is pulled vertically by a string (see Figure Q.6) and has an acceleration of magnitude $a > g$. Explain why the tension in the string is greater for an upward acceleration than for a downward acceleration.

FIGURE Q.6

EXAMPLE 5.7

Two mountaineers of masses m_1 and m_2 strangely happen to be hanging at rest on one ideal rope in equilibrium over a cliff, as shown in Figure 5.53. Find the magnitudes of the tensions in the two sections of rope.

Solution

Let the person on the bottom be the m_1-system. The second law force diagram for this system includes (see Figure 5.54)

1. the weight $\vec{w}_1$ of the person, of magnitude m_1g, acting in the downward direction (this is the gravitational force of the Earth on the person); and
2. the force $\vec{T}_1$ of the lower rope on the m_1-system, which has a magnitude equal to the magnitude of the tension in the lower rope.

Since the m_1-system is in equilibrium, the *right*-hand side of Newton's second law (Equation 5.1) is zero:

$$\vec{F}_{1\,total} = m_1\vec{a}_1$$
$$m_1g(-\hat{\jmath}) + T_1\hat{\jmath} = m_1(0 \text{ m/s}^2)$$
$$= 0 \text{ N}$$

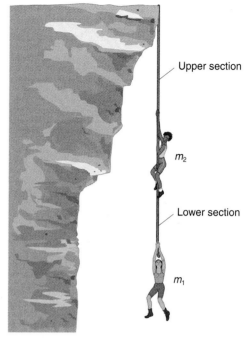

FIGURE 5.53

Thus the vector sum of the forces acting *on* the m_1-system (the *left*-hand side of the second law) must be zero. After transposing,

$$T_1 = m_1g$$

The tension in the lower rope is of magnitude m_1g.

There are two ways to find the tension in the upper rope.

Method 1

Call the upper person the m_2-system. The second law force diagram for this system includes the forces shown in Figure 5.55:

1. the weight $\vec{w}_2$ of the person, of magnitude m_2g, directed vertically down;
2. the force $\vec{T}_1$ of the lower rope on the m_2-system; this force is equal in magnitude to the tension in the lower rope; and
3. the force $\vec{T}_2$ of the upper rope on the m_2-system, which has a magnitude equal to the magnitude of the tension in the upper rope.

Since the mass of each rope is negligible, the tension in each rope has a constant magnitude along its respective length. Thus the magnitude of the force $\vec{T}_1$ of the lower rope on m_2 is

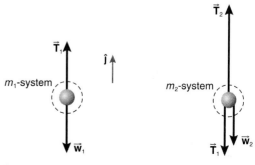

FIGURE 5.54   FIGURE 5.55

equal to the magnitude of the force of the lower rope on m_1, and both are equal to the magnitude of the tension in the lower rope. Since the m_2-system has zero acceleration, the right-hand side of the second law is zero:

$$\vec{F}_{2\,total} = m_2\vec{a}_2$$
$$T_2\hat{j} + m_2g(-\hat{j}) + T_1(-\hat{j}) = m_2(0 \text{ m/s}^2)$$
$$= 0 \text{ N}$$

Solving for T_2,

$$T_2 = m_2g + T_1$$

But T_1 is the magnitude of the tension in the lower rope, which is equal to m_1g (from our analysis of the m_1-system); thus

$$T_2 = m_2g + m_1g$$
$$= (m_1 + m_2)g$$

Method 2
Let both m_1 and m_2 be considered a *single* system called the $(m_1 + m_2)$-system. In this case the second law force diagram for the system includes the following forces (see Figure 5.56):

1. the weight $\vec{w}_{12}$ of the $(m_1 + m_2)$-system, which is of magnitude to $(m_1 + m_2)g$, directed vertically down; and
2. the force $\vec{T}_2$ of the upper rope on the $(m_1 + m_2)$-system, which is equal in magnitude to the tension in the rope leading to the top of the cliff.

Notice that the forces of the lower rope on m_1 and m_2 are *not* forces acting *on* the $(m_1 + m_2)$-system; these forces are *within* (or *internal* to) the system. Since the rope is of negligible mass, the tension in the rope between m_1 and m_2 is the same along its length; so the force of this rope acting on m_1 and the force of this rope acting on m_2 are of equal magnitude, point in opposite directions, and vector sum to zero. The internal forces can be considered as Newton's third law pairs within the larger $(m_1 + m_2)$-system.

The $(m_1 + m_2)$-system has zero acceleration, and so the right-hand side of Newton's second law, Equation 5.1, is zero:

$$\vec{F}_{total} = (m_1 + m_2)\vec{a}$$
$$(m_1 + m_2)g(-\hat{j}) + T_2\hat{j} = (m_1 + m_2)(0 \text{ m/s}^2)$$
$$= 0 \text{ N}$$

FIGURE 5.56

Note that the mass on the right-hand side of the second law is the mass of the *system*, here $m_1 + m_2$. Solving for T_2,

$$T_2 = (m_1 + m_2)g$$

You obtain the same result by either method. It *has* to come out that way: the physical world does not know, or care, which system you use to analyze the situation. But notice that with the $(m_1 + m_2)$-system, you gain *no* information about the tension in the lower rope *between* the two masses, since the forces this rope exerts on m_1 and m_2 are *internal* to the combined system.

STRATEGIC EXAMPLE 5.8

Two climbers of mass m_1 and m_2 are connected by an ideal rope and held in place by a force $\vec{F}$ directed up and parallel to an icy, frictionless slope that makes an angle θ with the horizontal (see Figure 5.57).

a. Consider each climber as a separate system and make second law force diagrams schematically indicating all forces acting on each system.
b. Determine the magnitude of the normal force of the surface on each climber, the magnitude of the tension in the rope between them, and the magnitude of the force $\vec{F}$ needed to hold the climbers in place at rest in terms of m_1, m_2, g, and θ.
c. Will the results of any of your calculations be different if the climbers are pulled up the incline at a constant velocity?

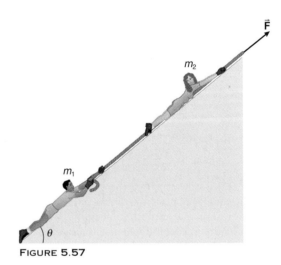

FIGURE 5.57

Solution
a. Let the climber of mass m_1 be the system. There are three forces acting on him (see Figure 5.58):
   1. the gravitational force of the Earth on him—his weight $\vec{w}_1$, of magnitude m_1g, directed vertically down;
   2. the normal force $\vec{N}_1$ of the surface on him, directed perpendicular to the surface; and
   3. the force of the rope on him, $\vec{T}_1$, equal in magnitude to the tension in the rope between them and directed parallel to and up the slope.

FIGURE 5.58

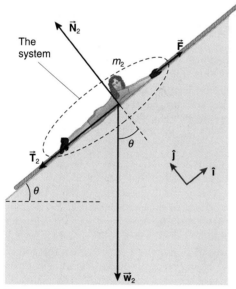

FIGURE 5.59

Let the climber of mass m_2 be another system. There are four forces acting on her (see Figure 5.59):

1. the gravitational force of the Earth on her—her weight $\vec{w}_2$, of magnitude to $m_2 g$, directed vertically down;
2. the normal force $\vec{N}_2$ of the surface on her, directed perpendicular to the surface;
3. the force of the lower rope on her, $\vec{T}_2$, equal in magnitude to the tension in the rope (and to the magnitude of $\vec{T}_1$, since the rope is ideal), directed parallel to and down the slope; and
4. the force $\vec{F}$ on m_2, directed parallel to and up the inclined plane, needed to hold the system at rest on the inclined plane.

b. The forces of the connecting rope on the two masses have the same magnitude since the rope is ideal; that is, $T_1 = T_2 = T$, the tension at any point along the length of the ideal rope between them. Introduce a coordinate system with axes parallel to and perpendicular to the plane, as shown in Figures 5.58 and 5.59. This coordinate choice is most convenient since any possible motion of the masses is parallel to the incline. The accelerations of both masses are zero since they are at rest, but this does not detract from the convenience of this choice for the coordinate system. With this coordinate choice, write the weight of each climber in Cartesian form as

$$\vec{w}_1 = (m_1 g \sin \theta)(-\hat{\imath}) + (m_1 g \cos \theta)(-\hat{\jmath})$$
$$= (-m_1 g \sin \theta)\hat{\imath} - (m_1 g \cos \theta)\hat{\jmath}$$

and

$$\vec{w}_2 = (m_2 g \sin \theta)(-\hat{\imath}) + (m_2 g \cos \theta)(-\hat{\jmath})$$
$$= (-m_2 g \sin \theta)\hat{\imath} - (m_2 g \cos \theta)\hat{\jmath}$$

The two normal forces are, in Cartesian form,

$$\vec{N}_1 = N_1 \hat{\jmath}$$
$$\vec{N}_2 = N_2 \hat{\jmath}$$

and the two forces of the connecting rope on each mass are

$$\vec{T}_1 = T\hat{\imath}$$
$$\vec{T}_2 = T(-\hat{\imath})$$

The force directed parallel to and up the inclined plane is

$$\vec{F} = F\hat{\imath}$$

Consider the mass m_1 as a system. The m_1-system is in equilibrium (zero acceleration). Since $a_x = 0$ m/s², Newton's second law of motion, Equation 5.2, becomes

$$F_{x\,\text{total}} = m_1 a_x$$
$$-m_1 g \sin \theta + T = m_1 (0 \text{ m/s}^2)$$
$$= 0 \text{ N}$$

After transposing, the magnitude of the tension in the rope is found to be

$$T = m_1 g \sin \theta$$

Likewise, since the m_1-system is not accelerating along the y-direction, $a_y = 0$ m/s² and the sum of the force components along the y-direction must also be zero. The second law of motion, Equation 5.2, becomes

$$F_{y\,\text{total}} = m_1 a_y$$
$$N_1 - m_1 g \cos \theta = m_1 (0 \text{ m/s}^2)$$
$$= 0 \text{ N}$$

After transposing, the normal force of the surface on mass m_1 is found to be

$$N_1 = m_1 g \cos \theta$$

Now apply Newton's second law of motion to the m_2-system. The m_2-system also is in equilibrium (zero acceleration); the sum of the force components *on* m_2 along the x-direction is zero since $a_x = 0$ m/s². Thus

$$F_{x\,total} = m_2 a_x$$
$$F - (m_2 g \sin \theta) - T = m_2(0 \text{ m/s}^2)$$
$$= 0 \text{ N}$$

Transposing,

$$F = m_2 g \sin \theta + T$$

Substituting for T, the magnitude of the force F is found to be

$$F = m_2 g \sin \theta + m_1 g \sin \theta$$
$$= (m_1 + m_2)g \sin \theta$$

Since m_2 also is not accelerating along the y-direction, $a_y = 0$ m/s² and the sum of the force components along the y-direction also is zero for mass m_2. The second law of motion, Equation 5.2, becomes

$$F_{y\,total} = m_2 a_y$$
$$N_2 - m_2 g \cos \theta = m_2(0 \text{ m/s}^2)$$
$$= 0 \text{ N}$$

After transposing, the normal force of the surface on the mass m_2 is found to be

$$N_2 = m_2 g \cos \theta$$

c. If the masses are pulled up the inclined plane at constant velocity, the system of two masses still is in equilibrium. The results we obtained, therefore, also apply to such a situation.

STRATEGIC EXAMPLE 5.9

Two masses m_1 and m_2 are connected by an ideal string passing over an ideal pulley, as shown in Figure 5.60. Mass m_2 is on an inclined frictionless surface.

a. Treat each mass as an individual system and make a second law force diagram indicating all forces acting on each mass.
b. Determine the acceleration of the system.
c. Find the magnitude of the tension in the ideal string connecting the two masses.

Solution

a. Choose m_1 to be a system. The second law force diagram for it has the two forces shown in Figure 5.61:
   1. the gravitational force of the Earth on m_1 (the weight $\vec{w}_1$ of m_1), of magnitude $m_1 g$, directed vertically downward; and
   2. the force $\vec{T}_1$ of the cord on m_1, equal in magnitude to the tension in the cord and directed vertically upward.

FIGURE 5.61

Choose m_2 to be another system. The second law force diagram for it includes the three forces shown in Figure 5.62:
   1. the gravitational force of the Earth on m_2 (the weight $\vec{w}_2$ of m_2), of magnitude $m_2 g$, directed vertically downward;
   2. the force $\vec{T}_2$ of the cord on m_2, which has a magnitude equal to that of the tension in the cord. Recall that the tension has the same magnitude everywhere along an ideal cord; the ideal pulley does not alter this situation but merely changes the direction in which the tension acts. Thus the force $\vec{T}_2$ on m_2 has the same *magnitude* as the force $\vec{T}_1$ of the cord on m_1; call their common magnitude T; and
   3. the normal force $\vec{N}$ of the surface on the mass m_2, directed perpendicular to the surface of the inclined plane.

b. To find the acceleration, apply Newton's second law to the motion of each mass considered as a separate system. But you do not know which way the system of masses will move! In such instances, *assume* a direction for the acceleration, say with m_1 falling vertically down. The actual direction that the system accelerates will emerge from the analysis.

Choose a coordinate system with $\hat{\imath}$ vertically down as indicated in Figure 5.61. Newton's second law, Equation 5.2, applied to the m_1-system implies that the vector sum of the forces in the

FIGURE 5.60

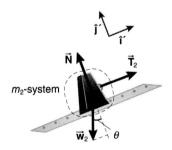

FIGURE 5.62

x-direction is equal to the mass of the system times its acceleration $a_x\hat{\imath}$:

$$m_1 g\hat{\imath} - T\hat{\imath} = m_1 a_x\hat{\imath}$$

Thus

$$m_1 g - T = m_1 a_x \qquad (1)$$

Notice that this equation has two unknowns, T and a_x. You need to find another equation relating T and a_x in order to find the solution. The remaining equation is found by applying Newton's second law to the m_2-system.

Choose a coordinate system with axes parallel and perpendicular to the incline, indicated in Figure 5.62, so the movement of m_2 in the positive x'-direction corresponds to m_1 moving in the positive x-direction. Write the weight $\vec{\mathbf{w}}_2$ in terms of its Cartesian components parallel and perpendicular to the plane:

$$\vec{\mathbf{w}}_2 = (m_2 g \sin\theta)(-\hat{\imath}') + (m_2 g \cos\theta)(-\hat{\jmath}')$$
$$= (-m_2 g \sin\theta)\hat{\imath}' - (m_2 g \cos\theta)\hat{\jmath}'$$

The acceleration of m_2 perpendicular to the plane is zero ($a_{y'} = 0$ m/s^2), and so the vector sum of the forces on m_2 in the direction perpendicular to the plane, according to Newton's second law, must vanish:

$$N\hat{\jmath}' - (m_2 g \cos\theta)\hat{\jmath}' = m_2 a_{y'}\hat{\jmath}'$$
$$= 0 \text{ N}$$

Thus you have

$$N = m_2 g \cos\theta$$

With the assumption made about the direction of the acceleration of m_1, the mass m_2 accelerates *up* the plane. The string is ideal and does not stretch, and so the acceleration of m_2 has the same *magnitude* as the acceleration of m_1. Look at the motion of m_2 parallel to the incline. The force $\vec{\mathbf{T}}_2$ is directed up the plane while a component of the weight ($m_2 g \sin\theta$) is directed down the plane. Apply Newton's second law to the m_2-system along the x'-direction:

$$T\hat{\imath}' - (m_2 g \sin\theta)\hat{\imath}' = m_2 a_{x'}\hat{\imath}'$$
$$T - m_2 g \sin\theta = m_2 a_{x'}$$

The acceleration components a_x and $a_{x'}$ have the same value since the string does not stretch; thus

$$T - m_2 g \sin\theta = m_2 a_x \qquad (2)$$

Now you can solve equations (1) and (2) simultaneously for T and a_x. First solve equation (1) for T:

$$T = m_1(g - a_x)$$

Then substitute for T into equation (2):

$$m_1(g - a_x) - m_2 g \sin\theta = m_2 a_x$$

Solving for a_x, you find

$$a_x = \frac{m_1 - m_2 \sin\theta}{m_1 + m_2} g \qquad (3)$$

Notice in equation (3) that the acceleration component a_x is *positive* if $m_1 > m_2 \sin\theta$. In other words, the system will move in the direction you assumed ($\hat{\imath}$ for m_1 and $\hat{\imath}'$ for m_2) if this inequality is satisfied. On the other hand if $m_1 < m_2 \sin\theta$, then the acceleration component a_x is *negative*. So the system accelerates in a direction *opposite* to the one you presumed to begin the calculation. Thus the *actual* direction of the acceleration emerges naturally from the analysis.

c. With the acceleration component in hand, you can determine the magnitude of the tension in the cord from either equation (1) or (2). Using equation (1), you find

$$T = m_1(g - a_x)$$

Substituting for a_x with equation (3),

$$T = m_1 g - m_1 \frac{m_1 - m_2 \sin\theta}{m_1 + m_2} g$$

After some algebraic simplification, you find that the tension in the rope is of magnitude

$$T = \frac{m_1 m_2(1 + \sin\theta)}{m_1 + m_2} g$$

Notice how you accounted for the forces by analyzing their components *consistent* with the directions of the forces indicated on the force diagram and the coordinate system(s) chosen.

EXAMPLE 5.10

Mexican Indians have an apparatus that, for spiritual purification, swings a person around a circle at constant angular speed ω, as shown in Figure 5.63. The length of each cable is ℓ and the radius of the circular hub is R.

a. Show that the angle θ that the cable makes with the vertical direction is independent of the mass of the patron and harness and is given by

$$\tan\theta = \frac{(R + \ell \sin\theta)\omega^2}{g}$$

b. If $R = 2.0$ m, $\ell = 3.0$ m, and the apparatus is spinning at 12.0 rev/min, find the angle θ.

FIGURE 5.63

Solution

a. Consider a person and harness to be a single system of mass m. A second law force diagram on the system on the left includes the two forces shown in Figure 5.64:

1. the weight $\vec{w}$ of the system, of magnitude mg, directed vertically down; and
2. the force $\vec{T}$ of the cable on the system, equal in magnitude to the tension in the cable, directed along the cable making an angle θ with the vertical direction.

There is no centripetal force, since there is no such force. However, there *is* a centripetal acceleration.

FIGURE 5.64

The coordinate system is fixed on the ground and does not rotate with the apparatus. We analyze the system when it is instantaneously on the *x*-axis as indicated in Figure 5.64.

The system is not accelerating in the vertical direction, so that $a_z = 0$ m/s² and the application of Newton's second law to this axis yields

$$F_{z\,total}\hat{k} = ma_z\hat{k}$$
$$(T\cos\theta)\hat{k} + mg(-\hat{k}) = m(0\ \text{m/s}^2)\hat{k}$$
$$= 0\ \text{N}$$
$$T\cos\theta = mg \qquad (1)$$

Along the *x*-direction, there is a force component $T\sin\theta$ directed toward $\hat{i}$ at the instant depicted in Figure 5.64. Since the system is in uniform circular motion, there is a centripetal acceleration of magnitude v^2/r, directed toward the center of the circle, toward $\hat{i}$ at this instant, where v is the speed of the system. Thus Newton's second law applied to the *x*-direction gives

$$F_{x\,total}\hat{i} = ma_x\hat{i}$$

$$(T\sin\theta)\,\hat{i} = m\frac{v^2}{r}\,\hat{i}$$

$$T\sin\theta = m\frac{v^2}{r} \qquad (2)$$

Divide equation (2) by equation (1); the magnitude of the tension is eliminated:

$$\frac{T\sin\theta}{T\cos\theta} = \frac{m\frac{v^2}{r}}{mg}$$

$$\tan\theta = \frac{v^2}{rg}$$

Note that the angle is independent of the mass of the system, so that all persons swing at the same angle with the vertical.

For circular motion, the speed v and the angular speed ω are related by Equation 4.34,

$$v = r\omega$$

Making this substitution into the expression for $\tan\theta$, you find

$$\tan\theta = \frac{r^2\omega^2}{rg}$$
$$= \frac{r\omega^2}{g}$$

From the geometry of Figure 5.63, $r = R + \ell\sin\theta$; thus

$$\tan\theta = \frac{(R + \ell\sin\theta)\omega^2}{g} \qquad (3)$$

b. The angular speed of the apparatus is 12.0 rev/min or

$$\omega = \frac{(12.0\ \text{rev/min})\,(2\pi\ \text{rad/rev})}{60\ \text{s/min}}$$
$$= 1.26\ \text{rad/s}$$

Since $R = 2.0$ m, $\ell = 3.0$ m, and $g = 9.81$ m/s², equation (3) becomes

$$\tan\theta = \frac{[2.0\ \text{m} + (3.0\ \text{m})\sin\theta]\,(1.26\ \text{rad/s})^2}{9.81\ \text{m/s}^2}$$
$$= 0.32 + 0.49\sin\theta$$

This is a transcendental equation for θ, not an uncommon outcome. There are two ways to find θ.

Method 1

A solution can be found graphically by making two graphs on the same sheet, one of $\tan\theta$ versus θ, the other of $0.32 + 0.49\sin\theta$ versus θ, as in Figure 5.65. The value of θ where the two graphs intersect is the value of θ that makes the sides of the equation equal.

Method 2

The angle θ also can be found by successive approximations, beginning with a value of θ that may be *reasonable*, such as 20°, then trying both higher and lower values of θ to see if they are closer to satisfying the equation, and walking-in on the solution this way.

Whichever method is chosen, you find

$$\theta = 29°$$

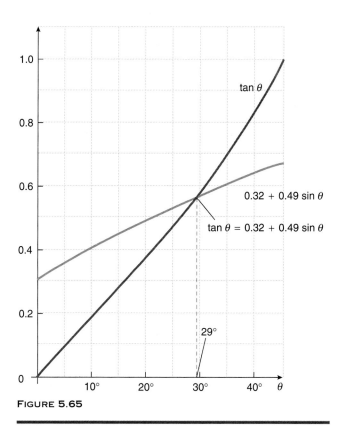

FIGURE 5.65

5.12 STATIC FRICTION

The obedience of matter is limited by friction.

Victor Hugo*

Frictional effects permeate the macroscopic world around us, and so we need to have a good understanding of them. As we noted in the introduction to this chapter, the existence of friction likely impeded the development of a consistent theory of motion. You should not think that frictional forces are always bad. Without friction, your car could never move, and neither could you, as you may know from attempting to accelerate from rest on an icy pavement. Friction is the proverbial two-edged sword: a blessing or a curse depending on the circumstances.

There are two general types of frictional forces: (1) a **static frictional force**, involving no motion (no sliding) at the places where a system stays in contact with a surface, such as the bottom of your hiking boots in contact with the trail, or the bottom of a tire in contact with the surface of a road; and (2) a **kinetic frictional force** that arises when one surface slides along a second surface, such as when your car wheels lock up and skid along the road, or when a moving automotive engine piston rubs against the cylinder containing it. Kinetic frictional forces also arise when a system moves through a fluid (such as air or water), as when you ride your bike at great speed.

A series of simple experiments will help you to see what is involved with static frictional effects.

Experiment 1: Imagine your calculus text lying at rest on your desk. The contact surface is *not* frictionless. The book is the

Les Misérables (1862) (A. L. Burt, New York), volume II, Jean Valjean, Book VI, The White Knight, IV, page 673.

system we want to analyze. We know of two forces on the text (see Figure 5.66):

1. the gravitational force of the Earth on the text, its weight $\vec{\mathbf{w}}$, of magnitude mg, directed vertically downward; and
2. the normal force $\vec{\mathbf{N}}$ of the surface on the text, directed perpendicular to the surface.

The second law of motion, Equation 5.1, becomes

$$\vec{\mathbf{F}}_{total} = m\vec{\mathbf{a}}$$
$$\vec{\mathbf{w}} + \vec{\mathbf{N}} = m(0\ \text{m/s}^2)$$
$$= 0\ \text{N}$$

Thus

$$\vec{\mathbf{N}} = -\vec{\mathbf{w}}$$

Since the book is not accelerating in the horizontal direction, the total force on it in the horizontal direction is zero; indeed, there are no forces at all in the horizontal direction.

Experiment 2: Now push slightly on the text in the horizontal direction with your finger. Since the motion in the vertical direction is independent of the motion in the horizontal direction, nothing changes in our previous analysis of the vertical motion (Experiment 1). You find the text *remains* in equilibrium despite the existence of the slight horizontal force you apply. The acceleration of the text along the horizontal direction is zero, since its velocity (equal to zero) is not changing with time. Newton's second law of motion implies that the vector sum of the forces *parallel* to the surface must be zero because the text is in equilibrium. One such horizontal force is the slight push $\vec{\mathbf{F}}$ by your finger (see Figure 5.67). Another force *must* exist parallel to the surface so that the vector sum of the two forces is zero. This other force is the **static frictional force** $\vec{\mathbf{f}}_s$.

FIGURE 5.66 Forces on your text when it is at rest on your desk.

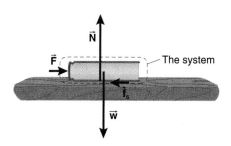

FIGURE 5.67 Forces on your text when you push it slightly in the horizontal direction.

The *total force* in the horizontal direction is zero, in keeping with our observation that the text is in equilibrium (zero acceleration). Therefore the static frictional force must be of the same magnitude as that of the force you exert on the text, but it must point in the opposite direction.

> The static frictional force always acts in a direction parallel to the surface regardless of the orientation of the surface with respect to the horizontal direction. The static fictional force is directed to prevent the system from slipping.

Note that the static frictional force of the surface on the text and the force you exert on the text are equal in magnitude and opposite in direction; but they are *not* a Newton's third law force pair, since they act on the *same* system (the text). The two forces are of equal magnitude and opposite direction because of zero acceleration and the *second* law of motion.

Experiment 3: Now push a bit harder with your finger. The text will still be in equilibrium. The static frictional force matches your push newton-for-newton. The vector sum of the two forces remains zero because the acceleration of the text, once again, is zero. Push a bit harder and the static frictional force grows to match it. The static frictional force is seemingly quite "intelligent"; it knows exactly how hard you are pushing and matches the push appropriately! The vector sum of the forces on the book acting parallel to the surface stays zero because the acceleration is zero.

Eventually, however, if you push hard enough, the text reaches a state where it is on the verge of slipping across the table. When the text is on the verge of slipping, and *only* then, the ratio between the magnitude of this greatest static frictional force and the magnitude of the normal force is defined to be a pure number called the **coefficient of static friction** μ_s. This definition is applied only when the system is on the verge of slipping:

$$\frac{f_{s\,max}}{N} \equiv \mu_s \qquad \text{(system ready to slip)}$$

This is usually written as

$$f_{s\,max} = \mu_s N \qquad \text{(system ready to slip)} \qquad (5.6)$$

> Equation 5.6 for $f_{s\,max}$ is *not* a vector equation. It is a *scalar* relationship between the *magnitudes* of the maximum static frictional force and the normal force; the two forces point in different directions. The normal force is directed perpendicular to the surface, while the static frictional force is parallel to the surface; the angle between the two forces $\vec{f}_{s\,max}$ and $\vec{N}$ is always 90°.*

The numerical value of the (dimensionless) coefficient of static friction μ_s depends on the specific materials that compose the two surfaces in contact, but to a first approximation it is independent of the area in contact. In other words, the coefficient

TABLE 5.3 Approximate Values of the Coefficient of Static Friction

Surfaces in contact	Coefficient of static friction μ_s when system is ready to slip
Glass on glass	0.9
Rubber on dry concrete	~1
Rubber on wet concrete	0.4
Steel on steel	0.6
Steel on steel (lubricated)	0.1
Wood on wood	0.3
Waxed ski on snow	0.1

of static friction for the text on the table has one value; the coefficient of static friction for the text on, say, a shag rug has quite another numerical value. Approximate numerical values of the coefficient of static friction for various surfaces in contact with each other are indicated in Table 5.3. Precise values must always be determined from experiments with the actual materials used. Example 5.11 illustrates a common experimental procedure for determining μ_s. The numerical value of the coefficient of static friction μ_s is the same *regardless* of the orientation of the surfaces with respect to the horizontal or vertical directions.

Recall that the static frictional force matched your pushes newton-for-newton up to the point where slipping was about to occur. The relationship given by Equation 5.6 is the *maximum* magnitude of the static frictional force, as indicated by the subscript.

The static frictional force always opposes *slippage* of surfaces in contact but may, under certain circumstances, actually be the force responsible for the *acceleration* of a system. Example 5.14 illustrates this perhaps counterintuitive idea.

Equation 5.6 for $f_{s\,max}$ is *not* a fundamental physical law on a par with, say, Newton's laws of motion. It is an empirical definition of the coefficient of static friction, used to model a complex molecular surface interaction in a straightforward way.

> **PROBLEM-SOLVING TACTIC**
>
> **5.10 Equation 5.6 is useful only when the system is ready to slip across the surface.** If the system is not on the verge of slipping, the magnitude f_s of the actual static frictional force is *less than* $f_{s\,max}$ and *cannot* be found using Equation 5.6.

QUESTION 7

In Experiment 2, if a graph is made of the magnitude of the static frictional force $\vec{f}_s$ versus the magnitude of your horizontal push $\vec{F}$ on the text, what is the slope of the graph? If the experiment were performed on an inclined surface, would the graph be the same?

QUESTION 8

For a given system in contact with a given surface, what would a graph of $f_{s\,max}$ versus N look like?

---

*It is possible to relate a vector in one direction to a vector pointing in another direction; we did this once via vector products, but that is a special case. The most general way to formulate this relationship is using matrices.

STRATEGIC EXAMPLE 5.11

A dictionary of mass m is placed at rest (and remains at rest) on a rough reading table surface inclined at an angle θ to the horizontal, as shown in Figure 5.68. The coefficient of static friction for the mass on the wood surface is μ_s.

FIGURE 5.68

a. Calculate the magnitude of the static frictional force on the dictionary.
b. What is the maximum angle θ for which the dictionary does not slip?

Solution

a. Consider the dictionary to be the system. There are three forces on the system (see Figure 5.69):
   1. the weight $\vec{w}$ (of magnitude mg), directed vertically down;
   2. the normal force $\vec{N}$ of the surface on the mass, directed perpendicular to the surface; and
   3. the static frictional force $\vec{f}_s$, parallel to the surface. If you do not know which direction the static frictional force points, either direction parallel to the surface (up or down the plane) can be chosen. The resulting analysis will indicate the right direction through the algebraic sign of the result. The second law force diagram then is shown in Figure 5.69.

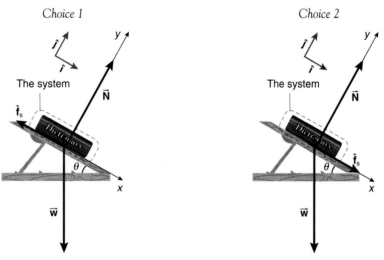

Choose the direction of $\vec{f}_s$ to be up the plane.   Choose the direction of $\vec{f}_s$ to be down the plane.

FIGURE 5.69

Choose the coordinate system indicated. There is nothing sacred about this choice, as we have seen on many occasions; but for inclined planes, it is usually convenient to choose coordinate axes parallel and perpendicular to the incline when the motion or potential motion of the system is up or down the inclined surface.

Write the forces in Cartesian form, *consistent* with the directions of the forces as indicated in the force diagram:
   1. The weight has components along both the x- and y-directions:

$$\vec{w} = (mg\sin\theta)\hat{\imath} + (mg\cos\theta)(-\hat{\jmath})$$

2. The normal force is

$$\vec{N} = N\hat{j}$$

3. The static frictional force is

$\vec{f}_s = f_s(-\hat{i})$	$\vec{f}_s = f_s\hat{i}$

If the dictionary remains at rest on the slope, its acceleration is zero. Newton's second law of motion then indicates that the vector sum of these forces must be zero. Sum the force vectors along the y-direction and apply Newton's second law with $a_y = 0 \text{ m/s}^2$:

$$F_{y\,total}\hat{j} = ma_y\hat{j}$$
$$N\hat{j} - (mg\cos\theta)\hat{j} = m(0 \text{ m/s}^2)\hat{j}$$
$$= 0 \text{ N}$$

Solving for N,

$$N = mg\cos\theta$$

Note that here the magnitude of the normal force N is less than mg.

The vector sum of the forces in the x-direction also must be zero according to Newton's second law of motion, because $a_x = 0 \text{ m/s}^2$:

$$F_{x\,total}\hat{i} = ma_x\hat{i}$$
$$= m(0 \text{ m/s}^2)\hat{i}$$
$$= 0 \text{ N}$$

$(mg\sin\theta)\hat{i} - f_s\hat{i} = 0 \text{ N}$	$(mg\sin\theta)\hat{i} + f_s\hat{i} = 0 \text{ N}$

or after transposing,

$f_s = mg\sin\theta$	$f_s = -mg\sin\theta$
	This means that the static frictional force $\vec{f}_s$, which we wrote as $\vec{f}_s = f_s\hat{i}$, consistent with the second law force diagram, is
	$$\vec{f}_s = (-mg\sin\theta)\hat{i}$$
	$$= (mg\sin\theta)(-\hat{i})$$
	which indicates that $\vec{f}_s$ actually is directed parallel to $(-\hat{i})$, or *up* the plane. Notice that the *magnitude* of the frictional force still is $mg\sin\theta$.

b. The expression for the magnitude of the static frictional force is valid for *any* angle θ as long as the mass is at rest. Notice that as θ increases, the magnitude of the static frictional force also increases, since $\sin\theta$ increases with increasing θ. Eventually, when the inclination angle θ is large enough, the system is on the verge of slipping. At this critical angle θ_{verge}, the static frictional force reaches its maximum magnitude:

$$f_{s\,max} = mg\sin\theta_{verge}$$

As the angle increases to θ_{verge}, the magnitude of the normal force *decreases* since $N = mg\cos\theta$. When the system is ready to slip, the corresponding magnitude of the normal force is $mg\cos\theta_{verge}$.

Divide the magnitude of the maximum static frictional force by the magnitude of the corresponding normal force when the system is ready to slip; you find

$$\frac{f_{s\,max}}{N} = \frac{mg\sin\theta_{verge}}{mg\cos\theta_{verge}}$$

$$= \tan\theta_{verge}$$

When the system is on the verge of slipping, the ratio of these force magnitudes is (by definition) the coefficient of static friction (Equation 5.6):

$$\frac{f_{s\,max}}{N} \equiv \mu_s$$

Equating the expressions for the ratio $f_{s\,max}/N$, you find that the coefficient of static friction is

$$\mu_s = \tan \theta_{verge} \qquad (1)$$

In other words, when the system is ready to slip, the tangent of the inclination angle is equal to the coefficient of static friction. Notice that if the angle θ_{verge} is greater than 45°, then the coefficient of static friction is greater than 1, which is entirely possible for some surfaces. Equation (1) is the basis for the experimental determination of the coefficient of static friction for two surfaces in contact.

STRATEGIC EXAMPLE 5.12

Take a dictionary of mass 5.00 kg and push it hard against a vertical wall with a horizontal force $\vec{F}$ of magnitude 200 N, so that the tome does not slip down the wall, as shown in Figure 5.70. The coefficient of static friction for the book in contact with the wall is $\mu_s = 0.300$.

a. Determine the magnitude of the normal force of the wall on the book.
b. Find the magnitude of the static frictional force on the book.
c. If the magnitude of the force $\vec{F}$ decreases until the book is ready to slip, what is the magnitude of $\vec{F}$ when slipping is imminent?

FIGURE 5.71

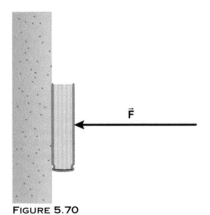

FIGURE 5.70

Solution
Consider the dictionary to be the system. There are four forces acting on the dictionary, shown in Figure 5.71:

1. the weight $\vec{w}$ of the book, of magnitude

$$mg = (5.00 \text{ kg})(9.81 \text{ m/s}^2)$$
$$= 49.1 \text{ N}$$

directed vertically down;
2. the horizontal force $\vec{F}$ that you exert on the dictionary;
3. the normal force $\vec{N}$ of the surface on the book, directed perpendicular to the surface; and

4. the static frictional force $\vec{f}_s$ on the book arising from its contact with the vertical wall. This force acts parallel to the surface to prevent the book from slipping down the wall; thus the static frictional force here acts vertically upward.

Introduce a coordinate system with horizontal and vertical axes, as indicated in Figure 5.71.

a. The book is in equilibrium and so the acceleration components a_x and a_y both are zero. Apply Newton's second law of motion, Equation 5.2, to the horizontal direction:

$$F_{x\,total}\hat{\imath} = ma_x\hat{\imath}$$
$$N\hat{\imath} + F(-\hat{\imath}) = m(0 \text{ m/s}^2)\hat{\imath}$$

Thus

$$N = F$$

The magnitude of the normal force is equal to the magnitude of the force you exert on the book, here 200 N. Notice that here the normal force of the surface on the book has *no* relationship whatever to the weight of the book, illustrating that the magnitude of the normal force must be determined on a case-by-case basis (Problem-Solving Tactic 5.8).

b. Apply Newton's second law of motion to the vertical direction:

$$F_{y\,total}\hat{\jmath} = ma_y\hat{\jmath}$$
$$mg(-\hat{\jmath}) + f_s\hat{\jmath} = m(0 \text{ m/s}^2)\hat{\jmath}$$

Thus

$$f_s = mg$$
$$= (5.00 \text{ kg})(9.81 \text{ m/s}^2)$$
$$= 49.1 \text{ N}$$

The magnitude of the static frictional force is equal to the magnitude of the weight. Notice that the magnitude of the static frictional force here is not found from Equation 5.6, because the book is not on the verge of slipping. Here Equation 5.6 would yield

$$f_{s \text{ max}} = \mu_s N$$
$$= (0.300)(200 \text{ N})$$
$$= 60.0 \text{ N}$$

well above the *actual* magnitude of the static frictional force of 49.1 N.

c. Now ease your push $\vec{F}$ on the book until the book *is* on the verge of slipping down the wall. What magnitude of force F_{min} is needed so the book is on the verge of slipping? As you ease your push, *the magnitude of the static frictional force acting on the book does not change* and still is equal in magnitude to the weight of the book (49.1 N). The book will be on the verge of slipping when the normal force has decreased to a magnitude N', where

$$f_{s \text{ max}} = \mu_s N'$$

The value of $f_{s \text{ max}}$ is 49.1 N, *not* 60.0 N—be sure you understand why! Substitute for $f_{s \text{ max}}$ and μ_s:

$$49.1 \text{ N} = 0.300 N'$$

Solve for N':

$$N' = 164 \text{ N}$$

In this example the normal force of the surface on the book is, at all times, equal in magnitude to that of the force you apply. The book is on the verge of slipping when the force you apply has a magnitude of 164 N.

EXAMPLE 5.13

The roadbeds underneath highways and railroad tracks typically are banked around high-speed curves, as shown in Figure 5.72. There is less wear on the tires, wheels, roadway, and rails if there is no static frictional force down the incline,

FIGURE 5.72 Curves on racetracks and highways are banked for a reason.

parallel to the surface. The ride inside the vehicle also is more comfortable, for there then is no tendency to slide across the seat. Aircraft also bank when turning for similar reasons.

a. Let the radius of the arc of a circular curve be r; your car is traveling at a speed v around the curve. What angle should the roadbed make with the horizontal to assure the least amount of wear? The angle is called the *banking angle* of the turn.

b. Calculate the banking angle for a car traveling at 100 km/h around a circular turn of radius 500 m.

Solution

a. Let the system be the mass of the wheel along with its share of the mass of the car and its contents; call this entire mass m. The forces acting on the system are the normal force of the surface and the weight, indicated in the second law force diagram (Figure 5.73). Remember that each perpendicular direction can be analyzed independently of the others. Here you are concerned only with the instantaneous motion in the plane of Figure 5.73. You can ignore any horizontal forces perpendicular to the plane of Figure 5.73.

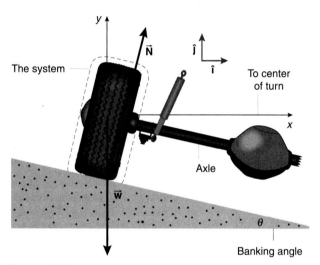

FIGURE 5.73

It is convenient to choose one of the coordinate axes to be in the same direction as the acceleration of the system. When you examined inclined planes, such as in Example 5.11, you chose the axes to be parallel and perpendicular to the incline, as in Figure 5.69, because any accelerations for that problem would be parallel to the incline. Here, however, the centripetal acceleration is toward the center of the curve; so choose coordinate axes that are horizontal and vertical. In this way, one coordinate axis is along the line of the acceleration. This is simpler than choosing the directions of the coordinate axes to be parallel and perpendicular to the slope. The preferred choice of axes is that which makes the problem simplest. Part of the skill of attacking a problem is making an appropriate choice for the axis directions, a skill you will develop with practice.

Since the car is moving around a curve, the vector sum of the forces in the x–y plane must provide the centripetal acceleration toward the center of the curve.

Express the normal force $\vec{\mathbf{N}}$ in terms of its two Cartesian components along the x- and y-axes. The normal force makes an angle with the vertical direction equal to the banking angle θ (since two angles whose sides are mutually perpendicular are equal). Hence

$$\vec{\mathbf{N}} = (N \sin \theta)\hat{\mathbf{i}} + (N \cos \theta)\hat{\mathbf{j}}$$

The weight is

$$\vec{\mathbf{w}} = -mg\,\hat{\mathbf{j}}$$

Analyze the motion along the x- and y-axes independently:

The y-coordinate of the wheel remains constant around the curve. Therefore there is no y-component to the acceleration ($a_y = 0$ m/s^2). Newton's second law of motion in this direction yields

$$F_{y\,\text{total}} = ma_y$$
$$N \cos \theta - mg = m(0 \text{ m/s}^2)$$
$$= 0 \text{ N}$$

Solving for N,

$$N = \frac{mg}{\cos \theta}$$

Along the x-direction, the acceleration is the centripetal acceleration toward the center of the turn, of magnitude v^2/r. Applying Newton's second law of motion to the x-direction gives

$$F_{x\,\text{total}} = ma_x$$

$$N \sin \theta = m\frac{v^2}{r}$$

Substituting for N,

$$\frac{mg}{\cos \theta} \sin \theta = m\frac{v^2}{r}$$

Divide by the mass and recall that

$$\frac{\sin \theta}{\cos \theta} = \tan \theta$$

So

$$g \tan \theta = \frac{v^2}{r}$$

Thus

$$\tan \theta = \frac{v^2}{rg}$$

Notice that the proper banking angle is *independent* of the mass.
b. Convert 100 km/h into m/s:

$$100 \text{ km/h} = \frac{100 \text{ km/h} \times 10^3 \text{ m/km}}{3600 \text{ s/h}}$$
$$= 27.8 \text{ m/s}$$

Substituting this into the equation for the tangent of the banking angle gives

$$\tan \theta = \frac{(27.8 \text{ m/s})^2}{(500 \text{ m})(9.81 \text{ m/s}^2)}$$
$$= 0.158$$

Thus

$$\theta = 8.98°$$

The same principle is used for banking turns on bobsled runs. Should you wish to go around a curve where there is *no* static friction, the curve must be banked so the component of the normal force directed toward the center of the turn is able to cause the centripetal acceleration. If a bobsled does not have the appropriate angle when executing a turn, there is trouble, which is why much practice is needed to be a good bobsled driver.

STRATEGIC EXAMPLE 5.14

A Reese's* Peanut Butter Cup® candy bar of mass m_{PBC} rests on a tray of mass m_{tray} that is on a horizontal frictionless surface (see Figure 5.74). The coefficient of static friction for the candy bar on the tray surface is μ_s. A constant horizontal force $\vec{\mathbf{F}}$ on the tray accelerates the tray and candy bar to the right.

FIGURE 5.74

a. Draw a second law force diagram indicating the forces acting on the tray and candy bar, considering them as a single system.
b. Determine the magnitude of the acceleration in terms of the applied horizontal force and appropriate masses.
c. Sketch another second law force diagram indicating the forces acting on the candy bar, considering it as a separate system. What force accelerates the candy bar?
d. What is the maximum magnitude of the acceleration if the candy bar does not slip on the tray?
e. Considering the tray as a separate system, draw yet another second law force diagram indicating the forces acting on the tray. Use this force diagram to determine the magnitude of the acceleration of the tray, and show that the acceleration is the same as the result obtained in part (b).

Solution
a. The forces acting on the tray and candy bar system are (see Figure 5.75)
  1. the gravitational force of the Earth on the tray and candy bar system; this is the combined weight $\vec{\mathbf{w}}$ of the tray and candy bar, of magnitude $(m_{\text{PBC}} + m_{\text{tray}})g$, directed vertically down;
  2. the normal force $\vec{\mathbf{N}}$ of the surface on the tray and candy bar system, directed vertically upward; and

---

*No known relation to the author. Sigh.

FIGURE 5.75

FIGURE 5.76

3. the horizontal force $\vec{\mathbf{F}}$ applied to the tray and candy bar system.

There is no frictional force on the tray and candy bar system since it is on a frictionless surface.

b. Use the coordinate system indicated in Figure 5.75. The acceleration of the system is found by applying Newton's second law of motion to the tray and candy bar system in the horizontal direction:

$$F\hat{\mathbf{i}} = (m_{PBC} + m_{tray})a_x\hat{\mathbf{i}}$$

The mass on the right-hand side of Newton's second law is the mass of the system, here the tray and candy bar. Solve for the acceleration component a_x:

$$a_x = \frac{F}{m_{PBC} + m_{tray}} \quad (1)$$

Note that in the vertical direction, there is no acceleration ($a_y = 0$ m/s²); thus Newton's second law applied to this direction yields

$$N\hat{\mathbf{j}} + (m_{PBC} + m_{tray})g(-\hat{\mathbf{j}}) = (m_{PBC} + m_{tray})(0 \text{ m/s}^2)\hat{\mathbf{j}}$$
$$= 0 \text{ N}$$

Solving for N,

$$N = (m_{PBC} + m_{tray})g$$

c. The forces on the candy bar, considering it as a separate system, are the following (see Figure 5.76):
   1. the gravitational force of the Earth on the candy bar (its weight $\vec{\mathbf{w}}_{PBC}$) of magnitude $m_{PBC}g$, directed vertically down;
   2. the normal force $\vec{\mathbf{N}}'$ of the surface of the tray on the candy bar, directed vertically up; and
   3. the static frictional force $\vec{\mathbf{f}}_s$ of the surface of the tray on the candy bar.

You may be uncertain which way along the horizontal direction the static frictional force of the tray on the candy bar is directed. Is it parallel or antiparallel to the horizontal acceleration? The static frictional force opposes slippage. The candy bar (if it slips) will slip to the left if the applied force $\vec{\mathbf{F}}$

to the right on the tray is great enough; therefore, the direction of $\vec{\mathbf{f}}_s$ is to the right.*

Now apply Newton's second law of motion to the candy bar system. In the vertical direction, there is no acceleration, so that $a_y = 0$ m/s² and the second law of motion becomes

$$N'\hat{\mathbf{j}} + m_{PBC}g(-\hat{\mathbf{j}}) = m_{PBC}(0 \text{ m/s}^2)\hat{\mathbf{j}}$$
$$= 0 \text{ N}$$

Thus

$$N' = m_{PBC}g \quad (2)$$

In the x-direction, Newton's second law implies

$$f_s\hat{\mathbf{i}} = m_{PBC}a_x\hat{\mathbf{i}}$$

Hence

$$f_s = m_{PBC}a_x \quad (3)$$

If the candy bar does not slip, its acceleration is identical to the acceleration of the tray, so that a_x is given by equation (1). Notice that the *static* frictional force on the candy bar causes its acceleration!

d. The maximum acceleration permitted is attained when the candy bar is ready to slip. The static frictional force then attains its maximum value, given by Equation 5.6:

$$f_{s\,max} = \mu_s N'$$

Using equation (2) for N',

$$f_{s\,max} = \mu_s m_{PBC}g$$

Using Newton's second law to determine the maximum acceleration,

$$f_{s\,max}\hat{\mathbf{i}} = m_{PBC}a_{x\,max}\hat{\mathbf{i}}$$

Substituting for $f_{s\,max}$, you find

$$\mu_s m_{PBC}g = m_{PBC}a_{x\,max}$$

---

*If you choose the reverse direction for $\vec{\mathbf{f}}_s$ in the sketch, the result for the component of $\vec{\mathbf{f}}_s$ will be negative, indicating that the direction of $\vec{\mathbf{f}}_s$ is in fact opposite to the direction you chose in the second law force diagram.

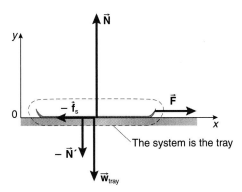

FIGURE 5.77

Thus

$$a_{x\,max} = \mu_s g$$

e. The forces acting on the tray, considering it as an individual system, are (see Figure 5.77)

1. the gravitational force of the Earth on the tray (the weight of the tray $\vec{w}_{tray}$), of magnitude $m_{tray}g$, directed vertically down;
2. the normal force $\vec{N}$ of the surface on the tray, directed vertically up;
3. the force of the candy bar on the tray, $-\vec{N}'$, of magnitude $m_{PBC}g$, directed vertically down; this force and the force $\vec{N}'$ of the tray (surface) on the candy bar [equation (2)] form a Newton's third law pair (notice that $\vec{N}'$ acts on the candy bar, *not* on the tray);
4. the horizontal force $\vec{F}$ exerted on the tray; and
5. the static frictional force $-\vec{f}_s$ of the candy bar on the top of the tray. This force is equal in magnitude, but opposite in direction to the static frictional force of the tray on the candy bar [equation (3)]; they form a Newton's third law pair. Thus the static frictional force of the candy bar on the tray is of magnitude $f_s = m_{PBC}a_x$ and is directed parallel to $-\hat{i}$.

Now apply Newton's second law of motion to the tray, considering it as an independent system. In the vertical direction we have, since $a_y = 0$ m/s²,

$$F_{y\,total}\hat{j} = m_{tray}a_y\hat{j}$$
$$m_{tray}g(-\hat{j}) + m_{PBC}g(-\hat{j}) + N\hat{j} = m_{tray}(0\ \text{m/s}^2)\hat{j}$$
$$= 0\ \text{N}$$

Solving for N,

$$N = (m_{PBC} + m_{tray})g$$

as we obtained before in part (a).

In the horizontal direction, Newton's second law implies

$$F\hat{i} + f_s(-\hat{i}) = m_{tray}a_x\hat{i}$$

But since $f_s = m_{PBC}a_x$, this becomes

$$F\hat{i} + m_{PBC}a_x(-\hat{i}) = m_{tray}a_x\hat{i}$$

Solving for a_x, you obtain

$$a_x = \frac{F}{m_{PBC} + m_{tray}}$$

the same as we found in part (a).

5.13 KINETIC FRICTION AT LOW SPEEDS

Solid surfaces in contact stick more effectively when at rest than when there is sliding motion parallel to the surfaces. Therefore, once a system begins sliding across a surface, the coefficient of friction changes from the static value μ_s to a smaller kinetic (motional) value μ_k. The ratio between the magnitude of the **kinetic frictional force** $\vec{f}_k$ and the magnitude of the normal force $\vec{N}$ defines the **coefficient of kinetic friction** μ_k:

$$\mu_k \equiv \frac{f_k}{N} \qquad \text{(system sliding)}$$

This is usually written as

$$f_k = \mu_k N \qquad \text{(system sliding)} \qquad (5.7)$$

As with the coefficient of static friction, numerical values for the coefficient of kinetic friction μ_k depend on the specific materials and are, to a first approximation, independent of the surface area in contact. When the system is sliding at low speeds,* the coefficient of kinetic friction is approximately constant, independent of speed. However, as the speed increases, the kinetic frictional force depends on the speed in complex ways that are typically determined empirically on a case-by-case basis. Approximate values for μ_k are tabulated in Table 5.4 for various surfaces.

PROBLEM-SOLVING TACTIC

5.11 The kinetic frictional force on a system by a surface always is directed opposite to the velocity of the system; the static frictional force on a system is directed opposite to the attempted slippage of the system.

*The quantitative meaning of a low speed depends on the specific surfaces in contact.

TABLE 5.4 Approximate Values of the Coefficient of Kinetic Friction[†]

Surfaces in contact	Coefficient of kinetic friction μ_k when system is moving at low speed
Glass on glass	0.4
Rubber on dry concrete	0.8
Rubber on wet concrete	0.3
Steel on steel	0.5
Steel on steel (lubricated)	0.05
Wood on wood	0.2
Waxed ski on snow	0.05

[†]For more precise values or for other surfaces, it is necessary to measure the coefficient of kinetic friction with the actual materials used. See Example 5.16 for a way to determine the coefficient of kinetic friction experimentally.

EXAMPLE 5.15

A 10.0 kg concrete block is sliding down an inclined plane at a construction site, as shown in Figure 5.78. The coefficient of kinetic friction for the block on the surface is 0.25. Determine the magnitude and direction of the kinetic frictional force on the block.

FIGURE 5.78

Solution
Let the 10.0 kg block be the system. Since the direction of the kinetic frictional force on a system always is opposite to its velocity, the kinetic frictional force on the block is directed parallel to the surface and up the inclined plane. The second law force diagram for the system includes the following forces (see Figure 5.79):

1. the weight $\vec{\mathbf{w}}$, of the magnitude mg, vertically down;
2. the normal force of the surface $\vec{\mathbf{N}}$, perpendicular to the surface; and
3. the kinetic frictional force $\vec{\mathbf{f}}_k$ antiparallel to the velocity, or up the plane.

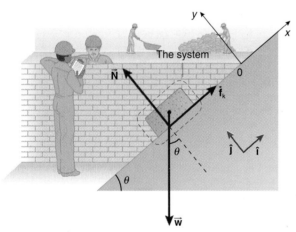

FIGURE 5.79

The magnitude of the kinetic frictional force is determined from Equation 5.7, and so you need to know the magnitude of the normal force. With the coordinate system indicated in Figure 5.79, the weight and normal forces are

$$\vec{\mathbf{w}} = (-mg \sin \theta)\hat{\mathbf{i}} - (mg \cos \theta)\hat{\mathbf{j}}$$
$$\vec{\mathbf{N}} = N\hat{\mathbf{j}}$$

The kinetic frictional force $\vec{\mathbf{f}}_k$ has only an x-component. The system is not accelerating along the y-direction—$a_y = 0$ m/s². Applying the second law of motion,

$$F_{y\,\text{total}} = ma_y$$
$$N - mg \cos \theta = m(0 \text{ m/s}^2)$$
$$= 0 \text{ N}$$

Hence

$$N = mg \cos \theta$$
$$= (10.0 \text{ kg})(9.81 \text{ m/s}^2) \cos 40.0°$$
$$= 75.1 \text{ N}$$

With Equation 5.7, the magnitude of the kinetic frictional force thus is

$$f_k = \mu_k N$$
$$= (0.25)(75.1 \text{ N})$$
$$= 19 \text{ N}$$

Since the kinetic frictional force is directed opposite to the velocity of the system, the force is directed up the plane:

$$\vec{\mathbf{f}}_k = (19 \text{ N})\hat{\mathbf{i}}$$

STRATEGIC EXAMPLE 5.16

A daring downhill skier is sliding down a steep slope that makes an angle θ with the horizontal direction, as shown in Figure 5.80. The mass of the fully equipped skier is m, the coefficient of kinetic friction for the waxed skis on the snowy slope is μ_k, the sky is clear, and 1156 fans are shivering.

FIGURE 5.80

a. Sketch a second law force diagram indicating all forces acting on the skier as he accelerates down the slope.
b. Calculate the magnitude of his acceleration.
c. If his teammate of mass m' follows him down the slope, is his acceleration the same?
d. For what slope angle θ is his acceleration zero?

Solution
Take the fully equipped skier to be the system.

a. The forces on the skier system are (see Figure 5.81)

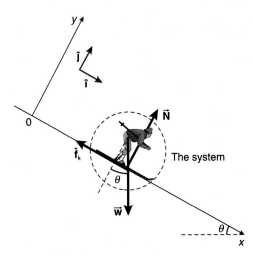

FIGURE 5.81

1. the gravitational force of the Earth on the skier (the weight $\vec{w}$ of the skier), of magnitude mg, directed vertically downward;
2. the normal force $\vec{N}$ of the slope on the skier, directed perpendicular to the slope; and
3. the kinetic frictional force $\vec{f}_k$ of the surface on the skier system, directed opposite to his velocity, so that $\vec{f}_k$ is parallel to and up the slope.

Psychological forces such as fear (I'm going down THAT??), panic (No way!), and peer pressure (I can't believe I got talked into this!) are not relevant to the physical problem.

Since the motion is down the slope, choose a set of coordinate axes parallel and perpendicular to the downward slope, as indicated in Figure 5.81.

b. The weight of the skier system is expressed in terms of its Cartesian components as

$$\vec{w} = (mg \sin \theta)\hat{i} - (mg \cos \theta)\hat{j}$$

The normal force of the surface on the skier system is

$$\vec{N} = N\hat{j}$$

and the kinetic frictional force is

$$\vec{f}_k = f_k(-\hat{i})$$

Apply Newton's second law to each coordinate direction separately:

Examine the motion along y. The y-coordinate of the skier remains constant as he zips down the slope. Therefore he is not accelerated along the y-direction, and so $a_y = 0$ m/s². Apply Newton's second law to the y-direction:

$$F_{y\,total}\hat{j} = ma_y\hat{j}$$
$$N\hat{j} + (mg \cos \theta)(-\hat{j}) = m(0 \text{ m/s}^2)\hat{j}$$
$$= 0 \text{ N}$$

Thus

$$N = mg \cos \theta$$

Apply Newton's second law of motion to the x-direction:

$$F_{x\,total}\hat{i} = ma_x\hat{i}$$
$$(mg \sin \theta)\hat{i} + f_k(-\hat{i}) = ma_x\hat{i}$$
$$mg \sin \theta - f_k = ma_x \qquad (1)$$

The magnitude of the kinetic frictional force is related to the magnitude of the normal force through the coefficient of kinetic friction (Equation 5.7):

$$f_k = \mu_k N$$

Substituting for N, you find

$$f_k = \mu_k mg \cos \theta$$

Substituting for f_k in equation (1) for the motion along the x-axis,

$$mg \sin \theta - \mu_k mg \cos \theta = ma_x$$
$$g(\sin \theta - \mu_k \cos \theta) = a_x$$

c. Notice that the acceleration is independent of the mass of the skier. So his teammate of mass m' will proceed down the slope with the same acceleration if his waxing is the same.
d. His acceleration is zero if

$$\sin \theta - \mu_k \cos \theta = 0$$

or

$$\sin \theta = \mu_k \cos \theta$$

Since $(\sin \theta)/(\cos \theta) = \tan \theta$, we have

$$\tan \theta = \mu_k \qquad \text{(system moving at constant speed down an incline)} \qquad (2)$$

If you use the numerical value for the coefficient of kinetic friction for waxed skis on snow in Table 5.4, the slope angle for zero acceleration is

$$\tan \theta = 0.05$$

Hence

$$\theta = 3°$$

This result is independent of the skier, since his mass did not enter into this calculation.

Notice from equation (2) that if a system is sliding down a slope at constant speed, the coefficient of kinetic friction is the tangent of the angle of the slope with the horizontal direction. This is a practical experimental way to determine the coefficient of kinetic friction.

5.14 KINETIC FRICTION PROPORTIONAL TO THE PARTICLE SPEED*

Drop a marble into a jug of molasses, and the marble accelerates during a short interval but quickly attains a constant speed. The

TABLE 5.5 Approximate Terminal Speeds of Various Sports Balls when Dropped in Air*

	Terminal speed (m/s)
baseball	42
basketball	20
golf ball	40
Ping-Pong ball	9
soccer ball	25
tennis ball	31
skydiver (unopened parachute) (hardly a sport ball!)	≈80

*Converted to SI units from a table in Peter J. Brancazio, *Sport Science* (Simon & Schuster, New York, 1984), page 334.

kinetic frictional force on the marble increases in magnitude with the speed of the marble until the total force on the marble and its acceleration are zero. Such effects are quite common for systems moving through a fluid (a gas or a liquid).

If a system is moving through a surrounding fluid, the magnitude of the kinetic frictional force on the system depends on its shape, the resistance of the medium to flow (the viscosity† of the fluid), and the speed of the system. The system is essentially sliding through the surrounding fluid. Experiments indicate that the faster the speed or the greater the viscosity, the larger the magnitude of the kinetic frictional force. This empirical relationship for the kinetic frictional force is written as

$$\vec{\mathbf{f}}_{\text{fluid}} = -K\eta\vec{\mathbf{v}} \qquad (5.8)$$

where η is the viscosity of the fluid (with SI units of $\text{N} \cdot \text{s/m}^2$), K is a geometric factor that depends on the shape of the particle (the SI units of K are m), and $\vec{\mathbf{v}}$ is the velocity of the particle. The minus sign indicates that the kinetic frictional force on the particle always is directed opposite to the velocity of the particle. For spherical objects, the geometrical factor K is found to be proportional to the radius of the sphere; the relationship is rewritten as

$$\vec{\mathbf{f}}_{\text{fluid}} = -kr\eta\vec{\mathbf{v}} \qquad \text{(spherical object)} \qquad (5.9)$$

where the radius of the spherical particle is explicitly indicated and a new dimensionless constant k is introduced.

As a particle freely falls from rest through a fluid, it accelerates because the downward gravitational force is greater than the upward force of kinetic friction (see Figure 5.82). However, as the speed of the particle increases, the kinetic frictional force increases in magnitude (see Figure 5.83). Eventually the magnitude of the kinetic frictional force comes to equal that of the gravitational force (Figure 5.84). The vector sum of the two forces is then zero, and so is the acceleration of the particle. The particle then is in equilibrium and falls at a constant speed known as the **terminal speed** v_{term}. The approximate terminal speeds of various sports balls falling in air are indicated in Table 5.5.

Frequently, the relationship between the kinetic frictional force and the particle speed is more complicated than the empirical one in Equations 5.8 and 5.9. Kinetic frictional forces

†We introduce the viscosity formally in Chapter 11.

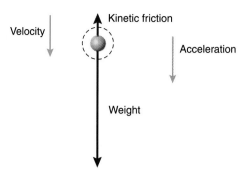

FIGURE 5.82 The forces on a particle falling through a fluid a short interval after release.

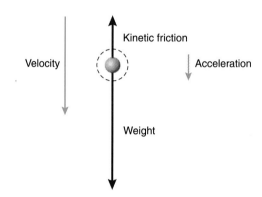

FIGURE 5.83 As the speed of the particle increases, so does the force of kinetic friction.

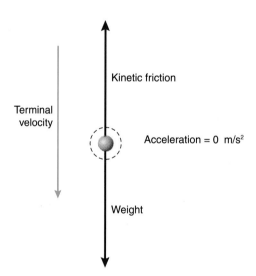

FIGURE 5.84 When the forces are of equal magnitude, the acceleration of the particle ceases.

approximately proportional to the *square of the speed* also are quite common.

All of the equations (Equations 5.7–5.9) for the kinetic frictional force under various circumstances are *empirical* equations; they are *not* fundamental, theoretical physical laws. The equations are used only because they are relatively simple approximations to complex situations. Microscopic theories of friction are quite complicated.

EXAMPLE 5.17 _____

A particle of mass m falls through a fluid near the surface of the Earth, and the *magnitude* of the kinetic frictional force acting on the particle is

$$f_{\text{fluid}} = K\eta v$$

a. Draw a sketch that indicates the directions and relative sizes of the gravitational force and the kinetic frictional force when the particle has reached its terminal speed.
b. Show that the terminal speed is

$$v_{\text{term}} = \frac{mg}{\eta K}$$

Solution

a. When the particle is falling at its terminal speed, the forces on the particle are (see Figure 5.85)
 1. the downward gravitational force $\vec{w}$, of magnitude mg; and
 2. the upward kinetic frictional force $\hat{\mathbf{f}}_{\text{fluid}}$, of magnitude $K\eta v$ and direction opposite to the downward velocity.

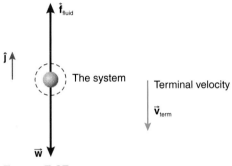

FIGURE 5.85

The two forces are of equal magnitude when the particle reaches its terminal speed.
b. When the particle reaches its terminal speed, it has zero acceleration. Choose $\hat{\mathbf{j}}$ to be the upward direction. Newton's second law of motion, Equation 5.2, becomes

$$F_{y\,\text{total}}\hat{\mathbf{j}} = ma_y\hat{\mathbf{j}}$$
$$K\eta v_{\text{term}}\hat{\mathbf{j}} - mg\hat{\mathbf{j}} = m(0 \text{ m/s}^2)\hat{\mathbf{j}}$$
$$= 0 \text{ N}$$

Hence

$$K\eta v_{\text{term}} = mg$$

Solving for v_{term}, you find

$$v_{\text{term}} = \frac{mg}{\eta K}$$

5.15 FUNDAMENTAL FORCES AND OTHER FORCES REVISITED*

The fundamental forces (see Section 5.2) of gravitation and electromagnetism underlie and are ultimately responsible for the host of forces encountered in applications in macroscopic physics and engineering. Forces such as pushes and pulls, tensions exerted by ropes or cables, friction, forces exerted by springs, levers, and what-not are essentially *contact forces*. However, when analyzed on a microscopic basis, *all* such contact forces are complex manifestations of the underlying fundamental electromagnetic and gravitational forces, which are noncontact forces. It is electromagnetic forces that dominate the interactions between the atoms and molecules and give rise to the tension in a rope and friction and normal forces. But it is extraordinarily difficult to describe in detail how common forces, such as tensions, friction, and normal forces, arise from the underlying electromagnetic force. This is precisely the reason that we bury the intractability of such a description; because it is simpler, we talk about these forces as if they had an independent existence of their own when we deal with macroscopic objects. This is a ruse, of course, but it is a very useful ruse for the solution to many practical problems with common macroscopic objects.

5.16 NONINERTIAL REFERENCE FRAMES*

Newton's laws of motion are applicable in inertial, but not noninertial, reference frames. In this section we see what the problem is for noninertial frames and how Newton's laws can, nonetheless, be salvaged for them. For simplicity and by way of example, we consider only a particular noninertial reference frame.

Consider a noninertial reference frame S' accelerating uniformly at a constant acceleration

$$\vec{\mathbf{a}}_0 = a_{x0}\hat{\mathbf{i}}$$

with respect to an inertial reference frame S along the common x-axes, as shown in Figure 5.86. Let x, y, and z represent the coordinates of a particle, the system whose motion is to be studied, in the inertial reference frame S and let x', y', and z' be the coordinates

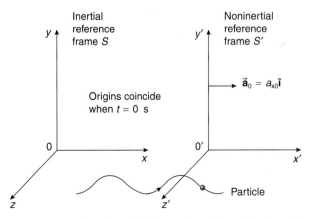

FIGURE 5.86 A noninertial reference frame S' accelerating with respect to inertial frame S.

of the same system in the noninertial reference frame S'. Since the constant acceleration of the noninertial frame has only an x-component, the transformation equations relating the coordinates of the particle in each reference frame are

$$x = x' + v_{x0}t + \frac{1}{2}a_{x0}t^2$$

$$y = y'$$

$$z = z'$$

The acceleration of the particle in the inertial reference frame S is

$$\vec{a} = \frac{d^2x}{dt^2}\hat{\imath} + \frac{d^2y}{dt^2}\hat{\jmath} + \frac{d^2z}{dt^2}\hat{k}$$

Substituting for x, y, and z from the transformation equations,

$$\vec{a} = \frac{d^2}{dt^2}\left(x' + v_{x0}t + \frac{1}{2}a_{x0}t^2\right)\hat{\imath} + \frac{d^2y'}{dt^2}\hat{\jmath} + \frac{d^2z'}{dt^2}\hat{k}$$

Differentiating to find the relationship between the accelerations gives

$$\vec{a} = \frac{d^2x'}{dt^2}\hat{\imath} + a_{x0}\hat{\imath} + \frac{d^2y'}{dt^2}\hat{\jmath} + \frac{d^2z'}{dt^2}\hat{k} \qquad (5.10)$$

Since the acceleration of the system in the S' reference frame is

$$\vec{a}' = \frac{d^2x'}{dt^2}\hat{\imath} + \frac{d^2y'}{dt^2}\hat{\jmath} + \frac{d^2z'}{dt^2}\hat{k}$$

Equation 5.10 becomes

$$\vec{a} = \vec{a}' + a_{x0}\hat{\imath}$$

or

$$\vec{a} = \vec{a}' + \vec{a}_0 \qquad (5.11)$$

Newton's second law is legitimate in the inertial frame S, and so an observer in this reference frame can say

$$\vec{F}_{total} = m\vec{a}$$

Substituting for $\vec{a}$ from Equation 5.11, we find

$$\vec{F}_{total} = m\vec{a}' + m\vec{a}_0 \qquad (5.12)$$

The term $m\vec{a}'$ is the right-hand side of Newton's second law *as written in the accelerated reference frame S'*. But notice that $m\vec{a}'$ is *not equal* to the vector sum of all the forces acting on the particle $\vec{F}_{total}$ because of the additional term, $m\vec{a}_0$, on the right-hand side of Equation 5.12. However, if you *rearrange* Equation 5.12 to read

$$\vec{F}_{total} - m\vec{a}_0 = m\vec{a}'$$

$$\vec{F}_{total} + (-m\vec{a}_0) = m\vec{a}'$$

then the *left*-hand side of the equation has an additional term equal to $-m\vec{a}_0$. Such a term is called a **pseudoforce**:

$$\vec{F}_{pseudo} \equiv -m\vec{a}_0$$

This is not a real force like the kinds mentioned earlier; the term is an accident of our choice of the accelerated reference frame S'.

By using such pseudoforce terms in an accelerated reference frame, Newton's second law can be rewritten as

$$\vec{F}_{real\ +\ pseudo} = m\vec{a}' \qquad (5.13)$$

where the $\vec{F}_{real\ +\ pseudo}$ term is interpreted as the *vector sum of the real and the pseudoforces* on the system. The pseudoforces arise *only* because of the acceleration of the noninertial (primed) reference frame. Without the pseudoforce term, the accelerated frame S' cannot properly describe the motion of the particle using Newton's laws. That is, in the noninertial reference frame S', you cannot say the acceleration of the system arises because of just the *real* forces acting on the system. The noninertial frame S' must also include the pseudoforces to make the patched-up second law applicable for properly describing the motion of the particle or system in the accelerated reference frame.

A specific example illustrates the difference between using Newton's second law in an inertial frame S and in a noninertial frame S'. Imagine a mass m suspended by a string from the ceiling of an airplane accelerating along a horizontal runway with an acceleration $\vec{a}_0$ with respect to the ground. We used such a plumb bob device as a crude accelerometer in Chapters 3 and 4 to detect the existence of horizontal accelerations. A passenger inside the plane is in a noninertial reference frame S'. The plumb bob is observed to be suspended in apparent equilibrium, with the string inclined to the vertical direction, as shown in Figure 5.87.

The greater the horizontal acceleration $\vec{a}_0$ of the plane, the greater the angle that the plumb bob makes with the vertical direction. The point is that an observer inside the airplane draws a second law force diagram of the real forces acting on the bob, as indicated in Figure 5.88. Once again there is the force $\vec{T}$ of the string on the bob and the weight $\vec{w}$ of the bob. The vector sum of the two real forces is *not zero*, since the forces are not collinear. The mass has *zero* acceleration according to the observer in the plane; it is just hanging there at rest with the string making an angle θ with the vertical direction! The observer inside the accelerating plane confronts a situation where the vector sum of

FIGURE 5.87 The plumb bob as observed by a passenger in an airplane accelerating along a horizontal runway.

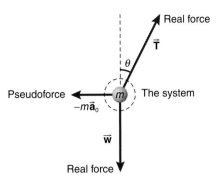

FIGURE 5.88 Second law force diagram according to an observer in the accelerating airplane.

the *real* forces is *not* zero but the observed acceleration *is* zero, since the bob is in apparent equilibrium according to the observer in the cabin of the plane. Hence

$$\vec{F}_{real} \neq m\vec{a}'$$

Newton's second law fails in the accelerated reference frame S'. However, we can make the second law applicable there by *adding* the *pseudoforce* $-m\vec{a}_0$ to the *left*-hand side of the second law. This is high-class fudging; but it does enable the observer inside the accelerating plane (the noninertial frame) to use a patched-up version of Newton's second law. The vector sum of the *real forces and pseudoforces* in the accelerated frame S' will be equal to the mass times the acceleration of the particle as observed in the accelerated frame:

$$\vec{F}_{real \,+\, pseudo} = m\vec{a}'$$

The real forces are $\vec{T}$ and $\vec{w}$, the pseudoforce is $-m\vec{a}_0$, and so according to an observer in the accelerating plane, the modified Newton's second law of motion is

$$\vec{T} + \vec{w} + (-m\vec{a}_0) = m\vec{a}'$$

Since the plumb bob stays right in front of the observer's nose, the bob is not accelerating relative to the observer in the accelerated frame—$\vec{a}' = 0$ m/s². The observer in the S' frame concludes that sum of the real and pseudoforces must be zero, consistent with the patched-up second law for noninertial frames. Thus, if pseudoforces are introduced into the *left*-hand side of Newton's second law, the second law can be used by observers in accelerated (i.e., noninertial) reference frames.

If the accelerating airplane and plumb bob are observed by an observer on the ground, as in Figure 5.89, Newton's second law works just fine with just the real forces, since the ground is an inertial reference frame S and no pseudoforces are needed or appropriate. The observer on the runway draws the second law force diagram indicated in Figure 5.90.

Since the plumb bob is accelerating *horizontally* along the runway, the *vertical* acceleration a_y is zero; therefore

$$
\begin{aligned}
F_{y\,total}\hat{\jmath} &= ma_y\hat{\jmath} \\
&= m(0 \text{ m/s}^2)\hat{\jmath} \\
&= 0 \text{ N}
\end{aligned}
$$

FIGURE 5.89 Observation of the plumb bob by an observer at rest on the runway.

FIGURE 5.90 Forces on the plumb bob according to an observer on the runway.

Hence this observer says the sum of the forces in the vertical direction is zero:

$$(T\cos\theta - mg)\hat{\jmath} = 0 \text{ N}$$

or, transposing,

$$T\cos\theta = mg \qquad (5.14)$$

In the horizontal direction the bob has acceleration (with the plane) $a_{x0}\hat{\imath}$, and so the second law in this direction is

$$
\begin{aligned}
F_{x\,total}\hat{\imath} &= ma_x\hat{\imath} \\
(T\sin\theta)\hat{\imath} &= ma_{x0}\hat{\imath} \qquad (5.15)
\end{aligned}
$$

The horizontal component of the force of the string on the bob causes the horizontal acceleration. Notice how the second law works as it should in the inertial frame S with just the real forces, whereas in the noninertial frame S' the second law works only with the introduction of the pseudoforce $-ma_0\hat{\imath}$ into *left*-hand side of the patched-up Newton's second law.

We now can see how the deflection angle of the plumb bob is a measure of the magnitude of the acceleration. Divide the magnitude of Equation 5.15 by Equation 5.14:

$$\tan \theta = \frac{a_{x0}}{g}$$

Hence the deflection angle θ is a direct measure of the acceleration component a_{x0}:

$$a_{x0} = g \tan \theta \qquad (5.16)$$

A measurement of the angle of deflection θ enables us to determine the magnitude of the acceleration. The direction of the deflection is opposite to that of the acceleration.

Spinning reference frames such as those attached to the surface of the rotating Earth also are (slightly) noninertial frames. Effects of pseudoforces (particularly one called the Coriolis force) are apparent in the dynamics of ballistic rockets, artillery shells, and even the circulations of air (winds) around high- and low-pressure weather patterns in the atmosphere of the Earth and on other planets as well. You will encounter such pseudoforces in more advanced courses in dynamics.

CHAPTER SUMMARY

The particle (or particles) whose motion is to be studied defines a *system*. In this chapter we considered systems with *constant mass*.

A system with an acceleration equal to zero is said to be in mechanical *equilibrium*.

Newton's three laws of motion are the pillars of classical dynamics and can be used in inertial reference frames.

Newton's first law of motion states: *any system in equilibrium remains in equilibrium unless compelled to change that state by a nonzero total force acting on the system.* A system in equilibrium has zero total force acting on it.

Newton's second law of motion states: *the vector sum $\vec{F}_{total}$ of all the forces acting on a system of constant mass m is equal to the product of m and its acceleration $\vec{a}$*:

$$\vec{F}_{total} = m\vec{a} \qquad (5.1)$$

The SI unit of force is the *newton* (N). A total force of magnitude 1 N acting a 1 kg mass will give it an acceleration of magnitude 1 m/s². So

$$1 \text{ N} = (1 \text{ kg})(1 \text{ m/s}^2)$$
$$= 1 \text{ kg·m/s}^2$$

Newton's third law of motion states: *if a system A exerts a force on another system B, then system B exerts a force on system A that is equal in magnitude but opposite in direction to the force of A on B:*

$$\vec{F}_{B \, on \, A} = -\vec{F}_{A \, on \, B} \qquad (5.3)$$

That is, forces always occur in pairs, but each member of the pair acts on a different system.

To apply Newton's second law to study the motion of a system, it is useful to draw a *second law force diagram* indicating all forces acting on the system. The vector sum of these forces is the total force $\vec{F}_{total}$ on the system and is equal to the mass of the system times its acceleration.

A *third law force diagram* indicates the two forces that form the pair of forces in Newton's third law of motion; these forces are of equal magnitude, point in opposite direction, and *act on different systems*.

The *weight* $\vec{w}$ of a system of mass m is the gravitational force on the system. The weight of a system of mass m near the surface of the Earth is of magnitude mg, where $g \approx 9.81$ m/s²:

$$w = mg \qquad (5.4)$$

The weight is directed vertically down.

The *normal force* $\vec{N}$ of a surface on a system always is directed *perpendicular* to the surface. The magnitude of the normal force must be determined from the specifics of the situation; there is no universal equation for the normal force.

A pull on an *ideal* string, rope, or cable is transmitted by the cord, undiminished in magnitude along its length, to objects at the other end of the cord. The force by any point along the string on its neighbors is called *tension*. The forces of the string on the objects at its ends (also called tension forces) are directed parallel to and toward the string. A frictionless and massless pulley changes the direction of a tension force, keeping its magnitude the same.

The *static frictional force* $\vec{f}_s$ of a surface on a system is directed *parallel to the surface* of contact and opposes slippage of the system. The static frictional force has a magnitude ranging from *zero* to a maximum value

$$f_{s \, max} = \mu_s N \qquad \text{(system ready to slip)} \qquad (5.6)$$

where N is the magnitude of the normal force of the surface on the system. *The maximum static frictional force occurs only when the system is on the verge of slipping.*

The *kinetic frictional force* $\vec{f}_k$ on a system moving at a low speed is of magnitude

$$f_k = \mu_k N \qquad \text{(system sliding)} \qquad (5.7)$$

and always is directed *opposite to the velocity of the system*. Here N is again the magnitude of the normal force of the surface on the object. For speeds that are not small, the kinetic frictional force depends on the particle speed in ways that must be determined from experiment.

The Newtonian laws of motion may be applied legitimately to systems in equilibrium or systems accelerating with respect to inertial reference frames as long as the speeds involved are much less than the speed of light ($\approx 3.00 \times 10^8$ m/s).

In noninertial reference frames (coordinate systems that are themselves being accelerated with respect to an inertial reference frame), Newton's second law is salvaged by adding *pseudoforces* to the real forces on the left-hand side of the law.

SUMMARY OF PROBLEM-SOLVING TACTICS

5.1 **(page 173)** It is essential to specify clearly the system whose motion is to be examined.

5.2 **(page 177)** The two sides of Newton's second law represent different things.

5.3 **(page 179)** Watch the little words. The words *of*, *on*, and *by* have important ramifications.

5.4 **(page 179)** Remember that the two forces in a third law force pair act on different systems.

5.5 **(page 179)** Decide what the systems are before you sort out third law force pairs.

5.6 **(page 182)** Identify the system and indicate the forces acting on it in a sketch (a second law force diagram).

5.7 **(page 182)** If the direction of a particular force is unknown, presume a direction.

5.8 **(page 184)** The normal force always is directed perpendicular to the surface, but the magnitude of the normal force is *not* always equal to the weight of the system.

5.9 **(page 189)** An ideal string exerts forces on the things at its ends that are directed toward and along the string; these forces are equal in magnitude.

5.10 **(page 198)** Equation 5.6, $f_{s\,max} = \mu_s N$, is useful only when the system is ready to slip across the surface.

5.11 **(page 205)** The kinetic frictional force on a system by a surface always is directed opposite to the velocity of the system; the static frictional force on a system is directed opposite to the attempted slippage of the system.

QUESTIONS

1. (page 176); 2. (page 180); 3. (page 181); 4. (page 184); 5. (page 185); 6. (page 191); 7. (page 198); 8. (page 198)

9. In an article about an automated pavement repair vehicle, a reporter wrote: "It carries just two people, but sports $400,000 of high-technology gear, including a 3-D infrared-laser-vision system to scan potholes, two computers to generate a 'filling algorithm,' a hose to scoop out debris, a hot-air lance to heat the hole and a robotic arm that plops precisely mixed rock and goo—'aggregate' and 'emulsion' to those in the trade—into the hole at a force of 60 mph."* Comment on the technical accuracy of this report.

10. Explain what may be wrong with the following statement: "An object at rest or moving at constant velocity has no forces on it."

11. A nonzero total force changes the velocity of a certain system but not its speed. Is this an oxymoron? What kind of motion is the mass experiencing?

12. (a) State Newton's second law of motion in words (no equations permitted). (b) State Newton's third law of motion in words (no equations permitted). (c) Explain why Newton's first flaw . . . err . . . law is more than just a special case of the second law.

13. Explain the physical principles involved in the spin cycle of a washing machine.

14. When you flick or shake a rug to get the dust out, what physical principles are involved?

15. Imagine a large mass and a small mass both resting on a horizontal frictionless surface. A constant horizontal force of magnitude F_0 is applied to each mass. What is the ratio of the magnitudes of the accelerations?

$$\frac{\text{magnitude of the acceleration of large mass}}{\text{magnitude of the acceleration of small mass}}$$

16. Two masses m_1 and m_2 are on a horizontal frictionless surface. Horizontal forces of various magnitudes are applied to each, and the corresponding resulting horizontal accelerations are measured. A graph is made plotting the magnitude of the horizontal force (in N) on each versus the magnitude of the consequent horizontal acceleration (in m/s^2). The results appear in Figure Q.16. Which graph represents the data for the greater mass? How did you reach your conclusion? Make measurements on the graph to determine the ratio of the two masses. If the two masses are glued together, sketch the corresponding graph for the system of mass $m_1 + m_2$.

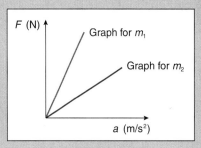

FIGURE Q.16

17. A system is on a horizontal frictionless table. A single horizontal force of magnitude F acts on the system and the resulting acceleration is measured. Forces of various magnitudes act on the system, and the resulting accelerations are measured. The magnitude of the horizontal force is plotted versus the magnitude of the corresponding acceleration. Is it possible for the graphs to appear as indicated in Figures Q.17? If so, what is the mass of the system? If it is not possible, explain why not.

18. Does Newton's second law imply anything about the relative directions of the total force acting on a system and the *velocity* of the system?

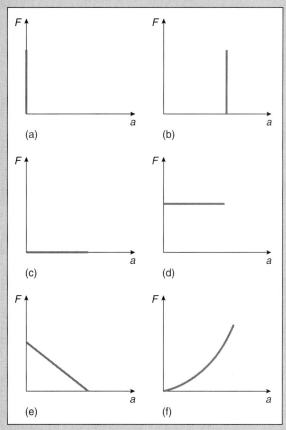

FIGURE Q.17

19. Drop a raw egg on a hard kitchen floor and you get a mess to clean up; drop the raw egg on a plush carpet and you likely can pick up the unbroken egg. Explain the physics underlying the two situations.

20. A marble is rolled around a partial circular barrier on a horizontal table as indicated in the Figure Q.20. Sketch the path of the marble after it reaches the end of the barrier, explaining the trajectory on the basis of Newton's laws of motion; be specific as to which law(s) you cite. Does your answer depend on whether the tabletop is considered frictionless or not?

FIGURE Q.20

21. A popular physics demonstration is to quickly yank a tablecloth out from under a set of dishes on a table. Explain the physical principles that are involved in such a demonstration.

22. A mass has one and only one force acting on it during a time interval Δt. The magnitude of the force is not equal to zero during the interval. Can the acceleration of the mass be zero at some instant during the interval? Can its velocity be zero at some instant during the interval? Explain.

23. An intrepid canoeist spies a nice spot to picnic on a flat rock in the middle of a beautiful stream. The canoeist blithely lunges out of the boat toward the rock and the canoe goes the other way. Explain why this happens on the basis of Newton's laws of motion.

24. The velocity of an automobile is $\vec{v}$ and its acceleration is $\vec{a}$. Describe the situations in which (a) $\vec{a} \neq 0$ m/s² but $\vec{v} = 0$ m/s; (b) $\vec{a} = 0$ m/s² but $\vec{v} \neq 0$ m/s; (c) $\vec{a} \neq 0$ m/s² and $\vec{v} \neq 0$ m/s, but $\vec{a} \cdot \vec{v} = 0$ m²/s³; (d) $\vec{a} \neq 0$ m/s² and $\vec{v} \neq 0$ m/s, but $\vec{a} \times \vec{v} = 0$ m²/s³. For each of (a)–(d), what is the direction of the total force acting on the automobile?

25. Two forces $\vec{F}_1$ and $\vec{F}_2$ act on a mass m. The forces are neither parallel nor antiparallel. What is the smallest number of additional forces needed to ensure that the particle is not accelerated? Specify the orientation(s) of these additional force(s) if the information permits. Repeat the problem for the case where three noncollinear forces $\vec{F}_1, \vec{F}_2$, and $\vec{F}_3$ act on the mass.

26. Newton's second law states that the total force on a mass m produces an acceleration of m that is parallel to the total force. Does this mean that the total force must be parallel (or antiparallel) to the velocity of the particle? Explain.

27. Describe a situation in which the total force on a mass is perpendicular to the velocity of the object.

28. *Imagine* that a particle with negative mass exists. What does Newton's second law imply about how such a hypothetical particle would respond to the application of a single force to it? Such behavior *never* has been observed in nature, which is why we believe negative masses do not exist.

29. A truck driver, wishing to save time, uses a forklift to place a massive load at the rear of a long trailer, but fails to secure the load with ropes before heading off (see Figure Q.29). Explain why and under what circumstances this could be dangerous to the rig, to the driver, and to vehicles in close proximity to the truck. Invoke appropriate laws of motion to explain your answer.

Load

FIGURE Q.29

30. Are seat belts really needed during a successful takeoff of an airplane? During level flight at constant velocity with no turbulence? During landings? Explain your answers on the basis of Newton's laws of motion, indicating which law(s), by number, you cite.

31. After a large repast, you check in with your (metric) bathroom scale and find that it reads 85 kg with you atop it. What is your mass? What is your weight?

32. If your weight on the surface of the Earth has a magnitude of 600 N, what is the magnitude and direction of the force that you exert *on the Earth*?

33. If a mass slides down a frictionless slope, is the magnitude of its acceleration ever as great as g (= 9.81 m/s²)? Justify your answer.

34. If a mass accelerates down an inclined plane, does the magnitude of the normal force of the plane on the mass have any influence on the magnitude of the acceleration (a) if the inclined surface is frictionless? (b) if the inclined surface is not frictionless? Explain your reasoning.

35. Toss a penny down a well. Sketch a second law force diagram indicating all forces acting on the penny after you release it. What is the acceleration of the penny? Make a third law force diagram indicating the forces acting on the penny and on the Earth. How large is the magnitude of the acceleration of the Earth compared with that of the penny?

36. Take a 0.200 kg ball and throw it upward with a speed of 2.00 m/s. After the ball leaves your hand, what is the acceleration of the ball when it is at each of the following places along its trajectory? (Neglect air friction.) (a) Halfway to its maximum height; (b) at its maximum height; (c) one-fourth of the way down from the peak of its trajectory.

37. Suppose you throw a bowling ball vertically upward. Umph! What is the qualitative relationship among the magnitude of the normal force of the floor on you, the magnitude of your weight, and the magnitude of the weight of the bowling ball during each of the following phases of the motion: (a) while you are throwing the ball upward but before the ball leaves your hands; (b) after the ball leaves your hands and is moving upward in flight; (c) while the ball is falling downward; (d) as you catch the ball on its return but before the ball stops moving. Indicate in a sketch the direction of the bowling ball's acceleration during each of (a)–(d). On the same sketches, indicate the direction of the total force acting on the bowling ball during each phase.

38. For the ball of the previous question, suppose you include the effect of air friction with a magnitude proportional to the speed of the ball. Choose ĵ to be directed vertically upward. Make a qualitative graph of the y-component of the frictional force on the ball as a function of time.

39. Near the surface of the Earth, is it possible to accelerate a system in the vertical direction with a magnitude greater than the magnitude of the local acceleration of gravity g (= 9.81 m/s²)? If so, explain how. If not, explain why not.

40. Crouch down in preparation for making a vertical jump upward off a horizontal floor. As you begin your jump, but before you leave the floor, is the normal force of the floor on you equal to your weight? Explain. Suppose you did the experiment on a bathroom scale. Explain what happens to the scale reading as you begin the jump (but have not yet left the scale).

41. Why can you only pull and not push with long ropes or cables?

42. A large bell is suspended from a thread as pictured in Figure Q.42. Another identical thread is attached to the clanger and is suspended below it. If the lower string is pulled down with a force that gradually increases its magnitude, the *upper* string eventually breaks. On the other hand, if the lower

FIGURE Q.42

string is yanked downward suddenly, the *lower* string breaks. Explain why.

43. To warm up for a tug-of-war, you and a friend each practice with a rope attached to a wall and discover you both individually exert a maximum horizontal force of magnitude 900 N on the rope. What is the force of the wall on the rope in each case? (Specify its magnitude and direction relative to the force each of you exerts on the rope.) Sketch the second law force diagram of the forces on the rope in this case. What is the magnitude of the tension in the rope? Now you and your friend take the rope and pull with forces of magnitude 900 N in opposite directions in a tug-of-war. Sketch the second law force diagram for the rope for this case. What is the tension in the rope during the tug-of-war? What, then, determines who wins a tug-of-war?

44. A plumb bob is suspended from the ceiling, pulled away from the vertical direction, released, and swings back and forth (see Figure Q.44). The string is cut at the instant when the mass is at these alternative choices of location: (a) the top point of the swing; (b) the bottom point of the swing; (c) a point about midway between the bottom and top of its swing. For each of these situations, answer the following questions: (i) What is the direction of the velocity of the mass at this instant if the velocity is not zero? (ii) What is the name of the total force on the mass after the string is cut? (iii) Sketch the trajectory of the mass after the string is cut.

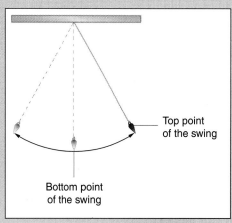

Top point of the swing

Bottom point of the swing

FIGURE Q.44

45. Explain several circumstances in which the static and kinetic frictional forces are (a) good; (b) bad.

46. Describe several situations in which the static frictional force on a mass produces an acceleration of this mass.

47. Will a drag racer accelerate faster if the tires spin on the pavement or if they do not? Explain. See Figure Q.47.

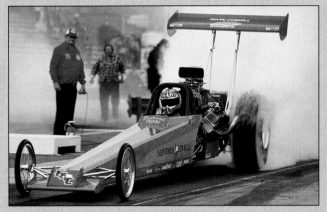

FIGURE Q.47

48. Aircraft tires have to be replaced less frequently if they are set spinning before touchdown than if they are not. Explain why.

49. What is the advantage of having an antilock braking system on a car? Explain your answer in terms of the coefficients of static and kinetic friction for the tires on the road.

50. You need to find a wax appropriate for cross-country skiing. For best performance, describe the characteristics of the coefficients of static and kinetic friction for the waxed skis on the snow.

51. Is it possible for the coefficients of static and kinetic friction to be greater than 1? How or why not?

52. Make a few simple measurements to determine the approximate value of the coefficient of static friction between a penny and the surface of this textbook page.

53. What observations imply that the kinetic frictional force of air on a system is a function of its speed?

54. Make appropriate measurements to estimate the terminal speed of an air-filled balloon falling in air.

55. Is it possible for a particle projected vertically *upward* in a resisting fluid to reach a terminal speed? Explain why or why not.

56. What are the SI units associated with the constant k in Equation 5.9 for the kinetic frictional force on a spherical object moving through a viscous fluid?

57. What is the instantaneous speed of the particles on a tire that are in contact with the road as the tire rolls along? What force ultimately is responsible for the forward acceleration of the tire? Under what circumstances is this force very small?

58. Do Newton's laws of motion legitimately apply in a reference frame attached to a merry-go-round when it is in motion at constant angular velocity? Explain.

59. Is it possible to apply Newton's laws of motion rigorously in a reference frame that is attached to an orbiting space laboratory? Why or why not?

60. You and a friend greet each other with a firm hand clasp. Describe the hand clasp from the viewpoint of Newton's third law (somewhat the way we used socks in Section 5.6).

61. Explain why the kinetic frictional force during the free-fall motion of a system never can have a magnitude greater than the weight of the system. Is the kinetic frictional force on an airplane in horizontal flight similarly restricted in magnitude? Explain why or why not.

62. You drive your car quite fast to the left around a circular turn and your passenger comments: "Holy cow, feel that centrifugal force!" On the basis of Newton's laws, explain what force the passenger senses. Is the total force on the passenger centrifugal (away from the center of the circle) or centripetal? Explain.

63. Is it possible for the coefficient of static friction for two surfaces to be less than the coefficient of kinetic friction for the same surfaces?*

---

*Question suggested to the author by Bryant E. Adams, a freshman.

PROBLEMS

Sections 5.3 Newton's First Law of Motion and a Qualitative Conception of Force
    5.4 The Concept of Force and Its Measurement
    5.5 Newton's Second Law of Motion
    5.6 Newton's Third Law of Motion
    5.7 Limitations to Applying Newton's Laws of Motion

1. Your car of mass 1.25×10^3 kg has an acceleration of magnitude 0.150 m/s^2. What is the magnitude of the total force on the car?

2. A system initially at rest has a constant total force on it of magnitude 295 N. After 3.00 s the speed of the system is 15.0 m/s. What is the mass of the system?

3. A 1.20×10^4 kg fighter jet has an acceleration of magnitude 2.0 g when taking off via a steam catapult from the flight deck of an aircraft carrier. See Figure P.3. What is the magnitude of the total force on the aircraft?

4. One force on a system of mass 5.60 kg is $(3.50 \text{ N})\hat{\imath} - (7.00 \text{ N})\hat{\jmath}$. The system has a constant velocity of $(2.50 \text{ m/s})\hat{\imath} + (3.00 \text{ m/s})\hat{\jmath}$. (a) What is the total force on the system? (b) If there is one other force acting on the system, find its magnitude and direction.

5. A 4.00 kg doohickey is subjected to the forces in the x–y plane shown in Figure P.5; they are the only forces acting on the doohickey. (a) What are the magnitude and direction (relative to the +x-axis) of the total force on the doohickey?

FIGURE P.3

FIGURE P.9

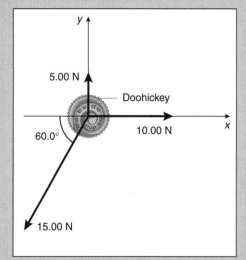

FIGURE P.5

vector in the direction of the acceleration in terms of the unit vectors of the coordinate system you introduced in part (a). Is the total force also directed along this unit vector?

•10. An incoming soccer ball of mass 0.430 kg bounces off the head of a striker. See Figure P.10. The incoming speed is 15.0 m/s and the outgoing speed is 18.0 m/s, with the two velocities in opposite directions. The ball is in contact with the cranium for 0.100 s. (a) Determine the magnitude of the acceleration of the ball during the brief interval of the head-shot. Assume the acceleration is constant. (b) What is the magnitude of the force of the ball on the player's head?

FIGURE P.10

(b) Determine the magnitude of the acceleration of the doohickey.

•6. A 2.50 kg system has an acceleration $\vec{a} = (4.00 \text{ m/s}^2)\hat{\imath}$. One of the forces acting on the system is $(3.00 \text{ N})\hat{\imath} - (6.00 \text{ N})\hat{\jmath}$. (a) Is this the total force on the system? (b) If not, find the other force acting on the system.

•7. A 1.50×10^3 kg car is pulling a 500 kg trailer with an acceleration of magnitude 0.150 m/s². (a) What is the magnitude of the total force on the trailer? (b) What is the magnitude of the total force on the car? (c) What is the magnitude of the total force on the car-and-trailer system?

•8. When the starting gun fires, a 60.0 kg sprinter accelerates with a constant acceleration for the first 10.0 m of the track and then runs with constant speed for the duration, completing the 100 m sprint in 10.0 s. What is the magnitude of the total force causing the acceleration of the sprinter?

•9. An 80 kg cubical poet on a horizontal frictionless surface is subjected to the horizontal critical forces shown in Figure P.9. These are the *only* horizontal forces on the poet. (a) Introduce a convenient coordinate system. (b) Find the total force on the poet as well as the magnitude of this force. (c) Determine the magnitude of the acceleration of the poet. (d) Find a unit

•11. Two forces act simultaneously on a 4.0 kg mass. One force is $\vec{F}_1 = (-3.0 \text{ N})\hat{\imath} + (6.5 \text{ N})\hat{\jmath}$ and the other has a magnitude 5.0 N directed at 240° counterclockwise from the +x-axis. Find the magnitude and direction of the acceleration.

•12. The following three forces act simultaneously (beginning when t = 0 s) and continuously on a 20 kg mass initially at rest. These are the *only* forces acting on the mass.

$$\vec{F}_1 = (3.0 \text{ N})\hat{\imath} + (2.0 \text{ N})\hat{\jmath}$$
$$\vec{F}_2 = (2.0 \text{ N})\hat{k}$$
$$\vec{F}_3 = (2.0 \text{ N})\hat{\imath} - (3.0 \text{ N})\hat{\jmath}$$

(a) Show that each force is perpendicular to the other two. (b) When $t = 3.0$ s, find the velocity and the speed of the mass.

•13. A point-particle politician of mass 90 kg, mostly pork, is subjected *only* to the political forces shown in Figure P.13. All forces are in the x–y plane. (a) Choose and clearly label an appropriate coordinate system to analyze the dynamics of the politician. (b) Determine the acceleration of the fat cat. (c) What is the magnitude of the acceleration?

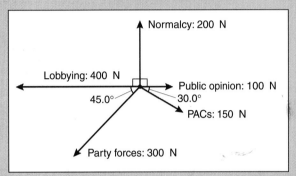

FIGURE P.13

•14. You are trying out a new Porsche of mass 1.50×10^3 kg. The car accelerates from a standing start to 130 km/h during 7.00 s along a level stretch of straight highway. Vroooom! Assume the acceleration is constant (not a particularly good approximation in fact). (a) Calculate the magnitude of the total force on the car causing the acceleration. (b) If the road is not level but inclined upward at a 5.0° angle to the horizontal, and the car still can reach the speed of 130 km/h during 7.00 s, is the magnitude of the total force on the car the same as in part (a)? Greater? Smaller? Explain your reasoning.

•15. A rather massive 400 kg walrus (*Odobenus rosmarus*) is frozen to the ice of a pond in Lake Wobegon and will not move, even when pulled on simultaneously by the following three horizontal forces:

3.00 kN	directed north
5.00 kN	directed east
4.00 kN	directed west

(a) Sketch the situation and introduce a convenient coordinate system. (b) Calculate the force exerted by the pond ice on the walrus.

Section 5.8 Do Inertial Reference Frames Exist?*

•16. (a) Calculate the magnitude of the centripetal acceleration of a point on the equator of the Earth due to its spin. (b) Compare the magnitude found in part (a) with the magnitude of the local acceleration due to gravity, g, near the surface of the Earth by forming the ratio:

$$\frac{a_{c \text{ spin}}}{g}$$

(c) Calculate the magnitude of the centripetal acceleration of the Earth (as a whole) in its orbital motion about the Sun. (d) Compare the magnitude of the orbital centripetal acceleration

with the magnitude of the local acceleration due to gravity g by forming the ratio:

$$\frac{a_{c \text{ orbital}}}{g}$$

(e) Compare the magnitudes of the spin and orbital centripetal accelerations by forming the ratio:

$$\frac{a_{c \text{ spin}}}{a_{c \text{ orbital}}}$$

Sections 5.9 Second Law and Third Law Force Diagrams
5.10 Weight and the Normal Force of a Surface

17. What is the mass of a fig (*Ficus carica*) with a weight of magnitude 1.00 N?

18. (a) What is the magnitude of the weight of a juicy 1.00 kg steak? (b) What is the magnitude of the force of a grill on the steak as you broil it (i.e., assume no fat loss)?

19. Your mass is 60.5 kg including the pizza just consumed while struggling with this physics homework at 2 A.M. early on a Sunday morning. That's dedication. (a) You now stand on a bunk to deliver a soliloquy. What is the gravitational force of the Earth on you? (b) You decide to jump off the bunk. What is the gravitational force of the Earth on you while you are in the air?

•20. (a) What is your mass in kilograms? (b) What is the magnitude of your weight on the surface of the Earth? (c) Jump off the floor. While you are in the air, is there a force acting on you? If so, what is the force as you go upward? reach the highest elevation? and fall? (d) If, on the way up, you let go of your wallet, will the wallet travel away from you as you continue up and then fall? (e) If (a)–(d) were asked on the surface of a planet where the local acceleration due to gravity was 4.50 m/s², which answers (if any) are different and what are the new answers?

•21. A 1.50 kg rock is in the middle of the bottom of an open bucket of mass 2.00 kg. The bucket and rock rotate in a *vertical* circle of radius 1.20 m at a constant speed of 4.00 m/s. (a) Make a second law force diagram indicating the forces on the rock when it is at the lowest point in its motion. (b) What is the magnitude of the total force on the rock when the bucket is at the lowest point of its circular path? (c) What is the force of the bucket on the rock at this lowest point? (d) At the highest point on the circular path, will the rock fall from the bottom of the bucket? Indicate your reasoning with an appropriate calculation and explanation.

•22. Muscles. On the ground you are just barely able to lift 100 kg. Inside an elevator, you discover that you can lift 110 kg (barely)! (a) What is the direction of the acceleration of the elevator relative to the ground? Sketch a second law force diagram showing the forces on the mass you lift in the elevator. (b) Calculate the magnitude of the acceleration of the elevator.

•23. You are sitting in the rear seat of an automobile that accelerates uniformly from rest to a speed of 120 km/h during 6.00 s along a straight stretch of level highway. The seat has a horizontal surface and a vertical seat back. Your mass is 75.0 kg. (a) Sketch a second law force diagram indicating all forces on you in the horizontal and vertical directions from the viewpoint of an observer at rest outside the car on the side of the highway. (b) Determine the magnitude of your acceleration. (c) What is the magnitude of the total force on you? In which direction does the total force act? (d) Identify the forces that are the Newton's third law pairs to the forces identified in (a) and sketch third law force diagrams indicating these forces and the systems on which they act.

•24. A car traveling at constant speed reaches the crest of a huge frost-heave (hill) that is a cylindrical cross section of radius 70 m as shown in Figure P.24. (a) At the instant when the car is at the crest of the hill, make a second law force diagram indicating schematically all forces acting on the car. (b) What is the maximum speed the car can travel without becoming airborne at the crest of the hill? Express your result in km/h.

FIGURE P.24

•25. (a) What inclination angle is needed for a frictionless inclined plane so that masses released on the plane experience an acceleration of magnitude precisely 1.00 m/s²? (b) If the mass is given an initial speed of 2.00 m/s down the plane, is its magnitude of acceleration the same as in part (a) after it is released? (c) If a 1.00 kg mass is placed at rest on this incline and released, what is the magnitude of the total force on the mass?

•26. A 51.0 kg crate is pushed up a frictionless 20.0° incline with a constant horizontal force (see Figure P.26). The magnitude of the acceleration of the crate is 0.100 m/s². (a) What are the magnitude and direction of the total force on the crate? (b) What is the magnitude of the weight of the crate? (c) What is the magnitude of the horizontal force applied? (d) What is the magnitude of the normal force of the surface on the crate?

FIGURE P.26

‡27. A chain of total mass m has n_0 identical links. The chain is suspended vertically from the top link and is at rest with no links touching the floor. (a) What is the mass of a single link? What is the magnitude of the weight of a single link? (b) What is the magnitude of the upward force on the link at the bottom of the chain (call this link number 1) by the next link up?

(c) What is the magnitude of the upward force on the link at the top of the chain (link number n_0)? (d) What is the magnitude of the upward force acting on link number n, counting from the bottom of the chain? (e) If the chain has an upward acceleration of magnitude a, recalculate your answers to parts (a)–(d).

‡28. A large Ferris wheel of radius 15.0 m is rotating wildly out of control at 4.00 rev/min. See Figure P.28. A student of mass 80.0 kg is becoming increasingly uncomfortable with the ride. On the other hand, you (of mass 50.0 kg) are rather enjoying the wild ride. To calm your companion's panic, you decide to explain soothingly the physics of the effects you both are experiencing on the Ferris wheel. You first point out that during the entire ride, there are only two forces acting on each of you: (1) your weight and (2) the normal force of the bench seat on your derriere. (a) When the chair is at the top of the circle, which of the two forces has the greater magnitude? What is the magnitude of the difference between your weight and the normal force of the seat on your derriere? (b) When the chair is at the bottom of the circle, which of the two forces is greater in magnitude? What is the difference in the magnitudes of the forces on you at this location? (c) When the chair is not at the top or bottom of the circle, the two forces are no longer antiparallel to each other, since the chair swivels. At what two places in the motion is the deviation of the direction of the normal force from the local vertical a maximum? At these locations, what is the angle θ between the direction of this force and the local vertical direction? (d) At what angular speed must the wheel rotate so you both are apparently "weightless" at the top of the spin? Quote the result in rev/min. Are you truly weightless at the top of the spin? Explain.

FIGURE P.28

Section 5.11 Tensions in Ropes, Strings, and Cables
Unless otherwise stated, assume all ropes, strings, and cables are ideal.

29. A fishing line will break if the magnitude of the tension in it exceeds 100 N. What is the greatest mass that can be hung vertically from the string?

30. A towing firm owned by Ima Wheeler is attempting to pull your car horizontally and frictionlessly out of an icy culvert. You and the car have a combined mass of 1355 kg. The cable attached to the car will break if its tension exceeds 4.00 kN.

What is the magnitude of the maximum acceleration that can be given to you and the car?

•31. Two masses m_1 and m_2 are connected by a string and are pulled across a horizontal frictionless surface by a horizontal force of magnitude F_0, as shown in Figure P.31. Find an expression for the magnitude of the tension in the cord connecting the masses.

FIGURE P.31

•32. A string will break if its tension exceeds 500 N. (a) Is it possible to hang a 35.0 kg stone on the string? (b) If part (a) indicates that the mass can be hung from the string, what is the *maximum* magnitude of acceleration vertically upward that can be given to the mass by the attached string without its breaking? (c) What maximum magnitude of acceleration vertically down can be imparted to the stone without breaking the string? (d) If the rock lies on a horizontal frictionless surface, what is the maximum horizontal acceleration that a horizontal pull with the string can give to the stone?

•33. A rope is hanging down from the ceiling of a gym. While standing on the floor, you grab the rope and pull down on the rope with a force of magnitude 100 N. Assume your mass is 70.0 kg. (a) What are the magnitude and direction of the force that the rope exerts on you? (b) Draw a second law force diagram indicating all forces acting on you. (c) For every force acting on you in part (b), state in words the corresponding force of a Newton's third law pair. (d) What is the magnitude of the gravitational force of the Earth on you? (e) What is the magnitude of the normal force that the floor exerts on you? (f) If you did this experiment while standing on a bathroom scale (calibrated in kilograms), the magnitude of which force can be determined from the scale reading? What is the scale reading?

•34. An intrepid mountain climber is resting in the awkward situation depicted in Figure P.34. (a) Draw a second law force diagram indicating schematically all forces acting on the climber. (b) Find the magnitude of the tensions in the two supporting ropes.

•35. A 200 kg meteorite is suspended in a science museum by two cables as shown in Figure P.35. (a) Sketch a second law force diagram considering the meteorite as the system. (b) Determine the magnitude of the forces the cables exert on the meteorite.

•36. A 1.20×10^4 kg helicopter is vertically lifting a 3.00×10^3 kg load (see Figure P.36). The upward acceleration is of magnitude 0.500 m/s². (a) Define an appropriate system and determine the magnitude of the tension in the vertical cable. (b) Define an appropriate system and determine the magnitude of the upward force of the air on the helicopter blades.

FIGURE P.34

FIGURE P.35

FIGURE P.36

•37. Two masses, m_1 and m_2, hang over an ideal pulley and the system is free to move (see Figure P.37). Such an arrangement is called an *Atwood's machine*. (a) Show that the acceleration $\vec{a}$ of the system of two masses has a magnitude

$$a = \frac{|m_2 - m_1|g}{m_2 + m_1}$$

(b) Show that the tension in the cord is of magnitude

$$\frac{2m_1m_2g}{m_1 + m_2}$$

•38. Two unfortunate frozen, cubical mountain climbers connected by a rope find themselves sliding over an icy (frictionless) precipice as indicated in Figure P.38. Rather than help them,

FIGURE P.37

FIGURE P.38

FIGURE P.39

FIGURE P.40

solve the following problems: (a) Consider the system to be the pair of climbers. Sketch a second law force diagram indicating the forces acting on this system. Write Newton's second law for this system to determine the magnitude and direction of the acceleration of the system. (b) Now sketch second law force diagrams indicating the forces acting on each climber considered as an individual system. (c) By considering the motion of either climber individually, find the magnitude of the tension in the rope.

•**39.** A 50.0 kg gymnast hangs on one side of a rope over an ideal pulley (see Figure P.39). The other end of the rope is attached to a 75.0 kg mass at rest on a scale on the floor. What is the magnitude of the minimum upward acceleration of the gymnast sufficient to have the scale indicate zero?

•**40.** Three 2.00 kg decoys on a frictionless table are connected in series by strings as indicated in Figure P.40. The final string passes over an ideal pulley at the edge of the table and suspends the mother of all decoys with a mass of 3.00 kg. The ducks are initially all at rest. (a) Which string has the least tension and why? (b) Which string has the greatest tension? Show that the greatest tension has a magnitude less than 29.4 N. (c) Do you

expect the magnitude of the acceleration of the ducky system to be less than, greater than, or equal to g? (d) Calculate the magnitude of the acceleration of the system of ducks.

•**41.** A string made from braided dental floss has a maximum permissible tension of 500 N and is used by a 60.0 kg inmate to escape prison (this actually happened!). The inmate starts at rest and slides vertically down the string for 8.0 m. (a) Make a second law force diagram indicating schematically all forces acting on the inmate in the descent. (b) Determine the magnitude of the acceleration of the inmate so the string barely does not break. Is this a minimum value or a maximum value? Explain. (c) What is the minimum speed of the former inmate at the end of the 8.0 m descent if the string does not break?

•**42.** A mass m_1 lies on a fixed, smooth (frictionless) cylinder. An ideal cord attached to m_1 passes over the cylinder to mass m_2 as pictured in Figure P.42. (a) Make a second law force diagram indicating schematically all forces acting on m_2. (b) Make a second law force diagram indicating schematically all forces acting on m_1. (c) Show that the system is in equilibrium if

$$\sin \theta = \frac{m_2}{m_1}$$

(d) Evaluate θ if $m_1 = 40$ kg and $m_2 = 10$ kg.

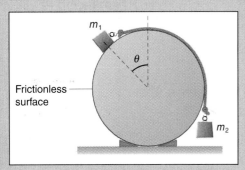

FIGURE P.42

•**43.** You are standing in an elevator that is accelerating vertically upward. (a) Sketch a second law force diagram indicating all forces acting on you. (b) Which force has the greatest magnitude? (c) Specify with a verbal description the forces that are the Newton's third law pairs to the forces delineated in part (a). Sketch corresponding third law force diagrams indicating the systems on which these forces act. (d) Repeat parts (a) and (b) for a situation in which the elevator is accelerating downward.

•**44.** A 500 kg elevator car is moving up in the world with an acceleration of 0.50 m/s² by means of a cable. (a) Make a second law force diagram indicating schematically all forces acting on the elevator car. (b) Determine the magnitude of the tension in the cable.

•**45.** A 100 kg oaf is in a 1.00×10^3 kg elevator car accelerating upward at 1.50 m/s². (a) Determine the magnitude of the tension in the cable supporting the elevator car. (b) What is the force of the floor of the elevator car on the oaf? Is this force equal in magnitude to that of the weight of the oaf? Explain why or why not.

•**46.** Two gargoyles, each of mass 2.00 kg, are hanging from the ceiling of an elevator by means of strings as indicated in Figure P.46. The elevator is moving downward with an acceleration of magnitude 2.0 m/s². (a) Choose an appropriate system, sketch an appropriate second law force diagram, and determine the magnitude of the tension in the lower string. (b) Choose an appropriate system, sketch an appropriate second law force diagram, and determine the magnitude of the tension in the upper string.

•**47.** Wahoo! You are swinging a mass m at speed v around on a string in a circle of radius r whose plane is 1.00 m above the ground (see Figure P.47). The string makes an angle θ with the vertical direction. (a) Make a second law force diagram in-

FIGURE P.46

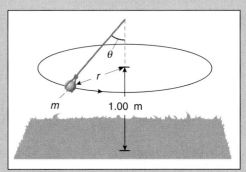

FIGURE P.47

dicating all forces acting on the mass at a particular instant in its motion. Indicate the direction to the center of its circular path. (b) What is the *direction* of the acceleration of the mass? (c) Apply Newton's second law to the horizontal and vertical directions to show that

$$\tan \theta = \frac{v^2}{rg}$$

(d) If the angle $\theta = 47.4°$ and the radius of the circle is 1.50 m, find the speed of the mass. (e) If the mass is 1.50 kg, what is the magnitude of the tension in the string? (f) The string breaks unexpectedly when the mass is moving exactly eastward. Where does the mass hit the ground?

•**48.** A 3.00 kg mass attached to a string is swung around, at constant speed, in a circle of radius 0.500 m by means of a massless string of length 2.00 m as indicated in Figure P.48. (a) Draw a second law force diagram schematically indicating the forces on the mass at some instant. (b) Is the total force on

FIGURE P.48

the mass zero? Explain. (c) Find the magnitude of the tension in the string. (d) What is the speed of the mass? (e) What is the time needed for the mass to complete one circular trajectory?

•49. The Dean of Deans (whose mass is 80.0 kg) steps into an express elevator car headed for a meeting of the Committee on Committees, the mother of all committees. The elevator initially is at rest. (a) Sketch a second law force diagram indicating schematically all forces on the Dean. (b) What is the magnitude of the gravitational force of the Earth on the Dean? (c) Find the magnitude of the force that the floor exerts on the Dean while the elevator car is at rest. In (d)–(h), the elevator car now gets under way with an upward acceleration of 1.00 m/s². (d) Sketch a second law force diagram indicating schematically all forces on the Dean. (e) What is the magnitude of the weight of the Dean during this acceleration? (f) Find the magnitude of the force of the floor on the Dean during this upward acceleration. (g) If the Dean were standing on a scale, what would the scale read during this acceleration? Express your result in newtons. How do you reconcile the scale reading with the answer to part (e)? In (h)–(k), as the appropriate floor is approached, the elevator car slows down with an acceleration of magnitude of 1.00 m/s². (h) Sketch a second law force diagram indicating schematically all forces on the Dean. (i) What is the magnitude of the gravitational force of the Earth on the Dean during this phase of the motion? (j) Calculate the magnitude of the force of the floor on the Dean during this phase of the motion. (k) If the Dean were standing on a scale, what would the scale read during this phase of the motion? Express your result in newtons.

•50. An igneous rock of mass 3.0 kg is moving in uniform circular motion in a horizontal plane 2.0 m above the ground as indicated in Figure P.50. The radius of the orbit is 0.80 m. (a) What is the magnitude of the force of the cord on the mass? (b) What is the magnitude of the acceleration of the rock? (c) What is the angular speed ω of the rock? (d) How many revolutions per second does the rock make? (e) If the cord breaks, how long will it take the rock to hit the ground?

FIGURE P.50

•51. You are headed up to the offices of the famous law firm, Dewey, Cheatam, and Howe, and take an elevator. Strangely, there is a scale on the floor of the elevator car and you stand on the scale. With the elevator car accelerating upward with an

acceleration of magnitude 0.981 m/s², the scale reads 600 N. (a) Sketch a second law force diagram indicating the forces acting on you. (b) Determine your mass. (c) If the total mass of the elevator car and its contents is 2.00×10^3 kg, what is the magnitude of the tension in the cable on which it hangs while the system is accelerating upward? (d) A loose 3.00 g screw falls out of the ceiling (while the elevator is accelerating upward) and falls to the floor 1.80 m below. Calculate the total force on the screw while it falls through the air.

•52. A *massive* cable is attached to a 30.0 kg mass on a frictionless surface and a horizontal force of magnitude F accelerates the arrangement as shown in Figure P.52. If the force of the cable on the mass has a magnitude 0.900F, what is the mass of the cable?

FIGURE P.52

Sections 5.12 Static Friction
      5.13 Kinetic Friction at Low Speeds
      5.14 Kinetic Friction Proportional to the
            Particle Speed*

53. If a horizontal force of magnitude 350 N is exerted on a 50 kg mushroom anchor, the anchor is on the verge of slipping across the surface of a dock. What is the coefficient of static friction for the anchor on the dock surface?

54. A 10.0 m long chain of mass 25 kg is dragged at constant velocity across a flat beach by means of a horizontal force of magnitude 270 N. What is the coefficient of kinetic friction for the chain on the beach surface?

55. You find that a horizontal force of magnitude 250 N is able to slide a 30 kg mass across a horizontal surface with an acceleration of magnitude 0.500 m/s². What is the coefficient of kinetic friction for the mass on the surface?

56. The coefficients of friction for a file cabinet on the floor are $\mu_s = 0.45$ and $\mu_k = 0.30$. You push horizontally with a force of magnitude 75 N on a 100 kg file cabinet and are unable to move it. What is the force of static friction by the floor on the cabinet?

57. A skier can coast directly down a 5° straight slope at constant velocity. What is the coefficient of friction for the skis and the snow surfaces? Is this the coefficient of static or kinetic friction?

58. An exit ramp with a circular curve of radius 200 m, banked at an angle of 20°, is covered with ice. At what speed (in km/h) could a car execute the turn successfully without slipping down or up the curve and hitting the guard rails?

•59. While executing a level, unbanked circular curve to the right of radius 200 m at constant speed 100 km/h, a passenger of mass 55 kg in your car discovers that she is ready to slip across the horizontal seat. What is the coefficient of static friction for the passenger on the seat?

•60. Your 2.50 kg physics text lies on a reading surface that is inclined to the horizontal at an angle of 35° as shown in Figure P.60. While enjoying a reverie about a fantastic physics lecture, you determine that a force of magnitude 20 N directed parallel to and up the plane is barely sufficient to slide the text up the incline at constant speed. Amazing stuff. (a) Draw a second law force diagram indicating schematically all forces acting on the text. (b) Determine the coefficient of friction for the text on the surface. (c) Is this the coefficient of static or kinetic friction?

35°

FIGURE P.60

•61. The static and kinetic coefficients of friction for a 50 kg mass on a table surface are $\mu_s = 0.20$ and $\mu_k = 0.15$. See Figure P.61. The pulley is ideal. The system is at rest in equilibrium. (a) For each mass, sketch a second law force diagram indicating schematically all forces that are acting on each mass. (b) How large can the mass m be and not move the system? What is the magnitude of the frictional force on the 50 kg mass when this situation exists? (c) If m is only half the maximum mass calculated in part (b), what is the magnitude of the frictional force on the 50 kg block? Indicate whether this is a force of static friction or a force of kinetic friction.

50 kg

m

FIGURE P.61

•62. You try to stand on a 30° slope on which the coefficients of static and kinetic friction for your shoes on the slope are 0.55 and 0.45 respectively. Your mass is 65.0 kg. (a) Can you stand motionless on one foot and/or two feet, or will you slip? (b) If you do not slip and slide down the plane, what is the magnitude of the static frictional force? If you do slip, find the magnitude of your acceleration. Will you continue to slide all the way down the slope?

•63. (a) A circular turn of radius 3000 m is being designed for high-speed trains traveling at speeds of 200 km/h. Calculate the appropriate banking angle of the turn for minimum wear on the track. (b) If trains execute the turn at speeds *slower* than the design speed, which rail experiences greater wear (the rail on the inside of the turn or the rail on the outside of the turn)? Explain your reasoning.

•64. Your car of mass 1250 kg makes a circular turn of radius 150 m while traveling at a constant speed of 35 km/h. The curve is

not banked but is horizontal. (a) What is the magnitude of the centripetal acceleration of the car? (b) What is the magnitude of the tangential acceleration of the car? (c) What magnitude force is required to keep the car going around the curve? What provides this force? (d) What is the minimum value of the coefficient of static friction required for the car to make the turn without skidding?

•65. You are on the indicated inclined plane shown in Figure P.65 and are connected to mass m by means of an ideal rope passing over an ideal pulley. Your mass is 70 kg and the coefficients of friction for you on the plane are $\mu_s = 0.40$ and $\mu_k = 0.35$. (a) If $m = 0$ kg, draw a second law force diagram indicating all forces acting on you. Will you slide down the plane? Explain your reasoning. (b) If $m \neq 0$ kg, draw a second law force diagram indicating all forces acting on you. Draw a similar second law force diagram for mass m. What mass m enables you to accelerate *up* the plane with an acceleration of magnitude 1.50 m/s²? (c) Under the conditions of part (b), determine the tension in the rope.

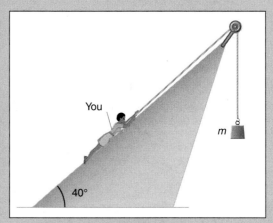

You

m

40°

FIGURE P.65

•66. Talus slopes are formed at the base of large cliffs because of the accumulation of rocks that break off the cliff and fall into a debris pile at the base. See Figure P.66. Sand piles also keep getting steeper until slippage of the particles occurs. If the coefficient of static friction between the particles is μ_s, find the angle θ that the talus slope makes with horizontal direction.

FIGURE P.66

•67. An old stereo turntable of 15.0 cm radius turns at 33.0 rev/min while mounted on a 30.0° incline as indicated in Figure P.67. (a) If a mass m can be placed anywhere on the turntable without slipping, where is the most critical place on the disk where slipping might occur? Why? (b) Calculate the least possible coefficient of friction that must exist if no slipping occurs. Is your answer independent of m? Is this a static or kinetic coefficient of friction?

FIGURE P.67

•68. A mass m is placed on a rough cylinder as shown in Figure P.68. The coefficient of static friction for the mass on the cylinder is 0.15. (a) Draw a second law force diagram indicating all forces acting on the mass. (b) Find the angle θ at which the mass is just ready to slip off the cylinder. (c) Is the angle different if the rough surface is a sphere?

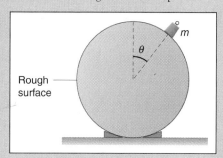

FIGURE P.68

•69. A 500 kg crate of apples is in the middle of the flat, horizontal bed of a pickup truck, but is not tied down. The coefficients of friction for the crate on the bed of the truck are $\mu_s = 0.35$ and $\mu_k = 0.25$. What is the magnitude of the maximum horizontal acceleration of the truck so that the crate does not slide across the bed?

•70. A 100 kg container is on a loading ramp inclined to the horizontal by 20.0° as shown in Figure P.70. The coefficients of friction for the crate on the slope are $\mu_s = 0.300$ and $\mu_k = 0.250$. (a) Draw a second law force diagram indicating the forces on the crate. (b) Will the mass slip down the ramp of its own accord? If so, what minimum magnitude force up the plane must be used to keep the crate in place? If not, what magnitude of force is needed down the plane to get the crate to slide?

FIGURE P.70

•71. A frustrated parent of mass 70 kg is dragging a child of mass 20 kg having a temper tantrum across level but rough ground. The parent exerts a force of magnitude 80 N on the child; the force is upward and forward at an angle of 45° to the horizontal (see Figure P.71). The parent and child are moving at constant speed of 1.5 m/s in a straight line homeward from a video arcade. (a) Draw a second law force diagram indicating all forces on the child. (b) Calculate the frictional force of the ground on the child. (c) Find the magnitude of the normal force of the surface on the child. Is the normal force equal in magnitude to that of the weight of the child? Explain why or why not. (d) Find the coefficient of friction for the child on the ground. Is this the coefficient of static or kinetic friction?

FIGURE P.71

•72. Two spherical students of identical 60 kg mass are dragged by Professor Mediocratus to a final exam in classics as indicated in Figure P.72. The coefficient of kinetic friction for the students on the floor is 0.30. The tension in the rope between the students is 300 N. (a) Choose an appropriate system and calculate the initial acceleration of the students. (b) Calculate the magnitude of the tension in the rope between the leading student and professor.

FIGURE P.72

•73. A fast merry-go-round is turning at a constant angular speed of 1.00 rad/s. Wheeee! You are standing on the merry-go-round minding a young sibling. The coefficient of static friction for your shoes on the merry-go-round is 0.80 and the coefficient of kinetic friction is 0.50. (a) What is the direction of the total force acting on you and why is it in this direction? (b) If you can stand 2.0 m from the axis of rotation, with what speed are you moving? (c) How far from the axis can you stand without slipping? (d) If you threw a ball straight up and parallel to the axis of rotation, describe the path of the ball (i) from the viewpoint of an observer on the ground, and (ii) from your viewpoint on the merry-go-round.

•74. A nerd of mass 50 kg sits in a chair of mass 5.0 kg on a platform of mass 100 kg that is accelerated off a stage in a straight line with an acceleration of magnitude 1.0 m/s². (a) What coefficient of static friction will keep the chair from slipping on the platform? (b) The chair exerts a frictional force on the platform, equal in magnitude but opposite in direction to the frictional force of the platform on the chair. How, then, can there be any acceleration of either one? Explain. (c) Is the platform an inertial reference frame? Explain.

•75. A large 5.0 kg bone rests on a level surface and the coefficients of friction for the bone on the surface are $\mu_s = 0.40$ and $\mu_k = 0.20$. A dog is pulling horizontally with a force of magnitude 30 N on the bone as indicated in Figure P.75. (a) Draw a second law force diagram indicating all forces acting on the bone. (b) Choose an appropriate coordinate system and indicate it clearly in a sketch. (c) What is the normal force on the bone? (d) What is the *maximum* magnitude of the static frictional force on the bone? (e) Will the bone accelerate? If so, determine its acceleration. If not, explain why it does not accelerate. (f) Determine the magnitude of the frictional force on the bone. (g) How many newtons of force must the dog use to give the bone an acceleration of magnitude of 1.0 m/s²?

FIGURE P.75

•76. A train consists of 50 identical hopper cars each with a mass of 2.00×10^4 kg. The train lacks a caboose; the hopper cars are numbered, with car number 1 being the first car behind the engines. A series of three locomotives at the head of the train provide a force on the first hopper car of magnitude 100 kN in the horizontal direction. The acceleration of the train has a magnitude 6.00×10^{-2} m/s² on a straight stretch of level track. Chug, chug, chug; I think I can, I think I can, I think I can do this problem. (a) What is the magnitude of the vertical force exerted by the track on *each* hopper car? (b) The frictional force exerted by the track on each car is the same; what is the magnitude of the frictional force on each car? (c) What is the magnitude of the force that the coupling at the rear of car 24 exerts on the coupling at the front of car 25? What is the magnitude of the force that the coupling at the front of car 25 exerts on the coupling at the rear of car number 24? Do these forces form a Newton's third law pair?

•77. A slapshot sends a hockey puck of mass m in a straight line with an initial speed v_0 across a vast expanse of ice. The coefficient of kinetic friction for the puck on the ice is μ_k. (a) Set up an appropriate coordinate system to analyze the problem. (b) Make a second law force diagram appropriate for the puck as it slides across the ice. (c) Find the velocity component as a function of time. (d) How long does it take the puck to come to rest? (e) Find an expression for the position of the puck as a function of time with respect to the origin chosen for your coordinate system. (f) How far does the puck travel before

it stops? (g) Evaluate the distance determined in part (f) if $v_0 = 25$ m/s and $\mu_k = 0.050$.

•78. An ideal physics text of mass 2.5 kg is placed on a rough inclined plane as indicated in Figure P.78. The coefficients of static and kinetic friction for the text on the plane are 0.30 and 0.20 respectively. An ideal string is attached to the text and passes over an ideal pulley to another mass m. (a) Sketch a second law force diagram indicating schematically all forces acting on the text. (b) Make a similar diagram for the mass m. (c) What is the maximum mass m that can be attached to the cord before the text starts to slip? (d) If m is 10.0 kg, determine the magnitude of the acceleration of the text.

FIGURE P.78

•79. A 20 kg mass is attached to a 30 kg mass by means of an ideal cord as indicated in Figure P.79. Both masses are on a level surface, and the coefficients of friction for the masses on the surface are $\mu_s = 0.40$ and $\mu_k = 0.20$. A dog is pulling horizontally with a force of magnitude 30 N as indicated. (a) Draw a second law force diagram indicating all forces acting on the 20 kg mass. (b) Draw a second law force diagram indicating all forces acting on the 30 kg mass. (c) Define a system consisting of *both* masses. Draw a second law force diagram indicating all forces acting on the combined system. (d) Will the system accelerate? If so, determine the magnitude of the acceleration. (e) What magnitude of force is needed for the system of two masses to have an acceleration of magnitude 1.2 m/s²?

FIGURE P.79

•80. A chicken feather and a fist-sized rock are dropped at the same time from a height 50 m in the air. (a) Why does one object reach the ground before the other? (b) Sketch qualitatively accurate graphs of the speeds of the two objects as functions of time. (c) Sketch qualitatively accurate graphs of the magnitudes of the accelerations of the two objects as functions of time.

•81. An automobile tire of mass m is spinning at angular speed ω_0 while held at rest a slight distance above a horizontal road. The tire then is released onto the roadway and skids away from its initial position (see Figure P.81). (a) Sketch a second law force diagram indicating all forces on the tire while it is skidding. (b) While the tire is skidding, name the force that accelerates the tire in the horizontal direction. (c) Show that while the tire is skidding, the horizontal acceleration is of magnitude $\mu_k g$, where μ_k is the coefficient of kinetic friction for the tire on the road.

FIGURE P.81

•82. Before pulling the rip cord on a parachute that fails to open, a skydiving physics student of mass 50.0 kg, falling vertically, somehow observes that air friction has caused the magnitude of her acceleration at some instant to be only $0.250g$. (a) Draw a second law force diagram indicating all forces acting on the unfortunate lass. (b) Calculate the magnitude of the frictional force acting on the student. (c) Sometime later, with the parachute now deployed but not open, she reaches a terminal speed (in more ways than one) of 150 km/h. What is her acceleration when the terminal speed is achieved? If the magnitude of the total kinetic frictional force at this speed is assumed to be proportional to the speed:

$$f_k = Kv$$

find the constant K and indicate its units.

•83. Amusement parks occasionally have a large cylindrical tube mounted to spin about the vertical axis of the cylinder as shown in Figure P.83. Patrons stand against the wall of the cylinder and the apparatus is set into rotation. At a certain critical angular speed, the floor on which the patrons are standing can be lowered and the patrons are left hanging on the wall (but on the verge of slipping) due to the force of static friction on them. Let R be the radius of the apparatus, m the mass of a patron, and μ_s the coefficient of static friction for a patron on the wall. (a) Consider a patron to be the system whose motion is to be analyzed. Make

a sketch of a second law force diagram indicating all forces acting on the patron. (b) Is there an acceleration in the vertical direction? What does this imply about the relationship between the static frictional force on the patron and the weight of the patron? Are these forces a third law force pair? (c) Is there an acceleration of the patron in the horizontal direction? How is this acceleration related to the speed of the patron? (d) How is the angular speed ω of the cylinder related to the tangential speed of the patron? If the patron is ready to slip, find the angular speed ω of the tube in terms of R, μ_s, and g. Express your result in rev/s.

‡84. Two masses, m_1 and m_2, are connected by an ideal cord as shown in Figure P.84. The cord passes over an ideal pulley. A student determines that when the surface is inclined at an angle θ to the horizontal, m_1 is just on the verge of slipping to the *right* up the incline. (a) Draw a second law force diagram indicating schematically all forces on mass m_2. (b) Draw a second law force diagram indicating schematically all forces on mass m_1. (c) Introduce appropriate coordinate system(s) and show that the coefficient of static friction for m_1 on the inclined plane is

$$\mu_s = \frac{|m_2 - m_1 \sin \theta|}{m_1 \cos \theta}$$

Just to be mean, the inclination angle of the inclined plane now is increased to angle ϕ so that the mass m_1 is just on the verge of slipping to the *left* down the incline. (d) Draw a second law force diagram indicating schematically all forces on m_2. (e) Draw a second law force diagram indicating schematically all forces on m_1. (f) Show that the coefficient of static friction for m_1 on the inclined plane is

$$\mu_s = \frac{|m_1 \sin \phi - m_2|}{m_1 \cos \phi}$$

Notice that if the suspended mass $m_2 = 0$ kg, then $\phi = \theta$ and this equation reduces to

$$\mu_s = \tan \phi = \tan \theta$$

in agreement with what was derived in Example 5.11. (g) Is this the *same* coefficient of friction as that determined in part (c)? Why or why not? (h) Show that the ratio of the two masses is given by the rather elegant expression

$$\frac{m_2}{m_1} = \frac{\tan \theta + \tan \phi}{\sec \theta + \sec \phi}$$

FIGURE P.83

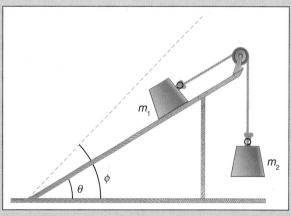

FIGURE P.84

‡85. For gung-ho students who have completed the previous problem, consider this: the mass ratio m_2/m_1 must be less than unity. Why? Likewise, the problem also indicates that the angle ϕ must be greater than the angle θ. For a *given* angle θ, then, the angle ϕ has a range from a minimum value of $\phi_{min} = \theta$ to a maximum value of $\phi_{max} = 90°$. (a) Show that when $\phi_{min} = \theta$, the mass ratio m_2/m_1 has a value equal to $\sin\theta$. (b) Show that when ϕ has its maximum value of $90°$, the mass ratio is equal to unity. Thus, for a given θ, the *possible* values of the mass ratio are located between $\sin\theta$ and 1. (c) Make a graph indicating the possible values of the mass ratio as a function of θ.

‡86. Assume the kinetic frictional force on a falling mass m is proportional to its speed v, with a proportionality constant β. Choose $\hat{\jmath}$ to be vertically downward. (a) Show that Newton's second law of motion yields

$$m\frac{dv_y}{dt} = mg - \beta v_y$$

(b) When the mass reaches its terminal speed, what is

$$\frac{dv_y}{dt}$$

(c) Show that the terminal speed is

$$v_{term} = \frac{mg}{\beta}$$

(d) If the mass is dropped from rest, show that

$$v_y(t) = v_{term}(1 - e^{-\beta t/m})$$

Sections 5.15 Fundamental Forces and Other Forces Revisited*
5.16 Noninertial Reference Frames*

•87. Calibrate the scale of the plumb bob accelerometer mentioned in Section 3.6. That is, mark the circular arc of the

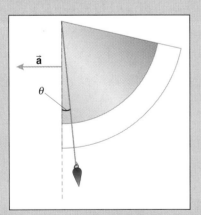

FIGURE P.87

accelerometer in Figure P.87 so that if the plumb bob is hung from the center of the circular arc, the scale on the circular arc indicates horizontal accelerations of magnitudes from 0 to 20.0 m/s². Is the scale linear with angle? Explain why or why not.

•88. For the plumb bob accelerometer referred to in Section 3.6 and calibrated in the previous problem: (a) What magnitude acceleration in the horizontal direction results in a deviation angle of the plumb bob from the vertical direction of 63.4°? (b) If an acceleration in the horizontal direction is of magnitude 98.1 m/s², what is the deviation angle θ of the plumb bob accelerometer?

•89. The equatorial radius of Jupiter is 7.19×10^4 km, more than ten times as great as that of the Earth. The local acceleration due to gravity near the surface* of Jupiter is of magnitude 24.5 m/s², two and a half times as great as that on the surface of the Earth. Jupiter rotates quite fast: a complete rotation in only 9 h 50 min. Calculate the ratio between the magnitudes of the centripetal acceleration on the Jovian equator and the local acceleration of gravity near the surface of Jupiter. Comment on whether the effects of pseudoforces can be neglected on Jupiter as much as they can be on the Earth in analyzing the motion of particles.

---

*Unlike the Earth, Jupiter does not have a well-defined surface.

INVESTIGATIVE PROJECTS

A. Expanded Horizons

1. Investigate historical ideas regarding motion and the causes of motion, in particular so-called *natural motion* proposed by the ancient Greek philosopher Aristotle, and how these ideas differed from those of Galileo and Newton.

   Arnold B. Arons, *Development of Concepts of Physics* (Addison-Wesley, Reading, Massachusetts, 1965), Chapter 2, pages 35–61.

   Giorgio De Santillana, *The Origins of Scientific Thought* (University of Chicago Press, Chicago, 1961).

   Herbert Butterfield, *The Origins of Modern Science 1300–1800* (MacMillan, New York, 1956).

   Allan Franklin, "Inertia in the Middle Ages," *The Physics Teacher*, 16, #4, pages 201–208 (April 1978).

   Lindsay Judson (editor), *Aristotle's Physics* (Oxford at the Clarendon Press, Oxford, 1991).

   Robert J. Whitaker, "Aristotle is not dead: student understanding of trajectory motion," *American Journal of Physics*, 51, #4, pages 352–357 (April 1983).

2. The remarkable forces involved in karate and other Asian martial arts are rather dramatic; they can exceed 3 kN in magnitude. Boards break and concrete blocks shatter because of the huge force of the human hand on them. Newton's third law implies the blocks and boards exert forces of the *same magnitude* on the hand! Investigate how the human hand can withstand these tremendous forces without injury.

   Jearl Walker, "The amateur scientist," *Scientific American*, 243, #1 (July 1980), pages 150–162; additional references are on page 162.

   Michael S. Field, Ronald E. McNair, and Stephen R. Wilk, "The physics of karate," *Scientific American*, 240, #4 (April 1979), pages 150–158; additional references are on page 190.

   Jearl D. Walker, "Karate strikes," *American Journal of Physics*, 43, #10, pages 845–849 (October 1975).

   Donn F. Draeger and Robert W. Smith, *Asian Fighting Arts* (Kodansha International, Tokyo, 1969).

3. Investigate the dynamics of particles moving at speeds *faster* than the so-called terminal speeds in a fluid medium such as air or water.

   Ernest Zebrowski Jr., "Superterminal velocities," *The Physics Teacher*, 27, #8, pages 618–619 (November 1989).

4. Plot the terminal speeds in Table 5.5 versus the cross-sectional area of the various sports balls. Formulate a hypothesis to account for the graph and any deviations of a ball from any trends apparent.

B. Lab and Field Work

5. Design and perform an experiment to determine the static coefficient of friction for two Velcro surfaces in contact. Also investigate whether or not μ_s is independent of the surface area in contact. Velcro was invented in 1948 by the Swiss engineer Georges deMestral.

   Judith Stone, "Velcro: the final frontier," *Discover*, 9, #5 (May 1988), pages 82–84.

6. Design an experiment to investigate the frictional characteristics of belts and capstans when the belt or rope is on the verge of slipping. Theory indicates that the frictional force is exponentially related to the total angle of contact between the belt and groove or between the rope and capstan; see if your experiments confirm the theory.

   George L. Hazelton, "A force amplifier: the capstan," *The Physics Teacher*, 14, #7, pages 432–433 (October 1976).

   William F. Riley and Leroy D. Sturges, *Engineering Mechanics: Statics* (John Wiley, New York, 1993), pages 400–402.

   Eugene Levin, "Friction experiments with a capstan," *American Journal of Physics*, 59, #1, pages 80–84 (January 1991).

   Clifford Bettis, "Capstan experiment," *American Journal of Physics*, 49, #11, pages 1080–1081 (November 1981).

7. Investigate the speed-dependent kinetic frictional effects of the air on the flight of baseballs, golf balls, tennis balls, and other balls used in various sports.

   Cliff Frohlich, "Aerodynamic drag crisis and its possible effect on the flight of baseballs," *American Journal of Physics*, 52, #4, pages 325–334 (April 1984).

   Peter J. Brancazio, "Looking into Chapman's homer: the physics of judging a fly ball," *American Journal of Physics*, 53, #9, pages 849–855 (September 1985).

   Joseph M. Zayas, "Experimental determination of the coefficient of drag of a tennis ball," *American Journal of Physics*, 54, #7, pages 622–625 (July 1986).

8. Design and perform an experiment to determine the coefficient of viscosity of air using spherical air-filled balloons falling at their terminal speeds.

9. Visit a local orthopedic medical practice and/or orthodontist and investigate the various ways in which traction is applied to various parts of the body and teeth. In particular, investigate the mechanics of the following traction devices: Bryant traction; Russell traction; and Sayre traction.

   Marian Williams and Herbert R. Lissner, *Biomechanics of Human Motion* (Saunders, Philadelphia, 1962).

C. Communicating Physics

10. Newton's laws had consequences that rippled well beyond the sciences, even into political and social arenas. Write a précis about the following article, discussing how Newton's mechanistic view of nature influenced the framing of the Constitution of the United States.

    John Patrick Diggins, "Science and the American experiment: how Newton's laws shaped the Constitution," *The Sciences*, 27, #6 (November–December 1987), pages 28–31.

11. Think about the following provocative statement: *science is always wrong*. Write an essay explaining why the statement is in some sense true and in other ways false. Most professional scientists submit their writing to colleagues for a critique before submission to a professional journal, after which the article typically is sent out for peer review to determine if it is acceptable for publication. If your instructor permits, let a colleague critique your essay before you submit it to your professor.

12. The conflict between science and religion arose around the time of Copernicus (16th century), and continued with the famous trial of Galileo (17th century) down to the present day. Science and religion clearly are very different intellectual endeavors. Nonetheless, in the spirit of good debate, think about and write an essay that explores how science and religion are *similar*. If your instructor permits, have a good discussion of the topic with your friends before writing the essay.

    Albert Einstein, *Ideas and Opinions* (Crown, New York, 1954), pages 36–54.

    Jacob Needleman, *A Sense of the Cosmos: The Encounter of Modern Science and Ancient Truth* (Open Court Publishers, Peru, Illinois, 1995).

CHAPTER 6
THE GRAVITATIONAL FORCE AND THE GRAVITATIONAL FIELD

. . . the Newtonian principle of gravitation is now more firmly established, on the basis of reason, than it would be were the government to step in, and make it an article of necessary faith. Reason and experiment have been indulged, and error has fled before them.

Thomas Jefferson (1743–1826)*

The gravitational force governs much of the great, large-scale natural drama of the universe. Gravitation determines many of the features of star formation, planetary systems, galaxies, and the universe as a whole. Gravitation was the first of the four fundamental and primordial forces (see Section 5.2) to be described quantitatively, and its characterization was another magnificent achievement for Newton. Newton's law of universal gravitation was the scientific culmination of the Renaissance, for it brought order and harmony to our understanding of the cosmos.

There are two likely reasons that the gravitational force was the first fundamental force discovered:

1. The force determines the structure of the solar system, a problem that intrigued scholars since the dawn of civilization.
2. The force is all-encompassing; it is apparent in virtually everything you do from simply dropping a pencil to lugging your books to the library for yet another long night of study. Gravity never ceases its action, but this very fact meant that for millennia the force was not recognized—it was all but impossible to imagine situations without it.

With ingenuity and inspiration, Newton realized that the gravitational influence of the Earth was weaker at the distance of the Moon but was nonetheless felt by it (see Section 6.1). Thus the gravitational force was not a purely local interaction but extended into space at least as far as the Moon. Occasionally we get more out of our thoughts and equations than we originally put into them: in a single stroke, Newton realized that the gravitational force likely explained the dynamics of the solar system and the universe itself. The gravitational force thus has a very long range: its effects are felt from microscopic to cosmological distances. Unlike the other fundamental forces of nature, gravitation is a truly *universal* force that acts, however weakly or strongly, on *every* particle in the universe as we mentioned in Section 5.2.

In this chapter we first see how Newton deduced the law of gravitation and will explore the nature and features of the gravitational force. We also examine some of its implications in the local environment of the Earth, within the solar system, and beyond. Lastly we investigate a convenient concept, the *gravitational field*, for describing how the gravitational force acts on masses that are not in direct physical contact with each other, such as astronomical masses separated by billions of kilometers or even light-years of empty space.

6.1 HOW DID NEWTON DEDUCE THE GRAVITATIONAL FORCE LAW?

After dinner, the weather being warm, we went into the garden and drank thea, under the shade of some appletrees, only he and myself. Amidst other discourse, he told me, he was just in the same situation, as when formerly, the notion of gravitation came into his mind. It was occasion'd by the fall of an apple, as he sat in a contemplative mood.
William Stukeley (1687–1765)[†]

Sir Isaac Newton is pictured on this British £1 note.

Newton realized that both a falling apple and the distant orbiting Moon accelerate toward the Earth. The force causing these accelerations he called the **gravitational force**. The ratio of the magnitudes of the accelerations of the apple and the Moon gave a clue about how the gravitational force depended on distance. Consideration of Newton's second and third laws of motion, coupled with experiments on falling objects, provided information about the mass dependence of the force. Here we examine these arguments.

Dependence on Distance

For simplicity, assume the Moon moves in uniform circular motion about the Earth, as shown in Figure 6.1.[‡] The Moon experiences a centripetal acceleration of magnitude

$$a_c = \frac{v^2}{r}$$

where v is the speed of the Moon in its orbit and r is the orbital radius, measured from the center of the Earth to the center of the Moon.

The assumption that the Moon is in uniform circular motion means the instantaneous speed v of the Moon at any instant is equal to the average speed during any interval, found by dividing the distance traversed by the corresponding time interval. Since the Moon makes one circumference of its orbit ($2\pi r$) in a time equal to its orbital **period** T, the average speed of the Moon in its orbit is

$$v = \frac{2\pi r}{T}$$

*(Chapter Opener) *Notes on the State of Virginia*, Query 17, "Religion" (Prichard and Hall, Philadelphia, 1788).

[†]*Memoirs of Sir Isaac Newton's Life* (1752), pages 19–20. Recent edition published by Taylor and Francis, London (1936).

[‡]A circle is a reasonable first approximation; a second approximation is an ellipse. The *actual* orbital path of the Moon is quite complicated because of the gravitational effects of the Sun and the major planets on the Moon. The effects of these other masses on the motion of the Moon was a problem that worried even Newton and "made his head ache and kept him awake so often that he would think of it no more" [attributed to Edmund Halley; see Roger R. Bate, Donald D. Mueller, and Jerry E. White, *Fundamentals of Astrophysics* (Dover, New York, 1971), page 322].

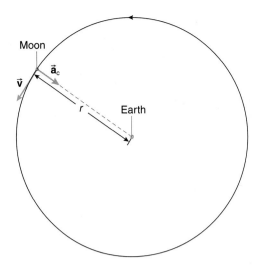

FIGURE 6.1 The Moon in a circular orbit experiences a centripetal acceleration.

Thus the magnitude of the centripetal acceleration of the Moon is

$$a_c = \frac{\left(\frac{2\pi r}{T}\right)^2}{r}$$

$$= \frac{4\pi^2 r}{T^2} \tag{6.1}$$

Both r and T are determined from astronomical observations and long-distance surveying techniques. The radius of the lunar orbit is 3.84×10^8 m and the orbital period of revolution is 27.3 days ($= 2.36 \times 10^6$ s). Upon substitution of these numerical values into Equation 6.1, the magnitude of the centripetal acceleration of the Moon is found to be

$$a_c = 2.72 \times 10^{-3} \text{ m/s}^2$$

or just less than 3 mm/s².

The magnitude of the acceleration due to gravity near the surface of the Earth is found, experimentally, to be about 9.81 m/s². The ratio of the magnitude of the acceleration due to gravity on the surface of the Earth to its magnitude at the distance of the lunar orbit is

$$\frac{a_{\text{near the surface of Earth}}}{a_{\text{at the distance of the Moon}}} = \frac{9.81 \text{ m/s}^2}{2.72 \times 10^{-3} \text{ m/s}^2}$$

$$\approx 3.60 \times 10^3$$

Hence the magnitude of the acceleration due to gravity at the distance of the Moon is about 1/3600 that on the surface of the Earth. Newton realized that the distance from the center of the Earth to the lunar orbit is about 60 times the radius of the Earth:

$$\frac{\text{radius of lunar orbit}}{\text{radius of the Earth}} = \frac{3.84 \times 10^8 \text{ m}}{6.37 \times 10^6 \text{ m}}$$

$$\approx 60$$

Note that $(60)^2 = 3600$.

Newton concluded that since the *acceleration* caused by the gravitational force, acting alone, evidently decreased in magnitude proportionally to the inverse square of the distance, the magnitude of the gravitational force must do likewise because, according to Newton's second law of motion, the force (acting alone) is proportional to the acceleration. That is,

$$F_{\text{grav}} \propto \frac{1}{r^2}$$

where r is the center-to-center distance between the Earth and the Moon.

Dependence on Mass

Since the only common physical property of the Earth and Moon is *mass*, it was reasonable to suppose that their masses also play a role in the magnitude of the gravitational force they exert on each other. Let M be the mass of the Earth and m that of the Moon. Newton's third law of motion states that if the Earth (as one system) exerts a force $\vec{F}_{\text{M on } m}$ on the Moon (another system), then the Moon exerts a force $\vec{F}_{m \text{ on M}}$ on the Earth of the *same magnitude*, but in the opposite direction:

$$\vec{F}_{\text{M on } m} = -\vec{F}_{m \text{ on M}} \tag{6.2}$$

Thus if the magnitude of the gravitational force depends on mass, it must involve the masses of *both* the Earth and Moon in a *symmetric* way, so that if the masses are interchanged in the gravitational force law, the magnitude of the force on each is unchanged. The force could depend on *either* the *sum* of the masses (M + m), *or* the *product* (Mm), or perhaps some power n of the sum or product:

$$(M + m)^n \quad \text{or} \quad (Mm)^n$$

since these forms are the same if the masses are interchanged in their algebraic position in the expression [to $(m + M)^n$ or $(mM)^n$]. Therefore the possibilities for the mass dependence of the magnitude of the gravitational force are

$$F_{\text{grav}} \propto (M + m)^n \quad \text{or} \quad F_{\text{grav}} \propto (Mm)^n$$

By appealing to the *second law of motion* and experiment, Newton ruled out all possibilities except the first power of the product of the masses. Why is that? The argument is the following: From the experiments of Galileo, experiments easily repeated today, it is known that the *acceleration* of *any* mass m produced by the *sole* action of the gravitational force of another given mass M (say, the mass of the Earth) is *independent* of the mass m. Rocks, socks, and blocks fall with the same acceleration during free-fall motion (in a vacuum). We can use this observation to determine the mass dependence of the gravitational force.

Assume the magnitude of the gravitational force of each mass on the other depends on, say, the *sum* of the masses to some power n, with a proportionality constant K. Then, when the gravitational force is the only force on m, the second law of motion applied to system m,

$$\vec{\mathbf{F}}_{grav} = m\vec{\mathbf{a}}_{grav}$$

yields, using the magnitudes of the vectors,

$$K(M + m)^n = ma_{grav}$$

Hence the magnitude of the acceleration of m is

$$a_{grav} = \frac{K(M + m)^n}{m}$$

This magnitude of acceleration depends on m no matter what the exponent n happens to be,* which contradicts experiment. Therefore the magnitude of the gravitational force cannot depend on the *sum* of the masses or a power of the sum.

On the other hand, assume the magnitude of the gravitational force of M on m depends on the *product* of the masses raised to a power n, with a proportionality constant K'. When the gravitational force is the only force on m, the second law of motion applied to m gives (again, using magnitudes)

$$F_{grav} = ma_{grav}$$
$$K'(Mm)^n = ma_{grav}$$

Solving for the magnitude of the acceleration of m,

$$a_{grav} = K'\frac{(Mm)^n}{m}$$
$$= K'M^n m^{n-1}$$

This magnitude of acceleration is independent of m *if and only if* n = 1. *Only* the simple product Mm means that the gravitational acceleration of m caused by mass M is independent of m, confirming experimental observations.

Therefore the gravitational force *must* depend on the simple product of the masses.

Now combine the distance and mass dependence into a single relation: the magnitude of the gravitational force must depend directly on the product of the masses and inversely on the square of their separation:

$$F_{grav} \propto \frac{Mm}{r^2}$$

Rewriting this relation with a constant of proportionality, G, yields

$$\boxed{F_{grav} = \frac{GMm}{r^2}} \tag{6.3}$$

The constant of proportionality G is the **universal constant of gravitation**. Equation 6.3 is called **Newton's law of universal gravitation**.

The gravitational force law is called an **inverse square law** because the exponent of r in the denominator is 2.

The inverse square nature of the force has remarkable consequences (see Section 6.15). Nature was kind to create the gravitational force as an inverse square law; gravitational physics would be very different and difficult with *any* other exponent for the power of r.

PROBLEM-SOLVING TACTIC

6.1 Watch your m & M's. Even though they have the same magnitude, be sure you understand which force is needed: the force of M on m, or the force of m on M. Use Equation 6.3 to find the magnitude of the gravitational force on each of the two pointlike systems of mass m and M. To write the forces as vectors, you will need to introduce a coordinate system.

6.2 NEWTON'S LAW OF UNIVERSAL GRAVITATION

Newton's law of universal gravitation describes the remarkable observation that any two pointlike particles in the universe exert a mutually attractive force on each other that is proportional to the product of their masses and inversely proportional to the square of the distance of their separation.

The *magnitude* of the gravitational force *on each particle* is given by Equation 6.3 as shown in Figure 6.2. The mutually attractive forces are directed along the line joining the particles and represent a Newton's third law force pair; Figure 6.2 is a third law force diagram.

The universal constant of gravitation G is a fundamental constant of nature whose value is determined in the laboratory (with difficulty) by directly measuring the slight force between two known masses separated by a known distance. The value of G, in SI units, is approximately

$$G \approx 6.67 \times 10^{-11} \ N \cdot m^2/kg^2$$

Newton mentions that even after he deduced Equation 6.3, he still did not know what gravity actually *is*; he could describe only what it *does*. Gravity itself was still just as much of a mystery after Newton as it was before him. *Why does*

---

*Of course if the exponent n = 0, then $(M + m)^n = 1$; but then the gravitational force would have no mass dependence at all.

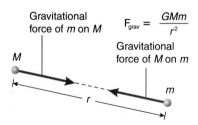

FIGURE 6.2 The gravitational force of two pointlike masses on each other.

gravity act this way and not some other way? *How* is the force transmitted between the masses across, say, the vacuum of space? Gravity itself (along with the other fundamental forces of nature) is an abstract, almost unfathomable concept that greatly intrigued Einstein and led him to develop his general theory of relativity.

Several aspects of the gravitational force need to be noted.

> Equation 6.3 yields the *magnitude* of the gravitational force of two *pointlike* masses on each other.

Newton worried about what happens if the objects have appreciable physical size; we will, too, in Sections 6.3 and 6.4. In fact, as long as the sizes of the particles are much less than their separation, Equation 6.3 describes the interaction quite well; nevertheless, as we will see, it is an approximation unless the masses are spherical in shape and have a spherically symmetric mass distribution.

> Equation 6.3 is not an independent law of motion (like Newton's three laws of motion); rather it is an analytical expression, based on reason and experiment, for the magnitude of a specific force: the gravitational force of each particle on the other.

> The gravitational force always appears in nature as an *attractive force*, directed along the line joining the two particles, as indicated in Figure 6.2.

While there are particles of *zero* mass, there are *no* particles with negative masses in nature.

QUESTION 1

If negative masses existed, what does Newton's second law of motion imply about their dynamics (the direction of the total force on a particle relative to the direction of its consequent acceleration)?

Looking ahead, the gravitational force contrasts strikingly with the electrical force (Chapter 16). Electrical forces are observed to be *either* attractive *or* repulsive. Therefore there must be two distinct types of electric charge, called positive charge and negative charge, names invented by Benjamin Franklin. The same kind of charges (like charges) *repel* each other; different kinds of charges (unlike charges) attract each other. Since there is only one kind of mass, the gravitational force must always be either attractive *or* repulsive; experiments indicate it is attractive, quite different from electricity. "Like" masses attract; there is no possibility for "unlike" masses in gravitation.

> The masses that appear in Equation 6.3, the law of universal gravitation, are called, formally, **gravitational masses**.

The mass m that appears on the right-hand side of Newton's second law of motion (Equation 5.1),

$$\vec{F}_{\text{total}} = m\vec{a}$$

is the *inertial mass* of a system, as we mentioned in Section 5.5. Newton was concerned about whether the gravitational mass and the inertial mass of a system were, in fact, the same or not. There is no *a priori* reason to think they are the same, since the second law of motion is a different type of statement than Equation 6.3 for the magnitude of the gravitational force. The second law indicates how a particle (via its inertial mass) responds when *any* total force whatsoever acts on the particle. The mass in the gravitational force law of Equation 6.3 provides a way to calculate the magnitude of a *particular* force: the gravitational force.

What do experiments indicate is the relationship between the inertial and gravitational masses? Newton's second law of motion states that the magnitude of the total force on a system and the magnitude of its acceleration are related by

$$F_{\text{total}} = m_{\text{inertial}}a$$

where we explicitly write a subscript indicating that the right-hand side of the second law involves the inertial mass of the system. If, say, the gravitational force of the Earth (with gravitational mass M_{grav}) is the only force acting on the system, then the left-hand side of the second law is the gravitational force, and the second law becomes

$$\frac{GM_{\text{grav}}m_{\text{grav}}}{r^2} = m_{\text{inertial}}a$$

where m_{grav} is the gravitational mass of the system whose inertial mass is m_{inertial}. So the acceleration of the system has a magnitude

$$a = \frac{GM_{\text{grav}}}{r^2}\frac{m_{\text{grav}}}{m_{\text{inertial}}} \tag{6.4}$$

The acceleration depends on the ratio of the gravitational and inertial masses. Experiments with the greatest precision indicate that *all* freely falling masses (in a vacuum) have the *same* acceleration, the so-called acceleration due to gravity, if the gravitational force is the only force acting on them. In a vacuum, feathers fall with the same gravitational acceleration as rocks. This observation implies that the ratio between the gravitational and inertial masses in Equation 6.4 must be equal to 1. The latest experiments indicate that the gravitational and inertial masses of the system are, in fact, equal to each other to better than 1 part in 10^{12}:

$$\frac{m_{\text{inertial}}}{m_{\text{grav}}} = 1.000\,000\,000\,000?$$

Equation 6.4 then indicates that the acceleration due to the gravitational force of the Earth alone is independent of the mass of the system (but depends on the mass of the Earth), in conformity with experiment. About 1915, Einstein elevated the experimental equality of the gravitational and inertial masses of a system to the status of a postulate called the **principle of equivalence**:

> The inertial and gravitational masses of a system are *exactly equal*:
> $$m_{\text{grav}} \equiv m_{\text{inertial}} \tag{6.5}$$

The universal gravitational constant G is the first of a number of fundamental physical constants of nature that we will encounter. Other such fundamental constants likely are familiar to you, among them the speed of light and the charge on the electron. The numerical value of G depends of course on the particular unit system used to measure the forces, the masses, and the distances in Equation 6.3. This is also true for other fundamental physical constants. For example, the speed of light is $2.997\,924\,58 \times 10^8$ m/s (exactly) in SI units, or about 186 000 miles per second in peculiarly American units, or more quaintly 1.80×10^{12} furlongs per fortnight; but the plethora of different unit systems is of our doing and does not make the fundamental constants any less fundamental.

Among the fundamental constants of physics, the gravitational constant G is known with the least precision (fewest significant figures) because of the weakness of the gravitational interaction. The smallness of G reflects the experimental observation that the magnitudes of the gravitational forces of common everyday objects on each other are extremely small; see Example 6.1. Therefore experiments to determine the value of G by directly measuring the gravitational force between two laboratory masses are notoriously difficult. There is precious little gravitational attraction between you and this text (perhaps in more ways than one), because the masses involved are so small; the mutual gravitational force of attraction on you and the book is of negligible magnitude compared with other forces on you and the text. Your *weight* (the gravitational force of the Earth on you) is substantial only because the mass of the Earth is huge.

Experiments to determine the numerical value of G by directly measuring the gravitational force of two laboratory masses on each other first were undertaken by Sir Henry Cavendish (1731–1810) in 1798, over a century after Newton. A modern-day version of the experimental apparatus, known as a Cavendish balance, is shown in Figure 6.3 along with a brief description of the procedure involved in the measurement.

There have been recent theoretical speculations about whether the fundamental constants of nature really are constant or change as the universe itself ages and evolves. The constant G has been of particular interest in these speculations, since the gravitational force dominates the large-scale dynamical aspects of the universe itself.* The fundamental philosophical question here is whether the laws of physics themselves might evolve or change along with inexorable changes in the universe as a whole. There is as yet *no* definitive experimental evidence to confirm these speculations about variations in G or in other fundamental constants. It seems that the physics "now" was the physics "then" and will be the physics in the future. Thank goodness. Just imagine how wacky the world and the universe would be if science changed with time (or were subject to change by Congress or your state legislature)! We like to seek out the unchanging statements and rules about the natural world. We are quite confident that the laws of physics, whatever they may be, are now and will be as they were in the beginning.

Experiments indicate that the gravitational force follows a **principle of linear superposition**.

FIGURE 6.3 The Cavendish balance used for the experimental determination of the numerical value of the universal gravitational constant G. The apparatus consists of a dumbbell arrangement of two small lead spheres, each of mass m, suspended from its center by a thin fiber so the dumbbell is free to turn in the horizontal plane. The arrangement is allowed to hang until it is at rest. Two other larger lead spheres, each of mass M, then are mounted near the small spheres and attract them, causing the dumbbell to rotate. Careful measurements of the rotation provide a way of directly calculating the force in newtons of each sphere on the other. In Section 6.3 we will see that a uniform spherical mass acts, for purposes of calculating its gravitational force on other masses located outside the sphere itself, as if it were a point mass located at the center of the sphere. By measuring the force, the masses of the spheres, and the separation between their centers, the only unknown quantity in Equation 6.3 is the universal gravitational constant G. The numerical value of G may thus be determined. There is a more sophisticated version of the experiment that involves switching the locations of the large spheres and then measuring the oscillation period of the dumbbell and its new equilibrium position. Although the procedure is straightforward, the experiment is not easy to perform; small vibrations in the laboratory induced by people walking up nearby stairs, motor vehicles lumbering by, and other nearby disturbances affect the apparatus and make data collection challenging.

The gravitational forces of two particles on each other are independent of forces arising from the presence of other masses.[†] The principle means that if, say, three particles are considered, the total gravitational force on one of them is the vector sum of the individual gravitational forces on that particle by the others, *taken one at a time*. In other words, the total gravitational force on, say, a particle of mass m is the vector sum of the gravitational force of particle M on m and the gravitational force of particle M′ on m, as shown in Figure 6.4. That is,

$$\vec{F}_{\text{total gravitational force on } m} = \vec{F}_{\text{gravitational force of M on } m}$$
$$+ \vec{F}_{\text{gravitational force of M' on } m}$$

Of course, the same applies to the total force on any one of the other particles: they are calculated by vector summing all the

---

*Edward R. Harrison, *Cosmology* (Cambridge University Press, Cambridge, England, 1981), pages 323–324, 336–337.

[†] The statement is true as long as the presence of other masses does not affect the shape of either of the given two masses.

FIGURE 6.4 The principle of linear superposition.

individual gravitational forces on it arising from the *other* particles present. Example 6.2 illustrates the procedure.

QUESTION 2

In T. S. Eliot's *Old Possum's Book of Practical Cats*, there is the couplet:

Macavity, Macavity, there's no one like Macavity,
He's broken every human law, he breaks the law of gravity
"Macavity, The Mystery Cat," lines 5–6

What is the difference between a human law and a law of nature? Is the phenomenon of gravitation a human or natural law? Is Equation 6.3 a human law, since it was deduced by Newton? Explain why Macavity's last feat is science fiction.

EXAMPLE 6.1

A 6.00 kg mass is located 0.500 m from a 3.00 kg mass, as shown in Figure 6.5. Assume they are both pointlike masses.

3.00 kg

FIGURE 6.5

a. Find the gravitational force of the 6.00 kg mass on the 3.00 kg mass.
b. What is the gravitational force of the 3.00 kg mass on the 6.00 kg mass?

Solution
a. The *magnitude* of the gravitational force that each exerts on the other is found using Equation 6.3:

$$F = \frac{GMm}{r^2}$$
$$= \frac{(6.67 \times 10^{-11} \text{ N·m}^2/\text{kg}^2)(6.00 \text{ kg})(3.00 \text{ kg})}{(0.500 \text{ m})^2}$$
$$= 4.80 \times 10^{-9} \text{ N}$$

Note the very small magnitude for the force: only a few nanonewtons.

The force of the 6.00 kg mass on the 3.00 kg mass is directed toward the 6.00 kg mass. Use the coordinate system indicated in Figure 6.6:

$$\vec{F}_{\text{on 3 kg mass}} = (4.80 \times 10^{-9} \text{ N})\hat{\imath}$$

FIGURE 6.6

b. The gravitational force of the 3.00 kg mass on the 6.00 kg mass has the same magnitude but is directed toward the 3.00 kg mass, as shown in Figure 6.6. The two forces form a Newton's third law pair of forces: they have the same magnitude, are in opposite directions, and act on different systems.

$$\vec{F}_{\text{on 6 kg mass}} = (4.80 \times 10^{-9} \text{ N})(-\hat{\imath})$$
$$= (-4.80 \times 10^{-9} \text{ N})\hat{\imath}$$

STRATEGIC EXAMPLE 6.2

Three masses are arranged as shown in Figure 6.7. Assume the masses are pointlike masses.

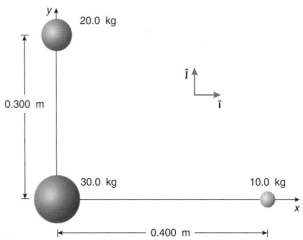

FIGURE 6.7

a. Find the gravitational force of the 30.0 kg mass on the 10.0 kg mass.
b. Find the gravitational force of the 20.0 kg mass on the 10.0 kg mass.
c. What is the *total* gravitational force on the 10.0 kg mass caused by the two other masses?
d. Find the magnitude of the total gravitational force on the 10.0 kg mass.
e. Find the angle that the total gravitational force on the 10.0 kg mass makes with î in Figure 6.7.

Solution

a. Let the 10.0 kg mass be the system. We consider the effect of the 30.0 kg mass on the 10.0 kg mass as if the 20.0 kg mass were not present, as shown in Figure 6.8. The *magnitude* of the gravitational force of the 30.0 kg mass on the 10.0 kg mass is found from Equation 6.3:

$$F_1 = \frac{GMm}{r^2}$$

$$= \frac{(6.67 \times 10^{-11} \ \text{N} \cdot \text{m}^2/\text{kg}^2)(30.0 \ \text{kg})(10.0 \ \text{kg})}{(0.400 \ \text{m})^2}$$

$$= 1.25 \times 10^{-7} \ \text{N}$$

Use the coordinate system indicated in Figure 6.8. The gravitational force $\vec{F}_1$ then is

$$\vec{F}_1 = (-1.25 \times 10^{-7} \ \text{N})\hat{\imath}$$

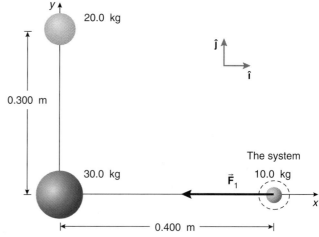

FIGURE 6.8

b. Now consider the effect of the 20.0 kg mass on the 10.0 kg mass as if the 30.0 kg mass were not present, as in Figure 6.9. Use the Pythagorean theorem to find the distance between the 20.0 kg and 10.0 kg masses:

$$[(0.300 \ \text{m})^2 + (0.400 \ \text{m})^2]^{1/2} = 0.500 \ \text{m}$$

The *magnitude* of the gravitational force of the 20.0 kg mass on the 10.0 kg mass is found from Equation 6.3:

$$F_2 = \frac{(6.67 \times 10^{-11} \ \text{N} \cdot \text{m}^2/\text{kg}^2)(20.0 \ \text{kg})(10.0 \ \text{kg})}{(0.500 \ \text{m})^2}$$

$$= 5.34 \times 10^{-8} \ \text{N}$$

Use the coordinate system indicated in Figure 6.9 to express this force in Cartesian form:

$$\vec{F}_2 = (5.34 \times 10^{-8} \ \text{N})[(\cos \theta)(-\hat{\imath}) + (\sin \theta)\hat{\jmath}]$$

The sine and cosine of θ are found from the geometry of the arrangement in Figures 6.8 and 6.9:

$$\cos \theta = 0.800 \quad \text{and} \quad \sin \theta = 0.600$$

Thus

$$\vec{F}_2 = (5.34 \times 10^{-8} \ \text{N})(-0.800\hat{\imath} + 0.600\hat{\jmath})$$

$$= (-4.27 \times 10^{-8} \ \text{N})\hat{\imath} + (3.20 \times 10^{-8} \ \text{N})\hat{\jmath}$$

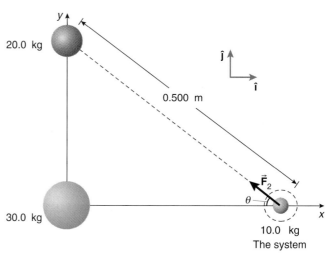

FIGURE 6.9

c. The *total* gravitational force on the 10.0 kg mass caused by the other two masses together is the *vector sum* of $\vec{F}_1$ and $\vec{F}_2$ according to the principle of superposition (see Figure 6.10).

$$\vec{F}_{total} = \vec{F}_1 + \vec{F}_2 = (-1.25 \times 10^{-7} \ \text{N})\hat{\imath}$$

$$+ [(-4.27 \times 10^{-8} \ \text{N})\hat{\imath} + (3.20 \times 10^{-8} \ \text{N})\hat{\jmath}]$$

$$= (-1.68 \times 10^{-7} \ \text{N})\hat{\imath} + (3.20 \times 10^{-8} \ \text{N})\hat{\jmath}$$

d. The magnitude of the total gravitational force on the 10.0 kg mass is the magnitude of $\vec{F}_{total}$:

$$F_{total} = [(-1.68 \times 10^{-7} \ \text{N})^2 + (3.20 \times 10^{-8} \ \text{N})^2]^{1/2}$$

$$= 1.71 \times 10^{-7} \ \text{N}$$

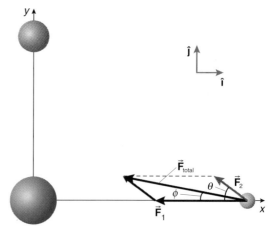

FIGURE 6.10

e. The angle ϕ in Figure 6.10 is found from the components of F_{total}:

$$\tan \phi = \left| \frac{F_{y\,\text{total}}}{F_{x\,\text{total}}} \right|$$

$$= \frac{3.20 \times 10^{-8} \text{ N}}{1.68 \times 10^{-7} \text{ N}}$$

$$= 0.190$$

So

$$\phi = 10.8°$$

The angle the total force makes with $\hat{\imath}$ is, therefore,

$$180° - \phi = 169.2°$$

6.3 GRAVITATIONAL FORCE OF A UNIFORM SPHERICAL SHELL ON A PARTICLE

Newton's law of universal gravitation describes the magnitude and direction of the force of two pointlike particles on each other. Newton was concerned about the gravitational effects of extended objects, because he was interested particularly in the gravitational influences of the Earth, Moon, Sun, and the planets, all objects of considerable size. Many of these celestial objects are roughly spherical in shape, and so it is of some interest to consider the gravitational effects of spherically symmetric systems. It seemed reasonable to think that a spherical mass of constant density, a so-called *uniform sphere*, could be treated as if it were a particle located at the center of the sphere. In fact, this is the case but it is not easy to show rigorously.

The case for a sphere is best considered by first examining the gravitational force of a uniform, thin, spherical shell of mass M on a pointlike particle of mass m. The results then can be extended to solid spheres.

Appendix A provides detailed proofs of the following two important gravitational shell theorems:

If a pointlike particle of mass m is located anywhere within a thin, uniform, hollow spherical shell of mass M, the total gravitational force of the shell on m is equal to zero (see Figure 6.11).

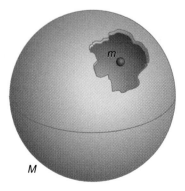

FIGURE 6.11 The gravitational force of a uniform spherical shell on a mass m inside it is zero.

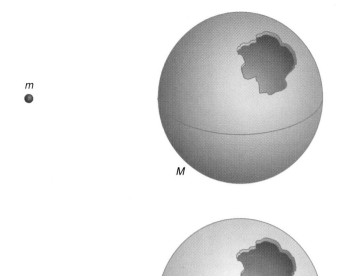

FIGURE 6.12 A mass m outside a uniform spherical shell experiences a gravitational force as if the mass of the shell were located at its center.

QUESTION 3

What is the gravitational force of m on the shell?

If a pointlike particle of mass m is located anywhere outside a thin, uniform, hollow spherical shell of mass M, the total gravitational force of the shell on m is the same as that of a pointlike mass M located at the center of the shell, as shown in Figure 6.12.

In order to prove the shell theorems, Newton had to invent the calculus. Ironically, he kept the results of his work to himself for years before Edmund Halley finally persuaded him to publish them.*

6.4 GRAVITATIONAL FORCE OF A UNIFORM SPHERE ON A PARTICLE

The gravitational shell theorems are used to determine the gravitational force of a uniform sphere of mass M on a pointlike particle of mass m. We consider two situations: (1) a point mass m located outside the sphere, and (2) a point mass m located within the volume enclosed by the sphere itself.

---

*Edmund Halley (1656–1743) was the first person to realize that comets orbited the Sun and periodically returned. He predicted that what we now call Comet Halley would return in 1758, but did not live to see his prediction vindicated. Comet Halley orbits the Sun in a highly elliptical orbit with a period of about 76 years and will next appear about the year A.D. 2061.

A Point Mass *m* Located Outside the Sphere

If the point mass *m* is located *outside* a uniform sphere of radius R and total mass M, divide the uniform sphere into a nested network of concentric, thin, uniform spherical shells, much like an onion, as shown in Figure 6.13. The magnitude of the gravitational force of any shell on *m* is calculated as if the mass of the shell were located at the center of the sphere.

> Therefore, for purposes of calculating the magnitude of the gravitational force of the sphere on *m*, the *entire* mass of the sphere acts as if it were concentrated at the sphere's center, as shown in Figure 6.14.

This result is wonderful!

Thus the magnitude of the gravitational force of the sphere on *m* is just the magnitude of the force between two pointlike particles as in Equation 6.3:

$$F = \frac{GmM}{r^2} \quad \begin{array}{l}\text{(}m\text{ outside a uniform sphere }or\text{ a spherically}\\ \text{symmetric mass distribution)}\end{array} \quad (6.6)$$

where *r* is the distance between *m* and the center of the sphere. The force of the sphere on *m* is directed toward the center of the sphere, since the gravitational force always is an attractive force.

You can use similar arguments for *nonuniform spheres* with a mass distribution that is *spherically symmetric*. The term spherically symmetric means that the density of the sphere depends *only* on the radial coordinate *r*, not on any angles such as the spherical coordinates θ and ϕ. We imagine breaking up the spherically symmetric mass distribution into thin shells, each of which is uniform; again the mass of each shell, whatever it is, acts as if its mass were concentrated at the center of the sphere. So the effect once again is just to have the entire mass of the sphere acting at the center. The Earth is *not* a uniform sphere; but the distribution of mass within the Earth is, to a good first approximation, spherically symmetric. Thus, to calculate the gravitational force of the Earth on particles outside the Earth, the entire mass of the Earth is imagined to be located at the center of the Earth. Other planets and stars also are spherical to a first approximation.*

---

*Jupiter and Saturn are the most oblate planets; the difference between their polar and equatorial diameters is about 10%. For the Earth, the difference is less than 1%. The Sun has no measurable difference between its polar and equatorial diameters.

Like the Earth, they also are not uniform spheres; but their mass distribution is approximately spherically symmetric.

> Thus the entire mass of a spherically symmetric mass distribution is imagined to be located at the center of the sphere for purposes of calculating the gravitational influence of the sphere on exterior particles such as planets, moons, and spacecraft.

Of course, according to Newton's third law of motion, the mass *m* exerts a force of the same magnitude on the sphere in the direction opposite to that of the force of the sphere on *m*; the two forces are a third law pair of forces.

A Point Mass *m* Located Within the Volume Enclosed by the Sphere Itself

Let the point mass *m* lie at some distance *r* from the center but *inside* a sphere of total mass M with a spherically symmetric mass distribution, as shown in Figure 6.15. Imagine the portion of the mass of the sphere that lies more distant from its center than *m* to be composed of a nested network of concentric uniform spherical shells, as indicated in Figure 6.15. Each individual shell exerts *zero* total gravitational force on *m*, since *m* is interior to each of these overlying shells.

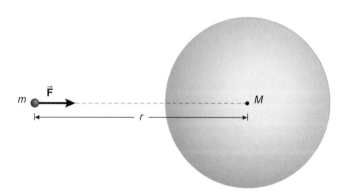

FIGURE 6.14 The entire sphere can be imagined as a pointlike particle of mass M located at the center of the sphere.

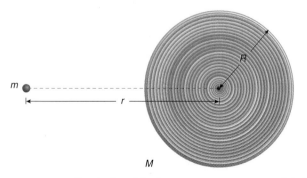

FIGURE 6.13 Divide the solid sphere into a concentric nest of shells.

FIGURE 6.15 Mass *m* inside a solid sphere; the overlying mass has no gravitational effect on *m*.

Thus the overlying portion of the sphere contributes *zero* total gravitational force on *m*.

Imagine the portion of the mass of the sphere that lies closer to the center than *m* also as a nested network of uniform spherical shells, as shown in Figure 6.16. In this case, *m* is outside the shells, and so the mass of these shells can be considered to be located at the center of the sphere for purposes of calculating the gravitational force on *m*.

Thus the magnitude of the total gravitational force of the entire sphere on *m* is due *only* to that portion of the total mass of the sphere that lies closer to the center than *m* itself.

If the sphere is uniform, the fraction of M that lies within *r* of the center is the ratio of the volume of this smaller sphere to the volume of the entire sphere:

$$\frac{\frac{4}{3}\pi r^3}{\frac{4}{3}\pi R^3} = \frac{r^3}{R^3}$$

So the magnitude of the total gravitational force of the sphere on *m* is the gravitational attraction between *m* and a mass $(r^3/R^3)M$ located at the center of the sphere. From Equation 6.3, this is just

$$F = \frac{Gm\left(\dfrac{r^3}{R^3}M\right)}{r^2}$$
$$= \frac{GmM}{R^3}r \qquad (m \text{ within a uniform sphere } only) \qquad (6.7)$$

Thus the magnitude of the gravitational force on *m* is directly proportional to the distance of *m* from the center of the sphere.

The magnitude of the gravitational force exerted by a uniform spherical mass on a particle *m* is plotted as a function of the radial distance of *m* from the center of the sphere in Figure 6.17. Notice that the magnitude of the force on *m* increases linearly with the distance *r* until *m* is at the surface of the sphere, at which point the magnitude of the force reaches its maximum;

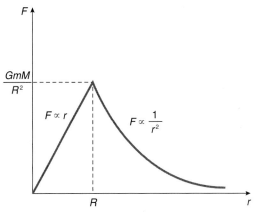

FIGURE 6.17 A schematic graph of the magnitude of the gravitational force on *m* versus its radial distance *r* from the center of a uniform sphere of radius *R* and mass M.

outside the sphere, the magnitude decreases as the inverse square of the distance: $1/r^2$.

If the spherical mass is *not* uniform, Equation 6.7 does not apply. However, if the mass distribution (i.e., the density) is *spherically symmetric*, one can still state that the gravitational force on *m* arises only from the material between *m* and the center of the sphere, but Equation 6.7 is inappropriate for such cases. As always, the equations of physics have strings attached to their use, and we must be aware of the limitations associated with every equation modeling physical phenomena.

QUESTION 4

Assume the Earth is a uniform sphere. Make qualitatively accurate graphs indicating how the mass and the magnitude of the weight of a rock vary as a function of its distance from the *center* of the Earth, from $r = 0$ m to $r \gg R_{\text{Earth}}$.

6.5 MEASURING THE MASS OF THE EARTH

It is hard to imagine how the mass of the Earth is determined, because the traditional technique for determining the mass of common objects by using a laboratory balance simply is inapplicable for the Earth itself. Here we see how the mass of the Earth and that of other astronomical bodies is found.

FIGURE 6.16 Only the mass of the sphere interior to *m* has a nonzero contribution to the gravitational force on *m*.

How to measure the mass of the Earth takes some thought.

The gravitational force on a particle of mass m was defined to be the weight of the particle. Model the Earth as a sphere, to a first approximation,* with a spherically symmetric mass distribution, so that we can use the results of the previous section to find the gravitational force of the Earth on m.

Let the mass m be outside the Earth but close to its surface. The distance r is the distance of m from the *center* of the Earth; r is essentially equal to the radius R of the Earth. Then the gravitational force of the spherical Earth on m is of *magnitude*

$$\frac{GmM}{R^2} \tag{6.8}$$

where M is the mass of the Earth. The force on m is directed, of course, toward the center of the Earth. Previously (in Chapter 5), when m is close to the surface of the Earth, we wrote the magnitude of the weight of a mass m as

$$w = mg \tag{6.9}$$

where g is the magnitude of the local acceleration due to gravity. Equations 6.8 and 6.9 are equivalent expressions for the *identical* force. The magnitudes must be the same:

$$mg = \frac{GmM}{R^2}$$

Hence

$$g = \frac{GM}{R^2} \tag{6.10}$$

The local acceleration due to gravity g is easily measured in the laboratory. A more precise experimental value of the universal gravitational constant is $G = 6.672 \times 10^{-11}$ N·m^2/kg^2 and the effective value of the radius of the Earth is 6.378×10^6 m. Therefore Equation 6.10 can be used to calculate the mass of the Earth. We find

$$\begin{aligned}
M &= \frac{gR^2}{G} \\
&= \frac{(9.81 \text{ m/s}^2)\ (6.378 \times 10^6)^2}{6.672 \times 10^{-11} \text{ N·m}^2/\text{kg}^2} \\
&= 5.98 \times 10^{24} \text{ kg}
\end{aligned}$$

This technique is quite versatile. If the magnitude a of the acceleration of a spacecraft is measured at a known distance r from the center of a planet or moon of mass M, then M is found using a generalization of the equation above:

$$M = \frac{ar^2}{G} \tag{6.11}$$

6.6 ARTIFICIAL SATELLITES OF THE EARTH

Stand on a high mountain and begin throwing rocks horizontally off the top. The rocks, of course, fall to the surface of the Earth.

*The shape is an oblate spheroid to a second approximation, since the polar and equatorial diameters of the Earth differ by 43 km. This distortion of its shape occurs because of the rotation of the Earth as we will see in Chapter 10.

Pitch the rocks with greater initial horizontal velocity and the rocks have a greater range before colliding with the surface of the Earth. Keep pitching the rocks with ever greater speed and the range keeps right on increasing. Neglect any frictional effects due to the air in the low atmosphere and any biological limitations on your throwing arm. Eventually, if the speed of the rock is sufficient, the projectile will circle the Earth, as shown in Figure 6.18.

The rock is falling but continuously *missing* the Earth! Figure 6.18 actually is from Newton's *Principia* and clearly indicates that Newton, in the late 17th century, anticipated the feasibility of artificial **satellites**[†] of the Earth.

[†]The word *satellite*, meaning a small secondary object (i.e., a "moon") in orbit about a larger body, first was used in this context by Johannes Kepler in 1611, referring to the four so-called "secondary planets" of Jupiter discovered by Galileo in 1610. Galileo called the Jovian moons the *sidera Medicaea*, or Medici stars, to curry favor with the Medici family that dominated Florence and Tuscany in Italy during the Renaissance. The four moons now are called the Galilean moons of Jupiter.

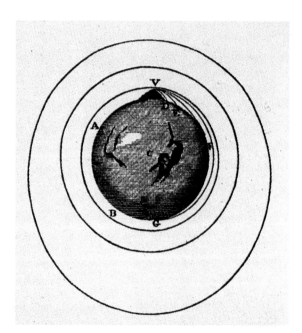

FIGURE 6.18 Launching a satellite of the Earth, as depicted by Newton in the *Principia*.

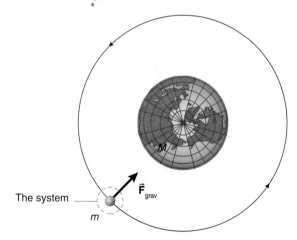

FIGURE 6.19 A satellite orbiting the Earth.

Consider a satellite in circular orbit about the Earth (as in Figure 6.19), well above the atmosphere, so that we can avoid the frictional effects of the air. The *only* force acting on the satellite system is the gravitational force of the Earth. The mass of the satellite typically is constant and the total force and acceleration are in the same direction (toward the center of the circular orbit). Newton's second law relates the total force to the acceleration; in terms of their magnitudes,

$$F_{\text{total}} = ma$$

Here F_{total} is the magnitude of the gravitational force of the Earth on the satellite, since this is the only force on it.

Since the satellite is in a circular orbit, the acceleration is the centripetal acceleration. The second law of motion thus becomes

$$\frac{GmM}{r^2} = m\,\frac{v^2}{r} \qquad (6.12)$$

The period T of the satellite is the time for one complete orbit. The average speed during any interval is equal to the constant instantaneous speed for uniform circular motion. So the speed of the satellite is the distance traveled, say the circumference of the circular orbit, divided by the time interval it takes to travel this distance, which is the period:

$$v = \frac{2\pi r}{T}$$

Substituting for v in Equation 6.12, we find

$$\frac{GmM}{r^2} = \frac{m\left(\dfrac{2\pi r}{T}\right)^2}{r}$$

Solving for T^2,

$$T^2 = \frac{4\pi^2}{GM}\,r^3 \qquad (6.13)$$

The square of the period of the satellite is proportional to the cube of the radius of its orbit. Equation 6.13 is a special simple case of **Kepler's third law of planetary motion**. Equation 6.13 is legitimate as long as the mass of the satellite is negligible compared with the mass of the planet, so that the acceleration of the mother planet is negligible when acted on by the third law counterpart to the gravitational force on the satellite. The approximation $m \ll M$ is excellent for any artificial satellite but is not necessarily a good approximation for natural satellites (the moons) of the planets. We will develop a more general statement of Kepler's third law of planetary motion in Section 6.11.

A few points to clarify:

1. One point is semantics. We defined the weight of a system of mass m to be the gravitational force on the system. A satellite certainly is subjected to the gravitational force of the Earth. Indeed, it is the gravitational force of the Earth on the satellite that gives the satellite its centripetal acceleration. The satellite is *not weightless*; indeed, it is the weight that keeps the satellite circling in its orbit! If the gravitational force somehow ceased its action on the satellite beginning at some instant, the satellite would proceed in a straight line tangent to the orbit. A satellite in orbit is in free-fall motion around the Earth, continually missing it. It is undergoing the same type of free-fall experienced when you jump off a low wall or a high cliff (though the latter is not recommended). If you do such a foolish thing while simultaneously standing on a scale (whose reading is a measure not of your weight but of the magnitude of the normal force of you on the scale), the scale will read zero because it too is in free-fall. You are normal-force-less, *not* weightless, when undergoing free-fall motion. Objects within a satellite (such as astronauts) experience normal-forcelessness because they are in free-fall motion around the Earth along with the satellite itself.

2. Any ordinary projectile, such as a baseball thrown from the surface of the Earth, actually is in orbit about the center of the Earth. The projectile is just *unable to complete the orbit* because the surface of the Earth gets in the way. You can have some fun with this idea: just wager to your friends that you can launch an Earth satellite with your bare hands. Guarantee them double if you lose the wager. Collect the bets. With the money safely in your pocket, simply toss a rock into the air. That's all there is to it; you have *launched* a satellite of the Earth. While the rock is in flight, it is a satellite of the Earth; there is no arguing with that fact. But... but... but, your friends may claim. But nothing. You never claimed that the satellite would *complete* an orbit; *that* is another matter! You simply claimed that you could *launch* it.

3. There is nothing sacred about the third rock from the Sun (the Earth!) in this argument about satellites. With the appropriate mass for M used in Equation 6.13, the arguments equally well apply to satellites or spacecraft in circular orbit about Jupiter, Mars, Venus, the Sun, or Superman's fictional home planet Krypton.

Astronauts in orbit are not weightless but normal-force-less.

What are *two* technical errors in the following journalistic description of a space shuttle mission in orbit about the Earth?

Their research, along with most of the nine other experiments on the shuttle, is designed to measure how various materials behave or are created in a weightless environment, outside Earth's gravitational pull.

Boston Globe (1 October 1988)

STRATEGIC EXAMPLE 6.3

a. What is the magnitude of the gravitational force of the Earth on an IRS surveillance satellite of mass 1040 kg that is in a grazing circular orbit 100 km above the *surface* of the Earth?

b. Find the magnitude of the acceleration of a satellite in such a grazing orbit. The order of magnitude of the result should not be unexpected. Explain why.

c. Determine the speed of a satellite in such an orbit.

d. Calculate the period of the satellite.

Solution

a. The magnitude of the gravitational force of the Earth on the satellite is found from Newton's law of universal gravitation, Equation 6.3:

$$F_{grav} = \frac{GMm}{r^2}$$

The distance r is the distance of the satellite from the *center* of the Earth, equal to 100 km *plus* the radius of the Earth. That is,

$$r = 0.100 \times 10^6 \text{ m} + 6.37 \times 10^6 \text{ m}$$
$$= 6.47 \times 10^6 \text{ m}$$

Let M be the mass of the Earth and m the mass of the satellite. Hence the magnitude of the force is

$$F_{grav} = \frac{(6.67 \times 10^{-11} \text{ N} \cdot \text{m}^2/\text{kg}^2)(5.98 \times 10^{24} \text{ kg})(1040 \text{ kg})}{(6.47 \times 10^6 \text{ m})^2}$$

$$= 9.91 \times 10^3 \text{ N}$$

b. Determine the magnitude of the centripetal acceleration from Newton's second law; use the magnitudes of the vectors:

$$F_{total} = ma$$
$$9.91 \times 10^3 \text{ N} = (1040 \text{ kg})a$$

So

$$a = 9.53 \text{ m/s}^2$$

The satellite is in free-fall motion and the magnitude of the acceleration due to gravity 100 km above the surface of the Earth should be slightly less than the value at the surface, 9.81 m/s².

c. The speed of the satellite is found from the magnitude of its centripetal acceleration:

$$a_c = \frac{v^2}{r}$$

where r is the radius of its orbit.

$$9.53 \text{ m/s}^2 = \frac{v^2}{6.47 \times 10^6 \text{ m}}$$

Solving for v,

$$v = 7.85 \times 10^3 \text{ m/s}$$
$$= 7.85 \text{ km/s}$$

The satellite is really moving fast!

d. The period is the time for one complete orbit.

Method 1

The satellite traverses the circumference of its circular orbit ($2\pi r$) at the constant speed v, and so the time it takes is

$$T = \frac{2\pi r}{v}$$

$$= \frac{2\pi \times 6.47 \times 10^6 \text{ m}}{7.85 \times 10^3 \text{ m/s}}$$

$$= 5.18 \times 10^3 \text{ s}$$

or 86.3 minutes.

Method 2

Use Kepler's third law of planetary motion, Equation 6.13:

$$T^2 = \frac{4\pi^2}{GM} r^3$$

$$= \frac{4\pi^2 (6.47 \times 10^6 \text{ m})^3}{(6.67 \times 10^{-11} \text{ N} \cdot \text{m}^2/\text{kg}^2)(5.98 \times 10^{24} \text{ kg})}$$

$$= 2.68 \times 10^7 \text{ s}^2$$

Taking the square root, you find

$$T = 5.18 \times 10^3 \text{ s}$$

6.7 KEPLER'S FIRST LAW OF PLANETARY MOTION AND THE GEOMETRY OF ELLIPSES

Almost a century before Newton, between about 1600 and 1620, Johannes Kepler (1571–1630) discovered three empirical laws of planetary motion that, for the first time, accurately described the heliocentric orbital motion of the planets in the solar system. Kepler made these discoveries by tediously analyzing the precise observational planetary position data collected painstakingly for 20 years during the late 16th century, before telescopes, by his mentor Tycho Brahe (1546–1601). All three of the planetary laws of Kepler were subsequently shown by Newton to be consequences of Newton's laws of motion and his law of universal gravitation. Kepler's laws, therefore, are not independent laws of motion. This illustrates how succeeding laws can build on and supplant earlier laws. We study Kepler's laws here to gain an appreciation for the beauty and structure of the solar system in which we reside. The laws also are applicable to any two-body gravitational interaction elsewhere in the cosmos, such as a

The planetary data secured by Tycho Brahe enabled Kepler to decipher the structure of the solar system.

moon or satellite in orbit around a mother planet, or the newly discovered planets orbiting stars other than the Sun.

> **Kepler's first law** of planetary motion states that the path of a planet orbiting the Sun is an ellipse with the Sun at one focus of the ellipse; the other focus is empty.

The empty focus perplexed and annoyed Kepler, but is not physically significant. Newton showed that the inverse square nature of the gravitational force law permits orbits that are *any* of the **conic sections** familiar to you from your study of analytic geometry: circles, ellipses, parabolas, and hyperbolas. More advanced courses in classical mechanics will show you how the conic section orbital shapes stem from the law of universal gravitation. Here we will look at only several characteristic features of the circular and elliptical orbits that are so common in orbital dynamics.

An **ellipse** is defined mathematically as the locus of points, the sum of whose distances from two fixed points, F_1 and F_2, known as the **foci**, is a constant.* Consider a planet in an elliptical orbit about the Sun, as in Figure 6.20. The Sun is at one focus of the

---

*Do not confuse the foci labels F_1 and F_2 with the magnitudes of forces.

ellipse, say F_1. Let the planet be at point P in its orbit. The distance of either focus from the geometric center of the ellipse is c.[†]

The definition of an ellipse implies that the sum of the distance r of the planet at P from the Sun and the distance s of the planet from the other (empty) focus is a constant. Call the constant $2a$, where the numerical factor of 2 is introduced for later convenience.[‡] So the definition of an ellipse implies that

$$r + s \equiv 2a \qquad (6.14)$$

Imagine the planet at the special point P_1 in Figure 6.20, located at one end of the **major axis** of the ellipse, closest to the Sun, called the **perihelion** point. Equation 6.14 then becomes

$$P_1F_1 + P_1F_2 = 2a \qquad (6.15)$$

Examine Figure 6.20; you can see that $P_1F_2 = P_1F_1 + 2c$, and so Equation 6.15 becomes

$$P_1F_1 + (P_1F_1 + 2c) = 2a \qquad (6.16)$$

But because of the symmetry of the ellipse, P_1F_1 is the same as P_2F_2, where P_2 is at the other end of the major axis; when at P_2, the planet is furthest from the Sun, at its **aphelion** point. Making this substitution into Equation 6.16, you get

$$P_1F_1 + (P_2F_2 + 2c) = 2a$$

The left-hand side of this equation is the sum of line segments that form the major axis of the ellipse (P_1P_2). Thus $2a$ is equal to the major axis of the ellipse, and the physical meaning of the constant a is *half the major axis*, known as the **semimajor axis** of the ellipse.

The **eccentricity** ε of the ellipse is defined to be the ratio of c to a:

$$\varepsilon \equiv \frac{c}{a} \qquad (6.17)$$

If the two foci of the ellipse are coincident, so $2c = 0$ m (or $c = 0$ m), then the eccentricity $\varepsilon = 0$ as well. The resulting geometrical figure is a circle. Thus a circle is an ellipse with zero eccentricity. At the other extreme, if c approaches a, then the ellipse becomes more and more elongated. When $c = a$, making $\varepsilon = 1$, the resulting geometrical figure is a parabola, which is not a closed curve.

The general equation of an ellipse in polar coordinates is found in the following way. Let $\vec{r}$ be the position vector of the planet in its orbit using the Sun as the origin, as shown in Figure 6.21. Let $\vec{c}$ be a vector from the Sun to the center of the ellipse. Let $\vec{s}$ be a vector from the second (empty) focus to the planet.

The rules of vector addition indicate that

$$\vec{r} = 2\vec{c} + \vec{s}$$

Thus

$$\vec{s} = \vec{r} - 2\vec{c}$$

[†] Do not confuse this distance c with the speed of light, also called c.

[‡] Do not confuse the constant a here with the magnitude of the acceleration. Here we will see that the constant a is the semimajor axis of the ellipse. It is usually clear from the context (if only from a dimensional analysis) whether a means the magnitude of the acceleration or a means the semimajor axis. Regrettably, there is a certain amount of unavoidable notational overlap in every technical field.

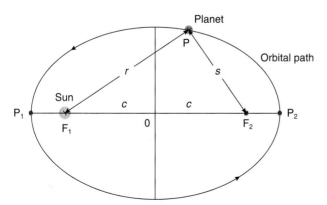

FIGURE 6.20 A planet in an elliptical orbit about the Sun.

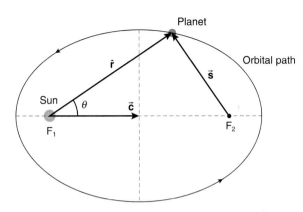

FIGURE 6.21 Position vector of a planet in an elliptical orbit.

Take the scalar product of $\vec{\mathbf{s}}$ with itself to find the square of its magnitude:

$$s^2 = \vec{\mathbf{s}} \cdot \vec{\mathbf{s}} = (\vec{\mathbf{r}} - 2\vec{\mathbf{c}}) \cdot (\vec{\mathbf{r}} - 2\vec{\mathbf{c}})$$
$$= r^2 + 4c^2 - 4\,\vec{\mathbf{r}} \cdot \vec{\mathbf{c}} \qquad (6.18)$$

But from the definition of an ellipse,

$$s + r = 2a$$

and so

$$s = 2a - r$$

From the definition of the eccentricity, Equation 6.17, $c = \varepsilon a$. Making these substitutions for s and c into Equation 6.18, we find

$$(2a - r)^2 = r^2 + 4(\varepsilon a)^2 - 4r(\varepsilon a)\cos\theta$$

where θ is the polar coordinate angle between $\vec{\mathbf{r}}$ and $\vec{\mathbf{c}}$, shown in Figure 6.21. After performing the indicated multiplications, we obtain

$$4a^2 - 4ra + r^2 = r^2 + 4\varepsilon^2 a^2 - 4r\varepsilon a\cos\theta$$

The quadratic terms in r subtract out. Solving this expression for r, we find

$$r = \frac{a(1 - \varepsilon^2)}{1 - \varepsilon\cos\theta} \qquad (6.19)$$

This is the equation for an ellipse in polar coordinates. The polar coordinate r is measured from the focus and θ is measured with respect to the major axis (see Figure 6.21).

EXAMPLE 6.4

When a planet is at position P_1 in Figure 6.22, the planet is at perihelion, its closest distance to the Sun. When the planet is at the position P_2, the planet is at aphelion, its greatest distance from the Sun.

a. What is the angle θ in the polar equation for an ellipse, Equation 6.19, when the planet is at aphelion?
b. Use the polar equation for the ellipse to show that the aphelion distance is related to the semimajor axis and the eccentricity of the ellipse by

$$r_{aph} = a(1 + \varepsilon)$$

c. What is the angle θ when the planet is at perihelion?

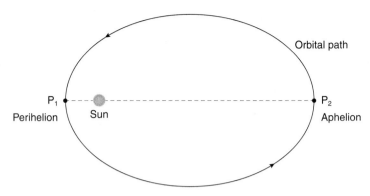

FIGURE 6.22

d. Show that the perihelion distance is related to the semimajor axis a and the eccentricity by

$$r_{peri} = a(1 - \varepsilon)$$

Solution
a. When the planet is at aphelion, point P_2, the angle θ is equal to $0°$, as shown in Figure 6.23.

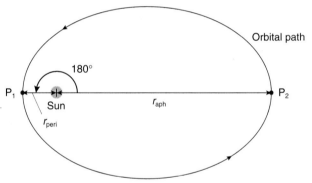

FIGURE 6.23

b. Substitute $\theta = 0°$ into the polar equation for an ellipse, Equation 6.19:

$$r_{aph} = \frac{a(1 - \varepsilon^2)}{1 - \varepsilon\cos 0°}$$
$$= \frac{a(1 - \varepsilon)(1 + \varepsilon)}{1 - \varepsilon}$$
$$= a(1 + \varepsilon)$$

c. When the planet is at perihelion, the angle $\theta = 180°$, as shown in Figure 6.23.
d. Substitute $\theta = 180°$ into the polar equation for an ellipse, Equation 6.19:

$$r_{peri} = \frac{a(1 - \varepsilon^2)}{1 - \varepsilon\cos 180°}$$
$$= \frac{a(1 - \varepsilon)(1 + \varepsilon)}{1 + \varepsilon}$$
$$= a(1 - \varepsilon)$$

Note that the *ratio* of the aphelion and perihelion distances depends only on the eccentricity:

$$\frac{r_{aph}}{r_{peri}} = \frac{a(1 + \varepsilon)}{a(1 - \varepsilon)}$$

$$= \frac{1 + \varepsilon}{1 - \varepsilon} \tag{1}$$

EXAMPLE 6.5

The eccentricity of the orbit of the Earth about the Sun is 0.0167 and the semimajor axis of the orbit is 1.496×10^8 km. Determine the aphelion and perihelion distances of the Earth from the Sun.

Solution
In Example 6.4, you showed that

$$r_{aph} = a(1 + \varepsilon)$$

and

$$r_{peri} = a(1 - \varepsilon)$$

Substituting numerical values for a and ε, you find

$$r_{aph} = (1.496 \times 10^8 \text{ km})(1 + 0.0167)$$
$$= 1.521 \times 10^8 \text{ km}$$

and

$$r_{peri} = (1.496 \times 10^8 \text{ km})(1 - 0.0167)$$
$$= 1.471 \times 10^8 \text{ km}$$

For cultural enrichment, in 1999 the Earth was at perihelion on 3 January and at aphelion on 6 July.

6.8 SPATIAL AVERAGE POSITION OF A PLANET IN AN ELLIPTICAL ORBIT*

The distance of a planet from the Sun varies between the extremes of the perihelion and aphelion distances. What is the average of all the possible spatial distances of the planet in its orbit?

Such a spatial average might be estimated by drawing n lines from the Sun to equally spaced positions of the planet along its orbital path, as in Figure 6.24. The n distances then are summed, and an estimate of the spatial average found by dividing the sum by n.[†] If you let the number of such lines increase indefinitely by letting the spacing Δs in Figure 6.24 approach zero, you obtain a true spatial average of all the possible positions of the planet in its orbit.

There is, however, an easier way to determine such a spatial average. The following imaginative procedure enables you to calculate the spatial average without undue effort.

Consider two possible positions of the planet on its orbit, points P and P′ in Figure 6.25, that are located *symmetrically* with respect to the **minor axis** of the ellipse. When the planet is at P its distance from the Sun is r, and when the planet is at

---

[†]In fact, this was the way Kepler found this average in the early 1600s.

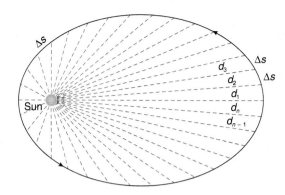

FIGURE 6.24 Lines from the Sun to points equally spaced along the orbital path.

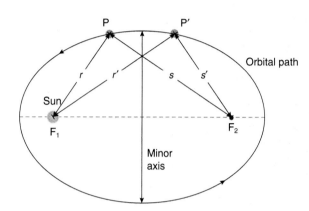

FIGURE 6.25 Locate two points P and P′ symmetrically with respect to the minor axis.

P′ its distance from the Sun is $r′$. The symmetrical placement of P and P′ means that $r′ = s$. Thus the average of these two distances becomes

$$\frac{r + s}{2}$$

But the definition of the ellipse, Equation 6.14, indicates that $r + s = 2a$. Thus the average is $2a/2$ or just a, the semimajor axis. But the points P and P′ were *any* two points on the orbit symmetrically placed. So the average of all such paired points also is just the semimajor axis.

> Thus the spatial average of the distances of the planet from the Sun is equal to the semimajor axis of the ellipse.

6.9 KEPLER'S SECOND LAW OF PLANETARY MOTION

Kepler also discovered another aspect of the orbital motions of the planets, now known as **Kepler's second law** (although chronologically, it actually was the first one he discovered).

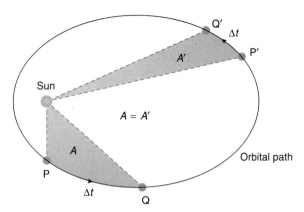

FIGURE 6.26 Kepler's second law of planetary motion.

A line from the Sun to a given planet sweeps out equal areas during equal time intervals.

If it takes the same time interval Δt for the planet to move between the positions at points P and Q and between P′ and Q′ in Figure 6.26, then the areas A and A′ also are equal.

This law implies that the speed of a planet when near its perihelion point (i.e., closest to the Sun) must be greater than when near its aphelion point (most distant from the Sun).

QUESTION 6

Imagine a line from the Sun to the Earth and another line from the Sun to Mars. One week elapses. Are the areas swept out by the lines equal? Explain your answer.

A postmark depicting Kepler's second law of planetary motion.

6.10 CENTRAL FORCES, ORBITAL ANGULAR MOMENTUM, AND KEPLER'S SECOND LAW*

A force that acts along the line between two particles is called a **central force**. The gravitational force $\vec{F}$ of the Sun on a planet is antiparallel to the position vector $\vec{r}$ locating the planet, using the Sun as an origin, as indicated in Figure 6.27. Thus the gravitational force is a central force. The magnitude of the *vector product* $\vec{r} \times \vec{F}$ must be zero, because the sine of the 180° angle between two antiparallel vectors is zero (sin 180° = 0):

$$|\vec{r} \times \vec{F}| = rF \sin 180°$$
$$= 0 \text{ N·m} \tag{6.20}$$

Notice that this result is true for *any central force*, regardless of the functional form of the force.

The central gravitational force is the only force on the planet. Newton's second law of motion indicates that

$$\vec{F}_{central} = m\vec{a}$$

where $\vec{a}$ is the acceleration of the planet. Substituting $m\vec{a}$ for the force in the vector product $\vec{r} \times \vec{F}$ and using Equation 6.20, you have

$$\vec{r} \times (m\vec{a}) = 0 \text{ N·m} \tag{6.21}$$

Look for a moment at the time derivative of the vector product $\vec{r} \times (m\vec{v})$:

$$\frac{d}{dt}[\vec{r} \times (m\vec{v})]$$

Remember that when differentiating you must preserve the order of the terms in the vector product and that m is constant. The differentiation proceeds as follows:

$$\frac{d}{dt}[\vec{r} \times (m\vec{v})] = \frac{d\vec{r}}{dt} \times (m\vec{v}) + \vec{r} \times \left(m\frac{d\vec{v}}{dt}\right) \tag{6.22}$$

However, since

$$\frac{d\vec{r}}{dt} = \vec{v} \quad \text{and} \quad \frac{d\vec{v}}{dt} = \vec{a},$$

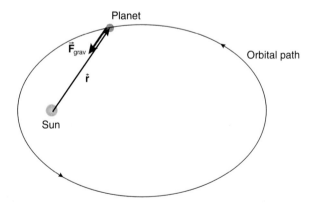

FIGURE 6.27 The gravitational force acts along the line connecting the two particles.

Equation 6.22 becomes

$$\frac{d}{dt}[\vec{r} \times (m\vec{v})] = \vec{v} \times (m\vec{v}) + \vec{r} \times (m\vec{a}) \qquad (6.23)$$

But the vector product of *any* vector with itself is zero, and so

$$\vec{v} \times (m\vec{v}) = m(\vec{v} \times \vec{v}) = 0 \ \text{kg·m}^2/\text{s}^2$$

and Equation 6.23 simplifies to

$$\frac{d}{dt}[\vec{r} \times (m\vec{v})] = \vec{r} \times (m\vec{a})$$

But $\vec{r} \times (m\vec{a}) = 0 \ \text{kg·m}^2/\text{s}^2$ from Equation 6.21, because $\vec{r}$ and $\vec{a}$ are antiparallel vectors for any central force. Thus we have

$$\frac{d}{dt}[\vec{r} \times (m\vec{v})] = 0 \ \text{kg·m}^2/\text{s}^2 \qquad (6.24)$$

> The product $m\vec{v}$ is defined to the **momentum $\vec{p}$** of the mass m:*
> $$\vec{p} \equiv m\vec{v} \qquad (6.25)$$

The momentum is a vector parallel to the velocity. Using this definition in Equation 6.24, we find

$$\frac{d}{dt}(\vec{r} \times \vec{p}) = 0 \ \text{kg·m}^2/\text{s}^2$$

If the time derivative of a vector is zero, the vector does not change with time and is a *constant vector*. Thus the vector product $\vec{r} \times \vec{p}$ does not change with time for *any central force*, acting alone, on m. If we think of a planet moving about the Sun in an elliptical orbit, the vector $\vec{r}$ changes with time (both in magnitude and direction), the vector $\vec{p}$ changes with time (since the velocity of the planet changes in both magnitude and direction in an elliptical orbit), but *the vector $\vec{r} \times \vec{p}$ does not change with time for a particle under the influence of a central force.* We say such a vector is **conserved**.

> The word *conserved* means the quantity is *constant at all times and locations.*

Thus $\vec{r} \times \vec{p}$ is quite a special vector; it occurs in many contexts in physics and is called the **orbital angular momentum $\vec{L}$**:[†]

> $$\vec{L} \equiv \vec{r} \times \vec{p} \qquad (6.26)$$

> The orbital angular momentum of a particle (or planet) under the influence of a central force is conserved; that is,
> $$\frac{d\vec{L}}{dt} = 0 \ \text{kg·m}^2/\text{s}^2$$

Since $\vec{L}$ is a constant vector for central force motion, and $\vec{L}$ is perpendicular to *both* $\vec{r}$ and $\vec{p}$ (from the definition of the vector product), *the motion of the particle under the influence of a central force must be confined to a plane.*

An interesting geometrical consequence of the constancy of the orbital angular momentum vector is Kepler's law of areas. To see this connection, consider a planet moving in an elliptical orbit, as in Figure 6.28. Let $d\vec{r}$ be the *change* in the position vector of the planet during a time interval dt. The position vector of the planet sweeps out a small differential area dA during this time. The small area swept out is essentially triangular in shape.

The geometric formula for the area of a triangle is

$$dA = \frac{1}{2}(\text{base})(\text{height})$$

The base of the skinny triangle is $r \, d\theta$ and the height is r; thus, the differential area is

$$dA = \frac{1}{2}(r \, d\theta) \, r$$

This expression is the same as the *magnitude* of the vector

$$\frac{1}{2}\vec{r} \times d\vec{r}$$

because the magnitude of this vector product is

$$\frac{1}{2} r \, dr \sin \phi$$

where ϕ is the angle between $\vec{r}$ and $d\vec{r}$ as indicated in Figure 6.28. But $dr \sin \phi$ is the same as the base of the little triangle ($r \, d\theta$). Thus we create a differential *area vector*

$$d\vec{A} = \frac{1}{2}\vec{r} \times d\vec{r}$$

whose magnitude is the differential area dA; its direction is *perpendicular* to the plane of the orbit from the vector product right-hand rule (since $\vec{r}$ and $d\vec{r}$ are in the plane of the orbit).

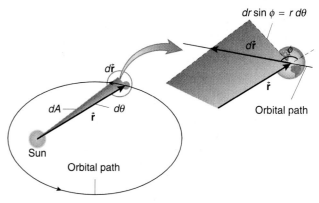

FIGURE 6.28 A line from the Sun to the planet sweeps out a small triangular area during a short interval dt.

---

*For now, we take this definition on an ad hoc basis; we introduce the momentum of a particle in a more pedagogical fashion in Chapter 9.

[†] We explore the orbital angular momentum in more detail in Chapter 10.

If we divide the differential area vector by the differential time dt, we obtain

$$\frac{d\vec{A}}{dt} = \frac{1}{2}\frac{\vec{r} \times d\vec{r}}{dt}$$

$$= \frac{1}{2}\vec{r} \times \frac{d\vec{r}}{dt}$$

$$= \frac{1}{2}\vec{r} \times \vec{v} \qquad (6.27)$$

The orbital angular momentum vector is $\vec{L} = \vec{r} \times \vec{p} = \vec{r} \times m\vec{v}$; hence, $\vec{r} \times \vec{v} = \vec{L}/m$ and Equation 6.27 becomes

$$\frac{d\vec{A}}{dt} = \frac{\vec{L}}{2m} \qquad (6.28)$$

Since $\vec{L}$ is a constant vector and m is just the (constant) mass of the planet, the area vector has a constant time derivative. That is, A itself must vary linearly with the time t. This means that a line from the Sun to the planet sweeps out equal areas during equal time intervals, which is Kepler's law of areas.

To avoid any misconceptions, note that the constant orbital angular momentum vector and the constant rate at which the area is swept out by the position vector to the planet are characteristic of the *individual planet* whose motion is being considered. The amounts of angular momentum of the various planets differ from each other because they have different orbits and different masses.

EXAMPLE 6.6

What is the direction of the orbital angular momentum vector $\vec{L}$ for the planetary motion depicted in the Figure 6.29? Is the magnitude or direction of $\vec{L}$ different at another position of the planet in its orbit?

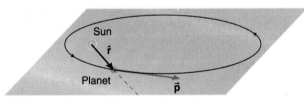

Plane of elliptical orbit

FIGURE 6.29

Solution
To find the direction of the angular momentum $\vec{L} = \vec{r} \times \vec{p}$, use the vector product right-hand rule. The vector $\vec{r} \times \vec{p}$ is directed perpendicular to the plane of the orbit of the planet as indicated in Figure 6.30.

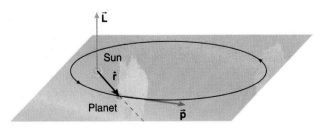

Plane of elliptical orbit

FIGURE 6.30

Since the angular momentum $\vec{L}$ is a constant vector, it has the same magnitude and direction for all positions of the planet in its orbit.

EXAMPLE 6.7

Consider a planet at its aphelion and perihelion points, as in Figure 6.31.

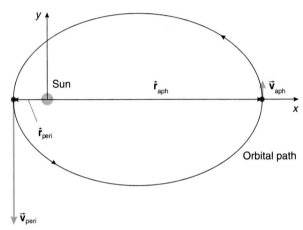

FIGURE 6.31

a. What is the angle between the position vectors and the respective velocity vectors of the planet at these locations?
b. By applying conservation of the orbital angular momentum of the planet at these two locations, show that the ratio of the *speeds* of the planet at the perihelion and aphelion locations is

$$\frac{v_{peri}}{v_{aph}} = \frac{1 + \varepsilon}{1 - \varepsilon}$$

Solution
a. The velocity vector always is tangent to the orbital path. At the aphelion and perihelion points in the orbit (but nowhere else in an elliptical orbit), the angle between $\vec{r}$ and $\vec{v}$ is 90°.
b. Since the orbital angular momentum vector is conserved, it must be the same at the perihelion and aphelion locations. Use the coordinate system indicated in Figure 6.31. The angular momentum of the planet of mass m when at aphelion is

$$\vec{L}_{aph} = \vec{r}_{aph} \times (m\vec{v}_{aph})$$
$$= mr_{aph}\hat{i} \times (v_{aph}\hat{j})$$
$$= mr_{aph}v_{aph}\hat{k}$$

Likewise, at perihelion

$$\vec{L}_{peri} = \vec{r}_{peri} \times (m\vec{v}_{peri})$$
$$= mr_{peri}(-\hat{i}) \times [v_{peri}(-\hat{j})]$$
$$= mr_{peri}v_{peri}\hat{k}$$

The angular momentum is a constant vector throughout the motion, and so

$$\vec{L}_{aph} = \vec{L}_{peri}$$
$$mr_{aph}v_{aph}\hat{k} = mr_{peri}v_{peri}\hat{k}$$

Divide by m and rearrange this to obtain the ratio you want:

$$\frac{v_{peri}}{v_{aph}} = \frac{r_{aph}}{r_{peri}} \tag{1}$$

Equation (1) in Example 6.4 showed that the ratio of the perihelion and aphelion distances is

$$\frac{r_{aph}}{r_{peri}} = \frac{1 + \varepsilon}{1 - \varepsilon}$$

Equation (1) here thus becomes

$$\frac{v_{peri}}{v_{aph}} = \frac{1 + \varepsilon}{1 - \varepsilon}$$

The speed at perihelion thus is faster than at aphelion.

6.11 NEWTON'S FORM FOR KEPLER'S THIRD LAW OF PLANETARY MOTION

According to Newton's law of universal gravitation, two spherically symmetric masses m and M (such as the Moon and the Earth) exert gravitational forces on each other directed along the line between their centers (see Figure 6.32). The gravitational forces on each are equal in magnitude and opposite in direction, in accordance with Newton's third law. Assume the gravitational force is the only force acting on each mass.

The gravitational force of M on m results in an acceleration of m according to Newton's second law. The magnitudes of the force and acceleration are related by

$$F_{on\ m} = ma$$

But the magnitude of the total force on m is the magnitude of the gravitational force of M on m, so that

$$\frac{GmM}{r^2} = ma$$

Consider also the gravitational force of m on M and the resulting acceleration a' of M. Once again, the magnitudes are related by the second law of motion:

$$F_{on\ M} = Ma'$$

The magnitude of the total force on M is the magnitude of the gravitational force of m on M, and so

$$\frac{GmM}{r^2} = Ma'$$

where a' is the acceleration of M.

Notice that the *same magnitude of force acts on each mass* (because of Newton's third law of motion) but the *magnitude of the acceleration of each mass is different* (as deduced from Newton's second law of motion, applied to each as an individual system):

Magnitude of acceleration of m:

$$a = \frac{GM}{r^2}$$

Magnitude of acceleration of M:

$$a' = \frac{Gm}{r^2}$$

These expressions imply that the magnitude of the acceleration of each mass is proportional to the *other* mass. Take the ratio of the magnitudes of the accelerations of the two masses:

$$\frac{\text{magnitude of acceleration of } m}{\text{magnitude of acceleration of } M} = \frac{a}{a'}$$

$$= \frac{\dfrac{GM}{r^2}}{\dfrac{Gm}{r^2}}$$

$$= \frac{M}{m} \tag{6.29}$$

Let m be a trivial mass compared with M, say the mass of a small rock compared with the mass of the Earth. The acceleration of the *rock* is proportional to the mass of the *Earth*. Since the mass of the Earth is so much larger than that of the small rock, the acceleration of the rock is significant. The rock exerts an *equal* gravitational *force* on the Earth; however, the acceleration of the *Earth* is proportional to the mass of the *rock*. Since the mass of the rock is so much smaller than that of the Earth, the acceleration of the Earth toward the rock is negligible. This is why we could neglect the gravitational effects *of* a satellite *on* the Earth in Section 6.6; the acceleration of the Earth is minuscule compared with that of the satellite. This argument also is precisely the reason we neglect any motion of the Earth when considering the motions of common objects subjected to the Earth's gravitational force.

However, if the two masses m and M are comparable in magnitude, the accelerations are comparable as well according to Equation 6.29. What happens then? Let the two masses be separated by a distance r, as in Figure 6.33. Imagine a point 0 on the line connecting the masses, and let r_1 be the distance of m from 0 and r_2 the distance of M from 0. Suppose 0 is the special point at which the products of the masses and their respective distances from 0 are equal; then 0 is called the **center of mass** point.

FIGURE 6.32 Two masses m and M exert gravitational forces on each other.

FIGURE 6.33 Two masses separated by a distance r.

FIGURE 6.34 Two masses released at rest.

The location of the center of mass point is defined by the equation*:

$$mr_1 = Mr_2 \qquad (6.30)$$

The center of mass point thus is located closer to the more massive object.[†]

If the two masses are released from rest, as in Figure 6.34, they accelerate toward each other along a straight line and eventually collide. Notice that if we differentiate Equation 6.30 twice with respect to time, we get the relationship between the magnitudes of the accelerations of the two objects (Equation 6.29, slightly rearranged): $ma = Ma'$. If we neglect the physical sizes of the two masses, they collide at the center of mass point of the two-body system.

So if the objects initially are at rest, and the only forces on them are their mutual gravitational forces, they will collide after some time interval. The collision occurs at the center of mass (if we neglect the different finite sizes of the two objects). When a little mass m is released near the Earth and falls, the center of mass of the system is effectively at the center of the Earth; but the collision occurs on the surface because of the size of the Earth.

A collision of the masses may be avoided if the objects initially are not at rest but are moving in a direction not coincident with the line between them. In such situations, Newton showed that both objects move in paths that

1. are similar in shape,
2. are one of the conic sections, and
3. describe motion about the center of mass point (which is fixed in space).

The special case of circular orbits about the center of mass point is the easiest to consider (see Figure 6.35), and so we examine it in some detail. Our objective is to find a relationship between the period T of the orbital motion and the separation r of the masses for such circular orbital motion of the masses about the center of mass point.

The magnitudes of the gravitational forces (1) of M on m and (2) of m on M are the *same* by Newton's third law. We assume both masses have appropriate spherical symmetry to use the point-mass gravitational force law. According to Newton's second law, the single force on each causes an acceleration in the

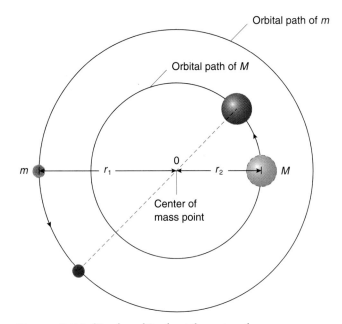

FIGURE 6.35 Circular orbits about the center of mass.

same direction as the force. Since the orbits are circular, the acceleration of each mass is a centripetal acceleration.

For the motion of m (use Newton's second law of motion with m as the system), the magnitudes of the force and acceleration are related via

$$\frac{GMm}{r^2} = ma$$

Substituting for a the magnitude of the centripetal acceleration of m, we obtain

$$\frac{GMm}{r^2} = m\frac{v_1^2}{r_1} \qquad (6.31)$$

where v_1 is the orbital speed of m in its orbit of radius r_1 about the center of mass.

Likewise, for the motion of M apply Newton's second law to M, with M as a separate system. The magnitudes of the force and acceleration are related by

$$\frac{GMm}{r^2} = Ma'$$

Again, substitute for a' the magnitude of the centripetal acceleration of M:

$$\frac{GMm}{r^2} = M\frac{v_2^2}{r_2} \qquad (6.32)$$

where v_2 is the speed of M in its orbit of radius r_2 about the center of mass.

Both masses have the same orbital period of revolution, since the center of mass must lie on the line between them.[‡] The time for

---

*The location of the center of mass for a collection of particles (not just two) will be considered in greater detail in Chapter 9.

[†]You can think of the center of mass point as the location of a "balance point" along the line between the two masses, just like the balance point of an imaginary (massless) see-saw with the two masses m and M on each end. To do this, one would need a third great mass equidistant from both of these masses to provide the gravitational forces to enable the balance to operate.

[‡]If the periods differed, eventually both masses could be on the same side of the center of mass, which is ridiculous. Why?

one orbital revolution is the period T. The constant speed of each mass is found from the distance traveled during one period (the circumference of the circular orbit) divided by the time it takes (one period); that is:

For the speed of m:

$$v_1 = \frac{2\pi r_1}{T}$$

For the speed of M:

$$v_2 = \frac{2\pi r_2}{T}$$

Substitute these expressions for the speeds in the second law equations of each mass. Equation 6.31 for m becomes

$$\frac{GMm}{r^2} = \frac{m}{r_1}\left(\frac{2\pi r_1}{T}\right)^2$$

or

$$\frac{GM}{r^2} = \frac{4\pi^2 r_1}{T^2} \qquad (6.33)$$

Equation 6.32 for M becomes

$$\frac{GMm}{r^2} = \frac{M}{r_2}\left(\frac{2\pi r_2}{T}\right)^2$$

or

$$\frac{Gm}{r^2} = \frac{4\pi^2 r_2}{T^2} \qquad (6.34)$$

Now solve Equations 6.33 and 6.34 for the radius of each orbit:

$$r_1 = \frac{GMT^2}{4\pi^2 r^2} \qquad \text{(orbital radius of } m) \qquad (6.35)$$

and

$$r_2 = \frac{GmT^2}{4\pi^2 r^2} \qquad \text{(orbital radius of M)} \qquad (6.36)$$

The orbital radius of each mass is proportional to the mass of the *other* object. The *smaller* mass m has the *larger* orbital radius about the center of mass point.

The separation distance r is the sum of the orbital radii:

$$r = r_1 + r_2$$

Use Equation 6.35 to substitute for r_1 and Equation 6.36 to substitute for r_2. The separation distance r thus is

$$r = \frac{GMT^2}{4\pi^2 r^2} + \frac{GmT^2}{4\pi^2 r^2}$$

After factoring, we have

$$r = \frac{G(M + m)T^2}{4\pi^2 r^2}$$

Solve for T^2:

$$T^2 = \frac{4\pi^2}{G(M + m)} r^3 \qquad \text{(circular orbits)} \qquad (6.37)$$

The square of the period is proportional to the cube of the center-to-center separation of the two masses. Equation 6.37 is **Newton's form for Kepler's third law** of planetary motion for *circular orbits* about the center of mass point. Notice that this law of Kepler is not an independent law; it is a natural consequence of the general Newtonian theory of motion and the law of universal gravitation. The other two empirical laws of planetary motion discovered by Kepler about 75 years before Newton also are not independent laws, but follow from Newton's general theory of motion and law of gravitation.

Rearrange Equation 6.30 to form the mass ratio:

$$\frac{m}{M} = \frac{r_2}{r_1}$$

As the mass ratio m/M becomes smaller and smaller, the ratio of the radii of the orbits of the two masses also decreases. In particular, as the mass ratio decreases, the orbital radius r_2 of M becomes smaller relative to the orbital radius r_1 of m, as shown in the sequence of sketches in Figure 6.36.

When m is *negligible* compared with M, the center of mass point is essentially located at the position of the center of the more massive object M. In this case m orbits M with an orbital radius $r_1 \approx r$, since $r_2 \approx 0$ m. The orbital motion of M is negligible because the center of mass point is essentially coincident with M itself.

When $m \ll M$, the constant of proportionality in Equation 6.37 becomes approximately

$$\frac{4\pi^2}{G(M + m)} \approx \frac{4\pi^2}{GM} \qquad \text{for } m \ll M$$

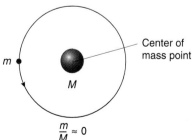

FIGURE 6.36 As the mass ratio m/M becomes smaller, the orbital radius of M decreases.

Then Newton's form of Kepler's third law, Equation 6.37, becomes

$$T^2 = \frac{4\pi^2}{GM} r^3 \qquad m \ll M, \text{ circular orbits} \qquad (6.38)$$

where r is the radius of the circular orbit of m about M. This is the result we derived in Section 6.6 when considering artificial satellites of the Earth, which satisfy the condition $m \ll M$.

Newton's form for Kepler's third law was derived here for the special case of *circular orbits*. When $m \ll M$ and the orbit of m about M is *elliptical*, Equation 6.38 has exactly the same form with r replaced by a, the semimajor axis (see Figure 6.37):*

$$T^2 = \frac{4\pi^2}{GM} a^3 \qquad m \ll M, \text{ elliptical orbit} \qquad (6.39)$$

Equation 6.39 is a more general expression for the Newtonian form of Kepler's third law of planetary motion.

Within the solar system, the mass of any planet m is considerably less than the mass M of the Sun,[†] so that the condition $m \ll M$ is well satisfied. The square of the period of a planet is proportional to the cube of the semimajor axis of its orbit, with a proportionality constant essentially independent of the planetary mass m. That is, for a mass m in an elliptical orbit of semimajor axis a about another much more massive object M, the square of the period T is proportional to the cube of the semimajor axis of the orbit (Equation 6.39). This equation certainly is applicable to *any* artificial satellite orbit about the Earth, or spacecraft or planetary orbits about the Sun, since the condition $m \ll M$ is well satisfied.

6.12 CUSTOMIZED UNITS

For the orbits of the planets or spacecraft about the Sun, the mass m of the planet or spacecraft is very small compared with the mass $M_\odot$ of the Sun. Equation 6.39 thus takes the form

$$T^2 = \frac{4\pi^2}{GM_\odot} a^3 \qquad m_{\text{orbiting object}} \ll M_\odot \qquad (6.40)$$

---

*We leave the proof of this statement to more advanced courses in dynamics.
[†]The most massive planet (Jupiter) has a mass that is somewhat less than 10^{-3} times the mass of the Sun, so that $m/M \approx 10^{-3}$. The mass of the Earth is on the order of 10^{-6} times the mass of the Sun.

where T is the period of the planet or spacecraft in its orbit about the Sun, a is the semimajor axis of the orbit, and $M_\odot$ is the mass of the Sun.

The semimajor axis of the orbit of the Earth about the Sun is defined to be one **astronomical unit** (AU):

$$1 \text{ AU} \approx 1.496 \times 10^{11} \text{ m}$$

The period of the Earth in its orbit is, of course, one year.[‡] If we measure the period, not in seconds, but in years, and measure the semimajor axis of the planetary orbit in astronomical units rather than meters, then Equation 6.40 has the following form when applied to the Earth:

$$(1 \text{ y})^2 = \frac{4\pi^2}{GM_\odot} (1 \text{ AU})^3$$

Thus

$$\frac{4\pi^2}{GM_\odot} = 1 \text{ y}^2/\text{AU}^3$$

Newton's form for Kepler's third law (Equation 6.40) then takes on the simple form:

$$T^2 = (1 \text{ y}^2/\text{AU}^3)a^3 \qquad (6.41)$$

T *must* be in y

a *must* be in AU

for *any* object in orbit about the Sun.

This legitimate trick of creating a simplified equation using *customized units* is employed in various branches of the sciences for equations that are frequently used. But once again, in using such equations (indeed, all equations), realize that you must be careful to know and consider any strings attached to their use.

PROBLEM-SOLVING TACTIC

6.2 **Be sure to express the period T in years and the semimajor axis in astronomical units when using Equation 6.41.** The proportionality constant on the right-hand side of Equation 6.41 is written explicitly as $1 \text{ y}^2/\text{AU}^3$ to remind you of the appropriate units to use. Equation 6.41 is *not valid* in SI units. Use Equation 6.39 when using SI units with T in s and a in m.

EXAMPLE 6.8

The orbital period of Jupiter is 11.87 y. Find the semimajor axis of the orbit of Jupiter.

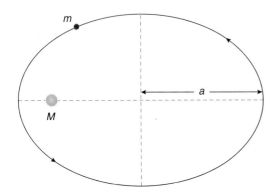

FIGURE 6.37 Elliptical orbit.

[‡] The true orbital period of the Earth about the Sun is the *sidereal year*, equal to approximately 365.256 37 days. The *calendar year*, also known as the *tropical year* or the year of the seasons, is slightly shorter and is approximately 365.242 199 days. We neglect the small difference between the two years in this text except where explicitly mentioned.

The largest planet in the solar system is Jupiter.

Solution

Method 1

Use the customized version of Kepler's third law, Equation 6.41:

$$T^2 = (1 \text{ y}^2/\text{AU}^3)a^3$$

where T is in y and a is in AU. Substituting $T = 11.87$ y, you find a in AU:

$$(11.87 \text{ y})^2 = (1 \text{ y}^2/\text{AU}^3)a^3$$

Solving for a,

$$a = 5.204 \text{ AU}$$

If desired, this result can be converted to meters using the conversion factor between AU and meters

$$1 \text{ AU} = 1.496 \times 10^{11} \text{ m}$$

Hence

$$a = (5.204 \text{ AU}) \times (1.496 \times 10^{11} \text{ m/AU})$$
$$= 7.785 \times 10^{11} \text{ m}$$

Method 2

Use Equation 6.39:

$$T^2 = \frac{4\pi^2}{GM} a^3$$

which necessitates SI units: T in *seconds* and M equal to the mass of the Sun in kg:

$$T = 11.87 \text{ y}$$
$$= 3.746 \times 10^8 \text{ s}$$
$$G = 6.67 \times 10^{-11} \text{ N·m}^2/\text{kg}^2$$
$$M_\odot = 1.99 \times 10^{30} \text{ kg}$$

The semimajor axis emerges in meters. Making the substitutions, you get

$$(3.746 \times 10^8 \text{ s})^2$$
$$= \frac{4\pi^2 a^3}{(6.67 \times 10^{-11} \text{ N·m}^2/\text{kg}^2)(1.99 \times 10^{30} \text{ kg})}$$

Solving for a,

$$a = 7.78 \times 10^{11} \text{ m}$$

This is more awkward computationally, but equally legitimate. The result can be converted to AU, if desired, using the conversion factor

$$1 \text{ AU} = 1.496 \times 10^{11} \text{ m}$$

STRATEGIC EXAMPLE 6.9

A far-out spacecraft is in an elliptical orbit about the Sun with the perihelion point at the orbit of the Earth and the aphelion point at the orbit of Saturn, as shown in Figure 6.38. Such orbits are used by spacecraft making direct trips to the outer planets of the solar system. Similar orbits are used to visit the inner planets, but the aphelion point then is at the orbit of the Earth and the perihelion point is at the orbit of the inner planet. Once the spacecraft is inserted into its orbit (which necessitates a large rocket at launch), the spacecraft proceeds unpowered along its orbit of the Sun.

The planet Saturn and its spectacular ring system.

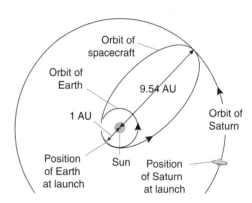

FIGURE 6.38

Consider the orbits of the Earth and Saturn to be circular. The radius of the orbit of Saturn is **9.54 AU**.

a. Find the time it takes for the spacecraft to go from the orbit of the Earth to that of Saturn.
b. Determine the eccentricity of the orbit of the spacecraft.

Solution

a. The major axis of the orbit of the spacecraft is

$$9.54 \text{ AU} + 1.00 \text{ AU} = 10.54 \text{ AU}$$

The semimajor axis a of the orbit of the spacecraft is then

$$a = \frac{10.54 \text{ AU}}{2}$$
$$= 5.27 \text{ AU}$$

Use Equation 6.41:

$$T^2 = (1 \text{ y}^2/\text{AU}^3)a^3$$
$$= (1 \text{ y}^2/\text{AU}^3)(5.27 \text{ AU})^3$$

Solving for T,

$$T = 12.1 \text{ y}$$

To travel from the Earth to Saturn, the spacecraft only traverses *half* its orbit, from one end of its semimajor axis to the other end. Hence the time of flight to Saturn is *half* the period of the spacecraft in its complete orbit, or

$$t_{\text{flight}} = 6.05 \text{ y}$$

This problem also illustrates the reason behind *launch windows*. When the spacecraft arrives at the orbit of Saturn, it would be nice if Saturn also was in the vicinity; it would be quite embarrassing if this were not the case and the planet was located elsewhere in its orbit! Therefore the spacecraft must be launched at such a time that when it arrives at the orbit of Saturn, the planet Saturn itself also is at that point in its orbit.

b. To determine the eccentricity ε of the orbit of the spacecraft, use the result of Example 6.4, knowing that the perihelion distance of the orbit of the spacecraft is 1.00 AU and its aphelion distance is 9.54 AU:

$$\frac{r_{\text{aph}}}{r_{\text{peri}}} = \frac{1 + \varepsilon}{1 - \varepsilon}$$

Solving for ε, you obtain

$$\varepsilon = 0.810$$

6.13 THE GRAVITATIONAL FIELD

It is inconceivable, that inanimate brute Matter should, without the Mediation of something else, which is not material, operate upon, and affect other Matter, without mutual Contact. . .

Isaac Newton (1642–1727)*

Newton clearly was troubled by the idea of the gravitational force magically acting over vast distances without any apparent physical mechanism for conveying the force between the masses. The force may as well have been transmitted by angels—or the devil. Such action-at-a-distance never really was resolved by him. The problem simply was ignored, for no conceptual framework existed to approach the problem adequately. Nonetheless, one could solve many problems of interest without worrying about such subtleties.

During the 19th century an idea emerged that attempted to explain action-at-a-distance via an intermediary known as a **field**. The concept of a field still is very useful for conveying the action of each of the fundamental forces (see Section 5.2). Thus it is important to consider the concept of a **gravitational field** in some detail; it serves as a paradigm for the treatment of the other fundamental forces, such as the electric and magnetic forces.

Consider the gravitational interaction between two point-like particles of mass m and M (or particles with appropriate spherical symmetry). Let the mass $m \ll$ M.[†] Mass m experiences a gravitational force from the presence of mass M located a

distance r away, as shown in Figure 6.39. The gravitational force of M on m is of magnitude

$$\frac{GmM}{r^2}$$

and is directed, of course, toward M.

Suppose we simply rearrange the expression for the magnitude of the gravitational force on m as

$$m\left(\frac{GM}{r^2}\right)$$

The stuff in the parentheses can be attributed to just M itself and the space surrounding M. In other words, we imagine M influencing space itself, setting up a "condition of space" or field at each point. Now if m is placed at a point in space, the *field of M at the point where m is placed* acts on m to produce the force on m. The field that M establishes at all points in space, by virtue of its mass, is called the gravitational field $\vec{g}$. It conveys the gravitational force. In other words, the very existence of M sets up or establishes the gravitational field at *all* points in space *whether the other mass m is present or not*. The gravitational field established by a point mass M at any point (say a distance r away from M) is a vector directed toward M and is of magnitude

$$\frac{GM}{r^2}$$

Let $\hat{\mathbf{r}}$ be a dimensionless unit vector, analogous to the familiar Cartesian $\hat{\mathbf{i}}$, $\hat{\mathbf{j}}$, and $\hat{\mathbf{k}}$ unit vectors, but pointing *radially away* from M, as shown in Figure 6.40. Then the gravitational field vector established in space by a point mass M is

$$\vec{g}(r) = -\frac{GM}{r^2}\hat{\mathbf{r}} \tag{6.42}$$

If mass m now is placed at a point in space where the gravitational field is $\vec{g}$, m experiences a force given by the product of m times the value of the gravitational field at the location where m is placed. That is,

$$\vec{F}_{\text{grav force on } m} = m[\text{gravitational field wherever } m \text{ is placed}]$$
$$= m\vec{g} \tag{6.43}$$

Equation 6.42 for the gravitational field of a point mass M is *not* the only kind of gravitational field. The specific analytical form for the gravitational field established by a mass M depends on how the mass M is shaped and distributed within its volume.

FIGURE 6.39 The gravitational force of M on m.

FIGURE 6.40 The unit vector $\hat{\mathbf{r}}$ points radially away from M, while the gravitational field of M points toward it.

---

*Letter to Richard Bentley, 25 February 1692(3), *Isaac Newton's Papers and Letters on Natural Philosophy*, edited by I. Bernard Cohen (Cambridge, Massachusetts, 1958), pages 300–309; quote is on page 302.

[†]This condition implies that the presence of m does *not* (via gravitational forces) affect the distribution or geometrical shape of the mass M, when M is not a pointlike mass.

The expression for $\vec{\mathbf{g}}$ in Equation 6.42 is just a specific example of a gravitational field: that established by a pointlike mass M. Other specific types of gravitational fields are listed in Table 6.1. *Whatever* the specific form of the gravitational field $\vec{\mathbf{g}}$ at a certain point in space, any mass m, placed at the point where the value of the gravitational field is $\vec{\mathbf{g}}$, experiences a gravitational force given by Equation 6.43. The gravitational field $\vec{\mathbf{g}}$ into which m is placed is established by *other* masses, not by m itself.

A few things to note:

1. Rearrange Equation 6.43 to find:

$$\vec{\mathbf{g}}_{\text{at the position of } m} = \frac{\vec{\mathbf{F}}_{\text{on } m}}{m}$$

Hence the gravitational field can be imagined as the *force per unit mass* (the force on one kilogram) at a given point in space.

Notice the similarity between Equation 6.43 and Newton's second law of motion when the gravitational force of M on m is the *only* force acting on m: $\vec{\mathbf{F}}_{\text{on } m} = m\vec{\mathbf{a}}$. The field has the same magnitude and direction as the acceleration caused by the total gravitational force on m when m is placed at the point in question.

Hence the gravitational field at a point in space is the same as the acceleration due to gravity at that location.

When near the surface of the Earth, the gravitational field has a magnitude 9.81 m/s² directed vertically down, but at other locations in space, the magnitude of the gravitational field of the Earth (or that due to other masses) is different.

PROBLEM-SOLVING TACTIC

6.3 Do not confuse the magnitude of the gravitational field $\vec{\mathbf{g}}(r)$ at any location in space with g, the magnitude of the local acceleration due to gravity near the surface of the Earth. The field $\vec{\mathbf{g}}(r)$ has a magnitude $g = 9.81$ m/s² only when near the surface of the Earth.

The gravitational field is a vector quantity since it is a force per unit mass (or, equivalently, the local acceleration due to gravity).

2. Since gravitational *forces* follow a principle of linear superposition, gravitational *fields* also must obey the same type of superposition rule because the field is a force per unit mass. Thus the total gravitational field established at some point in space by a number of masses is the *vector sum* of the individual gravitational fields at that point established by each mass acting individually, as shown in Figure 6.41.

The gravitational field established in space by an *extended mass* is found by dividing the extended mass into small pointlike differential masses dM (see Figure 6.42). The field $d\vec{\mathbf{g}}$ established at a point P in space by this differential mass dM is

$$d\vec{\mathbf{g}} = -\frac{G\,dM}{r^2}\,\hat{\mathbf{r}} \tag{6.44}$$

where r is the distance between dM and P and $\hat{\mathbf{r}}$ is a unit vector pointing radially away from dM.

TABLE 6.1 Gravitational Fields Established in Space by Various Mass Distributions

Mass distribution	Location	Gravitational field
Point mass M	gravitational field a distance r away from M	$\vec{\mathbf{g}}(r) = \dfrac{-GM}{r^2}\,\hat{\mathbf{r}}$
Uniform spherical shell	gravitational field *inside* the shell	$\vec{\mathbf{g}}(r) = 0$ (m/s²)
	gravitational field *outside* the shell a distance r away	$\vec{\mathbf{g}}(r) = \dfrac{-GM}{r^2}\,\hat{\mathbf{r}}$
Uniform sphere (either a uniform or a spherically symmetric distribution of mass)	gravitational field *outside* the sphere a distance r away	$\vec{\mathbf{g}}(r) = \dfrac{-GM}{r^2}\,\hat{\mathbf{r}}$
Uniform sphere	gravitational field *inside* the sphere a distance r from the center	$\vec{\mathbf{g}}(r) = \dfrac{-GM\,r}{R^3}\,\hat{\mathbf{r}}$
Gravitational field	near the surface of the Earth	(a constant vector) $g = 9.81$ m/s² directed vertically down
Gravitational field of a mass ring	a distance z from the plane of the ring along the symmetry axis of the ring	$\vec{\mathbf{g}}(z) = \dfrac{-GM\,z}{(R^2 + z^2)^{3/2}}\,\hat{\mathbf{k}}$

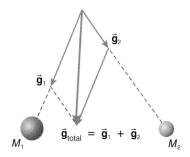

FIGURE 6.41 The principle of superposition applies to gravitational fields.

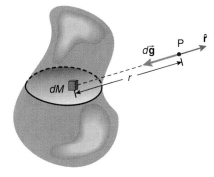

FIGURE 6.42 A differential mass establishes a differential gravitational field at a point in space.

The total gravitational field established by the extended mass at point P in space then is found by integrating Equation 6.44 over the physical extent of the mass distribution M:

$$\vec{g} = -G \int_{\text{mass distribution}} \frac{dM}{r^2} \hat{r} \qquad (6.45)$$

This integral is a vector integration and care must be taken when executing it. Example 6.12 illustrates how to perform such integrations. The gravitational field established by an extended mass is a sophisticated application of the principle of superposition. Since m cannot exert a gravitational force on itself, if m is placed *in* a gravitational field, the field affecting m is *not* the field established by m itself, but is the field established by all *other* masses.

3. The influence of a mass M *on* another mass m is thought of as being conveyed by the field. If M changes its location at some instant, the change in the field of M at a particular point in space occurs at a later time. Einstein's theory of general relativity predicts that such changes in the gravitational field propagate through space at the speed of light. Thus if, say, the Sun somehow disappeared at some instant, the gravitational field of the Sun at the position of the Earth would be unaffected for eight minutes, the light-travel-time from the Sun to the Earth, before ceasing. As mentioned earlier, we will not be concerned in this text with such propagation delays, but you should be aware that they exist.

4. If it looks as if we have merely taken one abstract idea for conveying the force, action-at-a-distance, and substituted the equally abstract idea of a field, you are right to some extent. The field solves the problem of instantaneous propagation, but it still does not answer the question of what gravity *is* or why it acts this way. But the field idea is a very useful abstraction.

One of the reasons for introducing the idea of a gravitational field in the space surrounding a mass M is that it enables us to talk about the gravitational influence of M itself without the bother of having another mass (say m) in the picture. In other words, with the gravitational field, we can begin to talk about the gravitational characteristics of M all by itself. A problem that initially had *two* masses (M and m) and *two* forces (of M on m and of m on M) now just has *one* mass (say M) and its very own gravitational field.*

QUESTION 7

Is it possible to measure directly the gravitational field, or is it determined only from its action on a mass placed in the field?

EXAMPLE 6.10

a. Find the gravitational field established by the Earth at a point in space located a distance from the center of the Earth equal to twice the Earth's radius.

_____
*Of course, m produces its own gravitational field too, which acts on the other mass M.

b. What is the acceleration due to gravity at this location?
c. What is the gravitational force on a 1.00 kg mass at this location?
d. What is the gravitational force of the Earth on a 500 kg space-cow chewing its cud at this unlikely location in the field?

Solution

a. The Earth is approximately spherical in shape, and so, from Table 6.1, the gravitational field established by the Earth in space is

$$\vec{g} = -\frac{GM}{r^2} \hat{r}$$

where M is the mass of the Earth (5.98×10^{24} kg) and r is the distance from the center of the Earth to the point where we want the field, here two Earth radii:

$$r = 2(6.37 \times 10^6 \text{ m})$$
$$= 12.7 \times 10^6 \text{ m}$$

Substituting into the expression for the field,

$$\vec{g} = -\frac{(6.67 \times 10^{-11} \text{ N} \cdot \text{m}^2/\text{kg}^2)(5.98 \times 10^{24} \text{ kg})}{(12.7 \times 10^6 \text{ m})^2} \hat{r}$$
$$= (-2.47 \text{ m/s}^2) \hat{r}$$

b. The gravitational field *is* the acceleration due to gravity at this location, and so the acceleration due to gravity at this location is

$$(-2.47 \text{ m/s}^2) \hat{r}$$

c. The gravitational force on a mass m at a location where the field is $\vec{g}$ is

$$\vec{F} = m\vec{g}$$
$$= (1.00 \text{ kg})(-2.47 \text{ m/s}^2) \hat{r}$$
$$= (-2.47 \text{ N}) \hat{r}$$

and is directed toward the center of the Earth. Notice that the gravitational force *on one kilogram* is numerically equal to the value of the field; thus the field can be considered to be the gravitational force per kilogram at the point.

d. The gravitational force of the Earth on a 500 kg space-cow at this location is the product of the mass of the cow and the gravitational field of the Earth at this point in space:

$$\vec{F} = m\vec{g}$$
$$= (500 \text{ kg})(-2.47 \text{ m/s}^2) \hat{r}$$
$$= (-1.24 \times 10^3 \text{ N}) \hat{r}$$

STRATEGIC EXAMPLE 6.11

a. What is the total gravitational field established by both the Earth and Moon at the point P in Figure 6.43?
b. What is the magnitude of the total gravitational field at this point?
c. What is the acceleration of a 1200 kg spacecraft at this location? What is the magnitude of this acceleration?

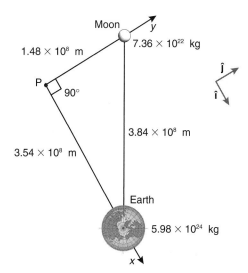

FIGURE 6.43

FIGURE 6.44

d. What is the total gravitational force on the spacecraft at this location? What is the magnitude of the total gravitational force on the spacecraft?

Solution

a. Consider both the Earth and Moon to be spherical masses. Use the principle of superposition to find the total field by first finding, individually, the field established by the Moon at this location (as if the Earth were not present) and the field established by the Earth at this location (as if the Moon were not present).

The gravitational field established by the Moon at this location is of magnitude

$$g_{Moon} = \frac{GM_{Moon}}{r^2}$$

where r is the distance between the center of the Moon and the point P, here 1.48×10^8 m. Substituting numerical values, you find

$$g_{Moon} = \frac{(6.67 \times 10^{-11} \text{ N·m}^2/\text{kg}^2)(7.36 \times 10^{22} \text{ kg})}{(1.48 \times 10^8 \text{ m})^2}$$
$$= 2.24 \times 10^{-4} \text{ m/s}^2$$

This field is directed toward the center of the Moon. Use the coordinate system indicated in Figure 6.44. The gravitational field established by the Moon at the point P is

$$\vec{g}_{Moon} = (2.24 \times 10^{-4} \text{ m/s}^2)\,\hat{j}$$

The magnitude of the gravitational field established by the Earth at the point P also is of the form

$$g_{Earth} = \frac{GM_{Earth}}{r^2}$$

where now r is the distance from the center of the Earth to P, here 3.54×10^8 m. Making the substitutions,

$$g_{Earth} = \frac{(6.67 \times 10^{-11} \text{ N·m}^2/\text{kg}^2)(5.98 \times 10^{24} \text{ kg})}{(3.54 \times 10^8 \text{ m})^2}$$
$$= 3.18 \times 10^{-3} \text{ m/s}^2$$

This field is directed toward the center of the Earth. Use the coordinate system in Figure 6.44; the gravitational field established by the Earth at point P is

$$\vec{g}_{Earth} = (3.18 \times 10^{-3} \text{ m/s}^2)\,\hat{i}$$

The *total* gravitational field at the point P is the *vector sum* of the two fields, as indicated in Figure 6.45:

$$\vec{g}_{total} = \vec{g}_{Earth} + \vec{g}_{Moon}$$
$$= (3.18 \times 10^{-3} \text{ m/s}^2)\,\hat{i} + (0.22 \times 10^{-3} \text{ m/s}^2)\,\hat{j}$$

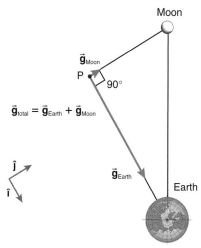

FIGURE 6.45

b. The magnitude of the total field at P is the magnitude of the vector $\vec{g}_{total}$:

$$g_{total} = [(3.18 \times 10^{-3} \text{ m/s}^2)^2 + (0.22 \times 10^{-3} \text{ m/s}^2)^2]^{1/2}$$
$$= 3.19 \times 10^{-3} \text{ m/s}^2$$

c. The total gravitational field at P *is* the acceleration due to gravity at this location:

$$\vec{g}_{total} = (3.18 \times 10^{-3} \text{ m/s}^2)\,\hat{i} + (0.22 \times 10^{-3} \text{ m/s}^2)\,\hat{j}$$

The magnitude of the acceleration of the spacecraft is the magnitude of the field:

$$g_{\text{total}} = 3.19 \times 10^{-3} \text{ m/s}^2$$

The acceleration is independent of the mass of the spacecraft.

d. The gravitational force on the spacecraft when it is at this location is found using Equation 6.43:

$$\vec{F} = m\vec{g}_{\text{total}}$$
$$= (1200 \text{ kg})[(3.18 \times 10^{-3} \text{ m/s}^2)\hat{i} + (0.22 \times 10^{-3} \text{ m/s}^2)\hat{j}]$$
$$= (3.82 \text{ N})\hat{i} + (0.26 \text{ N})\hat{j}$$

The magnitude of this force is

$$F = [(3.82 \text{ N})^2 + (0.26 \text{ N})^2]^{1/2}$$
$$= 3.83 \text{ N}$$

The magnitude of the force also could be found using

$$F = mg_{\text{total}}$$
$$= (1200 \text{ kg})(3.19 \times 10^{-3} \text{ m/s}^2)$$
$$= 3.83 \text{ N}$$

STRATEGIC EXAMPLE 6.12

A mass M is distributed uniformly in the shape of a circular ring of radius R, as shown in Figure 6.46. Such mass rings are used to model the individual rings of the planets Jupiter, Saturn, Uranus, and Neptune.

FIGURE 6.46

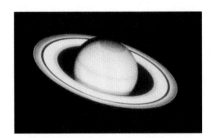

The concentric ring system of Saturn consists of a vast collection of orbiting rocks. The rings are not solid.

a. Find the gravitational field established by the ring along the symmetry axis of the ring at a distance z from the plane of the ring.

b. What is the gravitational field at the center of the ring?

Solution

a. The ring is an extended mass distribution; you need to use Equation 6.45 and perform a vector integration. On every meter of the circumference of the ring there is a mass λ, where

$$\lambda = \frac{M}{2\pi R}$$

The mass per meter λ is known as a linear mass density. A small differential length ds of the circumference of the ring, therefore, has a mass dM equal to the product of ds times the linear mass density λ:

$$dM = \lambda \, ds$$

This pointlike differential mass produces a gravitational field at the point P that is of magnitude

$$dg = G \frac{dM}{r^2}$$

and is directed toward the mass dM, as shown in Figure 6.47.

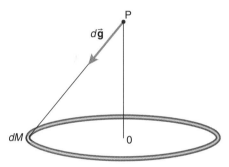

FIGURE 6.47

Another similar differential mass on the opposite side of the ring produces a similar field at P directed toward it (see Figure 6.48). Notice in Figure 6.48 that only the components of these fields parallel to $-\hat{k}$ survive a vector addition of the fields; the components parallel to the plane of the ring vector sum to zero. Hence the portion of the field produced by dM at P that survives the vector summation (integration) will be directed along $-\hat{k}$ and is

$$d\vec{g} = \left(G \frac{dM}{r^2} \cos \theta \right)(-\hat{k})$$

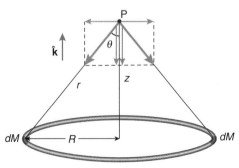

FIGURE 6.48

Substitute for dM and integrate over the extent of the ring. This is the total gravitational field of the ring at P:

$$\vec{\mathbf{g}} = G \int_{\text{ring}} \left(\frac{\lambda \, ds}{r^2} \cos \theta \right) (-\hat{\mathbf{k}})$$

When you integrate around the ring mass distribution, every mass element on the ring is the same distance r from the point P. Likewise $\cos \theta$ also is constant, since the angle θ is unchanged as you go around the ring. The linear mass density λ is constant too. Hence the integration becomes

$$\vec{\mathbf{g}} = -\frac{G\lambda \cos \theta}{r^2} \int_{\text{ring}} ds \, \hat{\mathbf{k}}$$

The integral of ds around the ring is the circumference of the ring, $2\pi R$. Hence the gravitational field of the ring at P is

$$\vec{\mathbf{g}} = -\frac{G\lambda \cos \theta}{r^2} 2\pi R \hat{\mathbf{k}}$$

Now substitute for $\lambda = M/(2\pi R)$. With the Pythagorean theorem you have

$$r^2 = R^2 + z^2$$

and geometrically (see Figure 6.48)

$$\cos \theta = \frac{z}{r} = \frac{z}{(R^2 + z^2)^{1/2}}$$

After these substitutions, the gravitational field established by the ring at P is found to be

$$\vec{\mathbf{g}}(z) = -\frac{GMz}{(R^2 + z^2)^{3/2}} \hat{\mathbf{k}} \tag{1}$$

Notice that for $z > 0$ m the field is directed toward the ring (parallel to $-\hat{\mathbf{k}}$). For $z < 0$ m the field *also* is directed toward the ring (parallel to $+\hat{\mathbf{k}}$).

b. There are two ways to find the gravitational field at the center of the ring.

Method 1
At the center of the ring where $z = 0$ m, equation (1) indicates the field is zero.

Method 2
Using symmetry considerations, you also can deduce that the field will be zero at the center of the ring since mass elements on opposite sides of the ring produce gravitational fields in opposing directions, as shown in Figure 6.49.

A mass placed at the center of the ring will experience zero gravitational force, since the field is zero at that location. Problem 67 asks you to show that the maximum magnitude of the field along the symmetry axis is at the point where $|z| = R/\sqrt{2}$.

6.14 THE FLUX OF A VECTOR*

The gravitational field has an interesting property associated with the masses that establish the field. To investigate this property of the gravitational field, it is necessary to introduce a new physical concept associated with a vector: its **flux**.

A differential surface area dS is a tiny area of differential size, indicated schematically in Figure 6.50. We use the letter S for the area here, meaning Surface. Since the differential area is arbitrarily small (the case with all differentials), the area is imagined to be *locally flat*, even if it is part of a larger curved surface.

We occasionally make a *vector* out of such a tiny differential area. Take the differential area as the *magnitude* of the differential area vector. In the plane of the differential area, there are infinitely many different directions that might be chosen and it is impossible to single out a unique direction for the area vector. But two unique directions are those lying along the perpendicular to the locally flat differential area, as shown in Figure 6.51. Either one of these perpendicular directions could serve as the direction of the differential area vector. However, if the differential area is part of a larger surface that also is a **closed surface**, then the convention is to choose the direction pointing to the *outside* of the enclosed volume, as in Figure 6.52.

What is a closed surface? Imagine submerging such a surface in water. A closed surface will not get wet on the inside; there are no holes. So, for example, an empty but sealed milk carton has a closed surface. An empty milk carton without its cap is not a closed surface. If the sealed carton has even a pin prick in it, the surface is not closed. A doughnut has a surface that is closed even though the doughnut is said to have a hole. We will encounter closed surfaces in a number of contexts.

FIGURE 6.50 A differential area dS.

FIGURE 6.51 Two directions are associated with the perpendicular to the differential area.

FIGURE 6.49

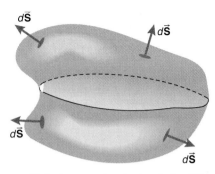

FIGURE 6.52 If the larger surface is closed, the differential area vector points to the outside of the enclosed volume.

Thus a differential area vector $d\vec{S}$ is locally flat, has a magnitude equal to the differential area in m², and is directed perpendicular to the surface in an outward direction if the larger surface is closed. If the larger surface is not closed, we are free to choose the direction of $d\vec{S}$ to be along either direction of the perpendicular to the differential surface area.

Now we tackle the idea of the flux of a vector, say, the flux of $\vec{g}$, the gravitational field vector, where $\vec{g}$ exists at the location of the differential area vector $d\vec{S}$. We first define what is meant by the flux in a mathematical sense and then explore its meaning to get a feeling for the terminology and a mental picture of what the flux represents.

The differential flux $d\Phi$ of the vector $\vec{g}$ through the differential area dS is defined to be the scalar product of $\vec{g}$ with $d\vec{S}$:

$$d\Phi \equiv \vec{g} \cdot d\vec{S} \qquad (6.46)$$

The value of $\vec{g}$ at the location of the differential area is used in this expression. We describe Equation 6.46 in the following way: the differential flux of the vector $\vec{g}$ *through* the differential area is $d\Phi$.

Note that the flux is a scalar quantity.

Use the definition of the scalar product of two vectors; the differential flux through the differential area is

$$d\Phi = g \, dS \cos \theta$$

where θ is the angle between the vector $\vec{g}$ and the differential area vector $d\vec{S}$, as shown in Figure 6.53. The differential flux $d\Phi$ may be positive, negative, or zero depending on the orientation of $\vec{g}$ relative to $d\vec{S}$.

Let's look at a few situations to get a feeling for the meaning of the flux. The word flux implies flow of something through the differential area dS.* While this has implications of fluids moving through the differential area dS rather than gravitational fields, the concept of flux finds wide use particularly with field vectors such as the gravitational field vector $\vec{g}$.†

If the angle between the vectors $\vec{g}$ and $d\vec{S}$ is 90°, as in Figure 6.54, then the cosine is zero and the differential flux of $\vec{g}$ through dS is zero:

$$d\Phi = g \, dS \cos 90° = 0 \ \text{m}^3/\text{s}^2$$

In this case the arrow representing the vector $\vec{g}$ does not pass through, or "thread," the little area if you imagine the differential area to be like the eye of a needle.

If the angle between the vectors $\vec{g}$ and $d\vec{S}$ is 0°, as in Figure 6.55, then the differential flux of the vector $\vec{g}$ through dS is

$$d\Phi = g \, dS \cos 0° = g \, dS$$

In this case, the arrow representing the vector $\vec{g}$ threads the little area dS "head-on." This is the geometry used when one really does try to thread a needle with a piece of thread.

If the angle between $\vec{g}$ and $d\vec{S}$ is somewhere between 0° and 90°, as in Figure 6.56, the differential flux has a value between its maximum value ($g \, dS$) and zero.

The largest negative value of the differential flux occurs when the angle between the two vectors is 180°, as in Figure 6.57. For this case,

$$d\Phi = \vec{g} \cdot d\vec{S} = g \, dS \cos 180°$$
$$= -g \, dS$$

---

*The word flux comes from the Lain *fluxus*, to flow. It originally was associated with the amount of fluid flowing per second through an area placed at various orientations to the direction of the fluid flow, but the concept has been generalized to encompass other vectors as well.
†We also will use the flux concept later with electric and magnetic fields.

FIGURE 6.54 Zero differential flux of $\vec{g}$ through dS.

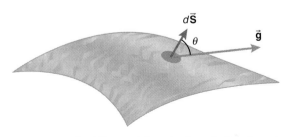

FIGURE 6.53 The differential flux of $\vec{g}$ through dS is the scalar product $\vec{g} \cdot d\vec{S}$.

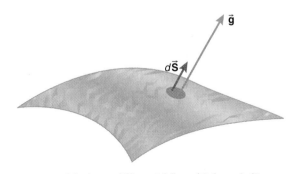

FIGURE 6.55 Maximum differential flux of $\vec{g}$ through dS.

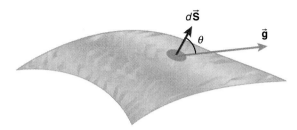

FIGURE 6.56 An intermediate case.

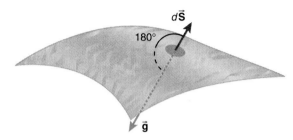

FIGURE 6.57 Largest negative differential flux of $\vec{g}$ through dS.

In this case the vector $\vec{g}$ threads the little area dS backward, or opposite to the direction of the differential area vector $d\vec{S}$.

Hence the differential flux of $\vec{g}$ through the differential area either can be positive, negative, or zero depending on the relative orientation of the vector $\vec{g}$ to $d\vec{S}$. The flux is a measure of the extent to which the vector $\vec{g}$ threads or flows through the differential area.

To find the total flux Φ of the vector $\vec{g}$ through a finite area S, we must integrate the differential flux over the entire area S:

$$\Phi = \int_S d\Phi = \int_S \vec{g} \cdot d\vec{S} = \int_S g\, dS \cos\theta \qquad (6.47)$$

This integration may be difficult or easy depending on the shape of the surface S and the analytical form of the field. As the differential area dS scans a large surface, the value of the vector $\vec{g}$ may change and the angle $\vec{g}$ makes with the local $d\vec{S}$ also may change, so that the integration can be difficult. Certain types of fields may have a high degree of symmetry, dictating surfaces through which the procedure of flux integration is tractable. We often *choose* the surface over which to calculate the flux, and it is best to make the job as easy as possible by choosing the surfaces dictated by the symmetry of the problem. The choice is accomplished by trying to select larger areas over which the vector $\vec{g}$ has constant magnitude and maintains a fixed orientation with respect to all the differential areas on the surface.

EXAMPLE 6.13 _____

a. Find the flux of the local gravitational field near the surface of the Earth through each of the five surfaces of the inclined plane shown in Figure 6.58.
b. What is the total flux through the entire closed surface?

Solution

a. Take $\hat{\jmath}$ to be vertically upward. Then the constant gravitational field near the surface of the Earth is

$$\vec{g} = -g\hat{\jmath}$$

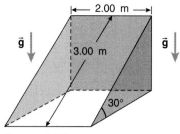

FIGURE 6.58

where g is 9.81 m/s² near the surface of the Earth. Since the larger surface is closed, all the differential area vectors must be chosen (by convention) to be directed to the outside of the closed surface.

Note that on each of the three vertical surfaces, every differential area vector $d\vec{S}$ is perpendicular to the direction of the field, as shown in Figure 6.59. Hence, at every $d\vec{S}$ on each vertical surface,

$$\vec{g} \cdot d\vec{S} = 0 \ \text{m}^3/\text{s}^2$$

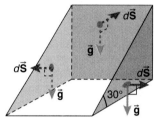

FIGURE 6.59

Hence there is no flux of $\vec{g}$ through the vertical parts of the larger closed surface:

$$\Phi_{\text{vert}} = 0 \ \text{m}^3/\text{s}^2$$

Note also that the lines and arrows representing $\vec{g}$ nowhere thread these vertical surfaces.

All the differential area vectors on the inclined surface are perpendicular to the inclined plane (see Figure 6.60). Since the larger area in question is closed, you have to choose the outward direction for all the $d\vec{S}$ on this inclined surface. Use the coordinate system indicated in Figure 6.60 to write the differential area vector as

$$d\vec{S} = dS\,[(-\sin\theta)\,\hat{\imath} + (\cos\theta)\,\hat{\jmath}]$$

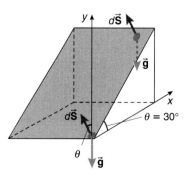

FIGURE 6.60

According to Equation 6.47, the flux of $\vec{\mathbf{g}}$ through the area S is

$$\begin{aligned}\Phi_{\text{incl}} &= \int_S \vec{\mathbf{g}} \cdot d\vec{\mathbf{S}} \\ &= \int_S (-g\hat{\mathbf{j}}) \cdot dS\,[(-\sin\theta)\,\hat{\mathbf{i}} + (\cos\theta)\,\hat{\mathbf{j}}] \\ &= -\int_S g\,dS\cos\theta\end{aligned}$$

Since g and $\cos\theta$ are constant over the entire area S, the flux becomes

$$\begin{aligned}\Phi_{\text{incl}} &= -gS\cos\theta \\ &= (-9.81 \text{ m/s}^2)(6.00 \text{ m}^2)\cos 30.0° \\ &= -51.0 \text{ m}^3/\text{s}^2\end{aligned}$$

The negative value of the flux indicates that the gravitational field vector here passes into the larger closed surface through the inclined surface.

On the bottom surface, all the differential area vectors are written as

$$d\vec{\mathbf{S}} = -dS\,\hat{\mathbf{j}}$$

since the outward direction from the closed surface must be chosen (see Figure 6.61). The flux of $\vec{\mathbf{g}}$ through this area then is

$$\begin{aligned}\Phi_{\text{btm}} &= \int_S \vec{\mathbf{g}} \cdot d\vec{\mathbf{S}} \\ &= \int_S (-g\hat{\mathbf{j}}) \cdot (-dS\,\hat{\mathbf{j}}) \\ &= \int_S g\,dS\end{aligned}$$

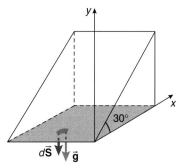

FIGURE 6.61

Since g is constant over the entire lower surface, the flux through this surface becomes

$$\begin{aligned}\Phi_{\text{btm}} &= gS \\ &= (9.81 \text{ m/s}^2)[(2.00 \text{ m})(3.00 \text{ m})\cos 30.0°] \\ &= 51.0 \text{ m}^3/\text{s}^2\end{aligned}$$

The positive result indicates that the gravitational field threads out of the larger closed surface through the bottom surface.

b. The total flux of the gravitational field is the sum of the fluxes through each of the parts of the larger closed surface:

$$\begin{aligned}\Phi_{\text{total}} &= \Phi_{\text{incl}} + \Phi_{\text{vert}} + \Phi_{\text{btm}} \\ &= -51.0 \text{ m}^3/\text{s}^2 + 0 \text{ m}^3/\text{s}^2 + 51.0 \text{ m}^3/\text{s}^2 \\ &= 0 \text{ m}^3/\text{s}^2\end{aligned}$$

Although the gravitational field vectors enter the closed surface through the inclined surface, they depart through the bottom, so that the total flux of the gravitational field of the Earth through the closed surface is zero.

With some reflection, you should be able to show that the flux of a constant vector through *any* closed surface is zero.

6.15 GAUSS'S LAW FOR THE GRAVITATIONAL FIELD*

Although one can calculate the flux of *any* vector through *any* surface, the flux of many vectors is not physically meaningful. For some vectors, however, the flux of the vector through a surface does have consequence; the gravitational field is one of those lucky vectors.

It will take some thoughtful study to find the total flux of *any* gravitational field through *any* closed surface; we will see, however, that the *result* is remarkably simple. The flux of the gravitational field through a closed surface involves calculating the integral over the surface of the scalar product of $\vec{\mathbf{g}}$ with a differential area element $d\vec{\mathbf{S}}$:

$$\Phi_{\substack{\text{clsd} \\ \text{surface S}}} = \int_{\substack{\text{clsd} \\ \text{surface S}}} \vec{\mathbf{g}} \cdot d\vec{\mathbf{S}}$$

The closed surface can be *any* closed surface we please to choose. It can be a real surface or a surface that we simply *imagine*.

To find the flux of the gravitational field through a closed surface, it first is convenient to extend the idea of angular measure to three dimensions. The radian measure of an angle is defined as the ratio between an arc length s of a circle of radius r and the radius itself (see Figure 6.62):

$$\theta \equiv \frac{s}{r}$$

where θ is measured at the center of the circle.

Since the circumference of a circle is $2\pi r$, there are

$$\frac{2\pi r}{r} = 2\pi \text{ rad}$$

subtended by a complete circle.

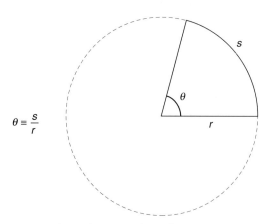

FIGURE 6.62 An angle measured in radians.

Instead of an arc on a circle, consider an area A on a sphere, as shown in Figure 6.63. The **solid angle** Ω measured at the center of the sphere, subtended by the area A is defined to be the ratio between the area A on the sphere and the square of the radius of the sphere:

$$\Omega \equiv \frac{A}{r^2} \qquad (6.48)$$

The units for measuring solid angles are **steradians** (sr). Steradians are dimensionless units, as are radians. Since the surface area of a sphere is $4\pi r^2$, the number of steradians subtended by the whole sphere is

$$\frac{4\pi r^2}{r^2} = 4\pi \ \text{sr}$$

A differential area dA on a sphere subtends a differential solid angle $d\Omega$ (see Figure 6.64), where

$$d\Omega = \frac{dA}{r^2}$$

If a differential area dS is *not* on a sphere of radius r, nor perpendicular to $\vec{r}$, the differential solid angle $d\Omega$ subtended by dS from a reference point 0 then is the quotient of the area of dS *projected* perpendicular to the radial direction (see Figure 6.65), divided by the square of the radial distance. The projected area dA involves the angle ϕ between the differential area vector $d\vec{S}$ and the radial direction. Indeed, $dA = d\vec{S} \cdot \hat{r}$.

Hence the solid angle $d\Omega$ subtended by dS is

$$d\Omega = \frac{dA}{r^2} = \frac{d\vec{S} \cdot \hat{r}}{r^2} = \frac{dS \cos \phi}{r^2} \qquad (6.49)$$

Now we have the tools to find the flux of any gravitational field through any closed surface.

Let a *closed surface* of *any arbitrary shape* surround (i.e., enclose) a particle of mass M, as in Figure 6.66. The mass M may be located anywhere within the closed surface. The flux of the gravitational field of M through this closed surface is

$$\Phi_{\substack{\text{clsd surface} \\ S \text{ encl M}}} = \int_{\substack{\text{clsd} \\ \text{surface } S}} \vec{g} \cdot d\vec{S}$$

Look at the integrand:

$$\vec{g} \cdot d\vec{S} = g \, dS \cos \theta$$

where $\vec{g}$ is the gravitational field vector created by mass M at the location of the differential area $d\vec{S}$. Here θ is the angle between the direction of $\vec{g}$ and the direction of the differential area vector $d\vec{S}$ as shown in Figure 6.66. A point mass M creates a gravitational field given by Equation 6.42. The *magnitude* of the gravitational field established by M is

$$g = \frac{GM}{r^2}$$

The differential area vector is normal to the surface and points *outward*, since the surface is a closed surface. From Figure 6.66, note that

$$\cos \theta = \cos(\pi - \phi) = -\cos \phi$$

FIGURE 6.65 A differential area dS not part of a sphere also subtends a differential solid angle $d\Omega$ from a reference point 0.

FIGURE 6.63 An area A on a sphere of radius R.

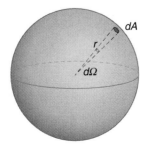

FIGURE 6.64 A differential area dA on a sphere subtends a solid angle $d\Omega$.

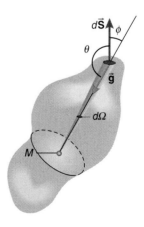

FIGURE 6.66 A closed surface surrounding a mass M.

Therefore the integrand in the flux integral is

$$g \, dS \cos \theta = \frac{GM}{r^2} \, dS \, (-\cos \phi)$$

$$= -GM \frac{dS \cos \phi}{r^2} \qquad (6.50)$$

But $(dS \cos \phi)/r^2$ is the differential solid angle $d\Omega$ subtended by the differential area dS as seen from M (Equation 6.49). Thus the integrand for the flux integral is

$$\vec{\mathbf{g}} \cdot d\vec{\mathbf{S}} = -GM \, d\Omega$$

Therefore the flux of the gravitational field of M through the entire closed surface is

$$\Phi_{\substack{\text{clsd surface} \\ S \text{ encl M}}} = \int_{\substack{\text{clsd} \\ \text{surface}}} \vec{\mathbf{g}} \cdot d\vec{\mathbf{S}} = -GM \int_{\substack{\text{clsd} \\ \text{surface}}} d\Omega$$

But the integral of the solid angle over the whole closed surface is the solid angle subtended by the whole surface as seen from M. Since the closed surface completely surrounds M, the solid angle subtended by the closed surface is *same* as the solid angle subtended by a sphere enclosing M—namely, 4π sr. Thus the integral of the solid angle over the whole closed surface is simply 4π sr. Hence the flux of the gravitational field $\vec{\mathbf{g}}$ through an arbitrarily shaped closed surface surrounding M is

$$\Phi_{\text{clsd surface encl M}} = -4\pi GM \qquad (6.51)$$

where we omit the steradian (sr) unit associated with the 4π, since it is dimensionless. Equation 6.51 is a remarkable result because it is independent of the surface shape *or* of where the mass M is located within the surface. If you study this derivation carefully (now, how else would you?), you will see that the simplicity of the result arises only because of the *inverse square* nature of the gravitational field: the r^2 term in Equation 6.50 came from the gravitational field vector of the pointlike mass M. We see how kind nature has been to formulate gravitation as an inverse square law!

What does the flux of the gravitational field mean? It is a measure of the extent or total amount that the gravitational field vector threads through the surface. In the case of a closed surface surrounding M, the gravitational field vector $\vec{\mathbf{g}}$ of M comes into the closed surface without any corresponding exits. The minus sign in Equation 6.51 for the flux indicates that the field threads into the closed surface rather than threading outward (exiting).

If some other mass M′ is located within the closed surface instead of M, its gravitational field produces a flux through the closed surface that is equal to

$$-4\pi GM'$$

because the flux through the surface is independent of the location of the mass within the surface. Thanks to the principle of superposition, if *both* M and M′ are within the surface, the total flux of the total field through the closed surface is the sum of the individual fluxes:

$$\Phi_{\substack{\text{total} \\ \text{gravitational} \\ \text{field}}} \text{(total through S)} = -4\pi GM - 4\pi GM'$$

So the total flux of the gravitational field through a closed surface is a measure of the total mass enclosed *anywhere* within the surface:

flux of the total gravitational field through a closed surface $= -4\pi G[\text{total mass enclosed by the surface}]$

or, more mathematically,

$$\int_{\substack{\text{clsd} \\ \text{surface } S}} \vec{\mathbf{g}} \cdot d\vec{\mathbf{S}} = -4\pi GM_{\text{within } S} \qquad (6.52)$$

We now address the question of what happens to the flux of the gravitational field through a closed surface if it does *not* have M inside, as in Figure 6.67. Something quite remarkable happens. Let this total surface be composed of two arbitrary parts A and B, as shown in Figure 6.68. Note the directions in Figure 6.68 of the outward differential area vectors on surface B. Next construct an *extension* of the surface to enclose M, as in Figure 6.69. You can consider the extension to be a closed surface composed of parts B and C. Note that the outward differential area vectors of area B, when considered as part of the extended closed surface enclosing M, are opposite to those in Figure 6.68.

FIGURE 6.67 A mass M outside a closed surface.

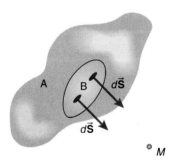

FIGURE 6.68 Divide the closed surface into two arbitrary parts, A and B.

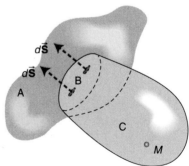

FIGURE 6.69 An extension surface C creates two closed surfaces: surface A with C and surface C with B.

Since we already know the value of the total flux threading a complete surface that encloses M, we have two cases here:

$$\Phi_A + \Phi_C = -4\pi GM \quad (6.53)$$

$$\Phi_C + (-\Phi_B) = -4\pi GM \quad (6.54)$$

The minus sign for Φ_B arises because the outward differential areas to surface B, when considered with surface C (forming a closed surface enclosing M), are opposite in direction to those when B is considered with surface A (forming the closed surface *not* enclosing M). Subtracting Equation 6.54 from Equation 6.53,

$$\Phi_A + \Phi_B = 0 \ \ \text{m}^3/\text{s}^2$$

Finally, since A plus B define our original empty closed surface of arbitrary shape, we have proved that the flux of the gravitational field through any arbitrarily shaped closed surface that excludes M vanishes. The gravitational field has no *net* threading of the closed surface if M is on the *outside* of the closed surface.

These are remarkable results. We summarize them here:

If M is *inside* the closed surface (whatever the shape of the surface or the location of M inside), the flux of the gravitational field through the surface is

$$\Phi_{\text{clsd surface, M inside}} = -4\pi GM$$

If M is anywhere *outside* the closed surface, the flux of the gravitational field through the closed surface is zero:

$$\Phi_{\text{clsd surface, M outside}} = 0 \ \ \text{m}^3/\text{s}^2$$

Both results can be summarized in the following way:

$$\Phi_{\text{clsd surface } S} = -4\pi GM_{\text{within } S} \quad (6.55)$$

This is known as **Gauss's law for the gravitational field**. The law is named for its discoverer, Carl Friedrich Gauss (1777–1855), whose almost unbelievably imaginative and productive contributions illuminated 19th-century physics, mathematics, and astronomy.

PROBLEM-SOLVING TACTIC

6.4 If you know the amount of mass within a closed surface S, the value of the flux of the gravitational field through the surface is found from the right-hand side of Gauss's law, Equation 6.55. It is not necessary to perform a complicated integral for the flux.

Despite poor and humble roots, Gauss was famous for his intellectual achievements when only in his mid-twenties.

STRATEGIC EXAMPLE 6.14

Consider the Earth–Moon system. The total gravitational field at any point in space is the vector sum of the gravitational field of the Earth and the gravitational field of the Moon at the point. Imagine a closed surface of arbitrary shape surrounding just the Earth, as indicated in Figure 6.70. What is the flux of the *total* gravitational field through this closed surface?

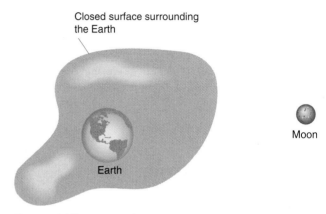

FIGURE 6.70

Solution

A direct calculation of the flux integral of the total gravitational field through this closed surface is formidable, because the total gravitational field of the Earth and Moon is a complicated function of position over the indicated surface. The angle the total gravitational field vector makes with differential area vectors on the imaginary closed surface also is a complicated function of position on the closed surface. So a brute force calculation of the flux of the *total* gravitational field is something to avoid like the plague. However, Gauss's law for the gravitational field, Equation 6.55, states that the numerical value of that complicated integral for the flux of the *total* gravitational field through this weird closed surface is very simply related to the mass enclosed within the closed surface. Here the only mass within the closed surface is the mass of the Earth, and so it is only the gravitational field established by the Earth that contributes to the flux of the *total* gravitational field through the indicated closed surface; the gravitational field established by the Moon contributes nothing to the flux because the Moon lies outside the indicated closed surface. So use Gauss's law for the gravitational field:

$$\Phi_{\text{clsd surface } S} = -4\pi GM_{\text{within } S}$$
$$= -4\pi GM_{\text{Earth}}$$

Substituting numerical values for G and M_{Earth}, you find

$$\Phi_{\text{clsd surface } S} = -4\pi(6.67 \times 10^{-11} \ \text{N·m}^2/\text{kg}^2)(5.98 \times 10^{24} \ \text{kg})$$
$$= -5.01 \times 10^{15} \ \text{N·m}^2/\text{kg}$$
$$= -5.01 \times 10^{15} \ \text{m}^3/\text{s}^2$$

CHAPTER SUMMARY

Newton's law of universal gravitation states that two pointlike masses m and M exert attractive forces on each other, called gravitational forces, of equal magnitude, given by

$$F = \frac{GMm}{r^2} \qquad (6.3)$$

where r is the separation of the masses and G is the *universal gravitational constant*

$$G \approx 6.67 \times 10^{-11} \text{ N·m}^2/\text{kg}^2$$

The gravitational force acts along the line between the two masses and so is a *central force*.

The *principle of linear superposition* states that when two or more masses exert gravitational forces on another mass m, the total gravitational force on m is the vector sum of the individual gravitational forces on m.

The total gravitational force of a uniform spherical shell of mass M on a mass m located anywhere *inside* the shell is equal to *zero*. If the mass m is located *outside* the shell, the effect of the shell on m is identical to that of a point mass, equal to the mass M of the shell, located at the center of the shell.

The total gravitational force of a spherically symmetric mass M on a pointlike mass m located *outside* the sphere is equivalent to that obtained by considering the sphere as a pointlike mass M located at the center of the sphere. If a mass m is located *inside* the sphere of radius R at a distance $r < R$, only that part of the mass of the sphere that is *closer* to the center than m is effective in exerting a gravitational force on m.

The experimental observation that *all* masses near the surface of the Earth experience the *same* gravitational acceleration (when the gravitational force is the only force on the system) implies that the gravitational mass of a system (the mass appearing in Newton's law of universal gravitation) is identical to the inertial mass of the system (the mass appearing on the right-hand side of Newton's second law of motion).

Kepler's empirical laws of planetary motion describe several characteristics of the orbital motion of a system of mass m around another mass M, where $m \ll M$. The laws of Kepler describe the orbital motion of planets about the Sun, of spacecraft in orbit about the Sun, of artificial satellites of the Earth, and of planetary moons as long as the mass of the moon is much less than the mass of the mother planet. Applied to the planets of the Sun, the laws are as follows:

Kepler's first law: The shape of the orbit of a planet about the Sun is an ellipse with the Sun at one focus of the ellipse; the second focus is empty.

Kepler's second law: A line from the Sun to a given planet sweeps out equal areas during equal time intervals. This law arises because the gravitational force is a *central force*—that is, one directed along the line between two particles.

Kepler's third law: The square of the period of a planet is proportional to the cube of the semimajor axis of its orbit.

Kepler's laws are consequences of Newton's general theory of motion and Newton's law of universal gravitation.

Two objects of comparable mass (for example, two stars, or a planet with a massive moon) both execute orbits about the center of mass of the two-mass system. The center of mass is located closer to the more massive of the pair. The orbits of the two masses have similar shapes, but the orbit of the more massive object of the pair is smaller in size.

For *circular orbits*, the *Newtonian form* for Kepler's third law is

$$T^2 = \frac{4\pi^2}{G(M + m)} r^3 \qquad (6.37)$$

where r is the distance between the two masses. The separation distance r is the sum of the radii of the circular orbits of the two masses about their *center of mass*.

For the planets of the Sun, for spacecraft in orbit about the Sun, or for spacecraft in orbit about the Earth or another very massive object (M), the orbital size and acceleration of the very massive object about the center of mass is negligible, since the center of mass is essentially coincident with the mass M itself. For these situations, the Newtonian form of Kepler's third law can be *simplified* (since m is *negligible* compared with M) and *generalized* to include elliptical orbits as well as circular orbits:

$$T^2 = \frac{4\pi^2}{GM} a^3 \qquad (m \ll M) \qquad (6.39)$$

where a is the semimajor axis of the orbit of m about the (stationary) mass M. A circle is an ellipse of zero eccentricity; its semimajor axis is identical to the radius of the circle.

For masses in orbit about the Sun, Kepler's third law may be simplified further to

$$T^2 = (1 \text{ y}^2/\text{AU}^3)a^3 \qquad (6.41)$$

where the period T must be expressed in years and the semimajor axis a in astronomical units (AU).

By definition, 1 AU is the semimajor axis of the orbit of the Earth:

$$1 \text{ AU} \approx 1.496 \times 10^{11} \text{ m}$$

A mass M establishes a *gravitational field* $\vec{g}$ in space. The gravitational field depends on the *shape* of M and how mass is *distributed* within its shape. The gravitational fields produced by various mass shapes and distributions are tabulated in Table 6.1. *The gravitational field at any location is identical to the acceleration due to gravity at that location.* The field at any location also may be thought of as the gravitational force per kilogram at that location in space.

If another mass m is placed at a location where the gravitational field has the value $\vec{g}$, then m experiences a gravitational force $\vec{F}$ equal to

$$\vec{F} = m\vec{g} \qquad (6.43)$$

The *flux* Φ of the gravitational field through any surface S is

$$\Phi = \int_S \vec{g} \cdot d\vec{S} \qquad (6.47)$$

The flux of the gravitational field through S is a measure of the extent to which the gravitational field vector threads the surface.

Gauss's law for the gravitational field states that the flux Φ of the gravitational field through any *closed surface* S is proportional to the mass enclosed within S:

$$\Phi_{\substack{\text{clsd} \\ \text{surface } S}} = -4\pi G M_{\text{within } S} \qquad (6.55)$$

SUMMARY OF PROBLEM-SOLVING TACTICS

6.1 **(page 234)** Watch your m & M's. Be sure you understand which force is needed: the force on m or the force on M.

6.2 **(page 254)** Be sure to express T in years and a in astronomical units when using Equation 6.41:

$$T^2 = (1 \ y^2/AU^3)a^3 \qquad (6.41)$$

6.3 **(page 257)** Do not confuse the magnitude of the gravitational field $\vec{g}(r)$ at *any* location in space with g, the magnitude of the local acceleration due to gravity near the surface of the Earth.

6.4 **(page 267)** If you know the amount of mass within a closed surface S, the value of the flux of the gravitational field through the surface is found from the right-hand side of Gauss's law, Equation 6.55:

$$\Phi_{\substack{\text{clsd} \\ \text{surface } S}} = -4\pi G M_{\text{within } S} \qquad (6.55)$$

It is then not necessary to perform the typically complicated integration for the flux.

QUESTIONS

1. (page 235); 2. (page 237); 3. (page 239); 4. (page 241); 5. (page 244); 6. (page 248); 7. (page 258)

8. Carefully explain why the gravitational force of the Earth on a mass m is *directly proportional* to m while the free-fall acceleration is *independent* of m.

9. The Ring Nebula in the constellation Lyra (see Figure Q.9) is a spherically symmetric expanding shell of gas of mass m ejected from a central star of mass M.* (a) What is the total gravitational force of the shell on the central star? (b) What is the total gravitational force of the central star on the shell? (c) Are these a Newton's third law pair of forces? (d) The shell is outside the star; why then is the force of the central star on the shell equal to zero?

FIGURE Q.9

10. Imagine yourself located somewhere in the empty interior of a hypothetical hollow planet. Describe the trajectory of a pea shot from a peashooter.

11. A reporter for a magazine once wrote[†]:

> Ah, but the worse is yet to come. Also locked up in a French vault is a block of metal that weighs exactly one kilogram.

Explain why the first sentence accurately describes the second sentence.

---

*The shell has the appearance of a ring because there is more material along our line of sight around the edges of the shell than along the line of sight to the center of the shell.
[†] *Smithsonian*, 24, #12 (March 1994), page 21.

12. A mass m is inside a uniform spherical shell of mass M. Another mass m_0 lies outside the shell as shown in Figure Q.12. (a) What is the gravitational force of the shell on m? (b) Is the gravitational force of m_0 on m also zero, since m lies inside the shell? In other words, does the shell act as a gravitational insulator, effectively *shielding* m from the influence of m_0?

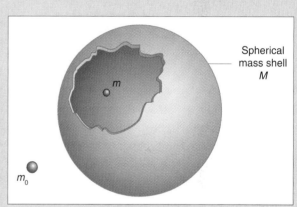

Spherical mass shell M

FIGURE Q.12

13. The Earth and the Moon exert gravitational forces of equal magnitude on each other. If a fraction of the mass of the Earth were transferred to the Moon, would the magnitude of their mutual gravitational force increase, decrease, or remain the same? Explain your reasoning. (See Problem 6.)

14. If the mass of the Earth remained constant but its radius doubled, would this affect the orbit of the Moon? Explain.

15. If the gravitational force on masses on the surface of the Earth were much stronger than it actually is (say, because the Earth had the same size but a larger mass density and total mass), discuss the influence this might have had on the evolution of human anatomy.

16. If gravity did not exist, could life be possible? Explain.

17. Despite massive backpacks and spacesuits, the Apollo astronauts were able to jump much higher on the Moon than on the Earth without all the accoutrements. Carefully explain why.

18. Consult an astronomy text to discover why the sidereal year is slightly longer than the tropical year as mentioned in the footnote in Section 6.12 on page 254.

19. A north–south line in the sky passing through your zenith is known as your *local celestial meridian*. The time between successive celestial meridian passages of the Sun is called an *apparent solar day*. Will the interval of an apparent solar day about the time the Earth is at perihelion (3 January) be longer, shorter, or the same as the apparent solar day about the time the Earth is at aphelion (6 July). Explain.

20. Comet Halley has an elliptical orbit of high eccentricity about the Sun. The comet is visible to the naked eye for only several months out of its 76 y period. Explain how Kepler's law of areas might account for this.

21. The Earth is closest to the Sun (at its perihelion distance) on 3 January and most distant from the Sun (at its aphelion distance) on 6 July. What, then, causes the seasons? Consult an astronomy text if necessary.

22. How does Kepler's law of areas account for the unequal length of the seasons? See the following table:

Northern hemisphere season	Southern hemisphere season	Approximate calendar dates	Length of the season
spring	fall	20 March–21 June	92.76 days
summer	winter	21 June–23 September	93.65 days
fall	spring	23 September–21 December	89.83 days
winter	summer	21 December–20 March	89.00 days
		Total	365.24 days

23. Which geometrical figure has the greater area: an ellipse with semimajor axis a and eccentricity $\varepsilon = 0$ (a circle!) or an ellipse of semimajor axis a and eccentricity $\varepsilon \neq 0$?

24. How does Kepler's law of areas explain why a planet has the greatest speed when it is at its perihelion distance?

25. The time it takes a planet to go from one end of the major axis of its orbit to the other end is half the orbital period of the planet. Explain why the time to go from one end of the *minor* axis to the other end is *not* half the period.

26. Two satellites are in circular orbit around the Earth, one at a radius r from the center of the Earth, the other at radius $2r$. What is the *ratio* of their periods? Speeds? Accelerations?

27. Earth satellites usually are launched to the east. Why is this more advantageous than launching them to the west?

28. Explain why the orbital planes of all the planets must pass through the center of the Sun.

29. Explain why an orbiting astronaut experiences what is colloquially but unfortunately called "weightlessness."

30. An orbiting astronaut is not, strictly speaking, weightless. But a geologist located in the empty interior of a hypothetical hollow planet, *is* truly weightless. Explain why.

31. Explain why it is not possible to place a satellite in an orbit that follows a path directly above either the Tropics of Cancer or Capricorn nor the Arctic or Antarctic Circles.

32. Explain why it is easier to determine the mass of a planet that has a moon than to find the mass of a planet with no moon. (Only Mercury and Venus lack any moons.)

33. If a planetary moon has appreciable mass compared with the mother planet, Newton's form for Kepler's third law (Equation 6.37) only enables you to determine the *sum* of the mass of the planet and the mass of the moon. What additional information is necessary to determine the masses of the individual objects?

34. For a planet in an elliptical orbit around the Sun, will the *temporal* average of the various positions of the planet from the Sun (the average of its distance from the Sun spaced at equal time intervals) be equal to the semimajor axis of the orbit? Explain why or why not. If not, will the temporal average likely be less than or greater than the semimajor axis?

35. Weather and communication satellites are placed in circular orbits such that they hover or stay above certain locations on the surface of the Earth. (a) Explain why the satellite does not fall down to the surface of the Earth. (b) What is the orbital angular speed of the satellite? (c) Could such a weather satellite be placed so that it hovered directly over the Big Apple (New York)? Why or why not?

36. The three laws of planetary motion discovered by Kepler concern the shape of an orbit, the variation of the speed around an orbit, and the period of the orbit. If the numerical value of the gravitational constant G were *different* from what it is, which of the Keplerian laws would be affected and which would be unaffected? Explain.

37. Newton showed that the possible shapes of the paths followed by objects subjected to (only) the gravitational force of the Sun are not only circles and ellipses, but also parabolas and hyperbolas. Kepler's *first* law is generalized to state that the Sun is located at either (i) the center of a circular path; (ii) the single focus of a parabolic path; (iii) one of the two foci for an elliptical path (the other focus being empty); or (iv) one focus of a hyperbolic path (the other focus and sheet of the hyperbola being empty). (a) Kepler's *second* law (the law of areas) was formulated by Kepler for elliptical (and circular) orbits. Does Kepler's second law *also* apply to parabolic and hyperbolic paths of objects about the Sun? Explain why or why not. (b) Kepler's *third* law concerns the orbital period. Is it meaningful to talk about a *period* associated with parabolic or hyperbolic paths? Explain.

38. Explain how a geologist might exploit *variations* in the gravitational field over the surface of the Earth to detect the presence of a large subterranean oil pool. (The density of oil is only a fraction of the density of rock.)

39. Show that the SI units for the flux Φ of the gravitational field are either $N \cdot m^2/kg$ or m^3/s^2.

40. If the flux of the gravitational field through a closed surface is zero, does this imply there is zero mass within the surface? Does such a zero flux through a closed surface imply the gravitational field *itself* is zero at every point on the closed surface? Explain.

41. *Imagine* that negative masses exist as well as positive masses. If a closed surface surrounded such a negative mass, what does Gauss's law for the gravitational field imply about the flux of the gravitational field through the closed surface? What does this mean in terms of the direction of the gravitational field of such a hypothetical negative mass? If a normal positive mass particle were placed in the gravitational field of such a hypothetical negative mass, what would be the direction of the gravitational force on the normal positive mass particle relative to the line between the two particles?

PROBLEMS

Sections 6.1 How Did Newton Deduce the Gravitational Force Law?
6.2 Newton's Law of Universal Gravitation

1. Two 2.50 kg pointlike masses are separated by 1.50 m. What is the magnitude of the gravitational force that each exerts on the other?

2. What is the magnitude of the gravitational force of a 1.00 kg young goshawk (*Accipiter gentilis*) on the Earth? In what direction does this force point?

3. A classics student of mass 70.0 kg is 0.50 m from a physics student of mass 50.0 kg. Calculate the approximate magnitude of the gravitational force that each exerts on the other. Explain why your result is only approximately equal to the magnitude of the actual gravitational force of each on the other.

•4. When the giant planet Jupiter ($M_{Jupiter}$ = 1.90 × 10²⁷ kg) is closest to the Earth, it is about 5.90 × 10⁸ km from us. (a) Calculate the magnitude of the gravitational force of Jupiter on a 70.0 kg astrologer when Jupiter is closest to the Earth. Assume both Jupiter and the astrologer are pointlike particles. So much for the influence of the planets on our daily lives. (b) What mass m placed 10.0 m from the astrologer would produce a gravitational force on the astrologer of the same magnitude as the force of Jupiter calculated in part (a)?

•5. The orbit of the Earth about the Sun is circular to a first approximation (the eccentricity of the elliptical orbit is only about ε = 0.0167). (a) Using Newton's second law of motion and the gravitational force law (among other things), determine the orbital speed of the Earth in terms of the mass of the Sun, the radius of the orbit, and other appropriate constants. Consider both the Earth and Sun to be pointlike particles and the Sun to be fixed in position. Evaluate the speed and express it in km/s. (b) If the Earth suddenly split gently into two halves of equal mass, what would happen to the orbital speed of each piece, assuming the radius of the orbit were unchanged?

‡ 6. Two masses m_1 and m_2 exert gravitational forces of equal magnitude on each other. Show that if the total mass $M = m_1 + m_2$ is fixed, the magnitude of the mutual gravitational force on each of the two masses is a maximum when they have equal mass.

‡ 7. A small mass m is on the surface of the Earth as indicated in Figure P.7. The mass is located along the line joining the centers of the Earth and Moon. (a) What is the magnitude of the gravitational force *of the Moon* on the mass m? (b) What is the magnitude of the gravitational force of the Moon on mass m if m were located at the center of the Earth? (c) The difference between the magnitude of the force of the Moon on m calculated in part (a) and that calculated in part (b) is known as the magnitude of the *tidal force* $\vec{F}_{tidal}$ on m due to the Moon; $\vec{F}_{tidal}$ is directed toward the Moon. Show that the magnitude of the tidal force on m at this location can be written approximately as

$$F_{tidal} \approx GMm \frac{2R}{d^3}$$

where M is the mass of the *Moon*. Assume $d \gg R$. The tidal force of the Moon on a mass m thus depends on (1) the size of

the object on which m finds itself (in this case, the radius R of the Earth), and (2) the inverse *cube* of the distance d between the centers of the Earth and Moon. (d) Show that if a similar calculation is made for the tidal force of the Sun on m, the result is approximately 2.2 times smaller than the tidal force of the Moon on m. So both the Moon and the Sun cause tidal forces on the Earth. These tidal forces cause the familiar ocean tides.

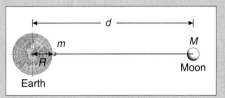

FIGURE P.7

‡ 8. In 1850 the French mathematician Édouard Roche showed that if a massive moon of substantial physical size ventured too close to a planet, the moon would be broken up by tidal forces. The moon must be of sufficient size and mass that its shape is determined by its self-gravitation rather than by other forces. Typically this means a generally spherical moon whose radius is several hundreds of kilometers. The distance of closest approach for such a moon is called the *Roche limit*. For a moon and planet of the *same density*, the Roche limit is equal to 2.455 planetary radii. A crude calculation can estimate the order of magnitude of the Roche limit using the approximations for the tidal forces in Problem 7 and crude assumptions about the shapes of the objects. Imagine a small mass m on the surface of a spherical moon of mass M' and radius R as indicated in Figure P.8. The moon is located a distance d from a planet of mass M and radius R_0. The magnitude of the tidal force on the mass m due to the *planet* is (using the results of Problem 7)

$$\frac{GMm\,2R}{d^3}$$

This force is directed toward the planet. The magnitude of the gravitational force of attraction of the moon for the mass m (the self-gravitational force) is

$$\frac{GM'm}{R^2}$$

by Newton's law of universal gravitation. This force is directed toward the moon. If the magnitudes of the two forces are set equal to one another, the moon is about to break up under the action of the tidal force. (a) Write the equation that gives this critical condition. (b) Show that the distance of separation d at which the tidal force on m is equal in magnitude to the force of the moon on m satisfies

$$d^3 = \frac{M\,2R^3}{M'}$$

(c) Show that if the spherical planet and moon have the *same density*, then

$$d^3 = 2R_0^3$$

or

$$d \approx 1.26R_0$$

where R_0 is the radius of the *planet*. This calculation means that if the moon ventures closer than 1.26 planetary radii to the planet of radius R_0, the moon will break up under the action of the tidal forces of the planet. This result disagrees with the exact calculation by Roche by about a factor of 2, which is not surprising because of the rather extreme assumptions made about the shapes of the objects and the tidal forces when the objects are so close together. While the calculation is crude, it does give an order of magnitude result. The rings of Saturn are within the Roche limit of the planet, whereas its first sizable moon (Mimas, with a radius of about 200 km) is outside the Roche limit.

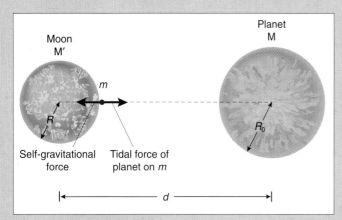

FIGURE P.8

Sections 6.3 Gravitational Force of a Uniform Spherical Shell on a Particle
6.4 Gravitational Force of a Uniform Sphere on a Particle

9. A pair of 10.0 cm diameter spheres of masses 4.00 kg and 0.400 kg are placed 10.1 cm apart, center-to-center. (a) What is the magnitude of the gravitational force of the 4.00 kg sphere on the 0.400 kg sphere? (b) What is the magnitude of the gravitational force of the 0.400 kg sphere on the 4.00 kg sphere? (c) If these forces are the *only* forces acting on the spheres, what is the magnitude of the acceleration of each sphere?

•10. A Cavendish balance (see Figure 6.3), used to determine the value of G, uses a small lead sphere of radius 0.500 cm and a larger lead sphere of radius 3.00 cm. The spheres are separated by 4.00 cm center-to-center. The density of lead is 11.0×10^3 kg/m³. (a) What is the magnitude of the gravitational force of each sphere on the other? (b) What is the magnitude of the weight of the smaller sphere? (c) What is the ratio of the magnitude of the weight of the smaller sphere to the magnitude of the gravitational force of the larger sphere on it? This ratio indicates why the Cavendish experiment is not easy to do.

•11. At what altitude above the surface of the Earth is the magnitude of the acceleration due to gravity 1.00% less than its value on the surface of the Earth?

•12. A mass m is inside a uniform spherical shell of mass M' and a mass M is outside the shell as shown in Figure P.12. What is the magnitude of the total gravitational force on m?

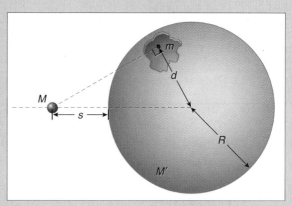

FIGURE P.12

•13. (a) A spherical asteroid has a radius R and average density ρ. (a) Show that the magnitude of the acceleration due to gravity on the surface of the asteroid is $(4/3)\pi GR\rho$. (b) The asteroid 107 Camilla has a diameter of 252 km and an average density of 4.5×10^3 kg/m³. What is the magnitude of the local acceleration due to gravity on the surface of 107 Camilla? (c) If you can high jump 2.0 m on the surface of the Earth, how high a jump could you make on 107 Camilla? How long would you be off the surface of the asteroid in making the jump?

•14. A certain neutron star has a radius of 10.0 km and a mass of 4.00×10^{30} kg, about twice the mass of the Sun. (a) Calculate the magnitude of the acceleration of an 80.0 kg student foolish enough to be 100 km from the center of the neutron star. (b) How many times g is the magnitude of this acceleration? (c) If the student is in a circular orbit of radius 100 km about the neutron star, use Newton's laws to find the orbital period.

•15. Imagine a spherical planet of radius R with an average density of 5.00×10^3 kg/m³; this is approximately the average density of the Earth. An 80.0 kg astronaut is on the surface of the planet. The magnitude of the gravitational force of the planet on the astronaut is only 1.00 N. What is the radius of the planet?

•16. (a) Given the mass and radius of the Earth, calculate the average density of the Earth. (b) If the mass of the Earth remained constant, but the radius of the Earth changed to increase the magnitude of the acceleration due to gravity on the surface of the Earth to twice its current value (i.e., to 19.6 m/s²), what would be the new radius of the Earth? Assume a uniform density.

‡17. Two identical, uniform spherical masses, each with mass m and radius R, are placed in contact with each other. (a) What must be the value of the density ρ so that the gravitational force of one mass on the other has a magnitude F? (b) If $F = 1.00$ N, do *any* of the natural elements have sufficient density for 1.00 cm radii spheres? Consult a reference that gives the densities of the natural elements. Surprisingly, the most dense of the 92 natural elements is not uranium ($\rho = 18.7 \times 10^3$ kg/m³) but osmium ($\rho = 22.5 \times 10^3$ kg/m³). (c) If the spheres are composed of osmium, what radius R is needed to produce a gravitational force of magnitude 1.00 N by one sphere on the other? What is the mass of each sphere in this case?

‡18. A mass m is at rest on the equator of a spinning, spherical planet. (a) Sketch a second law force diagram indicating all the forces acting on the mass m. (b) What happens, qualitatively, to the magnitude of the normal force of the surface on the mass as the spin angular speed of the planet increases?

(c) If the planet spins fast enough, the normal force of the surface on the mass vanishes and the centripetal acceleration is provided by the gravitational force alone. Under these circumstances, the gravitational force is barely able to keep the planet together. For a planet of average density ρ, show that the critical period of spin T_{crit} is

$$T_{\text{crit}} = \left(\frac{3\pi}{G\rho} \right)^{1/2}$$

Thus, to prevent a planet from breaking up,* the *rotational period* of the planet T must be greater than T_{crit}. Many moons within the solar system have average densities of about 3.0×10^3 kg/m^3. Calculate a numerical value for T_{crit} for this average density. No objects within the solar system rotate with periods this short: they would have come apart. (d) The same arguments can be used for stars as well as planets. Show that for a star of radius R, spinning at *frequency* ν (the number of rotations per second), the *minimum* mass that the star must have to be stable against rotational breakup is

$$M_{\text{min}} = \frac{4\pi^2}{G} R^3 \nu^2$$

Explain why this is a minimum mass. (e) Exotic stars known as neutron stars have radii of about 10 km. The neutron star at the center of the Crab Nebula in the constellation Taurus is spinning 33 times each second. Using the results of part (d), calculate the *minimum* mass for this neutron star. Express your result in terms of solar mass units: 1 solar mass = 1.99×10^{30} kg. Real neutron stars have masses between about 1.4 and 3.5 solar masses, considerably greater than the minimum mass calculated here.

Sections 6.5 Measuring the Mass of the Earth
6.6 Artificial Satellites of the Earth

• **19.** Calculate the period of a satellite of the Earth in a circular orbit with a radius 200 km greater than the mean radius of the Earth.

• **20.** Two satellites are in circular orbit about the Earth at radii r_1 and r_2. (a) Find the *ratio* between their orbital periods. (b) Evaluate this ratio if $r_2 = 2r_1$.

• **21.** A spacecraft on the surface of Mars discovers that a freely falling screw has an acceleration of magnitude 3.776 m/s^2. The radius of Mars is known to be 3.37×10^6 m. Use these data to calculate the mass of Mars. Express your result both in kilograms and as a fraction of the mass of the Earth.

• **22.** (a) Calculate the radius of the orbit of a satellite in circular orbit about the Earth with a period of one day. The planes of the orbits of such satellites typically are made coincident with the equatorial plane of the Earth and the motion is in the same directional sense that the Earth rotates (i.e., eastward). In this way the satellite remains above a fixed geographical position above the equator. Many such geosynchronous satellites are communication satellites, since they assist in transmitting radio, television, and telephone signals between continents. Three such satellites spaced 120° apart in approximately the same orbit constitute a global grid capable of

relaying messages virtually instantaneously over most of the Earth. (b) Find the orbital speed of the satellite. (c) What is the magnitude of the acceleration of the satellite?

• **23.** Two satellites are launched into circular equatorial orbits at essentially equal radii r about the Earth. One satellite moves east and the other west. Sketch the situation. (a) Will the satellites have identical periods? (b) At how many locations in your sketch will they pass each other? (c) Will the places where the satellites pass each other remain over the same locations on the surface of the Earth? (d) Which satellite (if either) will have the shorter time interval between successive passages over the same place on the equator of the Earth?

‡ **24.** On a clear day, you can see forever. . . . Well, not quite. Model the Earth as a sphere of radius R ($= 6370$ km). Seen from an altitude h above the surface of the Earth (see Figure P.24), the horizon lies a distance s away measured along the surface of the Earth. (a) Using the figure and your knowledge of geometry, show that

$$h = R(\sec \theta - 1)$$

where $\theta = s/R$. (b) For small angles θ,

$$\cos \theta = 1 - \frac{\theta^2}{2}$$

Also, by definition $\sec \theta = 1/\cos \theta$. Use these facts and the binomial expansion to show that, for small angles,

$$\sec \theta \approx 1 + \frac{\theta^2}{2}$$

(c) Use the results of parts (a) and (b) to show that an approximate expression for the distance s to the horizon is

$$s = (2Rh)^{1/2}$$

(d) Show that when at a height $h = 1.00$ km, the distance s to the horizon is approximately 113 km.

FIGURE P.24

Sections 6.7 Kepler's First Law of Planetary Motion and the Geometry of Ellipses
6.8 Spatial Average Position of a Planet in an Elliptical Orbit*

• **25.** The eccentricity of the orbit of the Earth about the Sun is $\varepsilon = 0.0167$. The semimajor axis of the orbit is 1.496×10^8 km.

---

*This is colloquially referred to as a breakup due to centrifugal pseudoforces.

(a) What is the separation of the foci of the orbit of the Earth? (b) Is the center of the orbit inside or outside the Sun (the radius of the Sun is 6.96×10^5 km). (c) What are the perihelion and aphelion distances of the Earth from the Sun? (d) In a scale model of the orbit of the Earth, let 10.0 cm represent the semimajor axis; what is the separation of the foci of the ellipse representing the orbit in this model?

•26. (a) On the same polar plot, make a series of accurate graphs of ellipses, all with a semimajor axis of 1.00 AU, appropriately scaled, with eccentricities ε between 0 and 0.900, in increments of 0.100. Include the orbits of Venus (0.723 AU), Mercury (0.387 AU), and Mars (1.52 AU), assuming they are circular. Locate the Sun on your graph. (b) Which elliptical orbits have perihelions closer to the Sun than Mercury? Do any of these orbits have aphelions beyond the orbit of Mars?

•27. Use the polar equation for an ellipse

$$r = \frac{a(1 - \varepsilon^2)}{1 - \varepsilon \cos \theta}$$

to describe the position r of a planet from the Sun (located at one of the foci of the ellipse). The value of r ranges between a minimum at the perihelion distance and a maximum at the aphelion distance. Show that when the planet is on the ends of the minor axis of the ellipse, the distance of the planet from the Sun is equal to the semimajor axis of the orbit, and that $\cos \theta = \pm \varepsilon$ at these positions.

•28. What is the eccentricity of the elliptical orbit of a planet whose aphelion distance is twice its perihelion distance?

•29. The eccentricity of the lunar orbit about the Earth is $\varepsilon = 0.055$. Calculate the ratio of the angle subtended by the diameter of the Moon at perigee (when it is closest to the Earth) to the angle subtended by the lunar diameter at apogee (when it is most distant from the Earth). Imagine yourself at the center of the Earth.

•30. A planet is in an elliptical orbit of semimajor axis a about the Sun. Use Equation 6.19 for the polar equation for the elliptical orbit:

$$r = \frac{a(1 - \varepsilon^2)}{1 - \varepsilon \cos \theta}$$

to prove that it is impossible for the planet to be located at a distance r from the Sun equal to half the aphelion distance unless $\varepsilon \geq 1/3$. Of course, ε also must be less than 1 for an elliptical orbit.

•31. The orbit of Pluto, with an eccentricity of 0.248, is the most eccentric of all the major planets. The semimajor axis of Pluto is 39.4 AU. (a) How far is the center of the elliptical orbit of Pluto from the center of the Sun? (b) Determine the perihelion and aphelion distances of Pluto. This should explain why Pluto occasionally is closer to the Sun than Neptune (most recently, between 1979 and 1999). The orbit of Neptune has a semimajor axis of 30.1 AU but an eccentricity of only 0.0086.

•32. The area A of an ellipse is

$$A = \pi ab$$

where a is the semimajor axis and b is the semiminor axis. (a) Use the geometrical properties of an ellipse to show that the area also can be expressed in terms of the eccentricity as

$$A = \pi a^2 (1 - \varepsilon^2)^{1/2}$$

(b) For a fixed semimajor axis, say equal to 1.00 AU, make a graph of the area of the ellipse as a function of its eccentricity.

•33. (a) Show that for an ellipse of eccentricity ε, the ratio of the major to minor axes is

$$\frac{\text{major axis}}{\text{minor axis}} = \frac{1}{(1 - \varepsilon^2)^{1/2}}$$

(b) What is the eccentricity of an ellipse if the ratio of the major axis to the minor axis is 1.01? 1.10 ?

‡34. Prove that the area of an ellipse of semimajor axis a and semiminor axis b is equal to πab.

‡35. Construct a Cartesian coordinate system with an origin at the geometrical *center* of an ellipse with the x- and y-axes oriented along the major and minor axes respectively. Show that the Cartesian equation for an ellipse

$$\frac{x^2}{a^2} + \frac{y^2}{b^2} = 1$$

(where a is the semimajor axis and b is the semiminor axis) is the same as the polar expression for an ellipse:

$$r = \frac{a(1 - \varepsilon^2)}{1 - \varepsilon \cos \theta}$$

where ε is the eccentricity of the ellipse and the origin for the polar coordinates r and θ is at one focus.

‡36. Prove that in the plane of any ellipse, the normal to the ellipse at any point P bisects the angle formed by lines from P to the two foci.

Sections 6.9 Kepler's Second Law of Planetary Motion
6.10 Central Forces, Orbital Angular Momentum,
and Kepler's Second Law*

37. The eccentricity of the orbit of the Earth is $\varepsilon_{\text{Earth}} = 0.0167$. Find the ratio of the speed of the Earth when at perihelion to its speed at aphelion.

•38. Calculate the surface area of the spherical Earth. How long a time is necessary for a line from the Sun to the center of the Earth to sweep out an area equal to the surface area of the Earth? Assume the orbit of the Earth is circular.

•39. A planet is in an elliptical orbit about the Sun with an eccentricity ε as indicated in Figure P.39. Let the semimajor and semiminor axes of the ellipse be a and b respectively. (a) Let the speed of the planet at perihelion (point P) be u, its speed at aphelion (point A) be v, and its speed at point C be V. At point C the planet is at its greatest distance from the major axis. By applying conservation of the orbital angular momentum of the planet at the points P, A, and C, show that

$$Vb = ua(1 - \varepsilon) = va(1 + \varepsilon)$$

(b) Use the expressions [from part (a)]

$$V = \frac{ua(1 - \varepsilon)}{b}$$

and

$$V = \frac{va(1 + \varepsilon)}{b}$$

to form an expression for V^2 and show that

$$V^2 = uv$$

This problem was inspired by a paper by Ira M. Freeman, "An interesting property of the Kepler ellipse," *American Journal of Physics*, 45, #6, pages 585–586 (June 1977).

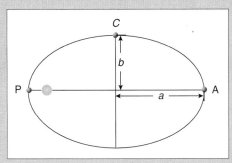

FIGURE P.39

•**40.** A planet of mass m is in an elliptical orbit about the Sun ($m \ll M_{Sun}$) with an orbital period T. Let L be the magnitude of the orbital angular momentum of the planet and let A be the area of the orbit. Show that

$$L = \frac{2mA}{T}$$

Sections 6.11 Newton's Form for Kepler's Third Law
of Planetary Motion
6.12 Customized Units

41. Most of the asteroids have orbits between Mars and Jupiter and have semimajor axes of about 2.8 AU. Determine the periods of the asteroids in these orbits.

42. The orbit of Pluto has a semimajor axis of 39.4 AU. Determine the period of Pluto. Pluto was discovered in 1930. In what year A.D. will Pluto first complete one orbital trip since its discovery?

43. Comet Swift-Tuttle (first discovered in 1862) has a period of about 130 y. It has been predicted that this comet may collide with the Earth in A.D. 2126; so . . . if you have nothing else to worry about . . . don't lose sleep; the prediction subsequently was found to be erroneous. What is the semimajor axis of the orbit of the comet in AU?

44. (a) What is the semimajor axis of a hypothetical planet of the Sun with a period equal to 10.0 y? (b) What is the period of a hypothetical planet of the Sun with a semimajor axis equal to 10.0 AU?

•**45.** The famous Comet Halley has a perihelion distance of 0.57 AU and a period of 75.6 y. (a) Determine the semimajor axis of its orbit. (b) What is the eccentricity of the orbit? (c) Find

the aphelion distance of the comet. (d) Determine the ratio of the maximum speed to the minimum speed of the comet.

•**46.** If the mass of the Sun were doubled, at what distance from the Sun (in AU) would a planet have an orbital period of 1.00 y?

•**47.** If the semimajor axis of an orbit about the Sun increases by a factor of 2, by what factor does the period change? Will the factor be the same for orbits about the Earth?

•**48.** The mass of the Sun is 3.30×10^5 times the mass of the Earth. Find the location of the center of mass of the Earth–Sun system and compare this with the radius of the Sun, 6.96×10^5 km.

•**49.** The spatial average distance between the Earth and the Moon, center-to-center, is about 3.84×10^8 m. The mass of the Earth is about 80 times the mass of the Moon. Determine the location of the center of mass of the Earth–Moon system: (a) relative to the center of the Earth; (b) relative to the *surface* of the Earth.

•**50.** What would be the orbital period of a rather hot grand piano in circular orbit about the Sun with a radius equal to the radius of the Sun?

•**51.** A planet is in a circular orbit about the Sun. Show that the square of the period of the planet is proportional to the 3/2 power of the *area* of the orbit.

•**52.** The orbit of Venus, with an eccentricity of 0.007, is the most circular of all the planets. The semimajor axis of the orbit is 0.723 AU. (a) How far is the center of the elliptical orbit of Venus from the center of the Sun? (b) Determine the perihelion and aphelion distances of Venus from the Sun. (c) Determine the period of the planet in its orbit.

•**53.** A peculiar triple star system is discovered in the constellation Physics Major (see Figure P.53). Two stars, called Ψ and Ψ^*, each of mass M/4, are at opposite points of the same circular orbit about a central star of mass M. (a) Determine the magnitude of the gravitational force of star Ψ^* on star Ψ. (b) Determine the magnitude of the gravitational force of the central star on star Ψ. (c) What is the magnitude of the total gravitational force on star Ψ? (d) Find an expression for the speed of star Ψ in terms of the radius r of its orbit and the period T of its orbital motion. (e) Find an expression for the square of the orbital period of star Ψ in terms of the various masses, the radius of the orbit, and the universal constant of gravitation G.

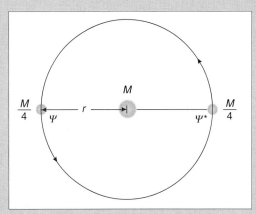

FIGURE P.53

•**54.** (a) Construct a graph of the square of the period (T^2) versus the cube of the semimajor axis (a^3) for the planets of the Sun.

Use the data located on the inside covers. (b) Is the graph linear? Why or why not? (c) What is the slope of the graph? (d) What does the graph represent or show?

•55. A spacecraft is placed in an elliptical orbit about the Sun. The perihelion distance is that of the Earth (1 AU) and its aphelion distance is the distance to the nearby star, Proxima Centauri, located approximately 2.7×10^5 AU from the Sun. (a) Sketch the situation. (b) Calculate the number of years it takes the spacecraft to make a one-way trip from its perihelion point to its aphelion point; neglect the gravitational influence of Proxima Centauri on the spacecraft. Anyone care to go along for the ride? (c) What is the approximate eccentricity of the orbit of the spacecraft?

•56. The time it takes for a mass m to fall from a distance d (expressed in AU) to, say, the Sun, can be approximated in the following way. Imagine m in a very eccentric elliptical orbit ($\varepsilon \approx 1$) about the Sun as indicated in Figure P.56. The semimajor axis a of the orbit then is approximately $a \approx d/2$. Use Kepler's third law with the customized units (of Section 6.12) to show that the time in years for *half* the orbit (aphelion to perihelion) to be completed is

$$t = \frac{T}{2} = \left(\frac{1}{2\sqrt{8}} \ \text{y/AU}^{3/2} \right) d^{3/2}$$

Estimate this time for an object located at the distance of the Earth from the Sun ($d = 1.00$ AU).

This problem was inspired by an article by John S. Dilsaver and Joseph R. Siler, "The solar swan dive," *The Physics Teacher*, 29, #1, pages 28–29 (January 1991).

Elliptical orbit of eccentricity $\varepsilon \lesssim 1$

FIGURE P.56

•57. Plaskett's binary (also known by the more exotic but less romantic name HD 47129), in the constellation Monoceros, consists of two blue-giant stars. Assume they are of equal mass and in circular orbits about their center of mass located midway between them. The orbital speed of each star has been found to be 250 km/s and the period of their orbit is 14.4 d. Find the mass of each star and express it as a multiple of the mass of our Sun ($M_{Sun} = 1.99 \times 10^{30}$ kg). These are among the most massive stars known.

•58. The orbits of the Earth and Venus are essentially circular. A spacecraft is in an orbit about the Sun with the perihelion point at the orbit of Venus and the aphelion point at the orbit of the Earth. The radius of the orbit of Venus is 0.723 AU. (a) Calculate the flight time for the spacecraft from the Earth to Venus. (b) Determine the eccentricity of the orbit of the spacecraft.

•59. The eccentricity of the orbit of Pluto is the largest of all the major planets: $\varepsilon_{Pluto} = 0.248$. The semimajor axis of Pluto is 39.4 AU. (a) Determine the perihelion and aphelion distances of Pluto from the Sun. (b) The eccentricity of the orbit of the planet Neptune is 0.0100 and its semimajor axis is 30.1 AU. Is

it possible for the planet Pluto to be closer to the Sun than Neptune? (c) Calculate the flight time in years for a spacecraft to go from the Earth to Pluto if the orbit of the spacecraft has its perihelion point at the orbit of the Earth and its aphelion point at the orbit of Pluto when Pluto is at its perihelion point. Assume the orbit of the Earth is circular. (d) Find the eccentricity of the orbit of the spacecraft.

‡60. A planet is in an elliptical orbit about the Sun with a semimajor axis a and eccentricity ε. The mass m of the planet is negligible with respect to the mass M of the Sun. Show that the square of the period of the planet is equal to the following:

$$\frac{4\sqrt{\pi}}{GM(1 - \varepsilon^2)^{3/4}} A^{3/2}$$

where A is the area of the elliptical orbit (see Problem 32 for an expression for the area of an ellipse).

‡61. The solar system as a whole is in orbit about the center of the Milky Way Galaxy. The distance of the Sun from the center of the galaxy is about 2.5×10^4 light-years.* Assume for simplicity that the orbit of the solar system is circular. Astronomers also have been able to determine, independently, that the speed of the solar system in its galactic orbit is about 230 km/s. (a) How many years does it take the solar system to complete one galactic orbit? (b) The age of the solar system is about 5.0×10^9 y. How many orbits about the galactic center has the solar system completed in its lifetime? (c) Assume a spherical distribution of matter between us and the center of the galaxy.[†] How much total mass exists between the orbit of the solar system and the center of the galaxy? The mass of the Sun is 1.99×10^{30} kg. Express your result both in kilograms and in units of solar masses. By using sampling techniques, astronomers can estimate the number of the stars (luminous matter) between the Sun and the center of the galaxy. The results are significantly less than the number deduced by arguments such as those used in this problem. This has convinced many astronomers that, in addition to stars, there is a significant amount of invisible, non-luminous, so-called *dark matter* in the galaxy. (d) What effect does material in the galaxy located at distances *beyond* the galactic orbit of the solar system have on the orbital motion of the solar system about the center of the galaxy?

‡62. One of the ancient Greek models of the solar system hypothesized an identical planet (called the antichthon [ἀντίχθων], or counterearth), orbiting the Sun in the *same* orbit as the Earth but on the opposite side of the Sun as in Figure P.62. Consider the orbit to be circular in shape and of radius r. (a) Sketch a second law force diagram indicating all the forces acting on the Earth if such a counterearth existed. (b) Using Newton's second law of motion, find an expression for the orbital period of the Earth in terms of the mass M of the Sun, the mass m of the counterearth, the radius of the orbit of the Earth and counterearth, and appropriate constants. (c) Does the presence of a counterearth mean that the period of the Earth in its orbit about the Sun is less or greater than the period of the Earth without the existence of a counterearth? (d) With a circular orbit, is it possible to see the counterearth at any time

---

*One light-year is the distance travelled by light during one year.
[†] In fact this is *not* a good assumption, but making it serves to give us an order of magnitude answer.

from the Earth, say during a total eclipse of the Sun? The Sun has a diameter about 100 times that of the Earth. Explain your reasoning. (e) If the Earth and counterearth were in the same *elliptical* orbit, would it be possible for the Earth at some time to catch a direct glimpse of the counterearth during a total solar eclipse? Explain your reasoning.

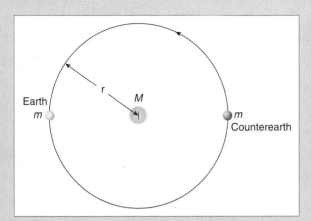

FIGURE P.62

‡63. The Sun exerts the dominant gravitational force on the Earth. The gravitational effects of the other planets on the motion of the Earth are small but measurable and are known as *perturbations*. Two distinct types of effects can take place: (1) permanent, long-term effects depending on the mere existence of another planet or the shape and location of the orbit of such a planet; these effects are known as *secular perturbations*; and (2) short-term effects depending on the detailed actual relative positions of the planets in their orbits; these effects are known as *periodic perturbations*. (a) Does the presence of the planets Mercury and Venus mean that the Earth orbits the Sun in less time or in more time than the time calculated assuming these planets did not exist? Is this effect a secular or a periodic perturbation? (b) Are the perturbations of *outer* planets on *inner* planets likely to be secular or periodic perturbations? Discrepancies between the observed and predicted motion of the planet Uranus, even after accounting for the effects of all the known planets, were used by John C. Adams (1819–1892) and Urbain J. J. LeVerrier (1811–1877) to predict the existence of a trans-Uranian planet (a planet beyond the orbit of Uranus). On the basis of their calculations and predictions, the planet Neptune was discovered on 23 September 1846 by Johann G. Galle (1812–1910) and Heinrich L. d'Arrest (1822–1875) on the first night they began the search! Ironically, discrepancies between the predicted and observed motion of the planet Mercury led LeVerrier to infer the existence of a planet closer to the Sun than Mercury, one that he even named: Vulcan. Despite numerous reported "sightings" of Vulcan, none has stood the test of time. Vulcan does not exist. The discrepancies in the motion of the planet Mercury were neatly and completely explained by Einstein as a crucial test of his general theory of relativity in 1914.

Section 6.13 The Gravitational Field

64. What are the magnitude and direction of the gravitational field of the Earth at a point $3R$ above the *surface* of the Earth, where R is the radius of the Earth?

•65. (a) At what point between the Earth and the Moon is their total gravitational field equal to zero? Express your answer in terms of the distance d between the centers of the Earth and Moon. This special point is known as the L_1 Lagrangian point. Hint:

$$\frac{\text{mass of Earth}}{\text{mass of Moon}} = 81.3$$

(b) If a mass at this location is moved slightly toward the Moon or toward the Earth, will the mass return to the L_1 point under the influence of only gravitational forces?

•66. (a) Find the magnitude of the total gravitational field at the point P in Figure P.66. (b) What is the magnitude of the acceleration experienced by a 4.00 kg salt lick at point P? (c) What is the magnitude of the total gravitational force on the salt lick if it is placed at P?

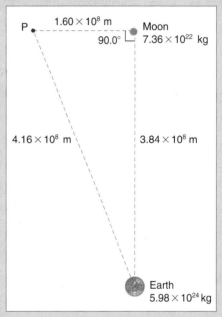

FIGURE P.66

•67. For the mass ring of Example 6.12, notice from Table 6.1 that when $z = 0$ m, the gravitational field is zero. Also at very large distances z from the ring, the field becomes vanishingly small. The magnitude of the gravitational field therefore has a maximum value somewhere along both the positive and negative z-axes. Show that the gravitational field has its maximum magnitude on the axis where $|z| = R/\sqrt{2}$.

•68. Let the mass of the Earth be $M_\oplus$ and let its radius be $R_\oplus$. (a) Show that the *ratio* between the magnitude of the gravitational field on the surface of the Earth and that on the surface of an object of radius R and mass M is

$$\frac{M}{M_\oplus}\left(\frac{R_\oplus}{R}\right)^2$$

This ratio is a pure number called the number of gs. (b) Evaluate the numerical value of this ratio for a white dwarf star that has a mass 1.50 times that of the Sun and a radius equal to the radius of the Earth. (c) Evaluate this ratio for a neutron star with a radius of 10.0 km and a mass 2.00 times that of the Sun.

69. Precise measurements of the gravitational field of the Earth indicate that it has a magnitude of 9.832 m/s² at the poles and 9.780 m/s² on the equator, indicating that the polar diameter of the Earth is slightly smaller than the equatorial diameter. Use these data, the value of G, the average radius of the Earth, and the mass of the Earth to estimate the difference between the equatorial and polar diameters of the Earth.

70. A mass M is in the shape of a thin uniform disk of radius R. Let the z-axis represent the symmetry axis of the disk as indicated in Figure P.70. The Milky Way Galaxy is modeled as such a mass disk to a first approximation. (a) Show that the gravitational field of the disk at a coordinate z along the symmetry axis of the disk is

$$\vec{g} = -\frac{2GM}{R^2}\left[1 - \frac{z}{(R^2 + z^2)^{1/2}}\right]\hat{k}$$

(b) Show that for $z \gg R$ the expression for $\vec{g}$ in part (a) becomes

$$\vec{g} \approx -\frac{GM}{z^2}\hat{k} \qquad (z \gg R)$$

That is, for distances very far from the disk along the symmetry axis, the gravitational field of the disk is essentially that of a point mass (since at great distances, the disk looks like a point). (c) Let σ be the surface mass density of the disk (the number of kilograms per square meter of the disk), so that

$$\sigma = \frac{M}{\pi R^2}$$

Show that as $z \to 0$ m along the positive z-axis, the gravitational field in part (a) approaches

$$\vec{g} \approx -2\pi G\sigma\hat{k}$$

This is a constant (uniform) field, directed downward. For very small z, the disk looks essentially like an infinite plane and resembles the situation on the surface of the Earth. On the surface of the Earth, the Earth appears to be a flat, infinite plane and we have a constant (uniform) gravitational field directed downward.

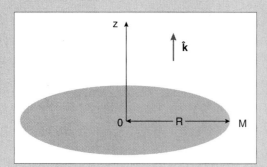

FIGURE P.70

71. The Earth is not, in fact, a sphere of uniform density. A high-density core is surrounded by a shell or mantle of lower-density material. Suppose we model a planet of radius R as indicated in Figure P.71. A core of density 2ρ and radius $3R/4$ is surrounded by a mantle of density ρ and thickness

R/4. Let M be the mass of the *core* and m be the mass of the *mantle*. (This is *not* an accurate model of the interior structure of the Earth but makes for an interesting and tractable problem.) (a) Show that the mass of the core can be expressed as

$$M = \frac{9}{8}\pi R^3\rho$$

(b) Show that the mass of the mantle can be expressed as

$$m = \frac{37}{48}\pi R^3\rho$$

(c) Show that the total mass of the planet (core + mantle) is

$$\frac{91}{48}\pi R^3\rho$$

(d) Show that the magnitude of the gravitational field at the surface of the planet is

$$g_{\text{surface}} = \frac{91}{48}\pi GR\rho$$
$$= 1.896\,\pi GR\rho$$

(e) Show that the magnitude of the gravitational field at the interface between the core and the mantle is

$$g_{\text{i}} = 2\pi GR\rho$$

Note that the magnitude of the field (the magnitude of the local acceleration of gravity) at the interface between the core and the mantle of the planet is greater than the surface field by about 5%.

This problem is based on the following two papers:
Donald G. Ivey, "Gravity in the real world," *The Physics Teacher, 30,* #4, pages 242–243 (April 1992).
Laurent Hodges, "Gravitational field strength inside the Earth," *American Journal of Physics, 59,* #10, pages 954–956 (October 1991).

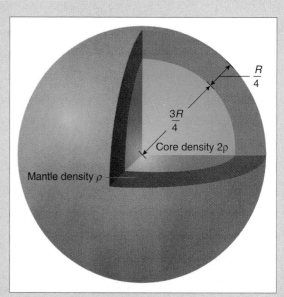

FIGURE P.71

Sections 6.14 The Flux of a Vector*
6.15 Gauss's Law for the Gravitational Field*

72. What is the flux of the gravitational field of mass m through the truncated cone shown in Figure P.72?

FIGURE P.72

•**73.** What is the flux of the total gravitational field of the Earth and Moon through a closed surface that encloses the Moon? If another surface encloses both the Earth and Moon, by how much is the flux of the gravitational field changed?

•**74.** (a) What is the flux of the total gravitational field caused by the Sun and all the planets through a closed surface that encloses only the Sun? (b) If the surface is expanded so that the entire solar system is enclosed by the surface, does the flux of the gravitational field change significantly?

•**75.** A thin uniform shell has a mass M and radius R as indicated in Figure P.75a. (a) Imagine a concentric spherical closed surface S of radius $r < R$. How much mass is within the surface S? What is the total flux of the gravitational field through this closed surface S? (b) Now imagine a closed surface of *any* shape, drawn *anywhere* inside the mass shell so long as it is totally within the mass shell as indicated in Figure P.75b. What is the mass within such a closed surface? What is the flux of $\vec{g}$ through any such surface? What does this imply about the value of the gravitational field on the closed surface within the shell? In this

way, Gauss's law for the gravitational field may be used to prove one of the gravitational shell theorems.

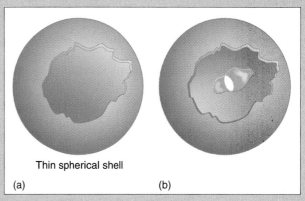

Thin spherical shell

(a)   (b)

FIGURE P.75

•**76.** Viewed from the center of a sphere, the sphere subtends an angle of 4π sr. If one marks out an area on the surface of a sphere that is one degree by one degree (see Figure P.76), the area subtends one *square degree*. Show that 4π sr is equivalent to approximately 41 253 square degrees.

FIGURE P.76

INVESTIGATIVE PROJECTS _____

A. Expanded Horizons

1. Investigate the relationship between the *inertial mass* (the mass that appears on the right-hand side of Newton's second law of motion) and the *gravitational mass* (the mass that appears in Newton's law of universal gravitation). The apparent equality of these masses is known as the principle of equivalence.
   Wolfgang Rindler, *Essential Relativity* (Springer-Verlag, New York, 1995).
   P. W. Worden Jr. and C. W. F. Everitt, Resource letter GI-1: Gravity and inertia," *American Journal of Physics*, 50, #6, pages 494–500 (June 1982); contains several references to this topic.
   Hans C. Ohanian, "What is the principle of equivalence," *American Journal of Physics*, 45, #10, pages 903–909 (October 1977).
   Albert Einstein and Leopold Infeld, *The Evolution of Physics* (Simon & Schuster, New York, 1967).

2. Investigate *Mach's principle*, the idea that the inertial mass of a system is caused by the existence of the rest of the universe.
   Wolfgang Rindler, *Essential Relativity* (Springer-Verlag, New York, 1995).
   J. David Nightingale, "Specific consequences of Mach's principle," *American Journal of Physics*, 45, #4, pages 376–379 (April 1977).
   J. Barbour and H. Pfester (editors), *From Newton's Bucket to Quantum Gravity* (Birkhauser Boston, Cambridge, Massachusetts, 1996).

3. Investigate experimental attempts to verify the inverse square law nature of the gravitational force law. That is, how close to exactly 2 is the exponent in the denominator of the expression for the magnitude of the gravitational force? Also research the consequences of *not* having the exponent exactly equal to 2.
   P. W. Worden Jr. and C. W. F. Everitt, Resource letter GI-1: Gravity and inertia," *American Journal of Physics*, 50, #6, pages 494–500 (June 1982); contains several references to this topic.
   G. T. Gillies, "Resource letter MNG-1: Measurements of Newtonian gravitation," *American Journal of Physics*, 58, #6, pages 525–534 (June 1990); this is an extensive guide to the gravitational literature.

4. During the 1970s and 1980s, there was speculation about the existence of a *fifth* fundamental force that acted like a weak, *repulsive* gravity. Investigate the state of these speculations.
   Allan Franklin, *The Rise and Fall of the Fifth Force* (American Institute of Physics, New York, 1993).
   A. P. French, "Is there a fifth fundamental force?" *The Physics Teacher*, 24, #5, pages 270–273 (May 1986).
   Robert Pool, "Was Newton wrong?" *Science*, 241, #4867, pages 789–790 (12 August 1988).

Clifford M. Will, "Is it twilight time for the fifth force?" *Sky and Telescope*, 80, #5, pages 472–479 (November 1990).

B. Lab and Field Work

5. Investigate experimental techniques used to determine the value of G, the universal constant of gravitation. In particular, use a *Cavendish balance* to measure G. To how many significant figures can G be determined in your experiment?

   Leybold-Heraeus, Instruction sheet 332 10, "Gravitation torsion balance," (1978) and an earlier version: Leybold physics leaflets, "Mechanics, Gravitation" #DC 531.51, a and b (1959); see also E. M. Purcell, "Gravitational torsion balance," *American Journal of Physics*, 25, #6, pages 393–394 (September 1957).

   Michael S. Saulnier and David Frisch, "Measurements of the gravitational constant without torsion," *American Journal of Physics*, 57, #5, pages 417–420 (May 1989).

   B. E. Clotfelter, "The Cavendish experiment as Cavendish knew it," *American Journal of Physics*, 55, #3, pages 210–213 (March 1987).

   Eric M. Rogers, *Physics for the Inquiring Mind* (Princeton University Press, Princeton, New Jersey, 1973), pages 336–339.

   Kirtley F. Mather and Shirley L. Mason, *Source Book in Geology* (McGraw-Hill, New York, 1939), pages 103–107; this is a reprint of the original experimental results of Cavendish.

   G. T. Gillies, "Resource letter MNG-1: Measurements of Newtonian gravitation," *American Journal of Physics*, 58, #6, pages 525–534 (June 1990); this is an extensive guide to the gravitational literature.

   Jesse W. Beams, "Finding a better value for G," *Physics Today*, 24, #5, pages 34–40 (May 1971).

   John Maddox, "Continuing doubt on gravitation," *Nature*, 310, page 723 (30 August 1984).

   George T. Gillies and Alvin J. Sanders, "Getting the measure of gravity," *Sky and Telescope*, 85, #4, pages 28–32 (April 1993).

6. The true orbital period of the Earth about the Sun, called the *sidereal year*, is slightly longer than the *calendar year* (also known as the tropical year) (see the footnote in Section 6.12). Investigate the reason for this difference. Design an extended experiment to measure these time intervals.

   Eric Chaisson and Steve McMillan, *Astronomy Today* (2nd edition, Prentice Hall, Upper Saddle River, New Jersey, 1996), pages 11–12, 21–22.

7. Design an experiment to measure the time interval between successive meridian passages of the Sun and the same interval for stars. Are the two intervals the same? What is the difference in the time intervals?

   Eric Chaisson and Steve McMillan, *Astronomy Today* (2nd edition, Prentice Hall, Upper Saddle River, New Jersey, 1996), page 10.

C. Communicating Physics

8. The fine structure of the gravitational field of the Earth and other planets can be inferred from slight deviations in satellite motions about the planets or, in the case of the Earth, from direct, precise measurements of the local field at various locations on the surface. Such data yield information about the interior of the planets. As part of a large grant proposal to fund such research, your supervisor assigns you the task of writing a layperson's introduction describing the use of such detailed knowledge of the gravitational field in geophysics and planetary physics.

Paul A. Bender, "Resource letter SEG-1: Solid-earth geophysics," *American Journal of Physics*, 44, #10, pages 903–911 (October 1976); this contains many references to these topics.

9. You have been commissioned by the local science museum to design a scale model of the solar system. The Sun is to be modeled as a sphere with a diameter of 1.00 m. Use the data on the inside covers to specify the diameters of the planets and their distances from the solar sphere. Report your results in tabular form.

10. Occasionally, an important discovery eludes even the best scientists. The planets Mercury, Venus, Earth, Mars, Jupiter, and Saturn have been known since antiquity. The planet Uranus was discovered accidentally in 1781 by William Herschel; the planet Neptune was discovered by Johann G. Galle and Heinrich L. d'Arrest in 1846, based on calculations and predictions by John C. Adams and Urbain J. J. LeVerrier; but Galileo inadvertently observed it on 28 December 1612. How did Galileo miss out on the discovery of Neptune when he observed it? Pluto was discovered in 1930 by Clyde Tombaugh.

    Charles T. Kowal and Stillman Drake, "Galileo's observations of Neptune," *Nature*, 287, pages 311–313, (25 September 1980).

    Stillman Drake and Charles T. Kowal, "Galileo's sighting of Neptune," *Scientific American*, 243, #6, pages 74–81, 246, (December 1980).

    Sky and Telescope, 60, #5, page 363, (November 1980).

CHAPTER 7
HOOKE'S FORCE LAW AND SIMPLE HARMONIC OSCILLATION

An oscillation is a motion that cannot make up its mind which way it wants to go.

An anonymous fourth grader

In 1660 Robert Hooke (1635–1703) described in a quantitative fashion the force that arises when a common, helically coiled spring is stretched or compressed. His empirical discovery now is called Hooke's force law. Working at Oxford University, Hooke was a contemporary of Newton at rival Cambridge University. Unfortunately, as a result of controversies regarding the priority of several discoveries,* the two gentlemen had a relationship filled with personal animosity, disrespect, and thinly veiled insults. Even great physicists are human.

In this chapter we study the force of a helical spring on a mass attached to it. If our concern about the force exerted by a simple, mundane spring on a mass appears to be a bit arcane and antiquated, in some respects it is just that. But many physical systems exhibit similar behavior and are *modeled* by a Hooke's force law to a first approximation. More advanced courses in physics and engineering return repeatedly to Hooke's force law as a foundation for understanding many complex oscillatory motions on both macroscopic and microscopic scales. For example:

1. If atoms are displaced slightly from their equilibrium positions in molecules, they experience springlike elastic restoring forces proportional to their distance from equilibrium in the molecule.
2. If you attempt to pull a quark out of a proton or other hadron, the quark experiences a restoring force that increases with distance to a crude first approximation.
3. If a beam is loaded with mass, the weight causes the beam to sag a distance that is proportional to the load.
4. If a rod is twisted, the restoring force is proportional to the twist.
5. If a steel cable on a bridge is stretched under tension, there is a restoring force proportional to the stretch.

---

*The discovery of gravitation was among them.

Robert Hooke wrote the first important treatise on observations with a microscope, coining the biological term *cell*.

Thus the relationship discovered by Hooke for simple springs has applications far beyond whatever he imagined during the 17th century. The force law is encountered in classical and quantum mechanics (for small excursions from stable equilibrium positions) as well as in many areas of engineering. As you pursue your career in the sciences, you will see Hooke's force law used to model many complex physical phenomena; so this is an important force to master even in an introductory course.

We begin by examining the force law discovered by Hooke and studying the one-dimensional oscillatory dynamics of a mass subjected to only such a force. We then investigate several different oscillatory phenomena that, at first glance, have little to do with masses attached to springs but, upon closer inspection, are camouflaged manifestations of Hooke's force law, revealing that it frequently is encountered in nature well beyond its original domain of a simple spring. Finally, conjuring up memories of your childhood experiences on a swing, we investigate realistic oscillations that either die out over time or continue because of the periodic application of additional forces.

7.1 HOOKE'S FORCE LAW

Springs have many applications, from ballpoint pen retractors and automobiles to the docking collar for the space shuttle when connecting with the Russian Mir spacecraft.

Consider a mass m attached to one end of a horizontally oriented spring, as shown in Figure 7.1. The other end of the spring is secured to a rigid, immovable wall. For simplicity, let the surface on which the mass m rests be frictionless. Choose $\hat{\jmath}$ to be vertically upward, as indicated in Figure 7.1. The position of the mass when the spring is neither stretched nor compressed is the **equilibrium position**, since there is zero acceleration and zero total force on m at this location.

> Take the origin of a Cartesian coordinate system to be at the equilibrium position of the mass, with the $+x$-axis ($\hat{\imath}$) corresponding to *stretching* the spring, as in Figure 7.1. This choice for the direction of $\hat{\imath}$ is the conventional choice, which we will use consistently throughout our treatment in this chapter and elsewhere in the text when considering springs.

Hidden springs are used for the space shuttle docking collar on the Mir spacecraft; what look like springs in this photo actually are coiled data cables.

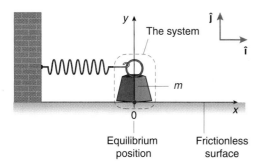

FIGURE 7.1 A mass on a frictionless surface is attached to a spring.

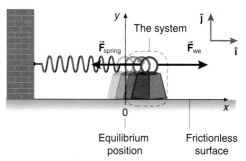

FIGURE 7.2 Horizontal forces on the mass when we stretch the spring with a force $\vec{\mathbf{F}}_{we}$.

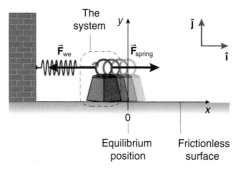

FIGURE 7.3 Horizontal forces on the mass when we compress the spring.

PROBLEM-SOLVING TACTIC

7.1 Always choose the origin to be at the equilibrium position with î corresponding to stretching the spring. The mass on the end of the spring, rather than the spring itself, is the system whose motion we will analyze. An **ideal spring** is one whose mass is negligible compared with the attached mass m. An ideal spring is, of course, an approximation to real springs, but many springs come close to this convenient idealization.

Consider all springs in this text to be ideal springs unless otherwise stated explicitly.

Two forces act on the mass in the vertical direction:

1. the upward force of the surface on the mass—the normal force; and

2. the downward gravitational force of the Earth on m—the weight of mass m.

These vertical forces, though, are irrelevant to motion in the horizontal direction since motions in perpendicular directions are independent. It is the motion parallel to the axis of the spring that we wish to examine.

We stretch or compress the spring horizontally with a force $\vec{\mathbf{F}}_{we}$ and hold the mass at rest at some position coordinate x. Since the origin was chosen at the equilibrium position of the mass, x also represents the x-component of the position vector of the system and measures the extent to which the spring is either stretched ($x > 0$ m) or compressed ($x < 0$ m). In the horizontal direction there are two forces acting on the system, shown in Figure 7.2:

1. the force $\vec{\mathbf{F}}_{we}$ that we exert on the mass to stretch the spring;

2. the force $\vec{\mathbf{F}}_{spring}$ of the spring on the mass.

Since the surface is frictionless, there is no frictional force parallel to the surface.

If we pull on the mass to stretch the spring, the force of the spring on m is directed opposite to our pull. Notice that the force of the spring on m is directed toward the equilibrium position. If instead of pulling on m and stretching the spring, we push on m

and compress the spring (see Figure 7.3), the force of the spring on m again is directed opposite to ours and once again attempts to restore m to the equilibrium position.

The force of the spring on the mass always is directed toward the equilibrium position whether the spring is stretched or compressed.

For this reason the force of the spring on m occasionally is called a **restoring force**, since it acts to restore the mass to its equilibrium position.

In the late 17th century, Robert Hooke discovered an empirical relationship between the stretch (or compression) of a spring and the force $\vec{\mathbf{F}}_{we}$ that we exert on m to hold it at rest at coordinate x. To stretch the spring, we clearly have to provide a force on m in the same direction as î; to compress the spring we need to push on m with a force directed along −î. Hooke discovered that a graph of the x-component of the force $\vec{\mathbf{F}}_{we}$ that we need to exert on m to keep it at a given value of x is directly proportional to x. In other words, the force component depends *linearly* on x, as shown in Figure 7.4.

Let the slope of such a linear plot be k. Hooke also noted that if we gradually decrease the magnitude of the x-component of the force we exert on m, the force component retraces its path on the graph; and so when the force we provide is zero, m is once again at the original equilibrium position. The graph thus has an intercept equal to zero. Furthermore, he discovered that the force on m to hold it at a given x-coordinate is, remarkably, *independent* of m and depends *only* on the particular spring used in the experiment.

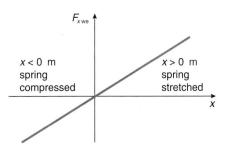

FIGURE 7.4 A graph of the *x*-component of $\vec{\mathbf{F}}_{we}$ (its only component) versus *x*.

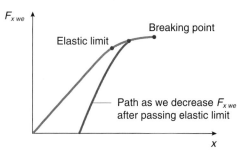

FIGURE 7.5 Stretch a spring too far and it is permanently deformed.

Since the equilibrium position of the mass on the spring is the origin of the Cartesian coordinate system chosen (Figure 7.1), we express Hooke's empirical discoveries by saying that the *x*-component of $\vec{\mathbf{F}}_{we}$ (its only component) is directly proportional to *x* itself; that is,*

$$\vec{\mathbf{F}}_{we} = kx\hat{\imath} \qquad (7.1)$$

where the proportionality constant *k*, the slope of the graph in Figure 7.4, is characteristic of the particular spring used in the experiment and is independent of the mass attached to the spring.

> The scalar constant *k* is called the **spring constant** and has the SI units of N/m.† The spring constant *k* represents the number of newtons of force needed to stretch (or compress) the spring one meter.

All springs have a positive spring constant. The larger the value of *k*, the harder it is for us to hold the mass on the spring at a given coordinate *x*. A spring with a large spring constant *k* is therefore a **stiff spring**.

If you ever have played with a small spring, such as that inside many retractable ballpoint pens, you are aware that if you stretch it too far, beyond what is called its **elastic limit**, the spring is permanently deformed and will not relax to its original equilibrium position if you subsequently remove the force you exert on *m*. If the spring is stretched still further, eventually it breaks. A schematic graph of the force component exerted versus the stretch or elongation of the spring from its relaxed position is shown in Figure 7.5.

The linear portion of the graph (including the linear section corresponding to compressing the spring) is called the **Hooke's law region**. The spring is said to be **elastic** if the stretch is confined to the Hooke's law region of the graph. In fact the graphs of many strangely behaving springlike forces, if repeatable, approach the linear shape if the departure from equilibrium is within small enough limits. Springs in special shapes like coils provide quite extensive ranges of linear behavior, and they are

examples of our ideal springs. We will only consider springs confined to the Hooke's law region.

Imagine the mass *held* at position *x*, as in Figure 7.6. The mass is not accelerated in the horizontal direction, and so the total force on the mass in the horizontal direction must be zero according to Newton's second law of motion. Thus, when $a_x = 0$ m/s², the vector sum of *the force that we exert* on the mass and *the force of the spring* on the mass must be zero:

$$\vec{\mathbf{F}}_{we} + \vec{\mathbf{F}}_{spring} = 0 \text{ N}$$

If we restrict ourselves to the Hooke's law region of the graph, and substitute for $\vec{\mathbf{F}}_{we}$ from Equation 7.1, we have

$$kx\hat{\imath} + \vec{\mathbf{F}}_{spring} = 0 \text{ N}$$

which states that the force of the spring on the mass in the Hooke's law region‡ is

$$\vec{\mathbf{F}}_{spring} = -kx\hat{\imath} \qquad (7.2)$$

Notice that if the spring is stretched, *x* is positive, and the force of the spring on *m* is toward $-\hat{\imath}$ (i.e., toward the equilibrium position). If the spring is compressed, *x* is negative, and the force of spring on *m* is toward $+\hat{\imath}$ (again, toward the equilibrium position). A graph of the *x*-component (the only component) of the force of the spring on *m* as a function of *x* is shown in Figure 7.7.

Notice that this graph has a negative slope equal to $-k$, the opposite of the slope of the graph of the component of $\vec{\mathbf{F}}_{we}$ versus *x* in Figure 7.4.

> Thus *the force of the spring* on *m*
>
> 1. is proportional to the amount of the stretch or compression of the spring and is always directed toward the equilibrium position; and
>
> 2. is independent of the mass *m* attached to the spring; the force depends only on the spring constant *k* and the stretch or compression of the spring *x*.

---

*Do not confuse the spring constant *k* (a scalar) with the Cartesian *unit vector* $\hat{\mathbf{k}}$.

†The spring constant also is known as the *force constant* of the spring. We will not use this terminology.

---

‡Problem 7 invites you to show that if the direction for $\hat{\imath}$ is chosen to correspond to *compressing* the spring, the Hooke's law expression for the force of the spring on *m still* is $\vec{\mathbf{F}}_{spring} = -kx\hat{\imath}$, unchanged in form.

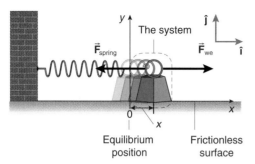

FIGURE 7.6 Stretch and hold the mass at position x.

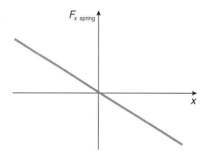

FIGURE 7.7 The x-component of $\vec{F}_{spring}$ (its only component) versus x.

PROBLEM-SOLVING TACTIC

7.2 For a mass attached to a spring, do not confuse the force we exert on the mass, $\vec{F}_{we} = kx\hat{\imath}$, with the force that the spring exerts on the mass, $\vec{F}_{spring} = -kx\hat{\imath}$. Be sure you know which force is appropriate in a problem.

QUESTION 1

When we pull on a mass attached to a horizontal spring on a frictionless surface and hold the mass at some position coordinate x, the force of the spring on m is equal in magnitude and opposite in direction to the force we exert on m. Are these forces a Newton's third law force pair? Explain why or why not.

EXAMPLE 7.1

A mechanic finds that a force of magnitude 150 N is sufficient to hold a 7.00 kg mass attached to a stiff spring 3.0 cm (stretched) from the equilibrium position. What is the force exerted by the spring on the mass when the spring is compressed 4.0 cm? The surface is frictionless.

Solution

In accordance with Problem-Solving Tactic 7.1, use the conventional coordinate system with $x = 0$ m as the equilibrium position with $\hat{\imath}$ corresponding to stretching the spring. Since the mechanic applies a force of magnitude 150 N to hold the mass 0.030 m (stretched) from its equilibrium position, use Equation 7.1,

$$\vec{F}_{mechanic} = kx\hat{\imath}$$

to find

$$(150 \text{ N})\hat{\imath} = k(0.030 \text{ m})\hat{\imath}$$

The spring constant k is

$$k = 5.0 \times 10^3 \text{ N/m}$$

Aware of Problem-Solving Tactic 7.2, do not confuse the force *the mechanic* exerts on the mass attached to the spring, $\vec{F}_{mechanic} = kx\hat{\imath}$, with the force that the *spring* exerts on the mass. The force of the spring is given by Equation 7.2:

$$\vec{F}_{spring} = -kx\hat{\imath}$$

When the spring is compressed 4.0 cm, $x = -0.040$ m, and so

$$\begin{aligned}\vec{F}_{spring} &= (-5.0 \times 10^3 \text{ N/m})(-0.040 \text{ m})\hat{\imath}\\ &= (2.0 \times 10^2 \text{ N})\hat{\imath}\end{aligned}$$

The force of the spring always is directed toward the equilibrium position. Since the spring is compressed, the force of the spring is parallel to $\hat{\imath}$.

EXAMPLE 7.2

A force of magnitude 2.50×10^3 N is able to hold a 10.0 kg mass attached to one end of a spring a distance of 10 cm from its equilibrium position. The spring is stretched.

a. What is the spring constant k of the spring?
b. What is the force of the spring on m?
c. What is the force needed to hold the mass at a location with the spring compressed by 10 cm? What then is the force of the spring on the mass at this location?

Solution

a. Take the equilibrium position of the mass as the origin for the x-coordinate with $\hat{\imath}$ corresponding to stretching the spring. The force needed to hold the mass at the coordinate $x = 0.10$ m then is, from Equation 7.1,

$$\vec{F}_{we} = kx\hat{\imath}$$
$$(2.50 \times 10^3 \text{ N})\hat{\imath} = k(0.10 \text{ m})\hat{\imath}$$

So the spring constant k is

$$k = \frac{2.50 \times 10^3 \text{ N}}{0.10 \text{ m}}$$

$$= 2.5 \times 10^4 \text{ N/m}$$

b. Since the mass is not accelerating, the force of the spring and the force we exert on the mass must vector sum to zero according to Newton's second law:

$$\vec{F}_{spring} + \vec{F}_{we} = 0 \text{ N}$$

So $\vec{F}_{spring}$ is opposite to the direction we pull and of the same magnitude as our pull. Since the force we applied was $(2.50 \times 10^3 \text{ N})\hat{\imath}$, the force of the spring on m is

$$\vec{F}_{spring} = (-2.5 \times 10^3 \text{ N})\hat{\imath}$$

This is parallel to $-\hat{\imath}$, because the force of the spring on the mass always is directed toward the equilibrium position and the spring was stretched.

c. When the spring is compressed the same distance, we need to provide a force

$$\vec{F}_{we} = (-2.5 \times 10^3 \text{ N})\hat{\imath}$$

to hold the mass at this location. Since the mass is not accelerating, the second law of motion implies

$$\vec{F}_{spring} + \vec{F}_{we} = 0 \text{ N}$$

The force of the spring on the mass thus is

$$\vec{F}_{spring} = (2.5 \times 10^3 \text{ N})\hat{\imath}$$

and, again, is directed toward the equilibrium position.

7.2 SIMPLE HARMONIC OSCILLATION

O damnéd vacillating state!
Alfred, Lord Tennyson (1809–1892)*

Many physical systems **oscillate**,[†] moving or swinging back and forth with a steady, repetitive rhythm, much like the tides. Masses attached to springs, playground swings, clock pendulums, tennis balls in a good rally, and professors pacing in front of a classroom all oscillate, or move back and forth. Here we examine a special kind of oscillation called **simple harmonic oscillation**. Not all oscillatory phenomena exhibit simple harmonic oscillation, but simple harmonic oscillation underlies the analy-

sis of almost every type of oscillation. It is important to know the characteristic features of simple harmonic oscillations and to be able to recognize whether an oscillatory motion is simple harmonic or not (which we study in Section 7.5). The oscillatory behavior of a mass on an ideal spring is the paradigm for simple harmonic oscillation.

Once again, we begin with a mass m attached to an ideal spring, lying on a horizontal frictionless surface. We pull the mass attached to the spring to the right a distance x_0 and then suddenly (instantaneously) release the mass at rest, as in Figure 7.8.

When we suddenly remove our force $\vec{F}_{we}$ on the mass, the force of the spring $\vec{F}_{spring}$ on m still exists, since the spring is stretched. From experience, we know what happens when our force is removed: the mass oscillates back and forth. Since the surface is frictionless and the spring is ideal, the oscillatory motion continues indefinitely. The oscillatory motion of m when the force of the spring is the *only* force on m in a given direction is simple harmonic oscillation, and the arrangement is called a **simple harmonic oscillator**. We describe this oscillatory motion of m in more detail using Newton's laws of motion. It typifies many cases of oscillation.

After the mass is released and located at coordinate x, the only force on the mass in the horizontal direction is the force of the spring on the mass, $\vec{F}_{spring} = -kx\hat{\imath}$. Newton's second law of motion, applied to the horizontal motion of m, then is

$$-kx\hat{\imath} = ma_x\hat{\imath} \tag{7.3}$$

The position vector of the mass is

$$\vec{r}(t) = x(t)\hat{\imath}$$

The x-component of the acceleration, a_x, is the second derivative of the x-component of the position vector with respect to time:

$$a_x = \frac{d^2x}{dt^2}$$

Make this substitution for the acceleration component in Newton's second law of motion for m. Equation 7.3 then becomes

$$-kx = m\frac{d^2x}{dt^2} \tag{7.4}$$

We want to discover how x varies with time. That is, we want to find the function $x(t)$ that satisfies this equation. Equation 7.4 is

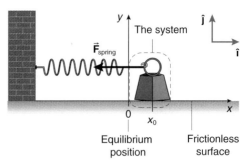

FIGURE 7.8 We pull a mass from the equilibrium position to coordinate x_0 and release the mass at rest.

Supposed Confessions of a Second-Rate Sensitive Mind Not in Unity with Itself, line 207. *The Poetical Works of Alfred Tennyson* (James R. Osgood and Company, Boston, 1877), page 285.

[†]The word comes from the Latin *oscillare*, meaning "to swing."

an example of a **differential equation**: an equation involving a function [in this case $x(t)$] and its derivatives (in this case just the second derivative). There are many types and kinds of differential equations; they are encountered frequently in physics and engineering. Many techniques exist for solving such equations, that is, for finding the function like our $x(t)$. You likely have not yet had a full-fledged course in these techniques for solving differential equations. But you will see that with your growing or existing knowledge of calculus, you can deduce a solution rather than use the formal techniques for solving such equations employed in higher-level mathematics, physics, and engineering courses.

The statement of Newton's second law for our problem, Equation 7.4, indicates that the second derivative of the (as yet) unknown function $x(t)$ is proportional to the function $x(t)$ itself. We need to find a function whose second derivative brings the *same* function back again, with perhaps a few constants thrown in for good measure. The mathematical functions that come to mind are the sines and cosines.* The first derivative of the sine is the cosine, but the second derivative of the sine brings back the sine again, apart from some purely numerical factors such as minus signs and other constants. Likewise the first derivative of the cosine is proportional to the sine, but the second derivative of the cosine brings back the cosine once again (apart from minus signs and constants).

Sine and cosine functions certainly oscillate as functions of their arguments, and so their use here may not be out of hand. Hence the solution we need for $x(t)$ likely is a sine or cosine function or some additive combination of them. We could try a solution for $x(t)$ such as

$$x(t) = a \cos \omega t + b \sin \omega t \qquad (7.5)$$

where ω, a, and b are some constants, independent of time, whose values are chosen to suit a given physical situation. Since t represents time, the constant ω must have the dimensions of inverse time, to make the argument of the trigonometric functions (the angle) dimensionless. Equation 7.5 will work just fine for these purposes. However, with no loss of generality, it is more convenient in the long run (although it is not apparent at this juncture) to use a single cosine function of the form

$$x(t) = A \cos(\omega t + \phi) \qquad (7.6)$$

where ω, A, and ϕ are three constants, independent of time, whose values must be determined by relating them to the oscil-

lator; ω and A are positive. Example 7.3 following shows that Equations 7.5 and 7.6 are equivalent expressions for $x(t)$.

PROBLEM-SOLVING TACTIC

7.3 Be careful! The quantity ωt is measured in radians, and so the constant angle ϕ, called the phase angle, also must be in radians. Thus radians are added to radians in the argument of the cosine.[†] Since the time t has the SI units of seconds and ωt is in radians, the units of ω are rad/s.

The quantity ω is called the **angular frequency** of the oscillation for reasons we will explore. It is customary *always* to indicate the SI units of the angular frequency ω explicitly as rad/s.

Recall that the maximum value of the cosine is 1; hence the maximum value of x in Equation 7.6 is A. The constant A in Equation 7.6 is called the **amplitude** of the oscillation, since it is the largest value that the x-coordinate of the oscillator can attain. Recall that x is the x-component of the position vector of the oscillator. Since the cosine can take on values only between (and including) -1 and $+1$, the range of x-coordinates visited at some time by the oscillator is twice the amplitude, or $2A$.

QUESTION 2 _____

A tennis ball is hit back and forth across the net by two excellent tennis players in fixed positions. Is the back-and-forth motion of the ball simple harmonic oscillation? Explain your reasoning.

EXAMPLE 7.3 _____

Show that Equation 7.6 is equivalent to Equation 7.5.

Solution

Use the trigonometric formula for the cosine of the sum of two angles,

$$\cos(\alpha + \beta) = \cos \alpha \cos \beta - \sin \alpha \sin \beta$$

to rewrite Equation 7.6:

$$x(t) = A \cos(\omega t + \phi)$$
$$= A \cos \omega t \cos \phi - A \sin \omega t \sin \phi$$

This is identical to Equation 7.5 if

$$a = A \cos \phi \qquad (1)$$

and

$$b = -A \sin \phi \qquad (2)$$

---

*Exponential functions involving the base of natural logarithms e also have this property. However, thanks to the Euler identity,

$$e^{i\theta} = \cos \theta + i \sin \theta$$

and its complex conjugate,

$$e^{-i\theta} = \cos \theta - i \sin \theta$$

the sines and cosines are combinations of exponentials with complex exponents. If you add the two complex exponential expressions, you find that

$$\cos \theta = \frac{e^{i\theta} + e^{-i\theta}}{2}$$

If the expressions are subtracted, you find that

$$\sin \theta = \frac{e^{i\theta} - e^{-i\theta}}{2i}$$

[†]In some disciplines, the angle ϕ occasionally is given in *degrees* for convenience. This is done frequently by electrical engineers. You must be aware that to perform the angular addition of ωt with ϕ, *both* must be expressed in the same angular units.

Divide equation (2) by equation (1) to eliminate A; this yields

$$\tan \phi = -\frac{b}{a}$$

Thus

$$\phi = \tan^{-1}\left(-\frac{b}{a}\right) \qquad (3)$$

Use the trigonometric identity

$$\sin^2 \phi + \cos^2 \phi = 1$$

and equations (1) and (2) for $\cos \phi$ and $\sin \phi$, to discover that

$$\frac{b^2}{A^2} + \frac{a^2}{A^2} = 1$$

Solve for A:

$$A = (a^2 + b^2)^{1/2} \qquad (4)$$

This shows that we can use

$$x(t) = A \cos(\omega t + \phi)$$

rather than

$$x(t) = a \cos \omega t + b \sin \omega t$$

without any loss of generality. Both expressions have two adjustable constants—A and ϕ, or a and b; so each expression will be equally able to be adjusted to match the starting conditions.

EXAMPLE 7.4 _____

Find the choices that need to be made for A and ϕ so that

$$x(t) = b \sin \omega t$$

can be written in the form

$$x(t) = A \cos(\omega t + \phi)$$

Solution
We use the trigonometric formula for the cosine of the sum of two angles,

$$\cos(\alpha + \beta) = \cos \alpha \cos \beta - \sin \alpha \sin \beta$$

to write

$$\begin{aligned} x(t) &= A \cos(\omega t + \phi) \\ &= A(\cos \omega t \cos \phi - \sin \omega t \sin \phi) \\ &= (A \cos \phi) \cos \omega t - (A \sin \phi) \sin \omega t \end{aligned}$$

Since this must be identical to $x(t) = b \sin \omega t$, we must have

$$A \cos \phi = 0 \text{ m} \qquad (1)$$

and

$$-A \sin \phi = b \qquad (2)$$

Equation (1) means we could have $A = 0$ m; but this choice is not appropriate since it implies $x(t) \equiv 0$ m for all times t. Hence, since $A \neq 0$ m, to satisfy equation (1), the angle ϕ can be $\pm \pi/2$ rad because $\cos(\pm \pi/2 \text{ rad}) = 0$. We must choose $\phi = -\pi/2$ rad to make A positive; then equation (2) implies that

$$-A \sin\left(-\frac{\pi}{2} \text{ rad}\right) = b$$

That is,

$$A = b$$

Hence $x(t) = b \sin \omega t$ is equivalent to

$$x(t) = b \cos\left(\omega t - \frac{\pi}{2} \text{ rad}\right)$$

_____

We have to see how to determine the constants ω, A, and ϕ in Equation 7.6. We know two things about the motion:

1. Equation 7.4 governs the motion at all times, since the equation stems from Newton's second law; and
2. We know how the mass was released, in our case, at position x_0 with zero speed.

Rearrange Equation 7.4:

$$m \frac{d^2x}{dt^2} + kx = 0 \text{ N}$$

Now divide by m:

$$\boxed{\frac{d^2x}{dt^2} + \frac{k}{m} x = 0 \text{ m/s}^2} \qquad (7.7)$$

This is the **equation for simple harmonic oscillation**. The equation must be satisfied by $x(t)$ at all times t, since it describes the dynamical behavior of the particle according to Newton's second law of motion. Take the function that we think is a solution, Equation 7.6,

$$x(t) = A \cos(\omega t + \phi)$$

and find its second derivative with respect to time (remember that A, ω, and ϕ are constants, independent of time). The first derivative of x with respect to time is the x-component (the only component) of the velocity:

$$\frac{dx}{dt} = -A\omega \sin(\omega t + \phi) \qquad (7.8)$$

The second derivative is the x-component (the only component) of the acceleration:

$$\begin{aligned} \frac{d^2x}{dt^2} &= -A\omega^2 \cos(\omega t + \phi) \\ &= -\omega^2 A \cos(\omega t + \phi) \\ &= -\omega^2 x(t) \end{aligned} \qquad (7.9)$$

Substitute the second derivative into Equation 7.7 for the simple harmonic oscillator:

$$-\omega^2 x(t) + \frac{k}{m} x(t) = 0 \ \text{m/s}^2$$

After factoring, we have

$$\left(-\omega^2 + \frac{k}{m}\right) x(t) = 0 \ \text{m/s}^2$$

This relation must be true at *any* time *t*. If $x(t)$ were *always* zero, we would not have much of a solution: there would be no oscillation. Indeed we know that $x(t)$ is not going to be zero at all times, since we stretched the spring and released it. Hence at all times *t* we must have

$$-\omega^2 + \frac{k}{m} = 0 \ \text{rad}^2/\text{s}^2$$

or

$$\boxed{\omega = \left(\frac{k}{m}\right)^{1/2}} \qquad (7.10)$$

The constant angular frequency ω is related to the spring constant *k* and the mass *m*. Recall that the units of ω always are expressed as rad/s. Notice that the differential equation describing the oscillation, Equation 7.7, also may be written cleanly using the angular frequency ω as

$$\boxed{\frac{d^2 x}{dt^2} + \omega^2 x = 0 \ \text{m/s}^2} \qquad (7.11)$$

The coefficient of the term involving *x* in the differential equation is the *square* of the angular frequency; we will use this fact in later examples of oscillatory motion.

> Any system of mass *m* whose position coordinate *x* satisfies Equation 7.7, or equivalently Equation 7.11, is executing simple harmonic oscillation.

We will return to investigate the meaning of ω after we discover more about the other constants we have to find: A and ϕ. The two constants A and ϕ are determined from the **initial conditions**. Such conditions also are called **boundary conditions**. Boundary conditions need not be the conditions when $t = 0$ s (which are called initial conditions). Boundary conditions also can be certain features of the motion at another specified time. In our case, the way in which the oscillation was initiated enables us to determine A and ϕ. For instance, we initiated the oscillation by stretching the spring to coordinate x_0 and releasing it with zero initial speed. Thus when $t = 0$ s, *x* must equal x_0. That is, from Equation 7.6,

$$x(0 \ \text{s}) = x_0 = A \cos[\omega(0 \ \text{s}) + \phi]$$
$$x_0 = A \cos \phi \qquad (7.12)$$

Since $x(t) = A \cos(\omega t + \phi)$, the *x*-component of the velocity (its only component) is

$$v_x(t) = \frac{dx(t)}{dt}$$
$$= -A\omega \sin(\omega t + \phi)$$

The mass was released when $t = 0$ s, at which time the velocity component was zero; thus,

$$0 \ \text{m/s} = -A\omega \sin[\omega(0 \ \text{s}) + \phi]$$
$$= -A\omega \sin \phi$$

The constant ω is not zero (earlier we found it was equal to $(k/m)^{1/2}$). Since A itself is not zero (if it *were* zero, our solution would be zero at all times, which is not the case), we must have $\phi = 0$ rad since $\sin(0 \ \text{rad}) = 0$.

Thus $\phi = 0$ rad, and, from Equation 7.12,

$$x_0 = A \cos \phi = A \cos(0 \ \text{rad})$$
$$= A \qquad (7.13)$$

PROBLEM-SOLVING TACTIC

7.4 The way the oscillator is released provides the information needed to determine the amplitude A and phase angle ϕ. Recall that the angular frequency ω depends only on the particular spring used (through the spring constant *k*) and the mass on the end of the spring; ω is *independent* of the initial conditions. See Example 7.6.

We have found that $x(t) = A \cos \omega t$, where $A = x_0$, the initial position of the mass *m* when it was released at rest. The position vector of the mass is $\vec{r}(t) = x(t)\hat{\imath}$. A graph of $x(t)$ is shown in Figure 7.9. We see that the range of *x*-coordinates visited by the oscillator is *twice* the amplitude or 2A. Note that when time $t = 0$ s, $x = A = x_0$, which is consistent with how we released the oscillator. Note also that when $\omega t = 2\pi$ rad, the mass is back at the position where it was released.

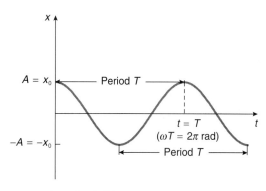

FIGURE 7.9 A graph of $x(t)$ versus *t*.

The time for the oscillator to make one complete cycle of its motion is the **period** T of the oscillation. Thus the period must satisfy the equation

$$\omega T = 2\pi \text{ rad}$$

Hence

$$T = \frac{2\pi \text{ rad}}{\omega} \qquad (7.14)$$

The **frequency** ν of the motion* is the number of cycles of the motion completed during one second. The SI units for frequency are s^{-1}; but since the frequency is so commonly used in discussing oscillatory motion, the frequency unit s^{-1} is renamed a hertz (Hz), after Heinrich Hertz (1857–1894), discoverer of radio waves (among other things). Since the period is the time for one complete oscillation, the frequency and period are reciprocals of each other:

$$\boxed{\nu = \frac{1}{T}} \qquad (7.15)$$

Notice also that since $T = (2\pi \text{ rad})/\omega$, the frequency ν is related to the angular frequency ω by

$$\nu = \frac{\omega}{2\pi \text{ rad}}$$

or equivalently,

$$\boxed{\omega = 2\pi\nu} \qquad (7.16)$$

where we suppress the radian units of the factor 2π because they are dimensionless.

The velocity component of the oscillator at any time is

$$v_x(t) = \frac{dx(t)}{dt}$$
$$= -A\omega \sin(\omega t)$$

A graph of the velocity component $v_x(t)$ is shown in Figure 7.10. Compare the graphs of $x(t)$ and $v_x(t)$. Notice that the *speed* reaches its maximum value ($A\omega$) when the oscillator is instantaneously at the *equilibrium* position $x = 0$ m.

The acceleration component $a_x(t)$ of the oscillator is found by differentiating the velocity component $v_x(t)$ with respect to time:

$$a_x(t) = \frac{dv_x(t)}{dt}$$

*The frequency of a system undergoing simple harmonic oscillation under the action of a Hooke's law force also is called the **natural oscillation frequency** of the oscillator.

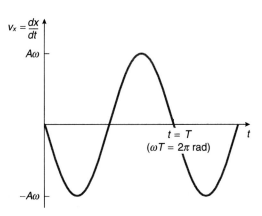

FIGURE 7.10 A graph of the velocity component $v_x(t)$ versus t.

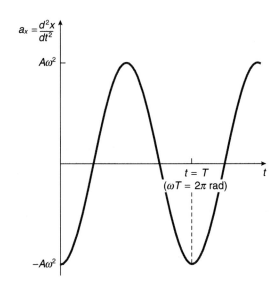

FIGURE 7.11 A graph of the acceleration component $a_x(t)$ versus t.

Since $v_x(t) = -A\omega \sin \omega t$,

$$a_x(t) = \frac{d}{dt}[-A\omega \sin(\omega t)]$$
$$= -A\omega^2 \cos(\omega t)$$

A graph of the acceleration component is shown in Figure 7.11. Notice that the acceleration component of a simple harmonic oscillator is *not constant*.

By comparing the graphs in Figures 7.9–7.11, we see that the magnitude of the maximum acceleration is $A\omega^2$. This occurs when the oscillator is instantaneously at its maximum distance from the equilibrium position (where $x = \pm A$) and its speed is zero. At these positions (where $x = \pm A$), the force of the spring on the mass is greatest, so that one expects the acceleration to be greatest at these positions too.

STRATEGIC EXAMPLE 7.5

A 0.500 kg passive, ruffed grouse (*Bonasa umbellus*) on a frictionless horizontal surface is attached to an ideal spring and is found to complete one oscillation every 2.00 s. The range of the oscillation is measured to be 0.40 m. The bird has zero speed when $t = 0$ s.

a. Determine the period T, the frequency ν, and angular frequency ω of the oscillation.
b. Find the spring constant k of the spring.
c. Determine the amplitude A and phase angle ϕ of the oscillation.
d. Find $x(t)$, where $x = 0$ m is the equilibrium position of the mass on the spring.
e. Determine $v_x(t)$ and the maximum speed.
f. Determine $a_x(t)$ and the magnitude of the maximum acceleration.

Solution

a. The bird mass executes one complete oscillation during 2.00 s. The period T of the oscillation is the time interval for one complete oscillation, and so

$$T = 2.00 \text{ s}$$

The frequency ν is the number of oscillations during one second, or the inverse of the period:

$$\nu = \frac{1}{T}$$

$$= \frac{1}{2.00 \text{ s}}$$

$$= 0.500 \text{ Hz}$$

The frequency ν *always* is expressed in hertz (Hz) in SI units.

The angular frequency ω is related to ν by Equation 7.16:

$$\omega = 2\pi\nu$$

$$= (2\pi \text{ rad})(0.500 \text{ Hz})$$

$$= 3.14 \text{ rad/s}$$

The angular frequency ω *always* is expressed in rad/s in SI units.

b. The angular frequency ω is related to the spring constant k and the mass m on the spring via Equation 7.10:

$$\omega = \left(\frac{k}{m}\right)^{1/2}$$

Solving for k, you find

$$k = m\omega^2$$

$$= (0.500 \text{ kg})(3.14 \text{ rad/s})^2$$

$$= 4.93 \text{ N/m}$$

The radian unit has no dimensions, and so k can be expressed in terms of N/m. The SI units for the spring constant *always* are expressed as N/m rather than kg/s^2. You should convince yourself that kg/s^2 are equivalent to N/m.

c. The range of the oscillation is 0.40 m. This is *twice* the amplitude A. Thus the amplitude A of the oscillation is

$$A = 0.20 \text{ m}$$

The mass has zero speed when $t = 0$ s; this information is used to determine the angle ϕ in $x(t) = A\cos(\omega t + \phi)$.

The velocity component is

$$v_x(t) = \frac{dx(t)}{dt}$$

$$= -A\omega \sin(\omega t + \phi)$$

When $t = 0$ s, $v_x(t) = 0$ m/s, because the oscillator was released at rest; hence

$$0 \text{ m/s} = -A\omega \sin(0 \text{ rad} + \phi)$$

This implies that

$$\phi = 0 \text{ rad}$$

d. With A and ϕ now determined, the position at any time is

$$x(t) = (0.20 \text{ m}) \cos[(3.14 \text{ rad/s})t]$$

e. The velocity component is

$$v_x(t) = \frac{dx(t)}{dt}$$

Use the expression for $x(t)$ in (d) to find

$$v_x(t) = (0.20 \text{ m})(-1) \sin[(3.14 \text{ rad/s})t](3.14 \text{ rad/s})$$

$$= (-0.63 \text{ m/s}) \sin[(3.14 \text{ rad/s})t]$$

Since the maximum value of the sine is 1, the maximum speed is 0.63 m/s; this is $A\omega$, according to Equation 7.8 and Figure 7.10.

f. The acceleration component $a_x(t)$ is the time derivative of $v_x(t)$:

$$a_x(t) = \frac{dv_x(t)}{dt}$$

$$= (-0.63 \text{ m/s}) \cos[(3.14 \text{ rad/s})t] (3.14 \text{ rad/s})$$

$$= (-2.0 \text{ m/s}^2) \cos[(3.14 \text{ rad/s})t]$$

Since the maximum value of the cosine is 1, the magnitude of the maximum acceleration is 2.0 m/s²; this is $A\omega^2$ (see Equation 7.9 and Figure 7.11).

STRATEGIC EXAMPLE 7.6

A 1.50 kg grapefruit is on a horizontal frictionless surface and attached to a spring with spring constant 75.0 N/m. With $x = 0$ m as the equilibrium position of the mass and $\hat{\imath}$ in the direction of stretching the spring, the moving mass is released when $t = 0$ s at $x = 0.200$ m with an initial velocity $(-0.500 \text{ m/s})\hat{\imath}$.

a. **Find the constants ω, A, and ϕ in Equation 7.6.**
b. **Make a graph of $x(t)$ versus t during two periods of the motion.**
c. **Find and plot $v_x(t)$ over the same domain for t.**
d. **Find and plot $a_x(t)$ over the same domain for t.**

Solution

a. The angular frequency depends only on m and k, not on how the mass is released. In particular, from Equation 7.10 you have

$$\omega = \left(\frac{k}{m}\right)^{1/2}$$

$$= \left(\frac{75.0 \text{ N/m}}{1.50 \text{ kg}}\right)^{1/2}$$

$$= 7.07 \text{ rad/s}$$

Since the oscillator is released when $t = 0$ s at the position $x = 0.200$ m, Equation 7.6 becomes

$$0.200 \text{ m} = A \cos \phi \qquad (1)$$

Differentiate Equation 7.6 to find the velocity component at any time:

$$v_x(t) = -A\omega \sin(\omega t + \phi)$$

According to the given information, when $t = 0$ s, the velocity component is -0.500 m/s, and so

$$-0.500 \text{ m/s} = -A\omega \sin \phi$$
$$0.500 \text{ m/s} = A\omega \sin \phi \qquad (2)$$

Divide equation (2) by equation (1) to eliminate A; you get

$$\frac{0.500 \text{ m/s}}{0.200 \text{ m}} = \frac{A\omega \sin \phi}{A \cos \phi}$$

$$2.50 \text{ s}^{-1} = \omega \tan \phi$$

Since $\omega = 7.07$ rad/s,

$$\tan \phi = 0.354$$

Thus

$$\phi = 0.340 \text{ rad}$$

Knowing both ω and ϕ, you find A using either equation (1) or (2). The result is

$$A = 0.212 \text{ m}$$

The position at any time, Equation 7.6, then is

$$x(t) = (0.212 \text{ m}) \cos[(7.07 \text{ rad/s})t + 0.340 \text{ rad}] \quad (3)$$

b. The frequency ν of the oscillation is related to ω by Equation 7.16; so

$$\nu = \frac{\omega}{2\pi}$$

$$= \frac{7.07 \text{ rad/s}}{2\pi \text{ rad}}$$

$$= 1.13 \text{ Hz}$$

The period T is the inverse of the frequency:

$$T = \frac{1}{\nu} = \frac{2\pi \text{ rad}}{7.07 \text{ rad/s}}$$

$$= 0.889 \text{ s}$$

A plot of $x(t)$, given by equation (3), encompassing two periods, from $t = 0$ s to 1.78 s is shown in Figure 7.12.

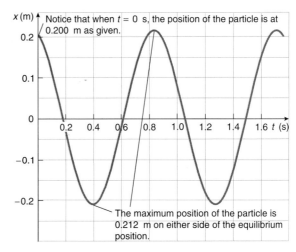

FIGURE 7.12

c. The velocity component at any time is found by differentiating Equation 7.6 with respect to time:

$$v_x(t) = \frac{dx(t)}{dt}$$

$$= -A\omega \sin(\omega t + \phi)$$

$$= (-1.50 \text{ m/s}) \sin[(7.07 \text{ rad/s})t + 0.340 \text{ rad}] \quad (4)$$

One could also differentiate equation (3) with respect to time and obtain the same result. A graph of this over two periods is shown in Figure 7.13.

d. The acceleration component at any time is the second derivative of Equation 7.6 with respect to time:

$$a_x(t) = -A\omega^2 \cos(\omega t + \phi)$$

$$= (-10.6 \text{ m/s}^2) \cos[(7.07 \text{ rad/s})t + 0.340 \text{ rad}]$$

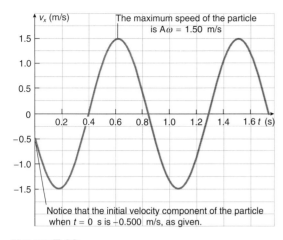

The maximum speed of the particle is $A\omega = 1.50$ m/s

Notice that the initial velocity component of the particle when $t = 0$ s is -0.500 m/s, as given.

FIGURE 7.13

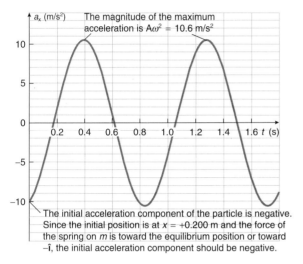

The magnitude of the maximum acceleration is $A\omega^2 = 10.6$ m/s^2

The initial acceleration component of the particle is negative. Since the initial position is at $x = +0.200$ m and the force of the spring on m is toward the equilibrium position or toward $-\hat{\imath}$, the initial acceleration component should be negative.

FIGURE 7.14

One also can obtain the same result by differentiating equation (4) once with respect to time. A graph of the acceleration component is shown in Figure 7.14.

7.3 A VERTICALLY ORIENTED SPRING

In the previous section we found that a mass m displaced from its equilibrium position on a spring oriented *horizontally*, on a *frictionless plane*, executes simple harmonic oscillation about the equilibrium position of the spring.

If the same spring now hangs by one end in the *vertical* direction, will a mass suspended from it still execute simple harmonic motion? When the spring is vertical, there are *two* vertical forces acting on the mass after it is released: the force

of the spring on m and the *constant* gravitational force of the Earth on m (its weight). How does the additional constant force affect the motion?

Let $x' = 0$ m correspond to the position of the bottom of the vertically oriented spring when *no* mass m is hung from it, as shown in Figure 7.15. The origin of the x'-axis is at this position $0'$. By convention, the positive x'-direction is taken to be in the direction of stretching the spring, and so $\hat{\imath}$ in Figure 7.15 is directed vertically *down*, as indicated.

A mass m now is carefully hung on the end of the spring. The spring stretches so that the mass is in equilibrium at coordinate x'_e, shown in Figure 7.16. There are two forces acting on the mass:

1. The gravitational force of the Earth on m, the weight $\vec{\mathbf{w}}$ of the mass. In terms of the coordinate system chosen,

$$\vec{\mathbf{w}} = mg\hat{\imath}$$

2. The force of the spring on the mass is

$$\vec{\mathbf{F}}_{spring} = -kx'_e\hat{\imath}$$

since the end of the spring was stretched to coordinate x'_e when the mass was hung on it.

$x' = 0$ m — $0'$

FIGURE 7.15 A vertical spring.

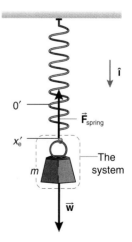

$0'$

$\vec{\mathbf{F}}_{spring}$

x'_e

m — The system

$\vec{\mathbf{w}}$

FIGURE 7.16 A mass on the vertical spring stretches it and is in equilibrium at coordinate x'_e.

The mass m is in equilibrium (zero acceleration, $a_{x'} = 0$ m/s²). Hence Newton's second law of motion becomes

$$F_{x' \text{ total}} = ma_{x'}$$
$$-kx'_e + mg = m(0 \text{ m/s}^2)$$

Transposing, we find

$$kx'_e = mg \qquad (7.17)$$

Now introduce a new x-axis with its origin 0 at the new equilibrium position x'_e. Pull the mass m down an additional distance x, *measured from the new equilibrium position*. Release the mass m with zero initial speed suddenly at this location. After release, there again are *two* forces acting on m, as shown in Figure 7.17.

1. The force of the spring on m. Since the total stretch of the spring is now $x'_e + x$, the (upward) force of the spring on m is

$$-k(x'_e + x)\hat{\imath}$$

2. The constant gravitational force of the Earth on m, its weight $\vec{w}$. The weight of the mass m is unchanged, and equal to

$$mg\hat{\imath}$$

Thus the *total force* acting on m in the vertical direction is

$$-k(x'_e + x)\hat{\imath} + mg\hat{\imath}$$

This is the left-hand side of Newton's second law. Newton's second law becomes

$$(-kx'_e - kx + mg)\hat{\imath} = ma_x\hat{\imath}$$

Using Equation 7.17 to substitute for kx'_e, we have

$$-mg - kx + mg = ma_x$$

and so

$$-kx = ma_x$$

Since

$$a_x = \frac{d^2x}{dt^2}$$

we have, after transposing,

$$m\frac{d^2x}{dt^2} + kx = 0 \text{ N}$$

Dividing by m, you may be surprised to find

$$\frac{d^2x}{dt^2} + \frac{k}{m}x = 0 \text{ m/s}^2$$

This is *identical* to the equation we found for the spring oscillating in the horizontal direction on a frictionless surface (Equation 7.7). The angular frequency for oscillations in the horizontal direction on a frictionless plane was (Equation 7.10)

$$\omega = \left(\frac{k}{m}\right)^{1/2}$$

and the angular frequency of the oscillation in the vertical direction is the *same*.

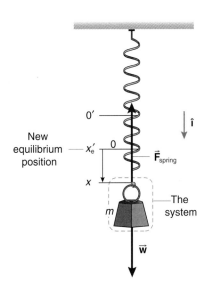

FIGURE 7.17 The mass is pulled down an additional distance x from the new equilibrium position and released at rest.

The only change when the spring is hung vertically is that the simple harmonic oscillation of m in the vertical direction is about the new equilibrium position of the mass when it is hung on the vertical spring. The constant force of the Earth on m has no effect on the angular frequency of the oscillation; it merely shifts the equilibrium position of the oscillation.

The frequency v of the oscillation and the period T of the oscillation are simply related to the angular frequency ω by Equations 7.15 and 7.16, as before:

$$\omega = 2\pi v = \frac{2\pi}{T}$$

Hence we can conclude:

> The frequency and period are unchanged by hanging the spring in the vertical direction.

QUESTION 3

A mass m undergoes vertical simple harmonic oscillation with frequency v on a spring located on the surface of the Earth. The system then is put into orbit about the Earth. Will the mass execute simple harmonic oscillation with the same frequency v while in orbit? Explain your reasoning.

EXAMPLE 7.7

An ideal spring with spring constant 150 N/m is hung vertically and has an unstretched length of 0.250 m. A 2.00 kg mass is hung from the spring. Find the equilibrium position of the mass with respect to the suspension point of the spring. If the mass is set into vertical oscillation, determine the angular frequency ω and the frequency v of oscillation about this equilibrium position.

Solution

In equilibrium, the spring has been slowly stretched until the upward force of the spring on the mass is equal in magnitude to the downward force of the Earth on the mass (its weight). The mass is unaccelerated in equilibrium.

With the $+x'$-axis downward (the direction that the spring will stretch when the mass is attached), let $x' = 0$ m correspond to the position of the bottom of the spring when no mass is attached, as in Figure 7.18a. Let x'_e be the new coordinate of the end of the spring when the mass is attached and in equilibrium, as in Figure 7.18b.

Newton's second law becomes

$$-kx'_e\hat{\imath} + mg\hat{\imath} = m(0 \text{ m/s}^2)\hat{\imath}$$

Hence

$$kx'_e = mg$$

and so

$$x'_e = \frac{mg}{k}$$

Substituting for m, g, and k, we have

$$x'_e = \frac{(2.00 \text{ kg})(9.81 \text{ m/s}^2)}{150 \text{ N/m}}$$

$$= 0.131 \text{ m}$$

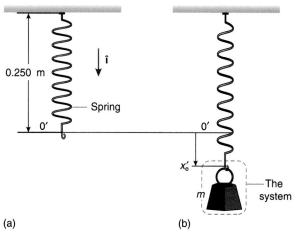

(a)   (b)

FIGURE 7.18

The mass is in equilibrium 0.131 m + 0.250 m = 0.381 m from the point of support of the spring. It is about this equilibrium position that the mass m will execute simple harmonic oscillation if the mass is, say, pulled down further and released.

The angular frequency of the oscillation ω is, from Equation 7.10,

$$\omega = \left(\frac{k}{m}\right)^{1/2}$$

$$= \left(\frac{150 \text{ N/m}}{2.00 \text{ kg}}\right)^{1/2}$$

$$= 8.66 \text{ rad/s}$$

The frequency of oscillation v is found using Equation 7.16:

$$v = \frac{\omega}{2\pi \text{ rad}}$$

$$= 1.38 \text{ Hz}$$

EXAMPLE 7.8

A mass m is on the end of an ideal spring (with a spring constant k) and is released at the equilibrium position ($x = 0$ m) when $t = 0$ s with an initial velocity $v_0\hat{\imath}$.

a. Determine the constants A, ω, and ϕ in the general solution for $x(t)$:

$$x(t) = A \cos(\omega t + \phi)$$

b. Make a graph of $x(t)$ versus t if

$$m = 0.200 \text{ kg}$$
$$k = 10.0 \text{ N/m}$$
$$v_0 = 0.500 \text{ m/s}$$

over the domain $0 \text{ s} \leq t \leq 2.0 \text{ s}$.

c. What is the period of the oscillation?

Solution

a. Since the particle is located at $x = 0$ m when $t = 0$ s, substitute these values into Equation 7.6, $x(t) = A \cos(\omega t + \phi)$. This yields

$$0 \text{ m} = A \cos \phi$$

Since $A \neq 0$ m, we could have $\phi = \pm \pi/2$ rad; choose $-\pi/2$ rad so A will be positive (see below).

We also are given the velocity of the mass when $t = 0$ s. The velocity component $v_x(t)$ at any time is

$$v_x(t) = \frac{dx(t)}{dt}$$
$$= -A\omega \sin(\omega t + \phi)$$

Since $\phi = -\pi/2$ rad and $v_x = v_0$ when $t = 0$ s, make these substitutions to obtain

$$v_0 = -A\omega \sin\left(-\frac{\pi}{2} \text{ rad}\right)$$

Hence

$$A = \frac{v_0}{\omega}$$

The solution for $x(t)$ thus is

$$x(t) = \frac{v_0}{\omega} \cos\left(\omega t - \frac{\pi}{2} \text{ rad}\right)$$

where $\omega = (k/m)^{1/2}$.

b. With the data provided,

$$\omega = \left(\frac{k}{m}\right)^{1/2}$$
$$= \left(\frac{10.0 \text{ N/m}}{0.200 \text{ kg}}\right)^{1/2}$$
$$= 7.07 \text{ rad/s}$$

Likewise

$$A = \frac{v_0}{\omega}$$
$$= \frac{0.500 \text{ m/s}}{7.07 \text{ rad/s}}$$
$$= 0.0707 \text{ m}$$

Therefore

$$x(t) = (0.0707 \text{ m}) \cos\left[(7.07 \text{ rad/s}) t - \frac{\pi}{2} \text{ rad}\right]$$

This is graphed in Figure 7.19.

c. The period of the oscillation is

$$T = \frac{2\pi \text{ rad}}{\omega}$$

where $\omega = 7.07$ rad/s, and so
$$T = 0.889 \text{ s}$$

7.4 CONNECTION BETWEEN SIMPLE HARMONIC OSCILLATION AND UNIFORM CIRCULAR MOTION

There is an interesting connection between simple harmonic oscillation and uniform circular motion (Sections 4.5 and 4.10). The position vector of a particle executing uniform circular motion at constant angular velocity $\vec{\omega}$ in a circle of radius r is Equation 4.28 with $\theta(t) = \omega t$, where ω is the magnitude of the angular velocity vector (the angular speed):

$$\vec{r}(t) = (r \cos \omega t)\hat{\imath} + (r \sin \omega t)\hat{\jmath}$$

Let the radius r of the circle be A, so that the position vector is (see Figure 7.20)

$$\vec{r}(t) = (A \cos \omega t)\hat{\imath} + (A \sin \omega t)\hat{\jmath}$$

FIGURE 7.19

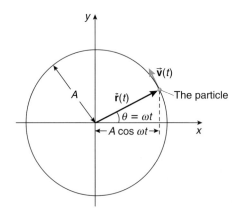

FIGURE 7.20 The position vector for uniform circular motion.

The x-component of the position vector is

$$x(t) = A \cos \omega t$$

This is the same description we found for the motion of the mass on the end of a spring executing *simple harmonic oscillation* at angular frequency ω and amplitude A with a phase angle $\phi = 0$ rad.

> We always can describe simple harmonic oscillation along a line as a component of a uniform circular motion position vector in a common plane of the motion.

If the light from a distant spotlight is directed vertically down on the particle executing uniform circular motion in a vertical plane, the *shadow* of the particle on a horizontal surface executes simple harmonic oscillation. The shadow represents the projection of the uniform circular motion of the particle onto a horizontal line; see Figure 7.21. It is because of this similarity that the constant

$$\omega = \left(\frac{k}{m} \right)^{1/2}$$

is called the angular frequency of the simple harmonic oscillation.

> • When referring to *simple harmonic oscillation*, ω is the *angular frequency* of the oscillation.
>
> • When referring to *uniform circular motion*, ω is the *angular speed* (the magnitude of the angular velocity).

QUESTION 4

A particle is executing uniform circular motion at constant speed in a vertical circle. Is the projection of the motion along a line in the plane of the circle but inclined at 45° to the horizontal simple harmonic oscillation?

FIGURE 7.21 A satellite in a circular orbit at constant speed. Viewed in the plane of the orbit, the motion is simple harmonic oscillation.

7.5 HOW TO DETERMINE IF AN OSCILLATORY MOTION IS SIMPLE HARMONIC OSCILLATION

Not every type of oscillatory motion is simple harmonic oscillation.

> Simple harmonic oscillation is a very particular kind of oscillation.

Oscillatory motion *is* simple harmonic oscillation *only* if either of two equivalent conditions is met:

1. If there is a total force on a system that is proportional to the position coordinate of the system measured from its equilibrium position with an effective spring constant k,

$$\vec{F}_{\text{total}} = -kx\hat{\imath}$$

   whether the force is from an actual spring or not, and if this force, or this force and a *constant* force along the same axis,* are the *only* forces acting in the x-direction.

2. If, in applying Newton's second law to a system, we find a differential equation *identical in form* to the differential equation for a simple harmonic oscillator (Equation 7.11),

$$\frac{d^2x}{dt^2} + \omega^2 x = 0 \text{ m/s}^2$$

   then the system will execute simple harmonic oscillation. The coefficient of the linear term x in the differential equation is the square of the angular frequency ω of the simple harmonic oscillation.

If the motion is found to be simple harmonic oscillation by virtue of one these conditions, then the angular frequency of the oscillation ω and the effective spring constant k are related by Equation 7.10:

$$\omega = \left(\frac{k}{m} \right)^{1/2}$$

and the period T, frequency ν, and angular frequency ω are related by Equations 7.15 and 7.16:

$$T = \frac{1}{\nu}$$

$$\omega = 2\pi\nu$$

We also can simply write down the expression for $x(t)$ (Equation 7.6),

$$x(t) = A \cos(\omega t + \phi)$$

without going through all the formalities again. The velocity component $v_x(t)$ and the acceleration component $a_x(t)$ then can be found by differentiating $x(t)$ once [for $v_x(t)$] and twice [for $a_x(t)$].

---

*Such as the constant gravitational force with a spring oriented in the vertical direction; see Section 7.3.

7.6 THE SIMPLE PENDULUM

A **simple pendulum** consists of a mass particle suspended on an ideal string, as shown in Figure 7.22. The pendulum swinging methodically in a grandfather's clock is a simple pendulum to a first approximation. A playground swing is another example. Is such oscillatory motion simple harmonic oscillation? Can you and your date "swing together" on side-by-side swings of the same length? How does the period of oscillation depend on the mass of the person in the swing and its length? Here we examine the physics of a simple pendulum.

If the mass is brought slightly to the side and released, the pendulum swings back and forth much like a swing. Let's look at this motion in more detail. The mass on the end of the string is the system. Draw a second law force diagram indicating all the forces acting on this system; there are two such forces, as shown in Figure 7.23:

1. the force $\vec{T}$ of the string on m (equal in magnitude to the tension in the string); and
2. the gravitational force of the Earth on m, its weight $\vec{w}$.

A coordinate system is introduced in Figure 7.23 with the origin at the equilibrium position of the pendulum at the bottom of its path. The force of the string on m is written in terms of its Cartesian components as*

$$\vec{T} = (-T \sin \theta)\hat{\imath} + (T \cos \theta)\hat{\jmath}$$

The weight is

$$\vec{w} = -mg\hat{\jmath}$$

Along the y-axis, Newton's second law is

$$F_{y\ total}\hat{\jmath} = ma_y\hat{\jmath}$$

where $F_{y\ total}\hat{\jmath}$ is the vector sum of the forces along the y-direction and a_y is the corresponding acceleration component. The vector sum of the forces along the y-direction is

$$(-mg + T \cos \theta)\hat{\jmath}$$

So for the y-direction, the Newton's second law of motion is

$$-mg + T \cos \theta = m \frac{d^2y}{dt^2} \qquad (7.18)$$

Geometrically, from Figure 7.23,

$$\cos \theta = \frac{\ell - y}{\ell}$$

---

*Do not confuse the notation for the tension force of the string $\vec{T}$ (a force), or its scalar magnitude T, with the *period* T of simple harmonic oscillation (a scalar time in seconds). These are the conventional symbols for each. It should be clear, if only from a dimensional analysis, whether T means a force magnitude or a time.

FIGURE 7.22 A simple pendulum.

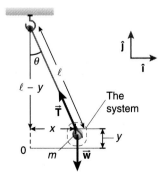

FIGURE 7.23 Forces on the system.

The small-amplitude oscillation of a swing is simple harmonic oscillation.

Thus Equation 7.18 for the y-motion is

$$-mg + T\frac{\ell - y}{\ell} = m\frac{d^2y}{dt^2}$$

In the x-direction Newton's second law is

$$F_{x\,total}\hat{\mathbf{i}} = ma_x\hat{\mathbf{i}}$$

where $F_{x\,total}\hat{\mathbf{i}}$ is the vector sum of the forces along the x-direction and a_x is the corresponding acceleration component. There is only one force along the x-direction:

$$(-T\sin\theta)\hat{\mathbf{i}}$$

Thus Newton's second law applied to the x-direction is

$$-T\sin\theta = m\frac{d^2x}{dt^2} \qquad (7.19)$$

Geometrically, from Figure 7.23, $\sin\theta = x/\ell$. Hence Equation 7.19 becomes

$$-T\frac{x}{\ell} = m\frac{d^2x}{dt^2} \qquad (7.20)$$

If the deviation of the pendulum from the equilibrium position is small (meaning that the angle θ made by the cord with the vertical direction is small), then the vertical motion is negligible compared with the horizontal motion along x. With this small-angle approximation, then, we neglect the acceleration in the vertical direction:

$$\frac{d^2y}{dt^2} \approx 0 \ \ \text{m/s}^2$$

Also, for small angles θ, $y \ll \ell$, and so the expression for the cosine of θ becomes

$$\cos\theta = \frac{\ell - y}{\ell} \approx 1$$

The equation for the vertical (y) motion (Equation 7.18) then becomes

$$-mg + T \approx 0 \ \ \text{N}$$

Hence

$$T \approx mg$$

for the magnitude of the force of the string on the mass.

Under these approximations, the equation in the horizontal direction, Equation 7.20, becomes (after substituting for T)

$$-mg\frac{x}{\ell} = m\frac{d^2x}{dt^2}$$

Notice that with this approximation, the force acting to restore the pendulum to its equilibrium position [$= (-mgx/\ell)\hat{\mathbf{i}}$] is proportional to the horizontal position of the mass from the equilibrium position. This is a Hooke's law type of force where the effective spring constant k is mg/ℓ.

The equation governing the motion of the pendulum in the horizontal direction becomes, after transposing,

$$\frac{d^2x}{dt^2} + \frac{g}{\ell}x = 0 \ \ \text{m/s}^2 \qquad (7.21)$$

which has the same form as the equation for the motion of a mass on the end of a spring (Equation 7.11). A pendulum swinging through small angles with respect to the vertical thus is essentially a simple harmonic oscillator. The solution for $x(t)$ is

$$x(t) = A\cos(\omega t + \phi)$$

The coefficient of the term involving x in Equation 7.21 is the square of the angular frequency of the oscillation. That is, comparing Equation 7.21 with Equation 7.11, we see that

$$\omega^2 = \frac{g}{\ell}$$

or

$$\omega = \left(\frac{g}{\ell}\right)^{1/2}$$

The period of the pendulum is, using Equations 7.15 and 7.16,

$$T = \frac{2\pi}{\omega}$$
$$= 2\pi\left(\frac{\ell}{g}\right)^{1/2} \qquad (7.22)$$

> The period of the pendulum is independent of the mass on its end and independent of the amplitude of the oscillation (as long as the amplitude is small compared with ℓ).

The greater the length ℓ of the pendulum, the longer the period T; the period is proportional to the square root of the length. The qualitative dependence of the period on the length is easily remembered if you think about the dynamical behavior of your childhood backyard swing: those of long length take more time to go back-and-forth than those of short length.

> A simple pendulum is a simple harmonic oscillator *only* to the extent you can neglect the vertical motion—that is, when the angle the pendulum makes with the vertical is small.

Hooke's law is a valid approximation for the simple pendulum as long as we stay within certain limits—here, small angles θ with the vertical. If θ becomes appreciable, the approximations we made are no longer valid. The motion of the pendulum, while still oscillatory, *is no longer simple harmonic oscillation* because the equations for x and y (Equations 7.18 and 7.19) then become more complicated. For large angles, the resulting force along x is not proportional to x, and so is not a Hooke's law force. We also cannot ignore the motion along the vertical direction y for large angles θ. For large angles θ, the solutions for the motion of the pendulum, $x(t)$ and $y(t)$, are complicated and beyond the scope of this course. Recall, also, that a spring can be approximated by a Hooke's force law *only* if we stay within the elastic limit of the spring. Our simple descriptions of nature have limits and are valid only to the extent we stay within them. Recognizing the existence of limits, as well as what the limits actually are, is a good lesson to learn in life as well as science.

QUESTION 5

A pendulum of length 1.00 m is hung inside a spacecraft orbiting the Earth every 100 min. What is the period of oscillation of the pendulum? Explain your answer.

EXAMPLE 7.9

Calculate the length of a swing whose period is 5.0 s for small oscillation amplitudes.

Solution

The expression for the period of a simple pendulum is Equation 7.22:

$$T = 2\pi \left(\frac{\ell}{g}\right)^{1/2}$$

Square this expression to find

$$T^2 = 4\pi^2 \frac{\ell}{g}$$

Solving for ℓ,

$$\ell = \frac{T^2 g}{4\pi^2}$$

Substituting for g and the period $T = 5.0$ s,

$$\ell = \frac{(5.0 \text{ s})^2 (9.81 \text{ m/s}^2)}{4\pi^2}$$

$$= 6.2 \text{ m}$$

Perhaps you were lucky enough to have a swing this long in your backyard!

EXAMPLE 7.10

You can measure g with a simple pendulum. Set up a simple pendulum of length 1.76 m in your room. With a stopwatch, you find that it executes 25 complete small oscillations during 66.4 s.

a. What value of g do these data imply?
b. If the oscillation amplitude were reduced to half its value, what would be the period?

Solution

a. Since 25 oscillations took 66.4 s, the time for one oscillation, the period, is

$$T = \frac{66.4 \text{ s}}{25}$$

$$= 2.66 \text{ s}$$

Since the period of the pendulum is related to its length and the value of g via Equation 7.22, solve that equation for g:

$$g = 4\pi^2 \frac{\ell}{T^2}$$

Substituting for ℓ and T, you have

$$g = 4\pi^2 \frac{1.76 \text{ m}}{(2.66 \text{ s})^2}$$

$$= 9.82 \text{ m/s}^2$$

b. The period is unchanged, because the period is independent of the amplitude for small oscillations.

EXAMPLE 7.11

What is the period of a simple pendulum whose length is 1.00 m?

Solution

The period T of a simple pendulum of length ℓ is, from Equation 7.22,

$$T = 2\pi \left(\frac{\ell}{g}\right)^{1/2}$$

Substituting numerical values for ℓ and g,

$$T = 2\pi \left(\frac{1.00 \text{ m}}{9.81 \text{ m/s}^2}\right)^{1/2}$$

$$= 2.01 \text{ s}$$

It is quite coincidental that the period of a one meter pendulum is very close to two seconds, making each swing from one side to the other take about one second. Thomas Jefferson proposed a standard unit of length based on a pendulum whose period was precisely 2 s, so that each swing had a duration of one second. Jefferson's standard was not adopted (see Problem 51).

EXAMPLE 7.12

a. Show that if the length of a simple pendulum is increased by a small amount $\Delta\ell$, the period of the pendulum increases by a small amount ΔT, where

$$\Delta T \approx \frac{\pi}{(g\ell)^{1/2}} \Delta\ell$$

b. If length of a pendulum 1.00 m long is increased by 1.0 cm, what is the change in the period?

Solution

a. The period and length of a pendulum are related by Equation 7.22:

$$T = 2\pi \left(\frac{\ell}{g}\right)^{1/2}$$

Take the derivative of T with respect to ℓ:

$$\frac{dT}{d\ell} = \frac{\pi}{(g\ell)^{1/2}}$$

Approximate the differentials with the small changes; that is, let $dT \approx \Delta T$ and $d\ell \approx \Delta \ell$, so that

$$\Delta T \approx \frac{\pi}{(g\ell)^{1/2}} \Delta \ell$$

b. For a 1.0 cm increase in the length of a 1.00 m pendulum, the increase in the period is

$$\Delta T = \frac{\pi (0.010 \text{ m})}{[(9.81 \text{ m/s}^2)(1.00 \text{ m})]^{1/2}}$$
$$= 0.010 \text{ s}$$

7.7 THROUGH A FICTIONAL EARTH IN 42 MINUTES

I'll put a girdle round the earth
In forty minutes.

William Shakespeare (1564–1616)*

In Chapter 6 we saw that as long as the mass distribution within the Earth is radially symmetric, the gravitational force of the Earth on particles on its surface or at distances from the center that are greater than the radius of the Earth is found by considering the Earth as a point particle in the gravitational force law. However, for a particle *within the Earth*, only that part of the mass of the Earth *closer* to the center than the particle produces a nonzero gravitational force on the particle.

The Earth is *not* a uniform sphere, although its mass distribution is approximately radially symmetric. The Earth has a dense nickel–iron core surrounded by material (the mantle) of lesser density. This means that the results we obtained for the magnitude of the gravitational force on a particle within a uniform sphere (Equation 6.7) cannot in fact be used for particles within the Earth. But what the heck; assume that the Earth *is* a uniform sphere, knowing full well that the assumption is only a crude first approximation. If we make this gross simplification, some rather interesting physics emerges for particles within this fictional model of the Earth.

Imagine you somehow fulfilled your childhood fantasy of digging a hole right through the center of the fictional Earth model (see Figure 7.24). A rock of mass m then is dropped into the hole. The rock falls down the hole until it reaches the center of the Earth, then goes up the hole on the opposite side.

Set up a coordinate system with its origin at the center of the fictional model Earth with an axis aligned with the hole, as in Figure 7.24. Neglect all air resistance (another unreasonable assumption) and assume the walls of the hole are frictionless.

The magnitude of the gravitational force on m is given by Equation 6.7, because the rock of mass m is within the uniform Earth. In accordance with our assumption about neglecting frictional effects, the gravitational force is the only force acting on the particle along the x-direction. With this choice of coordinates, the gravitational force of the Earth on m is

$$-\frac{GmM}{R^3} x\hat{\mathbf{i}} \qquad (7.23)$$

Notice that this force is proportional to the distance the mass m is from the origin and is directed toward the origin (at the center of the Earth) when x is positive *and* when x is negative. That is, the force is a Hooke's force law,

$$-kx\hat{\mathbf{i}} \qquad (7.24)$$

with an effective spring constant found by comparing Equation 7.23 with Equation 7.24:

$$k \equiv \frac{GmM}{R^3}$$

Hence the rock executes simple harmonic oscillation about the equilibrium position of m at the center of the fictional model Earth! Since we analyzed such motion in some detail, we immediately can write down the angular frequency of the oscillation, since it is related to the effective spring constant and the mass m on the *gravitational* spring. With Equation 7.10, we have

$$\omega = \left(\frac{k}{m}\right)^{1/2}$$
$$= \left(\frac{GM}{R^3}\right)^{1/2}$$

The period of the oscillation is found from Equations 7.15 and 7.16:

$$T = \frac{2\pi}{\omega}$$
$$= 2\pi \left(\frac{R^3}{GM}\right)^{1/2} \qquad (7.25)$$

which is independent of the mass of the rock dropped down the hole.

Another way to reach the same conclusion is to look at Newton's second law for the rock. Since the gravitational force is the only force acting on the rock, Newton's second law becomes

$$F_{x \text{ total}} = ma_x$$

$$-\frac{GmM}{R^3} x = m \frac{d^2x}{dt^2}$$

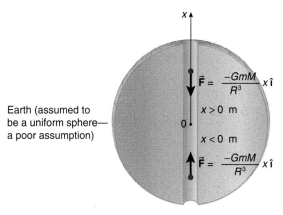

FIGURE 7.24 A frictionless hole through the center of the Earth.

The equation of motion for the rock is thus

$$\frac{d^2x}{dt^2} + \frac{GM}{R^3} x = 0 \text{ m/s}^2$$

which is identical in form to the simple harmonic oscillator equation (Equation 7.11). When the differential equation is in this form, the coefficient of x in the differential equation is ω^2. So once again we find for the angular frequency of the oscillation:

$$\omega = \left(\frac{GM}{R^3}\right)^{1/2}$$

If we substitute the numerical values for G, the radius R of the Earth, and the mass M of the Earth into Equation 7.25 for the period of the oscillation, we find

$$T = 2\pi \left(\frac{R^3}{GM}\right)^{1/2}$$

$$= 2\pi \left[\frac{(6.37 \times 10^6 \text{ m})^3}{(6.67 \times 10^{-11} \text{ N}\cdot\text{m}^2/\text{kg}^2)(5.98 \times 10^{24} \text{ kg})}\right]^{1/2}$$

$$= 5.06 \times 10^3 \text{ s}$$

or about 84.3 min. The particle can go *one-way* through the Earth in half the period, or just a tad over 42 min. This is the ultimate in express mail service: the mail delivers itself with no effort on our part other than dropping it down the letter slot. Of course there are a few minor engineering difficulties associated with constructing the slot through the center of the Earth, but . . . .

What happens if the frictionless tunnel dug through the fictional model Earth does *not* pass through the center of the Earth, as in Figure 7.25? The frictionless walls ensure that the rock slides freely in the hole. There are now two forces acting on the rock within the tunnel:

1. the gravitational force of the Earth on the rock; and
2. the normal force of the tunnel wall on the rock.

Set up the coordinate system indicated in Figure 7.25. The weight of the rock is written in terms of its Cartesian components as

$$-\frac{GmM}{R^3} r[(\cos\theta)\,\hat{\imath} + (\sin\theta)\,\hat{\jmath}]$$

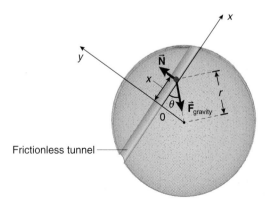

FIGURE 7.25 A frictionless tunnel between any two points on the surface of the Earth.

The rock is not accelerating in the y-direction, and so the vector sum of the two forces along y is zero according to Newton's second law:

$$\left[N - \frac{GmM}{R^3} r(\sin\theta)\right]\hat{\jmath} = 0 \text{ N}$$

There is only one force along the x-direction:

$$-\frac{GmM}{R^3} r(\cos\theta)\,\hat{\imath}$$

But from the geometry of Figure 7.25, $\cos\theta = x/r$ and the force in the x-direction is

$$-\frac{GmM}{R^3} r \frac{x}{r} \hat{\imath}$$

or

$$-\frac{GmM}{R^3} x\hat{\imath}$$

Notice that the force on m is proportional to its distance from the midpoint of the tunnel and is directed toward the origin at the center of the tunnel. Once again we have a Hooke's force law, and this is the only force on the mass in the x-direction. Therefore the mass executes simple harmonic oscillation about the midpoint of the tunnel. Indeed, the effective spring constant k for the problem is

$$k = \frac{GmM}{R^3}$$

just as with the tunnel through the center of the Earth. Therefore the period T of the oscillation is the same as for a tunnel through the center of the Earth, Equation 7.25. Hence the period of the motion through the tunnel is *independent of the path of the tunnel*. The period again is 84.3 min. A one-way trip through such a tunnel to *anywhere* on the Earth takes a little over 42 min. Such a rapid transit system need have only one timetable: 42 min to any- and everywhere. What is more, since the tunnel is frictionless, the trip uses no fuel.

7.8 DAMPED OSCILLATIONS*

The oscillations of the simple harmonic oscillator, described by Equation 7.6, continue indefinitely and do not decrease in amplitude with time. If you hang a mass on a spring, set it into oscillation, and plot the position of the mass as a function of time (see Figure 7.26), it is quite apparent that the oscillations do not continue indefinitely; they gradually decrease in amplitude and die out.

Even if we treat the spring as ideal, the decrease in amplitude arises because of frictional effects in the medium in which the mass is moving while oscillating. Such an oscillator is called a **damped harmonic oscillator**. Such oscillations are very common in the natural world and can be observed in everything from pendulums to diving boards.

Now imagine placing such an oscillator in thick, gooey molasses, pulling the mass from its equilibrium position, and then releasing the mass, as shown in Figure 7.27. The mass slowly returns to its equilibrium position without overshooting it, so

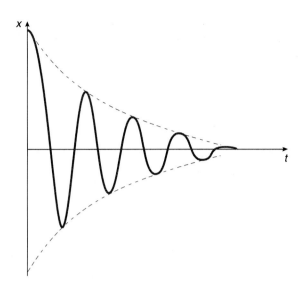

FIGURE 7.26 An oscillation that gradually dies out.

FIGURE 7.27 An oscillator in thick molasses.

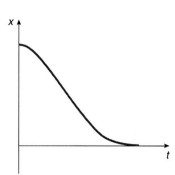

FIGURE 7.28 A plot of x versus t for an overdamped oscillator.

In a first and realistic approximation, the frictional effects are modeled by a frictional force proportional to but opposite in direction to the velocity of the mass:

$$\vec{f}_k = -\beta v_x \hat{i}$$

Newton's second law then becomes

$$-kx\hat{i} - \beta v_x \hat{i} = ma_x \hat{i} \qquad (7.26)$$

Since

$$v_x = \frac{dx}{dt}$$

and

$$a_x = \frac{d^2x}{dt^2}$$

Equation 7.26 becomes, after minor rearranging,

$$m\frac{d^2x}{dt^2} + \beta\frac{dx}{dt} + kx = 0 \text{ N} \qquad (7.27)$$

This is a more complicated differential equation than the simple harmonic oscillator equation. The mathematical technique for solving such an equation likely is unfamiliar to you, and so we will have to resort to some smoke and mirrors to find the function $x(t)$ that satisfies Equation 7.27. If you examine Figure 7.26, the damped oscillator has a sinusoidal-like oscillation, $\cos(\omega t + \phi)$, reminiscent of the simple harmonic oscillator, but whose amplitude decreases (decays) with time, perhaps as an exponential: $e^{-\alpha t}$ as an educated guess. One might, then, concoct a solution to Equation 7.27 of the form

$$x(t) = Ae^{-\alpha t}\cos(\omega t + \phi) \qquad (7.28)$$

In fact, this is just the thing to do. Differentiate Equation 7.28 to find

$$\frac{dx}{dt} \qquad \text{and} \qquad \frac{d^2x}{dt^2}$$

which will be rather messy looking expressions. Put these derivatives into Equation 7.27, and isolate the sine and cosine terms. After some tedious algebra, you should secure the following ungainly expression:

$$(m\alpha^2 - m\omega^2 - \beta\alpha + k)Ae^{-\alpha t}\cos(\omega t + \phi)$$
$$+ (2m\alpha\omega - \beta\omega)Ae^{-\alpha t}\sin(\omega t + \phi) = 0 \text{ N} \qquad (7.29)$$

that the mass is never really oscillating; this behavior is called **overdamped oscillation**. A graph of the position of the mass as a function of time appears as in Figure 7.28.

Suppose you mix water with the molasses to thin it, and repeat the experiment. When sufficient water is added to thin the molasses, the mass eventually reaches a state such that if more water is mixed, the mass, when released, will overshoot its equilibrium position before eventually coming back to rest at the equilibrium position. At this critical point of barely no overshoot, the oscillator is called a **critically damped oscillator**. If the molasses is thinned still more, the system behaves like the damped oscillator we described initially.

Here we look at such realistic oscillations in more detail. There are two forces on the mass:

1. the force of the spring,

$$\vec{F}_{\text{spring}} = -kx\hat{i}$$

where we use the standard coordinate system with positive x corresponding to stretching the spring; and
2. a kinetic frictional force.

Recall that this equation stemmed from Newton's second law, so it must be true at *any* and *every* instant t. The only way Equation 7.29 can be satisfied at all instants t is if *each* expression in parentheses is zero separately. That is,

$$2m\alpha\omega - \beta\omega = 0 \text{ kg/s}^2 \qquad (7.30)$$

and

$$m\alpha^2 - m\omega^2 - \beta\alpha + k = 0 \text{ kg/s}^2 \qquad (7.31)$$

Solve Equation 7.30 for α:

$$\alpha = \frac{\beta}{2m}$$

Use this in Equation 7.31, and then solve it for ω, obtaining

$$\omega = \left[\frac{k}{m} - \left(\frac{\beta}{2m}\right)^2\right]^{1/2} \qquad (7.32)$$

If there is no frictional force, $\beta = 0$ kg/s and the oscillation is simple harmonic with angular frequency $\omega_0 = (k/m)^{1/2}$. So we rewrite Equation 7.32 for the angular frequency of the damped oscillator as

$$\omega = \left[\omega_0^2 - \left(\frac{\beta}{2m}\right)^2\right]^{1/2} \qquad (7.33)$$

If the magnitude of the frictional force is small compared with that of the force of the spring, then β is small and the system oscillates with an angular frequency close to that of the undamped simple harmonic oscillator but with an amplitude that decreases with time. When the value of β is such that

$$\frac{\beta}{2m} = \omega_0$$

then the angular frequency ω in Equation 7.33 is zero and the oscillator is critically damped; the cosine term in Equation 7.28 then is constant, and so $x(t)$ decreases exponentially with time. If the resistive force is so great that $\beta/(2m) > \omega$, then the system is overdamped and $x(t)$ again decreases exponentially with time. For both the overdamped and critically damped conditions, the mass simply returns to and then remains at the equilibrium position without overshooting it.

7.9 FORCED OSCILLATIONS AND RESONANCE*

Think of a pendulum as a playground swing. Your little sister is on the swing and wants to be pushed. If you give her a small push and then wait for a minute or so before giving her another push, the amplitude of the motion of the swing will not be great and soon complaints will begin to emanate. For this situation, the frequency of your pushes (number of pushes per second) is small and the resulting amplitude of the swing is small. So you change strategies: now you execute a series of very fast pushes, say 10 pushes per second. Once again, the amplitude of the oscillation of the swing is not very large and your sister again starts to complain: "Come on, do it right!"

What is "right"? You push the swing every time it comes back to you. Under these circumstances the amplitude of the oscillation can grow to be quite large. What you have done is to apply a force to the oscillator at a frequency that matches the **natural oscillation frequency** of the oscillator. The natural oscillation frequency is the frequency of the swing when allowed to oscillate under the influence of *only* the Hooke's force law. The *additional* force that you exert on the oscillator (here by your occasional pushing) produces what is called **forced oscillation**. If the frequency of the applied force matches (is equal to) the natural oscillation frequency of the system, then a large-amplitude oscillation of the system results. Forced oscillation with matched frequencies produces **resonance**, large-amplitude oscillations. Resonance is a very general phenomenon with oscillators.

As another illustration of the idea of forced oscillation and resonance, fill a bathtub partially with water. Now use a paddle or a board to slosh the water back and forth in the tub, creating a "water oscillation." If the sloshing is done at too large or too small a frequency, only small-amplitude oscillations of the water are generated in the tub. But if the sloshing is done at just the right frequency (equal to the natural oscillation frequency of the water in the tub), then large-amplitude oscillations of the water are generated.

Any small child has the innate ability to find the right sloshing (forcing) frequency, with attendant parental consternation over the resulting resonance and flooded bathroom! The natural oscillation frequency of the water in the tub depends on the geometric shape of the tub, the depth of the water, and the presence of obstacles (people or rocks) in the tub itself. Since the frequency of the sloshing needed to create large water oscillations in the tub must match the natural oscillation frequency of the water in the tub, the effective frequency is different for different tubs and depths of water. In the Bay of Fundy,* a natural tub, the natural oscillation frequency of the water is close to the sloshing frequency of the *tides*; a large tidal rise and fall therefore exists in the Bay—a large tidal oscillation. The tidal rise and fall approaches 15 *meters*! Other bays have natural oscillation frequencies that are quite different from the tidal frequency, so that tidal oscillations of the water there are not as dramatic.

---

*The Bay of Fundy is located between New Brunswick and Nova Scotia.

The tidal frequency closely matches the natural oscillation frequency of the Bay of Fundy, resulting in large-amplitude tidal oscillations.

Complex structures such as bridges and buildings have many natural frequencies of oscillation. Resonance effects occur if a periodic (repetitive) force happens to coincide with one of the natural frequencies. Occasionally, disastrous effects result. Engineers must design structures to ensure that forces caused by winds, earthquakes, and the like do not induce or excite resonance effects in the structure. Tall buildings generally have lower natural frequencies of oscillation than do short buildings. Take two vertical meter sticks with one clamped at its end and the other at the midpoint; tweak them both. The one with the longer free length oscillates at a smaller natural frequency than does the one of shorter length. Earthquake motions are often closer to the natural frequencies of shorter buildings, and so these structures can be affected more severely than taller structures.

Forced oscillations and resonance effects are present in many aspects of the natural world and are involved electrically in everything from radios and TVs to lasers and medical imaging technologies. Detailed quantitative treatments of forced oscillations in a wide variety of systems (in mechanics, electromagnetism, optics, and quantum mechanics) will be encountered in more advanced courses in physics and engineering. Here we examine the behavior of such a forced oscillator in a semiquantitative way.

Assume there are three forces acting on a forced oscillator:

1. the applied force $F_{x\,\text{applied}}\hat{\imath}$;
2. frictional damping opposite in direction to the velocity; and
3. the Hooke's force law restoring force.

If, as before, we assume frictional damping proportional to the velocity, Newton's second law becomes

$$F_{x\,\text{applied}} - \beta v_x - kx = ma_x$$

or with a slight rearrangement,

$$m\frac{d^2x}{dt^2} + \beta\frac{dx}{dt} + kx = F_{x\,\text{applied}} \qquad (7.34)$$

In many situations, the applied force has an x-component that is sinusoidal:

$$F_{x\,\text{applied}} = F_0 \cos \omega t$$

The properly timed push you give your sister on a swing is *not* quite like this, since your push is applied only briefly near the peak amplitudes of the oscillation. Eventually, after a sinusoidal force is applied to the oscillator, the steady-state solution to Equation 7.34 has the form

$$x(t) = A(\omega)\cos(\omega t + \phi) \qquad (7.35)$$

where the amplitude is a function of the applied angular frequency. Using a procedure similar to that which we used for damped oscillations (substitution of the solution into the differential equation), and after a significant bit of differentiation and algebra, you will find that the amplitude depends on ω according to

$$A(\omega) = \frac{\dfrac{F_0}{m}}{\left[(\omega^2 - \omega_0^2)^2 + \left(\dfrac{\beta\omega}{m}\right)^2\right]^{1/2}} \qquad (7.36)$$

A qualitative graph of $A(\omega)$ versus ω (see Figure 7.29) confirms that the peak amplitude, as we know from our qualitative discussion, occurs when the angular frequency ω of the applied force is equal to the natural angular frequency ω_0 of the oscillator with no damping, which defines the *resonance* condition.

If there is no damping, $\beta = 0$ kg/s and the amplitude at resonance is infinite according to Equation 7.36. Since all oscillators in nature exhibit some damping, however small, an infinite amplitude never actually is encountered. As the damping increases, the amplitude of the oscillation at resonance decreases and the amplitude function becomes less sharply peaked.

Forced oscillations induced by earthquakes cause even more damage if they match the natural oscillation frequency of a building.

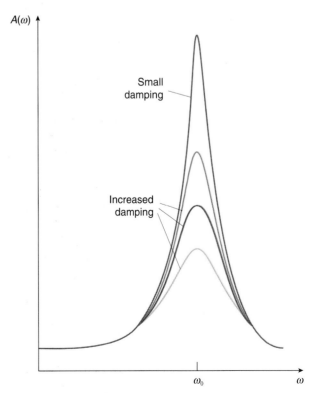

FIGURE 7.29 A graph of $A(\omega)$ versus ω.

CHAPTER SUMMARY

The force of a spring on an attached mass m is proportional to the position x of the mass m relative to its equilibrium position on the spring, taken to be $x = 0$ m:

$$\vec{F}_{\text{spring}} = -kx\hat{\imath} \qquad (7.2)$$

where k is a characteristic of the specific spring and is called the *spring constant*, expressed in N/m. By convention, $\hat{\imath}$ is chosen to be in the direction of stretching the spring. The force of the spring on m is independent of m and depends only on the stretch or compression of the spring and the spring constant. This empirical force law is known as *Hooke's force law* and is legitimate as long as the spring does not exceed its *elastic limit*.

If the mass is set into oscillation about its equilibrium position on the spring (whether the spring is horizontal or vertical), the position of the mass at any time $x(t)$ satisfies a differential equation of the form

$$\frac{d^2x}{dt^2} + \omega^2 x = 0 \ \text{m/s}^2 \qquad (7.11)$$

which is found from applying Newton's second law of motion to the mass m. Equation 7.11 is called the *simple harmonic oscillator equation*. Systems that obey Equation 7.11 are called *simple harmonic oscillators*. The solution of this differential equation is

$$x(t) = A\cos(\omega t + \phi) \qquad (7.6)$$

where A is the *amplitude* of the oscillation; its value is determined from the initial conditions of the oscillation; ω is the *angular frequency* of the oscillation in rad/s, given by

$$\omega = \left(\frac{k}{m}\right)^{1/2} \qquad (7.10)$$

and ϕ is a *phase angle* whose value is determined from the initial conditions of the oscillation.

Both the product ωt and ϕ are expressed in radians. The *period* of the oscillation T, its *frequency* ν, and the angular frequency ω are related by

$$\nu = \frac{1}{T} \qquad (7.15)$$

$$\omega = 2\pi\nu \qquad (7.16)$$

The velocity component $v_x(t)$ of the oscillator at any time is found by differentiating $x(t)$ with respect to time:

$$v_x(t) = \frac{dx(t)}{dt}$$
$$= -A\omega \sin(\omega t + \phi)$$

The acceleration component at any time is found by differentiating the velocity component with respect to time:

$$a_x(t) = \frac{dv_x(t)}{dt} = \frac{d^2x(t)}{dt^2} = -\omega^2 x(t)$$
$$= -A\omega^2 \cos(\omega t + \phi)$$

Hooke's law is important because such elastic forces are used to model many interactions in physics, engineering, and chemistry. Any situation in which you can recognize (a) that the *total* force on a system is proportional to the departure of the system from its equilibrium position—that is,

$$F_{x\,\text{total}} = -kx$$

or equivalently, (b) that in applying Newton's second law to the system you find a differential equation that has the form of Equation 7.11,

$$\frac{d^2x}{dt^2} + \omega^2 x = 0 \ \text{m/s}^2 \qquad (7.11)$$

is a system that you can conclude will execute simple harmonic oscillation. The angular frequency ω, the frequency ν, the period T, the position $x(t)$, the velocity component $v_x(t)$, and the acceleration component $a_x(t)$ are found using the equations just presented for these important parameters of the motion.

A simple pendulum is approximately a simple harmonic oscillator for small-angle oscillations with respect to the vertical direction. The period of a simple pendulum is independent of its mass and is proportional to the square root of its length ℓ:

$$T = 2\pi\left(\frac{\ell}{g}\right)^{1/2} \qquad (7.22)$$

An oscillator subjected to frictional forces is called a *damped harmonic oscillator* and the amplitude of its oscillation decreases with time. A *critically damped* or *overdamped* oscillator does not truly oscillate but merely returns to the equilibrium position without overshooting.

Large amplitudes result if a force is applied to a system at a frequency that matches a *natural oscillation frequency* of the system, a condition known as *resonance* under *forced oscillation*.

SUMMARY OF PROBLEM-SOLVING TACTICS

7.1   (page 283) Always choose the origin at the equilibrium position with $\hat{\imath}$ corresponding to stretching the spring.

7.2   (page 285) For a mass attached to a spring, do not confuse the force *we* exert on the mass, $\vec{F}_{\text{we}} = kx\hat{\imath}$, with the force that the *spring* exerts on the mass, $\vec{F}_{\text{spring}} = -kx\hat{\imath}$.

7.3   (page 287) Be careful! The quantity ωt is measured in radians, and so the constant angle ϕ, called the *phase angle*, also must be in radians.

7.4   (page 289) The way the oscillator is released provides the information to determine the amplitude A and phase angle ϕ.

7.5   (page 290) The SI units for the frequency ν are always given in hertz (Hz); the SI units for the angular frequency ω are always given in rad/s.

7.6   (page 290) If you know $x(t)$, then $v_x(t)$ is its first time derivative and $a_x(t)$ is its second time derivative. The maximum speed is $A\omega$. The magnitude of the maximum acceleration is $A\omega^2$.

QUESTIONS

1. (page 285); 2. (page 287); 3. (page 295); 4. (page 297); 5. (page 300)

6. A mass m is attached in turn to two different springs. Measurements are made of the x-component $F_{x\,we}$ of the force we need to exert on m to hold the mass at various positions along a meter stick scale beside each spring. Plots are made of the x-component of the force versus the position on the meter stick; see Figure Q.6. For each case, answer the following: (a) Does the spring obey Hooke's force law? Why or why not? (b) Draw the corresponding graph of the x-component of the *force of the spring* on m versus position on the meter stick. (c) If the spring does obey Hooke's force law, what needs to be done *to the data* to make the graph of the force component we exert on m look like Figure 7.4?

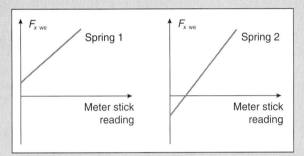

FIGURE Q.6

7. A mass m on a frictionless horizontal surface is attached to a horizontal spring. When we pull and hold the mass at some position coordinate x, where $x = 0$ m is the equilibrium position of the mass, our force on m and the force of the spring on m are of equal magnitude. If we gradually increase our pull on the mass, the mass gradually moves to a larger x-coordinate and the spring force on m gradually increases in magnitude, matching our force newton-for-newton. If we gradually decrease the magnitude of the force exerted on m, the spring force also decreases. Recall that the magnitude of the static frictional force also matches forces newton-for-newton (see Section 5.12); the magnitude of the static frictional force on m when it is on a rough surface also increases as we push harder, and decreases as we push less. Compare and contrast the force of a spring on m when it is on a frictionless surface to the static frictional force on m when it is on a rough surface. Explain why the static frictional force *cannot* produce simple harmonic oscillation of a mass.

8. A mass is attached to one end of a spring on a frictionless surface inclined at an angle θ to the horizontal (see Figure Q.8). When displaced from its equilibrium position, will the mass oscillate at the same angular frequency as it would if the spring were either horizontal (on a frictionless surface) or vertical? Justify your answer.

9. A mass is executing simple harmonic oscillation with amplitude A. What is the total distance traveled by the mass during one period of the oscillation? Does your answer depend

FIGURE Q.8

on the time instant from which the period of the oscillation is measured?

10. A mass is executing simple harmonic oscillation according to the equation

$$x(t) = A \cos \omega t$$

The period of the oscillation is T. Answer the following questions for each of the following time intervals: (i) 0 s $< t < T/4$; (ii) $T/4 < t < T/2$; (iii) $T/2 < t < 3T/4$; (iv) $3T/4 < t < T$. (a) Is the position vector parallel or antiparallel to the velocity? (b) Is the position vector parallel or antiparallel to the acceleration? (c) Is the velocity parallel or antiparallel to the acceleration?

11. A particle is executing simple harmonic motion according to the equation $x(t) = A \cos \omega t$. (a) What is the maximum speed of the particle? (b) What is the magnitude of the maximum acceleration of the particle? (c) At what positions is the *speed* of the particle *half* the maximum speed? (d) At what positions is the magnitude of the acceleration of the particle *half* the magnitude of the maximum acceleration?

12. One astronaut on the Moon has a spring with a known spring constant k while another astronaut has simply a string of known length ℓ. An interesting rock is found. Describe how the astronaut with the spring can determine the mass of the rock from observations and data secured with the spring without knowing the magnitude of the acceleration due to gravity on the Moon. Is it possible for the astronaut with the string to check the result? Explain how or why not.

13. You have a variety of masses available but only one spring. What can be done to double the maximum speed for simple harmonic oscillation at a fixed amplitude A?

14. The amplitude of a simple harmonic oscillation is tripled. Describe how this affects (a) the frequency of the oscillation; (b) the maximum speed of the oscillation; and (c) the magnitude of the maximum acceleration.

15. Make a list of several everyday examples of simple harmonic oscillation. Make a similar list of oscillating objects that do *not* exhibit simple harmonic oscillation.

16. A superball is dropped vertically and bounces up and down on the floor. Is the motion oscillatory? Is the motion damped simple harmonic oscillation? Explain your answers.

17. The x-component of the force needed to stretch or compress a mass m (attached to a peculiar spring) to a coordinate x (and hold it there) is plotted versus x with the results indicated in

Figure Q.17. The coordinate $x = 0$ m corresponds to the equilibrium position of the mass with $\hat{\imath}$ corresponding to stretching the spring. (a) *Describe* how this peculiar spring differs from an ordinary spring that obeys Hooke's law. (b) Think of a physical system that might behave this way.

FIGURE Q.17

18. Two springs are oriented vertically with the position of the bottom of each spring designated as $x = 0$ m and $\hat{\imath}$ directed vertically downward. Various masses are suspended from each spring in turn and the position of the bottom of the spring is noted when the mass is in equilibrium. The magnitude of the weight of each mass is plotted versus the corresponding position of the bottom of the spring from its massless equilibrium position. The resulting graphs appear in Figure Q.18. (a) Which spring has the greater spring constant k? (b) When the most mass is hung from each spring and the mass is set into vertical simple harmonic oscillation, which spring will cause the mass to oscillate with the largest frequency? With the greatest angular frequency? With the longest period? (c) With equal masses on each spring, does the information given provide a way to determine which spring gives the mass the greatest speed? The greatest acceleration? Explain your answer.

FIGURE Q.18

19. The mass of your car typically appears on its registration form. Push the car down quickly on one corner and estimate the period of oscillation. Knowing the mass and period, estimate the spring constant of each of the four springs, assuming they are equal.

20. If the length of a simple pendulum is tripled, by what factor does the period change?

21. A simple pendulum is hung from the ceiling inside an elevator at rest. Its period of oscillation is T_0. Describe qualitatively what happens to the period of the pendulum (if anything) when the elevator is: (a) moving upward at constant velocity; (b) moving downward at constant velocity; (c) moving upward with a constant acceleration; (d) moving downward with a constant acceleration.

22. The mass of a simple pendulum is doubled but the length of the pendulum is unchanged. By what factor does the period of the pendulum change?

23. The determination of longitude at sea necessitates an accurate clock. The first accurate marine chronometers were invented by John Harrison in the latter part of the 18th century. Several of his clocks are on display at the former Royal Observatory in Greenwich, England. Explain why a pendulum clock is rather useless as an accurate timekeeping device on a ship.

24. If the length of a simple pendulum increases by 1%, by what factor is the period of the pendulum changed? Does the period increase or decrease?

25. The period of a mass m oscillating on a spring is

$$2\pi \left(\frac{m}{k}\right)^{1/2}$$

and the period of a mass m on a simple pendulum executing small-angle oscillations is

$$2\pi \left(\frac{\ell}{g}\right)^{1/2}$$

Since both motions are simple harmonic, why does one period depend on the mass while the other does not?

26. A family moves its heirloom grandfather pendulum clock from Miami (at sea level) to Denver (elevation about 1500 m) without making adjustments in the length of the pendulum. If the temperature is the same in Miami and in Denver, will the clock run slow or fast in Denver? Should the length of the pendulum be slightly increased or decreased for it to keep accurate time?

27. We have studied several relationships involving force components or magnitudes of forces that have similar algebraic forms; among them are

$$F_{x\,total} = ma_x$$
$$f_{s\,max} = \mu_s N$$

$$f_k = \mu_k N$$
$$F_{x\,sprg} = -kx$$

$$w = mg$$
$$\vec{F}_{grav} = m\vec{g}$$

(a) Carefully explain what is meant by each of these expressions. (b) Each expression involves a particular constant (m, μ_s, μ_k, k); what are the SI units of each constant and what property of what objects does the constant quantify? (c) What are the *limitations* associated with *each* expression? (d) Is it possible to rank the expressions in the order of their importance or priority in physics? Justify your answer.

28. After a diver jumps off a diving board, the board oscillates at its natural frequency of oscillation. Qualitatively, how does the natural oscillation frequency of the diving board depend on the length of the board? (You can easily simulate the diving board by clamping a meter stick or ruler down to a table with part of its length hanging off the table and tweaking the free end of the stick. Vary the length of the free end of the stick and observe the effect on the frequency.)

29. Qualitatively, how would the natural frequencies of oscillation of a tall structure compare with those of a short structure? Earthquakes exert oscillating forces on all structures. To avoid significant structural damage, what should be the relationship between the shaking frequencies associated with earthquakes and the natural frequencies of oscillations of buildings?

30. Explain why automotive suspension systems typically are designed for critical damping of oscillations induced by road bumps.

31. The frequency of damped oscillatory motion is less than that of undamped oscillatory motion. Explain why this is reasonable in view of the forces acting on the system.

32. Why is it inadvisable to have large numbers of military troops marching in step across bridges?

33. A mass is attached to a peculiar spring. Every 10.0 s the x-component of the force we need to pull and hold m at a given coordinate x is suddenly changed; its value is quickly determined and then plotted versus x. The original position of the mass is at the coordinate $x = 0$ m when $t = 0$ s. The *chronological* sequence of data taken in the experiment appears in the table on the right.

Position coordinate x in cm of mass m	x-component of the force in N needed to act on m to keep m at the specified x-coordinate
0	0
2	4
5	16
10	28
15	34
20	40
15	38
10	37
5	34
0	30
−5	22
−10	0
−15	−30
−20	−40
−15	−38
−10	−37
−5	−34
0	−30
5	−22
10	0
15	30
20	40

(a) Make a graph of $F_{x\,we}$ versus x (in *meters*), indicating the path the mass follows *chronologically* on the graph. (b) Using the graph as a guide, *describe* what is happening during the experiment. (c) Does this spring obey Hooke's force law? Explain your answer. Such a peculiar spring exhibits what are called *hysteresis* effects. Can you think of a physical system that might behave in a way similar to this?

PROBLEMS

Section 7.1 Hooke's Force Law

1. A force of magnitude 5.00 N stretches a spring 10 cm. What is the spring constant of the spring?

2. You have a spring with a spring constant of 225 N/m. What magnitude force is needed to stretch the spring 5.0 cm?

•3. An inquisitive student hangs various masses from a stretchy cord and records the position of the end of the cord from the original position of its end, located at the 2.0 cm mark on a meter stick. The following data are secured:

Mass hung on the cord in kg	Position of the end of the cord in cm
0.100	7.7
0.200	10.1
0.300	11.9
0.400	13.4
0.500	14.8

(a) Make a graph of the magnitude of the force acting on the end on the cord versus the meter stick record of the *position* of the end of the cord. (b) Does the string obey Hooke's force law? Explain why or why not.

•4. An experiment is performed in which we measure the force needed to hold a 0.250 kg mass (attached to a spring) away from its equilibrium position on the spring. The equilibrium position of the mass on the spring is at $x = 0$ m and $x > 0$ m corresponds to stretching the spring. The mass is held at rest at various x-coordinates over the range -0.20 m $< x <$ $+0.20$ m. The spring constant is 10 N/m. On the same graph, accurately plot: (a) the x-component of the force of the spring on m; (b) the x-component of the force we exert on the mass; and (c) the x-component of the vector sum of the two forces on the mass versus x over the domain -0.20 m $< x <$ $+0.20$ m. Label each coordinate axis appropriately and be sure to distinguish among the graphs (a), (b), and (c).

•5. A horizontal spring is attached to a 15.5 kg mass on a rough horizontal surface with coefficients of friction $\mu_s = 0.25$ and $\mu_k = 0.20$. We pull the spring manually and find that when the spring is stretched 0.150 m, the mass remains at rest; but when the stretch is 0.151 m, the mass slips. What is the spring constant?

•6. A cage holds a 150 g mass, which is attached to one end of a spring. The other end of the spring is attached to the opposite end of the cage. The spring is stretched 4.0 cm as the mass is whirled at 25.0 rev/s in a circle of radius 6.0 cm about the center of the cage on a horizontal frictionless surface as indicated in Figure P.6. What is the spring constant?

FIGURE P.6

•7. Suppose, as is conventional, we take $x = 0$ m as the equilibrium position of a mass m attached to a spring. If $\hat{\imath}$ is chosen to be in the direction corresponding to *compressing* the spring, show that the force of the spring on m has the form

$$\vec{F}_{spring} = -kx\hat{\imath}$$

This expression for $\vec{F}_{spring}$ is the *same* form as when $\hat{\imath}$ corresponded to *stretching* the spring (Equation 7.2). Thus the direction chosen for $\hat{\imath}$ does *not* affect the results we obtained in this chapter.

•8. Show that if two identical springs, each with spring constant k, are connected in parallel as indicated in Figure P.8, the effective spring constant of the pair is $2k$.

FIGURE P.8

•9. Two springs with spring constants k_1 and k_2 are connected as shown in Figure P.9 with a mass m attached on a frictionless surface. We pull on m with a force $\vec{F}_{we}$ and hold m at rest. The mass m moves a distance x and point A moves a distance x' from their equilibrium positions. (a) By examining the forces on point A, show that

$$x' = \frac{k_2}{k_1 + k_2} x$$

(b) By examining the forces on m, show that if the system is modeled with a single spring stretching m a distance x, the effective spring constant of the single spring is

$$k_{eff} = \frac{k_1 k_2}{k_1 + k_2}$$

FIGURE P.9

•10. A spring with a spring constant k is cut transverse to its axis into two equal parts and then patched together again, so that the original spring can be modeled as two half-springs in series with each other (see the previous problem). Show that the spring constant of each half-spring is $2k$. Thus one way to create a stiffer spring (one with a larger spring constant) is to cut and remount the spring, so that it has fewer coils.

•11. A spring gun is used to project a 100 g mass horizontally across a frictionless surface. The spring is compressed 5.0 cm and has a spring constant of 30 N/m. What is the initial magnitude of the acceleration of the mass the instant the spring gun fires? Is the acceleration of the mass constant while it still is in contact with the spring?

Section 7.2 Simple Harmonic Oscillation

•12. A fleet-footed politician of mass 50.0 kg is observed oscillating back and forth on a slippery, frictionless issue according to the expression

$$x(t) = (6.0 \text{ m}) \cos\left[(8.00 \text{ rad/s})\, t + \frac{\pi}{4} \text{ rad}\right]$$

Physics can describe anything! (a) Find the amplitude A of the oscillation. (b) Find the angular frequency ω of the oscillation. (c) Find the frequency ν of the oscillation. (d) Find the period T of the oscillation. (e) Find the effective spring constant k associated with the oscillation. (f) Find the position x of the politician when $t = 2.00$ s. (g) Determine $v_x(t)$ and find the velocity component of the hack when $t = 2.00$ s. (h) Determine $a_x(t)$ and find the acceleration component of the pol when $t = 2.00$ s. (i) Make graphs of $x(t)$, $v_x(t)$, and $a_x(t)$ versus t during at least two periods of the motion, beginning when $t = 0$ s. Verify your answers to parts (f)–(h) from these graphs.

•13. A 2.00 kg mass executes simple harmonic oscillation with an amplitude of 5.00 cm. The maximum speed of the mass is 2.00 m/s. (a) Determine the angular frequency ω of the oscillation. (b) Find the frequency ν of the oscillation. (c) If the mass is at $x = 0$ m when $t = 0$ s, find $x(t)$. (d) What is the position of the mass when $t = 1.00$ s?

•**14.** The position of a mass undergoing simple harmonic oscillation is described by

$$x(t) = A \cos \omega t$$

(a) Qualitatively sketch the graph of $x(t)$ versus t during two periods of the motion. (b) Qualitatively sketch a graph of the velocity component $v_x(t)$ versus t during two periods of the motion. (c) Qualitatively sketch a graph of the acceleration component $a_x(t)$ versus t during two periods of the motion. On each graph indicate the value of the maximum position, speed, and magnitude of acceleration.

•**15.** A mass m rests on a block of mass M that is attached to a horizontal spring as indicated in Figure P.15. The block M rests on a frictionless surface. The coefficient of static friction for the contact surfaces of the two masses is μ_s and the coefficient of kinetic friction is μ_k. The system is set into simple harmonic oscillation with amplitude A. What spring constant ensures that m is on the verge of slipping? If m is then increased, will the mass slip or not? Explain.

FIGURE P.15

•**16.** The position of a mass m undergoing simple harmonic oscillation is described by $x(t) = A \cos \omega t$. Make a graph of the *distance* traveled by the mass as a function of time during one period of its motion.

•**17.** A 1.50 kg mass on a horizontal frictionless surface is attached to a horizontal spring with spring constant $k = 200$ N/m. The mass is in equilibrium at $x = 0$ m. The mass is released when $t = 0$ s at coordinate $x = 0.100$ m with a velocity $(2.00 \text{ m/s})\hat{i}$. (a) Find the constants A, ω, and ϕ in

$$x(t) = A \cos(\omega t + \phi)$$

(b) Find the period T of the oscillation. (c) Determine the maximum speed of the oscillation and the magnitude of the maximum acceleration. (d) Plot $x(t)$ during the time interval $t = 0$ s to $t = 2T$.

•**18.** A manufacturer claims that a device is capable of withstanding accelerations of magnitude at least equal to 50 times the magnitude of the local acceleration due to gravity g (= 9.81 m/s²) without destruction. As an engineer in a testing laboratory, you decide to test the claim by attaching the device to a spring and subjecting it to simple harmonic oscillation of amplitude 0.10 m. What oscillation frequency will subject the device to an acceleration of magnitude 50g at some times during the oscillation?

•**19.** A 500 g mass is undergoing simple harmonic consternation . . . oscillation (!) . . . that is described by the following equation for its position $x(t)$ from equilibrium:

$$x(t) = (0.50 \text{ m}) \cos\left[(6.0\pi \text{ rad/s}) t + \frac{\pi \text{ rad}}{6.0}\right]$$

(a) What is the amplitude A of the oscillation? (b) What is the angular frequency ω of the oscillation? (c) What is the frequency ν of the oscillation? (d) What is the period T of the oscillation? (e) What is the fall constant . . . rather, the spring constant k? (f) What is the position of the oscillator when $t = 0$ s? (g) At what time $t > 0$ s is the oscillator *first* at maximum distance from equilibrium? (h) What is the maximum speed of the oscillator? (i) What is the magnitude of the maximum acceleration of the oscillator?

•**20.** A mass is executing simple harmonic oscillation with a frequency of 9.0 Hz and amplitude 0.20 m about an equilibrium position at $x = 0$ m. When $t = 0$ s, the mass is located at the x-coordinate 0.10 m. (a) Find $x(t)$ and plot it versus t during a time interval of two periods beginning when $t = 0$ s. (b) Find $v_x(t)$ and plot it versus t during a time interval of two periods from $t = 0$ s. (c) Find $a_x(t)$ and plot it versus t during a time interval of two periods from $t = 0$ s.

•**21.** In Example 7.8, determine specific expressions for $v_x(t)$ and $a_x(t)$ and plot them versus t for the domain $0 \text{ s} \leq t \leq 2.0$ s as Figure 7.19 does for $x(t)$.

•**22.** A can of cola of mass $m = 0.250$ kg on a horizontal frictionless surface is attached to the end of a horizontal spring (with spring constant $k = 15.0$ N/m) and is released at the equilibrium position $x = 0$ m when $t = 0$ s with an initial velocity $v_0\hat{i} = (0.500 \text{ m/s})\hat{i}$. (a) Determine the constants A, ω, and ϕ in the general equation for $x(t)$:

$$x(t) = A \cos(\omega t + \phi)$$

(b) Determine the period T of the oscillation. (c) Find and plot $x(t)$, $v_x(t)$, and $a_x(t)$ versus t for the domain $0 \text{ s} \leq t \leq 2T$.

‡**23.** Real springs have mass. Thus when a mass m is attached to a spring of mass m_s and executes simple harmonic oscillation, part of the mass of the spring also is in oscillation. To account for the mass of the spring, let $M = m + \gamma m_s$, where γ represents the *fraction* of the mass of the spring that participates in the oscillation. The period of the oscillation T then is

$$T = 2\pi \left(\frac{M}{k}\right)^{1/2}$$

Square and substitute for M; you will find that

$$T^2 = \frac{4\pi^2}{k} m + \frac{4\pi^2}{k} \gamma m_s$$

A student uses a spring of mass 0.180 kg and measures the period of oscillation T when various masses m are suspended from the spring. The following data are secured:

Mass m in kg attached to spring	Measured period of oscillation in s
0.050	0.75
0.150	1.00
0.250	1.20
0.350	1.35
0.450	1.51
0.550	1.65
0.650	1.77
0.750	1.88
0.850	2.00
0.950	2.10

(a) Plot the *square* of the period (T^2) versus the mass m attached to the spring. In view of the foregoing expression for T^2, this graph should be linear. Find the best-fit straight line through the data. What is the slope of the graph? What is the intercept along the T^2-axis? (b) Use the slope and intercept to determine (i) the spring constant k of the spring and (ii) the fraction γ of the mass of the spring that participates in the oscillation.

‡24. The force of a spring on a mass m when it is a distance x from the equilibrium position at $x = 0$ m is

$$\vec{F}_{\text{spring}} = -kx\hat{i}$$

Suppose, instead, that the only force acting on mass m was a force given by

$$\vec{F} = kx\hat{i}$$

(a) Qualitatively describe the force $\vec{F}$ (its magnitude and direction) as the position of the mass moves toward increasing positive values of x. (b) Qualitatively describe the force $\vec{F}$ (its magnitude and direction) as the position of the mass moves toward increasingly negative values of x. (c) Use Newton's second law to show that the differential equation that describes the motion of the mass when $\vec{F}$ is the only force acting on it is

$$\frac{d^2x}{dt^2} - \frac{k}{m}x = 0 \text{ m/s}^2$$

(d) Does $x(t) = A\cos(\omega t + \phi)$ satisfy this differential equation? If so, what is the value of ω that emerges from this? (e) Does the expression

$$x(t) = Ae^{\omega t}$$

satisfy the differential equation? If so, what is ω? Qualitatively describe what happens to the particle for times $t > 0$ s if $x(t) = Ae^{\omega t}$. (f) Do you think that such a force actually exists in nature? Explain.

‡25. Two masses are undergoing simple harmonic oscillation according to the equations

$$x_1(t) = (0.150 \text{ m}) \cos\left[(0.250 \text{ rad/s})\, t + \frac{\pi}{6} \text{ rad}\right]$$

$$x_2(t) = (0.200 \text{ m}) \cos\left[(0.350 \text{ rad/s})\, t + \frac{\pi}{4} \text{ rad}\right]$$

(a) Find the period of each oscillation. (b) On the same graph, plot $x_1(t)$ and $x_2(t)$ versus t during $t = 0$ s to 60 s and determine the first six instants at which they have the same position. Is the interval between these instants the same? (c) By changing one of the amplitudes, see if the instants when the oscillators have the same position depend on the amplitudes.

‡26. Use the two oscillations in Problem 25 to make a graph of their superposition $x(t) = x_1(t) + x_2(t)$ during $t = 0$ s to 200 s. The result illustrates how more complex oscillations are imagined as composed of superpositions (sums and differences) of various simple harmonic oscillations with different amplitudes, frequencies, and phase angles. The process of determining the simple harmonic oscillations that constitute a complex oscillation is called *Fourier analysis*.

Section 7.3 A Vertically Oriented Spring

27. A 300 g mass is suspended from a vertical spring and the spring extends a distance of 0.15 m. (a) Determine the spring constant k. (b) Find the angular frequency ω and the period T at which the mass will oscillate on the spring.

•28. A medieval damsel in distress is on top of a tower of height h. An elastic bungee cord that obeys Hooke's law, with $k = 40.0$ N/m, hangs from the top of the tower and reaches halfway to the ground. The 60.0 kg damsel climbs down the cord and when she reaches the end amazingly finds that she is safely on the ground. What is the height of the tower from which she descended?

•29. A 2.00 kg mass is suspended from a vertical spring of spring constant 100 N/m and unstretched length 0.300 m. Find the position, relative to the fixed end of the spring, of the point about which the mass will make simple harmonic oscillations.

•30. A vertical spring is attached to the ceiling. An 8.00 kg frozen turkey is attached to the bottom of the spring and causes the spring to stretch an additional 10 cm. The turkey then is pulled down 4.0 cm beyond the equilibrium position and released at rest. Take the positive x-direction to be downward. (a) What is the spring constant? (b) If the oscillation is described by $x(t) = A\cos(\omega t + \phi)$, find A, ω, and ϕ. (c) What is the maximum speed of the turkey? (d) What is the magnitude of its maximum acceleration?

•31. A leprechaun of mass 70.0 kg sits on the end of a long diving board of negligible mass and the end is depressed 25 cm. (a) What is the effective spring constant k of this system if we model it as a simple spring? (b) At what frequency could this system oscillate? (c) What oscillation amplitude would allow the leprechaun to reach an upward acceleration equal in magnitude to that of the local acceleration due to gravity g?

•32. A 10.0 kg mass attached to a spring is dragged at constant speed up a rough surface ($\mu_s = 0.30$, $\mu_k = 0.25$) inclined at 10° to the horizontal as in Figure P.32. The spring is stretched

FIGURE P.32

8.0 cm. (a) What is the spring constant of the spring? (b) If the spring drags the mass at constant speed down the slope, what is the stretch of the spring?

•33. A 0.250 kg mass is suspended by a string from a 0.500 kg mass as shown in Figure P.33. The 0.500 kg mass is attached to a spring with a spring constant of 50.0 N/m. (a) What is the frequency of simple harmonic oscillation? (b) At what point in the oscillation is the tension in the string a minimum? (c) At what point in the oscillation is the tension in the string a maximum? (d) What amplitude oscillation ensures the tension in the string is zero at some instant during the motion?

FIGURE P.33

•34. A troll of mass m is placed on top of a piston executing simple harmonic motion in a vertical direction with an amplitude A. (a) At what position in the simple harmonic motion of the piston is the troll likely to leave the surface of the piston? (b) If the frequency of the simple harmonic oscillation of the piston is small, the troll on top of the piston is able to follow the oscillation. On the other hand, if the frequency of the simple harmonic oscillation is too large, the troll will bounce atop the piston. What is the maximum frequency of simple harmonic oscillation so that the troll barely remains continuously in contact with the piston at all times during the motion?

•35. A 1.50 kg mass is attached to a vertical spring. A student gradually pushes slowly up on the mass until the upward force of the student on the mass is equal to the weight of the mass. With the mass at rest, the student then suddenly removes the supporting hand and notes that the mass executes simple harmonic oscillation over a vertical range of 1.20 m. Determine the spring constant k and the frequency of oscillation ν.

•36. Two identical springs, each with spring constant k, are attached to two identical masses m with the lower mass suspended by its spring from the upper mass. The top of the upper spring is pulled upward in the vertical direction with a force sufficient to cause an acceleration of magnitude a. The unstretched length of each spring is ℓ_0. (a) What is the change in the length of the lower spring? (b) What is the change in the length of the upper spring?

•37. A 1.5 kg mass oscillates with an amplitude of 0.10 m on a vertical spring whose spring constant is 350 N/m. (a) What is the maximum speed of the mass? (b) What is the magnitude of the maximum acceleration of the mass?

•38. A 0.250 kg sugar plum fairy is attached to a vertically suspended spring whose spring constant is 2.00 N/m. The sugar plum fairy is pulled down 10 cm and released at rest. (a) Find $x(t)$. (b) At what position is the fairy when $t = 3.5$ s? (c) What is the velocity component of the fairy when $t = 3.5$ s? (d) What is the acceleration component of the fairy when $t = 3.5$ s?

‡39. A *massive spring* with mass m and spring constant k is horizontal and neither stretched nor compressed. Its length is measured to be ℓ_0. Show that the increase in the length of the spring, because of its weight, if it is hung in the vertical direction is

$$\frac{mg}{2k}$$

Hint: Generalize Problems 9 and/or 10 and consider the given spring to be a collection of n springs in series, each with spring constant nk, where n is an integer much greater than 1.

Section 7.4 Connection between Simple Harmonic Oscillation and Uniform Circular Motion

•40. A phonograph record with a diameter of 30.0 cm has a young cricket (*Acheta domestica*) sitting on its rim. The record and cricket are turning at 33.3 rev/min. The setting Sun casts a shadow of the cricket on a wall directly facing the Sun. (a) Does the shadow move in simple harmonic motion? (b) What is the amplitude of the motion of the shadow? (c) What is its period? (d) What is the maximum speed of the shadow? (e) What is the magnitude of the maximum acceleration of the shadow? (f) When the acceleration of the shadow is zero, is the velocity of the shadow also zero? (g) What are the forces acting on the cricket? Sketch an appropriate second law force diagram.

•41. A mass is executing uniform circular motion counterclockwise in a vertical circle of radius R at angular speed ω. Place the origin at the center of the circle, with the x-axis horizontal and in the plane of the circle with $\hat{\imath}$ to the right; choose the y-axis vertical with $\hat{\jmath}$ up. When $t = 0$ s, the particle is on the $+x$-axis. The motion is projected onto a line that passes through the center of the circle in the plane of the circle and is inclined at 30.0° to $\hat{\imath}$; see Figure P.41. Write the simple harmonic motion along the line in the standard form

$$A \cos(\omega t + \phi)$$

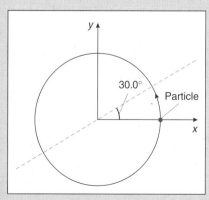

FIGURE P.41

•42. A 0.150 kg mass is attached to one end of a horizontal spring as indicated in Figure P.42. The mass lies on a frictionless surface. The unstretched length of the spring is 5.00 cm. The spring and mass then are whirled in a horizontal circle at an angular speed of 50.0 rad/s about a vertical axis through the other end of the spring. It is found that the mass extends a distance of 3.00 cm from its position when at rest. (a) Find the spring constant k of the spring. (b) A spotlight directed horizontally projects a shadow of m onto a wall. Is the motion of the shadow simple harmonic oscillation? If so, what is the angular frequency ω of the oscillation of the shadow? What is the effective spring constant of the simple harmonic oscillation of the shadow?

FIGURE P.42

•43. A line from the Sun to the star Regulus in the constellation Leo lies essentially in the plane of the orbit of the Earth about the Sun. Assume the orbit of the Earth is circular. Let the Earth be on this line when $t = 0$ s (about 21 February and again on 23 August). Write the projection of the motion of the Earth on this line in the form $x(t) = A \cos(\omega t + \phi)$ using SI units.

‡44. You have an unknown mass m, a spring whose spring constant is unknown, and a 1.000 kg standard mass. The mass m and the standard mass are attached to opposite ends of the spring. The spring is compressed on a frictionless horizontal surface and both masses released at rest when $t = 0$ s. You simultaneously determine the magnitude of the initial acceleration of m to be 9.00 m/s² while that of the standard 1.000 kg mass is 4.50 m/s² in the opposite direction. What is the unknown mass m?

Section 7.5 How to Determine if an Oscillatory Motion Is Simple Harmonic Oscillation

45. A beagle runs back and forth between two children, making one round trip every 10 s on a regular basis. (a) What is the period of the dog? (b) What is the frequency of the oscillation of the dog? (c) Is the motion of the dog simple harmonic oscillation or not? Explain.

•46. A mythical hero of mass m is strung between two springs as shown in Figure P.46. The spring constants are k_1 and k_2 as indicated. At the position $x = 0$ m, the hero is subjected to no force from either spring. The gods move the hero to position x. (a) What is the force of spring 1 on the hero? (b) What is the force of spring 2 on the hero? (c) What is the total force on the hero in the horizontal direction due to the two springs? (d) The horizontal surface is frictionless. When the gods release the hero at rest, apply Newton's second law to the hero and cast the result into the standard form for the differential equation for simple harmonic oscillation (a form like Equation 7.11).

FIGURE P.46

(e) Determine the period of oscillation of the hero from the equation obtained in part (d).

•47. A particle in a frictionless hemispherical bowl of radius R is moved slightly away from the bottom and released at rest. Show that particle executes simple harmonic oscillation about the bottom of the bowl and find the period of the oscillation.

‡48. A plumb bob of length ℓ is swung so that the mass on the end of the string moves in a horizontal circle at constant angular speed with the string making a constant, small angle θ with the vertical direction as shown in Figure P.48. Show that if the motion is viewed sideways in the plane of the horizontal circle, the motion is simple harmonic and of period

$$T = 2\pi \left(\frac{\ell \cos \theta}{g} \right)^{1/2}$$

FIGURE P.48

‡49. A piston moves inside a cylindrical tube and is connected by a tie-rod to the rim of a circular shaft that is turning at constant angular speed ω (see Figure P.49). Show that the oscillatory motion of the piston is *not* simple harmonic oscillation.

‡50. A spring with spring constant k is attached to a mass m that is confined to move along a frictionless rail oriented perpendicular to the axis of the spring as indicated in Figure P.50. The spring is initially unstretched and of length ℓ_0 when the mass is at the position $x = 0$ m in the indicated coordinate system.

FIGURE P.49

(a) If the mass m is pulled out a distance x along the rail, show that the new total length of the spring is

$$(\ell_0^2 + x^2)^{1/2}$$

(b) Show that the x-component of the force that the spring exerts on the mass is

$$F_x = -k[(\ell_0^2 + x^2)^{1/2} - \ell_0] \cos \theta$$

(c) Show that for *small* coordinates x along the rail, the horizontal force component on the mass m is approximately

$$F_x \approx -\frac{k}{2\ell_0^2} x^3$$

If the mass is released from the point x, oscillations occur but the oscillations are *not* simple harmonic oscillations. Indeed, the period of this oscillator is *inversely* proportional to the amplitude of the oscillator, not independent of the amplitude as is the case for simple harmonic oscillation. This is an example of a *nonlinear* or *anharmonic* oscillator.

This problem was inspired by a paper by Alan Cromer, "The x^3 oscillator," *The Physics Teacher*, 30, #4, pages 249–250 (April 1992).

FIGURE P.50

Section 7.6 The Simple Pendulum

51. Thomas Jefferson, as Secretary of State under George Washington, was asked to devise a decimal system of weights and measures. Jefferson proposed a standard of length based on a simple pendulum whose *half period* was *one second*, so that $T = 2.00$ s. A clock based on such a pendulum would tick once with each swing. If we take the magnitude of local acceleration due to gravity to be precisely 9.81 m/s^2, what length pendulum has a period of precisely 2.00 s? The length is very close to but not the same as a familiar standard of length, developed in France during the French Revolution.

52. The numerical value of the magnitude of the acceleration due to gravity g varies slightly with geographic location and altitude on the Earth. What value of g will give a simple pendulum with a length of precisely 1.000 m a period of 2.000 s? The largest values of g on the surface of the Earth are at the geographic poles, where $g \approx 9.83$ m/s^2, and the smallest values are on the equator, where $g \approx 9.78$ m/s^2.

53. In 1851, Jean Bernard Léon Foucault (1819–1868) suspended a very long pendulum (67.5 m) from the inside of the dome of the Panthéon in Paris to demonstrate the rotation of the Earth. (See chapter opening photo.) Such Foucault pendulums are a prominent feature of many science museums. (See Figure P.53.) What was the oscillation period of his pendulum?

FIGURE P.53

•54. The horizontal motion of a simple pendulum is given by the equation

$$x(t) = (0.050 \text{ m}) \cos\left[\left(\frac{\pi}{2.00} \text{ rad/s}\right) t + \frac{\pi}{6.00} \text{ rad}\right]$$

(a) What is the frequency ν of the oscillation? (b) What is the length of the pendulum?

•55. Before the advent of accurate clocks in the early 17th century, a crude instrument for measuring the pulse rate consisted of a pendulum whose length was adjusted until *half* its period matched the pulse rate of a patient. The length of the pendulum was recorded as a measure of the pulse rate. Construct a simple pendulum with a piece of string and a convenient mass. With the assistance of a buddy, adjust the length of the pendulum until its half period matches your pulse rate. From this length, determine your pulse rate in beats per minute, the now standard medical unit for expressing pulse rate.

•56. For a simple pendulum, at what angle from the vertical is the vertical rise 0.100 of the horizontal distance? See Figure P.56.

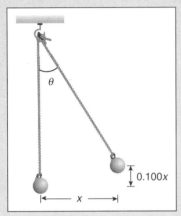

FIGURE P.56

•57. Two playground swings with lengths $\ell_1 > \ell_2$ have periods T_1 and T_2 respectively. They both are drawn aside together and released at the same instant to swing in small oscillations. After what time interval will they again be in step with each other?

•58. A simple pendulum that would have a period of 1.00 s on the surface of the Earth is carried to the Moon and set into oscillation on its surface. (a) Look up the radius and mass of the Moon and calculate the magnitude of the local acceleration due to gravity on the surface of the Moon. (b) Determine the period of oscillation of the pendulum on the lunar surface.

•59. If a pendulum 1.00 m long were somehow set into oscillation on the surface of a white dwarf star with a mass equal to that of the Sun and a radius equal to that of the Earth, determine what the frequency of oscillation would be.

‡60. Show that if a pendulum of length ℓ and period T is moved to a location where the magnitude of the local acceleration due to gravity differs by a small amount Δg, the change ΔT in the period of the pendulum is approximately

$$\Delta T \approx -\frac{T}{2g} \Delta g$$

‡61. The pendulum inside a grandfather clock has a *half* period of 1.0000 s at a location where the magnitude of the local acceleration of gravity is 9.800 m/s². The clock carefully is moved to another location at the same temperature and is found to run slow by 89.0 s per day. What is the magnitude of the local acceleration due to gravity at the new location?

‡62. A pendulum of length ℓ is released from rest when its string makes a small angle θ_0 with the vertical direction, as shown in Figure P.62. Find an expression for the angular velocity component of the pendulum as a function of time.

FIGURE P.62

Section 7.7 Through a Fictional Earth in 42 Minutes

63. Let your latitude be θ north and your longitude be ϕ west on the surface of the Earth. If you dig a vertical hole down through and beyond the center of the Earth, at what latitude and longitude does the hole emerge?

64. Determine the time for a mass m to make a one-way trip through a frictionless hole tunneled through a nonrotating neutron star with a mass twice that of the Sun and a radius of 10.0 km. Assume the star is a uniform sphere.

65. If you keep the mass of the Earth constant, what radius would permit a mass to make a one-way trip through a tunnel through the Earth in 1.00 min?

Section 7.8 Damped Oscillations*

66. What are the SI units of the damping constant β in Equation 7.26?

•67. A 0.950 kg mass hangs vertically from a spring that has a spring constant of 8.50 N/m. The mass is set into vertical oscillation and after 600 s you find that the amplitude of the oscillation is 1/10 that of the initial amplitude. What is the damping constant β associated with the motion?

•68. Show that if the position of a damped oscillator is given by

$$x(t) = Ae^{-\alpha t} \cos \omega t$$

where $\alpha = \beta/(2m)$, the time interval for the amplitude to decrease to half its initial value when $t = 0$ s is

$$\frac{2m}{\beta} \ln 2$$

•69. Hang a known mass from a string of known length to make a simple pendulum. Set the pendulum swinging in small-angle oscillations and time how long it takes for the amplitude to decrease to half its initial value. Use this measurement to determine the damping constant β for the pendulum.

Section 7.9 Forced Oscillations and Resonance*

70. What is the resonant frequency of a 0.500 kg mass attached to a spring with a spring constant $k = 10.0$ N/m?

71. What is the resonant frequency of a swing whose length is 3.50 m?

•**72.** The suspension system of your 1.20×10^3 kg car has a natural oscillation frequency of 2.0 Hz without functioning shock absorbers. The shock absorbers usually provide critical damping to the system. Along a worn section of concrete highway, you notice that the concrete joints, spaced 10.0 m apart, set up a violent oscillation of the car making it almost uncontrollable. At what speed are you driving?

•**73.** Country dirt roads frequently become rippled and when you drive over them at a certain speed, the bumps set up resonance in the car's suspension system, making control of the car problematic. If the ripples are 0.50 m apart and you find resonance occurs when you are driving at 25 km/h over them, what is the natural oscillation frequency of the suspension system on your car?

•**74.** At resonance, find the amplitude and maximum speed of a forced harmonic oscillator whose position coordinate $x(t)$ is given by Equation 7.35.

INVESTIGATIVE PROJECTS

A. Expanded Horizons

1. The superposition of two simple harmonic oscillations at right angles to each other produces some interesting patterns known as *Lissajous figures*, named for Jules Antione Lissajous (1822–1880) who first described them in 1855. Investigate this topic to determine how the various patterns are formed.

Thomas B. Greenslade Jr., "All about Lissajous figures," *The Physics Teacher*, 31, #6, pages 364–370 (September 1993).

William Fogg Osgood, *Mechanics* (Macmillan, New York, 1937), pages 182–190.

Forest K. Harris, *Electrical Measurements*, (John Wiley, New York, 1952), pages 644–647.

Gaylord P. Harnwell, *Principles of Electricity and Electromagnetism* (2nd edition, McGraw-Hill, New York, 1949), pages 256–258.

Brian A. Smith, "Lissajous figures," *Physics Education*, 16, #1, pages 38–46 (January 1981).

Mu-Shiang Wu and W. H. Tsai, "Corrections for Lissajous figures in books," *American Journal of Physics*, 52, #7, pages 657–658 (July 1984).

P. Jasselette and J. Vandermeulen, "More on Lissajous figures," *American Journal of Physics*, 54, #2, pages 182–183 (February 1986).

A. D. Crowell, "Motion of the Earth as viewed from the Moon and the Y-suspended pendulum," *American Journal of Physics*, 49, #5, pages 452–454 (May 1981).

Robert H. Romer, "A double pendulum 'art machine,'" *American Journal of Physics*, 38, #9, pages 1116–1121 (September 1970).

2. The superposition of two or more simple harmonic oscillations in the same plane with different amplitudes and frequencies results in complex oscillations (see Problem 26). Such superpositions form the basis for the *Fourier analysis* of complex, but periodic oscillations. Investigate this important area of more advanced analysis of oscillations.

Mary L. Boas, *Mathematical Methods in the Physical Sciences* (2nd edition, John Wiley, New York, 1983), Chapter 7, pages 297–335.

B. Lab and Field Work

3. The Lissajous patterns in Project 1 can be generated with an oscilloscope and a function generator or using a suitable software graphics package. Generate a set of such patterns with either or both of these methods.

4. The period of a simple pendulum of length ℓ for small oscillations is

$$T = 2\pi \left(\frac{\ell}{g}\right)^{1/2}$$

If the angle that a simple pendulum initially makes with the vertical is *not* small, the period of the pendulum no longer is independent of the amplitude and the equation for T becomes, approximately,

$$T \approx 2\pi \left(\frac{\ell}{g}\right)^{1/2} \left(1 + \frac{1}{16} \theta_0^2\right)$$

where θ_0 is the angle the pendulum initially makes with the vertical direction when released (at rest). When $\theta_0 = \pi/2$ rad, how much does the period differ from the small-angle result? Design and perform an experiment to see if you can detect the variation of T with θ.

Grant R. Fowles and George L. Cassiday, *Analytical Mechanics* (5th edition, Saunders, Fort Worth, 1993), page 107.

Keith R. Symon, *Mechanics* (3rd edition, Addison-Wesley, Reading, Massachusetts, 1971), pages 212–215.

5. Investigate the Foucault pendulum experiment (see Problem 53) used to detect the rotation of the Earth. If a suitable deep stairwell exists, set up a long Foucault pendulum using a large mass on its end. The wire supporting the pendulum must be freely gimballed with a device similar to the swivel used on fishing line. The pendulum should be released with as little perturbing force as possible by drawing the ball aside with a lightweight cloth sling tied with a strong but thin thread to a nearby wall. Break the string with a lighted match to set the pendulum into oscillation. Observe and measure the rate at which the plane of oscillation appears to rotate due to the rotation of the Earth. Compare your results with the predicted time period for the rotation of the plane of the pendulum:

$$\frac{24 \text{ h}}{\sin \theta}$$

where θ is your latitude.

Grant R. Fowles and George L. Cassiday, *Analytical Mechanics* (5th edition, Saunders, Forth Worth, 1993), pages 177–178.

6. Borrow an archery bow from a local sporting goods store or a friend, or use your own if you have one. Design and perform an experiment to determine if the draw of the bow obeys Hooke's force law.

W. C. Marlow, "Bow and arrow dynamics," *American Journal of Physics*, 49, #4, pages 320–333 (April 1981).

7. Design and perform an experiment to measure whether the vertical deflection of a diving board obeys Hooke's law over the common range of its deflection.

R. L. Page, "The mechanics of swimming and diving," *The Physics Teacher*, *14*, #2, pages 72–80 (February 1976).

C. Communicating Physics

8. The personal differences between Hooke and Newton illustrate that scientists are human, with all the complexities associated with the human species. Should you wish to investigate the sociological aspects of the relationship between these two great scientists on matters of the priority of discoveries, see the following references. Comment on the likelihood of whether it is possible to have such strong disagreements without them becoming personal in nature.

Gale E. Christianson, *In the Presence of the Creator* (The Free Press, New York, 1984); see the listing under Hooke in the index for extensive pages references.

A. Rupert Hall, *Isaac Newton: Adventurer in Thought* (Blackwell, Oxford, 1992); see the index under Hooke for page references.

Robert K. Merton, *On the Shoulders of Giants* (University of Chicago Press, Chicago, 1993).

Richard S. Westfall, *Never at Rest: A Biography of Isaac Newton* (Cambridge University Press, Cambridge, England, 1980).

CHAPTER 8
WORK, ENERGY, AND THE CWE THEOREM

Our fantastic civilization . . . [is one] obsessed with power, which
explains its whole world in terms of energy . . . .
*Henry Beston (1888–1968)**

W e have analyzed the dynamics of a system by considering the forces on it and using Newton's second law of motion. As you have learned, forces are vectors; therefore, you must carefully employ the rules for manipulating such force vectors to determine the resulting acceleration of the system. In this chapter, we will examine another valuable approach to dynamics, one that uses the *scalar* physical concepts of work and energy.

Our intuitive notions are a hindrance to understanding the concepts of work and energy. This is not surprising, for if you look up the word *work* in the *Oxford English Dictionary*, you will find more than eleven pages (in *very* small print) of various meanings of the word; almost a page is devoted to various meanings of the word *energy*. As we have seen in other contexts, we cannot let our everyday understanding of words interfere with their precise meaning in physics. Common ideas of what constitutes work in many cases bear little resemblance to the concept of work in the physical sciences. For example, when you say that you have "much work to do" to prepare for an examination, or when you say that a new computer "works fine," you are using the word in ways quite differently from the way you would use the term in physics. We will see that a system *cannot* be said to *have* work. *Work is not a property of a system* in the same sense that mass is a property; rather work is done *on* or *to* the system.

Our intuitive understanding of energy also differs from its meaning in physics. Colloquially, energy is associated with vigor. In physics, *energy is a property of a system*. There are many different types (denominations) of energy as we will see: kinetic energy, potential energy, mechanical energy, internal energy, and others, each having a precise meaning (much like one, five, ten, and twenty dollar bills all are denominations of money). A system *has* energy, much like you have an amount of money (and care little about the denominations of the currency). In this chapter we explore the important, close, but distinct relationship between work and energy.

8.1 MOTIVATION FOR INTRODUCING THE CONCEPTS OF WORK AND ENERGY

Aren't Newton's laws sufficient for every purpose? Why bother to introduce the intertwined concepts of work and energy? Their *raison d'être* is twofold:

1. Work and energy are defined as *scalar* quantities. On a practical level, scalars are easier to use than vectors in solving problems. We use the concepts of work and energy to reformulate Newton's second law of motion into a new, encompassing scalar equation quite different from the vector equation $\vec{\mathbf{F}}_{\text{total}} = m\vec{\mathbf{a}}$. We call the new scalar equivalent of Newton's second law the *CWE theorem*. The initials stem from its traditional name: the Classical Work–Energy

theorem.[†] The CWE theorem provides a powerful additional tool to understand the natural world.

2. The introduction of the scalar concepts of work and energy leads to a *totally new physical principle*, distinct from Newton's second law of motion: conservation of energy. There is no such corresponding conservation law for forces. The CWE theorem is a special case of energy conservation; a general statement of energy conservation must await our study of thermodynamics in Chapter 13.

We make the following assumptions:

> We restrict ourselves to systems that are represented as single particles with ignorable characteristics of size, internal structure, internal motions, and deformations; we also ignore thermal effects.

If you recall, these restrictions are synonymous with what we mean by the word particle.[‡][§] We also ignore internal complications such as engines or biochemical processes that necessarily lead to chemical and thermodynamic considerations involving heat transfer.

The forces on a physical system act both in *time* and *space*. The forces on a system may move with the system over *spatial distances* as the system travels from an initial to a final position. We explore one spatial aspect of forces in this chapter, leading us on a journey to the CWE theorem.[#] Forces also act during various *time intervals*, be they long or short; the temporal aspect of forces will be explored in Chapter 9.

Forces underlie changes in the velocity of a system, and so it is not surprising that the idea of the physical *work* done on a system is related to the concept of force and to changes in the position vector of the system. Since both are vector quantities, however, we invent a new scalar physical quantity called the *work done on a system by a force* using the mathematical concept of the scalar product.

In this chapter we explore how to calculate the work done on a system by any force and how the work done by various forces is related to changes in two kinds of energy: *potential energy* and *kinetic energy*. We then formulate the CWE theorem and use it to solve many problems that are much more difficult to attack using only Newton's second law of motion. Later in the chapter we explore the temporal aspects of the work done on a system, which gives rise to the concepts of *average power* and *instantaneous power*.

[†]In keeping with new trends in pedagogy, we prefer not to use the traditional term, because the theorem is not, strictly speaking, a universal statement about the concepts of work and energy. For our purposes, however, the name *CWE theorem* bridges the semantic gap between the traditional phrase and a necessary new term for this concept.

[‡]If you roll a cylinder down an inclined plane, the system (the cylinder) rotates. A car whose tires are rolling along a road also is a system with internal motions (the rotation of the tires and the innards of the engine). We do not consider such systems in this chapter except to the extent that such internal motions can be neglected and the system treated as a simple single particle.

[§]If you jump off the floor or push yourself off from a wall, you are a system that is changing its shape (i.e., is deformed). Analysis of such a system is not amenable to the techniques outlined in this chapter.

[#]Another spatial aspect of forces, a *vector* quantity known as torque, will be examined in Chapter 10.

(Chapter Opener) The Outermost House (Doubleday, Doran, New York, 1931), page 168.

8.2 THE WORK DONE BY ANY FORCE

Who first invented work . . .?

Charles Lamb (1775–1834)*

Moving day. You are loading a van with all your worldly possessions; there is a lot of real, physical work to be done. With a rope, pull a crate a distance s up a rough ramp, as in Figure 8.1. Let the crate be the system. There are several forces acting on it:

1. its weight $\vec{w}$;
2. the normal force $\vec{N}$ of the surface;
3. the force $\vec{T}$ of the rope on the crate; and
4. the kinetic frictional force $\vec{f}_k$.

> The differential **work** dW done by *any* force $\vec{F}$ on a system is defined to be the scalar product of $\vec{F}$ with the differential change in the position vector $d\vec{r}$ of the point of application of the force on the system:
>
> $$dW \equiv \vec{F} \cdot d\vec{r} \qquad (8.1)$$
>
> Since the force $\vec{F}$ acts *on* the system, we likewise say that the work done by $\vec{F}$ is work done *on* the system.

Since the system is represented as a particle, the change in the position vector of the point of application of the force is equivalent to the change in the position vector of the system itself.[†]

Notice that the definition of the differential work says *nothing* about the time interval dt during which the position vector of the system (i.e., point of application of the force) changes by $d\vec{r}$. Therefore the work of $\vec{F}$ on the system involves the force in purely a *spatial* context.

As in our initial example, many forces may act on the system. Hence you can calculate

- the work done by *each individual force* on the system; or
- the work done by *all the forces* on the system—that is, by the *total force* $\vec{F}_{\text{total}}$ on the system.

---

*Quoted from the *Examiner* in a letter by Charles Lamb to Bernard Barton, dated 11 September 1822; *The Letters of Charles Lamb*, edited by Alfred Ainger (Macmillan, London, 1900), volume 3, letter CCXV, page 64.

[†]This is not necessarily the case if the system undergoes rotation or is deformed; it is for this reason that we exclude such cases from consideration in this introduction to the concepts of work and energy.

First we see how to calculate the work done on the system by any individual force; then we can find the work done on the system by the total force $\vec{F}_{\text{total}}$ and investigate how the total work done on the system affects the system itself.

As the system moves along a path from an initial position to a final position (see Figure 8.2), the work done by a particular force $\vec{F}$ is the integral of its differential work along the specific path followed by the system. That is, the work done on the system by $\vec{F}$ as the system moves in space along a path from an initial position to a final position is

$$W = \int_i^f \vec{F} \cdot d\vec{r} \qquad (8.2)$$

Notice in Equation 8.2 that the lower limit of the definite integral is the initial position and the upper limit is the final position. We say W is the work done on the system by the force $\vec{F}$ over (or along) the path of travel from the initial to final position. The phrases *over the path* and *along the path* mean the same thing. The integral in Equation 8.2 is known as a **line integral**, although the path followed by the system need not be straight. The scalar product uses the value of $\vec{F}$ at the location of each differential $d\vec{r}$ along the path followed by the system as it moves from the initial to final positions. Since a force $\vec{F}$ may change its magnitude and/or direction at various points along the system's path, the integration in Equation 8.2 must be performed carefully. In Figure 8.1, for example, the kinetic frictional force, the normal force, and the weight are constant forces on the crate along the path on the ramp, but we might change the direction and magnitude of our pull as it moves along the path.

> **PROBLEM-SOLVING TACTIC**
>
> **8.1** The differential change $d\vec{r}$ in the position vector of the system is written as
>
> $$d\vec{r} = dx\,\hat{\imath} + dy\,\hat{\jmath} + dz\,\hat{k} \qquad (8.3)$$
>
> **whether the particle is moving toward increasing or decreasing values of the coordinates.** The limits of integration account for the actual direction the particle moves in going from its initial position (the lower limit) to its final position (the upper limit). Examples 8.1 and 8.2 illustrate this point. If the motion is purely along the direction you call x, then $d\vec{r} = dx\,\hat{\imath}$,[‡] whether the particle is moving toward increasing or decreasing values of x.

[‡]This is because $dy = dz = 0$ m when the motion is only along the x-axis.

FIGURE 8.1 Forces on a crate being pulled up a ramp.

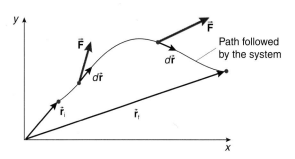

FIGURE 8.2 The work done by a force $\vec{F}$ is the integral of $\vec{F} \cdot d\vec{r}$ along the path followed by the system.

There are several important points to note about the definition of work:

> *Do not* think of work in the sense of simply a force times a distance. The orientation of the force vector $\vec{F}$ relative to the differential change in the position vector $d\vec{r}$ is critical.

For example, the work done by the normal force of the surface on the crate in Figure 8.1 is *not* Ns. Think of the differential work as the *scalar product* of the force and the differential change in the position vector. In Figure 8.1 the normal force $\vec{N}$ is perpendicular to $d\vec{r}$ at every point along the entire path from the initial to final positions of the crate along the ramp; thus, the scalar product $\vec{N} \cdot d\vec{r}$ is zero at each and every point along the entire path and *zero* work is done by this force.

> Work is a *scalar* quantity.

This is one reason for introducing the concept, though not the most important one. It is convenient to have a scalar quantity, so that we can use it without all the baggage that usually accompanies the manipulation of vectors.

The dimensions for work are those of force multiplied by distance or N·m in the SI unit system.

> Since the concept of work is used prolifically, the SI units, N·m, *when associated with work or energy,** are called *joules* (J).

The unit is named after James Prescott Joule[†] (1818–1889), a 19th-century wealthy English brewer-turned-scientist who spent much time formulating the ideas of work and energy in physics, while enjoying the financial and liquid fruits of his trade.

Because of the scalar product in the definition, the differential work dW done by a force $\vec{F}$ on a system as it changes its position vector by $d\vec{r}$ may be

1. positive, if the angle between $\vec{F}$ and $d\vec{r}$ is acute (less than 90°);

---

*In Chapter 10 we introduce a new vector concept called torque that also has the SI units of N·m but *never* is expressed in joules.

[†]Although Joule pronounced his name to rhyme with owl, the joule energy unit is now almost universally pronounced to rhyme with tool. We won't object if you prefer it Joule's way.

James Prescott Joule

2. negative, if the angle between $\vec{F}$ and $d\vec{r}$ is obtuse (greater than 90° but less than or equal to 180°); or
3. zero (if the angle between $\vec{F}$ and $d\vec{r}$ is equal to 90°).

The work done by a force over a finite path also may be positive, negative, or zero.

QUESTION 1

> If you push and hold your textbook very hard against the rigid wall of your room for ten minutes, is any work done by the force you exert on the book? Explain your answer.

EXAMPLE 8.1

You drag your textbook 0.75 m across your level desk with a constant force of magnitude 6.0 N, oriented at 30° to the horizontal direction.

a. What is the work done by the force you exert on the book?
b. What is the work done by the gravitational force of the Earth on the book?
c. What is the work of the normal force of the surface on the book?

Solution
The book is the system. Introduce a coordinate system and draw a second law force diagram indicating the forces on the book, as in Figure 8.3.

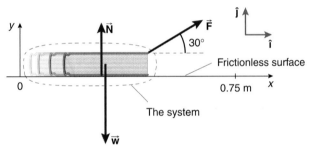

FIGURE 8.3

a. The constant force you exert on the book is

$$\vec{F} = (6.0 \text{ N})(\cos 30°)\hat{\imath} + (6.0 \text{ N})(\sin 30°)\hat{\jmath}$$
$$= (5.2 \text{ N})\hat{\imath} + (3.0 \text{ N})\hat{\jmath}$$

Since the book moves only along the *x*-axis, the differential change in the position vector of the book is written as

$$d\vec{r} = dx\,\hat{\imath}$$

The work done by this force then is

$$W = \int_{0 \text{ m}}^{0.75 \text{ m}} [(5.2 \text{ N})\,\hat{\imath} + (3.0 \text{ N})\,\hat{\jmath}] \cdot dx\,\hat{\imath}$$
$$= (5.2 \text{ N}) \int_{0 \text{ m}}^{0.75 \text{ m}} dx$$
$$= (5.2 \text{ N})(0.75 \text{ m} - 0 \text{ m})$$
$$= 3.9 \text{ J}$$

Notice that the component of the force perpendicular to the path followed by the book did no work. Only the component of the force along the direction of the path does work.

b. The weight $\vec{\mathbf{w}} = -mg\hat{\mathbf{j}}$ is perpendicular to $d\vec{\mathbf{r}} = dx\,\hat{\mathbf{i}}$ at each and every point along the path. Hence

$$\vec{\mathbf{w}} \cdot d\vec{\mathbf{r}} = -mg\hat{\mathbf{j}} \cdot dx\,\hat{\mathbf{i}}$$
$$= 0 \text{ J}$$

at each and every point along the path. The weight does zero work on the book along this path.

c. The normal force $\vec{\mathbf{N}} = N\hat{\mathbf{j}}$ also is perpendicular to $d\vec{\mathbf{r}} = dx\,\hat{\mathbf{i}}$ at each and every point along the path. Therefore

$$\vec{\mathbf{N}} \cdot d\vec{\mathbf{r}} = N\hat{\mathbf{j}} \cdot dx\,\hat{\mathbf{i}}$$
$$= 0 \text{ J}$$

at each and every point along the path. The normal force does zero work on the book along this path.

EXAMPLE 8.2

A 0.100 kg trinket is attached to a spring with spring constant 40.0 N/m and undergoes simple harmonic oscillation with an amplitude of 0.150 m before the glazed eyes of a dazed tourist. As usual, the equilibrium position is taken to be $x = 0$ m with $\hat{\mathbf{i}}$ in the direction of stretching the spring. What is the work done by the force of the spring on the trinket as it moves from the equilibrium position to $x = -0.150$ m?

Solution

The trinket is the system. The work done is the integral of $\vec{\mathbf{F}}_{spring} \cdot d\vec{\mathbf{r}}$ from the initial to final positions. The force of the spring on the trinket is given by Equation 7.2:

$$\vec{\mathbf{F}}_{spring} = -kx\hat{\mathbf{i}}$$

According to Problem-Solving Tactic 8.1, since the motion is purely along x, take $d\vec{\mathbf{r}} = dx\,\hat{\mathbf{i}}$ and let the limits of integration take care of the direction the system is moving along the x-axis:

$$W = \int_{i}^{f} (-kx\hat{\mathbf{i}}) \cdot dx\,\hat{\mathbf{i}}$$
$$= -k \int_{0 \text{ m}}^{-0.150 \text{ m}} x\, dx$$

After integrating, you have

$$W = -k\frac{x^2}{2} \Bigg|_{0 \text{ m}}^{-0.150 \text{ m}}$$
$$= -\frac{40.0 \text{ N/m}}{2}[(-0.150 \text{ m})^2 - (0 \text{ m})^2]$$
$$= -0.450 \text{ J}$$

Notice that the force of the spring and the differential change in the position vector were in opposite directions over the entire path; this accounts for the negative result for this work.

8.3 THE WORK DONE BY A CONSTANT FORCE

The weight of a system of constant mass is a constant vector no matter how the system moves about, as long as it remains close to

FIGURE 8.4 The weight is a constant vector as a system moves along any path near the surface of the Earth.

the surface of the Earth (see Figure 8.4). This is one of many situations that involve constant force vectors, and so it is useful to consider the special case of the work done on a system by a constant force as the system moves along a path that is *not* necessarily straight.

A constant force vector has both a constant magnitude *and* a constant direction.

For example, as you move a crate along a curved path, as in Figure 8.5, the weight of the crate *is* a constant force along the curved path, because its magnitude and direction are the same at all points along the path. However, the normal force and the kinetic frictional force are *not* constant forces because their magnitudes and directions change along this path.

To calculate the work done on the system by a constant force, begin with the general definition of the work, Equation 8.2:

$$W = \int_{i}^{f} \vec{\mathbf{F}} \cdot d\vec{\mathbf{r}}$$

It is useful to write the constant force $\vec{\mathbf{F}}$ in terms of its constant Cartesian components:

$$\vec{\mathbf{F}} = F_x\hat{\mathbf{i}} + F_y\hat{\mathbf{j}} + F_z\hat{\mathbf{k}}$$

FIGURE 8.5 The weight of the crate is a constant force along this surface and path, but the normal force and the kinetic frictional force are not.

We also express the change in the position vector $d\vec{\mathbf{r}}$ in its most general Cartesian form as (Equation 8.3)

$$d\vec{\mathbf{r}} = dx\,\hat{\mathbf{i}} + dy\,\hat{\mathbf{j}} + dz\,\hat{\mathbf{k}}$$

Making these substitutions into Equation 8.2, we get

$$W = \int_i^f [F_x\hat{\mathbf{i}} + F_y\hat{\mathbf{j}} + F_z\hat{\mathbf{k}}] \cdot [dx\,\hat{\mathbf{i}} + dy\,\hat{\mathbf{j}} + dz\,\hat{\mathbf{k}}]$$
$$= \int_i^f (F_x\,dx + F_y\,dy + F_z\,dz) \qquad (8.4)$$

Expressed in Cartesian form, the initial and final position vectors of the system are

$$\vec{\mathbf{r}}_i = x_i\hat{\mathbf{i}} + y_i\hat{\mathbf{j}} + z_i\hat{\mathbf{k}}$$
$$\vec{\mathbf{r}}_f = x_f\hat{\mathbf{i}} + y_f\hat{\mathbf{j}} + z_f\hat{\mathbf{k}}$$

Substitute the appropriate limits for the integrations into Equation 8.4. The work done is

$$W = \int_{x_i}^{x_f} F_x\,dx + \int_{y_i}^{y_f} F_y\,dy + \int_{z_i}^{z_f} F_z\,dz$$

Since the force components of a constant force are themselves constant, we have

$$W = F_x\int_{x_i}^{x_f} dx + F_y\int_{y_i}^{y_f} dy + F_z\int_{z_i}^{z_f} dz$$
$$= F_x(x_f - x_i) + F_y(y_f - y_i) + F_z(z_f - z_i) \qquad (8.5)$$

Let $\Delta\vec{\mathbf{r}}$ be the *change* in the position vector of the system:

$$\Delta\vec{\mathbf{r}} \equiv \vec{\mathbf{r}}_f - \vec{\mathbf{r}}_i$$
$$= (x_f - x_i)\hat{\mathbf{i}} + (y_f - y_i)\hat{\mathbf{j}} + (z_f - z_i)\hat{\mathbf{k}}$$

in Cartesian form.

The work done on the system by the constant force, Equation 8.5, is expressed succinctly as

$$\boxed{W = \vec{\mathbf{F}} \cdot \Delta\vec{\mathbf{r}} \qquad \text{(constant force } \vec{\mathbf{F}})} \qquad (8.6)$$

PROBLEM-SOLVING TACTIC

8.2 The work done by a constant force, Equation 8.6, can be used even if the path between the initial and final positions is not straight. However, take care: the force must be a *constant force*, with both a fixed magnitude and direction over the *entire path* followed by the system between its initial and final positions.

QUESTION 2

You push Victor Hugo's massive novel *Les Misérables* at constant speed along a path on your level desk. Describe a path along which the kinetic frictional force on the book is a constant force. Describe a path along which the kinetic frictional force is not constant.

EXAMPLE 8.3 ━━━━━━━━━━━━━━━

A force $\vec{\mathbf{F}} = (-30.0\,\text{N})\hat{\mathbf{i}} + (20.0\,\text{N})\hat{\mathbf{j}}$ acts on a 10.0 kg golf bag as it moves from $x = 10.00\,\text{m}$, $y = 0\,\text{m}$ to $x = 5.00\,\text{m}$, $y = 2.00\,\text{m}$

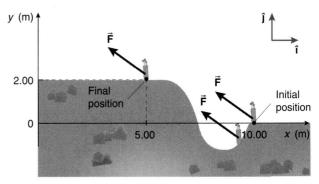

FIGURE 8.6

along the path indicated in Figure 8.6. Find the work done on the golf bag system by this force over this path.

Solution

Method 1
Notice that the force is a *constant vector*; it has the same magnitude and direction along the entire path followed by the system between the initial and final positions. Hence you can calculate the work by using Equation 8.6:

$$W = \vec{\mathbf{F}} \cdot \Delta\vec{\mathbf{r}} \qquad \text{(constant force } \vec{\mathbf{F}})$$

The initial and final position vectors of the system are

$$\vec{\mathbf{r}}_i = (10.00\,\text{m})\hat{\mathbf{i}}$$
$$\vec{\mathbf{r}}_f = (5.00\,\text{m})\hat{\mathbf{i}} + (2.00\,\text{m})\hat{\mathbf{j}}$$

Therefore the change in the position vector is

$$\Delta\vec{\mathbf{r}} = \vec{\mathbf{r}}_f - \vec{\mathbf{r}}_i$$
$$= [(5.00\,\text{m})\hat{\mathbf{i}} + (2.00\,\text{m})\hat{\mathbf{j}}] - (10.00\,\text{m})\hat{\mathbf{i}}$$
$$= (-5.00\,\text{m})\hat{\mathbf{i}} + (2.00\,\text{m})\hat{\mathbf{j}}$$

Use Equation 8.6 to find the work:

$$W = \vec{\mathbf{F}} \cdot \Delta\vec{\mathbf{r}}$$
$$= [(-30.0\,\text{N})\hat{\mathbf{i}} + (20.0\,\text{N})\hat{\mathbf{j}}] \cdot [(-5.00\,\text{m})\hat{\mathbf{i}} + (2.00\,\text{m})\hat{\mathbf{j}}]$$

Evaluating the various scalar products, you find

$$W = 150\,\text{J} + 40.0\,\text{J}$$
$$= 190\,\text{J}$$

Method 2
The definition of the work done on the system by any force over a path is given by Equation 8.2:

$$W = \int_i^f \vec{\mathbf{F}} \cdot d\vec{\mathbf{r}}$$

Notice that the force has an *x*-component directed toward decreasing values of *x* and the system is moving toward decreasing values of *x*. Thus the *x*-component of the force is directed parallel to the direction the system moves on the *x*-axis while traveling from the initial to the final position; you can anticipate that the work done by the *x*-component of the force will be positive. Since the *y*-coordinates of the system first decrease and then increase, it is perhaps not as clear what is the sign of the work done by the y-component of the force.

Since the system is moving in two dimensions, the differential change in the position vector $d\vec{r}$ is written as

$$d\vec{r} = dx\,\hat{\imath} + dy\,\hat{\jmath}$$

even though the particle is moving toward decreasing values of x and the value of y first decreases and then increases. The limits of integration account for all the changes in position. The work done by this force over this path is

$$W = \int_i^f [(-30.0 \text{ N})\,\hat{\imath} + (20.0 \text{ N})\,\hat{\jmath}] \cdot (dx\,\hat{\imath} + dy\,\hat{\jmath})$$

Evaluating the scalar products in the integrand, you obtain

$$W = (-30.0 \text{ N})\int_{10.00 \text{ m}}^{5.00 \text{ m}} dx + (20.0 \text{ N})\int_{0 \text{ m}}^{2.00 \text{ m}} dy$$

$$= (-30.0 \text{ N})\,x\,\Big|_{10.00 \text{ m}}^{5.00 \text{ m}} + (20.0 \text{ N})\,y\,\Big|_{0 \text{ m}}^{2.00 \text{ m}}$$

$$= (-30.0 \text{ N})(5.00 \text{ m} - 10.00 \text{ m})$$
$$+ (20.0 \text{ N})(2.00 \text{ m} - 0 \text{ m})$$

$$= 150 \text{ J} + 40.0 \text{ J}$$

$$= 190 \text{ J}$$

EXAMPLE 8.4

A satellite is in a circular orbit of radius r about the Earth. Calculate the work done by the gravitational force of the Earth on the satellite around one complete orbit.

Solution

The work done by the gravitational force around one complete orbital path is the integral of the scalar product of the gravitational force and the differential change in the position vector $d\vec{r}$ along the orbital path:

$$W = \int_{\substack{\text{one} \\ \text{orbital path}}} \vec{F}_{\text{grav}} \cdot d\vec{r}$$

Look at the integrand:

$$\vec{F}_{\text{grav}} \cdot d\vec{r}$$

As the satellite moves along its circular path, the gravitational force of the Earth on the satellite (which is directed radially toward the Earth) is perpendicular to $d\vec{r}$ at every point along the path, as shown in Figure 8.7. Thus the scalar product in the integrand is

$$\vec{F}_{\text{grav}} \cdot d\vec{r} = F_{\text{grav}}\, dr \cos 90°$$
$$= 0 \text{ J}$$

Thus the integrand is zero at every point along the path of the orbit. Therefore the work done by the gravitational force around the circular orbit is zero:

$$W = \int_{\substack{\text{one} \\ \text{orbital path}}} \vec{F}_{\text{grav}} \cdot d\vec{r}$$
$$= 0 \text{ J}$$

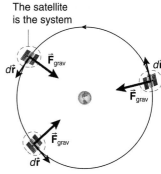

FIGURE 8.7

You might be tempted to use Equation 8.6 to solve this problem:

$$W = \vec{F} \cdot \Delta\vec{r} \qquad (\text{constant force } \vec{F})$$

However, while the *magnitude* of the force here is constant over the entire path, this force is *not a constant vector* since *its direction changes* along the path. Hence the criterion for using Equation 8.6 is *not* satisfied and it is incorrect to use it to calculate the work done on the system by $\vec{F}_{\text{grav}}$. Ironically, since we wanted to calculate the work done by $\vec{F}_{\text{grav}}$ around one complete orbital path, the change in the position vector $\Delta\vec{r}$ of the system here is *zero* because the initial and final position vectors are identical. With $\Delta\vec{r} = 0$ m, Equation 8.6 gives the correct answer, but the equation *cannot* be used legitimately, since $\vec{F}_{\text{grav}}$ is not a constant vector. Blindly using Equation 8.6 here for the work gives the right answer, but for the wrong reasons! This occasionally happens and illustrates that simply securing the right answer is not necessarily a good test of understanding.

QUESTION 3

Note that the work done by the gravitational force in Example 8.4 is *not* the magnitude of the force times the distance, $F_{\text{grav}}(2\pi r)$. Why?

8.4 THE WORK DONE BY THE TOTAL FORCE

Many systems have more than one force acting on them. The total force on a system is the vector sum of all the individual forces on it:

$$\vec{F}_{\text{total}} = \vec{F}_1 + \vec{F}_2 + \vec{F}_3 + \cdots$$

where $\vec{F}_1, \vec{F}_2, \vec{F}_3, \ldots$ are the individual forces on the system. The total work done on the system along some path is the work done by the total force:

$$W_{\text{total}} = \int_i^f \vec{F}_{\text{total}} \cdot d\vec{r}$$

Substitute the vector sum of the individual forces for $\vec{F}_{\text{total}}$:

$$W_{\text{total}} = \int_i^f (\vec{F}_1 + \vec{F}_2 + \vec{F}_3 + \cdots) \cdot d\vec{r}$$

$$= \int_i^f \vec{F}_1 \cdot d\vec{r} + \int_i^f \vec{F}_2 \cdot d\vec{r} + \int_i^f \vec{F}_3 \cdot d\vec{r} + \cdots$$

The integrals on the right-hand side of this expression are just the work done by the individual forces themselves. Thus

$$W_{\text{total}} = W_1 + W_2 + W_3 + \cdots$$

The work done by the total force acting on the system is the algebraic, scalar sum of the work done by the individual forces.

The work done by an individual force may be positive, negative, or zero, depending on the specific situation, and so W_{total} may be positive, negative, or zero.

STRATEGIC EXAMPLE 8.5

A crate of mass m containing a new lab instrument is dragged by enthusiastic physics students a distance s along a straight, level corridor from an elevator to a physics lab. The students maintain a constant pull of magnitude T newtons applied to the crate by means of a rope inclined to the horizontal at angle θ. The velocity of the crate along the straight path is *not* constant. The coefficient of kinetic friction for the crate on the floor is μ_k. Calculate the work done by each force on the crate and the work done by the total force on the crate.

Solution
Consider the crate to be the system and represent it schematically as a particle. A second law force diagram (see Figure 8.8) includes

1. the gravitational force of the Earth on the crate—the weight $\vec{w}$ of the crate;
2. the normal force $\vec{N}$ of the surface on the crate;
3. the force of the rope $\vec{T}$ on the crate; and
4. the kinetic frictional force $\vec{f}_k$.

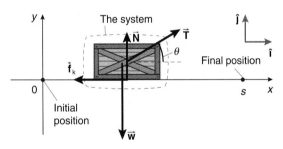

FIGURE 8.8

For convenience introduce a coordinate system with the origin at the initial position of the crate, so that $\vec{r}_i = 0$ m. The final position of the crate is $\vec{r}_f = s\hat{i}$. The change in the position vector of the crate thus is

$$\Delta\vec{r} = s\hat{i} - 0 \text{ m}$$
$$= s\hat{i}$$

The various forces on the crate are written in terms of their Cartesian components as

1. the weight: $\vec{w} = -mg\hat{j}$;
2. the normal force: $\vec{N} = N\hat{j}$; and
3. the force of the rope: $\vec{T} = (T\cos\theta)\hat{i} + (T\sin\theta)\hat{j}$.

Since the kinetic frictional force always is directed opposite to the velocity, this force is toward decreasing values of x:

4. the kinetic frictional force: $\vec{f}_k = -f_k\hat{i}$.

Each force is constant along the entire path, and so in each case you can use Equation 8.6 for the work done by a constant force:

$$W = \vec{F} \cdot \Delta\vec{r} \quad \text{(constant force } \vec{F}\text{)}$$

The work done by the normal force is

$$W_{\text{normal}} = N\hat{j} \cdot s\hat{i}$$
$$= 0 \text{ J}$$

The work done by the weight is
$$W_{\text{weight}} = -mg\hat{j} \cdot s\hat{i}$$
$$= 0 \text{ J}$$

The work done by the normal force $\vec{N}$ and the work done by the weight $\vec{w}$ are both zero because the horizontal path followed by the system is perpendicular to both of these forces at each and every point along the path.

The magnitude of the kinetic frictional force $\vec{f}_k$ is related to the magnitude of the normal force by the empirical equation

$$f_k = \mu_k N$$

The magnitude of the normal force is found by applying Newton's second law to the vertical direction, realizing that the system is not accelerating in this direction:

$$F_{y\,\text{total}} = ma_y$$
$$-mg + N + T\sin\theta = m(0 \text{ m/s}^2)$$

So

$$N = mg - T\sin\theta$$

Thus the kinetic frictional force is of magnitude

$$f_k = \mu_k(mg - T\sin\theta)$$

The work done by the kinetic frictional force is

$$W_{\text{frictn}} = (-f_k\hat{i}) \cdot s\hat{i}$$
$$= -f_k s$$
$$= -\mu_k(mg - T\sin\theta)s$$

Notice that the work done by the kinetic frictional force is *negative*, indicating that the force and the change in the position vector were in opposite directions.

The work done by the force of the rope on the system is

$$W_{\text{rope}} = [(T\cos\theta)\hat{i} + (T\sin\theta)\hat{j}] \cdot s\hat{i}$$
$$= Ts\cos\theta$$

Since the velocity of the crate is not constant, the total force on it is not zero. There are two ways you can find the work done by the total force.

Method 1
The work done by the *total* force is the *algebraic, scalar sum* of the work done by all the individual forces. Thus

$$W_{\text{total}} = W_{\text{weight}} + W_{\text{normal}} + W_{\text{frictn}} + W_{\text{rope}}$$

or

$$W_{total} = 0 \text{ J} + 0 \text{ J} + [-\mu_k(mg - T \sin \theta)s] + Ts \cos \theta$$
$$= Ts(\cos \theta + \mu_k \sin \theta) - \mu_k mgs$$

Method 2

The work done by the *total* force is

$$W_{total} = \int_i^f \vec{F}_{total} \cdot d\vec{r}$$

The total force is the vector sum of all the forces on the crate:

$$\vec{F}_{total} = \vec{w} + \vec{N} + \vec{T} + \vec{f}_k$$
$$= -mg\hat{j} + N\hat{j} + [(T \cos \theta)\hat{i} + (T \sin \theta)\hat{j}] + (-f_k\hat{i})$$
$$= (T \cos \theta - f_k)\hat{i} + (N + T \sin \theta - mg)\hat{j}$$

The total force is a constant vector along this path, and so you can use Equation 8.6 to find its work:

$$W_{total} = \vec{F}_{total} \cdot \Delta\vec{r}$$
$$= [(T \cos \theta - f_k)\hat{i} + (N + T \sin \theta - mg)\hat{j}] \cdot s\hat{i}$$
$$= (T \cos \theta - f_k)s$$

Since $f_k = \mu_k(mg - T \sin \theta)$, the total work is

$$W_{total} = [T \cos \theta - \mu_k(mg - T \sin \theta)]s$$
$$= Ts(\cos \theta + \mu_k \sin \theta) - \mu_k mgs$$

Notice that the y-component of the total force is everywhere perpendicular to the path and, thus, does no work; only the component of the total force along the path does nonzero work.

EXAMPLE 8.6 _____

A physics professor moves a mass m attached to a spring, stretching the spring very slowly along the entire path as m moves through a distance s along a frictionless surface from its equilibrium position (at $x = 0$ m), as shown in Figure 8.9. At any position the acceleration of the mass is essentially zero. Calculate the work done on the system of mass m:

a. by the force of the professor on m; and
b. by the force of the spring on m.
c. What is the work done by all the forces on the system?

FIGURE 8.9

Solution

Consider the mass m to be the system. Although the spring is deformed (stretched), the spring is not the system. Use the

coordinate system indicated in Figures 8.9 and 8.10. The forces on the system are as follows:

1. The force of the spring on m is, according to Equation 7.2,

$$\vec{F}_{spring} = -kx\hat{i}$$

2. Likewise, from Equation 7.1, the professor must exert a force

$$\vec{F}_{prof} = kx\hat{i}$$

on the system at any given position x.
3. The weight of the system, given by

$$\vec{w} = -mg\hat{j}$$

4. The normal force of the surface on the system, given by

$$\vec{N} = N\hat{j}$$

The force of the professor and that of the spring on the system are *not* constant forces over the path; you cannot use Equation 8.6 for a constant force but must calculate their work from the general expression for the work of any force, Equation 8.2.

FIGURE 8.10

Since the motion is one-dimensional, the differential change in the position vector $d\vec{r}$ is, from Equation 8.3,

$$d\vec{r} = dx\,\hat{i}$$

a. Assume that the spring remains within its elastic limit. The work done by the force of the professor on the mass is then

$$W_{prof} = \int_i^f \vec{F}_{prof} \cdot d\vec{r}$$
$$= \int_{0\text{ m}}^s kx\,\hat{i} \cdot dx\,\hat{i}$$
$$= \int_{0\text{ m}}^s kx\,dx$$
$$= \left. \frac{kx^2}{2} \right|_{0\text{ m}}^s$$
$$= \frac{ks^2}{2}$$

b. The force of the spring on m also is not a constant force. Hence the work done by this force on the system must be found using Equation 8.2, the general expression for the work of any force. The work is

$$
\begin{aligned}
W_{spring} &= \int_{i}^{f} \vec{F}_{spring} \cdot d\vec{r} \\
&= \int_{0\ m}^{s} (-kx\hat{\imath}) \cdot dx\,\hat{\imath} \\
&= -k \int_{0\ m}^{s} x\ dx \\
&= -k\frac{x^2}{2} \Big|_{0\ m}^{s} \\
&= -\frac{ks^2}{2}
\end{aligned}
$$

There are two other forces acting on the mass: the normal force $\vec{N}$ of the surface and the weight $\vec{w}$. These are *constant forces*. The work done by both of these forces is *zero* because both forces are perpendicular to the path followed by the system at each and every point along the path.

c. There are two ways to find the total work done by all the forces on the system:

Method 1

The work done by the total force is the *algebraic, scalar sum* of the work done on the system by the individual forces. The total work done by all the forces on the system is

$$
\begin{aligned}
W_{total} &= W_{spring} + W_{prof} + W_{weight} + W_{normal} \\
&= -\frac{ks^2}{2} + \frac{ks^2}{2} + 0\ J + 0\ J \\
&= 0\ J
\end{aligned}
$$

Method 2

Use Equation 8.2,

$$
W_{total} = \int_{i}^{f} \vec{F}_{total} \cdot d\vec{r}
$$

and note that the *total force* on the system is zero at all points along the path (because the system is not accelerating). Therefore the total work must be zero.

Does $W_{total} = 0$ J mean that no work is done by the professor pulling on the mass? No, that work is equal to the work done by the force of the professor on the system, $W_{prof} = ks^2/2$.

8.5 GEOMETRIC INTERPRETATION OF THE WORK DONE BY A FORCE

There is a geometric way to visualize the work done by a force on a system, which ties in well with your growing knowledge of calculus, where integrals can be viewed as areas under a curve. If a system moves from x_i to x_f along the *x*-coordinate axis and

$$
\vec{F} = F_x\hat{\imath} + F_y\hat{\jmath} + F_z\hat{k}
$$

is one of the forces on the system, the work done on the system by $\vec{F}$ is, according to the definition of work, Equation 8.2,

$$
\begin{aligned}
W &= \int_{i}^{f} \vec{F} \cdot d\vec{r} \\
&= \int_{i}^{f} (F_x\hat{\imath} + F_y\hat{\jmath} + F_z\hat{k}) \cdot dx\,\hat{\imath}
\end{aligned}
$$

After evaluating the scalar products, we find

$$
W = \int_{x_i}^{x_f} F_x\ dx
$$

Figure 8.11 is a schematic graph of the *x*-component of the force as a function of *x*.

> The work done on the system by the force component F_x as the system moves from x_i to x_f is the area under the curve between x_i and x_f.

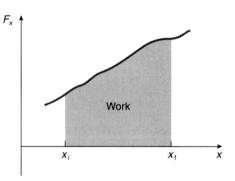

FIGURE 8.11   The work done by F_x is the area under the curve of F_x versus *x* between the initial and final *x*-coordinate positions.

Such areas may be positive or negative. Whether an area is positive or negative depends on two things: the direction the system is moving along the coordinate axis and whether the force component is positive or negative, as shown in Figure 8.12.

If the force

$$
\vec{F} = F_x\hat{\imath} + F_y\hat{\jmath} + F_z\hat{k}
$$

acts on a system while the system moves from position vector

$$
\vec{r}_i = x_i\hat{\imath} + y_i\hat{\jmath} + z_i\hat{k}
$$

to position vector

$$
\vec{r}_f = x_f\hat{\imath} + y_f\hat{\jmath} + z_f\hat{k}
$$

the work done by this force is

$$
\begin{aligned}
W &= \int_{i}^{f} (F_x\hat{\imath} + F_y\hat{\jmath} + F_z\hat{k}) \cdot (dx\,\hat{\imath} + dy\,\hat{\jmath} + dz\,\hat{k}) \\
&= \int_{x_i}^{x_f} F_x\ dx + \int_{y_i}^{y_f} F_y\ dy + \int_{z_i}^{z_f} F_z\ dz
\end{aligned}
$$

Each term is the area under the curve of the graph of that force component versus the corresponding coordinate.

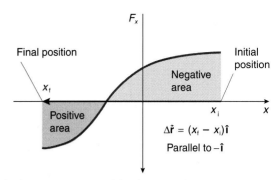

FIGURE 8.12 Positive or negative work (area) depends on both the sign of the force component and the direction of $\Delta\vec{r}$.

EXAMPLE 8.7

A force $\vec{F} = (3.00 \text{ N})\hat{i}$ acts on a system that moves along the x-axis from position $x_i = 3.00$ m to $x_f = 5.00$ m.

a. Graph F_x versus x.
b. Use the graph to find the work done by this force as the system moves between the two coordinates.

Solution
a. The x-component of the force is $F_x = 3.00$ N and is constant, and so a plot of F_x versus x is a horizontal line; see Figure 8.13.
b. The work done by F_x is the area under the curve between the two coordinates. This is the shaded rectangular area in Figure 8.13, which is

$$W = (3.00 \text{ N})(2.00 \text{ m})$$
$$= 6.00 \text{ J}$$

The area under the curve is positive since the system is moving parallel to $+\hat{i}$ and the force component is positive. We could anticipate a positive result since the force is in the same direction as the change in the position vector of the system.

FIGURE 8.14

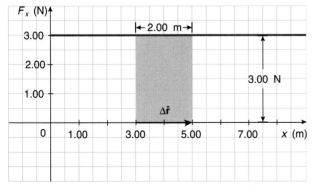

FIGURE 8.13

EXAMPLE 8.8

A force whose only component is F_x acts on a system; a graph of F_x versus x is shown in Figure 8.14. Use the graph to find the work done on the system by the force as the system moves along the x-axis from $x = -2.0$ m to $x = +10.0$ m.

Solution
The work done is the area under the graph between $x_i = -2.0$ m and $x_f = +10.0$ m. The area under the curve between -2.0 m and 1.0 m is *negative*, because the system is moving parallel to $\hat{i}$ and the force component for these values of x is negative. Since the region is triangularly shaped, the area is

$$\text{area} = \frac{1}{2} \text{(base)}\text{(height)}$$
$$= \frac{1}{2} (3.0 \text{ m})(-4.0 \text{ N})$$
$$= -6.0 \text{ J}$$

The area between $+1.0$ m and $+10.0$ m is positive, because the system is moving parallel to $\hat{i}$ and the force component is positive for these values of x. This region also is triangularly shaped, and so the area is

$$\text{area} = \frac{1}{2} (9.0 \text{ m})(12.0 \text{ N})$$
$$= 54 \text{ J}$$

The total area between $x_i = -2.0$ m and $x_f = +10.0$ m is the algebraic sum of these areas. Therefore the work W done by this force is

$$W = -6.0 \text{ J} + 54 \text{ J}$$
$$= 48 \text{ J}$$

EXAMPLE 8.9

A force has only an *x*-component, and a graph of F_x versus x is shown in Figure 8.15; the curves are arcs of circles. Use the graph to find the work done by this force on a system as it moves along the *x*-axis from $x_i = 10.00$ m to $x_f = -5.00$ m.

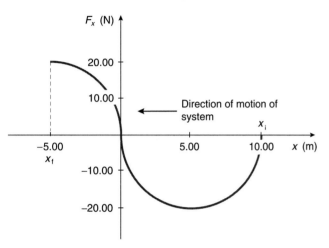

FIGURE 8.15

Solution

The work done by this force is the area under the curve between $x_i = 10.00$ m and $x_f = -5.00$ m. The semicircular area under the curve between $x = 10.00$ m and $x = 0$ m is positive, because the system is moving parallel to $-\hat{\imath}$ and the force component also is negative for these values of x. This semicircular area is equal to

$$\text{area} = \frac{1}{2}\pi(-20.0 \text{ N})(-5.00 \text{ m})$$
$$= 157 \text{ J}$$

The area under the curve between $x = 0$ m and $x = -5.00$ m is negative, because the system is moving parallel to $-\hat{\imath}$ but the force component is positive for these values of x. This area is equal to

$$\text{area} = \frac{1}{4}\pi(20.0 \text{ N})(-5.00 \text{ m})$$
$$= -78.5 \text{ J}$$

The total work is the algebraic sum of the areas:

$$W = -78.5 \text{ J} + 157 \text{ J}$$
$$= 78 \text{ J}$$

8.6 CONSERVATIVE, NONCONSERVATIVE, AND ZERO-WORK FORCES

Forces come in three work-related "flavors": conservative forces, nonconservative forces, and zero-work forces.

> A force is called a **conservative force**
>
> • if the work done by the force on any system as it moves between two points is *independent of the path* followed by the system between the two points; or equivalently,
>
> • if the work done by the force on any system *around any closed path is zero*. A closed path (round-trip path) is one that begins and ends at the same point in space.

The work done by a force around a *closed path* is written mathematically as

$$\oint \vec{\mathbf{F}} \cdot d\vec{\mathbf{r}}$$

where the elegant little circle through the integral sign is an indication that the integration is done around a *closed path*. If

$$\oint \vec{\mathbf{F}} \cdot d\vec{\mathbf{r}} = 0 \text{ J} \qquad (8.7)$$

for every closed path, then the force $\vec{\mathbf{F}}$ is conservative.

You will see the equivalence of these definitions as we consider several examples of conservative forces in the following section. We shall see that the work done by any conservative force on a system as it moves between two points depends only on the locations of the initial and final positions of the system.

> If the work done by a force around a closed path is *not zero*, or equivalently, if the work done by a force as a system moves between two points depends on the particular path taken between the two points, then the force is called a **nonconservative force.***

> Forces that *always do zero work* are called **zero-work forces**.

QUESTION 4

A simple pendulum swings back and forth once during its oscillation period. During one period, what work is done on the mass system attached to the string by the force of the string on the mass? Can you conclude that this force is conservative? Describe a physical situation in which the force of a string on a mass does nonzero work around a closed path.

*There are no corresponding "liberal forces" in the physics lexicon!

8.7 EXAMPLES OF CONSERVATIVE, NONCONSERVATIVE, AND ZERO-WORK FORCES

In this section we determine whether some of the common forces we considered in Chapters 5–7 are conservative, nonconservative, or zero-work forces.

The Local Gravitational Force on a Mass near the Surface of the Earth

The local gravitational force of the Earth on a system of mass m is its weight, of magnitude mg, where g is the magnitude of the gravitational field at the surface of the Earth (the magnitude of the acceleration due to gravity near the surface: $g = 9.81 \text{ m/s}^2$). This force is a constant vector. To see if this force is conservative, calculate the work done by the force on a system as it moves along an arbitrary path between two points A and B, as shown in Figure 8.16.

Set up a Cartesian coordinate system with $\hat{\mathbf{j}}$ directed up, as shown in Figure 8.16. The gravitational force on m is

$$\vec{\mathbf{F}} = -mg\hat{\mathbf{j}}$$

Other forces must also be involved in moving the particle. However, we focus on the gravitational force and the work it alone does on the system as it moves between the two points. Since the gravitational force is a *constant force* near the surface of the Earth, you can calculate the work done by this force using Equation 8.6 for the work done by a constant force:

$$W = \vec{\mathbf{F}} \cdot \Delta\vec{\mathbf{r}} \qquad \text{(constant force } \vec{\mathbf{F}}\text{)}$$

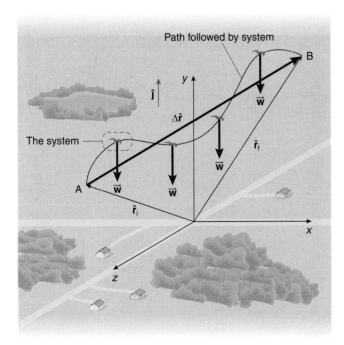

FIGURE 8.16 The weight is a constant vector along any path near the surface of the Earth.

where

$$\begin{aligned}
\Delta\vec{\mathbf{r}} &= \vec{\mathbf{r}}_\text{f} - \vec{\mathbf{r}}_\text{i} \\
&= (x_\text{f}\hat{\mathbf{i}} + y_\text{f}\hat{\mathbf{j}} + z_\text{f}\hat{\mathbf{k}}) - (x_\text{i}\hat{\mathbf{i}} + y_\text{i}\hat{\mathbf{j}} + z_\text{i}\hat{\mathbf{k}}) \\
&= (x_\text{f} - x_\text{i})\hat{\mathbf{i}} + (y_\text{f} - y_\text{i})\hat{\mathbf{j}} + (z_\text{f} - z_\text{i})\hat{\mathbf{k}}
\end{aligned}$$

Thus the work done by this force is

$$\begin{aligned}
W_\text{local} &= (-mg\hat{\mathbf{j}}) \cdot [(x_\text{f} - x_\text{i})\hat{\mathbf{i}} + (y_\text{f} - y_\text{i})\hat{\mathbf{j}} + (z_\text{f} - z_\text{i})\hat{\mathbf{k}}] \\
&= -mg(y_\text{f} - y_\text{i}) \qquad (\hat{\mathbf{j}} \text{ directed up}) \qquad (8.8)
\end{aligned}$$

Moreover, since the gravitational force is a constant force along *any* path between the initial and final positions, the result for W is independent of the path.

> Hence the local gravitational force is a conservative force.

There are several things to note about Equation 8.8:

a. The work is *negative* if the system *increases its elevation*—that is, if $y_\text{f} > y_\text{i}$. The negative answer can be anticipated since the gravitational force is *antiparallel* to the vertical change in the position vector over any path along which the system increases its elevation. The work is *positive* if the system *decreases its elevation* along a path—that is, if $y_\text{f} < y_\text{i}$; in this case, the gravitational force is *parallel* to the vertical change in the position vector. For any calculation of work, try to anticipate whether the result will be positive or negative on the basis of the relative directions of the force and the change in the position vector.

b. The work done by the local gravitational force depends *only* on the difference between the final and initial vertical positions of the particle; this will be important later (in Sections 8.8 and 8.9). The work done by the local gravitational force does *not* depend on differences in the horizontal coordinates of the system. The work done is *independent of the path* between the two points.

c. Notice that if the system is taken on a round trip around *any* closed path, the final and initial positions of the system are the same. In particular, $y_\text{f} = y_\text{i}$ and the work done by the local gravitational force is *zero*. Hence the work done by the local gravitational force around *any closed path* is zero. Thus we see explicitly that the two ways we defined a conservative force at the beginning of Section 8.6 here are equivalent.

Our results here can be generalized.

> Any force that is a constant vector over an arbitrary *closed path* is a conservative force.

The General Form for the Gravitational Force

Unlike the constant gravitational force near the surface of the Earth, the general form for the gravitational force is *not a constant force*. Thus we have to see if this force is conservative by calculating the work done as a system of mass m moves between two points using Equation 8.2 rather than Equation 8.6. If the work is independent of the path, the force is conservative.

Let's calculate the work done by the gravitational force on a system of mass m when it changes its position from point A (a distance r_i from mass M) to point B (a distance r_f from mass M), as shown in Figure 8.17. With an origin located at mass M, the gravitational force of M on m is, from Equation 6.6,

$$\vec{F}_{grav} = -\frac{GMm}{r^2}\hat{r}$$

where $\hat{r}$ is a unit vector pointing radially away from M, as shown in Figure 8.17.

How shall we get the system from A to B? Take the funky crooked path shown in Figure 8.18. Each segment of the path either is the arc of a circle centered on M or is along the radial direction itself.

The work done along each circular arc segment is zero, because the gravitational force is perpendicular to all the circular arc path segments $d\vec{r}$ at every point along them. So the only work done on the system by the gravitational force occurs when the system moves along the radial segments. We can accumulate all the contributions along the radial segments into a single integral since

$$\int_{r_i}^{r_1}\vec{F}\cdot d\vec{r} + \int_{r_1}^{r_2}\vec{F}\cdot d\vec{r} + \int_{r_2}^{r_3}\vec{F}\cdot d\vec{r} + \cdots + \int_{r_n}^{r_f}\vec{F}\cdot d\vec{r}$$

$$= \int_{r_i}^{r_f}\vec{F}\cdot d\vec{r}$$

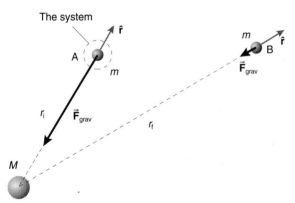

FIGURE 8.17 The gravitational force varies in magnitude and direction as you take m from A to B.

Thus the work done by the gravitational force in moving the system from A to B is

$$W = \int_{r_i}^{r_f}\left(-\frac{GmM}{r^2}\hat{r}\right)\cdot dr\,\hat{r}$$

In accordance with Problem-Solving Tactic 8.1, notice that we write $d\vec{r}$, the change in the position vector, as $dr\,\hat{r}$, because the limits of integration account for whether we are moving inward or outward in the radial direction.

Since the scalar product of a unit vector with itself is 1 ($\hat{r}\cdot\hat{r} = 1$), the work done is

$$W = -GmM\int_{r_i}^{r_f}\frac{dr}{r^2}$$

$$= -GmM\left(-\frac{1}{r}\right)\Big|_{r_i}^{r_f}$$

$$= -GmM\left(-\frac{1}{r_f} + \frac{1}{r_i}\right) \qquad (8.9)$$

The work done by the gravitational force depends only on the initial and final radial coordinate positions of the system of mass m relative to M. Note also that we could have chosen *any* other similarly segmented path to get from A to B with the same result.

A smooth path between the initial and final positions (see Figure 8.19) may be broken up into billions and billions of small radial and circular arc segments. Thus the argument made for the segmented path also applies to such smooth paths.* Therefore the work done by the gravitational force on the system as it moves from A to B is *independent of the path*.

> We can conclude that the general form for the gravitational force is a conservative force.

The work of the gravitational force depends only on the relative scalar radial position coordinates of A and B (and the constants

---

*This procedure is the standard limiting process in calculus.

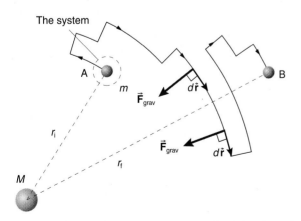

FIGURE 8.18 A path from A to B consisting of radial and circular segments.

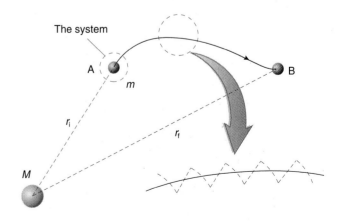

FIGURE 8.19 A smooth path can be broken up into many small radial and circular segments.

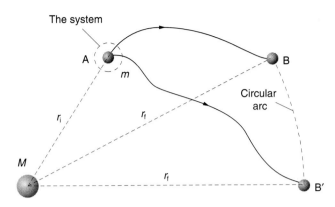

FIGURE 8.20 The work done by the gravitational force from A to B is the same as from A to B′ if B′ is the same distance from M as B.

FIGURE 8.21 A spring is stretched.

G, M, and m). The work done by the gravitational force on m in going from A to B′ in Figure 8.20 is the same as the work done in getting m from A to B because B′ has the same scalar radial coordinate as B.

If $r_f > r_i$, the work done by the gravitational force is *negative*. In this case the gravitational force is in the direction opposite to the outward radial change in the position vector of the system. On the other hand, if $r_f < r_i$, then the gravitational work is *positive*. In this case the gravitational force is in the same direction as the inward radial change in the position vector of the system.

A Central Force

Recall from Section 6.10 that a central force is a force that always is directed along the line connecting two particles. With some thought, the arguments used in the preceding subsection to show that the general form for the gravitational force is conservative can be used for *any central force*. Nonzero work is done by a central force only when the system changes its radial distance with respect to the source of the force; no work is done in moving perpendicular to the radial direction. Likewise, the work done by any central force depends only on the radial position coordinates of the initial and final points of the path.

Thus any central force is a conservative force.

Hooke's Force Law

We show here that Hooke's force law (see Chapter 7), important in simple harmonic oscillation, is conservative. In Figure 8.21, a system of mass m, attached to a spring characterized by spring constant k, is stretched from position x_i to position x_f, with $x = 0$ m corresponding to the equilibrium position of m on the spring (as usual).

We want to calculate the work done on the system by the force of the spring. In fact we already made such a calculation in Example 8.2. With the usual coordinate system for Hooke's law, the force of the spring on m is (from Equation 7.2)

$$\vec{F}_{spring} = -kx\hat{\imath}$$

This is *not* a constant force. Therefore to find the work we cannot use Equation 8.6 for the work of a constant force; we must use the general expression for the work of any force given by Equation 8.2:

$$W = \int_i^f \vec{F} \cdot d\vec{r}$$

Since the path is one-dimensional and along the x-axis, $d\vec{r}$ is

$$d\vec{r} = dx\,\hat{\imath}$$

because, as always, the limits of integration account for the specific direction the system moves along the axis. Thus the work done by the force of the spring on m is

$$
\begin{aligned}
W &= \int_{x_i}^{x_f} (-kx\hat{\imath}) \cdot (dx\,\hat{\imath}) \\
&= -k \int_{x_i}^{x_f} x\,dx \\
&= -k\frac{x^2}{2}\Bigg|_{x_i}^{x_f} \\
&= -k\frac{x_f^2}{2} + k\frac{x_i^2}{2} \quad\quad (8.10)
\end{aligned}
$$

Notice that the result depends only on the initial and final positions of the system attached to the spring. In particular, the spring does negative work on the system if $|x_f| > |x_i|$—that is, when the system is moving away from the equilibrium position, for then the force of the spring on the system and the change in its position vector are antiparallel. On the other hand, the work done by the spring on the system is positive if $|x_f| < |x_i|$—that is, when the force of the spring on the system is parallel to its change in position vector.

We also can calculate the work done by the force of the spring on the mass around a closed path by setting $x_f = x_i$ in Equation 8.10. The result is that the work done by the force of the spring on m around a closed path is zero.

Hence Hooke's force law is a conservative force.

The Kinetic Frictional Force

Though many forces are conservative, not all fit that criterion. A good example of a nonconservative force is the kinetic frictional force. Let's examine why this is the case.

To calculate the work done by the kinetic frictional force on a system of mass m around a closed path, move m along a straight-line distance s from point A to point B and then back to point A, as shown in Figure 8.22.

While other forces are involved in the motion, we are interested only in the work done by the kinetic frictional force. We use the specific form for the kinetic frictional force that is independent of the speed of the system; however, the nonconservative nature of the force does not depend on this choice for the specific form for the force.

Since the kinetic frictional force $\vec{\mathbf{f}}_k$ is constant along the path from A to B, the work done by it as the system moves from A to B is found by using Equation 8.6 for the work of a constant force:

$$W = \vec{\mathbf{f}}_k \cdot \Delta\vec{\mathbf{r}}$$

As the system moves from A to B, we anticipate that the kinetic frictional work will be *negative*, because the force is directed opposite to the change in the position vector of the system along the path from A to B. Use the Cartesian coordinate system indicated in Figure 8.22; the kinetic frictional force along the path from A to B then is

$$\vec{\mathbf{f}}_k = -f_k\hat{\mathbf{i}}$$

where f_k is the magnitude of the kinetic frictional force. The change in the position vector of the system is

$$\Delta\vec{\mathbf{r}} = s\hat{\mathbf{i}}$$

The work done by the kinetic frictional force in going from A to B is

$$W = (-f_k\hat{\mathbf{i}}) \cdot s\hat{\mathbf{i}}$$
$$= -f_k s$$

Now calculate the work W' done by the kinetic frictional force as the mass is moved from B back to A, thus completing a closed path. For this case, we *also* anticipate a negative result for the frictional work, because the kinetic frictional force once again is directed opposite to the change in the position vector of the system along the path from B back to A. In moving from B back to A, the kinetic frictional force is constant along this path and is

$$\vec{\mathbf{f}}_k = f_k\hat{\mathbf{i}}$$

since it is directed parallel to $\hat{\mathbf{i}}$ as we move from B back to A. The change in the position vector of the system now is

$$\Delta\vec{\mathbf{r}} = -s\hat{\mathbf{i}}$$

FIGURE 8.22 Move a mass from A to B and back to A.

So the work done by the kinetic frictional force from B back to A along this path is

$$W' = (f_k\hat{\mathbf{i}}) \cdot (-s\hat{\mathbf{i}})$$
$$= -f_k s$$

The *total work* done by the kinetic frictional force around the *closed path* from A to B and back to A again then is the sum of the work along each segment:

$$W_{\text{clsd path}} = W + W'$$
$$= -f_k s - f_k s$$
$$= -2f_k s$$

The work done around the closed path is *not zero*.

> Although we specifically considered only the speed-independent kinetic frictional force, *all* kinetic frictional forces are nonconservative.

If you are tempted to calculate the work done by the kinetic frictional force around a closed path using Equation 8.6 for the work done by a constant force, keep the following in mind. While the frictional force has a constant *magnitude* over the entire closed path, the force does *not* have a constant direction, and so it is *not* a constant vector over the *entire* closed path. This is quite unlike the constant local gravitational force, which is a constant vector over any closed path near the surface of the Earth and so is conservative.

QUESTION 5

Can you say that the work done by the kinetic frictional force *always* is negative?

Zero-Work Forces

Certain forces always do zero work and hence are called zero-work forces. Among them are the following:

a. The *static frictional force*. There is a subtle aspect to the work done by the static frictional force on a system. The static frictional force, by its very definition, involves no motion at the point(s) in contact; thus, the work done by static frictional forces is zero.
b. The *normal force* of a rigid surface. The normal force that a surface exerts on a system is perpendicular to the surface. If the system moves parallel to the surface, the normal force is perpendicular to the change in the position vector and so always does zero work. On the other hand, if the system moves perpendicular to the surface, thus leaving the surface (such as in jumping), the force exists only as long as the system is in contact with the surface (there is no motion at the point of contact); and so the normal force still does zero work.

8.8 THE CONCEPT OF POTENTIAL ENERGY

In Section 8.7, we showed that several common forces are conservative forces:

a. the *gravitational force* (in all its forms, whether the local gravitational force or the more general gravitational force);

b. *Hooke's force law* associated with springs and simple harmonic oscillation;

c. *any central force*; and

d. any force that is *constant over any and every closed path*.

We associate the word *work* with an *action* that is applied to the system externally. The action is applied by forces acting on the system as it moves from one place to another. This action is *transferred* to the system across our imaginary boundary that defines the system. It is useful and convenient to associate this action with the *change* in some *property* associated with the system itself. This property of the system is called its **energy**.

To exploit the position dependence of the work done by *conservative forces* on a system, a position-dependent property of the system called its **potential energy** is used. The work done by each conservative force has associated with it a change in a particular potential energy function.

> We define the work done on a system by a specific conservative force as the *negative* of the change in a potential energy function associated with the force. *Changes* always are taken as final minus initial values:
>
> $$W_{consrv} \equiv -(PE_f - PE_i)$$
> $$\equiv -\Delta PE \qquad (8.11)$$

The reason for the seemingly pesky negative sign in the definition will become apparent in Section 8.13. For now, just swallow hard and take the minus sign as part of the definition.

Several things to note immediately:

> *Each* conservative force has its own associated potential energy function. *Only* conservative forces have potential energy functions associated with them. Nonconservative forces do *not* have potential energy functions associated with them.

> Since the work done by a conservative force depends only on the initial and final positions of the particle, the change in the potential energy function also depends only on these positions. *Therefore potential energy functions are functions of the position of the system in space.*

> The work done by a conservative force on the particle *is the* negative of the change in the potential energy of the particle; the work done by the conservative force and the change in the associated potential energy function are not different things, they are the *same* thing.

> Potential energies (and all other energies) have the same dimensions as work (the product of a force and a distance); the SI units are joules (J).

Let's resurrect some of the results of the previous section regarding the work done by specific conservative forces and see what potential energy function is associated with each conservative force on a system.

8.9 THE GRAVITATIONAL POTENTIAL ENERGY OF A SYSTEM NEAR THE SURFACE OF THE EARTH

We discovered in Section 8.7 (Equation 8.8) that the work done by the gravitational force near the surface of the Earth in moving a system of mass m from $\vec{r}_i$ to $\vec{r}_f$ depends only on the difference in the elevation of the system along the y-coordinate axis (with $\hat{j}$ directed vertically up):

$$W_{local} = -mg(y_f - y_i)$$
$$= -(mgy_f - mgy_i) \qquad (\hat{j} \text{ directed up}) \quad (8.12)$$

We can choose the origin for the vertical y-axis at *any* convenient elevation; it does not have to be on the ground (see Figure 8.23).

According to Equation 8.11, the work done by a conservative force is defined to be the negative of the change in the associated potential energy function of the system:

$$W_{local} \equiv -(PE_f - PE_i)$$

Comparing Equation 8.12 with Equation 8.11, we identify the final potential energy of the system with the term mgy_f and its initial potential energy with mgy_i.

> Since these points can be located anywhere near the surface of the Earth, a system of mass m located at the vertical coordinate y has a gravitational potential energy of mgy:
>
> $$PE_{local} = mgy \qquad (\hat{j} \text{ directed up}) \quad (8.13)$$

We say the system *has* potential energy by virtue of its *position in space* at vertical coordinate y near the surface of the Earth.

Note several things about this local gravitational potential energy:

a. The local gravitational potential energy of the system is a *scalar* property of the system that is a function of its *position in space*.

b. The system has zero local gravitational potential energy at the elevation where we choose the y-coordinate to be zero; the potential energy is zero on the ground *only* if we choose the origin for the y-coordinate axis to be on the ground.

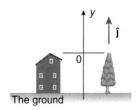

FIGURE 8.23 Choose $\hat{j}$ to be vertically up; you can choose the origin to be at any convenient elevation.

8.3 Equation 8.13 for the local gravitational potential energy must have ĵ directed up, since that is the choice we made in deriving the equation; but you can choose $y = 0$ m to be at any elevation you want. Thus you choose the elevation where the local gravitational potential energy is zero when choosing the origin for the y-coordinate axis. Specify this choice clearly in a sketch.

c. The local gravitational potential energy of a system is positive, negative, or zero depending on whether the y-coordinate of the system is positive, negative, or zero respectively.
d. *Changes* in the potential energy are equal to the negative of the work done on the system by the local gravitational force.
e. *If* the gravitational force is the *only* force acting on the system *and* the system is released *at rest*, then the system will move initially in a direction to lower its gravitational potential energy. That is, if m is released at rest some distance above the ground, so that the weight is the only force acting on m, then the local gravitational force moves m (doing positive work on m) in such a direction that it decreases the gravitational potential energy of the system. Since $PE = mgy$, with ĵ up, the system decreases its potential energy by decreasing its y-coordinate; cutting through the technical jargon, this argument is a physicist's way of saying m *falls*.

QUESTION 6

Three students are having a disagreement about the gravitational potential energy of a cola bottle at rest on a desk one meter above the floor. One says the potential energy of the bottle is zero, another argues it is positive, and the third says it is negative. Explain the details about how they each may be right within the context of the choices they make for a coordinate system.

STRATEGIC EXAMPLE 8.10

A 6.00 kg medicine ball is located 2.50 m above the ground.

a. **Find the change in the gravitational potential energy of the ball if it is moved to a point located 0.50 m above the ground.**
b. **What is the work done by the gravitational force on the ball over this change in position?**
c. **Does the work done by the gravitational force depend on the specific path used to get from one point to the other?**

Solution
a. Since the ball is near the surface of the Earth, use the local form for the gravitational potential energy, Equation 8.13:

$$PE = mgy \quad (\text{ĵ directed up})$$

You need to choose an elevation for the origin for the y-coordinate; this choice must be clearly indicated in a sketch. We show here the consequences of two choices for the y-coordinate origin to convince you that the *same* change in the potential energy emerges regardless of where the origin for y is chosen.

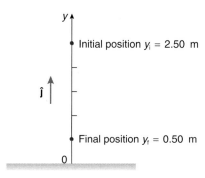

FIGURE 8.24

Method 1
Choose the elevation of the origin to be at ground level, as indicated in Figure 8.24. The initial y-coordinate of the mass then is $y_i = 2.50$ m and the initial potential energy of the mass is

$$PE_i = mgy_i$$
$$= (6.00 \text{ kg})(9.81 \text{ m/s}^2)(2.50 \text{ m})$$
$$= 147 \text{ J}$$

The final y-coordinate of the mass now is $y_f = 0.50$ m, and so the final potential energy of the mass is

$$PE_f = mgy_f$$
$$= (6.00 \text{ kg})(9.81 \text{ m/s}^2)(0.50 \text{ m})$$
$$= 29 \text{ J}$$

The *change* in the gravitational potential energy then is

$$\Delta PE \equiv PE_f - PE_i$$
$$= 29 \text{ J} - 147 \text{ J}$$
$$= -118 \text{ J}$$

Method 2
Choose the elevation of the origin to be at the 2.50 m elevation above the ground, as in Figure 8.25. Then the initial y-coordinate of the mass is $y_i = 0$ m, and the initial potential energy of the mass is

$$PE_i = mgy_i$$
$$= (6.00 \text{ kg})(9.81 \text{ m/s}^2)(0 \text{ m})$$
$$= 0 \text{ J}$$

When the mass is finally at the elevation 0.50 m above the ground, the y-coordinate of the mass is $y_f = -2.00$ m and its potential energy is

$$PE_f = mgy_f$$
$$= (6.00 \text{ kg})(9.81 \text{ m/s}^2)(-2.00 \text{ m})$$
$$= -118 \text{ J}$$

FIGURE 8.25

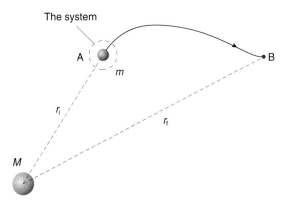

FIGURE 8.26 A mass m is moved from r_i to r_f in the gravitational field of another mass M.

The *change* in the potential energy of the mass then is

$$\Delta PE = PE_f - PE_i$$
$$= -118 \text{ J} - 0 \text{ J}$$
$$= -118 \text{ J}$$

Notice that the change in the potential energy of the system is the *same* for both coordinate choices. It is only *changes* in the potential energy that are physically significant.

b. By definition, the work done by the gravitational force on m is the *negative* of the change in the gravitational potential energy of m:

$$W_{grav} \equiv -\Delta PE$$
$$= -(-118 \text{ J})$$
$$= 118 \text{ J}$$

You could anticipate a positive result for the work done by the gravitational force, since the force is in the same direction as the vertical change in the position vector of the mass.

c. Since the gravitational force is a conservative force, the work done by the gravitational force is *independent* of the specific path used to get the system between the two points.

8.10 THE GENERAL FORM FOR THE GRAVITATIONAL POTENTIAL ENERGY

In Section 8.7, we found (Equation 8.9) that the work done by the gravitational force in moving a system of mass m from a position with radial coordinate r_i to a position with radial coordinate r_f (both measured with an origin at another mass M; see Figure 8.26) is

$$W_{grav} = -GmM \left(-\frac{1}{r_f} + \frac{1}{r_i} \right)$$

This can be rewritten even more awkwardly as

$$W_{grav} = -\left[\left(-\frac{GmM}{r_f} \right) - \left(-\frac{GmM}{r_i} \right) \right] \quad (8.14)$$

According to Equation 8.11, the work done by a conservative force is the negative of the change in the potential energy of the system:

$$W_{grav} \equiv -(PE_f - PE_i)$$

Compare Equation 8.14 with Equation 8.11 and identify

$$PE_f \text{ with } -\frac{GmM}{r_f}$$

and

$$PE_i \text{ with } -\frac{GmM}{r_i}$$

Therefore a system of mass m that finds itself at some position r from a mass M can be described in several ways:

• the system is at some position r in the gravitational field of M;

• there is a gravitational force of M on the system m.

We also can say that the system of mass m, because of its position relative to M, has a gravitational potential energy given by

$$PE_{grav} = -\frac{GmM}{r} \quad (8.15)$$

Note that:

a. The potential energy is a *scalar* property of the system that is a function of its position in space with respect to M.

b. The position coordinate r is raised to the *first* power in the denominator; r is raised to the *second* power in the expression for the magnitude of the gravitational *force* on m (and M), and in the expression for gravitational *field* of M.

PROBLEM-SOLVING TACTIC

8.4 When using Equation 8.15 for the gravitational potential energy, choose the origin of the coordinate system to be at M; no other choice for the origin should be made. This is the conventional choice for such two-particle gravitational interactions.

Note that with this choice of coordinates, the gravitational potential energy of m is *always negative*. The place where the gravitational potential energy is *zero* is at infinity ($r = \infty$ m), *not* the origin for r. The gravitational potential energy of m increases as its distance (r) from M increases: the potential energy becomes less negative as r increases.

If m is released *at rest* from some point and the gravitational force is the *only* force acting on m, then it moves in response to the force in a direction that *decreases* its gravitational potential energy. To do this, m falls toward M, decreasing r and making the potential energy of m more negative.

> Any *central force* will have a potential energy function that depends only on the spatial coordinate r:
>
> $$PE = PE(r) \qquad \text{(any central force)} \qquad (8.16)$$

EXAMPLE 8.11

A 500 kg satellite is initially in a circular orbit with a radius equal to twice the radius of the Earth. The satellite is moved to a circular orbit with a radius equal to three times the radius of the Earth.

a. Find the gravitational potential energy of the satellite in each orbit.
b. Find the change in the gravitational potential energy of the satellite.
c. Find the work done by the gravitational force as the satellite is moved between the two locations. Does this work depend on the path used to take the satellite between the two orbits?

Solution

a. Since the satellite is *not* located near the surface of the Earth, you *cannot* use the local expression for the gravitational potential energy, Equation 8.13. Rather you must use the more general expression of Equation 8.15:

$$PE_{grav} = -\frac{GmM}{r}$$

The radius of the Earth is $R_{\oplus} = 6.37 \times 10^6$ m. (The subscript $\oplus$ is the astronomical symbol for the Earth.) The satellite initially is located where

$$r_i = 2R_{\oplus} = 1.27 \times 10^7 \text{ m}$$

The mass of the Earth is $M_{\oplus} = 5.98 \times 10^{24}$ kg. Make these substitutions into the expression for the potential energy:

$$PE_i = -\frac{GmM}{r_i}$$
$$= -\frac{(6.67 \times 10^{-11} \text{ N}\cdot\text{m}^2/\text{kg}^2)(500 \text{ kg})(5.98 \times 10^{24} \text{ kg})}{1.27 \times 10^7 \text{ m}}$$
$$= -1.57 \times 10^{10} \text{ J}$$

Since the orbit of the satellite is circular, the gravitational potential energy of the satellite is the same at all positions

in its orbit. When the satellite is moved to the circular orbit at $r = 3R_{\oplus}$, or

$$r_f = 1.91 \times 10^7 \text{ m}$$

the gravitational potential energy is

$$PE_f = -\frac{GmM}{r_f}$$
$$= -\frac{(6.67 \times 10^{-11} \text{ N}\cdot\text{m}^2/\text{kg}^2)(500 \text{ kg})(5.98 \times 10^{24} \text{ kg})}{1.91 \times 10^7 \text{ m}}$$
$$= -1.04 \times 10^{10} \text{ J}$$

Once again, the gravitational potential energy is constant at all points in the new circular orbit.

b. The *change* in the gravitational potential energy is

$$\Delta PE = PE_f - PE_i$$
$$= -1.04 \times 10^{10} \text{ J} - (-1.57 \times 10^{10} \text{ J})$$
$$= 5.3 \times 10^9 \text{ J}$$

c. The work done by the gravitational force on the mass is the *negative* of the change in its potential energy:

$$W_{grav} \equiv -\Delta PE$$
$$= -5.3 \times 10^9 \text{ J}$$

You could have anticipated the negative result, since here the gravitational force on the mass is directed *opposite* to the radial direction that the mass moved.

The work done by the gravitational force is *independent* of the path used, since the gravitational force is a conservative force.

8.11 THE RELATIONSHIP BETWEEN THE LOCAL FORM FOR THE GRAVITATIONAL POTENTIAL ENERGY AND THE MORE GENERAL FORM*

The most general form for the gravitational potential energy of a system of mass m located outside the Earth at a distance r from its center is given by Equation 8.15:

$$PE_{grav} = -\frac{GmM}{r}$$

where M is the mass of the Earth. Near the surface of the Earth, we use another expression for the gravitational potential energy, Equation 8.13:

$$PE_{local} = mgy \qquad (\hat{\jmath} \text{ directed up})$$

where g is the magnitude of the local acceleration due to gravity near the surface of the Earth, 9.81 m/s². The two forms for the potential energy must be related to each other, since the force involved in both situations is the gravitational force of the Earth on the system. Here we show the connection between the two expressions for the gravitational potential energy.

For a system of mass m near the surface of the Earth, we write the distance r of the system from the *center* of the Earth as

$$r = R + y$$

where R is the distance from the center of the Earth to the origin of the Cartesian coordinate system we choose. As you can see in Figure 8.27, the distance R is equal to the radius of the Earth $R_\oplus$ plus the distance h that the origin of our coordinate system is above the surface of the Earth:

$$R = R_\oplus + h$$

Near the surface of the Earth, $h \ll R_\oplus$ and so

$$R \approx R_\oplus$$

In fact R is fixed once the origin for the y-coordinate axis is chosen. Substitute $r = R + y$ into the general expression for the gravitational potential energy of m (Equation 8.15):

$$PE_{grav} = -\frac{GMm}{R + y}$$

We can show how this is related to PE_{local}. To accomplish this, factor R out of the denominator:

$$PE_{grav} = -\frac{GMm}{R\left(1 + \dfrac{y}{R}\right)}$$

and rewrite this as

$$PE_{grav} = -\frac{GMm}{R}\left(1 + \frac{y}{R}\right)^{-1} \qquad (8.17)$$

Now we use the **binomial theorem** to expand the quantity in brackets. This theorem states that

$$(1 + x)^n = 1 + nx + \frac{n(n-1)}{2!}x^2 + \frac{n(n-1)(n-2)}{3!}x^3 + \cdots \qquad (8.18)$$

The binomial expansion implies that

$$\left(1 + \frac{y}{R}\right)^{-1} = 1 - \frac{y}{R} + \left(\frac{y}{R}\right)^2 - \cdots$$

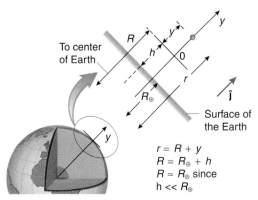

FIGURE 8.27 A coordinate system near the surface of the Earth.

However, since y is much less than R, we can neglect powers of (y/R) higher than the first power and keep just the first two terms of the expansion. Thus

$$\left(1 + \frac{y}{R}\right)^{-1} \approx 1 - \frac{y}{R}$$

Substituting this into Equation 8.17 for the potential energy, we get

$$PE_{grav} = -\frac{GMm}{R}\left(1 - \frac{y}{R}\right)$$
$$= -\frac{GMm}{R} + \frac{GM}{R^2}my$$

The first term on the right-hand side is a *constant* (since R is fixed). This constant represents the potential energy of m at the location chosen for the y-axis origin when using Equation 8.15. In the second term on the right-hand side, the quantity

$$\frac{GM}{R^2} \approx \frac{GM}{R_\oplus^2}$$

is g, the magnitude of the local acceleration due to gravity near the surface of the Earth (as we found in Section 6.5). Thus the gravitational potential energy is

$$PE_{grav} = -\frac{GMm}{R} + mgy$$
$$= -\frac{GMm}{R} + PE_{local}$$

Remember that only *changes* in the potential energy are physically significant; so the constant term,

$$-\frac{GMm}{R}$$

subtracts out when *differences* in potential energy are taken. Thus, for regions near the surface of the Earth, as long as the y-coordinate of the mass does not become large with respect to the radius of the Earth, you can use mgy as the gravitational potential energy with $\hat{\jmath}$ chosen in the vertically upward direction.

8.12 THE POTENTIAL ENERGY FUNCTION ASSOCIATED WITH HOOKE'S FORCE LAW

We discovered in Section 8.7 (Equation 8.10) that the work done by the force of a spring on a system of mass m as the system moves from x_i to x_f is

$$W_{spring} = -k\frac{x_f^2}{2} + k\frac{x_i^2}{2}$$

or with a slight rearrangement,

$$W_{spring} = -\left(k\frac{x_f^2}{2} - k\frac{x_i^2}{2}\right) \qquad (8.19)$$

The work done by a conservative force is the negative of the change in the potential energy (Equation 8.11):

$$W_{\text{spring}} \equiv -(PE_f - PE_i)$$

Compare Equation 8.19 with Equation 8.11 and identify the various potential energy terms.

The potential energy function associated with a system of mass m attached to a spring when the system is located at coordinate x (where $x = 0$ m is the equilibrium position of m) is

$$PE_{\text{spring}} = \frac{1}{2}kx^2 \qquad (8.20)$$

A few things to note about this potential energy function:

a. Once again we see that the potential energy is a scalar property of the system that is a function of its position in space.
b. The potential energy of this system *is independent of its mass*; the potential energy depends only on the spring constant k and the location of the system relative to the equilibrium position.
c. The potential energy is zero at the equilibrium position of the system ($x = 0$ m), where we chose the origin of our coordinate system.
d. Since the spring constant k is always positive, and $x^2 \geq 0$ m^2, this potential energy function is never negative.
e. The work done by the force of the spring on the system is the negative of the *change* in the potential energy of the system.
f. If the mass m is released *at rest* at some coordinate x and the spring force is the only force in the x-direction acting on m, then m moves initially in a direction to decrease its potential energy. In doing this, m approaches the equilibrium position.

Table 8.1 summarizes the potential energy function associated with each conservative force.

TABLE 8.1 **Summary of Potential Energy Functions**

(1) Gravitational potential energy when the system of mass m is always near the surface of the Earth:
$$PE = mgy \quad (\hat{\jmath}\text{ up}) \qquad (8.13)$$
(you choose the elevation of the y-coordinate origin)

(2) General form for the gravitational potential energy of a system of mass m located a distance r from another pointlike mass M or an appropriate sphere:
$$PE = -\frac{GmM}{r} \qquad (8.15)$$
(origin for r must be chosen at M)

(3) The potential energy of a system of mass m attached to a spring with spring constant k when m is at coordinate x:
$$PE = \frac{1}{2}kx^2 \qquad (8.20)$$
(where $x = 0$ m is the equilibrium position of m)

EXAMPLE 8.12 _____

A 2.00 kg mass on a frictionless horizontal surface is attached to a spring with a spring constant $k = 75.0$ N/m. The mass is initially at its equilibrium position ($x = 0$ m) and is moved to the coordinate $x = -0.30$ m.

a. Find the potential energy of the mass at each location.
b. Find the change in the potential energy of the mass.
c. Find the work done by the force of the spring on m when it is moved between these two locations.

Solution

a. The potential energy function for a system of mass m on a spring is given by Equation 8.20:

$$PE = \frac{1}{2}kx^2$$

When the mass is initially at the coordinate $x_i = 0$ m, the potential energy is

$$PE_i = \frac{1}{2}(75.0 \text{ N/m})(0 \text{ m})^2$$
$$= 0 \text{ J}$$

When the mass is at the coordinate $x_f = -0.30$ m, the potential energy is

$$PE_f = \frac{1}{2}(75.0 \text{ N/m})(-0.30 \text{ m})^2$$
$$= 3.4 \text{ J}$$

b. The *change* in the potential energy is

$$\Delta PE = PE_f - PE_i$$
$$= 3.4 \text{ J} - 0 \text{ J}$$
$$= 3.4 \text{ J}$$

c. The work done by the force of the spring on m is equal to the *negative* of the change in the potential energy of m:

$$W_{\text{spring}} \equiv -\Delta PE$$
$$= -3.4 \text{ J}$$

You could anticipate the negative result, because the force of the spring on m is opposite to the direction that m was moved from $x = 0$ m to $x = -0.30$ m.

8.13 THE CWE THEOREM

Energy is Eternal Delight

William Blake (1757–1827)*

The CWE theorem is a special case of an important physical principle that is the *raison d'être* for introducing energy concepts: conservation of energy. This central theorem of classical mechanics stems from Newton's second law of motion.

Recall from Section 8.1 that we consider pointlike, nondeformable, single-particle systems of constant mass with no

---

The Marriage of Heaven and Hell, Plate 4, The voice of the Devil. In *William Blake's Writings*, edited by Gerald E. Bentley Jr., Volume 1, Engraved and Etched Writings (Oxford at the Clarendon Press, Oxford, England, 1978), page 78.

internal motions, and ignore thermal effects. If a system has a nonzero total force acting on it, the system experiences a nonzero acceleration according to Newton's second law of motion. Accelerations imply changes in the velocity of the system.

The dynamics of the system are governed by Newton's second law of motion:

$$\vec{F}_{total} = m\vec{a}$$
$$= m\frac{d\vec{v}}{dt}$$

where $\vec{F}_{total}$ is the vector sum of all the forces acting on the system and $\vec{v}$ is its velocity. Let's calculate the total work W_{total} done by the total force $\vec{F}_{total}$ on the system as it moves from some initial position to some final position. According to the general definition of work, Equation 8.2, W_{total} is

$$W_{total} = \int_i^f \vec{F}_{total} \cdot d\vec{r}$$

Use Newton's second law for $\vec{F}_{total}$:

$$W_{total} = m\int_i^f \frac{d\vec{v}}{dt} \cdot d\vec{r} \qquad (8.21)$$

The velocity $\vec{v}$ is the time rate of change of the position vector:

$$\vec{v} = \frac{d\vec{r}}{dt}$$

So the differential change $d\vec{r}$ in the position vector can be written as

$$d\vec{r} = \vec{v}\, dt$$

Then substituting $\vec{v}\, dt$ for $d\vec{r}$ in Equation 8.21 changes the integration from one over position to one over time and then to one over the *speed*:

$$W_{total} = m\int_{i\,time}^{f\,time} \frac{d\vec{v}}{dt} \cdot \vec{v}\, dt$$
$$= m\int_{i\,speed}^{f\,speed} d\vec{v} \cdot \vec{v} \qquad (8.22)$$

To perform the integration on the right-hand side of Equation 8.22, look at the following derivative just for the fun of it:

$$\frac{d}{dt}(\vec{v} \cdot \vec{v})$$

Use the product rule to differentiate the scalar product:

$$\frac{d}{dt}(\vec{v} \cdot \vec{v}) = \frac{d\vec{v}}{dt} \cdot \vec{v} + \vec{v} \cdot \frac{d\vec{v}}{dt}$$
$$= 2\frac{d\vec{v}}{dt} \cdot \vec{v}$$

Hence

$$d\vec{v} \cdot \vec{v} = \frac{1}{2} d(\vec{v} \cdot \vec{v}) \qquad (8.23)$$

The left-hand side of Equation 8.23 is the integrand in Equation 8.22. After this substitution, Equation 8.22 becomes

$$W_{total} = m\int_{i\,speed}^{f\,speed} \frac{d(\vec{v} \cdot \vec{v})}{2}$$

Notice that $\vec{v} \cdot \vec{v} = v^2$, the square of the speed of the particle. Hence the integrand is a perfect differential:

$$W_{total} = \frac{1}{2} m\int_{i\,speed}^{f\,speed} d(v^2)$$
$$= \frac{1}{2} mv^2 \Big|_{i\,speed}^{f\,speed}$$
$$= \frac{1}{2} mv_f^2 - \frac{1}{2} mv_i^2 \qquad (8.24)$$

The right-hand side of this expression is defined to be the change in the classical translational **kinetic energy** of the system of mass m:

$$\Delta KE \equiv KE_f - KE_i$$
$$= \frac{1}{2} mv_f^2 - \frac{1}{2} mv_i^2$$

> The kinetic energy is a scalar property of a system that is proportional to the square of its speed v:*
>
> $$\boxed{KE \equiv \frac{1}{2} mv^2} \qquad (8.25)$$

The classical translational kinetic energy of the system typically is called, more simply, the kinetic energy of the system.

A few things should be noted about the kinetic energy of a system of mass m:

1. The kinetic energy is a *scalar* property of the system.
2. Since $m \geq 0$ kg and $v^2 \geq 0$ m²/s², the kinetic energy of a system is *never negative*.
3. Equation 8.25 for the kinetic energy of the system can be used as long as the speed v of the system is *small compared with the speed of light* (although this is not a fact that arises from our analysis). When the speed of the system becomes an appreciable fraction of the speed of light (universally called c), the expression for the kinetic energy is more elaborate than Equation 8.25, but reduces to it for speeds $v \ll c$.[†]
4. Since the SI units for work are joules, the SI units for the kinetic energy also are joules (J).

---

*The special significance of the quantity mv^2 first was recognized by Christian Huygens (1629–1695) in 1668, before Newton published his *Principia*. Wilhelm von Leibnitz (1646–1716), co-inventor of calculus, also recognized the significance of this quantity and gave it the name *vis viva* (meaning "living force").

[†]The reason a modification of the kinetic energy is needed for so-called relativistic speeds will have to await our treatment of relativistic dynamics in Chapter 25.

We rewrite Equation 8.24 as

$$W_{total} = \Delta KE \qquad (8.26)$$

This statement is the **CWE theorem**. It states that the effect of the work done by all the forces acting on a point-particle system is to change the value of the kinetic energy property of the system.

Think of the total work done on a system as a way of transferring energy to the system via the action of forces on the system. In other words, *work is a form of energy transfer*. A system does *not* "*have*" an amount of work by virtue of its position in space or time. We *cannot* say "the work *of* the system"; we *do* say the "work done *on* a system" by forces as it moves from one location to another.

If the system has a speed v, its kinetic energy is $(1/2)mv^2$. Thus a system *can* be said to *have* kinetic energy if $v \neq 0$ m/s. As a result of the energy transfer by means of the work done on the system by all the forces on the system, the CWE theorem states that the system changes its kinetic energy.

If nonzero total work is done *on* a system, the kinetic energy of the system changes; energy is transferred *to* the system. Such energy transfer may be positive, negative, or zero. If the total work W_{total} done *on* the system is positive (> 0 J), the energy transfer is positive and the system *increases* its kinetic energy (since $\Delta KE > 0$ J). If the total work W_{total} done on the system is negative, the energy transfer is negative and the kinetic energy change is negative—the system *loses* kinetic energy.

Several aspects of the CWE theorem should be noted:

1. It is critical to realize that the left-hand side of Equation 8.26, W_{total}, is the work done by *all the forces* acting on the system, not just some of the forces. Knowing what the forces are on the system and drawing a second law force diagram to indicate these forces still are essential.
2. Since the CWE theorem was derived from Newton's second law of motion (a vector equation), the CWE theorem is, in a sense, a scalar equivalent of the second law of motion.

Some forces are conservative forces. It is useful to separate the work done by *all* the forces acting on the system, W_{total}, into two categories:

1. the total work done by all *conservative* forces on the system: W_{consrv}; and
2. the total work done by all *other* forces on the system: W_{other}.

Then the CWE theorem, Equation 8.26, is rewritten as

$$W_{other} + W_{consrv} = \Delta KE$$

The work done by a conservative force was defined to be the *negative* of the change in the potential energy function associated with that conservative force (see Equation 8.11):

$$W_{consrv} \equiv -\Delta PE$$

If more than one conservative force acts on the system, the work done by *each* conservative force is the negative of the change in *each* corresponding potential energy function. Thus W_{consrv} gathers together all the (negative) changes in the potential

energy functions. Substitute for the work done by all the conservative forces acting on the system into the CWE theorem:

$$W_{other} + (-\Delta PE) = \Delta KE$$

Now you can finally see why the negative sign was introduced into the definition of the work done by conservative forces ($= -\Delta PE$). Transposing the changes in the potential energy functions to the right-hand side of the CWE theorem, we find

$$W_{other} = \Delta KE + \Delta PE \qquad (8.27)$$

or more explicitly,

$$W_{other} = (KE_f - KE_i) + (PE_f - PE_i)$$

Now collect the terms associated with the initial and final positions of the system:

$$W_{other} = (KE_f + PE_f) - (KE_i + PE_i)$$

The CWE theorem then becomes

$$W_{other} = \Delta(KE + PE) \qquad (8.28)$$

In other words, the work done by the "other" forces on the system (that is, by forces that we do not classify as conservative forces) is equal to the change in the sum of the kinetic and potential energies of the system between the initial and final positions.

The sum of the kinetic and potential energies of a system ($KE + PE$) at any point in space is called the **total mechanical energy** E of the system. So the work done by the "other" forces is equal to the change in the total mechanical energy of the system:

$$W_{other} = \Delta E \qquad (8.29)$$

PROBLEM-SOLVING TACTIC

8.5 Equations 8.26, 8.27, 8.28, and 8.29 are all equivalent expressions for the CWE theorem. The theorem is an extraordinarily useful result. Remember it. More importantly, use it. In terms of importance, the CWE theorem ranks right up there with Newton's second law of motion. It is one of the reasons that the physical concept of work is such a useful invention.

A frequent, special case of the general CWE theorem arises if either of the following situations arise:

a. if the forces acting on the system are all conservative forces; or
b. if the work done by all "other" forces acting on the system is zero—that is, $W_{other} = 0$ J.

For *either* case, the work done by nonconservative and zero-work forces (i.e., "other" forces) is zero, so that the left-hand side of the CWE theorem is zero in Equations 8.27, 8.28, and 8.29.

Rewriting Equation 8.28 with the left-hand side equal to zero puts the CWE theorem into a more restricted form:

$$0 \text{ J} = \Delta(KE + PE) \qquad (W_{\text{other}} = 0 \text{ J}) \qquad (8.30)$$
$$= \Delta E$$

In such situations [specified in (a) and (b)], the total mechanical energy of the system is conserved or unchanged throughout the motion; this *restricted version* of the CWE theorem is called **conservation of mechanical energy**.

QUESTION 7

If the total work done on a system is zero, what (if anything) can be said about the speed of the system?

EXAMPLE 8.13

A meteoroid of mass 2.09×10^4 kg is moving at a speed of 1.20×10^4 km/h; this meteoroid is a rock only about a meter in diameter. Meteoroids of about this mass occasionally collide with the atmosphere of the Earth, creating spectacular fireballs.

a. Find the kinetic energy of the meteoroid.
b. What mass m, moving at a speed of 10.0 m/s ($\approx$ 36 km/h), has the same kinetic energy?

Solution

a. First convert the speed from km/h to m/s to use proper SI units for the speed:

$$1.20 \times 10^4 \text{ km/h} = \frac{(1.20 \times 10^4 \text{ km/h})(10^3 \text{ m/km})}{3.600 \times 10^3 \text{ s/h}}$$
$$= 3.33 \times 10^3 \text{ m/s}$$

The kinetic energy of the meteoroid then is

$$KE = \frac{1}{2} mv^2$$
$$= \frac{1}{2}(2.09 \times 10^4 \text{ kg})(3.33 \times 10^3 \text{ m/s})^2$$
$$= 1.16 \times 10^{11} \text{ J}$$

This is a rather substantial amount of kinetic energy! The explosion of one metric ton (1 metric ton = 10^3 kg) of TNT*

---

*Trinitrotoluene ($C_7H_5N_3O_6$).

A meteoroid entering the atmosphere of the Earth.

produces about 4.9×10^9 J, and so the kinetic energy of this meteoroid has an energy equivalent to the explosion of about 24 metric tons of TNT.

b. A mass m moving at a speed of 10.0 m/s has kinetic energy

$$KE = \frac{1}{2} mv^2$$

Substituting 1.16×10^{11} J on the left-hand side and $v = 10.0$ m/s on the right-hand side, you have

$$1.16 \times 10^{11} \text{ J} = \frac{1}{2} m (100 \text{ m}^2/\text{s}^2)$$

Solving for m,

$$m = 2.32 \times 10^9 \text{ kg}$$

This is 2.32 *million* metric tons, about the mass of twenty fully loaded oil supertankers.

EXAMPLE 8.14

An American bald eagle (*Haliaeetus leucocephalus*) of mass 4.00 kg has a velocity $\vec{v} = (5.00 \text{ m/s})\hat{\imath} - (12.00 \text{ m/s})\hat{\jmath}$. What is the kinetic energy of the bird?

Solution

To determine the kinetic energy, we need the *speed* of the bird, that is, the magnitude of the velocity vector. Evaluating $v^2 = \vec{v} \cdot \vec{v}$, you find

$$v^2 = [(5.00 \text{ m/s})\hat{\imath} - (12.00 \text{ m/s})\hat{\jmath}] \cdot [(5.00 \text{ m/s})\hat{\imath} - (12.00 \text{ m/s})\hat{\jmath}]$$
$$= (5.00 \text{ m/s})^2 + (-12.00 \text{ m/s})^2$$
$$= 169 \text{ m}^2/\text{s}^2$$

Hence the kinetic energy is

$$KE = \frac{1}{2} mv^2$$
$$= \frac{1}{2}(4.00 \text{ kg})(169 \text{ m}^2/\text{s}^2)$$
$$= 338 \text{ J}$$

EXAMPLE 8.15

A car is moving at speed v_i. By what factor must its speed increase to *double* its kinetic energy?

Solution

Let the mass of the system be m and let its initial speed be v_i. Its initial kinetic energy then is

$$KE_i = \frac{1}{2} mv_i^2$$

When its speed is increased to v_f, its kinetic energy is

$$KE_f = \frac{1}{2} mv_f^2$$

and this must equal $2\,\text{KE}_i$. That is,

$$\frac{1}{2}\,mv_f^2 = 2\left(\frac{1}{2}\,mv_i^2\right)$$

Hence

$$v_f^2 = 2v_i^2$$

or

$$v_f = \sqrt{2}\,v_i$$

Therefore, to double the kinetic energy, the speed of the car must be increased by the factor $\sqrt{2} \approx 1.41$. For example, increasing the speed from 60 km/h to 85 km/h *doubles* the kinetic energy of the car, which is why car crashes rapidly become more dangerous as the speed of the car increases.

STRATEGIC EXAMPLE 8.16

Beginning at rest, a stale bagel of mass 0.250 kg slides down a frictionless slope of straight-line length 2.00 m inclined at an angle of 30.0° to the horizontal before entering the awaiting mouth of a domestic pig (*Sus scrofa*); see Figure 8.28. The slope has a frictionless bump partway down.

FIGURE 8.28

a. Sketch a second law force diagram indicating all the forces acting on the bagel during the motion.
b. Identify which forces are conservative (if any).
c. What work is done by "other" (i.e., nonconservative and zero-work) forces?
d. Apply the CWE theorem to determine the speed of the bagel at the bottom of the slope.

Solution

a. The forces acting on the bagel during the slippery slide are shown in Figure 8.29:

   1. the gravitational force of the Earth on the bagel—that is, the weight $\vec{w}$ of the bagel; and
   2. the normal force $\vec{N}$ of the surface on the bagel.

The surface is frictionless so there is no frictional force.

b. The gravitational force is a conservative force. The work done by the gravitational force on the bagel is accounted for by a gravitational potential energy term in the CWE theorem. The work done by the normal force must be accounted for in the W_{other} term on the left-hand side of the CWE theorem.
c. The work done by the normal force is the work done by "other" forces on the system, since it is a zero-work force:

$$W_{\text{normal}} = \int_i^f \vec{N} \cdot d\vec{r}$$

The normal force is perpendicular to the surface at every point along the path of the bagel, even over the bump. Hence the integrand is zero at every point along the path and the normal force does *zero work*; it is a zero-work force.

d. The presence of the bump makes it difficult, if not impossible, to apply Newton's second law, since the shape of the bump is not known. Use the CWE theorem instead. The appropriate potential energy function to use is the one associated with the gravitational force on a mass near the surface of the Earth:

$$\text{PE}_{\text{local}} = mgy \qquad (\hat{j} \text{ directed up})$$

You can choose the elevation for the y-axis origin to be anywhere you like for this potential energy function. Choose the origin to be at the bottom of the incline, as shown in Figure 8.30. The horizontal placement of the vertical y-axis is not crucial. At the initial position of the bagel at the top of the incline, the potential energy of the bagel is

$$\text{PE}_i = mgy_i$$

Geometrically,

$$y_i = (2.00 \text{ m}) \sin 30.0°$$
$$= 1.00 \text{ m}$$

Hence

$$\text{PE}_i = (0.250 \text{ kg})(9.81 \text{ m/s}^2)(1.00 \text{ m})$$
$$= 2.45 \text{ J}$$

The initial kinetic energy of the bagel is zero since the initial speed is zero:

$$\text{KE}_i = \frac{1}{2}\,mv_i^2$$
$$= 0 \text{ J}$$

At the bottom of the incline the y-coordinate of the bagel is zero, and so the final potential energy of the bagel is

$$\text{PE}_f = mgy_f$$
$$= (0.250 \text{ kg})(9.81 \text{ m/s}^2)(0 \text{ m})$$
$$= 0 \text{ J}$$

FIGURE 8.29

FIGURE 8.30

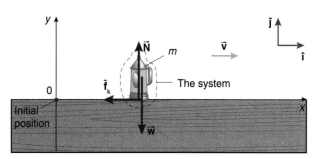

FIGURE 8.31

Let v be the unknown speed of the bagel at the bottom of the incline. The kinetic energy of the bagel at the bottom of the incline is

$$KE_f = \frac{1}{2} mv^2$$
$$= \frac{1}{2} (0.250 \text{ kg}) v^2$$
$$= (0.125 \text{ kg}) v^2$$

Making the appropriate substitutions into the CWE theorem,

$$W_{other} = \Delta(KE + PE)$$
$$= (KE_f + PE_f) - (KE_i + PE_i)$$

you get

$$0 \text{ J} = [(0.125 \text{ kg})v^2 + 0 \text{ J}] - (0 \text{ J} + 2.45 \text{ J})$$

Solving for v^2,

$$v^2 = \frac{2.45 \text{ J}}{0.125 \text{ kg}}$$
$$= 19.6 \text{ m}^2/\text{s}^2$$

The speed v thus is

$$v = 4.43 \text{ m/s}$$

STRATEGIC EXAMPLE 8.17

At the Longbranch saloon, a full beer stein of mass 2.00 kg is slid horizontally (and without rotation) across the bar with an initial speed of 3.00 m/s. If the stein travels a distance 1.53 m before stopping without losing a drop of amber liquid, find the coefficient of kinetic friction μ_k for the stein on the bar surface.

Solution

Our strategy is to apply the CWE theorem to the problem using the beer stein as the system. We find the changes in the kinetic and potential energies of the stein and set this equal to the work done by the "other" forces in the CWE theorem. The forces acting on the stein as it slides across the bar are (see Figure 8.31)

1. the gravitational force of the Earth on the stein, its weight $\vec{w}$ of magnitude mg, directed vertically down;

2. the normal force $\vec{N}$ of the surface of the bar on the stein, directed vertically up; and
3. the kinetic frictional force $\vec{f}_k$ on the stein, directed horizontally in the direction opposite to its velocity.

The normal force and the weight are perpendicular to the path of the stein at every point on its route. Thus both of these forces do zero work on the stein. Since the gravitational force is conservative, the work done by the gravitational force is equal to the negative of the change in the gravitational potential energy of the stein. The appropriate gravitational potential energy is mgy (with $\hat{j}$ directed up). The change in the gravitational potential energy of the stein is zero because its vertical coordinate is the same at both the initial and final positions. (The vertical coordinate y of the stein is in fact constant everywhere along the path.) Thus

$$\Delta PE = 0 \text{ J}$$

The change in the kinetic energy is

$$\Delta KE = KE_f - KE_i$$

The final speed of the stein is zero, so that

$$KE_f = \frac{1}{2} mv_f^2 = 0 \text{ J}$$

The initial kinetic energy is

$$KE_i = \frac{1}{2} mv_i^2$$
$$= \frac{1}{2} (2.00 \text{ kg})(3.00 \text{ m/s})^2$$
$$= 9.00 \text{ J}$$

Hence the change in the kinetic energy of the stein is

$$\Delta KE = KE_f - KE_i$$
$$= 0 \text{ J} - 9.00 \text{ J}$$
$$= -9.00 \text{ J}$$

Thus the CWE theorem becomes

$$W_{other} = \Delta KE + \Delta PE$$
$$= -9.00 \text{ J} + 0 \text{ J}$$
$$= -9.00 \text{ J} \tag{1}$$

The nonconservative force here is the kinetic frictional force on the stein. The magnitude of the kinetic frictional force $\vec{f}_k$ is related to the magnitude of the normal force $\vec{N}$ by

$$f_k = \mu_k N$$

Since there is no acceleration of the stein in the vertical direction, $a_y = 0$ m/s^2. Applying Newton's second law of motion to the y-direction,

$$F_{y\,total} = ma_y$$
$$N - mg = m(0 \text{ m/s}^2)$$

and so

$$N = mg$$

Thus the kinetic frictional force on the stein is of magnitude

$$f_k = \mu_k N$$
$$= \mu_k mg$$

Use the coordinate system indicated in Figure 8.31; the kinetic frictional force is

$$\vec{f}_k = -\mu_k mg\hat{\imath}$$

Since the kinetic frictional force is a *constant vector* over the *entire* one-way straight line path of the stein, you can use Equation 8.6 to find the work of the kinetic frictional force:

$$\begin{aligned} W &= \vec{F} \cdot \Delta\vec{r} \qquad (\vec{F} \text{ constant}) \\ &= \vec{f}_k \cdot (\vec{r}_f - \vec{r}_i) \\ &= -\mu_k mg\hat{\imath} \cdot [(1.53 \text{ m})\hat{\imath} - (0 \text{ m})\hat{\imath}] \\ &= (-1.53 \text{ m})\mu_k mg \end{aligned}$$

Substituting for m and g, you find

$$W_{frictn} = -30.0\mu_k \text{ J}$$

The work done by the kinetic frictional force is negative since the force is antiparallel to the change in the position vector of the system.

The work done by the kinetic frictional force is the left-hand side of the CWE theorem since this is the work done by nonconservative forces, the "other" forces, acting on the system. Equation (1) then becomes

$$(-30.0 \text{ J})\mu_k = -9.00 \text{ J}$$

Solving for μ_k,

$$\mu_k = 0.300$$

STRATEGIC EXAMPLE 8.18

A particle of mass m is at a height h_0 above the ground at the top of a smooth, frictionless slide and plane, as in Figure 8.32. The mass is released at rest, slides down the incline, and encounters a spring (with spring constant k). Determine the maximum extent to which the spring is compressed before it hurls the mass back in the direction whence it came.

Solution

Choose a coordinate system with its origin at the uncompressed end of the spring, as shown in Figure 8.33. Let mass m be the system. You could break the problem into two parts:

a. the descent down and then along the smooth slide, and
b. the encounter with the spring;

and use the CWE theorem to examine each part separately; you might try this as an exercise. However, it is easier to solve the problem with a *single* application of the CWE theorem. Take the initial position of the system of mass m to be when it is at the top of the incline and the final position when it is (instantaneously) at rest with the spring at maximum compression.

The forces acting on the system are as follows:

1. The gravitational force of the Earth on the system (the weight $\vec{w}$ of the system). This is a conservative force, and so the work done by this force is accounted for in the CWE theorem (Equation 8.28) by the change in the gravitational potential energy term.
2. The normal force $\vec{N}$ of the incline and horizontal surface on the system. This force changes its magnitude and direction as the system goes down the incline and onto the horizontal surface. Nonetheless, the normal force at all points along the path of the system does *zero work*, since it is at all points perpendicular to the path of the system; as always, it is a zero-work force.
3. The force of the spring on the system when it encounters the spring. This is a conservative force, and the work done on the system by this force is accounted for by the change in a spring potential energy term in the CWE theorem.

FIGURE 8.32

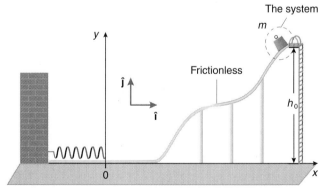

FIGURE 8.33

In applying the CWE theorem here, there are two potential energy terms, one associated with the gravitational force and one with the force of the spring on the system. Hence the theorem is

$$W_{\text{other}} = (\text{KE} + \text{PE}_{\text{grav}} + \text{PE}_{\text{spring}})_f - (\text{KE} + \text{PE}_{\text{grav}} + \text{PE}_{\text{spring}})_i$$

The term W_{other} is zero, since no work is done by the normal force and this is the only force on the system "other" than conservative forces. Since the system was released at rest, the initial kinetic energy is zero:

$$\text{KE}_i = 0 \text{ J}$$

The initial gravitational potential energy is mgy_i. With the coordinate system indicated in Figure 8.33, $y_i = h_0$, and you have

$$\text{PE}_{\text{grav i}} = mgh_0$$

The initial potential energy of the system associated with the spring is zero since the spring is not (yet) compressed:

$$\text{PE}_{\text{sprg i}} = 0 \text{ J}$$

At the final position of the system, with the spring under maximum compression, the system is instantaneously at rest, and so

$$\text{KE}_f = 0 \text{ J}$$

The final gravitational potential energy of the system, mgy_f, also is zero since $y_f = 0$ m:

$$\text{PE}_{\text{grav f}} = 0 \text{ J}$$

The spring is compressed to coordinate $x < 0$ m, and so the potential energy of the system because of the spring is

$$\text{PE}_{\text{sprg f}} = \frac{1}{2}kx^2$$

Substituting into the CWE theorem,

$$W_{\text{other}} = (\text{KE} + \text{PE}_{\text{grav}} + \text{PE}_{\text{spring}})_f - (\text{KE} + \text{PE}_{\text{grav}} + \text{PE}_{\text{spring}})_i$$

$$0 \text{ J} = \left(0 \text{ J} + 0 \text{ J} + \frac{1}{2}kx^2\right) - (0 \text{ J} + mgh_0 + 0 \text{ J})$$

Solve for x. The negative root must be chosen because the spring is compressed:

$$x = -\left(\frac{2mgh_0}{k}\right)^{1/2}$$

8.14 THE ESCAPE SPEED

The law of gravity says no fair jumping up without coming back down.

Anonymous fourth grader

Throw a ball vertically upward and it comes back down again. If the ball is thrown upward with a greater speed, it travels higher and spends a longer time in flight, but eventually it comes back

to haunt us again. However, if thrown vertically upward with sufficient speed, equal to or greater than what is called the **escape speed**, the mass will never return. So if you *really* want to get rid of something, just project it vertically upward with a speed exceeding the escape speed and your troubles are over.

The escape speed is the minimum speed needed to project a mass m upward from the surface of another mass M (such as a planet, asteroid, or star) so that m never returns.

The purpose of the huge rocket engines for interplanetary space probes is to provide the "kick" (the initial kinetic energy) needed to propel the spacecraft to the escape speed, so that it may escape from the Earth. A few meteorites from Mars, found in Antarctica, escaped from the red planet during an ancient collision of a small asteroid with Mars, propelling them into space, eventually to land on the Earth.

The CWE theorem can determine the escape speed. Throw a system of mass m radially outward from the surface of a spherically symmetric planet or star of mass M and radius R, as shown in Figure 8.34. For simplicity, disregard any kinetic frictional forces from the atmosphere as the mass m escapes, although this

A rocket must propel the payload of an interplanetary space probe to a speed at least equal to the escape speed. This is the Mars Global Surveyor Mission launched 11 November 1996.

This meteorite reached the Earth by being blasted from Mars by an impact that gave the rock a speed greater than the escape speed from Mars.

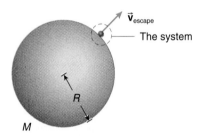

FIGURE 8.34 A mass m is propelled upward from a spherical mass M.

assumption is a bit unrealistic on planets with an atmosphere. While in its unpowered flight, the only force on the system is the gravitational force of the central mass M; there is no work done by nonconservative forces and the *left*-hand side of the CWE theorem, W_{other}, is zero:

$$W_{\text{other}} = 0 \text{ J}$$

We want to find the minimum possible launch speed necessary for mass m to escape from M; this is the escape speed v_{escape}. We launch mass m with an initial kinetic energy

$$\text{KE}_i = \frac{1}{2} m v_{\text{escape}}^2$$

Since m is under the gravitational influence of mass M, the mass m initially also has gravitational potential energy:

$$\text{PE}_i = -\frac{GMm}{R}$$

where we have to use the general expression for the gravitational potential energy, Equation 8.15, since m is not going to be confined to the local vicinity of the surface of M. Indeed, that's the point: we want to project m from M to great distances.

If m just barely escapes from M, the final speed of m at an infinite distance away is zero; thus the final kinetic energy of m is zero:

$$\text{KE}_f = 0 \text{ J}$$

Since m can go (effectively) infinitely far away from M after it escapes, $r_f = \infty$ m, and the final gravitational potential energy of m,

$$\text{PE}_f = -\frac{GMm}{r_f}$$

approaches zero as $r_f \to \infty$ m:

$$\text{PE}_f = 0 \text{ J}$$

Apply the CWE theorem:

$$W_{\text{other}} = \Delta(\text{KE} + \text{PE})$$
$$= (\text{KE} + \text{PE})_f - (\text{KE} + \text{PE})_i$$

$$0 \text{ J} = (0 \text{ J} + 0 \text{ J}) - \left[\frac{1}{2} m v_{\text{escape}}^2 + \left(-\frac{GMm}{R} \right) \right]$$

Hence

$$\frac{1}{2} m v_{\text{escape}}^2 = \frac{GMm}{R} \tag{8.31}$$

Solve for the escape speed v_{escape}:

$$v_{\text{escape}} = \left(\frac{2GM}{R} \right)^{1/2} \tag{8.32}$$

Notice that the escape speed does *not* depend on the mass m of the system propelled upward: it is the *same* speed for all masses escaping from M.

Systems (such as you, me, baseballs, and footballs) with speeds less than the escape speed of the Earth have kinetic energies that are too small to escape—fortunately!

Examine Equation 8.31. In order to escape, the initial kinetic energy of m must at least be equal to the absolute value of its initial gravitational potential energy. If not, the mass m is "bound" to M and will return to the surface when thrown upward. With this in mind, the absolute value of the gravitational potential energy of m, the quantity

$$\left| -\frac{GMm}{R} \right|$$

is called the **binding energy** of m when it is at rest on the surface of M. More generally, the binding energy of a system at any location is the additional energy the system needs to barely escape (with final speed zero) to an infinite distance away.

The escape speed from other planets or stars can be found in terms of the escape speed from the Earth by taking the ratio of the two escape speeds. The escape speed from a planet or star of mass M and radius R is

$$v_{\text{escape}} = \left(\frac{2GM}{R} \right)^{1/2}$$

The escape speed from the Earth (symbolized with the subscript $\oplus$) is

$$v_\oplus = \left(\frac{2GM_\oplus}{R_\oplus} \right)^{1/2}$$

Take the *ratio* of the two expressions; this eliminates G and numerical constants:

$$\frac{v_{escape}}{v_\oplus} = \left(\frac{\dfrac{M}{M_\oplus}}{\dfrac{R}{R_\oplus}} \right)^{1/2}$$

$$= \left(\frac{\text{mass ratio}}{\text{radius ratio}} \right)^{1/2} \tag{8.33}$$

Thus the relative magnitudes of the escape speeds can be found if the mass and radius ratios are known.

EXAMPLE 8.19

Calculate the escape speed from the surface of the Earth.

Solution
The escape speed from the surface of the Earth is given by Equation 8.32:

$$v_{escape} = \left(\frac{2GM_\oplus}{R_\oplus} \right)^{1/2}$$

where $M_\oplus$ is the mass of the Earth and $R_\oplus$ is its radius. Substituting numerical values for these quantities and for G, you get

$$v_{escape} = \left[\frac{2 \times (6.67 \times 10^{-11} \ \text{N·m}^2/\text{kg}^2)(5.98 \times 10^{24} \ \text{kg})}{6.37 \times 10^6 \ \text{m}} \right]^{1/2}$$

$$= 1.12 \times 10^4 \ \text{m/s}$$

or 11.2 km/s. Thus to send a spacecraft off the Earth to, say, Mars, a minimum speed of 11.2 km/s is needed.

EXAMPLE 8.20

Jupiter is 318 times as massive as the Earth; the radius of the giant planet is 10.9 times that of the Earth. Find the escape speed from Jupiter in terms of the escape speed from the Earth.

Solution
The mass ratio is

$$\frac{M_{Jupiter}}{M_\oplus} = 318$$

The radius ratio is

$$\frac{R_{Jupiter}}{R_\oplus} = 10.9$$

Use Equation 8.33; the ratio of the escape speeds is

$$\frac{v_{escape}}{v_\oplus} = \left(\frac{318}{10.9} \right)^{1/2}$$

$$= 5.40$$

Hence the escape speed from Jupiter is 5.40 times that from the Earth. Since the escape speed from the Earth is 11.2 km/s (see Example 8.19), it takes a minimum speed of 5.40 × 11.2 km/s =

60.5 km/s to escape from Jupiter. Don't get too close or you'll never come back! Escape speeds for other planets and even stars can be calculated using similar ratio techniques.

EXAMPLE 8.21

a. Calculate the binding energy of a 1.00 kg mass on the surface of the Earth.
b. How much kinetic energy is needed for this mass to escape from the Earth?

Solution
a. The binding energy is the absolute value of the gravitational potential energy of the mass on the surface of the Earth. Here you *cannot* use the local expression for the gravitational potential energy (mgy); rather you must use the more general expression for the gravitational potential energy, given by Equation 8.15:

$$PE = - \frac{GM_\oplus m}{R_\oplus}$$

where $M_\oplus$ is the mass of the Earth, $R_\oplus$ is the radius of the Earth, and m is the 1.00 kg mass whose binding energy we are asked to find. Making these substitutions, you find

$$|PE| = \frac{(6.67 \times 10^{-11} \ \text{N·m}^2/\text{kg}^2)(5.98 \times 10^{24} \ \text{kg})(1.00 \ \text{kg})}{6.37 \times 10^6 \ \text{m}}$$

$$= 6.26 \times 10^7 \ \text{J}$$

b. The binding energy is equal to the amount of kinetic energy needed to escape (barely) from the Earth; therefore, the 1.00 kg mass needs a kinetic energy of 6.26×10^7 J to escape (barely) from the surface of the Earth. Since this example involved a 1.00 kg mass, the binding energy of a mass m is

$$\text{binding energy} = m(6.26 \times 10^7 \ \text{J/kg})$$

This large amount of energy is why NASA puts a premium on minimizing the mass of payloads: it costs real money for the energy. It pays to be an astronaut of small mass!

8.15 BLACK HOLES*

Confinement to the Black Hole . . . to be reserved for cases of Drunkenness, Riot, Violence, or Insolence to Superiors
1844 Regulations and Orders of the Army 121; from the *Oxford English Dictionary.*

*. . . — black it stood as Night
Fierce as ten Furies, terrible as Hell*
John Milton (1608–1674)[†]

During the 18th century, John Michell (1724?–1793) realized that the escape speed from a sufficiently massive and compact star might exceed the speed of light (only measured as finite during the early part of that century), producing what we now call a **black hole**.

[†] *Paradise Lost*, Book II, lines 670–671.

The following argument for the possible existence of black holes is based on Newtonian mechanics. The expression for the escape speed from a spherically symmetric mass M of radius R is given by Equation 8.32:

$$v_{\text{escape}} = \left(\frac{2GM}{R} \right)^{1/2} \qquad (8.32)$$

If the mass M somehow is compressed into a smaller spherical volume, decreasing its radius R as in Figure 8.35, the escape speed increases.

> If the mass M is compressed to such a small size that the escape speed *exceeds* the speed of light, the mass M forms a black hole.

Not even light can escape from a black hole, since $v_{\text{escape}} > c$. The critical radius to which a mass M must be compressed for the escape speed to *equal* the speed of light is known as the **Schwarzschild radius** R_s of the mass, named for Karl Schwarzschild (1873–1916), who first deduced the expression from Einstein's then new theory of general relativity. The Schwarzschild radius of a mass M is found from the equation

$$v_{\text{escape}} = \text{speed of light} \equiv c$$

Substitute for v_{escape} from Equation 8.32:

$$\left(\frac{2GM}{R_s} \right)^{1/2} = c$$

Solving for the Schwarzschild radius R_s, we find

$$R_s = \frac{2GM}{c^2} \qquad (8.34)$$

In principle, anything can become a black hole if the radius R of the mass M is less than R_s. So if *you* want to turn into a black hole, just curl yourself up into a size smaller than your Schwarzschild radius R_s. Needless to say, thanks to the large value of the speed of light (in the denominator of the expression for R_s), the Schwarzschild radius for ordinary masses is *very* small (see Problem 25). However, for sufficiently massive stars at the end of their lifetimes, when the star exhausts its nuclear fusion fuel supplies, no known force can prevent a complete gravitational collapse of the core of the star and a black hole forms. If you use

the result of Example 8.22 (below), a black hole with a mass of three times that of the Sun has a Schwarzschild radius of about 9 km. Venture nearer to a black hole than the Schwarzschild radius and you can kiss the rest of the universe goodbye*; you never can get out, since the escape speed from the black hole is greater than the speed of light.[†‡] *Supermassive* black holes, with masses millions of times greater than the mass of the Sun, are thought to be located at the center of many galaxies, including our own Milky Way Galaxy.

A word of caution. We used classical nonrelativistic dynamics to "derive" the expression for the Schwarzschild radius, Equation 8.34. This derivation is hocus-pocus. We have noted on several occasions that Newtonian dynamics is restricted to speeds much smaller than the speed of light. Ironically, the complicated, full-blown relativistic calculation for the Schwarzschild radius yields precisely the same result as Equation 8.34. This is quite fortuitous for our purposes here, since such coincidences between nonrelativistic and relativistic equations are rare. The classical derivation is an illustration of getting the right answer, but with incorrect physics; the end rarely justifies the means, but here we could not resist!

EXAMPLE 8.22

Calculate the Schwarzschild radius of a hypothetical black hole with the mass equal to that of the Sun.

Solution
The Schwarzschild radius of a black hole of mass M is given by Equation 8.34:

$$R_s = \frac{2GM}{c^2}$$

Substituting for the mass of the Sun, the speed of light, and the universal gravitational constant, you get

$$R_s = \frac{2 \times (6.67 \times 10^{-11} \ \text{N·m}^2/\text{kg}^2)(1.99 \times 10^{30} \ \text{kg})}{(3.00 \times 10^8 \ \text{m/s})^2}$$

$$= 2.95 \times 10^3 \ \text{m}$$

or about 3 km. For reasons that are not readily explainable at this level, *real* black holes must have masses *greater* than about three times the mass of the Sun. The Sun, therefore, will *not* become a black hole when it exhausts the nuclear fuel in its core about A.D. 5×10^9. Real black holes that result from the evolution of stars have Schwarzschild radii of at least about $3 \times 3 \ \text{km} = 9 \ \text{km}$.

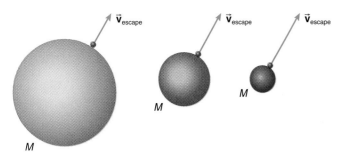

FIGURE 8.35 Decrease the radius of a mass M and the escape speed increases.

---

*Actually, by the time you reach the Schwarzschild radius you will be dissociated into your constituent particles by the enormous gravitational tidal forces in the vicinity of the hole; you reach the Schwarzschild radius as a hailstorm of little particles.

[†] There is an upper limit to the speed of any mass m: the speed of light. The reason the speed of light is an upper bound will become apparent when we study relativity in Chapter 25.

[‡] In the process of entering a black hole, there are frictional forces that cause material to lose kinetic and potential energy, with the result that the infalling material is captured by the hole. Once inside the black hole, there is no way to escape; it is impossible for matter to attain the escape speed, since it is greater than the speed of light.

8.16 LIMITATIONS OF THE CWE THEOREM: TWO PARADOXICAL EXAMPLES*

We have restricted ourselves to systems that can be represented as simple pointlike particles with no internal motions, deformations, or thermal effects. Here we consider two examples of systems that do not meet these criteria to show that the CWE theorem, useful as it is, is not the last word on energy considerations.[†]

Case 1: A Car Accelerates from Rest

Take your car out for a spin. The car, initially at rest, accelerates to speed v_0 along a horizontal section of straight highway. What does the CWE theorem have to say about this phenomenon? The car is a complex system with internal moving parts (the engine and tires) and thermal effects (the combustion of the fuel). The entire car moves through some distance, but the part of the tires in contact with the road surface is always at rest along the path followed by the car (unless skidding occurs).

Choose a coordinate system with $\hat{\jmath}$ vertically upward and $\hat{\imath}$ horizontal and in the direction the car is moving, as shown in Figure 8.36. The origin is at the point where the car begins to accelerate. The forces acting on the car are as follows (see Figure 8.36):

1. The gravitational force of the Earth on the car, its weight $\vec{w}$, directed vertically downward:

$$\vec{w} = -mg\hat{\jmath}$$

   The work done by the conservative gravitational force here is zero, since $\vec{w}$ is everywhere perpendicular to the horizontal path followed by the car.

2. The normal force $\vec{N}$ of the road surface on each tire of the car, directed perpendicular to the surface, so that

$$\vec{N} = N\hat{\jmath}$$

   These normal forces, as always, also do zero work because they are perpendicular to the path followed by the car at every point along its path.

3. The frictional force of the road surface on the drive wheels of the car is the static frictional force because the points of contact of the tires with the road are *at rest*:

$$\vec{f}_s = f_s\hat{\imath}$$

_____
[†] The full story is revealed in Chapter 13.

It is this force that causes the car to accelerate. However, since there is no motion at the point of application of the static frictional force of the road on the tires, the static frictional force does zero work.

Each of the forces on the car does *zero work*. The change in the kinetic energy of the car is

$$\Delta KE \equiv KE_f - KE_i$$
$$= \frac{1}{2}mv_0^2 - 0 \text{ J}$$
$$= \frac{1}{2}mv_0^2$$

Apply the CWE theorem to the car:

$$W_{total} = \Delta KE$$
$$0 \text{ J} = \frac{1}{2}mv_0^2$$

The left-hand side of the theorem apparently is zero but the right-hand side is not! Certainly there is something wrong with applying the CWE theorem to the motion, which should be unsettling to you. Applying Newton's second law of motion to the system *is* appropriate, since it is the static frictional force that accelerates the car. Without the static frictional force on the car, it simply would not change its velocity or accelerate (on an icy, frictionless road, this is indeed what happens).

Case 2: Push Off from a Wall

Another example of a problem arising from the CWE theorem is the following. Imagine yourself on a frictionless horizontal surface (such as an ice rink) next to a vertical wall, as shown in Figure 8.37. Push off from the wall with your arms, attaining a speed v in the horizontal direction. What does the CWE theorem have to say about this situation?

This is an example of a system (you) that *deforms* as your arms compress and extend during the time they are in contact with the wall; the system is not a simple pointlike particle. There are three forces on you during the push-off, shown in Figure 8.37:

1. your weight $\vec{w}$, directed vertically down;
2. the normal force $\vec{N}_{rink}$ of the surface of the ice rink on you, directed vertically upward; and
3. the normal force $\vec{N}_{wall}$ of the wall on you, directed horizontally.

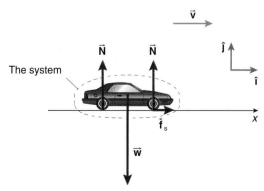

FIGURE 8.36 Forces on the car.

FIGURE 8.37 Forces on you when you push off from a wall.

The weight and the normal force of the rink each do zero work, because they are perpendicular to the change in the position vector of the system at every point along the horizontal path during the push-off. The normal force of the wall also does zero work, since the force exists only at the point of contact between you and the wall and there is no motion at the point of application of this normal force during the push-off.

Apply the CWE theorem:

$$W_{\text{total}} = \Delta KE$$
$$= KE_f - KE_i$$
$$0 \text{ J} = \frac{1}{2} mv^2 - 0 \text{ J}$$

Once again, the CWE theorem yields a ridiculous result. Just as in Case 1, there is something fishy going on here with the CWE theorem.

There is *nothing wrong with applying Newton's second law* to these cases. But yet there clearly *is* something incomplete with the CWE theorem. The clue to both examples is that something, somewhere is quite literally getting warm. Thermodynamic effects cannot be ignored.

One way to resolve or patch up* the CWE theorem for these cases is to imagine or invent a new kind of fictional work, called the **pseudowork** done by the static frictional force in Case 1 and the normal force of the wall in Case 2. In other words, separate the left-hand side of the CWE theorem, W_{total}, into two contributions—the *real work done* by forces and *pseudowork* done by forces that do no real work:

$$W_{\text{total}} \equiv W_{\text{real}} + W_{\text{pseudo}}$$

In this way, the CWE theorem becomes

$$W_{\text{real}} + W_{\text{pseudo}} = \Delta KE$$

The pseudowork of the forces that do no real work is found by integrating the force over the path followed by the system as if it were a pointlike particle. For example, the pseudowork of the static frictional force in Case 1 is

$$W_{\text{pseudo}} = \int_i^f \vec{\mathbf{f}}_s \cdot d\vec{\mathbf{r}}$$

For Case 1, the pseudowork of the frictional force is equal to the change in the kinetic energy of the car. For Case 2, the pseudowork of the normal force of the wall is equal to the change in the kinetic energy of the skater.

The source of the apparent pseudowork of these forces lies *within the system* itself. For the car, the energy that changes its kinetic energy ultimately comes from chemical reactions inside the engine. For the skater, the energy comes from biochemical reactions that drive the muscles.

The paradoxes arising from the CWE theorem are resolved by treating the system as a more complex system than the particle-systems we typically have considered. A more general statement about energy conservation than the CWE theorem is needed; we study this more general statement in Chapter 13.

---

*Another way to describe this is high-class fudging.

8.17 THE SIMPLE HARMONIC OSCILLATOR REVISITED

A system of mass m attached to an ideal spring on a horizontal frictionless surface undergoes simple harmonic oscillation. The only force on the system that does nonzero work is the force of the spring, the Hooke's law conservative force.

The work done by the force of the spring is accounted for by the change in a potential energy term in the CWE theorem. For a simple harmonic oscillator system, the *left*-hand side of the CWE theorem, W_{other} in Equation 8.28, is zero and the theorem takes the form

$$0 \text{ J} = \Delta(KE + PE)$$

The total mechanical energy E is the sum of the kinetic and potential energies. Thus the CWE theorem indicates that the total mechanical energy E of a simple harmonic oscillator is *constant* or *conserved*. That is, $0 \text{ J} = \Delta E$, and so E is constant. Since the simple harmonic oscillator is such an important system in physics, we want to calculate the total mechanical energy of *any* simple harmonic oscillator, be it a mass m attached to a spring or an atom executing simple harmonic oscillation about its equilibrium position in a molecule.

Recall that for *any* simple harmonic oscillator (Chapter 7) the position x of the oscillator as a function of time is given by Equation 7.6,

$$x(t) = A \cos(\omega t + \phi)$$

where ω is the angular frequency [$= (k/m)^{1/2}$ for a mass m attached to a spring with spring constant k]; A is the amplitude of the oscillation and ϕ is a phase angle. Both A and ϕ depend on how the oscillator is initially released. The position $x = 0$ m corresponds to the equilibrium position of the oscillator.

The potential energy of the oscillator at any position x is (according to Equation 8.20)

$$PE(t) = \frac{1}{2} kx(t)^2$$
$$= \frac{1}{2} kA^2 \cos^2(\omega t + \phi)$$

The potential energy varies with time.

To find the kinetic energy, we need to know the speed of the oscillator. There is only one velocity component, v_x, and this is

$$v_x(t) = \frac{dx(t)}{dt}$$
$$= -A\omega \sin(\omega t + \phi)$$

Hence the kinetic energy of the system is

$$KE(t) = \frac{1}{2} mv_x(t)^2$$
$$= \frac{1}{2} mA^2 \omega^2 \sin^2(\omega t + \phi)$$

But since $\omega = (k/m)^{1/2}$, the kinetic energy is

$$KE(t) = \frac{1}{2} kA^2 \sin^2(\omega t + \phi)$$

The kinetic energy also varies with time.

The total mechanical energy E of the oscillator is the sum of the kinetic and potential energies:

$$E = KE(t) + PE(t)$$

Substitute for the kinetic and potential energy of the harmonic oscillator; after factoring, we find

$$E = \frac{1}{2}kA^2[\cos^2(\omega t + \phi) + \sin^2(\omega t + \phi)]$$

But the sum of the squares of the sine and cosine of the same angle always is equal to 1. Thus the total mechanical energy of the oscillator is

$$E = \frac{1}{2}kA^2 \tag{8.35}$$

Since the time t does not appear in this expression, the total mechanical energy E of a simple harmonic oscillator is independent of time, confirming what the CWE theorem has to say about the system: $\Delta E = 0$ J, and so E is *constant*.

> The total mechanical energy of a simple harmonic oscillation is conserved; only the mix or proportion of the energy that is kinetic or potential energy changes with time.

If the oscillator is released at $x = A$ when $t = 0$ s, then the phase angle $\phi = 0$ rad and potential and kinetic energy of the oscillator are

$$PE(t) = \frac{1}{2}kA^2 \cos^2(\omega t)$$

$$KE(t) = \frac{1}{2}kA^2 \sin^2(\omega t)$$

with their sum equal to $E = (1/2)kA^2$.

The potential energy, kinetic energy, and total mechanical energy of this oscillator are plotted versus time in Figure 8.38. If the phase angle $\phi \neq 0$ rad, the potential and kinetic energy graphs are shifted from the previous case, but their sum is still the constant $E = (1/2)kA^2$. This is illustrated in Figure 8.39 for the case when $\phi = \pi/6$ rad.

Since

$$\omega^2 = \frac{k}{m}$$

the total mechanical energy of the oscillator also can be expressed in terms of the angular frequency ω:

$$E = \frac{1}{2}m\omega^2A^2$$

If we convert from the angular frequency ω (in rad/s) to the frequency ν (in Hz) using

$$\omega = 2\pi\nu$$

the total mechanical energy of the oscillator also can be written as

$$E = 2\pi^2m\nu^2A^2 \tag{8.36}$$

> The total mechanical energy of a simple harmonic oscillator is proportional to the product of the square of the frequency and the square of the amplitude.

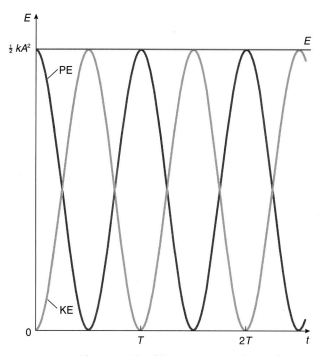

FIGURE 8.38 The potential and kinetic energies of a simple harmonic oscillator as functions of time if $\phi = 0$ rad; note that their sum, the total mechanical energy, is constant.

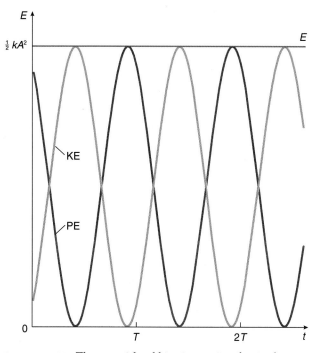

FIGURE 8.39 The potential and kinetic energies of a simple harmonic oscillator as functions of time if $\phi = \pi/6$ rad; note that their sum, the total mechanical energy, is constant.

Since the total mechanical energy of the oscillator is fixed, we can find the speed of the oscillator as a function of position x by writing

$$E = \text{KE} + \text{PE}$$

Substituting, we get

$$\frac{1}{2} kA^2 = \frac{1}{2} mv^2 + \frac{1}{2} kx^2$$

Solve for v (take the positive square root since the speed is a positive scalar):

$$v(x) = \left[\frac{k}{m} \left(A^2 - x^2 \right) \right]^{1/2} \qquad (8.37)$$

Remember the foregoing results apply to *any* simple harmonic oscillation with angular frequency ω and amplitude A.

EXAMPLE 8.23

A 3.00 kg mass is attached to a spring with spring constant $k = 300$ N/m and is executing simple harmonic motion with an amplitude of 0.40 m.

a. Find the total mechanical energy of the oscillator.
b. Make a graph of the *speed* v of the oscillator as a function of its position x.

Solution

a. The total mechanical energy of the oscillator is given by Equation 8.35:

$$E = \frac{1}{2} kA^2$$

The amplitude $A = 0.40$ m and $k = 300$ N/m, and so

$$E = \frac{1}{2} (300 \text{ N/m})(0.40 \text{ m})^2$$

$$= 24 \text{ J}$$

b. The speed v as a function of position x is found from Equation 8.37:

$$v(x) = \left[\frac{k}{m} \left(A^2 - x^2 \right) \right]^{1/2}$$

$$= \left\{ \frac{300 \text{ N/m}}{3.00 \text{ kg}} \left[(0.40 \text{ m})^2 - x^2 \right] \right\}^{1/2}$$

$$= (10.0 \text{ s}^{-1}) (0.16 \text{ m}^2 - x^2)^{1/2}$$

The domain of x is -0.40 m $\leq x \leq 0.40$ m. The function $v(x)$ is parabolic and is plotted in Figure 8.40. On the other hand, the velocity component $v_x(t)$ (plotted in Chapter 7) is sinusoidal. Notice in Figure 8.40 that the maximum speed of the simple harmonic oscillator occurs at the equilibrium position $x = 0$ m, and zero speed occurs at maximum amplitude.

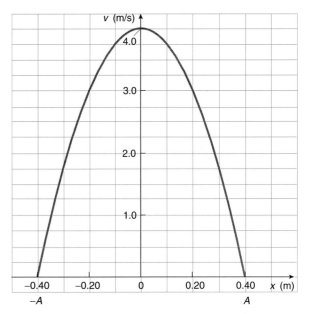

FIGURE 8.40 The speed of a simple harmonic oscillator as a function of position.

8.18 THE AVERAGE AND INSTANTANEOUS POWER OF A FORCE

I sell here, Sir, what all the world desires to have—POWER.
Matthew Boulton (1728–1809)*

As part of a work-study project, you have been hired to lift a number of bricks from the ground up to a height where a mason is laying bricks for a new science building at your university. You could pick up the bricks a few at a time and lift them, or you could take a nearby forklift and lift the whole collection of bricks at one time and get the job done more quickly. The forces that you exert or the forklift exerts on the bricks do the same amount of work on the collection of bricks, but there clearly is a difference between the two situations.

The difference involves the time interval required to do the work on the brick pile. The time rate at which work is done is called **power**. Since work is a way in which energy is transferred from external agents to the system, power represents the time rate at which energy is transferred to the system. Let's take a look at this definition from a quantitative point of view.

The **average power** P_{ave} transferred by a force to a system is defined to be the ratio of the work done by the force on the system to the time interval Δt during which the work is accomplished:

$$P_{\text{ave}} = \frac{\text{work done on system by } \vec{\mathbf{F}} \text{ during time interval } \Delta t}{\Delta t}$$

$$(8.38)$$

*Quoted when speaking to Boswell of his engineering works (22 March 1776). In James Boswell, *Life of Samuel Johnson*, edited by George Birkbeck Hill (revised and enlarged edition by Lawrence F. Powell, Oxford at the Clarendon Press, Oxford, England, 1979), volume 2, page 459.

For the brick problem, the average power in the two situations is clearly different. The forklift accomplished the same work on the bricks during a shorter time than you could; it did the job using more average power. The forklift can do more work during a given time interval than you can.

The dimensions of power are those of work divided by time, or J/s using SI units. Since the concept of power is so important in many contexts, the SI unit of power, J/s, is called a **watt** (W), after James Watt (1736–1819), the inventor of the steam engine. Although you probably are more familiar with the power unit through its use in electrical contexts (e.g., 100 W light bulbs), the unit is a general one in the SI system.*

> The time rate at which a force $\vec{\mathbf{F}}$ does work is called its **instantaneous power**.

The instantaneous power (typically called simply the **power**) transferred to a system by a force doing work on the system is found in the following way. The differential work dW done by a force $\vec{\mathbf{F}}$ when the system changes its position vector by $d\vec{\mathbf{r}}$ is, by definition (Equation 8.1),

$$dW = \vec{\mathbf{F}} \cdot d\vec{\mathbf{r}}$$

The differential change in the position vector $d\vec{\mathbf{r}}$ is related to the instantaneous velocity $\vec{\mathbf{v}}$:

$$\vec{\mathbf{v}} = \frac{d\vec{\mathbf{r}}}{dt}$$

Rearranging,

$$d\vec{\mathbf{r}} = \vec{\mathbf{v}} \, dt$$

---

*Another unit of power that may be familiar is the *horsepower* (hp). The horsepower is an antiquated English unit (with all the quaint peculiarities of that abominable system of weights and measures) introduced by James Watt to measure the efficiency of his newly invented steam engines in the early 19th century in terms of a "standard horse." For those interested, the conversion factor is 1 hp = 746 W ≈ 0.75 kW. We shall not use the horsepower unit in this text.

A forklift is more powerful than you because it can do more work per unit time.

Substituting for $d\vec{\mathbf{r}}$ in the expression for the differential work done by $\vec{\mathbf{F}}$, we find

$$dW = \vec{\mathbf{F}} \cdot \vec{\mathbf{v}} \, dt$$

The power P is

$$\boxed{P \equiv \frac{dW}{dt} = \vec{\mathbf{F}} \cdot \vec{\mathbf{v}}} \qquad (8.39)$$

Power is a *scalar* quantity. Think of the power provided by a force as a measure of the time rate of doing work by that force on the system. Notice that if the force is perpendicular to the velocity, the scalar product vanishes and the instantaneous power of the force is zero.

Take the definition of the instantaneous power of a force:

$$P = \frac{dW}{dt}$$

Rearrange:

$$dW = P \, dt$$

Finally, integrate between time t_i and time t_f:

$$W = \int_{t_i}^{t_f} P \, dt \qquad (8.40)$$

This can be interpreted geometrically in the following way. If a graph is made of the instantaneous power $P(t)$ versus t, as in Figure 8.41, the work done by the force during the interval from t_i to t_f is the area under the graph between the two times.

This complements the geometric interpretation of work, discussed in Section 8.5, where we saw that the work done by a force component F_x between *positions* x_i and x_f can be interpreted as the area under a graph of the force component F_x versus x between the two positions.

It is important to mention that the word power (as well as the words work, force, and energy) is used in many contexts *outside* the sciences that convey meanings unrelated to the concept in physics and engineering. Colloquial usages typically convey an *incorrect* impression of what is meant by power in physics (e.g., political power, power lunches, power play). As a scientist or engineer, you will recognize these uses differ from the technical meaning of power.

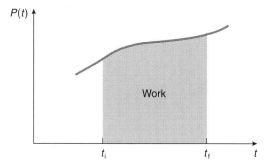

FIGURE 8.41 The work is the area under the graph of $P(t)$ versus t.

A system of mass m executes a closed path during a time interval Δt; one of the forces acting on the system is a conservative force. What is the average power of the conservative force during the time interval?

EXAMPLE 8.24

One hundred 2.00 kg bricks are to be lifted up 1.50 m from a pallet. A mason accomplishes the task during 10.0 min and a forklift does the same task in 3.00 s. Calculate the average power of the mason and that of the forklift.

Solution

The system is the 100 bricks. The forces on the bricks are

1. the gravitational force of the Earth on the bricks; and
2. the force of the mason on the bricks (for the case of the mason lifting the bricks), or (for the case of the forklift) the force of the forklift on the bricks.

The work done by the gravitational force in the process is accounted for by the change in the potential energy term in the CWE theorem. The work done by the mason or forklift must be accounted for on the *left*-hand side of the CWE theorem, since these forces have no potential energy function associated with them. Assume that the kinetic energy of the bricks is unchanged; they are picked up at rest and delivered at rest:

$$\Delta KE = 0 \text{ J}$$

Choose the coordinate system indicated in Figure 8.42 with $\hat{\jmath}$ up and the origin at the bricks on the pallet. Since $y_i = 0$ m and $y_f = 1.50$ m, the change in the potential energy of the bricks is

$$\Delta PE = PE_f - PE_i$$
$$= mgy_f - mgy_i$$
$$= (100 \text{ bricks})(2.00 \text{ kg/brick})(9.81 \text{ m/s}^2)(1.50 \text{ m}) - 0 \text{ J}$$
$$= 2.94 \times 10^3 \text{ J}$$

Apply the CWE theorem:

$$W_{\text{mason or forklift}} = \Delta KE + \Delta PE$$
$$= 0 \text{ J} + 2.94 \times 10^3 \text{ J}$$
$$= 2.94 \times 10^3 \text{ J}$$

FIGURE 8.42

The mason accomplished the work on the bricks during 10.0 min or $\Delta t_{\text{mason}} = 600$ s, and so the average power of the mason is

$$P_{\text{ave mason}} = \frac{W_{\text{mason}}}{\Delta t_{\text{mason}}}$$
$$= \frac{2.94 \times 10^3 \text{ J}}{600 \text{ s}}$$
$$= 4.90 \text{ W}$$

The forklift also did the same work (actually more since it raised the pallet too), but during a time interval of 3.00 s, and so the average power of the forklift is

$$P_{\text{ave forklift}} = \frac{W_{\text{forklift}}}{\Delta t_{\text{forklift}}}$$
$$= \frac{2.94 \times 10^3 \text{ J}}{3.00 \text{ s}}$$
$$= 980 \text{ W}$$

EXAMPLE 8.25

What is the (instantaneous) power provided by the gravitational force on a particle of mass m in *circular* orbit about the Sun (of mass M); see Figure 8.43.

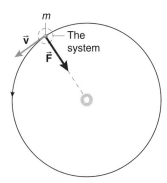

FIGURE 8.43

Solution

The instantaneous power is given by Equation 8.39:

$$P = \vec{F} \cdot \vec{v}$$

Note that the gravitational force is perpendicular to the velocity of m at every point in the circular orbit. Hence, at any instant, $\vec{F} \cdot \vec{v} = 0$ W, and the instantaneous power at all times and places around the circular orbit is equal to zero.

EXAMPLE 8.26

Hydroelectric power plants use falling water as a source of energy to turn generators to produce electricity. Consider a mass m of water released at rest that falls through a distance h.

a. Find the average power of the gravitational force during the interval of the fall.
b. Find the instantaneous power of the gravitational force the instant just before impact.

Solution

a. Take m to be the system. To find the average power, you need to know the work done on the system by the gravitational force and the time interval Δt during which this work was done. The time interval of the fall is found from the one-dimensional motion equations at constant acceleration (Chapter 3). Choose $\hat{\jmath}$ up and with the y-origin at bottom of the fall, as shown in Figure 8.44. Then the one-dimensional motion equation for constant acceleration,

$$y = y_0 + v_{y0}t + \frac{1}{2}a_yt^2$$

has

$$y_0 = h$$
$$v_{y0} = 0 \text{ m/s}$$
$$a_y = -g$$

At impact $y = 0$ m, so that

$$0 \text{ m} = h + \frac{1}{2}(-g)t^2$$

Solving for t (and taking the positive root, since the impact occurs after the mass is released),

$$t = \left(\frac{2h}{g}\right)^{1/2}$$

The work done by the gravitational force during the fall is the negative of the change in the gravitational potential energy. Use the coordinate system indicated in Figure 8.44:

$$\begin{aligned} W_{\text{grav}} &= -\Delta\text{PE} \\ &= -(\text{PE}_f - \text{PE}_i) \\ &= -(0 \text{ J} - mgh) \\ &= mgh \end{aligned}$$

The average power of the gravitational force during the interval of the fall then is

$$\begin{aligned} P_{\text{ave}} &= \frac{W_{\text{grav}}}{t} \\ &= \frac{mgh}{\left(\dfrac{2h}{g}\right)^{1/2}} \\ &= \frac{m}{\sqrt{2}}\sqrt{h}\,g^{3/2} \end{aligned}$$

b. Use the CWE theorem to calculate the speed of the mass just before impact. For variety, choose another coordinate system, the one shown in Figure 8.45.

During the fall the only force acting on the mass is the gravitational force. The work done by the gravitational force is accounted for by the potential energy term in the CWE theorem. The left-hand side of the theorem is zero, because there are no other forces acting on the mass during the fall. The initial kinetic energy is zero, since the mass was released from rest. The initial gravitational potential energy also is zero, because the mass was released at a point where its y-coordinate is zero. The final kinetic energy of the particle at the instant of impact is

$$\text{KE}_f = \frac{1}{2}mv^2$$

The final gravitational potential energy is equal to mgy_f where $y_f = -h$:

$$\begin{aligned} \text{PE}_f &= mg(-h) \\ &= -mgh \end{aligned}$$

The outflow from a hydroelectric power facility.

FIGURE 8.44

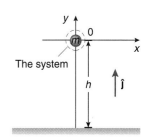

FIGURE 8.45

Therefore the CWE theorem assumes the form

$$W_{\text{other}} = \Delta(\text{KE} + \text{PE})$$

$$0 \text{ J} = (\text{KE} + \text{PE})_f - (\text{KE} + \text{PE})_i$$

$$= \left[\frac{1}{2} mv^2 + (-mgh) \right] - (0 \text{ J} + 0 \text{ J})$$

Hence

$$v = (2gh)^{1/2}$$

To calculate the power you need vector expressions for both the force and the velocity. With the coordinate system in Figure 8.45, the gravitational force on m is

$$\vec{\mathbf{F}}_{\text{grav}} = -mg\hat{\mathbf{j}}$$

and the velocity vector just before impact is

$$\vec{\mathbf{v}} = (2gh)^{1/2}(-\hat{\mathbf{j}})$$

Thus the instantaneous power P provided by the gravitational force is given by Equation 8.39:

$$P = \vec{\mathbf{F}}_{\text{grav}} \cdot \vec{\mathbf{v}}$$
$$= -mg\hat{\mathbf{j}} \cdot [-(2gh)^{1/2}\hat{\mathbf{j}}]$$
$$= \sqrt{2h} \, mg^{3/2}$$

Substitute representative numerical values to get a feeling for the quantity of this power: say $m = 2.00$ kg and $h = 30.0$ m. Then

$$P = [2(30.0 \text{ m})]^{1/2}(2.00 \text{ kg})(9.81 \text{ m/s}^2)^{3/2}$$
$$= 476 \text{ W}$$

8.19 THE POWER OF THE TOTAL FORCE ACTING ON A SYSTEM

It is the total force that causes the acceleration of a system. In this section we will see that the power of the total force dictates the time rate of change of the kinetic energy of the system.

The kinetic energy of a system of mass m when it has a speed v is given by Equation 8.25:

$$\text{KE} = \frac{1}{2} mv^2$$

Since $v^2 = \vec{\mathbf{v}} \cdot \vec{\mathbf{v}}$, the kinetic energy may be rewritten as

$$\text{KE} = \frac{1}{2} m \, \vec{\mathbf{v}} \cdot \vec{\mathbf{v}}$$

The time rate at which the kinetic energy of the system changes is the time derivative:

$$\frac{d}{dt}(\text{KE}) = \frac{d}{dt}\left(\frac{1}{2} m \, \vec{\mathbf{v}} \cdot \vec{\mathbf{v}} \right)$$

Consider a system of constant mass (we have considered only such systems). Differentiate the scalar product on the right-hand side with respect to time, obtaining

$$\frac{d}{dt}(\text{KE}) = \frac{1}{2} m \left(\frac{d\vec{\mathbf{v}}}{dt} \cdot \vec{\mathbf{v}} + \vec{\mathbf{v}} \cdot \frac{d\vec{\mathbf{v}}}{dt} \right)$$

The two terms on the right-hand side are equal. Thus

$$\frac{d}{dt}(\text{KE}) = m \frac{d\vec{\mathbf{v}}}{dt} \cdot \vec{\mathbf{v}} \qquad (8.41)$$

The *total force* acting on the system is, by Newton's second law,

$$\vec{\mathbf{F}}_{\text{total}} = m \frac{d\vec{\mathbf{v}}}{dt}$$

Substitute this in Equation 8.41 to obtain the time rate of change of the kinetic energy:

$$\frac{d}{dt}(\text{KE}) = \vec{\mathbf{F}}_{\text{total}} \cdot \vec{\mathbf{v}}$$

The right-hand side of this equation is the (instantaneous) power of the *total force*, i.e., the instantaneous total power.

The instantaneous *total power* is equal to the time rate of change of the kinetic energy of the system:

$$\frac{d}{dt}[\text{KE}(t)] = P_{\text{total}}(t) \qquad (8.42)$$

QUESTION 9

A graph is made of the kinetic energy of a system versus t. What is the physical significance of the slope of the graph?

EXAMPLE 8.27

A spaceship of essentially constant mass m, initially moving at velocity $v_{x0}\hat{\mathbf{i}}$ is briefly subjected to a constant acceleration $a_x\hat{\mathbf{i}}$ by its rocket motors. Find the power of the total force on the spacecraft during its acceleration.

Solution

The spacecraft is the system. The velocity component of the system as a function of time is found from the one-dimensional kinematic equations for a constant acceleration (Chapter 3):

$$v_x(t) = v_{x0} + a_x t$$

Method 1

The power is the scalar product of the total force and the velocity:

$$P = \vec{\mathbf{F}}_{\text{total}} \cdot \vec{\mathbf{v}}$$

The total force is found using Newton's second law:

$$\vec{\mathbf{F}}_{\text{total}} = m a_x \hat{\mathbf{i}}$$

Hence the power is

$$P(t) = m a_x \hat{\mathbf{i}} \cdot (v_{x0} + a_x t)\hat{\mathbf{i}}$$
$$= m a_x (v_{x0} + a_x t)$$

Method 2

The power of the total force is the time rate of change of the kinetic energy of the system. The kinetic energy of the system is

$$KE(t) = \frac{1}{2} m \left[v_x(t) \right]^2$$

$$= \frac{1}{2} m (v_{x0} + a_x t)^2$$

$$= \frac{1}{2} m (v_{x0}^2 + 2 v_{x0} a_x t + a_x^2 t^2)$$

Take the time derivative of the kinetic energy to find the power of the total force:

$$P_{\text{total}}(t) = \frac{d KE(t)}{dt}$$

$$= m(v_{x0} a_x + a_x^2 t)$$

$$= m a_x (v_{x0} + a_x t)$$

8.20 MOTION UNDER THE INFLUENCE OF CONSERVATIVE FORCES ONLY: ENERGY DIAGRAMS*

It is convenient to conceptualize the motion of a system subjected only to conservative forces with an **energy diagram**. Such diagrams are used extensively in contemporary physics, particularly in quantum mechanics. It is useful to explore them first in a classical context.

Consider a system that satisfies *either* of the following conditions:

a. the *only* forces acting on a particle are conservative forces; or
b. if there are "other" forces acting, the work done by all these other forces sums to zero—so $W_{\text{other}} = 0$ J.

Under *either* of these conditions, the CWE theorem of Equation 8.28 reduces to Equation 8.30:

$$0 \text{ J} = \Delta(KE + PE)$$

The total mechanical energy of the system is conserved, or unchanged during the motion. If the potential energy of the particle changes, the kinetic energy compensates, so that the sum of the two energies remains constant.

Let's see what this means from both an algebraic and graphical perspective. Substitute the kinetic energy, $KE = mv^2/2$, into the definition of the total mechanical energy:

$$E = \frac{mv^2}{2} + PE$$

Now solve for the speed:

$$v = \left[\frac{2}{m} (E - PE) \right]^{1/2} \tag{8.43}$$

For the speed of the particle to be a *real number*, we must have

$$E \geq PE$$

at all times and places during the motion. If $E < PE$, the speed is an *imaginary number*; while this is mathematically legitimate, it is not physically permissible because any physical measurement of the speed yields a real number.

> Thus the total mechanical energy of the system must at all times and places be greater than or equal to its potential energy.

Look at several specific examples to see what this condition on E means.

Simple Harmonic Oscillator

We saw in Section 8.17 that the total mechanical energy E of a simple harmonic oscillator is constant. The potential energy of a simple harmonic oscillator was found in Section 8.12 to be (Equation 8.20)

$$PE = \frac{1}{2} k x^2$$

where $x = 0$ m represents the equilibrium position of the oscillator.

Make a graph of the potential energy as a function of x (see Figure 8.46). The graph is parabolic in shape because of the quadratic dependence on x. Plot the specific value of the total mechanical energy E of the given simple harmonic oscillator on the same diagram. Since E is fixed, the graph of E is a horizontal line at the appropriate number of joules on the vertical axis. The line representing E intercepts this potential energy curve at two places, whose x-coordinates are called **turning points**.

The possible x-coordinates of the oscillator must lie between the turning points, since it is only for these values of x that $E \geq PE$, leading to real values of the speed according to Equation 8.43. Thus the x-coordinate of the oscillator is bounded (confined) by the turning points. For values of x outside these bounds, E is less than the PE, and at those positions the speed of the oscillator becomes imaginary according to the speed equation (Equation 8.43). Thus, for a given E in Figure 8.46, portions of the x-axis are prohibited. We say the particle is **bound**, because it is confined to a finite region of the x-axis between the turning points.

The maximum value of x is the amplitude of the motion; the turning points are at $x = \pm A$. At any permitted position coordinate x, the potential and kinetic energies sum to the total mechanical energy as indicated in Figure 8.46. When the x-coordinate of the oscillator is at either turning point, the potential energy has increased so that it is equal to the (fixed) value of the total energy E; the kinetic energy of the oscillator there is zero and the system is instantaneously at rest. At the two turning points, the oscillator has its maximum amplitude and is reversing direction.

Suppose the oscillator is initially at rest at the equilibrium position, $x = 0$ m. The potential energy is zero, since $x = 0$ m; the kinetic energy is zero, since $v = 0$ m/s; and the total mechanical energy is zero, since $E = KE + PE = 0$ J $+ 0$ J. The oscillator will not move, because the two turning points are both at $x = 0$ m in Figure 8.46. We mentioned previously that if a conservative force is the only force on a system and the system is released at rest, the system will tend to move *initially* in a direction to lower its potential energy. With the particle initially at $x = 0$ m and released at rest, the potential energy already is as

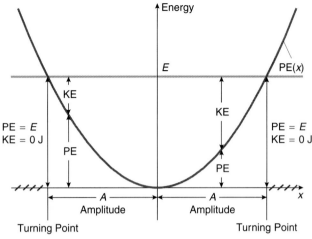

FIGURE 8.46 A graph of PE(x) versus x for a simple harmonic oscillator.

low as it can be, and so no motion will ensue. We also know that if $x = 0$ m, the spring exerts no force on the particle and, therefore, the particle will not change its velocity.

On the other hand, when the oscillator is pulled to $x = A$ and then released at rest, the total mechanical energy of the oscillator is

$$E = \frac{1}{2} kA^2$$

This corresponds to releasing the oscillator at one of the turning points. The oscillator moves initially in a direction to lower its potential energy—that is, toward decreasing values of $|x|$. The motion is bounded by the two turning points at $x = \pm A$.

In this one-dimensional situation, the differential work done by the force of the spring on the system is

$$\begin{aligned} dW &= \vec{\mathbf{F}}_{spring} \cdot d\vec{\mathbf{r}} \\ &= F_x \hat{\mathbf{i}} \cdot dx\,\hat{\mathbf{i}} \\ &= F_x\,dx \end{aligned} \quad (8.44)$$

The differential work done by a conservative force on the system is the *negative* of the differential change in the potential energy of the system (from the definition of the potential energy):

$$dW = -d(PE)$$

After substituting for dW, Equation 8.44 becomes

$$-d(PE) = F_x\,dx$$

and the x-component of the force is

$$F_x = -\frac{d}{dx}(PE) \quad (8.45)$$

The component of the force of the spring is the negative of the slope of the potential energy curve.

The potential energy function for Hooke's force law is

$$PE = \frac{1}{2} kx^2$$

We can find the force component by taking the negative spatial derivative of the potential energy function:

$$\begin{aligned} F_x &= -\frac{d}{dx}\left(\frac{1}{2} kx^2\right) \\ &= -kx \end{aligned}$$

That is,

$$\begin{aligned} \vec{\mathbf{F}}_{spring} &= F_x \hat{\mathbf{i}} \\ &= -kx\hat{\mathbf{i}} \end{aligned}$$

as before.

At the bottom of the curve, the slope is zero and so the force component is zero. When $x > 0$ m, the slope is positive, so the expression for the force component F_x is negative, indicating that the force is directed toward $-\hat{\mathbf{i}}$. When $x < 0$ m, the slope is negative, and the force component becomes positive or directed along $+\hat{\mathbf{i}}$.

QUESTION 10

A graph of the acceleration component a_x of a simple harmonic oscillator is made as a function of x. What is the shape of the curve?

The Gravitational Force

The gravitational force is a conservative force. If the gravitational force is the only force on a system, the CWE theorem becomes

$$\begin{aligned} W_{other} &= \Delta(KE + PE) \\ 0 \text{ J} &= \Delta(KE + PE) \end{aligned}$$

The total mechanical energy $E = KE + PE$ of the system is conserved.

The most general form for the potential energy function of a system of mass m located a distance r from another pointlike or spherical mass M is, from Equation 8.15,*

$$PE = -\frac{GmM}{r}$$

---

*The masses m and M must be either point masses or spheres with spherically symmetric mass distributions. The center-to-center position of m and M must be greater than the sum of the radii of the spheres to use this expression for the potential energy. A different form for the potential energy (see Problem 20) is used for values of r less than the radius of the sphere.

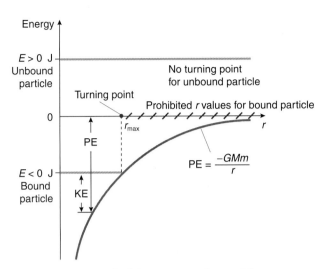

FIGURE 8.47 A graph of the gravitational potential energy as a function of r.

A graph of this potential energy as a function of r is shown in Figure 8.47. Recall that the radial coordinate r is always greater than or equal to zero, and so only positive values of r are plotted. According to Equation 8.43, the total mechanical energy E must be greater than or equal to the potential energy in order to have a real number value for the speed for the system.

For the value of the total mechanical energy $E < 0$ J indicated in Figure 8.47, the indicated region along the r-axis is prohibited, since for these r-coordinates the total mechanical energy E is less than the potential energy and the speed is imaginary. There is a turning point where the line representing E intersects the potential energy curve. What this means physically is that if E is fixed, there are many possible orbits of m about M (ellipses with various eccentricities), but all with an upper bound on r at the turning point.

At any given permitted r, the values of the kinetic and potential energies are schematically indicated in Figure 8.47. The fixed energy E is the sum of the kinetic and potential energies. Notice also that if the total mechanical energy E is greater than zero, there are no turning points and no bound on the motion of the particle: any r value from zero to infinity is permitted. Particles with $E \geq 0$ J are called **unbound** particles; their orbits are straight lines and the conic sections known as parabolas (for $E = 0$ J) and hyperbolas (for $E > 0$ J) for inverse-square-law forces. For negative total energies ($E < 0$ J), the motion is bounded (there is a maximum value that r can attain). Particles with negative total energies are bound particles, because the radial coordinate has an upper bound. The orbits of particles with negative total energies are circles (with M at the center), ellipses (with M at one focus), and straight lines (with M on the line) for attractive inverse-square-law forces.

Once again the differential work done by the force is

$$dW = \vec{F} \cdot d\vec{r}$$

Write the central gravitational force of M on m as

$$\vec{F} = F_r \hat{r}$$

(where $F_r < 0$ N, since gravitation is an attractive force). The gravitational force (like any central force) has *only* a radial component.

The only work done by the gravitational force on m occurs when m moves in the radial direction—that is, when $d\vec{r} = dr\,\hat{r}$:

$$dW = F_r \hat{r} \cdot dr\,\hat{r}$$
$$= F_r\,dr \qquad (8.46)$$

But the differential work done by a conservative force is the negative of the differential change in the potential energy:

$$dW = -d(\mathrm{PE})$$

This substitution for dW in Equation 8.46 gives

$$-d(\mathrm{PE}) = F_r\,dr$$

and so the radial component (and only component) of the force is

$$F_r = -\frac{d(\mathrm{PE})}{dr} \qquad (8.47)$$

Here again the force component is the negative of the slope of the potential energy function. Use the potential energy function,

$$\mathrm{PE} = -\frac{GmM}{r}$$

and take the negative of its derivative with respect to r to obtain the radial force component F_r:

$$F_r = -\frac{d}{dr}\left(-\frac{GmM}{r}\right)$$

We find

$$F_r = -\frac{GmM}{r^2}$$

The force on m then is

$$\vec{F} = F_r \hat{r}$$
$$= -\frac{GmM}{r^2}\,\hat{r}$$

Voilà. Figure 8.47 has a positive slope throughout its extent, so F_r is negative, indicating that the gravitational force always is an attractive force.

Central Forces

The gravitational force is a central force. Any central force has only a radial force component:

$$\vec{F} = F_r \hat{r}$$

Likewise, the potential energy function associated with a central force depends only on r:

$$PE = PE(r)$$

> The force component F_r is the negative of the slope of the graph of the potential energy versus r.

Nothing in the derivation of Equation 8.46,

$$F_r = -\frac{d(PE)}{dr}$$

depends on anything other than having the potential energy be a function of r alone: $PE = PE(r)$.

The force is repulsive ($F_r > 0$ N) where the slope of the potential energy curve is negative; the force is attractive ($F_r < 0$ N) where the slope is positive. Again the motion is restricted to regions where $E \geq PE$ to keep the speed of the system real rather than imaginary.

EXAMPLE 8.28

The gravitational potential energy of a mass m near the surface of the Earth is (Equation 8.13)

$$PE = mgy \qquad (\hat{\jmath} \text{ directed up})$$

a. Make a schematic graph of the potential energy as a function of y for both positive and negative values of y.
b. What is the value of the y-intercept of the graph?
c. What is the slope of this graph?
d. How is the slope related to the gravitational force on m?
e. Show that the gravitational force

$$\vec{F} = F_y \hat{\jmath}$$

has a component F_y that is

$$F_y = -\frac{d(PE)}{dy}$$

Solution

a. The graph of the potential energy as a function of y is shown in Figure 8.48.
b. The y-intercept of the graph is zero.
c. The slope of the graph is the coefficient of y in the equation $PE = mgy$. Thus the slope is mg.
d. The *negative* of the slope is the gravitational force component F_y on m.
e. Since

$$F_y = -\frac{d(PE)}{dy}$$

substitute for the potential energy:

$$F_y = -\frac{d(mgy)}{dy}$$

$$= -mg$$

Hence the force is

$$\vec{F} = F_y \hat{\jmath}$$

$$= -mg\hat{\jmath}$$

Since $\hat{\jmath}$ is vertically upward, the gravitational force on m is directed vertically downward.

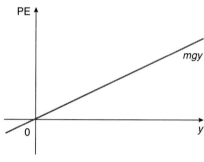

FIGURE 8.48

STRATEGIC EXAMPLE 8.29

A stone of mass 0.500 kg is thrown vertically upward with an initial speed of 6.00 m/s from a point at an elevation of 2.00 m above the ground. Choose the $\hat{\jmath}$ to be up with an origin at the ground.

a. Find the total mechanical energy of the stone at launch.
b. Does the total mechanical energy of the stone change during its flight?
c. Make a detailed graph of the potential energy of the stone as a function of y.
d. Plot the total mechanical energy of the stone on the potential energy graph.
e. Locate the turning point on the graph and explain its physical significance.
f. Locate the initial y-coordinate of the stone on the graph and indicate on the graph what represents the initial kinetic and potential energies of the stone.

Solution

a. The total mechanical energy of the stone at launch is the sum of its initial kinetic and potential energies:

$$E = KE + PE$$

$$= \frac{1}{2}mv_i^2 + mgy_i$$

Since the y-axis origin is on the ground,

$$y_i = 2.00 \text{ m}$$

Also you are given $v_i = 6.00$ m/s. Hence

$$E = \frac{1}{2}(0.500 \text{ kg})(6.00 \text{ m/s})^2$$
$$+ (0.500 \text{ kg})(9.81 \text{ m/s}^2)(2.00 \text{ m})$$
$$= 9.00 \text{ J} + 9.81 \text{ J}$$
$$= 18.81 \text{ J}$$

b. Since only the gravitational force acts on the stone during its flight, $W_{\text{other}} = 0$ J in the CWE theorem, and so the theorem becomes

$$0 \text{ J} = \Delta(\text{KE} + \text{PE})$$

The total mechanical energy of the stone *does not change* throughout the flight.

c. The potential energy of the stone as a function of y is

$$\text{PE} = mgy$$
$$= (0.500 \text{ kg})(9.81 \text{ m/s}^2)y$$
$$= (4.91 \text{ J/m})y$$

This potential energy function is graphed in Figure 8.49.

d. The total mechanical energy E is a horizontal line on the graph with a y-axis intercept of 18.81 J.

e. From the graph, the turning point is the y-coordinate where the line representing E intersects the potential energy curve. This intersection occurs at about $y_{\text{turn pt}} \approx 3.8$ m. At the turning point, $E = \text{PE}$, so that the KE of the stone there is zero. The stone is instantaneously at rest at this coordinate, and so this is the y-coordinate of the maximum height of the stone. Since we chose the y-axis origin at the ground, the maximum height of the stone is about 3.8 m.

The stone cannot have y-coordinates greater than the turning point, and so y > 3.8 m is a prohibited region.

f. The initial y-coordinate of the stone was $y_i = 2.00$ m. The initial potential energy is represented by the vertical distance

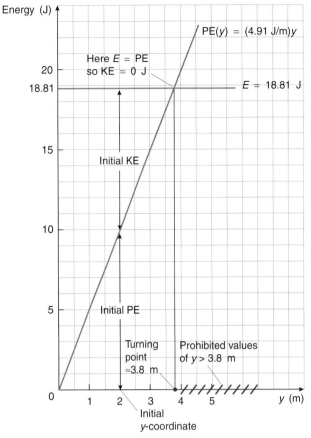

FIGURE 8.49

on the graph between the horizontal axis (y) at y = 2.00 m and the line representing the potential energy. Since $E = \text{PE} + \text{KE}$, the initial kinetic energy is the vertical distance on the graph between the potential energy (at y = 2.00 m) and the line representing E.

CHAPTER SUMMARY

The *work W* done by any force $\vec{\mathbf{F}}$ on a system of mass m is

$$W = \int_i^f \vec{\mathbf{F}} \cdot d\vec{\mathbf{r}} \qquad (8.2)$$

where the integral is taken along the path followed by the system between its initial and final positions. The SI units of work are N·m, called *joules* (J).

If a force is a *constant force* (i.e., a constant vector), along a path (whether straight or curved), then the work done by such a force on a system is

$$W = \vec{\mathbf{F}} \cdot \Delta\vec{\mathbf{r}} \qquad \text{(constant force } \vec{\mathbf{F}}) \qquad (8.6)$$

where $\Delta\vec{\mathbf{r}}$ is the *change* in the position vector of the system:

$$\Delta\vec{\mathbf{r}} = \vec{\mathbf{r}}_f - \vec{\mathbf{r}}_i$$

A system *cannot* be said to *have* work, so that *work is not a property of a system.* On the other hand, *energy is a property of a system.* Changes in the energy of a system are related to the work done on the system by the forces acting on it.

Forces are classified as *conservative, nonconservative,* or *zero-work* forces. The work done by a *conservative force* on a system as it moves between two positions is *independent of the path* followed by the system between the two points. Equivalently, the work done by a conservative force around any closed path is zero. If the work done by a force around an arbitrarily chosen closed path is not equal to zero, the force is a *nonconservative force.* Certain forces always do zero work and are known as *zero-work forces.* The normal force of a rigid surface on a system and the static frictional force are zero-work forces.

The work done by a conservative force is defined to be equal to the negative of the change in a potential energy function associated with the conservative force:

$$W_{consrv} \equiv -\Delta PE \qquad (8.11)$$

The potential energy of a system depends on its *location* in space with respect to a reference position where PE = 0 J.

Important potential energy functions to know are those associated with the following conservative forces (on a system of mass m):

a. For the *gravitational force* on the mass m when it is always near the surface of the Earth, the potential energy of m is

$$PE_{local} = mgy \qquad (\hat{\jmath} \text{ directed up})$$

where $y = 0$ m may be placed at any convenient elevation.

b. For the more general gravitational force of M on m, the potential energy of m when it is located a distance r from M is

$$PE_{grav} = -\frac{GmM}{r} \qquad (8.15)$$

where the origin for r is at M and the zero of potential energy always is where $r = \infty$ m.

c. For Hooke's force law associated with a mass m on a spring, the potential energy of m is

$$PE_{spring} = \frac{1}{2} kx^2 \qquad (8.20)$$

where $x = 0$ m is the equilibrium position of m and k is the spring constant.

The *CWE theorem* states that the work, W_{total}, done by *the total force* on a system of mass m is equal to the change in the kinetic energy of the system:

$$W_{total} = \Delta KE \qquad (8.26)$$

Since the forces acting on a system are either conservative forces, nonconservative ("other") forces, or zero-work forces, the total work done by the total force is broken up into two contributions:

$$W_{other} + W_{consrv} = \Delta KE$$

Since $W_{consrv} \equiv -\Delta PE$, the CWE theorem also can be put into the form

$$W_{other} = \Delta KE + \Delta PE$$
$$= \Delta(KE + PE) \qquad (8.28)$$

The sum of the kinetic and potential energies of a system is called the *total mechanical energy E* of the system, and so the CWE theorem has yet another variation:

$$W_{other} = \Delta E \qquad (8.29)$$

If $W_{other} = 0$ J, then the total mechanical energy E of the system is *conserved*, or constant. The condition $W_{other} = 0$ J is met if either

(1) *all* the forces on the system are conservative or zero-work forces, or (2) the total work done by "other" (e.g., nonconservative) forces on the system is zero.

The *escape speed* from a spherical mass M of radius R is

$$v_{escape} = \left(\frac{2GM}{R}\right)^{1/2} \qquad (8.32)$$

A simple harmonic oscillator has a total mechanical energy E that is constant and equal to

$$E = \frac{1}{2} kA^2 \qquad (8.35)$$

or

$$E = 2\pi^2 m v^2 A^2 \qquad (8.36)$$

The average power P_{ave} of a force $\vec{F}$ on a system is defined to be

$$P_{ave} \equiv \frac{\text{work done by } \vec{F} \text{ during a time interval } \Delta t}{\Delta t} \qquad (8.38)$$

The instantaneous power P of a force $\vec{F}$ on a system is

$$P = \vec{F} \cdot \vec{v} \qquad (8.39)$$

where $\vec{v}$ is the velocity of the system. The SI unit for power is the J/s or a *watt* (W).

The (instantaneous) *total power* P_{total} of the total force on a system is the time rate of change of the kinetic energy of the system:

$$P_{total} = \frac{d(KE)}{dt} \qquad (8.42)$$

Interesting information about a system for which $W_{other} = 0$ J can be obtained from an *energy diagram* in which the potential energy of the system is graphed as a function of its position. The CWE theorem implies that the speed of the system is

$$v = \left[\frac{2}{m}\left(E - PE\right)\right]^{1/2} \qquad (8.43)$$

To have a real (rather than imaginary) value for v, the (constant) total mechanical energy E must satisfy

$$E \geq PE$$

The coordinates on such energy diagrams where $E = PE$ are called *turning points*. Coordinates where $E < PE$ are prohibited.

At any point on such a potential energy graph, the *negative of the slope* is the force component associated with that potential energy function:

$$F_x = -\frac{d(PE)}{dx} \qquad (8.45)$$

SUMMARY OF PROBLEM-SOLVING TACTICS

8.1 **(page 321)** The differential change $d\vec{r}$ in the position vector of the system is written as

$$d\vec{r} = dx\,\hat{\imath} + dy\,\hat{\jmath} + dz\,\hat{k} \qquad (8.3)$$

whether the particle is moving toward increasing or decreasing values of the coordinates. The limits of integration account for the actual direction the particle moves in going

from its initial position (the lower limit) to its final position (the upper limit).

8.2 **(page 324)** The work done by a constant force, Equation 8.6, can be used even if the path between the initial and final positions is not straight.

8.3 **(page 336)** Equation 8.13 for the local gravitational potential energy must have $\hat{\jmath}$ directed up, since that is the choice we made in deriving the equation; you can choose $y = 0$ m to be at any elevation you want.

8.4 **(page 337)** When using Equation 8.15 for the gravitational potential energy, choose the origin of the coordinate system to be at M; no other choice for the origin should be made.

8.5 **(page 342)** Equations 8.26, 8.27, 8.28, and 8.29 are all equivalent expressions for the CWE theorem.

QUESTIONS

1. (page 322); 2. (page 324); 3. (page 325); 4. (page 330); 5. (page 334); 6. (page 336); 7. (page 343); 8. (page 356); 9. (page 358); 10. (page 360)

11. Can the normal force of a surface on a system ever do nonzero work?

12. Is the kinetic frictional force a constant force if the path followed by a system is not a straight-line path? Explain.

13. A raindrop is falling through the air at a terminal speed. (Refer back to Section 5.14.) What forces act on the raindrop? What is the relationship between the amounts of work done on the raindrop by these forces?

14. The normal force of a surface and the static frictional force both are zero-work forces, but typically for different reasons. Explain why each is a zero-work force.

15. A student slowly lifts a book from the floor up to a library shelf without changing the kinetic energy of the book. Newton's second law implies that the force of the Earth on the book (its weight) and the force of the student on the book are of equal magnitude and opposite direction. Are these a Newton's third law pair of forces? Explain. The total force on the book is zero, indicating zero total work was done. Yet certainly *some* work was done, since the book is now on the shelf. Resolve this dilemma.

16. A system of mass m lies on a rough table. The system is dragged at a constant low speed between two points A and B on the table. The kinetic frictional force on the system is independent of its low speed. The path is *not* a straight line. Carefully present arguments to show that the work done by the kinetic frictional force of magnitude f_k is equal to

$$W_{A \text{ to } B} = -f_k s$$

where s is the total length of the path taken between the two points. Thus the work done by the kinetic frictional force depends on the specific path taken between the two points, and so it is not a conservative force.

17. The orbital path of a satellite is an ellipse (see Figure Q.17). The work done by the gravitational force around one orbit is

$$W = \int_{\substack{\text{one} \\ \text{orbital path}}} \vec{F}_{\text{grav}} \cdot d\vec{r}$$

(a) At what points along the orbit is the integrand zero? (b) Is the *integrand* zero at *every* point along the orbital path? (c) Is the work done by the gravitational force over one complete orbital path zero? Explain why or why not.

18. A mass m is located somewhere inside a thin, uniform, hollow spherical shell. Give a convincing argument to show that the gravitational potential energy of m is constant, independent of its position within the shell.

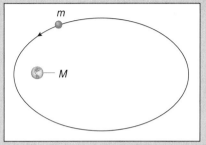

FIGURE Q.17

19. A student applies the CWE theorem and surprisingly discovers from a calculation that the kinetic energy of the system is negative at some location. What does a negative kinetic energy imply about the speed of the system? It is *certain* that the student made an error in applying the CWE theorem. Why?

20. By what factor must the speed of a system of mass m be increased to *triple* its kinetic energy?

21. A legislator argues that a statewide speed limit should not be increased from 100 km/h to 140 km/h because "the force of a collision (with a stationary object) doubles." Does it? Explain your reasoning.

22. A rock is dropped from rest. An identical rock is thrown downward with an initial speed of 5.00 m/s. Is the work done by the total force on each rock over the first meter of its flight (after their release) the same? Find the change in the speed of each rock after they have traveled a distance of one meter from the point of their release. The gravitational force did the same work on each; why are the changes in the speeds different? Are the changes in the kinetic energy of both rocks over the one meter path the same? Explain.

23. A ball at rest is dropped from your hand and falls to the floor. Use the CWE theorem to explain why the ball never can bounce up to a height greater than that from which it was released. How can the ball be released from your hand so it *does* bounce to a higher elevation than the point of release?

24. Let $\Delta \vec{v} = \vec{v}_f - \vec{v}_i$. Explain why the change in the kinetic energy is *not*

$$\frac{1}{2} m (\Delta v)^2$$

25. Is the force $\vec{F} = (1.00 \text{ N})\hat{i} - (1.00 \text{ N})\hat{j}$ a conservative force? Explain why or why not.

26. A physics student throws this (!) textbook off a cliff in one of the ways shown in Figure Q.26. He wants the impact speed to be as large as possible. Arg! If the *initial* speed of the text is the same along each trajectory, along which is the impact speed the greatest or does it matter? Justify your answer.

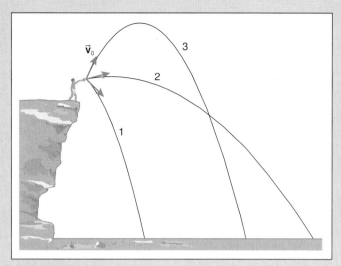

FIGURE Q.26

27. Consider an inertial reference frame S' moving at constant velocity $v\hat{\imath}$ with respect to another inertial reference frame S, with their respective coordinate axes parallel to each other, and their x-axes coincident as shown in Figure Q.27. A system of mass m is moving in reference frame S' with an initial velocity $v_0\hat{\imath}'$. (a) What is the kinetic energy of the system as measured in S'? (b) What is the kinetic energy of the system as measured in reference frame S? (c) We showed in Chapter 4 that the acceleration of a system is the *same* in both inertial frames, so that both frames will measure the *same* total force acting on the system. If the total force is not zero, will the two reference frames agree on the total work done by the total force acting on the system during a one second interval? Why or why not?

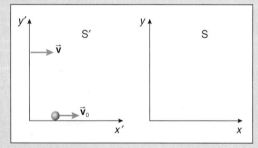

FIGURE Q.27

28. Can the CWE theorem be applied in noninertial reference frames? Explain.

29. Say the work done by all the nonconservative forces on a system is negative as it moves between two points. What happens (qualitatively) to the total mechanical energy of the system? Describe a physical situation that illustrates this scenario.

30. The nonconservative forces on a system are found to do positive work on the system. What happens (qualitatively) to the total mechanical energy of the system? Describe a physical situation that illustrates this scenario.

31. A rough sketch of the position x of a mass executing simple harmonic oscillation as a function of time is shown in Figure

Q.31. (a) Make a corresponding rough sketch of the velocity component v_x as a function of time. (b) Make a corresponding rough sketch of the acceleration component a_x as a function of time. (c) Make a corresponding rough sketch of the force component F_x as a function of time. (d) Make a corresponding rough sketch of the kinetic energy as a function of time. (e) Make a corresponding rough sketch of the potential energy as a function of time.

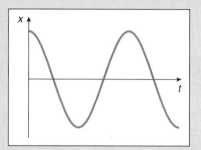

FIGURE Q.31

32. For an elliptical orbit of a planet about the Sun, name several physical quantities that are *constant* at all locations in the orbit and several physical quantities that depend on where the planet is located in its orbit.

33. A baseball of mass m is tossed vertically up into the air and caught at the same place on the way down. Ignoring the effects of air friction, what is the work done by the gravitational force on the ball over the path followed by the ball? If you consider air friction effects, does this change the answer to the previous question? If air friction *cannot* be neglected, what can be said (if anything) about the speed of the ball as it is caught compared with the launch speed? Justify your answer via the CWE theorem.

34. For a particle of mass m in an elliptical orbit about the Sun, is the instantaneous power of the gravitational force on m equal to zero at *every* point along the orbit? If not, are there points where it *is* equal to zero? If so, what points? Is the average power of the gravitational force over one period of the orbit equal to zero? Over what portions of the orbit is the work done by the gravitational force positive? Negative?

35. A log of mass 5.00 kg is dropped from rest off a cliff of height 30.0 m. What is the *initial* power of the gravitational force on the log?

36. A graph is made of the instantaneous power $P(t)$ versus time t. What is the physical significance of the area under the graph between $t = 0$ s and some other time $t > 0$ s?

37. Are the colloquial uses of the words work, energy, and power synonymous? What is meant by each term in physics? Think of several other scientific words that have different meanings in a colloquial context.

38. A graph of the potential energy of an α-particle (the nucleus of a helium atom) in the immediate vicinity of a massive nucleus has the schematic form indicated in Figure Q.38. If the total energy E of the α-particle is that indicated in the

sketch, qualitatively describe the motion of the α-particle (i) if it is located in the region to the left of the hump on the diagram, and (ii) if it is located to the right of the hump. Explain why it is (classically) impossible for the α-particle to be in the region between r_1 and r_2. One of the strange features of quantum mechanics is that it is possible for the α-particle to "tunnel" through the region where it is (classically) forbidden to be; among other things, this phenomenon describes radioactivity.

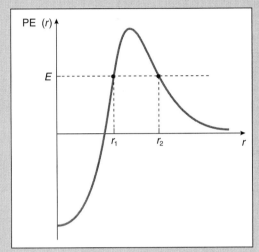

FIGURE Q.38

39. A satellite is in *circular* orbit about the Earth. How is the satellite represented on an energy diagram like Figure 8.47? How is a satellite in an *elliptical* orbit represented on such a diagram?

40. Only conservative forces act on a certain system. The total potential energy of the system is plotted as a function of a co-ordinate x; Figure Q.40 is obtained. The total mechanical energy of the system also is indicated. What can be said about the motion of the system?

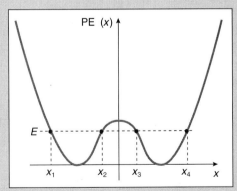

FIGURE Q.40

41. The strong force, also known as the nuclear force, is one of the fundamental forces of nature, as mentioned in Section 5.2. The strong force is conservative, which means that there is a potential energy associated with it. A nuclear particle such as a neutron has a potential energy that is a function of its position relative to the center of the nucleus. Write the strong force as

$$\vec{F}_{strong} = F_r \hat{r}$$

The potential energy of a neutron at a distance r from the center of a nucleus has the general schematic dependence on r indicated in Figure Q.41; in the figure R is the radius of the nucleus (about 10^{-15} m). (a) For distances $r > R + \Delta r$, what is the magnitude and direction of the strong force? (b) For distances $R < r < R + \Delta r$, what can be said about the magnitude and direction of the strong force? (c) For distances $R - \Delta r < r < R$, what can be said about the magnitude and direction of the strong force? (d) For distances $r < R - \Delta r$, what can be said about the magnitude and direction of the strong force? (e) Describe how the strong force on the neutron is quite *different* from the gravitational force of the nucleus on the neutron.

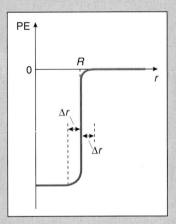

FIGURE Q.41

42. For forced oscillations, give a convincing argument to explain why the power transferred to the oscillator by a sinusoidal applied force is a maximum if the applied force component has the same frequency and phase angle as the velocity component.

43. Rewrite the following passage using proper physical terminology:

Master Andry raised his eyes, seemed for a moment to be measuring the height of the column, the weight of the rascal, mentally multiplying that weight by the square of the velocity, and was silent.

Victor Hugo, *Notre-dame de Paris*, Volume I, Book I, Chapter I, "The Great Hall" (Estes and Lauriat, Boston, 1892), page 11.

PROBLEMS

Sections 8.1 Motivation for Introducing the Concepts of Work and Energy
8.2 The Work Done by Any Force
8.3 The Work Done by a Constant Force
8.4 The Work Done by the Total Force
8.5 Geometric Interpretation of the Work Done by a Force*

1. A force $\vec{\mathbf{F}} = (15.0\text{ N})\hat{\imath}$ is one of the forces on a pumpkin (*Cucurbita pepo*) of mass 10.0 kg. For no apparent reason, calculate the work done by this force on the pumpkin for each of the changes in its position vector in Figure P.1.

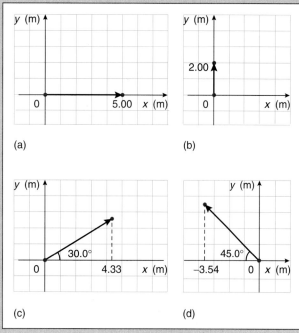

(a)   (b)

(c)   (d)

FIGURE P.1

2. A constant force $\vec{\mathbf{F}} = (20.0\text{ N})\hat{\imath}$ pulls a 5.00 kg monster zucchini along the *x*-coordinate axis from $x = 2.00$ m to $x = 5.00$ m. Find the work done by $\vec{\mathbf{F}}$ on the squash over this path.

3. A force $\vec{\mathbf{F}} = (5.00\text{ N})\hat{\jmath}$ acts on a 2.0 kg bundle of homework problems while it moves along the *y*-axis from the coordinate $y_1 = 4.00$ m to $y_2 = -2.00$ m. (a) Graph F_y versus *y*. (b) Find the work done by this force on the bundle between the two points.

4. A force whose only component is F_x acts on 1.50 kg system; a graph of F_x versus *x* appears in Figure P.4. Find the work done on the system by this force while the system moves along the *x*-axis from $x = -2.00$ m to $x = +10.00$ m.

5. A force $\vec{\mathbf{F}} = -(6.00\text{ N/m})x\hat{\imath}$ acts on a 3.0 kg boxful of nuts and bolts as it moves along the *z*-axis between the initial position $z_1 = 9.00$ m and final position $z_2 = 2.00$ m. (a) Make a graph of F_x versus *x*. (b) Find the work done by this force as the system moves between the two positions.

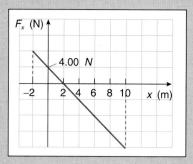

FIGURE P.4

6. A 5.00 kg watermelon (*Citrullus vulgaris*) moves from $x = 8.00$ m, $y = 2.00$ m, $z = 0$ m to $x = -4.00$ m, $y = 1.00$ m, $z = 0$ m by moving along the *curved* path shown in Figure P.6. One of the forces acting on the watermelon system is $\vec{\mathbf{F}} = (60.0\text{ N})\hat{\imath} - (40.0\text{ N})\hat{\jmath} + (25.0\text{ N})\hat{\mathbf{k}}$. Calculate the work done on the system by this force as the system moves from its initial position to final position.

FIGURE P.6

7. A 0.450 kg quahog clam (*Venus mercenaria*) moves from $x = -2.00$ m, $y = 1.00$ m to $x = 3.00$ m, $y = 1.00$ m along the path indicated in Figure P.7. One of the forces acting on the clam system is $\vec{\mathbf{F}} = (20.0\text{ N})\hat{\imath} + (10.0\text{ N})\hat{\jmath}$. What is the work done on the system by this force over the path followed by the clam?

FIGURE P.7

8. A force $\vec{\mathbf{F}} = (-30.0\text{ N})\hat{\imath} + (20.0\text{ N})\hat{\jmath}$ acts on a 10.0 kg log as it moves from $x = 10.0$ m to $x = 5.0$ m along the *x*-coordinate axis. Find the work done on the log by this force over this path.

9. A force on a system has only an *x*-component F_x; a graph of this component as a function of position *x* is shown in Figure P.9. Calculate the work done by this force if the system moves along the *x*-axis from $x = 0$ m to $x = 10.0$ m.

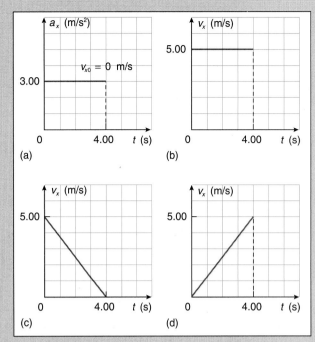

FIGURE P.9

FIGURE P.13

10. Your professor is having her office cleaned and asks you to slide a 175 kg filing cabinet 6.5 m across a rough floor in a straight line. You push with a horizontal force of magnitude 280 N and do the job for brownie points. How much work do you do?

•11. The force $\vec{F} = (1.00 \text{ N})\hat{\imath} - (1.00 \text{ N})\hat{\jmath}$ is one of the forces on a 4.00 kg system located at the origin. (a) Sketch the direction of the force on a suitable coordinate grid. (b) Indicate a path that the system may follow for 2.00 m so that $\vec{F}$ does *zero* work on the system along the path. If the system returns to the origin along this path, is the work done by this force zero along the return trip? Explain your reasoning. (c) Indicate a 2.00 m long path that the system may follow so that $\vec{F}$ does the maximum amount of *positive work* on the system along the path. What is the work done by the force along this path? If the system returns to the origin along this path, does the force do positive work along the return trip? Explain. (d) Indicate a 2.00 m long path that the system may follow so that $\vec{F}$ does the maximum amount of *negative work* on the system along the path. What is the work done by this force along this path? If the system returns to the origin along this path, does the force do negative work along the return trip? Explain.

•12. A professor moves a 3.00 kg mass attached to a spring (with $k = 10.0$ N/m) very slowly along a straight path from $x = -0.10$ m to $x = +0.15$ m; see Figure P.12. At any position, the acceleration of the mass is essentially zero. Calculate the work done on the mass m: (a) by the force of the professor on m; and (b) by the force of the spring on m. (c) What is the work done by the sum of the two forces on the system?

FIGURE P.12

•13. Each graph in Figure P.13 describes the one-dimensional motion of a 5.00 kg system during a 4.00 s interval. For each case, what is the work done on the system by the total force on the system during the interval?

•14. Show that the work done by the gravitational force in moving a system of mass m at a distance r_i from the center of a uniform sphere (of mass M) to a distance r_f is

$$W = -\frac{GmM}{R^3}\left(\frac{r_f^2}{2} - \frac{r_i^2}{2}\right)$$

where both r_i and r_f are less than the radius R of the sphere (see Figure P.14). What does this imply about the work done by this form of the gravitational force around a closed path?

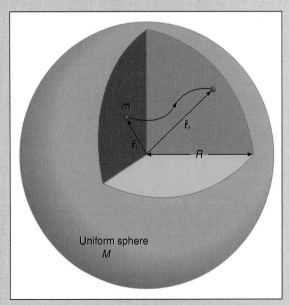

FIGURE P.14

Sections 8.6 Conservative, Nonconservative, and Zero-Work Forces
 8.7 Examples of Conservative, Nonconservative, and Zero-Work Forces
 8.8 The Concept of Potential Energy
 8.9 The Gravitational Potential Energy of a System near the Surface of the Earth
 8.10 The General Form for the Gravitational Potential Energy
 8.11 The Relationship between the Local Form for the Gravitational Potential Energy and the More General Form*
 8.12 The Potential Energy Function Associated with Hooke's Force Law

•**15.** A force $\vec{F} = (2.00 \text{ N})\hat{i} + (1.00 \text{ N})\hat{j}$ is one force on a system as it executes the rectangular path shown in Figure P.15. (a) Find the work done on the system by this force if the path is traversed in the clockwise sense. (b) Find the work done by this force if the loop is traversed in the counterclockwise sense. (c) Is this force conservative?

FIGURE P.15

•**16.** A force $\vec{F} = -kx\hat{i} - ky\hat{j}$ is one of the forces on a particle of mass m. The quantity k is constant. This force represents a two-dimensional Hooke's force law. (a) Is this force conservative? (b) What is an appropriate potential energy function for this force?

•**17.** Show that if the distance of a particle of mass m from another mass M increases by a small amount Δr, the change in the gravitational potential energy, ΔPE, of m is approximately

$$\Delta \text{PE} \approx \frac{GMm}{r^2}\, \Delta r$$

•**18.** How high above the surface of the Earth must a mass be located so its gravitational potential energy is only 1.00% of its value on the surface?

•**19.** A 3.00 kg mass is on a rough horizontal table. The coefficient of kinetic friction for the mass on the table surface is $\mu_k = 0.25$. The mass is pushed slowly between points A and B via the two different routes shown in Figure P.19. Find the work done by the kinetic frictional force along each path. The work is not the same along each path; hence the work done by the kinetic frictional force depends on the path. Thus the kinetic frictional force is not conservative.

•**20.** (a) Show that the gravitational potential energy of a system of mass m located inside a uniform sphere of radius R and mass M can be taken to be of the form

$$\text{PE} = \frac{1}{2} \frac{GmM}{R^3}\, r^2$$

FIGURE P.19

You can make the calculation easier by thinking of the result of the harmonic oscillator problem; why? (b) Where is this potential energy function equal to zero?

•**21.** If the position of a mass m attached to a spring changes by a small amount Δx, show that the change in the potential energy ΔPE of the mass is approximately

$$\Delta \text{PE} \approx kx\, \Delta x$$

•**22.** A system of mass m has a gravitational potential energy mgy, where $y > 0$ m and $\hat{j}$ is directed vertically up. If the mass then is attached to a horizontal spring of spring constant k, by how much must the spring be extended or compressed so that the potential energy of the mass on the spring is equal to its gravitational potential energy?

Sections 8.13 The CWE Theorem
 8.14 The Escape Speed
 8.15 Black Holes*
 8.16 Limitations of the CWE Theorem: Two Paradoxical Examples*

23. The velocity of a system of mass 3.00 kg is measured to be

$$(3.00 \text{ m/s})\hat{i} - (2.00 \text{ m/s})\hat{j} + (5.00 \text{ m/s})\hat{k}$$

when $t = 3.00$ s. Later, when $t = 6.00$ s, the velocity is found to be

$$(-4.00 \text{ m/s})\hat{i} + (6.00 \text{ m/s})\hat{j} - (8.00 \text{ m/s})\hat{k}$$

(a) What is the change in the velocity of the system? (b) What is the change in the kinetic energy of the system? (c) Why is the change in the kinetic energy *not* equal to

$$\frac{1}{2} m(\Delta v)^2$$

(d) What is the work done by all the forces on the system?

24. A spacecraft with a mass of 1.20×10^3 kg is to be sent to Mars from the surface of the Earth. What minimum kinetic energy is required to send the spacecraft on its way? (Consider only the gravitational influence of the Earth on the spacecraft.)

25. Show that the Schwarzschild radius for a 75 kg physics student is about $R_s \approx 10^{-25}$ m. This is *ten* orders of magnitude smaller than the size of a typical atomic nucleus. Teeny tiny for sure. So it is *quite* impossible to implode yourself to form a black hole, thank goodness.

•**26.** Starting from rest at point (1) in Figure P.26, a 4.00 kg tooth fairy slides down the indicated frictionless surface. The zero of gravitational potential energy is chosen to be at the elevation indicated. Take the fairy to be the system. At each of the points labeled (1), (2), (3), and (4), specify the values of:

(a) the total mechanical energy of the system; (b) the potential energy of the system; (c) the kinetic energy of the system; and (d) the speed of the system.

FIGURE P.26

•27. A 0.75 kg upside down bicycle helmet slides 8.0 m down the rough incline indicated in Figure P.27 at a constant small speed of 0.005 m/s. Use the CWE theorem to determine the coefficient of kinetic friction for the helmet on the surface.

FIGURE P.27

•28. A 100 kg warthog (*Phacochoerus aethiopics*) squeals with delight as it slides from rest down the greased (frictionless) slope and plane in Figure P.28a. The hog encounters a spring that compresses 150 cm before bringing the hog to rest for an instant. (a) What is the spring constant of the spring? (b) The hog is sent back in the opposite direction by the spring. To what height will the hog rebound up the slope? (c) Suppose the hog begins anew at the 5.00 m height but now encounters a 2.00 m stretch of level rough ground on the way to the spring and back (see Figure P.28b). The coefficient of kinetic friction for the hog on the rough ground is 0.30. How much will the spring now compress? To what height will the wild bacon rise on the rebound?

•29. Latter-day infidels have succeeded in constructing a modern Tower of Babel with a height equal to the radius of the Earth. To celebrate the accomplishment, the barbarians are celebrating with a sacrifice to appease the gravitational gods. The peter-principled Dean of Deans (*Maximus incompetus*) is dropped from the tower with zero initial speed and plummets toward the Earth. Neglect all histrionics and atmospheric effects. Determine the speed at impact with the surface of the Earth.

•30. It is thought that *supermassive* black holes exist at the center of many galaxies. Such black holes have masses of 10^6 times that of the our Sun, or more. Calculate the Schwarzschild radius of a black hole with a mass of 1.0×10^6 solar masses and compare the radius with the average distance of the Earth from the Sun (1 AU = 1.496×10^{11} m).

•31. A system of two identical springs (each with spring constant k) is used to accelerate a mass m from rest to as great a speed as possible on a frictionless surface. Various arrangements of the springs are possible: (i) use just one spring; (ii) use two springs in series; (iii) use two springs in parallel. See Figure P.31. Regardless of which design is used, the spring system can be set (cocked) the *same* distance x_0 from the equilibrium position of m. (a) Find the maximum potential energy of m in each spring system. (b) Find an expression for the maximum speed of m as it leaves each system. (c) To give m the greatest possible speed, which arrangement should be used (or does it make any difference)?

FIGURE P.28

FIGURE P.31

•**32.** The ancient Greek god Zeus decides to drop in for an unannounced inspection. Let the unknown mass of Zeus be $m > 0$ kg. He begins his fall at zero speed from an infinite distance and free-falls, under the influence of the gravitational force of the Earth alone, all the way down to Athens (latitude 38°N, longitude 24°E) on the surface of the Earth. At what speed does Zeus arrive? How does this speed compare with the escape speed from the Earth?

•33. (a) Calculate the orbital speed of a hypothetical, frictionless satellite in circular orbit several meters above the surface of a perfectly spherical Earth (a so-called "grazing" orbit). (b) What is the *ratio* of the orbital speed of the satellite to its escape speed from the surface?

•**34.** (a) What is the binding energy of a 250 kg satellite of the Earth in a circular orbit with a radius equal to twice the radius of the Earth? (b) What additional energy must be given to the satellite to have it escape from the gravitational influence of the Earth?

•**35.** From the surface of the Earth, Superman throws a baseball vertically upward with a speed equal to *half* the escape speed. Neglecting air frictional effects, what is the maximum height of the ball? Express your result in terms of multiples of the radius of the Earth.

•**36.** An 80.0 kg sphere is suspended by a wire of length 25.0 m from the ceiling of a science museum as indicated in Figure P.36. A *horizontal* force of magnitude F is applied to the ball, moving it very slowly at constant speed until the wire makes an angle with the vertical direction equal to 35°. (a) Sketch a second law force diagram indicating all the forces on the ball at any point along the path. (b) Is the force needed to accomplish the task constant in magnitude along the path followed by the ball? (c) Use the CWE theorem to find the work done by the force $\vec{F}$.

FIGURE P.36

•37. A very fast sprinter can run with a speed of about 10 m/s. A spherical asteroid has a density equal to the density of the Earth, about 5.3×10^3 kg/m³. What is the maximum radius for the asteroid so the escape speed from it does not exceed 10 m/s?

•**38.** Uh oh. You carelessly forget to take the car keys from the ignition, and your car (of mass 1.50×10^3 kg) is in neutral with the emergency brake off. As you rush to another exciting physics lecture, your car starts from rest and rolls away in the sloping parking lot; gravity never rests. You turn around and notice to your horror that the car, after descending 2.00 m in vertical height and 50.0 m along the slope of the lot, is moving

at 4.00 m/s just before it crashes into the very expensive car of your professor. Consider the car to be a particle; consider yourself in deep trouble. (a) What is the change in the kinetic energy of the car? (b) What is the change in the gravitational potential energy of the car? (c) What is the work done by forces other than the gravitational force in the process?

•**39.** A 60.0 kg engine block head is held against a spring ($k = 4.00 \times 10^3$ N/m) while it is compressed horizontally through a distance of 1.50 m. The block head is released at rest when $t = 0$ s and slides horizontally over a smooth surface. Along the way, the block head encounters a 2.00 m rough stretch of ground characterized by $\mu_k = 0.200$ and $\mu_s = 0.300$, as shown in Figure P.39. After passing over the rough ground, another smooth (frictionless) section is encountered as well as another spring ($k' = 3.00 \times 10^3$ N/m). This spring, of course, reverses the direction of the block head. (a) What is the initial total mechanical energy of the block head when $t = 0$ s? (b) How much work is done by the kinetic frictional force on the block head during *each* passage over the rough ground? (c) What is the maximum compression of the spring on the right when the block head encounters it for the first time? (d) How many times will the block head pass *completely* over the rough stretch of ground before coming to a halt?

FIGURE P.39

•**40.** A hiker pushes a 100 kg boulder off a cliff as pictured in Figure P.40. (This is not a particularly cool thing to do!) The rock falls vertically for 30.0 m and then slides down a talus slope. At the bottom of the slope and at a speed of 2.00 m/s, the rock passes some astonished park rangers who subsequently set off in hot pursuit of the dastardly hiker. (a) How much work was done by the kinetic frictional force as the rock slid down the slope to the point where it passed the rangers? (b) Assuming the kinetic frictional force is of constant magnitude, find the coefficient of kinetic friction for the boulder on the slope.

FIGURE P.40

•41. Consider the orbit of the Earth (mass m) about the Sun (mass M) to be circular and of radius r. In terms of G, m, M, and r, find (for the sheer joy of it): (a) the potential energy of the Earth; (b) the orbital speed of the Earth; and (c) the orbital kinetic energy of the Earth. (d) What is the *ratio* of the kinetic energy to the potential energy of the Earth? The result you obtain for this ratio is a consequence of a general theorem in (advanced) mechanics known as the *virial theorem*.

•42. A 1.00×10^3 kg surveillance satellite is placed in a circular orbit with an altitude of 910 km above the *surface* of the Earth. (a) Calculate the magnitude of the gravitational field of the Earth at the orbital radius of the satellite. (b) Calculate the magnitude of the gravitational force of the Earth on the satellite in its orbit. (c) Determine the speed of the satellite in its orbit in m/s. (d) What is the period of the satellite in hours? (e) Approximately how many revolutions around the Earth does the satellite make per day? (f) What is the kinetic energy of the satellite? (g) What is its gravitational potential energy? (h) What is the *ratio* of the kinetic energy of the satellite to its gravitational potential energy?

•43. Hi ho, hi ho, it's off to work we go. . . . A student flings a 2.00 kg physics tome down a rough inclined plane with an initial speed of 2.00 m/s as indicated in Figure P.43. Amazingly, as luck would have it, the book stops right at the bottom of the incline, just before the precipice. (a) What is the work done by the gravitational force on the book? (b) What is the work done by the normal force of the plane on the book? (c) What is the change in the kinetic energy of the book? (d) What is the change in the gravitational potential energy of the book? (e) What is the value of the coefficient of kinetic friction for the book on the incline?

2.00 m/s

10.0 m

15°

FIGURE P.43

•44. Consider the orbit of the Earth about the Sun to be circular. Let m be the mass of the Earth, M the mass of the Sun, and r the radius of the orbit of the Earth. (a) Derive an expression for the orbital speed v of the Earth in terms of M, r, and appropriate physical constants. (b) The Greek Titan, Atlas, screeches the Earth to a halt in its orbit and then kicks it with a speed *just* sufficient to expel the Earth entirely from the solar system. Neglecting the influence of the other planets, derive an expression for the minimum speed v_{escape} the Earth needs to escape to infinity from its former orbit at distance r. (c) What is the *ratio* of this escape speed to the orbital speed?

•45. You have been assigned the exciting task of designing a new roller coaster. The track is to include a loop-the-loop of radius R as shown in Figure P.45. To a first approximation, ignore all frictional effects. The approach to the loop is made by starting the roller coaster vehicle (of mass m) at zero speed down an incline of height h as indicated. (a) What is the speed of the vehicle at the beginning of the loop (point A)? (b) What is the speed of the

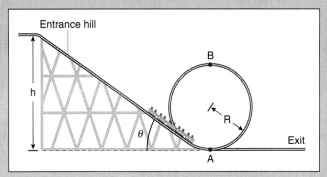

Entrance hill

h

B

R

θ

A

Exit

FIGURE P.45

vehicle at point B if the force of the track on the vehicle at this point is zero? (c) Find the ratio h/R so that the force of the track on the vehicle at point B just becomes zero. (d) What is the speed of the vehicle as it exits the loop? (e) What is the total work done by the gravitational force on the vehicle from the top of the incline to the exit from the loop? (f) Which of your answers to parts (a)–(e) depend on the mass m of the vehicle?

•46. A huge 10.0 kg otter (*Lutra canadensis*) is released at rest at the top of a frictionless inclined plane as shown in Figure P.46. The otter slides onto a horizontal surface with a coefficient of kinetic friction $\mu_k = 0.20$ for the otter tummy on the surface. (a) What is the speed of the otter as it reaches the bottom of the incline? (b) How far will it slide along the horizontal surface before coming to rest?

Frictionless

5.00 m

25.0°

Rough

FIGURE P.46

•47. A popular recreation for the fearless is jumping from tall structures while attached to an elastic bungee cord to ensure that less than catastrophic consequences result from the fall. Model the bungee cord as an ideal spring. A student of mass m drops off a tower of height h, tethered to the bungee cord. The

FIGURE P.47

elasticity of the bungee cord is chosen so that its effective spring constant is $k = 2mg/h$. The jump is begun at zero speed at altitude h with the bungee cord fastened vertically above her, barely taut (with no slack), but under no initial tension. See Figure P.47. (a) Choose a convenient coordinate system to analyze the problem. (b) What is the initial total mechanical energy of the student? (c) Show that the student will arrive at the ground with zero kinetic energy. Whew! (d) At what height above the ground is the speed of the student a maximum? (e) Will the *same* bungee cord be safe to use for a jump from the same tower (of height h) for a person with mass $m' < m$? With mass $m' > m$? Explain your reasoning.

•48. A spring gun is used in the lab to project a 0.070 kg mass from rest to a horizontal speed of 7.0 m/s. The spring is compressed a distance of 5.0 cm before release. Assume frictional forces are zero. (a) What is the kinetic energy of the projectile as it emerges from the gun? (b) What is the potential energy of the mass on the fully compressed spring? (c) What is the spring constant of the spring?

•49. A projectile is launched at initial speed v_0 at an angle θ with the horizontal direction. Neglect air friction and assume the trajectory is confined to a region near the surface of the Earth. Use the CWE theorem to show that the maximum vertical height of the projectile at the top of its flight path is

$$\frac{v_0^2 \sin^2 \theta}{2g}$$

•50. A train-forming railyard is designed so that boxcars of mass 3.00×10^4 kg, moving at an initial speed of 1.50 m/s, roll down an incline of height 2.00 m and across a level section of track before coupling at a speed of 0.500 m/s with other boxcars. See Figure P.50. Consider the boxcars to be particles, neglecting the rolling motion of the wheels. (a) What is the work done by the gravitational force on each boxcar? (b) What is work done by the normal force of the track on each boxcar? (c) What is the work done by other forces on each boxcar in the process?

FIGURE P.50

•51. A spacecraft of mass m is in a circular orbit about the Earth at a distance r from the center of the Earth (of mass M). (a) Use Newton's second law of motion to show that the speed v of the satellite satisfies

$$v^2 = \frac{GM}{r}$$

(b) Show that the total mechanical energy E of the spacecraft is

$$E = -\frac{1}{2}\frac{GMm}{r}$$

NASA wants to *change* the orbit of the spacecraft to one with an orbital radius of $2r$. (c) What is the total mechanical energy of the spacecraft in the new orbit? (d) How much energy must be given to the spacecraft to change its orbit from r to $2r$?

•52. (a) Use Equation 8.32 for the escape speed from a uniform spherical object of radius R and mass M to show that the escape speed from the surface of the Sun is about 1/500 the speed of light. (b) Assume M to be a uniform sphere. Show that the expression for the escape speed can be written in terms of the density ρ of the sphere in the following way:

$$v_{\text{escape}} = R\left(\frac{8\pi}{3}G\rho\right)^{1/2}$$

In 1784 John Michell realized that if the radius of the Sun were increased by a factor of 500 while keeping its density constant, that "all light emitted from such a body would be made to return to it, by its power of gravity,"[*] thus anticipating the idea of a black hole.

•53. A satellite of mass m is in an elliptical orbit about a planet of mass M ($m \ll M$), as shown in Figure P.53. The only force acting on the satellite is the gravitational force of M. The speed of the satellite at A is v_1 and its speed at B is v_2. (a) What is the potential energy of the satellite at point A? At point B? (b) What is the magnitude of the orbital angular momentum of the satellite at point A? At point B? (If necessary, see Section 6.10 for the meaning of orbital angular momentum.) Recall that one consequence of Kepler's second law of planetary motion, and indeed of all central force motion, is that the orbital angular momentum of the satellite is conserved: the magnitude (and direction) of the orbital angular momentum vector is the same at both A and B. (c) Use the CWE theorem and conservation of orbital angular momentum to show that the total mechanical energy E of the satellite can be written as

$$E = -\frac{GMm}{2a}$$

where $a = (r_1 + r_2)/2$ is the semimajor axis of the elliptical orbit. This result indicates that the total mechanical energy of

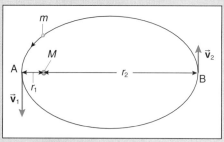

FIGURE P.53

---

[*]See Simon Schaffer, "John Michell and black holes," *Journal for the History of Astronomy*, 10, Part 1, #27, pages 42–43 (February 1979). Notice how the word "power" is *not* used by Michell in the way we now define the power of a force.

the orbit of a satellite (or planet about the Sun) depends only on the semimajor axis of the orbit and not other factors such as the eccentricity ε.

•**54.** A spring with $k = 500$ N/m is suspended vertically as indicated in Figure P.54. A pan of negligible mass is on the end of the spring. Let the position of the pan in this configuration be taken as the origin of a y-coordinate axis with $\hat{\jmath}$ directed up. (a) A 2.00 kg mass is dropped (with zero initial speed) into the pan from a position 0.500 m above the pan. What is the initial total mechanical energy of the mass? (b) After the mass is in the pan, show that the mass is instantaneously at rest where $y = -0.241$ m and $y = +0.163$ m. (c) If the 2.00 kg mass is simply placed (rather than dropped) into the pan, show that the mass is in equilibrium in the pan where $y = -0.039$ m. Notice that when the mass is dropped into the pan, the oscillation is symmetric about the position of the pan determined in part (c) and of amplitude 0.202 m. Thus, if a mass is hung vertically from a vertical spring, the mass undergoes simple harmonic oscillation about the new equilibrium position of the mass on the spring, as you learned in Section 7.3.

+y

0.500 m

0

FIGURE P.54

•**55.** The many fragments of Comet Shoemaker–Levy 9 plunged into the atmosphere of Jupiter in July 1994 with huge amounts of kinetic energy. Consider a fragment modeled as a spherical ice mass 1.00 km in diameter; the density of ice is 917 kg/m³. The incident speed of the fragment is 50 km/s. (a) What is the mass of the fragment? (b) What is its kinetic energy? (c) One metric ton (1000 kg) of the explosive TNT* has an energy yield of about 4.9×10^9 J. What is the incident kinetic energy of the comet fragment expressed in equivalent metric tons of TNT? (d) If such a comet fragment were approaching the Earth, comment on whether nuclear missiles could appreciably change the kinetic energy of the fragment.

*Trinitrotoluene ($C_7H_5N_3O_6$).

•**56.** Jupiter has a mass of 1.90×10^{27} kg and rotates once every 9.84 h. A 1200 kg spacecraft is placed into an equatorial circular orbit about Jupiter with the same rotational period, so that the satellite remains above a fixed point on the Jovian equator. (The giant planet has no solid surface.) (a) Use Newton's second law to determine the radius of the orbit of the satellite. (b) What is the orbital speed of the satellite? (c) What is the total mechanical energy of the satellite? (d) To what speed would the satellite need to be propelled to escape from this orbit to a great distance from Jupiter?

‡**57.** A system of mass m is traveling at a large initial speed v_0. A small amount of work ΔW_{total} is done by all the forces acting on the system. Begin with the CWE theorem in the form

$$W_{total} = \Delta KE$$

and show that the change in the speed Δv of the system is approximately given by

$$\Delta v \approx \frac{\Delta W_{total}}{mv_0}$$

Use this equation to estimate the change in the speed of a 10.0 kg mass initially traveling at 30.0 m/s when 1.00 J of work is done on it by the total force on the system.

‡**58.** A system of mass m is undergoing vertical motion near the surface of the Earth with the gravitational force of the Earth the *only* force on the system. For such vertical motion near the surface of the Earth, the total mechanical energy of a system of mass m is

$$E = KE + PE$$

$$= \frac{1}{2} m \left(\frac{dy}{dt}\right)^2 + mgy$$

where $\hat{\jmath}$ is vertically up. The system has a constant vertical (downward) acceleration and the position of the system at any time is

$$y = y_0 + v_{0y}t - \frac{1}{2} gt^2$$

Use this in the foregoing expression for E to show that E is independent of time (that is, the total mechanical energy is conserved) and to find E in terms of y_0, v_{y0}, m, and g. The result should not be unexpected.

‡**59.** We all likely have enjoyed the childhood experience of dropping a marble down a set of hardwood steps. The tap, tap, tap of the marble as it works its way down the steps is a delight to almost any child—or adult. Imagine a marble dropped down a set of steps of step-height h. The rebounding marble bounces to a height equal to the height h of the previous step as indicated in Figure P.59. (Use this figure to help solve part (b) only.) The *coefficient of restitution* ε of the bouncing marble is defined to be

$$\varepsilon = \frac{\text{speed immediately after the bounce}}{\text{speed immediately before the bounce}}$$

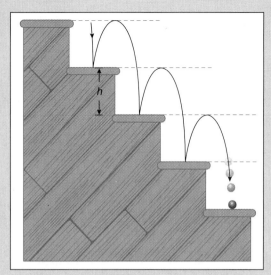

FIGURE P.59

(a) If the coefficient of restitution is zero, what is happening?
(b) Neglect the horizontal speed of the marble. Show that the coefficient of restitution for the situation indicated in Figure P.59 is

$$\varepsilon = \left(\frac{1}{2}\right)^{1/2}$$

(c) Sketch what happens to the marble if its coefficient of restitution is equal to 1.
The inspiration for this problem came from Robert F. Kingsbury, *Elements of Physics* (Van Nostrand, Princeton, New Jersey, 1965), page 172.

•60. The rules governing the construction and composition of tennis balls are quite stringent. When dropped from rest vertically from a height of 2.54 m, the ball must, after bouncing from a concrete floor, rebound to a height of at least 1.35 m, but no more than 1.47 m. What is the range of permissible coefficients of restitution of tennis balls? (See Problem 59 for a definition of the coefficient of restitution.)

‡61. A mass executing simple harmonic oscillation is described by

$$x(t) = A \cos \omega t$$

(a) Find the velocity component v_x of the oscillator as a function of time. (b) The potential energy of the oscillator,

$$PE = \frac{1}{2} kx^2$$

is time dependent. The average value of a time-dependent function $f(t)$ during an interval of time T is found using the following recipe:

$$f_{ave} = \frac{1}{T} \int_0^T f(t)\ dt$$

Find the time average value of the potential energy of the oscillator during one period of the oscillation. (c) The kinetic energy of the oscillator is

$$KE = \frac{1}{2} mv^2 = \frac{1}{2} mv_x^2$$

This also is time dependent. Find the time average value of the kinetic energy of the oscillator during a period of the oscillation. (d) Show that the time average value of the potential energy of the oscillator is equal to the time average value of the kinetic energy of the oscillator. That is, show that

$$PE_{ave} = KE_{ave}$$

This result is an application of a general theorem in (advanced) mechanics known as the *virial theorem*.

‡62. A particle of mass m is dropped (from rest) from a height h in the uniform gravitational field near the surface of the Earth as shown in Figure P.62. For convenience, the indicated coordinate system has been chosen to analyze this problem. (a) For review, show that the y-component of the velocity is

$$v_y = -gt$$

and the y-coordinate of the particle is

$$y(t) = h - \frac{1}{2} gt^2$$

where g is the magnitude of the local acceleration due to gravity (here $g = 9.81$ m/s^2). Show also that the time of flight of the particle is

$$T = \left(\frac{2h}{g}\right)^{1/2}$$

(b) The potential energy of the particle is time dependent. The following equation gives the average value of a time-dependent function $f(t)$:

$$f_{ave} = \frac{1}{T} \int_0^T f(t)\ dt$$

Use this equation to show that the time average value of the potential energy, during the time of flight T, is

$$PE_{ave} = \frac{2}{3} mgh$$

(c) The kinetic energy of the particle depends on the time. Show that the time average value of the kinetic energy of the particle, averaged over the time of flight, is

$$KE_{ave} = \frac{1}{3} mgh$$

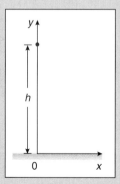

FIGURE P.62

The results of parts (b) and (c) indicate that

$$KE_{ave} = \frac{1}{2} PE_{ave}$$

which is a consequence of a theorem in (advanced) mechanics known as the *virial theorem*.

Section 8.17 The Simple Harmonic Oscillator Revisited

63. (a) What is the total mechanical energy of a 2.00 kg mass executing simple harmonic oscillation of amplitude 0.30 m on a spring of spring constant 40 N/m? (b) What is the period of the oscillation?

•64. For the simple harmonic oscillator of Problem 63, what are the kinetic and potential energies of the mass system when it is 0.10 m from its equilibrium position?

•65. A 4.00 kg mass is attached to a spring and executes simple harmonic oscillation with a period of 1.50 s. The total mechanical energy of the system is 12.0 J. (a) What is the spring constant? (b) Determine the amplitude of the oscillation.

•66. A simple harmonic oscillator has an amplitude A. At what coordinate(s) from the equilibrium position (as origin) is the kinetic energy equal to the potential energy?

•67. When a mass executing simple harmonic oscillation has a position equal to half the amplitude, what fraction of the total mechanical energy of the oscillator is kinetic?

•68. One end of a spring is placed on the floor with the axis of the spring vertical. The length of the spring is 60 cm. A 2.00 kg physics text is placed on the spring at rest and the spring compresses to three-fourths of its original length. The text then is raised slightly and dropped (from rest), compressing the spring now to one-fourth of its original length. Find the height above the floor from which the text was released.

Sections 8.18 The Average and Instantaneous Power of a Force
8.19 The Power of the Total Force Acting on a System

•69. A 2.00 kg mass falls from rest a distance of 100 m near the surface of the Earth. (a) What is the average power of the gravitational force during the time of the fall? (b) What is the instantaneous power of the gravitational force at the instant just before impact?

•70. You are playing with your little sister on a swing. At first, you pull the swing back and release her (without pushing—i.e., at rest) at an elevation that is a distance d above the bottom of the path of the swing. She is not content. So you then release her (again without pushing) a distance $2d$ above the low point of the motion. (a) By what factor will the maximum speed of the swing increase? (b) What is the power provided by the force of the ropes on the swing when the swing has its maximum speed? (c) What can be said (if anything) about the power provided by the force of the ropes on the swing at other points along the path of the motion?

•71. A mass m is dropped at rest from a height h when $t = 0$ s. Show that the instantaneous power of the gravitational force on m at any instant t is

$$P(t) = mg^2 t$$

•72. When $t = 0$ s, a mass m is projected at speed v_0 vertically upward. (a) Show that the instantaneous power of the gravitational force at any instant t is

$$P(t) = mg^2 t - mg v_0$$

(b) Sketch a graph of $P(t)$ versus t, labeling appropriate points. (c) At what point in the motion is the power zero? (d) What prevents the power from increasing indefinitely? (e) What is the physical significance of the area under the $P(t)$ versus t graph?

•73. An 80.0 kg premedical student decides to liven things up by sliding (from a standing start) down a slippery (i.e., frictionless) 25° mud slope as indicated in Figure P.73. At the end of the slope, the student encounters a horizontal stretch of terrain where the coefficient of kinetic friction for the student's derriere on the turf is $\mu_k = 0.20$. (a) Set up an appropriate coordinate system clearly indicating the origin and the direction of the coordinates axes. (b) Write an expression for the total mechanical energy of the premed at the top of the slide. (c) What is the speed of the premed as she leaves the incline? (d) How far will she travel along the rough horizontal plane before coming to a stop? (e) What is the total work done by the normal force of the surface on the student (i) during the descent and (ii) along the horizontal segment of the path? (f) What is the total work done by the kinetic frictional force on the student along the horizontal section? (g) What is the power of the kinetic frictional force as the student commences the horizontal segment? Is this power constant over the entire horizontal segment? Explain your reasoning.

FIGURE P.73

•74. The position coordinate of a 2.00 kg mass executing simple harmonic oscillation is described by the equation

$$x(t) = (0.50 \text{ m}) \cos[(7.00 \text{ rad/s})t]$$

where $x = 0$ m is the equilibrium position. (a) What is the spring constant? (b) When the mass is at the maximum distance from the equilibrium position, what is the power of the spring force? (c) When the mass goes through its equilibrium position, what is the power of the spring force? (d) Break up the oscillation into four successive parts:

Part 1: from $x = A$ to $x = 0$ m
Part 2: from $x = 0$ m to $x = -A$
Part 3: from $x = -A$ to $x = 0$ m
Part 4: from $x = 0$ m to $x = A$

During what parts of the motion is the power of the spring force positive? Explain why. (Do not evaluate the power specifically.)

During what parts of the motion is the power of the spring force negative? Explain why. (Do not evaluate the power specifically.)

•75. For each case shown in Figure P.13 (Problem 13), answer the following questions. (a) Find an expression for the kinetic energy as a function of time that is valid during the four second interval. (b) What is the average power of the total force on the system? (c) Find $P(t)$ at any instant during the four second interval. Is the instantaneous power of the total force constant during the four second interval? Explain why or why not.

•76. A passenger elevator car with a total mass of 1.20×10^3 kg is zipping upward at a constant speed of 4.00 m/s ferrying health inspectors to the famous catering firm of Cook, Wells, Dunn, and Burns. (a) What is the kinetic energy of the elevator car? (b) What is the force of the cable on the elevator car? (c) What is the power of the force of the cable? (d) What is the power of the gravitational force on the elevator car? (e) What is the power of the total force on the elevator car?

•77. A 1.00×10^3 kg sports car on a level road accelerates from rest to a speed of 150 km/h during 8.00 s. Consider the car to be a particle. What is the average power of the total force on the car during this time interval?

•78. A mass m is attached to a spring of spring constant k and is oscillating horizontally at angular frequency ω on a frictionless surface. The mass is released at rest when $t = 0$ s at the position $x = A$. (a) What is the position x of the oscillator as a function of time? (b) What is the velocity component

$$v_x = \frac{dx}{dt}$$

as a function of time? (c) What is the instantaneous power of the spring force when $t = 0$ s? Explain your answer. (d) What is the instantaneous power of the spring force when the oscillator is at position $x = 0$ m? Explain your answer. The results of parts (c) and (d) imply that the oscillator reaches a peak power sometime during the first quarter of a complete oscillation. (e) Show that the instantaneous power $P(t)$ of the oscillator force is

$$P(t) = \frac{k\omega A^2}{2} \sin(2\omega t)$$

(f) Show that the power is at a *maximum* for the first time when t is equal to one-eighth the period of oscillation. (g) What is the average power of the oscillator force during one period of the oscillation? (h) Make graphs of $F_{x\ \text{sprg}}(t)$, $P(t)$ [the result of part (e)], and $v_x(t)$ versus time during one period of the oscillation if the mass is 1.00 kg, the spring constant is 49.0 N/m, and the amplitude of the oscillation of 0.100 m. What conclusions can be drawn from the graphs?

•79. A particle of mass m, initially moving with a velocity $v_{x0}\hat{\imath}$, is subjected when $t \geq 0$ s to a constant acceleration $a_x\hat{\imath}$. (a) Find the kinetic energy as a function of time, $KE(t)$. (b) Determine the power of the total force by evaluating $\vec{F}_{\text{total}} \cdot \vec{v}$ and show that this is the same as the time rate of change of the kinetic energy, confirming Equation 8.42.

Section 8.20 Motion Under the Influence of Conservative Forces Only: Energy Diagrams*

•80. A 3.00 kg brick is hurled vertically upward at an initial speed of 5.00 m/s from a height of 20.0 m. Choose $\hat{\jmath}$ up and place the y-coordinate origin at ground level. (a) Find the total mechanical energy E of the brick. (b) Make a graph of the potential energy of the brick as a function of y. (c) Plot the total mechanical energy E on the graph. (d) For a general y-coordinate, indicate on the graph what represents the kinetic and potential energies of the brick. (e) Is there a turning point? (f) What y-coordinates are prohibited for the brick?

•81. The potential energy function of a conservative force acting on a system moving in one dimension along the x-axis is described by

$$PE = C|x|$$

where C is a positive constant. The conservative force is the only force on the system. (a) What are the SI units of C? (b) What is the x-component of the force on the system? (c) Make a schematic graph of the potential energy as a function of x. (d) Let the total mechanical energy E of the system be greater than zero. Sketch a representative value for $E > 0$ J on your graph of the potential energy function. Do turning points exist? If so, how many? Indicate prohibited values of x on the graph. (e) Describe the motion of the system. Is the motion simple harmonic oscillation? In what ways is the motion similar to simple harmonic oscillation? In what ways is it different?

•82. A pendulum of mass m and length ℓ makes an angle θ with the vertical direction. (a) Find an expression for the gravitational potential energy of m as a function of θ so that the potential energy is zero when $\theta = 0$ rad. (b) Make a sketch of this potential energy function in units of $mg\ell$ as a function of θ over the range

$$-\frac{\pi}{2} \text{ rad} \leq \theta \leq \frac{\pi}{2} \text{ rad}$$

(c) Let PE_{max} be the maximum potential energy of m over this range of θ. Let the total mechanical energy E of m be constant and have a value of $E = PE_{\text{max}}/2$. What are the turning points of a pendulum with such a value for the total mechanical energy?

‡83. The total force on a system of mass m varies with position according to

$$\vec{F} = F_0 \cos\left(\frac{2\pi x}{\ell}\right)\hat{\imath}$$

(a) Calculate the work done on the system by this force as the system moves from $x = 0$ m to $x = \ell/2$. (b) Calculate the work done on the system by this force as the system moves from $x = \ell/2$ to $x = 0$ m. (c) Is the force conservative? If so, proceed with the following questions. If not, you have the rest of the afternoon off. (d) Show that the following potential energy function is appropriate for this force:

$$PE(x) = -\frac{F_0\ell}{2\pi} \sin\left(\frac{2\pi x}{\ell}\right)$$

(e) Sketch the potential energy function over the domain $-\ell \leq x \leq \ell$. (f) The system is released at $x = 0$ m with zero speed. What is the total mechanical energy E of the system? Plot this value of E on the energy diagram of part (c). (g) Describe the motion of the system, calling particular attention to whether the system is bounded in its motion by turning points (and if so at what x values), and whether the system is oscillatory (and if so, whether the oscillation is simple harmonic or not). (h) If the total mechanical energy of the system somehow is decreased slightly to $E' < E$, what happens to the motion of the system? (i) For what values of E is the motion unbounded?

INVESTIGATIVE PROJECTS

A. Expanded Horizons

1. Investigate the relationship between *physical work* as we have defined it in physics and *isometric* work done by muscles when, for example, your arms hold a set of barbells at rest over your head or you push hard against an unyielding wall.
   Bruce Abernethy, Vaughan Kippers, Laurel Traeger Mackinnon, Robert J. Neal, and Stephanie Hanrahan, *The Biophysical Foundations of Human Movement* (Human Kinetics Publishers, Champaign, Illinois, 1997).
   Colin J. Pennycuick, *Newton Rules Biology* (Oxford University Press, Oxford, England, 1992).
   Theodore C. Ruch and Harry D. Patton, *Physiology and Biophysics* (W. B. Saunders, Philadelphia, 1965).

2. Investigate the notion of *scaling* and *mechanical similarity* and how these ideas can be applied to (a) gravitational potential energy and (b) the potential energy of a simple harmonic oscillator to deduce many of the features of motion under these forces. Also investigate notions of scaling in biology.
   Lev D. Landau and Evgenii M. Lifshitz, *Mechanics* (Addison-Wesley, Reading, Massachusetts, 1960), pages 22–24.
   Herbert Lin, "Fundamentals of zoological scaling," *American Journal of Physics*, 50, #1, pages 72–81 (January 1982).
   William H. Press, "Man's size in terms of fundamental constants," *American Journal of Physics*, 48, #8, pages 597–598 (August 1980).

3. Several problems in this chapter (Problems 41, 61, and 62) referred to a general theorem in mechanics known as the *virial theorem*. Find out what this theorem says in a general context and connect it to the results of these problems.
   Lev D. Landau and Evgenii M. Lifshitz, *Mechanics* (Addison-Wesley, Reading, Massachusetts, 1960), pages 23–24.
   Jerry B. Marion and Stephen T. Thornton, *Classical Dynamics of Particles and Systems* (Saunders, Philadelphia, 1995).
   Peter Kleban, "Virial theorem and scale transformations," *American Journal of Physics*, 47, #10, pages 883–888 (October 1979).

4. Investigate the possibility that the universe itself is a black hole. We might be enjoying life inside one!
   William J. Kaufmann III, *Universe* (3rd edition, W. H. Freeman, New York, 1990), page 546.

5. During the 19th century, Hermann von Helmholtz developed a theory that the Sun derived its energy from a gradual self-gravitational collapse now known as Helmholtz contraction. Investigate this theory and estimate the lifetime of the Sun if its energy were derived from this source. The energy source of the Sun actually is hydrogen fusion. Yet the mechanism of Helmholtz contraction is important in the *formation* of stars from the collapse of huge gaseous nebulae as well as in the late stages of stellar evolution.
   Elske v. P. Smith and Kenneth C. Jacobs, *Introductory Astronomy and Astrophysics* (W. B. Saunders, Philadelphia, 1973), pages 366–367.

B. Lab and Field Work

6. Determine the spring constant of a spring that is available in your physics laboratory (or the departmental attic!). Use the spring to design two harmonic oscillators, one with 1.0 J and the other with 10.0 J of total mechanical energy.

7. With appropriate equipment, design and perform an experiment to measure the average power of the force a weightlifter exerts on a set of barbells when lifting them from the floor to overhead.

C. Communicating Physics

8. Look up the words *work* and *energy* in the *Oxford English Dictionary* and read through the plethora of their meanings in the English language. If you are multilingual, consider whether these words are as laden with meaning in other languages. Write an essay comparing, contrasting, and discussing the various meanings and how they differ from their meanings in physics, either in English or another language.

9. As you read the newspaper or news magazines, begin a collection of technical bloopers consisting of misuses of words like force, work, energy, and power. See how many you can find during, say, the next month.

CHAPTER 9
IMPULSE, MOMENTUM, AND COLLISIONS

**Or by collision of two bodies grinde
The Air attrite to Fire**

John Milton (1608–1674)*

Collisions occur all around us. Whether marbles or pool, bowling or golf, baseball or football, tennis or paddleball, collisions are involved in some of our most cherished recreational sports. Cars occasionally collide with each other, sometimes with disastrous consequences on the occupants, but almost always on their wallets (though not the wallets of their lawyers). On a grand astronomical scale, the spectacular collisions of the pieces of Comet Shoemaker–Levy 9 with Jupiter during the summer of 1994 were extraordinarily violent.

It is likely that a stupendous collision of a large asteroid (about the size of Mars) with the Earth early in its history gave rise to the Moon. The collision of a smaller asteroid (about 10 km in diameter) with the Earth about 65 million years ago likely caused the mass extinctions that wiped out the dinosaurs.[†] Collisions of asteroids with the Earth are not only a thing of the past: they inevitably will occur again. Even whole galaxies, each with hundreds of billions of stars, collide with each other.[‡]

At the opposite extreme, collisions are ubiquitous in nuclear, atomic, and molecular physics. Atoms and molecules collide incessantly in gases. Physicists even use collisions of particles to fathom the ultimate nature of matter and to glean the characteristics of the fundamental forces of nature. There simply is no avoiding collisions, and so it is important to understand the physics of collision phenomena. We shall see that the concept of *momentum* is particularly useful for this purpose.[§]

In Chapter 5 we saw how Newton's second law relates the total force on a system to its acceleration. In this chapter we see that more generally, Newton's second law of motion relates the total force to the time rate of change of the system's momentum. In Chapter 8 we examined how the work done by the total force

---

*(Chapter Opener) *Paradise Lost*, Book X, lines 1072–1073.

[†] The Chicxulub impact crater from this collision spans land and sea on the Yucatan Peninsula in Mexico.

[‡] A dwarf galaxy presently is colliding with our own Milky Way Galaxy, but on the opposite side of the galactic center.

[§] We introduced the concept of momentum briefly in Section 6.10. Here we introduce momentum in a more formal way.

The collisions of the pieces of Comet Shoemaker–Levy 9 with Jupiter produced this series of huge dark blotches in the atmosphere of the giant planet.

---

acting during motion over a *distance* changes the *kinetic energy* of the system. Here we see that if we change our viewpoint and examine how the total force acts during a *time interval*, the *momentum* of the system changes.

Some forces on a system always are present, whereas others act only for limited time intervals. For example, in the vicinity of the Earth, the gravitational force of the Earth on a system of mass m never ceases its attraction, even for an instant. Although this force is ever present, it may not be the only force on the system. Other forces on the system may come and go as we or circumstances dictate. For example, you push your car only rarely, one hopes. The normal force of a surface on a system exists only as long as the system is in contact with the surface. Likewise, in the collision of a baseball with a bat, the force of the bat on the ball is present only during the very short interval they are in physical contact. The *impulse–momentum theorem* relates the change in the momentum of a system to the total force on it and the time interval during which this total force acts.

Until now our attention has focused on systems of just one or two distinct mass particles. We treated the system as a single particle of negligible size (were it a grand piano, a massive dictionary, or a subatomic particle) and examined its dynamical motion as a whole. We ignored the distribution of mass within the system itself. Toward the end of this chapter, we begin to consider systems either of *many* particles or of masses of *finite* size composed of an essentially continuous distribution of atoms and/or molecules. We begin to ask to what extent the physical size and composition of the system affect its motion. The dynamics of complex systems of many particles is simplified considerably by inventing the concept of the *center of mass*, a special point in space whose motion typifies a more complex system.[#] The concept of the center of mass will prove useful in the next chapter as well, where we investigate rotational motion in detail.

9.1 MOMENTUM AND NEWTON'S SECOND LAW OF MOTION

There is something qualitatively different about getting a 10 kg baby stroller moving from rest to a speed of 2 m/s and getting a 15 000 kg truck moving from rest to the same speed. It is doubtless easier for you to change the velocity of the baby stroller than that of the truck. The distinction between the situations has to do with their **momentum**.

The momentum $\vec{\mathbf{p}}$ of a mass m moving at a velocity $\vec{\mathbf{v}}$ is[¶]

$$\vec{\mathbf{p}} \equiv m\vec{\mathbf{v}} \tag{9.1}$$

The momentum is thus a vector quantity and its direction is always parallel to the velocity vector. The SI units of momentum are kg·m/s.

[#] We introduced the idea of the center of mass of a two-particle system in Chapter 6 when studying orbital motion. Here we will generalize to systems of many particles.

[¶] Equation 9.1 defines the *classical momentum* of a system, valid as long as the speed of the system is much less than the speed of light—that is, $v \ll c$. For systems having speeds that are significant fractions of the speed of light, the definition of the momentum must be modified. The reasons for the modification are examined in Chapter 25.

Systems with the same velocity but different masses have different momenta, as with the baby stroller and truck. Also we can agree from experience that a larger force is needed to *change* the momentum of the truck than the momentum of the baby stroller during the same time interval (e.g., bring them both from rest to velocity $\vec{\mathbf{v}}$ during the same time interval). So forces are involved in changing the momentum of a system.

We introduced Newton's second law of motion in Equation 5.1 as

$$\vec{\mathbf{F}}_{total} = m\vec{\mathbf{a}}$$

where $\vec{\mathbf{F}}_{total}$ is the vector sum of all the forces acting on a system of mass m and $\vec{\mathbf{a}}$ is the system's acceleration. Actually Newton formulated the second law of motion using the *momentum*.* The momentum of a single-particle system changes at a rate equal to the total force on it:

$$\vec{\mathbf{F}}_{total} = \frac{d\vec{\mathbf{p}}}{dt}$$

Now think of the total mass of a system being accelerated as composed of lots of different masses, perhaps fastened together or perhaps not, and each with its own distinct momentum. The vector sum of all the individual momenta is the total momentum $\vec{\mathbf{p}}_{total}$ of the entire system of masses. Each individual momentum changes at a rate equal to the total force on that piece. Hence the grand total momentum $\vec{\mathbf{p}}_{total}$ of the whole system changes at a rate equal to the vector sum $\vec{\mathbf{F}}_{total}$ of all the forces acting on all the pieces. This result,

$$\boxed{\vec{\mathbf{F}}_{total} = \frac{d\vec{\mathbf{p}}_{total}}{dt}} \qquad (9.2)$$

holds true for any system, be it a single point-particle mass, a huge boulder, or a wild collection of pieces flying helter-skelter in all sorts of directions.

> Equation 9.2 is the most general statement of Newton's second law of motion.

For systems of *constant* mass, Equations 5.1 and 9.2 for the second law are equivalent. To confirm this, substitute the definition of the momentum $\vec{\mathbf{p}}$ from Equation 9.1 into the right-hand side of Equation 9.2; we obtain Equation 5.1:

$$\begin{aligned}
\vec{\mathbf{F}}_{total} &= \frac{d\vec{\mathbf{p}}_{total}}{dt} \\
&= \frac{d}{dt}(m\vec{\mathbf{v}}) \\
&= m\frac{d\vec{\mathbf{v}}}{dt} \quad \text{(constant } m) \\
&= m\vec{\mathbf{a}} \quad \text{(constant } m)
\end{aligned}$$

The effects of the total force acting during a time interval on even systems of varying mass are investigated using Equation 9.2, since it is the most general statement of Newton's second law of motion.

*Newton called it the *state of motion*, a term no longer used.

We focus our attention here not on the individual forces on the system, but rather on the total force $\vec{\mathbf{F}}_{total}$, because it is the total force that determines the rate at which the total momentum of the system changes. For example, if a baby stroller and a truck both accelerate from rest to velocity $\vec{\mathbf{v}}$ during the same time interval, their momenta both change, but by very different amounts. Clearly, during the same time interval, it will take a total force of greater magnitude to change the momentum of the truck than that of the baby stroller because, according to Equation 9.2, the total force is equal to the time rate of change of the momentum.

QUESTION 1

Two masses have equal kinetic energies. Under what circumstances (if any) are the momenta of the particles the same?

EXAMPLE 9.1

An 18.2 kg stone is used in the Scottish game of curling. If the stone has a speed of 5.00 m/s and is moving toward increasing values of a coordinate *x*, what is the momentum of the stone?

Players coax the stone used in the game of curling.

Solution

The velocity of the stone is $(5.00 \text{ m/s})\hat{\mathbf{i}}$; so its momentum is (from Equation 9.1)

$$\begin{aligned}
\vec{\mathbf{p}} &= m\vec{\mathbf{v}} \\
&= (18.2 \text{ kg})(5.00 \text{ m/s})\hat{\mathbf{i}} \\
&= (91.0 \text{ kg·m/s})\hat{\mathbf{i}}
\end{aligned}$$

EXAMPLE 9.2

A 600 kg African black rhinoceros (*Diceros bicornis*) is charging a stalled safari vehicle filled with terrified tourists. The rhino has a velocity $\vec{\mathbf{v}} = (5.00 \text{ m/s})\hat{\mathbf{i}} + (6.00 \text{ m/s})\hat{\mathbf{j}}$ with respect to a convenient Cartesian coordinate system with $\hat{\mathbf{i}}$ north and $\hat{\mathbf{j}}$ west, laid out by a wildlife biologist.

a. Find the momentum of the rhino and the magnitude of its momentum.
b. If a baseball of mass 0.145 kg has momentum of the same magnitude, what is its speed?

Solution

a. Use Equation 9.1. The momentum is

$$\vec{p} = m\vec{v}$$
$$= (600 \text{ kg})[(5.00 \text{ m/s})\hat{\imath} + (6.00 \text{ m/s})\hat{\jmath}]$$
$$= (3.00 \times 10^3 \text{ kg}\cdot\text{m/s})\hat{\imath} + (3.60 \times 10^3 \text{ kg}\cdot\text{m/s})\hat{\jmath}$$

The magnitude of this vector is

$$p = [(3.00 \times 10^3 \text{ kg}\cdot\text{m/s})^2 + (3.60 \times 10^3 \text{ kg}\cdot\text{m/s})^2]^{1/2}$$
$$= 4.69 \times 10^3 \text{ kg}\cdot\text{m/s}$$

b. The speed of the baseball is found from the magnitude of its momentum:

$$p = mv$$

So

$$4.69 \times 10^3 \text{ kg}\cdot\text{m/s} = (0.145 \text{ kg})v$$

Solving for v, you find

$$v = 3.23 \times 10^4 \text{ m/s}$$
$$= 32.3 \text{ km/s}$$

This is almost three times its escape speed! A *real* fastball.

9.2 IMPULSE–MOMENTUM THEOREM

The longer the total force acts on a system, the longer the system is subjected to an acceleration and the greater the change in its velocity. Changes in velocity imply changes in momentum, according to the definition of the momentum, Equation 9.1. Here we find the relationship between the total force, the interval during which it acts, and the resulting change in the momentum of the system.

Let $\vec{F}_{\text{total}}$ be the total force on a system during a time interval Δt. Of course there is no guarantee that the total force is constant during the interval; the total force may well be time dependent. For example, the force of a bat on a baseball is not of constant magnitude during the brief interval they are in contact. Likewise the gravitational force on a planet in an elliptical orbit changes its magnitude and direction as the planet assumes different positions in space at various times, as shown in Figure 9.1. So we must account for situations where the total force is time dependent, $\vec{F}_{\text{total}}(t)$.

Rearrange the general form of Newton's second law, Equation 9.2, into the differential form

$$\vec{F}_{\text{total}}(t) \ dt = d\vec{p}$$

Integrate over the time interval between t_i and t_f. Due to the action of the total force, the total momentum of the system changes from $\vec{p}_i$ to $\vec{p}_f$ during the interval:

$$\int_{t_i}^{t_f} \vec{F}_{\text{total}}(t) \ dt = \int_{\vec{p}_i}^{\vec{p}_f} d\vec{p}$$

The integration on the left-hand side cannot be performed without full knowledge of how $\vec{F}_{\text{total}}(t)$ varies with time. On the

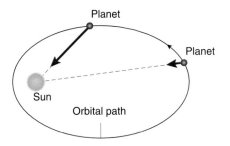

FIGURE 9.1 The gravitational force of the Sun on a planet in an elliptical orbit changes its magnitude and direction as the planet assumes different positions in its orbit.

other hand, the integration on the right-hand side *can* be performed easily: the integrand is a perfect differential. Hence we have

$$\int_{t_i}^{t_f} \vec{F}_{\text{total}}(t) \ dt = \vec{p}_f - \vec{p}_i \qquad (9.3)$$

The left-hand side of this relationship, the integral of the total force over the interval Δt, is called the **impulse** $\vec{I}$ of the total force.* Since force is a vector quantity, impulse is also a vector quantity[†]:

$$\boxed{\vec{I} \equiv \int_{t_i}^{t_f} \vec{F}_{\text{total}}(t) \ dt} \qquad (9.4)$$

This definition is a vector equation and so represents three equations, one for each of the Cartesian components of the impulse. That is,

$$I_x = \int_{t_i}^{t_f} F_{x \text{ total}}(t) \ dt$$

with similar equations for the y- and z-components. The SI units of impulse are N·s, which are equivalent to kg·m/s, the units for momentum.

The right-hand side of Equation 9.3 is the change in the momentum $\Delta\vec{p}$ of the system.

Hence Equation 9.3 grandly states that the impulse of the total force acting on the system is equal to the change in the total momentum of the system:

$$\boxed{\vec{I} = \Delta\vec{p}} \qquad (9.5)$$

The total impulse vector $\vec{I}$ has the *same* direction as the *change* in the momentum vector $\Delta\vec{p}$ and they both have identical magnitudes. Equation 9.5 is known as the **impulse–momentum theorem**.

As we have seen, to evaluate the integral for the impulse, defined by Equation 9.4, the time dependence of the force must

---

*The name impulse arises from the Latin *impulsus*, the past participle of *impellere*, to impel (here, to change the momentum of a system).

[†]We also could calculate the impulse provided by each individual force acting on the system using a similar integration. The total impulse then is the vector sum of the impulses of all the individual forces.

be known; this integration is thus often difficult to perform. This difficulty certainly is the case for a bat hitting a baseball, for example. However, if the total force is constant, it can be factored out of the integral and the integration for the impulse can be done explicitly (as we will see in Example 9.3).

If the way a component of the total force on the system varies with time is known and a graph is made of that component versus t, as in Figure 9.2, the corresponding component of the impulse of the total force between t_i and t_f is the area under the force component curve between the two times. Such areas may be positive or negative, of course, and the corresponding impulse component is likewise positive or negative.

Think of the collision of a bat with a ball, and take the ball to be the system. The force of the bat on the ball during the collision typically is much greater than the weight of the ball (see Example 9.4). Thus, to a good approximation, the total force on the ball during the collision is just the force of the bat on the ball. This force increases and decreases dramatically over the brief interval of the collision as indicated schematically in Figure 9.3. Here, for convenience, we represent the total force as if it has only one component. Forces that exist only for short intervals are called **impulsive forces**, in keeping with the colloquial meaning of the word impulse.

Often it is useful to calculate a time-averaged force component, say $F_{x\,ave}$; it provides an estimate of the order of magnitude of the actual force component during the interval Δt. By definition the average force component $F_{x\,ave}$, during the

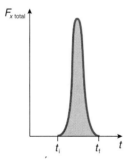

FIGURE 9.2 The x-component of the impulse during Δt is the area under the graph of $F_{x\,total}$ versus t between t_i and t_f.

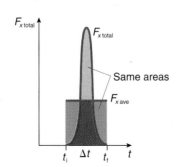

FIGURE 9.3 The area (component of the impulse) under the graph of the actual force component versus t is equal to the area under the average force component.

same time interval Δt, has the same area under its curve as does the actual total force component $F_{x\,total}$:

$$F_{x\,ave}\,\Delta t \equiv \int_{t_i}^{t_f} F_{x\,total}\,dt \equiv I_x \qquad (9.6)$$

The impulse–momentum theorem (Equation 9.5) indicates the x-component of the total impulse is equal to the x-component of the change in the momentum of the particle. Therefore the expression for the x-component of the average total force becomes

$$F_{x\,ave} = \frac{\Delta p_x}{\Delta t} \qquad (9.7)$$

This equation enables us to calculate the average total force *if the* interval during which it exists can be estimated in some way.

QUESTION 2

How can a small force on a system produce a large impulse? Conversely, how can a large force on a system produce a small impulse?

STRATEGIC EXAMPLE 9.3

A pickled egg of mass 0.250 kg of the extinct dodo bird (*Raphus cucullatus*) falls from rest off a museum shelf and plunges 2.00 m to the floor. Humpty Dumpty!

a. Use Equation 9.4 to calculate the impulse of the gravitational force of the Earth on the egg during the interval of the fall.
b. Determine the speed of the egg at the instant just before impact, and from this calculate the change in the momentum of the egg during its fall.
c. The egg is brought to a halt by the floor during 0.010 s. Find the impulse of the total force on the egg during the crash. Find the time-averaged total force on the egg during the crash and compare the magnitude of this force with the magnitude of the weight of the egg.

Solution

a. The gravitational force of the Earth on the egg (its weight) is a constant force near the surface of the Earth. Use the coordinate system indicated in Figure 9.4. The gravitational force is $-mg\hat{\jmath}$. The time of fall is found from the one-dimensional motion equation for a constant acceleration:

$$y = y_0 + v_{y0}t + \frac{1}{2}a_y t^2$$

Here

$$y_0 = 2.00 \text{ m}$$
$$v_{y0} = 0 \text{ m/s}$$
$$a_y = -9.81 \text{ m/s}^2$$

The impact occurs where $y = 0$ m at time t. Making these substitutions,

$$0 \text{ m} = 2.00 \text{ m} + (0 \text{ m/s})\,t + \frac{1}{2}(-9.81 \text{ m/s}^2)\,t^2$$

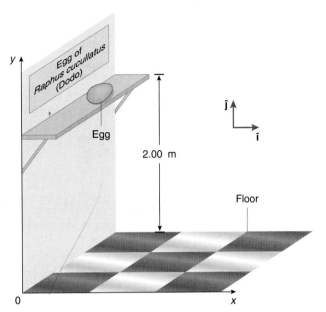

FIGURE 9.4

Solving for t, taking the positive root, you have

$$t = 0.639 \text{ s}$$

The integral for the impulse, Equation 9.4, then is

$$\vec{\mathbf{I}} = \int_{0 \text{ s}}^{t} (-mg\hat{\jmath}) \, dt$$

Since the force is constant, you find

$$\vec{\mathbf{I}} = -mg(t - 0 \text{ s})\hat{\jmath}$$

Substituting for m, g, and $t = 0.639$ s for the time of fall, you find the impulse to be

$$\vec{\mathbf{I}} = (-0.250 \text{ kg})(9.81 \text{ m/s}^2)(0.639 \text{ s})\hat{\jmath}$$
$$= (-1.57 \text{ kg} \cdot \text{m/s})\hat{\jmath}$$

b. The velocity component at impact is found using

$$v_y = v_{y0} + a_y t$$
$$= 0 \text{ m/s} + (-9.81 \text{ m/s}^2)(0.639 \text{ s})$$
$$= -6.27 \text{ m/s}$$

The final momentum of the egg is thus

$$\vec{\mathbf{p}}_f = m\vec{\mathbf{v}}_f$$
$$= mv_y\hat{\jmath}$$
$$= (0.250 \text{ kg})(-6.27 \text{ m/s})\hat{\jmath}$$
$$= (-1.57 \text{ kg} \cdot \text{m/s})\hat{\jmath}$$

The initial momentum of the egg is zero, because it began its fall with zero speed:

$$\vec{\mathbf{p}}_i = (0 \text{ kg} \cdot \text{m/s})\hat{\jmath}$$

The change in the momentum of the egg is

$$\Delta\vec{\mathbf{p}} = \vec{\mathbf{p}}_f - \vec{\mathbf{p}}_i$$
$$= (-1.57 \text{ kg} \cdot \text{m/s})\hat{\jmath} - (0 \text{ kg} \cdot \text{m/s})\hat{\jmath}$$
$$= (-1.57 \text{ kg} \cdot \text{m/s})\hat{\jmath}$$

Notice that the change in the momentum of the egg is equal to the impulse, consistent with Equation 9.5, the impulse–momentum theorem.

c. During the brief interval of the crash, take the initial time to be when the egg first makes contact with the floor and the final time to be when the egg stops. This time interval is given as $\Delta t = 0.010$ s. The impulse of the total force cannot be found by integrating $\vec{\mathbf{F}}_{\text{total}}(t)$ over the time interval, because the details of how the total force on the egg varies with time during the crash interval are unknown. However, the impulse–momentum theorem states the total impulse $\vec{\mathbf{I}}$ is equal to the change in the momentum of the system, and this we can calculate.

For this phase of the problem, the initial momentum of the egg is the momentum at the instant of contact:

$$\vec{\mathbf{p}}_i = m\vec{\mathbf{v}}_{\text{impact}}$$
$$= (0.250 \text{ kg})(-6.27 \text{ m/s})\hat{\jmath}$$
$$= (-1.57 \text{ kg} \cdot \text{m/s})\hat{\jmath}$$

The final momentum is zero, since the egg is then at rest:

$$\vec{\mathbf{p}}_f = (0 \text{ kg} \cdot \text{m/s})\hat{\jmath}$$

The change in the momentum of the egg is

$$\Delta\vec{\mathbf{p}} = \vec{\mathbf{p}}_f - \vec{\mathbf{p}}_i$$
$$= (0 \text{ kg} \cdot \text{m/s})\hat{\jmath} - (-1.57 \text{ kg} \cdot \text{m/s})\hat{\jmath}$$
$$= (1.57 \text{ kg} \cdot \text{m/s})\hat{\jmath}$$

Since $\vec{\mathbf{I}} = \Delta\vec{\mathbf{p}}$, the impulse of the total force during the crash is

$$\vec{\mathbf{I}} = (1.57 \text{ kg} \cdot \text{m/s})\hat{\jmath}$$

The average force during the crash is found using the vector version of Equation 9.6:

$$\vec{\mathbf{F}}_{\text{ave}} = \frac{\vec{\mathbf{I}}}{\Delta t}$$
$$= \frac{(1.57 \text{ kg} \cdot \text{m/s})\hat{\jmath}}{0.010 \text{ s}}$$
$$= (1.6 \times 10^2 \text{ N})\hat{\jmath}$$

This average total force is the vector sum of the force of the floor on the egg (the normal force) and the gravitational force of the Earth on the egg (its weight). The magnitude of the average total force is 1.6×10^2 N and the magnitude of the weight of the egg is mg, here 2.45 N. The ratio of the magnitude of the average total force during the crash to the magnitude of the weight is thus

$$\frac{1.6 \times 10^2 \text{ N}}{2.45 \text{ N}} = 65$$

Notice that the magnitude of the average total force on the egg during the crash is much greater than the magnitude of its weight.

EXAMPLE 9.4

A baseball of mass 0.145 kg arrives at home plate with a horizontal speed 20.0 m/s. The batter hits a line drive. Thwack! Immediately after the ball leaves the bat, the speed of the ball is 50.0 m/s in the opposite horizontal direction.

a. Calculate the impulse imparted to the ball by the total force on the ball.
b. Estimate the magnitude of the average total force on the ball if the collision of the ball with the bat lasts for 1.00×10^{-3} s.
c. Compare the magnitude of the average total force with the magnitude of the weight of the ball.

Solution

a. Diagram the situation and introduce the coordinate system shown in Figure 9.5. The ball is the system. The change in the momentum of the ball is

$$\Delta \vec{p} = \vec{p}_f - \vec{p}_i$$
$$= m\vec{v}_f - m\vec{v}_i$$
$$= (0.145 \text{ kg})(50.0 \text{ m/s})(-\hat{i}) - (0.145 \text{ kg})(20.0 \text{ m/s})\hat{i}$$
$$= (-10.2 \text{ kg·m/s})\hat{i}$$

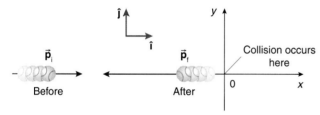

FIGURE 9.5

According to the impulse–momentum theorem, Equation 9.5, the impulse of the total force is equal to the change in the momentum, and so

$$\vec{I} = (-10.2 \text{ kg·m/s})\hat{i}$$

b. Use the vector version of Equation 9.6 to find the average total force:

$$\vec{F}_{ave} = \frac{\vec{I}}{\Delta t}$$
$$= \frac{(-10.2 \text{ kg·m/s})\hat{i}}{1.00 \times 10^{-3} \text{ s}}$$
$$= (-1.02 \times 10^4 \text{ N})\hat{i}$$

c. Take the ratio of the magnitude of the average total force to the magnitude of the weight of the ball:

$$\frac{F_{ave}}{mg} = \frac{1.02 \times 10^4 \text{ N}}{(0.145 \text{ kg})(9.81 \text{ m/s}^2)}$$
$$= 7.17 \times 10^3$$

Would you agree that the magnitude of the average total force on the ball during the collision is *much* greater than that of the weight of the ball?

EXAMPLE 9.5

An incoming horizontal pitch at speed v is hit. The baseball of mass m leaves the bat in the vertical direction at speed $2v$.

a. What is the impulse of the total force on the ball?
b. What is the magnitude of the impulse?

Solution

a. The ball is the system. Use the coordinate system indicated in Figure 9.6. The initial momentum of the ball is

$$\vec{p}_i = mv\hat{i}$$

and its final momentum is

$$\vec{p}_f = m(2v)\hat{j}$$

FIGURE 9.6

The change in momentum of the ball is then

$$\Delta \vec{p} = \vec{p}_f - \vec{p}_i$$
$$= 2mv\hat{j} - mv\hat{i}$$
$$= mv(-\hat{i} + 2\hat{j})$$

According to the impulse–momentum theorem, Equation 9.5, the impulse of the total force on a system is equal to the change in the momentum of the system, and so

$$\vec{I} = mv(-\hat{i} + 2\hat{j})$$

Notice that the direction of the impulse is *not* necessarily along the direction of either momentum vector; the impulse is in the direction of the *change* in the momentum.

b. The magnitude of the impulse is

$$I = mv[(-1)^2 + 2^2]^{1/2}$$
$$= \sqrt{5}\,mv$$

9.3 THE ROCKET: A SYSTEM WITH VARIABLE MASS*

An infamous editorial blunder:

> . . . *after the rocket quits our air and really starts on its longer journey [to the Moon] its flight would neither be accelerated nor maintained by the explosion of the charges it then might have left. To claim that it would be is to deny a fundamental law of dynamics. . . . That Professor Goddard [inventor of the liquid fueled rocket] . . . does not know the relation of action to reaction, and of the need to have something better than a vacuum against which to react—to say that would be absurd. Of course he only seems to lack the knowledge ladled out daily in high schools.*
>
> *New York Times (1920)**

The liftoff of a rocket propelling a payload into orbit is both a spectacular sight and a technological triumph. Most of the mass of the rocket on the launch pad is fuel. During the powered phase of the launch, the mass of the rocket system changes dramatically as exhaust gases are expelled furiously at great speed. A rocket can propel itself in the vacuum of outer space by expelling gases at great speed in a direction opposite to that of the rocket itself. Everything needed for the combustion of its fuel is carried within the rocket. Jet airplanes use the same physical principle of expelling gases at high speed but are restricted to lower altitudes to use atmospheric oxygen for the combustion of the fuel. Jets are more complex systems than rockets because they accumulate mass with their intakes and expel even more mass (combustion by-products) at great speed. Here we examine the physics of rockets, our first encounter with systems of variable mass.

For simplicity, consider a single-stage rocket moving in one dimension, say, along the x-axis parallel to $\hat{\imath}$; see Figure 9.7. Let the system be the rocket with any remaining fuel. Call the mass of the system $m(t)$, where we explicitly indicate that the mass varies with time. Its corresponding velocity relative to a fixed (inertial) coordinate system is $v_x(t)\hat{\imath}$.

Assume that the exhaust gases are ejected at constant speed v_{exh} *relative to the rocket*. The velocity of the rocket and the velocity of the exhaust gases are in opposite directions. Hence the velocity of the exhaust gases relative to the fixed inertial coordinate system is

$$v_x(t)\hat{\imath} - v_{\text{exh}}\hat{\imath} = [v_x(t) - v_{\text{exh}}]\hat{\imath}$$

At an instant t the momentum of the rocket (with the remainder of its unburned fuel) is

$$m(t)v_x(t)\hat{\imath} \tag{9.8}$$

During a short differential time interval dt, the mass of the rocket *changes* by dm. The mass change dm is *negative* since the rocket loses

*Editorial, "A Severe Strain on Credulity," 13 January 1920, page 12, column 5.

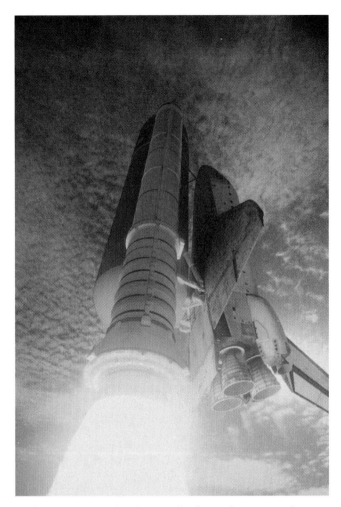

Rockets are systems that dramatically change their mass with time.

mass; this does *not* imply that mass is negative: dm is the *change* in the mass of the rocket. The mass of the exhausted gases is $-dm$ (which is positive since dm itself is negative). At the end of the interval dt, the mass of the rocket is $m(t) + dm$ (remember that dm is a negative quantity) and its velocity is $[v_x(t) + dv_x]\hat{\imath}$. Hence the momentum of the rocket and residual fuel in its tanks is (see Figure 9.8)

$$[m(t) + dm][v_x(t) + dv_x]\hat{\imath}$$

The momentum of the fuel expelled during dt is

$$(-dm)[v_x(t) - v_{\text{exh}}]\hat{\imath}$$

FIGURE 9.7 The rocket system at instant t.

FIGURE 9.8 The rocket system and expelled fuel at instant $t + dt$.

At the end of the interval dt the momentum of the total system is the vector sum

$$[m(t) + dm][v_x(t) + dv_x]\hat{\imath} + (-dm)[v_x(t) - v_{exh}]\hat{\imath}$$
$$= [m(t)v_x(t) + m(t)\,dv_x + dm\,v_x(t) - dm\,v_x(t) + v_{exh}\,dm]\hat{\imath}$$
$$= [m(t)v_x(t) + m(t)\,dv_x + v_{exh}\,dm]\hat{\imath} \qquad (9.9)$$

where we neglect the term $dm\,dv_x\,\hat{\imath}$ involving the product of two differentials, since as differentials approach zero, their product vanishes even faster. The differential change $dp_x\,\hat{\imath}$ in the momentum of the system is, therefore, the difference between Equations 9.9 and 9.8:

$$dp_x\,\hat{\imath} = [m(t)\,dv_x + v_{exh}\,dm]\hat{\imath}$$

The impulse–momentum theorem (Equation 9.5) states that the differential change in the x-component of the momentum must equal the x-component of the differential impulse of the total force on the system during the interval dt:

$$[m(t)\,dv_x + v_{exh}\,dm]\hat{\imath} = F_{x\,total}\,dt\,\hat{\imath}$$

Hence

$$F_{x\,total} = m(t)\frac{dv_x}{dt} + v_{exh}\frac{dm}{dt} \qquad (9.10)$$

The x-component of the acceleration of the rocket is

$$\frac{dv_x}{dt}$$

Equation 9.10 can be rewritten as

$$F_{x\,total} - v_{exh}\frac{dm}{dt} = m(t)\frac{dv_x}{dt} \qquad (9.11)$$

Equation 9.11 is known as the **rocket equation**. Since the rocket is losing mass,

$$\frac{dm}{dt} < 0 \ \text{kg/s}$$

and the term

$$-v_{exh}\frac{dm}{dt}$$

actually is a force component in the same direction as the acceleration component. The term

$$v_{exh}\left|\frac{dm}{dt}\right| \qquad (9.12)$$

is called the **thrust** of the rocket and arises because the mass of the rocket is not constant. The thrust can be increased by increasing the speed v_{exh} of the exhaust gases relative to the rocket or by increasing the rate at which the rocket loses mass.

For a launch near the surface of the Earth, $F_{x\,total}$ must include the weight of the rocket (which varies with time during the flight) and speed-dependent kinetic frictional forces caused by air resistance. At launch the thrust must be greater than the weight of the rocket for it to leave the launch pad.

If we restrict ourselves to an isolated rocket, far away from any other large masses like the Earth, or to a direction perpendicular to gravitational forces, $F_{x\,total} = 0$ N and the rocket equation, 9.11, becomes

$$-v_{exh}\frac{dm}{dt} = m(t)\frac{dv_x}{dt}$$

Rearranging, we find

$$-\frac{dm}{m} = \frac{dv_x}{v_{exh}} \qquad (9.13)$$

Let the initial speed of the rocket be zero when its mass is m_0. Integrate Equation 9.13:

$$-\int_{m_0}^{m}\frac{dm}{m} = \int_{0\text{ m/s}}^{v_x}\frac{dv_x}{v_{exh}}$$

Since the speed of the exhaust gases is constant, we have

$$\ln m_0 - \ln m = \frac{v_x}{v_{exh}}$$

Hence

$$v_x = v_{exh}\,\ln\left(\frac{m_0}{m}\right) \qquad (9.14)$$

Separate the initial mass of the rocket into payload and fuel:

$$m_0 = m_{payload} + m_{fuel}$$

When all fuel is exhausted, the mass of the rocket is just $m_{payload}$. Equation 9.14 then indicates that the maximum speed of the rocket is

$$v_x = v_{exh}\,\ln\left(\frac{m_{fuel} + m_{payload}}{m_{payload}}\right)$$
$$= v_{exh}\,\ln\left(\frac{m_{fuel}}{m_{payload}} + 1\right) \qquad (9.15)$$

From this result we can see that to attain a large final speed, the fuel to payload mass ratio must be large. Since in deriving Equation 9.15 we ignored the presence of forces such as the weight of the rocket and kinetic friction, you *cannot* use simply Equation 9.15 for satellite launches from the surface of the Earth.

EXAMPLE 9.6

A rocket motor expels exhaust gases at a rate of 1.5×10^3 kg/s with a speed of 3.0 km/s. What is the thrust of the motor?

Solution

The thrust is the product of the speed of the exhaust gases relative to the rocket and the absolute value of the rate of mass loss:

$$v_{exh}\left|\frac{dm}{dt}\right| = (3.0 \times 10^3 \ \text{m/s})(1.5 \times 10^3 \ \text{kg/s})$$
$$= 4.5 \times 10^6 \ \text{N}$$

EXAMPLE 9.7 ———————————

Well away from any other masses, a rocket with a fuel to payload mass ratio of $m_{fuel}/m_{payload} = 4.0$ is fired. The exhaust gases leave the rocket at a speed of 2.0 km/s relative to the rocket. If it is initially at rest with respect to an inertial reference frame, what is the speed of the rocket in this frame when the fuel is exhausted?

Solution

The problem meets the conditions attached to the use of Equation 9.15. Thus

$$v_x = v_{exh} \ln \left(\frac{m_{fuel}}{m_{payload}} + 1 \right)$$
$$= (2.0 \times 10^3 \text{ m/s})(\ln 5.0)$$
$$= 3.2 \times 10^3 \text{ m/s}$$

The speed is 3.2 km/s.

9.4 CONSERVATION OF MOMENTUM

If the total force acting on a system is zero, the system is said to be mechanically **isolated**. For such isolated systems, Equation 9.2 becomes

$$0 \text{ N} = \frac{d\vec{p}_{total}}{dt}$$

The zero time derivative implies that $\vec{p}_{total}$ must be a constant vector.

When $\vec{p}_{total}$ is constant, we say the system exhibits **conservation of momentum.***

Nonzero forces of one particle within the system on other particles within it may exist; these are called **internal forces**. Such internal forces may change the momentum of the individual particles within the system, but the *total* momentum of the system remains fixed; it is just reapportioned among these particles in different ways. Since the internal forces of one particle on another within the system always are paired according to Newton's third law of motion, they are of equal magnitude and opposite direction; so the internal forces vector sum to zero and cannot affect the total momentum. Only external forces from outside the system (i.e., a nonzero total force) can change its total momentum.

We shall see that conservation of momentum, like the CWE theorem, is a very important physical theorem. In particular, in the next section we see that the total momentum is conserved in collisions if the system you define is large enough to include all the particles participating in the collision.

———————————
*The expression really should be called conservation of the *total* momentum $\vec{p}_{total}$ of the system, since the *individual* particle momenta may change; it is the *total* momentum of the system (the vector sum of the individual particle momenta that make up the system) that is unvarying.

9.5 COLLISIONS

We already have considered a few examples of collisions as examples of impulsive forces. We now present a systematic, general approach to collision phenomena.

If a system has zero total force on it ($\vec{F}_{total} = 0$ N), then the time rate of change of the total momentum of the system also is zero according to Newton's second law of motion, expressed in Equation 9.2:

$$0 \text{ N} = \frac{d\vec{p}_{total}}{dt}$$

This implies that the total momentum of the system must be a constant vector; that is, the total momentum of the system is conserved.

Take *all* the particles involved in a collision as a single system, as shown in Figure 9.9. The total momentum of the system just before the collision is the vector sum of the individual momenta of the particles; for a two-particle collision, this is

$$\vec{p}_{total \ bfr} \equiv \vec{p}_{1 \ bfr} + \vec{p}_{2 \ bfr}$$

Likewise the total momentum of the system after the collision is the vector sum of the individual momenta just after the collision; for a two-particle collision, this is

$$\vec{p}_{total \ aft} \equiv \vec{p}_{1 \ aft} + \vec{p}_{2 \ aft}$$

If the total force on the system is zero, conservation of momentum means that the total momentum of the system the instant before the collision is equal to the total momentum of the system the instant after the collision; that is,

$$\vec{p}_{total \ bfr} = \vec{p}_{total \ aft} \qquad (9.16)$$

In a collision, the *individual momenta* of the particles *do change*, but the *total momentum of the system of colliding particles does not.*

Why do the individual particle momenta change? Consider the collision between a baseball and a bat. Take the ball as a single-particle system. During the collision, there are two forces on the ball: (1) the large force of the bat on the ball and (2) the comparatively small weight of the ball (the gravitational force of the Earth on the ball). During the brief time interval of the collision, the force of the bat on the ball causes a much greater impulse on the ball than the impulse of the weight. Hence, to an excellent approximation, the change in the momentum of the ball is determined by the impulse of the bat; we can essentially

The system for a two-particle collision

FIGURE 9.9 Take the collection of all the particles in a collision as the system.

neglect the impulse of the gravitational force. In any case, the total force on the ball (considered as a single-particle system) during the collision is not zero; so according to Newton's second law of motion (applied to the single-particle ball system), the momentum of this system (the ball) is not conserved.

It is only by considering *all* the particles involved in a collision as a single system that Newton's second law leads to conservation of the total momentum of the system.

> Conservation of the total momentum is exact as long as the vector sum of all the forces on the system is zero.

Conservation of the total momentum can be applied to two-particle collisions or two-million-particle collisions as long as the system is defined to include all the particles participating in the collision.

Occasionally there are nonzero total forces on a system of all particles involved in a collision. For example, when two balls collide in midair, the gravitational force of the Earth on the two balls (their weight) is, of course, ever present. However, during the typically brief time interval of the collision, the gravitational forces usually are negligible compared with the forces of the particles on each other arising from the collision. Hence, during the collision, the dynamics of each particle is dominated by the forces involved in the collision. These internal collision forces are paired according to Newton's third law and vector sum to zero when you consider the system to be all the particles involved in the collision. Conservation of the total momentum thus is an excellent approximation for analyzing the dynamics of the system of colliding particles during the brief time interval between the instant before and the instant after the collision. Once the brief collision is over, however, the dynamics of each particle is determined by the gravitational force on each ball, since the forces on each associated with the collision no longer exist.

It is kinetic energy considerations that enable us to classify collisions into two broad regimes.

Elastic Collisions

> If the total kinetic energy of the particles is conserved, then the collision is called an **elastic collision**.

In other words, for elastic collisions there is no change in the total kinetic energy of the system of particles:

$$\Delta KE_{total} = 0 \text{ J}$$

Thus

$$KE_{total\,aft} = KE_{total\,bfr}$$

where $KE_{total\,aft}$ is the (scalar) sum of the scalar kinetic energies of the particles just after the collision and $KE_{total\,bfr}$ is the sum of the kinetic energies of the particles just before the collision.

If the kinetic energy is conserved (unchanged) before and after the collision, then the CWE theorem, Equation 8.26, becomes

$$W_{total} = \Delta KE_{total}$$
$$= 0 \text{ J}$$

Thus the work done by the total force acting during the collision is zero according to the CWE theorem. The forces involved in the collision therefore must be either conservative forces (such as Hooke's-law elastic forces) or forces doing zero work.

During the brief time interval of the collision, the kinetic energy is *not* conserved; some or all of the initial total kinetic energy of the particles is stored during the brief collision as the potential energy of various conservative forces (such as Hooke's law elastic forces, or electrical potential energy on a microscopic scale). But as the time interval of collision ends, this potential energy all is dumped back into kinetic energy. No loss of kinetic energy arises if we look at the kinetic energy the instant just before the collision and the kinetic energy the instant just after the collision. There are no thermal effects, and no work is done by nonconservative forces during an elastic collision. Collisions between macroscopic particles are rarely, if ever, elastic but can come close to that ideal in certain cases.

Inelastic Collisions

> If the total kinetic energy of the particles involved in a collision is *not* conserved, the collision is called an **inelastic collision**. Collisions of macroscopic particles typically are inelastic. *If the particles stick together after the collision*, the collision is called a **completely inelastic collision**.

The famous and important CWE theorem of Chapter 8 *cannot* account for this lost kinetic energy of the system. Why? The forces involved in the collision are paired according to Newton's third law (they are internal to the system of particles involved in the collision). The total force on the system of particles during the collision thus is zero. If the total force is zero, the work done by the total force is zero. Hence the left-hand side of the CWE theorem,

$$W_{total} = \Delta KE_{total}$$

is zero but the right-hand side is not! The loss of kinetic energy typically manifests itself as thermal energy (what physicists call the thermodynamic internal energy, as we will see in Chapter 13).* The CWE theorem of classical mechanics *neglects* such thermal effects. We return to this question once again in Chapter 13 to see how the problem finally is resolved.

> The things to remember about collisions are the following:
>
> a. The total momentum of the system is conserved in *both* elastic and inelastic collisions as long as
>   - the system is defined to include all the particles participating in the collision, and
>   - the total external force on the system is either zero or can be neglected compared with forces present during the collision itself.
> b. The total kinetic energy is conserved *only* for elastic collisions.
> c. A collision that is not elastic is called inelastic.
> d. If the particles stick together after the collision, it is called a completely inelastic collision.

---

*Some energy also may appear initially as sound waves.

PROBLEM-SOLVING TACTIC

9.1 None of the many specialized equations for various types of specific collisions enumerated in the examples is worth remembering; just remember the general points we have highlighted. Such specialized equations come with so many strings attached to them, restricting their usage, that by the time you pick your way through the strings to find out if a particular collision problem fits the special circumstances appropriate for the equation, you likely could derive the equation from the general principles just outlined. The memorization of equations is not what physics is all about.*

We look at several examples to illustrate the approach used to solve collision problems. *The results we obtain are not as important as the method used to set up the problems*, since each individual collision problem must be analyzed from scratch.

QUESTION 3

Two masses, one initially at rest, collide. Is it possible for *both* masses to be at rest after the collision? Explain.

EXAMPLE 9.8

Two particles of masses m and $2m$ approach each other with equal speeds along a line on a frictionless surface and make a completely inelastic collision in one dimension, as shown in Figure 9.10. After the collision the composite particle moves along the line of their approach. Find the velocity of the composite particle.

Frictionless surface

Before   After

FIGURE 9.10

Solution

Define the system to be the two particles undergoing the collision. Introduce the coordinate system shown in Figure 9.11.

There are external forces acting on the system, namely the respective weights of the particles and the normal forces of the surface on them. However, the vector sum of these external forces acting on the system is zero by Newton's second law, because there is no acceleration along the vertical direction. The forces that each particle exerts on the other during the collision are *internal forces*; they vector sum to zero according to Newton's third law. Since the total force on the system is zero, the total momentum of the system is conserved: the total vector momentum of the system of particles is constant.

The individual momenta of the particles before the collision are

$$\vec{\mathbf{p}}_{1\,bfr} = mv\hat{\imath}$$

---

*In an introductory physics course, quite frankly, it may *seem* like you are awash in equations; but there are in fact only a few fundamental physical principles. A good understanding of the latter makes it unnecessary to memorize many specialized equations because you can figure them out! This is part of what makes physics fun! (?)

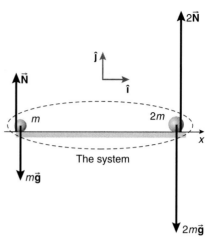

FIGURE 9.11

and

$$\vec{\mathbf{p}}_{2\,bfr} = 2mv(-\hat{\imath})$$
$$= -2mv\hat{\imath}$$

The total momentum of the system before the collision is thus

$$\vec{\mathbf{p}}_{total\,bfr} = \vec{\mathbf{p}}_{1\,bfr} + \vec{\mathbf{p}}_{2\,bfr}$$
$$= mv\hat{\imath} - 2mv\hat{\imath}$$
$$= -mv\hat{\imath}$$

Since the total momentum of this system is conserved, the momentum after the collision must be the same as the momentum before the collision. Let $\vec{\mathbf{v}}'$ be the velocity of the composite particle system after the collision. Since the mass of the system is $m + 2m = 3m$, the total momentum of the system after the collision is, by the definition of momentum,

$$\vec{\mathbf{p}}_{total\,aft} = 3m\vec{\mathbf{v}}'$$

Applying momentum conservation,

$$\vec{\mathbf{p}}_{total\,aft} = \vec{\mathbf{p}}_{total\,bfr}$$
$$3m\vec{\mathbf{v}}' = -mv\hat{\imath}$$

Solving for $\vec{\mathbf{v}}'$,

$$\vec{\mathbf{v}}' = -\frac{v}{3}\hat{\imath}$$

The minus sign indicates the velocity of the composite particle actually is opposite to that pictured in Figure 9.10. Notice that if we were to define the system to be only, say, the particle on the left (of mass m) (see Figure 9.11), then the momentum of *that* system is *not* conserved. This particle does suffer a change in momentum during the collision; the change arises from the force of the other particle (of mass $2m$) on it, a force that is now *external* to this single-particle system.

STRATEGIC EXAMPLE 9.9

Two particles make a one-dimensional, head-on elastic collision on a frictionless surface, as shown in Figure 9.12. One particle, of mass m_2, initially is at rest at the origin of the coordinate system introduced to analyze the problem.

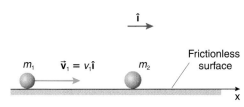

FIGURE 9.12

a. **Find the velocity components v_1' and v_2' of the particles after the collision.**
b. **Discuss the results if $m_2 \gg m_1$.**
c. **Discuss the results if $m_1 \gg m_2$.**

Solution

a. The system is the two particles involved in the collision. The total vector momentum before collision is

$$\vec{P}_{\text{total bfr}} = m_1 v_1 \hat{\imath} + m_2 (0 \text{ m/s}) \hat{\imath}$$
$$= m_1 v_1 \hat{\imath}$$

The total vector momentum after the collision is

$$\vec{P}_{\text{total aft}} = m_1 v_1' \hat{\imath} + m_2 v_2' \hat{\imath}$$

Since the total force on the system of two particles is zero, the total momentum is conserved:

$$\vec{P}_{\text{total bfr}} = \vec{P}_{\text{total aft}}$$

That is,

$$m_1 v_1 \hat{\imath} = m_1 v_1' \hat{\imath} + m_2 v_2' \hat{\imath}$$
$$m_1 v_1 = m_1 v_1' + m_2 v_2' \qquad (1)$$

Since the collision is elastic, the total kinetic energy also is conserved. That is, the total kinetic energy before the collision is equal to the total kinetic energy after the collision:

$$KE_{\text{total bfr}} = KE_{\text{total aft}}$$

or

$$\frac{1}{2} m_1 v_1^2 = \frac{1}{2} m_1 v_1'^2 + \frac{1}{2} m_2 v_2'^2$$

Simplifying slightly,

$$m_1 v_1^2 = m_1 v_1'^2 + m_2 v_2'^2 \qquad (2)$$

That's it for the physics: we conserved momentum and also kinetic energy, since *this* collision is an *elastic* collision. The problem now is mathematical. This is typically the case in collision problems: the mathematics is more tedious than the physics. We tune up our algebraic prowess. Kramer's rule for solving sets of simultaneous *linear* equations is of no use, because the kinetic energy equation is nonlinear in the velocity components. We resort to brute force algebraic techniques or algebraic software.

The algebraic solution proceeds as follows: Take equation (2) stemming from the conservation of kinetic energy and transpose the term in particle 1 from the right to the left side, yielding

$$m_1 (v_1^2 - v_1'^2) = m_2 v_2'^2$$

Now factor the difference of squares:

$$m_1 (v_1 + v_1')(v_1 - v_1') = m_2 v_2'^2 \qquad (3)$$

Make a similar transposition in the momentum component equation (1):

$$m_1 (v_1 - v_1') = m_2 v_2' \qquad (4)$$

Now divide equation (3) by equation (4), assuming $v_1 - v_1' \neq 0$ m/s. (If $v_1 = v_1'$, the problem then is quite uninteresting, since it means that particle m_1 has the same velocity after the collision. In such a case, no collision actually takes place: m_1 misses m_2.) The division yields

$$v_1 + v_1' = v_2' \qquad (5)$$

Now substitute for v_2' in the right-hand side of equation (4):

$$m_1 (v_1 - v_1') = m_2 v_2'$$
$$= m_2 (v_1 + v_1')$$

Solve for v_1':

$$v_1' = \frac{m_1 - m_2}{m_1 + m_2} v_1$$

Then use equation (5) to find v_2':

$$v_2' = v_1 + v_1'$$
$$= v_1 + \frac{m_1 - m_2}{m_1 + m_2} v_1$$

Place things over a common denominator, and you finally get

$$v_2' = \frac{2m_1}{m_1 + m_2} v_1$$

Whew.

b. Suppose $m_2 \gg m_1$. This is like a BB pellet or a Ping-Pong ball colliding head-on with a stationary bowling ball. Such elastic collisions of particles with immobile objects will be important when we examine the kinetic theory of gases in Chapter 14. The velocity components after the collision are then

$$v_1' = \frac{m_1 - m_2}{m_1 + m_2} v_1$$
$$\approx -v_1 \qquad \text{since } m_2 \gg m_1$$

and

$$v'_2 = \frac{2m_1}{m_1 + m_2}$$

$$\approx 0 \text{ m/s} \quad \text{since } m_2 \gg m_1$$

The incident particle m_1 merely reverses direction and m_2 is unmoved by the whole affair. No surprises here.

c. On the other hand suppose $m_1 \gg m_2$, so that the mass of the incident particle is huge with respect to that of the stationary target particle sitting placidly at the origin. This is like a cannonball colliding head-on with a stationary BB pellet. In this case the expressions just derived indicate that the velocity components of the respective particles after the collision are

$$v'_1 = \frac{m_1 - m_2}{m_1 + m_2} v_1$$

$$\approx v_1 \quad \text{since } m_1 \gg m_2$$

and

$$v'_2 = \frac{2m_1}{m_1 + m_2} v_1$$

$$\approx 2v_1 \quad \text{since } m_1 \gg m_2$$

The huge incident mass essentially is unaffected by the encounter and the tiny target mass m_2 zips off at *twice* the speed of the incident behemoth! What does this case tell you about the speed of a baseball relative to the speed of the bat when the ball is hit from its position on a "T-ball" stand?

STRATEGIC EXAMPLE 9.10

Two particles, each with speed v, make a completely inelastic collision. The collision is not head-on, as shown in Figure 9.13.

a. Find the velocity of the composite particle after the collision, and the angle ϕ that this velocity makes with the x-axis in Figure 9.13.
b. What is the change in the total kinetic energy of the particles?

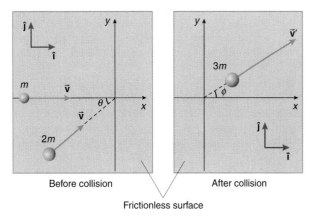

Before collision   **After collision**

Frictionless surface

FIGURE 9.13

Solution

Consider the system to be the two particles involved in the collision.

a. Momentum is conserved in collisions, so that the *total* momentum before the collision is equal to the *total* momentum after the collision. Remember that momentum is a *vector*, so that

$$\vec{\mathbf{P}}_{\text{total bfr}} = \vec{\mathbf{P}}_{\text{total aft}}$$

Use the coordinate system indicated in Figure 9.13. The *individual* momenta of the particles before the collision are

$$\vec{\mathbf{P}}_{1 \text{ bfr}} = mv\hat{\imath}$$

$$\vec{\mathbf{P}}_{2 \text{ bfr}} = (2mv \cos \theta)\hat{\imath} + (2mv \sin \theta)\hat{\jmath}$$

After the collision the composite particle has mass $3m$; let its unknown velocity be $\vec{\mathbf{v}}'$. The total momentum of the system after the collision is the momentum of this single composite particle:

$$\vec{\mathbf{P}}_{\text{total aft}} = 3m\vec{\mathbf{v}}'$$

Conserve the total momentum:

$$\vec{\mathbf{P}}_{\text{total aft}} = \vec{\mathbf{P}}_{\text{total bfr}}$$
$$3m\vec{\mathbf{v}}' = mv\hat{\imath} + [(2mv \cos \theta)\hat{\imath} + (2mv \sin \theta)\hat{\jmath}]$$

The mass m is common to every term and may be divided out. Solving for $\vec{\mathbf{v}}'$, you find

$$\vec{\mathbf{v}}' = \frac{1}{3} v \left[\left(1 + 2 \cos \theta \right) \hat{\imath} + \left(2 \sin \theta \right) \hat{\jmath} \right]$$

The angle ϕ that the velocity of the composite particle makes with the x-axis is found from the ratio of the components of the velocity $\vec{\mathbf{v}}'$ indicated in Figure 9.13:

$$\tan \phi = \frac{v'_y}{v'_x}$$

$$= \frac{2 \sin \theta}{1 + 2 \cos \theta}$$

$$\phi = \tan^{-1} \left(\frac{2 \sin \theta}{1 + 2 \cos \theta} \right)$$

b. The total kinetic energy after the collision is the kinetic energy of the single composite particle of mass $3m$:

$$KE_{\text{total aft}} = \frac{1}{2} (3m) v'^2$$

Substituting for v', you have

$$KE_{\text{total aft}} = \frac{1}{2} (3m) \frac{v^2}{9} \left[\left(1 + 2 \cos \theta \right)^2 + \left(2 \sin \theta \right)^2 \right]$$

After a bit of algebraic simplification, you find

$$KE_{\text{total aft}} = \frac{mv^2}{6} (1 + 4 \cos \theta + 4 \cos^2 \theta + 4 \sin^2 \theta)$$

Since $\sin^2\theta + \cos^2\theta = 1$, the expression in the bracket can be simplified, yielding

$$KE_{\text{total aft}} = \frac{mv^2}{6}(5 + 4\cos\theta)$$

The kinetic energy before the collision is the sum of the kinetic energies of the two individual particles involved in the collision:

$$KE_{\text{total bfr}} = \frac{mv^2}{2} + \frac{(2m)v^2}{2}$$

$$= \frac{3mv^2}{2}$$

Thus the *change* in the kinetic energy is

$$KE_{\text{total aft}} - KE_{\text{total bfr}} = \frac{mv^2}{6}(5 + 4\cos\theta) - \frac{3mv^2}{2}$$

$$= \frac{mv^2}{6}(5 + 4\cos\theta - 9)$$

$$= -\frac{mv^2}{6}4(1 - \cos\theta)$$

The change of kinetic energy is negative. Kinetic energy was lost in the collision since $KE_{\text{total aft}} < KE_{\text{total bfr}}$. In Chapter 13 we will see what happens to this lost kinetic energy.

EXAMPLE 9.11

Two small, hard spheres of *equal mass* collide elastically, as shown in Figure 9.14. One sphere is at rest before the collision. If the spheres do not collide head-on, after the collision they emerge making angles θ and ϕ with the incident direction as indicated. Find the angle β between the paths followed by the masses after the collision. That is, find the *sum* of the angles θ and ϕ.

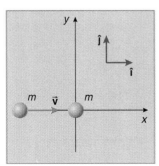

Before collision   After collision

FIGURE 9.14

Solution

As with many things, there is an easy way and a hard way to solve this problem.

Method 1

Let us see what is involved if we tackle the problem in a straightforward fashion by conserving momentum and, since the collision

is elastic, kinetic energy as well. With the choice of coordinates indicated in Figure 9.14, the total momentum of the particles before the collision is

$$\vec{\mathbf{p}}_{\text{total bfr}} = mv\hat{\mathbf{i}} + m(0 \text{ m/s})\hat{\mathbf{i}}$$

$$= mv\hat{\mathbf{i}}$$

After the collision the total momentum is

$$\vec{\mathbf{p}}_{\text{total aft}} = \vec{\mathbf{p}}_{1\text{ aft}} + \vec{\mathbf{p}}_{2\text{ aft}}$$

$$= (mv_1\cos\theta)\hat{\mathbf{i}} + (mv_1\sin\theta)\hat{\mathbf{j}}$$

$$+ (mv_2\cos\phi)\hat{\mathbf{i}} - (mv_2\sin\phi)\hat{\mathbf{j}}$$

where v_1 and v_2 are the speeds of the particles after the collision. Momentum conservation means that

$$\vec{\mathbf{p}}_{\text{total bfr}} = \vec{\mathbf{p}}_{\text{total aft}}$$

Two vectors are equal to each other if and only if all of their respective components are each equal to each other. Equate the x-components of the total momenta:

$$mv = mv_1\cos\theta + mv_2\cos\phi$$

$$v = v_1\cos\theta + v_2\cos\phi \qquad (1)$$

Equate the y-components:

$$0 \text{ kg}\cdot\text{m/s} = mv_1\sin\theta - mv_2\sin\phi$$

so

$$v_1\sin\theta = v_2\sin\phi \qquad (2)$$

Since the collision is elastic, the kinetic energy also is conserved. That is,

$$\frac{mv^2}{2} = \frac{mv_1^2}{2} + \frac{mv_2^2}{2}$$

which reduces to

$$v^2 = v_1^2 + v_2^2 \qquad (3)$$

This essentially ends the physics of the problem. We now have a mathematical problem, and a rather messy one at that. You have to eliminate the speeds and find the sum of the two angles: $\beta = \theta + \phi$. You have three equations to work with: equations (1) and (2) from conservation of momentum, and equation (3) from conservation of kinetic energy. If you like algebraic morasses, this is a good one to curl up with and wade through when you are in a good mood. This approach is cumbersome but legitimate, what a physicist likely might call correct, but inelegant.

Method 2

An easier way to do the problem is, amazingly, not to get so involved. In particular write momentum conservation in vector form *without* reference to a particular coordinate choice:

$$m\vec{\mathbf{v}} = m\vec{\mathbf{v}}_1 + m\vec{\mathbf{v}}_2$$

$$\vec{\mathbf{v}} = \vec{\mathbf{v}}_1 + \vec{\mathbf{v}}_2$$

Take the scalar product of this equation with itself:

$$\vec{\mathbf{v}} \cdot \vec{\mathbf{v}} = (\vec{\mathbf{v}}_1 + \vec{\mathbf{v}}_2) \cdot (\vec{\mathbf{v}}_1 + \vec{\mathbf{v}}_2)$$

So

$$v^2 = v_1^2 + v_2^2 + 2\,\vec{v}_1 \cdot \vec{v}_2 \qquad (4)$$

But the conservation of kinetic energy equation [equation (3)] indicates that

$$v^2 = v_1^2 + v_2^2$$

Comparing equation (4) with this equation indicates that the scalar product $\vec{v}_1 \cdot \vec{v}_2$ must be zero:

$$\vec{v}_1 \cdot \vec{v}_2 = 0 \ \text{m}^2/\text{s}^2$$

Hence

$$v_1 v_2 \cos \beta = 0 \ \text{m}^2/\text{s}^2$$

Since neither v_1 nor v_2 is zero, $\cos \beta$ must be zero, which means $\beta = 90°$. The two velocity vectors are perpendicular to each other! After the collision the particles emerge with an angle $\beta = 90°$ between their paths. Wow. That was fast and the result is quite remarkable!

This example illustrates that occasionally the straightforward way of solving a problem may not necessarily be the easiest; some thought and inspiration occasionally yield imaginative solutions. Physicists (and mathematicians too) would call the vector shortcut quite elegant, even beautiful, because of its compactness and clarity compared with the algebraic drudgery of the more straightforward but mathematically cumbersome procedure of Method 1.

9.6 DISINTEGRATIONS AND EXPLOSIONS

The mental image we have of explosions and disintegrations is of many particles flying off in all directions. *BOOM!* The investigations into the tragic explosion and disintegration of TWA Flight 800 in the summer of 1996 and the bombing of Pan Am Flight 103 in 1988 attempted to discover what happened by extrapolating backward from the distribution of the recovered pieces.

For explosions and disintegrations the original system, considered as a single particle, suddenly breaks up into many particles zipping off in all directions. Disintegrations, such as those that occur in the radioactive decay of certain nuclei, are completely inelastic collisions in reverse; a video of the events, run backward, would look like a completely inelastic collision. Explosions also can be considered as inelastic collisions in reverse, to the extent that the mass of the explosive itself can be ignored.*

Define the system to be all particles involved in the disintegration or explosion. The *internal* forces that exist by and on the particles that compose the system during the short time interval of the disintegration or explosion typically are *much* greater than *external* forces on the particles. External forces, such

---

*This assumption is *not*, in many cases, a good approximation for many explosions. Certainly in much ordnance the mass of the explosive is much greater than the mass of the casing.

as gravitational forces of the Earth on the particles (their weight), typically are negligible by comparison. Since the internal forces are paired according to Newton's third law and vector sum to zero, the total force on the system of particles is zero (or negligible) during the disintegration or explosion. As a consequence, the total momentum of the system is conserved, to an excellent approximation. In other words, the total momentum of the system at the instant before the disintegration or explosion is equal to the total momentum of all the particles at the instant immediately after the event:

$$\vec{p}_{\text{total bfr}} = \vec{p}_{\text{total aft}}$$

The total kinetic energy of the system is *not conserved*, just as for any inelastic collision. The source of the resulting kinetic energy of the particles comes from chemical, mechanical, or nuclear sources associated with the disintegration or the explosive material. Energy considerations for explosions are accounted for by a very general statement of energy conservation to be introduced later, in Chapter 13.

EXAMPLE 9.12

A 15.0 kg object moving vertically upward at speed 10.0 m/s suddenly explodes into three pieces of mass 2.0 kg, 3.0 kg, and 10.0 kg. Immediately after the breakup, the 2.0 kg mass is found by astute, hawk-eyed physics students to be moving vertically upward at a speed of 20.0 m/s and the 3.0 kg chunk is discovered to be moving horizontally at a speed of 5.0 m/s. Find the velocity and speed of the 10.0 kg piece.

Solution

Conserve the momentum of the system of particles before and after the disintegration. Use the coordinate system indicated in Figure 9.15. The total momentum of the system before the disintegration is the momentum of the single particle of mass 15.0 kg:

$$\begin{aligned}\vec{P}_{\text{total bfr}} &= (15.0 \ \text{kg})(10.0 \ \text{m/s})\hat{\jmath} \\ &= (150 \ \text{kg}\cdot\text{m/s})\hat{\jmath}\end{aligned}$$

The total momentum of the system after the explosion is the vector sum of the momenta of the three particles. The momentum of the 2.0 kg mass is

$$\begin{aligned}\vec{p}_{1 \text{ aft}} &= (2.0 \ \text{kg})(20.0 \ \text{m/s})\hat{\jmath} \\ &= (40 \ \text{kg}\cdot\text{m/s})\hat{\jmath}\end{aligned}$$

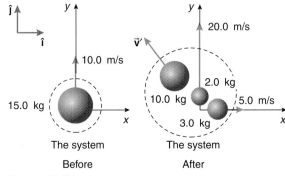

FIGURE 9.15

The momentum of the 3.0 kg chunk is

$$\vec{p}_{2\,\text{aft}} = (3.0 \text{ kg})(5.0 \text{ m/s})\hat{\imath}$$
$$= (15 \text{ kg·m/s})\hat{\imath}$$

Let $\vec{v}'$ be the velocity of the 10.0 kg piece. Its momentum after the disintegration then is

$$\vec{p}_{3\,\text{aft}} = (10.0 \text{ kg})\vec{v}'$$

The total momentum after the disintegration thus is

$$\vec{p}_{\text{total aft}} = (15 \text{ kg·m/s})\hat{\imath} + (40 \text{ kg·m/s})\hat{\jmath} + (10 \text{ kg})\vec{v}'$$

Conserve momentum:

$$\vec{p}_{\text{total bfr}} = \vec{p}_{\text{total aft}}$$
$$(150 \text{ kg·m/s})\hat{\jmath} = (15 \text{ kg·m/s})\hat{\imath} + (40 \text{ kg·m/s})\hat{\jmath} + (10 \text{ kg})\vec{v}'$$

Solving for $\vec{v}'$, you obtain

$$\vec{v}' = (-1.5 \text{ m/s})\hat{\imath} + (11.0 \text{ m/s})\hat{\jmath}$$

The speed of the 10.0 kg is the magnitude of its velocity vector, or

$$v' = [(-1.5 \text{ m/s})^2 + (11.0 \text{ m/s})^2]^{1/2}$$
$$= 11.1 \text{ m/s}$$

9.7 THE CENTRIPETAL ACCELERATION REVISITED*

There is an interesting way to think of the centripetal acceleration of a particle in circular motion that stems from our study of collisions and the impulse–momentum theorem. Imagine a particle of mass m moving at speed v on a frictionless surface inside a fixed horizontal circular hoop of radius r, as shown in Figure 9.16.

Assume the collisions of the particle with the circular hoop are elastic. Since the kinetic energy is conserved in elastic collisions and the hoop is fixed, the particle maintains a constant speed before and after each such collision. The magnitude of the momentum of the particle is therefore constant but, clearly, its direction is changed with each collision with the hoop. Thus *the momentum of the particle is not conserved in its collision with the hoop.* (See Question 33.) Viewed from the center of the circle, successive collisions are separated by an angle θ, as indicated in Figure 9.16.

Examine the collision at point A; use the coordinate system indicated in Figure 9.16 to write the momentum of the particle before and after the collision at point A:

$$\vec{p}_{\text{bfr}} = \left[mv \cos\left(\frac{\theta}{2}\right) \right]\hat{\imath} + \left[mv \sin\left(\frac{\theta}{2}\right) \right]\hat{\jmath}$$

$$\vec{p}_{\text{aft}} = \left[mv \cos\left(\frac{\theta}{2}\right) \right]\hat{\imath} - \left[mv \sin\left(\frac{\theta}{2}\right) \right]\hat{\jmath}$$

The change in the momentum $\Delta\vec{p}$ of the particle is

$$\Delta\vec{p} \equiv \vec{p}_{\text{aft}} - \vec{p}_{\text{bfr}}$$
$$= \left[-2mv \sin\left(\frac{\theta}{2}\right) \right]\hat{\jmath}$$

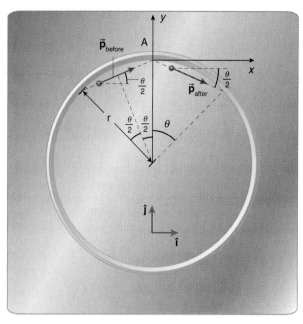

FIGURE 9.16 Geometry for a collision of the particle with the hoop.

Notice that the change in the momentum of the particle is directed toward the center of the circle. The impulse–momentum theorem (Equation 9.5) implies that the impulse of the force of the hoop on the particle also must be directed toward the center of the circle.

The *magnitude* of the change in the momentum of the particle is

$$\Delta p = 2mv \sin\left(\frac{\theta}{2}\right)$$

The hoop exerts an impulse on the particle that, according to the impulse–momentum theorem, is equal to the change in the momentum of the particle. Use Equation 9.6, which expresses the impulse as the product of the magnitude of the average force $\vec{F}_{\text{ave}}$ and the time interval of the collision Δt, in the impulse–momentum theorem:

$$F_{\text{ave}} \Delta t = \Delta p$$
$$= 2mv \sin\left(\frac{\theta}{2}\right) \qquad (9.17)$$

The time interval between successive collisions, $\Delta t'$, is the distance between successive collisions divided by the speed of the particle. The distance between successive collisions is, using the geometry of Figure 9.16, equal to $2r \sin(\theta/2)$. Hence the time interval between successive collisions is

$$\Delta t' = \frac{2r}{v} \sin\left(\frac{\theta}{2}\right) \qquad (9.18)$$

Now imagine the particle moving so the time interval between successive collisions, $\Delta t'$, approaches the time interval associated with the collision of the particle with the hoop, Δt; the particle then is essentially in continuous contact with the hoop and in circular motion. Use Equation 9.18 for $\Delta t'$ to substitute for Δt in Equation 9.17; we get

$$F_{ave} \frac{2r}{v} \sin\left(\frac{\theta}{2}\right) = 2mv \sin\left(\frac{\theta}{2}\right)$$

or

$$F_{ave} = m \frac{v^2}{r}$$

This we recognize as Newton's second law applied to the particle: the force of the hoop on the particle is the total force on the particle in the horizontal plane, and so it must be equal to the mass m times the acceleration. Thus we can identify v^2/r as the magnitude of the centripetal acceleration of the particle. Newton made this argument in his *Principia*.[†]

9.8 AN ALTERNATIVE WAY TO LOOK AT FORCE TRANSMISSION*

When we studied gravitation in Chapter 6, the gravitational field was introduced as a way of understanding how the force mysteriously is transmitted even over long, empty distances between masses not in physical contact with each other. The idea of a field also is very useful for electromagnetism, as we will see in Chapters 16 and 20.

Our study of collisions can give us an inkling of an alternative way to visualize force transmission: via the exchange of **mediating particles**. The mediating particles associated with the four fundamental forces were listed in Table 5.2. The example we examine next illustrates how a repulsive force can be conveyed by particle exchange.

Consider a series of collision experiments involving a 100.0 kg astronaut and a 1000 kg spacecraft both initially floating freely at rest in outer space, well away from any other masses such as the Earth, other planets, and the Sun. In this way the astronaut and spacecraft are isolated in space. We neglect the small gravitational attraction of the astronaut and spacecraft on each other.

Experiment 1: One-Particle Exchange

Imagine the 100.0 kg astronaut to be 2.00 m from the spacecraft. The astronaut initially has zero velocity relative to the spacecraft, as shown in Figure 9.17. A launcher on the spacecraft sends out a 1.00 kg canister to the astronaut at a speed of, say, 2.00 m/s.

The astronaut catches the canister 1.00 s after launch. The catch can be viewed as a completely inelastic collision of the astronaut and the canister. Define the system to be the canister and the astronaut; momentum is conserved in the collision. The momentum of the astronaut before the collision is zero, because the astronaut had zero velocity:

$$\vec{p}_{ast\ bfr} = (0\ kg \cdot m/s)\hat{\imath}$$

[†]Newton's argument is in Book I, Proposition IV. The author is indebted to Arnold Arons, A *Guide to Introductory Physics Teaching* (Wiley, New York, 1990), pages 138–140 for bringing this argument to his attention.

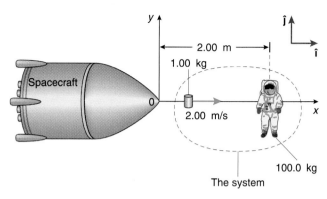

FIGURE 9.17 A canister is sent from the spacecraft to the astronaut.

The momentum of the canister before the collision is (with the choice of coordinate axes in Figure 9.17)

$$\vec{p}_{can\ bfr} = (1.00\ kg)(2.00\ m/s)\hat{\imath}$$

Therefore the total initial momentum of the astronaut–canister system before the collision is

$$\vec{p}_{total\ bfr} = \vec{p}_{can\ bfr} + \vec{p}_{ast\ bfr}$$
$$= (2.00\ kg \cdot m/s)\hat{\imath} + (0\ kg \cdot m/s)\hat{\imath}$$

Since the astronaut caught and held the canister, the total momentum of the single, composite particle after the collision is

$$\vec{p}_{total\ aft} = (100.0\ kg + 1.00\ kg)\vec{v}$$
$$= (101.0\ kg)\vec{v}$$

where $\vec{v}$ is the common velocity of the astronaut and the canister after the collision.

Momentum conservation in the collision means that the total momentum before the collision is equal to total momentum after the collision:

$$\vec{p}_{total\ bfr} = \vec{p}_{total\ aft}$$
$$(2.00\ kg \cdot m/s)\hat{\imath} = (101.0\ kg)\vec{v}$$

So the velocity $\vec{v}$ of the astronaut and canister after the collision is

$$\vec{v} = \frac{2.00\ kg \cdot m/s}{101.0\ kg}\hat{\imath}$$
$$= (0.0198\ m/s)\hat{\imath}$$

A graph of the velocity component of the astronaut as a function of time is shown in Figure 9.18. The velocity component of the astronaut is zero until the canister touches the hand of the astronaut. After the short interval of the collision, the canister and astronaut have the same velocity component.

Experiment 2: Two-Particle Exchange

Instead of the previous collision experiment, suppose the spacecraft successively threw *two* canisters each of mass 0.500 kg, half that of the single canister in Experiment 1, at the same speed of 2.00 m/s. Each of the canisters now has *half* of the momentum of the original larger 1.00 kg canister. The astronaut catches each smaller canister in turn. The total change in the momentum of

FIGURE 9.18 Velocity component of the astronaut as a function of time for the one-canister experiment.

FIGURE 9.19 Velocity component of the astronaut as a function of time for the two-canister experiment.

FIGURE 9.20 Velocity component of the astronaut as a function of time for the n-canister experiment.

the astronaut is the *same* as in the first experiment; the momentum just changes via two smaller impulses rather than in one larger impulse. So the final velocity of the astronaut is the same as in the first (single-canister) experiment. You should be able to verify this statement by considering the two collisions of the canisters with the astronaut.

Suppose the two canisters were thrown so they were caught successively at times 0.500 s and 1.00 s by the astronaut. A plot of the velocity component of the astronaut as a function of time is shown in Figure 9.19.

Experiment 3: n-Particle Exchange

Now divide the original 1.00 kg canister into n smaller canisters, each of mass $(1.00 \text{ kg})/n$. Each is thrown at the astronaut successively, each at speed 2.00 m/s, and all caught within one second. The astronaut experiences n successive inelastic collisions and the velocity and momentum of the astronaut change in n small steps (see Figure 9.20).

Recall that the acceleration component is the slope of the corresponding velocity component versus t graph. If n is very large, then the acceleration is almost continuous (and constant)

during the one second interval. The astronaut experiences something more like a steady push rather than a series of jerky impulses from the collisions. The astronaut likely would say a *force* was pushing on her.

According to the most general statement of Newton's second law, Equation 9.2, the total force on a system is equal to the time rate of change of the total momentum of the system:

$$\vec{F}_{total} = \frac{d\vec{p}_{total}}{dt}$$

Let's calculate the average force on the astronaut (considered as a single-particle system) during the one second time interval. The momentum change of the astronaut during the one second interval was

$$\Delta \vec{p} = (2.00 \text{ kg·m/s})\hat{\imath}$$

The impulse on the astronaut is, from the vector version of Equation 9.6,

$$\vec{I} = \vec{F}_{ave} \Delta t$$

According to the impulse–momentum theorem, the impulse is equal to the change in the momentum:

$$\vec{F}_{ave}(1.00 \text{ s}) = (2.00 \text{ kg·m/s})\hat{\imath}$$
$$\vec{F}_{ave} = (2.00 \text{ N})\hat{\imath}$$

Experiment 4: Effect on the Spacecraft

Now let's look at the other side of the picture: the spacecraft that was projecting the large canister or smaller canisters toward the astronaut. Examine the first of the experiments we performed. The spacecraft initially is at rest before it emits the canister. The emission of the canister also can be viewed as a collision, since if we play a video of the event backward in time, the time-reversed event has all the character of a completely inelastic collision.

Define the system to be the spacecraft and the emitted canister. The spacecraft throws out a 1.00 kg canister at a speed of 2.00 m/s. Initially the spacecraft and canister each have zero momentum since both are not moving. The total momentum of the spacecraft–canister system before the collision is zero:

$$\vec{p}_{total \ bfr} = (0 \text{ kg·m/s})\hat{\imath}$$

After the emission, however, the canister has a momentum of

$$(2.00 \text{ kg·m/s})\hat{\imath}$$

and the spacecraft recoils with a velocity $\vec{v}'$ and momentum

$$(999 \text{ kg})\vec{v}'$$

The total momentum after the emission is

$$\vec{p}_{total \ aft} = (2.00 \text{ kg·m/s})\hat{\imath} + (999 \text{ kg})\vec{v}'$$

Conservation of momentum says the total momentum before the interaction must be equal to the total momentum after the interaction. So

$$(0 \text{ kg·m/s})\hat{\imath} = (2.00 \text{ kg·m/s})\hat{\imath} + (999 \text{ kg})\vec{v}'$$

and the velocity of the recoiling spacecraft is

$$\vec{\mathbf{v}}' = \frac{-2.00 \ \text{kg} \cdot \text{m/s}}{999 \ \text{kg}} \ \hat{\imath}$$

$$= (-0.002\ 00 \ \text{m/s}) \ \hat{\imath}$$

The momentum change of the recoiling spacecraft has exactly the same magnitude as the momentum change of the astronaut, but the two vectors point in opposite directions.

We could analyze the other experiments (which transferred momenta by using many particle canisters) for their effects on the spacecraft as well but you likely can see that the game is essentially the same. For the *n*-particle exchange, we attribute the change in the momentum of the astronaut to the impulse of the canisters on the astronaut (thanks to the reception or *absorption* of the many little particles); the momentum change of the spacecraft also can be attributed to an impulse on the spacecraft (thanks to the *emission* of many little particles). The two impulses and the two average forces are of the same magnitude but are oppositely directed. This equality is equivalent to Newton's third law of motion.

These experiments have several important features that touch on many major issues in contemporary physics. Since the exchanged particles travel at finite speeds, forces take time to propagate from one object to another. In other words, when the spacecraft emits the exchanged particles (the canisters), the astronaut does not feel the effects until they are received some finite time later. Such propagation delays are important in many areas of physics. These effects bothered Newton himself, but he was unable to come to grips with the problem appropriately.

The astronaut and spacecraft move away from each other as a result of the particle exchange, much like a *repulsive force*. This was easy to see. However, using classical physics, it is quite difficult to imagine how an attractive force is conveyed by means of particle exchange. Somehow nature manages to do it, but we lack the conceptual framework to describe such interactions with simple classical models. Quantum mechanics is another game entirely. So here we are in an *introductory* physics course and find ourselves face-to-face with a difficult question we cannot answer easily. Maybe you can figure it out. No one else has been able to do this as yet.* Think about it. This problem should convince you that there is a difficult underlying subtlety about forces and their effects.

We will not pursue further this view of force transmission via particle exchange, but this example should serve to make you aware that there are various ways to model the way forces are transmitted. In this text, we will be content only to examine the effects of the forces themselves on the motion of particles and the field mechanism of force transmission; that is trouble enough for now!

9.9 THE CENTER OF MASS

How do we deal neatly with a collection of masses at different locations as a single system? The most representative location of the system is called the **center of mass** point. The center of

mass of an extended system is the point whose dynamics typifies the system as a whole when it is treated as a particle. We will find that this concept leads to useful simplifications of the dynamics of an extended system of many particles. For example, when a high-board diver or gymnast executes a complicated maneuver, their center of mass follows the trajectory you would calculate when treating the system as a simple particle (see Figure 9.21).

How do you locate the center of mass point of an extended system? Think of this analogy. The center of population of the United States is not just the average of the positions of all the towns, because some towns have more people than others; Dixville Notch, New Hampshire, is not Los Angeles! In the averaging, a town of 10 000 persons should count ten times as much as a town of 1000. And so it is with the concept of the center of mass of a collection of particles. Each particle carries an importance proportional to its mass. The sum of the products of the mass of each particle and its position vector gives the grand total mass m_{total} times the position vector of the center of mass (see Figure 9.22):

$$m_1\vec{\mathbf{r}}_1 + m_2\vec{\mathbf{r}}_2 + m_3\vec{\mathbf{r}}_3 + \cdots \equiv m_{\text{total}}\vec{\mathbf{r}}_{\text{CM}}$$

We write this equation more compactly as

$$\sum_i m_i\vec{\mathbf{r}}_i \equiv m_{\text{total}}\vec{\mathbf{r}}_{\text{CM}} \qquad (9.19)$$

where $\vec{\mathbf{r}}_i$ is the position vector of the *i*th particle (of mass m_i) with respect to the origin of the chosen coordinate system.

Rearranging Equation 9.19 slightly to isolate $\vec{\mathbf{r}}_{\text{CM}}$, we see that the position vector of the center of mass is

$$\vec{\mathbf{r}}_{\text{CM}} \equiv \frac{1}{m_{\text{total}}} \sum_i m_i\vec{\mathbf{r}}_i \qquad (9.20)$$

Since two vectors are equal only if all their respective components are equal, the equation defining the location of the center

FIGURE 9.21 When executing this complicated maneuver, the center of mass of this gymnast follows a parabolic path after leaving the apparatus.

---

*There is a simplistic (but unsatisfactory) way of imagining an attractive force arising from particle exchange: two individuals on ice skates are each holding a basketball (the mediating particle). If the two individuals snatch or grab the basketballs from each other, the individuals likely will be drawn toward each other on the ice. But this crude model necessitates almost immediate contact between the particles to be attracted, whereas attractive forces in physics can be exerted on particles over vast distances.

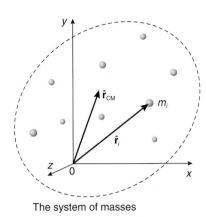

The system of masses

FIGURE 9.22 The vector $\vec{r}_{CM}$ locates the position of the center of mass.

FIGURE 9.23 The center of mass of a uniform rectangular plate is at its geometric center.

of mass is really three scalar equations, one for each (Cartesian) component of the center of mass:

$$x_{CM} \equiv \frac{1}{m_{total}} \sum_i m_i x_i$$

$$y_{CM} \equiv \frac{1}{m_{total}} \sum_i m_i y_i \qquad (9.21)$$

$$z_{CM} \equiv \frac{1}{m_{total}} \sum_i m_i z_i$$

The center of mass is a *point in space*; there may or may not be any mass actually located at the center of mass.

It is useful to consider common objects as essentially continuous distributions of matter; this is an excellent approximation for macroscopic objects, because atoms are so small and close together. For such essentially continuous distributions of matter, the discrete summations (Equations 9.21) defining the components of the position vector of the center of mass become integrals over the physical extent of the object:

$$x_{CM} \equiv \frac{1}{m_{total}} \int x \, dm$$

$$y_{CM} \equiv \frac{1}{m_{total}} \int y \, dm \qquad (9.22)$$

$$z_{CM} \equiv \frac{1}{m_{total}} \int z \, dm$$

where dm is the differential amount of mass in a differential volume dV located at the position vector whose components are x, y, and z. Let ρ be the mass density at the location of the volume element dV; then

$$dm = \rho \, dV$$

If the system is an extended mass of *uniform composition* (i.e., of constant density ρ), then the geometric *symmetry* of the system (if it has any!) often can be exploited to locate the center of mass quickly—without the agony of performing the summations or integrations indicated in Equations 9.21 or 9.22. For example:

1. A uniform rectangular plate has its center of mass at the geometric center of the plate (see Figure 9.23). We can see this as follows. If we choose to locate the origin of a coordinate system at the geometric center of the plate, then any bit of mass dm

with position vector $\vec{r}$ has a symmetrically placed counterpart located at $-\vec{r}$. The two terms sum to zero: $dm \, \vec{r} + dm \, (-\vec{r}) = 0 \text{ kg·m}$. The mass of the entire plate can be accounted for this way, and so the summations in Equations 9.21 or, really, the integrals in Equations 9.22 are zero. The position vector of the center of mass is zero; hence the origin chosen at the geometric center of the plate is coincident with the center of mass point, since its position vector is zero.

2. Similar symmetry arguments are used to show that the center of mass of

   • a uniform spherical shell,
   • a uniform sphere, or
   • a sphere with a spherically symmetric mass distribution

   each is located at the geometric center of the sphere.

Occasionally the center of mass can be found using several legitimate tricks of the trade:

Trick 1: Segmented mass method

a. Segment the extended mass into regular geometric pieces and locate the center of mass of each piece using symmetry considerations.

b. Find the fraction of the total mass represented by each segment.

c. Then treat the collection of mass segments as a collection of pointlike particles each located at the center of mass of the respective segment. The center of mass of the entire system then is found using Equation 9.20 for a collection of particles. See Example 9.14 (Method 1) for an illustration of this procedure.

Trick 2: Strategic subtraction method

Another way is to create voids or holes by judicious subtraction of regularly shaped pieces. See Example 9.14 (Method 2) for an illustration of this technique.

Trick 3: An experimental method

If the system is irregular in shape or not of uniform composition, the analytic techniques we have outlined become intractable in a practical sense. In such situations the location of the center of mass can be determined *experimentally* by suspending the object freely from two arbitrary points. For example, consider an irregular two-dimensional (planar) object. When the object is freely suspended, the mass will hang with its center of mass on the vertical line through the pivot, and below it, as shown in Figure 9.24.

This position minimizes the gravitational potential energy of the system. A vertical line is drawn on the object indicating that the center of mass is located somewhere along the line. The object then is suspended freely from any other point and another such line drawn, as in Figure 9.25. The intersection of the two lines locates the center of mass in the plane of the object. If the mass is not planar, extrapolate the lines drawn outside the object to their point of convergence within the mass.

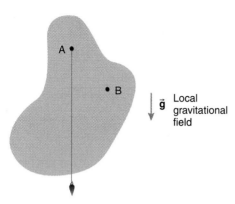

FIGURE 9.24 The center of mass lies on the vertical line through the pivot and below it.

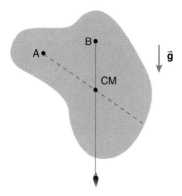

FIGURE 9.25 Suspend the system from another point.

QUESTION 4

Where is the location of the center of mass of the ring system surrounding the planet Saturn? Where is the center of mass of a hula hoop?

EXAMPLE 9.13

Find the location of the center of mass of the system of three particles shown in Figure 9.26.

FIGURE 9.26

Solution

Introduce a coordinate system as in Figure 9.27. The position vectors of the three masses are then

$$\vec{r}_1 = \ell\hat{i}$$
$$\vec{r}_2 = \ell\hat{j}$$
$$\vec{r}_3 = (0 \text{ m})\hat{j}$$

The position vector of the center of mass is found using Equation 9.20. Since there are three masses, there are three terms in the summation:

$$\vec{r}_{CM} = \frac{1}{m_{total}}(m_1\vec{r}_1 + m_2\vec{r}_2 + m_3\vec{r}_3)$$

where m_{total} is the total mass of the system, $4m$. Substituting the masses and their respective position vectors,

$$\vec{r}_{CM} = \frac{1}{4m}\left[2m\ell\hat{i} + m\ell\hat{j} + m(0 \text{ m})\hat{j}\right]$$
$$= \frac{\ell}{4}(2\hat{i} + \hat{j})$$

This position is indicated in Figure 9.27. For this system, there is no mass located at the center of mass.

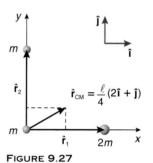

FIGURE 9.27

STRATEGIC EXAMPLE 9.14

Find the location of the center of mass of the uniform plate of mass m, shown in Figure 9.28.

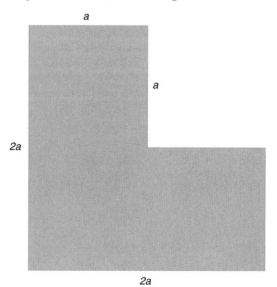

FIGURE 9.28

Solution

Method 1: The segmented mass method
Consider the system to be composed of three uniform square plates each with sides a, as shown in Figure 9.29. By symmetry, the center of mass of each segment is at the geometric center of each square plate. Since the plates all are of equal area, the mass of each plate segment is $m/3$.

FIGURE 9.29

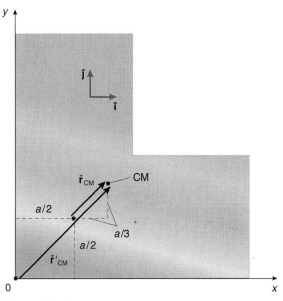

FIGURE 9.30

Choose the origin of the coordinate system to be at the center of the square on the lower left, as indicated in Figure 9.29. The position vector of the center of mass of each of the segments then is

$$\vec{\mathbf{r}}_1 = a\hat{\mathbf{j}}$$
$$\vec{\mathbf{r}}_2 = (0 \text{ m})\hat{\mathbf{i}}$$
$$\vec{\mathbf{r}}_3 = a\hat{\mathbf{i}}$$

The position vector of the center of mass of the three-particle system then is, from Equation 9.20,

$$\vec{\mathbf{r}}_{CM} = \frac{1}{m_{total}} \sum_i m_i \vec{\mathbf{r}}_i$$

$$= \frac{1}{m_{total}} (m_1 \vec{\mathbf{r}}_1 + m_2 \vec{\mathbf{r}}_2 + m_3 \vec{\mathbf{r}}_3)$$

$$= \frac{1}{m} \left[\frac{m}{3} a\hat{\mathbf{j}} + \frac{m}{3} (0 \text{ m})\hat{\mathbf{i}} + \frac{m}{3} a\hat{\mathbf{i}} \right]$$

$$= \frac{a}{3} (\hat{\mathbf{i}} + \hat{\mathbf{j}})$$

The position of the center of mass with respect to the lower left corner means $a/2$ must be added to each component in the last line, as shown in Figure 9.30. Hence, measured from the corner, the position vector of the center of mass is

$$\vec{\mathbf{r}}'_{CM} = \left(\frac{a}{2} + \frac{a}{3} \right) \hat{\mathbf{i}} + \left(\frac{a}{2} + \frac{a}{3} \right) \hat{\mathbf{j}}$$

$$= \frac{5}{6} a\hat{\mathbf{i}} + \frac{5}{6} a\hat{\mathbf{j}}$$

Method 2: The strategic subtraction method
Consider the original system to be what is left when a square plate, of sides $2a$, has a square with sides a cut from it, as shown in Figure 9.31. The mass of the large square is $(4/3)m$, while that

of the piece to be cut away is $m/3$. The location of the center of mass of each square plate is at its respective geometric center. Choose the origin of the coordinate system to be at the lower left corner, as shown in Figure 9.31.

The position vector of the center of mass of the larger plate thus is

$$\vec{\mathbf{r}}_1 = a\hat{\mathbf{i}} + a\hat{\mathbf{j}}$$

while that of the smaller plate is

$$\vec{\mathbf{r}}_2 = \frac{3}{2} a\hat{\mathbf{i}} + \frac{3}{2} a\hat{\mathbf{j}}$$

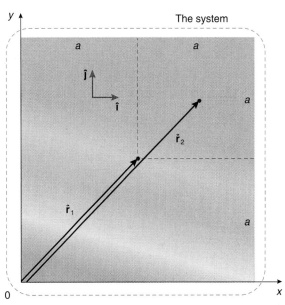

FIGURE 9.31

The problem has been reduced to a two-mass system, except that we *subtract* the contribution of the smaller square from that of the larger square:

$$\vec{\mathbf{r}}_{\text{CM}}' = \frac{1}{m_{\text{total}}}(m_1\vec{\mathbf{r}}_1 - m_2\vec{\mathbf{r}}_2)$$

$$= \frac{1}{\left(\dfrac{4}{3}m - \dfrac{m}{3}\right)}\left[\frac{4}{3}m(a\hat{\mathbf{i}} + a\hat{\mathbf{j}}) - \frac{m}{3}\left(\frac{3}{2}a\hat{\mathbf{i}} + \frac{3}{2}a\hat{\mathbf{j}}\right)\right]$$

$$= \frac{5}{6}a\hat{\mathbf{i}} + \frac{5}{6}a\hat{\mathbf{j}}$$

9.10 DYNAMICS OF A SYSTEM OF PARTICLES

With the concept of the center of mass, there is a great simplification in store for us. Here we want to see what happens to a system of many particles when forces act on them. It is important, of course, to decide first what particles constitute our chosen system.

The location of the center of mass of a system of particles is defined by Equation 9.20:

$$\vec{\mathbf{r}}_{\text{CM}} = \frac{1}{m_{\text{total}}}\sum_i m_i\vec{\mathbf{r}}_i$$

where m_{total} is the total mass of the collection of particles. Multiplying this equation by m_{total}, we have

$$\sum_i m_i\vec{\mathbf{r}}_i = m_{\text{total}}\vec{\mathbf{r}}_{\text{CM}}$$

To see what happens to the position vectors as functions of time, differentiate with respect to time (assume that the mass of each particle is constant):

$$\sum_i m_i\frac{d\vec{\mathbf{r}}_i}{dt} = m_{\text{total}}\frac{d\vec{\mathbf{r}}_{\text{CM}}}{dt}$$

The time derivatives of the position vectors are the respective velocities of the particles, and so we have

$$\sum_i m_i\vec{\mathbf{v}}_i = m_{\text{total}}\vec{\mathbf{v}}_{\text{CM}} \tag{9.23}$$

The momentum of each individual particle is $\vec{\mathbf{p}}_i = m_i\vec{\mathbf{v}}_i$, so that the summation on the left-hand side of Equation 9.23 is the vector sum of the individual particle momenta, which is the total momentum $\vec{\mathbf{p}}_{\text{total}}$ of the system of particles:

$$\vec{\mathbf{P}}_{\text{total}} = m_{\text{total}}\vec{\mathbf{v}}_{\text{CM}} \tag{9.24}$$

Equation 9.24 means that the total momentum of the system of particles is the same as the product of the mass of the entire system, m_{total}, and the velocity of the center of mass point, $\vec{\mathbf{v}}_{\text{CM}}$.

What a nice simplification!

Notice that if the center of mass of the system is not moving—that is, if $\vec{\mathbf{v}}_{\text{CM}} = 0$ m/s—then the *vector sum* of the individual particle momenta also is zero, even if the individual momenta themselves are not zero. For example, consider the case of an air-filled balloon at rest: the individual atoms each are moving and, therefore, have individual momenta $\vec{\mathbf{p}}_i$; yet the vector sum of all the $\vec{\mathbf{p}}_i$ is zero because the center of mass of the air in the stationary balloon is at rest.

How do we handle all the forces acting on a many-particle system? If Equation 9.23 is differentiated with respect to time, we find

$$\frac{d}{dt}\left(\sum_i m_i\vec{\mathbf{v}}_i\right) = \frac{d}{dt}\left(m_{\text{total}}\vec{\mathbf{v}}_{\text{CM}}\right) \tag{9.25}$$

Look at a typical particle within the system, say the *i*th particle. According to Newton's second law of motion, the product

$$m_i\frac{d\vec{\mathbf{v}}_i}{dt}$$

must be equal to the vector sum of *all* the forces acting on *this* particle, $\vec{\mathbf{F}}_{i\,\text{total}}$. Let's include what is happening to all the particles at once. Equation 9.25 then becomes

$$\sum \vec{\mathbf{F}}_{i\,\text{total}} = m_{\text{total}}\frac{d\vec{\mathbf{v}}_{\text{CM}}}{dt} \tag{9.26}$$

The forces on the *i*th particle arise from two distinct sources:

1. Forces on the *i*th particle by the *other* particles of the system. Call the vector sum of these forces on the *i*th particle $\vec{\mathbf{F}}_{i\,\text{total int}}$ since they arise from other particles *within* (i.e., *internal* to) the system of particles under consideration.
2. Forces on the *i*th particle arising from sources *outside* the system itself. Call the vector sum of these forces on the *i*th particle originating from outside (i.e., *external* to) the system $\vec{\mathbf{F}}_{i\,\text{total ext}}$.

Classifying the forces in this way in Equation 9.26, we have

$$\sum_i\left(\vec{\mathbf{F}}_{i\,\text{total int}} + \vec{\mathbf{F}}_{i\,\text{total ext}}\right) = m_{\text{total}}\frac{d\vec{\mathbf{v}}_{\text{CM}}}{dt} \tag{9.27}$$

The first summation on the left-hand side of this equation,

$$\sum_i\vec{\mathbf{F}}_{i\,\text{total int}}$$

gives the grand sum of all the forces by each particle on each of the other particles of the system. Fortunately, as Newton's third

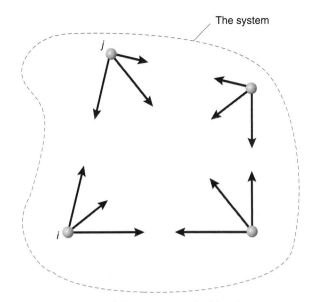

FIGURE 9.32 Internal forces occur in third law force pairs.

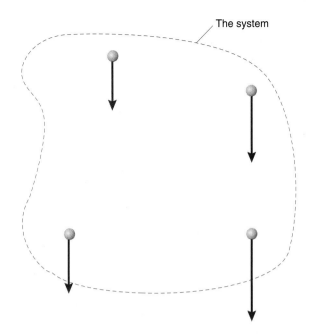

FIGURE 9.33 External forces on the system may not sum to zero.

law of motion states, all forces occur in pairs of equal magnitude and opposite direction (though the pair of forces act on different objects). Hence the force of particle j on particle i is equal in magnitude and opposite in direction to the force of particle i on particle j, as shown in Figure 9.32.

Since *both* these forces are in the sum of the *internal* forces, they will vector sum to zero. *All* of the internal forces are so paired; thus, the vector sum of all the internal forces is zero:

$$\sum_i \vec{F}_{i\,\text{total int}} = 0 \text{ N} \qquad \text{(because of Newton's third law)}$$

This certainly helps simplify things!

The corresponding vector sum of the applied *external* forces on the particles of the system need *not* be zero, since only *one* member of each external force third law force pair acts on the ith particle. For example, the force of the Earth on the ith particle (its weight) is present in the sum, but the force of the ith particle *on* the external Earth is *not* in the sum if the Earth is not part of the system you chose. See Figure 9.33. There is no Newton's third law pairwise summation of the external forces since only *one* member of the external force pairs is a force on a particle of the system.

Incorporating these results into Equation 9.27, we have

$$\sum_i \vec{F}_{i\,\text{total ext}} = m_{\text{total}} \frac{d\vec{v}_{\text{CM}}}{dt}$$

Let $\vec{F}_{\text{total}}$ be the vector sum of all the *external* forces on *all the particles of the system*; the overall motion of the system has a simple description:

$$\boxed{\vec{F}_{\text{total}} = m_{\text{total}} \frac{d\vec{v}_{\text{CM}}}{dt}} \qquad (9.28)$$

The vector sum of the total external forces on the system produces an acceleration of the center of mass of the system inversely proportional to the total mass of the system. This is why the center of mass concept is so useful: the motion of the entire system is determined by imagining the system as a *single particle* of mass m_{total} located at the center of mass point.

Since $m_{\text{total}}\vec{v}_{\text{CM}}$ is the total momentum $\vec{p}_{\text{total}}$ of the system, we can write Equation 9.28 as

$$\vec{F}_{\text{total}} = \frac{d\vec{p}_{\text{total}}}{dt} \qquad (9.29)$$

which is just Newton's second law for the system of particles treated as a *single particle* located at the center of mass point.

QUESTION 5

A handful of marbles is pitched into the air simultaneously. What is the acceleration of the center of mass of the system of marbles while they are in the air?

9.11 KINETIC ENERGY OF A SYSTEM OF PARTICLES

The center of mass also provides a useful simplification when finding the total kinetic energy of a system of many moving particles. Choose an origin 0 at some convenient point in space and locate a typical particle in the system by means of a position vector $\vec{r}_i$ originating at 0 and another position vector $\vec{r}_i'$

originating at the center of mass, as shown in Figure 9.34. The two vectors are related by

$$\vec{r}_i = \vec{r}_{CM} + \vec{r}'_i \qquad (9.30)$$

where $\vec{r}_{CM}$ locates the center of mass with respect to the origin 0.

Differentiate the position vectors with respect to time to find the relationship between the velocities:

$$\vec{v}_i = \vec{v}_{CM} + \vec{v}'_i \qquad (9.31)$$

The kinetic energy of the *i*th particle as measured in the reference frame with its origin at 0 is

$$\frac{1}{2} m_i v_i^2$$

The total kinetic energy of the whole system is the scalar sum of the kinetic energies of each of the particles:

$$KE_{total} = \sum_i \frac{1}{2} m_i v_i^2$$

$$= \sum_i \frac{1}{2} m_i \vec{v}_i \cdot \vec{v}_i$$

Substituting for $\vec{v}_i$ using Equation 9.31, we obtain

$$KE_{total} = \sum_i \frac{1}{2} m_i (\vec{v}_{CM} + \vec{v}'_i) \cdot (\vec{v}_{CM} + \vec{v}'_i) \qquad (9.32)$$

But

$$(\vec{v}_{CM} + \vec{v}'_i) \cdot (\vec{v}_{CM} + \vec{v}'_i) = v_{CM}^2 + v_i'^2 + 2\,\vec{v}_{CM} \cdot \vec{v}'_i$$

The total kinetic energy thus is

$$KE_{total} = \sum_i \frac{1}{2} m_i\left(v_{CM}^2 + v_i'^2 + 2\vec{v}_{CM} \cdot \vec{v}'_i\right)$$

$$= \frac{1}{2}\sum_i m_i v_{CM}^2 + \sum_i \frac{1}{2} m_i v_i'^2 + \vec{v}_{CM} \cdot \sum_i m_i \vec{v}'_i \qquad (9.33)$$

In the first summation, the sum of all the masses is the total mass m_{total} of the system of particles and there is only one v_{CM}. The second summation is the kinetic energy of the system of particles with respect to the center of mass. For the third sum, consider the following. Since each $\vec{r}'_i$ is measured with respect to the center of mass, Equation 9.20 implies

$$\sum_i m_i \vec{r}'_i = 0 \ kg \cdot m$$

If we differentiate this with respect to time, we find

$$\sum_i m_i \vec{v}'_i = 0 \ kg \cdot m/s$$

and so the third summation in Equation 9.33 is zero. The total kinetic energy of the system of particle is then

$$KE_{total} = \frac{1}{2} m_{total} v_{CM}^2 + \sum_i \frac{1}{2} m_i v_i'^2 \qquad (9.34)$$

Therefore the total kinetic energy of the system is the sum of

a. the kinetic energy of the center of mass, as if the total mass of the system were located all at that point; and
b. the kinetic energy of the particles with respect to (wrt) a reference frame with its origin at the center of mass.

That is,

$$KE_{total} = KE_{of\ CM} + KE_{wrt\ CM} \qquad (9.35)$$

EXAMPLE 9.15

An 8.00 kg runaway snowboard is approaching you at a speed of 12.00 m/s while a 4.00 kg small sled is coming at you from the opposite direction at a speed of 3.00 m/s on a frozen lake at the base of a ski area; see Figure 9.35.

a. What is the total kinetic energy of the two masses?
b. What is the velocity of their center of mass with respect to you?
c. What is the velocity of each mass with respect to the center of mass?

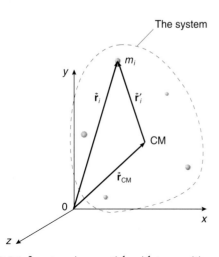

FIGURE 9.34 Locate a given particle with two position vectors.

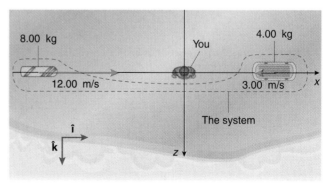

FIGURE 9.35

d. What is the kinetic energy of the two masses when viewed from the center of mass point?

e. Show that the total kinetic energy you found in part (a) is equal to the sum of the kinetic energy of the center of mass point and the kinetic energy with respect to the center of mass point, as predicted by Equation 9.35.

Solution

a. You determine the total kinetic energy by taking the sum of the kinetic energy of each mass individually:

$$KE_{total} = \frac{1}{2}(8.00 \ kg)(12.00 \ m/s)^2 + \frac{1}{2}(4.00 \ kg)(3.00 \ m/s)^2$$
$$= 594 \ J$$

b. The position vector of the center of mass of the two particles is (from Equation 9.20)

$$\vec{r}_{CM} = \frac{1}{m_1 + m_2}(m_1\vec{r}_1 + m_2\vec{r}_2)$$

If you differentiate this equation with respect to time, you find an expression for the velocity of the center of mass in terms of the velocities of the particles:

$$\vec{v}_{CM} = \frac{1}{m_1 + m_2}(m_1\vec{v}_1 + m_2\vec{v}_2)$$

Using the coordinate system indicated in Figure 9.35,

$$\vec{v}_{CM} = \frac{(8.00 \ kg)(12.00 \ m/s)\,\hat{\imath} + (4.00 \ kg)(-3.00 \ m/s)\,\hat{\imath}}{8.00 \ kg + 4.00 \ kg}$$
$$= (7.00 \ m/s)\,\hat{\imath}$$

c. Since the 8.00 kg mass is moving parallel to $\hat{\imath}$ at speed 12.00 m/s and the center of mass is also moving parallel to $\hat{\imath}$ but at speed 7.00 m/s, the velocity of the 8.00 kg mass with respect to the center of mass is $(5.00 \ m/s)\hat{\imath}$. Since the 4.00 kg is moving parallel to $-\hat{\imath}$ at speed 3.00 m/s while the center of mass is approaching it at speed 7.00 m/s, the velocity of the 4.00 kg mass with respect to the center of mass is $(-10.00 \ m/s)\hat{\imath}$.

d. The kinetic energy of the two particles with respect to the center of mass is calculated using their speeds with respect to the center of mass, found in part (c):

$$\frac{1}{2}(8.00 \ kg)(5.00 \ m/s)^2 + \frac{1}{2}(4.00 \ kg)(10.00 \ m/s)^2 = 300 \ J$$

e. You determine the kinetic energy of the center of mass point by imagining the mass of the whole system at that point, moving with speed v_{CM}:

$$KE_{CM} = \frac{1}{2}(8.00 \ kg + 4.00 \ kg)(7.00 \ m/s)^2$$
$$= 294 \ J$$

Equation 9.35 computes the total kinetic energy of the two masses as the sum of the kinetic energy of the center of mass

point (which you just found) and the kinetic energy of the particles with respect to the center of mass (found in part (d)):

$$KE_{total} = 294 \ J + 300 \ J$$
$$= 594 \ J$$

This is the same as the kinetic energy you found directly in part (a); hence Equation 9.35 correctly determines the total kinetic energy.

9.12 THE VELOCITY OF THE CENTER OF MASS FOR COLLISIONS*

When various masses collide in a complicated manner, does the velocity of the center of mass before and after the collision change? Perhaps surprisingly, the velocity of the center of mass is conserved (i.e., unchanged) if the external forces on the system of colliding particles during the collision are negligible compared with the internal collision forces. Let's now show why this is true.

Consider a system of two particles of masses m_1 and m_2 moving with velocities $\vec{v}_{1 \ bfr}$ and $\vec{v}_{2 \ bfr}$, respectively, and subjected to zero (or negligible) total external force. Use the coordinate system indicated in Figure 9.36.

The position vector of the center of mass of the two-particle system is, using Equation 9.20,

$$\vec{r}_{CM} = \frac{1}{m_1 + m_2}(m_1\vec{r}_1 + m_2\vec{r}_2)$$

The velocity of the center of mass is found by differentiating the equation for $\vec{r}_{CM}$ with respect to time:

$$\vec{v}_{CM} = \frac{d\vec{r}_{CM}}{dt}$$
$$= \frac{1}{m_1 + m_2}(m_1\vec{v}_{1 \ bfr} + m_2\vec{v}_{2 \ bfr})$$

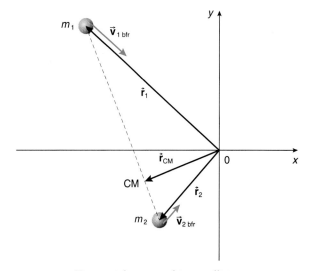

FIGURE 9.36 Two particles approaching a collision.

Rearrange this slightly to find

$$(m_1 + m_2)\vec{v}_{CM} = m_1\vec{v}_{1\,bfr} + m_2\vec{v}_{2\,bfr} \qquad (9.36)$$

The terms on the right-hand side of Equation 9.36 are the momenta of the individual particles; the term on the left-hand side of the equation is the *momentum of the center of mass*, if you imagine the entire mass of the system located at that point.

If the two particles undergo a collision and emerge with velocity vectors $\vec{v}_{1\,aft}$ and $\vec{v}_{2\,aft}$, the total momentum after the collision is

$$m_1\vec{v}_{1\,aft} + m_2\vec{v}_{2\,aft}$$

But we know that momentum is conserved in all collisions where external forces are negligible compared with the internal forces arising from the collision. Therefore

$$m_1\vec{v}_{1\,bfr} + m_2\vec{v}_{2\,bfr} = m_1\vec{v}_{1\,aft} + m_2\vec{v}_{2\,aft}$$

Using this equality in Equation 9.36, we find

$$(m_1 + m_2)\vec{v}_{CM} = m_1\vec{v}_{1\,bfr} + m_2\vec{v}_{2\,bfr}$$
$$= m_1\vec{v}_{1\,aft} + m_2\vec{v}_{2\,aft}$$

> This means that the velocity of the center of mass is the same before and after the collision.

Although we derived this result considering just a two-particle collision, the result is true for any number of colliding particles because conservation of momentum, the key step, is true for all collisions (with negligible external forces).

QUESTION 6

If the velocity of the center of mass is *unchanged* by a collision, why, then, do two colliding cars invariably come to rest after a collision even if their brakes are not used?

9.13 THE CENTER OF MASS REFERENCE FRAME*

There is great flexibility in choosing the origin for a coordinate system. A reference frame with its origin at the center of mass point of a system is called the **center of mass reference frame**. If the motion of a system is described from the vantage point of the center of mass reference frame, we find that the total momentum of the system is *zero*.

To see that this is the case, recall from Equation 9.20 that the position vector of the center of mass is

$$\vec{r}_{CM} = \frac{1}{m_{total}} \sum_i m_i\vec{r}_i$$

where m_{total} is the mass of the entire system of particles. If we choose the origin for the coordinate system to be at the center of mass itself, then the position vector of the center of mass point is zero; that is, $\vec{r}_{CM} = 0$ m, or

$$0 \text{ m} = \frac{1}{m_{total}} \sum_i m_i\vec{r}_i'$$

where the $\vec{r}_i'$ locate each particle with respect to the center of mass as the origin. Since $m_{total} \neq 0$ kg, the summation must be zero:

$$0 \text{ kg} \cdot \text{m} = \sum_i m_i\vec{r}_i'$$

Differentiating this equation with respect to time, we find

$$0 \text{ kg} \cdot \text{m/s} = \sum_i m_i\vec{v}_i'$$

where $\vec{v}_i'$ is the velocity of the ith particle relative to the center of mass frame. Each term in the sum is the momentum $\vec{p}_i'$ of the respective particle in the center of mass frame, and so the summation is the total momentum $\vec{p}_{total\,wrt\,CM}$ of the system when measured with respect to the center of mass; that is,

$$0 \text{ kg} \cdot \text{m/s} = \sum_i \vec{p}_i'$$
$$= \vec{p}_{total\,wrt\,CM}$$

> Thus the total momentum of the system is zero when measured with respect to the center of mass. For this reason, the center of mass reference frame, with its origin at the center of mass, also is known as a **zero momentum reference frame**.

Whether such a reference frame is an inertial (nonaccelerated) reference frame depends on the absence of any appreciable external force.

QUESTION 7

Since the total momentum of a system is zero in its center of mass reference frame, why is the total kinetic energy *not* zero (unless the particles all are at rest in this reference frame)?

STRATEGIC EXAMPLE 9.16

Two particles are sliding on a frictionless ice rink toward an impending collision, as shown in Figure 9.37. Thanks to some Velcro, the particles make a one-dimensional, completely inelastic collision, as indicated in Figure 9.38.

a. Find the velocity of the center of mass point according to an observer at rest on the ice rink.
b. Find the momentum of each particle with respect to the center of mass reference frame.
c. What is the total momentum of the two particles in the center of mass reference frame?
d. Find the velocity of the composite particle in the laboratory reference frame.

FIGURE 9.37

FIGURE 9.38

Solution

a. The position vector of the center of mass point in the lab is found from Equation 9.20:

$$\vec{r}_{CM} = \frac{1}{m_{total}} \sum_i m_i \vec{r}_i$$

The velocity of the center of mass point is found by differentiating this with respect to time:

$$\vec{v}_{CM} = \frac{1}{m_{total}} \sum_i m_i \vec{v}_i$$

Using the coordinate system indicated in Figure 9.37, this becomes

$$\vec{v}_{CM} = \frac{\left(3.00 \ \text{kg}\right)\left(10.00 \ \text{m/s}\right)\hat{i} + \left(5.00 \ \text{kg}\right)\left(2.00 \ \text{m/s}\right)\left(-\hat{i}\right)}{3.00 \ \text{kg} + 5.00 \ \text{kg}}$$

$$= (2.50 \ \text{m/s}) \,\hat{i}$$

The center of mass point is moving parallel to $\hat{i}$ at a constant speed of 2.50 m/s as seen by the observer at rest on the ice rink.

b. Since the 3.00 kg particle is moving to the right at a speed of 10.00 m/s and the center of mass point is moving to the right at a lower speed of 2.50 m/s, the speed of the 3.00 kg particle with respect to the center of mass point is the difference, or 7.50 m/s, also to the right, as shown in Figure 9.39. Thus the

FIGURE 9.39

velocity of the 3.00 kg particle with respect to the center of mass point is $(7.50 \ \text{m/s})\hat{i}$.

The momentum of the 3.00 kg particle in the center of mass reference frame then is

$$\vec{p}_1' = (3.00 \ \text{kg})(7.50 \ \text{m/s})\hat{i}$$
$$= (22.5 \ \text{kg·m/s})\hat{i}$$

The 5.00 kg particle is moving to the left at a speed of 2.00 m/s while the center of mass point is moving to the right at a speed of 2.50 m/s. Hence the speed of the 5.00 kg particle with respect to the center of mass is the sum, or 4.50 m/s. Viewed from the center of mass, the 5.00 kg particle is moving to the left—that is, toward $-\hat{i}$. Hence the velocity of the 5.00 kg particle in the center of mass reference frame is $(-4.50 \ \text{m/s})\hat{i}$, as shown in Figure 9.39. The momentum of the 5.00 kg particle in the center of mass reference frame is

$$\vec{p}_2' = (5.00 \ \text{kg})(-4.50 \ \text{m/s})\hat{i}$$
$$= (-22.5 \ \text{kg·m/s})\hat{i}$$

c. The total momentum of the system with respect to the center of mass point is

$$\vec{p}_{total \ wrt \ CM} = \vec{p}_1' + \vec{p}_2'$$
$$= (22.5 \ \text{kg·m/s})\hat{i} + (-22.5 \ \text{kg·m/s})\hat{i}$$
$$= 0 \ \text{kg·m/s}$$

This calculation shows explicitly that the total momentum of the system in its center of mass reference frame is zero.

d. The velocity $\vec{v}$ of the composite particle in the laboratory is found by conserving the total momentum of the system in the ice rink reference frame:

$$\vec{P}_{total \ bfr} = \vec{P}_{total \ aft}$$
$$(3.00 \ \text{kg})(10.00 \ \text{m/s})\hat{i} + (5.00 \ \text{kg})(2.00 \ \text{m/s})(-\hat{i})$$
$$= (3.00 \ \text{kg} + 5.00 \ \text{kg})\vec{v}$$

Solving for $\vec{v}$, you find

$$\vec{v} = (2.50 \ \text{m/s})\hat{i}$$

This is the same as the velocity of the center of mass, calculated in part (a). Why? Since the collision is completely inelastic, the center of mass point is coincident with the composite particle itself. Therefore, after the collision, the velocity of the particle and the velocity of the center of mass point must be the same: they are the same point!

CHAPTER SUMMARY

The classical *momentum* $\vec{p}$ of a particle of mass m is defined to be

$$\vec{p} \equiv m\vec{v} \tag{9.1}$$

where $\vec{v}$ is its velocity. This relationship is valid as long as the speed of the particle is much less than the speed of light.

The most general statement of Newton's second law of motion is

$$\vec{F}_{total} = \frac{d\vec{p}_{total}}{dt} \tag{9.2}$$

where $\vec{p}_{total}$ is the total momentum of the system.

The *impulse* $\vec{I}$ imparted by a force $\vec{F}(t)$ during a time interval Δt is the integral of the force over the time interval:

$$\vec{I} \equiv \int_{t_i}^{t_f} \vec{F}(t)\, dt \tag{9.4}$$

$$= \vec{F}_{ave}\, \Delta t \tag{9.6}$$

where $\Delta t \equiv t_f - t_i$.

The *impulse–momentum theorem* states that impulse of the *total force* on a system is equal to the change in the total momentum of the system:

$$\vec{I} = \Delta\vec{p} \tag{9.5}$$

If the total force $\vec{F}_{total}$ on a system of one or more particles is equal to zero, then the total momentum of the system is a *constant vector*; that is, the total momentum is *conserved*. This is known as *conservation of momentum*. The individual momenta of the particles of the system may change because of the existence of forces internal to the system, but the total momentum of the system does not change.

The total momentum of a system of particles is conserved in collisions, disintegrations, and explosions; that is,

$$\vec{p}_{total\, bfr} = \vec{p}_{total\, aft} \tag{9.16}$$

Collisions are classified in the following way:

1. *Elastic collisions*: The total kinetic energy of the system of particles is conserved; that is,

$$KE_{total\, bfr} = KE_{total\, aft}$$

2. *Inelastic collisions*: The total kinetic energy of the system of particles is *not* conserved; that is,

$$KE_{total\, bfr} \neq KE_{total\, aft}$$

A collision is *completely inelastic* if all the particles stick together after the collision.

The position vector $\vec{r}_{CM}$ of the *center of mass* point of a system of particles is determined from

$$\vec{r}_{CM} = \frac{1}{m_{total}} \sum_i m_i \vec{r}_i \tag{9.20}$$

where m_{total} is the total mass of the system,

$$m_{total} = \sum_i m_i$$

and $\vec{r}_i$ is the position vector of the ith particle (of mass m_i) with respect to the origin of a coordinate system.

For a system of one or more particles, the total force on the system changes the velocity of the center of mass:

$$\vec{F}_{total} = m_{total} \frac{d\vec{v}_{CM}}{dt} \tag{9.28}$$

The total kinetic energy of a system of particles is the sum of the kinetic energy of the center of mass (imagining the total mass of the system to be located at that point) and the kinetic energy of the particles with respect to the center of mass:

$$KE_{total} = KE_{of\, CM} + KE_{wrt\, CM} \tag{9.35}$$

$$= \frac{1}{2} m_{total} v_{CM}^2 + \sum_i \frac{1}{2} m_i v_i'^2 \tag{9.34}$$

If a coordinate system has its origin at the center of mass point, the reference frame is called the *center of mass reference frame*. The total momentum of a system of particles in its center of mass reference frame is zero, and so such a reference frame also is known as a *zero momentum reference frame*. The center of mass reference frame is an inertial frame if the total (external) force on the system is zero.

In collisions, disintegrations, and explosions, the velocity of the center of mass point is the same immediately before and after the collision, disintegration, or explosion.

SUMMARY OF PROBLEM-SOLVING TACTICS

9.1   **(page 392)** None of the many specialized equations for various types of specific collisions is worth remembering; just remember the general points highlighted on pages 390–391.

QUESTIONS

1. (page 383); 2. (page 385); 3. (page 392); 4. (page 402); 5. (page 405); 6. (page 408); 7. (page 408)

8. Is it ever possible for the momentum to be *antiparallel* to the velocity vector? Explain why or why not.

9. A large mass M and a small mass m have equal kinetic energies. Which has the greater momentum?

10. Compare and contrast the colloquial meanings of the words *momentum* and *impulse* with their meanings in physics.

11. What condition must be satisfied for you to say the total momentum of a system is conserved?

12. The forces involved in automobile collisions can be lethal if too great. Explain how air bags are able to decrease the magnitudes of average forces in collisions and thereby save lives.

13. What is wrong with the physics in the editorial statement quoted at the beginning of Section 9.3?

14. When a rock is dropped, its momentum is not conserved. Explain why.

15. Is it possible for a system of two moving masses to collide, leaving both at rest? Explain.

16. If the total kinetic energy of a system is conserved, what can be said about the work done by the total force on the system?

17. It *is* possible to conserve the total momentum of a system but not its total kinetic energy (e.g., in inelastic collisions). Is it possible to conserve the total kinetic energy of a system but *not* the total momentum? Explain.

18. In an elastic collision, is the potential energy of the system conserved? Is the potential energy of the system the same at all instants during the time interval of the collision?

19. The total momentum of a system is conserved if the total force on the system is zero. If the total force on the system is zero and remains zero, we also can say that there is a "total force conservation law." Why is this "total force conservation law" *not* as useful as momentum conservation in dynamical situations?

20. A physicist observes an unstable particle at rest. Suddenly the particle breaks up (we say *decays*) into at least two fragments, perhaps more. The paths and velocity vectors of two of the resulting fragments are noted and sketched in Figure Q.20. In which cases (if any) can the physicist be certain of the existence of at least one other additional, unobserved particle associated with the decay? Explain your reasoning. For each case, what can be said (if anything) about the relative masses of the fragments?

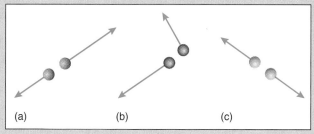

(a)   (b)   (c)

FIGURE Q.20

21. A piece of bubble gum is thrown at and sticks to a wall. Is the momentum of the gum conserved? What happens to the momentum of the gum?

22. In a collision, the occupants of a massive car are more likely to survive than the occupants of a compact car of small mass. Give a critique of why this is so. What can be said about the magnitude of the force that each car exerts on the other at each instant during the collision?

23. Explain why a single particle at rest cannot disintegrate (or decay) into *two* particles, one of which is at rest. Is a decay into three particles, with one at rest, possible?

24. If a single particle at rest disintegrates (or decays) into two particles, explain why the two particles *must* go off in opposite directions.

25. A small compact car collides head-on with a huge trailer truck moving in the opposite direction. What is the relationship between the force of the car on the truck and the force of the truck on the car during the collision? Explain your answer.

26. How could you design a rifle to minimize its recoil when it is fired?

27. Explain the physics associated with the executive toy known as Newton's cradle, pictured in Figure Q.27.

FIGURE Q.27

28. A mass m moving at speed v collides elastically with an identical mass at rest. Compare the initial kinetic energy associated with this collision with one in which the two masses approach each other, both initially with speed v.

29. Is it possible for a mass to have nonzero momentum yet zero total mechanical energy? Is it possible for a mass to have nonzero total mechanical energy yet zero momentum? Explain your answers.

30. A radioactive nucleus at rest emits a β^- particle (an electron). The direction of the emitted electron and that of the recoiling so-called daughter nucleus are found experimentally to be *not* in opposite directions. Explain why this observation means that an *additional particle* must have been emitted in the decay process. The third particle is known as an antineutrino.

31. Explain the underlying physical principles behind jet engines and rocket propulsion.

32. Is it possible for high jumpers to pass *over* the bar while their centers of mass pass *under* the bar? Explain your answer, perhaps with a series of sketches.

33. In the argument of Section 9.7, since the particle makes elastic collisions with the circular hoop, why is the momentum of the particle *not conserved* in each collision?

34. While meditating. . . . A small impulse applied to a roll of toilet paper can unwind the entire roll. A large impulse rips off a single sheet. Explain.

35. Must there be mass at the center of mass?

36. What configuration of planets in the solar system places the center of mass of the solar system at the greatest distance from the Sun? (The planets of the solar system are more or less confined to within a few degrees of the orbital plane of the Earth, known as the ecliptic plane.)

37. Sketch the shape of several objects whose center of mass lies outside the object itself.

38. An imaginary sphere of radius R is barely circumscribed around a massive object of arbitrary shape. No part of the mass lies outside the circumscribed sphere. Is the center of mass of the object *guaranteed* to be within the sphere? Explain why or why not.

39. Stand up straight with both feet slightly spread apart. If you raise one foot off the floor, what must happen to the rest of your torso to prevent yourself from falling?

40. Explain why it is not possible to stand next to a wall with one foot in front of the other *and* with the sides of both feet touching the wall.

41. A traditional time-keeping hourglass is placed on a sensitive scale to determine its mass (see Figure Q.41a). The hourglass now is inverted (see Figure Q.41b). Has the mass of the hourglass changed? Will the scale indicate the same value as in Figure Q.41a? Explain.

42. If the motion of a system is described using the center of mass reference frame, is the total momentum of the system zero even if the center of mass point is undergoing a nonzero acceleration as seen in the laboratory? Justify your answer.

(a)

(b)

FIGURE Q.41

PROBLEMS

Section 9.1 Momentum and Newton's Second Law of Motion

1. A rock-filled dump truck with a mass of 2.0×10^4 kg is creeping along at a speed of only 1.0 km/h. How fast would you have to run to have the same magnitude of momentum as that of the truck? Assume your mass is 60 kg. Express your speed in km/h.

2. A 100 g bullet has a speed of 400 m/s. How fast would you have to run to have the same magnitude of momentum? Assume your mass is 60 kg.

3. (a) The α-particle (a helium nucleus with mass 6.68×10^{-27} kg) emitted in the radioactive decay of $^{238}_{92}$U has a speed of 1.4×10^7 m/s. Consider this speed to be small compared with the speed of light (it is only about 5% of the speed of light). What is the magnitude of the momentum of the α-particle? (b) What is the speed of a 1.00 g mass that has the same magnitude of momentum? (c) How long would it take the 1.00 g mass to traverse 1.00 m at this speed? Express your result in years.

4. (a) Show that the kinetic energy of a system of mass m can be written in terms of the magnitude of its momentum as

$$KE = \frac{p^2}{2m}$$

(b) Show that the kinetic energy of the system also can be written as

$$KE = \frac{pv}{2}$$

where v is the speed of the system. The first of these expressions is used quite frequently; the second expression almost never is seen because p is a function of v as well.

5. How fast must a 60.0 kg physics student run so she has the same magnitude of momentum as a 1.00×10^3 kg automobile moving at 1.00 km/h?

6. Your car of mass 1.20×10^3 kg has a horizontal speed of 100 km/h in a northwest direction. A Cartesian coordinate grid has $\hat{\imath}$ east and $\hat{\jmath}$ north, as in Figure P.6. (a) What is the momentum of the car? (b) What is the magnitude of the momentum?

FIGURE P.6

7. A 50 caliber machine gun* fires bullets, each with a mass of about 45 g, at a speed of about 8.0×10^2 m/s. (a) What is the magnitude of the momentum of each bullet? (b) How fast would a 70 kg soldier have to run to have the same magnitude of momentum?

---

*The caliber number gives the diameter of the bullet in hundredths of an inch. A 50 caliber bullet has a diameter of 0.50 inch.

•8. A 2.73 kg bag of bread flour accidentally is dropped from rest from a height 2.00 m above the floor of a supermarket. What is the magnitude of the momentum of the bag the instant before its impact with the floor?

•9. For breakfast, a herring gull (*Larus argentatus*) flying horizontally at speed 15.0 m/s drops a tasty cockleshell (*Dinocardium robustum*) of mass 0.200 kg from a height of 20.0 m above a flat rocky beach as in Figure P.9. What is the magnitude of the momentum of the bivalve the instant before it strikes the beach?

FIGURE P.9

Section 9.2 Impulse–Momentum Theorem

10. Show that the SI units for impulse, N·s, are the same as those for momentum, kg·m/s, and therefore that the impulse–momentum theorem has the same SI units on each side of the equation.

11. A fire hose discharges water at a rate of 15.0 kg/s with a speed of 20 m/s as in Figure P.11. What magnitude of force is needed to hold the nozzle stationary?

FIGURE P.11

•12. Golf pro Guy Boros whacks a golf ball of mass 45.0 g. The ball leaves the tee at an angle of 45.0° to the horizontal and with a speed of 60.0 m/s. (a) Choose an appropriate coordinate system. (b) What impulse was given to the ball? (c) If the club is in contact with the ball for 1.00 ms, what is the magnitude of the average force of the club on the ball? See Figure P.12.

FIGURE P.12

•13. An ostrich (*Struthio camelus*) egg of mass 1.00 kg is dropped from rest from a height of 10.0 m. (a) Choose a convenient co-ordinate system and indicate it clearly in a diagram. (b) What is the sum of the kinetic and potential energies of the egg at a height of 5.0 m? (c) Does the momentum of the egg change at a constant rate with time as it falls, or not? Explain. (d) Find the impulse imparted to the egg by the gravitational force during the fall to the instant before impact.

•14. A supertanker of mass 1.0×10^5 metric tons traveling at 25 km/h runs aground on a reef and is brought to a complete stop in 8.0 s. Crunch. Call out the environmental lawyers. What is the magnitude of the average horizontal force of the reef on the tanker? (1 metric ton = 1000 kg)

•15. A golf club hits a golf ball of mass 0.045 kg and gives it a speed of 50 m/s. Estimate the magnitude of the average force of the club on the ball if the collision lasts for 1.0 ms.

•16. A battleship at rest fires a 400 mm diameter cylindrically symmetric shell with a mass of 1.00×10^3 kg. BOOM! The barrel is aimed off the stern and elevated at an angle of 30.0° to the horizontal. The shell emerges from the gun with a speed of 600 m/s. (a) What impulse is given initially to the shell in the firing? (b) What impulse is given to the battleship in the firing? (c) If the ship has a mass of 5.0×10^4 metric tons, what is the recoil speed of the battleship? Assume the ship can move both horizontally and vertically.

•17. The time dependence of a force component $F_x(t)$ on a system is indicated in Figure P.17. (a) What is the impulse of the force during the first 2.00 s? (b) What is the average force during the first 2.00 s? (c) What is the impulse of the force during the first 4.00 s? (d) What is the average force during the first 4.00 s?

FIGURE P.17

•18. (a) During the time interval from $t = 0$ s to 4.0 s, a total force whose only component is $F_x(t)$ acts on a moonshine jug system; see Figure P.18. Find the impulse of the total force. (b) What is the change in the momentum of the system during the same time interval?

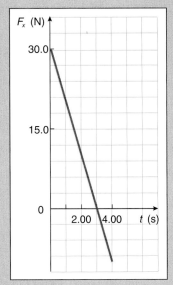

FIGURE P.18

•19. A 15 g cricket (*Acheta domestica*) leaps through a 2.20 m horizontal distance. The initial velocity of the cricket makes a 30° angle with the horizontal direction. If it takes the cricket 0.010 s to leave the ground, what is the magnitude of the impulse of the total force on the cricket that gives it the speed needed for the jump?

•20. A cockroach (*Periplaneta americana*) of mass 0.0100 kg runs off a precipice at speed 1.50 m/s and falls to the ground 2.00 m below as shown in Figure P.20. (a) Indicate a convenient coordinate system. (b) What is the initial momentum of the roach? (c) What is the momentum of the roach at the instant just before impact? (d) What is the *change* in the momentum of the roach? (e) What force provides the impulse to change the momentum of the roach? (f) Calculate the impulse of this force using

$$\vec{\mathbf{I}} = \int_{t_i}^{t_f} \vec{\mathbf{F}} \, dt$$

and show that the result is the same as the change in the momentum calculated in part (d).

•21. There is an old adage that goes: "It's not the fall that hurts, it's the sudden stop." Estimate the magnitude of the acceleration experienced by a 60 kg painter during the sudden stop associated with a fall off a scaffold at an elevation of 8.0 m. Assume the painter is brought to rest in 0.15 s.

•22. Marbles of mass *m* are thrown at speed *v* at a rigid wall as shown in Figure P.22. Assume the collision of each marble

FIGURE P.20

with the wall is elastic. (a) What is the momentum change of each marble? (b) What is the impulse of the force of the wall on each marble? (c) What is the impulse of the force of a marble on the wall? (d) If the marbles are thrown at a rate of *n* marbles each second, what is the average force of the marbles on the wall? (e) If the marbles hit the wall in a randomly distributed manner over an area A, what is the average force per square meter of the marbles on the wall? The magnitude of the force per unit area is called the *pressure*.

FIGURE P.22

•23. Repeat Problem 22 but consider the marbles and wall to be covered with contact cement, so that the collisions of the marbles with the wall are completely inelastic collisions.

•24. Repeat Problem 22 but assume that the marbles encounter and rebound from the wall at an angle *θ* with respect to a normal line as shown in Figure P.24. Assume the collisions with the wall are elastic.

Section 9.3 The Rocket: A System with Variable Mass*

25. What fuel to payload ratio is need for a rocket to achieve a final speed equal to that of its exhaust gases if the rocket is launched at rest well away from any other large masses?

26. The exhaust gases leave a certain rocket with a speed of 2.5 km/s with respect to the rocket. What fuel to payload ratio is needed for the rocket to achieve a final speed equal to the escape speed from the Earth if the rocket is initially at rest well away from any other masses (including the Earth)? Will the fuel to

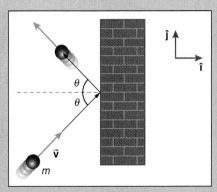

FIGURE P.24

(d) Integrate this equation to show that

$$v = \left(1 - \frac{m_0}{m}\right)v_0$$

‡29. A massive anchor chain of length ℓ is held vertical from its top link so that its lowest link is barely in contact with the horizontal ground. The chain is dropped. When the top link has fallen a distance y (see Figure P.29), show that the magnitude of the normal force of the ground on the chain is $3mgy/\ell$.

payload ratio have to be greater or the same if the rocket actually is launched from the Earth to achieve the escape speed? Explain your answer.

•27. The Boeing 747-400 has the maximum, fully loaded takeoff mass of any civilian aircraft: a whopping 3.956×10^5 kg. A newly developed jet engine for this plane can produce 333 kN of thrust. Using four such engines, what is the magnitude of the maximum acceleration of the fully loaded aircraft? Neglect effects of air resistance on the plane, the mass of air needed by each engine per second for combustion, as well as the (comparatively small) mass loss during takeoff.

‡28. A fire hose fixed to a support delivers a horizontal stream of water at a speed v_0 and a mass flow rate of β kilograms per second. (a) The water stream is directed against a wall along a perpendicular to the wall and is brought to rest. What is the magnitude of the force of the water on the wall? (b) What is the momentum per unit length of the stream? (c) The stream now is directed into a hole in the back of a tank car (see Figure P.28) that is free to roll from rest (with no frictional work done). The tank car has an initial mass m_0. Let the total mass of the tank car be m (tank car plus accumulated water). Let v be the x-component of the velocity of the tank car at time t. Show that

$$m\, dv = (v_0 - v)\, dm$$

FIGURE P.29

Sections 9.4 Conservation of Momentum
        9.5 Collisions
        9.6 Disintegrations and Explosions
        9.7 The Centripetal Acceleration Revisited*
        9.8 An Alternative Way to Look at
           Force Transmission*

30. You are running at speed v_0 to another dynamic physics lecture and unfortunately collide in a completely inelastic, one-dimensional manner with a buddy of equal mass at rest. What fraction of the initial kinetic energy is lost in the collision?

•31. Two particles, of masses m and $2m$, are incident as indicated in Figure P.31. The particles collide and, thanks to the wonders of contact cement, stick together after the collision. (a) What type of collision is this? (b) Introduce a Cartesian coordinate system to analyze the problem; use the impact point as the origin and clearly indicate the directions of the positive coordinate axes. (c) What is the total momentum of the system of two particles before the collision? (d) What is the velocity

For part (a)   For part (c)

FIGURE P.28

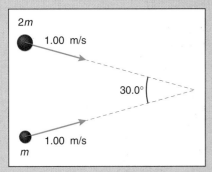

FIGURE P.31

vector of the composite particle after the collision? (e) What is the speed of the composite particle after the collision? (f) What is the angle that the velocity vector of the composite particle makes with the original direction of m after the impact? (g) What is the change in the kinetic energy of the system?

•**32.** Trouble has come to River City. Pool. A hustler takes a cue and sends the cue ball at speed v_0 slamming into the 6 ball initially at rest (see Figure P.32). The collision is elastic and the 6 ball clunks into a pocket of the table. The two balls have equal mass. Treat the balls as particles, neglecting any rotation. Show that the speed of the rebounding cue ball is $0.500v_0$ and that the speed of the 6 ball after the collision is $0.865v_0$.

Before

After

FIGURE P.32

•**33.** A 1.00×10^3 kg car is moving toward increasing values of x along the x-axis and strikes a stationary 4.00×10^3 kg truck. CRASH! Immediately after the collision, the car is motionless and the truck is moving with the velocity of $(2.00 \text{ m/s})\hat{\imath}$.

Assume only collision forces are relevant. (a) What was the original velocity of the car? (b) Was the collision elastic? Support your answer with a calculation. (c) Was the momentum of the *car* conserved? Explain.

•**34.** Two pucks slide on a frictionless, horizontal airtable. One puck has a mass of 0.200 kg and a velocity of $(3.00 \text{ m/s})\hat{\imath} - (4.50 \text{ m/s})\hat{\jmath}$. The second puck has mass 0.250 kg and velocity $(-7.10 \text{ m/s})\hat{\imath}$. The pucks collide, leaving the first puck at rest. (a) Find the velocity of the second puck after the collision. (b) What is the change in the kinetic energy of the system of two pucks?

•**35.** A quarterback of mass 100 kg, running at speed 10.0 m/s toward increasing values of x, runs (stupidly) directly into the arms of a 150 kg tackler moving in the opposite direction at 5.0 m/s. The impact occurs in one dimension in midair and lasts for 0.20 s, after which the players are locked together. (a) Is the momentum of *each individual* conserved in the collision? Explain. (b) What is the velocity of the composite mass after the collision? (c) What is the impulse imparted to the quarterback by the tackler? (d) What is the magnitude of the average force exerted by the tackler on the quarterback? Compare this with the magnitude of the weight of the quarterback.

•**36.** A cart with a mass of 400 kg is sliding across a frictionless ice pond with a speed of 10.0 m/s. A horrendous downpour dumps a mass m of rain (falling vertically) into the cart, slowing it to 8.00 m/s. Define an appropriate system. (a) Is the horizontal component of the momentum of your system conserved in this process? Explain. (b) Find the mass of rain that fell into the cart.

•**37.** Two cars of equal mass $m = 1.00 \times 10^3$ kg are approaching an intersection at speeds of 60.0 km/h and 40.0 km/h as indicated in Figure P.37. Neither driver yields the right of way and the two jerks collide. The cars stick together on impact. (a) What kind of a collision is this? (b) What is the total momentum of

FIGURE P.37

the system of two cars before the collision? (c) What is the total momentum after the collision? (d) What is the speed of the crumpled composite car immediately after the collision? (e) What was the change in the kinetic energy of the system of two cars in the collision?

•**38.** A 10.0 g bullet is fired into and remains in a 3.00 kg block of wood located 1.00 m from the edge of a frictionless table as indicated in Figure P.38. The block slides along the table surface, falls off the table, and lands 2.50 m from the base of the table (of height 75.0 cm). (a) Does the table move? Explain why or why not. (b) Determine the speed of the bullet the instant before it hits the block.

FIGURE P.38

•**39.** Five equal masses are free to slide along a horizontal friction-less surface as indicated in Figure P.39. Initially four of the masses are at rest and the fifth is approaching the others at a speed v. At each collision the particles stick together. (a) What is the total momentum of the system of five masses before the first collision? After the last collision? (b) Calculate the speed of the composite particle after each collision in turn. (c) Find the change in the kinetic energy of the system of five masses, taking the final kinetic energy as that after the last collision and the initial KE as that before the first collision.

FIGURE P.39

•**40.** Thanks to front and side air bags, a speeding sports car of mass 1.50×10^3 kg makes a nonfatal, completely inelastic collision with a truck of mass 3.00×10^3 kg [which includes its load of turkeys (*Meleagris gallopavo*)] traveling at speed 25.0 m/s. After the collision the vehicle pair is moving (with uninjured drivers) at an angle of 70.0° with respect to the initial direction of the car (as indicated in Figure P.40) along with some *very* agitated gobblers. (a) Establish a coordinate system to analyze the problem, clearly labeling the positive direction of the axes. (b) Determine the *speed* of the smashed car–truck system immediately after the collision. (c) Determine the initial speed of the car.

•**41.** Your 1050 kg car is skidding across a frictionless ice pond in northern Minnesota with a velocity 22 m/s east. Unfortunately another car of mass 900 kg is sliding at speed 18 m/s directed 30° south of east and collides with and sticks to your car. Crunch! Now cry. (a) After the collision, what is the velocity of the vehicles? (b) What is the change in the kinetic energy of the two-car system as a result of the collision?

FIGURE P.40

•**42.** The muzzle speed of a bullet can be determined using a device called a ballistic pendulum, shown in Figure P.42. A bullet of mass m moving at speed v encounters a large mass M hanging vertically as a pendulum at rest. The mass M absorbs the bullet. The hanging mass (now consisting of M + m) then swings to some height h above the initial position of the pendulum as shown. (a) Show that the initial speed v' of the pendulum (with the embedded bullet) after impact is

$$v' = \frac{mv}{M + m}$$

(b) Show that the muzzle speed v of the bullet is

$$v = \frac{M + m}{m} (2gh)^{1/2}$$

(c) If h = 10.0 cm, M = 2.50 kg, and m = 10.0 g, find v.

FIGURE P.42

•**43.** There has been increasing concern about the inevitability of a collision between the Earth and an incident asteroid. (a) If a vertically incoming asteroid has a speed of 50 km/s and makes a completely inelastic collision with the Earth, estimate the mass of an asteroid that will give the Earth a recoil speed of 1.0 mm/s. (b) If the density of the asteroid is 5.0×10^3 kg/m^3 and it is spherical in shape, what is its diameter?

•**44.** A collection of identical pool balls (each of mass 0.165 kg) is at rest on a pool table. A ball of unknown mass is incident upon the collection at speed 1.50 m/s. As a result of the elastic collision, three of the identical pool balls (initially at rest) move off as shown in Figure P.44. All the other balls, including the incident ball, are at rest. Treat the balls as particles. (a) What is the speed of the ball labeled A? (b) What is the mass of the unknown ball?

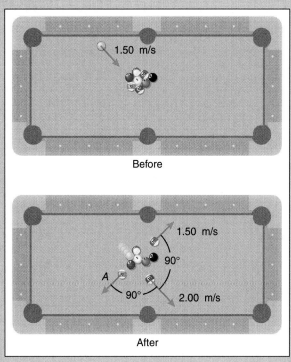

Before

After

FIGURE P.44

•**45.** You are tooling along in a 1.00×10^3 kg sports car in a dense fog. Uh oh. You suddenly collide into the rear end of an 1.80×10^3 kg behemoth that was moving at 15.0 m/s in the same direction. Your cars stick together and move at 20.0 m/s immediately after the collision. Consider the motion of all objects in this problem to be in a straight line with only collision forces relevant. (a) Were you exceeding the legal speed limit of 30.0 m/s (roughly 110 km/h) before the crash? (b) An instant after the foregoing collision, the pair of cars strikes yet another car (of mass 1.50×10^3 kg) head-on that had been traveling at 40.0 m/s in the opposite direction. All the cars stick together. What is the speed of the system of three mangled cars just after this collision?

•**46.** A Cadillac of mass 2250 kg going east collides with a Geo of mass 1200 kg headed north at 25.0 km/h on a level highway. The cars remain tangled and slide to rest after traveling 25.0 m in a straight line at an angle of 20.0° north of east. The coefficient of kinetic friction for the tires on the road is 0.20. How fast was the Cadillac traveling before the unfortunate collision?

•**47.** Two particles collide at the origin of a frictionless horizontal x–y plane. Prior to the collision, the mass $m_1 = 3.50$ kg was moving along the x-axis toward increasing values of x with speed 4.00 m/s, while mass $m_2 = 5.00$ kg was moving along the y-axis toward increasing values of y at a speed 1.50 m/s, as shown in Figure P.47. After the collision, m_1 is at rest at the

origin. (a) Find the velocity of m_2 after the collision. (b) Is the collision elastic or inelastic?

FIGURE P.47

•**48.** A coin of mass m is released from rest on a frictionless plane inclined to the horizontal at angle θ. After sliding a distance d to the bottom of the incline, the coin encounters a rough horizontal surface. The coefficient of kinetic friction for the coin on the surface is μ_k. The coin slides a short distance s and then collides completely inelastically with a coin of mass M that is initially at rest. Find an expression for each of the following in terms of the given quantities. Indicate explicitly what physical principles (such as the CWE theorem, conservation of momentum, etc.) you use for each part. (a) What is the speed of the coin of mass m at the bottom of the incline? (b) What is the speed of the coin of mass m just before the collision? (c) What is the speed of the coins immediately after the collision?

•**49.** A block of mass m is released at rest on a frictionless inclined plane as indicated in Figure P.49. The horizontal floor is rough (the coefficients of friction for m on the floor surface are $\mu_s = 0.40$ and $\mu_k = 0.25$). The mass m slides 0.20 m and makes a completely inelastic collision with another mass M that is initially at rest. The composite particle has a speed 0.25 m/s immediately after the collision. Find the ratio M/m of the two masses.

FIGURE P.49

$50.** A system of two small spheres of identical mass m collide obliquely as shown in Figure P.50. The collision is inelastic. Initially one of the masses is at rest. The momenta of the particles immediately *after* the collision are $\vec{p}_1'$ and $\vec{p}_2'$. The angle between the paths of the masses after the collision is β. Show that the change in the kinetic energy of the system is

$$\Delta KE = -\frac{p_1'p_2'}{m}\cos\beta$$

Notice that for an *elastic* collision, $\Delta KE = 0$ J and $\beta = 90°$ as we found in Example 9.11. (Hint: Write the momentum conservation equation in vector form and the kinetic energies in terms of the magnitudes of the momenta. Use a technique similar to the short and elegant Method 2 in Example 9.11.)

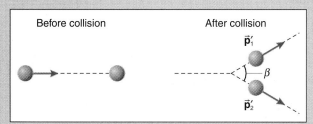

FIGURE P.50

⊞51. The curved inclines of two identical wedge-shaped masses M smoothly merge with a horizontal, frictionless surface as indicated in Figure P.51. The curved surfaces of the inclines also are frictionless. Another mass m is released at rest from a height h on one of the curved, smooth surfaces of the wedges. (a) Consider first the motion of m from its initial height h down to the horizontal plane. Is the total mechanical energy of m and M conserved in the descent? If so, write an equation representing this. Is the total momentum of the system of m and M conserved in the descent? If so, write an appropriate equation expressing this conservation when m reaches the horizontal surface. (b) After m reaches the horizontal plane, it then encounters the right-hand wedge and ascends it to some height h'. Is the total mechanical energy of the system of m and the right-hand wedge conserved in the ascent? If so, write an equation expressing this relationship. Is the total momentum of m and the right-hand wedge M conserved in the ascent? If so, write an equation expressing this relationship. (c) Show that the height h' to which m ascends the right-hand wedge is

$$h' = \frac{M^2}{(M + m)^2}\,h$$

(d) What mass ratio M/m results in $h' = h/4$?
This problem was inspired by a problem in *Quantum*, 3, #2 (November–December 1992), pages 27, 55–56.

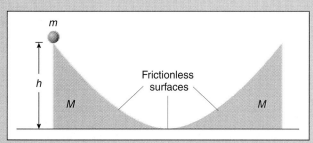

FIGURE P.51

⊞52. A particle of mass M at rest disintegrates into *three* particles. Two of the particles (of masses m and M′) move off with speeds v and v' respectively and directions indicated in Figure P.52. Show that the angle ϕ, locating the trajectory of the third particle, is given by the expression

$$\tan\phi = \frac{mv}{M'v'}\sec\theta - \tan\theta$$

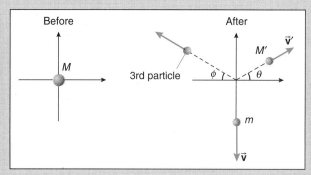

FIGURE P.52

⊞53. For elastic collisions, show that the relative speed of approach before the collision (the speed of one particle with respect to the other particle) is equal to the relative speed of recession after the collision.

⊞54. A ball of large mass M and a ball of small mass m are dropped from the same height h above the floor as shown in Figure P.54a. Neglect the (small) radii of the balls in this problem. (a) Show that if no mechanical energy is lost in the bounce of the large ball from the floor, then the speed with which the large ball rebounds is

$$v = (2gh)^{1/2}$$

(b) The instant after the large mass rebounds from the floor, the small mass [which is still momentarily traveling downward at the speed v given in the answer to part (a)] makes an elastic collision with the large ball rising at the same speed v. Imagine yourself on the large ball just after it rebounds from the floor (see Figure P.54b). What is the speed of approach of the small ball when viewed by an observer on the large ball? (c) Relative to an observer on the large ball, what is the speed at which the small ball moves *after* its elastic collision with the large ball? (d) Show that with respect to the *ground*, the small ball rebounds from this collision with a speed $3v$. (e) Show that the height to which the small ball rebounds is $9h$. This problem can be used to explain why interplanetary space probes occasionally make encounters with planets in maneuvers known colloquially as slingshot effects or gravitational pumping. Imagine a spacecraft traveling at speed v and at a distance r from a planet traveling at speed v' in the opposite direction, as in Figure P.54c. The spacecraft makes an "elastic collision" with the planet by being whipped around the planet by the gravitational interaction. The relative speed of approach of the spacecraft (viewed from the *planet*) is $v' + v$. The speed at which the spacecraft recedes from the encounter, when at the same distance as in Figure P.54d, also is $v' + v$ with respect to the *planet*. But the speed of the spacecraft relative to the reference frame of Figure P.54e is $v' + v$ plus the speed v' of the planet itself—that is,

$$(v' + v) + v'$$

or

$$2v' + v$$

The change in the speed of receding spacecraft is twice the speed of the planet! This example also qualitatively illustrates the extraordinary speeds of matter ejected from supernova explosions of massive stars.

This problem was inspired by G. Stroink, "Superball problem," *The Physics Teacher*, 21, #7, page 466 (October 1983).

FIGURE P.54

FIGURE P.56

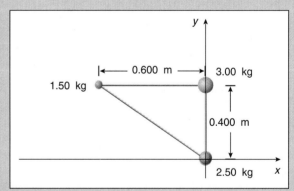

FIGURE P.57

Section 9.9 The Center of Mass

55. A 3.0 kg sack of sugar is located at coordinates $x = 3.0$ m, $y = -4.0$ m. A 4.0 kg sack of industrial grade coffee is located at $x = -2.0$ m, $y = 2.2$ m. Determine the position vector of the center of mass of the two-particle system.

56. For amusement (?), find the location of the center of mass of the collections of particles shown in Figure P.56.

•**57.** Three particles are held in a rigid triangle by thin rods of negligible mass, as shown in Figure P.57. Find the coordinates of the center of mass of the system of three particles.

•**58.** The doohickey shown in Figure P.58 is constructed from a uniform plate of steel and is to be used in a sophisticated ornamental doodad. Find the center of mass of the doohickey in *two different ways* to be sure that your calculations are correct.

•**59.** Find the center of mass of a system of identical particles distributed at the vertices of a regular tetrahedron with sides of length ℓ.

•**60.** A uniform circular plate of material of radius R has an off-center hole cut out as indicated in Figure P.60. The total mass of the plate with the hole is m. The radius of the hole is $R/2$. Find the location of the center of mass of the plate.

FIGURE P.58

FIGURE P.60

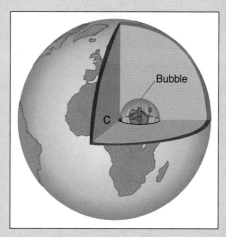

FIGURE P.61

*61. For millennia Hell was thought to be located inside the Earth. (Many students now think it is located at their university during finals.) Suppose we model the quaint Hell-inside-the-Earth cosmology with a uniform sphere of radius R that has an off-center, hellish, spherical bubble of radius $R/4$ as indicated in Figure P.61. The Devil orders you to find the position vector of the center of mass of the system using the center of the large sphere C as the origin of your coordinate axis; choose the x-axis to be directed from the origin to the center of the bubble representing Hell.

*62. An open-topped, thin-walled, cubical box with sides of length ℓ is constructed from sheet metal. Determine the location of the center of mass of the box.

*63. A collection of six uniform planks each of mass m and length ℓ are stacked according to the scheme indicated in Figure P.63. (a) Show that the center of mass of the *top five planks* is located directly over the right edge of the bottom plank. The system will not topple over, even though the top plank is lying totally beyond the edge of the table! (Note that only *five* planks really are needed to accomplish this feat.) You might like to try this stacking experimentally with, say, six meter sticks. It is a good idea to keep them separated by slightly less than the indicated offset distances as a safety precaution in building the stack. (b) The stable stacking can be continued with more planks. If a seventh plank were added at the *bottom* of the stack, what would be its offset relative to the plank directly above it? The general rule governing the offsets is that the right edge of the bottom plank is directly under the center of mass of the collection of planks on top of it.

FIGURE P.63

Section 9.10 Dynamics of a System of Particles

*64. An 80.0 kg ΦBK student stands inside a 20.0 kg canoe that can slide freely on a frictionless water surface; see Figure P.64. Both are initially at rest. The student then runs like crazy at a speed of 3.0 m/s with respect to the canoe along its length (ignore the thwarts of the canoe). (a) Describe and explain what happens to the center of mass of the student–canoe system. (b) How fast is the student moving relative to the water? How fast is the canoe moving relative to the water?

FIGURE P.64

•65. A father (of mass 70 kg) faces his young daughter (of mass 20 kg) on a level roller rink. Dad pushes his daughter and she rolls backward at a speed of 2.5 m/s. Determine the speed of her Dad and the direction of his motion relative to hers.

FIGURE P.65

•66. A 2.00 kg mass and a 3.00 kg mass are used to compress opposite ends of an ideal spring with spring constant 1.50×10^3 N/m on a frictionless table; see Figure P.66. The spring is compressed 40.0 cm from its unstretched length and released with both masses initially at rest. Define the system to be the two masses and the ideal spring. (a) Explain why the momentum of the system is conserved. (b) Explain why the total mechanical energy of the system is conserved. (c) Determine the speeds of the masses as they leave the spring.

FIGURE P.66

Section 9.11 Kinetic Energy of a System of Particles

•67. A 5.00 kg dingbat is approaching a stationary 3.00 kg Cheshire cat at a speed of 10.0 m/s as shown in Figure P.67. (a) What is the total kinetic energy of the system of two particles? (b) What is the velocity of the center of mass? (c) What is the kinetic energy of the center of mass? (d) Find the velocities of the two particles with respect to the center of mass. (e) Calculate the kinetic energy of the two particles with respect to the center of mass point. (f) Use your calculations to check Equation 9.35.

FIGURE P.67

‡68. Two barred owls (*Syrnium varium*) are dive-bombing a small opossum (*Didelphis virginiana*) at rest, as shown in Figure P.68. (a) What is the total kinetic energy of the system of two owls? (b) What is the velocity of the center of mass of the owls? (c) What is the kinetic energy of the center of mass of the owls? (d) Find the velocities of the two owls with respect to their center of mass. (e) Calculate the kinetic energy of the two owls with respect to their center of mass. (f) Use your calculations to check Equation 9.35.

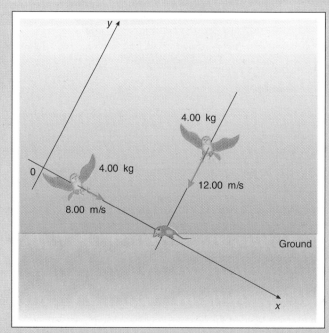

FIGURE P.68

Sections 9.12 The Velocity of the Center of Mass for Collisions*

9.13 The Center of Mass Reference Frame*

•69. A loaded coal car with a mass of 120 metric tons traveling at 2.0 m/s collides and couples with an empty, stationary coal car with a mass of 10 metric tons. (a) Determine the final speed of the two cars. (b) What is the change in the kinetic energy of the system of two cars in the collision? (c) What is the speed of the center of mass of the system of two cars throughout the process? (1 metric ton = 10^3 kg)

•70. A 20 kg projectile is launched over level ground near the surface of the Earth with a speed of 50 m/s and at an angle of 60° with the horizontal. At the top of its trajectory, the projectile suddenly breaks into two pieces, each of mass 10 kg. One of the 10 kg pieces is instantaneously at rest after the explosion (before falling). (a) Introduce an appropriate coordinate system and find the velocity and speed of the remaining 10 kg piece. (b) Sketch the trajectory of the center of mass point from launch to the instant of the breakup, and from the instant of the breakup to the instant when the center of mass point strikes the ground. Place appropriate horizontal distances on your sketch. (c) How far apart are the two 10 kg pieces when they hit the ground?

•71. A mass m is at rest. It disintegrates into *two* particles of mass m_1 and m_2 such that $m = m_1 + m_2$. (a) Is kinetic energy conserved in this process? (b) Show that the two particles m_1 and m_2 *must* move off in *opposite* directions in order to conserve momentum. This problem has implications in radioactivity and other types of particle decay. If the initial particle is at rest and decays into two particles, the two decay particles move off in opposite directions. In β-decay of a radioactive nucleus at rest, the recoiling (so-called daughter) nucleus and the emitted electron do *not* move off in opposite directions. This is a strong clue that β-decay produces not two particles but three (or perhaps more). The third particle is called a neutrino. Neutrinos finally were detected in the 1950s. Neutrinos are very difficult to detect because they interact extraordinarily weakly with matter.

INVESTIGATIVE PROJECTS

A. Expanded Horizons

1. Investigate methods used to reduce injuries from automobile accidents. Include aspects of automotive design, the use of seat belts, air bags, energy absorbing bumpers, collapsible front ends and steering columns, and so forth.

   M. A. Paulo, "An occupant restraint primer," *Forensic Accident Investigation: Motor Vehicles*, edited by Thomas L. Bohan and Arthur C. Damask (Michie Butterworth, Charlottesville, Virginia, 1995).

   James S. Baker and Lynn B. Fricke, *Traffic Accident Investigation Manual*, volume I (Northwestern University Traffic Institute, Evanston, Illinois, 1986).

   Lynn B. Fricke, *Traffic Accident Reconstruction*; volume 2 of the *Traffic Accident Investigation Manual* (Northwestern University Traffic Institute, Evanston, Illinois, 1990).

2. A mass m makes an elastic collision with a rigid wall. Will the collision also be elastic if the wall is moving at speed v_{wall} along the direction of its normal line? Will the angle the ball makes with the normal line as it rebounds be the same, less than, or greater than the angle at which it approached the moving wall?

B. Lab and Field Work

3. Design and perform an experiment to estimate the order of magnitude of the impulse delivered to a nail when hit by a hammer as it is driven into a 2 × 4 or other board.

4. Design and perform an experiment to estimate the magnitude of the impulse and average force used by a karate or tae kwon do practitioner when they shatter a board with their hand or foot.

   George A. Amann and Floyd T. Holt, "Karate demonstration," *The Physics Teacher*, 23, #1, page 40 (January 1985).

   S. R. Wilk, R. E. McNair, and M. S. Feld, "The physics of karate," *American Journal of Physics*, 51, #9, pages 783–790 (September 1983).

   Jearl D. Walker, "Karate strikes," *American Journal of Physics*, 43, #10, pages 845–849 (October 1975).

5. Your physics department may have an airtable and force transducers that electronically measure the magnitude of forces. Use this equipment to demonstrate that in a two-particle collision between two different masses, the magnitude of the force that each exerts on the other during the collision is the same. Compare this magnitude of force with that of the weight of each mass.

6. If a ball in a game of billiards hits the edge of the table at an angle θ with respect to a horizontal normal line to the edge, design and perform an experiment to see if the ball rebounds at the same angle with the normal line.

   R. D. Edge, "The errant pool balls," *The Physics Teacher*, 20, #1, pages 50–52 (January 1982).

7. Design a system that can stop a raw egg without breaking it when the egg is dropped from a height of about 5 m. Estimate the impulse and total force on the egg during the portion of its motion when its velocity is decreasing with time.

8. See how many meter sticks you can stack according to the scheme in Problem 63 to determine how far out from the table the top meter stick can project from the table.

9. Physicists occasionally are called as expert witnesses in accident investigations and other aspects of forensic science. Ask the head of your physics department if any local faculty members have been involved in such activities; you also might acquire names of local physicists from neighboring legal firms. Write a report about an actual investigation.

   Arthur L. Robinson, "Mastering the physics of disaster," *Physics Today*, 47, #12, pages 47–48 (December 1994).

   P. K. Tao, *Experiments, Devices, and Techniques: The Physics of Traffic Accident Investigation* (Oxford University Press, Hong Kong, 1987).

   Tim Folger, "Road scholar," *Discover*, 12, #8, pages 62–65 (August 1991).

   Vincent Lytle, "After the crash," *Discover*, 17, #5, pages 98–101 (May 1996).

   See the last two references for Project 1 as well.

10. A favorite recreational pastime at physics picnics is a water balloon toss. Two people toss a water-filled balloon back and

forth, gradually moving apart until one of them either misses or the water balloon breaks while being caught. (a) Secure a balloon from an appropriate source, fill it partially with water, and tie it shut. Determine its mass. By placing various weights on the filled balloon on a suitably protected flat surface, estimate the magnitude of the force needed to break the balloon. (b) Making reasonable assumptions, estimate the maximum separation between the balloon tossers when the water balloon breaks while being caught.

C. Communicating Physics

11. Interview one or more members of the physics department at your college or university to learn more about force transmission via the exchange of mediating particles. You likely will want to seek out theoretical physicists. See if they can explain to you how an attractive force might be transmitted by such particle exchange. Write a report about your findings appropriate for the audience of your campus newspaper. Perhaps you can become their science reporter!

12. Suppose an asteroid with a diameter of 1.0 km and an average density of 5.0×10^3 kg/m^3 is detected approaching the Earth at a relative approach speed of 30 km/s. The estimated impact time is only one week away. As an intern for the scientific advisor to the President of the United States, you are asked to evaluate whether firing one, several, or even the entire nuclear arsenal of the United States (~10,000 warheads with an average energy yield of 100 metric kilotons of TNT) might be effective in deflecting the asteroid sufficiently for it to miss the Earth. Organize a technical answer to the question, stating clearly any assumptions you make in coming to a conclusion whether the scheme is hopeless. The energy yield of TNT is about 4.9×10^9 J per metric ton (see Example 8.13).

CHAPTER 10
SPIN AND ORBITAL MOTION

Surely the devil of rotations is afoot.
James Watt (1736–1819) *

Rotational motion is apparent everywhere. Wheels and gears turn. Ice skaters and ballet dancers impress us with their pirouettes. Gymnasts and hot-dog skiers perform amazing flips in flight. Helicopter rotors, lawn mower and automobile engines, turbines, and jet engines all exhibit rotational motion. The planets spin while simultaneously orbiting the Sun. Electrons in atoms also "spin" and "orbit" the nucleus.

In Chapter 4 we studied the *kinematics* of circular rotational motion and introduced the angular velocity and angular acceleration vectors. In this chapter we study the *dynamics* of spin and orbital motion; we will see that it has many parallels with rectilinear dynamics. Kinematic similarities were apparent in Chapter 4, where the description of circular motion was seen to be a kind of generalized one-dimensional motion involving angular coordinates rather than rectilinear coordinates. The kinematic equations used to describe rotation were similar to those used to describe motion in one dimension. In this chapter, we will see further parallels and find a relationship that is the rotational analog of Newton's second law of motion.

We begin by examining the orbital motion of a single pointlike particle. We use this as a paradigm for a more complex spinning system of many particles: a symmetric rigid body. Finally, we look into simultaneous spin and orbital motion, along the way investigating several special examples of rotational motion. These include the precession of spinning tops and the Earth, synchronous rotation, rolling motion, and wheels. We conclude by noting two conditions necessary for the equilibrium of rigid bodies.

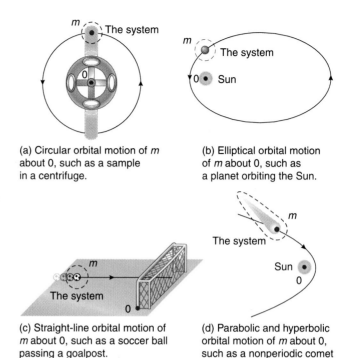

(a) Circular orbital motion of *m* about 0, such as a sample in a centrifuge.

(b) Elliptical orbital motion of *m* about 0, such as a planet orbiting the Sun.

(c) Straight-line orbital motion of *m* about 0, such as a soccer ball passing a goalpost.

(d) Parabolic and hyperbolic orbital motion of *m* about 0, such as a nonperiodic comet about the Sun.

FIGURE 10.1 Orbital motion of *m* about the point 0.

10.1 THE DISTINCTION BETWEEN SPIN AND ORBITAL MOTION

We use the word **spin** to describe rotational motion of a system about an axis through its center of mass.

Thus we say the Earth spins on its axis, or a CD spins in a stereo system. A **rigid body** is a system composed of many pointlike particles that maintain fixed distances from each other at all times. Each pointlike particle of a spinning rigid body system executes circular motion about an axis through the center of mass.

If the center of mass of the system *also* is moving in space from the perspective of a particular reference frame, we say the center of mass is executing **orbital** motion with respect to the origin of the coordinate system.

The motion of the center of mass need not be circular to be called orbital motion, as shown in Figure 10.1. Orbital motion describes *motion of the center of mass*, even though its path need *not* be a closed path in many cases.

A traditional example of orbital motion is the annual motion of the Earth about the Sun; this orbit is elliptical in shape. In many cases we can consider the orbit of the Earth to be

circular because the eccentricity of the orbit is small ($\varepsilon = 0.016$). The Earth also spins once a day about an axis through its center of mass. The actual spin period with respect to the distant stars is 23 h 56 min 4.10 s. Thus the Earth exhibits both spin and orbital motion, as shown in Figure 10.2.

For the Earth, the spin angular velocity $\vec{\omega}_{spin}$ is quite different from the orbital angular velocity $\vec{\omega}_{orbit}$ in both magnitude and direction. The Earth spins once in about 24 h, and so its angular speed of spin is

$$\omega_{spin} = \frac{2\pi \text{ rad}}{(24 \text{ h})(3600 \text{ s/h})}$$
$$= 7.3 \times 10^{-5} \text{ rad/s}$$

It takes the Earth 365.25 days to complete one orbit of the Sun, and so its average orbital angular speed is

$$\omega_{orbit} = \frac{2\pi \text{ rad}}{(365.25 \text{ d})(86\,400 \text{ s/d})}$$
$$= 1.9910 \times 10^{-7} \text{ rad/s}$$

The plane of the equator of the Earth is inclined to the plane of its orbit about the Sun by about 23.5°; hence, the two angular velocity vectors also have an angle between them of 23.5°.

Rolling wheels also exhibit both spin and orbital motion. Skidding, locked wheels exhibit no spin, but do have orbital motion with respect to the ground because the center of mass of each wheel is moving.

In this chapter, we begin our study of rotational motion by examining the orbital motion of a single pointlike particle; later,

---

*(Chapter Opener) Quoted in Henry W. Dickinson and Rhys Jenkins, *James Watt and the Steam Engine* (Oxford at the Clarendon Press, Oxford, England, 1927), page 159.

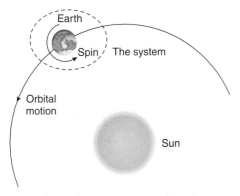

FIGURE 10.2 The Earth spins on an axis through its center of mass and simultaneously orbits the Sun.

we shall see that the single particle also may represent the center of mass point of a more complex system. We then examine the spinning motion of a system of particles about an axis coincident with a symmetry axis through the center of mass, when the center of mass itself is at rest (so there is no orbital motion). Finally, we examine what happens when a system undergoes simultaneous spin and orbital motion.

QUESTION 1

Consider two stars of equal mass in circular orbit about their center of mass point as in Figure Q.1. Consider the stars to be particles and ignore any spin associated with each. (a) If you consider the system to be *one* of the stars, is the system undergoing orbital or spin motion? (b) If you consider the system to be *both* stars, is the system undergoing orbital or spin motion?

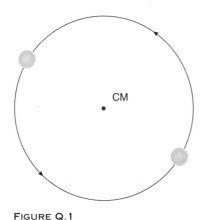

FIGURE Q.1

10.2 THE ORBITAL ANGULAR MOMENTUM OF A PARTICLE

One of the principle parameters associated with rotational motion (in all its hideous variations) is **angular momentum**. We briefly encountered the orbital angular momentum of a particle in Section 6.10 where we found the orbital angular momentum of each planet was constant, a manifestation of Kepler's empirical second law of

planetary motion.* In this section we develop the concept of angular momentum more leisurely; in the next few sections we discover why it is so important for spin and orbital dynamics.

Recall from Chapter 9 that a pointlike particle of mass m moving with velocity $\vec{v}$ has a (classical) momentum $\vec{p}$ given by (Equation 9.1)

$$\vec{p} \equiv m\vec{v}$$

The velocity of the particle is measured with respect to a reference frame with an origin 0 at some convenient point of your choosing, as in Figure 10.3.

The momentum of a particle is important because, as we saw in the last chapter, the most general statement of Newton's second law of motion relates the total force on the particle to the time rate of change of its momentum; angular momentum has a similar importance.

Whatever the path or trajectory of the particle, be it a straight line, a curved path, or a closed orbital path, the **orbital angular momentum $\vec{L}$** of the particle at any position with respect to a reference point (such as the origin of your coordinate choice) is

$$\boxed{\vec{L} \equiv \vec{r} \times \vec{p}} \qquad (10.1)$$

Since the position vector $\vec{r}$ is measured with respect to the origin at 0, we say the orbital angular momentum of the particle is taken *about* the point 0. The dimensions of the orbital angular momentum are those of distance times those of the momentum; using SI units, m(kg·m/s) = kg·m²/s.

We shall see in this chapter that the total angular momentum of a system is closely connected to a rotational analog of Newton's second law of motion. Bear with us (i.e., hang in there) as we develop this unifying and useful idea.

QUESTION 2

A certain particle has nonzero momentum $\vec{p}$ and nonzero orbital angular momentum $\vec{L}$. The angle between $\vec{r}$ and $\vec{p}$ is neither 0° nor 180°. What is the angle between the vectors $\vec{p}$ and $\vec{L}$?

---

*A line from the Sun to a given planet sweeps out equal areas during equal times. [Section 6.10 was optional; if you skipped it, there is *no* need to go back and read it now (unless you want to!).]

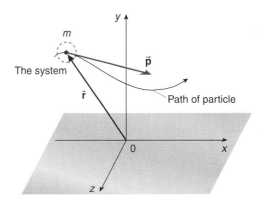

FIGURE 10.3 The momentum of a particle and its position vector with respect to the origin at 0.

EXAMPLE 10.1 ────────────

A small 8.00 kg Australian wild dog called a dingo (*Canis dingo*) is running with a velocity $(-6.0 \text{ m/s})\hat{j}$ along the y-axis as indicated in Figure 10.4. Find the momentum and the orbital angular momentum of the dingo with respect to a terrified rabbit (*Lepus timidus*) located at the origin in Figure 10.4.

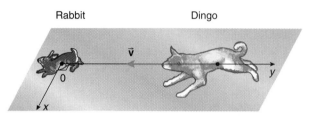

FIGURE 10.4

Solution
The dingo is the system. The momentum of the dingo is

$$\vec{p} = m\vec{v}$$
$$= (8.00 \text{ kg})(-6.0 \text{ m/s})\hat{j}$$
$$= (-48 \text{ kg·m/s})\hat{j}$$

The orbital angular momentum of the dingo is found from Equation 10.1:

$$\vec{L} = \vec{r} \times \vec{p}$$

The position vector of the dingo when it is at coordinate y on the y-axis is

$$\vec{r} = y\hat{j}$$

The orbital angular momentum of the dingo with respect to the rabbit at the origin thus is

$$\vec{L} = y\hat{j} \times (-48 \text{ kg·m/s})\hat{j}$$
$$= 0 \text{ kg·m}^2/\text{s}$$

regardless of the y-coordinate of the dingo.

EXAMPLE 10.2 ────────────

A 2.00 kg partridge (*Perdix perdix*) is flying at a constant velocity $\vec{v} = (5.00 \text{ m/s})\hat{i}$ along a straight line past a pear tree at the origin,

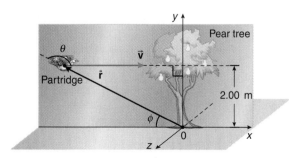

FIGURE 10.5

as shown in Figure 10.5. Find the orbital angular momentum of the partridge with respect to the base of the pear tree as an origin.

Solution

Method 1
The momentum $\vec{p}$ of the partridge is given by Equation 9.1:

$$\vec{p} = m\vec{v}$$
$$= mv\hat{i}$$

When the partridge is a distance r from the pear tree, the position vector of the bird is

$$\vec{r} = (-r\cos\phi)\hat{i} + (r\sin\phi)\hat{j}$$

Hence the orbital angular momentum is

$$\vec{L} = \vec{r} \times \vec{p}$$
$$= [(-r\cos\phi)\hat{i} + (r\sin\phi)\hat{j}] \times mv\hat{i}$$
$$= -(r\sin\phi)mv\hat{k}$$

Geometrically, from Figure 10.5, $r\sin\phi = 2.00$ m; this is the perpendicular distance between the line of motion and the origin. Hence the orbital angular momentum is

$$\vec{L} = -(2.00 \text{ m})(2.00 \text{ kg})(5.00 \text{ m/s})\hat{k}$$
$$= (-20.0 \text{ kg·m}^2/\text{s})\hat{k}$$

Notice (perhaps with some surprise!) that the orbital angular momentum here is *constant* and independent of the location of the partridge along the straight-line path.

Method 2
You also could determine the orbital angular momentum by first finding its magnitude and then its direction. The magnitude of the orbital angular momentum is

$$L = |\vec{r} \times \vec{p}|$$
$$= rp\sin\theta$$

where θ is the angle between the vectors $\vec{r}$ and $\vec{p}$ (with their tails at a common point). Geometrically, from Figure 10.5, $\theta = \pi - \phi$, where we measure the angles in radians, so that

$$L = rp\sin(\pi - \phi)$$
$$= rp(\sin\pi\cos\phi - \cos\pi\sin\phi)$$
$$= rp\sin\phi$$
$$= (r\sin\phi)(mv)$$

From the geometry of Figure 10.5, $r\sin\phi = 2.00$ m. So the orbital angular momentum has a magnitude

$$L = (2.00 \text{ m})(2.00 \text{ kg})(5.00 \text{ m/s})$$
$$= 20.0 \text{ kg·m}^2/\text{s}$$

The vector product right-hand rule indicates that $\vec{L}$ must be directed as shown in Figure 10.6: parallel to $-\hat{k}$. Hence

$$\vec{L} = (-20.0 \text{ kg·m}^2/\text{s})\hat{k}$$

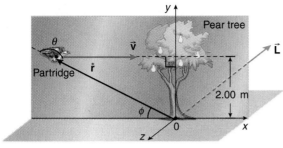

FIGURE 10.6

10.3 THE CIRCULAR ORBITAL MOTION OF A SINGLE PARTICLE

Here we consider both uniform and nonuniform circular motion of a single particle, since this will form the basis for the spinning rotational motion of more complex, many-particle systems such as rigid bodies. The *circular* orbital motion of a planet about the Sun is an example of circular motion that is uniform (i.e., with constant angular speed). The circular orbital motion of a sample in a centrifuge is another example of circular orbital motion, although *not* necessarily at constant angular speed.

Let r be the radius of the circular motion of the single particle, as shown in Figure 10.7. Choose the center of the circle as the origin. The direction of the angular velocity vector is found from the circular motion right-hand rule given in Chapter 4.

The orbital angular momentum of the particle is, from Equation 10.1,

$$\vec{L} = \vec{r} \times \vec{p}$$

where $\vec{r}$ is the position vector of the particle and $\vec{p}$ is its momentum. We say the particle has angular momentum $\vec{L}$ about the point chosen as origin, here the center of the circle. Thanks to the vector product right-hand rule, the orbital angular momentum vector of the particle is in the same direction as the orbital angular velocity vector of the particle. Both $\vec{L}$ and $\vec{\omega}$ are perpendicular to the plane of the circular motion, as shown in Figure 10.7.

Since the orbital angular momentum and orbital angular velocity vectors are parallel, perhaps they are related to each other in some other way. Substitute for the momentum of the particle into the orbital angular momentum:

$$\vec{L} = \vec{r} \times m\vec{v}$$
$$= m\vec{r} \times \vec{v}$$

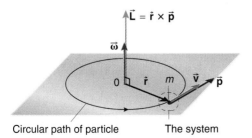

FIGURE 10.7 Circular motion of a particle and the vectors associated with it.

For circular motion $\vec{r}$ is perpendicular to $\vec{v}$; hence, the magnitude of the angular momentum is

$$L = mrv \sin 90°$$
$$= mvr \qquad (10.2)$$

The velocity $\vec{v}$ and angular velocity $\vec{\omega}$ in circular motion are related according to Equation 4.34:

$$\vec{v} = \vec{\omega} \times \vec{r}$$

For circular motion $\vec{\omega}$ is perpendicular to $\vec{r}$, and so the speed is

$$v = \omega r \sin 90°$$
$$= \omega r \qquad (10.3)$$

We use Equation 10.3 to substitute for v in Equation 10.2:

$$L = m(\omega r)r$$
$$= mr^2 \omega$$

But we know $\vec{L}$ is parallel to $\vec{\omega}$, so that we can write

$$\vec{L} = mr^2 \vec{\omega} \qquad (10.4)$$

> The orbital angular momentum of the particle is proportional to the orbital angular velocity; the proportionality constant is called the **moment of inertia** I of the particle.* The moment of inertia of the particle about the axis through the center of the circle is, therefore,
>
> $$I \equiv mr^2 \quad \text{(orbital motion of a particle } only) \quad (10.5)$$
>
> The SI units of the moment of inertia are $\text{kg} \cdot \text{m}^2$.

Notice that the moment of inertia of the particle depends on *both* its mass and its location with respect to the axis of rotation. The moment of inertia does *not* depend on the angular velocity vector.

Equation 10.4 for the orbital angular momentum of the particle about the origin can be rewritten using the moment of inertia as

$$\vec{L} = I\vec{\omega} \qquad (10.6)$$

For *uniform* circular motion, the angular velocity is constant; thus, according to Equation 10.6, the angular momentum is constant as well. For *nonuniform* circular motion, where there exists a nonzero angular acceleration $\vec{\alpha}$, the angular velocity vector is not constant; therefore, the angular momentum is not constant either.

We want to see what causes angular accelerations and, consequently, changes in the angular momentum. Forces cause changes in the momentum via Newton's second law of motion. Forces also are involved in causing changes in the angular momentum, but in a more subtle way.

If the total force on the particle is directed toward the center of the circle, the angular acceleration $\vec{\alpha}$ is zero. This is uniform

---

*Do not confuse the (scalar) moment of inertia I with the magnitude of the vector impulse $\vec{I}$ of a force, discussed in Chapter 9.

circular motion with a purely centripetal acceleration. Note that the vector product of $\vec{r}$ and $\vec{F}_{total}$ is zero in this case, because the two vectors are antiparallel (see Figure 10.8).

We saw in Chapter 4 that nonzero angular acceleration means the particle experiences a *total acceleration* that is *not* directed toward the center of the circle; the total acceleration then has both centripetal and tangential components. Thus, from Newton's second law of motion, the total force is *not* directed toward the center of the circle for nonuniform circular motion, as shown in Figure 10.9. The particle in circular motion can either speed up or slow down in its circular motion if the total force on the particle is not directed toward the center of the circle. This condition also implies that the vector product of $\vec{r}$ with $\vec{F}_{total}$ is *not* zero.

Hence the vector product of $\vec{r}$ and $\vec{F}_{total}$ is an indicator of whether $\vec{F}_{total}$ produces an angular acceleration or not. To account for the effect of a force in causing angular accelerations, we introduce a new physical concept: the **torque** of a force.* One can calculate the torque of an individual force or of the total force.

Whatever the force $\vec{F}$, its torque is defined to be

$$\vec{\tau} \equiv \vec{r} \times \vec{F} \qquad (10.7)$$

where $\vec{r}$ is the *position vector of the point of application of the force with respect to the chosen origin* for the coordinate system (see Figure 10.10). If the motion is circular, the origin is always taken to be the center of the circle.

The SI units of torque are N·m. While N·m are equivalent to joules (J), we *always* quote the SI units of torque as N·m, reserving joules specifically for work or energy (scalar quantities).

---

*The word comes from the Latin *torquis*, a twisted neck chain worn by ancient Britons, Gauls, and Germans.

FIGURE 10.8 In uniform circular motion, the position vector of a particle is antiparallel to the total force on it.

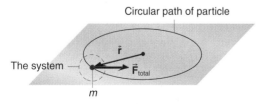

FIGURE 10.9 In nonuniform circular motion, the position vector of a particle and the total force on it are not antiparallel.

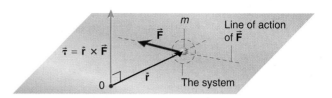

FIGURE 10.10 The torque $\vec{\tau}$ is perpendicular to both the position vector $\vec{r}$ and the force $\vec{F}$.

The torque is a vector quantity and, because of its definition as a vector product, its direction always is perpendicular to both $\vec{r}$ and $\vec{F}$; see Figure 10.10. The line along which the force lies is called the **line of action** of the force.

The torque is a measure of how effective the force is in causing an angular acceleration of the system about the point 0; a nonzero torque means there will be a nonzero angular acceleration *if the force $\vec{F}$ acts alone.*

The torque is zero if $\vec{r}$ and $\vec{F}$ are either parallel or antiparallel, as in Figure 10.11. In such cases the force $\vec{F}$ has no ability to cause an angular acceleration of the system about the point 0.

PROBLEM-SOLVING TACTIC

10.1 If the line of action of a force goes through a point 0, the force has zero torque about that point. The torque is a maximum when the two vectors $\vec{r}$ and $\vec{F}$ are perpendicular to each other, as in Figure 10.12. In this case the force $\vec{F}$ is most effective in causing angular acceleration of the system about 0.

The magnitude of the torque is

$$\tau = rF \sin \theta$$

where θ is the angle between the vectors $\vec{r}$ and $\vec{F}$ (when their tails are at a common point). The quantity $r \sin \theta$ is the perpendicular distance from the point 0 to the line of action of $\vec{F}$. This

FIGURE 10.11 A force that has zero torque about the point 0.

FIGURE 10.12 The torque of $\vec{F}$ about 0 is a maximum when $\vec{F}$ is perpendicular to $\vec{r}$.

perpendicular distance is called the **moment arm** of the force, as shown in Figure 10.13.*

We shall see that the torque of the total force on the particle is closely related to changes in the angular momentum of the particle, in much the way the total force is related to changes in the momentum of the particle via Newton's second law of motion. To show this we first must find the torque of the total force $\vec{\mathbf{F}}_{\text{total}}$ on the particle. Calculate the torque of the total force about the origin of your coordinate system:

$$\vec{\boldsymbol{\tau}}_{\text{total}} = \vec{\mathbf{r}} \times \vec{\mathbf{F}}_{\text{total}}$$

Since we say the total force acts on the particle, we also say that the total torque acts on the particle. Substitute

$$\frac{d\vec{\mathbf{p}}}{dt}$$

for the total force, according to Newton's second law of motion:

$$\vec{\boldsymbol{\tau}}_{\text{total}} = \vec{\mathbf{r}} \times \frac{d\vec{\mathbf{p}}}{dt} \qquad (10.9)$$

As an aside, note that

$$\frac{d}{dt}(\vec{\mathbf{r}} \times \vec{\mathbf{p}}) = \frac{d\vec{\mathbf{r}}}{dt} \times \vec{\mathbf{p}} + \vec{\mathbf{r}} \times \frac{d\vec{\mathbf{p}}}{dt} \qquad (10.10)$$

The first term on the right-hand side of Equation 10.10 is zero because the vector product of any vector with itself is zero:

$$\frac{d\vec{\mathbf{r}}}{dt} \times \vec{\mathbf{p}} = \vec{\mathbf{v}} \times m\vec{\mathbf{v}}$$
$$= m\vec{\mathbf{v}} \times \vec{\mathbf{v}}$$
$$= 0 \ \text{kg}\cdot\text{m}^2/\text{s}^2$$

Hence Equation 10.10 becomes

$$\frac{d}{dt}(\vec{\mathbf{r}} \times \vec{\mathbf{p}}) = \vec{\mathbf{r}} \times \frac{d\vec{\mathbf{p}}}{dt} \qquad (10.11)$$

We use Equation 10.11 in Equation 10.9 for the total torque and obtain

$$\vec{\boldsymbol{\tau}}_{\text{total}} = \frac{d}{dt}(\vec{\mathbf{r}} \times \vec{\mathbf{p}})$$

*The term *lever arm* also is used synonymously with the term moment arm.

FIGURE 10.13 The moment arm of a force.

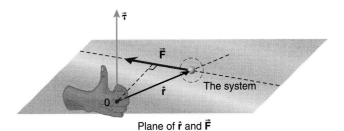

FIGURE 10.14 If you place your right-hand thumb along the direction of the torque, the right-hand fingers will indicate the sense that the force will tend to accelerate (or rotate) the system about 0.

But $\vec{\mathbf{r}} \times \vec{\mathbf{p}}$ is the orbital angular momentum $\vec{\mathbf{L}}$ of the particle about the origin. Making this substitution, we discover the following:

Since the angular momentum is proportional to the angular velocity via Equation 10.6, Equation 10.12 also can be written as

$$\vec{\boldsymbol{\tau}}_{\text{total}} = \frac{d}{dt}(I\vec{\boldsymbol{\omega}})$$

The moment of inertia I is constant as long as the particle is of fixed mass and is in circular orbital motion. Hence

$$\vec{\boldsymbol{\tau}}_{\text{total}} = I\frac{d\vec{\boldsymbol{\omega}}}{dt}$$

But

$$\frac{d\vec{\boldsymbol{\omega}}}{dt}$$

is the angular acceleration $\vec{\boldsymbol{\alpha}}$.

Hence we find that

$$\vec{\tau}_{\text{total}} = I\vec{\alpha} \quad \text{(constant } I\text{)} \quad (10.13)$$

Equation 10.13 is the rotational counterpart of $\vec{F}_{\text{total}} = m\vec{a}$. Equation 10.13 can be used only when the moment of inertia I of the particle is constant, just as $\vec{F}_{\text{total}} = m\vec{a}$ only can be used if m is constant.

QUESTION 3

If $\vec{\omega} = 0$ rad/s at some instant, does this imply $\vec{\tau}_{\text{total}} = 0$ N·m at that instant? If $\vec{\alpha} = 0$ rad/s² at some instant, does this imply that $\vec{\tau}_{\text{total}} = 0$ N·m at that instant?

EXAMPLE 10.3

A mechanic tightens a bolt on a V-8 engine head with a wrench. He applies a tangential force 20 cm from the bolt, as shown in Figure 10.15. The bolt will break if the applied torque about the axis of the bolt exceeds 80 N·m in magnitude. What is the maximum magnitude of force that the mechanic can apply without breaking the bolt?

FIGURE 10.15

Solution

Method 1
The magnitude of the torque is

$$\tau = |\vec{r} \times \vec{F}|$$
$$= rF \sin\theta$$

where $\vec{r}$ is the vector from the point about which the torque is taken (here the axis of the bolt) to the point of application of the force $\vec{F}$ (here 20 cm from the bolt); θ is the angle between $\vec{r}$ and $\vec{F}$—here $\theta = 90°$. Since the magnitude of the torque cannot exceed 80 N·m, you have

$$80 \text{ N·m} = (0.20 \text{ m})F \sin 90°$$

Solving for F,

$$F = 4.0 \times 10^2 \text{ N}$$

Method 2
The moment arm of the force is the perpendicular distance from the origin (here at the center of the circle) to the line of action

FIGURE 10.16

of the force, as shown in Figure 10.16. Here the moment arm is the radius of the circle. The magnitude of the torque is the product of the magnitude of the force and its moment arm:

$$\tau = F[\text{moment arm}]$$
$$80 \text{ N·m} = F(0.20 \text{ m})$$

Solving for F,

$$F = 4.0 \times 10^2 \text{ N}$$

For safety many pieces of machinery, such as snow blowers and jet engine mounts have bolts, called shear pins, that are designed to break (or shear off) if they experience forces that exceed some specified maximum magnitude.

EXAMPLE 10.4

A huge centrifuge is used to subject astronauts to large accelerations. It consists of two 1500 kg chambers on opposite ends of a rigid boom of length 20 m, much like a dumbbell. The chamber and boom system is brought from rest to an angular speed of 12 rev/min during 1.5 min by means of a force of constant magnitude applied 2.0 m from the center of mass and perpendicular to the boom, as shown in Figure 10.17. Consider the chambers to be pointlike masses at the end of the boom; neglect the mass of the boom.

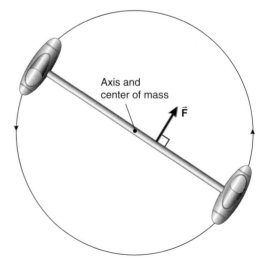

FIGURE 10.17

a. Find the torque of the force.
b. Find the magnitude of the force.

Solution

Each chamber is a particle in circular motion. Since the moment of inertia is a scalar and both particles have the same angular speed, the total moment of inertia of the two-particle system is

$$I = mr^2 + mr^2$$
$$= 2mr^2$$
$$= 2(1500 \text{ kg})(10 \text{ m})^2$$
$$= 3.0 \times 10^5 \text{ kg·m}^2$$

Choose the z-axis to be parallel to the angular velocity vector, whose direction is found from the circular motion right-hand rule; $\hat{\mathbf{k}}$ is perpendicularly out of the page in Figure 10.17. The final angular velocity component of the system is

$$\omega_z = \frac{(12 \text{ rev/min})(2\pi \text{ rad/rev})}{60 \text{ s/min}}$$
$$= 1.3 \text{ rad/s}$$

Use Equation 4.49 to find the constant angular acceleration component; the initial angular velocity component is 0 rad/s:

$$\omega_z = \omega_{z0} + \alpha_z t$$
$$1.3 \text{ rad/s} = 0 \text{ rad/s} + \alpha_z(90 \text{ s})$$

Solving for α_z,

$$\alpha_z = 1.4 \times 10^{-2} \text{ rad/s}^2$$

The constant angular acceleration is

$$\vec{\alpha} = \alpha_z \hat{\mathbf{k}}$$
$$= (1.4 \times 10^{-2} \text{ rad/s}^2)\hat{\mathbf{k}}$$

a. The torque is in the same direction as the angular acceleration, according to Equation 10.13:

$$\vec{\tau} = I\vec{\alpha}$$
$$= I\alpha_z \hat{\mathbf{k}}$$
$$= (3.0 \times 10^5 \text{ kg·m}^2)(1.4 \times 10^{-2} \text{ rad/s}^2)\hat{\mathbf{k}}$$
$$= (4.2 \times 10^3 \text{ N·m})\hat{\mathbf{k}}$$

b. The force is applied perpendicular to the boom and 2.0 m from the axis, so that

$$\tau = |\vec{\mathbf{r}} \times \vec{\mathbf{F}}|$$
$$= rF \sin 90°$$
$$= rF$$

where $\vec{\mathbf{r}}$ locates the point of application of $\vec{\mathbf{F}}$ with the axis as an origin; here $r = 2.0$ m. Solving for the magnitude of the force,

$$F = \frac{\tau}{r}$$
$$= \frac{4.2 \times 10^3 \text{ N·m}}{2.0 \text{ m}}$$
$$= 2.1 \times 10^3 \text{ N}$$

EXAMPLE 10.5

A hammer-throw athlete whirls a 7.30 kg mass in a circle of radius 1.20 m at a constant speed of 8.00 m/s, barely above the ground, as shown in Figure 10.18.

FIGURE 10.18

a. Does the mass have spin and/or orbital angular momentum?
b. Determine the orbital angular momentum of the particle with respect to the center of the circle.

Solution

a. The mass has orbital angular momentum because the center of mass of the single particle is moving. Since the particle also rotates once during each revolution, it also has some spin angular momentum.
b. There are several ways to calculate the orbital angular momentum $\vec{\mathbf{L}} = \vec{\mathbf{r}} \times \vec{\mathbf{p}}$. Here $\vec{\mathbf{r}}$ locates the particle with respect to the center of the circle, since you want the angular momentum about this point.

Method 1

Find the magnitude of angular momentum using the magnitude of the vector product $\vec{\mathbf{L}} = \vec{\mathbf{r}} \times \vec{\mathbf{p}}$:

$$L = rp \sin \theta$$
$$= rmv \sin \theta$$

In this case the angle between $\vec{\mathbf{r}}$ and $\vec{\mathbf{p}}$ is always 90°. Substituting for r, m, v, and θ, you obtain

$$L = (1.20 \text{ m})(7.30 \text{ kg})(8.00 \text{ m/s}) \sin 90°$$
$$= 70.1 \text{ kg·m}^2/\text{s}$$

The direction of $\vec{\mathbf{L}}$ is found by applying the vector product right-hand rule to $\vec{\mathbf{r}} \times \vec{\mathbf{p}}$; see Figure 10.19.

Method 2

Use Equation 10.6, $\vec{\mathbf{L}} = I\vec{\boldsymbol{\omega}}$, to find the angular momentum. The moment of inertia of a point particle is

$$I = mr^2$$

The angular velocity $\vec{\boldsymbol{\omega}}$ is found from the linear velocity using Equation 4.34:

$$\vec{\mathbf{v}} = \vec{\boldsymbol{\omega}} \times \vec{\mathbf{r}}$$

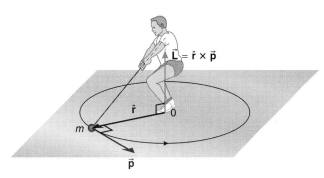

$$\vec{L} = \vec{r} \times \vec{p}$$

FIGURE 10.19

Since $\vec{\omega}$ is perpendicular to $\vec{r}$ in circular motion, the magnitudes of these vectors are related by

$$v = \omega r$$

or

$$\omega = \frac{v}{r}$$

Notice that the angular speed is the magnitude of the component of the velocity perpendicular to $\vec{r}$ (which is just v itself for circular motion) divided by r. Thus the magnitude of the orbital angular momentum $\vec{L}$ is

$$
\begin{aligned}
L &= I\omega \\
&= mr^2 \frac{v}{r} \\
&= mrv \\
&= (7.30 \text{ kg})(1.20 \text{ m})(8.00 \text{ m/s}) \\
&= 70.1 \text{ kg} \cdot \text{m}^2/\text{s}
\end{aligned}
$$

The direction of $\vec{L}$ is the same as the direction of $\vec{\omega}$, since $\vec{L} = I\vec{\omega}$. The direction of $\vec{\omega}$ is found using the circular motion right-hand rule: wrap the fingers of your right hand around the center of the circle in the sense the particle is moving; the extended right-hand thumb indicates the direction of $\vec{\omega}$. Thus $\vec{\omega}$ is perpendicular to the plane of the circle and so is $\vec{L}$, as shown in Figure 10.20.

FIGURE 10.20

STRATEGIC EXAMPLE 10.6

A 3.00 kg mass on a frictionless horizontal surface is attached to a string of length 50.0 cm that is fixed at one end but can slide around the axis. The mass initially is at rest and the string is taut. For times $t \geq 0$ s, a force of magnitude 5.00 N is applied tangentially as indicated in Figure 10.21.

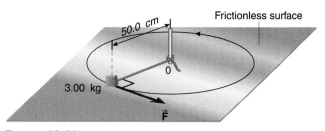

FIGURE 10.21

a. What are the forces acting on the mass?
b. What is the torque of the total force on the mass?
c. What is the moment of inertia of the mass about the fixed end of the string?
d. What is the angular acceleration of the mass?
e. Is the angular momentum of the mass constant? Explain your answer.
f. What is the angular momentum of the mass when $t = 3.00$ s?

Solution
Consider the 3.00 kg mass as a particle that is the system.

a. The forces on the system are (see Figure 10.22)
   1. the gravitational force of the Earth on the system (its weight $\vec{w}$), directed vertically down;
   2. the normal force $\vec{N}$ of the surface on the system, directed perpendicular to the surface, here vertically upward;
   3. the applied tangential force $\vec{F}$ of constant magnitude 5.00 N; and
   4. the force $\vec{T}$ of the string on the system, directed toward the center of the circle.

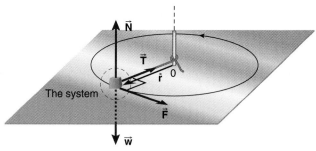

FIGURE 10.22

b. There is no acceleration in the vertical direction, and so the vector sum of the weight and normal force is zero. The vector sum of the weight and the normal force produces zero torque since the vector sum of these forces is zero and they are colinear. Since the tension is directed toward the center of the circle, antiparallel to $\vec{r}$, the torque of the tension also is zero:

$$\vec{r} \times \vec{T} = 0 \text{ N} \cdot \text{m}$$

The total torque on the mass about the axis of its rotation thus is equal to the torque of the tangential force. Examine Figure 10.23; the torque of this force is given by

$$\vec{\tau}_{total} = \vec{r} \times \vec{F}$$

where $\vec{r}$ is the vector from the origin to the point where $\vec{F}$ is applied. Using the vector product right-hand rule, the direction of the torque is perpendicular (upward in Figure 10.23) to the plane of the motion.

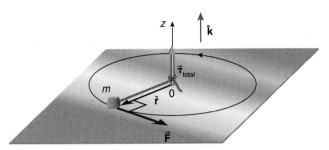

FIGURE 10.23

The magnitude of the torque is

$$\tau_{total} = rF \sin 90°$$
$$= (0.500 \text{ m})(5.00 \text{ N})(1)$$
$$= 2.50 \text{ N·m}$$

With the coordinate system in Figure 10.23, the total torque is

$$\vec{\tau}_{total} = (2.50 \text{ N·m})\hat{k}$$

c. The moment of inertia of the 3.00 kg particle is found using Equation 10.5 for a pointlike mass in orbital motion:

$$I = mr^2$$
$$= (3.00 \text{ kg})(0.500 \text{ m})^2$$
$$= 0.750 \text{ kg·m}^2$$

d. The angular acceleration of the particle is found from Equation 10.13,

$$\vec{\tau}_{total} = I\vec{\alpha}$$
$$(2.50 \text{ N·m})\hat{k} = (0.750 \text{ kg·m}^2)\vec{\alpha}$$

Solving for $\vec{\alpha}$ and simplifying the units,

$$\vec{\alpha} = (3.33 \text{ rad/s}^2)\hat{k}$$

e. Equation 10.12 implies that the orbital angular momentum of the mass is *not constant* since the total torque is not zero.

f. To determine the angular momentum when $t = 3.00$ s, use Equation 10.6,

$$\vec{L} = I\vec{\omega}$$

You have to find the angular velocity when $t = 3.00$ s. The angular acceleration of the mass is constant, so that you can use Equation 4.49 for circular motion with a constant angular acceleration:

$$\omega_z(t) = \omega_{z0} + \alpha_z t$$

When $t = 0$ s, the mass had no angular speed, so that $\omega_{z0} = 0$ rad/s. From part (d), you see that $\alpha_z = 3.33$ rad/s^2. Hence, when $t = 3.00$ s, the angular velocity component is

$$\omega_z = 0 \text{ rad/s} + (3.33 \text{ rad/s}^2)(3.00 \text{ s})$$
$$= 9.99 \text{ rad/s}$$

The angular velocity then is

$$\vec{\omega} = \omega_z\hat{k}$$
$$= (9.99 \text{ rad/s})\hat{k}$$

The orbital angular momentum then is

$$\vec{L} = I\vec{\omega}$$
$$= (0.750 \text{ kg·m}^2)(9.99 \text{ rad/s})\hat{k}$$
$$= (7.49 \text{ kg·m}^2/\text{s})\hat{k}$$

10.4 NONCIRCULAR ORBITAL MOTION

Equation 10.12,

$$\vec{\tau}_{total} = \frac{d\vec{L}}{dt}$$

also can be applied to noncircular orbital motion, such as a particle moving in a straight line or the familiar case of the elliptical orbital motion of a planet about the Sun. For the latter case, take the Sun as the origin of a coordinate system and consider the planet as a particle, as in Figure 10.24. Recall from Chapter 6 that the gravitational force is a *central force*—that is, it acts along the line connecting the two particles. The total torque on the planet is zero: $\vec{r}$ is always along the same line as the gravitational force $\vec{F}$, so that $\vec{r} \times \vec{F} = 0$ N·m.

Thus Equation 10.12 becomes

$$0 \text{ N·m} = \frac{d\vec{L}}{dt}$$

That is, the orbital angular momentum of the planet is constant, which is equivalent to Kepler's second law of planetary motion as we demonstrated in Section 6.10.

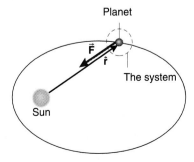

FIGURE 10.24 The gravitational force on a planet is a central force with zero torque.

If the planet is in a circular orbit (Venus very nearly fits this case), then the moment of inertia of the planet about the Sun ($I = mr^2$) is constant and the angular velocity is too. On the other hand, if the planet or comet is moving in an orbit that is elliptical, parabolic, or hyperbolic, the moment of inertia of the particle changes (since r changes). However, the magnitude of the angular velocity changes to compensate, and so the angular momentum of the particle, given by Equation 10.6,

$$\vec{\mathbf{L}} = I\vec{\boldsymbol{\omega}}$$

remains constant (as it must because the total torque on the planet is zero). Since the angular velocity of a planet in an elliptical orbit changes, the angular acceleration $\vec{\boldsymbol{\alpha}}$ of the planet is *not* zero. However, you *cannot* use Equation 10.13,

$$\vec{\boldsymbol{\tau}}_{\text{total}} = I\vec{\boldsymbol{\alpha}}$$

for such planets. The total torque on the planet *is* zero, because the gravitational force is a central force. The left-hand side of Equation 10.13 is zero, but the right-hand side is not! Clearly this result is ridiculous. You cannot use Equation 10.13 for such noncircular orbital motion because the equation assumes that the moment of inertia of the particle is constant. For noncircular motion I is *not* constant, because the distance of the particle from the origin changes with time. Just as $\vec{\mathbf{F}}_{\text{total}} = m\vec{\mathbf{a}}$ *cannot* be used for a system with varying mass (the more general equation

$$\vec{\mathbf{F}}_{\text{total}} = \frac{d\vec{\mathbf{p}}}{dt}$$

must be used instead), $\vec{\boldsymbol{\tau}}_{\text{total}} = I\vec{\boldsymbol{\alpha}}$ *cannot* be used for a system with a time-varying moment of inertia (the more general equation

$$\vec{\boldsymbol{\tau}}_{\text{total}} = \frac{d\vec{\mathbf{L}}}{dt}$$

must be used instead). Note the satisfying parallelism.

10.5 RIGID BODIES AND SYMMETRY AXES

You boil it in sawdust: You salt it in glue:
You condense it with locusts and tape:
Still keeping one principal object in view —
To preserve its symmetrical shape

Lewis Carroll (1832–1898)*

We now know how to calculate the angular momentum and the torque of the total force on a single-particle system as well as its consequent angular acceleration. In this section we consider more complex systems containing *many* particles.

A rigid body system is not deformable but maintains its shape. With a few exceptions,[†] we restrict ourselves to multi-

---

The Hunting of the Snark, Fit the Fifth, The Beaver's Lesson, lines 93–96 (Macmillan, New York, 1927), page 38.

[†] The exceptions we consider are either (a) single-particle systems, (b) swarms of unconnected single particles, or (c) systems that deform under the action of *internal forces* in such a way that the system at all times still is symmetric about the axis of symmetry. For example, a figure skater who gradually brings in both arms or a star that gradually expands or contracts are deformable systems but, at any instant, are symmetric about a symmetry axis.

FIGURE 10.25 A symmetric rigid body has particles located symmetrically with respect to an axis of symmetry.

particle systems that are rigid bodies. A perfectly rigid body is an idealization, since all objects deform to a greater or lesser degree under the action of various forces, as we will see in Chapter 11. Nonetheless, the idea of a rigid body is extremely useful for simplifying the dynamics of rotational motion. For simplicity we also restrict ourselves to rigid bodies that possess at least one **axis of symmetry**.

> An axis of symmetry can be imagined as a line in space such that for each pointlike particle m_i making up part of the rigid body, located at position vector $\vec{\mathbf{r}}_{i\perp}$, there is a corresponding pointlike particle of equal mass located at position vector $-\vec{\mathbf{r}}_{i\perp}$; see Figure 10.25.

Both position vectors are measured *perpendicular* to the axis of symmetry (hence the subscript $\perp$) and have the *axis* as origin. Such a symmetry axis always passes through the center of mass of the system.

We shall see in the next section that the assumption of a rigid body with at least one axis of symmetry greatly simplifies the dynamical description of rotational motion. A *general* description of the rotational motion of *un*symmetric rigid bodies and nonrigid bodies is beyond the scope of an introductory physics course, which may be a relief to you!

QUESTION 4

Sketch one of the symmetry axes through the center of mass of a uniform, square plate.

10.6 SPIN ANGULAR MOMENTUM OF A RIGID BODY

Systems rotating about an axis through the center of mass exhibit spinning motion or, more simply, *spin*. The rotational motion of a CD, a windmill pinwheel, an ultracentrifuge rotor, or even a common (empty) clothes dryer all are examples of symmetric rigid bodies undergoing spin. Each pointlike particle of the rigid body system undergoes circular orbital motion with the same angular velocity.

We consider the spinning motion of a system of many pointlike particles of a *rigid body* with at least one axis of symmetry, as in Figure 10.25. The symmetry axis passes through the center of mass point of the system. For the time being we restrict ourselves to situations in which two conditions hold:

1. *The center of mass of the system is at rest.*

We make this restriction because spin is the simplest kind of rigid body rotational motion and the kinematics of spin is an extension of the kinematic description of circular motion that we encountered in Chapter 4.

2. Also, for simplicity, we assume *the rotational axis of the system coincides with the symmetry axis of the system.*

Later we relax both of these restrictions, when we have developed the conceptual tools to do so.

All the pointlike particles that make up the rigid body have the *same* angular velocity $\vec{\boldsymbol{\omega}}$. The direction of the angular velocity vector is found using the familiar circular motion right-hand rule of Chapter 4. We shall see that the total angular momentum of such a system is proportional to the angular velocity vector, just as for single-particle orbital motion (Equation 10.6).

Take the fixed center of mass point to be the origin of a coordinate system (see Figure 10.26). A pointlike particle of mass m_i (that is part of the rigid body system) has an angular momentum $\vec{\mathbf{L}}_i$ that is the vector product of the position vector $\vec{\mathbf{r}}_i$ of the particle with its momentum $\vec{\mathbf{p}}_i$:

$$\vec{\mathbf{L}}_i \equiv \vec{\mathbf{r}}_i \times \vec{\mathbf{p}}_i$$
$$= \vec{\mathbf{r}}_i \times m_i \vec{\mathbf{v}}_i$$

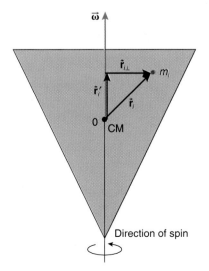

FIGURE 10.26 **The relationship between the position vectors $\vec{\mathbf{r}}_i$, $\vec{\mathbf{r}}_{i\perp}$, and $\vec{\mathbf{r}}_i'$.**

The *total* angular momentum $\vec{\mathbf{L}}$ of the spinning system is the vector sum of the orbital angular momenta of all the pointlike particles that make up the system:

$$\vec{\mathbf{L}} = \sum_i \vec{\mathbf{L}}_i$$
$$= \sum_i m_i \vec{\mathbf{r}}_i \times \vec{\mathbf{v}}_i \qquad (10.14)$$

Each particle of the rigid body executes *circular* motion about the rotational axis. Let $r_{i\perp}$ be the radius of the circle for the ith particle, as shown in Figure 10.26. The position vector $\vec{\mathbf{r}}_{i\perp}$ locates the particle *with respect to the axis of rotation* rather than the center of mass point.

The velocity $\vec{\mathbf{v}}_i$ of the ith particle is related to the angular velocity $\vec{\boldsymbol{\omega}}$ and $\vec{\mathbf{r}}_{i\perp}$ according to Equation 4.34 (modified slightly for the new notation here):

$$\vec{\mathbf{v}}_i = \vec{\boldsymbol{\omega}} \times \vec{\mathbf{r}}_{i\perp}$$

Also note in Figure 10.26 that we can write $\vec{\mathbf{r}}_i$ as the vector sum of $\vec{\mathbf{r}}_i'$ and $\vec{\mathbf{r}}_{i\perp}$, where $\vec{\mathbf{r}}_i'$ locates the tail of $\vec{\mathbf{r}}_{i\perp}$ with respect to the center of mass point:

$$\vec{\mathbf{r}}_i = \vec{\mathbf{r}}_i' + \vec{\mathbf{r}}_{i\perp}$$

Making these substitutions for $\vec{\mathbf{v}}_i$ and $\vec{\mathbf{r}}_i$ into Equation 10.14 for the spin angular momentum, we have quite an algebraic mouthful:

$$\vec{\mathbf{L}} = \sum_i m_i (\vec{\mathbf{r}}_i' + \vec{\mathbf{r}}_{i\perp}) \times (\vec{\boldsymbol{\omega}} \times \vec{\mathbf{r}}_{i\perp})$$

Expand this into two summations:

$$\vec{\mathbf{L}} = \sum_i m_i \vec{\mathbf{r}}_i' \times (\vec{\boldsymbol{\omega}} \times \vec{\mathbf{r}}_{i\perp}) + \sum_i m_i \vec{\mathbf{r}}_{i\perp} \times (\vec{\boldsymbol{\omega}} \times \vec{\mathbf{r}}_{i\perp}) \quad (10.15)$$

Now stare carefully at Figure 10.26. The vector product $\vec{\boldsymbol{\omega}} \times \vec{\mathbf{r}}_{i\perp}$ is directed perpendicularly into the page. The vector product of $\vec{\mathbf{r}}_i'$ with $\vec{\boldsymbol{\omega}} \times \vec{\mathbf{r}}_{i\perp}$ thus is parallel to $-\vec{\mathbf{r}}_{i\perp}$ and, since the vectors are all mutually perpendicular, is of magnitude $r_i' \omega r_{i\perp}$. Therefore the first summation on the right-hand side of Equation 10.15 can be rewritten as

$$-\sum_i m_i r_i' \omega \vec{\mathbf{r}}_{i\perp}$$

Likewise, by using the vector product right-hand rule, the term $\vec{\mathbf{r}}_{i\perp} \times (\vec{\boldsymbol{\omega}} \times \vec{\mathbf{r}}_{i\perp})$ in the second summation of Equation 10.15 is directed parallel to $\vec{\boldsymbol{\omega}}$ and is of magnitude $r_{i\perp}^2 \omega$, since the vectors are mutually perpendicular. Therefore the second summation in Equation 10.15 can be written as

$$\sum_i m_i r_{i\perp}^2 \vec{\boldsymbol{\omega}}$$

In this way Equation 10.15 can be rewritten as

$$\vec{\mathbf{L}} = -\sum_i m_i r_i' \omega \vec{\mathbf{r}}_{i\perp} + \sum_i m_i r_{i\perp}^2 \vec{\boldsymbol{\omega}} \qquad (10.16)$$

This expression looks rather ghastly, and might mirror your feelings if your eyes have glazed over. Because of the presence of the first summation in Equation 10.16, the spin angular momentum vector is *not* necessarily parallel to the angular velocity vector! This is bad news for simplicity. The vector $\vec{L}$ is an involved vector summation of the angular velocity $\vec{\omega}$ and a host of other vectors $\vec{r}_{i\perp}$. Herein lies the complexity of rotational motion for extended objects of *arbitrary* shape. The rotation of an oddly shaped object about *any* axis of rotation is beyond the scope of this text. By restricting ourselves to *symmetric* rigid bodies, we can spare ourselves much complication and still produce quite useful results.

For symmetric rigid bodies with the rotational axis coincident with a symmetry axis, the first summation in Equation 10.16 is *zero*. Why? Since the object is symmetric about the rotational axis, for every particle located at position vector $\vec{r}_{i\perp}$ with respect to the axis, there is an identical particle located at $-\vec{r}_{i\perp}$ as in Figure 10.25. Hence, when you sum over all the particles, the terms of the first summation pairwise add to zero. Indeed this is the very reason that we choose to consider only the rotation of rigid bodies that have an axis of symmetry! For unsymmetrical systems, the first summation is *not* zero and the dynamical description of their rotation is much more complicated, because the angular momentum vector then is not parallel to the angular velocity vector.*

Therefore, for symmetric rigid bodies spinning about an axis coincident with the symmetry axis, the total spin angular momentum of the system, Equation 10.16, reduces to

$$\vec{L} = \sum_i m_i r_{i\perp}^2 \vec{\omega}$$

Since the angular velocity is a common factor of every term in the summation, we have

$$\vec{L} = \left(\sum_i m_i r_{i\perp}^2 \right) \vec{\omega}$$

The summation in parentheses is the *moment of inertia* I_{CM} of the rigid body about a rotational axis through the center of mass and parallel to the symmetry axis of the rigid body. That is, by definition

$$I_{CM} = \sum_i m_i r_{i\perp}^2 \qquad (10.17)$$

The moment of inertia of a rigid body thus depends on not only its mass but also how this mass is distributed with respect to the axis of rotation. The greater the distance of a mass point from the axis of rotation, the greater its contribution to the moment of inertia. In Section 10.8 we will see how to calculate the moment of inertia of several common symmetric rigid bodies; some values will be tabulated there in Table 10.1. The moment of inertia of a symmetric rigid body about a given symmetry axis is a characteristic of the particular rigid body shape, its mass, and that symmetry axis. It is quite possible for a rigid body to have more than one moment of inertia I_{CM}, but each one then is appropriate for a *different* symmetry axis.

---

*For unsymmetric rigid bodies, the moment of inertia is not a scalar but a second rank tensor. Such tensors are studied in more advanced courses in dynamics.

For symmetric rigid bodies rotating about one of their symmetry axes, the spin angular momentum vector of the system is *parallel* to the spin angular velocity vector; the scalar moment of inertia I_{CM} relates the two:

$$\vec{L}_{spin} = I_{CM} \vec{\omega}_{spin} \qquad (10.18)$$

This equation is identical in form to Equation 10.6 for a *single* particle undergoing orbital motion, with an appropriate value for the moment of inertia.

10.7 THE TIME RATE OF CHANGE OF THE SPIN ANGULAR MOMENTUM

We want to see if the time rate of change of the spin angular momentum is equal to the total torque on the system, which is the case for a single particle. The total spin angular momentum of a system is the vector sum of the (circular orbital) angular momenta of all the individual particles that make up the system. As in the previous section, take the origin of a coordinate system to be at the stationary center of mass point of the system, as shown in Figure 10.27.

The spin angular momentum is

$$\vec{L} = \sum_i \vec{L}_i$$
$$= \sum_i \vec{r}_i \times \vec{p}_i$$

where $\vec{r}_i$ is the position vector of the particle with momentum $\vec{p}_i$. Substituting $m_i \vec{v}_i$ for the momentum, we have

$$\vec{L} = \sum_i m_i \vec{r}_i \times \vec{v}_i$$

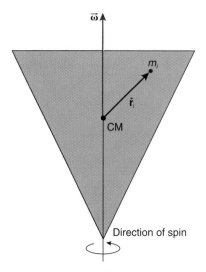

FIGURE 10.27 Use an origin at the center of mass.

The time rate of change of the spin angular momentum is

$$\frac{d\vec{L}}{dt} = \sum_i \frac{d}{dt}\left(m_i\vec{r}_i \times \vec{v}_i\right)$$

Differentiate the vector product, remembering to preserve the order of the terms. We find

$$\frac{d\vec{L}}{dt} = \sum_i m_i\left(\frac{d\vec{r}_i}{dt} \times \vec{v}_i + \vec{r}_i \times \frac{d\vec{v}_i}{dt}\right)$$

Since

$$\frac{d\vec{r}_i}{dt} = \vec{v}_i$$

the first term in the bracket is $\vec{v}_i \times \vec{v}_i$, which is zero; you might explain why. This leaves you with

$$\frac{d\vec{L}}{dt} = \sum_i m_i\left(\vec{r}_i \times \frac{d\vec{v}_i}{dt}\right)$$
$$= \sum_i \vec{r}_i \times \frac{d\vec{p}_i}{dt}$$

According to Newton's second law of motion, the time rate of change of the momentum of the *i*th particle is equal to the total force $\vec{F}_{i\,\text{total}}$ on the *i*th particle. Hence

$$\frac{d\vec{L}}{dt} = \sum_i \vec{r}_i \times \vec{F}_{i\,\text{total}}$$

Each term in the sum is the total torque on each particle of the system.

The total force acting on any particle of the system can be written as the sum of two types of forces: forces originating from outside the system itself (*external* forces such as externally applied pushes or pulls) and forces originating from other particles within the system itself (*internal* forces). Let's look first at the torques produced by the internal forces between any two pointlike particles; the results can then be generalized easily to any number of such particles.

Let $\vec{F}_{1\,\text{int}}$ be the force on particle 1 due to particle 2 and let $\vec{F}_{2\,\text{int}}$ be the force on particle 2 due to particle 1. According to Newton's third law of motion, these forces have the same magnitude but point in opposite directions. However, there is nothing in the third law that indicates the specific direction along which these mutual forces act. *Central* internal forces act along the line between the particles,* as shown in Figure 10.28. For central internal forces between the particles, the total internal torque produced by the force of each particle on the other is

$$\vec{r}_1 \times \vec{F}_{1\,\text{int}} + \vec{r}_2 \times \vec{F}_{2\,\text{int}}$$

where the vectors $\vec{r}_1$ and $\vec{r}_2$ locate the points of application of the forces with respect to the center of mass point. Since $\vec{F}_{2\,\text{int}} = -\vec{F}_{1\,\text{int}}$

Noncentral internal forces do *not* act along the line between the particles. Although noncentral forces exist, they are more appropriately considered in advanced and specialized areas of physics. We will not consider them.

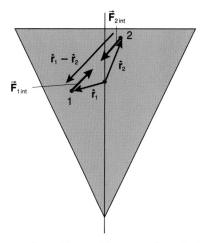

FIGURE 10.28 Central internal forces act along the line connecting the particles.

according to Newton's third law, we can eliminate $\vec{F}_{2\,\text{int}}$; so the sum of the internal torques of these forces is

$$\vec{r}_1 \times \vec{F}_{1\,\text{int}} + \vec{r}_2 \times (-\vec{F}_{1\,\text{int}})$$

or

$$(\vec{r}_1 - \vec{r}_2) \times \vec{F}_{1\,\text{int}}$$

Notice, in Figure 10.28, that the vector $\vec{r}_1 - \vec{r}_2$ points along the line between the two particles. However, the force $\vec{F}_{1\,\text{int}}$ also is directed along this line (as is $\vec{F}_{2\,\text{int}}$), since it is a central force. Thus the vector product is zero:

$$(\vec{r}_1 - \vec{r}_2) \times \vec{F}_{1\,\text{int}} = 0 \text{ N·m}$$

Therefore the *internal* torques between *every* pair of particles in the system *vanish*. Consequently the sum of the torques of the internal forces vanishes if the internal forces are central forces.

> The absence of internal torques means that the total torque on a system is caused only by forces external to the system itself.

> Thus the time rate of change of the spin angular momentum of the system is equal to sum of the torques of the external forces on the system (the total torque on the system due to external forces):
>
> $$\vec{\tau}_{\text{total ext}} = I_{\text{CM}}\frac{d\vec{\omega}}{dt}$$
>
> If the moment of inertia I_{CM} is constant, use Equation 10.18, $\vec{L} = I_{\text{CM}}\vec{\omega}$, and we have
>
> $$\vec{\tau}_{\text{total ext}} = \frac{d\vec{L}}{dt} \qquad (10.19)$$
>
> or in terms of the angular acceleration,
>
> $$\vec{\tau}_{\text{total ext}} = I_{\text{CM}}\vec{\alpha} \qquad (10.20)$$

After all this, you can rejoice because Equations 10.19 and 10.20 are similar to the earlier Equations 10.12 and 10.13 for a single particle executing orbital motion about a fixed point.

10.8 THE MOMENT OF INERTIA OF VARIOUS RIGID BODIES

Since the moment of inertia is so important to the dynamics of rotational motion, we must know how to calculate or find it for many common rigid bodies. In this section we see how this is accomplished.

Point Particle

The moment of inertia of a particle of mass m located a distance r from the axis of rotation (see Figure 10.29) we found to be given by Equation 10.5:

$$I = mr^2 \qquad \text{(point particle)}$$

Collection of Point Particles

Since the moment of inertia is a scalar quantity, the moment of inertia of a collection of particles all with the same angular velocity (see Figure 10.30) is the sum of the moments of inertia of the individual particles about their common axis of rotation:

$$I = \sum_i m_i r_{i\perp}^2 \qquad (10.21)$$

where $r_{i\perp}$ is the *perpendicular* distance of each mass m_i from the axis of rotation.

A Rigid Body

The moment of inertia of a rigid body is found by

a. breaking up the rigid body into a collection of differential masses dm, as in Figure 10.31;

b. finding the moment of inertia of each differential mass (a pointlike particle) about the axis of rotation:

$$r_\perp^2 \, dm$$

where $r_\perp$ is the distance of dm from the axis of rotation; and then

c. integrating this expression over the geometrical extent of the rigid body:

$$I = \int_{\text{object}} r_\perp^2 \, dm \qquad (10.22)$$

If the mass is distributed along essentially one dimension (thin rods), this integration is a one-dimensional integral. Example 10.7 illustrates this procedure. The integral is a two-

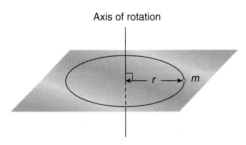

FIGURE 10.29 A pointlike particle located a distance r from an axis of rotation.

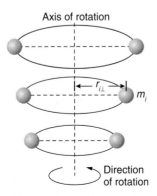

FIGURE 10.30 A collection of pointlike particles rotating about a common axis.

TABLE 10.1 Moment of Inertia I_{CM} of Some Common Symmetric Rigid Bodies About an Axis Coincident with the Symmetry Axis Through the Center of Mass

Thin cylindrical shell or hoop:

$$I_{\text{CM}} = mR^2$$

Solid cylinder:

$$I_{\text{CM}} = \frac{1}{2} mR^2$$

Thick cylindrical shell or hoop:

$$I_{\text{CM}} = \frac{1}{2} m(R_1^2 + R_2^2)$$

Rectangular plate:

$$I_{\text{CM}} = \frac{1}{12} m(a^2 + b^2)$$

Long thin rod:

$$I_{\text{CM}} = \frac{1}{12} m\ell^2$$

Sphere:

$$I_{\text{CM}} = \frac{2}{5} mR^2$$

Thin spherical shell:

$$I_{\text{CM}} = \frac{2}{3} mR^2$$

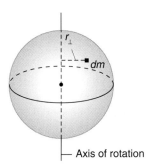

FIGURE 10.31 A solid, rigid body is considered as a collection of small differential mass elements dm.

dimensional integral if the object is planar. The integral is a three-dimensional volume integral if the object is three-dimensional (spheres, cubes, and so forth).

The moment of inertia is an intrinsically *positive quantity* that is the product of a purely numerical factor times the mass and the square of some characteristic dimension of the object. The moments of inertia of some common symmetrical rigid bodies are tabulated in Table 10.1.

The moments of inertia of the rigid bodies in Table 10.1 about axes *parallel* to, but not coincident with, the indicated axis will be discussed in Section 10.14.

PROBLEM-SOLVING TACTIC

10.3 If the rigid body can be segmented into several symmetrically shaped pieces, such as in Figure 10.32, the moment of inertia of the object as a whole is the sum of the moments of inertia of all the segmented pieces provided the moments are about the same axis.

QUESTION 5

Explain why it is possible for some rigid bodies to have more than one moment of inertia I_{CM}. Give an example of such a rigid body.

EXAMPLE 10.7

Calculate the moment of inertia of a thin rod of mass m and length ℓ about a symmetry axis through the center of mass and perpendicular to the length of the rod, as shown in Figure 10.33.

Solution

The moment of inertia of an extended, continuous rigid body is found by integrating the moment of inertia of a differential mass dm over the object:

$$\int r_{\perp}^2 \; dm$$

where $r_{\perp}$ is the distance of dm from the axis of rotation (see Figure 10.34; here $r_{\perp} = x$.

The thin rod has a mass per unit length, λ, that is given by

$$\lambda \equiv \frac{m}{\ell}$$

Thus the differential amount of mass dm on a differential length dx of the rod is

$$dm = \lambda \; dx$$

(a)

(b)

FIGURE 10.32 Segment a more complicated rigid body into smaller, symmetrical shapes whose moments of inertia are known. (a) Rigid body consisting of two symmetrically shaped pieces, a cylinder and a square plate. (b) Determine the moment of inertia of each symmetrical piece:

$$I_{\text{cyldr}} = \frac{1}{2}\, mR^2$$

$$I_{\text{plate}} = \frac{m'a^2}{6}$$

The total moment of inertia is the sum $I_{\text{cyldr}} + I_{\text{plate}}$.

FIGURE 10.33

FIGURE 10.34

The differential bit of moment of inertia of this pointlike mass about the axis of rotation is

$$dI = x^2 \, dm$$
$$= x^2 \lambda \, dx$$

Integrate this over the extent of the rod, from $x = -\ell/2$ to $x = \ell/2$, to find the moment of inertia of the rod I:

$$I = \lambda \int_{-\ell/2}^{\ell/2} x^2 \, dx$$

$$= \lambda \left. \frac{x^3}{3} \right|_{-\ell/2}^{\ell/2}$$

$$= \frac{m\ell^2}{12}$$

STRATEGIC EXAMPLE 10.8

A satellite is placed in a high-altitude orbit for precisely measuring large distances across the globe. The satellite consists of a uniform, solid sphere of Invar of mass 905 kg and radius 0.300 m. For stability it is to be set spinning about an axis through its center of mass. The final angular speed of the sphere is to be 300 rev/min. The sphere begins at rest and reaches its final angular speed in 30.0 s with a constant angular acceleration.

a. Determine the angular acceleration of the sphere.
b. Determine the total torque acting on the sphere that causes the angular acceleration.
c. If the total torque is provided by a force applied tangential to and along the equator of the sphere by small thrusters of negligible mass, what is the magnitude of the force required?

Solution

Introduce the coordinate system shown in Figure 10.35.

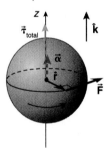

FIGURE 10.35

a. The final angular speed is

$$\frac{(300 \ \text{rev/min})(2\pi \ \text{rad/rev})}{60 \ \text{s/min}} = 31.4 \ \text{rad/s}$$

For a constant angular acceleration we have, from Equation 4.49,

$$\omega_z(t) = \omega_{z0} + \alpha_z t$$

Here ω_{z0} is zero, since the sphere begins at rest; also $\omega_z = 31.4$ rad/s when $t = 30.0$ s. Making these substitutions, you have

$$31.4 \ \text{rad/s} = 0 \ \text{rad/s} + \alpha_z (30.0 \ \text{s})$$

Solving for α_z,

$$\alpha_z = 1.05 \ \text{rad/s}^2$$

Hence the angular acceleration is

$$\vec{\boldsymbol{\alpha}} = (1.05 \ \text{rad/s}^2)\hat{\mathbf{k}}$$

The angular acceleration vector is sketched in Figure 10.35.

b. The moment of inertia of a solid sphere about its symmetry axis is found in Table 10.1:

$$I_{\text{CM}} = \frac{2}{5} mR^2$$
$$= (0.400)(905 \ \text{kg})(0.300 \ \text{m})^2$$
$$= 32.6 \ \text{kg} \cdot \text{m}^2$$

The total torque on the sphere is found from Equation 10.20:

$$\vec{\boldsymbol{\tau}}_{\text{total}} = I_{\text{CM}} \vec{\boldsymbol{\alpha}}$$
$$= (32.6 \ \text{kg} \cdot \text{m}^2)(1.05 \ \text{rad/s}^2)\hat{\mathbf{k}}$$
$$= (34.2 \ \text{N} \cdot \text{m})\hat{\mathbf{k}}$$

The total torque vector also is sketched in Figure 10.35.

c. The total torque is

$$\vec{\boldsymbol{\tau}}_{\text{total}} = \vec{\mathbf{r}} \times \vec{\mathbf{F}}$$

Since the force is applied tangentially to and along the equator of the sphere, $\vec{\mathbf{r}}$ is perpendicular to $\vec{\mathbf{F}}$ and

$$\vec{\boldsymbol{\tau}}_{\text{total}} = rF\hat{\mathbf{k}}$$
$$(34.2 \ \text{N} \cdot \text{m})\hat{\mathbf{k}} = (0.300 \ \text{m})F\hat{\mathbf{k}}$$

The magnitude of the total force thus is

$$F = 114 \ \text{N}$$

10.9 THE KINETIC ENERGY OF A SPINNING SYSTEM

A spinning rigid body system has many pointlike particles in motion. The kinetic energy associated with the spinning rigid body is the sum of the kinetic energies of all its constituent particles. At first glance, this may appear to be a complicated summation, but in this section we find that the result is a rather elegant expression for the kinetic energy associated with the spin, known as the **kinetic energy of rotation**.

Figure 10.36 depicts a rigid body system spinning about an axis coincident with a symmetry axis through the center of mass point of the system. The total kinetic energy of the system is the sum of the kinetic energy of each pointlike particle:

$$KE_{\text{rot}} = \sum_i \frac{1}{2} m_i v_i^2 \qquad (10.23)$$

Each particle is moving in a circle of radius $r_{i\perp}$ with angular velocity $\vec{\boldsymbol{\omega}}$; the angular velocity is the *same* for all the particles because the system is a rigid body. For such circular spinning motion, the velocity of the ith particle is found from Equation 4.34, as we have seen before:

$$\vec{\mathbf{v}}_i = \vec{\boldsymbol{\omega}} \times \vec{\mathbf{r}}_{i\perp}$$

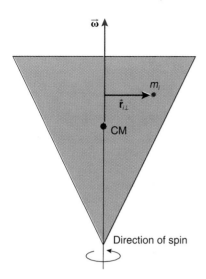

FIGURE 10.36 A rigid body spinning about a symmetry axis through its center of mass point.

where $\vec{r}_{i\perp}$ is the perpendicular position vector of the ith particle from the axis of rotation. Since $\vec{\omega}$ is perpendicular to $\vec{r}_{i\perp}$, the speed of the ith particle is

$$v_i = \omega r_{i\perp}$$

Substituting for v_i in Equation 10.23 for the kinetic energy, we find

$$KE_{rot} = \sum_i \frac{1}{2} m_i \omega^2 r_{i\perp}^2$$

$$= \sum_i \frac{1}{2} m_i r_{i\perp}^2 \omega^2$$

Since the angular speed ω is the same for all the particles, ω^2 can be factored out of each term in the summation:

$$KE_{rot} = \frac{1}{2} \left(\sum_i m_i r_{i\perp}^2 \right) \omega^2$$

Perhaps you recognize the expression in parentheses as the moment of inertia I_{CM} of the system relative to the symmetry axis passing through the center of mass (Equation 10.17).

> The rotational kinetic energy about the center of mass is
>
> $$KE_{rot} = \frac{1}{2} I_{CM} \omega^2 \qquad (10.24)$$

This has the same basic form as the kinetic energy

$$\frac{1}{2} mv^2$$

of a single particle of mass m, with the moment of inertia taking the place of the mass and the angular speed replacing the speed.

QUESTION 6

A sphere, a cylinder, and a thin-walled cylindrical pipe all have the same mass and radius and are spinning at the same angular speed about a symmetry axis through their respective centers of mass (for each cylinder, the symmetry axis coincident with the axis of the cylinder). Which has the greatest rotational kinetic energy? Which has the least?

EXAMPLE 10.9

Massive flywheels are used to store significant quantities of energy. Consider a flywheel disk of mass 150.0 kg and radius 25.0 cm spinning at 1000 rev/min about an axis perpendicular to the disk and passing through its center of mass, as shown in Figure 10.37.

FIGURE 10.37

a. Find the rotational kinetic energy of the flywheel system.
b. At what speed would a 1.00 kg particle have the same kinetic energy?

Solution

a. The moment of inertia of a disk is, from Table 10.1,

$$I_{CM} = \frac{1}{2} mR^2$$

$$= \frac{1}{2} (150.0 \text{ kg})(0.250 \text{ m})^2$$

$$= 4.69 \text{ kg} \cdot \text{m}^2$$

The angular speed of the disk is

$$\omega = \frac{1000 \text{ rev/min}}{60 \text{ s/min}} (2\pi \text{ rad/rev})$$

$$= 104.7 \text{ rad/s}$$

The kinetic energy associated with the spin then is found from Equation 10.24:

$$KE_{rot} = \frac{1}{2} I_{CM} \omega^2$$

$$= \frac{1}{2} (4.69 \text{ kg} \cdot \text{m}^2)(104.7 \text{ rad/s})^2$$

$$= 2.57 \times 10^4 \text{ J}$$

b. A particle of mass m moving with speed v has kinetic energy

$$KE = \frac{1}{2} mv^2$$

If a 1.00 kg particle has 2.57×10^4 J of kinetic energy, its speed is determined from

$$2.57 \times 10^4 \text{ J} = \frac{1}{2} (1.00 \text{ kg}) v^2$$

Solving for v,

$$v = 227 \text{ m/s}$$

10.10 SPIN DISTORTS THE SHAPE OF THE EARTH*

To a first approximation, the Earth has a spherical shape. A completely fluid, *nonspinning* Earth would assume a spherical shape under its own self-gravitation. Careful measurements, however, indicate that the equatorial diameter of the Earth is about 43 km greater than its polar diameter. This means that the Earth is oblate and has an equatorial bulge or equatorial deviation from spherical shape. The Earth is more nearly shaped like an **oblate ellipsoid** rather than a sphere. An oblate ellipsoid is a solid figure obtained by rotating an ellipse about its minor axis; see Figure 10.38. (A **prolate ellipsoid** is the figure obtained by rotating an ellipse about its major axis, as shown in Figure 10.39.)

Newton was the first to realize that a spinning, fluid Earth assumes the shape of an oblate ellipsoid. In this section, we show why the spinning Earth cannot be spherical in shape but must become oblate.

Consider a system of mass m lying at rest on the surface of a stationary, *nonspinning*, spherical Earth. The mass experiences two forces (see Figure 10.40):

1. the gravitational force of the Earth on the system, its weight $\vec{w}$; and
2. the normal force $\vec{N}$ that the surface exerts on the system, directed perpendicular to the surface.

Since the mass is not accelerated, the two forces must be equal in magnitude and opposite in direction. Their vector sum is zero in accordance with Newton's second law of motion.

On the other hand, since the Earth really is spinning, it cannot remain spherical because the mass m on the surface is moving around a circular path and so must experience a centripetal

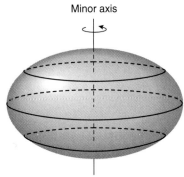

FIGURE 10.38 An oblate ellipsoid.

Minor axis

FIGURE 10.39 A prolate ellipsoid.

Major axis

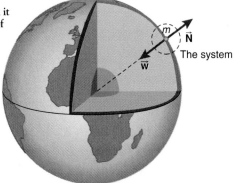

FIGURE 10.40 Forces on m when it is on the surface of a spherical, nonspinning Earth.

m $\vec{N}$

The system

$\vec{w}$

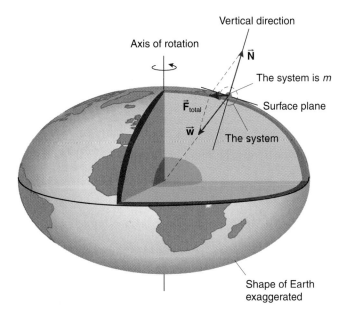

Vertical direction

Axis of rotation

$\vec{N}$

The system is m

$\vec{F}_{\text{total}}$

Surface plane

$\vec{w}$

The system

Shape of Earth exaggerated

FIGURE 10.41 Forces on a mass m on the surface of the spinning Earth.

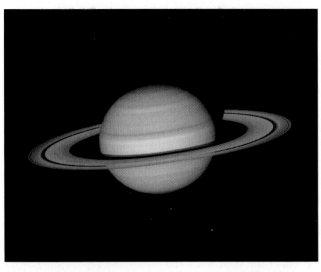

The giant planet Saturn spins so fast that it is noticeably oblate.

acceleration toward the axis of rotation. A force is needed to provide this acceleration. The shape of the Earth distorts so that the vertical direction (a direction perpendicular to the surface of a liquid on the surface of the Earth) no longer passes through the center of the Earth (except for locations at the poles and the equator), as shown in Figure 10.41.

The weight $\vec{\mathbf{w}}$ is no longer directed opposite to the normal force on a spinning Earth.

The shape of the Earth distorts so the vector sum of the two forces produces a total force directed toward the axis of rotation, which then provides the centripetal acceleration associated with the rotational motion of m about the axis. The faster the spin, the greater the distortion needed to provide the necessary force for the greater centripetal acceleration of the mass m. The giant planets Jupiter and Saturn spin very fast (Jupiter once in 9.84 h and Saturn once in 10.2 h) and have noticeable oblateness in telescopic images; the differences between the equatorial and polar diameters of these planets is almost 10%.

The reason the Earth distorts into an oblate ellipsoid rather than a prolate ellipsoid can be seen in Figure 10.42. The prolate ellipsoid shape would provide a vector sum of $\vec{\mathbf{w}}$ and $\vec{\mathbf{N}}$ that is directed *away* from the axis of rotation. As you know, however, for circular motion, the total force must be directed *toward* the center of the circle; a prolate ellipsoid shape cannot provide such a force.

The oblate shape of the Earth implies that the surface distance corresponding to a degree of latitude[†] is smaller near the

---

[†] Latitude is the angle between the local vertical direction (a direction perpendicular to the surface of liquid on the Earth's surface) and the perpendicular direction to the axis of spin. Equivalently, it is the angle between the local zenith and the place where the celestial equator (the plane of the Earth's equator projected onto the sky or celestial sphere) intersects the local celestial meridian (a north–south line through the zenith).

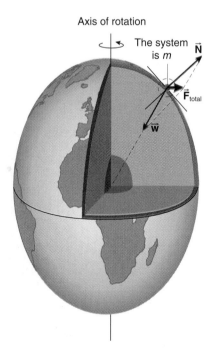

FIGURE 10.42 A mass m on a hypothetical prolate-ellipsoid-shaped planet.

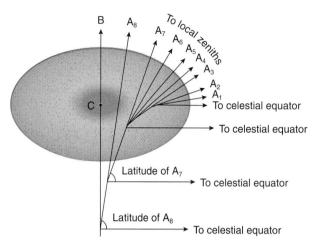

FIGURE 10.43 The distance corresponding to a degree of latitude is smaller near the equator than near the poles.

equator than near the poles, as shown in Figure 10.43. The reverse is true on a prolate ellipsoid. On the Earth, a degree of latitude represents 111.70 km near the pole and 110.57 km near the equator. The first convincing evidence for this difference came with the careful measurements made during several French expeditions to various parts of the Earth in the 1730s. The measurements were performed by Pierre Bouguer (1698–1758), Charles Marie de la Condamine (1701–1744), Louis Godin (1704–1760), and Pierre Louis Moreau de Maupertius (1698–1759).

10.11 THE PRECESSION OF A RAPIDLY SPINNING TOP*

One of the more remarkable physical systems in classical dynamics is a spinning top, familiar to us from the halcyon days of our childhood. A spinning top defies our preconceived notions of what should happen. It is wonderful to see how the Newtonian formulation of mechanics is able to account for the motions of such a marvelous toy. The same principles are used in the design of gyroscopic systems for navigating airplanes, spacecraft, and ships.

A simple top has an axis of symmetry passing through the center of mass. Several forces act on the top:

1. the weight $\vec{\mathbf{w}}$ (of magnitude mg), which can be thought of as acting on the top at its center of mass (see Figure 10.44); and
2. the force of the surface on the top at the fixed contact point between the top and the surface (the vector sum of the usual normal force of the surface and the static frictional force of the surface); this force is not shown in Figure 10.44.

The normal force of the surface and the static frictional force at the contact point produce zero torque about the contact point taken as an origin because their lines of action pass through the origin. Only the weight produces a torque about the point of contact:

$$\vec{\boldsymbol{\tau}}_{\text{total}} = \vec{\mathbf{r}} \times \vec{\mathbf{w}}$$

The vector $\vec{\mathbf{r}}$ is the vector from the origin (the point of contact) to the point of application of the weight (the center of mass of the top).

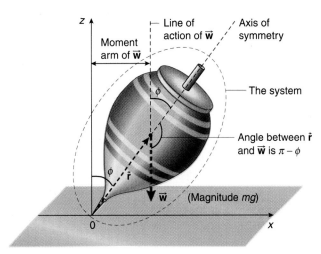

FIGURE 10.44 Only the weight has a nonzero torque about the point of contact of the top with the surface.

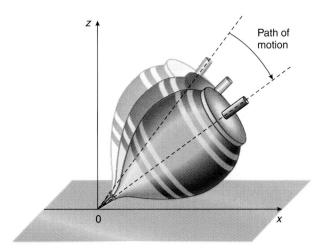

FIGURE 10.45 A nonspinning top falls over.

The magnitude of the total torque on the top is, from Figure 10.44,

$$\tau_{\text{total}} = r(mg)\sin(\pi - \phi)$$
$$= mgr\sin\phi$$

where ϕ is the angle between the symmetry axis of the top and the vertical direction.

The vector product always is perpendicular to the vectors forming the product, and so the direction of the torque is perpendicular to both $\hat{\mathbf{r}}$ and $\vec{\mathbf{w}}$. The direction of the torque is perpendicular to the plane formed by the symmetry axis of the top and the local vertical direction. *The torque on the top always is perpendicular to this plane regardless of the orientation of the top with respect to the particular coordinate system* of Figure 10.44.

According to Equation 10.19, the torque is equal to the time rate of change of the angular momentum:

$$\vec{\tau}_{\text{total}} = \frac{d\vec{\mathbf{L}}}{dt}$$

Hence the *differential change* $d\vec{\mathbf{L}}$ in the angular momentum vector is in the same direction as the torque—that is, in a direction perpendicular to the plane of the symmetry axis and the local vertical direction.

We consider two cases.

A Nonspinning Top

In this case the top initially has zero angular velocity and zero angular momentum, since $\vec{\mathbf{L}} = I\vec{\boldsymbol{\omega}}$. The *change* in the angular momentum is in the direction of the torque according to Equation 10.19. The torque is directed along $\hat{\mathbf{j}}$ when the top is in the position indicated in Figures 10.44 and 10.45.

Because of the torque, the *change* in the angular momentum (initially zero) also is along $\hat{\mathbf{j}}$, perpendicular to the plane of the symmetry axis and the local vertical direction. This is precisely what happens when the top simply falls over and hits the table. In Figure 10.45 the top rotates in a clockwise sense as it falls over, and so the top develops a growing angular velocity vector

oriented along $\hat{\mathbf{j}}$ according to the circular motion right-hand rule. The angular momentum vector grows from zero and is parallel to the growing angular velocity vector, because $\vec{\mathbf{L}} = I\vec{\boldsymbol{\omega}}$. Both are in the same direction as the torque (perpendicular to the plane of the symmetry axis and the local vertical). Nothing surprising here: the top falls over because of the torque caused by its weight.

A Rapidly Spinning Top

In this case the top has a *nonzero* initial angular momentum vector directed along the symmetry axis of the top; see Figure 10.46. The angular momentum is related to the angular velocity through the moment of inertia I_{CM} of the top:

$$\vec{\mathbf{L}}_{\text{spin}} = I_{\text{CM}}\vec{\boldsymbol{\omega}}_{\text{spin}}$$

This relationship, though, is peripheral to understanding what happens to the top as a result of the torque. All we need to know is that there is an angular momentum vector directed along the symmetry axis. The torque is unchanged regardless of whether the top is spinning.

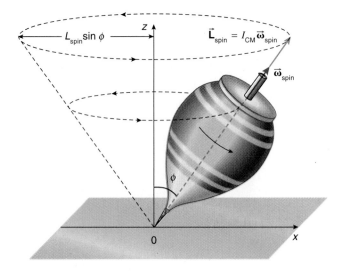

FIGURE 10.46 A rapidly spinning top.

But the *effect* of the torque now is quite different. The *change* in the angular momentum vector is (as always) in the *same* direction as the torque. The torque forces the *existing* angular momentum vector to move in a direction perpendicular to the plane of the symmetry axis and the local vertical direction! Since the change in the angular momentum vector is perpendicular to the angular momentum vector itself, the effect of the torque is to change the *direction* of the *existing* angular momentum vector but not its magnitude! The angular momentum vector is tied to the symmetry axis of the top, and so the top does *not* fall down: it moves in a direction perpendicular to the plane of the symmetry axis and the local vertical. As the top moves, the direction of the torque vector changes, because it is always perpendicular to the plane of the symmetry axis and the local vertical. The torque and the change in the angular momentum vector remain perpendicular to $\vec{L}_{spin}$ itself, and so the axis of the top moves (we say *precesses*) counterclockwise about the z-axis of Figure 10.46. The angular momentum vector $\vec{L}_{spin}$ and the top trace out a precessional cone.

We can calculate the angular speed of the precession. Notice the spin angular momentum vector $\vec{L}_{spin}$ and the change $d\vec{L}$ in this vector in Figure 10.47. From the geometry, the magnitude of the change is

$$dL = L_{spin} \sin \phi \, d\theta$$

Solving for $d\theta$,

$$d\theta = \frac{dL}{L_{spin} \sin \phi}$$

The angular speed of the precession is

$$\omega_{precs} = \frac{d\theta}{dt}$$

or upon substituting for $d\theta$,

$$\omega_{precs} = \frac{1}{L_{spin} \sin \phi} \frac{dL}{dt}$$

But

$$\frac{dL}{dt}$$

is equal to the magnitude of the torque ($= mgr \sin \phi$). With this substitution, the precessional angular speed is

$$\omega_{precs} = \frac{mgr \sin \phi}{L_{spin} \sin \phi}$$
$$= \frac{mgr}{L_{spin}}$$

The faster the spin given to the top, the greater L_{spin} and *the slower* or smaller the precessional angular speed. The precessional angular velocity is parallel to $\hat{k}$ in Figures 10.46 and 10.47.

Since the top has a precessional angular velocity, the center of mass of the top is in motion, and so the top also has a small amount of *orbital* angular momentum (directed along $\hat{k}$ in Figures 10.46 and 10.47). Thus the *total* angular momentum of the top,

$$\vec{L}_{total} = \vec{L}_{orbit} + \vec{L}_{spin}$$

is not, in a strict sense, coincident with the symmetry axis of the top; $\vec{L}_{spin}$ is directed along the symmetry axis while $\vec{L}_{orbit}$ is along $\hat{k}$. The torque is equal to the time rate of change of the *total* angular momentum. However, if the spin of the top is large, the precessional angular speed is small and, therefore, the magnitude of the orbital angular momentum of the center of mass also will be small compared with the magnitude of the spin angular momentum about the center of mass:

$$L_{orbit} << L_{spin}$$

Therefore the total angular momentum is essentially $\vec{L}_{spin}$ and our analysis of the dynamics is a good approximation. An exact analysis, particularly for slowly spinning tops, is more involved.

QUESTION 7

What happens to the direction of precession of a top if the spin of the top is in the direction opposite to that indicated in Figure 10.46?

10.12 THE PRECESSION OF THE SPINNING EARTH*

The Earth spins on its axis about once a day. The rotational axis of the Earth defines the geographic north and south poles. The spin angular momentum vector of the Earth is directed along the symmetry axis, as shown in Figure 10.48. You can think of the Earth as a huge spinning top, like the one we examined in the previous section.

The Earth is not precisely a sphere; it has a middle-aged spread (an equatorial bulge) caused by its rotation, as was discussed in Section 10.10. The equatorial diameter of the Earth is about 43 km larger than the polar diameter, and the Earth is roughly shaped like an oblate ellipsoid.

The orbital plane of the Earth in its motion about the Sun, known as the plane of the **ecliptic**, is *different* from the equatorial

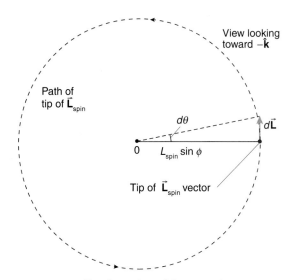

FIGURE 10.47 Top-down view of the precession.

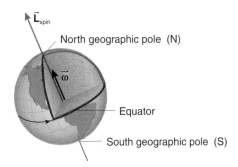

FIGURE 10.48 The spinning Earth.

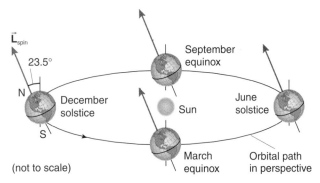

FIGURE 10.49 The orbital plane of the Earth (the ecliptic) is different from the equatorial plane.

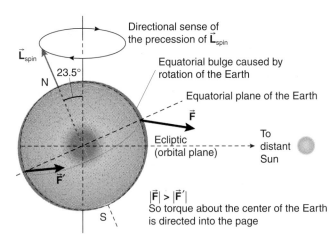

FIGURE 10.50 The force on the parts of the bulge closer to the Sun is of slightly greater magnitude than that on parts further from the Sun.

plane of the Earth, as shown in Figure 10.49. The two planes are inclined to each other by about 23.5°, an angle known as the **obliquity of the ecliptic**.

Consider the effect of the Sun on the equatorial bulge of the Earth. The gravitational attraction of the Sun on the part of the equatorial bulge in the hemisphere facing the Sun is slightly greater than that on the part of the bulge in the hemisphere opposite the Sun because of their slightly different distances from the Sun, stemming from the finite size of the Earth. The force on the parts of the bulge closer to the Sun and that on the parts further from the Sun also point in slightly different directions, as shown in Figure 10.50.

The forces of the Sun on the equatorial bulge produce torques about the center of the Earth. The resulting *total* torque is directed perpendicularly into the page of Figure 10.50.

Notice that if the Earth were spherical or if the angle between the equatorial plane and the ecliptic plane were zero, there would be no torque. Examining Figure 10.49, you can see that the torque is greatest at the times of the solstices (in June and December) and least at the times of the equinoxes (in March and September), so that it has a semiannual periodicity associated with it. The torque tries to force the equatorial bulge (and the Earth with it) into the plane of the ecliptic and align the polar axis of the Earth perpendicular to this plane. This is precisely what would occur if the Earth were not spinning.

The Earth *is* spinning, however, and so the *change* in the spin angular momentum vector (which is in the *same* direction

as the *torque*) is *perpendicular* to the spin angular momentum vector of the Earth, because the torque is perpendicular to $\vec{\mathbf{L}}_{spin}$. As a result the rotational axis of the Earth and the spin angular momentum vector maintain a *fixed* angle with the ecliptic plane and slowly change their orientation in space, tracing out a precessional cone. The spinning Earth precesses. The speed of the precessional motion is quite small, 50.262 arc seconds (= 0.013 962°) per year, completing the precessional circle in 25 785 y.

The motion is similar to that of a spinning top, with one important difference. Set a top spinning counterclockwise (as seen from above) with its axis inclined to the vertical, as shown in Figure 10.51. The gravitational force of the Earth on the top does not cause the top to fall over (which is what would occur if the top were not spinning). The top precesses, the axis of the top tracing out a cone in space about the vertical direction. In the

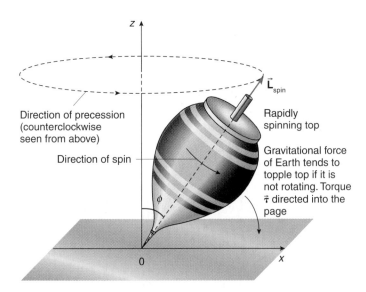

FIGURE 10.51 A counterclockwise spinning top precesses in the counterclockwise direction.

case depicted here, the precessional motion of the top also is *counterclockwise* about the vertical direction (when viewed from above), since the force is tending to topple the top, not to make it vertical or to "right" the top.

The precession of the spinning Earth is similar except that the precessional motion of the axis is in the *opposite direction*—

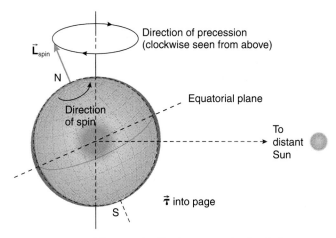

Direction of precession
(clockwise seen from above)

$\vec{L}_{spin}$

N

Direction
of spin

Equatorial plane

To
distant
Sun

$\vec{\tau}$ into page

S

FIGURE 10.52 The spinning Earth precesses in the clockwise sense when viewed from above the Northern Hemisphere.

that is, *clockwise* (when seen from above) about a line perpendicular to the plane of the ecliptic; see Figure 10.52. In this case the torque tends to "right" the Earth, not topple it or tip it over as was the case for the top.

The precession of the rotational axis of the Earth has interesting observational consequences for astronomers on Earth. The point toward which the rotational axis of the Earth points in the heavens in the Northern Hemisphere is known as the **North Celestial Pole**, or NCP. The brightest star nearest the position of the NCP customarily is called the **pole star** (the north star for Northern Hemisphere observers). Precession means that the identity of the pole star changes with time. The north star is now Polaris (α Ursae Minoris). About 5000 years ago, when the Great Pyramid of Khufu was built in Egypt, the north star was Thuban (α Draconis).* The bright star Vega (α Lyrae) will be the north star for the class of A.D. 14 000 at your college or university. The path traced out by the North Celestial Pole in the sky over the 25 785-year precessional period is indicated in the star map of the northern sky shown in Figure 10.53.

---

*The sloping shaft into the interior of the Great Pyramid of Khufu (c. 2900 B.C.) at Giza (near modern Cairo) was oriented toward the pole star and so is parallel to the spin axis of the Earth. The shaft still points toward the pole star, but now it is Polaris, not Thuban.

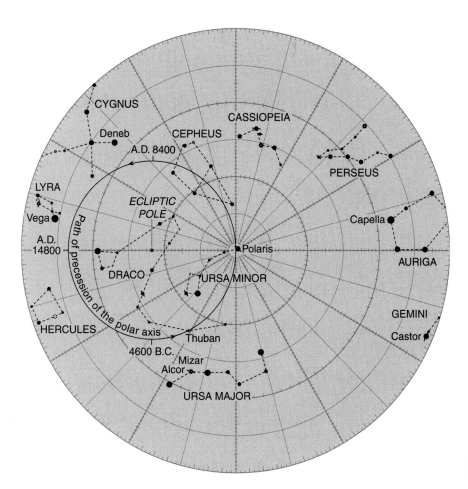

CYGNUS

Deneb

CEPHEUS

CASSIOPEIA

A.D. 8400

PERSEUS

LYRA

ECLIPTIC
POLE

Vega

Capella

A.D.
14800

Polaris

AURIGA

Path of precession of the polar axis

DRACO

URSA MINOR

GEMINI

Castor

HERCULES

Thuban

4600 B.C.

Mizar

Alcor

URSA MAJOR

FIGURE 10.53 The precessional path of the polar axis of the Earth during the 25 785-year precessional period.

The ancient Greek astronomer Hipparchus (c. 125 B.C.) discovered the precession of the rotational axis of the Earth. The NCP moves only about 1° around the precessional circle during an average human lifetime of 70 years (much longer than the average lifetime at the time of Hipparchus!). The discovery gives you a feeling for how diligently ancient cultures observed the sky!

QUESTION 8

Imagine yourself at the north geographic pole of the Earth. As you look vertically up into the sky, does the slow precessional path of the axis of the spinning Earth trace out a clockwise or counterclockwise path during 25 785 y? Is your answer the same if you stand at the south geographic pole?

10.13 SIMULTANEOUS SPIN AND ORBITAL MOTION

There are many examples of systems that undergo both spin and orbital motion. The system spins about a symmetry axis through the center of mass point and, simultaneously, the center of mass is in motion. The tires of your car spin while their center of mass moves (orbital motion). A beautiful football pass has both spin and orbital motion. Each planet spins about its axis and orbits the Sun. The moons of the planets simultaneously spin on their axes while orbiting the mother planet.

A spinning system has rotational kinetic energy as we discovered in Section 10.9. The kinetic energy associated with the spin is (Equation 10.24 with appropriate subscripts added)

$$KE_{rot} = \frac{1}{2} I_{CM} \omega_{spin}^2$$

But there is additional kinetic energy associated with the motion of the center of mass. The kinetic energy associated with the motion of the center of mass is

$$KE = \frac{1}{2} m v_{CM}^2$$

or equivalently,

$$KE = \frac{1}{2} (m r_\perp^2) \omega_{orbit}^2$$

We showed in Section 9.11 that the total kinetic energy is the sum of the rotational kinetic energy about the center of mass and the kinetic energy associated with the motion of the center of mass itself:

$$KE_{total} = KE_{rot} + \frac{1}{2} m v_{CM}^2 \qquad (10.25)$$

Likewise, for a system with spin and orbital motion, the total angular momentum has two contributions. The spin angular

momentum $\vec{L}_{spin}$ of the system is proportional to the spin angular velocity (Equation 10.18):

$$\vec{L}_{spin} = I_{CM} \vec{\omega}_{spin}$$

where I_{CM} is the moment of inertia of the system about the appropriate symmetry axis through the center of mass point, and $\vec{\omega}_{spin}$ is the spin angular velocity of the system about this axis. For orbital motion, the system is imagined to be a point particle, with a mass equal to the total mass m of the entire system, located at the center of mass point. The orbital angular momentum of the system is

$$\vec{L}_{orbit} = \vec{r} \times \vec{p}_{CM}$$

where $\vec{p}_{CM}$ is the momentum of the center of mass,

$$\vec{p}_{CM} = m \vec{v}_{CM}$$

(here $\vec{v}_{CM}$ is the velocity of the center of mass). Equivalently the orbital angular momentum can be found from Equation 10.4 (with appropriate subscripts added):

$$\vec{L}_{orbit} = m r_\perp^2 \vec{\omega}_{orbit}$$

where $\vec{\omega}_{orbit}$ is the angular velocity of the center of mass about an axis and $r_\perp$ is the perpendicular distance of the center of mass point from the axis of orbital motion.

The total angular momentum of the system is the vector sum of the spin and orbital angular momenta:

$$\begin{aligned} \vec{L}_{total} &= \vec{L}_{spin} + \vec{L}_{orbit} \\ &= I_{CM} \vec{\omega}_{spin} + m r_\perp^2 \vec{\omega}_{orbit} \end{aligned} \qquad (10.26)$$

EXAMPLE 10.10

A 3.00 kg disk of radius 0.100 m is spinning at 15.0 rev/s while simultaneously zipping across a horizontal (frictionless) ice surface at a speed of 5.00 m/s, as shown in Figure 10.54.

4.00 m

30.0°

0

$\vec{v}$

Frictionless ice surface

FIGURE 10.54

a. What is the spin angular momentum of the disk?
b. What is the orbital angular momentum of the disk about the point 0?
c. What is the total angular momentum of the disk?
d. Find the total kinetic energy of the disk.

Solution

a. The moment of inertia of a disk about the symmetry axis perpendicular to the disk and through the center of mass is found from Table 10.1 to be

$$I_{CM} = \frac{1}{2} mR^2$$

$$= \frac{1}{2} (3.00 \text{ kg})(0.100 \text{ m})^2$$

$$= 1.50 \times 10^{-2} \text{ kg} \cdot \text{m}^2$$

The spin angular momentum is, from Equation 10.18,

$$\vec{L}_{spin} = I_{CM} \vec{\omega}_{spin}$$

Use the circular motion right-hand rule and the coordinate system indicated in Figure 10.55 to discover that the direction of the spin angular velocity vector is along $+\hat{k}$. Hence

$$\vec{L}_{spin} = I_{CM} \omega_{spin} \hat{k}$$

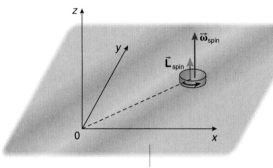

FIGURE 10.55

The spin angular speed is

$$\omega_{spin} = (15.0 \text{ rev/s})(2\pi \text{ rad/rev})$$

$$= 94.2 \text{ rad/s}$$

Hence the spin angular momentum is

$$\vec{L}_{spin} = I_{CM} \omega_{spin} \hat{k}$$

$$= (1.50 \times 10^{-2} \text{ kg} \cdot \text{m}^2)(94.2 \text{ rad/s}) \hat{k}$$

$$= (1.41 \text{ kg} \cdot \text{m}^2/\text{s}) \hat{k}$$

b. The orbital angular momentum of the disk about the point 0 is

$$\vec{L}_{orbit} = \vec{r} \times \vec{p}$$

Its magnitude is

$$L_{orbit} = rp \sin \theta$$

$$= (4.00 \text{ m})(3.00 \text{ kg})(5.00 \text{ m/s}) \sin 30.0°$$

$$= 30.0 \text{ kg} \cdot \text{m}^2/\text{s}$$

The direction of the orbital angular momentum is the direction of $\vec{r} \times \vec{p}$. From the vector product right-hand rule, this is toward $-\hat{k}$. Hence

$$\vec{L}_{orbit} = (-30.0 \text{ kg} \cdot \text{m}^2/\text{s}) \hat{k}$$

c. The total angular momentum is the vector sum of the spin and orbital angular momenta (see Figure 10.56):

$$\vec{L}_{total} = \vec{L}_{spin} + \vec{L}_{orbit}$$

$$= (1.41 \text{ kg} \cdot \text{m}^2/\text{s}) \hat{k} + (-30.0 \text{ kg} \cdot \text{m}^2/\text{s}) \hat{k}$$

$$= (-28.6 \text{ kg} \cdot \text{m}^2/\text{s}) \hat{k}$$

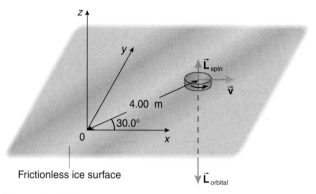

FIGURE 10.56

d. The total kinetic energy is the sum of the kinetic energy of rotation plus the kinetic energy associated with the translation of the center of mass:

$$KE_{total} = \frac{1}{2} I_{CM} \omega_{spin}^2 + \frac{1}{2} mv_{CM}^2$$

$$= \frac{(1.50 \times 10^{-2} \text{ kg} \cdot \text{m}^2)(94.2 \text{ rad/s})^2}{2}$$

$$+ \frac{(3.00 \text{ kg})(5.00 \text{ m/s})^2}{2}$$

$$= 66.6 \text{ J} + 37.5 \text{ J}$$

$$= 104.1 \text{ J}$$

10.14 SYNCHRONOUS ROTATION AND THE PARALLEL AXIS THEOREM

In many cases the orbital angular velocity $\vec{\omega}_{orbit}$ is different from the spin angular velocity $\vec{\omega}_{spin}$. For example, the spin angular velocity of the Earth $\vec{\omega}_{spin}$ is significantly greater than its orbital angular velocity $\vec{\omega}_{orbit}$ about the Sun; the two vectors also point in different directions. Other planets and some moons are similar in these respects.

Occasionally, however, a system has *equal* spin and orbital angular velocities:

$$\vec{\omega}_{spin} = \vec{\omega}_{orbit}$$

In this case:

1. the two vectors are parallel; and

2. the spin angular speed is equal to the orbital angular speed.

If these conditions are satisfied, the system is said to undergo **synchronous rotation**.

A good example of synchronous rotation is the motion of the Moon about the Earth*: as the Moon orbits the Earth, the Moon always keeps the *same* hemisphere facing the Earth.[†]

In synchronous rotation, as the system completes a revolution with respect to the orbital axis, the system *also* completes a spin revolution about an axis through its center of mass, as shown in Figure 10.57. If $\vec{\omega}_{\text{spin}}$ is equal to $\vec{\omega}_{\text{orbit}}$, call the common angular velocity $\vec{\omega}$. Then the total angular momentum of the system is, according to Equation 10.26,

$$\vec{L}_{\text{total}} = I_{\text{CM}}\vec{\omega} + md^2\vec{\omega}$$
$$= (I_{\text{CM}} + md^2)\vec{\omega} \qquad (10.27)$$

Write the total angular momentum as

$$\vec{L}_{\text{total}} = I\vec{\omega} \qquad (10.28)$$

where I is the moment of inertia of the system about an axis parallel to the symmetry axis through the center of mass point, but not coincident with this symmetry axis. Compare Equations 10.27 and 10.28.

> We see that the moment of inertia I of the system about an axis parallel to the symmetry axis of the system and separated from it by a perpendicular distance d is
>
> $$I = I_{\text{CM}} + md^2 \qquad (10.29)$$
>
> Equation 10.29 is known as the **parallel axis theorem**.

---

*Tidal gravitational forces of the Earth on the Moon caused the motion of the Moon to have this characteristic. For similar reasons, many (but not all) of the moons of the planets also exhibit synchronous rotation.

[†] The spin axis and the orbital axis are not quite parallel (they have an angle of about 6° between them), but for simplicity we assume they are parallel here. The 6° angle and the elliptical orbit of the Moon enable observers on the Earth to see approximately 59% of the lunar real estate as the Moon traverses its orbital path.

This theorem applies to any rigid body in synchronous rotation about an axis that is parallel to the symmetry axis through the center of mass. The parallel axis theorem can be applied to a system *only* when the spin and orbital angular velocities of the system are the same—that is, for synchronous rotation.

10.15 ROLLING MOTION WITHOUT SLIPPING

A rolling stone gathers no moss

               Publius Syrus (c. 1st century B.C.)[‡]

The rolling motion of wheels is such a common feature of our technological society that we take it for granted. No one, least of all a college student, wants to be without wheels. In this section we consider rolling motion in some detail, for it is not as trivial as it may seem.

When a symmetric rigid body system undergoes rolling motion, it is simultaneously spinning about its center of mass while the center of mass is undergoing translation. Notice in Figure 10.57 that if the radius of the orbit of point A is zero, the system is *rolling without slipping* about the orbital rotational axis. No slipping means there is no sliding at the point of contact of the system with the orbital rotational axis. The wheels of a car that is moving forward while its tires are spinning madly, leaving behind black stripes on the road, is an example of rolling motion *with slipping*; we commonly call it skidding. Likewise, when drag racers spin the tires of a car, the tires are rolling with slipping at the point of contact between the tires and the road.

> Rolling motion (without slipping) is a special case of synchronous rotation.

The spin and orbital axes are parallel and separated by the radius of the circularly shaped system undergoing rolling motion. The

---

[‡] *The Moral Sayings of Publius Syrus, a Roman Slave*, translated by Darius Lyman (Andrew J. Graham, New York, 1862), page 48 (saying #524).

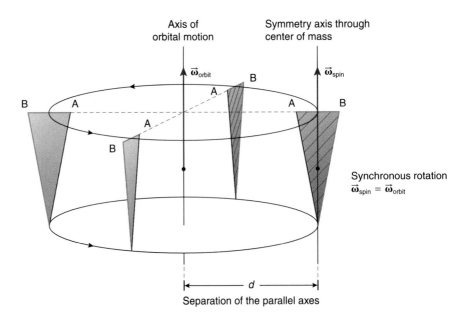

FIGURE 10.57 In synchronous rotation, a system keeps the same side facing the orbital axis.

moment of inertia I, found from the parallel axis theorem, then represents the moment of inertia of the system about the point of rolling contact (e.g., the point of contact between a tire and the road).

It is preferable for tires to roll without slipping on pavement, for then the appropriate coefficient of friction is the static value, rather than the smaller coefficient of kinetic friction. Tires rolling *with* slipping are skidding (such as on icy pavements), which can be quite dangerous for the occupants of the vehicle, not to say for the vehicle itself. Antilock braking systems are designed to ensure that tires roll without slipping during braking.

If a symmetric rigid body system is rolling without slipping, there are relationships between

1. the distance the center of mass travels and the corresponding angle through which the system rotates about the symmetry axis through the center of mass;
2. the speed of the center of mass and the angular speed of rotation; and
3. the magnitude of the acceleration of the center of mass and the magnitude of the angular acceleration of the system.

We want to determine what these relationships are.

Consider a round rigid body system such as a wheel, sphere, or other system with a circular cross section, as shown in Figure 10.58. After one complete rotation, the center of the system moves through a distance equal to the circumference of its circular cross section. In other words the center moves through a distance s, where

$$s = 2\pi R$$

If the system rotates through an angle θ (in radians), as in Figure 10.59, then the center moves through a distance

$$s = R\theta \tag{10.30}$$

FIGURE 10.58 One complete revolution of a rolling system translates the center of mass by the circumference of the circular cross section.

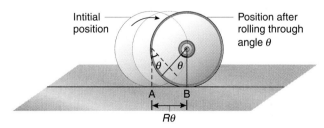

FIGURE 10.59 Rolling through an angle θ translates the center of mass by a distance $R\theta$.

Differentiating Equation 10.30 with respect to time, we find

$$\left| \frac{ds}{dt} \right| = R \left| \frac{d\theta}{dt} \right|$$

But

$$\left| \frac{ds}{dt} \right| = v_{\text{center}}$$

the speed of the center of the system. Likewise

$$\left| \frac{d\theta}{dt} \right| = \omega$$

the angular speed of the system about the rotational axis through the center of the circular cross section.

Thus the rolling condition is equivalent to the following relation between the translational speed of the center and the angular speed:

$$v_{\text{center}} = R\omega \tag{10.31}$$

Differentiation of Equation 10.31 relates the magnitude of the linear acceleration of the center of the wheel, $\vec{a}_{\text{center}}$, to the magnitude of the angular acceleration $\vec{\alpha}$ about the axis through the center of the wheel:

$$a_{\text{center}} = R\alpha \tag{10.32}$$

Equations 10.30, 10.31, or 10.32 imply that the object is rolling without slipping; therefore they are known as **rolling constraints**.

Consider an origin at 0 in Figure 10.60. The position vector $\vec{r}$ of any point P within or on the circular cross section is the sum of the vector $\vec{r}_{\text{center}}$ locating the center of the circle and the position vector $\vec{r}'$ locating the point P with respect to the center of the circle:

$$\vec{r} = \vec{r}_{\text{center}} + \vec{r}'$$

Differentiating this equation with respect to time, we find

$$\vec{v} = \vec{v}_{\text{center}} + \vec{v}' \tag{10.33}$$

where $\vec{v}_{\text{center}}$ is the velocity of the center of the wheel and $\vec{v}'$ is the velocity of the point P with respect to the center of the wheel. *For points on the circumference of the wheel*, the magnitude of $\vec{v}'$ is the same as the magnitude of $\vec{v}_{\text{center}}$ because of the rolling

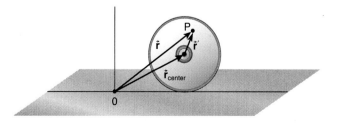

FIGURE 10.60 Geometry for describing rolling motion.

constraints. The velocity $\vec{v}$ obtained by performing the vector addition in Equation 10.33 for various points on the wheel is shown in Figure 10.61. Notice that the point of the wheel in contact with the surface is at rest while the point at the top of the wheel is moving with a velocity $2\vec{v}_{center}$.

Assume the center is also the center of mass. As we learned in Section 9.11, the *kinetic energy* of a rolling system has two contributions:

1. the kinetic energy associated with the motion of the center of mass:

$$KE_{trans} = \frac{1}{2}mv_{CM}^2$$

where v_{CM} is the speed of the center of mass; and

2. the kinetic energy associated with the spin about the center of mass:

$$KE_{rot} = \frac{1}{2}I_{CM}\omega^2$$

where I_{CM} is the moment of inertia of the object about the center of mass and ω is the angular speed of the rotation (spin) about the center of mass.

Thus the total kinetic energy is

$$KE_{total} = KE_{trans} + KE_{rot}$$
$$= \frac{1}{2}mv_{CM}^2 + \frac{1}{2}I_{CM}\omega^2$$

Use the rolling constraint, Equation 10.31, for v_{CM}:

$$KE_{total} = \frac{1}{2}mR^2\omega^2 + \frac{1}{2}I_{CM}\omega^2$$
$$= \frac{1}{2}(mR^2 + I_{CM})\omega^2$$

The parallel axis theorem, Equation 10.29, enables us to write the total kinetic energy as

$$KE_{total} = \frac{1}{2}I\omega^2 \qquad (10.34)$$

PROBLEM-SOLVING TACTIC

10.4 If you use the parallel axis theorem to determine the moment of inertia I of the system, the total kinetic energy of the system due to *both* its spin and orbital motion is given by Equation 10.34:

$$KE_{total} = \frac{1}{2}I\omega^2 \qquad (10.34)$$

QUESTION 9

An object of mass m, circular cross section (of radius R), and moment of inertia I_{CM} is *rolling* without slipping along a horizontal surface. (a) What is the *ratio* of the kinetic energy of translation to the kinetic energy of rotation? That is, find

$$\frac{KE_{trans}}{KE_{rot}}$$

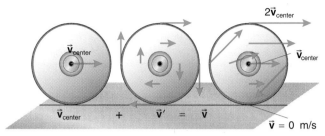

FIGURE 10.61 The velocity of the center, the velocity of points relative to the center, and the velocity with respect to the ground.

(b) Does the ratio depend on the speed of the object?
(c) Which of the objects listed in Table 10.1 with circular cross sections has the *greatest* value for this ratio? For which (if any) is the ratio equal to 1?

STRATEGIC EXAMPLE 10.11

Release at rest a spherical marble of mass m and radius R and let it roll down an inclined plane of vertical height h on your desk. Determine the speed of the marble at the base of the incline.

Solution

Consider the marble to be the system. The forces acting on the sphere are the following:

1. Its weight $\vec{w}$ of magnitude mg. The work done by the gravitational force is accounted for in the CWE theorem through the change in the gravitational potential energy of the center of mass of the sphere.
2. The normal force $\vec{N}$ of the surface on the sphere. As always, this force does zero work, because the normal force is directed perpendicular to the change in the position vector of the sphere at all points along its path.
3. The frictional force on the sphere by the surface. For rolling motion without slipping, the point of contact between the sphere and the surface is at rest, so that the frictional force is the *static* frictional force. This force does zero work.

The CWE theorem here takes the form

$$0 \text{ J} = \Delta(KE + PE)$$
$$= (KE + PE)_f - (KE + PE)_i$$

At the top of the incline the KE is initially zero:

$$KE_i = 0 \text{ J}$$

Take the origin of a coordinate system to be at ground level, as shown in Figure 10.62. The initial gravitational PE of the center of mass of the sphere is

$$PE_i = mg(h + R)$$

At the bottom of the incline, the gravitational PE of the center of mass of the sphere is

$$PE_f = mgR$$

The final KE may be found in two (equivalent) ways.

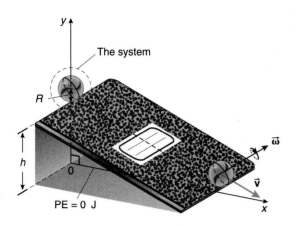

FIGURE 10.62

Method 1

Think of the KE as having two contributions: the motion of the center of mass and the spin about the center of mass. The KE of the center of mass is

$$\frac{1}{2} m v^2$$

and the rotational KE about the center of mass is

$$\frac{1}{2} I_{\text{CM}} \omega^2$$

where I_{CM} is the moment of inertia of a sphere about an axis passing through its center, found in Table 10.1:

$$I_{\text{CM}} = \frac{2}{5} m R^2$$

That is,

$$\text{KE}_\text{f} = \frac{1}{2} m v^2 + \frac{1}{2} \left(\frac{2}{5} m R^2 \right) \omega^2$$

Use the rolling constraint of Equation 10.31, for ω, and you have

$$\text{KE}_\text{f} = \frac{1}{2} m v^2 + \frac{1}{2} \left(\frac{2}{5} m R^2 \right) \left(\frac{v}{R} \right)^2$$

$$= \frac{7}{10} m v^2$$

Method 2

Realize that the sphere is instantaneously rotating about the point of contact between the sphere and the plane. With this view, the total KE then is found from Equation 10.34 where I is the moment of inertia of the sphere about an axis at the point of contact. This moment of inertia is obtained from the parallel axis theorem of Equation 10.29:

$$I = \frac{2}{5} m R^2 + m R^2$$

$$= \frac{7}{5} m R^2$$

Hence

$$\text{KE}_\text{f} = \frac{1}{2} \left(\frac{7}{5} m R^2 \right) \omega^2$$

Use the rolling constraint of Equation 10.31 for ω, and you obtain

$$\text{KE}_\text{f} = \frac{7}{10} m v^2$$

Having found an expression for the final KE, apply the CWE theorem:

$$
\begin{aligned}
0 \text{ J} &= (\text{KE} + \text{PE})_\text{f} - (\text{KE} + \text{PE})_\text{i} \\
&= \left(\frac{7}{10} m v^2 + m g R \right) - \left(0 \text{ J} + m g [h + R] \right) \\
&= \frac{7}{10} m v^2 - m g h
\end{aligned}
$$

Solving for v,

$$v = \left(\frac{10}{7} g h \right)^{1/2}$$

10.16 Wheels*

The wheel is come full circle

William Shakespeare (1564–1616)[†]

It is difficult to imagine life without wheels, since they form the basis of many machines. Trains, trucks, cars, carts, Roman chariots, roller skates, roller blades, or roller bearings: wheels are a symbol of intelligent, technological life. The invention of the wheel certainly was one of the paramount discoveries in the history of technology. Here we form a crude model of an ideal wheel to see how the great advantage arises in using wheels to assist us in doing physical work.

A Stationary Wheel

First consider a wheel of mass M and radius R with a smaller hub of radius r, as shown in Figure 10.63. The wheel is free to turn about its center, which is fixed in position.

A mass m hangs from an ideal string that is wound around the central hub. Another ideal string is wrapped around the perimeter on which we pull with a force $\vec{\mathbf{F}}$. For simplicity we neglect any friction in the axle bearing.

If the wheel has zero angular acceleration, the total torque on the wheel must also be zero. Choose the center of the wheel as the point about which to calculate the total torque. The force $\vec{\mathbf{F}}'$ of the axle on the wheel and the weight $\vec{\mathbf{w}}'$ of the wheel itself (of magnitude Mg) produce zero torque about this point, because their lines of action pass through the point about which we are

---

[†] *King Lear*, Act V, Scene 3, line 176.

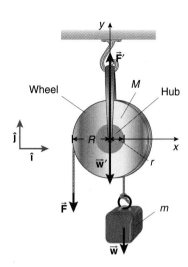

FIGURE 10.63 A wheel that can turn about its fixed center.

calculating the torque. The torques supplied by the hanging mass and the force $\vec{\mathbf{F}}$ are

$$\text{torque of } \vec{\mathbf{F}}: \quad -R\hat{\mathbf{i}} \times (-F\hat{\mathbf{j}}) = RF\hat{\mathbf{k}}$$
$$\text{torque of the weight of } m: \quad r\hat{\mathbf{i}} \times (-mg\hat{\mathbf{j}}) = -rmg\hat{\mathbf{k}}$$

The total torque must sum to zero, since there is zero angular acceleration:

$$(RF - rmg)\hat{\mathbf{k}} = 0 \text{ N·m}$$

Thus

$$RF = rmg$$

Rearranging this slightly, we find

$$F = \frac{r}{R}mg \qquad (10.35)$$

The magnitude of the force $\vec{\mathbf{F}}$ is *less* than the magnitude of the weight hanging from the hub by the factor r/R. The quantity r/R is known as the **mechanical advantage** of the wheel.*

Now let the force $\vec{\mathbf{F}}$ slowly raise the weight through one turn of the wheel, so that the kinetic energy of the system is unchanged throughout the process. Since the force $\vec{\mathbf{F}}$ is in the same direction as the change in the position vector of its point of application (the end of the rope), the work done by this constant force is

$$F(2\pi R)$$

The work done by the gravitational force is accounted for in the change in the potential energy of the mass m. Since the mass m is raised a distance $2\pi r$ when the wheel turns through one revolution, the *change* in the gravitational potential energy of the mass is

$$mg(2\pi r)$$

---

*The term mechanical advantage also is used for other simple machines such as the lever, inclined plane, and screw.

The CWE theorem states that

$$W_{\text{other}} = \Delta KE + \Delta PE$$

The work done by "other" forces is the work done by $\vec{\mathbf{F}}$. The change in the kinetic energy is zero (because the mass was raised at constant speed). Thus we have

$$F(2\pi R) = 0 \text{ J} + mg(2\pi r) \qquad (10.36)$$

> The work we do is the same as if we simply raised the mass m a distance $2\pi r$ without the advantage of the wheel.

> The real advantage of the wheel, then, is *not* in the work done on m, but rather that the force $\vec{\mathbf{F}}$ is smaller than that force used when the mass is raised without the assistance of the wheel.

The price we pay for using a smaller force $\vec{\mathbf{F}}$ is that it must move through a greater distance ($2\pi R$) to accomplish the same amount of work as raising the mass directly through a distance $2\pi r$.

Notice that the result of the CWE theorem (Equation 10.36) also indicates that

$$F = \frac{r}{R}mg$$

as we found from considering the torques (Equation 10.35).

A Rolling Wheel

You likely know from personal experience that it is far easier to pull a mass when it is on wheels than to drag it. Here we consider a simplified model to explain why this is the case.

Consider a rather idealized case of a steel wheel (of radius R) moving on a steel rail, as on a railroad track. Let the steel wheel have an axle of radius r, around which the wheel can slip, as shown in Figure 10.64. Let the track be straight and horizontal.

Imagine the wheel to be locked, so that it cannot roll or spin. A horizontal force $\vec{\mathbf{F}}_{\text{drag}}$ is needed to drag the wheel along at constant speed. The kinetic frictional force on the locked wheel depends on the coefficient of kinetic friction μ_k (here for steel sliding on steel) and the normal force of the track on the wheel. The magnitude of the normal force of the track on the wheel here is equal to the magnitude of the weight of the wheel and its fraction of the load of the rail car; say this is of magnitude mg. If we drag the locked wheel through a distance

FIGURE 10.64 A rolling wheel.

equal to the circumference of the wheel, $2\pi R$, the work done by the force $\vec{F}_{drag}$ is then

$$W_{drag} = F_{drag}(2\pi R)$$

The work done by the kinetic frictional force is *negative* because the change in the position vector is opposite to the direction of the force:

$$W_{frictn} = -(\mu_k mg)2\pi R$$

Now we apply the CWE theorem,

$$W_{other} = \Delta KE + \Delta PE$$

There is no change in either the kinetic energy or the gravitational potential energy, and so we have

$$W_{drag} + W_{frictn} = 0 \text{ J} + 0 \text{ J}$$

or

$$F_{drag}(2\pi R) - \mu_k mg(2\pi R) = 0 \text{ J}$$

Solving for F_{drag},

$$F_{drag} = \mu_k mg \qquad (10.37)$$

The force $\vec{F}_{drag}$ is equal in magnitude to that of the *kinetic* frictional force on the steel wheel by the steel track.

Now allow the wheel to *roll* without slipping and pull the wheel through one revolution. The frictional force on the wheel at the rail now is the *static* frictional force, which does zero work because there is no motion at the point of contact. There only is steel-on-steel *kinetic* friction at the place where the wheel slides around the axle.* The kinetic frictional force certainly is no bigger than it was before:

$$\mu_k mg$$

since the axle must bear the weight of the load of mass m (the force likely is less since the axle does not need to bear the weight of half the wheel). The work done by the kinetic frictional force as the wheel turns through one revolution now is effectively[†]

$$W_{frictn} = -\mu_k mg(2\pi r)$$

where r is the radius of the *axle*. This work is negative because the kinetic frictional force opposes the motion. The work done by the applied horizontal force $\vec{F}_{roll}$ after the wheel moves through one revolution is

$$W_{roll} = F_{roll}(2\pi R)$$

where R is the radius of the *wheel*. The CWE theorem states that

$$W_{other} = \Delta KE + \Delta PE$$

Since there is no change in either the kinetic or potential energies, the CWE theorem becomes

$$W_{roll} + W_{frictn} = 0 \text{ J} + 0 \text{ J}$$
$$F_{roll}(2\pi R) - \mu_k mg(2\pi r) = 0 \text{ J}$$

---

*For simplicity we assume no roller bearings are used.

[†] This is an approximation because the frictional force in this case is *distributed* around the entire circumference of the axle, which makes the actual calculation of the work more complicated than this simplified argument.

Solving for the magnitude of the rolling force,

$$F_{roll} = \frac{r}{R}\mu_k mg$$

We use Equation 10.37 to cast this in the form

$$F_{roll} = \frac{r}{R}F_{drag} \qquad (10.38)$$

> The magnitude of the force needed to roll the wheel is less than that needed to drag it by the factor r/R, the mechanical advantage of the wheel.

It should be emphasized that Equation 10.38 is *not* a general relationship, but was derived for this simplified model of a horizontally rolling wheel. Equation 10.38 implies that the magnitude of the rolling force approaches zero for an axle with zero radius; this limit, of course, cannot be reached in practice.

Although this argument is somewhat oversimplified, it does explain, at least semiquantitatively, why it is easier to roll rather than drag things. Many other factors come into play in modeling the behavior of real wheels. Among these are (1) the use of roller bearings with no sliding contacts, which make F_{roll} even less than what we calculated here; (2) real wheels do not make point contact with the surface but deform and create a small area of contact, called the footprint of the wheel, easily visible with your car tires after a rain; and (3) kinetic energy is converted into thermodynamic internal energy by deformation forces, which is one reason that freely rolling wheels (those with no force $\vec{F}_{roll}$ pulling them) eventually come to a stop. These factors serve to complicate the analysis of real wheels, and make for fertile fodder for thoughtful analysis of engineering applications in more advanced courses.

10.17 TOTAL ANGULAR MOMENTUM AND TORQUE

Newton's second law of motion states that the total force acting on a system is equal to the time rate of change of its momentum. We found there is a corresponding relation between the total torque and the time rate of change of the *spin* angular momentum (Equation 10.19), as well as for a single particle in *orbital* motion (Equation 10.12). In this section we will see that the *same* relationship exists when *both* spin and orbital motion exist, provided certain conditions are met.

A system of particles is shown in Figure 10.65. The origin at 0 is located in an inertial reference frame. Another reference point is at the point P, which may or may not be the center of mass of the system of particles. Point P is located with respect to 0 by means of the position vector $\vec{r}_P$. Each particle of mass m_i in the system is located either via the position vector $\vec{r}_i$ (with respect to origin 0) or the position vector $\vec{r}_i'$ (with respect to the reference point P).

These position vectors are related to each other by the vector sum

$$\vec{r}_i = \vec{r}_P + \vec{r}_i'$$

Rewrite this as

$$\vec{r}_i' = \vec{r}_i - \vec{r}_P \qquad (10.39)$$

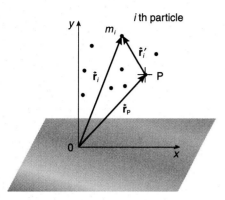

FIGURE 10.65 A particle is located via position vectors with respect to 0 and P.

The total angular momentum $\vec{\mathbf{L}}_P$ of the system about the point P is the vector sum of the individual angular momenta of the particles about the point P. That is,

$$\vec{\mathbf{L}}_P = \sum_i m_i \vec{\mathbf{r}}_i' \times \vec{\mathbf{v}}_i' \qquad (10.40)$$

where $\vec{\mathbf{v}}_i'$ is the velocity of m_i with respect to the reference point P. Now differentiate Equation 10.39 with respect to time:

$$\frac{d\vec{\mathbf{r}}'}{dt} = \frac{d\vec{\mathbf{r}}_i}{dt} - \frac{d\vec{\mathbf{r}}_P}{dt}$$
$$\vec{\mathbf{v}}_i' = \vec{\mathbf{v}}_i - \vec{\mathbf{v}}_P \qquad (10.41)$$

where $\vec{\mathbf{v}}_i$ is the velocity of m_i with respect to the origin 0 and $\vec{\mathbf{v}}_P$ is the velocity of the reference point P with respect to origin 0. Substituting for $\vec{\mathbf{r}}_i'$ using Equation 10.39 and for $\vec{\mathbf{v}}_i'$ using Equation 10.41, the expression for the angular momentum about the point P (Equation 10.40) becomes

$$\vec{\mathbf{L}}_P = \sum_i m_i (\vec{\mathbf{r}}_i - \vec{\mathbf{r}}_P) \times (\vec{\mathbf{v}}_i - \vec{\mathbf{v}}_P) \qquad (10.42)$$

Taking a cue from Newton's second law of motion, suppose we find the time rate of change of the angular momentum about the point P; that is, differentiate Equation 10.42 with respect to t. You have to be careful to preserve the order of the terms in the vector product:

$$\frac{d\vec{\mathbf{L}}_P}{dt} = \sum_i m_i \left[\frac{d}{dt} (\vec{\mathbf{r}}_i - \vec{\mathbf{r}}_P) \right] \times (\vec{\mathbf{v}}_i - \vec{\mathbf{v}}_P)$$
$$+ \sum_i m_i (\vec{\mathbf{r}}_i - \vec{\mathbf{r}}_P) \times \frac{d}{dt} (\vec{\mathbf{v}}_i - \vec{\mathbf{v}}_P)$$

The derivatives of the various position vectors are the various velocities. We find

$$\frac{d\vec{\mathbf{L}}_P}{dt} = \sum_i m_i (\vec{\mathbf{v}}_i - \vec{\mathbf{v}}_P) \times (\vec{\mathbf{v}}_i - \vec{\mathbf{v}}_P)$$
$$+ \sum_i (\vec{\mathbf{r}}_i - \vec{\mathbf{r}}_P) \times m_i \frac{d\vec{\mathbf{v}}_i}{dt}$$
$$- \sum_i (\vec{\mathbf{r}}_i - \vec{\mathbf{r}}_P) \times m_i \frac{d\vec{\mathbf{v}}_P}{dt} \qquad (10.43)$$

Since the vector product of any vector (here $\vec{\mathbf{v}}_i - \vec{\mathbf{v}}_P$) with itself is zero, the first summation vanishes, which is partial relief from the complexity.

In the second summation, the term

$$m_i \frac{d\vec{\mathbf{v}}_i}{dt}$$

can be written as

$$\frac{d\vec{\mathbf{p}}_i}{dt}$$

where $\vec{\mathbf{p}}_i$ is the momentum of particle m_i. In the third summation, the term

$$\frac{d\vec{\mathbf{v}}_P}{dt}$$

is the acceleration $\vec{\mathbf{a}}_P$ of the reference point P. Make these changes in Equation 10.43, which tidies it up a bit:

$$\frac{d\vec{\mathbf{L}}_P}{dt} = \sum_i (\vec{\mathbf{r}}_i - \vec{\mathbf{r}}_P) \times \frac{d\vec{\mathbf{p}}_i}{dt} - \sum_i (\vec{\mathbf{r}}_i - \vec{\mathbf{r}}_P) \times m_i \vec{\mathbf{a}}_P \quad (10.44)$$

From Newton's second law, the time rate of change of the momentum of a particle,

$$\frac{d\vec{\mathbf{p}}_i}{dt}$$

is equal to the total force $\vec{\mathbf{F}}_{i\,\text{total}}$ acting on it. Thus Equation 10.44 becomes

$$\frac{d\vec{\mathbf{L}}_P}{dt} = \sum_i (\vec{\mathbf{r}}_i - \vec{\mathbf{r}}_P) \times \vec{\mathbf{F}}_{i\,\text{total}} - \sum_i (\vec{\mathbf{r}}_i - \vec{\mathbf{r}}_P) \times m_i \vec{\mathbf{a}}_P$$

$$(10.45)$$

Using Equation 10.39 for the position vectors, Equation 10.45 becomes

$$\frac{d\vec{\mathbf{L}}_P}{dt} = \sum_i \vec{\mathbf{r}}_i' \times \vec{\mathbf{F}}_{i\,\text{total}} - \sum_i \vec{\mathbf{r}}_i' \times m_i \vec{\mathbf{a}}_P$$

The first summation is the *total torque* $\vec{\boldsymbol{\tau}}_P$ of the system of particles about the reference point P. Hence

$$\frac{d\vec{\mathbf{L}}_P}{dt} = \vec{\boldsymbol{\tau}}_P - \sum_i \vec{\mathbf{r}}_i' \times m_i \vec{\mathbf{a}}_P$$

Rearrange the terms in the summation slightly to obtain

$$\frac{d\vec{\mathbf{L}}_P}{dt} = \vec{\boldsymbol{\tau}}_P - \sum_i m_i \vec{\mathbf{r}}_i' \times \vec{\mathbf{a}}_P$$

Since the acceleration $\vec{\mathbf{a}}_P$ of the reference point P is independent of the particle locations and masses (P was just a *reference point*), the summation can be written as

$$\frac{d\vec{\mathbf{L}}_P}{dt} = \vec{\boldsymbol{\tau}}_P - \left(\sum_i m_i \vec{\mathbf{r}}_i' \right) \times \vec{\mathbf{a}}_P$$

The second term on the right-hand side is zero if any one of the following three conditions on the reference point P is met:

1. *The acceleration $\vec{\mathbf{a}}_P$ of the reference point P is zero.* If this is the case, the reference point P is in an inertial reference frame (just like the origin 0).

2. *The reference point P is the center of mass of the system of particles.* If this is the case, the summation vanishes because of the definition of the location of the center of mass (Equation 9.20):

$$\vec{\mathbf{r}}'_{\text{CM}} = \frac{1}{m_{\text{total}}} \sum_i m_i \vec{\mathbf{r}}'_i$$

That is, if P is the center of mass then $\vec{\mathbf{r}}'_{\text{CM}} = 0$ m, and so we have

$$\sum_i m_i \vec{\mathbf{r}}'_i = 0 \ \text{kg} \cdot \text{m}$$

3. *The acceleration $\vec{\mathbf{a}}_P$ of the point P is in the same direction as the vector*

$$\sum_i m_i \vec{\mathbf{r}}'_i$$

This is equivalent to saying that the acceleration $\vec{\mathbf{a}}_P$ is parallel or antiparallel to the vector (from P) locating the center of mass of the system.

In summary, if the reference point P

a. is in an inertial reference frame, or
b. is the center of mass of the system, or
c. has an acceleration parallel (or antiparallel) to a vector (from P) locating the center of mass,

then the time rate of change of the angular momentum of the system (about P) is equal to the total torque (about P):

$$\frac{d\vec{\mathbf{L}}_P}{dt} = \vec{\boldsymbol{\tau}}_P \tag{10.46}$$

Now that is a welcome simplification in dealing with a horde of masses.

For a *rigid body*, the total torque is related to the angular acceleration via Equation 10.20:

$$\vec{\boldsymbol{\tau}}_{\text{total}} = I\vec{\boldsymbol{\alpha}}$$

Thus Equation 10.46 also can be written as

$$\frac{d\vec{\mathbf{L}}_P}{dt} = I\vec{\boldsymbol{\alpha}} \tag{10.47}$$

STRATEGIC EXAMPLE 10.12

Wind a string around a rigid body with a circular cross section of radius R and mass m, as shown in Figure 10.66. Fix the string at one end and release the mass at rest. Determine the magnitude of the downward acceleration of the mass as it falls. Let I_{CM} be the moment of inertia about the symmetry axis through the center of mass.

Solution
Let the rigid body be the system. The system falls under the influence of two forces, shown in Figure 10.67:

1. its weight $\vec{\mathbf{w}}$ of magnitude mg; and
2. the force $\vec{\mathbf{T}}$ of the string on the system.

FIGURE 10.66

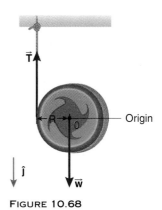

The system

FIGURE 10.67

Origin

$\hat{\jmath}$

FIGURE 10.68

Method 1
Use the center of mass of the rigid body as the origin or the point about which to calculate the torque of the forces. With this choice (see Figure 10.68), the weight has zero torque since its line of action passes through the chosen origin. The torque of the force of the string is of magnitude TR, since the moment arm of the tension is the radius of the rigid body.

Thus the total torque on the system about the point at its center is of magnitude TR. The magnitude of the torque about the center of the disk is equal to the moment of inertia of the disk *about this point* times the magnitude of the angular acceleration $\vec{\boldsymbol{\alpha}}$ of the disk about this point according to Equations 10.46 and 10.47:

$$\tau = I\alpha$$

The moment of inertia of the disk to use here is the moment of inertia I_{CM} of the system about the point where we calculated the torque. This is because I_{CM} is moment of inertia of the object about its symmetry axis through the center of mass point. Since the mass rolls off the string, the magnitude of the angular acceleration $\vec{\boldsymbol{\alpha}}$ and the magnitude of the downward acceleration $\vec{\mathbf{a}}$ are related by the rolling constraint of Equation 10.32:

$$\alpha = \frac{a}{R}$$

Hence Equation 10.47 becomes

$$TR = I_{CM}\frac{a}{R} \tag{1}$$

To find T, look at the translational motion of the system as a whole and apply Newton's second law. Taking $\hat{\mathbf{j}}$ down, the total force component in the downward direction is

$$mg - T$$

and this must equal the mass of the object times the acceleration component a in this direction:

$$mg - T = ma$$

Thus

$$T = m(g - a)$$

Substituting this expression for T into equation (1), you obtain

$$m(g - a)R = I_{CM}\frac{a}{R}$$

Solving for a, you obtain

$$a = \frac{m}{m + \dfrac{I_{CM}}{R^2}}g$$

or

$$a = \frac{mR^2}{mR^2 + I_{CM}}g$$

Notice that the greater the moment of inertia, the smaller the downward acceleration.

Method 2
Choose the point on the rim of the rigid body where the string last is in contact with the object as the origin about which to calculate the torque, as shown in Figure 10.69. With this choice, the torque of the force $\vec{\mathbf{T}}$ of the string is zero since its line of action passes through the chosen origin. Thus the total torque on

FIGURE 10.69

the object is the torque of the weight. The magnitude of the torque provided by the weight is mg times its moment arm R, or

$$mgR$$

The magnitude of the total torque about the origin is equal to the moment of inertia of the object *about this point* times the magnitude of the angular acceleration of the disk about this point, or

$$\tau = I\alpha \tag{2}$$

In this case the moment of inertia I must be obtained from the parallel axis theorem because the rotational axis through the point about which we calculated the torque is parallel to but not coincident with the symmetry axis through the center of mass. So, by Equation 10.29,

$$I = I_{CM} + mR^2$$

Since the system is rolling off the string, the magnitude of the angular acceleration $\vec{\boldsymbol{\alpha}}$ here is found using the rolling constraint, Equation 10.32:

$$\alpha = \frac{a}{R}$$

where a is the magnitude of the downward acceleration of the system. Substituting for the magnitude of the torque, the moment of inertia, and magnitude of the angular acceleration into equation (2), you find

$$mgR = \left(mR^2 + I_{CM}\right)\frac{a}{R}$$

Solving for a,

$$a = \frac{mR^2}{mR^2 + I_{CM}}g$$

10.18 CONSERVATION OF ANGULAR MOMENTUM

Conservation laws are important in physics not only because they indicate that some features of a system are unchanged as other factors associated with it do change, but also because such conservation laws provide a convenient tool for solving practical problems. Newton's second law of motion,

$$\vec{\mathbf{F}}_{\text{total}} = \frac{d\vec{\mathbf{p}}_{\text{total}}}{dt}$$

states that if the total force on a system is zero, then

$$0 \text{ N} = \frac{d\vec{\mathbf{p}}_{\text{total}}}{dt}$$

and the total momentum $\vec{\mathbf{p}}_{\text{total}}$ of the system is a constant vector. This statement of *conservation of momentum* is one of the great conservation laws of physics; we used it extensively in analyzing collision phenomena in Chapter 9. Likewise the rotational counterpart of Newton's second law is, as we have seen,

$$\vec{\boldsymbol{\tau}}_{\text{total}} = \frac{d\vec{\mathbf{L}}_{\text{total}}}{dt}$$

If the total torque on a system is zero, then

$$0 \text{ N·m} = \frac{d\vec{L}_{total}}{dt}$$

and then the total angular momentum $\vec{L}_{total}$ of the system must be a *constant vector*. This is **conservation of angular momentum**.

This conservation law also is of great importance and utility in physics.

EXAMPLE 10.13

An ice skater with a moment of inertia of 0.600 kg·m^2 about a vertical axis is spinning with an angular speed of 12.6 rad/s. The skater brings her arms in closer to the axis, decreasing her moment of inertia to 0.240 kg·m^2. Find her new angular speed.

Solution

Take the skater to be the system. Although the system is neither rigid nor frigid, the forces involved in the deformation of the system are internal forces. The two external forces on the skater, her weight and the normal force of the ice surface on her, are of equal magnitude and in opposite directions (but are *not* Newton's third law forces; why not?). Take the center of mass of the skater as the point about which to calculate the torques. The two external forces produce zero torque on the skater because their lines of action pass through the point about which we are taking the torque, the center of mass. Hence the total torque on the system is zero and the total angular momentum of the system is conserved:

$$\vec{L}_i = \vec{L}_f$$

The initial angular momentum of the skater is

$$\vec{L}_i = I_i \vec{\omega}_i$$

and, by the circular motion right-hand rule, is directed upward along the vertical direction, as shown in Figure 10.70.

The final angular momentum of the skater is

$$\vec{L}_f = I_f \vec{\omega}_f$$

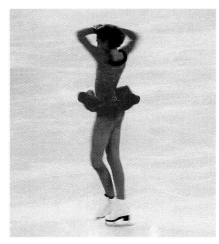

Figure skaters change their angular speed of spin by changing their moment of inertia to exploit conservation of angular momentum.

Conservation of angular momentum for the skater means that

$$I_i \vec{\omega}_i = I_f \vec{\omega}_f$$
$$(0.600 \text{ kg·m}^2)(12.6 \text{ rad/s})\hat{k} = (0.240 \text{ kg·m}^2)\vec{\omega}_f$$

Solving for $\vec{\omega}_f$,

$$\vec{\omega}_f = (31.5 \text{ rad/s})\hat{k}$$

Her new angular speed is 31.5 rad/s.

STRATEGIC EXAMPLE 10.14

You (mass m) are incident at velocity $\vec{v}_0$ along a straight line tangent to a disk-shaped merry-go-round of mass M and radius R that is spinning with an angular speed ω_0 about a fixed axle; see Figure 10.71. You jump on and hold on to the rim of the disk.

a. Find the impulse provided by the axle to the system consisting of you and the merry-go-round.
b. Find the final angular speed of the merry-go-round after you jump on.
c. What incident speed v_0 is sufficient for you to bring the merry-go-round to a halt?

Solution

Let the system be you and the merry-go-round.

a. The collision of you with the merry-go-round is completely inelastic, and so we do *not* expect the kinetic energy of the

FIGURE 10.70

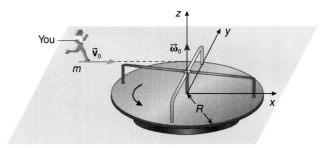

FIGURE 10.71

system to be conserved. If the axle of the disk were not fixed, the momentum of the system would be conserved. But because the axle is fixed, the system has no horizontal velocity after the collision. The momentum of the system thus is not conserved. An external force changes the momentum of the system. The force of you on the merry-go-round and the force of the merry-go-round on you are internal forces to the system; they form a Newton's third law pair (and sum to zero). However, the axle provides a horizontal external force on the system during the collision. The impulse provided by the force of the axle of the system equals the change in its momentum. The momentum of the system initially is due only to you; the merry-go-round has zero momentum because its center of mass is at rest. Use the coordinate system indicated in Figure 10.71. The initial momentum of the system is

$$mv_0\hat{\imath}$$

The final momentum of the system is zero. The change in the momentum of the system is equal to the impulse in accordance with the impulse–momentum theorem of Chapter 9. Thus the impulse provided by the axle on the disk is

$$
\begin{aligned}
\vec{I} &= \Delta\vec{p} \\
&= \vec{p}_f - \vec{p}_i \\
&= (0 \text{ kg·m/s})\hat{\imath} - mv_0\hat{\imath} \\
&= -mv_0\hat{\imath}
\end{aligned}
$$

b. Let's see if you can conserve the *angular momentum* of the system consisting of you and the merry-go-round. An external force (the force of the axle) acts on the system in the horizontal direction; however, if you use the center of the disk as an origin, the force of the axle on the system produces *zero* torque about this point, because the line of action of the force passes through the point about which you take the torque. Thus, if you measure the angular momentum of the system about the center of the disk, there are no external torques on the system. Thus Equation 10.46,

$$\vec{\tau}_{\text{total}} = \frac{d\vec{L}_{\text{total}}}{dt} \qquad (10.46)$$

becomes

$$0 \text{ N·m} = \frac{d\vec{L}_{\text{total}}}{dt}$$

which means $\vec{L}_{\text{total}}$ is a constant vector. The total angular momentum of the system about the center of the disk *is* conserved:

$$\vec{L}_{\text{total f}} = \vec{L}_{\text{total i}}$$

The total angular momentum of the system initially consists of the vector sum of the orbital angular momentum of you and the spin angular momentum of the merry-go-round. The initial orbital angular momentum of you about the center of the disk can be found in two ways.

Method 1

Use the definition of the angular momentum of a point particle, Equation 10.1:

$$\vec{L}_{\text{you}} = \vec{r} \times \vec{p}$$

where $\vec{p}$ is your momentum: $\vec{p} = mv_0\hat{\imath}$ with the coordinate system indicated in Figure 10.72. The magnitude of $\vec{L}_{\text{you}}$ is equal to the magnitude of $\vec{r}$ times the magnitude of $\vec{p}$ times the sine of the angle θ between the two vectors:

$$L_{\text{you}} = r(mv_0)\sin\theta$$

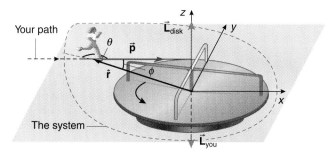

FIGURE 10.72

Geometrically,

$$
\begin{aligned}
r\sin\theta &= r\sin(180° - \phi) \\
&= r\sin\phi \\
&= R
\end{aligned}
$$

which is the radius of the disk. So

$$L_{\text{you}} = mv_0R$$

The direction of your angular momentum is found using the usual vector product right-hand rule: this vector product is directed along $-\hat{k}$. Thus

$$\vec{L}_{\text{you}} = -mv_0R\hat{k}$$

Method 2

One might think of using the relationship in Equation 10.6,

$$\vec{L} = I\vec{\omega}$$

where I is your moment of inertia about the axis:

$$I = mr^2$$

The only trouble is that your distance from the axis is *changing*, so that your moment of inertia about this axis *also* is changing. The distance to the axis is decreasing, so that your moment of inertia is decreasing. Your angular velocity about the axis *also* is changing: when you are very far from the axis your angular velocity is small, and as you approach the merry-go-round your angular velocity about the axis increases. The two effects compensate for each other, but it is more involved to calculate the angular momentum using $\vec{L} = I\vec{\omega}$ at a general position for the incoming particle (you). However, the instant before you collide with the disk, you can use Equation 10.6 relatively easily. At this location your instantaneous moment of inertia (of the you-particle) about the axis is mR^2, since you are then a distance R from the axis. Your instantaneous angular speed about the axis is equal to

v_0/R. The instantaneous angular velocity then is (using the coordinate system indicated in Figure 10.71 or 10.72)

$$-\frac{v_0}{R}\hat{k}$$

In this way your orbital angular momentum about the axis is found to be

$$\vec{L}_{you} = I\vec{\omega}$$
$$= mR^2\left(-\frac{v_0}{R}\right)\hat{k}$$
$$= -mv_0R\hat{k}$$

Having found your initial angular momentum, the initial spin angular momentum of the merry-go-round is found from Equation 10.18:

$$\vec{L} = I_{CM}\vec{\omega}$$

Use the circular motion right-hand rule to determine the direction of $\vec{\omega}$. The angular velocity of the disk is directed along $+\hat{k}$. So the angular velocity of the disk is

$$\vec{\omega} = \omega_0\hat{k}$$

The moment of inertia of the disk-shaped merry-go-round is found in Table 10.1; we neglect the rails:

$$I_{CM} = \frac{1}{2}MR^2$$

Thus the initial angular momentum of the disk is

$$\vec{L}_{disk} = \frac{1}{2}MR^2\omega_0\hat{k}$$

Hence the total angular momentum of the system initially is

$$\vec{L}_i = \vec{L}_{disk} + \vec{L}_{you}$$
$$= \left(\frac{1}{2}MR^2\omega_0 - mv_0R\right)\hat{k}$$

Now you have to find the *final* angular momentum of the system. You can do this two ways.

Method 1

The merry-go-round now is spinning at an unknown angular velocity $\omega\hat{k}$. Thus the final spin angular momentum of the merry-go-round is found using Equation 10.18 once again:

$$\frac{1}{2}MR^2\omega\hat{k}$$

You are now on the rim of the merry-go-round, also rotating at the same angular velocity $\omega\hat{k}$. Your orbital angular momentum can be found from Equation 10.6:

$$\vec{L}_{you} = I_{you}\vec{\omega}$$

Since I_{you} is just mR^2, your orbital angular momentum is

$$\vec{L}_{you} = mR^2\omega\hat{k}$$

The final total angular momentum of the system is the vector sum of your angular momentum and that of the merry-go-round:

$$\vec{L}_f = \left(\frac{1}{2}MR^2 + mR^2\right)\omega\hat{k}$$

Method 2

You also can find the final total angular momentum of the system by considering the system to be a composite of a disk, of moment of inertia

$$\frac{1}{2}MR^2$$

and a point particle (you!) on its rim (of moment of inertia mR^2). Since both objects are rotating about the same axis with the same angular velocity, the total moment of inertia of the system is the sum of the moments of inertia of its constituent pieces:

$$I_{total} = \frac{1}{2}MR^2 + mR^2$$

The final total angular momentum is

$$\vec{L}_f = I_{total}\vec{\omega}$$
$$= \left(\frac{1}{2}MR^2 + mR^2\right)\omega\hat{k}$$

which is what we obtained in Method 1.

The total angular momentum of the system is conserved:

$$\vec{L}_f = \vec{L}_i$$

so

$$\left(\frac{M}{2} + m\right)R^2\omega\hat{k} = \left(\frac{1}{2}MR^2\omega_0 - mv_0R\right)\hat{k}$$

From this you can find the final angular velocity component ω:

$$\omega = \frac{\frac{1}{2}MR^2\omega_0 - mv_0R}{\left(\frac{M}{2} + m\right)R^2} \tag{1}$$

c. If your speed v_0 is sufficiently large, the collision can bring the disk to rest. The final angular velocity component of the disk then is zero; therefore the numerator in equation (1) must be zero. This means that

$$\frac{1}{2}MR^2\omega_0 = mv_0R$$

Thus your initial speed v_0 that is sufficient to halt the rotation of the merry-go-round is

$$v_0 = \frac{MR\omega_0}{2m}$$

10.19 CONDITIONS FOR STATIC EQUILIBRIUM

Buildings, towers, bridges, and jungle gyms must be stable structures for safety, as well as to avoid lawsuits from lean and hungry packs of trial attorneys. What are some of the principles involved in a safe design? Certainly such structures should neither accelerate nor begin to rotate because of forces acting on them. Here we find there are but two conditions for static equilibrium that form the basis for the engineering analysis of all static structures.

Forces cause changes in momentum; torques cause changes in angular momentum. The two fundamental equations that govern the dynamics of a physical system are

Newton's second law of motion

$$\vec{F}_{total} = \frac{d\vec{p}_{total}}{dt} \quad \text{(most general form)}$$

$$= m\vec{a} \quad \text{(for systems of constant mass)}$$

and its rotational counterpart

$$\vec{\tau}_{total} = \frac{d\vec{L}_{total}}{dt} \quad \text{(most general form)}$$

$$= I\vec{\alpha} \quad \text{(for systems with constant moment of inertia } I)$$

We now can extend the concept of single-particle equilibrium, first introduced in Chapter 5, to that of many-particle systems. If the acceleration of the center of mass of the system is zero, and the angular acceleration of the system about a symmetry axis through its center of mass also is zero, the system is in **static equilibrium**.

Hence the **two conditions for static equilibrium** of a system are

1. the total force on the system must be zero: $\vec{F}_{total} = 0$ N; and

2. the total torque on the system must be zero: $\vec{\tau}_{total} = 0$ N·m.

Therefore the center of mass of a system in static equilibrium has a constant velocity: $\vec{v}_{CM}$ is constant. The angular velocity about a symmetry axis through the center of mass point also is constant: $\vec{\omega}$ is constant. The velocity of the center of mass and the angular velocity about the center of mass need not be zero; the important thing for static equilibrium is that both are *constant* vectors.

The acceleration $\vec{a}$ and angular acceleration $\vec{\alpha}$ both must be zero for a system to be in static equilibrium.

As we have seen, to calculate the torque on a system, we typically introduce a coordinate system and calculate the torque of the various forces with respect to the origin of that coordinate system. The motion of the system, however, could care less about where we choose the origin of our coordinate system. Thus we can choose *any point* as the origin about which to calculate the torques of the various forces. For static equilibrium, the vector sum of the torques of the forces must vanish *regardless* of where we choose to place the origin for the purpose of calculating the torques. This gives us great freedom when analyzing a problem.

PROBLEM-SOLVING TACTIC

10.5 If the lines of action of several forces pass through a particular point, that is the point about which to calculate torques for static equilibrium problems. The torques of the forces whose lines of action pass through the point are zero, because their moment arms are zero. This choice of a point about which to calculate torques is not obligatory, of course, but it does make the calculations easier to do.

QUESTION 10

Stand up straight with both feet spread about 0.5 m apart. If you raise one foot off the floor, explain what must happen to your center of mass to prevent yourself from falling over.

EXAMPLE 10.15

Two scales are positioned beneath the ends of a uniform board of mass M and length ℓ_0. A person of mass m is standing partway between the scales, a distance ℓ from one end, as shown in Figure 10.73. The board and person are at rest. What magnitude of force does each scale exert on the board?

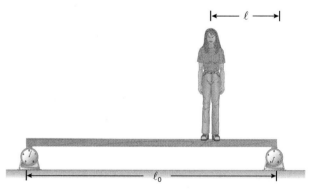

FIGURE 10.73

Solution

Let the board be the system. The forces on the board are (see Figure 10.74)

1. the force of the person acting down on the surface of the board, equal in magnitude and direction to the weight $\vec{w}$ of the person.

2. the weight $\vec{w}'$ of the board, of magnitude Mg; and

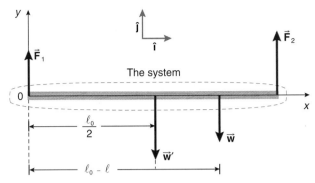

FIGURE 10.74

3. the forces of the two scales on the board. Call these forces $\vec{F}_1$ and $\vec{F}_2$, directed as shown in Figure 10.74. Introduce the coordinate system indicated in that figure.

The first condition of equilibrium states that the vector sum of the forces must be zero:

$$F_1\hat{j} + F_2\hat{j} + mg(-\hat{j}) + Mg(-\hat{j}) = 0 \text{ N}$$

or

$$F_1 + F_2 = (m + M)g \qquad (1)$$

Equation (1) alone is not sufficient to answer the question, since we have two unknowns: the force components F_1 and F_2. The second condition of equilibrium enables us to write another equation. The second condition of equilibrium states that the total torque about any point must vanish. Choose the origin of the coordinate system, located at the left end of the board, as the point about which to take the torques. The torque of the force $\vec{F}_1$ then is zero, since its line of action passes through the point about which we are taking the torque. The torque of the force of the person on the board is

$$\begin{aligned}\vec{\tau}_{\text{person}} &= (\ell_0 - \ell)\hat{i} \times (-mg)\hat{j} \\ &= -(\ell_0 - \ell)mg\hat{k}\end{aligned}$$

The torque of the weight of the board is

$$\begin{aligned}\vec{\tau}_{\text{board}} &= \frac{\ell_0}{2}\hat{i} \times (-Mg)\hat{j} \\ &= -\frac{\ell_0}{2}Mg\hat{k}\end{aligned}$$

The torque of the force $\vec{F}_2$ is

$$\begin{aligned}\vec{\tau}_{\text{of } F_2} &= \ell_0\hat{i} \times F_2\hat{j} \\ &= \ell_0 F_2\hat{k}\end{aligned}$$

The vector sum of the torques must be zero since the system obviously has zero angular acceleration. The vector sum is

$$\ell_0 F_2\hat{k} - (\ell_0 - \ell)mg\hat{k} - \frac{\ell_0}{2}Mg\hat{k} = 0 \text{ N·m}$$

Solving for F_2, you find

$$F_2 = \left(1 - \frac{\ell}{\ell_0}\right)mg + \frac{Mg}{2}$$

Using equation (1) to find F_1, you find that

$$F_1 = \frac{\ell}{\ell_0}mg + \frac{Mg}{2}$$

CHAPTER SUMMARY

The rotational motion of a system about an axis through its center of mass is called *spin*. If the center of mass is moving with respect to a reference frame, the center of mass is executing *orbital motion* with respect to the origin of the coordinate system. Orbital motion need not be circular orbital motion.

The *orbital angular momentum* $\vec{L}$ of a particle about a point 0 is

$$\vec{L} \equiv \vec{r} \times \vec{p} \qquad (10.1)$$

where $\vec{r}$ is the position vector of the particle with respect to 0 and $\vec{p}$ is the momentum of the particle.

The orbital angular momentum of a particle of mass m executing circular orbital motion is related to its angular velocity by

$$\vec{L} = I\vec{\omega} \qquad (10.6)$$

where I is the *moment of inertia* of the point particle about the center of the circle:

$$I = mr^2 \quad \text{(single particle only)} \qquad (10.5)$$

For a symmetric rigid body undergoing spin about a symmetry axis through the center of mass, we also have

$$\vec{L}_{\text{spin}} = I_{\text{CM}}\vec{\omega}_{\text{spin}} \qquad (10.18)$$

where I_{CM} is the moment of inertia of the rigid body about the symmetry axis through its center of mass point. The moment of inertia is defined to be

$$I_{\text{CM}} = \sum_i m_i r_{i\perp}^2 \qquad (10.17)$$

where $r_{i\perp}$ is the perpendicular distance of the point mass m_i from the axis of rotation. The moments of inertia I_{CM} for various symmetric rigid bodies are listed in Table 10.1 on page 440.

For a rigid body undergoing simultaneous spin and orbital motion, the total angular momentum of the system is the vector sum of the spin and orbital angular momenta. The orbital angular momentum is that of a point particle with a mass equal to the total mass of the system located at the center of mass point.

For a system undergoing *synchronous rotation*, the total angular momentum also is proportional to its angular velocity:

$$\vec{L}_{\text{total}} = I\vec{\omega} \qquad (10.28)$$

where I then is found from the *parallel axis theorem*.

The *torque* $\vec{\tau}$ of any force $\vec{F}$ is defined to be

$$\vec{\tau} \equiv \vec{r} \times \vec{F} \qquad (10.7)$$

where $\vec{r}$ is the position vector that locates the point of application of the force with respect to a chosen origin.

The total torque of a system about some point is equal to the time rate of change of the total angular momentum of the system about the same point:

$$\vec{\tau}_{\text{total}} = \frac{d\vec{L}_{\text{total}}}{dt} \tag{10.12}$$

Equation 10.12 is the rotational analog of Newton's second law:

$$\vec{F}_{\text{total}} = \frac{d\vec{P}_{\text{total}}}{dt}$$

If the moment of inertia I is constant, the rotational analog of Newton's second law also may be written as

$$\vec{\tau}_{\text{total}} = I\,\vec{\alpha} \qquad \text{(constant } I) \tag{10.13}$$

where $\vec{\alpha}$ is the angular acceleration. This is analogous to

$$\vec{F}_{\text{total}} = m\vec{a} \qquad \text{(constant } m)$$

The kinetic energy associated with the spin of a rigid body is

$$KE_{\text{rot}} = \frac{1}{2} I_{\text{CM}} \omega^2 \tag{10.24}$$

The total kinetic energy is the sum of the kinetic energy of the center of mass point and the rotational kinetic energy of the system associated with the spin about the symmetry axis through the center of mass:

$$KE_{\text{total}} = KE_{\text{rot}} + \frac{1}{2} m v_{\text{CM}}^2 \tag{10.25}$$

If the system is either rolling without slipping or undergoing synchronous rotation, its moment of inertia about an axis parallel to the symmetry axis through the center of mass point can be found from the *parallel axis theorem*:

$$I = I_{\text{CM}} + md^2 \tag{10.29}$$

where d is the perpendicular distance between the symmetry axis and the parallel rotation axis. The total kinetic energy of the system then is

$$KE_{\text{total}} = \frac{1}{2} I\omega^2 \tag{10.34}$$

For a system with circular cross section of radius R that is rolling without slipping, the *rolling constraints* are

$$s = R\theta \tag{10.30}$$
$$v = R\omega \tag{10.31}$$
$$a = R\alpha \tag{10.32}$$

where s is the distance through which the center of mass of the system moves, θ is the angle through which the system turns while rolling, v is the speed of the center of mass point, ω is the angular speed about the symmetry axis through the center of mass point, a is the magnitude of the acceleration of the center of mass point, and α is the magnitude of the angular acceleration of the system about the symmetry axis through the center of mass point.

If the total torque on a system about some point is zero, the total angular momentum of the system about the same point is conserved. This is *conservation of angular momentum*.

A system is in *static equilibrium* if its acceleration $\vec{a}$ and angular acceleration $\vec{\alpha}$ both are zero. Newton's second law and its rotational counterpart mean that the *two conditions for the static equilibrium* of a rigid body are

$$\vec{F}_{\text{total}} = 0 \text{ N}$$

and

$$\vec{\tau}_{\text{total}} = 0 \text{ N·m}$$

SUMMARY OF PROBLEM-SOLVING TACTICS

10.1 (page 430) If the line of action of a force goes through a point 0, the force has zero torque about that point.

10.2 (page 431) The magnitude of the torque is the product of the magnitude of the force and its moment arm; the direction of the torque is determined from the vector product right-hand rule.

10.3 (page 441) If a rigid body can be segmented into several symmetrically shaped pieces, the moment of inertia of the object as a whole is the sum of the moments of inertia of all the segmented pieces provided the moments are about the same axis.

10.4 (page 454) If you use the parallel axis theorem to determine the moment of inertia I of the system, the total kinetic energy of the system due to *both* its spin and orbital motion is given by Equation 10.34:

$$KE_{\text{total}} = \frac{1}{2} I\omega^2 \tag{10.34}$$

10.5 (page 464) If the lines of action of several forces pass through a particular point, that is the point about which to calculate torques for static equilibrium problems.

QUESTIONS

1. (page 427); 2. (page 427); 3. (page 432); 4. (page 436); 5. (page 441); 6. (page 443); 7. (page 447); 8. (page 450); 9. (page 454); 10. (page 464)

11. A particle is moving in a straight line. Sketch the situation. Indicate a point about which the particle has zero orbital angular momentum. Indicate a point about which the particle has nonzero orbital angular momentum.

12. Show that the SI units of angular momentum kg·m²/s are equivalent to J·s. Later in the course (in Chapter 26), we will see that the SI units of an important fundamental constant called Planck's constant, always symbolized as h, also are J·s: $h = 6.626 \times 10^{-34}$ J·s. This suggests that Planck's constant is in some way connected to angular momentum; we also will see that on the level of elementary

particles, the angular momentum is quantized in integral or half-integral units of $h/(2\pi)$. Those particles with spin angular momentum that are half-integral units of $h/(2\pi)$ are called *fermions*; those with integral units are called *bosons*. The combination $h/(2\pi)$ occurs so frequently in quantum physics that it is given its own special symbol, $\hbar$, pronounced "h-bar": $\hbar \equiv h/(2\pi)$.

13. Give a convincing argument to show that the center of mass point must lie on the symmetric axis (if one exists) of a rigid body with uniform density.

14. Sketch the shape of an object that has two distinct symmetry axes with different moments of inertia about each axis.

15. How many symmetry axes does a solid sphere have?

16. Sketch and indicate the number of symmetry axes through the center of mass point for each rigid body depicted in Figure Q.16. The objects are planar and of uniform composition.

FIGURE Q.16

17. The rotation of the Sun was discovered by Galileo in the early 17th century by observing sunspots. Suppose you notice several sunspots aligned as indicated in Figure Q.17a. Several weeks later, after the Sun has executed one rotation, the sunspots are observed as in Figure Q.17b. Can you conclude that the Sun is a rigid body? What assumption must be made to reach your conclusion?

18. A disk is mounted so it can rotate about a horizontal axis as indicated in Figure Q.18. (a) Indicate how *two* forces, of *equal magnitude*, can act on the rim of the disk so that the total force on it is *zero* but the total torque on it is *not* zero. (b) Indicate how *two* forces, of *equal magnitude*, can act on the rim of the disk so that the total force on it is *not* zero but the total torque on it *is* zero.

19. If a system is rotating, can you conclude that there is a nonzero total torque on the system? Explain your answer.

20. With respect to a fixed reference frame attached to the Sun (but not rotating with the Sun) (see Figure Q.20), is your speed on the Earth greater at noon or at midnight?

FIGURE Q.17

FIGURE Q.18

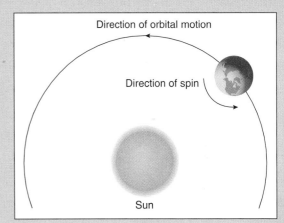

FIGURE Q.20

21. Two solid cylinders of the same radius and length are made of materials of different densities. The cylinders are released at rest at the top of an inclined plane. Figure out whether both cylinders reach the bottom of the incline simultaneously. If not, does the cylinder with the greater density reach the bottom first or second? Explain your reasoning.

22. The Mississippi River ("Big Muddy") carries large amounts of sediment south into the Gulf of Mexico. What does this do (if anything) to the moment of inertia of the Earth? Will this affect the length of the day? Explain your answers.

23. After making appropriate measurements, determine the approximate length of the record groove on an old-fashioned LP record.

24. Explain why the moment of inertia of an object is a minimum if the rotational axis passes through the center of mass of the object.

25. Consider a thin uniform disk of mass m and radius R and two possible axes of rotation: (i) an axis perpendicular to the disk through its center; (ii) an axis coincident with a diameter of the disk. Without making a detailed calculation, determine about which axis the moment of inertia of the disk is greater.

26. A rigid body rotates about a symmetry axis through its center of mass. Two points on the object lie at different radii from the center such that $r_1 > r_2$. (a) Which point (if either) has the greater angular speed? (b) Which point (if either) has the greater linear speed? (c) Which point (if either) has the greater centripetal acceleration?

27. Two interlocking gears of radii r_1 and r_2 have n_1 and n_2 teeth respectively. What is the ratio of the angular speeds of the two gears?

28. Two disks have the same thickness and radii, but are manufactured from materials of different densities. Which, if either, has the greater moment of inertia? Explain your reasoning.

29. The planar solids in Figure Q.29 have equal mass and the same thickness and height. Which has the greatest moment of inertia? Which has the smallest moment of inertia?

FIGURE Q.29

30. Explain how it is experimentally possible to distinguish between a raw egg and a hard-boiled egg without breaking the shell. Both eggs are at the same temperature.

31. When next in the weight room, take a barbell and spin it about the two axes indicated in Figure Q.31. Explain why it is easier for you to spin the barbell about the axis in Figure Q.31a than about the axis in Figure Q.31b.

FIGURE Q.31

32. Why do divers tuck themselves into a ball-like shape when executing several turns in a dive? As the diver straightens out before entering the water (see Figure Q.32), does the diver actually stop rotating?

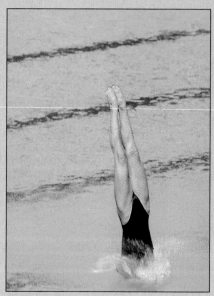

FIGURE Q.32

33. If the tread of a retreaded tire fails and breaks away from the tire itself, explain why it is likely that the failure occurs when the weakest part of the tread is at the top of the rolling motion of the tire.

34. A thin pillar on a level frictionless surface initially is upright but then falls. Sketch the trajectory of the center of mass of the pillar.

35. A flywheel is mounted in a special suitcase and set spinning at a large angular speed about a horizontal axis. A skycap picks up the suitcase and begins walking with it. Explain what happens when the skycap attempts to turn a corner with the suitcase.

36. A bicycle racer clearly wants to use a bicycle with a small mass. Explain why for a given frame it is beneficial to use tires with small moments of inertia.

37. Explain why tightrope walkers carry a long pole. (See Figure Q.37.) Why is the pole carried horizontally and not vertically?

FIGURE Q.37

38. A sphere, cylinder, and hoop all have the same radius and are made from the same material. They all are set rolling from rest from the same position down an inclined plane. Is there sufficient information to determine which wins the race to the bottom? Explain your reasoning. Does your answer change if the three objects also have the same mass?

39. A right-handed quarterback jumps vertically upward to make a critical pass. Describe and explain the motions of his lower body as he makes the pass.

40. What is the *ratio* of the (average) angular speeds of the second and minute hands of a clock to that of the hour hand? What is the direction of each of their angular velocity vectors?

41. Why do helicopters have at least two rotors, usually set at right angles to each other? (See Figure Q.41.)

FIGURE Q.41

42. In designing a car for a soapbox derby, is it advisable to design the wheels with large or small moments of inertia? Explain your answer.

43. Is it possible to have $\vec{F}_{total} = 0$ N but $\vec{\tau}_{total} \neq 0$ N·m? If so, describe such a situation.

44. Is it possible to have $\vec{F}_{total} \neq 0$ N but $\vec{\tau}_{total} = 0$ N·m? If so, describe such a situation.

45. When you buy a new tire for your car, you should be offered the option of having the tire balanced. (a) Why is this a good idea? (b) Tire stores offer the option of a static balance or a (more expensive) dynamic balance. Why is a dynamic balance better?
See Richard C. Smith, "Static vs. spin balancing of automobile wheels, *American Journal of Physics*, **40**, #1, pages 199–201 (January 1972).

46. There is essentially zero torque on a spinning ice skater; hence according to

$$\vec{\tau}_{total} = \frac{d\vec{L}}{dt}$$

her angular momentum is constant or conserved. If the skater brings in her extended arms closer to her body, the angular speed increases dramatically, indicating the presence of a nonzero angular acceleration. The left-hand side of

$$\vec{\tau}_{total} = I_{CM}\vec{\alpha}$$

is zero, but the right-hand side is not. Resolve this dilemma with a clear explanation.

47. After studying hard, you and a friend decide to go cycling using identical bicycles. Your friend has a greater mass than you. If you both begin together from rest and coast down the same hill, which of you will reach the bottom of the incline first? Or will it be a tie? Justify your answer.

48. A twin engine airplane has propellers or jet engines on each side or wing. Why is it advantageous to have the engines spinning in opposite directions?

49. Since wheels are so advantageous, why did they not evolve in biological evolution?

50. Name a 20th-century technological discovery that might rival the invention of the wheel for its importance in technology.

51. The drive wheels of steam locomotives (see Figure Q.51) do *not* have a symmetric distribution of mass about the axle, unlike modern railroad locomotives. Explain why.

52. What elements of physics are involved in the design of *mobiles*? (See Figure Q.52.) Formulate a rule indicating where to place a vertical wire to hold all masses located below the wire in their desired locations in the mobile.

53. Is a particle in uniform circular motion in static equilibrium? Explain your answer.

FIGURE Q.51

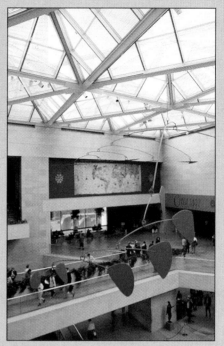

FIGURE Q.52

PROBLEMS

Sections 10.1 The Distinction between Spin and Orbital Motion
10.2 The Orbital Angular Momentum of a Particle
10.3 The Circular Orbital Motion of a Single Particle
10.4 Noncircular Orbital Motion

1. What is the angular momentum of the fully loaded brown pelican (*Pelecanus occidentalis*) in Figure P.1 with respect to the flying fish (*Cypselurus heterurus*) at point Q? Is the angular momentum orbital or spin angular momentum?

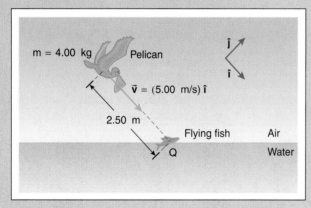

FIGURE P.1

2. Calculate the orbital angular momentum of a mute swan (*Cygnus olor*) of mass m about you (located at the point P), when the bird is at positions (a) and (b) in Figure P.2.

FIGURE P.2

- 3. What is the magnitude of the orbital angular momentum about the center of the Earth of a 1000 kg satellite in a circular orbit 100 km above the *surface* of the Earth?

- 4. A fantastic slapshot gives a hockey puck (of mass 0.170 kg) a speed of 40.0 m/s in a straight line toward a petrified goalie as indicated in Figure P.4. Consider the ice to be frictionless. (a) Calculate the momentum of the puck. (b) Is the momentum of the puck conserved along its trajectory? Why or why not? (c) What is the orbital angular momentum of the puck about the goalie when the puck is a distance of 2.0 m from the goalie? (d) Calculate the orbital angular momentum of the puck about the goalie when the puck is closest to him.

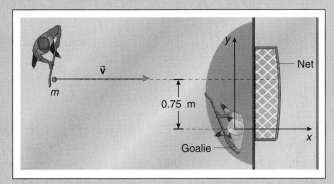

FIGURE P.4

•5. A planet is in an *elliptical* orbit about the Sun. The orbit has a semimajor axis a and an eccentricity ε. (a) Show that the ratio of the aphelion and perihelion distances is

$$\frac{r_{aph}}{r_{peri}} = \frac{1 + \varepsilon}{1 - \varepsilon}$$

(b) Show that the ratio of the moment of inertia of the planet (with respect to the Sun) at aphelion to that at perihelion is

$$\frac{I_{aph}}{I_{peri}} = \left(\frac{1 + \varepsilon}{1 - \varepsilon}\right)^2$$

•6. A planet of mass m is in circular motion about the Sun of mass M (where $M \gg m$). Since the mass of the Sun is so much larger than that of the planet, consider the Sun to be the origin and essentially fixed in position. Let r be the radius of the orbit of the planet and let T be its period. (a) Show that the magnitude of the orbital angular momentum $\vec{L}$ of the planet can be written as

$$L = \frac{2\pi m r^2}{T}$$

(b) In what direction is the orbital angular momentum vector? (c) Use the result of part (a) and the Newtonian form for Kepler's third law (Equation 6.37) to show that the magnitude of the orbital angular momentum also can be written in the form

$$L = m[G(M + m)r]^{1/2}$$
$$\approx m(GMr)^{1/2}$$

Thus the magnitude of the orbital angular momentum of the planet is proportional to the square root of the radius of the orbit.

•7. Two wheels of diameter 0.76 m are attached to opposite ends of an axle of length 2.0 m. The wheels roll around a circular track of inside radius 10 m. (a) Through what angle around the circular track must the axle assembly move so that the outer wheel makes one revolution more than the inner wheel? (b) What is the ratio of the angular speeds of spin of the two wheels? The differing angular speeds of the wheels is the reason for the differential in the drivetrain of a car or a truck.

‡8. Imagine a huge Ferris wheel of radius 40.0 m spinning at angular speed ω. A small mass m emerges from the wheel with a horizontal trajectory and is placed into circular orbit about the Earth (see Figure P.8). Clearly, we neglect air frictional effects! (a) What is the angular speed of the Ferris wheel if it is located at the north pole of the Earth? (b) If the Ferris wheel is located on the equator of the Earth and oriented so the particle is launched traveling *east*, is your answer to the question different? Why or why not? What is the new angular speed of the Ferris wheel if it is different from that calculated in part (a)? (c) If the Ferris wheel is located on the equator but the particle is launched traveling *west*, is your answer different? If so, find the new angular speed of the Ferris wheel.

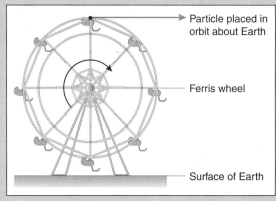

FIGURE P.8

Sections 10.5 Rigid Bodies and Symmetry Axes
10.6 Spin Angular Momentum of a Rigid Body
10.7 The Time Rate of Change of the Spin Angular Momentum
10.8 The Moment of Inertia of Various Rigid Bodies
10.9 The Kinetic Energy of a Spinning System

9. Imagine a rigid disk of radius r spinning at 1.00 rev/s. At what distance from the center of the disk is the tangential speed of a particle equal to the speed of light $c = 3.00 \times 10^8$ m/s? When we study relativity in Chapter 25, we shall see that it is not possible for material particles to travel at the speed of light; hence the disk cannot remain rigid.

•10. Calculate the total angular momentum of the system of particles pictured in Figure P.10 about the origin at 0.

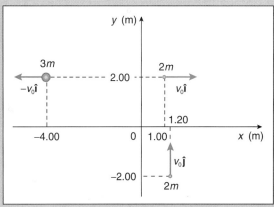

FIGURE P.10

•11. Calculate the moment of inertia of each particle system in Figure P.11 about the indicated axis of rotation.

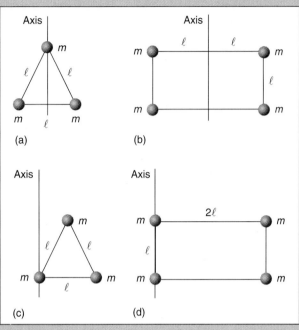

FIGURE P.11

•12. Find the center of mass of the collection of mass points in Figure P.12 and then find the moment of inertia of the system about an axis through the center of mass and parallel to the y-axis. Is the system of particles symmetric about this axis through the center of mass?

FIGURE P.12

•13. A uniform disk of mass m and radius R has an additional rim of mass m as well as four symmetrically placed masses, each of mass $m/4$, fastened at positions $R/2$ from the center as shown in Figure P.13. What is the total moment of inertia of the disk about an axis perpendicular to the disk through its center?

FIGURE P.13

•14. The total mass of the bolt shown in Figure P.14 is 200 g. (a) What is the mass of the rectangular plate that constitutes the head of the bolt? (b) What is the moment of inertia of the head of the bolt about the indicated symmetry axis? (c) What is the mass of the body of the bolt? (d) What is the moment of inertia of the body of the bolt about the indicated symmetry axis? (e) What is the total moment of inertia of the bolt about the indicated symmetry axis?

FIGURE P.14

•15. The moment of inertia of a uniform sphere of mass m and radius R about an axis through the center of mass is (see Table 10.1)

$$I_{\text{CM uniform sphere}} = \frac{2}{5} mR^2 = 0.400mR^2$$

The moment of inertia of a spherical shell of the same mass and radius is

$$I_{\text{CM spherical shell}} = \frac{2}{3} mR^2 = 0.667mR^2$$

Spherical planets and moons are neither uniform spheres nor spherical shells. Write the moment of inertia of such spherical masses about an axis through their center of mass as

$$I_{\text{CM}} = \beta mR^2$$

(a) What is the maximum value for β? Is there a minimum value for β? (b) As β decreases from the value 0.400 (the value for a uniform sphere), what does this imply about the mass distribution within the planet? Is it likely that a spherical planet or moon has value of β much larger than 0.400? What would this imply about the structure of the planet or moon? (c) Models of the mass distribution of the Earth, Moon, and Mars (inferred from detailed tracking of orbiting spacecraft) indicate that the numerical value of β is

for the Moon	0.391
for Mars	0.377
for the Earth	0.331

What conclusions can be drawn from this data about the comparative interior structure of these celestial objects?

•16. Consider a two-dimensional *planar* object as indicated in Figure P.16. Let the x- and y-axes be two perpendicular axes in the plane of the figure as indicated and let I_x and I_y be the moments of inertia of the system about each of these axes respectively.* Show that the moment of inertia I_z of the system about the z-axis, perpendicular to the plane of the object, is

$$I_z = I_x + I_y$$

This result is known as the *perpendicular axis theorem* and applies only to *planar* masses. Explain why this restriction is necessary.

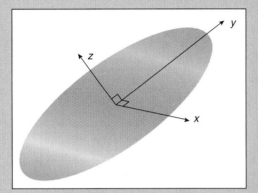

FIGURE P.16

•17. Use the perpendicular axis theorem of the previous problem to show that the moment of inertia of a circular hoop of mass m and radius R about the axis indicated in Figure P.17 is

$$\frac{1}{2} mR^2$$

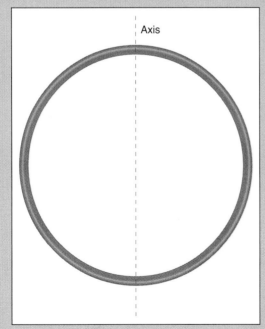

FIGURE P.17

•18. Professorial dreams. Two students are fighting over a physics text and exerting forces of equal magnitude F on it as indicated in Figure P.18. The mass of the text is 2.50 kg and its dimensions are 20 cm by 30 cm. Assume it can be modeled as a uniform and unusually dense rectangular plate. (a) What is the magnitude of the total torque about the center of mass of the text? (b) What is the magnitude of the initial angular acceleration of the text about its center of mass?

FIGURE P.18

•19. Assume the Sun is a uniform, rigid sphere (it is spherically shaped but actually is not of uniform density nor a rigid body). The Sun rotates once every 27 days.[†] (a) Calculate the rotational kinetic energy of the Sun. (b) The luminosity (power output) of the Sun is 3.826×10^{26} W, a rather powerful light bulb! For how many years could the Sun shine at its present luminosity if it were radiating away its rotational kinetic energy? The energy source of the Sun is not the rotational kinetic energy but rather the nuclear fusion of hydrogen nuclei, creating helium.

•20. Massive spinning flywheels (disks) can be used for storing energy. Consider a flywheel with a diameter of 1.00 m and a mass of 500 kg. A constant force of magnitude F is applied tangentially to the rim of the flywheel to accelerate it from rest to 4.00×10^3 rev/min during 3.00 min. (a) What is the magnitude of the angular acceleration of the flywheel during this interval? (b) What magnitude of torque is necessary to cause this angular acceleration? (c) What magnitude F of force is needed? (d) What is the rotational KE of the flywheel at the end of this acceleration? (e) What average power did the accelerating mechanism provide to the flywheel? (f) From what height h would you have to drop a mass of 500 kg for it to achieve a KE just before impact that is equal to the rotational KE of the flywheel?

•21. The total mass of your new sports car is 1.500×10^3 kg. The four wheels each have a mass of 35.0 kg and a radius of 25.0 cm.

---

*The quantities I_x, I_y, and I_z are *not* the components of a vector since the moment of inertia is a scalar, not a vector.

[†] The Sun actually exhibits *differential rotation*: the rotational period depends on solar *latitude* or the distance from the solar equator. The rotational period is about 25 days on the solar equator and about 35 days near the poles.

To a first approximation, consider the wheels as uniform disks. (a) When the car is zipping along a straight section of highway at 150 km/h attempting to elude the flashing blue lights visible in the rearview mirror, what is the total kinetic energy of the car? (b) What fraction of the total kinetic energy is the rotational kinetic energy of the wheels?

•22. Space stations have been proposed to accommodate the surplus population of the Earth. The initial design is for a hollow, uniform, cylindrical space station of diameter 3.00 km, length 10.00 km, and total mass of 1.20×10^{10} metric tons. (See Figure P.22.) The space station is to be spun about the symmetry axis coincident with the axis of the cylindrical shape. (a) What angular speed of rotation ω is needed to simulate the magnitude of the local acceleration due to gravity (9.81 m/s^2) for objects on the perimeter of the space station? Express your result in rev/min. (b) What is the rotational kinetic energy of the space station? (c) Small, thrusting rockets mounted tangential to the circular cross section are to set the space station in rotational motion. Starting from rest, the spacecraft reaches the angular speed calculated in part (a) after 1.00 y. Each rocket is capable of exerting a force of magnitude 1000 N continuously over the year. How many thrusting rockets are needed?

FIGURE P.22

•23. A Maytag repairman models the washer drum of a washing machine as a thin cylindrical shell of mass 30.0 kg, radius 25.0 cm, and height 50.0 cm together with a uniform disk (the bottom of the drum) of mass 10.0 kg of the same radius, as indicated in Figure P.23. The washer is 3/4 filled with water (of density $\rho = 1.00 \times 10^3$ kg/m^3) as it begins its spin cycle. The drain fails to open and the machine accelerates the load from rest to 10.0 rev/s during 60.0 s. Assume that the mass of water retains a cylindrical shape during the spin cycle (the top

surface actually assumes the shape of a paraboloid of revolution). (a) Find the moment of inertia of the tub. (b) Find the moment of inertia of the water. (c) Find the total moment of inertia of the tub and water system. (d) Calculate the magnitude of the angular acceleration. (e) What magnitude of torque is provided by the motor? (f) What is the kinetic energy of the system when rotating at 10.0 rev/s?

Direction of spin

FIGURE P.23

‡24. Show that the moment of inertia of a thick cylindrical shell of mass m, whose inner and outer radii are R_1 and R_2 respectively (see Figure P.24), about its symmetry axis is

$$\frac{1}{2} m \left(R_1^2 + R_2^2 \right)$$

Explain why this moment of inertia cannot be found by taking the moment of inertia of a disk of radius R_2 and subtracting that of a disk of radius R_1.

R_2 Symmetry axis

R_1

FIGURE P.24

‡25. At the center of the famous Crab Nebula in the constellation Taurus lies a compact (radius $R = 10$ km) but massive neutron star that spins with a period T of only 0.0333 s. Astronomers

have discovered that the period of the neutron star is gradually increasing at the rate of 3.65×10^{-8} s/d. Assume the star has a mass m of twice that of the Sun ($M_{Sun} = 1.99 \times 10^{30}$ kg). (a) Show that the rotational kinetic energy of the system is

$$KE_{rot} = \frac{4\pi^2}{5} \frac{mR^2}{T^2}$$

(b) Evaluate the rotational kinetic energy. (c) Use the result of part (a) to show that the rate of kinetic energy loss can be expressed as

$$\frac{d(KE)}{dt} = -\frac{2\,KE}{T}\frac{dT}{dt}$$

(d) Evaluate this rate of kinetic energy loss (power!) and compare it with the power output of the Sun (3.83×10^{26} W).

‡26. A mass m is attached to a rigid rod of negligible mass as in Figure P.26. The system is pivoted at point 0 and rotates about the indicated z-axis with angular velocity $\vec{\boldsymbol{\omega}}$, maintaining a fixed angle θ with the axis. (a) Show that at the instant pictured the angular momentum of m about the pivot 0 is

$$\vec{L} = (-mr^2\omega\sin\theta\cos\theta)\hat{i} + (mr^2\omega\sin^2\theta)\hat{k}$$

(b) Note that $\vec{L}$ is *not* parallel to $\vec{\boldsymbol{\omega}}$. Explain why. Is $\vec{L}$ parallel to $\vec{\boldsymbol{\omega}}$ at any instant during the motion?

This problem was inspired by the article by Herman Erlichson, "Angular momentum and angular velocity," *The Physics Teacher, 32*, #5, pages 274–275 (May 1994).

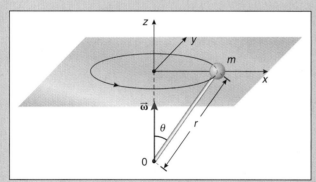

FIGURE P.26

Sections 10.11 The Precession of a Rapidly Spinning Top*
10.12 The Precession of the Spinning Earth*

•27. A thin disk of mass m has a moment of inertia I about a symmetry axis perpendicular to the plane of the disk. The disk is spinning with a large angular speed ω. The disk is suspended from a swivel that permits the symmetry axis of the disk (a rod of negligible mass) to turn freely while keeping the position of an end of the axis fixed; see Figure P.27. The disk slowly precesses in a horizontal circle. (a) In a sketch, indicate the direction of the angular momentum vector of the disk. (b) What force provides a torque on the disk about the point of support? What is the direction of the total torque on the

disk about the point of support? (c) When viewed from above, does the disk precess in the clockwise or counterclockwise sense? (d) Show that the angular speed of the precession is

$$\frac{mgb}{I\omega}$$

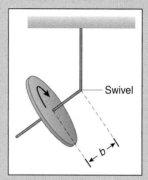

FIGURE P.27

•28. The planet Venus spins in the sense opposite to the spin of the Earth and has a spin period of 243 days; see Figure P.28. The Earth has a spin period of one day. The radius of Venus is 6.05×10^3 km, close to that of the Earth (6.37×10^3 km). The rotational axis of Venus makes an angle of about 3.0° with a perpendicular to the plane of its orbit about the Sun (for the Earth, the corresponding angle is about 23.5°). (a) Do you expect the oblateness of Venus to be greater than, less than, or about the same as that of the Earth? Explain your answer. (b) Will the axis of Venus precess? If so, is the directional sense of the precession the same as that of the Earth's axis or in the opposite sense? Explain your answer. (c) Will the precessional period likely be greater or less than that of the Earth (for the Earth it is 25 785 years)? Explain your answer.

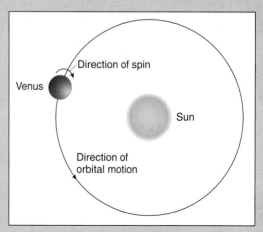

FIGURE P.28

Sections 10.13 Simultaneous Spin and Orbital Motion
10.14 Synchronous Rotation and the Parallel
Axis Theorem
10.15 Rolling Motion without Slipping
10.17 Total Angular Momentum and Torque
10.18 Conservation of Angular Momentum

29. Calculate the moment of inertia of each mass in Figure P.29 about the indicated axis.

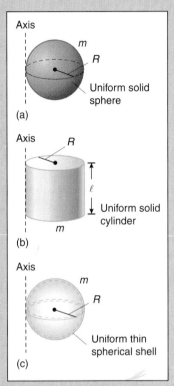

FIGURE P.29

•30. A solid spherical bowling ball has a mass of 8.5 kg and a radius of 10 cm. How much work is it necessary to do on the ball to get it to a speed of 6.0 m/s? Assume the sphere starts at rest and eventually is rolling without slipping on a horizontal surface.

•31. A sphere has a diameter equal to the length of an edge of a cube of the same mass. Both are released at rest at the top of an inclined plane with a smooth and rough path down the incline. The sphere rolls without slipping down the rough surface and the cube slides without friction down the smooth surface. (a) Which (if either) has the greater kinetic energy at the bottom of the incline? Justify your answer with a quantitative argument. (b) Which (if either) has the greater speed at the bottom of the incline? Justify your answer with a quantitative argument.

•32. A rod of length ℓ and mass m is free to rotate in a vertical plane about an axis at one end. The rod is released at rest in the horizontal orientation as shown in Figure P.32. (a) What is the moment of inertia of the rod about this axis?

(b) What is the initial torque of the gravitational force on the rod about this axis? (c) Specify any other forces that act on the rod as it rotates. Calculate their torques about the axis. (d) Specify the initial angular acceleration of the rod. (e) What is the torque of the gravitational force when the rod is vertically oriented? What is the angular acceleration when the rod has this orientation? (f) Use the CWE theorem to find the angular velocity of the rod when it has a vertical orientation. (g) Find the angular momentum of the rod when it has a vertical orientation.

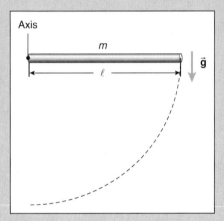

FIGURE P.32

•33. As a first approximation, consider the Earth to be spherical in shape with radius 6.37×10^3 km. (a) Calculate the angular speed of the spin of the Earth. (b) Calculate the magnitude of the spin angular momentum of the Earth. (c) Rivers such as the Mississippi carry large quantities of sediment from northern latitudes to more southern latitudes. Over long periods of time, what *qualitative effect* [if any] will such sediment transport toward the equator have on the length of a day? (No calculation is intended; simply explain what change takes places in the length of a day (if any) and why a change does or does not occur.) (d) Calculate the orbital angular speed of the Earth in its revolutionary motion about the Sun. Assume the orbit is circular. (e) Calculate the magnitude of the orbital angular momentum of the Earth about the Sun. (f) What is the ratio of the magnitudes of the orbital and spin angular momenta of the Earth?

•34. A frictionless human-powered merry-go-round has a radius of 4.0 m and a moment of inertia of 4.00×10^4 kg·m². Your kid sister (mass 30 kg) is located 3.0 m from the axis and asks you to give her a ride. You oblige by pushing tangentially along the rim of the merry-go-round with a force of magnitude 250 N. (a) What is the total moment of inertia of the system composed of the merry-go-round and your sister? (b) How long will it take you to get the system moving from rest until you reach a speed of 1.00 m/s? (c) Show the *ratio* of the kinetic energy of the system to the magnitude of its angular momentum is proportional to t, and so increases with time.

•35. A uniform plank of mass 100 kg is mounted so it can rotate about a horizontal axis as indicated in Figure P.35. Two physics students with masses 60.0 kg and 80.0 kg are attached to the ends of the plank as indicated. Consider the students as particles. (a) Calculate the moment of inertia of the system of the plank and two students about the indicated axis. (b) What is the magnitude of the total torque acting on the system about the indicated axis arising from gravitational forces? (c) Determine the magnitude of the initial angular acceleration of the system, assuming it is released initially at rest. (d) Is the angular acceleration constant as the plank turns? (e) Determine the angular speed of rotation of the system when the plank is vertical. (Assume unrealistically the students do not fall off!) (f) What is the magnitude of the angular acceleration of the system when the plank is vertical?

FIGURE P.35

•36. A disk of mass m and radius R is free to turn about a fixed, horizontal axle. The disk has an ideal string wrapped around its periphery from which another mass m (equal to the mass of the disk) is suspended, as indicated in Figure P.36. (a) What is the magnitude of the acceleration of the falling mass? (b) What is the magnitude of the angular acceleration of the disk?

FIGURE P.36

•37. A thin, hollow sphere of mass m is filled with a liquid of mass $m/2$. The filled sphere is spinning freely with no friction at angular speed ω_0 about an axis through the center of mass point of the sphere. The hollow sphere and all parts of the fluid have the same angular speed ω_0. A leak develops along the axis of the sphere (see Figure P.37) and all the fluid leaks out. What is the angular speed of the now hollow sphere?

FIGURE P.37

•38. Two uniform spheres of identical mass and radius are placed on inclined planes at the same height h and inclination angle θ. One plane is rough and causes one sphere to roll down the plane; the other is frictionless, and so the sphere on it slides down the incline. (a) Find the ratio of the kinetic energies of the two spheres at the bottom of the inclines:

$$\frac{KE_{slide}}{KE_{roll}}$$

(b) Which sphere takes the greater time to reach the bottom of the incline or do they take the same time? (c) Find the ratio of the times it takes the spheres to reach the bottom of the inclines:

$$\frac{t_{slide}}{t_{roll}}$$

(d) Find the ratio of their speeds at the base of the inclines:

$$\frac{v_{slide}}{v_{roll}}$$

•39. Oops. You just lost your marbles. One marble in particular has a mass m and radius R and rolls without slipping down a 30° incline. (a) Show that the magnitude of the acceleration of the marble is independent of its mass and equals 0.36g. (b) If the marble began at rest, how far does it go during the 3.0 s it takes you to recoup all your other marbles?

•40. A cylinder is rolled without slipping from rest by a constant force of magnitude F_0 in the horizontal direction applied by a string wrapped around the cylinder as indicated in Figure P.40.

FIGURE P.40

(a) If a point P on the horizontal section of the string moves a distance x, how far does the center of the cylinder move? (b) If P moves a distance x, how much work is done by the force? (c) After P moves the distance x, what is the speed v of the center of mass of the cylinder?

•41. A solid sphere, a spherical shell, a disk, and a thin cylindrical shell all have the same radius R and the same mass m. They all are situated at rest at the top of an inclined plane of height h as shown in Figure P.41. When t = 0 s, they all begin to roll, without slipping, down the plane. (a) Write the moment of inertia of each object as

$$I = \beta m R^2$$

with the appropriate value of β for each as found in Table 10.1. Show that the speed of the objects at the bottom of the incline can be written as

$$v = \left(\frac{2gh}{1 + \beta} \right)^{1/2}$$

(b) At the bottom of the incline the kinetic energy of each object has two contributions: (1) the kinetic energy of the center of mass and (2) rotational kinetic energy. Rank the objects in the order of largest to smallest values of the rotational and translational kinetic energy that they have at the bottom of the incline. (c) Indicate the order in which the objects arrive at the base of the incline.

FIGURE P.41

•42. Two hoops of radii R_1 and R_2 have identical masses. The centers of the hoops are traveling at the same speed v parallel to a level surface along which the hoops are rolling. (a) What is the ratio of the magnitudes of the momenta of the two hoops? (b) What is the ratio of the magnitudes of the angular momenta of the hoops taken about their respective centers of mass? (c) What is the ratio of the kinetic energies of the hoops? (d) If the hoops roll without slipping to a stop by going up an incline, what is the ratio of the change in the heights attained by the center of mass point of each hoop?

•43. You are driving a car of mass m at a constant speed v around a horizontal circular track of radius r. On this track the magnitude of the frictional force of the road on the tires of the car can be as great as half the magnitude of the weight of the car before slipping occurs. (a) Specify the forces that act on the car. (b) What is the direction of the total force that acts on the car? (c) Is the momentum of the car conserved? Explain. (d) What is the maximum possible speed of the car such that slipping does not occur?

•44. During World War II, the British mounted an unorthodox and successful night air attack on the Mohne Dam in Nazi Germany. Large, almost cylindrical, horizontal bombs were dropped at a speed of 360 km/h, at low altitude (18 m), and about 550 m from the dam itself. Much like flat stones, the bombs skipped across the lake behind the dam (about eight skips!), came to rest just behind the dam itself, sank, and exploded; the resulting shock wave destroyed the dam. The mass of each bomb was about 4.0×10^3 kg and its dimensions were approximately 0.70 m in radius with a length of 2.1 m. When released, each bomb had a spin about its horizontal cylindrical symmetry axis of 500 rev/min. (a) Find the moment of inertia of each bomb, assuming it is of uniform density. (b) Calculate the work needed to get each bomb from an initial angular speed of zero to the given angular speed. (c) Comment on the likely reason for spinning the bombs before they were dropped.

A fascinating account, complete with other aspects of the physics underlying the mission, can be found in an article by Arthur Stinner, "Physics and the dambusters," *Physics Education*, 24, #5, pages 260–267 (September 1989). The mission was recounted in a book by Paul Brickhill, *The Dambusters* (Ballantine, New York, 1951) and made into a movie in 1954.

•45. A 200 kg uniform beam hangs from a hinge and is supported by a cable as shown in Figure P.45. Time inexorably takes its toll and the cable snaps. (a) What is the magnitude of the total torque about the hinge the instant after the cable breaks? (b) What is the magnitude of the initial angular acceleration of the beam? (c) Is the angular acceleration of the beam constant as it swings down? Explain why or why not. (d) What is the angular speed of the beam just before it crashes into the vertical wall?

•46. An ice skater spinning with an angular speed of 1.50 rev/s is holding two 3.00 kg masses out at a radius of 0.80 m from her axis of symmetry using her two arms (of negligible mass). The *total* moment of inertia of the skater and the two 3.00 kg masses is 6.00 kg·m². (a) Find the moment of inertia of the skater herself. (b) If the two 3.00 kg masses are brought in to

FIGURE P.45

(a)

(b)

FIGURE P.48

a radius of 0.25 m, what is the new angular speed of the skater in rev/s? (c) If the two masses were dropped with her arms left extended, instead of being pulled in, would her spin rate change? If so, to what value in rev/s?

•47. An empty flat-bottomed cylindrical tub with vertical sides is freely spinning about its symmetry axis at 180 rev/min. The bottom of the tub has a mass of 5.0 kg and a radius of 25 cm, while the cylindrical sides have a mass of 6.0 kg. Sand gradually and gently is shoveled uniformly into the tub. Neglect any initial kinetic energy of the sand. As the tub fills, the sand assumes the same angular speed as the tub. Assume the sand in the tub has a cylindrical volume. (a) What is the moment of inertia of the tub about its symmetry axis? (b) What is the initial kinetic energy of the tub? (c) What mass of sand must be added to the tub to reduce the angular speed to 60 rev/min? (d) What is the kinetic energy of the rotating tub–sand system when its angular speed is 60 rev/min? (e) What fraction of the initial kinetic energy of the tub–sand system is part (d)?

•48. A mass m is hung from an ideal cord wrapped around the periphery of a disk of mass M and radius R as indicated in Figure P.48a. The disk is free to turn on a fixed horizontal axle with negligible friction. (a) Sketch a second law force diagram indicating all the forces acting on m. (b) What is the magnitude of the total torque on the disk about its axis? (c) Apply Newton's second law of motion to m and the rotational counterpart of Newton's second law to the spin of the disk to show that the downward acceleration of the mass m is of magnitude

$$a = \frac{1}{1 + \dfrac{M}{2m}} g$$

(d) What must be the ratio of M to m so the magnitude of the downward acceleration of m is only 0.010g? (e) If such a device is used as a fire escape on a building (see Figure P.48b) for a person of mass 80 kg, what must be the mass of the disk to satisfy the conditions of part (d)? Neglect the mass of the bucket compared with that of the person. If the person is to hit

the ground with a speed no greater than 2.0 m/s, what is the maximum height of a building that can employ such a system?

•49. Use conservation of angular momentum to find the ratio between the angular speeds of a planet at perihelion and aphelion in terms of the eccentricity of its orbit. That is, find

$$\frac{\omega_{aph}}{\omega_{peri}}$$

•50. After getting a flat tire fixed, you are hurriedly rolling the tire back to your car, simmering because you now are at least an hour late for a lavish party and your dress is, well, a sight to be seen. Model the tire (no pun intended) as a uniform disk of mass m and radius R; we better not model you. Beginning at rest at the base of an incline, you roll the tire up the incline to a vertical height h so the tire has a speed v when you reach your car (see Figure P.50). Way to go! What work is done by your constant force on the tire to accomplish this? And now you are sweaty too.

FIGURE P.50

•51. A skater is rotating at an angular speed of 1.00 rad/s on smooth ice with both arms and legs outstretched (as indicated in Figure P.51). When the arms and legs are returned to their normal positions, the angular speed of the skater increases to

10.0 rad/s. (a) Is the angular momentum of the skater conserved in this change? Explain. (b) By what multiplying factor does the magnitude of the angular momentum of the skater change? (c) By what factor does the moment of inertia of the skater change? (d) Is kinetic energy of the skater conserved in this change? (e) By what multiplying factor does the kinetic energy of the skater change?

•52. A log rolling politician is (to first order) in the shape of a cylinder of mass 60.0 kg and radius 20.0 cm. An ideal cord is wrapped around the circumference of the good-ol'boy with one end attached to the ceiling of the dome of the Capitol as in Figure P.52. The politician is released at rest and becomes totally unraveled due to the lack of a spin doctor. (a) Sketch a second law force diagram indicating the forces on the politician. (b) What is the magnitude of the downward acceleration of the chap? (c) What is the magnitude of the angular acceleration? (d) What is the angular speed 1.50 s after release? (e) What is the speed of the center of mass 1.50 s after release?

FIGURE P.52

•53. Consider the Earth–Moon system. Both objects spin about axes through their respective centers of mass; both also orbit the center of mass of the Earth–Moon system. Assume that

both the Earth and Moon are spherical in shape and that their orbits about the center of mass are circular in shape. The orbital period of the Earth and Moon about the center of mass is known as the sidereal period of the Moon and is 27.322 d. The Moon also is in synchronous rotation; that is, its period of rotation is the same as the period of its orbital motion about the center of mass of the Earth–Moon system. (This equality is not coincidental but came about because of tidal forces of the Earth acting on the Moon over the long history of the Earth–Moon system.) The rotational period of the Earth is 23.933 h. (a) Find the location of the center of mass of the Earth–Moon system. (b) Calculate the magnitude of the orbital angular momentum of the Moon about the center of mass of the Earth–Moon system. (c) Calculate the magnitude of the orbital angular momentum of the Earth about the center of mass of the Earth–Moon system. (d) Calculate the magnitude of the spin angular momentum of the Moon. (e) Calculate the magnitude of the spin angular momentum of the Earth. Notice that most of the total angular momentum of the Earth is *spin* angular momentum while most of total angular momentum of the Moon is *orbital* angular momentum. For the Earth–Moon system, the gravitational tidal forces of the Earth on the Moon and of the Moon on the Earth are *internal forces*. To a first approximation, there are no external torques on the Earth–Moon system, and so the total angular momentum of the Earth–Moon system is conserved. The tidal force of the Moon on the Earth produces an internal torque on the Earth whose effect is to decrease the angular speed of the spin of the Earth, thus lengthening the day.* This, of course, decreases the spin angular momentum of the Earth, the dominant contribution to the angular momentum of the Earth. To keep the *total* angular momentum of the Earth–Moon system constant, the angular momentum of the Moon must increase. Since most of the angular momentum of the Moon is orbital angular momentum, the orbital angular momentum of the Moon increases. The effect of this increase in the orbital angular momentum of the Moon is to increase the distance between the Earth and the Moon by about 3.8 centimeters per year. [See Problem 6, part (c).] Tidal forces thus gradually slow the rotation of the Earth (lengthening the day by several thousandths of a second per century) and increase the distance between the Earth and the Moon (by several meters per century). Both effects have been measured directly! Eventually the duration of an Earth day will be equal to the orbital period of the Moon, and both will be about 53 (current) days long [see Arthur Eisenkraft and Larry D. Kirkpatrick, "When days are months," *Quantum, 1, #2*, page 45 (November–December 1990), for a way to calculate this time period].

•54. A kid of mass 30 kg is at the center of a playground merry-go-round that is rotating (on frictionless bearings) at 20 rev/min. The mass of the merry-go-round is 150 kg and its radius is

---

*This is because the fast rotational speed of the Earth carries the tidal bulges of the Earth ahead of the Earth–Moon line. The lagging Moon exerts gravitational forces on these tidal bulges that tend to slow or brake the rotation of the Earth.

2.00 m. Model the merry-go-round as a disk and the kid as a point particle. The kid slowly crawls to the rim. Find the new angular speed of the merry-go-round in revolutions per minute.

•55. A uniform beam of length ℓ is in a vertical position with its lower end on a rough surface that prevents this end from slipping. The beam topples. Show that at the instant before impact with the floor, the angular speed of the beam about its fixed end is

$$\omega = \left(\frac{3g}{\ell}\right)^{1/2}$$

•56. (a) Model a star as a rigid uniform spherical mass of radius R_0 spinning with a period T_0. (In fact, such a model of a star like the Sun as a rigid body is *not* very realistic, but we will go ahead with it anyway.) If the star collapses under gravitational forces to a radius R and loses no mass nor angular momentum, what is the new period of rotation of the star? (b) Consider a star with twice the mass of the Sun, yet with an initial radius that is equal to the radius of the Sun and with a initial period of rotation of 25.0 d. If this star collapses and becomes a neutron star with a radius of 10.0 km, what is the new rotational period of the star? Assume (unrealistically) the star satisfies the conditions of the model in part (a) of this problem. Such small neutron stars are created by the gravitational collapse of massive stars when they exhaust their options for generating energy via nuclear fusion reactions in their cores. In fact, in the formation of neutron stars via supernova explosions, a significant percentage of the mass of the star is lost to the system, perhaps as much as 95%. This means, of course, that the initial star had to have a mass considerably greater than what we assumed in part (b). As a result of the mass loss and the concomitant loss of the angular momentum associated with the expelled material, neutron stars do not spin as rapidly as the calculation in part (b) would indicate. Nonetheless, neutron stars (which always have masses between about 1.5 and 3.5 times the mass of the Sun, and radii of only about 10 km) *do* rotate quite quickly: the neutron star in the middle of the Crab Nebula in the constellation Taurus (the bull) rotates 30 times each second!

•57. A solid sphere and a hollow spherical shell have equal masses and radii. (a) Without consulting a table or making a detailed calculation, which object has the larger moment of inertia? Explain your reasoning. (b) How could you distinguish one from the other by rolling them (without slipping) down an inclined plane? Justify your answer with an appropriate technical argument.

•58. A 20 kg pointlike mass and a 30 kg pointlike mass are separated by 0.80 m, connected by a string under tension, and spinning clockwise at 8.0 rad/s on a frictionless frozen pond. (a) Determine the location of the center of mass of the system. (b) What is the total angular momentum of the system about the center of mass point? (c) What is the total kinetic energy of rotation about the center of mass point? (d) If the connecting string slips and becomes 1.6 m long, what is the new angular speed of the particles about the center of mass point? (e) If the string breaks, what happens to the kinetic energy and to the angular momentum?

•59. Model a Ferris wheel by means of point masses $m = 100$ kg connected together by thin rigid rods of negligible total mass as shown in Figure P.59. (a) Determine the moment of inertia of the Ferris wheel about its axis. The Ferris wheel initially is not moving. Another mass $M = 200$ kg now is dropped from a height $h = 50.0$ m above the ground, lands in the indicated bucket, and sticks there. (b) What is the magnitude of the orbital angular momentum of the falling mass M about the axis of the Ferris wheel the instant before M hits the bucket? (c) Determine the angular speed of the system the instant after the mass M hits the bucket.

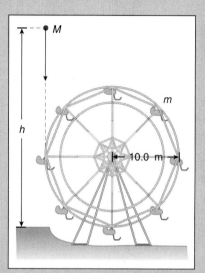

FIGURE P.59

•60. A girl of mass 30.0 kg, initially running tangentially at a speed of 5.00 m/s, jumps onto a disk-shaped playground merry-go-round initially at rest as shown in Figure P.60. The mass of the merry-go-round is 100 kg and its radius is 1.50 m. Find the angular speed of the merry-go-round after the girl jumps on.

FIGURE P.60

•61. You spy a common house mouse (*Mus musculus*) of mass m at rest on the rim of a lazy Susan (also at rest). The mass of the lazy Susan is M and its radius is R. The fleet-footed rodent is frightened into running clockwise around the rim of the lazy Susan at speed v_0 relative to the surface of the lazy Susan. In what rotational sense does the lazy Susan rotate? What is the angular speed of the lazy Susan as measured by an observant and terrified cook in the kitchen? Assume frictionless bearings.

•62. A varsity-sized tarantula (*Eurypelma californica*) of mass 100 g is hanging above a stereo turntable. The platter is of mass 2.00 kg and radius 15.0 cm and is spinning with no friction at an antiquated 33.0 rev/min with the motor off. The spider decides to make the turntable into a merry-go-round, drops from a height of 2.00 mm, and sticks to the rim of the turntable. (a) Calculate the magnitude of the initial spin angular momentum of the platter of the turntable. (b) What is the total moment of inertia of the platter and spider system when the spider is riding the merry-go-round? (c) Determine the new angular speed of the platter in rev/min.

•63. A neutron star with a mass of twice that of the Sun and a radius of 10 km rotates 30 times each second. Neutron stars realistically can be modeled as uniform, solid stars. (a) Calculate the rotational kinetic energy of the star. The result you obtain is greater than the entire energy output of the Sun during its 10 billion year lifetime! (b) Calculate the magnitude of the spin angular momentum of the neutron star. (c) Suppose the star shrinks 1.0 m in radius. What is the *ratio* of the new period of rotation to the initial period of rotation? Radio astronomers can detect changes in the rotational period of such stars corresponding to a mere 1 mm change in radius. Such changes occur occasionally in the rotational periods of neutron stars and are thought to be caused by "starquakes" that slightly change the radius of the star.

•64. The planet Venus has the following characteristics:

Rotational (spin) period	243 d
Radius	6051 km
Mass	4.87×10^{24} kg
Orbital period	225 d
Orbital radius	1.08×10^8 km

The spin is in the sense opposite to the orbital motion, as shown in Figure P.64. Consider the planet to be spherical in shape and, for simplicity, of uniform composition. The orbit of Venus is essentially a circle (the eccentricity of the orbit is only 0.007). The axis of rotation is essentially perpendicular to the plane of its orbit (the inclination of the axis to a perpendicular to its orbital plane is only 3.39°; we neglect this inclination in the present problem). (a) Find the magnitude of the spin angular momentum of the planet. (b) Find the magnitude of the orbital angular momentum of the planet. (c) Find the magnitude of the total angular momentum of the planet. Is most of the angular momentum orbital or spin angular momentum? (d) Find the total kinetic energy of the planet. Is most of the kinetic energy associated with the spin or with the orbital motion?

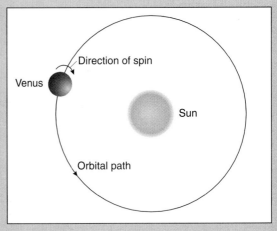

FIGURE P.64

•65. The Sun rotates on its axis about once every 27 days. (a) Calculate the magnitude of the spin angular momentum of the Sun. (b) The spin angular momentum of the Sun represents only about 2% of the *total* angular momentum of the solar system. The planets have 98% of the total angular momentum of the solar system (in both orbital angular momentum associated with their orbital motion and the orbital motion of their moons and spin angular momentum associated with the spin of the various planets and moons). Using the result of part (a), what is the magnitude of the total angular momentum of the solar system? (c) Assume that we could somehow transfer the angular momentum of the planets and their moons to the Sun. If the Sun contained 100% of the total angular momentum of the solar system, in other words, if we increased the magnitude of the spin angular momentum of the Sun by a factor of 50, what would be the rotational period of the Sun? When the solar system was formed about five billion years ago from the gravitational collapse of a giant nebula, conservation of angular momentum indicates that the Sun should have been spinning much faster than the result you obtained in part (c). This means that angular momentum was lost in the collapse of the primitive solar nebula. The missing angular momentum was carried off by mass that was lost to the system during the early epochs of the formation of the solar system.

•66. A disk with moment of inertia I_1 is rotating with initial angular speed ω_0; a second disk with moment of inertia I_2 initially is not rotating (see Figure P.66). The arrangement is much like a LP record ready to drop onto an unpowered, freely spinning turntable. The second disk drops onto the first and friction between them brings them to a common angular speed ω. Show that

$$\omega = \frac{I_1}{I_1 + I_2} \omega_0$$

‡67. A hula hoop of mass m and radius R is hung over a peg as indicated in Figure P.67. (a) What is the moment of inertia of the

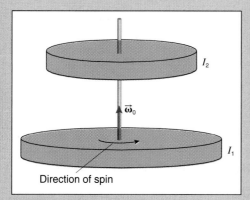

FIGURE P.66

hula hoop about the peg? (b) If the hoop is moved so that a line from the peg to the center of the hoop makes an angle θ with the vertical direction, what is the torque on the hoop about the peg? (c) For small angles θ, will the system execute simple harmonic oscillation? If so, what is the period of the oscillation?

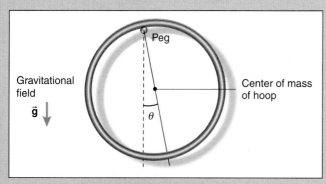

FIGURE P.67

‡68. A wheel of radius R is rolling at constant speed v_0 along a level road. (a) Find an expression for the speed v of a particle on the rim of the wheel as a function of the angle θ in Figure P.68. Test the expression to show that it gives the appropriate values of v when $\theta = 0°$ and when $\theta = 180°$. (b) Make a graph of v versus θ.

FIGURE P.68

‡69. One way to determine the moment of inertia of a disk of radius R and mass M in the laboratory is to suspend a mass m from it to give the disk an angular acceleration. The mass m is attached to a string that is wound around a small hub of radius r, as shown in Figure P.69. The mass and moment of inertia of the hub can be neglected. The mass m is released and takes a time t to fall a distance h to the floor. The total torque on the disk is due to the torque of the tension in the string and the small but unknown frictional torque $\vec{\tau}_{\text{frictn}}$ of the axle on the disk. Once the mass m hits the floor, the disk slowly stops spinning during an additional time t' because of the sole influence of the small frictional torque on the disk. Show that the moment of inertia of the disk is given by

$$I = \frac{mr^2\left(\dfrac{gt^2}{2h} - 1\right)}{1 + \dfrac{t}{t'}}$$

where m is the falling mass and r is the radius of the hub. In an experiment the value of I measured can be compared with the theoretical value of the moment of inertia of a disk of mass M and radius R:

$$\frac{1}{2}MR^2$$

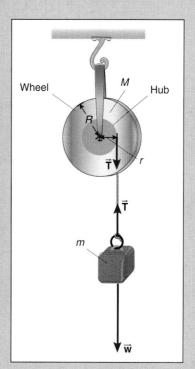

FIGURE P.69

‡70. A marble of radius r rolls without slipping off a fixed globe of radius R, beginning from rest at the top of the globe (see

FIGURE P.70

Figure P.70). Show that the angle θ at which the marble leaves the globe is given by

$$\theta = \cos^{-1}\left(\frac{10}{17}\right)$$

71. The mass of the polar ice caps on the Earth is on the order of $m = 6 \times 10^{18}$ kg. The Antarctic ice cap has a mass of $0.9m$; the Arctic ice cap has a mass of $0.1m$. Model the polar ice caps as two point masses coincident with the rotational axis of the Earth as in Figure P.71a. Should sufficient global warming occur, the ice caps will melt and the resulting liquid water will spread over the ocean areas of the Earth, raising sea levels around the globe. Assume the Earth is a uniform sphere. (a) Make the naive assumption that the water from the ice caps is spread *uniformly* over the surface of the Earth (i.e., ignore continental land forms), thus increasing its effective radius (see Figure P.71b). If the ice caps mass were spread uniformly over the surface of the Earth, by how much would the present radius of the Earth (6.37×10^3 km) increase? Express your result in meters. What does this bode for owners of ocean waterfront property? (b) The redistributed mass changes the moment of inertia of the Earth. If all the polar ice melts according to the assumptions in part (a), what is the increase in the rotational period of the Earth? Express your result in seconds.

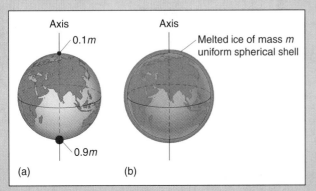

FIGURE P.71

Section 10.19 Conditions for Static Equilibrium

72. A uniform beam of mass m and length ℓ is hung from two cables, one at the end of the beam and the other $5\ell/8$ of the way to the other end as shown in Figure P.72. Determine the magnitudes of the forces the cables exert on the beam.

FIGURE P.72

•73. A uniform ladder of mass 20 kg and length 5.0 m is leaning against a smooth (frictionless) wall. The coefficient of static friction for the ladder on the ground is $\mu_s = 0.30$. What is the minimum angle θ that the ladder can make with the ground without slipping?

•74. A massive 250 kg door is hung from two hinges in the portico of a cathedral as indicated in Figure P.74. Each hinge exerts both a horizontal and a vertical force on the door to keep it in equilibrium. The door is of uniform composition. (a) Sketch a second law force diagram indicating the forces acting on the door. (b) Find the horizontal and vertical components of the forces that each hinge exerts on the door.

FIGURE P.74

•75. An unsharpened pencil of length ℓ and mass m is leaning against a smooth (i.e., frictionless) vertical surface. The pencil is approximately uniform in composition. The static coefficient of friction for the pencil on the horizontal surface is μ_s. (a) Draw a second law force diagram indicating all the forces acting on the pencil. (b) What is the smallest angle θ that the pencil can make with the horizontal direction without slipping? (c) You might try this situation experimentally on your desk using a protractor to measure the minimum angle θ as a way to find the coefficient of static friction. To satisfy the conditions of a frictionless vertical surface, is it better to have the eraser end of the pencil on the horizontal or on the vertical surface?

• **76.** An extended object of mass 5.00 kg is subjected *only* to the forces shown in Figure P.76. (a) What is the total force acting on the mass? Will the object accelerate? (b) Calculate the torque of each force about the point P in Figure P.76. (c) What is the total torque acting on the object? (d) Is the object in equilibrium?

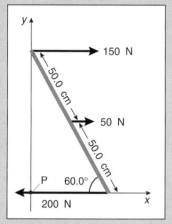

FIGURE P.76

• **77.** Uh oh. You (mass 70.0 kg) decide to tempt fate. A massive uniform plank of length 4.00 m and mass 100 kg lies flat but extends 1.50 m beyond the edge of the physics building as shown in Figure P.77. What is the maximum distance from the edge of the building that you can walk along the plank without disaster?

FIGURE P.77

• **78.** While hiking, you notice a dead tree trunk 4.00 m long resting on level ground. Always prepared to do some interesting physics, you discover experimentally that a force of 900 N is barely sufficient to lift one end while a force of 450 N barely is sufficient to lift the other end. What is the mass of the tree trunk?

• **79.** A crude model of an arm and its biceps muscle is shown in Figure P.79. Let the mass of the arm and biceps muscle be negligible (the classic 40.9 kg weakling). (a) What magnitude of force must the biceps muscle provide for the arm to hold a 10 kg mass in the orientation shown? (b) The mechanical advantage of a lever such as this is defined as the ratio between the magnitude of the force of the load (the weight of the 10 kg mass) and the magnitude of the force exerted by the muscle. Determine the mechanical advantage of this lever system. The arm is a very inefficient lever, since its mechanical advantage is so small. The great advantage of the system arises in the ability of the arm to move freely and throw objects at high speed. If the biceps were attached to the lower arm close to the hand, the mechanical advantage of the muscle would increase (we could hold greater masses in the given orientation) but our throwing abilities (controlled by the triceps on the back of the upper arm) would be severely compromised; no fastballs!

FIGURE P.79

INVESTIGATIVE PROJECTS

A. Expanded Horizons

1. Investigate the dynamics associated with throwing a bola, a traditional South American hunting weapon.

   D. L. Mathieson, "Wrap up rotational motion by throwing a bola," *The Physics Teacher*, 30, #3, pages 180–181 (March 1992).

2. Yo-yos are fascinating examples of spin and orbital motion. Investigate the physics of the yo-yo.

   William Boudreau, "Cheap and simple yo-yos," *The Physics Teacher*, 28, #2, page 92 (February 1990).

   Wolfgang Bürger, "The yo-yo: a toy flywheel," *American Scientist*, 72, #2, pages 137–142 (March–April 1984).

   Edward Zuckerman, "Quest for the perfect yo-yo," *Science Digest*, 93, #7, pages 54–55, 60 (July 1985).

3. Investigate the use of an ultracentrifuge to determine molecular masses of complex molecules.

   I. W. Richardson and Ejler B. Neergaard, *Physics for Biology and Medicine* (Wiley-Interscience, New York, 1972), pages 158–160.

4. Investigate the dynamics associated with boomerangs.

   Vernon Barger and Martin Olsson, *Classical Mechanics: A Modern Perspective* (2nd edition, McGraw-Hill, New York, 1995), pages 195–202.

   Allen L. King, "Project boomerang," *American Journal of Physics*, 43, #9, pages 770–773 (September 1975).

   Henk Vos, "Straight boomerang of balsa wood and its physics," *American Journal of Physics*, 53, #6, pages 524–527 (June 1985).

   Michael Hanson, "The flight of the boomerang," *The Physics Teacher*, 28, #3, pages 142–147 (March 1990).

   Jacques Thomas, "Why boomerangs boomerang," *New Scientist*, 99, #1376, pages 838–843 (22 September 1983).

5. Investigate the physics associated with riding a bicycle.

   J. Lowell and H. D. McKell, "The stability of bicycles," *American Journal of Physics*, 50, #12, pages 1106–1112 (December 1982).

   David E. H. Jones, "The stability of the bicycle," *Physics Today*, 23, #4, pages 34–40 (April 1970).

   Thomas B. Greenslade Jr., "More bicycle physics," *The Physics Teacher*, 21, #6, pages 360–363 (September 1983).

   Daniel Kirshner, "Some nonexplanations of bicycle stability," *American Journal of Physics*, 48, #1, pages 36–38 (January 1980).

6. Falling chimneys tend to break at characteristic locations along their length before they hit the ground. Investigate the causes of this phenomenon.

   Ernest L. Madsen, "Theory of the chimney breaking while falling," *American Journal of Physics*, 45, #2, pages 182–184 (February 1977).

   Arthur Taber Jones, "The falling chimney," *American Journal of Physics*, 14, #4, page 275 (July–August 1946).

   Albert A. Bartlett, "More on the falling chimney," *The Physics Teacher*, 14, #6, pages 351–353 (September 1976); this reference has a bibliography on the subject.

7. Investigate the static equilibrium and dynamics of tightrope walking.

   Robert B. Prigo, "Classroom tightrope walking," *American Journal of Physics*, 50, #5, pages 471–473 (May 1982).

8. Investigate the physics associated with "pumping" a swing.

   William B. Case and Mark A. Swanson, "The pumping of a swing from the seated position," *American Journal of Physics*, 58, #5, pages 463–467 (May 1990).

   Stephen M. Curry, "How children swing," *American Journal of Physics*, 44, #10, pages 924–926 (October 1976).

9. Measurements of the moment of inertia are common in physics laboratories. On the other hand, it is quite another story to *measure* the moment of inertia of the Moon or a planet such as Venus or Mars; they are not easily tinkered with in the lab! Measurements of such moments of inertia enable astronomers and geologists to model the interiors of the moons and planets. How might the moments of inertia of moons and planets be measured?

   William B. Hubbard, *Planetary Interiors* (Van Nostrand Reinhold, New York, 1984).

   Ralph Snyder, "Two-density model of the Earth," *American Journal of Physics*, 54, #6, pages 511–513 (June 1986).

10. Investigate the physics of the rotating, orbiting space colony stations proposed by Gerald K. O'Neill.

    Gerald K. O'Neill, "The colonization of space," *Physics Today*, 27, #9, pages 32–40 (September 1974).

11. High angular speed flywheels are used as energy storage devices in hybrid race cars. Investigate the state of this technology.

    William B. Scott, "Satellite control concepts bolster civil, defense systems," *Aviation Week and Space Technology*, 142, pages 43, 46 (6 March 1995).

B. Lab and Field Work

12. Many science museums have a dramatic pendulum in their entrance foyers called a Foucault pendulum. The slow rotation of the plane of oscillation of the pendulum is experimental proof that the Earth rotates. Investigate the dynamics of a Foucault pendulum to explain why this is the case. Imagine such a pendulum at the north pole, for simplicity. Construct a Foucault pendulum in a large open stairwell and measure the period associated with the apparent rotation of its plane of oscillation. See also Project 5 in Chapter 7.

    Charles Kittel, Walter D. Knight, and Malvin S. Ruderman, *Mechanics*, Berkeley Physics Course, volume 1 (McGraw-Hill, New York, 1965), pages 77–79.

    H. Richard Crane, "The Foucault pendulum as a murder weapon and a physicist's delight," *The Physics Teacher*, 28, #5, pages 264–269 (May 1990).

    Anthony P. French, "The Foucault pendulum," *The Physics Teacher*, 16, #1, pages 61–62 (January 1978).

    H. Richard Crane, "Foucault pendulum wall clock," *American Journal of Physics*, 63, #1, pages 33–39 (January 1995).

13. Several basic aspects of rotational motion can be demonstrated easily with common materials. Create and explain several of the demonstrations mentioned in the article listed here.

    George Lehmberg, "Investigating rotational inertia using coffee cans," *Science Teacher*, 60, #3, pages 68–69 (March 1993).

14. Design and perform an experiment to measure the moment of inertia of an automobile tire when mounted on its hub. Problem 69 may be of use in this regard. Specify and discuss the assumptions you make about the tire.

    Richard C. Smith, "General physics and the automobile tire," *American Journal of Physics*, 46, #8, pages 858–859 (August 1978).

15. Plan and make suitable astronomical observations to measure the rotational period of the Earth with respect to the stars and with respect to the Sun. Explain the difference in the values you obtain for these two time periods.

C. Communicating Physics

16. Take two quarters (or any two identical coins). Orient the two coins so they are both heads up with both profiles of George Washington upright as indicated in Figure I.16a. Roll one of the coins, say the one on the left, around *half* the circumference of the other coin as indicated. It is amazing to discover that the profile of the rolling coin is *not* upside down but right

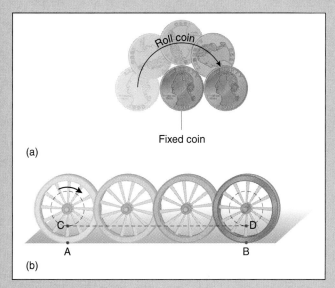

(a)

(b)

FIGURE I.16

side up after this maneuver. Explain why. This phenomenon is related to another curious aspect of rolling motion. Suppose a wheel (see Figure I.16b) rolls through one complete turn, carrying it from point A to point B. The distance between A and B is, therefore, equal to the circumference of the wheel. Notice, however, that point C, along the radius from the center of the wheel *also* executes one revolution as the wheel moves from A to B. Since the distance AB is equal to the distance CD, "we are confronted with the evident absurdity that the circumference of the small circle is equal to the circumference of the large circle." Resolve this paradox.

See James R. Newman, *The World of Mathematics* (Simon & Schuster, New York, 1956), volume 3, pages 1937–1939; the quotation is from this work, page 1938.

17. Call or visit a tire retailer and determine the manufacturer of a dynamic tire balancing machine. Communicate with engineers at the manufacturer to learn the physics associated with dynamic balancing of tires. Write a summary of your findings appropriate for an audience of your peers.

Richard C. Smith, "Static vs spin balancing of automobile wheels," *American Journal of Physics*, 40, #1, pages 199–201 (January 1972).

18. The tippe top is a toy that has fascinated even great physicists like Niels Bohr and Wolfgang Pauli, perhaps even you. Demonstrate and explain its peculiar dynamics.

Richard J. Cohen, "The tippe top revisited," *American Journal of Physics*, 45, #1, pages 12–17 (January 1977).

Ivars Peterson, "Topsy-turvy top," *Science News*, 146, page 108 (13 August 1994).

CHAPTER 11
SOLIDS AND FLUIDS

Archimedes' lost principle: When a body is immersed in water, the phone will ring.

Don Simanek*

Who has not marveled at the wonders of the sea, the dynamics of hurricanes and tornados, the thrill of a white-water river adventure, or our ability to build structures of great size and beauty such as the Golden Gate Bridge, the grand arch in St. Louis, the World Trade Center, a giant jumbo jet, or the space shuttle?

Solids and fluids (liquids and gases) are the tangible part of the natural world in which we live. The study of the mechanical and dynamical properties of materials is essential for the advancement of our technological prowess as well as for an appreciation of the raw elements of nature and our environment. In this chapter we investigate some of the basic classical mechanical properties of solids and fluids.

11.1 STATES OF MATTER

We typically classify macroscopic matter (large collections of atoms or molecules) into one of four **states of matter**: solids, liquids, gases, and plasmas. These states are difficult to define precisely, since some materials, even common ones, defy such elementary characterizations and exhibit characteristics of two or more of them. Physicists and engineers have developed branches of their disciplines devoted to studying the static and dynamical behavior of each of the various states of matter. Thus you hear of disciplines such as fluid mechanics, solid mechanics, condensed matter physics, plasma physics, and materials science. The essential physics underlying them all is the same: forces, the response of the system to those forces, and energy considerations.

A **solid** usually is thought of as a semirigid collection of atoms or molecules that maintains a definite shape and volume. Bricks, baseball bats, and boulders are examples of typical solids.

---

*(Chapter Opener) *The Physics Teacher*, 27, #7, page 536 (October 1989).

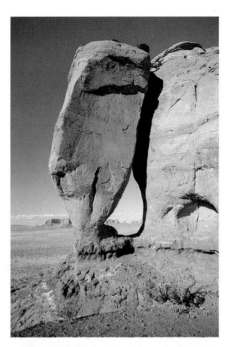

A rock is a solid. Duh! But when heated to a high temperature, rocks can exhibit the flowlike properties of a liquid or even vaporize.

A **liquid** typically is imagined as a collection of atoms or molecules that has a density comparable to that of solids, maintains a definite volume, but also flows and assumes the shape of any container, though the liquid may not necessarily fill it. The milk in a partially filled container illustrates the properties of a typical liquid.

A **gas** typically is imagined as a loose collection of atoms or molecules that also exhibits flow characteristics but completely fills a container of any shape or volume in which it is placed. Gases are more easily compressed or expanded than solids and liquids.

Another unusual state of matter is a **plasma**. Plasmas exist typically at temperatures high enough for some electrons to be liberated from their parent atoms. A plasma is a gas- or liquid-like collection of relatively independent electrons and positive ions that is, overall, electrically neutral. Matter is found in the plasma

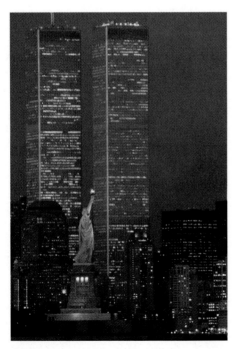

The World Trade Center in New York City with a spectacular 19th-century structural gift from the French in the foreground.

Milk is a liquid, but on another level, it is a colloidal suspension of small, hydrophobic particles.

These balloons are filled with helium gas. The gas liquefies at the chilly temperature of a few degrees above the absolute zero of temperature.

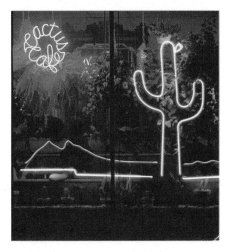

Plasmas exist inside neon lights.

Particulate materials like sand and grains exhibit properties of both solids and liquids.

Even solids like this glacier exhibit properties of liquids, such as flow.

state inside the Sun and other stars (but not all stars). Plasmas also exist inside neon lights and laser tubes that employ gases such as neon, carbon dioxide, xenon, argon, or gas mixtures such as helium with neon.

Such simple characterizations of materials are not always valid. Water, for example, can exist as a solid, liquid, gas, or combinations of them depending on the temperature and pressure; it is the same with many other materials. Materials such as talcum powder, dry sand, cement, or harvested grains are composed of small macroscopic solid particles that are able to flow and can take the shape of their container, similar to liquids (think of an hourglass). Likewise ice, typically thought of as a solid, also can flow, albeit slowly, as it does in large glacial formations.

Asphalt, typically thought of as a solid, also can flow slowly, particularly on hot days, creating annoying ripples in the pavement. Materials that flow (typically gases and liquids) usually are called **fluids**. So most matter is not as easily categorized as you might have thought.

In all these states of matter, *internal* forces exist among the particles of the material. The magnitude or strength of these internal forces of interaction among the particles is greatest in solids, less in liquids, and smallest in gases. To a large extent, the response of the system to applied *external* forces determines whether a substance is properly treated as a solid, liquid, or gas.

11.2 STRESS, STRAIN, AND YOUNG'S MODULUS FOR SOLIDS

Stretch a cord too tightly, and it will be likely to break.
Publius Syrus (c. 1st century B.C.)*

The stress and strain of the academic year likely are by now taking their toll on you. In this section we see yet another example of how physics takes common words from our everyday discourse and uses them in a different, technical context.

---

*The Moral Sayings of Publius Syrus, a Roman Slave, translated by Darius Lyman (Andrew J. Graham, New York 1862), page 52 (saying #570).

Perhaps even as a child you noticed that when you sit on a rope swing, the ropes stretch slightly. Likewise, if a bulldozer pulls on a wire rope attached to a massive tree stump, the wire rope stretches. What factors influence the amount of the stretch?

Rigid bodies in fact deform when subjected to forces. The rigid bodies that we presumed in Chapter 10 when studying spin and orbital motion are idealizations that, strictly speaking, actually do not exist. We now seek to characterize the deformations of materials under forces.

Consider a long, thin, essentially one-dimensional solid such as a rope or wire, as in Figure 11.1. The rope is hung vertically from a fixed support. We neglect the small weight of the rope. A force of magnitude F is applied to stretch the rope along its length.

An applied force that attempts to stretch the rope is called a **tensile force**.* Experimental observations indicate that a tensile force of magnitude F stretches the rope by an amount $\Delta \ell$. The original experiments investigating such stretching, made by Robert Hooke during the 17th century, discovered that the elongation $\Delta \ell$ of a wire or rope is proportional to the magnitude of the applied force $\vec{\mathbf{F}}$. That is,

$$F = k \, \Delta \ell \qquad (11.1)$$

where k is a proportionality constant that depends on the specific material composition of the rope and on geometric factors such as its length and cross-sectional area. The constant k has the SI units N/m and is effectively the spring constant of the material, because of the similarity of Equation 11.1 to the force we exert to stretch a Hooke's force law spring (Equation 7.1). So perhaps it is no surprise that a solid is a large collection of atoms or molecules whose interactions are modeled (as a first approximation) by imaginary Hooke's law springs.

As in most situations, there are limits. If the applied tensile force becomes too large, the rope or wire is permanently deformed (stretched) and will not return to its original length when the force is removed, much like a spring under similar circumstances. The rope or wire then is said to be inelastic. A schematic graph of the magnitude of the applied tensile force versus the elongation $\Delta \ell$ is shown in Figure 11.2.

The linear region of the graph is called the **elastic region** and it is in this region that Equation 11.1 is valid. The constant k

*A force in the opposite direction is called a compressional force; but it is hardly possible to apply such a force along a long thin rope.

in Equation 11.1 is the slope of the linear region. The curved region is the **inelastic region**. Eventually, if the applied tensile force becomes too great, the rope breaks; this is very similar to the behavior of springs we investigated in Chapter 7. Just as with springs, a large value of k (or spring constant) means a very stiff rope or wire, one that is difficult to stretch. Steel wires can be subjected to large forces before reaching the elastic limit; fishing line reaches its elastic limit with the application of relatively small forces.

The elongation depends not only on the magnitude of the force applied, but also on the composition of the rope or wire and its geometric dimensions. Wires of the same material but of different lengths ℓ and different cross-sectional areas A have different elongations when subjected to the same tensile force $\vec{\mathbf{F}}$. Experiments indicate that for a tensile force of a given magnitude, the longer the rope (of fixed cross section), the greater the elongation $\Delta \ell$; in contrast, for a given length ℓ, the greater the cross-sectional area, the less the elongation $\Delta \ell$. These experimental observations are summarized by saying that $\Delta \ell$ is proportional to F and ℓ, but inversely proportional to the cross-sectional area A. That is,

$$\Delta \ell = \frac{1}{E} \frac{F\ell}{A} \qquad (11.2)$$

This is the same as Equation 11.1 with

$$k = \frac{EA}{\ell}$$

For reasons of convenience, the constant of proportionality in Equation 11.2 is written as $1/E$. The larger the value of E, the stiffer the wire or rope, meaning the smaller the deformation $\Delta \ell$ for a given force of magnitude F. By incorporating these geometric considerations, the constant E is left dependent *only* on the material from which the wire is formed. The constant E is called **Young's modulus** of the material.

The magnitude of the applied tensile force $\vec{\mathbf{F}}$ divided by the cross-sectional area A of the rope or wire is called the **tensile stress** on it:

$$\text{tensile stress} \equiv \frac{F}{A}$$

FIGURE 11.1 A force of magnitude F stretches a rope or wire.

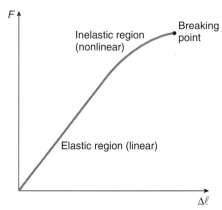

FIGURE 11.2 A schematic graph of the magnitude of the tensile force versus the elongation it produces.

The change in length of the rope $\Delta\ell$ divided by the length of the rope ℓ is called the **tensile strain**:

$$\text{tensile strain} \equiv \frac{\Delta\ell}{\ell}$$

Take the ratio of the tensile stress to the strain:

$$\frac{\text{tensile stress}}{\text{tensile strain}} = \frac{F/A}{\Delta\ell/\ell}$$

Using Equation 11.2, we find that the ratio of the tensile stress to tensile strain is Young's modulus E:

$$E \equiv \frac{\text{tensile stress}}{\text{tensile strain}}$$

$$= \frac{F/A}{\Delta\ell/\ell} \qquad (11.3)$$

Values of Young's modulus for various materials are found in Table 11.1. Note that the SI units for Young's modulus are the same as for tensile stress, N/m^2, since the strain is dimensionless.

In three dimensions, there are other ways in which forces can be applied to materials to cause deformation. For example, consider a solid rectangular parallelopiped, as in Figure 11.3. If a pair of forces, each of magnitude F, is applied to the material over and parallel to opposite surfaces of areas A and separation ℓ, as in Figure 11.4, the material is subjected to what is called a **shear stress**.

The applied forces typically are distributed over the area A rather than at a single point. The shear stress then is F/A. A shear stress deforms a solid initially in the shape of a rectangular parallelopiped into a *non*rectangular parallelopiped. The deformation $\Delta\ell$ divided by the separation ℓ of the two surfaces is known as the **shear strain**.

The ratio of the shear stress to the shear strain is the **shear modulus** G of the material:

$$G \equiv \frac{\text{shear stress}}{\text{shear strain}} = \frac{F/A}{\Delta\ell/\ell} \qquad (11.4)$$

FIGURE 11.3 A solid rectangular parallelopiped.

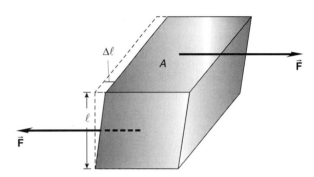

FIGURE 11.4 A shear stress applied to a solid material.

Values of the shear modulus depend on the material and are tabulated in Table 11.1. The SI units for the shear modulus are N/m^2, since the strain is dimensionless.

Although solids and liquids typically are thought of as incompressible, when subjected to compressive forces per unit area, they deform (or are squeezed) into smaller volumes (even if slightly). Let such a material be subjected to uniform squeezing from all directions, as in Figure 11.5. The magnitude of the applied compressive force per unit area is the **pressure** (it could also be considered a volume stress). Let the volume of the material originally be V_i and its change ΔV. The change in the volume is

$$\Delta V = V_f - V_i$$

TABLE 11.1 Moduli of Various Materials

Material	Young's modulus $(N/m^2 \equiv Pa)$	Shear modulus $(N/m^2 \equiv Pa)$	Bulk modulus $(N/m^2 \equiv Pa)$
Aluminum	7.0×10^{10}	2.4×10^{10}	7.0×10^{10}
Boron carbide	35×10^{10}		
Brass	9.0×10^{10}	3.5×10^{10}	6.1×10^{10}
Copper	11×10^{10}	4.2×10^{10}	14×10^{10}
Gold	7.9×10^{10}		
Kevlar	$6.2\text{--}12.4 \times 10^{10}$		
Lead	1.5×10^{10}	0.5×10^{10}	
Mercury	—	—	2.7×10^{10}
Nylon	2×10^{9}		
Platinum	17×10^{10}	6.4×10^{10}	
Silicon carbide	41×10^{10}		
Silver	7.8×10^{10}	2.6×10^{10}	
Steel	20×10^{10}	8.1×10^{10}	16×10^{10}
Tungsten	35×10^{10}	15×10^{10}	20×10^{10}
Water	—	—	0.20×10^{10}

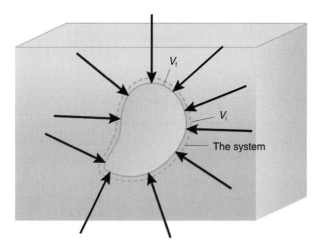

FIGURE 11.5 Squeezing a solid or liquid changes its volume slightly.

and, since the material is compressed into a smaller volume, the change in the volume is *negative*. The ratio of $\Delta V / V_i$ is called the **volume strain**. The ratio of the pressure (volume stress) to the volume strain is the **bulk modulus** B of the material:

$$B \equiv -\frac{\text{pressure}}{\text{volume strain}} \qquad (11.5)$$

The minus sign is introduced so the bulk modulus B is positive. The bulk modulus for various materials also is tabulated in Table 11.1. The SI units for the bulk modulus are the same as those for pressure: N/m^2. The inverse of the bulk modulus is called the **compressibility** β:

$$\beta = \frac{1}{B} \qquad (11.6)$$

Notice that all three moduli are defined as the ratio of an applied stress to the resulting strain in the material, and all three moduli have the SI units of N/m^2.

If the applied stress of this course does not result in much resulting strain on your part, then your "learning modulus" for physics is quite large. Of course, the value of this learning modulus varies with the specific material too (be it you or the topic in physics)!

QUESTION 1

Would you expect Young's modulus for rubber to be greater than or less than that for copper? Make a few crude measurements to determine the order of magnitude of Young's modulus for the elastic material used to make rubber bands.

EXAMPLE 11.1

A wire of diameter 0.617 mm and length 0.984 m is suspended vertically from a rigid support. The wire is subjected to a tensile force by hanging an 8.00 kg mass from its end. Careful measurement* indicates that the wire stretches 1.30 mm. Determine Young's modulus of the material.

---

*Such small distances can be measured precisely using an optical lever. You might ask your laboratory instructor about using such a neat device.

Solution

Young's modulus is the ratio of the tensile stress to the tensile strain. The tensile stress is the magnitude of the force per unit cross-sectional area of the wire:

$$\text{tensile stress} = \frac{F}{A}$$

The magnitude of the applied tensile force is the magnitude of the weight of an 8.00 kg mass:

$$w = mg$$
$$= (8.00 \text{ kg})(9.81 \text{ m/s}^2)$$
$$= 78.5 \text{ N}$$

The cross-sectional area of the wire of diameter d is

$$A = \pi r^2 = \pi \frac{d^2}{4}$$
$$= \pi \frac{(6.17 \times 10^{-4} \text{ m})^2}{4}$$
$$= 2.99 \times 10^{-7} \text{ m}^2$$

Hence the tensile stress is

$$\text{tensile stress} = \frac{F}{A}$$
$$= \frac{78.5 \text{ N}}{2.99 \times 10^{-7} \text{ m}^2}$$
$$= 2.63 \times 10^8 \text{ N/m}^2$$

The tensile strain is

$$\text{tensile strain} = \frac{\Delta \ell}{\ell}$$
$$= \frac{1.30 \times 10^{-3} \text{ m}}{0.984 \text{ m}}$$
$$= 1.32 \times 10^{-3}$$

Therefore Young's modulus is

$$E = \frac{\text{tensile stress}}{\text{tensile strain}}$$
$$= \frac{2.63 \times 10^8 \text{ N/m}^2}{1.32 \times 10^{-3}}$$
$$= 1.99 \times 10^{11} \text{ N/m}^2$$

The wire likely is made of steel (see Table 11.1).

11.3 FLUID PRESSURE

Scuba divers can only go so deep before the effects of increasing pressure become quite apparent. The concept of pressure is central to an understanding of fluids. We introduced the idea in our discussion of the bulk modulus in the previous section.

The effects of fluid pressure are apparent when scuba diving.

FIGURE 11.6 The fluid on the left of a differential area exerts a differential force on the fluid to the right of the area.

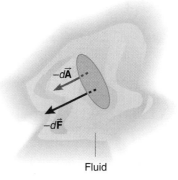

FIGURE 11.7 The fluid on the right of a differential area exerts a differential force on the fluid to the left of the area.

Consider a differential area $d\vec{\mathbf{A}}$ at any location and with any orientation in a fluid, as in Figure 11.6. The differential force $d\vec{\mathbf{F}}$ exerted by one part of the fluid on another part of the fluid perpendicularly across the area $d\vec{\mathbf{A}}$ is proportional to the size of the differential area. We write this as

$$d\vec{\mathbf{F}} = P\,d\vec{\mathbf{A}}$$

where the quantity P is a *scalar* called the pressure; it has the dimensions of force per unit area. In Figure 11.6 the fluid to the left of the area exerts a force on the fluid to the right of the area. Likewise, in Figure 11.7 the fluid to the right of the area exerts a force on the fluid to the left of the area. These two forces are in opposite directions, are of equal magnitude, and form a Newton's third law force pair. In terms of the magnitudes of the quantities,

$$dF = P\,dA$$

or

$$P = \frac{dF}{dA} \qquad (11.7)$$

Experimental observations indicate that this relationship is true regardless of the orientation of the differential area in the fluid.

> The pressure is a scalar quantity that is the magnitude of the force per unit area acting on *any* differential area, regardless of its orientation within the fluid.

Hence the fluid exerts a force on the walls of its container that is perpendicular to its walls at all points; from Newton's third law, we see that the walls exert a force of equal magnitude in the opposite direction on the fluid.

The SI unit of pressure is N/m². Since pressure is such a central concept in studying fluids, the SI unit of pressure is defined to be a **pascal** (Pa):

$$1\ \text{N/m}^2 \equiv 1\ \text{Pa} \qquad (11.8)$$

The unit of pressure is named after the 17th-century French scientist Blaise Pascal (1623–1662) who did important early work investigating fluids.

There are several other common, *convenient* units of pressure. A historically convenient unit for measuring pressure is the *atmosphere* (atm),* but it is not an SI unit. The atmosphere unit of pressure stems from the magnitude of the average weight per horizontal square meter of the atmosphere of the Earth at sea level. An atmosphere of pressure is defined to be

$$1\ \text{atm} \equiv 1.013\,25 \times 10^5\ \text{Pa} \qquad (11.9)$$

The *actual* pressure in our atmosphere at any time may differ from this standard atmosphere because of meteorological factors and elevation. We shall use the pressure unit atm occasionally for convenience, much as we do grams and centimeters.

---

*In banking, an "atm" means automated teller machine!

You may encounter other common pressure units in other technical contexts, but we will not use them in this text (except on rare occasions) because they are not SI units.

1. The *torr** is defined to be the pressure difference between the top and bottom of a column of mercury exactly 1 mm in height when at sea level. Torr, therefore, also are referred to as *millimeters of mercury* (mm Hg)

$$1 \text{ torr} \equiv 1 \text{ mm Hg} \tag{11.10}$$

Experimentally, a column of mercury 76.0 cm in height corresponds to one atmosphere pressure. Hence the conversion to the SI pressure unit Pa is found from the identity:

$$1.013\,25 \times 10^5 \text{ Pa} \equiv 1 \text{ atm} \equiv 760 \text{ torr} \equiv 760 \text{ mm Hg} \tag{11.11}$$

The pressure unit mm Hg commonly is encountered in meteorological work and is frequently mentioned in weather reports. Your blood pressure also is commonly measured in mm Hg.

2. In many areas of engineering, the **gauge pressure** is used. The gauge pressure is the actual pressure P minus whatever the actual pressure of the atmosphere (≈ 1 atm) is at the location of the gauge, because the other side of the gauge mechanism is exposed to the atmosphere:

$$\text{gauge pressure} \equiv P - 1 \text{ atm} \tag{11.12}$$

The actual pressure P then is called the **absolute pressure**. Gauge pressure is the pressure indicated by a common tire pressure gauge, which is why it is occasionally a useful unit of convenience in physics and engineering. A totally flat tire has a gauge pressure equal to zero. The absolute pressure P of a thoroughly flat tire is the *actual* value of the atmospheric pressure at the location of the tire, which is approximately 1 atm.

11.4 STATIC FLUIDS

It is the pressure that limits the depth to which submarines can dive. Mountain climbers likewise know that as they ascend high peaks, the pressure of the atmosphere decreases, making it difficult for their bodies to secure adequate oxygen. Commercial airplanes are equipped with systems to increase the cabin pressure above the ambient outside air pressure at high altitudes, so that passengers can breathe normally. Air travelers are reminded before every flight of emergency procedures to follow in the event of a sudden decrease of cabin pressure. Here we discover how pressure varies with altitude in a fluid.

Consider a fluid at rest near the surface of the Earth. We want to figure out how the pressure in the fluid changes either with height or depth from a reference level. We assume that any changes in height or depth are small when compared with the radius of the Earth, so that the magnitude of the acceleration due to gravity g at the surface can be taken as constant.[†] Choose a

---

*The unit named after Evangelista Torricelli (1608–1647), the inventor of the mercury barometer; see Example 11.3.

[†] A change in altitude of approximately 32 km changes the magnitude of the local acceleration due to gravity by only about 1%. You should be able to show this!

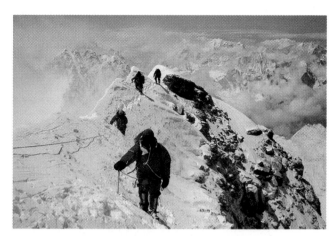

High-altitude climbing requires the use of supplemental oxygen because of the low atmospheric pressure.

coordinate system with $\hat{\jmath}$ directed vertically up. Examine the static equilibrium of a differentially thin slice of the fluid of area A and vertical thickness dy; see Figure 11.8. Let $P(y)$ be the pressure at a height y. When y is increased slightly to $y + dy$, let the pressure at the new location be written as $P(y) + dP$.[‡]

There are several forces acting on the thin slice (see Figure 11.9):

1. The weight of the fluid in the thin slice. Let ρ be the density of the fluid at this height. The mass of material in the slice is its density times the volume of the slice: $\rho(A\,dy)$. Thus the weight of the slice is of magnitude

$$(\rho\,A\,dy)g$$

where g is the magnitude of the local acceleration due to gravity (the magnitude of the local gravitational field). With the coordinate system chosen in Figures 11.8 and 11.9, the weight is directed along $-\hat{\jmath}$. Thus the weight is

$$(-\rho\,A\,dy\,g)\hat{\jmath}$$

[‡] The pressure actually decreases as we go up, but this will emerge naturally from our analysis.

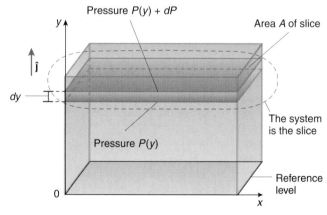

FIGURE 11.8 The pressure on either side of a thin slice of fluid.

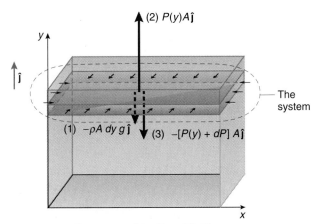

FIGURE 11.9 Forces on a thin slice of fluid.

2. The pressure at the bottom of the slice is $P(y)$. The magnitude of the force exerted by the rest of the fluid on the bottom of the slice is $P(y)A$ and is directed upward $(+\hat{\jmath})$. Thus the force on the bottom of the slice is

$$P(y)A\hat{\jmath}$$

3. The pressure at the top of the slice is $P(y) + dP$, so that the force of the rest of the fluid on the top of the slice is directed downward and is

$$[P(y) + dP]A(-\hat{\jmath})$$

4. Forces also are exerted horizontally by the rest of the fluid on the tiny lateral area of the thin slice, as shown in Figure 11.9, but these horizontal forces pairwise sum to zero and can be omitted from further consideration.

Since the slice is in mechanical equilibrium, the sum of the forces acting in the vertical direction must be zero according to the first condition of equilibrium. Thus

$$F_{y\,\text{total}}\hat{\jmath} = 0 \text{ N}$$
$$\{P(y)A - [P(y) + dP]A - \rho A\, dy\, g\}\hat{\jmath} = 0 \text{ N}$$

Rearranging, we get

$$-A\, dP - \rho A\, dy\, g = 0 \text{ N}$$

Thus the pressure changes with height according to the following differential equation:

$$\frac{dP}{dy} = -\rho g \qquad (11.13)$$

The minus sign means that the *pressure decreases with increasing height* in the fluid: a small positive height change dy results in a small negative value for the pressure change dP. For finite changes in height, we have to solve this differential equation to find the pressure.

Two situations arise according to whether the fluid is quite incompressible (most liquids) or quite compressible (gases).

Incompressible Liquids

The bulk modulus of most liquids is quite large (see Table 11.1), and so to a first approximation liquids can be considered to be

incompressible. With this assumption the density of the liquid is constant, regardless of height or depth. Equation 11.13 then can be integrated directly to find the pressure at any elevation y in the liquid:

$$dP = -\rho g\, dy$$

Let the pressure at the origin (i.e., the reference level), $y = 0$ m, be P_0 and the pressure at elevation y be $P(y)$. The definite integrals are

$$\int_{P_0}^{P(y)} dP = -\int_{0\,\text{m}}^{y} \rho g\, dy$$

Performing the integrations, we find

$$P(y) - P_0 = -\rho g(y - 0 \text{ m})$$

This can be rearranged into the following form:

$$P(y) = P_0 - \rho g y \qquad (\hat{\jmath}\text{ up}) \qquad (11.14)$$

This equation means that as we increase height in the liquid, the pressure decreases. Or equivalently, pressure increases in direct proportion to depth in an incompressible liquid.

> Since Equation 11.14 depends only on the vertical coordinate y, the pressure is the same at all points at the same depth or elevation regardless of their horizontal location, as shown in Figure 11.10.

PROBLEM-SOLVING TACTIC

11.1 You have the freedom to choose where to place the origin of the coordinate system; you simply need to have $\hat{\jmath}$ directed up to use Equation 11.14. The place where you choose the origin is called the *reference level* and the pressure at this level (where $y = 0$ m) is P_0. The pressure at *any* coordinate y within the liquid then is $P(y)$, whether the coordinate y is above the origin (positive values of y) or below the origin (negative values of y).

FIGURE 11.10 The pressure is the same at any horizontal level within a connected fluid.

QUESTION 2

When designing a dam it is *not* the surface area of the lake or its volume of water behind the dam that is the critical design element, but rather the depth of the water at the dam site and the vertical projection of the surface area of the dam face. Explain why.

STRATEGIC EXAMPLE 11.2

Find the fluid pressure in units of atmospheres at the depth of the ill-fated, allegedly unsinkable *Titanic*. The huge ship sank after hitting an iceberg on its maiden voyage from Southampton, England, to New York on 14 April 1912 with great loss of life. The ocean liner is located approximately 3.00×10^3 m below the surface of the northern Atlantic Ocean at latitude 41.8°N and longitude 50.2°W. Sea water has a density of 1.025×10^3 kg/m³.

The *Titanic* embarking on its ill-fated maiden voyage in 1912.

Solution

Use Equation 11.14 for the pressure at any coordinate y in a liquid:

$$P(y) = P_0 - \rho g y \qquad (\hat{\jmath} \text{ up})$$

Method 1

Choose the origin of the coordinate system (the reference level in the fluid) to be at the location of the ship at the bottom of the sea, as shown in Figure 11.11. Then P_0 is the pressure at this level; you are trying to find the value of P_0. Solve Equation 11.14 for P_0:

$$P_0 = P(y) + \rho g y \qquad (1)$$

The pressure at the surface of the sea (at coordinate y) is $P(y)$. The water surface is at coordinate $y = 3.00 \times 10^3$ m and the pressure $P(y)$ at this coordinate location is atmospheric pressure, 1.0 atm. The term $\rho g y$ is

$$\rho g y = (1.025 \times 10^3 \text{ kg/m}^3)(9.81 \text{ m/s}^2)(3.00 \times 10^3 \text{ m})$$
$$= 3.02 \times 10^7 \text{ Pa}$$

This can be converted to atmospheres by using the conversion

$$1 \text{ atm} = 1.013 \times 10^5 \text{ Pa}$$

Hence

$$\rho g y = 298 \text{ atm}$$

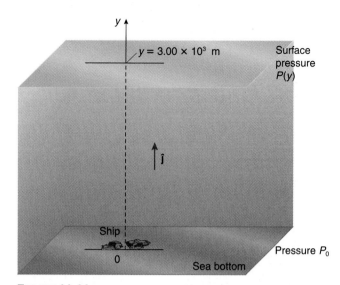

FIGURE 11.11

Since $P(y)$ is 1.0 atm, equation (1) is

$$P_0 = 1.0 \text{ atm} + 298 \text{ atm}$$
$$= 299 \text{ atm}$$

Method 2

Choose the surface of the sea as the origin of the coordinate system, as shown in Figure 11.12. In this case the reference pressure P_0 is 1.0 atm. We want the pressure $P(y)$ at the coordinate y, and so Equation 11.14,

$$P(y) = P_0 - \rho g y$$

is employed with $y = -3.00 \times 10^3$ m, the value of the y-coordinate at the location of the ship.

Thus

$$P = P_0 - (1.025 \times 10^3 \text{ kg/m}^3)(9.81 \text{ m/s}^2)(-3.00 \times 10^3 \text{ m})$$
$$= P_0 + 3.02 \times 10^7 \text{ Pa}$$

FIGURE 11.12

Converting this pressure to atmospheres, and substituting 1.0 atm for P_0, you get

$$P = 1.0 \ \text{atm} + 298 \ \text{atm}$$
$$= 299 \ \text{atm}$$

EXAMPLE 11.3

A common device for measuring the pressure in the surrounding atmosphere is the mercury barometer. It is constructed by filling a uniform tube and a pan partially with mercury; see Figure 11.13. When the tube is inverted with its open end in the pan, some of the mercury flows out of the tube, as shown in Figure 11.14.

FIGURE 11.13

FIGURE 11.14

The closed end of the tube essentially is a vacuum.* The pressure acting on the top surface of the mercury in the inverted column thus is zero to an excellent approximation. The height of the mercury column is measured to be 76.0 cm. Find the atmospheric pressure.

Solution
Method 1
Using Problem-Solving Tactic 11.1, choose a coordinate system with the origin at the elevation of the pool of mercury, as in Figure 11.15. The origin locates the reference level in the fluid. You must have $\hat{\jmath}$ up, consistent with the choice made in deriving Equation 11.14.

The pressure at the level of the pool of mercury open to the air is atmospheric pressure. Since this is our reference level, this pressure is P_0 in Equation 11.14:

$$P(y) = P_0 - \rho g y \qquad (\hat{\jmath} \ \text{up})$$

The pressure at the top of the column of mercury is 0 Pa since there is essentially a vacuum above the surface at that point. So $P(y) = 0$ Pa. Thus

$$0 \ \text{Pa} = P_0 - \rho g y$$

---

*The only pressure in this region is from the small vapor pressure of mercury, which we neglect here.

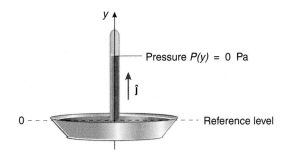

FIGURE 11.15

Hence

$$P_0 = \rho g y$$

You must take care to use consistent SI units. The density of mercury is $\rho = 13.6 \times 10^3 \ \text{kg/m}^3$ and of course $g = 9.81 \ \text{m/s}^2$. The observed height of the mercury column is 76.0 cm or 0.760 m. Substituting these numerical values, you find

$$P_0 = (13.6 \times 10^3 \ \text{kg/m}^3)(9.81 \ \text{m/s}^2)(0.760 \ \text{m})$$
$$= 1.01 \times 10^5 \ \text{kg/(m·s}^2)$$

The units kg/(m·s^2) are equivalent to Pa since

$$\text{Pa} = \text{N/m}^2 = \frac{\text{kg·m}}{\text{s}^2 \cdot \text{m}^2} = \frac{\text{kg}}{\text{m·s}^2}$$

Thus

$$P_0 = 1.01 \times 10^5 \ \text{Pa}$$

Method 2
Choose the origin of the coordinate system to be at the top of the column, as shown in Figure 11.16. The pressure P_0 in Equation 11.14 is the pressure at the level of the origin in the fluid. This is now at the top of the column where the vacuum exists, and so $P_0 = 0$ Pa. Hence Equation 11.14 becomes

$$P(y) = P_0 - \rho g y \qquad (\hat{\jmath} \ \text{up})$$
$$P(y) = 0 \ \text{Pa} - \rho g y$$
$$= -\rho g y$$

FIGURE 11.16

The pressure $P(y)$ is the pressure at the coordinate y. But the y-coordinate of the surface of the mercury pool is $y = -0.760$ m. Hence you obtain

$$P = -(13.6 \times 10^3 \ \text{kg/m}^3)(9.81 \ \text{m/s}^2)(-0.760 \ \text{m})$$
$$= 1.01 \times 10^5 \ \text{Pa}$$

A Gaseous Atmosphere

Equation 11.13 also may be applied to gases such as the atmosphere of the Earth. Gases are easily compressed. For such situations, the density ρ of the gas is not constant but is a function of height $\rho(y)$, which is why climbers find it more difficult to breathe at high altitudes and may need supplementary oxygen tanks. Make a crude (and unrealistic) assumption that the temperature of the atmosphere is independent of height; the pressure at any height then is found to be proportional to the density. Once again, call the pressure P_0 where $y = 0$ m and the density of the gas there ρ_0. Then

$$P_0 \propto \rho_0$$

Likewise at height y, where the pressure is $P(y)$ and density is $\rho(y)$, we have

$$P(y) \propto \rho(y)$$

Dividing these two proportions, we find

$$\frac{P(y)}{P_0} = \frac{\rho(y)}{\rho_0}$$

or

$$\rho(y) = \rho_0 \, \frac{P(y)}{P_0}$$

We substitute this expression into Equation 11.13 for ρ and solve for dP; we find

$$dP = -\rho_0 \, \frac{P(y)}{P_0} \, g \, dy$$

If we now divide by $P(y)$, the equation can be integrated:

$$\frac{dP}{P} = -\frac{\rho_0}{P_0} \, g \, dy$$

We integrate the left-hand side from P_0 to $P(y)$ with the corresponding limits for y on the right-hand side:

$$\int_{P_0}^{P(y)} \frac{dP}{P} = -\frac{\rho_0}{P_0} \, g \int_{0 \text{ m}}^{y} dy$$

$$\ln P(y) - \ln P_0 = -\frac{\rho_0}{P_0} \, gy$$

$$\ln \frac{P(y)}{P_0} = -\frac{\rho_0}{P_0} \, gy$$

We put this into exponential form:

$$P(y) = P_0 \exp\left(-\frac{\rho_0}{P_0} \, gy\right) \tag{11.15}$$

The pressure, here the magnitude of the weight per unit area of the *remaining* gases overhead, decreases exponentially with height in the atmosphere.

Note that since an exponent must be a pure number with no dimensions, the quantity

$$\frac{P_0}{\rho_0 g}$$

must have the dimensions of a length, or m in SI units. This characteristic length is called the **scale height** in the atmosphere:

$$\text{scale height} \equiv \frac{P_0}{\rho_0 g} \tag{11.16}$$

When the height y above the origin equals the scale height, the pressure is e^{-1} of the pressure at the position of the origin, where $y = 0$ m. In fact the scale height also depends on the temperature, though this certainly is not apparent since we assumed the temperature of the atmosphere is constant in deriving Equations 11.15 and 11.16.

EXAMPLE 11.4

The surface atmospheric pressure of the Earth is approximately $P_0 = 1.013 \times 10^5$ Pa and the surface air density is about $\rho_0 = 1.29$ kg/m³. Calculate the scale height of the atmosphere measured from the surface of the Earth.

Solution
Use Equation 11.16, choosing the origin to be at the surface of the Earth:

$$\text{scale height} \equiv \frac{P_0}{\rho_0 g}$$

$$= \frac{1.013 \times 10^5 \text{ Pa}}{(1.29 \text{ kg/m}^3)(9.81 \text{ m/s}^2)}$$

$$= 8.00 \text{ km}$$

You might convince yourself that the units indeed check out. This result means that at a height of 8.00 km, the pressure is $e^{-1} = 0.368$ of the surface atmospheric pressure. By way of comparison, the peak of Mt. Everest is approximately 8.85 km above sea level.

EXAMPLE 11.5

The scale height of the atmosphere of the Earth is 8.00 km (see Example 11.4). What is the ratio of the atmospheric pressure in Denver (altitude 1.6 km above sea level) to that in Seattle (at sea level)?

Solution
Choose the origin to be at sea level, so that the pressure at sea level is P_0 in Equation 11.15. Substituting the scale height of the atmosphere, Equation 11.15 becomes

$$P(y) = P_0 \exp\left(\frac{-y}{8.00 \text{ km}}\right)$$

where the altitude y must be in km, and so the exponent is a pure number. Since Denver has an altitude of $y = 1.6$ km above sea level, you get

$$\frac{P(y)}{P_0} = \exp\left(\frac{-1.6 \text{ km}}{8.00 \text{ km}}\right)$$

$$= 0.82$$

The atmospheric pressure in Denver is only about 82% that in Seattle, which is why people (especially visiting athletes) typically have to adjust to the reduced pressure there.

11.5 PASCAL'S PRINCIPLE

Notice in Equation 11.14 that if we increase $P(y)$ by some amount for a liquid *enclosed* in and totally filling a container, say by squeezing on it, then P_0 also must increase by the same amount to maintain the equality; the converse also is true.

> Hence, if we increase the pressure on the surface of an enclosed liquid, the pressure increases by the same amount at all points throughout the liquid and at the walls of its container, independent of the shape of the fluid container.

This result, discovered by Pascal in the 17th century, is called **Pascal's principle**.

A common application of Pascal's principle is in hydraulic lifts and machinery. A force of magnitude F_1, applied to a piston of area A_1 (see Figure 11.17), increases the pressure in the entire liquid by F_1/A_1. Hence, at a second piston of area A_2, the pressure *increase* is the *same*, even if it is at a different elevation in the fluid. The pressure increase at the second piston is F_2/A_2 where F_2 is the force on the second piston and A_2 is its area. Thus, according to Pascal's principle,

pressure increase at piston 1 = pressure increase at piston 2

$$\frac{F_1}{A_1} = \frac{F_2}{A_2}$$

Hence

$$F_2 = \frac{A_2}{A_1} F_1 \qquad (11.17)$$

If $A_2 > A_1$, a small-magnitude force F_1 applied to a small piston results in a large-magnitude force F_2 at the other, larger piston. The second piston may be used to lift a car (the hydraulic lift common in many automotive service shops), move a blade or large bucket on a bulldozer or front loader, or allow you to stop your car by pushing with your foot on the hydraulic brake piston with a force of small magnitude. In each case a small-magnitude force on a small piston results in tremendous forces on a larger piston.

Blaise Pascal discovered the principle underlying hydraulic machinery. He also is noted for his writings on religion, philosophy, and literature.

FIGURE 11.17 Apply a force of magnitude F_1 to a piston of area A_1.

EXAMPLE 11.6

A service station hydraulic lift has a large piston of radius 15 cm and a small piston of radius 2.5 cm. What magnitude of force increase must be applied to the small piston to lift a car of mass 2.00×10^3 kg?

Solution
The weight of the car is of magnitude

$$mg = (2.00 \times 10^3 \text{ kg})(9.81 \text{ m/s}^2)$$
$$= 1.96 \times 10^4 \text{ N}$$

Use Equation 11.17:

$$F_2 = \frac{A_2}{A_1} F_1$$

$$1.96 \times 10^4 \text{ N} = \frac{\pi (0.15 \text{ m})^2}{\pi (0.025 \text{ m})^2} F_1$$

Solving for F_1,

$$F_1 = 5.4 \times 10^2 \text{ N}$$

11.6 ARCHIMEDES' PRINCIPLE

Have you ever wondered why some things float while others sink? Throw a penny into the water and it sinks like a rock, but a huge steel aircraft carrier floats. Why is that? What factors determine the answer to this old question? The answer lies with Archimedes' principle, formulated over two millennia ago.

The story goes that Archimedes (c. 287–212 B.C.) was given the problem of determining whether a new crown of King Hieron II was made of pure gold or of a cheaper gold–silver alloy (known as electrum). This was one of the first known problems in consumer testing. Since the crown was irregularly shaped and Archimedes understandably was prohibited from cutting or drilling into it, the problem was not an easy one to solve. Archimedes allegedly thought of a way to solve the problem while entering a communal bath and noting either the

Archimedes formulated one of the oldest principles of classical physics. He died at the hands of an unknown Roman soldier during the capture of Syracuse in 212 B.C.

overflow of the water or the apparent "lightness" of his limbs in the water. Legend has it that he ran streaking stark naked from the bath, shouting *Eureka! Eureka!* ("I have found it!"). Appropriately enough, the word eureka now is the state motto of California, a state whose citizens are known for their uninhibited inventiveness.

Archimedes' principle states that a system submerged or floating in a fluid has a *buoyant force* acting on it whose magnitude is equal to that of the weight of the fluid displaced by the system. As its name implies, the buoyant force acts on the system in the vertically upward direction.

For us in these latter days, Archimedes' principle is not a new fundamental law of physics, nor is the buoyant force a new, fundamental force of nature. The principle is merely a consequence of one of the conditions for static equilibrium that, in turn, stem from Newton's laws of motion. To show this, consider an arbitrarily shaped region of a fluid at rest, as in Figure 11.18. The vector sum of all the forces acting on the fluid within this volume must be zero because the fluid is in mechanical equilibrium. The forces on this region of the fluid arise from two sources:

1. the *weight* $\vec{\mathbf{w}}$ of the region of fluid (that is, the gravitational force of the Earth on the fluid); and
2. the force $\vec{\mathbf{F}}_{\text{buoy}}$ arising from the *pressure of the rest of the fluid* on this region of fluid.

Since these forces vector sum to zero for a fluid in equilibrium, they must be of equal magnitude and in opposite directions. In other words, the force exerted by the rest of the fluid on this region is equal in magnitude to the weight of the fluid inside the region, but is directed upward.

Now imagine replacing the fluid in the region with an object of exactly the same shape; we consider this new object to be the system, as in Figure 11.19. The force caused by the pressure exerted by the rest of the fluid on this region is the same as before: it could care less what was actually inside the region. The magnitude of the force exerted by the rest of the fluid on this region we found to be equal to that of the weight of the fluid within the region. So the system that replaces (we say *displaces*) the fluid has an upward buoyant force acting on it equal in magnitude to that of the weight of the fluid displaced by the system.

The foregoing argument and conclusion are the same regardless of the shape of the region or the object. Hence any system immersed completely in a fluid has at least two forces acting on it in the vertical direction: the upward buoyant force and the downward weight of the system itself. If the magnitude of the weight of the system is greater than that of the buoyant force on it, as in Figure 11.20, then the total force on the system is in the downward direction and the system sinks. If the magnitude of the buoyant force is greater than that of the weight of the system, as in Figure 11.21, then the total force on the system is in the upward direction and the system rises in the fluid. If the two forces are equal in magnitude, as in Figure 11.22, the total force in the vertical direction is zero. Thus, when the system is released, it stays at the vertical location where it is placed, a situation called **neutral buoyancy**.

To find an expression for the buoyant force on a system, let the fluid density be ρ_{fluid} and the average density of the

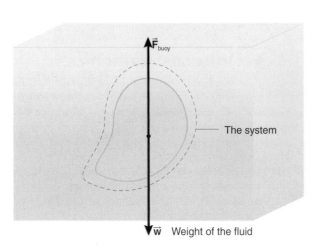

FIGURE 11.18 The forces on a region of fluid at rest.

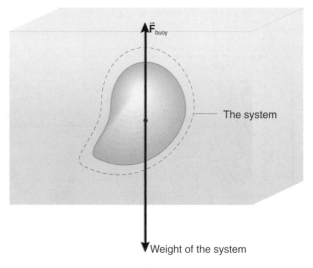

FIGURE 11.19 Forces on a submerged object with the same shape as the region of fluid in Figure 11.18.

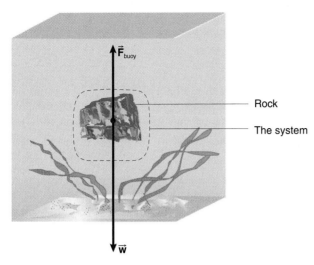

FIGURE 11.20 A system sinks if the magnitude of its weight is greater than the magnitude of the buoyant force.

FIGURE 11.21 If the magnitude of the buoyant force is greater than the magnitude of the weight, the system rises in the fluid.

system be ρ_{system}. Let the total volume of the system be V. Then the mass of the system is

$$\rho_{system}V$$

and its weight is of magnitude

$$\rho_{system}Vg$$

Let V′ be the volume of the system *immersed in the fluid*; if the system is totally submerged, V′ = V. The weight of the fluid displaced by the system is the weight of a region of fluid of volume V′.

Thus the buoyant force on the system is of magnitude

$$F_{buoy} = \rho_{fluid}V'g \qquad (11.18)$$

and is directed vertically upward.

When a system floats, the system experiences neutral buoyancy. Of course, the system does not literally float *on* the

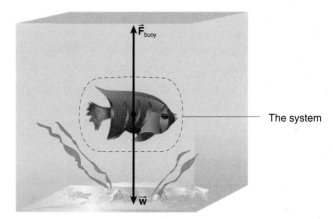

FIGURE 11.22 When the magnitude of the buoyant force is equal to the magnitude of the weight, neutral buoyancy occurs.

surface, although we probably drew childhood pictures of floating boats this way! See Figure 11.23. You know better now. When placed in the fluid, the system sinks into the fluid until the magnitude of the buoyant force on the system (equal to the magnitude of the weight of the liquid displaced by the system) is equal to the magnitude of the weight of the system, as shown in Figure 11.24. That is, the system floats when

$$\rho_{fluid}V'g = \rho_{system}Vg$$

or

$$\rho_{fluid}V' = \rho_{system}V$$

Since V′ can at most be equal to V, a system will float in any liquid with a density greater than the average density of the system itself.

FIGURE 11.23 Typical child's drawing of a floating system.

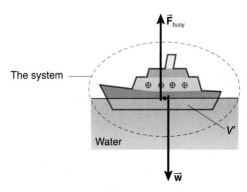

FIGURE 11.24 Forces on a floating system.

For neutral buoyancy or for floating partially submerged, the system thus has an upward buoyant force acting on it (exerted by the rest of the fluid) equal in magnitude to the downward weight of the system itself. The two forces then vector sum to zero and the system floats—that is, it is in vertical equilibrium.

Applications of Archimedes' principle are ubiquitous. Swim bladders in fish, gas-filled chambers in the nautilus shell (*Nautilus pompilius*), and the act of stone swallowing in crocodiles all control the buoyancy of these marine species. Submarines and hot air balloons regulate their altitude in their respective fluids by controlling the magnitude of their weight vis-à-vis that of the buoyant force. Archimedes' principle of displacement also is used to determine the volume of irregular objects. In medicine, the principle is used to assess the amount of swelling in limbs as well as in hydrotherapy for arthritic and other patients.

QUESTION 3

The buoyant force on a system and the weight of the fluid displaced by the system are of equal magnitudes but point in opposite directions. Do these forces form a Newton's third law force pair? Explain why or why not. If they are not a third law force pair, specify the third law force pairs of these forces.

EXAMPLE 11.7

A rock is suspended from a spring scale in air and is found to have a weight of magnitude w; see Figure 11.25a. The rock then is submerged completely in water while attached to the scale, which is found to give a new reading of w_{submrg}; see Figure 11.25b. Find the density of the rock ρ_{rock} in terms of the scale readings and the density of water ρ_{water}.

(a)   (b)

FIGURE 11.25

Solution

Let the system be the rock whose volume is V; its density is given as ρ_{rock}. Then the magnitude of its weight $\vec{w}$ is its mass ($\rho_{rock}V$) times the magnitude of the local gravitational field (g):

$$w = \rho_{rock}Vg$$

When the rock is submerged in the water while attached to the scale, the rock experiences an additional upward buoyant force with a magnitude equal to that of the weight of the water

displaced by the system. Equation 11.18 states that the buoyant force has a magnitude

$$F_{buoy} = \rho_{water}V'g$$

with $V' = V$, since the rock is totally submerged. The magnitude of the total downward force on the system when it is in the water is the new scale reading, w_{submrg}. Thus the new scale reading is the difference between the magnitude of the weight of the rock and the magnitude of the buoyant force on it:

$$w_{submrg} = w - F_{buoy}$$
$$= \rho_{rock}Vg - \rho_{water}Vg$$

Form the ratio

$$\frac{w_{submrg}}{w} = \frac{\rho_{rock}Vg - \rho_{water}Vg}{\rho_{rock}Vg}$$
$$= \frac{\rho_{rock} - \rho_{water}}{\rho_{rock}}$$

Solving for the density of the system (i.e., the rock),

$$\rho_{rock} = \frac{w}{w - w_{submrg}}\,\rho_{water}$$

EXAMPLE 11.8

A cylindrical object with a weight of magnitude w is placed in two different liquids; in each liquid, it floats with the axis of the cylinder vertical.* When it is placed in an unknown liquid of density ρ, it sinks to a depth h and floats; when placed in water (of known density ρ_{water}), it sinks to a depth h_{water} and floats. What is the density ρ of the unknown liquid in terms of the two depths and the density of water? This example illustrates the operation of a hydrometer, used to measure the relative density of a fluid.

Solution

A floating object sinks until it displaces a volume of liquid sufficient to create a buoyant force of magnitude equal to that of the weight of the object itself. For a cylindrical tube floating with its long axis vertical, the volume of the liquid displaced is proportional to the depth of the tube in the liquid. Indeed, if the cylinder is of cross-sectional area A and sinks to a depth h_{water} when placed vertically in water (see Figure 11.26), the volume of liquid displaced is Ah_{water}. The buoyant force on the floating cylinder thus is of magnitude

$$\rho_{water}(Ah_{water})g$$

and this is equal to the magnitude of the weight of the cylinder since it is floating.

When the cylinder is allowed to float in a liquid of density ρ, it sinks to a depth h, and the buoyant force on the cylinder now has magnitude

$$\rho Ahg$$

---

*The cylinder will need to be supported along its side to avoid capsizing. The long-axis vertical floating position is unstable; the reason for this will be examined in Section 11.7.

FIGURE 11.26

The raising of the remains of the hull of the *Mary Rose* outside Portsmouth, England.

which also is equal to that of the weight of the cylinder. Since the buoyant forces on the floating cylinder in water and in the unknown liquid are both equal in magnitude to that of the weight of the cylinder, these buoyant forces themselves have equal magnitudes:

$$\rho A h g = \rho_{water} A h_{water} g$$

Solving for the density of the unknown liquid,

$$\rho = \rho_{water} \frac{h_{water}}{h}$$

In this way the density of the unknown liquid can be found in terms of the density of water. Of course another liquid of known density also can be used instead of water.

11.7 THE CENTER OF BUOYANCY*

In July 1545 the great warship of King Henry VIII, the *Mary Rose*, was sailing forth from Portsmouth harbor in southern England in a light breeze on its way to engage a superior French fleet poised for invasion. The *Mary Rose* quite unexpectedly capsized after listing slightly and quickly sank. In horror, the king, his court, and the wife of the captain all witnessed the disaster, which took place only about a mile offshore (oops! make that about 1.6 km). Approximately 700 sailors and soldiers perished; there were only about a dozen survivors. The vessel had been newly outfitted and refitted with massive guns and armament; unfortunately the additions made the vessel unstable and it capsized when slightly rocked.[†] More recently, on 28 September 1994, the ferry *Estonia* lost its bow doors during a fierce storm in the Baltic Sea; sea water rushed in and quickly made the ferry unstable. It capsized and sank in minutes with the loss of over 900 lives; only a few passengers and crew survived.

These tragic sinkings illustrate that for floating objects to be in *stable* equilibrium, any torques about the center of mass arising from a tilting or rocking from the equilibrium position must be directed to restore or rotate the system back to its original upright floating position. If not, the system will capsize.

The stability of a system that is floating (i.e., in neutral buoyancy) depends on the relative positions of the points of application of the weight of the system and of the buoyant force. These points of application are different.

The weight acts at the center of mass point of the system. The position of the center of mass depends on the specific shape and distribution of the mass of the system. On the other hand, the buoyant force acts at the position of the center of mass of the *fluid* imagined to be in the *hole* created by the system in the fluid, a point known as the **center of buoyancy**, as shown in Figure 11.27.

FIGURE 11.27 The point of application of the weight is the center of mass C; the buoyant force acts at the center of buoyancy B.

[†]The remains of the vessel and many nautical and personal artifacts were recovered during the early 1980s and now are displayed in a fascinating museum in Portsmouth.

For a system at rest in equilibrium in neutral buoyancy, the weight and buoyant forces are collinear, having the same line of action; thus the center of mass and the center of buoyancy lie along the same vertical line. This condition ensures that there is zero torque about the center of mass of the system, and so no spin results, fulfilling the second condition for static equilibrium. But what happens if the system is tilted slightly? Under what conditions will the system return to its original, upright, floating position rather than capsize? The answer depends on whether the system is completely submerged or floating, and on the shape of the submerged hull.

Totally Submerged Systems

For simplicity, consider a system, say a submarine, with a circular cross section, as shown in Figure 11.28. The submarine is not symmetrically loaded and we represent this with a large mass at the bottom; the center of mass of the submarine is located below the geometric center of the circular cross section, say at point C. The center of buoyancy B is located at the center of mass of the fluid imagined to be in the *hole* created in the fluid by the submarine. In this case that point is at the center of the circle, point B, since the fluid is of constant density. The center of mass thus lies *below* the center of buoyancy.

If the submarine is tilted slightly, as in Figure 11.29, the buoyant force then produces a torque about the center of mass of the submarine that tends to rotate the submarine back to its original equilibrium position. Thus the original orientation of the submarine is stable.

Suppose, on the other hand, that the center of mass of the submarine is located *above* the elevation of the center of buoyancy, as in Figure 11.30. If the submarine is tilted, then in this case the torque of the buoyant force about the center of mass accentuates the tilt and the system is in unstable equilibrium; the submarine will roll over.

> Therefore, for a submerged system to be in stable equilibrium, the center of mass must be *below* the elevation of the center of buoyancy.

Floating Systems

The stability of a floating system is more complicated because the center of buoyancy may be located at different positions relative to the center of mass, depending on the shape of the system *and* the specific position in which it is floating. For example, Figure 11.31 depicts a square cross section of a level ship that has its center of mass C *above* the center of buoyancy B.

According to what we learned about submerged systems, you might think this is an unstable configuration that would result in a capsize. But if the ship is tilted slightly, as in Figure 11.32, the center of mass at C remains in the same location on the ship but the center of buoyancy *shifts* in the direction of the tilt (to the right in Figure 11.32). The triangular shaped region DAE, which was part of the *original* buoyant volume, is transferred to the new buoyant volume GAF. Since the center of buoyancy is at the center of mass of the *hole* created by the system in the fluid, the center of buoyancy must move to the right. Thus, in Figure 11.32, we see that the torque of the buoyant force about the center of mass point will *return* the ship to its original upright position; the ship is in stable equilibrium.

So what condition then determines whether a floating ship is in stable equilibrium or not?

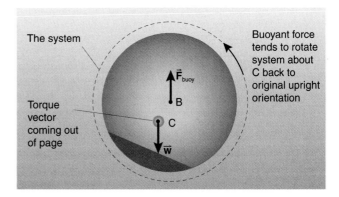

FIGURE 11.29 The torque of this buoyant force about the center of mass C will rotate the system back to the equilibrium position.

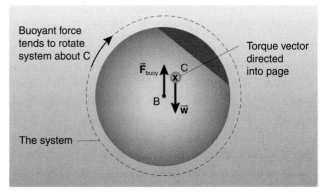

FIGURE 11.30 If the center of mass C of the submerged system is above the center of buoyancy B, the torque of the buoyant force about the center of mass will cause the system to roll directly toward equilibrium.

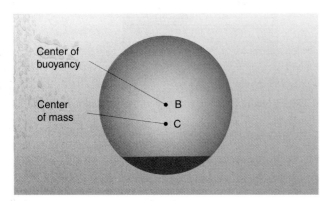

FIGURE 11.28 The center of mass C of this submerged system lies below the center of buoyancy B.

FIGURE 11.31 A cross section of a ship with its center of mass C above the center of buoyancy B.

FIGURE 11.33 When a horizontal pole is tilted, the metacenter M is above the center of mass C.

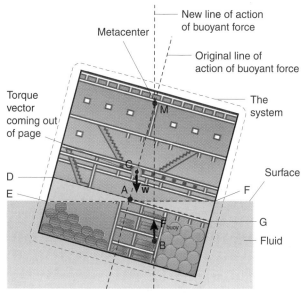

Torque of buoyant force about C rights the ship.

FIGURE 11.32 When tilted the center of buoyancy B shifts in the direction of the tilt.

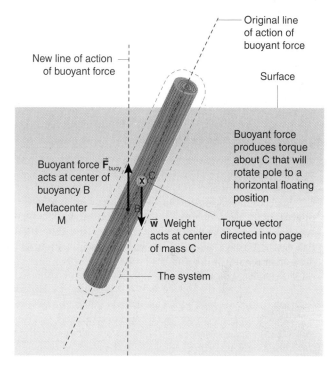

FIGURE 11.34 When a pole is vertical, the metacenter M lies below the center of mass C.

The point of intersection of the *original* line of action of the buoyant force (a line imagined to rotate with the ship when it is rocked) with the *new* line of action of the buoyant force (a vertical direction) when the system is tilted is known as the **metacenter**; see Figure 11.32.

If the elevation of the metacenter is *above* the elevation of the center of mass of the ship, the system will return to its original orientation when tilted and the floating equilibrium position is stable. If the elevation of the metacenter is *below* that of the center of mass, the system is in unstable equilibrium because the torque of the buoyant force then accentuates the tilt; the system will roll over (capsize).

This is seen in the following example.

A long, uniform, cylindrical wooden telephone pole floats with its long axis horizontal rather than vertical. When the long pole is horizontal and is tilted, as shown in Figure 11.33, we see that the location of the metacenter is above the center of mass, indicating the system will return to its original configuration.

Suppose, on the other hand, that the telephone pole is floating with its long axis vertical, as shown in Figure 11.34. If the system is tilted slightly, the location of the center of buoyancy shifts to the left of the center of mass (because the long axis of the pole is vertical). The metacenter now is *below* the center of mass. The torque of the buoyant force about the center of mass point now accentuates the tilt. The system does not return to its original vertical configuration but assumes the stable horizontal floating configuration. Thus the vertical floating position of the pole is unstable.

Hence we see that naval architects must be sure that the following condition is satisfied for stable floating equilibrium:

> A floating system is in stable equilibrium if the metacenter is located above the center of mass of the system—say, a ship— for every possible roll of the ship in heavy seas. The same is true for submerged systems.

For a sailboat and for other boats experiencing high winds, the problem is more complicated, since we must also account for the torque on the boat caused by forces arising from the wind on the sails and/or ship.

Needless to say, these principles were not known in the time of Henry VIII (who lived over a century before Newton), much to the peril of the unfortunate souls aboard the *Mary Rose*.

QUESTION 4

Given enough depth, a sinking ship (such as the *Titanic*) likely will settle on the bottom in an upright position even if it first capsizes while sinking. Explain why.

11.8 SURFACE TENSION*

Water striders (*Hygrotrechus conformis*) literally can walk on water. Many other water insects have the same ability. Such insects are not floating. They are able to execute such feats not because of Archimedes' principle but because of surface tension. Surface tension also is responsible for the spherical shape of small water droplets. More domestically, if you are careful, with a pair of tweezers perhaps you can gently lay a needle from your mending kit on top of the surface of a glass of water.

The molecules inside a liquid have molecular neighbors on all sides; molecules at the surface do not. As a result, if you imagine trying to pull a surface molecule out of a liquid, the molecular bonds with neighboring molecules are stretched and the given molecule experiences a force much like that of a stretched spring. Correspondingly, if a surface molecule is depressed slightly into the liquid, the molecule also experiences a restoring force. The surface of the liquid thus behaves much like an elastic rubber sheet. Small water droplets are spherical in shape because the restoring forces act to minimize the surface area of the drop.

If you gently pull upward on a needle lying on the surface of a glass of water using a calibrated, sensitive spring, you will find that the magnitude of the force needed is proportional to the length ℓ in contact with the liquid. The length of needle edge in contact with the liquid is *twice* the length of the needle because there is liquid on both sides of the needle. When the needle is on the verge of being pulled from the surface, the magnitude of the upward force (indicated by the extension of the spring) is equal to the magnitude of the weight of the needle (mg) plus that of the lifted liquid (which we neglect in comparison to the weight of the needle in a crude first approximation)[†] plus the magnitude $\gamma\ell$ of a *surface tension* force. The quantity γ is the magnitude of the force per unit length in contact with the liquid, known as the **coefficient of surface tension**. Values of γ for representative liquids are given in Table 11.2.

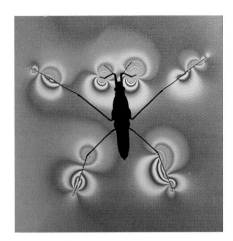

A water strider walking on water.

A more accurate method for measuring the coefficient of surface tension is the Wilhelmy slide method.[‡] A microscope slide (or other thin flat plate) is suspended from a spring or other sensitive balance and the magnitude of the weight w of the slide in air is noted. The slide is lowered slowly until it makes contact with the fluid; the scale reading then increases to w'. The two scale readings are related by

$$w' - w = \gamma p$$

where p is the perimeter of the slide and γ is the coefficient of surface tension.

QUESTION 5

Given the data in Table 11.2, why might water striders have trouble walking on acetone?

EXAMPLE 11.9

A thin, circular wire ring of diameter 4.0 cm and total mass 0.70 g is gently pulled vertically from a water surface by means

---

[†]Typically metal needles or rings are used. Since the density of metals is much greater than that of most liquids, neglecting the weight of the lifted liquid is a reasonable, although crude, first approximation. This extraction method (and that of Example 11.9) typically yields coefficients of surface tension that may have uncertainties of about 25%.

[‡]The method was invented in 1863 by Ludwig Wilhelmy (1812–1864).

TABLE 11.2 Coefficient of Surface Tension at 20 °C

Liquid	γ (N/m)
Acetone	0.024
Ammonia	0.020
Benzene	0.029
Glycol	0.048
Methanol	0.023
Water	0.073

of a sensitive spring with spring constant $k = 0.75$ N/m, as shown in Figure 11.35. When the spring is stretched 3.4 cm from its equilibrium extension in air with *no ring attached*, the ring is on the verge of being pulled free from the water surface. Find the coefficient of surface tension of water. Neglect the mass of the lifted water.

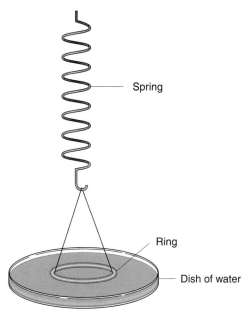

Spring

Ring

Dish of water

FIGURE 11.35

Solution

Let the wire ring be the system. When the ring is on the verge of breaking free from the surface, the magnitude of the upward force of the spring on the ring (kx) is equal to the sum of the magnitude of the weight of the ring (mg) and the magnitude of the force of surface tension ($\gamma\ell$):

$$kx = mg + \gamma\ell$$

The length in contact with the water surface is *twice* the circumference of the ring since there is fluid on both sides of the ring:

$$\ell = 2(2\pi r)$$

where r is the radius of the ring. Hence the coefficient of surface tension is

$$\gamma = \frac{kx - mg}{4\pi r}$$

$$= \frac{(0.75 \text{ N/m})(3.4 \times 10^{-2} \text{ m}) - (7.0 \times 10^{-4} \text{ kg})(9.81 \text{ m/s}^2)}{4\pi(2.0 \times 10^{-2} \text{ m})}$$

$$= \frac{0.026 \text{ N} - 0.0069 \text{ N}}{0.25 \text{ m}}$$

$$= \frac{0.019 \text{ N}}{0.25 \text{ m}}$$

$$= 0.076 \text{ N/m}$$

11.9 CAPILLARY ACTION*

The forces between a molecule of a liquid and other molecules of the liquid are called **cohesive forces**. Forces between a molecule of a liquid and other materials such as the side of a container are known as **adhesive forces**.

The **contact angle** θ_{cont} between the vertically upward direction of the wall of a container and the surface of the liquid at the wall (see Figure 11.36) gives a measure of the relative magnitudes of the cohesive and adhesive forces. When the adhesive forces are greater in magnitude than are the cohesive forces, we say the liquid **wets** the surface. In such circumstances the liquid climbs the wall of the container and produces a concave **meniscus**, like that of water around the edges of a measuring cup. The contact angle is less than 90° when wetting occurs. For water and glass, the contact angle is close to 0°.

If the cohesive forces have magnitudes greater than those of the adhesive forces, no wetting occurs and the meniscus is convex. The contact angle is greater than 90° when no wetting occurs. Liquid mercury in glass has this characteristic; the contact angle is about 140°.

When wetting occurs, surface tension will draw a liquid up into a thin tube inserted into the liquid, a phenomenon called **capillary action**; see Figure 11.37. The liquid rises into the tube until the vertical component of the surface tension

Water in a glass container has a concave meniscus with a contact angle close to 0°. On the other hand, liquid mercury produces a convex meniscus with an obtuse contact angle.

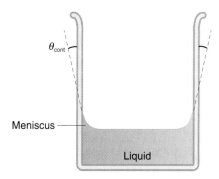

θ_{cont}

Meniscus

Liquid

FIGURE 11.36 The contact angle between a liquid and the wall of its container.

FIGURE 11.37 Capillary action.

force has a magnitude equal to that of the weight of the liquid in the column:

$$\gamma \ell \cos \theta_{cont} = mg \qquad (11.19)$$

If the column has a circular cross section of radius r, then

$$\ell = 2\pi r$$

Measure the height h of liquid in the column to the bottom of the meniscus. Neglect the small amount of liquid in the concave surface of the liquid. Then the mass of liquid of density ρ in a liquid column of height h is approximately

$$m = \pi r^2 h \rho$$

Substituting for ℓ and m in Equation 11.19 and solving for h, we find

$$h = \frac{2\gamma \cos \theta_{cont}}{\rho g r} \qquad (11.20)$$

Capillary action is what draws kerosene up into the wick of an emergency lamp or melted wax up to the flame of a candle. Capillary action also is used in the analysis of biological molecules using techniques called chromatography.

EXAMPLE 11.10

A glass tube of inside diameter 1.0 mm is placed vertically into a glass of water. How high will the water rise into the tube due to capillary action?

Solution
Assume the contact angle between the water surface and the glass wall is 0°. Use the value of the coefficient of surface tension for water given in Table 11.2. The density of water is about 1.00×10^3 kg/m³. Using Equation 11.20,

$$h = \frac{2\gamma \cos \theta_{cont}}{\rho g r}$$

$$= \frac{2\,(0.073 \text{ N/m}) \cos 0°}{(1.00 \times 10^3 \text{ kg/m}^3)(9.81 \text{ m/s}^2)(1.0 \times 10^{-3} \text{ m})/2}$$

$$= 0.030 \text{ m}$$

$$= 3.0 \text{ cm}$$

11.10 FLUID DYNAMICS: IDEAL FLUIDS

The motions of real fluids are quite complex, as the observation of a fast flowing river indicates. Turbulence, eddies, and viscous

Fuel is drawn up to the flame by capillary action in both candle strings and kerosene lamp wicks.

frictional effects make whitewater enthusiasts happy, but complicate the study of the kinematics and dynamics of real fluids. Many of these complications are best left for advanced courses in fluid mechanics.

Thus we initially confine ourselves to the kinematics and dynamics of **ideal fluids** that have the following characteristics:

1. No viscous, frictional effects are present. Thus the fluid experiences frictionless flow, analogous to the ideal of frictionless surfaces we commonly assumed in our study of mechanics.
2. The flow is steady; that is, the velocity of the fluid at every point in space is time independent. However, the velocity may have a *spatial* dependence. Different regions of the fluid may be moving at different velocities, but wherever we are, the velocity of the fluid at that point in space, whatever its value, does not vary with time.
3. The density of the fluid is constant in space and time; this is called **incompressible flow**.
4. No eddies or rotational flow are present; this condition is termed **irrotational (nonturbulent) flow**. The center of mass of a small pinwheel set loose at any point in the fluid would move, but the pinwheel itself will not spin if the flow is irrotational.

Fluid flow can exhibit complex motions, which make whitewater sports fun and exciting—and occasionally dangerous.

11.11 EQUATION OF FLOW CONTINUITY

A small particle of a flowing, incompressible fluid follows a path known as a **streamline**. For steady flow the flow rate and streamlines are static (not changing with time). Streamlines are like the paths followed by cars on a multilane freeway when the cars maintain fixed separation distances (incompressible traffic!).

Consider the set of streamlines in Figure 11.38, which delineate the motion of a fluid. The set of streamlines form a **flow tube** that may be either part of a larger volume of surrounding fluid or may represent flow inside a physical pipe. At a particular place along the flow tube, consider a section of the fluid of cross-sectional area A and differential length ds (see cross section #1 in Figure 11.38). The amount of mass dm in this small volume is the product of the density ρ and the volume $A\,ds$:

$$dm = \rho A\,ds \qquad (11.21)$$

A differential time interval dt later, the fluid has flowed just past cross section #1, as shown in Figure 11.39. Take Equation 11.21 and divide by dt, obtaining

$$\frac{dm}{dt} = \rho A \frac{ds}{dt}$$
$$= \rho A v$$

where v is the speed of the fluid. For steady flow the rate at which mass passes any given location in the flow tube must be constant. That is, if mass dm passes cross section #1 during a time dt, then the same amount of mass dm also must pass cross section #2 further along the flow tube during dt; see Figure 11.40.

FIGURE 11.38 A small amount of fluid in a flow tube.

Flow tube

Streamlines

ds

The system

Cross section #1

Cross section #1

FIGURE 11.39 Fluid flow past a cross section of a flow tube.

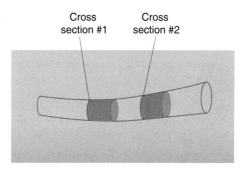

Cross section #1   Cross section #2

FIGURE 11.40 The same amount of fluid mass must flow past cross sections #1 and #2 along a flow tube during dt.

If this were not the case, then the mass would be either increasing or decreasing, leading to accumulations or losses of mass during an interval of time. Thus

$$\frac{dm}{dt}$$

must be constant for steady flow. Since

$$\frac{dm}{dt} = \rho A v$$

the product $\rho A v$ must be constant at any point along the flow tube.

Therefore, at any two points along the flow tube,

$$\rho_1 A_1 v_1 = \rho_2 A_2 v_2$$

For incompressible fluids the density is constant, and so

$$A_1 v_1 = A_2 v_2 \qquad (11.22)$$

This equation is known as the **equation of flow continuity**.

Equation 11.22 represents conservation of mass: the amount of mass per second flowing past all cross-sectional areas of a flow tube must be the same.

For such incompressible fluids, the equation of flow continuity means that the **flow rate** $Q = Av$, the volume of fluid flowing past any cross-sectional area in the flow tube during one second, is constant:

$$Q_1 = Q_2$$

The volume flow rate is measured in m³/s in SI units, or liters/s using units of convenience. One liter is a cube exactly 10 cm on a side, so that 1 liter = 10^{-3} m³.

11.12 BERNOULLI'S PRINCIPLE FOR INCOMPRESSIBLE IDEAL FLUIDS

Your experience using a garden hose with and without a nozzle (or your thumb) indicates that even an ideal fluid filling and moving through a smooth pipe can race or creep depending on changes in the cross-sectional area of the pipe, the pressure inside the pipe, and changes in elevation of the pipe itself. Here we discover the connecting principle, first formulated by Daniel

Bernoulli (1700–1782), that lets us understand the flow throughout the pipe.

Consider an incompressible ideal fluid (i.e., one with steady flow and no viscous or rotational effects), which permits us to ignore any complications from thermal effects associated with turbulent fluid flow. The work done by "other" (nonconservative) forces acting on the fluid results only in changes in the mechanical energy of the fluid, and so we can use the faithful CWE theorem of Chapter 8. Recall that this theorem states that the work done by the "other" (i.e., nonconservative) forces acting on the system, W_{other}, is equal to the change in the mechanical energy of the system (Equation 8.27):

$$W_{other} = \Delta KE + \Delta PE$$

Examine a section of fluid in a flow tube, as in Figure 11.41. Let the pressure at cross section #1 in the tube be P_1 and let the cross-sectional area of the flow tube here be A_1. Let y_1 be the average height of the tube (the height of the midpoint of the tube) above some reference level. The speed of the flow here is v_1. At cross section #2 along the flow tube, let the corresponding quantities be P_2, A_2, y_2, and v_2.

The fluid to the left of cross section #1 exerts a force $\vec{F}_1$ on the section of fluid between cross sections #1 and #2. If the fluid undergoes a change in its position vector $\Delta\vec{r}_1$, then the work done by this force is

$$\vec{F}_1 \cdot \Delta\vec{r}_1$$

Since $\vec{F}_1$ is parallel to $\Delta\vec{r}_1$, the scalar product reduces to simply

$$F_1 \Delta r_1$$

The fluid to the right of cross section #2 exerts a force $\vec{F}_2$ on the fluid located between the two cross sections #1 and #2. The fluid at cross section #2 experiences a change in position vector $\Delta\vec{r}_2$, which may be different from $\Delta\vec{r}_1$ because of the different cross-sectional areas of the flow tube. The work done by the force $\vec{F}_2$ is

$$\vec{F}_2 \cdot \Delta\vec{r}_2$$

But here the two vectors are antiparallel, so that the scalar product is

$$F_2 \Delta r_2 \cos 180°$$

or

$$-F_2 \Delta r_2$$

The work done by other pressure forces on the fluid between these two points is zero because these forces are perpendicular to the changes in the position vectors along the flow tube. These other forces are caused either by the rest of the fluid (surrounding the flow tube) or by the walls of the constraining pipe. So, gathering these results, the CWE theorem here becomes

$$F_1 \Delta r_1 + (-F_2 \Delta r_2) = \Delta KE + \Delta PE$$

The magnitude of the force $\vec{F}_1$ on the fluid at cross section #1 is the average pressure P_1 at this point times the cross-sectional area A_1;* and likewise at cross section #2: $F_2 = P_2 A_2$. With these substitutions the CWE theorem becomes

$$P_1 A_1 \Delta r_1 - P_2 A_2 \Delta r_2 = \Delta KE + \Delta PE$$

Notice that $A_1 \Delta r_1$ is the volume of the fluid pushed past cross section #1; likewise $A_2 \Delta r_2$ is the volume of fluid pushed past cross section #2. Since the fluid is incompressible, these volumes must be the same by the principle of flow continuity. Call this volume of fluid ΔV; so the CWE theorem becomes

$$(P_1 - P_2)\,\Delta V = \Delta KE + \Delta PE \qquad (11.23)$$

Fluid flow means that some fluid enters the region (between cross sections #1 and #2) at cross section #1 and some leaves this region at cross section #2. The same amount of fluid exists between the two cross sections as before, because of flow continuity. Thus the total *change* in the mechanical energy of the entire amount of fluid between the two cross sections takes place at cross sections #1 and #2. The *change* in the kinetic energy of the system is the kinetic energy of a volume ΔV of fluid at cross section #2 minus the kinetic energy of the same volume of fluid at cross section #1. That is,

$$\Delta KE = \frac{1}{2}(\rho\,\Delta V)v_2^2 - \frac{1}{2}(\rho\,\Delta V)v_1^2$$

The change in the gravitational potential energy of the fluid is the potential energy of the fluid in volume ΔV when at cross section #2 minus that at cross section #1:

$$\Delta PE = (\rho\,\Delta V)gy_2 - (\rho\,\Delta V)gy_1$$

With these substitutions, Equation 11.23 becomes

$$(P_1 - P_2)\,\Delta V = \frac{1}{2}\rho\,\Delta V\,v_2^2 - \frac{1}{2}\rho\,\Delta V\,v_1^2$$
$$+ \rho\,\Delta V\,gy_2 - \rho\,\Delta V\,gy_1$$

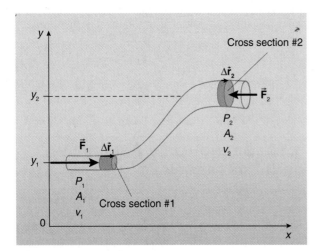

FIGURE 11.41 Fluid flow at two different cross sections along a flow tube.

---

*Since the pipe or flow tube has some finite size in the vertical dimension, there is a variation in pressure over the cross-sectional area; hence we let P_1 and P_2 be the *average* pressures over the respective cross-sectional areas.

The volume change ΔV divides out of the equation. Shuffling the terms involving cross section #1 to one side of the equation and those involving #2 to the other yields

$$P_1 + \frac{1}{2}\rho v_1^2 + \rho g y_1 = P_2 + \frac{1}{2}\rho v_2^2 + \rho g y_2 \quad (11.24)$$

Equation 11.24 means that the quantity

$$P + \frac{1}{2}\rho v^2 + \rho g y$$

has the same value at every point along the flow tube! This conservation law, essentially a consequence of the CWE theorem applied to the fluid, is known as **Bernoulli's principle** for incompressible flow of an ideal fluid.

Let's look at a few special applications.

Static Incompressible Fluids

If the fluid is not moving, both v_1 and v_2 are zero and Bernoulli's principle reduces to

$$P_1 + \rho g y_1 = P_2 + \rho g y_2$$

Call cross section #1 the reference level where $y_1 = 0$ m, so that the pressure here is P_0. Then let cross section #2 have a pressure P at an elevation y. So the previous equation becomes

$$P_0 + \rho g(0 \text{ m}) = P + \rho g y$$

or

$$P = P_0 - \rho g y$$

which is identical to what we secured before (Equation 11.14) for static incompressible fluids.

Incompressible Fluids Moving in a Horizontal Pipe

For this case (see Figure 11.42), there is no change in the gravitational potential energy of the fluid. Although some fluid in the larger-diameter pipe lowers its gravitational potential energy, an equivalent amount raises its gravitational potential energy, and so the net change is zero. Another way to look at this is that the mean elevation of both pipes is the same. So Bernoulli's principle becomes

$$P_1 + \frac{1}{2}\rho v_1^2 = P_2 + \frac{1}{2}\rho v_2^2 \quad (11.25)$$

FIGURE 11.42 A horizontal pipe with two different cross-sectional areas.

The equation of flow continuity for incompressible fluid flow, Equation 11.22, implies that the speed of flow in the small-diameter pipe must be greater than that in the larger-diameter pipe. Since $v_2 > v_1$, Bernoulli's principle (as expressed in Equation 11.25) then indicates that the pressure $P_2 < P_1$.

That is, if the speed of the fluid *increases*, the pressure *decreases*.

Solve the flow continuity equation (Equation 11.22) for v_2:

$$v_2 = \frac{A_1}{A_2} v_1$$

Substitute for v_2 in Equation 11.25:

$$P_1 + \frac{1}{2}\rho v_1^2 = P_2 + \frac{1}{2}\rho \left(\frac{A_1}{A_2} v_1\right)^2$$

This can then be solved for v_1:

$$v_1 = \left[\frac{2(P_1 - P_2)}{\rho\left(\dfrac{A_1^2}{A_2^2} - 1\right)}\right]^{1/2}$$

This means that by measuring the pressure difference between the two regions and knowing the cross-sectional areas of the two pipes, the speed of the fluid v_1 in the pipe can be determined (for a fluid of known density). A device that enables you to make such a measurement is known as a **Venturi meter**.

Lift

Bernoulli's principle also can explain, at least qualitatively, how an airplane wing develops an upward force called **lift** to oppose the downward force of gravitation. To a first approximation, the air flow around a wing is incompressible. Wings are designed so that the flow *over* the wing is faster than the flow *under* the wing, as shown in Figure 11.43.

This means that the pressure in the region above the wing is less than the pressure in the region under the wing. Thus the upward force of the air acting on the underside of the wing is greater in magnitude than that of the downward force of the air on the upper surface. As a result, there is a net upward force on the wing: lift. One aspect of aeronautical engineering is to design wing surfaces and cross sections that maximize the lift for a given

FIGURE 11.43 A wing is designed so the air flow over the wing is faster than the flow under it.

speed. The problem is not as easy as Bernoulli's principle makes it out to be (since we assumed, unrealistically, air flows like an ideal fluid), but at least you can get an inkling of how airplanes manage to get off the ground and stay aloft. Winging it (wing design) is not easy!

QUESTION 6

Explain the operation of a soda straw. Could such a straw be used to draw liquids on the Moon?

EXAMPLE 11.11

A desperado sneaks into town and at the train depot shoots a small hole in a full, open water tank at a depth h below the water surface, as shown in Figure 11.44. What is the speed of the water as it exits the hole?

FIGURE 11.44

Solution

Choose a coordinate system with the origin at the level of the hole, as shown in Figure 11.45. Use Bernoulli's principle at the top of the tank and the hole:

$$P_{\text{top}} + \frac{1}{2}\rho v_{\text{top}}^2 + \rho g y_{\text{top}} = P_{\text{hole}} + \frac{1}{2}\rho v_{\text{hole}}^2 + \rho g y_{\text{hole}}$$

Since the hole is small, the vertical speed of the fluid at the top of the tank is negligible compared with the speed of the fluid exiting the hole, so take $v_{\text{top}} = 0$ m/s. Since both the top of the tank and the hole are open to the atmosphere, $P_{\text{top}} = P_{\text{hole}} \approx 1.00$ atm $= 1.01 \times 10^5$ Pa. With the coordinate system chosen, $y_{\text{top}} = h$ and $y_{\text{hole}} = 0$ m. With these substitutions, you have

$$\rho g h = \frac{1}{2}\rho v_{\text{hole}}^2$$

Solving for v_{hole}, you find

$$v_{\text{hole}} = (2gh)^{1/2}$$

FIGURE 11.45

Notice that the *horizontal* speed of the liquid emerging from the hole is the same as the *vertical* speed it would have as a result of a free-fall from rest through a height h. This observation is known as **Torricelli's law**, after Evangelista Torricelli (1608–1647), who also invented the barometer (see Example 11.3).

11.13 NONIDEAL FLUIDS*

Slow as molasses in January

Southern folk saying

Turn an open bottle of soda pop upside down and the fluid quickly runs out all over you and your textbook in just a few seconds. On the other hand, a comparable bottle filled with molasses will empty itself much more slowly if turned upside down, but still creates a real mess. We say the molasses is more *viscous* than the soda pop. The **coefficient of viscosity** of a fluid is a measure of its resistance to flow. Here we will see how this coefficient of viscosity can be measured.

Consider a thin fluid layer, as in Figure 11.46. A plate of area A, separated by a small distance Δy from the bottom of the

FIGURE 11.46 Fluid flow velocities between a moving and fixed surface.

layer of fluid, is dragged across the fluid at constant speed v. Experiments indicate that the fluid immediately adjacent to the moving plate moves essentially at the speed of the plate, while the fluid immediately adjacent to the bottom of the layer has essentially zero speed. Therefore there exists a speed change Δv across the thin fluid layer of thickness Δy.

Plane layers of fluid parallel to the plate slide over each other and exert friction-like, viscous forces on each other. The existence of the viscous forces means that an applied horizontal force of constant magnitude F is needed to drag the upper moving plate at constant speed, much as a constant applied force is needed to drag an object along a rough surface at constant speed because of friction. Experiments indicate that for many fluids, the magnitude F of the applied force needed to drag the moving plate at constant speed v is proportional to the speed and to the area of the plate but inversely proportional to the separation of the plates as long as the separation distance Δy is small. This empirical dependence is written as

$$F = \eta \, \frac{A \, \Delta v}{\Delta y}$$

where $\Delta v = v - 0$ m/s $= v$. The proportionality constant η depends on the particular fluid used and is called the coefficient of viscosity of the fluid. Thus the coefficient of viscosity η is

$$\eta = \frac{F}{A} \, \frac{\Delta y}{\Delta v}$$

or

$$\eta = \frac{F/A}{\Delta v / \Delta y} \qquad (11.26)$$

The numerator of Equation 11.26 is the shear stress applied to the fluid, while the denominator, the variation in the speed of the fluid over a small distance measured perpendicular to the direction of the flow, is called the speed gradient.[†]

Inspecting Equation 11.26, the SI units for the coefficient of viscosity[‡] are $(N/m^2)/[(m/s)/m] = N \cdot s/m^2 = Pa \cdot s$. Table 11.3 indicates the value of the coefficient of viscosity for various fluids under various conditions.

11.14 Viscous Flow*

When a fluid flows through a long horizontal pipe of constant cross-sectional area, as in Figure 11.47, Bernoulli's equation (Equation 11.24) implies that pressure is constant along the length of the pipe. For real fluids, however, the pressure decreases along the direction of the flow, as shown in Figure 11.48. This has significant ramifications in the firefighting profession if long lengths of hose are used, or in providing water to a house at the end of a long supply pipe.

[†]The term gradient arises by making an analogy with a mountain slope. We say there is a steep gradient if the slope makes a considerable angle with the horizontal.

[‡]An old unit for the viscosity that still appears in some reference manuals is the poise. The conversion is

$$1 \text{ Pa} \cdot \text{s} = 10 \text{ poise}$$

The poise unit was named after Jean Louis Marie Poiseuille (1799–1869), a French physician *and* physicist.

TABLE 11.3 Coefficient of Viscosity of Various Liquids

Liquid	Coefficient of viscosity η (Pa·s)
Benzene (20 °C)	0.65×10^{-3}
Ethanol (20 °C)	1.2×10^{-3}
Glycerin (20 °C)	1.41
Mercury (20 °C)	1.6×10^{-3}
Methanol (20 °C)	0.60×10^{-3}
Molasses (20 °C)	~100
Water (0 °C)	1.8×10^{-3}
Water (20 °C)	1.0×10^{-3}
Water (100 °C)	0.28×10^{-3}
Glacial ice	12×10^{11}
Blood (whole) (37 °C) [average values]*	
Newborns	5.5×10^{-3}
Men	4.6×10^{-3}
Women	4.0×10^{-3}
Sweat*	0.93×10^{-3}
Tears*	1.3×10^{-3}
SAE 10 motor oil (30 °C)	200×10^{-3}

Geigy Scientific Tables (1981); all other data from *Handbook of Chemistry and Physics* (34th edition).

The pressure drop occurs because of *viscous forces*, exerted by adjacent layers of the fluid on each other as they rub or flow past each other. The flow is not uniform across the cross section of the pipe. For a fluid confined to a pipe, the fluid flows with greatest speed at the center of the pipe and essentially zero speed at the walls. Fluid flow in a straight river is similar: the current is greatest at midstream and decreases as you approach the banks.

Experimental observations indicate that the absolute value of the pressure drop $|\Delta P|$ along the direction of flow is proportional to the volume flow rate Q (in m³/s), the length ℓ of the pipe, and the coefficient of viscosity η of the fluid and further depends the shape of the cross section of the pipe:

$$|\Delta P| = KQ\eta\ell$$

where K is a geometrical factor depending on the cross-sectional shape of the pipe.

For circular pipes, Jean Louie Marie Poiseuille was able to show that[§]

$$|\Delta P| = \frac{8}{\pi r^4} Q\eta\ell$$

now known as **Poiseuille's law** for simple fluids undergoing steady, nonturbulent flow and whose viscosity does not depend on the speed of the flow. The very strong dependence on the radius of the pipe has implications in medicine. If the radius of an artery gradually decreases with time due to arteriosclerosis, the absolute value of the pressure drop along an artery increases dramatically. To maintain the same flow rate the heart (a biological pump) must do more work per unit time (power), a condition that may result in heart failure.

[§]A derivation of this relationship can be found in Leonard Leyton, *Fluid Behavior in Biological Systems* (Oxford, 1975), pages 198–199.

FIGURE 11.47 For an ideal fluid, the pressure is constant along the length of a pipe.

FIGURE 11.48 The pressure of a real fluid decreases gradually along the length of a pipe.

In fact blood is not a simple fluid of constant viscosity. Blood consists of a variety of distinct ingredients, among them biconcave disklike red blood cells (erythrocytes, approximately 8 μm in diameter and 2 μm thick) that make up about 50% of the total volume of the blood. The red blood cells are suspended in a yellowish-clear, cell-free plasma fluid.* The red

blood cells are randomly oriented at low speeds but the planes of the disks line up at higher flow speeds. As a result the viscosity of blood at higher speeds is smaller than at lower speeds. Flow speeds in the human circulatory system range from about 0.1 m/s in major arteries (1–10 mm in diameter) down to about 10^{-3} m/s in capillaries (about 8 μm in diameter). Other blood constituents, such as the white blood cells and platelets, make up only a small volume fraction and do not significantly affect the flow properties.

---

*Blood plasma is *not* the same as the state of matter a physicist calls a plasma.

CHAPTER SUMMARY

The *tensile stress* applied to a rope or wire of cross-sectional area A is the magnitude of the tensile force divided by the cross-sectional area:

$$\text{tensile stress} = \frac{\text{magnitude of the tensile force}}{\text{cross-sectional area}}$$

The *tensile strain* of a rope or wire of length ℓ, when subjected to a tensile force, is the change in length $\Delta\ell$ divided by ℓ itself:

$$\text{tensile strain} = \frac{\Delta\ell}{\ell}$$

Young's modulus E for a solid material is the tensile stress divided by the tensile strain:

$$E \equiv \frac{\text{tensile stress}}{\text{tensile strain}} \qquad (11.3)$$

The *pressure P* in a fluid is the magnitude of the force per unit area on any surface in any orientation within or surrounding the fluid. The pressure is a scalar quantity whose SI unit is the *pascal* (Pa), equal to one newton per square meter (N/m^2).

The pressure P at any vertical coordinate y in a static liquid of constant density ρ near the surface of the Earth is

$$P = P_0 - \rho g y \qquad (\hat{\jmath} \text{ up}) \qquad (11.14)$$

where P_0 is the pressure at the elevation where the coordinate y is chosen to be zero (the origin); g is the magnitude of the local acceleration due to gravity near the surface of the Earth (g = 9.81 m/s^2).

Pascal's principle states that if the pressure is increased at some point in an enclosed liquid, the pressure increases by the same amount at all points within the liquid.

Archimedes' principle states that if a system is partially or totally submerged in a fluid, the fluid exerts an upward buoyant force on the system whose magnitude is equal to that of the weight of the fluid displaced by the system. The buoyant force acts at the *center of buoyancy*, located at the center of mass of the fluid imagined to fill the hole created in the fluid by the system.

A submerged system is in stable equilibrium if the elevation of its center of mass is below that of the center of buoyancy.

If a floating system is tilted slightly from its equilibrium position, the intersection of the new line of action of the buoyant force with the original line of action of the buoyant force is known as the *metacenter*. A floating (or submerged) system is in stable equilibrium if the elevation of the metacenter is above that of the center of mass of the system.

Surface tension forces account for capillary action.

An *ideal fluid* is nonviscous and its flow is steady and irrotational (nonturbulent). The flow rate is the product of the cross-sectional area A of a flow tube and the fluid speed v at that location. If the fluid also is incompressible, the flow rate Q of such an ideal fluid is constant:

$$A_1 v_1 = A_2 v_2 \qquad (11.22)$$

which is known as the *equation of flow continuity*. Such ideal incompressible flow also satisfies *Bernoulli's principle*:

$$P_1 + \frac{1}{2} \rho v_1^2 + \rho g y_1 = P_2 + \frac{1}{2} \rho v_2^2 + \rho g y_2 \quad (11.24)$$

which follows from applying the CWE theorem to the fluid in a flow tube.

The *viscosity* of a real fluid is a measure of its resistance to flow. The *coefficient of viscosity* is the shear stress applied to a fluid divided by the speed gradient (the variation of fluid speed in a direction perpendicular to its flow).

SUMMARY OF PROBLEM-SOLVING TACTICS

11.1 (page 497) You have the freedom to choose where to place the origin of the coordinate system when using Equation 11.14:

$$P(y) = P_0 - \rho g y \qquad (\hat{j} \text{ up}) \qquad (11.14)$$

You simply need to have $\hat{j}$ directed up. You can choose any convenient elevation for the origin. The pressure where y = 0 m is P_0.

QUESTIONS

1. (page 494); 2. (page 498); 3. (page 504); 4. (page 508); 5. (page 508); 6. (page 514)

7. To demonstrate the effect of increasing pressure with depth, Torricelli took a sturdy wine cask, which, when filled with water, held fast as in Figure Q.7a. He then connected a thin tube to it (as in Figure Q.7b) and filled it to a great height, bursting the cask to the astonishment of a surrounding audience, but not to Torricelli—or to you. Explain why. Why did he use water and not wine?

8. Mercury is a very toxic and hazardous material, yet barometers invariably use mercury rather than water as the working substance. What are two reasons that such water barometers are not used in practice?

9. In Example 11.4 we found that the scale height of the atmosphere measured from the surface of the Earth is 8.00 km. If you place an origin on top of Mt. Everest, is the numerical value of the scale height the same? Explain your answer.

10. Using the technique of Example 11.8, how can the density of the unknown liquid be determined if the cylinder *sinks* in the unknown liquid?

11. A rock of volume V is suspended as indicated in Figure Q.11a. The weight of the rock and its support is equal to the weight of the water and the beaker on the opposite pan of the balance. Describe what happens to the balance if the rock is lowered into the water as indicated in Figure Q.11b.

12. A swimmer floating in the Dead Sea (located between Jordan and Israel) floats higher out of the water than in the nearby Mediterranean Sea. What does this imply about the relative densities of the water in the two seas?

FIGURE Q.7

13. Typical large icebergs are made of frozen fresh water. Look up the density of freshwater ice and that of sea water in a reference such as *The Handbook of Chemistry and Physics*. Explain why approximately 90% of the mass of an iceberg lies submerged. This gives rise to the colloquial expression about seeing only the "tip of the iceberg."

14. If you make the observation that very little of the human body floats out of the water when doing a dead-man's float, what is the approximate average density of the human body?

15. Explain the operation of a perfume atomizer.

Rock

(a)

Rock

(b)

FIGURE Q.11

FIGURE Q.23

FIGURE Q.24

16. Explain why airplanes take off going *into* the wind rather than *with* the wind. When landing, do planes head into or with the wind?

17. Two cubes are of identical dimensions. One is made of granite and the other of lead. When submerged in water, what is the relationship between the buoyant forces on each cube?

18. To carry large panes of glass, suction cups are used to attach handles to the glass surface. Explain the operation of a suction cup.

19. A woman of mass 50 kg wears spiked high-heel shoes. Making reasonable estimates, calculate the pressure of each heel on the floor and compare it with that exerted by the feet of an elephant.

20. Explain why a ship made of steel (which is much denser than water) can float.

21. You are relaxing peacefully on a foam float or boogie board in a fraternity hot tub and note the level of the water in the tank carefully. Now you slip into the water and float beside the board. What happens to the level of the water in the tank? Explain your answer.

22. In a shower with the water on full blast, the bottom of the shower curtain is drawn slightly into the shower rather than hanging vertical (as it will when the water is off). Explain why.

23. Using a blowdryer, you can suspend a ping-pong ball in midair. Explain the physics. See Figure Q.23.

24. Tall, old farm silos were constructed of vertical timbers with steel bands wrapped around the periphery. Explain why the spacing of the bands is greatest at the top of the silo and decreases as the bottom of the silo is approached. See Figure Q.24.

25. Is Archimedes' principle applicable in a spacecraft orbiting the Earth? Explain carefully.

26. One pail of water is filled to the brim with water. Another identical pail of the same mass has a wooden toy ship floating in it and it also is filled to the brim with water. Both pails are weighed. What is the relationship between the weights of the two pails?

27. An oil barge floats with its bottom 3.0 m below the surface of a calm ocean. As the barge is towed from the ocean up a large river, will the barge float at the same level, higher in the water, or lower in the water? Explain.

28. After your election to the prestigious ΦBK academic honor society, your mother shows you her cherished ΦBK key. She cannot remember if the key is solid gold or gold-plated steel. With your growing knowledge of physics, describe an experiment to determine from which material the key was made.

29. A smoky fire is burning in a fireplace. Explain why the smoke is drawn up the chimney better if there is a wind blowing horizontally across the top of the chimney.

30. Astronomy books often state (ridiculously) that if there were an ocean big enough (there really isn't), the planet Saturn

could float in it. What does this imply about the density of Saturn?

31. Explain why logs float with their long axis horizontal rather than vertical.

32. Under what circumstances will a damaged and sinking ship capsize before sinking?

33. Why do sailboats and other vessels have massive keels?

34. Kayaks tip over easily; rowboats do not. Explain why.

35. Modern aircraft carriers look decidedly top heavy, yet do not capsize even in heavy seas. Explain how this is accomplished. See Figure Q.35.

FIGURE Q.35

36. Describe a procedure and the equipment needed to determine the volume of an irregularly shaped object by immersing it in a liquid. What assumption must be made about the object for the procedure to be valid?

37. Examine a hydraulic car jack (common at many service stations) and explain the physics associated with its operation.

38. When pitching curve balls in baseball, does the pitcher give the ball a spin about a horizontal or vertical spin axis? Explain your answer.

39. When hitting golf balls, is it desirable to give the ball a spin around a vertical or a horizontal axis? Does this give greater range to the ball?

40. Is Equation 11.14,

$$P = P_0 - \rho g y$$

valid in an orbiting spacecraft with a suitably adjusted value for g? Explain your answer.

41. Describe an experimental procedure that will determine whether the crown of King Hieron II was made of pure gold.

42. A cylindrical steel barrel is of diameter 0.50 m and height 1.0 m. Describe how you could cause the air around the barrel to collapse it.

43. A jumbo jet is cruising at an altitude of 10.7 km. The cabin pressure is that at sea level. Is there a buoyant force on the jet? Explain.

44. Explain the physics involved in the operation of a water pistol.

45. Why is it so easy to get a painful paper cut?

46. What principle is involved in squeezing the toothpaste out of a tube?

47. Take a pin and insert it through the center of an index card. Insert the pin in the end of a spool of thread as shown in Figure Q.47. The purpose of the pin is simply to keep the index card centered on the spool. Now blow hard through the opposite end of the spool. Explain why the index card is *not* blown off the other end; indeed, the harder you blow, the more difficult it is to move the index card from the other end.

FIGURE Q.47

48. Cork, of course, floats easily on water. A shipment of 10^6 corks is loaded into a freighter for transport to a winery. Does the boat float higher or lower in the water as a result of its highly buoyant cargo? Explain the reasoning behind your answer.

49. If the value of the local acceleration due to gravity g is increased, will the depth to which a ship sinks into the water before floating also change? If so, will the ship sink to a greater or lesser depth? Justify your answer.

50. Explain why the canvas top of a speeding convertible bulges outward.

51. Because of the viscosity of air, spinning baseballs, tennis balls, golf balls, and Ping-Pong balls drag some of the air close to the ball around with it. Left-handed and right-handed pitchers spin the ball in different directions around a vertical axis for a curve ball. What direction is the angular velocity vector of the ball for each pitcher? Which way will the ball curve for each pitcher as the pitcher looks at home plate? If a golf shot hooks to the right, discuss the likely direction of the spin angular velocity vector of the ball.

52. Blood consists of fluid plasma in which various cells are carried. Blood near the walls has a smaller speed than blood near the center of an artery or vein. Explain why this speed profile tends to push large red blood cells toward the center of the arteries and veins.

53. After you wax your car, and it inevitably rains within the hour, do the water droplets on it indicate that cohesive or adhesive forces dominate? What about when there is no wax on your car finish?

PROBLEMS

Sections 11.1 States of Matter
11.2 Stress, Strain, and Young's Modulus for Solids

1. A vertical steel wire with a diameter of 1.0 mm is 2.00 m long. What mass should be hung from the wire to stretch it 1.0 mm?

2. A stool in a diner is supported by a solid steel column of diameter 4.0 cm and length 65 cm. When a 90 kg trucker sits on the chair with his feet off the floor, determine the compression of the column. Neglect the mass of the seat of the stool.

3. A swing is suspended by means of two nylon ropes from a tree limb as indicated in Figure P.3. The ropes have a diameter of 1.0 cm and are each 8.0 m long. When a passing 25 kg tyke gets on the swing, the ropes stretch 1.0 cm. Use this information to find Young's modulus for the nylon ropes.

8.0 m

FIGURE P.3

4. A small grouper fish (*Epinephelus striatus*) of mass 2.5 kg hangs from a vertical 1.50 m length of fishing line and stretches it 1.2 cm. The diameter of the line is 0.50 mm. What is Young's modulus for the line?

•5. A 2.00 m long bungee cord of 8.0 mm diameter stretches 250 cm when an 80 kg college-level turkey is suspended from a hook on one end of the cord. Find Young's modulus for the cord.

•6. The well-known civil engineering firm of Rivers, Rhodes, and Waters is designing a walkway that is to be suspended by means of four steel rods of diameter 2.0 cm and length 3.00 m. The stretch of the rods is not to exceed 0.50 cm under any circumstances. What is the maximum additional mass that the 1000 kg walkway can carry to meet the design specifications? Assume that the load is uniformly distributed over the walkway so that each rod carries an equal share of the load.

•7. A 1.00 m³ volume of water is subjected to a pressure of 1.08×10^8 Pa. What is the change in the volume of the water, expressed in cm³? This pressure is approximately the pressure at the deepest part of the ocean in the Mariana Trench (approximately 11 km deep) off the southeastern coast of Japan.

•8. What pressure increase will decrease the volume of a sample of water by 1.0%?

‡9. A mass m is suspended from a fixed support by means of a wire of diameter d and length ℓ made from material whose Young's modulus is E. Show that the frequency ν of small vertical oscillations is

$$\nu = \left(\frac{Ed^2}{16\pi\ell m} \right)^{1/2}$$

Sections 11.3 Fluid Pressure
11.4 Static Fluids
11.5 Pascal's Principle

10. (a) What is the pressure at the surface of a swimming pool? (b) What is the pressure at the bottom of a filled swimming pool with a depth of 2.50 m? (c) What is the gauge pressure at the top and at the bottom?

11. At what depth does a diver experience a gauge pressure of 1.00 atm?

12. The Mariana Trench in the Pacific Ocean off the southeastern coast of Japan has water to approximately 11 km deep. Calculate the pressure in atm at the bottom of the trench.

•13. Assume your mass is 60.0 kg and your two bare footprints have a total area of about 4.0×10^{-2} m². How high a water column produces a gauge pressure on the floor equal to the pressure you exert on the floor?

•14. Normal atmospheric pressure is 1.00 atm = 760 mm Hg. A meteorologist finds that a storm decreases the pressure by 25 mm Hg. What is the atmospheric pressure inside the storm in pascals?

•15. A vertical water pipe extends from the bottom to the top of the World Trade Center in the Big Apple (New York City), a distance of about 420 m. Calculate the difference in pressure in atm between the bottom and the top of the water pipe.

•16. Hydraulic equipment makes use of Pascal's principle to create what is called a hydraulic lever, to magnify or increase forces; see Figure P.16. Use the CWE theorem to show that if the place where a force of constant magnitude F_1 is applied moves slowly through a distance d_1, the place where force F_2 (also of constant magnitude) is applied moves through a distance d_2 given by

$$d_2 = \frac{F_1}{F_2} d_1$$

Assume the fluid is incompressible.

•17. Imagine the atmosphere of the Earth to be incompressible and therefore of constant density (similar to a liquid) equal to its value at sea level (about 1.30 kg/m³). (a) What total depth of atmosphere would yield a pressure at sea level equal to the present value of 1.00 atm? (b) Mt. Everest has a height of about 8.85 km. Would Mt. Everest poke through this uniform atmospheric blanket? If so, by how much?

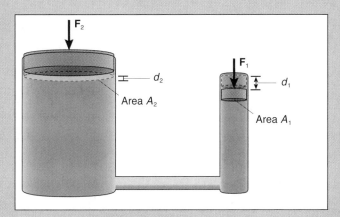

FIGURE P.16

•18. With a quantitative calculation, explain why even with the best efforts, it is not possible to use a soda straw longer than about 10.3 m for sipping water. How is this length determined?

•19. In 1654 Otto von Guericke, the inventor of the air pump, made a spectacular demonstration of his new device before the Emperor Ferdinand III. Von Guericke took two hollow hemispheres of copper that could be fitted snugly to a gasket between them (see Figure P.19). He then used his air pump to evacuate the air from inside the sphere formed from the two hemispheres so that the atmospheric pressure held them together, preventing them from falling apart. Thirty horses, fifteen pulling on each hemisphere, were unable to tear the hemispheres apart. Assume the hemispheres used were 1.00 m in diameter and that Guericke was able to completely evacuate the interior. (a) Find the magnitude of the force of the atmosphere on each hemisphere. (b) What maximum pull per horse would still be incapable of separating the hemispheres? Hemispheres that can be so evacuated now are known as *Magdeburg hemispheres* after the town of von Guericke's birth and where he was burgomaster. Von Guericke's invention of the air pump was such a sensation that it is now immortalized as a constellation called *Antlia*, or Air Pump, in the southern celestial hemisphere.

•20. You and five classmates are drinking by means of straws from a large tankard of a beverage derived from hops and other assorted spices and ingredients. The beverage has essentially the density of water. The level of liquid in the various straws is as indicated in Figure P.20. The levels are stationary and the liquid is not in motion. The levels of liquid in the straws are lifted or depressed the same distance d relative to the open liquid surface. The distance d corresponds to a pressure difference of 1.00×10^{-2} atm. (a) Calculate the value of d. (b) The pressures in atmospheres at various locations are denoted by the symbols P_1, P_2, P_3, and so forth. Indicate what the following pressures or pressure differences are (in units of atmospheres); the *signs* are important:

$$P_1 =$$
$$P_2 - P_1 =$$
$$P_3 =$$
$$P_5 - P_4 =$$
$$P_5 - P_6 =$$
$$P_7 - P_1 =$$
$$P_8 - P_1 =$$

•21. A hydraulic jack is filled with an oil of density 0.80 times that of water ($\rho_{water} = 1.00 \times 10^3$ kg/m³). The input piston has a cross-sectional area that is 1/10 that of the output piston; see Figure P.21. In its operation: (a) Calculate the ratio of the pressures at the output and input pistons. (b) Calculate the ratio of the magnitudes of the forces at the output and input pistons.

FIGURE P.19 A drawing of the Guericke demonstration. It depicts an incorrect number of horses pulling on each hemisphere.

FIGURE P.20

FIGURE P.21

•22. The gauge pressure of the air in an automobile tire is 2.08×10^5 Pa.[†] The tire supports one-fourth of the weight of a car of mass 1.50×10^3 kg. What is the area of the tire that is in contact with the road? This area is called the footprint of the tire. As a tire goes flat, explain what must happen to the footprint of the tire.

•23. A barber chair is supported by a piston that is 6.0 cm in diameter. The chair and its occupant have a total mass of 100 kg. (a) What pressure is required to raise the chair and its occupant? (b) If the pressure is produced by a plunger whose diameter is 1.5 cm, what force is needed to accomplish the task?

•24. A cylindrical steel barrel has a diameter of 0.50 m and a height of 1.00 m. Calculate the magnitude of the maximum force that air pressure can exert on the entire barrel from the outside.

•25. The Italian scientist Evangelista Torricelli originally made a barometric device such as examined in Example 11.3 with water instead of mercury. What height of water column corresponds to one atmosphere pressure?

‡26. A right-circular cone of radius R and height h (see Figure P.26) is filled with a fluid whose density is ρ. The top of the cone is exposed to the atmosphere. (a) Determine the magnitude of the weight of the fluid in the cone. (b) What is the gauge pressure at the bottom of the fluid? (c) What is the magnitude of the force of the fluid on the container's bottom caused by the gauge pressure? By how much is this greater than the magnitude of the weight of the fluid in the cone? Explain and resolve this paradoxical result, first noted by Simon Stevin (1548–1620). This problem was inspired by the article by Alpha E. Wilson, "The hydrostatic paradox," *The Physics Teacher*, 33, #8, pages 538–539 (November 1995).

[†]This corresponds to 30 pounds per square inch (psi). Common tire pressure gauges in the United States indicate the gauge pressure in pounds per square inch.

FIGURE P.26

Sections 11.6 Archimedes' Principle
11.7 The Center of Buoyancy*

27. A 5.0 kg stone with a density of 5.3×10^3 kg/m³ is placed on a scale when totally submerged in water. What does the scale indicate (in kg)?

28. A boulder of granite of mass 200 kg is weighed in air and then when submerged in a pool of water. What is the difference in the readings of the scale used to perform the measurements? ($\rho_{granite} = 2.7 \times 10^3$ kg/m³.)

29. A cube of wood 15.0 cm on an edge floats in water (density 1.00×10^3 kg/m³) with 2.5 cm of a vertical edge above the surface. (a) Draw a second law force diagram indicating all the forces acting on the cube. (b) Determine the density of the wood.

30. A block with a density of 550 kg/m³ and a volume of 0.600 m³ is held completely under water by means of a rope. Determine the magnitude of the force of the cord on the block.

•31. What is the magnitude of the buoyant force on a cube of iron 0.100 m on a side when it is totally submerged in a pool of mercury? The density of iron is 7.9×10^3 kg/m³ and that of mercury is 13.6×10^3 kg/m³.

•32. A goldsmith alleges that a beautiful bracelet is made of links of pure gold. The mass of the bracelet is easily determined to be 85.00 g. You are a bit suspicious of the pure gold claim and decide to check it out. Quite by chance you happen to have a small graduated cylinder in your briefcase. By submerging the bracelet in water, you determine the bracelet has a volume of 6.1 cm³. Should you patronize the merchant or call the fraud squad? Explain the reasoning that leads to your conclusion. (Gold has a density of 19.3×10^3 kg/m³.)

•33. An olive is perched atop a cubical ice cube 3.0 cm on a side so that the cube barely floats in a martini; see Figure P.33. Consider the martini to be quite dilute (essentially water). What is the mass of the olive?

FIGURE P.33

•**34.** An object of density ρ (> ρ_{water}) and volume V contains a cavity of air of volume V'. Volume V does *not* include V'; see Figure P.34. How big must V' be (in terms of V, ρ_{water}, and g) so the object barely floats in water? This is the principle underlying the swim bladders of many fish.

FIGURE P.34

•**35.** The 40.0 kg steel ball of a Foucault pendulum is suspended in air as shown in Figure P.35. (a) Draw a second law force diagram indicating the forces on the ball. (b) Determine the magnitude of the tension in each cord. (c) If the system were submerged in water, draw a second law force diagram indicating the forces on the ball and determine the tension in each cord. Do you need to know the density of steel to do this calculation?

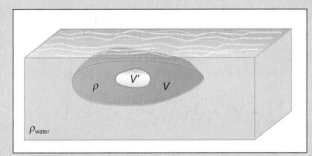

FIGURE P.35

•**36.** Model a prone human being as a rectangular box of dimensions 15 cm × 20 cm × 180 cm with the density of water, as indicated in Figure P.36a. (a) If the model then is placed in a huge pool of mercury, to what depth will the model sink? This may give you an indication of the extent to which you can float on mercury rather than on water. (b) If the model is placed in the mercury pool in the orientation indicated in Figure P.36b, to what depth will the model sink?

•**37.** A rectangular barge has a depth of 1.00 m, a floor area of 40.0 m², and a mass of 1.00×10^4 kg. (a) Will the barge float on water? Prove it. (b) If the barge will float, what is the maximum mass of cargo that the barge can carry (barely) without sinking?

FIGURE P.36

•**38.** One way of moving a large undersea rock or mooring is to fasten a closed, empty oil drum to the rock while at low tide as in Figure P.38a. As the tide comes in, the drum is drawn down into the water (see Figure P.38b), increasing the buoyant force acting on the system of the rock and drum. If the drum is of sufficient size (or if multiple drums are used), they are capable of lifting the rock or mooring, so that it can easily be floated to another location and released. Suppose a cylindrical drum of radius 25 cm, length 100 cm, and mass 20 kg is used. What is the maximum mass of a rock (of density 5.0×10^3 kg/m³) that the drum can lift from the bottom? Assume the density of sea water is the same as that of pure water.

FIGURE P.38

•**39.** A solid student in the shape of a rectangular parallelopiped of dimensions 30.0 cm × 20.0 cm × 2.000 m floats horizontally in a calm hot tub filled with warm water; 3.00 cm of the parallelopiped (along the 20.0 cm side) is above the water surface. Find the average density of the student.

•**40.** A cubical block of a solid material is suspended from a spring balance over a vat of water as indicated in Figure P.40a. The block and scale are gradually lowered into the vat until the block is completely submerged. The scale reading as a function of distance x is graphed in Figure P.40b. (a) What is the magnitude of the weight of the cube in air? (b) How far was the block lowered before it came into contact with the water? (c) What is the length of a side of the cube? (d) What is the density of the cube in kg/m³? (e) What is the maximum magnitude of the buoyant force on the cube? (f) What is the magnitude of the weight of the water displaced by the object?

FIGURE P.40

•**41.** A cork of volume 3.00×10^{-5} m³ and density 120 kg/m³ floats on water. No surprises here. (a) Sketch a second law force diagram indicating the forces acting on the cork as it floats. (b) What fraction of the volume of the cork is submerged as it floats? (c) The cork now is tied, using an ideal string, to a suspended piece of aluminum (of density 2.70×10^3 kg/m³) of such mass that the cork is just completely submerged without sinking. What mass of aluminum is needed? (d) What is the magnitude of the tension in the string connecting the cork and aluminum when the cork is just barely completely submerged?

•**42.** Archimedes, prove your stuff. A royal crown (cola?) of pure gold is suspended in air by a thread and the tension in the thread is of magnitude T. When the crown is totally immersed in water, the magnitude of the tension in the thread is $0.872T$. (a) Sketch a second law force diagram indicating the forces acting on the crown when suspended in air. (b) Sketch a second law force diagram indicating the forces acting on the crown when suspended in water. (c) Is the crown made of pure gold or of another metal with just a gold plating? (The density of pure gold is 19.3×10^3 kg/m³.) Justify your answer.

•**43.** You are strapped to a large floating cushion, enjoying the pool at a luxury resort in Tahiti. The float is submerged to half its depth. A friend (?) approaches from underneath and playfully upsets the cushion, flipping you and it over. You are still strapped to the cushion and now are in danger of drowning. Is the float still submerged to half, less than half, or more than half its thickness? Indicate your reasoning.

•**44.** A sample of pure gold is found to have a scale weight when submerged in water that is 95% of its weight in air. From this information, calculate the ratio of the density of gold to the density of water (this ratio is called the *specific gravity* of the substance).

•**45.** Liquid mercury has a density of 13.6×10^3 kg/m³. A heavy metal band floats on mercury with 80% of its volume above the surface. Find the average density of the heavy metal band.

•**46.** A huge, 8.00 m long, uniform log of wood (density 700 kg/m³) of mass 1000 kg is forced to float vertically in a clear mountain lake because of a massive iron ball attached to one end. Only 40 cm of the log remains above the surface of the lake. What mass of iron is attached to the end?

•**47.** You are idly fishing while sitting on the bank of a river. An empty soda can floats by and you decide to practice physics. The can has a mass of 20.0 g and a volume of 200 cm³. You intend to just barely sink the can by adding an appropriate mass of sand while the can has the orientation shown in Figure P.47. (a) Is it necessary to know the density of sand to determine the mass of sand to add? Explain. (b) Find the minimum mass of sand needed to sink the can, without tipping it over. Recover the can so you don't litter!

FIGURE P.47

•**48.** You (mass 70.0 kg) decide to take off—literally—with a helium balloon of mass 30.0 kg. The densities of air and helium are $\rho_{air} = 1.321$ kg/m³ and $\rho_{helium} = 0.179$ kg/m³. What volume of helium is needed to levitate you and the balloon? Is it reasonable to neglect your volume in making this calculation?

•**49.** A conical waxed paper drinking cup of negligible mass has a diameter equal to its height (see Figure P.49). The cup holds 0.040 kg of beach sand of density 2.00×10^3 kg/m³. How deep will the conical cup float in water?

FIGURE P.49

‡50. A cube of wood (of density ρ) with horizontal and vertical edges of length ℓ is floating on water of density ρ_{water}. (a) Show that the cube sinks to a depth x_0 given by

$$x_0 = \frac{\rho}{\rho_{water}} \ell$$

(b) An additional downward force of magnitude F now is applied that sinks the cube an *additional* distance x. Show that the magnitude of the force F is

$$F = \rho_{water} \ell^2 g x$$

Notice that the magnitude of F is proportional to the additional distance x. (c) Show that if the applied force is removed, the cube will begin simple harmonic oscillation with a period T equal to

$$T = 2\pi \left(\frac{\rho}{\rho_{water}} \frac{\ell}{g} \right)^{1/2}$$

Sections 11.8 Surface Tension*
11.9 Capillary Action*

51. A water strider (*Hygrotrechus conformis*) has a mass of 1.00 g and six legs. Assume the legs are identical. For each leg, what is the minimum length of leg that must be in contact with a water surface so the insect can walk on water? Such water striders actually have hairs on the ends of their legs that significantly increase the length actually in contact with the water surface.

52. Humans have two feet, of course. Measure the approximate perimeter of one foot. If your mass is 60.0 kg, what would the perimeter of your foot have to be if you could just barely stand on water because of surface tension? This is a rather large shoe size!

•53. Measure the inside diameter of a thin drinking straw. Place the straw vertically in a glass of water and measure the height to which the water is drawn by capillary action. Use this data to calculate the surface tension of water.

•54. A pipette of uniform inside diameter of 0.500 mm is inserted into a beaker of water. To what height is the water drawn by capillary action?

•55. A water surface is located 2.0 cm below the rim of a cup. What is the inside diameter of a circular tube that can be inserted into the water so that water is drawn by capillary action to the same height as the rim of the cup?

Sections 11.10 Fluid Dynamics: Ideal Fluids
11.11 Equation of Flow Continuity
11.12 Bernoulli's Principle for Incompressible
Ideal Fluids

•56. To quench a ravenous thirst, a notorious desperado shoots a hole 2.00 m below the level of the water in a tank as indicated in Figure P.56. What is the speed of the water as it emerges from the tank?

•57. A wind is blowing parallel to one face of a low building as indicated in Figure P.57. How large a wind speed is needed to cause a pressure difference of 10% between the inside and outside of the building? Such pressure differences are the

FIGURE P.56

principal cause of the damage to buildings during tornados; the buildings literally explode because of the pressure difference between the inside and outside!

FIGURE P.57

•58. Water flows at speed 1.0 m/s through a level pipe with a circular cross-sectional area of 10 cm²; the water continues into a pipe of twice the diameter as shown in Figure P.58. Assume the water is an ideal fluid. (a) What is the volume flow rate expressed in liters/s? (b) What is the pressure difference between the two sections of the pipe? (c) Which section of pipe has the fluid at higher pressure?

FIGURE P.58

•59. Two horizontal pipes of diameters 4.00 cm and 0.500 cm are coupled as indicated in Figure P.59. A pail collects the outflow and you measure 5.0 liters = 5.0×10^{-3} m³ of water flowing out of the pipe every minute. (a) Determine the speed of the flow in each pipe. (b) What is the difference in pressure of the fluid in the two pipes? (c) Which pipe has the fluid at higher pressure?

•60. The pressure in a uniform pipe of flowing water is enough to hold up 4.00 m of water in an open, vertical tube as shown in Figure P.60. Assume that water is an ideal fluid. The pipe leads 3.00 m upstairs and empties into the open air through an

FIGURE P.59

opening with an area half that of the uniform pipe. (a) At the exit opening of the pipe, what is the pressure? (b) What is the pressure at the point where the vertical tube is connected to the pipe? (c) What is the exit speed of the water from the pipe?

FIGURE P.60

•61. Beer is flowing in a brewery through a horizontal pipe of circular cross section and (inside) diameter 5.0 cm. The pressure in this pipe is 9.60×10^5 Pa. The pipe tapers to (inside) diameter 2.0 cm and pressure 5.50×10^5 Pa. What is the volume flow rate in the pipe in m^3/s? Assume the density of beer is that of water (1.00×10^3 kg/m^3).

•62. The funnel in Figure P.62 is filled with water. You initially have an index finger covering the outlet of the funnel to prevent the water from draining out. Assume that water is an ideal fluid. (a) What is the gauge pressure at the bottom of the funnel? Now remove your thumb, so water begins to drain out of the funnel. (b) What is the ratio of the speed of the fluid exiting the funnel to the downward speed of the surface of the fluid at the top of the funnel? (c) What is the initial exit speed of the fluid from the funnel? (d) What is the initial rate at which the water exits the funnel when the funnel is full? Express your result in liters/s.

•63. The plumbing contractor Flood, Leeks, and Drain uses a 5.0 cm diameter, high-pressure pipe to supply a building. The pressure inside the pipe is 6.0 atm and the speed of the water flowing past the inlet valve is 3.00 m/s. On a floor 5.0 m above the inlet pipe, the single pipe has only a 2.5 cm diameter; see Figure P.63. Assume that water is an ideal fluid. (a) What is the speed of the water in the smaller pipe? (b) What is the pressure in the smaller pipe at the 5.0 m height? (c) If the smaller pipe ruptures, propelling water vertically upward as in the figure, to what further height h will the water rise? Neglect air resistance.

•64. Just after its completion, the glass-walled skyscraper known as the John Hancock tower in Boston (see Figure P.64) was faced with an embarrassing problem: its windows kept falling

FIGURE P.62

FIGURE P.63

out on windy days! The building soon became known as the "plywood palace," since at one point plywood covered a significant percentage of the exterior of the building as a temporary replacement for the lost windows. Needless to say, the 2.00 m × 3.00 m glass panes were quite a safety hazard as they cascaded to the pavement below. Fortunately, no casualties resulted from the flying panes. If a wind with a speed of 60.0 km/h is blowing parallel to the face of one of the windows, what is the magnitude of the net outward force on each window caused by the pressure difference between the inside and outside of the building? Compare the magnitude of this force to the magnitude of the weight of a 70.0 kg insurance salesperson.

•65. The architectural firm of Steele, Woods, and Glass is designing a decorative fountain for the new terminal building of U.S. Snarelines at Kennedy International Airport in New York. The fountain is fed by a water pipe of 10.0 cm inside diameter; water

FIGURE P.64

will be projected vertically as indicated in Figure P.65. Assume that water is an ideal fluid. (a) At what speed must the water be projected from the pipe to reach a height of 20.0 m? (b) How many liters per second of water will emerge from the pipe if it is fed directly into the air with no nozzle? (c) If the fountain is gravity-fed from the indicated elevated, open reservoir, what must be the altitude of the top of the water level in the reservoir measured with respect to the nozzle of the fountain?

FIGURE P.65

•**66.** The diameter of the aorta emerging from the human heart is approximately 2.0 cm. Assume the aorta is of circular cross section and that blood is an ideal fluid. The heart pumps about 5.0 liters of blood per minute at an average pressure of 100 mm Hg (= 1.33×10^4 Pa). (a) Determine the approximate speed of the blood in the aorta. (b) The blood is distributed to billions of capillaries in the circulatory system. Assume the average diameter of a capillary is about 7.0 μm. If there are, say, 5×10^9 capillaries, what is the average speed of the blood in the capillaries?

•**67.** A Pitot tube can be used to determine the speed of an airplane relative to the surrounding air. The device consists of a U-shaped tube containing a fluid with a density ρ; see Figure P.67. One end of the U-shaped tube (A) opens to the air at the side of the plane; the other end (B) is open to the air in the direction

the plane is flying so the air is stagnant (at zero speed) at this location. Show that the speed of the plane can be expressed in terms of the difference in the heights h of the fluid in the tube as

$$v = \left(2gh \, \frac{\rho}{\rho_{\text{air}}} \right)^{1/2}$$

where ρ_{air} is the density of air. Why will this not serve to measure the speed of a satellite in Earth orbit?

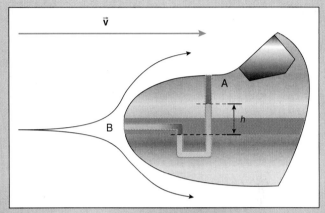

FIGURE P.67

•**68.** An airplane of mass 1.00×10^3 kg has a total wing area of 30 m². (a) What pressure difference between the upper and lower sides of the wing is needed to keep the plane in level flight? (b) If the speed of the air relative to the underside of the wing is 300 km/h, what is the speed of the air over the wing? (The density of air is about 1.30 kg/m³.)

‡**69.** A perfume dispenser consists of a tube that dips into the liquid. Air blown across the top of the tube causes the perfume to rise in the tube and be dispersed out the nozzle (see Figure P.69). The device is called an aspirator. The tube is 6.0 cm long with its top 5.0 cm above the level of the liquid perfume (of density 900 kg/m³). The density of air is 1.30 kg/m³. What is the air speed needed to raise the perfume to the top of the tube? Neglect capillary action. Paint sprayers also employ similar physical principles.

FIGURE P.69

‡70. One of the more remarkable devices involving fluids is the siphon (see Figure P.70). The device is used to transfer fluid out of a container that (for one reason or another) cannot be easily tipped. Let the density of the liquid be ρ and let atmospheric pressure be P_0. The tube (that is called the siphon) is filled with the liquid. One end of the filled tube then is inserted into the tank and the other end held at a level below that of the liquid in the tank as shown. Liquid will flow (seemingly all by itself) out of the tank until the level of the tank reaches the level of the output orifice. (a) Calculate the speed at which the liquid leaves the orifice shown in Figure P.70. (b) Let A be the point at which the siphon has its greatest elevation h_1 above the level of the fluid in the tank. What is the pressure at the point A in the siphon? (c) What is the maximum height $h_{\max}$ that A can be above the surface of the liquid? (Hints: When A is at maximum height, the speed of the fluid in the siphon is zero. What is the smallest pressure that can exist at point A?) (d) If the fluid is water, what is the maximum height $h_{\max}$ for the point A?

Sections 11.13 Nonideal Fluids*
11.14 Viscous Flow*

•71. A rectangular metal plate has dimensions of 10 cm by 20 cm. A thin film of oil of thickness 0.20 mm separates the plate from a horizontal surface. An ideal string is attached to the plate (see Figure P.71) and passes over an ideal pulley to a mass m. When $m = 125$ g, the metal plate moves at a constant speed of 5.0 cm/s across the horizontal surface. Determine the coefficient of viscosity of the oil.

FIGURE P.70

FIGURE P.71

INVESTIGATIVE PROJECTS

A. Expanded Horizons

1. Investigate the aspects of Bernoulli's principle applicable to sailing.
   Rachel P. Gray, "Investigation of the airflow around a sail," *Physics Education*, 21, #1, pages 10–13 (January 1986).
   Paul Jackson, "What limits the speed of a sailboat," *The Physics Teacher*, 18, #3, pages 224–225 (March 1980).
   George C. Goldenbaum, "Equilibrium sailing velocities," *American Journal of Physics*, 56, #3, pages 209–215 (March 1988).
   Sabinus H. Christensen, "Sailing into the wind," *The Physics Teacher*, 17, #7, pages 416, 418 (October 1979).

2. Investigate the operation of a hydraulic ram, a simple device to pump water uphill to, say, a storage reservoir.
   H. Richard Crane, "The hydraulic ram: how to make water go uphill," *The Physics Teacher*, 25, #4, pages 245–247 (April 1987).

3. Investigate the application of Bernoulli's principle to lifting forces in aerodynamics.
   Klaus Weltner, "Bernoulli's law and aerodynamic lifting force," *The Physics Teacher*, 28, #2, pages 84–86 (February 1990).
   Klaus Weltner, "A comparison of explanations of the aerodynamic lifting force," *American Journal of Physics*, 55, #1, pages 50–54 (January 1987).

4. Investigate the physics associated with the circulatory system.
   Alan Van Heuvelen, "Physics of the circulatory system," *The Physics Teacher*, 27, #8, pages 590–596 (November 1989).
   L. J. Bruner, "Cardiovascular simulation for the undergraduate physics laboratory," *American Journal of Physics*, 47, #7, pages 608–611 (July 1979).

5. At a bend in a flowing river, the depth of the river along the outer bank (the outside of the curve) is deeper than along the inner bank (the inside of the curve). Investigate how this arises.
   Robert W. Pohl, *Physical Principles of Mechanics and Acoustics* (Blackie and Sons, London, 1951), pages 189–190.
   Albert Einstein, *Ideas and Opinions* (Crown, New York, 1954), pages 249–253.

6. By consulting appropriate sources, find out how the SAE (Society of Automotive Engineers) rating of the viscosity of various motor oils is related to the SI unit of viscosity.

B. Lab and Field Work

7. Design and conduct appropriate experiments to determine the elastic limits and breaking points of various types of fishing line.
   Graig A. Spolek, "The mechanics of flycasting: the fly line," *American Journal of Physics*, 54, #9, pages 832–836 (September 1986).

8. Silk has extraordinary mechanical properties. Many insects secrete silks, among them, of course, silkworms (*Bombyx mori*) and spiders. The golden orb-weaving spider (*Nephila clavipes*) secretes silk that has attracted much interest in attempts to design and manufacture artificial silks. Design and perform an experiment to determine Young's modulus, the elastic limit, and the breaking point for spiderweb silk.
   Richard Lipkin, "Artificial spider silk," *Science News*, 149, pages 152–153 (9 March 1996).
   Somdev Tyagi and Arthur E. Lord Jr., "Simple and inexpensive apparatus for Young's modulus measurement," *American Journal of Physics*, 48, #3, pages 205–206 (March 1980).

9. Investigate the strength of bone material to determine its Young's modulus and the forces necessary to fracture various

bones in the human body such as a femur, tibia, or fibula. Are bones intrinsically stronger under tension or compression? Do most fractures occur because of tensile forces, compressive forces, or shear forces? Use chicken bones to measure the Young's modulus of a readily available bone material; does it make a difference if the bone is fresh or dry?

John R. Cameron and James G. Skofronick, *Medical Physics* (Wiley, New York, 1978), pages 46–55.

10. Design and perform experiments to measure the terminal speed of a small spherical bead falling through liquids such as motor oils or cooking oils with various coefficients of viscosity. See if you can determine if there is a correlation between the terminal speed and the coefficient of viscosity. You might want to use small air bubbles rising through the fluid instead of falling beads.

Metin Yersel, "A simple demonstration of terminal velocity," *The Physics Teacher*, 29, #6, pages 334–335 (September 1991).

Jim Nelson, "About terminal velocity," *The Physics Teacher*, 22, #4, pages 256–257 (April 1984).

11. Use the Wilhelmy slide method (see Section 11.8) to measure the coefficient of surface tension for several liquids. Compare the results with the simple pulled ring method of Example 11.9.

R. D. Edge, "Surface tension," *The Physics Teacher*, 26, #9, pages 586–587 (December 1988).

S. Y. Mak and K. Y. Wong, "The measurement of surface tension by the method of direct pull," *American Journal of Physics*, 58, #8, pages 791–792 (August 1990).

12. Devise and perform an experiment to measure the coefficient of viscosity of various motor oils at room temperature and near 0 °C.

13. Investigate how capillary action is used in paper chromatography, which has applications in chemistry and biology.

Richard J. Block, *Paper Chromatography* (Academic Press, New York, 1958).

14. The physics of granular materials (like sand and food grains) is quite different from the physics of fluids. Use vertical pipes with convenient diameters to investigate how the pressure at the bottom of a pipe varies with the depth of dry sand in it and with the diameter of the pipe. Present your results in graphical form. In what way is the pressure of such granular materials *not* like that of a fluid? What might account for your results?

Heinrich M. Jaeger, Sidney R. Nagel, and Robert P. Behringer, "The physics of granular materials," *Physics Today*, 49, #4, pages 32–38 (April 1996).

15. Design an experiment to measure the pressure decrease along the length of a copper water pipe of small diameter.

C. Communicating Physics

16. When your blood pressure is measured, two numbers are quoted—for example, 120/80. What do these numbers mean and represent? What is the physics underlying an instrument used to measure your blood pressure? Make your explanation appropriate for the patients of a general practitioner.

17. Investigate and explain the operation of an automotive battery hydrometer; consider your audience to be the employees of an automotive repair facility.

Michael Davis, "A straw hydrometer," *The Physics Teacher*, 25, #3, page 184 (March 1987).

18. Consult the department of civil engineering at your university and/or appropriate engineering materials resources to determine the advantage of using prestressed concrete for bridge decks, floors, and ceilings. Explain why prestressed concrete is not used for walls and columns. Consider the audience for your explanation to be fellow classmates in an introductory physics course.

CHAPTER 12
WAVES

**Life is a *wave* which in no two consecutive moments of its
existence is composed of the same particles.**
John Tyndall (1820–1893) *

We all like to make waves. Nature does too. At the beach, the light we detect when viewing the sea, the characteristic sounds we hear of the surf, and the ocean waves in which we frolic illustrate three very different types of waves. More sobering, we experience with terror the motions of seismic waves produced by earthquakes. The technological marvels of radio waves bring us music, entertainment, and information from around the globe, while their natural counterparts inform radio astronomers about cataclysmic events in the distant cosmos. Microwaves have meant new convenience in our kitchens and dorms. You read this page by detecting waves of light. It is difficult to imagine the practice of medicine without X-rays and gamma rays.[†] Ultrasound waves enable obstetricians to view developing fetuses in the womb and other medical specialists to practice lithotripsy. These examples indicate the wide variety of wave phenomena in our natural and technological environments. We will see in Chapter 27 that the waves in quantum mechanics take on a decidedly unfamiliar, almost mystical cast. So the study of waves is an essential aspect of physics, many other sciences, and even our very culture.

We begin this chapter with an examination of wave phenomena in general. Then we investigate a special class of waves called sinusoidal (harmonic) waves that form the mathematical building blocks for modeling complex, realistic wave phenomena such as acoustic waves from oboes, pianos, and mellophoniums or seismic waves from earthquakes. We also examine several aspects of waves on strings as well as sound waves, including the acoustic Doppler effect. Finally, we see what happens when two or more waves of the same physical type are present simultaneously at the same location; interesting and unusual physical effects result from their superposition.

12.1 WHAT IS A WAVE?

Undulate around the world, serenely arriving, arriving. . .
Walt Whitman (1819–1892)[‡]

You all know what a wave is . . . at least until you have to write a paragraph defining one, illustrating once again that physics is as much language as mathematics. In the sciences it is often quite difficult to generalize or explain precisely what is meant by our observations using the deep and common pool of words from the well of our linguistic heritage.

We need a paradigm for a wave that encompasses the many varieties of waves that occur in nature. If you think of graceful ocean swells, or the ripples generated when you drop a pebble into a calm pond, it is apparent that water waves are an oscillating disturbance that moves, we say *propagates*, across the water surface. When such waves make contact with you or a canoe floating in the water, it is evident that they give or transfer energy and momentum to you or the canoe. A wave involves energy and momentum transport at a characteristic speed that has little direct relationship to the speeds of the entrained par-

---

*[*](Chapter Opener) *Fragments of Science* (Hooper, Clarke, Chicago, 1900), Chapter XXVI, "Vitality," page 379.*

[†]*Radio waves, microwaves, infrared light, visible light, ultraviolet light, X-rays, and gamma rays all are electromagnetic waves. We will study electromagnetic waves in Chapter 21.*

[‡]*Leaves of Grass, Memories of President Lincoln, When lilacs last in the dooryard bloom'd, stanza 14, line 29.*

Experimenting with the right wave can be exciting!

ticles. In a water wave well away from the beach, the dominant motion of an individual water particle is up and down, perpendicular to the direction the water wave is moving. Such waves are much more complex as they "break" near a beach!

As we know, energy and momentum also can be transferred by the movement of a particle, such as when a pitcher throws a baseball to the catcher. Energy and momentum can be transferred by one, two, or a whole stream of particles; but such particle streams are not what we imagine as waves.

A propagating wave disturbance may involve any manner of physical quantity. A **mechanical wave** is a propagating disturbance in and of a material (such as a solid, liquid, or gas) from its equilibrium configuration. The term mechanical wave indicates its close association with matter. These waves involve oscillatory motions of the atoms and/or molecules that make up the material, without requiring net transport of the material itself over long distances in the direction of motion of the wave. Such mechanical waves represent propagating oscillations from equilibrium, be they departures from equilibrium positions (e.g., water waves and acoustic waves), or pressure or density oscillations (also acoustic waves). An ocean wave predominantly moves water up and down with little or no average motion of the water along the direction of travel of the wave. A huge locomotive engine generates mechanical waves (vibrations and sound waves) in the steel rails that propagate speedily along their length; although the particles in the rail move slightly, the rail itself does not progressively move as do the waves.

Other types of waves, capable of propagating even through a vacuum, involve propagating oscillations of *fields* of various kinds: gravitational fields (gravity waves), or electric and magnetic fields (electromagnetic waves—i.e., light). Such waves are not mechanical waves.

Generalizing from these observations, we characterize a **classical wave** as a propagating disturbance that transfers energy and momentum at its own characteristic speed from one region of space to another with little if any mass transfer.

Mechanical waves and waves of oscillating fields are both classical waves because they satisfy a general **classical wave equation**, which we will develop in Section 12.6. In this chapter we confine our study principally to mechanical waves. We see how electromagnetic waves arise in Chapter 21. Later, when we study quantum mechanics in Chapter 27, we shall see that the peculiar waves associated with quantum mechanics satisfy a different wave equation and are not themselves directly observable. The waves in quantum mechanics are not classical waves. We also shall see in that now distant chapter that there is a subtle, mysterious, almost oxymoronic connection between particles and waves called the wave–particle duality: ultimately (and surprisingly) we cannot have waves without particles, nor particles without waves!

QUESTION 1

The word wave is used in many colloquial contexts. Examples include:

- Don't make waves.
- Wave goodbye.
- Whew! What a heat wave we had this summer!
- Waves of troops invaded.
- Her hair has beautiful waves.
- Flag waving children greeted the President.

Perhaps you can think of others. In what ways does the use of the word wave in these contexts differ from the meaning of the word in physics? Check out the meaning of the word *wave* in the *Oxford English Dictionary*; you may be surprised at the many and varied uses of the word!

12.2 LONGITUDINAL AND TRANSVERSE WAVES

There are two different kinds of "pure" waves. In this section we contrast them.

Longitudinal Waves

Take a slinky and stretch it across your desk, holding one end in your hand with the other end fixed to a wall. Quickly push or compress the end you hold along the axis of the slinky, as in Figure 12.1, without returning your hand back to its original position. The single **compression** or pulse propagates down the axis of the slinky at a speed called the *speed of the wave*.

If you now quickly extend the end you hold, as in Figure 12.2, again without returning your hand to its original position, an anti-compression pulse called a **rarefaction** is transmitted down the axis of the slinky. Multiple successive compressions and extensions by your hand produce multiple compressions and rarefactions that move along the axis of the slinky. For such disturbances, the oscillation (jiggling) of the particles of the slinky is along the line of propagation of the wave.

> Waves whose oscillation or jiggling is along the line the wave propagates are called **longitudinal waves**.

A given particle of the slinky has a maximum speed when the coils are most compressed or rarefied. This is characteristic of longitudinal mechanical waves: particle speeds are a maximum at the instants of maximum compression and rarefaction. This observation is *not* obvious, so do not worry about it for now; we will explore this aspect of longitudinal waves in greater detail in Section 12.13.

FIGURE 12.1 Compress an end of a slinky and a compression pulse propagates along it.

FIGURE 12.2 Extend your end of the slinky quickly and a rarefaction propagates along the slinky.

Common sound waves in air or other gases are longitudinal waves that can be viewed as propagating density or pressure variations that result from longitudinal motions of the particles that make up the medium through which the sound wave travels. Regions of increased density (or pressure) are compressions; regions of reduced density (or pressure) are rarefactions. Such *variations* or oscillations in density (or pressure) about the ambient or mean value are positive for compressions and negative for rarefactions. The density (or pressure) itself is never negative. The density (or pressure) variations are correlated to the particle motions in ways that will be discussed in Section 12.13.

Transverse Waves

As another experiment, take a thin clothesline or climbing rope and stretch it taut across your room. If you quickly move your hand up and down once, perpendicular to the axis of the rope, the disturbance you generate propagates along the rope; see Figure 12.3. But in this case the disturbance (the jiggling—i.e., the motion of the particles that make up the rope) is perpendicular to the direction the wave disturbance propagates.

Quick motion of hand

FIGURE 12.3 A propagating pulse on a rope.

Move your hand up and down several times quickly, as in Figure 12.4, and you produce multiple oscillations that propagate along the rope.

> Waves whose oscillation or jiggling is perpendicular to the line along which the wave propagates are called **transverse waves**.

Waves on a rope or string are transverse waves.

> The plane in which the jiggling of a transverse wave disturbance occurs is called the **plane of polarization** of the transverse wave; see Figure 12.4.

Only transverse waves have planes of polarization; we say such waves are **plane polarized**. Light also is a transverse wave, even though it is not a mechanical wave but a wave of oscillating electric and magnetic fields.*

Hence the primary physical distinction between longitudinal and transverse waves is the direction of the jiggling or oscillations of the particles:

Transverse wave: the oscillations are perpendicular to the line along which the wave propagates.

---

*We explore the properties of light waves in greater detail in Chapter 24, including the evidence that suggests that light is indeed a wave (it is *not* obvious that this is the case) and what observations (polarization) enable us to conclude that it is a transverse wave. The polarization of ordinary light is not apparent because typical sources emit many waves with many different planes of polarization present simultaneously.

Longitudinal wave: the oscillations are along the line along which the wave propagates.

Earthquakes produce *both* longitudinal and transverse mechanical waves, called **seismic waves**, that propagate from the **focus**, the source of the quake. The **epicenter** is the place on the surface of the Earth directly above the focus of the quake. The longitudinal wave is called a **P-wave**, where the P means *p*ressure wave. The transverse wave is called an **S-wave**, where the S means *s*hear wave. The S- and P-waves travel at different speeds, on the order of 3 km/s and 5 km/s respectively.† Because of this difference, seismometers detect the arrival of the P-wave before the arrival of the S-wave, as in Figure 12.5. The difference in the time of arrival enables us to estimate the distance from the seismometer to the site of the earthquake that generated the waves. The S-waves do not penetrate the outer core of the Earth, since it is fluid and cannot sustain shear or transverse forces.

Do *not* get the impression that all waves are either longitudinal or transverse! Certain types of waves exhibit *both* characteristics simultaneously, and so are neither purely transverse nor purely longitudinal in character. Surface water waves are good examples of this mixed type of wave. If you carefully follow the jiggling of water particles as a water wave passes in deep water well away from the shore,‡ the particles generally move up and down but also exhibit smaller motions parallel and antiparallel to the direction of propagation of the wave, as shown in Figure 12.6.

---

†The speeds vary depending on the density of the rock through which they travel.

‡This can be done by using a water suspension of small particles (dirty water!) and illuminating the dirt particles with a strong light.

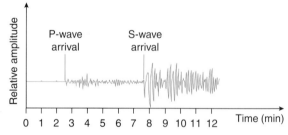

FIGURE 12.5 A record of a moderate earthquake in Chile on 3 September 1993, made by a seismograph at Washington and Lee University in Virginia; note the different arrival times of the P- and S-waves. The time $t = 0$s is not when the earthquake occurred in Chile but here is an arbitrary choice for the time axis.

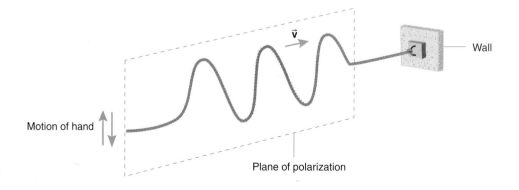

FIGURE 12.4 The plane of polarization of a transverse wave.

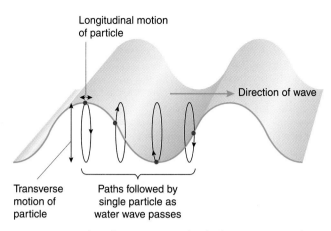

Longitudinal motion
of particle

Direction of wave

Transverse
motion of
particle

Paths followed by
single particle as
water wave passes

FIGURE 12.6 A surface water wave has both a transverse and longitudinal character.

FIGURE 12.7 A waveform of a water wave shows its shape at an instant of time.

Under certain circumstances, even light exhibits both transverse and longitudinal characteristics (for example, when microwaves propagate in certain ways inside pipes called wave guides).

Grab the end of a taut rope whose other end is tied to a wall. Whirl the end you are holding around in a circle in a plane perpendicular to the axis of the rope. Is the wave on the rope transverse, longitudinal, or neither?

12.3 WAVEFUNCTIONS, WAVEFORMS, AND OSCILLATIONS

Waves are closely related to oscillations.* Both words imply undulatory (back-and-forth) motion but there is a distinction. Unlike the oscillatory behavior we studied in Chapter 7, waves are *propagating* oscillatory disturbances.

We denote the propagating wave disturbance by the Greek letter Ψ, called psi (pronounced like the English word "sigh"), whatever the wave may represent physically. This way we can avoid being too parochial and can, without bias, think of Ψ as representing *any* kind of specific wave disturbance. We call Ψ the **wavefunction** since it represents the wave disturbance. Since a wave propagates through space and time, Ψ must be a function of space and time: $\Psi(x, y, z, t)$.

If you "freeze" the wave in time, say by taking an imaginary picture of it at time instant t_0, the resulting spatial pattern of the disturbance $\Psi(x, y, z, t_0)$ is called a **waveform**. A photograph taken after you shake the end of a rope or a photograph of a water wave depicts the waveforms of the respective waves (see Figure 12.7).

On the other hand, if you imagine a specific point in space (x_0, y_0, z_0) and examine the behavior there as a function of time, we have a *dynamical picture* of the *oscillatory motion* at that point (like a video camera viewing one location, without panning the camera). For example, we can take a video of the oscillatory motion of a given point on a rope you shake. Likewise we could take a video of your canoe lazily bobbing up and down as a water wave passes, since the boat responds to the oscillatory behavior of the water particles at its location.

12.4 WAVES PROPAGATING IN ONE, TWO, AND THREE DIMENSIONS

Waves can propagate in one or more spatial dimensions. A longitudinal wave on a slinky is a good example of a one-dimensional wave. Let the x-axis be the line along which the wave propagates. Then the wave disturbance Ψ is a function of the position along the x-axis as well as the time: $\Psi(x, t)$.[†] Look at the waveform at some instant, as shown in Figure 12.8. The **wave train** (also called the **pulse train**) is an indication of the spatial region over which the waveform is nonzero at an instant. The **carrier wavelength** is the distance between successive compressions or successive rarefactions measured on the waveform.

A transverse wave on a taut rope is another example of a one-dimensional wave; see Figure 12.9. The wave propagates along a line, the x-axis, coincident with the undisturbed rope and the disturbance Ψ is a measure of the position, perpendicular to the x-axis,[‡] where a particle of the rope is located. Again Ψ

---

*An oscillation is any kind of back-and-forth motion. Recall that simple harmonic oscillation (Chapter 7) is a very specific type of oscillation. Not all oscillations are simple harmonic oscillations.

[†]Here is a good example of why we call the wave disturbance Ψ. For such a longitudinal compression wave, the disturbance is along the same direction the wave propagates, say x. By using Ψ to represent the disturbance, we avoid likely notational confusions such as $x(x, t)$.

[‡]We can measure this perpendicular distance along the y-axis, so $\Psi(x, t) \equiv y(x, t)$, but we prefer to call all wave disturbances Ψ.

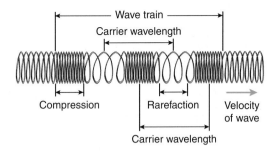

FIGURE 12.8 The waveform of a compressional wave on a slinky.

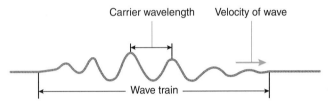

FIGURE 12.9 The waveform of a wave on a rope.

depends on both the position x along the rope and t: $\Psi(x, t)$. The wave train is the distance over which the waveform is nonzero. The carrier wavelength is the distance between successive corresponding points of the wave along the waveform.

Drop a rock into a calm pond and the resulting surface water wave is two-dimensional, since it propagates in all directions across the surface of the water (see Figure 12.10). In this case we have $\Psi = \Psi(x, y, t)$. The wave train is the radial distance between the leading and trailing edges of the waveform—that is, the region over which the disturbance is nonzero. Taking a cue from water waves, we call the peaks of a waveform **crests**, and the minima **troughs**. The **carrier wavelength** is the distance between successive peaks, troughs, or other corresponding successive points on the ripples. The carrier wavelength may vary along the wave train, particularly for complex waveforms.

Scream and shout and the resulting acoustic wave in the air is three-dimensional, since it propagates in all spatial directions;

thus, $\Psi = \Psi(x, y, z, t)$. The wave train is the distance over which the sound is nonzero at some instant, and the carrier wavelength is the distance between successive compressions or rarefactions on the waveform.

With the slinky, if you compress the end once, a single, one-dimensional, longitudinal, compressional pulse propagates down the axis. A one-dimensional transverse pulse is generated if you shake a taut rope side-to-side once. The sound of a rifle shot is approximately a three-dimensional, longitudinal, acoustic wave pulse. A tsunami closely approximates a surface water wave pulse. Waveforms with wave trains of finite extent are called **wavepackets**.

If the waveform is essentially random (see Figure 12.11), then the wave is called **noise**, whether the wave be acoustic or not. Noise has no single, well-defined carrier wavelength. The waveforms of the sound of a waterfall or the crackle of static from your radio are good examples of waveforms with no definitive carrier wavelength.

A crest or trough or any locus of connected points where Ψ has a constant value defines a region of **constant phase**, known as a **wavefront**, shown in Figure 12.12. Directed lines drawn perpendicular to wavefronts and parallel to the velocity of the wave are called **rays**.

QUESTION 3

Make a quick sketch of the waveform of a water swell on an otherwise calm ocean, well away from the shore. Indicate what is meant by a crest, trough, wavefront, and ray.

FIGURE 12.11 The waveform of noise.

FIGURE 12.10 A two-dimensional surface water waveform.

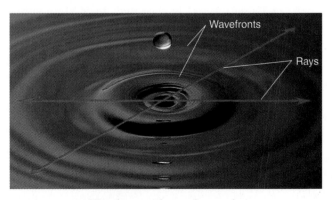

FIGURE 12.12 Wavefronts and rays of a waveform.

12.5 ONE-DIMENSIONAL WAVES MOVING AT CONSTANT VELOCITY

In this section we seek to find the way the wavefunction $\Psi(x, t)$ depends more specifically on x and t for a one-dimensional wave propagating along the x-axis at constant velocity.

Imagine a waveform of Ψ with a wavefront crest of maximum size located at $x = 0$ m when $t = 0$ s, as shown in Figure 12.13. We always can arrange the coordinate system and begin measuring time when this largest peak is at the origin, so that there is no loss of generality in imposing this restriction. This largest wavefront crest of this disturbance Ψ occurs where the argument of Ψ is zero, that is, where the stuff inside the functional parentheses is all zeros. Let's keep track of this wavefront as the wave moves at constant velocity. Tracking a particular wavefront means following a point of constant phase; the speed v of the wave is thus called a **phase speed**.

At a later instant t the wavefront has moved along the x-axis, say in the direction of increasing x, a distance equal to vt, so that the wavefront we are following is now at coordinate $x = vt$, as shown in Figure 12.14. Thus we have

$$x - vt = 0 \text{ m}$$

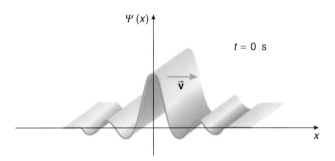

FIGURE 12.13 A large wavefront crest at $x = 0$ m when $t = 0$ s.

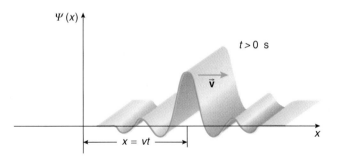

FIGURE 12.14 The wavefront has moved to coordinate $x = vt$ when the time is t.

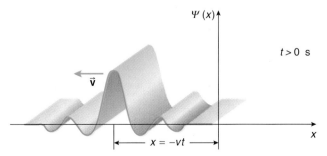

FIGURE 12.15 A wave moving toward decreasing values of x.

From a mathematical viewpoint, if $x - vt$ holds a fixed value, here 0 m, this value specifies a particular wavefront of the disturbance.

The wave is moving at speed v *along* $\hat{\imath}$; the wavefront is thus at the coordinate $x = vt$ at time t, and $x - vt$ is therefore constant for a given wavefront. Thus we describe the disturbance $\Psi(x, t)$ more specifically as a function of the quantity $x - vt$:

$$\Psi(x - vt)$$

In this way, whenever the argument $x - vt$ of the function is equal to zero, we have $\Psi(0$ m$)$ and have located this chosen wavefront of the disturbance.

> Therefore a one-dimensional wave propagating at constant speed v in the direction of *increasing* values of x is described by the wavefunction
>
> $$\Psi(x - vt) \qquad (12.1)$$

If you want to describe the same disturbance moving in the direction of *decreasing* values of x, as in Figure 12.15, the wavefront will have moved out along the negative x-axis for times $t > 0$ s and the wavefront is at coordinate $x = -vt$. So

$$x + vt = 0 \text{ m}$$

locates the chosen wavefront that originally was at the origin when $t = 0$ s.

> Hence the wavefunction
>
> $$\Psi(x + vt) \qquad (12.2)$$
>
> represents a one-dimensional disturbance moving at constant speed v in the direction of *decreasing* values of x.

PROBLEM-SOLVING TACTIC

12.1 The sign between x and vt determines the direction Ψ is moving along the x-axis. When the sign is negative, $\Psi(x - vt)$, the wave is moving toward *increasing* values of x; when the sign is positive, $\Psi(x + vt)$, the wave is moving toward *decreasing* values of x.

12.6 THE CLASSICAL WAVE EQUATION FOR ONE-DIMENSIONAL WAVES*

It is quite remarkable that all one-dimensional classical waves propagating with constant velocity satisfy a single equation known as the **classical wave equation**. In this section we formulate this special equation.

One-dimensional waves propagating at constant phase speed v in the direction of increasing values of x are described by a wavefunction $\Psi(x - vt)$. Notice that Ψ is a function of two variables: x and t. The calculus you have learned to date likely involved just one independent variable (the ubiquitous "x" in calculus courses). When there are two independent variables (such as x and t here), you can take derivatives of Ψ with respect to either x or t. When differentiating with respect to x, the other variable t is treated as a constant, and vice versa. Such derivatives are called **partial derivatives**.* The same familiar rules of differentiation apply, but the derivative symbol is changed from the familiar

$$\frac{d}{dx}$$

to the more stylish

$$\frac{\partial}{\partial x}$$

If Ψ is differentiated with respect to x, treating t as a constant, we call the result the partial derivative of Ψ with respect to x and write it as

$$\frac{\partial \Psi}{\partial x}$$

We could also differentiate Ψ with respect to the other independent variable, t, in which case you treat x as a constant and apply the usual differentiation rules. This is the partial derivative of Ψ with respect to t, written

$$\frac{\partial \Psi}{\partial t}$$

We know that a wave Ψ propagating in the direction of increasing values of x is a function of the quantity $x - vt$:

$$\Psi(x - vt)$$

Simply from a mathematical standpoint, we might suspect that there is a relationship between the partial derivatives of Ψ with respect to x and those of Ψ with respect to t. Let's see if we can discover what that relationship is. For convenience, we define a parameter u to be

$$u \equiv x - vt \qquad (12.3)$$

To differentiate Ψ partially with respect to x, we use the chain rule from calculus: differentiate Ψ with respect to u and then multiply the result by the derivative of u with respect to x:

$$\frac{\partial \Psi}{\partial x} = \frac{\partial \Psi}{\partial u} \frac{\partial u}{\partial x}$$

From Equation 12.3 we see the partial derivative of u with respect to x is just 1:

$$\frac{\partial u}{\partial x} = 1$$

(remember that t is treated as a constant in partial differentiations with respect to x). The wave speed v is a constant for the class of waves with which we are concerned. So we have

$$\frac{\partial \Psi}{\partial x} = \frac{\partial \Psi}{\partial u} 1$$

---

*There is nothing really "partial" about the differentiation process, however, since the same differentiation rules of calculus are used for either choice of the variable.

For good measure we take the second partial derivative of Ψ with respect to x (again using the chain rule), getting

$$\frac{\partial^2 \Psi}{\partial x^2} = \left[\frac{\partial}{\partial u} \left(\frac{\partial \Psi}{\partial x} \right) \right] \frac{\partial u}{\partial x}$$

$$= \left[\frac{\partial}{\partial u} \left(\frac{\partial \Psi}{\partial u} \right) \right] 1$$

$$= \frac{\partial^2 \Psi}{\partial u^2} \qquad (12.4)$$

Now we look at the partial derivatives of Ψ with respect to t (in which case we treat x as a constant). Once again, we use the chain rule:

$$\frac{\partial \Psi}{\partial t} = \frac{\partial \Psi}{\partial u} \frac{\partial u}{\partial t} \qquad (12.5)$$

But from Equation 12.3 we have

$$\frac{\partial u}{\partial t} = -v$$

Equation 12.5 thus becomes

$$\frac{\partial \Psi}{\partial t} = -v \frac{\partial \Psi}{\partial u}$$

Since

$$\frac{\partial \Psi}{\partial u} = \frac{\partial \Psi}{\partial x}$$

we could stop here and say that the desired relationship between the partial derivatives is

$$\frac{\partial \Psi}{\partial t} = -v \frac{\partial \Psi}{\partial x} \qquad (12.6)$$

Unfortunately, if we did stop here, we would find a *different* relationship between the partial derivatives for a wave traveling in the opposite direction, toward decreasing values of x! (See Problem 12.) This is mathematically and physically inelegant (and becomes even more of a problem when considering waves in two and three dimensions). So we push on and see if we can find a relationship between the second partial derivatives.

Taking the second partial derivative of Ψ with respect to t, applying the chain rule once again, we get

$$\frac{\partial^2 \Psi}{\partial t^2} = \frac{\partial}{\partial t} \left(\frac{\partial \Psi}{\partial t} \right) = \frac{\partial}{\partial u} \left(\frac{\partial \Psi}{\partial t} \right) \frac{\partial u}{\partial t}$$

$$= \frac{\partial}{\partial u} \left(-v \frac{\partial \Psi}{\partial u} \right)(-v) = v^2 \frac{\partial^2 \Psi}{\partial u^2}$$

Equation 12.4 enables us to write this as

$$\frac{\partial^2 \Psi}{\partial t^2} = v^2 \frac{\partial^2 \Psi}{\partial x^2}$$

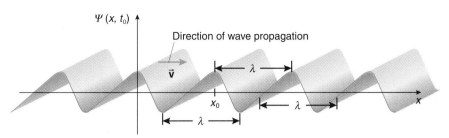

$\Psi(x, t_0)$

Direction of wave propagation

$\vec{v}$

λ

x_0

λ

λ

x

FIGURE 12.16 The wavelength λ of a periodic waveform.

With some slight rearrangement, we find the *classical wave equation* for one-dimensional waves is

$$\frac{\partial^2 \Psi}{\partial x^2} - \frac{1}{v^2} \frac{\partial^2 \Psi}{\partial t^2} = 0 \qquad (12.7)$$

Such an equation is called a partial differential equation for Ψ, since it involves the partial derivatives of the function. This particular partial differential equation is very famous in physics and engineering and is given its own name: the classical wave equation for one-dimensional waves. Any classical real wave Ψ, propagating at constant velocity $\vec{v}$ in one dimension (along the x-axis), must satisfy this relationship among its partial derivatives.*

12.7 PERIODIC WAVES

Many waves, such as ocean swells, periodically repeat themselves. A **periodic wave** is a wave with characteristic, constant length and time scales. The waveform of a periodic wave repeats itself along a characteristic, constant scale of length known as the **wavelength** λ. The wavelength λ is the distance between successive corresponding points along the waveform, as shown in Figure 12.16. In SI units, λ is expressed in meters. Note that Figure 12.16 is a graph of the *waveform* at a particular instant t_0 of time, plotting Ψ as a function of x.

On the other hand, if you fix your attention at any given position x_0, the *oscillatory* behavior of a periodic wave passing that position has a constant **period** T, as shown in Figure 12.17. In SI units, the period is measured in seconds. Note that the graph in Figure 12.17 depicts the *dynamic*, oscillatory behavior of Ψ at a fixed position as a function of time.

---

*The peculiar waves encountered in quantum mechanics are complex waves (complex in the mathematical sense: they involve $i = \sqrt{-1}$). We will see in Chapter 27 that the waves in nonrelativistic quantum mechanics do not satisfy the classical wave equation, but another wave equation known as the Schrödinger equation.

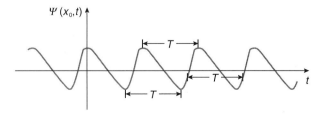

$\Psi(x_0, t)$

T

T

T

t

FIGURE 12.17 The period T of a periodic wave.

The period T is the interval of time needed to execute one complete oscillation. The **frequency** ν is the number of oscillations made during exactly one second. The SI unit of frequency is the hertz (Hz). The frequency is the reciprocal of the period:

$$\nu = \frac{1}{T} \qquad (12.8)$$

Let the constant phase speed of the wave be v. The wave moves through a distance equal to the wavelength during the time interval of one period. Since the distance is the product of the speed and time interval, we have

$$\lambda = vT \qquad (12.9)$$

Rewriting Equation 12.9 in terms of the frequency instead of the period (using Equation 12.8), we obtain

$$v = \nu\lambda \qquad (12.10)$$

Equation 12.10 is valid for any periodic wave moving at constant speed.

A particularly important class of periodic waves is **sinusoidal waves** (also known as *harmonic waves*); their waveform has the shape of the sine or cosine trigonometric functions. The oscillatory disturbance at any position then is simple harmonic oscillation, also described by sinusoidal functions. Such sinusoidal waves are idealizations, since they have infinite wave trains and exist at all times. No real wave, of course, exists over all space for all time. Nonetheless, sinusoidal waves are important because they form the building blocks for the analysis of more complex waves, as we will see in Section 12.21. We look at sinusoidal waves in greater detail in the following section.

STRATEGIC EXAMPLE 12.1

A periodic wave traveling at a constant speed 10.0 m/s toward increasing values of x has the waveform at the instant $t = 0$ s sketched in Figure 12.18.

a. What is the wavelength λ of the periodic wave?
b. Sketch the corresponding oscillatory behavior, Ψ versus t, at a fixed position, say the origin.
c. What is the period T of the periodic wave?
d. What is the frequency ν of the periodic wave?

Solution

a. The wavelength λ is the distance between successive corresponding points along the waveform. This is 6.0 m.
b. Consider the origin in Figure 12.18. Notice that as the wave moves to the right at constant speed, the time to reach the peak

FIGURE 12.18

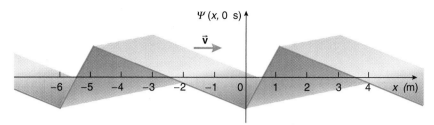

will be five times as long as the time to fall to the trough of the wave. Since the wave is moving at speed 10 m/s to the right in Figure 12.18, it will take 0.50 s for the first peak to the left of the origin in Figure 12.18 to reach the origin; 0.10 s later, the first trough to the left of the origin in Figure 12.18 reaches the origin. Hence the oscillatory behavior at the origin has the appearance shown in Figure 12.19.

FIGURE 12.19

c. The period T of the wave is the time between successive corresponding points on the graph of Ψ versus t. The period is 0.60 s.
d. There are two ways to find the frequency.

Method 1

From Equation 12.8, the frequency is the inverse of the period:

$$\nu = \frac{1}{T}$$
$$= \frac{1}{0.60 \text{ s}}$$
$$= 1.7 \text{ Hz}$$

Method 2

Since you know the speed v and the wavelength λ, you can use Equation 12.10 to find the frequency ν:

$$v = \nu\lambda$$

Solving for the frequency,

$$\nu = \frac{v}{\lambda}$$
$$= \frac{10 \text{ m/s}}{6.0 \text{ m}}$$
$$= 1.7 \text{ Hz}$$

EXAMPLE 12.2 ━━━━━━━━

A periodic wave, traveling toward *decreasing* values of x at speed 4.0 m/s, produces oscillatory behavior at the origin of the x-axis depicted in Figure 12.20.

a. What is the period of the wave?
b. What is the frequency of the wave?

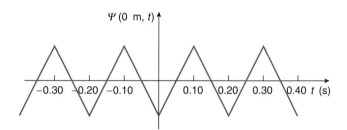

FIGURE 12.20

c. What is the wavelength of the wave?
d. Sketch the waveform, Ψ versus x, when $t = 0$ s.

Solution
a. The period of the wave is the time between successive corresponding points on the Ψ versus t graph. This time is $T = 0.20$ s.
b. The frequency of the wave is the inverse of the period:

$$\nu = \frac{1}{T}$$
$$= \frac{1}{0.20 \text{ s}}$$
$$= 5.0 \text{ Hz}$$

c. The wavelength can be found in two equivalent ways.

Method 1

The wave travels at speed v through a distance equal to one wavelength during one period. Hence the wavelength is

$$\lambda = vT$$
$$= (4.0 \text{ m/s})(0.20 \text{ s})$$
$$= 0.80 \text{ m}$$

Method 2

Since you know the frequency and the speed, you can find the wavelength using Equation 12.10:

$$v = \nu\lambda$$

Solving for the wavelength,

$$\lambda = \frac{v}{\nu}$$
$$= \frac{4.0 \text{ m/s}}{5.0 \text{ Hz}}$$
$$= 0.80 \text{ m}$$

d. The oscillatory pattern at the origin, given in Figure 12.20, indicates that Ψ increases linearly with time from a trough when $t = 0$ s to a peak when $t = 0.10$ s. (Notice that the waveform is moving toward negative values of x.) The wave then decreases to a trough after another 0.10 s. Since the wave is moving at constant speed, the waveform of Ψ must have linear segments of equal length. Since the wavelength is 0.80 m and, when $t = 0$ s, Ψ is at a trough at the origin, a graph of Ψ versus x has the appearance shown in Figure 12.21.

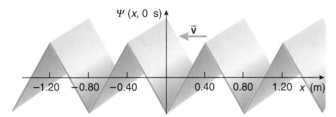

FIGURE 12.21

12.8 SINUSOIDAL (HARMONIC) WAVES

Since any one-dimensional wave traveling at constant speed v toward increasing values of x has a wavefunction depending on the combination of variables $x - vt$, a **sinusoidal wave** traveling in the direction of increasing x will have the form

$$\Psi(x, t) = A \cos[k(x - vt)] \qquad (12.11)$$

where we have to determine the meaning and physical significance of k. Since the cosine can be taken only of angles (which have units but no dimensions), the product of k with x is dimensionless; hence k itself must have the dimensions of inverse length.

> By convention the angular argument of the cosine function— that is, $k(x - vt)$—is expressed in radian units of angular measure, *not* degrees.

Although we use the cosine function in Equation 12.11, the wave still is called a sinusoidal wave because the sine and cosine functions have the same general shape and characteristics; they simply are shifted by $\pi/2$ rad with respect to each other. One could just as well use a sinusoidal wavefunction Ψ of the form $A \sin[k(x - vt)]$. However, by convention, we will use Equation 12.11 for sinusoidal waves.

Let's examine the waveform and oscillatory characteristics of a sinusoidal wave.

Waveform of a Sinusoidal Wave

Take a "snapshot" of the wave when $t = 0$ s and make a graph of the waveform of Ψ as a function of x. That is, when $t = 0$ s, we have

$$\Psi(x, 0 \text{ s}) = A \cos(kx) \qquad (12.12)$$

A graph of this waveform as a function of x is a cosine function, as shown in Figure 12.22.

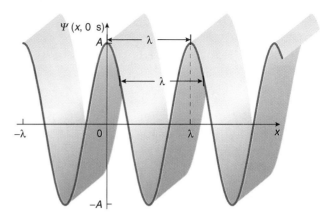

FIGURE 12.22 The waveform of the sinusoidal wave given by Equation 12.11 when $t = 0$ s.

Since the maximum value of a cosine function is 1, the maximum value of Ψ is equal to A, known as the **amplitude** of the sinusoidal wave. The amplitude has the same dimensions as Ψ itself (*not* necessarily a distance, since Ψ represents whatever particular physical wave is under discussion). Notice that the variation of Ψ from trough to peak, the so-called peak-to-peak amplitude, is 2A.

A sinusoidal waveform has a fixed wavelength. Notice that a shift in the value of x by λ in either direction along the x-axis yields the same value again for Ψ. For example, at $x = 0$ m the value of Ψ is A according to Equation 12.12. When $x = \pm\lambda$, the value of Ψ is again A. The argument of the cosine function changes by 2π rad as x changes from 0 m to λ; in other words, when $x = \lambda$ the angle (or argument) of the cosine is 2π rad. That is, from Equation 12.12

$$k\lambda = 2\pi \text{ rad}$$

The constant k therefore is

$$k = \frac{2\pi \text{ rad}}{\lambda} \qquad (12.13)$$

The constant k is known as the **angular wavenumber**.* Physically, the angular wavenumber represents the number of wavelengths in a distance of exactly 2π meters.†

---

*The angular wavenumber also commonly is called the *propagation vector*. The latter term bears some explanation. The propagation vector k for a one-dimensional wave is a scalar; it represents the magnitude of a propagation vector. For waves in two or three dimensions, k takes on vector properties, which is the reason that it is called the propagation vector. In fact, the product kx in Equations 12.11 and 12.12 is the one-dimensional scalar product of a vector $\vec{k}$ with the position vector $\vec{r}$:

$$\vec{k} \cdot \vec{r} = kx \qquad \text{(in one dimension)}$$

So we have a case of trickle-down terminology for the one-dimensional case. To avoid calling k a vector when it is a scalar for a one-dimensional situation, we use the terminology *angular wavenumber*.

†The reciprocal of the wavelength, $1/\lambda$, without the factor of 2π, is called the *wavenumber*. The wavenumber is the number of wavelengths in a distance of exactly one meter.

The Oscillatory Behavior of a Sinusoidal Wave

You can monitor the oscillatory behavior of the wave passing a particular location, say $x = 0$ m. The wavefunction Ψ then is a function of time. Equation 12.11 becomes

$$\Psi(0 \text{ m}, t) = A \cos(-kvt) \qquad (12.14)$$

Since the cosine of a negative angle is the same as the cosine of the positive angle $[\cos(-\theta) = \cos(\theta)]$, Equation 12.14 becomes

$$\Psi(0 \text{ m}, t) = A \cos(kvt)$$

whose graph also is sinusoidal, as shown in Figure 12.23. Notice that the graph is periodic in time. The constant period T of the wave is the time between any two successive corresponding points on the graph shown in Figure 12.23.

As t increases from $t = 0$ s to $t = T$, the argument of the cosine goes through 2π rad. Hence

$$kvT = 2\pi \text{ rad} \qquad (12.15)$$

The period is the time for Ψ to execute a complete oscillation at any given location. The inverse of the period is the frequency ν of the wave. Since the period is the time for one complete oscillation, the frequency is the number of oscillations completed during one second. Substituting $1/\nu$ for T in Equation 12.15,

$$\frac{kv}{\nu} = 2\pi \text{ rad}$$

But from Equation 12.13, $k = (2\pi \text{ rad}/\lambda)$, so that

$$\frac{2\pi \text{ rad}}{\lambda} \frac{v}{\nu} = 2\pi \text{ rad}$$

and we regain Equation 12.10, valid for any periodic wave:

$$v = \nu\lambda$$

Note that the product

$$kv = \frac{2\pi}{\lambda} v\lambda = 2\pi\nu = \omega$$

The product $2\pi\nu$ is called the angular frequency ω of the wave from the analogous quantity in circular motion.

The sinusoidal wavefunction of Equation 12.11 can be written in many equivalent forms:

$$\Psi(x, t) = A \cos[k(x - vt)]$$
$$= A \cos(kx - kvt)$$

$$\Psi(x, t) = A \cos(kx - \omega t) \qquad (12.16)$$

$$= A \cos\left[2\pi\left(\frac{x}{\lambda} - vt\right)\right]$$

$$= A \cos\left[2\pi\left(\frac{x}{\lambda} - \frac{t}{T}\right)\right]$$

To avoid notational overload, we choose Equation 12.16 as the standard form for a sinusoidal wave because of its notational simplicity.

The sinusoidal wavefunction contains an incredible amount of information within its rather humble mathematical form.

PROBLEM-SOLVING TACTIC

12.2 To find the value of $\cos(kx - \omega t)$ in Equation 12.11 or 12.16, be sure your calculator is set for radian measure of angle, not degrees.

STRATEGIC EXAMPLE 12.3

A sinusoidal wave representing the transverse oscillations on a string has the wavefunction

$$\Psi(x, t) = (0.40 \text{ m}) \cos[(6.00 \text{ rad/m})x - (10.0 \text{ rad/s})t]$$

a. What is the amplitude of the wave?
b. What is the wavelength of the wave?
c. Find the frequency of the wave.
d. Find the speed of the wave.
e. What is the value of Ψ at the position $x = 5.00$ m when $t = 2.00$ s?
f. Plot the waveform of Ψ as a function of x over the domain $x = 0$ m to $x = 3.00$ m when $t = 2.00$ s. Indicate the wavelength λ on your graph.
g. Plot the oscillation at $x = 5.00$ m as a function of t over the domain $t = 0$ s to $t = 3.00$ s. Indicate the period T on your graph.

Solution

a. The amplitude is the coefficient of the cosine term. Thus the amplitude is 0.40 m.

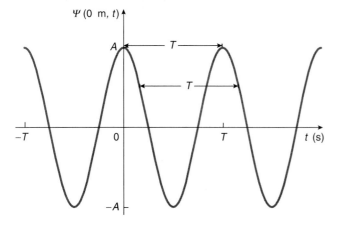

FIGURE 12.23 The oscillatory behavior of a sinusoidal wave at a fixed location, here the origin where $x = 0$ m.

b. The coefficient of x in the argument of the cosine is the angular wavenumber k. Thus

$$k = 6.00 \text{ rad/m}$$

But according to Equation 12.13, k is $(2\pi \text{ rad})/\lambda$; hence, you have

$$\frac{2\pi \text{ rad}}{\lambda} = 6.00 \text{ rad/m}$$

Solving for the wavelength λ,

$$\lambda = \frac{\pi}{3.00} \text{ m}$$

$$= 1.05 \text{ m}$$

c. The coefficient of t in the argument of the cosine is the angular frequency ω. Thus

$$\omega = 10.0 \text{ rad/s}$$

But the angular frequency ω is related to the frequency ν by

$$\omega = 2\pi\nu$$

Therefore

$$10.0 \text{ rad/s} = 2\pi\nu$$

Solving for the frequency,

$$\nu = \frac{5.00}{\pi} \text{ s}^{-1}$$

$$= 1.59 \text{ Hz}$$

d. According to Equation 12.10, the speed of the wave is the product of the frequency and the wavelength. Thus

$$v = (1.59 \text{ Hz})(1.05 \text{ m})$$
$$= 1.67 \text{ m/s}$$

e. Use the given wavefunction and substitute for x and t:

$$\Psi(5.00 \text{ m}, 2.00 \text{ s}) = (0.40 \text{ m}) \cos[(6.00 \text{ rad/m})(5.00 \text{ m})$$
$$- (10.0 \text{ rad/s})(2.00 \text{ s})]$$
$$= (0.40 \text{ m}) \cos(10.0 \text{ rad})$$
$$= (0.40 \text{ m})(-0.839)$$
$$= -0.34 \text{ m}$$

f. Make the substitution $t = 2.00$ s into the given wavefunction, which yields the waveform

$$\Psi(x, 2.00 \text{ s}) = (0.40 \text{ m}) \cos[(6.00 \text{ rad/m})x - 20.0 \text{ rad}]$$

A graph of this appears in Figure 12.24 with the wavelength λ indicated.

g. Making the substitution $x = 5.00$ m into the given wavefunction yields the oscillatory behavior at this location:

$$\Psi(5.00 \text{ m}, t) = (0.40 \text{ m}) \cos[30.0 \text{ rad} - (10.0 \text{ rad/s})t]$$

A graph of this is shown in Figure 12.25 with the period T indicated.

FIGURE 12.24

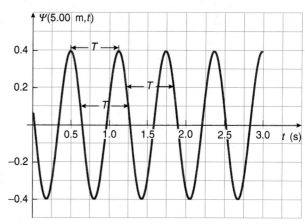

FIGURE 12.25

EXAMPLE 12.4 ⎯⎯⎯⎯⎯⎯⎯⎯⎯⎯⎯⎯⎯

Show that a sinusoidal wavefunction of the form

$$\Psi(x, t) = A \cos\left(kx - \omega t - \frac{\pi}{2} \text{ rad}\right)$$

is equivalent to

$$\Psi(x, t) = A \sin(kx - \omega t)$$

Solution
Use the double-angle trigonometric formula

$$\cos(\alpha - \beta) = \cos\alpha\cos\beta + \sin\alpha\sin\beta$$

and the given wavefunction with $\alpha = kx - \omega t$ and $\beta = \pi/2$ rad to obtain (omitting the radian units on $\pi/2$)

$$\Psi(x, t) = A\left[\cos\left(kx - \omega t\right)\cos\frac{\pi}{2} + \sin\left(kx - \omega t\right)\sin\frac{\pi}{2}\right]$$

$$= A \sin(kx - \omega t)$$

This means that a sinusoidal wave of the form

$$\Psi(x, t) = A \sin(kx - \omega t)$$

can be written in the cosinusoidal form by subtracting $\pi/2$ rad from the argument of the sine.

12.9 WAVES ON A STRING

Waves on a string or wire under tension are involved in the rich musical texture of all stringed instruments, and so an understanding of the physics associated with such waves can add to your appreciation of the guitar, banjo, violin, or piano that you might play.

When you stretch a string taut and send a transverse wave along it by wagging your hand back and forth, you can easily observe that the speed of the wave increases if you increase the tension in the cord. If you use a cord of greater mass per unit length, the speed decreases. Let's use our knowledge of physics to see how the wave speed depends on the tension and mass per unit length of the cord.

A transverse wave Ψ propagating along a string represents a departure of sections of the string from their equilibrium positions along the straight original position of the string. Such transverse departures from the equilibrium position typically are small compared with the length of the string; we will consider only such small-amplitude waves.

Place a uniform string under a tension of magnitude T. Let μ be the mass per unit length of the string, found by dividing the total mass m of the string by its total length ℓ:

$$\mu = \frac{m}{\ell}$$

If the end of the string suddenly is moved slightly to the side, a kink in the string propagates along the string at speed v. We want to examine the transverse motion of the bit of string caused by the propagating kink. Consider a differential length dx of the string just encompassing the kink during a small time interval dt, so that

$$dx = v\, dt$$

The mass dm of this small section of the string is the product of the length dx and the mass per unit length μ:

$$dm = \mu\, dx$$
$$= \mu v\, dt$$

Let the transverse component of the velocity of the string element be v_y, where we choose the y-axis to be perpendicular to the length of the string in the direction of motion of the mass element of the string. During the interval dt, the change in the y-component of the momentum of the mass element is

$$d\vec{\mathbf{p}} = (\mu v\, dt)v_y\hat{\mathbf{j}} - (0 \text{ kg·m/s})\hat{\mathbf{j}}$$
$$= (\mu v\, dt)v_y\hat{\mathbf{j}}$$

The forces acting on this bit of string are the tension exerted by other parts of the string and the weight of the string segment.

The strings of this Bösendorfer grand piano are under great tension. This piano also is pictured in Question 5 of Chapter 1 (page 12).

Since the segment of string is small, we neglect the weight compared with the tension; we then have the second law force diagram shown in Figure 12.26. Our results, therefore, will not depend on whether the axis of the string is horizontal, vertical, or inclined at some angle to the horizontal, which is fortunate for musicians! Since the tension has a component $T \sin \theta$ in the transverse direction, the impulse of the tension in the y-direction during the interval dt is

$$(T \sin \theta)\, dt\, \hat{\mathbf{j}}$$

The impulse–momentum theorem from Chapter 9 states that the total impulse is equal to the change in the momentum. For the y-direction we have

$$(T \sin \theta)\, dt\, \hat{\mathbf{j}} = (\mu v\, dt)v_y\hat{\mathbf{j}}$$
$$T \sin \theta = \mu v v_y \qquad (12.17)$$

From Figure 12.27 we have

$$\tan \theta = \frac{v_y}{v}$$

$$\frac{\sin \theta}{\cos \theta} = \frac{v_y}{v} \qquad (12.18)$$

Using Equation 12.18 for $\sin \theta$ in Equation 12.17, we find

$$(T \cos \theta)\frac{v_y}{v} = \mu v v_y$$

FIGURE 12.26 A propagating kink and the forces on the small mass element at the kink.

FIGURE 12.27 The velocity components at the kink.

Isolating the wave speed v, we obtain

$$v^2 = \frac{T}{\mu} \cos \theta$$

Since the motion of the string to the side was small, the angle θ also is small, in which case $\cos \theta \approx 1$. Hence, to a good approximation, we have

$$v^2 = \frac{T}{\mu}$$

The wave speed v is

$$v = \left(\frac{T}{\mu} \right)^{1/2} \qquad (12.19)$$

Notice that the greater the tension, the greater the speed; and the greater the mass per unit length, the smaller the speed of the wave on the string, both in keeping with our qualitative observations for a stretched cord.

In fact, Equation 12.19 is a special case of a general relationship for the speed of mechanical waves in material media. The speed is proportional to the square root of the magnitude of some stiffness quantity related to a force divided by an inertial quantity related to the mass:

$$v \propto \left(\frac{\text{magnitude of a force factor}}{\text{mass factor}} \right)^{1/2}$$

For example, Young's modulus E is the ratio between tensile stress and tensile strain in a material (see Equation 11.3). Young's modulus is a force factor. A corresponding mass factor for a bulk material ought to be the density ρ. A dimensional analysis of the ratio of Young's modulus to the density gives the square of a speed; in SI units,

$$\frac{[E]}{[\rho]} = \frac{\text{N/m}^2}{\text{kg/m}^3} = \frac{(\text{kg} \cdot \text{m/s}^2)/\text{m}^2}{\text{kg/m}^3} = \text{m}^2/\text{s}^2$$

In fact, a rigorous derivation finds that the speed of sound waves in solids is

$$v_{\text{solid}} = \left(\frac{E}{\rho} \right)^{1/2} \qquad (12.20)$$

In a similar way, the speed of sound waves in liquids is related to the bulk modulus B:

$$v_{\text{liquid}} = \left(\frac{B}{\rho} \right)^{1/2} \qquad (12.21)$$

EXAMPLE 12.5

A nylon clothesline of length 5.0 m and mass 0.076 kg is subjected to a tension of magnitude 5.1 N. If the end of the rope is flicked slightly to the side, what is the speed of the pulse that travels along the rope?

Solution
The mass per unit length of the rope is

$$\mu = \frac{m}{\ell}$$
$$= \frac{0.076 \text{ kg}}{5.0 \text{ m}}$$
$$= 0.015 \text{ kg/m}$$

Use Equation 12.19 to find the speed:

$$v = \left(\frac{T}{\mu} \right)^{1/2}$$
$$= \left(\frac{5.1 \text{ N}}{0.015 \text{ kg/m}} \right)^{1/2}$$
$$= 18 \text{ m/s}$$

EXAMPLE 12.6

The bulk modulus of water is 0.20×10^{10} N/m² and its density is 1.00×10^3 kg/m³. What is the speed of sound in water?

Solution
The speed of sound in liquids is related to the bulk modulus and the density through Equation 12.21:

$$v_{\text{liquid}} = \left(\frac{B}{\rho} \right)^{1/2}$$
$$= \left(\frac{0.20 \times 10^{10} \text{ N/m}^2}{1.00 \times 10^3 \text{ kg/m}^3} \right)^{1/2}$$
$$= 1.4 \times 10^3 \text{ m/s}$$

12.10 REFLECTION AND TRANSMISSION OF WAVES

Echoes are reflected sound waves; when you hear traffic noise through the wall of your room, the sound waves were transmitted through the wall. So longitudinal waves can be reflected and transmitted at boundaries between two media.

Likewise, take a rope and tie one end to a rigid wall. If you send a pulse down the rope toward the wall, the pulse will be reflected from the wall and return toward you. Notice that if the pulse incident on the wall is an "up" pulse, the reflected pulse is a "down" pulse, as shown in the series of sketches in Figure 12.28.

We say the pulse experiences a **change of phase** of π rad (=180°) on reflection. The phase change is caused by the force of the wall on the rope. Imagine the particle of the rope at the wall. As the wave pulse impinges on the wall, the wave tries to force the rope upward, and so the rope exerts an upward force on the wall. Thanks to Newton's third law, the wall exerts a force of the same magnitude in the opposite direction; hence the wall exerts a downward force on the rope, which produces the reflected pulse in the downward direction.

On the other hand, if we tie the rope to a ring that is free to slide on a frictionless, vertical post, an incident "up" pulse is reflected as an "up," as shown in the series of sketches in Figure 12.29, quite unlike the reflection from the rigid wall. There is no phase change on reflection.

If we tie the rope to another rope with greater mass per unit length and send a pulse down the rope, as in Figure 12.30(a), we discover that a pulse is transmitted to the second rope *and* a pulse is reflected. Both the transmitted and reflected pulses have smaller amplitudes than the original pulse, indicating that the energy in the original wave is somehow divided between the transmitted and reflected waves. Since the ropes have the same tension, but different masses per unit length, Equation 12.19 implies that the speed of the pulse along the rope with greater mass per unit length is slower. The characteristic length of the pulse also is smaller on this rope. An incident harmonic wave has a smaller wavelength on the rope of greater mass per unit length.

Such simultaneous reflection and transmission of a wave occurs whenever there is a change in the medium in which a wave propagates. Sound waves are both reflected and transmitted at walls and windows, and so you can hear a neighbor's conversation or stereo, much to their consternation or yours.* Light waves are both reflected and transmitted at boundaries between two transparent media such as air and water.

The simultaneous reflection and transmission of a wave is quite unlike the corresponding case for a particle; an incident particle is *either* totally reflected *or* totally transmitted, not both simultaneously.

Question 4

A surface water wave impinges on a sturdy sea wall. Does the reflected wave experience a phase change at the wall?

FIGURE 12.29 No phase change on reflection.

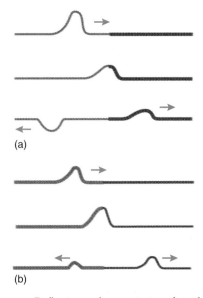

(a)

(b)

FIGURE 12.30 Reflection and transmission of a pulse at the boundary between two ropes with different masses per unit length. (a) A change of phase occurs when reflecting from a thicker rope. (b) No change of phase occurs when reflecting from a thinner rope.

---

*Intelligence agencies have developed ways to detect (from some distance) the small vibrations of window panes induced by the chatter within.

FIGURE 12.28 An incident pulse reflects from a rigid wall.

12.11 ENERGY TRANSPORT VIA MECHANICAL WAVES

Mechanical waves involve masses in motion, and so it should come as no surprise that energy is associated with waves. What is unusual about repetitive mechanical waves is that energy is transported from one point to another along the direction of the wave velocity without the net transport of mass between the two points: there are no particle streams.

Harmonic acoustic waves involve longitudinal simple harmonic oscillations of the particles that compose the medium. Likewise, for transverse harmonic waves on string, the mass particles of the string execute simple harmonic motion. We saw in Chapter 7 that simple harmonic oscillation of a mass m has a total mechanical energy E proportional to the square of the amplitude of the oscillation:

$$E = \frac{1}{2} k A^2 \qquad (12.22)$$

where k is the effective spring constant, equal to $m\omega^2$ via Equation 7.10. For a differential amount of mass dm, there is a differential amount of mechanical energy dE, and we write Equation 12.22 in the form

$$dE = \frac{1}{2} dm \, \omega^2 A^2 \qquad (12.23)$$

Let dm be the mass of a small length dx of the string, so that

$$dm = \mu \, dx$$

where μ is the mass per unit length of the string. Hence the total mechanical energy on the length dx is

$$dE = \frac{1}{2} \mu \, dx \, \omega^2 A^2 \qquad (12.24)$$

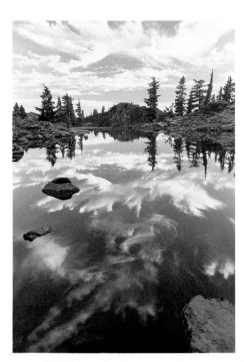

Light waves are both reflected and transmitted at the boundary between two transparent materials.

Since the wave is traveling in the x-direction at speed v,

$$v = \frac{dx}{dt}$$

so

$$dx = v \, dt$$

Substituting for dx in Equation 12.24, we find

$$dE = \frac{1}{2} \mu v \, dt \, \omega^2 A^2 \qquad (12.25)$$

The time rate of energy transport,

$$\frac{dE}{dt}$$

is the *power P* transmitted (transferred) by the wave.

Hence the power is

$$P = \frac{1}{2} \mu \omega^2 A^2 v \qquad (12.26)$$

The power transfer is proportional to the square of the angular frequency (equivalently, to the square of the frequency ν, since $\omega = 2\pi\nu$), to the square of the amplitude, and to the propagation speed of the wave.

The power transfer of any classical wave has the same dependence on frequency, amplitude, and speed, although we will not show this explicitly.

QUESTION 5

To double the power of a wave, by what factor must its amplitude increase if other factors are unchanged?

EXAMPLE 12.7

A nylon clothesline of length 5.0 m and mass 76 g has a tension of magnitude 50 N. One end of the string is wiggled back and forth in simple harmonic motion with a frequency of 10 Hz, producing a transverse harmonic wave of amplitude 2.0 cm on the string. What is the power transferred from one end of the string to the other?

Solution

To apply Equation 12.26, you first need to find the speed of the waves on the string from Equation 12.19:

$$v = \left(\frac{T}{\mu} \right)^{1/2}$$

$$= \left[\frac{50 \text{ N}}{(0.076 \text{ kg})/(5.0 \text{ m})} \right]^{1/2}$$

$$= 57 \text{ m/s}$$

The angular frequency ω of the oscillation is

$$\omega = 2\pi\nu$$

$$= 2\pi(10 \text{ Hz})$$

$$= 63 \text{ rad/s}$$

From Equation 12.26, the power is

$$P = \frac{1}{2} \mu \omega^2 A^2 v$$

$$= \frac{1}{2} \frac{0.076 \text{ kg}}{5.0 \text{ m}} (63 \text{ rad/s})^2 (0.020 \text{ m})^2 (57 \text{ m/s})$$

$$= 0.69 \text{ W}$$

12.12 WAVE INTENSITY

The **intensity** of a wave is the average power transmitted by the wave through one square meter oriented perpendicular to the direction the wave is propagating; see Figure 12.31.

The SI units of intensity are, therefore, watts per square meter (W/m^2).

Generalizing from Equation 12.26, the intensity of a wave is proportional to the square of its amplitude and the square of its frequency.

For waves moving out isotropically* in three dimensions, the power is distributed over the *area* of a sphere of radius r centered on the source, as shown in Figure 12.32. Since the area

---

*The word isotropic means "the same in all directions."

FIGURE 12.31 The intensity is the average power through one square meter perpendicular to the direction of propagation of the wave.

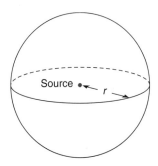

FIGURE 12.32 The power of a pointlike wave source is distributed uniformly over the area of sphere centered on the source.

of sphere is $4\pi r^2$, the intensity of the waves at a distance r from the source is

$$I = \frac{P}{4\pi r^2} \tag{12.27}$$

The intensity decreases with the inverse square of the distance from the source, and so the intensity of a *spherical* wave follows an inverse square law.

EXAMPLE 12.8

A source emitting waves isotropically in three dimensions has an intensity I_0 at a distance r_0 from the source. At what distance from the source is the intensity $0.100I_0$?

Solution
Since the wave is propagating isotropically in three dimensions, the intensity follows the inverse square law of Equation 12.27:

$$I_0 = \frac{P}{4\pi r_0^2} \tag{1}$$

Likewise at distance r, the intensity is

$$I = \frac{P}{4\pi r^2}$$

where the power P of the source is the same. Since you want $I = 0.100I_0$, you have

$$0.100 \, I_0 = \frac{P}{4\pi r^2}$$

Using equation (1) to substitute for P, you have

$$0.100 \, I_0 = \frac{1}{4\pi r^2} 4\pi r_0^2 I_0$$

Solving for r, you obtain

$$r = 3.16 r_0$$

Hence if you increase the distance from the source by a factor of 3.16, the intensity decreases by a factor of 10.0.

12.13 WHAT IS A SOUND WAVE?*

We say a **sound** exists in an elastic material if a propagating disturbance is a variation in the ambient density caused by a bulk shift in the positions of the particles of the material away from their nominal equilibrium positions. The disturbance Ψ can represent either the density variation or the departure of the particles from their erstwhile positions. For sound propagating in gases, the density variation also can be interpreted as a variation in pressure from the ambient pressure.

When you think of sound, you likely consider sound propagating through air (a gas), but it equally well can propagate through other media such as liquids and solids. It is difficult to imagine a gas such as air as an elastic material; if you swing your hand through the air, you do not feel much of a countervailing restoring force. However, take a tire pump, raise the plunger, and then hold a finger tightly over the end of the hose. When you press on the plunger, a restoring force is quite apparent: the air has elastic-like properties. Sound waves in air or other gases represent such rapid oscillations in density that the inertial properties of the air make it quite elastic. The propagating variations in density are correlated with the bulk particle motions, much like the longitudinal compressional waves along the axis of a stretched slinky. But, as we will see, the density wave and particle wave disturbances are not in phase (in step) with each other.

The air particles are randomly moving on a microscopic basis but, for clarity, we ignore these random motions here.* We represent the particles of the air as a uniform distribution of fixed dots that represents the ambient density, as shown in Figure 12.33.

Imagine a membrane (such as the diaphragm of a speaker) or wall executing rapid, back-and-forth, simple harmonic motion with period T in the gas. As the wall moves from its equilibrium position to the right, it accelerates the air particles next to it, compressing them because the particles beyond them have not yet shifted their positions. The compression is represented by a crowding of dots near the wall in Figure 12.34, indicating that the density is greater than the ambient density in this region. Dots further from the wall are as yet undisturbed, and the density there remains at its ambient value.

---

*We will examine the microscopic motions of the particles in a gas in greater detail in Chapter 14. Here we are interested in bulk motions of the particles of the gas.

In addition to their microscopic, random motion momentum, the particles also now have some momentum to the right, transferred to them by the moving wall. The momentum is transferred to other air particles to the right by collisions, and in this way, the initial compression propagates through the gas.

The motion of the wall in the opposite direction causes a rarefaction in the distribution of particles near it. A rarefaction has a density less than the ambient value, indicated by the increased spacing between the dots shown in Figure 12.35, causing them to accelerate in the opposite direction. As the wall oscillates back and forth, the process is repeated faithfully, and successive compressions and rarefactions propagate out into the gas, as shown in Figure 12.36. The particle motions are longitudinal—that is, along the line the wave is propagating.

Let's see how the variations in the ambient density correlate with the positions of the particles right and left of their original positions, and with the velocity components of the particles. Since the wall is executing simple harmonic oscillation and the gas is

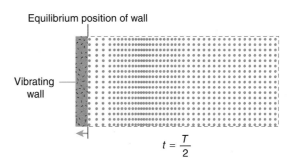

Equilibrium position of wall

Vibrating wall

$$t = \frac{T}{2}$$

FIGURE 12.35 The formation of a rarefaction.

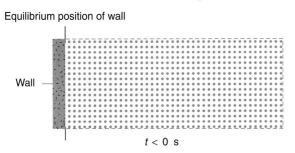

Equilibrium position of wall

Wall

$$t < 0 \text{ s}$$

FIGURE 12.33 Undisturbed air is represented as a uniform distribution of fixed dots.

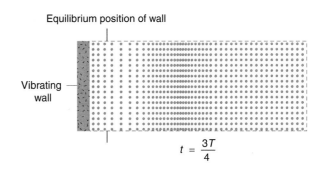

Equilibrium position of wall

Vibrating wall

$$t = \frac{3T}{4}$$

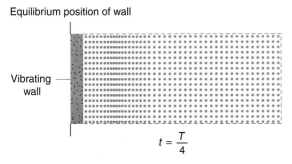

Equilibrium position of wall

Vibrating wall

$$t = \frac{T}{4}$$

FIGURE 12.34 The shift in the membrane causes a bunching of the dots near it.

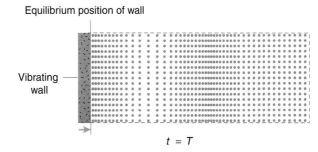

Equilibrium position of wall

Vibrating wall

$$t = T$$

FIGURE 12.36 Successive compressions and rarefactions propagate into the gas.

considered elastic, the particles also execute simple harmonic oscillations about their original nominal equilibrium positions. Recall that in simple harmonic oscillation of a particle, the speed is zero at greatest amplitude; a maximum speed exists when passing through the equilibrium position. The maximum density occurs at the center of the compression, where the particles are at their equilibrium position but are moving with greatest velocity to the right in our pictorial representation of the rightward traveling sound wave. The particle at the center of the compression must be moving faster than the ones on its right and left; we represent this by velocity arrows with different lengths in Figure 12.37. In this way the particle at the center of the compression is catching up to the particle on its right while pulling away from the particle on its left. The result is that the compression moves to the right.

Likewise in a rarefaction, where the variation in density has its greatest value below the ambient value, the particle at the center of the rarefaction must be moving faster to the left than the ones to its left and right, as indicated in Figure 12.38. In this way the given particle is catching up to the particle on its left and pulling away from the particle on its right. As a result the rarefaction also propagates to the right.

Thus we see that the velocity component of the particles is in phase (in step) with the density variation: greatest compression occurs at the same times and places where the velocity of the particles is greatest to the right, while the center of a rarefaction occurs where the velocity of the particles is greatest to the left.

We summarize the information about the relationship among the variations in density above (+) and below (−) the ambient value, the velocity component to the right (+) or left (−), and the variation in position of the particles from their ambient positions, taking + to the right and − to the left in the graphical representation of their respective waveforms shown in Figure 12.39. We reach several conclusions from this analysis. In the regions where the waves exist,

- the density variation and the longitudinal velocity component of the particles have their maxima and minima at those locations where the particles are at their former equilibrium positions; and

- the particles have maximum departure from equilibrium where the density is the ambient value (zero variation in density).

Center of compression is at equilibrium position of particle's simple harmonic oscillation with particle moving to the right

Magnified view of region around center of compression

FIGURE 12.37 The velocities of particles at and on either side of the maximum compression.

Center of rarefaction is at equilibrium position of particle's simple harmonic oscillation with particle moving to the left

Magnified view of region around center of rarefaction

FIGURE 12.38 The velocities of particles at and on either side of a rarefaction.

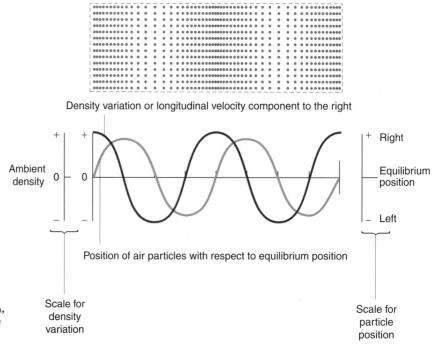

Density variation or longitudinal velocity component to the right

Position of air particles with respect to equilibrium position

FIGURE 12.39 Graphs of the density variation, velocity component, and position (relative to the equilibrium position) of the particles.

Hence the density wave disturbance Ψ_{density} is 90° (= $\pi/2$ rad) out of phase with the particle position wave disturbance Ψ_{position}.

12.14 SOUND INTENSITY AND SOUND LEVEL*

The Sounds of Silence

Paul Simon and Art Garfunkel[†]

From the sound of gently rustling leaves to the incredible roar of a space shuttle launch, the ear is a very sensitive detector of sound wave intensities. The **threshold for hearing** of the human ear for acoustic waves is an intensity of only about 10^{-12} W/m² at a frequency of about 1 kHz.[‡] At the other extreme, the intensity of sound waves that causes deafening pain (such as encountered frequently at rock concerts or a shuttle launch) is about a factor of 10^{12} greater than the threshold of hearing and is quite independent of frequency.

The enormous range of intensities detectable by the human ear makes the introduction of a more practical measure of sound level a matter of convenience. Logarithms are useful for this purpose. If a quantity α increases to $10^{12}\alpha$, the change in the value of the logarithm is only a factor of 12, rather than 10^{12}:

$$\log 10^{12}\alpha - \log \alpha = 12$$

The threshold of hearing intensity is defined to be the intensity

$$I_0 \equiv 10^{-12} \text{ W/m}^2$$

and is used as a reference for sound intensity. The **sound level** β of a sound intensity I is defined to be

$$\beta \equiv (10 \text{ dB}) \log \frac{I}{I_0} \tag{12.28}$$

The ratio I/I_0 is used since logarithms are exponents and must be dimensionless. The units for the sound level β are **decibels** (dB) one-tenth of the larger, but rarely used unit called a bell (B) named after the famous inventor of the telephone, Alexander Graham Bell (1847–1922).[#] From Equation 12.28 with $I = I_0$, the threshold of hearing thus has a sound level of 0 dB.

Each *factor* of 10 increase in sound intensity I produces an *additive* change in sound level of 10 dB; see Table 12.1.

PROBLEM-SOLVING TACTIC

12.3 Logarithms are exponents, so the expression

$$Y = \log X$$

means, after taking the antilogarithms,

$$X = 10^Y$$

Therefore when you take the antilogarithms of Equation 12.28, you have, after first dividing by 10 dB,

$$\frac{I}{I_0} = 10^{\beta/(10 \text{ dB})} \tag{12.29}$$

The human ear is not equally sensitive to sounds at all frequencies, as indicated in the graph in Figure 12.40. The ears of other lifeforms have different sensitivities to sounds, as you likely know if you have a dog or cat.

QUESTION 6

Dogs have very sensitive ears, capable of detecting sound intensities less than the threshold for human hearing. What sound levels does this correspond to in dB?

TABLE 12.1 Approximate Sound Levels and Sound Intensities

Sound	Sound level β (dB)	Sound intensity I (W/m²)
Threshold of hearing (definition)	0	10^{-12}
Cat purring	10	10^{-11}
Whisper at 1 m	20	10^{-10}
Normal conversation	60	10^{-6}
Urban traffic	70	10^{-5}
Subway train	90	10^{-3}
Chain saw at 1 m	100	10^{-2}
Loud rock concert	120	1
Threshold of pain	120	1
Jet engine at 10 m	150	10^3
Space shuttle launch at 50 m	200	10^8

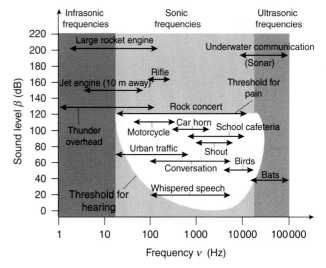

FIGURE 12.40 The frequency dependence of the threshold of hearing and of pain for a typical human ear.

[†]Title of song on album titled *Sounds of Silence* CS9269 Columbia Records CL2469 (Eclectic Music Company 1965).

[‡]The threshold intensity is a function of frequency; see Figure 12.40.

[#]The decibel unit also can be used for other types of waves if a reference intensity I_0 is chosen for those waves. The unit also is used in electronics for comparing quantities other than wave intensity.

EXAMPLE 12.9

If a sound level increases by 3.00 dB, by what factor does the sound intensity increase?

Solution

Let β be the sound level with sound intensity I. Then, according to Equation 12.28,

$$\beta = (10\ dB)\log\frac{I}{I_0}$$

Dividing by 10 dB and then taking antilogarithms, you have

$$\frac{I}{I_0} = 10^{\beta/(10\ dB)} \tag{1}$$

Let I' be the sound intensity when the sound level is $\beta + 3.00$ dB; hence,

$$\beta + 3.00\ dB = (10\ dB)\log\frac{I'}{I_0}$$

Dividing by 10 dB and then taking the antilogarithms, you have

$$\frac{I'}{I_0} = 10^{(\beta+3.00\ dB)/(10\ dB)} \tag{2}$$

Dividing equation (2) by equation (1),

$$\frac{I'}{I} = \frac{10^{(\beta+3.00\ dB)/(10\ dB)}}{10^{\beta/(10\ dB)}}$$
$$= 10^{0.300}$$
$$= 2.00$$

Therefore an increase in sound level of 3.00 dB increases the intensity by a factor of 2.00.

12.15 THE ACOUSTIC DOPPLER EFFECT*

If not by name, the qualitative aspects of the **acoustic Doppler effect** likely are familiar to you. If a source of sound (siren, train whistle, or screaming Mimi,[†] etc.) is approaching at high speed, you hear a higher frequency than when the source is at rest. Conversely, if the source is receding from you, a lower frequency is heard. The change in frequency is quite apparent as the source whooshes by you. The effect is qualitatively the same if the source is stationary and the observer is approaching the source (higher frequency heard) or receding from it (lower frequency heard).

A quantitative description of the acoustic Doppler effect first was devised in 1842 by Johann Christian Doppler (1803–1853). To verify the theory, Buys Ballot at Utrecht used trumpeters moving on a railroad flatcar and stationary observers who had perfect pitch.[‡] The experiment must have made a curious sight—and sound.

As Doppler himself speculated, the effect is a general one for all types of waves (sound, light, water, and other types of waves), although the quantitative description of the changes in frequency depends on the nature of the particular wave phenomena. The quantitative expressions for the frequency change are different for sound and for light. The distinctions arise because, unlike sound, light waves do not need a medium through which to travel, as we will see in Chapters 22 and 25. Here we examine the acoustic Doppler effect. The Doppler effect for electromagnetic waves (light) is best approached using relativity and will be examined when we tackle that fascinating aspect of physics in Chapter 25.

[†]Heroine of the opera *La Bohème* by Puccini (1896).

[‡]Perfect pitch is the ability to distinguish and name musical notes (frequencies) simply by hearing them.

Johann Christian Doppler first quantified the acoustic effect that now bears his surname.

The Doppler effect first was tested using moving and stationary musicians!

Let v be the speed of sound. The speed of sound in air varies with temperature and is about 343 m/s at 20 °C. At other temperatures t_{celsius}, the speed of sound in air is given approximately by the following empirical equation:

$$v = 331 \text{ m/s} + [0.60 \text{ m/(s·°C)}]t_{\text{celsius}}$$

The speed of sound in various other materials is tabulated in Table 12.2.

We need to consider motion of the source, the observer, and the medium through which the sound waves travel. For convenience, the speeds will be measured with respect to a fixed reference frame, say the ground. Although this seems to imply the existence of a preferred reference frame, we use this device as a means of keeping the conceptual development clear; it is easier to apply the resulting equations with this approach.

Motion of the Observer with the Source and Medium at Rest

Let a source emit sound of frequency v. The waves propagate toward an observer (see Figure 12.41). The lines represent circles of constant phase (wavefronts) and are spaced one wavelength apart; they propagate toward the observer at a speed v through the medium. If the observer is at rest, v wavefronts (crests) are received each second because the frequency is the number of such crests or wavelengths that are received during each second.

If the observer is moving with a speed v_{obs} toward the stationary source, as in Figure 12.42, the observer moves a distance $v_{\text{obs}}t$ during a time interval t. During one second, then,

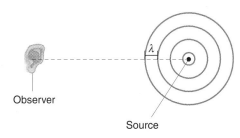

FIGURE 12.41 Wave propagating from a stationary source toward a stationary observer.

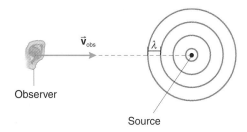

FIGURE 12.42 Observer moving toward a stationary source increases the effective frequency.

the observer moves a distance equal to $v_{\text{obs}}(1 \text{ s})$. There are $v_{\text{obs}}(1 \text{ s})/\lambda = v_{\text{obs}}/\lambda$ crests in this distance. So the observer receives an additional v_{obs}/λ crests each second because of motion toward the source.

Hence the total number of wavefronts the observer collects during the one second interval (which is the frequency v' of the waves detected by the observer) is

$$v' = v + \frac{v_{\text{obs}}}{\lambda}$$

Factoring the right-hand side,

$$v' = v\left(1 + \frac{v_{\text{obs}}}{\lambda v}\right) \qquad (12.30)$$

But the product of the wavelength and frequency is the speed of the wave (Equation 12.10), here the speed of sound v. Thus Equation 12.30 becomes

$$v' = v\left(1 + \frac{v_{\text{obs}}}{v}\right)$$

$$= v\left(\frac{v + v_{\text{obs}}}{v}\right) \qquad (12.31)$$

If the observer is moving away from the source, fewer wavefronts are received during each second. Indeed, instead of gaining v_{obs}/λ additional crests each second, this same number is lost; the number of wavefronts received by the observer during one second is thus

$$v' = v - \frac{v_{\text{obs}}}{\lambda}$$

TABLE 12.2 Approximate Value of the Speed of Sound in Various Materials

Material	Speed (m/s)
Gases	
Air (0 °C)	331
Air (20 °C)	343
Nitrogen (20 °C)	351
Oxygen (20 °C)	328
Liquids	
Ethyl alcohol (20 °C)	1.17×10^3
Mercury (20 °C)	1.45×10^3
Petroleum (15 °C)	1.33×10^3
Water (fresh) (20 °C)	1.48×10^3
Water (sea) (20 °C)	1.52×10^3
Solids	
Aluminum	6.42×10^3
Brass	4.25×10^3
Glass (crown glass)	5.66×10^3
Gold	3.24×10^3
Granite	6.0×10^3
Silver	3.60×10^3
Steel	6.10×10^3
Wood (oak)	4.10×10^3

Myer Kutz (editor), *Mechanical Engineer's Handbook* (John Wiley, New York, 1986), pages 692–693.

Hence the frequency detected by the receding observer is

$$v' = v\left(1 - \frac{v_{\text{obs}}}{v}\right)$$

$$= v\left(\frac{v - v_{\text{obs}}}{v}\right) \qquad (12.32)$$

We can combine Equations 12.31 and 12.32 into the following form:

$$v' = v\left(\frac{v \pm v_{\text{obs}}}{v}\right) \qquad (12.33)$$

where we use the + sign if the observer is approaching the source (higher frequency heard), and the − sign if the observer is receding from the source (lower frequency heard).

> **PROBLEM-SOLVING TACTIC**
>
> **12.4** The sign to use in Equation 12.33 can be determined as long as you remember that the frequency v' is higher than v if the observer is approaching the source (so the + sign must be used) and less than v when receding (use the − sign).

Boaters are quite aware of similar Doppler-like effects when riding with or against the waves.

Motion of the Source with the Observer and Medium at Rest

Once again, let v be the frequency emitted by the sound source, λ its wavelength, and v the speed of sound. Equation 12.10 still is valid: $v = v\lambda$. Let the speed of the source toward the observer be v_{source}.

The motion of the source toward an observer results in a decrease in (shortening of) the wavelength. Once a crest (wavefront) is emitted, the wavefront travels at speed v through the medium toward the observer independent of the motion of the source. If T is the period of the sound waves, the next crest is emitted after the source has traveled a distance $v_{\text{source}}T$ toward the observer. So instead of being a distance λ apart, the crests are a distance $\lambda - v_{\text{source}}T$ apart, as shown in Figure 12.43. Thus the effective wavelength λ' of the sound that reaches the observer is

$$\lambda' = \lambda - v_{\text{source}}T$$

The frequency v' of the sound reaching the observer is v/λ'. Hence

$$v' = \frac{v}{\lambda - v_{\text{source}}T}$$

$$= \frac{v}{\lambda - \dfrac{v_{\text{source}}}{v}}$$

$$= \frac{vv}{\lambda v - v_{\text{source}}}$$

$$= v\left(\frac{v}{v - v_{\text{source}}}\right) \qquad (12.34)$$

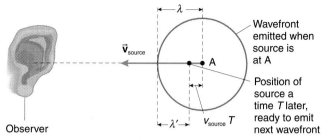

FIGURE 12.43 Motion of the source toward the observer decreases the effective wavelength of the sound.

If the source is receding from the observer, the wavelength is increased because of the motion of the source. In this case the wavelength λ' of the sound reaching the observer is

$$\lambda' = \lambda + v_{\text{source}}T$$

In a manner similar to that just shown, we find the frequency v' the observer detects is

$$v' = v\left(\frac{v}{v + v_{\text{source}}}\right) \qquad (12.35)$$

For the motion of the source, Equations 12.34 and 12.35 can be combined into the single expression:

$$v' = v\left(\frac{v}{v \mp v_{\text{source}}}\right) \qquad (12.36)$$

where we use the − sign for an approaching source (higher frequency heard by the observer), and the + sign for a receding source (lower frequency heard by the observer).

> **PROBLEM-SOLVING TACTIC**
>
> **12.5** The sign to choose in Equation 12.36 can be determined by whether the observer hears a higher or lower frequency (as the source approaches or recedes, respectively).

> The equations for the motion of the observer, Equation 12.33, and for the motion of the source, Equation 12.36, can be combined into a single equation:
>
> $$v' = v\left(\frac{v \pm v_{\text{obs}}}{v \mp v_{\text{source}}}\right)$$
>
> If the source is not moving, $v_{\text{source}} = 0$ m/s and we have Equation 12.33 for the motion of the observer. If the observer is not moving, $v_{\text{obs}} = 0$ m/s and we have Equation 12.36 for the motion of the source. If both are moving, we put in both speeds and choose the signs, as if only one or the other was moving (see Example 12.10).

Motion of the Medium (a Wind) with the Source and Observer at Rest

Let the medium be moving along the line from the source to the observer with a speed v_{med}, as shown in Figure 12.44. The

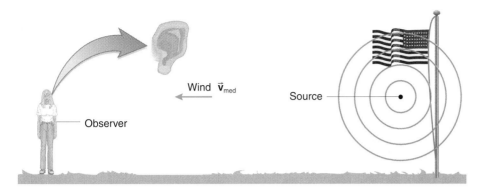

motion of the medium past the source and observer (at rest relative to each other) is a wind. In this case the wavefront crests are spaced further apart as they leave the source; the wavelength is increased.

This might lead you to conclude that the frequency heard by the observer decreases, but this is *not* the case. Although the distance between the wavefronts crests is longer, the wavefronts are carried to the observer at a faster speed: $v + v_{med}$. The two effects exactly compensate for each other, and so the observer receives exactly the same number of crests per second as with no wind.

Likewise, if the wind is blowing from the observer to the source, the distance between the crests of the waves decreases, which leads you to think you would hear a higher frequency. But the speed the crests move toward the observer also decreases (and is $v - v_{med}$). The two effects again exactly compensate for each other and the observer hears no change in frequency.

> Thus the wind, acting alone (with the source and observer at rest with respect to each other), has *no effect* on the frequency and $v' = v$.

Motion of the Observer and/or Source with a Wind

The effect of the wind is to change the speed of the waves propagating from the source to the observer. Ironically, if either the source or observer is moving, the wind *does* have an effect on the frequency detected! That is, if the wind is blowing in a direction from the source to the observer, we replace v with $v + v_{med}$; on the other hand, if the wind is blowing from the observer and toward the source, we replace v with $v - v_{med}$.

> Therefore the most general equation for the acoustic Doppler effect is
>
> $$v' = v \left(\frac{v \pm v_{med} \pm v_{obs}}{v \pm v_{med} \mp v_{source}} \right) \qquad (12.37)$$

Notice that if the source and observer are at rest, $v_{obs} = v_{source} = 0$ m/s, and then the wind speed v_{med} has no effect: $v' = v$. On the other hand, if v_{obs} and/or v_{source} is not zero, then v_{med} *does* have an effect on the frequency detected by the observer.

PROBLEM-SOLVING TACTIC

12.6 To use the general equation for the Doppler effect (Equation 12.37), there are a few things to keep in mind. These conditions may remind you of the infamous IRS 1040 tax form in view of their written complexity:

a. If there is a wind blowing from the source toward the observer, then the + sign is used for the v_{med} term since the effective speed of the waves increases; if the wind is blowing from the observer toward the source, the − sign is used for the v_{med} term since the effective speed decreases. This accounts for the change in the speed of the waves relative to a stationary source and observer.

b. If the source is moving away from the observer (thought of as being fixed), then the frequency heard by the observer is lower and we need to use the + sign in the v_{source} term in the denominator. Likewise, if the source is moving toward the observer (thought of as at rest), then the frequency heard by the observer is higher and we need to use the − sign in the v_{source} term in the denominator.

c. If the observer is moving toward the source (thought of as being at rest), then the observer hears a higher frequency and so the + sign must be used with the v_{obs} term in the numerator. On the other hand, if the observer is moving away from the source (thought of as being at rest), then the frequency heard by the observer is lower and so the − sign must be used with the v_{obs} term in the numerator.

Our treatment of the acoustic Doppler effect involved motion of the source, observer, and medium along the line connecting the source and observer. If the motions are not along this line, then it is the component of the motion along this line that is operative in the acoustic Doppler effect.

QUESTION 7

An observer is moving with respect to a source of sound yet *no* Doppler effect is detected (no change in frequency is heard). Describe the motion of the observer with respect to the source.

EXAMPLE 12.10

A 261.6 Hz source (the famous middle C), an observer, and a wind are moving as indicated in Figure 12.45 with respect to the ground. What frequency is heard by the observer?

FIGURE 12.45

Solution

Since the wind is blowing in a direction from the observer to the source, use the minus sign attached to the v_{med} term in Equation 12.37 with v_{med} as 5.0 m/s since the effective speed of sound is decreased. The source is moving away from the observer (considered to be fixed); thus you use the plus sign with the v_{source} term in the denominator, since the frequency heard by the observer is lower than normal. The observer is moving away from the source (considered to be fixed); thus you use the minus sign with the v_{obs} term in the numerator of Equation 12.37, since a receding observer results in a decrease in perceived frequency. Now the frequency heard by the observer can be found from Equation 12.37:

$$v' = (261.6 \text{ Hz}) \left(\frac{343 \text{ m/s} - 5.0 \text{ m/s} - 20.0 \text{ m/s}}{343 \text{ m/s} - 5.0 \text{ m/s} + 10.0 \text{ m/s}} \right)$$

$$= 239 \text{ Hz}$$

An observer-musician would say the note is not middle C but between B♭ and B.

12.16 SHOCK WAVES*

As with most equations in physics, Equation 12.37 has limitations. Consider a situation in which

a. the medium is at rest, so that $v_{med} = 0$ m/s;
b. an observer is at rest, so that $v_{obs} = 0$ m/s; and
c. a source is moving toward the observer, so that the minus sign in front of term for v_{source} must be used in Equation 12.37 for the general acoustic Doppler effect.

Under these conditions, the general equation for the Doppler effect becomes

$$v' = v \left(\frac{v}{v - v_{source}} \right) \qquad (12.38)$$

As the speed of the source increases, the denominator decreases and the frequency v' heard by the observer increases. But notice that as v_{source} approaches the speed of sound v, Equation 12.38 diverges (when $v_{source} = v$) and then yields negative frequency values for v when $v_{source} > v$. These results mean that the speed

of the source must be less than the speed of sound in the medium for Equations 12.37 and 12.38 to apply.

As v_{source} approaches v, the distance between the wavefronts ahead of the source approaches zero (the wavelength approaches zero) and the source is able to keep pace with all the wavefronts in the forward direction, as indicated in Figure 12.46.

What happens if the speed of the source exceeds the speed of sound v in the medium? If you blindly use Equation 12.38, it yields a negative answer for the frequency: an unphysical result. The limitation to Equation 12.38 certainly is not apparent mathematically, since there is nothing mathematically wrong with the negative result for the frequency if $v_{source} > v$. Something new must happen physically when the speed of the source exceeds the speed of the wave in the surrounding medium: the source escapes from inside its propagating wavefront before the next one is emitted, as shown in Figure 12.47.

In this case the envelope of the propagating wavefronts forms a cone, known as the **Mach cone,*** with its apex at the

---

*The term is named for Ernst Mach (1836–1916), an Austrian physicist.

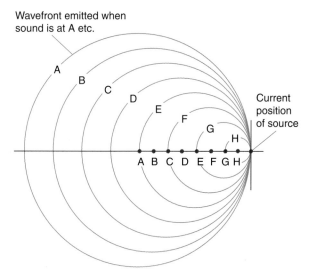

FIGURE 12.46 When the source speed approaches the speed of sound, the wavelength in the forward direction approaches zero.

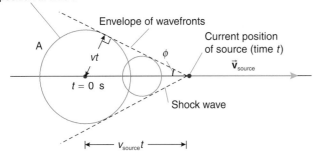

Wavefront emitted when $t = 0$ s
when source is at current
position at time t

Envelope of wavefronts

A

vt

ϕ

Current position
of source (time t)

$\vec{v}_{source}$

$t = 0$ s

Shock wave

$v_{source}t$

Supersonic regime $v_{source} > v$

FIGURE 12.47 A source exceeding the speed of sound escapes
from the wavefront before the next one is emitted.

Cerenkov radiation (light) is formed when charged particles travel
faster than the speed of light in a material.

source. If the wavefront labeled A in Figure 12.47 is emitted
when $t = 0$ s, then the radius of its circular wavefront at time t
is vt, since the wavefront moves at the speed v through the
medium. The source traveling at speed v_{source} moves a distance
$v_{source}t$ during the same time interval. Thus the half-angle ϕ of
the Mach cone (half the apex angle of the cone itself) is found
from the trigonometric relation

$$\sin \phi = \frac{v}{v_{source}} \qquad (v_{source} > v) \qquad (12.39)$$

The envelope of the wavefronts forms a **shock wave**.

Such acoustic shock waves commonly are called **sonic
booms**, familiar to us from the passage of supersonic aircraft.
Cracking a whip is another way to generate a sonic boom. When
the whip cracks, its tip is briefly traveling at a speed greater than
the speed of sound in air. Notice that since $v_{source} \geq v$ in Equation
12.39, the largest angle ϕ for the shock wave is 90°: when
$v_{source} = v$, $\sin \phi = 1$, corresponding to the situation depicted in
Figure 12.46. As the speed of the source increases past v, the half-
angle ϕ of the Mach cone decreases from 90°.

The **Mach number** is the ratio of the speed of the source to
the speed of sound in the medium:

$$\text{Mach number} = \frac{v_{source}}{v} \qquad (12.40)$$

Note that the half-angle ϕ of the Mach cone is the sine of the
reciprocal of the Mach number:

$$\sin \phi = \frac{1}{\text{Mach number}} \qquad (\text{Mach numbers} \geq 1) \qquad (12.41)$$

Of course, this equation is legitimate only for Mach numbers
greater than or equal to one. (Why? What is $\sin \phi$ for Mach
numbers less than 1?)

Shock waves also are apparent in other wave phenomena as
well. When electrically charged particles such as electrons or
protons travel through a material faster than the speed of light in
that medium, an unusual type of shock wave called Cerenkov

Bow waves (wakes) are quite complex waves.

radiation is formed: light. Such Cerenkov radiation is visible in
nuclear reactors that use water as a radiation shield.

The angle of the Mach cone has a strong dependence on the
speed. Observations of the waves produced by boats, swimmers,
or ducks show that their V-shaped patterns have an apex angle
of about 40° over a fairly wide range of speeds. Such bow waves
are *not* shock waves caused by motion faster through the water
than a fixed propagation speed. These bow waves or **wakes** are
quite complex physical waves that arise when wave speed varies

with wavelength, a phenomenon called **dispersion**. You might recall that small-wavelength ripples on a pond propagate more slowly than large-wavelength ocean waves.

12.17 DIFFRACTION OF WAVES

When waves pass through an opening whose size is comparable to the wavelength, the waves spread out on the other side of the opening, as shown in Figure 12.48.

> The spreading out of waves after they pass through or around such apertures or obstacles is called **diffraction**.

Diffraction is a characteristic feature of any wave. If the opening is made larger, as in Figure 12.49, the diffraction becomes less pronounced, except near the edges of the opening. When the opening is much larger than the wavelength, diffraction is minimal and the waves pass essentially straight through the opening, following the path of the incident rays.

If a person is talking outside your open door but is not visible, you still can easily hear the conversation. Hence audible sound waves must have wavelengths on the order of magnitude

FIGURE 12.48 Diffraction of waves at an aperture; the size of the aperture is comparable to the wavelength.

FIGURE 12.49 Diffraction is minimal when the opening is much larger than the wavelength.

of the width of the door. For middle C with a frequency of 261.6 Hz, the wavelength is found using Equation 12.10:

$$v = \nu\lambda$$
$$\lambda = \frac{v}{\nu}$$
$$= \frac{343 \text{ m/s}}{261.6 \text{ Hz}}$$
$$= 1.31 \text{ m}$$

This is about the order of magnitude of a door opening. The sound waves diffract, which is why you can hear them around corners. Visible light waves, however, travel in straight lines through openings such as doors or keyholes, producing well-defined shadows behind the objects. Hence the wavelengths of visible light waves must be quite small compared with the sizes of common objects. We will study light diffraction effects in greater detail in Chapter 24. Here we are content to note that diffraction effects give a clue to the size of the wavelength of any wave.

QUESTION 8
You may have noticed that behind a small hill AM radio stations can be received more easily than FM radio stations. What does this observation imply about the relative sizes of the wavelengths of these radio waves?

12.18 THE PRINCIPLE OF SUPERPOSITION

And spend her strength with over-matching waves.
William Shakespeare (1564–1616)*

In this section we see what happens when and where two or more waves of the same type[†] meet. For example, if you drop two pebbles into a pond simultaneously at slightly different locations, two wave trains are produced that propagate across the water surface and overlap. Likewise, if you and a friend hold opposite ends of a taut rope and each flick your respective ends in the same plane to generate two wave pulses, the two waves propagate along the rope in opposite directions; see Figure 12.50. Where and when they reach each other, the multiple waves combine to produce a more complicated disturbance. If the amplitudes are not too large, the total wave disturbance Ψ_{total} at any point x and time t is the sum of the individual wave disturbances:

$$\Psi_{\text{total}}(x, t) = \Psi_1(x, t) + \Psi_2(x, t) + \Psi_3(x, t) + \cdots \quad (12.42)$$

Wave disturbances that follow this linear **principle of superposition** are called linear waves and are the only types of waves we will consider.[‡]

---

*Henry VI, Part III, Act I, Scene 4, line 21.
[†]Two (or more) water waves, waves on a rope, sound waves, light waves, and so on. For transverse waves, the waves must have the same plane of polarization. If not, the waves must be summed like vectors, not scalars.
[‡]If the amplitudes are large, then nonlinear superposition of the waves occurs. Such nonlinear superposition is beyond the scope of an introductory course.

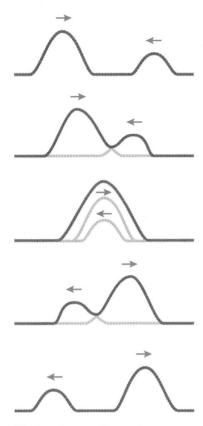

FIGURE 12.50 The interference of two pulses on a rope.

When two or more similar types of waves exist simultaneously at a point in space, we say the waves **interfere** with each other. If the resulting wave disturbance at a point is greater than that produced by any of the waves acting alone, we say the waves exhibit **constructive interference**. If the resulting disturbance is less than that produced by any alone, the waves exhibit **destructive interference**.

Superposition implies that on a string two wave pulses in the same plane moving in opposite directions proceed past each other without being changed in any way except where and when the two waves overlap in space and time; see Figure 12.50. Perhaps you also have observed such constructive and destructive interference of water waves while boating on a busy waterway.

In the next two sections we examine two special kinds of wave interference using the principle of superposition.

12.19 STANDING WAVES

. . . there shall thy proud wave be stayed.

Book of Job*

When you relax by plucking a guitar or banjo, or tickle the ivories on a piano, the waves you generate on the strings have rather remarkable properties. The plucking of a string that is fixed at both ends produces at least two harmonic (sinusoidal)

---
*Chapter 38, verse 11.

waves of the same amplitude that travel in opposite directions. The superposition of the two waves and their reflections from the fixed ends results in a rather astounding total wave disturbance.

Let the x-axis be along the length of the string. A harmonic (sinusoidal) wave of amplitude A moving toward increasing values of x has the form

$$\Psi_1(x, t) = A \cos(kx - \omega t)$$

A similar harmonic wave traveling toward decreasing values of x has the form

$$\Psi_2(x, t) = A \cos(kx + \omega t)$$

The resulting disturbance at any x and t is, according to the principle of superposition,

$$\Psi_{\text{total}}(x, t) = \Psi_1(x, t) + \Psi_2(x, t)$$
$$= A[\cos(kx - \omega t) + \cos(kx + \omega t)] \quad (12.43)$$

That's all the physics. To see what is going on, we dust off a trigonometric identity that indicates how to add two cosine functions together:

$$\cos \alpha + \cos \beta = 2 \cos\left(\frac{\alpha + \beta}{2}\right) \cos\left(\frac{\alpha - \beta}{2}\right) \quad (12.44)$$

We apply this identity to the sum of the cosines in Equation 12.43:

$$\Psi_{\text{total}}(x, t) = 2A \cos\left[\frac{(kx - \omega t) + (kx + \omega t)}{2}\right]$$
$$\times \cos\left[\frac{(kx - \omega t) - (kx + \omega t)}{2}\right]$$
$$= 2A \cos(kx) \cos(-\omega t) \quad (12.45)$$

Since the cosine of a negative angle is the same as the cosine of the positive angle, we rewrite Equation 12.45 in the following form:

$$\Psi_{\text{total}}(x, t) = 2A \cos(kx) \cos(\omega t) \quad (12.46)$$

This is a remarkable type of total wave disturbance. For certain positions, namely where the product kx is an *odd* integral multiple of $\pi/2$ rad, the wave disturbance $\Psi_{\text{total}}(x, t)$ vanishes for all times t! Let n be an odd integer. Hence if

$$kx = n\frac{\pi}{2} \text{ rad} \quad (n \text{ odd integer}) \quad (12.47)$$

the total disturbance is zero at all times, since

$$\cos\left(n\frac{\pi}{2} \text{ rad}\right) = 0 \quad (n \text{ odd integer})$$

The locations where the total wave disturbance is zero at all times are called **nodes**. Since the angular wavenumber k is $(2\pi \text{ rad})/\lambda$, Equation 12.47 implies that the nodes are located at those positions that satisfy the following condition:

$$\frac{2\pi \text{ rad}}{\lambda} x_{\text{node}} = n\frac{\pi}{2} \text{ rad} \quad (n \text{ odd integer})$$

Thus

$$x_{\text{node}} = n\frac{\lambda}{4} \quad (n \text{ odd integer})$$

By substituting n and then $n + 2$ into this expression, we can show the following:

> Successive nodes are separated in space by *half* a wavelength.

The maximum amplitude of the resulting superposition of the two waves occurs where the $\cos(kx)$ factor in Equation 12.46 is equal to unity. The amplitude at these locations is then $2A$, or twice what each individual wave could contribute. The positions where the amplitude of the resulting motion has its maximum are called **antinodes**. The antinodes occur where kx is an *even* integral multiple of $\pi/2$ rad, since then the $\cos(kx)$ term becomes

$$\left| \cos\left(n\frac{\pi}{2} \right) \right| = 1 \qquad (n \text{ even integer})$$

That is, the antinodes occur where

$$kx_{\text{antinode}} = n\frac{\pi}{2} \qquad (n \text{ even integer})$$

Since $k = 2\pi/\lambda$, the antinodes are located where

$$x_{\text{antinode}} = n\frac{\lambda}{4} \qquad (n \text{ even integer})$$

> Note that successive antinodes along the string also are separated by half a wavelength.

The resulting disturbance has the appearance shown in Figure 12.51, where the double-ended arrows indicate the amplitude of the oscillations at that position. The string oscillates in a stationary envelope pattern and has lost all indication of the underlying traveling waves!

> The disturbance resulting from the superposition of two waves of equal amplitude and frequency traveling in opposite directions is called a **standing wave**.

Notice that the amplitude of the resulting standing wave disturbance depends on the location x. At any given position x, the oscillation is simple harmonic oscillation, varying sinusoidally with time [thanks to the $\cos(\omega t)$ term] with amplitude

$$2A \cos(kx)$$

Plucked guitar and banjo strings illustrate standing waves on strings; their dynamical behavior is fun to watch using a strobe light. Since the string is tied down or fixed at both ends, there must be nodes at the extremes of the string. Let ℓ be the distance between the fixed ends of the string.* Since the nodes of a standing wave are separated by half a wavelength, the longest wavelength λ_1 that can be established on the string satisfies the condition

$$\ell = \frac{\lambda_1}{2}$$

or

$$\lambda_1 = 2\ell$$

Since the speed of the wave is the product of the frequency times the wavelength,

$$v = \nu\lambda$$

the lowest frequency ν_1 of the vibrating string is

$$\nu_1 = \frac{v}{\lambda_1}$$

Using Equation 12.19 for the wave speed, the lowest frequency is found to be

$$\nu_1 = \frac{1}{2\ell}\left(\frac{T}{\mu} \right)^{1/2}$$

---

*For stringed instruments, ℓ may be different from the total length of the string, used to determine the mass per unit length μ, because the string may be wrapped around pegs, pass over bridges, or be shortened by a finger fixing one end, such as in instruments like guitars, banjos, and violins.

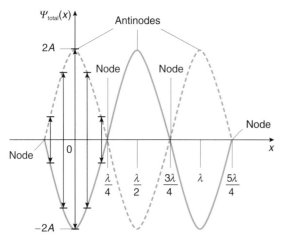

FIGURE 12.51 A graph of a standing wave on a string. The amplitude is a function of position along the string.

The nodes and antinodes of a standing wave on a string fixed at both ends are quite apparent in this photograph.

where T is the magnitude of the tension in the string. The lowest frequency is called the **fundamental frequency** of the string or, equivalently, the first harmonic (whence the subscript 1). For low fundamental frequencies, an instrument must have long strings with considerable mass per unit length like the bass strings on a piano or double bass. *Tuning* a stringed instrument involves adjusting the magnitude T of the tension in each string to obtain precisely the desired note or frequency.

Standing waves of other frequencies, with different amplitudes, also are permitted simultaneously on the string. Since the fixed ends must always be nodes, the only condition that must be met is that the distance ℓ between the fixed ends of the string must be an integral number of half wavelengths:

$$\ell = n \frac{\lambda_n}{2}$$

Thus the permitted wavelengths are given by

$$\lambda_n = \frac{2\ell}{n} \qquad (n \text{ is a positive integer}) \qquad (12.48)$$

The frequencies of these vibrations are then

$$\nu_n = \frac{v}{\lambda_n}$$

Using Equation 12.48 for λ_n and 12.19 for v, we get

$$\nu_n = n \frac{1}{2\ell} \left(\frac{T}{\mu} \right)^{1/2} \qquad (12.49)$$

When $n = 1$, we have the fundamental frequency (i.e., the first harmonic). The second harmonic corresponds to $n = 2$; the third harmonic to $n = 3$; and so forth. See Figure 12.52. So the integer n numbers the various harmonics.

The frequencies of the higher harmonics are integral multiples of the fundamental frequency ν_1.

Not all frequencies are present on the string—only certain special, characteristic frequencies.

The frequencies that may exist on the string are known as the **eigenfrequencies** of the system, from the German word *eigen*, meaning *characteristic*.

The collection of allowed frequencies and their respective amplitudes is called the **frequency spectrum** of the system.

For a vibrating string, the frequency spectrum consists of the fundamental and integral multiples of it. The amplitudes of the various eigenfrequencies may vary, producing the unique sound associated with various instruments. Figure 12.53 indicates the high and low frequency range of many instruments.

As a practical matter, the transverse standing waves on strings generally produce little sound: the coupling between the string and the air is very small because the longitudinal cross section of the string is tiny. The vibrating string alone cannot transfer energy effectively to the air to cause an acoustic wave with an appreciable sound level. Hence the vibration of the

Two-dimensional standing waves are produced on a vibrating drumhead.

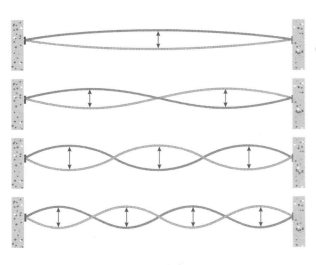

Fundamental (first harmonic) ν_1	$n = 1$	$\lambda_1 = \frac{2\ell}{1}$
Second harmonic $2\nu_1$	$n = 2$	$\lambda_2 = \frac{2\ell}{2}$
Third harmonic $3\nu_1$	$n = 3$	$\lambda_3 = \frac{2\ell}{3}$
Fourth harmonic $4\nu_1$	$n = 4$	$\lambda_4 = \frac{2\ell}{4}$

FIGURE 12.52 Standing wave pattern of the fundamental and several harmonics on a string fixed at both ends.

FIGURE 12.53 Frequency ranges of various musical instruments.

string must be made more effective to create louder (more intense) sounds. The amplification can be done mechanically via sounding boards (in pianos and harpsichords), open-backed membranes (in banjos), resonant sound boxes (violins, guitars, etc.), or electronically (electric guitars and synthesizers). The sounding boards and boxes have much greater surface area and couple the vibration to the air with greater efficiency than the naked string, much like the vibrating wall in Section 12.13. Musical synthesizers are capable of closely simulating electronically the frequency spectrum of many musical instruments without the need for the strings, reeds, and air columns associated with acoustic instruments.

Sound waves in organ pipes are examples of resonant, longitudinal acoustic standing waves. Consider a pipe closed at one end by water whose level can be adjusted by means of a reservoir, as shown in Figure 12.54. A tuning fork that vibrates at frequency ν is held over the open end of the pipe. The fork emits sound waves of wavelength λ. The sound waves propagate down the pipe and are reflected from the water surface. The reflected and incident waves superimpose to form a longitudinal standing wave pattern in the pipe.

If the water level in the pipe is adjusted so that maximum sound is heard, a condition of **resonance** is established in the column. At resonance, there is a node of the standing longitu-

FIGURE 12.54 A way to generate acoustic standing waves in a pipe closed at one end.

dinal vibratory particle position waves at the water surface and an antinode at the open end. The distance between nodes (or between antinodes) is $\lambda/2$, and so the distance between a node and an antinode is $\lambda/4$. Therefore there must be an odd number of quarter wavelengths in the length of the resonant air column:

$$\ell = n\frac{\lambda_n}{4} \qquad (n \text{ an odd integer}$$

Thus the permitted wavelengths for a closed pipe are

$$\lambda_n = \frac{4\ell}{n} \qquad (n \text{ an odd integer})$$

The longest resonant wavelength corresponds to $n = 1$ and produces the smallest (lowest) frequency, known as the fundamental frequency of the pipe:

$$\nu_1 = \frac{v}{\lambda_1}$$
$$= \frac{v}{4\ell}$$

Higher odd integers correspond to odd-numbered harmonics of the fundamental; not all harmonics are possible. Closed organ pipes are closed at one end like the pipe in Figure 12.54. They produce the fundamental frequency and the associated odd-numbered harmonics of the fundamental, as shown in Figure 12.55.

A pipe open at both ends, as shown in Figure 12.56, has a fundamental frequency at resonance corresponding to antinodes of the longitudinal vibratory particle position standing waves at each end of the pipe with a single node in the middle. The length of the pipe then is equal to $\lambda/2$. Since $v = \nu\lambda$, the fundamental frequency is

$$\nu_1 = \frac{v}{2\ell}$$

The standing wave pattern with two nodes between antinodes at the ends has a wavelength equal to the length of the pipe. The corresponding frequency is twice the fundamental and so this is the second harmonic. A standing wave pattern with three nodes between the antinodes at the end is the third harmonic. Thus *all* harmonics now are possible.

The wavelengths of the harmonics of an open pipe are given by

$$\lambda_n = \frac{2\ell}{n} \qquad (n \text{ any positive integer})$$

Open organ pipes, open at both ends, are constructed this way.

Since it is possible to have all the harmonics of the fundamental with open pipes rather than with closed pipes (in which only the odd-numbered harmonics are possible), open-piped instruments produce richer musical tones than instruments with closed pipes.

QUESTION 9

Examine Equation 12.46. Are there instants t when the wave disturbance is zero for every x? If so, how many times during one period of the wave disturbance? Sketch such a waveform.

Note: The acoustic waves are *longitudinal* standing waves, not transverse waves (which are indicated for clarity only).

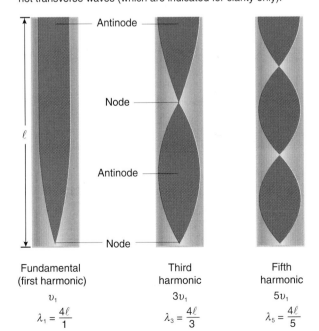

FIGURE 12.55 A pipe closed at one end, its fundamental frequency, and several harmonics. Only odd harmonics are possible.

Note: The acoustic waves are *longitudinal* standing waves, not transverse waves (which are indicated for clarity only).

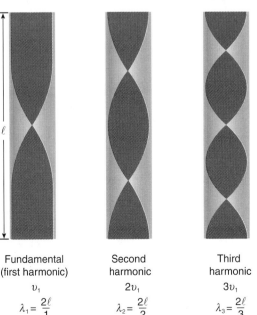

FIGURE 12.56 A pipe open at both ends, its fundamental frequency, and several harmonics. All harmonics are possible.

QUESTION 10

Why do the strings for the lowest-frequency notes of a piano have wire wrapped tightly around them?

EXAMPLE 12.11

A banjo string with a length 0.585 m between its fixed ends has a mass per unit length of 5.74×10^{-5} kg/m. It is tuned to play the note A_4 (with a frequency of 440 Hz) as the fundamental.

a. What is the tension in the string?
b. What is the wavelength of the fundamental?
c. What is the frequency of the second harmonic?
d. What is the distance between the nodes of the second harmonic?

Solution

a. The frequency, tension, length, and mass per unit length are related by Equation 12.49 with $n = 1$ for the fundamental:

$$v = \frac{1}{2\ell}\left(\frac{T}{\mu}\right)^{1/2}$$

Solving for T, you find

$$
\begin{aligned}
T &= 4\ell^2 v^2 \mu \\
&= 4(0.585 \text{ m})^2 (4.40 \times 10^2 \text{ Hz})^2 (5.74 \times 10^{-5} \text{ kg/m}) \\
&= 15.2 \text{ N}
\end{aligned}
$$

b. The wavelength of the fundamental can be found in two ways.

Method 1

The fundamental has nodes only at each fixed end of the string. The distance between nodes is half a wavelength. Hence the wavelength of the fundamental is twice the distance between the fixed ends, or 1.17 m.

Method 2

First find the speed of the waves on the string via Equation 12.19:

$$
\begin{aligned}
v &= \left(\frac{T}{\mu}\right)^{1/2} \\
&= \left(\frac{15.2 \text{ N}}{5.74 \times 10^{-5} \text{ kg/m}}\right)^{1/2} \\
&= 5.15 \times 10^2 \text{ m/s}
\end{aligned}
$$

Then, since the frequency is known, use Equation 12.10 to find the wavelength:

$$
\begin{aligned}
v &= \nu\lambda \\
\lambda &= \frac{v}{\nu} \\
&= \frac{5.15 \times 10^2 \text{ m/s}}{440 \text{ Hz}} \\
&= 1.17 \text{ m}
\end{aligned}
$$

c. The frequency of the second harmonic is twice the fundamental frequency, or 880 Hz.

Optical standing waves exist in laser cavities.

Note how the bass strings on a piano are wrapped with wire.

d. The distance between nodes of the second harmonic is half a wavelength. For the second harmonic, the wavelength is equal to the distance between the fixed ends of the string. Therefore the distance between nodes is $(1/2)(0.585 \text{ m}) = 0.293 \text{ m}$; hence, there is a node at the middle of the string.

EXAMPLE 12.12

You are designing a closed organ pipe to play the first A above middle C, with a frequency of 440 Hz.

a. **What is the length of the pipe?**
b. **What are the frequencies of the next two permitted harmonics?**

Solution

a. For a closed organ pipe, the length of the pipe is one-fourth of the wavelength of the fundamental frequency:

$$\ell = \frac{\lambda}{4}$$

The speed of sound is the product of the frequency and the wavelength:

$$v = \nu\lambda$$

Hence the length of the pipe is

$$\ell = \frac{v}{4\nu}$$
$$= \frac{343 \text{ m/s}}{4\,(440 \text{ Hz})}$$
$$= 0.195 \text{ m}$$

b. Since only odd-numbered harmonics can exist in a closed organ pipe, the next higher frequencies are 3 and 5 times the fundamental, or 1.32×10^3 Hz and 2.20×10^3 Hz.

12.20 WAVE GROUPS AND BEATS*

Those of you who play musical instruments likely know that tuning such instruments frequently involves listening for very low frequency **beats** between two acoustic frequencies that are nearly the same. One frequency is adjusted until the beating disappears, indicating that the two notes have the same frequency. Here we see how such beats arise for sound and other types of waves and oscillations.

Consider the superposition of two waves with different frequencies traveling in the *same* direction. For simplicity, we assume the waves have equal amplitudes. Let k_1 and ω_1 be the angular wavenumber and angular frequency of one wave traveling toward increasing values of x:

$$\Psi_1(x, t) = A \cos(k_1 x - \omega_1 t)$$

Since the wave is a function of $x - vt$, the speed can be determined by factoring out k_1 in the argument of the cosine:

$$\Psi_1(x, t) = A \cos\left[k_1\left(x - \frac{\omega_1}{k_1}t\right)\right]$$

Thus the speed of the wave is

$$v_1 = \frac{\omega_1}{k_1}$$

This is just a fancy way of writing Equation 12.10, the familiar $v = \nu\lambda$, since

$$v_1 = \frac{\omega_1}{k_1} = \frac{2\pi\nu_1}{2\pi/\lambda_1} = \nu_1\lambda_1$$

Let k_2 and ω_2 be the angular wavenumber and angular frequency of another wave of the same amplitude A, traveling in the same direction as the first wave (toward increasing values of x):

$$\Psi_2(x, t) = A \cos(k_2 x - \omega_2 t)$$

The speed of this second wave is

$$v_2 = \frac{\omega_2}{k_2}$$

The speeds of the individual waves are known as phase speeds. The phase speeds of the individual waves may or may not be the same. Whether they are the same or not depends on the particular kind of wave being considered and the medium through which the waves are propagating.

> If the speeds of waves of different frequency are *not* the same, we say there is **dispersion**; dispersion means that the speeds of individual waves depend on the particular wavelength (or frequency).[†]

It is not necessary for the superimposed waves to have the same amplitudes, but this condition makes the effects of such a superposition easier to discuss both from an analytical and from a physical viewpoint. The principle of superposition implies that the resulting wave disturbance is

$$\Psi_{\text{total}}(x, t) = \Psi_1(x, t) + \Psi_2(x, t)$$
$$= A[\cos(k_1 x - \omega_1 t) + \cos(k_2 x - \omega_2 t)] \quad (12.50)$$

[†] Light waves traveling in a vacuum exhibit no dispersion; all light wavelengths travel with the same speed, c, the speed of light in vacuum. On the other hand, in material media (such as glass and water), the speed of light is a function of wavelength and dispersion is therefore present. This simply means that the index of refraction of a material (perhaps familiar to you from a high school physics course and a concept we will introduce in Chapter 23) depends on the wavelength; among other things, when light passes through a prism, the component colors (wavelengths) of the light are separated.

We use the trigonometric identity for the sum of two cosines (Equation 12.44):

$$\cos \alpha + \cos \beta = 2 \cos\left(\frac{\alpha + \beta}{2}\right) \cos\left(\frac{\alpha - \beta}{2}\right)$$

Applying this identity to the two cosine terms in Equation 12.50, we find

$$\Psi_{\text{total}}(x, t) = 2A \cos\left[\frac{\left(k_1 x - \omega_1 t\right) + \left(k_2 x - \omega_2 t\right)}{2}\right]$$
$$\times \cos\left[\frac{\left(k_1 x - \omega_1 t\right) - \left(k_2 x - \omega_2 t\right)}{2}\right]$$

Collecting similar terms in the arguments of the cosines,

$$\Psi_{\text{total}}(x, t) = 2A \cos\left[\frac{\left(k_1 + k_2\right)}{2} x - \frac{\left(\omega_1 + \omega_2\right)}{2} t\right]$$
$$\times \cos\left[\frac{\left(k_1 - k_2\right)}{2} x - \frac{\left(\omega_1 - \omega_2\right)}{2} t\right]$$

This is a rather ungainly expression! We can tidy it up a bit by calling

$$\frac{k_1 + k_2}{2} \equiv k$$

which is the average value of the two angular wavenumbers. Likewise we call

$$\frac{\omega_1 + \omega_2}{2} \equiv \omega$$

which is the average value of the angular frequencies of the two waves. In a similar fashion, we call

$$|k_1 - k_2| = \Delta k$$

and

$$|\omega_1 - \omega_2| = \Delta \omega$$

With these cosmetic changes, the superposition of the two waves becomes somewhat prettier:

$$\Psi_{\text{total}}(x, t) = 2A \cos(kx - \omega t) \cos\left[\left(\frac{\Delta k}{2}\right) x - \left(\frac{\Delta \omega}{2}\right) t\right]$$
$$(12.51)$$

So what is the meaning of all this? *Both* cosine terms represent waves, since they both have the proper general form of Equation 12.16 for a wave traveling toward increasing values of *x*. Let's look at each harmonic wave individually. The term

$$\cos(kx - \omega t)$$

is a wave much like the two individual waves in the superposition. This wave term has an angular wavenumber *k* equal to the average of the angular wavenumbers of the component waves and an angular frequency equal to the average of the angular frequencies of the two waves superimposed. If these angular frequencies and angular wavenumbers are only slightly different from each other, ω and *k* are essentially equal to the corresponding quantities of either component wave. The speed of this wave is found from

$$v = \frac{\omega}{k}$$

Substituting for ω and *k*,

$$v = \frac{\dfrac{\omega_1 + \omega_2}{2}}{\dfrac{k_1 + k_2}{2}} = \frac{\omega_1 + \omega_2}{k_1 + k_2} \qquad (12.52)$$

We factor this expression in the following awkward way:

$$v = \frac{\omega_1\left(1 + \dfrac{\omega_2}{\omega_1}\right)}{k_1\left(1 + \dfrac{k_2}{k_1}\right)} \qquad (12.53)$$

If the frequencies and wavelengths of the individual waves are only slightly different from each other, then $\omega_1 \approx \omega_2$ and $k_1 \approx k_2$, and Equation 12.53 reduces to

$$v \approx \frac{\omega_1}{k_1} = v_1$$

Here *v* is essentially the phase speed of *either* wave in the superposition.

The other factor resulting from the superposition is

$$2A \cos\left[\left(\frac{\Delta k}{2}\right) x - \left(\frac{\Delta \omega}{2}\right) t\right] \qquad (12.54)$$

and also is in the form of a traveling wave. It has a lower angular frequency than the two component waves because $\Delta\omega/2 < \omega_1$ or ω_2. This wave represents a slowly varying amplitude factor for the superposition, which travels at its own **group speed** v_{group}. We write the factor now as

$$2A \cos\left[\frac{\Delta k}{2}\left(x - \frac{\Delta \omega}{\Delta k} t\right)\right] \qquad (12.55)$$

and can identify this group speed as the coefficient of *t*:

$$v_{\text{group}} = \frac{\Delta \omega}{\Delta k}$$

In the limit of small differential differences between the frequencies and wavelengths of the constituent waves, we get

$$v_{\text{group}} = \frac{d\omega}{dk} \qquad (12.56)$$

where ω is related to k via the phase speed equation:

$$v_{\text{phase}} = \frac{\omega}{k}$$

or

$$\omega = k v_{\text{phase}}$$

We substitute for ω in Equation 12.56 for the group speed:

$$v_{\text{group}} = \frac{d}{dk}(k v_{\text{phase}})$$

$$= v_{\text{phase}} + k\frac{dv_{\text{phase}}}{dk} \qquad (12.57)$$

If the phase speed is independent of wavelength (i.e., independent of k, since $k = 2\pi/\lambda$)—that is, if there is *no* dispersion—then the phase and group speeds are equal since

$$\frac{dv_{\text{phase}}}{dk} = 0 \ \text{m}^2/\text{s}$$

On the other hand, if there is dispersion, the phase speed depends on wavelength (or k, since $k = 2\pi/\lambda$), and the phase and group speeds are different.

We write the amplitude term in the standard form of a traveling wave:

$$2A\cos(k_A x - \omega_A t)$$

The angular wavenumber is k_A and ω_A is the angular frequency of the wave. From Equation 12.54 we get

$$k_A = \frac{\Delta k}{2}$$

and

$$\omega_A = \frac{\Delta\omega}{2}$$

Let's see what these mean. Call λ_A the long overall wavelength of this amplitude factor wave. Since the angular wavenumber is related to the wavelength by

$$k_A = \frac{2\pi \ \text{rad}}{\lambda_A}$$

we can relate the wavelength λ_A to the wavelengths of the individual waves involved in the superposition:

$$k_A = \frac{\Delta k}{2} = \frac{|k_1 - k_2|}{2}$$

or

$$\frac{2\pi}{\lambda_A} = \frac{\left|\dfrac{2\pi}{\lambda_1} - \dfrac{2\pi}{\lambda_2}\right|}{2}$$

Solving for λ_A, we find

$$\lambda_A = \frac{2\lambda_1\lambda_2}{|\lambda_2 - \lambda_1|}$$

If the two wavelengths superimposed are not very different from each other, the wavelength λ_A is much larger than either λ_1 or λ_2.

Also the angular frequency ω_A of the long-wavelength wave is related to the angular frequencies of the two waves superimposed:

$$\omega_A = \frac{\Delta\omega}{2} = \frac{|\omega_1 - \omega_2|}{2}$$

In terms of the frequencies, we have

$$\nu_A = \frac{|\nu_1 - \nu_2|}{2}$$

If the frequencies of the two waves superimposed are not very different from each other, the frequency ν_A is much lower than the frequencies of either of the waves superimposed.

To see what the entire superposition looks like as a function of time, we consider a fixed position and then plot the resulting disturbance as a function of time. For convenience, we choose the position $x = 0$ m. Then the superimposed wave of Equation 12.51 becomes the oscillation

$$\Psi_{\text{total}}(0 \ \text{m}, t) = 2A\cos(-\omega t)\cos\left(-\frac{\Delta\omega}{2}t\right)$$

Since the cosine of a negative angle has the same value as the cosine of the positive angle, we have

$$\Psi_{\text{total}}(0 \ \text{m}, t) = 2A\cos(\omega t)\cos\left(\frac{\Delta\omega}{2}t\right) \qquad (12.58)$$

This expression is plotted in Figure 12.59 in Example 12.13 for two waves with frequencies of 8.000 Hz and 8.250 Hz.

Notice that the high-frequency cosine term

$$\cos(\omega t)$$

is modulated by a low-frequency *envelope* given by the term

$$\cos\left(\frac{\Delta\omega}{2}t\right)$$

The envelope is of maximum amplitude when this cosine term has the value ± 1. These maxima are known as **beats**. Since the cosine assumes the values ± 1 during each cycle, the beats occur at *twice* the frequency ν_A of the wave itself. The *beat frequency* ν_B thus is

$$\text{beat frequency} \equiv \nu_B \equiv 2\nu_A = 2\left(\frac{|\nu_1 - \nu_2|}{2}\right)$$

$$= |\nu_1 - \nu_2|$$

The beat frequency is equal to the absolute value of the difference in frequencies of the two waves involved in the superposition.

As mentioned at the beginning of this section, the beating between two waves of slightly different frequency is, among other things, the method used by musicians to tune instruments.

The note in question is played along with a standard reference tone or note. If the frequencies of the two sound waves are not identical, a trained ear can detect the small beat frequency oscillations that are a direct measure of the difference in the frequencies of the two notes. If the two notes differ in frequency by, say, 2 Hz, then the beat frequency is 2 Hz. The frequency of the note on one instrument is adjusted (by changing the tension in a string or other means for other instruments) until no more beats are heard. At that point the difference in the two frequencies is zero and the two notes are in tune with each other.

The beating phenomenon between waves or oscillations of slightly different frequencies is manifested in many aspects of nature, not just acoustics. In other words, it is a general characteristic of wave or other periodic phenomena. For example, Figure 12.57 is a graph of the time interval between successive full moons (called the synodic period of the Moon) versus time (or the calendar date) over about a 30-year period. The graph suggests a beating between two slightly different frequencies associated with the Sun–Earth–Moon system. A detailed analysis of the graph* indicates that the periods associated with the beating frequencies are 366 days (about a year) and 411 days (the time it takes the Sun to return to the same position along the major axis of the lunar orbital ellipse).

EXAMPLE 12.13

Two harmonic waves with frequencies 8.000 Hz and 8.250 Hz are superimposed. The waves each have amplitude A. Consider the position where $x = 0$ m.

a. **On the same graph, plot $\Psi_1(0\text{ m}, t)$ and $\Psi_2(0\text{ m}, t)$ as functions of time over the domain $t = 0$ s to 8.00 s.**
b. **On a separate graph, plot $\Psi_{\text{total}}(0\text{ m}, t)$ versus t over the domain $t = 0$ s to 8.00 s.**
c. **What is the beat frequency?**
d. **What is the beat period?**

Solution
a. Each harmonic wave has the form

$$\Psi(x, t) = A\cos(kx - \omega t)$$

---

*See Ronald Lane Reese, George Y. Chang, and David L. Dupuy, "The oscillation of the synodic period of the Moon: a 'beating' phenomenon," *American Journal of Physics*, 57, #9, pages 802–807 (September 1989).

At the origin, $x = 0$ m, so that

$$\Psi(x, t) = A\cos(-\omega t)$$
$$= A\cos(\omega t)$$

Since $\omega = 2\pi\nu$, the two oscillations at the origin are

$$\Psi_1(0\text{ m}, t) = A\cos[2\pi(8.000\text{ Hz})t]$$
$$= A\cos[(50.27\text{ rad/s})t]$$

and

$$\Psi_2(0\text{ m}, t) = A\cos[2\pi(8.250\text{ Hz})t]$$
$$= A\cos[(51.84\text{ rad/s})t]$$

These functions are plotted on the same grid in Figure 12.58.
b. At the origin, the superimposed waves have the form of Equation 12.51, with $x = 0$ m:

$$\Psi_{\text{total}}(0\text{ m}, t) = 2A\cos(\omega t)\cos\left(\frac{\Delta\omega}{2}t\right) \quad (1)$$

Here ω is the average of two angular frequencies:

$$\omega = \frac{\omega_1 + \omega_2}{2}$$
$$= \frac{2\pi(\nu_1 + \nu_2)}{2}$$
$$= 51.05\text{ rad/s}$$

and $\Delta\omega$ is the difference in the two angular frequencies:

$$\Delta\omega = \omega_1 - \omega_2$$
$$= 2\pi(\nu_2 - \nu_1)$$
$$= 1.571\text{ rad/s}$$

Hence

$$\frac{\Delta\omega}{2} = 0.7855\text{ rad/s}$$

Equation (1) thus takes on the specific form

$$\Psi_{\text{total}}(0\text{ m}, t) = 2A\cos[(51.05\text{ rad/s})t]\cos[(0.7855\text{ rad/s})t]$$

This oscillation is graphed in Figure 12.59. Notice how the resulting oscillation in Figure 12.59 is large when the

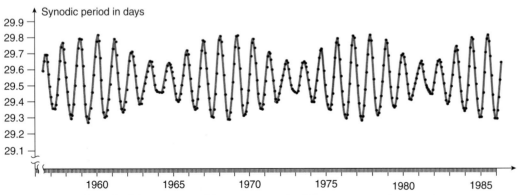

FIGURE 12.57 A graph of the time interval between successive full moons versus time reveals a beating of two frequencies. Other frequencies are present too, but have smaller amplitudes.

FIGURE 12.58

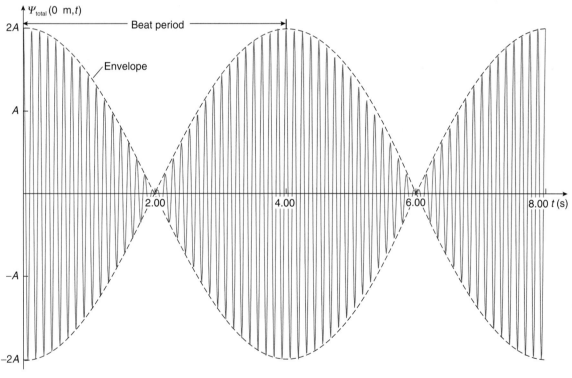

FIGURE 12.59

two oscillations in Figure 12.58 are in phase ("in step") with each other. The resulting oscillation is small when the two oscillations are out of phase ("out of step") with each other.

c. The beat frequency is the difference between the two superimposed frequencies, or 0.250 Hz. In Figure 12.59 the beat frequency is the number of "bumps" of the envelope during an interval of one second.

d. The beat period is the inverse of the beat frequency, or 4.000 s. This is the time interval between the "bumps" of the envelope shown in Figure 12.59.

12.21 FOURIER ANALYSIS AND THE UNCERTAINTY PRINCIPLES*

Ambiguous undulations . . . .

Wallace Stevens (1879–1955)[†]

Sinusoidal waves are like Legos. From sinusoidal waves, any manner of more complex periodic waves can be built. More complicated periodic waves may have waveforms such as those

[†]From the poem *Sunday Morning*, stanza VIII, line 14. In *The Collected Poems of Wallace Stevens* (Alfred A. Knopf, New York, 1954), pages 66–70; quotation is on page 70.

shown in Figure 12.60 (along with their corresponding oscillatory behavior) or many others.

It was Jean Baptiste Joseph Fourier (1768–1830) who realized that *any* periodic oscillation (or waveform) can be represented by a suitable superposition (perhaps with an infinite number of terms) of sinusoidal oscillations (or sinusoidal waveforms) that are harmonics of the fundamental frequency of the oscillation (or harmonics of the fundamental angular wavenumber of the periodic waveform). His work created a branch of mathematics now called **Fourier analysis**.

To see how combinations of sinusoidal functions are used to represent more complicated periodic functions, we examine a particular example of such a superposition. To this end, we look at the oscillation caused by a complicated periodic wave at a particular position. We also could look at the waveform at a particular time. The oscillatory behavior and waveform are complementary aspects of wave motion, as we have seen. The oscillation viewed at a fixed position will be sufficient to present the general ideas underlying Fourier analysis. The analysis is similar in both cases; by examining periodic oscillatory behavior, we can generalize the results so they apply to periodic waveforms as well.

Periodic but nonharmonic oscillatory behavior is imagined as a superposition of harmonic oscillations of various amplitudes.

> The frequencies in a Fourier representation of a periodic oscillation always are harmonics of the fundamental frequency of the periodic oscillation.

The periodic oscillation shown in Figure 12.61 is known, not surprisingly, as a sawtooth. The graph represents the oscillation produced by a sawtooth-shaped traveling wave at a particular place, say $x = 0$ m. The methods used to find the Fourier representation of complicated oscillations such as this one are slightly beyond the level of this course.* Here we will pluck the mathematical result out of thin air, just to show you how such a superposition leads to a representation of more complicated wave shapes.

The results of a Fourier analysis of the sawtooth oscillation indicate that it is represented by the superposition of an infinite number of progressively smaller sinusoidal oscillations that are harmonics of the fundamental frequency (or an-

---

*The methods are standard recipes that involve integrations of the given periodic function with sinusoidal functions.

Fourier was almost executed during the French Revolution. He went to Egypt with Napoleon and contracted a disease (likely myxedema) that eventually proved fatal.

gular frequency). In Figure 12.61 the period of the sawtooth is 2.00 s; its fundamental frequency is the inverse of the period: 0.500 Hz. The fundamental angular frequency is 2π times the fundamental frequency or

$$(2\pi \text{ rad/s})(0.500 \text{ Hz}) - \pi \text{ rad/s}$$

The representation of the sawtooth oscillation involves harmonics of the fundamental angular frequency, integral multiples of π rad/s:

$$\Psi(0 \text{ m}, t) = \frac{8}{\pi^2} \left\{ \cos(\pi t) + \frac{1}{9} \cos(3\pi t) \right.$$
$$+ \frac{1}{25} \cos(5\pi t) + \frac{1}{49} \cos(7\pi t) + \cdots$$
$$\left. + \frac{1}{(2n+1)^2} \cos\left[(2n+1)\pi t\right] + \cdots \right\} \quad (12.59)$$

where n is an integer greater than or equal to zero and each π has the units of rad/s. For this particular case, only odd-numbered harmonics contribute. The right-hand side of Equation 12.59 is called the **Fourier series** of the sawtooth periodic oscillation.

To show you that the series begins to approximate the sawtooth oscillation as more and more terms are superimposed, Figures 12.62–12.64 are graphs of successively more terms in the

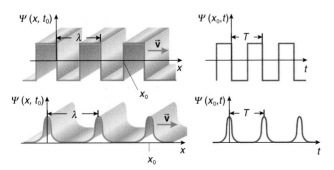

FIGURE 12.60 More complicated periodic waveforms and oscillations.

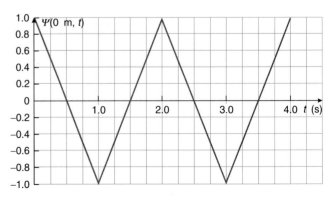

FIGURE 12.61 A sawtooth oscillation.

series on the right-hand side of Equation 12.59. Figure 12.62 is just the first term on the right-hand side of Equation 12.59. Figure 12.63b is the superposition of the first two terms. Figure 12.64b is the superposition of the first three terms. And we could continue in the same manner. As more and more terms are included, the resulting superposition approaches that of the desired sawtooth oscillation. The amplitudes and frequencies of the terms used to create the more complicated oscillation make up the frequency spectrum of the oscillation.[†]

The frequency spectrum of any periodic oscillation is discrete: only integral multiples of the fundamental frequency are present. For the sawtooth, the discrete spectrum contains only the odd-numbered harmonics.

The Fourier representation of a sawtooth waveform, rather than a sawtooth oscillation, has the same form as Equation 12.59, with x replacing t and k replacing ω. In this case we have a spectrum of angular wavenumbers that are integral multiples of the fundamental angular wavenumber of the waveform. The fundamental angular wavenumber is related to the fundamental wavelength of the periodic waveform via $k = 2\pi/\lambda$.

> Although this is a particular example, the important thing to realize is that *any* periodic function can be represented by a Fourier sum of harmonic, sinusoidal functions having a discrete spectrum. This is why the study of sinusoidal waveforms and oscillations is so important: it is the foundation for the quantitative treatment of more complicated waveforms and oscillations.

Even nonperiodic functions, such as a single pulse oscillation or pulse waveform, shown in Figure 12.65, can be represented by

---

[†]We saw in Section 12.19 how the complex sounds of various musical instruments have a frequency spectrum that consists of integral multiples of the first harmonic. The frequency spectrum indicates the amplitudes and frequencies of the various terms of the Fourier superposition that represents the oscillation

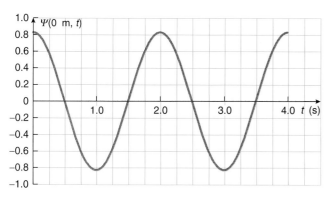

FIGURE 12.62 The first term in the Fourier series of a sawtooth.

(a)

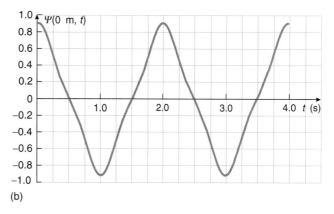

(b)

FIGURE 12.63 (a) The first two terms in the Fourier series of a sawtooth oscillation and (b) their sum.

(a)

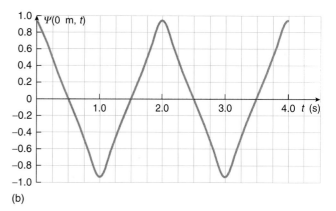

(b)

FIGURE 12.64 (a) The first three terms in the Fourier series of a sawtooth oscillation and (b) their sum.

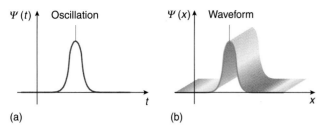

F!GURE 12.65 (a) A pulse oscillation and (b) a pulse waveform.

sinusoidal functions. Such nonperiodic oscillations and waveforms are imagined as periodic functions with a very long fundamental period T or fundamental wavelength λ, as shown in Figure 12.66. A long fundamental period T implies a small fundamental frequency ν and small fundamental angular frequency ω since

$$\nu = \frac{1}{T}$$

$$\omega = (2\pi \text{ rad})\,\nu$$

Hence the harmonics of the fundamental angular frequency become more closely spaced in angular frequency as the period T increases. Indeed, as $T \to \infty$ s, the fundamental angular frequency $\omega \to 0$ rad/s, and so the Fourier series sum becomes a continuous sum, an integral,* and the frequency and angular frequency spectrum are *continuous* rather than discrete.

One finds in Fourier analysis that if the time interval Δt over which the pulse is nonzero is finite, the angular frequency spectrum also is nonzero over a finite range: $\Delta \omega = \omega_{max} - \omega_{min}$. The superposition of terms is completely destructive except during the interval of the pulse. In other words, the frequencies are out of phase at the beginning of the pulse, come into phase at the peak of the pulse, and are out of phase at the end of the pulse. This condition is closely met if the number of periods associated with ω_{max} during the interval Δt of the pulse is about one more than number of periods associated with ω_{min} during the same interval.[†] The number of periods T during an interval Δt is

$$\frac{\Delta t}{T} = \Delta t\,\frac{\omega}{2\pi \text{ rad}}$$

for an oscillation with angular frequency ω. So the condition regarding the relative number of periods can be expressed as

$$\frac{\Delta t}{2\pi \text{ rad}}\left(\omega_{max} - \omega_{min}\right) \approx 1$$

---

*The integral is known as the *Fourier transform.*
[†]This may not be obvious to you, so don't fret it.

Simplifying, we have

$$\Delta t\,\Delta \omega \approx 2\pi \text{ rad} \tag{12.60}$$

> The shorter the duration of the pulse, the greater the spread of angular frequencies in its Fourier representation, and vice versa.

For a pure harmonic oscillation, the interval Δt over which it is nonzero is very long (strictly speaking, it is infinite), and so the spread of the angular frequency spectrum $\Delta \omega$ is very small: namely, the oscillation is represented by a single angular frequency.

Similar arguments also can be made for a pulse waveform: a long fundamental wavelength λ implies a small fundamental angular wavenumber k since

$$k = \frac{2\pi \text{ rad}}{\lambda}$$

Thus the harmonics of the fundamental angular wavenumber become more closely spaced as λ increases. As $\lambda \to \infty$ m, $k \to 0$ rad/m and the Fourier series sum becomes an integral.[‡] The spectrum of angular wavenumbers becomes continuous. Since k and x in the waveform play the roles of ω and t in the oscillation, an argument similar to that made for the oscillatory behavior means that we have a relation analogous to Equation 12.60 for x and k. If the waveform is nonzero over a wavetrain of length Δx, the range of angular wavenumbers Δk in its Fourier representation satisfies

$$\Delta x\,\Delta k \approx 2\pi \text{ rad} \tag{12.61}$$

Equations 12.60 and 12.61 are called **uncertainty relations** and are inexorably linked with all wave phenomena. In Chapter 27 (an introduction to quantum mechanics) the notions of waves and particles become intertwined, and the uncertainty relations limit the precision with which we can simultaneously measure the position and momentum of a particle, or measure its energy. The word "uncertainty" is used is the sense of "indeterminable," not capable of being fixed or measured precisely. In the context of quantum mechanics, the uncertainty relations are called **Heisenberg uncertainty principles**, after Werner Heisenberg (1901–1976), who first recognized their importance in measurement processes in quantum mechanics.

[‡]This integral also is called a Fourier transform.

FIGURE 12.66 A periodic function with a very long period or wavelength approximates a pulse oscillation or pulse waveform.

CHAPTER SUMMARY

A *classical wave* is a propagating disturbance that transfers energy and momentum at its own characteristic speed from one region of space to another with little if any mass transfer. Particle streams also transfer energy and momentum but are not classical waves. A *mechanical wave* is a propagating disturbance in and of a material. Other types of classical waves, capable of propagating even through a vacuum, involve propagating fields.

Waves in which oscillations are along the line that the wave is propagating are called *longitudinal waves*. Sound waves in gases are longitudinal, as are compressional waves on a slinky or spring. *Transverse waves* have oscillations perpendicular to the velocity of the wave. Waves on strings under tension as well as light waves are transverse waves. Some waves may be both longitudinal and transverse.

A wave is represented by a *wavefunction* Ψ whose value indicates the size of the disturbance, whatever it may involve physically. Any one-dimensional wave propagating at constant speed v toward increasing values of x is represented by a wavefunction that is a function of the quantity $x - vt$:

$$\Psi(x - vt) \tag{12.1}$$

A wave propagating toward decreasing values of x is represented by a wavefunction of the form

$$\Psi(x + vt) \tag{12.2}$$

There are two complementary ways to view a wave. The *waveform* of the wavefunction is found by fixing $t = t_0$ and plotting the wavefunction Ψ as a function of x:

$$\Psi(x, t_0)$$

This graph is a spatial representation of the shape of the wavefunction. The *oscillatory* behavior of the wave at a fixed position x_0 is found by plotting the wavefunction Ψ as a function of time t at that position:

$$\Psi(x_0, t)$$

Any classical one-dimensional wave traveling at constant velocity $\vec{v}$ along the x-axis satisfies the *classical wave equation*:

$$\frac{\partial^2 \Psi}{\partial x^2} - \frac{1}{v^2}\frac{\partial^2 \Psi}{\partial t^2} = 0 \tag{12.7}$$

Periodic waves have a well-defined (constant) *wavelength* λ and *period* T. The wavelength is the distance between corresponding successive points on the waveform. The period is the time for one complete oscillation. The *frequency* ν is the inverse of the period and is the number of oscillations completed during an interval of exactly one second. For a wave traveling at speed v, the wave moves a distance equal to a wavelength during one period. Hence the speed is the product of the frequency and the wavelength:

$$v = \nu\lambda \tag{12.10}$$

Sinusoidal (harmonic) waves, propagating toward increasing values of x, have the form

$$\Psi(x, t) = A\cos(kx - \omega t) \tag{12.16}$$

where k is the *angular wavenumber*:

$$k = \frac{2\pi \text{ rad}}{\lambda} \tag{12.13}$$

and ω is the angular frequency:

$$\omega \equiv (2\pi \text{ rad})\nu$$

The angular wavenumber is the number of wavelengths in a distance of 2π meters.

The speed v of a transverse wave on a string with mass per unit length μ and a tension of magnitude T is

$$v = \left(\frac{T}{\mu}\right)^{1/2} \tag{12.19}$$

The power P transferred by a wave is proportional to the wave speed v, the square of the amplitude A, and the square of the angular frequency. For transverse waves on a stretched string, the power transferred by the wave is

$$P = \frac{1}{2}\mu\omega^2 A^2 v \tag{12.26}$$

The *intensity* I of a wave is the average power transmitted through one square meter oriented perpendicular to the direction the wave is propagating. For a wave propagating in three dimensions, the intensity a distance r from a point source of waves follows an inverse square law:

$$I = \frac{P}{4\pi r^2} \tag{12.27}$$

where P is the power of the source.

The *sound level* β in *decibels* (dB) of an acoustic wave of intensity I is

$$\beta = (10 \text{ dB})\log\frac{I}{I_0} \tag{12.28}$$

where I_0 is the average human *threshold of hearing*; $I_0 \equiv 10^{-12}$ W/m^2.

Motion of a source and/or observer changes the observed frequency of acoustic waves, a phenomenon known as the *acoustic Doppler effect*. If the source frequency is ν, the frequency ν' detected is

$$\nu' = \nu\left(\frac{v \pm v_{\text{med}} \pm v_{\text{obs}}}{v \pm v_{\text{med}} \mp v_{\text{source}}}\right) \tag{12.37}$$

where v is the speed of sound, v_{med} is the wind speed (use the positive sign for a wind blowing from the source toward the observer, use the negative sign for the opposite direction), v_{obs} is the speed of the observer, and v_{source} is the speed of the source. If the distance between source and observer is decreasing (increasing), the detected frequency is greater (less) than the source frequency and you should use the top (bottom) signs in front of v_{obs} and v_{source} in Equation 12.37.

If the speed of a source of sound v_{source} is greater than the speed of sound v in the medium, a *shock wave* (sonic boom) is formed. The half-angle ϕ of the *Mach cone* defining the shock wave is found from

$$\sin\phi = \frac{v}{v_{\text{source}}} \qquad (v_{\text{source}} > v) \tag{12.39}$$

The *Mach number* is

$$\text{Mach number} = \frac{v_{\text{source}}}{v} \tag{12.40}$$

Two or more waves together produce a total disturbance that is the sum of the individual wave disturbances:

$$\Psi_{\text{total}}(x, t) = \Psi_1(x, t) + \Psi_2(x, t) + \cdots$$

This is the linear *principle of superposition*. We consider only waves that follow such a linear principle of superposition. You should be aware that not all waves do so.

Standing waves are produced by the superposition of two sinusoidal waves of equal amplitude and frequency propagating in opposite directions; if the waves are transverse, they must have the same plane of polarization. The *nodes* of a standing wave are the locations where there is zero wave disturbance at all times. Nodes are separated in space by half a wavelength. *Antinodes* are places of maximum disturbance and also are separated by half a wavelength.

Standing waves on a string fixed at both ends have frequencies given by

$$\nu_n = n \frac{1}{2\ell}\left(\frac{T}{\mu}\right)^{1/2} \quad (n \text{ is an integer}) \quad (12.49)$$

where n is an integer that numbers the harmonics, μ is the mass per unit length of the string, T is the magnitude of the tension of the string, and ℓ is the distance between the fixed ends of the string. The frequency corresponding to $n = 1$ is called the *fundamental frequency* or, equivalently, the *first harmonic*.

Two similar waves of different frequency propagating in the same direction produce *beats*. The beat frequency is equal to the absolute value of the difference between the frequencies of the two waves.

Sinusoidal (harmonic) waves are used to describe more complex periodic waves through *Fourier analysis*. The Fourier analysis leads to two *uncertainty relations*:

$$\Delta t\, \Delta\omega \approx 2\pi \text{ rad} \quad (12.60)$$

$$\Delta x\, \Delta k \approx 2\pi \text{ rad} \quad (12.61)$$

SUMMARY OF PROBLEM-SOLVING TACTICS

12.1 (page 537) The sign between x and νt determines the direction $\Psi(x, t)$ is moving along the x-axis. When the sign is negative, $\Psi(x - \nu t)$, the wave is moving toward *increasing* values of x; when the sign is positive, $\Psi(x + \nu t)$, the wave is moving toward *decreasing* values of x.

12.2 (page 542) To find the value of $\cos(kx - \omega t)$ in Equation 12.11,

$$\Psi(x, t) = A\cos(kx - \omega t) \quad (12.11)$$

be sure your calculator is set for radian measure of angle, not degrees.

12.3 (page 551) Logarithms are exponents, so that the expression $Y = \log X$ means $X = 10^Y$. Therefore the sound level

$$\beta = (10 \text{ dB})\log\frac{I}{I_0}$$

means

$$\frac{I}{I_0} = 10^{\beta/(10 \text{ dB})}$$

12.4 (page 554) The sign to use in Equation 12.33,

$$\nu' = \nu\left(\frac{v \pm v_{\text{obs}}}{v}\right) \quad (12.33)$$

can be determined as long as you remember that the frequency ν' is greater than ν if the observer is approaching the source (so the + sign must be used) and less than ν if receding (use the − sign).

12.5 (page 554) The sign to use in Equation 12.36

$$\nu' = \nu\left(\frac{v}{v \mp v_{\text{source}}}\right) \quad (12.36)$$

can be determined by whether the observer hears a higher or lower frequency (as the source approaches or recedes, respectively).

12.6 (page 555) To use the general equation for the Doppler effect (Equation 12.37), recall the conventions associated with various signs on page 555.

QUESTIONS

1. (page 533); 2. (page 535); 3. (page 536); 4. (page 546); 5. (page 547); 6. (page 551); 7. (page 555); 8. (page 558); 9. (page 563); 10. (page 564)

11. Fans at various sporting events like to make "the wave" as shown in Figure Q.11. Is this a classical wave as we define it in physics? Explain.

12. A line of cars is zooming along a freeway at a constant speed of 100 km/h. One car brakes suddenly and a traffic pulse ripples along the line of cars. Is this traffic pulse a wave? What factors determine its speed?

13. If a radio wave (a light wave) and a sound wave have the same wavelength, which has the greatest frequency?

14. A flash of lightning is observed; t_0 seconds later, a crash of thunder is heard. Given that the speed of sound is about 340 m/s, devise an *approximate* expression for the distance in *kilometers* to the lightning flash. If you wish, devise an approximate expression for the distance in miles.

15. A transverse wave of amplitude A is propagating along a rope. Imagine a small segment of the rope. At what positions in its transverse motion is the magnitude of its acceleration greatest? At what positions is the acceleration of zero magnitude? At what positions is its speed a maximum? At what positions is its speed zero?

16. If you double the tension in a string, by what factor does the speed that a wave propagates on the string increase?

17. When you drop a rock into a calm pond, the circular water waves decrease in amplitude as they radiate outward. Explain why.

And like circles on the water, which, as they grow fainter, expand.
Herman Melville (1819–1891)*

_____
Moby Dick (University of California Press, Berkeley, 1979), Chapter 110, page 486.

FIGURE Q.11

18. In the acoustic Doppler effect, the effect of an approaching and then receding sound source is a transition from a higher to lower frequency as time passes. Explain why a transition *from* a lower *to* a higher pitch never is observed for a source with a constant velocity.

19. Examine a guitar or banjo and consider one of its strings under a constant tension. Measure the lengths of the string corresponding to two frets an octave apart and see if the frequencies found using Equation 12.49 (for $n = 1$) differ by a factor of two (an octave).

20. Pianos come in various sizes, from baby grands to concert grands. As you progress to lower-frequency notes, at what note are the strings wire-wrapped? Is it the same note for all piano sizes? Does your answer depend on whether the piano is made by Steinway, Bösendorfer, Yamaha, or another manufacturer?

21. The high-frequency string on a standard guitar is tuned to the E above middle C and has a frequency of 329.63 Hz.* The lowest-frequency string is tuned to the E two octaves lower. If the strings of a guitar all were made of wire with the same mass per unit length, strung with the same magnitude tension, what would be the ratio of the longest to shortest string?

22. When you tune a guitar or banjo, what parameter in Equation 12.49 are you changing?

23. When you use different frets along a guitar or banjo string, what parameter in Equation 12.49 are you changing?

24. Standing waves are generated on a string fixed at both ends. The distance between the fixed ends is ℓ. You observe a node located

a distance $\ell/3$ from each fixed end as shown in Figure Q.24. What harmonic of the fundamental frequency is oscillating on the string?

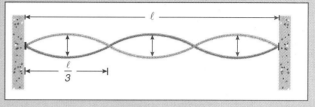

FIGURE Q.24

25. Do the locations of the nodes and antinodes of a standing wave move?

26. Describe an experiment to see if there is dispersion of sound waves in air, that is, to see whether the speed of sound is independent of frequency.

27. Pianos are strung with many strings under considerable tension. A frame is used on which to string the strings. Explain why the transition from wood frames to steel frames, pioneered in the mid-1800s by the now famous Steinway piano factory, revolutionized piano making technology, enabling musicians to have their pianos tuned much less frequently.

28. Consult an appropriate book on acoustics to distinguish between the musical concepts of pitch, loudness, and tone.

29. Two-dimensional standing waves exist on vibrating surfaces such as drum heads. Describe an experimental procedure that can be used to visually locate the nodes and antinodes on the surface if it is horizontal. (See the picture on page 561.)

30. A space shuttle in orbit releases a rocket from its cargo bay and moves 200 m away before the rocket engine is ignited. Should the crew wear hearing protectors? Explain.

31. A xylophone is an instrument with metal bars of various lengths that produce the notes of the musical scale. A vibraphone is similar except that open-ended tubes of appropriate lengths are placed beneath each metal bar to enhance the sound. If you strike a xylophone key, the sound will persist for almost a minute. If you strike the same note on a vibraphone, the sound is much louder than with a xylophone but persists for only a small fraction of the interval for a xylophone. Explain.

32. The strings of a properly tuned guitar play the notes E_2, A_2, D_3, G_3, B_3, and E_4 as fundamentals. Use the frequency chart in Figure 12.53 on page 562 to determine the ratio of the frequencies of the notes to the lowest frequency as a reference.

33. Measure the mass and total length of each string on a guitar to determine its mass per unit length μ. Suppose each string is adjusted to have the same magnitude tension. What is the ratio of the frequencies of the fundamental notes on each string to that of the lowest frequency?

34. In the same way as a boat is set into oscillation by a passing water wave, a sound wave induces oscillations in objects in its path. An opera singer "tweaks" the rim of a wine glass and listens to its ringing. She then sings a note of the same frequency. Explain how the singer can shatter the glass if she sings loud enough near the glass. See Figure Q.34.

---

*In guitar music, this E typically is written for convenience as the E located one octave higher on the treble clef.

In the case of a cock putting its head into an empty utensil of glass where it crowed so that the utensil thereby broke, the payment must be in full. . . .

The Babylonian Talmud*

FIGURE Q.34

35. What observational evidence can you cite that indicates sound is a wave rather than a stream of particles?

36. What observational evidence indicates that radio waves really are waves and not streams of particles?

37. If you inhale helium and then talk, your voice has a falsetto tone that sounds like a chipmunk. Why?

38. Does the speed of sound increase, decrease, or remain the same as a function of altitude in the atmosphere of the Earth? Explain the reasons for your answer.

39. As you fill a thermos with coffee, the frequency of the sound you hear changes. Does the frequency increase or decrease? Explain why.

40. Write a brief answer to the old question that goes like this: A tree falls in a forest. Is there a sound if no one is there to hear it?

41. If the sound intensity doubles, what is the increase in the sound level to the nearest tenth of a decibel? Do you need to know the original sound level to make this calculation?

42. Two sources of sound have the same frequency. Under what circumstances could you hear beats between them?

43. From observations of water wave ripples on a pond and water waves on the open ocean, do surface water waves exhibit dispersion or no dispersion? If so, does the speed increase or decrease with wavelength?

44. For standing acoustic waves, what is the distance between a node of the pressure variation wave and a node of the longitudinal particle position wave? Use wavelength as the unit of measurement.

PROBLEMS

For problems where it is germane, assume the speed of sound in air is 343.0 m/s unless otherwise indicated.

Sections 12.1 What Is a Wave?
 12.2 Longitudinal and Transverse Waves
 12.3 Wavefunctions, Waveforms, and Oscillations
 12.4 Waves Propagating in One, Two, and Three Dimensions
 12.5 One-Dimensional Waves Moving at Constant Velocity
 12.6 The Classical Wave Equation for One-Dimensional Waves*
 12.7 Periodic Waves

1. At a fireworks celebration on the fourth of July, you notice the flash from a cherry bomb and 0.50 s later hear the sound of the acoustic pulse from the explosion. The speed of sound waves in air is about 343 m/s while that of light is about 3.0×10^8 m/s. What was the distance between you and the explosion?

•2. A transverse wave pulse on a string has the wavefunction

$$\Psi(x, t) = \frac{0.250 \text{ m}^3}{2.00 \text{ m}^2 + [x - (5.00 \text{ m/s}) t]^2}$$

(a) What is the speed of the pulse? (b) On the same plot, graph the waveform of the pulse at the instants when $t = 0$ s,

1.00 s, and 2.00 s to show that the pulse is moving toward increasing values of x at the speed you found in part (a). Let the domain of your graph be $x = -10.0$ m to 20.0 m. (c) What is the maximum value of the disturbance? (d) If you change the wavefunction to

$$\Psi(x, t) = \frac{0.250 \text{ m}^3}{2.00 \text{ m}^2 + [x + (5.00 \text{ m/s}) t]^2}$$

show with another series of plots of the waveform when $t = 0$ s, 1.00 s, and 2.00 s that the pulse is moving toward decreasing values of x. Let the domain of this plot be $x = -20.0$ m to 10.0 m.

•3. One-dimensional waves propagating toward increasing values of x are functions of the quantity $x - vt$, so we write $\Psi = \Psi(x - vt)$. Consider the following functions of $x - vt$ where A, α, and v are constants: (i) $A[x - vt]$; (ii) $\log[\alpha(x - vt)]$; (iii) $Ae^{-\alpha(x - vt)^2}$. (a) When $t = 0$ s, sketch the waveform of these functions. (b) At $x = 0$ m, sketch the dynamical behavior of these functions (if any) for $t \geq 0$ s. (c) Two of these functions cannot represent realistic waves; one can. Explain which and why.

•4. An asymmetric transverse pulse on a taut string has the waveform shape indicated in Figure P.4 when $t = 0$ s, where Ψ represents the transverse position of the string in meters, measured from its equilibrium position. The value of Ψ represents the position of a particle of the string along the y-axis, perpendicular to the velocity of the wave. The velocity of the pulse is $(2.0 \text{ m/s})\hat{\imath}$. (a) For the position $x = 6.0$ m in Figure P.4,

---

**The Soncino Talmud*, translated by Isidore Epstein (Soncino Press, London, 1935), *Seder Nezikin*, Volume I, Chapter 2, page 87.

sketch the corresponding oscillatory behavior on a Ψ versus t graph. (b) At the instant shown in Figure P.4 ($t = 0$ s), make a sketch of the transverse velocity component of the particle as a function of position x along the string. (c) For the position $x = 6.0$ m, sketch the transverse velocity component as a function of time. (d) From your graphs, at what value of the transverse position Ψ does a particle of the string have a velocity component equal to zero? (e) When $t = 0$ s, at what positions along the x-axis is the transverse speed of a particle on the string equal to zero?

This problem was inspired by Arnold Arons, *A Guide to Introductory Physics Teaching* (Wiley, New York, 1990), page 203.

FIGURE P.4

•5. An earthquake generates two principal types of waves known as S-waves (for shear waves) and P-waves (for pressure waves) that propagate away from the focus of the quake. The P-waves travel at a speed of about 5.0 km/s whereas the S-waves poke along at only about 3.0 km/s. A seismometer can detect these waves at sites remote from the focus (or origin) of the quake. The P-waves arrive at a seismometer before the S-waves from the same quake (see Figure 12.5). The time interval between the arrival of the waves is an indication of the distance of the seismometer from the focus of the quake. (a) If the P-waves arrive at the seismometer 4.00 min before the S-waves, what is the distance to the focus of the quake? (b) The previous part determined the distance to the focus of the quake but not the specific location. Why not? (c) By using several seismometers at widely separated locations it is possible to pinpoint the specific location of the quake; indicate how.

•6. A periodic transverse wave moving at 4.00 m/s toward decreasing values of x has the waveform indicated in Figure P.6 when $t = 0$ s. (a) What is the wavelength? Indicate the wavelength on the waveform. (b) Draw a graph of the disturbance at $x = 0$ m as a function of time. Indicate the period of the wave on this graph. What is the period in seconds? (c) Use the results of part (b) to determine the frequency and confirm that the speed is the product of the frequency and wavelength as Equation 12.10 states. (d) What is the maximum transverse speed?

FIGURE P.6

•7. While bottom fishing, you notice that your anchored boat completes 60 complete up-and-down motions during one minute because of the passage of a wave. You estimate the speed of the waves to be 7.0 m/s. Another nearby anchored boat is located in a direction parallel to the velocity of the waves. You notice that the other boat is in a wave trough while yours is at a crest, and vice versa. At what possible distances from you is the other boat?

•8. A perfectly elastic ball bounces on a rigid floor and continually rebounds to a height of 1.20 m. Each time the ball strikes the floor, a sensor emits a brief beep. (a) Are the intervals between successive beeps the same? (b) What is the period of the acoustic beeps? (c) What is the wavelength associated with the periodic acoustic beeps?

•9. A periodic wave moving with a speed of 3.00 m/s toward increasing values of x passes the position $x = 0$ m and the disturbance at that location is graphed as a function of time in Figure P.9. (a) Indicate the period of the wave on the graph. (b) What is the frequency of the wave? (c) At the instant when $t = 0$ s, make a graph of the waveform as a function of x and indicate the wavelength of the wave on your graph. (d) Verify that the speed of the wave is the product of the frequency and wavelength.

FIGURE P.9

•10. Repeat Problem 9 but with the wave moving toward decreasing values of x.

•11. Show that the sinusoidal wave function

$$\Psi(x, t) = A \cos(kx - \omega t)$$

satisfies the classical wave equation (Equation 12.7).

•12. (a) Show that a wavefunction of the form

$$\Psi(x + vt)$$

representing a wave traveling at constant speed v in a direction toward decreasing values of x has the following relationship between its first partial derivatives:

$$\frac{\partial \Psi}{\partial t} = +v \frac{\partial \Psi}{\partial x}$$

Notice that this result is different from the relationship for a wave traveling in the direction of increasing values of x (Equation 12.6). (b) Show that $\Psi(x + vt)$ satisfies the same classical wave equation (Equation 12.7) as a wave $\Psi(x - vt)$ traveling in the opposite direction, toward increasing values of x.

‡13. An unusual wave is described by the rather formidable wavefunction:

$$\Psi(x, t) = Ae^{-[\alpha(x - vt)]^2} + Ae^{-[\alpha(x + vt)]^2}$$

Use the following values for the parameters A, α, and v:

$$A = 3.00 \text{ m}$$
$$\alpha = 0.500 \text{ m}^{-1}$$
$$v = 1.00 \text{ m/s}$$

Make a series of graphs of the waveform of this wavefunction over the domain $x = -10.00$ m to $x = +10.00$ m at each of the following times: $t = -6.00$ s; $t = -3.00$ s; $t = 0.00$ s; $t = 3.00$ s; $t = 6.00$ s. (b) What is the meaning of the constant A on the graph? (c) Describe how Ψ changes as time advances during the interval from $t = -6.00$ s to 6.00 s. Note the successive positions of a particular crest of the wave. The series of waveforms illustrates the principle of linear superposition, discussed in greater detail in Section 12.18. (d) Investigate the effect on the graphs of changing the parameter α to 1.00 m^{-1} and v to 2.00 m/s. Make the plots over the domain $x = -15.00$ m to $x = +15.00$ m for the same five instants as in part (a).

Section 12.8 Sinusoidal (Harmonic) Waves

14. What is the approximate frequency of a sinusoidal sound wave that has a wavelength equal to the straight-line distance between your ears?

•**15.** Alas, after a sybaritic festival, the cheap upright piano in your fraternity house is found upright at the bottom of the house swimming pool. You decide to play Handel's *Water Music* but first test the sound of middle C (261.6 Hz). The speed of sound in water is 1.48×10^3 m/s. What is the wavelength of the sound wave corresponding to middle C in the pool? What assumptions are needed to make the determination?

•**16.** A transverse wave on a string is described by the wavefunction

$$\Psi(x, t) = (0.300 \text{ m}) \cos[(12.57 \text{ rad/m})x - (251.3 \text{ rad/s})t]$$

Find: (a) the amplitude of the wave; (b) the angular wavenumber of the wave; (c) the wavelength of the wave; (d) the angular frequency of the wave; (e) the frequency of the wave; (f) the period of the wave; (g) the speed of the wave; (h) the wave disturbance at $x = 3.00$ m when $t = 1.50 \times 10^{-3}$ s.

•**17.** An aging hippie is making good vibes that are described by the wavefunction

$$\Psi(x, t) = (6.00 \times 10^{-3} \text{ Pa}) \cos[(8.06 \text{ rad/m})x - (2.76 \times 10^3 \text{ rad/s})t]$$

(a) What is the amplitude of the wave? (b) What is the angular wavenumber of the wave? (c) How many wavelengths will fit into a distance of 2π meters? (d) What is the wavelength of the waves? (e) What is the angular frequency of the wave? (f) What is the frequency of the wave? (g) What is the period of the oscillation? (h) What is the speed of the wave? (i) Plot the waveform when $t = 1.00$ s. Choose an appropriate domain for x to show several wavelengths. (j) Plot the oscillation produced by the wave at the position $x = 2.00$ m as a function of time. Choose an appropriate domain to show several oscillations. (k) What is the wave disturbance Ψ at $x = 1.00$ m when $t = 0$ s? Explain the physical significance of the negative pressure value of Ψ.

•**18.** Figure P.18 depicts the waveform of a traveling sinusoidal wave at two instants, when $t = 0$ s and 0.250 s later. The wave is traveling toward increasing values of x. Find $\Psi(x, t)$.

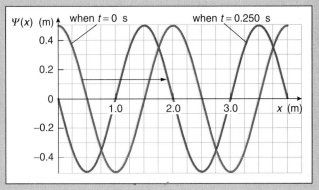

FIGURE P.18

•**19.** A very slow acoustic sales pitch emitted by a traveling wave salesperson has the wavefunction

$$\Psi(x, t) = (2.00 \times 10^{-3} \text{ Pa}) \cos[(8.00 \text{ rad/m})x - (6.00 \text{ rad/s})t]$$

For the Better Business Bureau, find: (a) the amplitude of the wave; (b) the angular wavenumber of the wave; (c) the angular frequency of the wave; (d) the frequency of the wave; (e) the speed of the wave; and (f) the direction the wave is moving. (g) Just to be ornery, find the value of the disturbance at $x = 2.00$ m when $t = 1.50$ s.

•**20.** Duck season has arrived. A Harlequin duck (*Histrionicus histrionicus*) is bobbing vertically up and down due to a passing water wave. The duck bobs through a total vertical distance of 10.0 cm and completes 7.00 cycles during 10.0 s according to your Mickey Mouse watch. You also note that a particular crest of the wave travels the 9.00 m from the duck to your canoe during 5.00 s. (a) What is the frequency of the water wave? (b) What is the speed of the wave? (c) What is the wavelength of the waves? (d) Write a mathematical expression for the wavefunction $\Psi(x, t)$ assuming it is a sinusoidal wave with the duck located at the origin for x and at a peak of the wave when $t = 0$ s. Express all distances in meters and times in seconds.

•**21.** You like to make waves and succeed in generating small, sinusoidal waves that propagate across a liquid surface at a speed of 0.600 m/s. The ripple crests are 0.200 m apart and are 1.00 cm in height (from trough to peak). (a) Find the frequency of the waves. (b) Write an equation for the vertical position Ψ as a function of horizontal position x and time t if the origin in the vertical direction is at the water surface when it is calm. Use SI units. A peak is at the origin when $t = 0$ s.

•**22.** By careful measurements, you determine that the lecture sound wave propagating from your professor to your ears is described by the following wavefunction:

$$\Psi(x, t) = (1.60 \times 10^{-3} \text{ Pa}) \cos[(11.0 \text{ rad/m})x + (3.77 \times 10^3 \text{ rad/s})t + 10.0 \text{ rad}]$$

Your professor is suitably impressed with your experimental prowess, but now asks you to find: (a) the wavelength; (b) the frequency, (c) the speed; and (d) the direction of propagation of the wave.

•23. During a hairy calculation by the famous classicist I. M. Pompous Mediocritus, an astute physics student detects a "hand wave" propagating across the blackboard. The hand wave is described by the wavefunction

$$\Psi(x, t) = (3.00 \text{ obfuscations}) \cos[(18.0 \text{ rad/m})x - (36.0 \text{ rad/s})t]$$

(a) Determine the wavelength and frequency of the hand wave. (b) Determine the speed of the hand wave (otherwise known as the "sleight of hand").

Section 12.9 Waves on a String

24. Transverse waves with a speed of 25 m/s are to be generated on a 6.50 m length of clothesline of mass 100 g. What magnitude of tension is needed?

25. A string has a tension of magnitude 50 N and transverse waves propagate along its length with a speed of 35 m/s. What magnitude of tension is needed to have the waves propagate with speed 45 m/s?

•26. A 14 gauge copper wire has a diameter of 0.2032 cm. Transverse waves on the wire are to travel at a speed of 125 m/s. What magnitude tension is needed? The density of copper is $8.50 \times 10^3 \text{ kg/m}^3$.

•27. A rope with linear mass density μ_1 is spliced to another rope with linear mass density μ_2; one end of the latter rope is attached rigidly to a wall. You pull on the end of the first rope with a tension of magnitude T, directed along the rope. What is the ratio of the speeds v_1/v_2 of transverse waves on the two strings?

•28. Two separate strings have linear mass densities μ_1 and μ_2. Transverse waves travel at equal speeds along each string. What is the ratio of the magnitudes of the tensions T_1/T_2 in the strings?

•29. You and a buddy hold opposite ends of a rope of length of 10.00 m and mass 156 g. You exert a force of magnitude 40.0 N on your end of the rope, directed along the rope; the rope is stationary. A brief interval Δt after you initiate a pulse down the rope, your buddy does the same. The two pulses meet 1.00 m from the center of the rope. What was the time interval Δt between initiation of the pulses?

•30. When $t = 0$ s, a pulse on a string is described by the equation

$$\Psi(x, 0 \text{ s}) = \frac{1.00 \text{ m}^3}{2.50 \text{ m}^2 + x^2}$$

(a) Sketch $\Psi(x, 0 \text{ s})$ as a function of x over the domain $-10.00 \text{ m} \le x < 10.00 \text{ m}$. (b) The pulse moves parallel to $-\hat{\imath}$ with a speed of 8.00 m/s. Find $\Psi(x, t)$.

Sections 12.10 Reflection and Transmission of Waves
12.11 Energy Transport via Mechanical Waves
12.12 Wave Intensity

31. You clap your hands once at the top of a deep well shaft and hear the echo 0.40 s later. How far below the top of the well is the water surface?

•32. A river canyon is 150 m across as shown in Figure P.32. You are located on the river surface and clap your hands once to generate a brief acoustic pulse. You detect a time interval of 0.408 s between the twin echoes recorded on tape, analyzed when back home (?) in a physics lab. Where were you relative to the canyon walls? (Neglect multiple echoes.)

River

FIGURE P.32

•33. The ancient Great Pyramid at Giza, Egypt, built to be the tomb of Khufu, the first pharaoh of the Fourth Dynasty (reigning from 2789–2767 B.C.), had smooth polished stone on each side. Most of the fine stone facing was stripped away by scavengers during the subsequent millennia for other building projects near and in Cairo. The smooth facing remains only near the peak. The tomb also was plundered; so much for eternally resting in peace. The faces of the pyramid now reveal huge blocks of stone that form large steps as shown in Figure P.33. Imagine yourself located on a nearby sand mound on a line that is perpendicular to the side of one face of the pyramid. You generate a very short acoustic pulse by shouting a short "Yip!", setting off a small firecracker, or shooting a starter's gun. The pulse propagates to and is reflected from each step. About how many reflected pulses per second do you hear? You may be able to do a similar experiment on a local scale using stepped bleachers at your university stadium.

You

2.0 m

1.5 m

FIGURE P.33

•34. You are scuba diving 3.00 m below the surface of the Caribbean Sea. Fun! A support boat is located directly above you. A bell on the boat is located 1.5 m above the water line.

(a) How long does it take the sound of the bell to propagate to you? (b) If the sound reaches the ear of the skipper in the same time, how far is the skipper from the bell? See Table 12.2 (on page 553) for appropriate values of the speed of sound.

●35. The intensity of solar light waves impinging on the Earth is measured by a spacecraft to be 1.36 kW/m². This intensity is called the *solar constant*, and its rather significant value is what motivates much research into tapping such solar energy. Consider the Sun to be a point source of light. The Earth is located 149.6×10^6 km from the Sun. Use this information to calculate the *solar luminosity*, the power output of the Sun.

●36. An acoustic source emits sound waves with a power of 10.0 W in all directions. What is the intensity of the sound waves at distances of 5.00 m and 25.0 m from the source?

●37. The inside of a long straight tunnel has a strip of fluorescent lights along its length; see Figure P.37. The lights are each ℓ meters long and have a power of P watts of light. What is the light intensity at a distance r from the lamps, measured perpendicular to the axis of lights? Neglect reflections from the walls of the tunnel.

FIGURE P.37

Sections 12.13 What Is a Sound Wave?*
12.14 Sound Intensity and Sound Level*

38. A speaker emits 50 W of power in acoustic waves that propagate away from the source uniformly in all directions. (a) What is the intensity of the sound waves 20 m from the source? (b) What is the sound level in dB at this location?

39. An acoustic source has an output sound level of 96 dB at a distance of 2.00 m from the source. What is the intensity of the sound?

●40. If the sound level increases by 5.0 dB, by what factor does the sound intensity increase?

●41. A sound source that emits acoustic waves in all directions has a sound level of 120 dB at a distance of 5.0 m from the source. (a) What is the power of the source? (b) At what distance is the sound level 90 dB?

●42. A sound level meter indicates that a single trombone player produces a sound level of 75 dB, measured a certain distance away. (a) If two identical trombonists now are used, both the same distance away, what sound level will the meter indicate? (b) If 76 trombones lead the big parade, what sound level will they produce if they all are approximately the same distance from the sound level meter?

●43. As a precaution at a rock concert or in other noisy environments, you wear ear protectors that reduce the sound level by 20 dB. What percentage of the intensity of the sound is blocked?

●44. Professional screamers (a.k.a. small siblings and/or other whiners) can attain sound levels approaching 100 dB, measured 2.0 m away. (a) What is the corresponding sound intensity? (b) Assuming (unrealistically) there is no absorption of sound waves by the air, how far away is the sound level of the scream equal to the threshold of hearing?

●45. A rock concert is given in an auditorium with front row seats 6.0 m from the speakers. The rear row seats are 30.0 m from the speakers. If the sound level is 120 dB in the front row, what is the sound level in the rear row? Assume (unrealistically) that the speakers propagate sound uniformly in all directions.

●46. An experiment called the Acoustic Thermometry of Ocean Climate Project (ATOC) attempted to measure global warming through its effect on long-term changes in ocean water temperatures. The experiment involved measuring the propagation times of sound waves through sea water over very large distances, upward of 2×10^4 km. The speed of sound is a function of temperature, so that the travel times of sound waves to distant receivers vary as ocean temperatures change. The long distances give good average values for ocean temperatures that were monitored over time. Low-frequency sound waves (≈ 57 Hz) were used to minimize interference with marine mammal communication frequencies among other things. The powerful acoustic source was placed deep in the ocean (~ 1 km) to avoid dangerous sound levels for nearby fish and marine mammals and to minimize interference with their communications, but the experiment nonetheless sparked environmental controversy and was abandoned. The source emitted sound waves in all directions. The power of the source was about 1.0×10^5 W. At what distance from the source did the sound level become 100 dB? Assume no absorption of the sound waves by sea water; this assumption is unrealistic.

Sections 12.15 The Acoustic Doppler Effect*
12.16 Shock Waves*

●47. Surface water waves with a wavelength of 3.0 m travel across the surface of Lake Superior with a speed of 7.0 m/s. A ship is moving with a speed of 2.5 m/s. Both speeds are measured with respect to the bottom of the lake. What is the frequency of the waves according to the ship if it is moving: (a) parallel to the velocity of the waves? (b) antiparallel to the velocity of the waves? (c) perpendicular to the velocity of the waves?

●48. A collection of standing passengers is spaced uniformly 1.50 m apart along a moving sidewalk at a busy air terminal. The speed of the sidewalk is 2.0 m/s. An executive is running at speed 5.0 m/s to catch a flight. (a) What is the number of people per second that move past a coffee shop beside the walkway? (b) If the executive on the walkway runs parallel to the velocity of the walkway, how many people per second does she pass?

●49. After a particularly hairy literature exam covering the many works by the famous author Anonymous, you run in a straight line from the examination room at constant speed while screaming. Your scream normally has a frequency of 2.200×10^3 Hz but your fellow sacrificial lambs, who remain in the room enduring the exasperating examination, somewhat out of tune hear a frequency that differs from this frequency by 20 Hz. (a) What frequency is heard by your classmates? (b) Determine the speed of your departure.

•50. A policewoman is chasing a speeder along a straight interstate highway. Both the speeder and trooper are moving at 150.0 km/h. The siren on the police cruiser has a frequency of 2.000×10^3 Hz. What frequency is detected: (a) by the policewoman? (b) by the speeder? (c) by a doomed, dumb possum (*Didelphis marsupialis*) watching the cars approach from a vantage point in the middle of the road?

•51. The submarine *U. S. S. Blowfish* lies peacefully at rest just off Tahiti. A fast yellow submarine is moving directly toward the *Blowfish* at a speed of 75.0 km/h using sonar (sound waves in water) of frequency 40.0 kHz to detect the *Blowfish*. The *Blowfish* detects the sonar from the yellow submarine. (a) What frequency does the *Blowfish* detect? Use 1.520×10^3 m/s for the speed of sound in sea water. (b) If the yellow submarine was stationary and the *Blowfish* was moving directly toward it at a speed of 75.0 km/h, what frequency would the *Blowfish* detect?

•52. The Doppler effect was first tested experimentally by using moving and stationary musicians. A violinist plays middle C with a frequency of 261.6 Hz. (a) If the violinist is placed on an open railway carriage moving directly away from you (at rest), will the note you hear be sharp or flat? Sharp means a higher frequency; flat means a lower frequency. (b) If you also are playing a violin at the frequency 261.6 Hz, at what speed must the railway carriage travel so you hear a beat frequency (see Section 12.20) of 2.0 Hz from the two violins?

•53. An ambulance siren emits a wail at 1200 Hz. Responding to an accident, the ambulance is rushing to the scene at a speed of 130.0 km/h. For the usual reasons, a lawyer is chasing the ambulance at a speed of 140.0 km/h. (a) What frequency is heard by the paramedic in the ambulance? (b) What frequency does the barrister hear? (c) After the lawyer passes the ambulance, what frequency does she hear?

•54. You have a source that emits sound waves with a frequency of 440.0 Hz. (a) At what speed must the source be set in motion either toward or away from you so that you hear a frequency of 439.0 Hz? (b) With the speed calculated in part (a), determine the frequency at which you hear the harmonic of 880.0 Hz.

•55. A new jet engine, developed by musically inclined aeronautical engineers, emits sound precisely at a frequency of 440.0 Hz; this thing really hums! A plane equipped with such engines is directly approaching you at 250 km/h on its takeoff roll along a runway, roars past, and recedes into the distance without turning. (a) What is the frequency you hear as the plane approaches? (b) What is the frequency you hear as the plane recedes? (c) Is the frequency *difference* on approach equal to the frequency *difference* as it recedes? Explain your answer.

•56. An audio transmitter and receiver are mounted side by side as shown in Figure P.56. The transmitter emits sound of frequency v. A distant flat plate approaches the instrument at speed v_{target} and the receiver detects sound waves of frequency v' reflected from the target. The difference in frequency between the emitted and detected frequencies can be used as a motion detector.

(a) Show that

$$\frac{v'}{v} = \frac{v + v_{target}}{v - v_{target}}$$

(b) In many cases $v_{target} \ll v$. For such cases, show that

$$\frac{v'}{v} \approx 1 + 2\frac{v_{target}}{v}$$

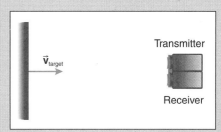

FIGURE P.56

•57. A motion detector (see Problem 56) emits an inaudible sound wave with a frequency of 30.0 kHz. A dog trots toward the stationary detector with a speed of 1.00 m/s. What is the frequency detected by the motion detector?

•58. A jet fighter in level flight passes directly overhead at a speed corresponding to Mach number 1.30. The sonic boom is heard by you on the ground 12.0 s later. What is the altitude of the plane? Assume (unrealistically) the speed of sound does not change with altitude.

•59. After taking measurements from Figure P.59, which shows the shock wave created by a bullet moving in air, determine the speed and Mach number of the bullet.

FIGURE P.59

•60. Some fighter planes can attain speeds of Mach number 2.5 at an altitude of 15.0 km. The speed of sound at that altitude is less than at sea level and is about 2.9×10^2 m/s.* (a) What is the

*The speed of sound as a function of altitude depends on the season of the year as well. See Allan D. Pierce, *Acoustics* (McGraw-Hill, New York, 1981), pages 389 and 395 for profiles of the speed of sound as a function altitude and season.

half-angle of the Mach cone formed by the plane? (b) What is the speed of the plane in km/h?

•61. The Concorde supersonic passenger jet attains speeds of about 2150 km/h at a cruising altitude of 19.8 km. The speed of sound at that altitude is about 2.9×10^2 m/s. (a) What is the Mach number of the aircraft at its cruising altitude? (b) What is the half-angle of the Mach cone formed by the plane?

‡62. A source of sound of frequency v (perhaps your brother singing a note from a favorite aria) is perched on the rim of a merry-go-round of radius R spinning with frequency v_0. You are located at rest a sufficient distance away that all points on the rim are essentially in the same direction to you; see Figure P.62. Find an expression for the frequency v' you detect as a function of time if the source is at the position shown in Figure P.62 when $t = 0$ s.

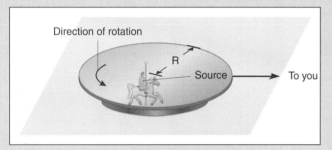

Direction of rotation

R

Source

To you

FIGURE P.62

Section 12.17 Diffraction of Waves

63. The entrance to a marina is approximately 10 m in width, as shown in Figure P.63. What is the order of magnitude of the range of wavelengths of water waves that can pass into the marina with minimal diffraction effects?

10 m

FIGURE P.63

•64. If the wavelength emitted by a source such as a loudspeaker is much greater than the size of the source (say, at least ten times greater), the waves propagate outward uniformly in all directions and the source is effectively a point source. Conversely, if the wavelength is quite small compared with the size of the source (say, at least ten times smaller), the source is directional and the waves propagate outward in the

direction the speaker faces. Two speakers are mounted in a console as shown in Figure P.64. One speaker has a diameter of 25 cm and the other 10 cm. (a) For what range of frequencies is each speaker a point source? (b) For what range of frequencies is each a directed source?

FIGURE P.64

Sections 12.18 The Principle of Superposition
12.19 Standing Waves
12.20 Wave Groups and Beats*

65. A hollow, gas-filled glass cylinder has a piston at each end as shown in Figure P.65. The bottom of the tube has a sprinkling of finely ground cork particles littered along its length. One piston is connected to a oscillator that moves it back and forth at frequency v, generating sound waves of the same frequency inside the tube. The position of the other piston is adjusted until standing sound waves exist in the tube. The bits of cork particles collect in small piles at the nodes of the standing waves. Let ℓ be the distance between the nodes. Show that the speed of sound v in the gas-filled tube is

$$v = 2\ell v$$

This procedure, called *Kundt's method*, was first used by August Adolph Kundt (1839–1894) in 1866 to measure the speed of sound in various gases. Your physics department may have such an apparatus in its attic.

To oscillator

Clamp

FIGURE P.65

66. Two strings, call them A and B, have equal mass. String A has twice the length of string B. (a) If string A has a tension of magnitude twice that of string B, what is the ratio of their fundamental frequencies v_A/v_B? (b) What fraction of the tension of string B must string A have to produce a fundamental frequency half that of string B?

67. The base string for the low A (27.5 Hz) on a Model 225 Bösendorfer piano has a total length 2.00 m and a total mass of 0.50 kg; the distance between the ends of the clamped string is 1.90 m. What tension is needed for the properly tuned note?

•**68.** Repeat Problem 13 part (a) but with

$$\Psi(x, t) = Ae^{-[\alpha(x-vt)]^2} - Ae^{-[\alpha(x+vt)]^2}$$

for the wavefunction. Explain the meaning of each term in the wavefunction in terms of your graphs.

•**69.** (a) Describe the physical difference between a wave described by the wavefunction

$$\Psi_1(x, t) = (3.00 \times 10^{-2} \text{ m}) \cos[(9.00 \text{ rad/m})x - (4.00 \text{ rad/s})t]$$

and one whose wavefunction is

$$\Psi_2(x, t) = (3.00 \times 10^{-2} \text{ m}) \cos[(9.00 \text{ rad/m})x + (4.00 \text{ rad/s})t]$$

(b) If these two waves are superimposed, will a standing wave pattern be formed? If so, what is the distance between the nodes of the standing wave? If no standing wave is formed, explain why. (c) If a standing wave is formed, determine the x-coordinates of the first three nodes going from the origin toward increasing values of x.

•**70.** A wave is described by the wavefunction

$$\Psi(x, t) = (2.00 \times 10^{-2} \text{ m}) \cos[(18.0 \text{ rad/m})x - (24.0 \text{ rad/s})t]$$

(a) What wave, when added to the given wave, will form a standing wave pattern? (b) For such a standing wave, what is the distance between the nodes of the wave? (c) For such a standing wave pattern, what is the distance between the antinodes? (d) What is the distance between a node and the nearest antinode?

•**71.** The middle C strings on a piano typically have a length of about 0.60 m. Middle C has a frequency of 261.6 Hz. (a) If the same kind of wire and the same magnitude tension is used to string the C located four octaves higher (the highest note on the piano), what length string is needed? In practice, this string is too short to use for this note. A longer string with less mass per unit length is used in order to accommodate other aspects involved in the construction of a piano. (b) If the same kind of wire and same magnitude of tension as the middle C string are used to produce the lowest note on the common piano (the A with a frequency of about 27.5 Hz), what string length is needed? In practice, this string is much too long to make an affordable (let alone an aesthetic looking) piano even if you had the huge rooms of the Schönbrunn palace in Vienna* to accommodate the instrument.

•**72.** An open organ pipe has a fundamental frequency of 440 Hz. A closed organ pipe has a fundamental frequency that is the second harmonic of the open organ pipe. What are the lengths of the two pipes?

_____

*This palace is where Mozart played for Maria Teresa.

•**73.** An instrument has a string of length ℓ that produces a fundamental frequency v. A finger shortens the length of the string to produce higher frequencies. If the string is shortened to a length $\alpha\ell$, where α is a pure number less than or equal to 1, what is the frequency of the fundamental corresponding to the new length?

•**74.** The strobe lights on top of two tall radio towers both have a period of about 1.00 s. When turned on simultaneously, a keen observer notes that during an 800 s interval, the lights gradually fall out of synchronization and then come back into flashing synchronously. What is the interval ΔT between the periods of the two beacons?

•**75.** Two wavefunctions have the following descriptions:

$$\Psi_1(x, t) = (2.00 \times 10^{-3} \text{ Pa}) \cos[(1.00 \text{ rad/m})x - 1.00 \text{ rad/s})t]$$

$$\Psi_2(x, t) = (2.00 \times 10^{-3} \text{ Pa}) \cos[(1.01 \text{ rad/m})x - 1.03 \text{ rad/s})t]$$

(a) Find the speed of each wave. (b) If the waves are superimposed, find the beat frequency and the approximate value of the group speed. (c) Explain what is meant by the phase speed and the group speed.

❚**76.** Two waves with slightly different frequencies combine to form beats. (a) Show that the group speed, given by Equation 12.57, also can be expressed as

$$v_{\text{group}} = -\lambda^2 \frac{dv}{d\lambda}$$

(b) If the waves exhibit no dispersion, that is, if all frequencies and wavelengths travel at the same constant speed $v_{\text{phase}} = v\lambda$, show that the phase and group speeds are identical. (c) For waves that exhibit no dispersion, make a rough sketch of a graph of the frequency v versus λ.

Section 12.21 Fourier Analysis and the Uncertainty Principles*

•**77.** The waveform of a periodic traveling wave when $t = 0$ s has the Fourier series

$$\Psi(x, 0 \text{ s}) = \frac{4.00 \text{ m}}{\pi} \left\{ \frac{\sin\left[(1.00 \text{ rad/m}) x\right]}{1} \right.$$

$$+ \frac{\sin\left[(3.00 \text{ rad/m}) x\right]}{3}$$

$$+ \frac{\sin\left[(5.00 \text{ rad/m}) x\right]}{5} + \cdots$$

$$\left. + \frac{\sin\left[\left([2n + 1] \text{ rad/m}\right) x\right]}{2n + 1} + \cdots \right\}$$

where n is a positive integer. (a) Make a graph of the sum of the first six terms of the Fourier series. Choose a domain for x to show several wavelengths of the periodic waveform. (b) As more and more terms are included in such a graph, what periodic waveform does the Fourier series represent? (c) What is the wavelength of the periodic waveform represented by the Fourier series?

•**78.** A traveling wave produces the following function for the oscillation at $x = 0$ m:

$$\Psi(0 \text{ m}, t) = \frac{6.00 \text{ m}}{\pi} \left\{ \sin\left[(5.00 \text{ rad/s}) t\right] + \frac{\sin\left[(10.00 \text{ rad/s}) t\right]}{2} \right.$$
$$\left. + \cdots + \frac{1}{n} \sin\left[n(5.00 \text{ rad/s}) t\right] + \cdots \right\}$$

where n is a positive integer. (a) Make a graph of the sum of the first ten terms of the series. (b) As more and more terms are included in such a plot, what periodic oscillation does the Fourier series represent? (c) What is the period of the oscillation?

•**79.** Digital communication systems use short-duration electronic pulses to transmit data. One such system can transmit as many as 10^8 pulses per second. (a) If the pulses do not overlap, what is the maximum duration of a single pulse? (b) According to the uncertainty relations, what range of frequencies is involved in a Fourier representation of such pulses?

INVESTIGATIVE PROJECTS

A. Expanded Horizons

1. Investigate the use of ultrasonic technology for medical applications.

   Peter N. T. Wells, "Medical ultrasonics," *IEEE Spectrum*, 21, #12, pages 44–51 (December 1984).

   Gilbert B. Devy and Peter N. T. Wells, "Ultrasound in medical diagnostics," *Scientific American*, 238, #5, pages 98–112, 172 (May 1978).

   Stephen J. Riederer, "Resource letter MI-1: Medical imaging," *American Journal of Physics*, 60, #8, pages 682–693 (August 1992).

   Russell K. Hobbie, "Resource letter MP-1: Medical physics," *American Journal of Physics*, 53, #9, pages 822–829 (September 1985).

   Hylton B. Meire and Pat Farrant, *Basic Ultrasound* (John Wiley, New York, 1995).

   Steven Webb, *The Physics of Medical Imaging* (Institute of Physics, Philadelphia, 1988).

2. Investigate how bats use sound to detect flying insects.

   Gerhard Neuweiler, "How bats detect flying insects," *Physics Today*, 33, #8, pages 34–40 (August 1980).

   Don E. Wilson and Merlin D. Tuttle, *Bats in Question: The Smithsonian Answerbook* (Aubrey Books, Silver Spring, Maryland, 1997).

3. Noise: it is everywhere, it seems. Investigate the nature of acoustic noise and efforts to combat environmental noise from various sources.

   Thomas D. Rossing, "Resource letter ENC-1: Environmental noise control," *American Journal of Physics*, 46, #5, pages 444–454 (May 1978); this contains many references to these topics.

   John I. Shonle, "Resource letter PE-1: Physics and the environment," *American Journal of Physics*, 42, #4, pages 267–273 (April 1974); this contains references to these topics.

   Sukhbir Mahajan, Michael Shea, D. L. Robinson, and Dave Mathes, "Environmental noise: an undergraduate research project," *American Journal of Physics*, 45, #10, pages 987–990 (October 1977).

   A. E. Lord Jr., "Research projects involving inner city high school students," *American Journal of Physics*, 43, #7, pages 602–605 (July 1975).

   Anthony A. Silvidi and Louis J. Kristopson, "Noise measurements in a university: an open-ended student experiment," *American Journal of Physics*, 41, #7, pages 909–913 (July 1973).

4. Investigate the structure and function of the various parts of the human ear. What are the minimum and maximum motions of the ear membrane at audible frequencies? Are there different forms of ears among animals and fish? Can plants hear or respond to acoustic waves?

   T. S. Littler, *Physics of the Ear* (Pergamon Press, Oxford, 1965).

   Neal P. Rowell, "The response of the ear," *The Physics Teacher*, 18, #7, pages 531–532 (October 1980).

   D. W. Kammer and J. A. Williams, "Some experiments with biological applications for the elementary laboratory," *American Journal of Physics*, 43, #6, pages 544–547 (June 1975).

   C. Daniel Geisler, *From Sound to Synapse: Physiology of the Mammalian Ear* (Oxford University Press, New York, 1998).

5. Investigate an area of the acoustics of music that interests you. There are many aspects to musical acoustics, including the perception of sound, the physics of various musical instruments, and the physical principles underlying the design of concert auditoriums.

   Thomas D. Rossing, "Resource letter MA-2: Musical acoustics," *American Journal of Physics*, 55, #7, pages 589–601 (July 1987); this contains an extensive bibliography to many of these areas.

   John I. Shonle, "Implementing a course on the physics of music," *American Journal of Physics*, 44, #3, pages 240–243 (March 1976).

   Neville H. Fletcher and Thomas D. Rossing, *The Physics of Musical Instruments* (Springer-Verlag, New York, 1998).

   Arthur H. Benade, *Fundamentals of Musical Acoustics* (Dover, Mineola, New York, 1990).

6. Investigate how seismic waves are used to glean information about the structure of the interior of the Earth and earthquakes.

   Paul A. Bender, "Resource letter SEG-1: Solid-earth geophysics," *American Journal of Physics*, 44, #10, pages 903–911 (October 1976); this contains references to these topics.

   Fred Carrington, "The Northridge earthquake—a giant physics laboratory," *The Physics Teacher*, 32, #4, pages 212–215 (April 1994).

   A. J. Berkhout, *Applied Seismic Wave Theory* (Elsevier, New York, 1987).

B. Lab and Field Work

7. Design and perform an experiment to determine the range of frequencies and minimum sound levels to which your ears respond. Is there a difference between your left and right ears? Investigate how the range of frequencies to which the ear responds changes with age from infancy to old age.

   Thomas D. Rossing, *The Science of Sound* (Addison-Wesley, Reading, Massachusetts, 1982), Chapter 6.

8. Using a ripple tank and other necessary equipment, conduct a series of experiments to measure the dispersion of surface water waves, the variation of the speed of the waves with wavelength.

9. Write a proposal for an experiment to determine if audible sound waves exhibit dispersion. Your proposal should include a statement of the objective of the experiment, a detailed discussion of the methods to be used, and the reasons for any equipment needed to perform the experiment. If your proposal is acceptable to your professor and/or laboratory instructor, proceed to perform the experiment. Check to see if such experiments have been made before at the undergraduate laboratory level by checking such journals as the *American Journal*

of Physics, *The Physics Teacher*, and the "Amateur Scientist" section of *Scientific American*. Do you think it reasonable to assume that ultrasonic sound waves (those with frequencies higher than we can hear) and infrasonic sound waves (those with frequencies lower than what we can hear) also exhibit (or do not exhibit) dispersion?

10. Use a sound level meter and an experienced musician (perhaps yourself) to measure the approximate sound levels for music marked *pp*, *p*, *mf*, *f*, *ff*, and so on at a standard distance from the instruments. To what extent do your results depend on the instrument?

11. Conduct a series of experiments to determine if the acute angle that the crests of a bow wave make with the line along which a boat travels is a function of the speed of the boat.

12. Contact a piano manufacturer to secure samples of the wire used to string the 88 notes on the standard piano keyboard. Measure the mass per unit length of the various strings. Some notes are strung with multiple strings. Use the frequencies indicated in Figure 12.53 for the various notes to estimate the magnitude of the total tension that the piano frame must withstand due to all the strings.
Anders Askenfelt (editor), *Acoustics of the Piano* (Swedish Academy of Music, Stockholm, 1990).

13. Measure the mass and total length of the E_2 string on a guitar (the lowest-note string). (a) Determine its mass per unit length μ. The frequency of this note is 82.4 Hz. (b) Calculate the tension in the string when it is properly tuned. (c) Calculate the speed of the waves on the string.

14. Carefully measure the positions of the frets along one string of a guitar, banjo, mandolin, or another fretted string instrument. Calculate the various fundamental frequencies that can be produced from the string in terms of the fundamental frequency of the unfretted string. (See Problem 73.)

15. Use a sound level meter to survey the sound levels at various locations on your campus including (a) a class bell while ringing, (b) a car horn, (c) a fume hood in a chemistry lab, (d) cafeteria chatter, (e) a crowded reception hall, (f) a rock concert, and (g) a passing bus, train, or subway. Do any sound levels you survey exceed the threshold for pain?

C. Communicating Physics

16. A group of physicians has commissioned you to write a brief but articulate article for a lay audience (their patients) about the history and physical principles that underlie the common medical stethoscope. Your article will replace the outdated magazines in their waiting room, so you will need to make your writing clear and engaging.
John R. Cameron and James G. Skofronick, *Medical Physics* (Wiley, New York, 1978), pages 263–265.

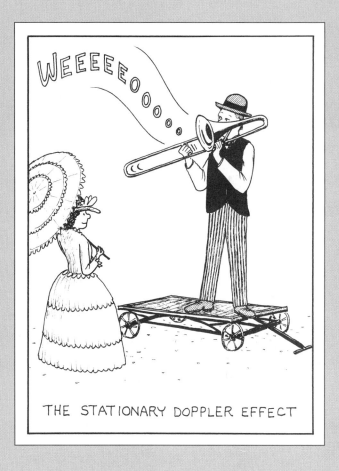

CHAPTER 13

TEMPERATURE, HEAT TRANSFER, AND THE FIRST LAW OF THERMODYNAMICS

Thermodynamics is a funny subject. The first time you go through it, you don't understand it at all. The second time you go through it, you think you understand it, except for one or two points. The third time you go through it, you know you don't understand it, but by that time you are so used to the subject, it doesn't bother you anymore.

Arnold Sommerfeld (1868–1951) *

Among other things, you can now describe the flight of a baseball and understand the mechanics of what happens when it is hit by a bat. In considering the dynamics of a system, the paramount concerns were changes in physical quantities with time. The fundamental principles of mechanics or dynamics are Newton's laws of motion. The CWE theorem, conservation of momentum, and conservation of angular momentum also stem, in one way or another, from Newton's laws of motion.

In our study of dynamics, we ignored or neglected thermal effects. Such an approach is tantamount to assuming that everything in the universe is at the same, constant temperature. But such a universe is a strange one indeed. In it many things we take for granted would not be possible: the stars and the seasons, engines and refrigerators, even life itself could not exist! Thus, to complement our study of mechanics, we now change our perspective and study *thermodynamic*[†] effects on a system.

In thermodynamics, one of the central parameters is *temperature* rather than time. Effects (such as thermal expansion) and processes (such as heat transfer) that cause or arise from changes or differences in temperature are important aspects of thermodynamics. But what is temperature? Is it the same as heat? In a word: *No!* In this chapter, we will first explore the distinct meanings of these words and then investigate aspects of a system that are affected by them. The fundamental principles of thermodynamics are known as the *laws of thermodynamics*; we encounter two of them in this chapter, reserving others for Chapter 15.

For the study of thermodynamics, we must introduce a new fundamental unit of the SI system, one which is as hallowed as the now familiar kilogram, second, and the meter: the SI unit for quantifying temperature, the *kelvin* (K).[‡]

13.1 SIMPLE THERMODYNAMIC SYSTEMS

A **thermodynamic system** is defined as a collection of many particles such as atoms and/or molecules.[§] A cup of coffee, a can of cola, the gas in a laser discharge tube, and even *you* are examples of thermodynamic systems. In this text, with few exceptions, we consider only **simple thermodynamic systems**.

A simple thermodynamic system is a system that is macroscopic, homogeneous, isotropic, uncharged, chemically inert, and experiences no change in its total mechanical energy. The system is sufficiently large that surface effects can be neglected. No electric or magnetic fields are present, and gravitational fields are irrelevant.

Gravitational effects are irrelevant if the system experiences no variation in the gravitational field over the physical dimensions of the system and experiences no changes in its gravitational po-

tential energy. For simplicity, we omit chemical, electrical, magnetic, and gravitational effects from this introductory treatment of thermodynamics. The thermodynamics of *complex* systems, those that are not simple, is best deferred (gladly, you might feel) to more advanced courses.[#] While a cup of coffee and a can of cola at rest are *simple* thermodynamic systems, the gas in a laser discharge tube and *you* are not.

QUESTION 1

Explain why *you*, even at rest, are not a simple thermodynamic system. Is a corpse a simple thermodynamic system?

13.2 TEMPERATURE

Some Like It Hot[¶]

*If you don't like the heat, get out of the kitchen.***

"Whew! It's *hot*!" is a common remark in midsummer; in contrast, a winter day in Maine or Minnesota might elicit the opposite comment: "Brrrr! It's *cold*!" Of course, what you are referring to is the temperature. Just as we all have an intuitive sense of the meaning of time, we all have an intuitive feeling for the meaning of temperature. To determine what is hot and cold, we could use our fingers (carefully!), or we could rely on the numerical readings of an appropriate thermometer, such as the common mercury household or laboratory thermometer. For our purposes early in this chapter, this understanding of the temperature as a quantitative measure of the hotness or coldness of a system will suffice. However, a precise definition of temperature, like a precise definition of time, requires some clear thinking to formulate, which we will do in Section 13.5.

Colloquially the word *heat* is often confused with *temperature*; in physics and the sciences, the two words are *not* synonymous.

The next section investigates the distinction between these terms.

13.3 WORK, HEAT TRANSFER, TEMPERATURE, AND THERMAL EQUILIBRIUM

. . . my bones are burned with heat.

The Book of Job[††]

When you catch a soda can tossed to you by a buddy, the can loses kinetic energy because of the (negative) work you do on it. When you place the warm can in the refrigerator to cool, we often hear it said that the can "loses heat" as it cools. How can we distinguish between heat and work and between heat and temperature?

---

[*](Chapter Opener) A statement (possibly apocryphal) attributed to Arnold Sommerfeld; quoted in Stanley W. Angrist and Loren G. Hepler, *Order and Chaos: Laws of Energy and Entropy* (Basic Books, New York, 1967), page 215.

[†] The word stems from the Greek θερμός (*thermos*), meaning "hot."

[‡] The unit formerly was called the *degree kelvin*, but the word degree and its symbol (°) now are considered superfluous when using the kelvin unit.

[§] By many particles, we mean the number of particles in the system is huge.

[#]We do consider an occasional complex system (in Section 13.14 for example, and in several problems) to illustrate the connection between dynamics and thermodynamics.

[¶]The title of a 1959 motion picture starring Marilyn Monroe, produced and directed by Billy Wilder.

[**]This remark is attributed to Harry S. Truman, 33rd president of the United States (the remark evidently was first quoted in *Time* magazine, 28 April 1952 (volume LIX, #17, page 19). Truman attributed the origin of the expression to Major General Harry Vaughan.

[††]Chapter 30, verse 30.

Work

When we examined the process of forces doing work on a system in Chapter 8, we saw that it was possible to change the kinetic energy of a system thanks to the work done *on* it by mechanical forces. The process of *energy transfer* to a system via work was neatly summarized in the CWE theorem. The work done by the total force is positive if $\Delta KE > 0$ J and negative if $\Delta KE < 0$ J. Work done by forces on a system is one way of transferring energy from external agents *to* a system. We say work is done *on* the system. You can think of work as that (energy) which is transferred from external agents to the system by the total force acting on it over a distance. But when a system is in mechanical equilibrium, we *cannot* say that the system "has" an amount work in the same way that the system *can* be said to have a definite amount of kinetic energy. There is no instantaneous value for the work; there is no mathematical work function associated with a system in mechanical equilibrium by virtue of its position in space or speed. However, there *is* an instantaneous value of the kinetic energy of a system in mechanical equilibrium: the by now well-known kinetic energy functions

$$KE_{trans} \equiv \frac{1}{2} mv^2$$

and

$$KE_{rot} \equiv \frac{1}{2} I\omega^2$$

In the same way, the potential energy of a system is a function of its position at a point in space.

In our study of the CWE theorem in Chapter 8, we noted that the work done on the system by the total force is equal to the change in the kinetic energy of the system. Changes in kinetic energy arise because of the process we call the work of the total force.

Hence work can change the state of mechanical equilibrium of the system. Later in the chapter we shall see that work *also* can change something we call the *internal energy* of system, quite apart from the change in its kinetic energy.

Heat and Heat Transfer

Imagine that you and a friend have different amounts of money in your wallet, say $25 and $15 respectively. To equalize the riches, you generously hand $5 to your friend as a donation, so that you each have the same wealth, $20. Once the money is transferred from you to your friend, the donation no longer exists (except in your mind—and for tax purposes). The donation can be considered to be both the *amount* transferred as well as the act (or *process*) of donating. Now suppose you have two systems with different temperatures in contact with each other.

Heat is defined to be that (energy) which is transferred between the two systems simply because they are at different temperatures.

To emphasize that heat is only involved in a process that occurs between two systems at different temperatures, rather than being a characteristic property of either system, *we will use the term* **heat transfer** *instead of the single word* heat. Heat transfer (like your donation) can be considered to be both the *amount* of energy transferred as well as the *process* itself. Work has this same dichotomous meaning.

The *process* of heat transfer continues until both systems have the same temperature. As with your monetary donation, once the systems have the same temperature, the process of heat transfer no longer exists—it is zero; the energy (the *amount* of heat transfer) manifests itself in other ways, which we will discuss in due course in Section 13.13. Heat transfer is essentially a microscopic atom-to-atom transfer of energy that commonly is not macroscopically apparent to the eye, because no macroscopic motion typically is involved. Thus, when you place a warm soda can in the refrigerator, energy is exchanged between the warm can and the cool air inside the refrigerator since they are initially at different temperatures; the process continues until they have the same temperature. But this energy exchange is not readily visible on a macroscopic basis: the can just sits passively on the shelf.

The energy exchange cools the can and may temporarily warm the inside of the refrigerator. The energy exchanged in this process is what we call a heat transfer. The heat transfer is *from* the warm can *to* the surrounding cool air. The energy exchange continues until a common temperature is reached. Once this common temperature is attained, there is no more energy exchange, no further heat transfer (process) from one system (the can) to the other (the air inside the refrigerator)—in other words, *zero* heat transfer. Nonetheless, both now have the same finite temperature; thus the concept of temperature is quite distinct from heat transfer.

Contrast heat transfer with what happens when a force on the can does work on it and changes its kinetic energy. You can easily see the effects of this energy exchange reflected in the

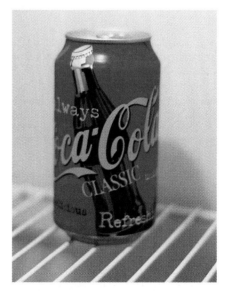

Using your eyes, you cannot see the heat transfer taking place between a warm soda can and the cool air within a refrigerator.

changing speed of the can. Thus macroscopic *work* is distinctly different from *heat transfer*.

It is not necessary to have direct physical contact between the systems to have heat transfer; for example, there is heat transfer between the Sun and the Earth through the vacuum of space (via a mechanism called *radiation*, as we will see in Section 13.11).

> If it is *possible* for two systems to exchange energy via heat transfer (whether or not they actually do), the systems are said to be in **thermal contact** with each other.

> If two systems at the *same* temperature are placed in thermal contact with each other, there is zero heat transfer between them. Such systems then are said to be in **thermal equilibrium** with each other.

By definition, systems in thermal equilibrium with each other have the *same temperature*. In Chapter 15, we shall see that the common temperature is *never* 0 K, yet the heat transfer rate becomes *zero* when they reach the common temperature. If the combined system is isolated from its surroundings (i.e., if the system is well **insulated**), the common temperature does not change with time. Thermal equilibrium in thermodynamics is analogous to static equilibrium in mechanics.

Heat transfer exists when systems at different temperatures are placed in thermal contact with each other. Heat transfer occurs *from* the hotter system (the one at the higher temperature) *to* the cooler system (the one at the lower temperature). The reason heat transfer occurs from the system at the higher temperature to the system at the lower temperature, and not the other way around, is a subject we will examine in due course (in Chapter 15). For now, take this as an incontrovertible experimental fact; like guano, it just happens that way.

> Just as you *cannot* say that a system in mechanical equilibrium has an amount of work, you *cannot* say a thermodynamic system in thermodynamic equilibrium has an amount of heat.

There is *no* mathematical heat function associated with a system in thermodynamic equilibrium. As stated earlier, heat transfer from one system to another system is nonzero *only* if there is a temperature difference between the systems.

The dimensions of heat transfer are the same as those for work, since both represent energy transferred to (or from) a system. Hence heat transfer is expressed in joules (J) in SI units.

Historically, several other energy units are associated with heat transfer, though rarely with work. We will not use these other units in this text, precisely because they are antiquated and are not SI units. We mention these traditional units, however, because they occur frequently in the literature that you may encounter in your scientific careers.

1. One of these units is the **calorie** (cal), originally defined as the amount of energy needed to be transferred to precisely one gram of pure liquid water to raise its temperature by exactly one kelvin. The precise amount of energy transfer needed to accomplish the one kelvin change in temperature depends on the original temperature of the water. Hence, to avoid the problem of having a unit tied to a specific substance, the calorie now is defined to be exactly 4.186 J:

$$1 \text{ cal} \equiv 4.186 \text{ J} \quad \text{(an exact conversion)}$$

2. The food **Calorie** (Cal), of concern to sports enthusiasts and anyone on a diet, is the kilocalorie (kcal) or 4186 J:

$$1 \text{ Cal} \equiv 1 \text{ kcal} \equiv 4186 \text{ J} \quad \text{(exact conversion)}$$

Note that the food Calorie (Cal) is spelled and abbreviated with a capital C.

3. Another common energy unit associated with heat transfer is the **British thermal unit** (Btu) in the English system of units. The Btu unit frequently is used in engineering applications, particularly in the United States. The British thermal unit is the energy transfer needed to raise the temperature of one pound* of water one degree fahrenheit (°F). If needed, the appropriate conversion factor is

$$1 \text{ Btu} = 1055 \text{ J}$$

You may have noticed when purchasing an air conditioning unit that its cooling capacity is quantified in Btu.

QUESTION 2

Cookbooks often include baking instructions such as "cook at moderate heat" or "bake at high heat." Is this terminology correct from the standpoint of a physical scientist? Carefully explain your answer.

QUESTION 3

Two systems at different temperatures will eventually assume the same temperature when brought into thermal contact. Specify another physical quantity that becomes the same when two systems are combined or connected together.

13.4 THE ZEROTH LAW OF THERMODYNAMICS

If a can of cola and a bottle of milk each have the same temperature as the inside of your refrigerator, then the milk and cola have the same temperature as well. This common, almost trivial observation is called the **zeroth law of thermodynamics**[†]:

> If systems A and B each are in thermal equilibrium with a third system C, then A and B are in thermal equilibrium with each other.

---

*Actually, the English unit of mass is not the pound but the *slug*; the pound is the English unit for force. What you think of as one pound of sugar has a mass of approximately 1/32 slug, no matter where it may be located. You can see why we try to avoid the English system of units in the sciences!

[†]The importance of this observation for the formal structure of thermodynamics was recognized only long after the first and second laws of thermodynamics were formulated, yet this law must logically precede them. Hence, to avoid a confusing renumbering of the laws, this observation about systems in thermal equilibrium is called the zeroth law of thermodynamics after a suggestion by R. H. Fowler and E. A. Guggenheim [*Statistical Thermodynamics* (Cambridge University Press, Cambridge, England, 1939), page 56].

FIGURE 13.1 If vectors $\vec{A}$ and $\vec{B}$ each are perpendicular to $\vec{C}$, this need not mean that $\vec{A}$ and $\vec{B}$ are perpendicular to each other.

The law really stems directly from the definition of thermal equilibrium, but the law is *not* as obvious as it may seem. For example, suppose we return to the now familiar vectors of Chapter 2. If vectors $\vec{A}$ and $\vec{B}$ each are perpendicular to a vector $\vec{C}$, this fact does *not* necessarily mean that $\vec{A}$ is perpendicular to $\vec{B}$: see Figure 13.1.

Likewise, you know sociologically that if both Daniel and Andrew like Anna, and Anna likes them both, this does not necessarily mean that Daniel likes Andrew, particularly if they both are trying to court her! Thus the zeroth law of thermodynamics is not immediately obvious; it is a generalization made from countless experiments. No experimental systems ever have been observed to violate or disagree with the zeroth law of thermodynamics.

13.5 THERMOMETERS AND TEMPERATURE SCALES

Very high and very low temperature extinguishes all human sympathy and relations. It is impossible to feel affection beyond 78° or below 20° of Fahrenheit; human nature is too solid or too liquid beyond these limits.

Sydney Smith (1771–1845)*

The measurement of temperature and the definition of an appropriate temperature scale are not as simple as they might seem initially. When you think of temperature, you think about the reading indicated on a common thermometer. But what exactly *is* a thermometer and what does its reading tell you?

Types of Thermometers

Constructing a reliable and accurate **thermometer** (any device for measuring temperature) relies on using some physical property of matter, known as a **thermometric property**, that changes with temperature. The physical property might be a volume of mercury whose changes with temperature are indicated by changes in the length of a thin, uniform column of mercury connected to the volume, as in common household and laboratory thermometers.

Another useful thermometer of significance, but likely less well known to you, is the **constant volume gas thermometer**, pictured in Figure 13.2. The device keeps a gas at constant volume by means of a column of liquid. The thermometer can measure temperature in the following way. When brought into thermal contact with a system, if the temperature of the gas increases, the pressure increases inside the constant volume. The pressure increase manifests itself as an increase in the height of the liquid column needed to maintain the gas at constant

FIGURE 13.2 A constant volume gas thermometer.

volume. Thus the column height of the liquid provides a way to quantify the temperature.

Another convenient thermometer uses electrical resistance as a thermometric property; *platinum resistance thermometers* and common medical thermometers operate in this fashion. Still other thermometers use an electrical potential difference (known colloquially as a voltage) as the thermometric property; such a device is called a *thermocouple*. An example of a device that uses such electrical thermometers is shown in Figure 13.3.

Calibrating Thermometers

Typically any thermometer, such as the common mercury thermometer, is calibrated by bringing it successively into thermal contact with a large thermodynamic system at two different temperatures, known as **fixed points**. The process proceeds as follows:

1. The thermometer is brought into thermal contact with a massive ice–water slurry at one atmosphere pressure. When the length of the mercury column in the thermometer remains constant, the thermometer is in thermal equilibrium

FIGURE 13.3 Weather balloons use thermometers based on electrical thermometric properties so that data can be easily transmitted to ground-based observers.

---

*Letter to an unidentified woman, July 1836; in Evert A. Duyckinck, *Wit and Wisdom of the Rev. Sydney Smith* (Armstrong, New York, 1882), page 404.

with the slurry, and the location of the top of the mercury column can be arbitrarily marked as zero, as in Figure 13.4a.

2. Then the thermometer is brought into thermal contact with a huge mass of boiling water, also at one atmosphere pressure, and the new location of the top of the mercury column is noted and conveniently labeled 100, as in Figure 13.4b.

3. Now divide the change in the length of the uniform mercury column into 100 equally spaced divisions and you have created what we call the celsius scale; see Figure 13.4c. The boiling point of water at one atmosphere pressure is 100 °C and its normal freezing point is 0 °C.* The celsius scale is a *convenient* temperature scale and we shall use it occasionally.

The same procedure can be undertaken to calibrate other thermometers, such as the constant volume gas thermometer discussed earlier. All thermometers calibrated this way will indicate that the freezing point of water is 0 °C and the boiling point of water is 100 °C at a pressure of one atmosphere, because these are the fixed calibration points.

A significant question arises: if a mercury thermometer indicates that the temperature of a system is, say, 37 °C, does a constant volume gas thermometer (or another type of thermometer) *also* indicate 37 °C? Precision experiments indicate that the thermometers *differ slightly* in their temperature readings of a system with a temperature that is not at one of the fixed points. The discrepancy increases for temperatures increasingly removed from the calibration points. This observation raises two new questions: (1) Which thermometer indicates the correct temperature? and (2) How can you be sure that the actual temperature of the system is independent of the particular means or substance used to determine the temperature?

---

*The scale is named after Anders Celsius (1701–1744), a Swedish astronomer who originally proposed having 100 degrees between the boiling and freezing points of water. Curiously, the original proposal of Celsius had the scale inverted from current practice, so that the zero was at the boiling point and 100 was at the freezing point. The celsius scale also was known in the past as the centigrade scale (meaning 100 gradations or divisions between the normal freezing and boiling points of water); the term celsius now is used rather than the word centigrade.

Experiments indicate that the pressure P inside a constant volume gas thermometer, particularly for a dilute gas, is a linear function of the temperature T it indicates (after calibration at the fixed points) as long as the gas is not near its liquefaction temperature. We express this linear relationship between the temperature and pressure in this type of thermometer as

$$T = aP + b \qquad (13.1)$$

The constants a and b are determined by calibrating the thermometer at the two fixed points at a known pressure. Equation 13.1 indicates that a graph of T versus P is linear with a slope equal to a and an intercept b, as shown in Figure 13.5, which presents the results of using three different dilute gases inside the bulb of a constant volume gas thermometer.

Further experiments indicate that the temperature of a system indicated by the constant volume gas thermometer is *independent* of the particular gas used in the calibrated thermometer as long as the gas is dilute and far from its liquefaction temperature. This means that the *intercept b* in Equation 13.1 is the *same* for all gases. The temperature intercept corresponds to $P = 0$ Pa and is found by extrapolating experimental data with a variety of gases to a temperature lower than the liquefaction points of all gases. One finds, on the celsius scale, that the intercept b is equal to -273.15 °C.

Precisely calibrating a thermometer is difficult. The freezing and boiling points of water are experimentally difficult to reproduce because they depend on the pressure on the water surface. A common pressure cooker exploits the fact that the boiling temperature increases with the pressure on the surface of the water, decreasing the cooking time. On the other hand, climbers know that water boils at a lower temperature at high altitude (lower pressure) than at sea level, which increases the cooking time.

To avoid these difficulties, a *single-point* calibration of a thermometer is made. Liquid water, ice, and water vapor all coexist simultaneously in thermal equilibrium at a unique temperature and pressure called the **triple point** of water. The triple point of water has a temperature defined to be 0.01 °C; the pressure of the mixture at the triple point is experimentally found to be 611.73 Pa.

(a) Step 1
Ice-water slurry

(b) Step 2
Boiling water

(c) Step 3
Label scale

FIGURE 13.4 One way to calibrate a thermometer.

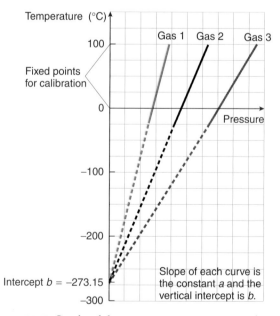

FIGURE 13.5 Graphs of the temperature versus pressure for different dilute gases used in a constant volume gas thermometer.

The boiling temperature of water is lower at a high altitude than at sea level because of the reduced pressure on the water surface; therefore, longer cooking times are necessary.

The **ideal gas temperature scale** is defined by setting the temperature of the triple point of water to be 273.16 K, in which case the intercept in Figure 13.5 is 0 K, called **absolute zero**.

Thus one **kelvin** is defined to be 1/273.16 of the temperature of the triple point of water.*

A degree on the celsius scale is the same size as a degree on the kelvin scale, but the zeros of each scale are different.

The zero of the kelvin scale is shifted by 273.15 degrees, and so the normal freezing point of water (at a pressure of exactly one atmosphere) is 273.15 K (= 0 °C) and the triple point is 0.01 °C. That is, a temperature T in kelvin on the ideal gas temperature scale is

$$T = t_{celsius} + 273.15 \text{ K} \qquad (13.2)$$

where $t_{celsius}$ is the temperature on the celsius scale.

Temperature changes on the celsius and kelvin scales are the same because the degrees are the same size, but the two scales *never* indicate precisely the same numerical reading for the temperature of a system because of the different locations of the zero on each scale.

The ideal gas temperature scale means that if the gas in a constant volume gas thermometer has a pressure P when it is in thermal equilibrium with a system at temperature T, the temperature T is found from

$$T = \frac{273.16 \text{ K}}{P_3} P \qquad (13.3)$$

where P_3 is the pressure of the gas in the *same* thermometer when in thermal equilibrium with water at its triple point. The pressure P_3

is *not* the pressure of the triple point of water (611.73 Pa), but the pressure of the gas in the constant volume gas thermometer when it is in thermal equilibrium with a system of water, ice, and water vapor at its triple point. The pressure P_3 in Equation 13.3 depends on the specific amount of gas in the constant volume gas thermometer, but not the specific kind of gas.

The lowest temperatures that can be measured with a constant volume gas thermometer are on the order of a few kelvin, in which case the dilute gas used is helium, since it has the lowest liquefaction temperature of any gas.

When we study the second law of thermodynamics in Chapter 15, we shall see that it is possible to define an **absolute temperature scale** that uses the kelvin and does *not* depend on any thermometric property of *any* substance; the absolute temperature scale also is coincident with the ideal gas temperature scale over the latter's wide range of applicability.

The absolute temperature scale thus is a truly thermodynamic temperature scale and is used to quantify any temperature, including those close to the absolute zero of temperature (0 K). There is no upper limit on its range.[†]

Choosing the Right Measuring Device

From a practical standpoint, the particular thermometer chosen to measure the temperature of a system depends on the situation. What would be an appropriate thermometer for one task is completely inappropriate for another. For example, since mercury freezes at −39 °C at 1 atm pressure, a mercury thermometer is useless for measuring temperatures lower than that at the same pressure.

There are other considerations involved in the selection of an appropriate thermometer. For example, the ubiquitous laboratory mercury column thermometer can be used to determine your internal body temperature with no problem. However, it would be ridiculous to use such a large thermometer to measure the internal temperature of, say, a tiny baby mouse. Typically, the mass of the thermometer used to measure the temperature should be *much smaller* than the mass of the thermodynamic system whose temperature is to be measured, because otherwise the thermometer will change the temperature of the system.[‡]

We want the *rate* of heat transfer to the thermometer to be large, but the actual amount of heat transfer to be small, so that the temperature of the system is essentially unaffected. For instance, suppose you have an isolated thimbleful of hot water and an isolated bucketful of hot water. An ordinary mercury thermometer can measure the temperature of the water in the bucket accurately, as the mass of the thermometer is much less than the mass of the water (the system). But the same thermometer is inappropriate for measuring the temperature of the water in the thimble, since here the mass of the thermometer is comparable to or even greater than the mass of the system. If the amount of heat transfer is large,[§] then the thermometer ceases

---

*The kelvin unit is named for Lord Kelvin (William Thomson) (1824–1907), who first suggested (in 1854!) defining a temperature scale with only one fixed point. This procedure was finally sanctioned by the International Committee on Weights and Measures a century later in 1954.

[†]The temperature of the universe a mere 10^{-43} s after the Big Bang is estimated to have been 10^{32} K. Hot indeed!

[‡] The thermometer and system can be of comparable mass if the combined system is in thermal contact with a thermodynamic reservoir (see Section 13.10 for the meaning of the term reservoir).

[§] We clarify this relationship further in Section 13.10.

to act as a negligible probe: it significantly affects the thermodynamic system itself.

The question of the effect of a measuring instrument itself on the behavior of the system is one that occurs in many aspects of physics. In classical physics, the effects of a measuring instrument on a system are, in principle at least, able to be calculated, though occasionally with great difficulty. In the quantum mechanics of small systems such as atoms, molecules, and their constituent particles, the act of measuring unavoidably alters the system but, in contrast to classical physics, the alteration in quantum mechanics is in ways that may *not* be able to be calculated, even *in principle*.

QUESTION 4

The temperature inside the core of the Sun is about 1.5×10^7 K. Could you also say, legitimately, that the temperature was 1.5×10^7 °C? Explain your answer.

13.6 TEMPERATURE CONVERSIONS BETWEEN THE FAHRENHEIT AND CELSIUS SCALES*

In the United States another temperature scale is in common use: the **fahrenheit scale**, first introduced by the German scientist Daniel Gabriel Fahrenheit (1686–1736). The fahrenheit scale is defined in terms of the same two fixed points of water used by Celsius: the freezing point of water is 32 °F and its boiling point is 212 °F, both at one atmosphere of pressure. Thus there are $212 - 32 = 180$ fahrenheit degrees between the boiling and freezing points of water.

The basic distinction between the celsius and fahrenheit scales can be thought of in the following way. On a typical thermometer, one can note the column length at which water freezes and where water boils. With the celsius system, the freezing point was labeled 0 °C and boiling point 100 °C, so that the length between the points contained 100 celsius degrees. The divisions on the Celsius thermometer were introduced *after* locating these two fixed points on the thermometer.

On the other hand, Fahrenheit labeled his scale another way. The peculiar numbers for the freezing and boiling points of water came about because Fahrenheit set the zero of his scale at the coldest temperature he could obtain with a freezing mixture of salt and water. Following a suggestion of Newton, he originally took the temperature of the human body to be 12° on his scale, but this resulted in degrees that were too large for practical use. For finer graduations, the 12° interval was divided into eighths and the markings were renumbered, so that the human body temperature became $8 \times 12° = 96°$. Fahrenheit then discovered that pure water froze near the 32° mark and boiled near the 212° mark on his revised scale, with 180° between them (representing the "opposite" properties of water); he then took these as the two fixed points for his scale. With the recalibration of the values associated with the two fixed points, the average human body temperature (no longer a fixed point) became not 96 °F but 98.6 °F.

Since there are 180 fahrenheit degrees and 100 celsius degrees between the boiling and freezing points of water, the celsius degree is larger in size than a fahrenheit degree by the factor 180 °F/100 °C = 9 °F/5 °C. Thus, to convert celsius temperatures to fahrenheit, we do the following: (1) multiply the celsius temperature by the factor 9 °F/5 °C to get a larger number of

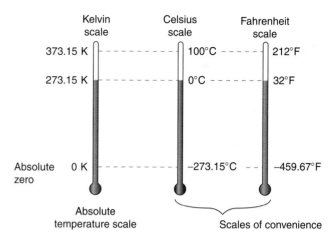

FIGURE 13.6 Schematic comparison of the kelvin (absolute), celsius, and fahrenheit temperature scales.

degrees; (2) then add 32 °F to the result to account for the difference in the location of the zeros of the two scales. That is,

$$t_{\text{farenheit}} = \frac{9 \text{ °F}}{5 \text{ °C}} t_{\text{celsius}} + 32 \text{ °F} \qquad (13.4)$$

To make the conversion from fahrenheit to celsius, we solve Equation 13.4 for t_{celsius}:

$$t_{\text{celsius}} = \frac{5 \text{ °C}}{9 \text{ °F}} \left(t_{\text{farenheit}} - 32 \text{ °F} \right) \qquad (13.5)$$

We think of this latter conversion in the following way: subtract 32 °F from the fahrenheit temperature to readjust the zero location, and multiply the result by 5 °C/9 °F to obtain a smaller number of the larger celsius degrees.

Conversion of degrees celsius to the absolute temperature T in kelvin is accomplished through Equation 13.2:

$$T = t_{\text{celsius}} + 273.15 \text{ K}$$

Corresponding temperatures on the kelvin (absolute), celsius, and fahrenheit scales are shown schematically in Figure 13.6.

PROBLEM-SOLVING TACTIC

13.1 If you recall that the sizes of the degrees on the celsius and fahrenheit scales differ by the factor 9 °F/5 °C or 5 °C/9 °F, depending on which way the comparison is made, you should be able to deduce either Equation 13.4 or 13.5 without memorizing them.

EXAMPLE 13.1

The average normal human body temperature is about 98.6 °F. What is this temperature on the celsius and kelvin scales?

Solution

The human body temperature is 66.6 fahrenheit degrees above the freezing point of water: 98.6 °F − 32.0 °F. Since the celsius degree is a larger size degree, fewer of them quantify this temperature difference. Therefore multiply by the factor 5 °C/9 °F to find the temperature on the celsius scale:

$$\left(\frac{5}{9} \text{ °C/°F}\right)(66.6 \text{ °F}) = 37.0 \text{ °C}$$

Since the zero of the kelvin scale is 273.15 celsius or kelvin degrees lower than the zero on the celsius scale, the temperature on the kelvin scale is found by adding 273.15 to the celsius temperature. Thus the average normal human body temperature is

$$310.2 \text{ K}$$

13.7 THERMAL EFFECTS IN SOLIDS AND LIQUIDS: SIZE

Most substances expand when their temperature is raised and contract when cooled.* Such thermal effects on the length of materials must be considered in many areas of engineering. Bridge designers must allow for the expansion of steel caused by seasonal changes in temperature; typically, this is done by means of expansion joints, such as the one pictured in Figure 13.7.

The common household thermostat, used to regulate heating and cooling systems, contains a coil of material known as a bimetallic strip; see Figure 13.8. The strip is composed of two different metal strips back-to-back in thermal contact. Different metals expand by different amounts for the same temperature changes, and so the strip will bend or curl as the temperature is raised or lowered. The curling is used to make or

___

*A common exception to the rule is rubber. If a rubber band is chilled while stretched and then placed in a warm environment, you can easily observe that the band contracts as its temperature rises.

FIGURE 13.8 In a thermostat, a coiled bimetallic strip expands and contracts with temperature changes, making or breaking an electrical circuit.

break an electrical contact and thereby act as a temperature-sensitive switch.

Old railroad track rails were relatively short to permit many small expansion joints (see Figure 13.9), giving rise to the familiar clickity-clack of the wheels, romanticized in song and legend. The development of specialty steels with small thermal expansion properties permitted the introduction of continuous weld rails, thus eliminating the necessity for many expansion joints and the rather annoying clickity-clack. All high-speed rail systems use such materials.

Changes in the Length of a Solid

Consider an essentially one-dimensional solid of length ℓ at temperature T, shown in Figure 13.10. If the temperature changes by ΔT, experiments indicate that the change in the length $\Delta \ell$ of the rod is directly proportional to the temperature change as well as to the length ℓ of the rod.

FIGURE 13.7 A common thermal expansion joint on a New Hampshire bridge (in need of some maintenance).

FIGURE 13.9 A thermal expansion joint between two railroad rails.

FIGURE 13.10 Thermal expansion of an essentially one-dimensional material.

We express these experimental facts as

$$\Delta \ell = \alpha \ell \, \Delta T \qquad (13.6)$$

The quantity α is known as the **coefficient of linear expansion** of the material.

Those materials that *contract* when their temperature increases have *negative* coefficients of linear expansion (see Table 13.1 for rubber); for these materials if $\Delta T > 0$ K, then $\Delta \ell < 0$ m.

Solving Equation 13.6 for α, we find

$$\alpha = \frac{\Delta \ell / \ell}{\Delta T}$$

We can interpret α as the fractional change in length per kelvin (or equivalently, per celsius degree). The coefficient of linear expansion *at a particular temperature* is found by taking the limit of Equation 13.6 as ΔT approaches zero:

$$\alpha = \lim_{\Delta T \to 0} \frac{1}{\ell} \frac{\Delta \ell}{\Delta T}$$

$$= \frac{1}{\ell} \frac{d\ell}{dT} \qquad (13.7)$$

For most solids and liquids, the value of α is not sensitive to the pressure but may have a temperature dependence. Average values of the coefficient of linear expansion for various materials over a range of temperatures are tabulated in Table 13.1. For all but the most precise work, the average values of α are sufficiently accurate to use when finding the change in length of a substance

TABLE 13.1 Approximate Coefficients of Linear Expansion of Some Elements and Materials at Approximately Room Temperature (293 K)

Element or material	α (K^{-1})
Aluminum	24×10^{-6}
Brass (65% Cu–35% Zn)	21×10^{-6}
Cast iron (ASTM A-47)	12×10^{-6}
Concrete	9.9×10^{-6}
Copper	17×10^{-6}
Ethyl alcohol	3.7×10^{-4}
Glass (ordinary)	9×10^{-6}
Glass (pyrex)	3×10^{-6}
Gold	14×10^{-6}
Invar (Ni–Fe alloy*)	1.3×10^{-6}
Lead	29×10^{-6}
Mercury	61×10^{-6}
Quartz (fused)	0.5×10^{-6}
Rubber (natural)	-620×10^{-6}
Silver	19×10^{-6}
Stainless steel (AISI 302)	17×10^{-6}
Steel (structural ASTM A36)	12×10^{-6}
Titanium	8.6×10^{-6}
Water	
Liquid	69×10^{-6}
Ice	51×10^{-6}

*Invar was designed for a very low thermal expansion coefficient. The name is contracted from the word *invariant*.

using Equation 13.6. Notice that the units attached to the coefficient of linear expansion are K^{-1} or, equivalently, °C^{-1} since the kelvin and celsius degrees are of equal size.

EXAMPLE 13.2

The Golden Gate Bridge, spanning the entrance to San Francisco Bay, has a span of 1.28×10^3 m between its two massive support towers. Calculate the resulting change in the length of the steel span for a seasonal temperature change of perhaps 30 K. Assume the value of α for structural steel in Table 13.1.

Solution
The coefficient of linear expansion for structural steel is, from Table 13.1,

$$\alpha_{\text{steel}} = 12 \times 10^{-6} \text{ K}^{-1}$$

Substitute into Equation 13.6 to find the change in length:

$$\Delta \ell = \alpha \ell \, \Delta T$$
$$= (12 \times 10^{-6} \text{ K}^{-1})(1.28 \times 10^3 \text{ m})(30 \text{ K})$$
$$= 0.46 \text{ m}$$

The expansion joints on the bridge deck are quite large!

Changes in Area of Solids

When pouring concrete for the vast expanses of tarmac and runways at airports, provision must be made for the thermal expansion of the concrete from winter to summer. To accomplish this, the concrete is poured in square or rectangular slabs, with a pliable expansion board around the perimeter (see Figure 13.11).

Temperature-induced changes in area are found by applying linear expansion to each of the two dimensions of the area. For example, consider a rectangular area of dimensions ℓ_1 and ℓ_2, forming an area A, as shown in Figure 13.12. The area is

$$A = \ell_1 \ell_2 \qquad (13.8)$$

When the temperature changes, the change in the area can be found by differentiating Equation 13.8 with respect to the temperature T:

$$\frac{dA}{dT} = \ell_1 \frac{d\ell_2}{dT} + \frac{d\ell_1}{dT} \ell_2$$

FIGURE 13.11 To allow for the thermal expansion in area, large expanses of concrete are poured in smaller slabs surrounded by pliable expansion boards.

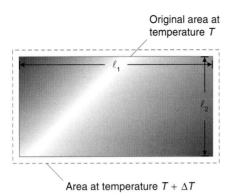

FIGURE 13.12 Thermal expansion of a rectangular area.

Using Equation 13.7 for each of the derivatives, we find

$$\frac{dA}{dT} = \ell_1 \alpha \ell_2 + \alpha \ell_1 \ell_2$$
$$= 2\alpha \ell_1 \ell_2$$

If α is constant over the temperature range of interest, this equation can be integrated:

$$\int_{A_i}^{A_f} dA = 2\alpha \ell_1 \ell_2 \int_{T_i}^{T_f} dT$$
$$\Delta A = (2\alpha)A \, \Delta T \qquad (13.9)$$

The change in the area is proportional to the area and to the change in the temperature.

> The **coefficient of area expansion** is 2α—that is, twice the coefficient of linear expansion.

Equation 13.9 can be used for areas of arbitrary shape and is not restricted to rectangular areas.

One of the curious aspects of area expansion is that for an area with a hole in it, such as a metal washer, the area of the *hole* expands with temperature as if the hole were completely filled with the material. This rather perplexing result can be demonstrated by considering a flat plate of the material with no hole in

it, as in Figure 13.13. A circle is drawn on the plate; it has an area A at temperature T.

If the temperature of the system now is increased to $T + \Delta T$, so that the plate expands, the circular area A increases according to Equation 13.9. Thus the *periphery* of the circular area moves further from the center of the circle, as shown in Figure 13.14. Hence, whether the circle is filled with material or not (i.e., is a hole in the larger plate), its periphery expands in the same way.

EXAMPLE 13.3

A structural steel plate of dimensions 2.00 m by 3.00 m is initially at temperature 20 °C. The temperature of the plate is raised to 50 °C. What is the increase in the area of the plate?

Solution
The area of the plate at 20 °C is

$$A = (2.00 \text{ m})(3.00 \text{ m})$$
$$= 6.00 \text{ m}^2$$

The change in the area of the plate is found via Equation 13.9:

$$\Delta A = (2\alpha_{\text{steel}})A \, \Delta T$$

The temperature change is $\Delta T = 30$ K, and from Table 13.1,

$$\alpha_{\text{steel}} = 12 \times 10^{-6} \text{ K}^{-1}$$

Making these substitutions into Equation 13.9, you find
$$\Delta A = 2(12 \times 10^{-6} \text{ K}^{-1})(6.00 \text{ m}^2)(30 \text{ K})$$
$$= 4.3 \times 10^{-3} \text{ m}^2$$
$$= 43 \text{ cm}^2$$

Changes in Volume of Solids and Liquids

Gasoline is sold by volume. Because of thermal expansion, a given mass of gasoline has a larger volume when warm than when cold. Hence, when possible, a smart consumer will purchase gas at the coolest time of the day.

FIGURE 13.13 A flat area with no hole, but with a circular area marked on it.

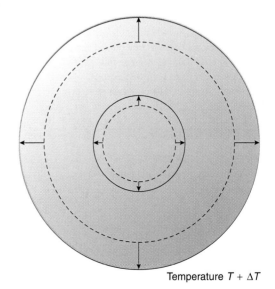

FIGURE 13.14 The periphery of the area expands if the temperature increases.

Think of the volume of a rectangular parallelopiped of sides ℓ_1, ℓ_2, and ℓ_3:

$$V = \ell_1 \ell_2 \ell_3$$

If a temperature change occurs, the change in volume can be found by differentiating this equation with respect to T:

$$\frac{dV}{dT} = \ell_1 \ell_2 \frac{d\ell_3}{dT} + \ell_1 \frac{d\ell_2}{dT} \ell_3 + \frac{d\ell_1}{dT} \ell_2 \ell_3$$

Using Equation 13.7 for the various derivatives, we find

$$\frac{dV}{dT} = \ell_1 \ell_2 (\alpha \ell_3) + \ell_1 (\alpha \ell_2) \ell_3 + (\alpha \ell_1) \ell_2 \ell_3$$
$$= (3\alpha) \ell_1 \ell_2 \ell_3$$

If α is constant over the temperature range of interest, this equation can be integrated to obtain

$$\int_{V_i}^{V_f} dV = 3\alpha \ell_1 \ell_2 \ell_3 \int_{T_i}^{T_f} dT$$

$$\Delta V = (3\alpha) V \, \Delta T \qquad (13.10)$$

Thus the change in volume is proportional to the volume and the change in temperature. Although we derived Equation 13.10 for the specific shape of a rectangular parallelopiped, the result is quite a general one and can be used for volumes of any shape.

The **coefficient of volume expansion** is three times the coefficient of linear expansion.

As for areas, voids or holes in a volume of solid material expand as if the void or hole were filled with the material.

Liquids also change their volume with temperature changes. Indeed, their coefficients of volume expansion (see Table 13.2) are much greater than for solids. Thus the effects of changes in temperature on the volume of liquids are quite substantial. This can be easily observed by filling a soda bottle to the brim with hot water; when the system cools to room temperature, the level of liquid in the bottle is noticeably below the rim. The decrease in the volume of the glass bottle is much less than the decrease in the volume of the liquid, because the coefficient of volume expansion for glass is much less than for most liquids.

Water has unusual thermal expansion properties that may explain why life likely developed first in an aqueous environment. You probably have noticed when swimming and diving that warmer water lies near the surface with cooler water at the bottom. The temperature of the water decreases with depth. Why, then, does ice form at the top of ponds and not at the bottom?

Most materials contract when cooled, thus decreasing their volume. Since the mass remains constant, a decrease in volume increases the density of these materials as they are cooled. Liquid water does this too. A graph of the *volume* of 1000 kg of water as function of temperature is shown in Figure 13.15.

As the temperature decreases, the volume decreases, causing the density of the water to increase. This is why cooler, more dense water sinks, and warmer, less dense water rises to the top of a pond or lake. The volume is a minimum at about 4 °C.

Below 4 °C, however, the volume *increases* as the freezing point (0 °C) is approached, causing the very cold water to migrate to the surface of the liquid. Thus, when water freezes at 0 °C, ice forms at the top rather than the bottom of ponds and lakes. Once the ice forms, the ice acts to insulate the water below from temperature changes above its surface. In this way, water remains in the liquid state below the ice layer on top. This is fortunate for aquatic life. Consider the alternative scenario: *if* water continued to contract until it froze, then ice would form at the *bottom* of ponds and lakes and the entire mass of water might freeze during the winter, along with the aquatic life in it. Once frozen solid, it is difficult to melt the entire mass again with solar energy incident only on top. When spring and summer come, the ice at the top might melt, but the ice further down could remain frozen all year. Fortunately this scenario does not happen, because the volume of water actually increases when cooled below 4 °C.

QUESTION 5

A bimetallic strip is made of two adjacent materials with different coefficients of linear expansion. The strip is straight at a temperature T. The temperature now is raised by an amount $\Delta T > 0$ K and the strip curves. Is the material with the greater coefficient of linear expansion on the inside or outside of the bend?

QUESTION 6

To get a greater mass of petrol, should you fill the tank of your car when the temperature is cold or hot?

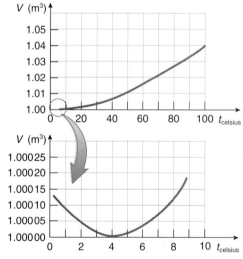

FIGURE 13.15 A graph of the volume of 1000 kg of water versus temperature.

TABLE 13.2 Approximate Coefficients of Volume Expansion for Various Liquids at Approximately Room Temperature

Liquid	3α (K^{-1})
Acetone	14.9×10^{-4}
Benzene	12.0×10^{-4}
Ethyl alcohol	11.2×10^{-4}
Mercury	1.83×10^{-4}
Petroleum (density 846 kg/m³)	9.6×10^{-4}
Water	2.1×10^{-4}

Liquid water may exist below the ice-covered surface of Europa, a moon of Jupiter.

EXAMPLE 13.4

A petrol station takes delivery of 10 m³ of gasoline (1.0 × 10⁴ liters) at a temperature of 0 °C. If the delivery of the same mass of petroleum were made when its temperature was 30 °C, what would be the increase in the volume of liquid delivered?

Solution

Using Table 13.2, petroleum has a coefficient of volume expansion of about 9.6×10^{-4} K⁻¹. The change in the volume of the petrol is found via Equation 13.10:

$$\Delta V = (3\alpha)V\,\Delta T$$

Making the appropriate substitutions,

$$\Delta V = (9.6 \times 10^{-4}\ \text{K}^{-1})(10\ \text{m}^3)(30\ \text{K})$$
$$= 0.29\ \text{m}^3$$
$$= 2.9 \times 10^2\ \text{liters}$$

Since petroleum is sold by volume, not mass, such dramatic temperature changes in the volume have instigated vituperative arguments between petroleum wholesalers and retailers about the timing of deliveries.

EXAMPLE 13.5

The (structural) steel petrol tank of your car has a capacity of 80 liters of petrol when the tank and petrol are at 0 °C.

a. What is the change in the volume of the tank when its temperature is increased to 30 °C?
b. What is the change in the volume of the petrol when its temperature is increased to 30 °C?
c. At which temperature will you get the greater mass of petrol? How much greater?

Solution

a. Note that the petrol tank is a three-dimensional solid with a large hole removed. The change in the volume of the steel tank is

$$\Delta V_{\text{tank}} = (3\alpha_{\text{steel}})V\,\Delta T$$
$$= 3(12 \times 10^{-6}\ \text{K}^{-1})(80\ \text{liters})(30\ \text{K})$$
$$= 0.086\ \text{liters}$$

b. When petrol is warmed from 0 °C to 30 °C, the change in the volume of petroleum is, using the data in Table 13.2,

$$\Delta V_{\text{petrol}} = (9.6 \times 10^{-4}\ \text{K}^{-1})(80\ \text{liters})(30\ \text{K})$$
$$= 2.3\ \text{liters}$$

c. The mass of petroleum you get when the tank is filled at 30 °C is less than the mass you get at 0 °C. The difference in mass is equal to the mass of about

$$2.3\ \text{liters} - 0.086\ \text{liters} = 2.2\ \text{liters}$$

of 30 °C petrol.

13.8 THERMAL EFFECTS IN IDEAL GASES

In 1660 Robert Boyle (1627–1691) empirically discovered a remarkable property of dilute gases. If the temperature of a fixed amount of a gas is kept constant, its pressure is inversely proportional to the volume: the smaller the volume, the greater the pressure. We usually express this by saying that the product of the pressure and volume is constant:

$$PV = \text{constant} \qquad \text{(at constant temperature)} \qquad (13.11)$$

This approximate empirical relationship is characteristic of all gases at low densities and is now known as **Boyle's law**. If the experiments are performed at another temperature, the numerical value of the constant changes in a way that is directly proportional to the absolute temperature T. In other words,

$$PV = DT \qquad (13.12)$$

where the constant D depends on the amount of gas used in the experiments. Equation 13.12 implies that at constant volume, the pressure of the gas is proportional to the absolute temperature, a result we noted before in formulating Equation 13.3 for the constant volume gas thermometer.

Let's see if we can determine how the constant D in Equation 13.12 depends on the amount of gas present. Take two identical samples of the same gas that have the same volume, temperature, and pressure. Equation 13.12 implies that the constant D must be the same for both individual samples. If the two samples now are brought together, experimentally we find that we have a combined sample with *twice* the volume at the *same* temperature and pressure. Hence, if we apply Equation 13.12 to the combined sample, the constant D must double; this implies that the constant D is directly proportional to the *amount* of gas present. If N is the number of molecules in the sample of gas, we write $D = kN$, and so Equation 13.12 becomes

$$PV = NkT \qquad (13.13)$$

Experiments indicate that k is independent of the particular gas used as well as the amount of the gas. The constant k is known as **Boltzmann's constant** and has the value

$$k = 1.380\,66 \times 10^{-23}\ \text{J/K} \qquad (13.14)$$

in SI units. It is named after Ludwig Boltzmann (1844–1906), a pioneer in the theoretical development of thermodynamics during the latter part of the 19th century.

The amount of a gas typically is expressed in moles (mol). The mass of one mole of a substance is called its molar mass M.* The molar mass in *grams per mole* is found by summing the molar masses of the atoms that make up the molecule. Oxygen gas is composed of the oxygen molecule, O_2, with two oxygen atoms. The molar mass of the oxygen atom (O) is 16 g, and so the molar mass of the oxygen molecule is 16 g + 16 g = 32 g. For carbon dioxide (CO_2), we add the molar mass of carbon (12 g) to that of O_2 (32 g), getting 44 g.

For *n* moles, the mass *m* of the sample is

$$m = nM \qquad (13.15)$$

The molar mass of several gases is given in Table 13.3.

In Chapter 1 we learned that one mole is Avogadro's number N_A of particles (be they atoms or molecules), where

$$N_A = 6.022\ 137 \times 10^{23}\ \text{particles/mol} \qquad (13.16)$$

Let *n* be the number of moles of a substance; hence, the number of particles N is equal to nN_A:

$$N = nN_A$$

In this way, we can rewrite Equation 13.13 as

$$PV = nN_A kT \qquad (13.17)$$

The product of Avogadro's number and Boltzmann's constant is known as the **universal gas constant** R:

$$R \equiv N_A k \qquad (13.18)$$
$$= (6.022\ 137 \times 10^{23}\ \text{particles/mol})(1.380\ 66 \times 10^{-23}\ \text{J/K})$$
$$= 8.314\ 52\ \text{J/(mol·K)} \qquad (13.19)$$

Equation 13.17 commonly is written as

$$PV = nRT \qquad (13.20)$$

Equation 13.20 (or its equivalent forms, Equation 13.13 or 13.17) is known as the **ideal gas law**.

We shall have more to say about the characteristic features of an ideal gas in Chapter 14.

___
*The molar mass also is known as the *molecular weight* (particularly in chemistry); in this context the word weight is *not* the force we call weight in physics. We will not use the term molecular weight.

TABLE 13.3 Molar Masses of Some Common Gases†

Gas	Molar mass	
	grams/mole	*kilograms/mole*
Oxygen (O_2)	32	0.032
Nitrogen (N_2)	28	0.028
Carbon dioxide (CO_2)	44	0.044
Carbon monoxide (CO)	28	0.028
Hydrogen (H_2)	2	0.002
Helium (He)	4	0.004

†Molar masses given on the periodic table of the elements reflect the relative natural abundances of the isotopes of each element and so may be given with a precision of fractions of a gram per mole.

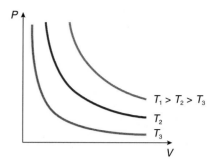

FIGURE 13.16 Isotherms of an ideal gas are hyperbolic in shape.

Notice that at constant temperature, a graph of the pressure versus the volume (called a **P–V diagram**) of an ideal gas results in a hyperbolic curve:‡ Such curves at constant temperature are called **isotherms**; see Figure 13.16.

According to the ideal gas law, Equation 13.20, the ratio

$$\frac{PV}{nT}$$

should be constant, independent of pressure and equal to the gas constant R. The behavior of *real gases* at pressures up to several atmospheres is closely approximated by the ideal gas law, as shown in Figure 13.17.

Any gas at a temperature of 0 °C (273.15 K) and 1 atm pressure (1.013×10^5 Pa) is said to be at **standard conditions**.

PROBLEM-SOLVING TACTIC

13.2 Be careful with the units when using the ideal gas law. To use the value of R = 8.314 52 J/(mol·K), the pressure P must be in pascals, V in cubic meters, and n in moles.

Different units for P and V (such as expressing P in atmospheres and V in liters, typically done for convenience) mean the gas constant has other numerical values than the one we have quoted; this is frequently the case in chemistry, where the gas

___
‡Recall from analytic geometry that a mathematical expression of the form

$$xy = \text{constant}$$

is a hyperbola.

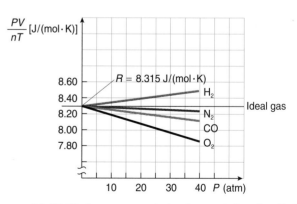

FIGURE 13.17 Real gases approach the characteristics of an ideal gas for pressure up to several atmospheres.

constant has values that change like a chameleon as various units are used. In physics, we typically stick to SI units, in which case the units for pressure and volume should be converted to standard SI units (pascals and cubic meters) when doing most problems.

> ### PROBLEM-SOLVING TACTIC
>
> **13.3 When using the ideal gas law, be sure to express the temperature in kelvin (not degrees celsius).** If you (incorrectly!) use negative or zero celsius degrees in the ideal gas law, you discover that the product of the pressure and volume (both intrinsically positive quantities) yields a negative or zero result. Such a result is physically absurd and should indicate that something you did is amiss.

EXAMPLE 13.6

Determine the volume of 1.00 mol of an ideal gas at a pressure of 1.000 atm and a temperature of 0 °C.

Solution

To use the ideal gas law, Equation 13.20, the temperature *must* be expressed in kelvin, following Problem-Solving Tactic 13.3, and so $T = 273.15$ K. Problem-Solving Tactic 13.2 indicates you need to convert 1.000 atm pressure to SI units: 1.013×10^5 Pa. Solve the ideal gas law for V:

$$V = \frac{nRT}{P}$$

Substituting the proper SI values, you get

$$V = \frac{(1.00 \ \text{mol})\,[8.315 \ \text{J}/(\text{mol} \cdot \text{K})]\,(273.15 \ \text{K})}{1.013 \times 10^5 \ \text{Pa}}$$

$$= 2.24 \times 10^{-2} \ \text{m}^3$$

$$= 22.4 \ \text{liters}$$

EXAMPLE 13.7

The temperature of a 4.00 liter sample of an ideal gas initially at standard conditions is increased 50.0 °C and its volume reduced to 3.00 liters. Determine the new pressure of the gas.

Solution

The ideal gas law, Equation 13.20, states that

$$PV = nRT$$

Rearrange this into the form

$$\frac{PV}{T} = nR$$

Since the right-hand side is constant for a given sample of gas, the ratio

$$\frac{PV}{T}$$

must be same for the initial and final states of the gas. That is,

$$\frac{P_i V_i}{T_i} = \frac{P_f V_f}{T_f}$$

Solve for P_f:

$$P_f = \frac{V_i}{V_f}\frac{T_f}{T_i}P_i$$

Notice that the answer depends on the volume ratio. Here you do not need to convert to cubic meters to determine this ratio. However, the ratio of the temperatures *must* be calculated using *absolute temperatures* in kelvin. The final pressure emerges in the same units you use for the initial pressure. Since the initial pressure is in atmospheres, the final pressure will be in atmospheres and you do not need to convert to pascals. (You might ask yourself why the temperature ratio must be converted to SI units while the volume and pressure ratios do not).

The ratio of the initial to final volumes is (4.00 liters)/(3.00 liters). The initial temperature is 273.15 K and the final temperature is 323.2 K. The initial pressure is 1.00 atm. Hence

$$P_f = \frac{4.00 \ \text{liters}}{3.00 \ \text{liters}}\frac{323.2 \ \text{K}}{273.15 \ \text{K}}(1.00 \ \text{atm})$$

$$= 1.58 \ \text{atm}$$

EXAMPLE 13.8

a. How many moles are in a 0.100 kg sample of carbon dioxide?
b. How many molecules are in the sample?

Solution

a. From Table 13.3, the molar mass of carbon dioxide, CO_2, is 44 g/mol or 0.044 kg/mol. Hence the number of moles in the sample of 0.100 kg is

$$n = \frac{0.100 \ \text{kg}}{0.044 \ \text{kg/mol}}$$

$$= 2.3 \ \text{mol}$$

b. Each mole has Avogadro's number of molecules, and so the number of molecules N is

$$N = nN_A$$

$$= (2.3 \ \text{mol})(6.02 \times 10^{23} \ \text{molecules/mol})$$

$$= 1.4 \times 10^{24} \ \text{molecules}$$

13.9 CALORIMETRY

Place a few ice cubes in a warm soft drink and in a few minutes you have a nice cool treat while you study. The soda is cooled and some or all of the ice melts. This observation is so common that we take it for granted, but the observation illustrates an important aspect about heat transfer between two systems initially at different temperatures.

When heat transfer occurs to or from a system, several things typically happen:

1. The temperature of the system may change (the soda cools off).
2. A *change of state* may occur such as from a solid to a liquid (the ice melts) or from a liquid to a gas (such as boiling water into

water vapor). Such changes of state are examples of **phase transitions*** and take place at a constant temperature, characteristic of the material and its environment (e.g., pressure).

Calorimetry experiments probe both of these aspects. In some such experiments (such as when ice cubes are added to soda at room temperature), one thermodynamic system at a cool temperature is placed in thermal contact with another thermodynamic system at a higher temperature; heat transfer then occurs between the two systems until the combined system reaches a common temperature and is in thermodynamic equilibrium.[†]

> The heat transfer *from* the warmer system is *to* the cooler system. By convention, heat transfer *from* a system is considered *negative*; heat transfer *to* a system is considered *positive*.

We typically ensure, by means of insulation, that the heat transfer of the combined system to the environment is negligible (for instance, by using an insulated cup). For such *isolated systems*, the total heat transfer to the combined system is zero.

Given the initial temperatures and masses of, say, the ice and the soda pop, we want to be able to predict the final temperature of the mixture when they are combined (or placed in thermal contact with each other). To do this, we have to see how heat transfer changes the temperature of a system or its phase.

Temperature Change

Put a thimbleful of water on a hot stove burner for 10 s and the water increases in temperature dramatically. Place a huge pot of water on the same hot stove burner for 10 s and the temperature increase of the water is quite small. From such observations, we see that the temperature change of a system in response to heat transfer is inversely proportional to the mass of the system. That is, for a differential heat transfer dQ to a system, the differential temperature change dT that the system experiences is directly proportional to the heat transfer and inversely proportional to the mass of the system. In other words,

$$dT \propto \frac{dQ}{m}$$

This relationship traditionally is written as

$$dQ \propto m\,dT$$

Different materials, all with the same mass, have different dT for a given dQ. Thus the proportionality constant needed to make the relationship above an equality depends on the specific material used in the experiment. We write

$$dQ = mc\,dT$$

where c is the **specific heat** of the substance. The specific heat *at* a specified temperature is defined as

$$c = \frac{1}{m}\frac{dQ}{dT} \tag{13.21}$$

The dimensions for c must therefore be those of energy divided by the product of a mass and temperature. In SI units, the specific heat is expressed in joules per kilogram-kelvin [J/(kg·K)].

If the specific heat is independent of temperature, then Equation 13.21 can be integrated to obtain

$$Q = mc\,\Delta T \tag{13.22}$$

If the amount of material is given in moles, rather than in kilograms, then the molar specific heat must be used. Equation 13.22 then becomes

$$Q = nc_{\text{molar}}\,\Delta T \tag{13.23}$$

where n is the number of moles of material in the system. Analogous to Equation 13.21, the molar specific heat is defined as

$$c_{\text{molar}} \equiv \frac{1}{n}\frac{dQ}{dT} \tag{13.24}$$

The molar specific heat is given in SI units of joules per mole-kelvin [J/(mol·K)].

> You can think of the specific heat (or molar specific heat) as the heat transfer to one kilogram (or one mole) of the material needed to raise its temperature by one kelvin.

> **PROBLEM-SOLVING TACTIC**
>
> **13.4 Be careful in looking up the specific heat of a substance in tables in technical literature; other common energy and mass units also are used.** Check the units carefully.

Table 13.4 lists the specific heat of several common substances and materials.

TABLE 13.4 Approximate Specific Heat of Various Substances at Approximately Room Temperature (20 °C ≈ 293 K)

Substance	J/(kg·K)	J/(mol·K)
Aluminum	896	24.2
Brass	385	
Copper	386	24.5
Gold	132	26.8
Ice (−20 °C)	2050	36.9
Iron	481	24.6
Lead	128	26.4
Mercury	138	27.9
Silver	233	24.9
Water (15 °C)	4186	75.3

The molar specific heat of many metallic elemental solids is approximately $3R = 24.9$ J/(mol·K). The reason for this is explained in Chapter 14.

---

*Phase transitions also are possible without a change of state. For example, when a solid crystalline structure (such as quartz) changes the form of its crystalline structure, the material undergoes a phase transition with no change of state, since both the initial and final structures are solids.

[†] Other types of calorimetric experiments transfer heat at a known rate and measure the corresponding changes in temperature as a function of time. Such devices are called scanning calorimeters.

A change in temperature of a system is always taken as

$$\Delta T = T_f - T_i$$

When $T_f > T_i$, then $\Delta T > 0$ K and, according to Equations 13.22 or 13.23, $Q > 0$ J. If the final temperature of the system is less than the initial temperature, then $Q < 0$ J.

Especially for gases, the value of the specific heat depends on *how* the heat transfer occurs: experimentally, it could be accomplished at constant pressure or at constant volume.* That is, we need to distinguish between *two* distinct specific heats. When the heat transfer is at constant volume, we designate the specific heat as

$$c_V = \frac{1}{m}\left(\frac{dQ}{dT}\right)_V \qquad (13.25)$$

where the subscripts indicate that the process takes place at *constant volume*. When the heat transfer is at constant pressure, we write

$$c_P = \frac{1}{m}\left(\frac{dQ}{dT}\right)_P \qquad (13.26)$$

where the subscripts indicate the process takes place at *constant pressure*. The molar specific heats are similar:

$$c_{\text{molar } V} = \frac{1}{n}\left(\frac{dQ}{dT}\right)_V \qquad (13.27)$$

$$c_{\text{molar } P} = \frac{1}{n}\left(\frac{dQ}{dT}\right)_P \qquad (13.28)$$

The specific heat of solids and liquids typically is determined at constant pressure (one atmosphere), because it is very difficult to keep these phases of matter at constant volume.† The specific heat at constant volume for most solids and liquids typically is within a few percent of the specific heat at constant pressure (but it *is* different). Unless extraordinary precision is needed, one can neglect the difference. This will be done in this text (for solids and liquids, but *not* gases), but you should be aware of the distinction.

On the other hand, with gases it is quite easy experimentally to keep them under either constant pressure or constant volume; the distinction is important for gases, since the two specific heats differ substantially, as we will see in the next chapter.

The specific heat of a substance usually varies with the temperature, so that the value of the specific heat at one temperature is not necessarily the same as at another. If this is the case, Equations 13.22 or 13.23 are not appropriate and we must revert to the more general defining Equations 13.21 and 13.24 and integrate them, accounting for the variation in the specific heat with temperature.

> In this text we will consider the specific heat to be independent of temperature as long as the substance does not change phase; hence, Equations 13.22 and 13.23 may be used.

The temperature dependence of the specific heat—that is, the way the value of the specific heat changes with temperature—is of great interest because it is a *macroscopic* manifestation of changes in the material on the *microscopic* level. Dramatic changes in the value of the specific heat of a material are indications of significant changes on the microscopic level. For example, note in Table 13.4 that the specific heat of ice is about half that of liquid water. The specific heat changes dramatically through the phase transition. This is characteristic of many types of phase transitions.

QUESTION 7

Explain why regions of land near large bodies of water such as oceans have a more moderate climate than inland areas at the same latitude.

EXAMPLE 13.9

The temperature of 4.50 kg of brass is raised from 20.0 °C to 150.0 °C. What heat transfer to the brass is needed to accomplish this, assuming no heat transfer to the surroundings?

Solution

Use Equation 13.22:

$$Q = mc\,\Delta T$$

The specific heat of brass c_{brass} is found from Table 13.4 to be

$$c_{\text{brass}} = 385 \ \text{J/(kg·K)}$$

The change in the temperature of the brass is

$$\Delta T = T_f - T_i$$
$$= 130.0 \ \text{K}$$

Making these substitutions into Equation 13.22, you get

$$Q = (4.50 \ \text{kg})[385 \ \text{J/(kg·K)}](130.0 \ \text{K})$$
$$= 2.25 \times 10^5 \ \text{J}$$

Changes of Phase

Place a pot of water on the stove and bring the water from room temperature to the boiling point. A remarkable thing happens once the boiling point is reached. Additional heat transfer to the boiling water does *not* increase the temperature of the water. Rather, the heat transfer goes into changing the *phase* of the water from a liquid to a gas.‡ Converting water at 100 °C to steam at 100 °C involves considerable heat transfer, but no change in the temperature. Likewise, melting ice at 0 °C to water at 0 °C also involves considerable heat transfer to the ice system, but no change in its temperature.

The heat transfer needed to change the phase of one kilogram of a substance is called the **latent heat**§ L of the phase transition. If the phase change is from a solid to a liquid (or vice versa), the latent heat is called the **latent heat of fusion** L_f. If the phase change is from a liquid to a gas (or vice versa), the latent heat is called the **latent heat of vaporization** L_v. The SI units of

---

*It also could be accomplished without controlling either pressure or volume.
†Extraordinary pressures are needed to keep the system at constant volume because the forces causing thermal expansion are quite large.

---

‡The energy is needed to weaken the atomic or molecular attractions between the particles of the system as it changes from a liquid to a gas (or, in the case of melting, from a solid to a liquid).
§The word comes from the Latin *latere*, meaning "hidden."

TABLE 13.5 Latent Heats of Fusion and Vaporization and Phase Transition Temperatures for Various Materials at 1.00 atm Pressure

Material	Melting point (K)	Latent heat of fusion L_f (J/kg)	Boiling point (K)	Latent heat of vaporization L_v (J/kg)
Aluminum	933	3.96×10^5		
Carbon dioxide*			195	5.73×10^5
Copper	1356	2.05×10^5	2839	47.26×10^5
Ethyl alcohol	159	1.09×10^5	351	8.79×10^5
Gold	1336	0.64×10^5	3081	17.01×10^5
Helium			4.2	0.24×10^5
Lead	601	0.247×10^5	2023	8.58×10^5
Nitrogen	63	0.257×10^5	77	1.99×10^5
Oxygen	55	0.138×10^5	90	2.13×10^5
Silver	1234	1.05×10^5	2436	23.23×10^5
Water	273.15	3.335×10^5	373.15	22.57×10^5

*At 1 atm, carbon dioxide has no liquid state; it sublimates directly from a solid to a gas. The values quoted are for this transition.

Handbook of Chemistry and Physics (73rd edition).

latent heat are J/kg, or else J/mol if the molar quantity of material is used rather than the mass. As always, it is necessary to check the units carefully when using values of the latent heats found in the technical literature. Table 13.5 lists the latent heats of fusion and vaporization associated with several phase transitions.

Thus, to change the phase of a mass m, the heat transfer to the mass is

$$Q = mL \qquad (13.29)$$

where L is the appropriate latent heat associated with the particular phase transition in J/kg. If you use moles instead of kilograms, then the appropriate equation is

$$Q = nL_{molar} \qquad (13.30)$$

where n is the number of moles of material, and L_{molar} is in J/mol.

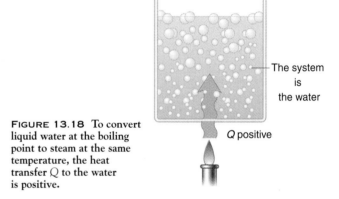

FIGURE 13.18 To convert liquid water at the boiling point to steam at the same temperature, the heat transfer Q to the water is positive.

PROBLEM-SOLVING TACTIC

13.5 There are no ready signals in Equations 13.29 and 13.30 to indicate whether Q should be positive or negative. Hence we impose the same sign convention as with temperature changes and Equations 13.22 and 13.23: if the heat transfer is *to* the system to change the phase, then Q is positive. If the heat transfer is *from* the system, then Q is negative.

To convert water to steam, heat transfer $Q = mL_v$ must accrue *to* the water (where L_v is the latent heat of vaporization associated with this phase transition), and so Q is *positive* ($Q = mL_v$); see Figure 13.18. As steam condenses into water, the heat transfer is *from* the (steam) system, and so Q is *negative* for the steam system ($Q = -mL_v$); see Figure 13.19.

Likewise, when ice at 0 °C melts to liquid water at 0 °C, the heat transfer Q to the ice system is positive ($Q = mL_f$), where L_f is the latent heat associated with this phase transition. When liquid water at 0 °C freezes into ice at 0 °C, the heat transfer is from the liquid water system, and so Q is negative ($Q = -mL_f$).

Not all phase transitions involve latent heats. Those that do are called **first-order phase transitions**. Phase transitions that have zero latent heat ($L = 0$ J/kg) are called **second-order phase**

FIGURE 13.19 To condense steam at 100 °C into water at the same temperature, the heat transfer Q is from the steam, and so Q is negative.

transitions. Many structural phase transitions in crystals (changes in the microscopic crystal structure of a material) are second-order phase transitions with no measurable latent heat associated with the change of crystalline phase.

EXAMPLE 13.10

a. A mass of 3.00 kg of water is boiling. What heat transfer to the water is needed to completely vaporize the water?
b. If a 1.00 kW heater is used to supply this heat transfer, how long will the process take?

Solution

a. Use Equation 13.29, with the latent heat of vaporization of water found in Table 13.5:

$$L_{v \text{ water}} = 22.57 \times 10^5 \text{ J/kg}$$

Hence the heat transfer Q to the water is

$$
\begin{aligned}
Q &= mL_{v \text{ water}} \\
&= (3.00 \text{ kg})(22.57 \times 10^5 \text{ J/kg}) \\
&= 6.77 \times 10^6 \text{ J}
\end{aligned}
$$

b. A 1.00 kW heater supplies energy at a rate of 1.00×10^3 J/s. To supply 6.77×10^6 J will take a time

$$
\begin{aligned}
t &= \frac{6.77 \times 10^6 \text{ J}}{1.00 \times 10^3 \text{ J/s}} \\
&= 6.77 \times 10^3 \text{ s} \\
&= 1.88 \text{ h}
\end{aligned}
$$

EXAMPLE 13.11

How much ice at -10.0 °C must be added to 4.00 kg of water at 20.0 °C to cause the resulting mixture to reach thermal equilibrium at 5.0 °C? Assume no heat transfer to the surrounding environment, so that heat transfer occurs only between the water and ice.

Solution

The problem is a matter of accounting: the heat transfer to the ice system is from the warm water system; thus, the total heat transfer to (or from) the combined ice and water system is zero.

Let m be the mass of ice needed. The specific heat of ice is found from Table 13.4. Use Equation 13.22. To warm the ice from -10 °C to 0 °C takes a heat transfer to the ice system of

$$
\begin{aligned}
Q_1 &= mc_{\text{ice}} \Delta T \\
&= m[2.050 \times 10^3 \text{ J/(kg·K)}][0 \text{ °C} - (-10.0 \text{ °C})] \\
&= m(2.050 \times 10^4 \text{ J/kg})
\end{aligned}
$$

Notice that since only the *change* in temperature is involved, you can find the change in temperature using the celsius temperatures (the temperature *change* is the same on both kelvin and celsius scales).

To melt the ice at 0 °C to water at 0 °C is a phase transition involving a latent heat of fusion $L_f = 3.335 \times 10^5$ J/kg, from Table 13.5. Thus the heat transfer to the ice is

$$
\begin{aligned}
Q_2 &= mL_f \\
&= m(3.335 \times 10^5 \text{ J/kg}) \\
&= m(33.35 \times 10^4 \text{ J/kg})
\end{aligned}
$$

To warm the newly formed liquid water to the final temperature of 5.0 °C involves an additional heat transfer to the mass m of

$$
\begin{aligned}
Q_3 &= mc_{\text{water}} \Delta T \\
&= m[4186 \text{ J/(kg·K)}](5.0 \text{ °C} - 0.0 \text{ °C}) \\
&= m(2.1 \times 10^4 \text{ J/kg})
\end{aligned}
$$

Thus, to warm the ice, melt the ice, and then warm the resulting liquid to 5.0 °C, the total heat transfer to the ice is

$$
\begin{aligned}
Q_1 + Q_2 + Q_3 &= m(2.050 \times 10^4 \text{ J/kg}) + m(33.35 \times 10^4 \text{ J/kg}) \\
&\quad + m(2.1 \times 10^4 \text{ J/kg}) \\
&= m(37.6 \times 10^4 \text{ J/kg})
\end{aligned}
$$

In cooling the 4.00 kg of water at 20.0 °C to 5.0 °C, the heat transfer from the water is

$$
\begin{aligned}
Q_4 &= mc \, \Delta T \\
&= (4.00 \text{ kg})[4186 \text{ J/(kg·K)}](5.0 \text{ °C} - 20.0 \text{ °C}) \\
&= -25.1 \times 10^4 \text{ J}
\end{aligned}
$$

where the minus sign indicates that the heat transfer is from the warm water.

Since the combined system of the ice and water is isolated from the environment, the total heat transfer to or from the isolated system is zero. Thus

$$
Q_1 + Q_2 + Q_3 + Q_4 = 0 \text{ J}
$$
$$
m(37.6 \times 10^4 \text{ J/kg}) + (-25.1 \times 10^4 \text{ J}) = 0 \text{ J}
$$

Solving for m, you find

$$m = 0.668 \text{ kg}$$

STRATEGIC EXAMPLE 13.12

One (1.00) liter of boiling water at 100.0 °C is poured into an insulated container that contains 20.0 kg of ice at 0.0 °C. Determine the final temperature of the system and, if any ice is left, how much ice remains.

Solution

Assume the density of water does not change with temperature (a reasonable first approximation). One (1.00) liter of water (1.00×10^{-3} m³) has a mass of 1.00 kg, since the density of water is 1.00×10^3 kg/m³.

Let t_{celsius} be the final temperature of the combined water and ice system in celsius degrees.

The heat transfer to the ice to melt it all is

$$
\begin{aligned}
Q_1 &= mL_f \\
&= (20.0 \text{ kg})(3.335 \times 10^5 \text{ J/kg}) \\
&= 6.67 \times 10^6 \text{ J}
\end{aligned}
$$

To warm the melted ice to the final temperature t_{celsius} takes heat transfer

$$
\begin{aligned}
Q_2 &= mc_{\text{water}} \Delta T \\
&= (20.0 \text{ kg})[4186 \text{ J/(kg·K)}](t_{\text{celsius}} - 0 \text{ °C}) \\
&= (8.37 \times 10^4 \text{ J/K})t_{\text{celsius}}
\end{aligned}
$$

To cool the hot water to temperature t_{celsius} takes

$$
\begin{aligned}
Q_3 &= mc_{\text{water}} \Delta T \\
&= (1.00 \text{ kg})[4186 \text{ J/(kg·K)}](t_{\text{celsius}} - 100.0 \text{ °C}) \\
&= (4.19 \times 10^3 \text{ J/K})(t_{\text{celsius}} - 100.0 \text{ °C})
\end{aligned}
$$

Notice that this heat transfer is negative because the final temperature of the originally hot water is less than the initial temperature.

The total heat transfer to or from the isolated, combined water and ice system is zero. Thus

$$Q_1 + Q_2 + Q_3 = 0 \text{ J}$$
$$6.67 \times 10^6 \text{ J} + (8.37 \times 10^4 \text{ J/K})t_{\text{celsius}}$$
$$+ (4.19 \times 10^3 \text{ J/K})(t_{\text{celsius}} - 100.0 \text{ °C}) = 0 \text{ J}$$

Solving for t_{celsius}, you obtain

$$t_{\text{celsius}} = -71.1 \text{ °C}$$

This is a ridiculous answer! There is no way that the final temperature of the system can be less than the coldest material in the problem. This result means that not all the ice melts. This calculation reminds us to question whether the result obtained is reasonable. This answer is not!

We have to begin anew. Since not all the ice melts, the final temperature of the system must be 0 °C. Let m be the mass of the ice that does melt. Then to melt this amount of ice takes

$$Q_1 = mL_f$$
$$= m(3.335 \times 10^5 \text{ J/kg})$$

The heat transfer from the hot water in cooling from 100 °C to 0 °C is

$$Q_2 = mc_{\text{water}} \Delta T$$
$$= (1.00 \text{ kg})[4186 \text{ J/(kg·K)}](0.0 \text{ °C} - 100.0 \text{ °C})$$
$$= -4.19 \times 10^5 \text{ J}$$

The total heat transfer to or from the combined ice and water must be zero, because the combined system is isolated:

$$Q_1 + Q_2 = 0 \text{ J}$$
$$m(3.335 \times 10^5 \text{ J/kg}) + (-4.19 \times 10^5 \text{ J}) = 0 \text{ J}$$

Solving for m, you get

$$m = 1.26 \text{ kg}$$

So 1.26 kg of ice melts. Thus the amount of ice left is

$$20.0 \text{ kg} - 1.26 \text{ kg} = 18.7 \text{ kg}$$

13.10 RESERVOIRS

If a thimbleful of boiling water is brought into thermal contact with a swimming pool full of cool water, the temperature change of the water in the pool is negligible and the thimble of water cools to the initial temperature of the pool. Likewise, place mom's all-American apple pie to bake in a hot oven and the pie eventually warms up to the temperature of the oven. A warm six-pack placed in a cool refrigerator eventually has the same temperature as the inside of the refrigerator. These examples illustrate what is meant by the term **reservoir** in thermodynamics. One meaning of the word reservoir is "a great supply of something"; we will use this definition for thermodynamic purposes.

A reservoir is a special thermodynamic system. We use a reservoir to bring other systems that are in thermal contact with it to a fixed temperature: the temperature of the reservoir. The large pool of water is a good reservoir for the thimble; the inside of an oven and a refrigerator are other types of reservoirs, maintained at

a fixed temperature by electrical means. Ovens and refrigerators may be called *dynamic* reservoirs, whereas large masses of material, such as the pool, may be called *passive* reservoirs. The distinction between active and passive reservoirs is not important.

> A system in thermal contact with a reservoir experiences heat transfer to or from the system until it has the same temperature as the reservoir. The temperature of the reservoir does not change, despite positive or negative heat transfer to it.

A passive reservoir can be designed as follows. Consider a large mass m of material at a temperature T; let c be the specific heat of the material. If heat transfer Q occurs to the mass (from another system at a different temperature placed in thermal contact with the reservoir), the temperature of the reservoir material changes by an amount ΔT, where

$$Q = mc \, \Delta T$$

according to Equation 13.22. The temperature change of the reservoir material thus is

$$\Delta T = \frac{Q}{mc} \tag{13.31}$$

The product in the denominator of this expression, mc, is called the **thermal mass** of the material. If the thermal mass of the material is sufficiently large, then the temperature change of the reservoir material can be made arbitrarily small. A passive thermodynamic reservoir is a thermodynamic system whose thermal mass is so large that its temperature change is negligible despite heat transfer *to* the reservoir system ($Q > 0$ J) or heat transfer *from* the reservoir system ($Q < 0$ J).

In a strict sense, such ideal passive reservoirs do not exist, since they are never of infinite thermal mass. But from a practical standpoint, there is no problem if the reservoir has a sufficiently large thermal mass (see Example 13.13). Therefore, a passive thermodynamic reservoir has a huge thermal mass. A good passive reservoir must have a thermal mass much larger than the thermal mass of any system placed in thermal contact with it.

> The word reservoir in thermodynamics means either (a) a huge thermal mass or (b) a region (such as the inside of an oven or a refrigerator) that maintains a fixed temperature despite heat transfer to or from it from other thermodynamic systems.

Experimentally, refrigeration and heating units, whose temperatures can be adjusted over a wide range, are used to provide dynamic reservoirs of small total thermal mass.

When the temperature of a system is measured with an appropriate thermometer, the system should act as a thermodynamic reservoir as far as the thermometer is concerned, since we do not want the presence of the thermometer to affect the temperature of the system itself.

EXAMPLE 13.13

A passive reservoir is needed. A heat transfer to the reservoir of 1.50×10^4 J must not change its temperature by more than 2.00×10^{-3} K. How many kilograms of water are needed to ensure a passive reservoir of adequate thermal mass?

Solution

Use Equation 13.22 and solve for *m*:

$$m = \frac{Q}{c_{\text{water}}\, \Delta T}$$

The specific heat of water is found in Table 13.4; substituting all values,

$$m = \frac{1.50 \times 10^4 \text{ J}}{[4186 \text{ J/(kg·K)}](2.00 \times 10^{-3} \text{ K})}$$
$$= 1.79 \times 10^3 \text{ kg}$$

This represents about 1.79 m³ of water, which is a cube of water about 1.21 m on a side.

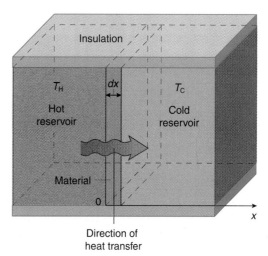

FIGURE 13.20 A material in thermal contact with two reservoirs.

13.11 MECHANISMS FOR HEAT TRANSFER*

Heat transfer between systems at different temperatures occurs by the mechanisms of (1) conduction, (2) convection, and (3) radiation.

Conduction

Hold one end of a silver spoon and put the other end into a hot liquid, such as a delicious soup simmering on a stove. You quickly become aware that the end of the spoon you are holding becomes hot. This illustrates heat transfer by **conduction**. Yet do the same experiment with a wooden spoon of identical dimensions and you know the end you hold does not become hot, even after many minutes of stirring the soup. Hence heat transfer by conduction depends critically on the material bridging or connecting the warmer and cooler regions.

Imagine two reservoirs at slightly different temperatures connected by means of a thin piece of material of uniform cross-sectional area, as pictured in Figure 13.20. The material is in thermal contact with each reservoir.

Heat transfer occurs *through* the material from the hot reservoir to the cold reservoir. The energy is transported from atom to atom through the material by means of atomic vibrations and/or collisions. There is no bulk, macroscopic motion of the material itself. This mechanism of heat transfer is conduction.

Let *dT* be the differential difference in temperature between the warmer and cooler reservoirs, respectively, and let *A* be the area of the material in thermal contact with each reservoir. Let the length of material connecting the reservoirs be *dx*, with the *x*-axis directed from the warmer to the cooler reservoir. Experiments indicate that the heat transfer per second

$$\frac{dQ}{dt}$$

called the **heat flow** (expressed in J/s ≡ W), between the reservoirs is directly proportional to the temperature difference *dT* and the area of thermal contact *A*, and inversely proportional to the length of material *dx*. It makes a great deal of difference whether the material is copper or Styrofoam; thus the propor-

tionality constant *k* depends on the material; *k* is called the **thermal conductivity** of the material. We summarize these experimental observations with the following equation:

$$\frac{dQ}{dt} = -kA\frac{dT}{dx} \qquad (13.32)$$

Why the minus sign? The heat flow is from the warmer to the cooler reservoirs, so that as *x increases*, the temperature *decreases* through the material. Thus

$$\frac{dT}{dx}$$

is *negative*; we want the heat flow

$$\frac{dQ}{dt}$$

from the hot reservoir to the cold reservoir to be positive, since this is the direction in which energy is transferred.[†] The minus sign in Equation 13.32 thus makes the heat flow positive. Notice also that large values of the thermal conductivity mean a large heat flow (a good thermal conductor). Values of the thermal conductivity *k* for various materials are tabulated in Table 13.6.

The way the temperature changes with position in the material,

$$\frac{dT}{dx}$$

is known as the **temperature gradient**. Equation 13.32 indicates that the heat flow through a material is proportional to the temperature gradient, the area in contact, and the thermal conductivity of the material.

[†]The reason the heat transfer is in this direction and not the other way (from the cold to the hot region) will be examined in Chapter 15.

TABLE 13.6 Thermal Conductivity k of Various Materials at Approximately Room Temperature and 1 atm

Material	k [W/(m·K)]
Air	0.026
Aluminum	237
Asbestos	0.08
Concrete	1.4
Copper	401
Gold	317
Glass (window)	0.78
Hydrogen	0.18
Ice	2.1
Iron	80.2
Lead	35.3
Mercury	8.18
Nitrogen	0.026
Oxygen	0.027
Sheetrock (dry wall)	0.13
Silver	429
Stainless steel	15
Water	0.613

Myer Kutz (editor), *Mechanical Engineer's Handbook* (John Wiley, New York, 1998).

Let the temperatures of the two reservoirs be T_H and T_C and the length of material through which heat transfer occurs be d, as shown in Figure 13.21.

For a steady-state situation, the temperature at any position x inside the material must remain constant. This means the heat flow

$$\frac{dQ}{dt}$$

past any point in the material must be independent of the position x in the material. If the heat flow were not constant at all points x, then the temperature at any point inside the material would be time dependent and we would not have a steady-state situation. So, for a steady-state situation, in a material of uniform cross section, both

$$\frac{dQ}{dt}$$

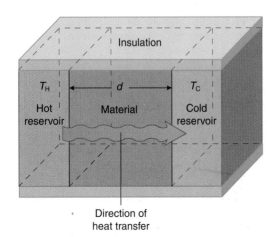

FIGURE 13.21 Two reservoirs connected by a material of length d.

and

$$\frac{dT}{dx}$$

are independent of x. Write the constant temperature gradient as

$$\frac{dT}{dx} \equiv -S \qquad (13.33)$$

where S is a constant. The derivative of T with respect to x is the same everywhere; this means that the temperature decreases linearly with x as we go through the material from the hot to the cold reservoir. We can see this mathematically by integrating Equation 13.33; rearranging it slightly, we have

$$dT = -S \, dx$$

Integrating this between corresponding limits,

$$\int_{T_H}^{T} dT = -S \int_{0\text{ m}}^{x} dx$$

or

$$T - T_H = -S[x - 0\text{ m}]$$
$$= -Sx$$

Solving for T, we find

$$T = T_H - Sx$$

This shows that the temperature decreases linearly with x through the material from the hot to cold regions. Solving for the temperature gradient S, we obtain

$$S = \frac{T_H - T}{x}$$

When $x = d$, the length of the material connecting the reservoirs, the temperature T is the temperature of the cold reservoir, T_C; hence

$$S = \frac{T_H - T_C}{d} \qquad (13.34)$$

Equation 13.32 for the heat flow becomes, using Equation 13.33:

$$\frac{dQ}{dt} = -kA \frac{dT}{dx}$$
$$= kAS$$

Substituting for the temperature gradient S using Equation 13.34, we get

$$\frac{dQ}{dt} = kA \frac{T_H - T_C}{d} \qquad (13.35)$$

Dividing both the numerator and denominator of the right-hand side of this equation by the thermal conductivity k, we obtain

$$\frac{dQ}{dt} = A \frac{T_H - T_C}{d/k}$$

The quantity d/k is known as the **R-value** R of the material, and represents a thermal resistance to heat flow*:

$$R \equiv \frac{d}{k} \qquad (13.36)$$

---

*It is called the R-value because it is the thermal Resistance. Do not confuse the R-value with the universal gas constant.

Note that the smaller the thermal conductivity, the larger the R-value, and the smaller the heat flow between the reservoirs. The R-value incorporates the length (thickness) of the material connecting the two reservoirs; thus if the thickness d of material is doubled, the R-value doubles and the heat flow is halved. Building materials typically are rated for their insulating effect—that is, their resistance to heat flow—by quoting their R-value; the higher the R-value, the greater the insulating effect of the material and the smaller the heat flow between the hot and cold reservoirs. Thus

$$\frac{dQ}{dt} = A\,\frac{T_H - T_C}{R} \qquad (13.37)$$

The approximate R-values of some materials are summarized in Table 13.7.

Materials in Series

In building construction, several layers of different materials typically make up the wall separating the inside and outside of a building. Think of the inside and outside as two reservoirs at different temperatures. We want to determine the effect of such layering on the heat flow between the reservoirs.

Let two layers of material of different thermal conductivities and thicknesses connect the two reservoirs, as shown in Figure 13.22. Such a layered system represents materials in **thermal series**, since the heat flow must unavoidably be through each material in succession (or series) to get from the hotter to the cooler reservoir. Let the R-values of the materials be

$$R_1 = \frac{d_1}{k_1} \quad \text{and} \quad R_2 = \frac{d_2}{k_2}$$

When a steady-state situation is reached, the heat flow through each material is the same. Let T be the temperature of the

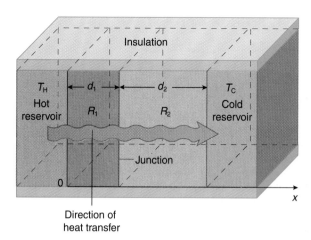

FIGURE 13.22 Two reservoirs connected by two materials in series.

junction between the two materials. The heat flow through the first material is

$$\left(\frac{dQ}{dt}\right)_1 = A\,\frac{T_H - T}{R_1} \qquad (13.38)$$

and that through the second material is

$$\left(\frac{dQ}{dt}\right)_2 = A\,\frac{T - T_C}{R_2} \qquad (13.39)$$

Since the heat flows are equal in the steady state, we have

$$A\,\frac{T_H - T}{R_1} = A\,\frac{T - T_C}{R_2}$$

We can solve for the temperature T at the interface between the materials. After a bit of algebra (famous last words!), we find that

$$T = \frac{1}{R_1 + R_2}\left(R_2 T_H + R_1 T_C\right)$$

Now we use Equation 13.38 and substitute for the temperature T at the interface, since this temperature is not typically of interest in determining the heat flow between the two reservoirs. The resulting expression for the heat flow involves only the temperatures of the two reservoirs, the area of thermal contact, and the R-values of the two materials. After making the substitution for the junction temperature T, the heat flow becomes (after a bit more algebraic simplification)

$$\frac{dQ}{dt} = A\,\frac{T_H - T_C}{R_1 + R_2}$$

Notice that the effect of an additional piece of material between the reservoirs is to increase the total R-value in the denominator of the heat flow equation.

TABLE 13.7 Approximate R-Values of Selected Building Materials

Material	R-value ($\mathrm{m^2 \cdot K \cdot s/J = m^2 \cdot K/W}$)
Glass (single-pane window glass) (thickness 3.2×10^{-3} m)	0.21
Brick (thickness 1.0×10^{-1} m)	0.93
Fiberglass batting (thickness 8.9×10^{-2} m)	2.53
(thickness 1.5×10^{-1} m)	4.37
Air space (thickness 8.9×10^{-2} m)	0.23
Pine board (thickness 1.9×10^{-2} m)	0.22
Plywood (thickness (1.2×10^{-2} m)	0.14
Polyurethane foam (thickness 2.54×10^{-2} m)	2.3
Sheetrock (dry wall) (thickness 1.3×10^{-2} m)	0.10

In the United States, the R-value of a material typically is given in the antiquated English system of units. In the English unit system, the temperature difference is expressed in degrees fahrenheit, the area in square feet, and the R-value is in the singularly awkward units of feet$^2 \cdot$°F$\cdot$h/Btu. The heat flow then emerges in Btu/h rather than watts. To convert the R-values in SI units to the English system, multiply each value by

$$5.68\;\frac{\mathrm{feet^2 \cdot {}^\circ F \cdot h/Btu}}{\mathrm{m^2 \cdot K/W}}$$

Thus, if other layers of material are added in series between the two reservoirs, the effective (or total) R-value of the composite is the sum of the R-values of the individual layers:

$$R_{\text{total}} = R_1 + R_2 + R_3 + R_4 + \cdots \quad \text{(layers in series)} \qquad (13.40)$$

The heat flow is then found from

$$\frac{dQ}{dt} = A\,\frac{T_H - T_C}{R_{total}} \qquad (13.41)$$

Materials in Parallel

If two different materials *each* connect the hot and cold reservoirs, as in Figure 13.23, then the materials are said to be connected thermally in **parallel**.

For example, both windows and walls connect the inside and outside of a building; the windows and walls are thermally in parallel. We neglect any heat transfer along the interface between the two materials. The heat flow through material 1 is

$$\left(\frac{dQ}{dt}\right)_1 = A_1\,\frac{T_H - T_C}{R_1} \qquad (13.42)$$

where A_1 is the area of this material in thermal contact with the reservoirs. Likewise, for the second material,

$$\left(\frac{dQ}{dt}\right)_2 = A_2\,\frac{T_H - T_C}{R_2} \qquad (13.43)$$

where A_2 is the area of this material in thermal contact with the reservoirs.

The total heat flow from the hot to the cold reservoir then is the sum of the individual heat flows:

$$\left(\frac{dQ}{dt}\right)_{total} = \left(\frac{dQ}{dt}\right)_1 + \left(\frac{dQ}{dt}\right)_2 \qquad (13.44)$$

Substituting from Equations 13.42 and 13.43, we find

$$\left(\frac{dQ}{dt}\right)_{total} = A_1\left(\frac{T_H - T_C}{R_1}\right) + A_2\left(\frac{T_H - T_C}{R_2}\right)$$
$$= (T_H - T_C)\left(\frac{A_1}{R_1} + \frac{A_2}{R_2}\right) \qquad \text{(materials in parallel)} \qquad (13.45)$$

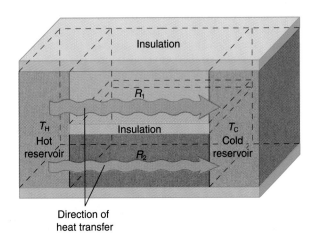

FIGURE 13.23 Two reservoirs connected by two materials in parallel.

EXAMPLE 13.14

A single-pane glass window of dimensions 1.00 m × 1.50 m connects a room at temperature 20 °C with outside air on a blistering summer day (temperature 37 °C). Find the heat flow through the window.

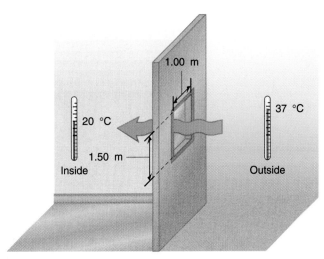

Heat transfer through a window.

Solution

The hot reservoir is the outside air and the cold reservoir is the inside of the room, maintained at its temperature by an air conditioning system. Thus $T_H - T_C = 17$ K. From Table 13.7, we see that the R-value for a single-pane window is

$$R = 0.21 \ \text{m}^2\cdot\text{K/W}$$

Use Equation 13.37 for the heat flow:

$$\frac{dQ}{dt} = A\,\frac{T_H - T_C}{R}$$

where A is the area of the window, here 1.00 m × 1.50 m = 1.50 m².

Making the substitutions, you have

$$\frac{dQ}{dt} = (1.50 \ \text{m}^2)\,\frac{17 \ \text{K}}{0.21 \ \text{m}^2\cdot\text{K/W}}$$
$$= 1.2 \times 10^2 \ \text{W}$$

EXAMPLE 13.15

A wall of a house consists of sheetrock (dry wall of thickness 1.3×10^{-2} m), fiberglass insulation (thickness 8.9×10^{-2} m), and plywood (thickness 1.2×10^{-2} m) separating a room from the outdoors. The wall is 2.00 m high and 5.00 m long and is windowless. If the inside temperature is 20 °C and the outside temperature is -25 °C on a cold winter day, what is the heat flow through the wall?

Heat transfer through a composite wall.

Solution

The materials are layered in series, and so the total R-value of the layers is the sum of the individual R-values of the material, each found from Table 13.7:

$$R_{total} = R_{dry\ wall} + R_{fiberglass} + R_{plywood}$$
$$= 0.10\ m^2 \cdot K/W + 2.53\ m^2 \cdot K/W + 0.14\ m^2 \cdot K/W$$
$$= 2.77\ m^2 \cdot K/W$$

The temperature difference is

$$T_H - T_C = 45\ K$$

and the wall area is

$$A = (2.00\ m)(5.00\ m)$$
$$= 10.0\ m^2$$

Using Equation 13.41 for the heat flow, you get

$$\left(\frac{dQ}{dt}\right)_{total} = A\frac{T_H - T_C}{R_{total}}$$
$$= \frac{(10.0\ m^2)(45\ K)}{2.77\ m^2 \cdot K/W}$$
$$= 1.6 \times 10^2\ W$$

EXAMPLE 13.16

A wall of a dormitory cell has overall dimensions 2.00 m by 3.50 m. A single-pane window of dimensions 0.75 m by 1.20 m is in the wall. The effective R-value of the opaque wall is 2.10 m² · K/W. If the inside temperature is 20 °C and the outside air temperature is −10 °C, determine the heat flow through the entire wall.

Solution

The wall and window connect the two reservoirs in parallel, and so the total heat flow is the sum of the heat flow through the window with the heat flow through the wall. You could use Equation 13.45 directly, but let's look at what is happening with

each parallel conductor separately and add the heat flows. The temperature difference is $T_H - T_C = 30$ K. The area of the window is

$$A_{window} = (0.75\ m)(1.20\ m)$$
$$= 0.90\ m^2$$

The R-value of the single-pane window is found from Table 13.7 to be

$$R_{window} = 0.21\ m^2 \cdot K/W$$

The heat flow through the window thus is, from Equation 13.37,

$$\left(\frac{dQ}{dt}\right)_{window} = A_{window}\frac{T_H - T_C}{R_{window}}$$
$$= (0.90\ m^2)\frac{30\ K}{0.21\ m^2 \cdot K/W}$$
$$= 1.3 \times 10^2\ W$$

The area of the wall itself (not including the window) is

$$(2.00\ m)(3.50\ m) - (0.75\ m)(1.20\ m) = 6.10\ m^2$$

The heat flow through the wall is found using Equation 13.37:

$$\left(\frac{dQ}{dt}\right)_{wall} = A_{wall}\frac{T_H - T_C}{R_{wall}}$$
$$= (6.10\ m^2)\frac{30\ K}{2.10\ m^2 \cdot K/W}$$
$$= 87\ W$$

The total heat flow is the sum of the heat flow through the wall and that through the window:

$$\frac{dQ}{dt} = 1.3 \times 10^2\ W + 87\ W$$
$$= 2.2 \times 10^2\ W$$

EXAMPLE 13.17

The air temperature above a calm pond with water at 0 °C near its surface suddenly plummets from 0 °C to −10 °C.

a. Show that the rate at which the thickness x of ice increases with time is

$$x = \left(\frac{2k\ \Delta T}{L_f \rho_{ice}}t\right)^{1/2}$$

where ΔT is the temperature difference $T_H - T_C$, ρ_{ice} is the density of ice, and L_f the latent heat of fusion of ice.
b. How long will it take to form a layer 1.0 cm thick? The density of ice is 920 kg/m³.

Solution

a. Let A be an area on the surface of the pond. In a differential thickness $|dx|$ of ice, there is a differential mass dm, where

$$dm = \rho_{ice}A\ |dx|$$

To solidify this mass of ice, the heat transfer from the water is

$$dQ = -L_f \, dm$$
$$= -L_f \rho_{ice} A \, |dx|$$

Let x be the thickness of the ice. The heat flow is upward through the ice from the water to the air. The additional ice formation occurs at the underside of the ice and proceeds downward. Choose the $+x$-direction downward; so dQ, which flows upward, passes through $-dx$ (i.e., toward negative x). Thus, to eliminate the absolute value signs, we write

$$dQ = -L_f \rho_{ice} A \, (-dx)$$
$$= L_f \rho_{ice} A \, dx$$

The heat flow is given by Equation 13.35:

$$\frac{dQ}{dt} = kA \frac{\Delta T}{x}$$

Substituting for dQ from the preceding, you have

$$L_f \rho_{ice} A \frac{dx}{dt} = \frac{kA \, \Delta T}{x}$$

or

$$x \, dx = \frac{k \Delta T}{L_f \rho_{ice}} \, dt$$

Integrating this expression,

$$\int_{0\,m}^{x} x \, dx = \frac{k \, \Delta T}{L_f \rho_{ice}} \int_{0\,s}^{t} dt$$

Solving for x, you obtain

$$x = \left(\frac{2k \, \Delta T}{L_f \rho_{ice}} t \right)^{1/2}$$

b. The temperature difference ΔT is 10 K. The thermal conductivity k of ice is (see Table 13.6)

$$k = 2.1 \ \text{W/(m·K)}$$

The heat of fusion for water is found in Table 13.5:

$$L_f = 3.335 \times 10^5 \ \text{J/kg}$$

The density of ice is

$$\rho_{ice} = 920 \ \text{kg/m}^3$$

Making these substitutions, you have

$$x = (3.7 \times 10^{-4} \ \text{m/s}^{1/2}) \sqrt{t}$$

Substitute $x = 0.010$ m and solve for t:

$$0.010 \ \text{m} = (3.7 \times 10^{-4} \ \text{m/s}^{1/2}) \sqrt{t}$$

or

$$t = 7.3 \times 10^2 \ \text{s}$$
$$= 12 \ \text{min}$$

Convection

Heat transfer resulting from the movement or circulation of material is called **convection**. For example, in a hot water tank the water in immediate contact with the hot heating element expands and becomes less dense than the surrounding fluid. Buoyant forces then cause the hot water to rise in the tank; cooler liquid takes its place and a circulation of water is established in the tank. Such circulation is called a **convection current**. Likewise, the air in contact with a flame or a hot radiator expands and becomes less dense; buoyant forces make it rise, and convection currents can be established. During the summer near the beach, hot air over inland areas rises, drawing cooler air inward from across the water surface, producing on-shore breezes in the afternoon. Such winds are a large-scale example of a convection current of air.

Heat transfer by convection can be *passive* (such as in the examples just mentioned) or *forced* with fans or pumps. Quantitative descriptions of convection involve considerations of fluid dynamics that are beyond the scope of an introduction to thermodynamics.

Radiation

Bask in sunlight at poolside and you experience heat transfer by **radiation**. The same radiation mechanism of heat transfer is involved in common heat lamps used to keep food warm in your college cafeteria and fast food restaurants while awaiting voracious student appetites. You likely have experienced the romantic effects of radiative heat transfer curled up in front of a roaring fire with a good physics text for company, or more likely with a boon companion.

> While both conduction and convection require a material medium for heat transfer, radiation does not.

It is well known that we receive energy from the Sun despite the virtually total absence of any material between the Sun and the Earth. The solar energy is transported by (electromagnetic) radiation: light in the most general sense of the term. The particles of light, called **photons**, carry the energy from an emitter (such as the Sun) to an absorber (your body frying beside the pool).

Every object emits energy by radiation.

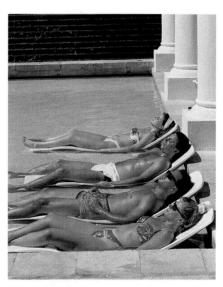

Radiative heat transfer between the Sun and poolside Sun worshipers.

The rate at which a system emits radiation is found experimentally to be proportional to the *fourth* power of its absolute temperature.

This remarkable result, discovered by Josef Stefan (1835–1893) in 1879, has deep historical significance in the history of physics. It was through a detailed study of the physics of such radiation that Max Planck (1858–1947) discovered (in 1901) a totally new and unanticipated fundamental constant of nature, known formally as the quantum of action, that now bears his name (Planck's constant). Planck's discovery was the opening salvo in the revolution of physics known as *quantum mechanics* (Chapters 26 and 27).

The rate at which energy is *radiated* by a system also depends directly on its surface area A. The heat flow via radiation is written as

$$\frac{dQ}{dt} = -eA\sigma T^4 \tag{13.46}$$

As before, the radiative heat flow

$$\frac{dQ}{dt}$$

is expressed in joules/second (J/s) or watts (W). The constant e in Equation 13.46, called the **emissivity** of the surface,* is a pure number (with no dimensions) between 0 and 1. The emissivity quantifies the relative effectiveness of the surface as a radiator. A perfect radiator has an emissivity $e = 1$; such a system is known as a *blackbody radiator* or, more simply, a **blackbody**.[†] The remaining constant in Equation 13.46, σ, is known as the **Stefan–Boltzmann constant**; experiments indicate it has the numerical value[‡]

$$\sigma = 5.670 \times 10^{-8} \ \text{W/(m}^2 \cdot \text{K}^4)$$

The minus sign in Equation 13.46 indicates that the system is emitting rather than absorbing the radiation, in keeping with our convention that heat transfer from a system is negative. Equation 13.46 is known as **Stefan's law**.

A system also absorbs radiation. The radiation absorbed by a system is proportional to the fourth power of the absolute temperature of its *surroundings*. We write the heat flow to a system as

$$\frac{dQ}{dt} = +aA\sigma T^4$$

where a is a pure number between 0 and 1 indicating the relative ability of the surface to absorb radiation from its surroundings. A surface with $a = 1$ is a perfect absorber.

If you were to place a hot object in a surrounding region at a cooler temperature and insulate the resulting system, as in

FIGURE 13.24 A hot object in a region of cooler surrounding temperature and insulated from everything else.

Figure 13.24, heat transfer would occur only between the hot object and the surrounding cooler region. The hot object radiates energy and cools, while the surroundings absorb radiant energy and increase in temperature.

Eventually, the object and its surroundings reach the same temperature T and the combined system (object and its surroundings) is in thermal equilibrium. At this point, the object must be radiating an amount of energy equal to the energy it is absorbing. When the combined system is in thermal equilibrium, the total heat flow (via radiation and absorption) is zero. That is, at thermal equilibrium,

$$\text{heat flow radiated} + \text{heat flow absorbed} = 0 \ \text{W}$$
$$-eA\sigma T^4 + aA\sigma T^4 = 0 \ \text{W}$$

Hence

$$e = a$$

In other words, the effectiveness of an object as a radiator (e) is the *same* as its effectiveness as an absorber (a). Good radiators are good absorbers; poor radiators are poor absorbers. We really need only one constant to describe both absorption and radiation; the constant chosen is e, the emissivity.

If an object is at a temperature T_{object} that is different from that of its surroundings, $T_{\text{surroundings}}$, then the object *radiates* energy at the rate

$$-eA\sigma T^4_{\text{object}}$$

and *absorbs* energy at the rate

$$+eA\sigma T^4_{\text{surroundings}}$$

The total heat flow via radiative heat transfer to the object from its surroundings then is

$$\frac{dQ}{dt} = eA\sigma\left(T^4_{\text{surroundings}} - T^4_{\text{object}}\right) \tag{13.47}$$

---

*Do *not* confuse the emissivity e with the base of natural logarithms (the pure number $e = 2.71828\ldots$) or with the magnitude of the charge on the proton and electron ($e = 1.602 \times 10^{-19}$ coulombs).

[†] The object may not appear black in color at all. An excellent approximation to a blackbody radiator is a closed chamber at temperature T with a small hole drilled into it, through which the radiation is emitted.

[‡] The Stefan–Boltzmann constant can be expressed in terms of other fundamental physical constants (in particular, the Boltzmann constant k, Planck's constant h, and the speed of light c). Thus the Stefan–Boltzmann constant is not truly a fundamental physical constant on a par with h, c, or G.

QUESTION 8

Before the advent of microwave ovens, a common household trick when baking potatoes in a standard oven was to insert a (clean) nail through each potato while baking. Explain how this helps to bake the potatoes faster.

EXAMPLE 13.18 _____

The surface temperature of the spherical Sun is 5.8×10^3 K and its radius is 6.96×10^8 m. Consider it as a blackbody radiator. What is the heat flow from the Sun via radiation?

Solution
Heat flow via radiation is found using Stefan's law, Equation 13.46:

$$\frac{dQ}{dt} = -eA\sigma T^4$$

For a blackbody radiator, the emissivity $e = 1$. The surface area of the Sun is that of a sphere of radius r:

$$A = 4\pi r^2$$
$$= 4\pi (6.96 \times 10^8 \text{ m})^2$$
$$= 6.09 \times 10^{18} \text{ m}^2$$

Make these substitutions into Stefan's law and use the value of the Stefan–Boltzmann constant $\sigma = 5.670 \times 10^{-8}$ W/(m²·K⁴); you find

$$\frac{dQ}{dt} = (-6.09 \times 10^{18} \text{ m}^2)[5.670 \times 10^{-8} \text{ W/(m}^2 \cdot \text{K}^4)]$$
$$\times (5.8 \times 10^3 \text{ K})^4$$
$$= -3.9 \times 10^{26} \text{ W}$$

The Sun is a rather powerful radiator! The minus sign indicates that the heat transfer per second (the heat flow) is *from* the Sun. The absolute value of the heat flow from the Sun via radiation is known as the *solar luminosity*.

EXAMPLE 13.19 _____

An object at an absolute temperature T_{object} finds itself in a region where the absolute temperature is only slightly greater:

$$T_{\text{surroundings}} = T_{\text{object}} + \Delta T$$

where ΔT is small compared with either T_{object} or $T_{\text{surroundings}}$. Show that the total heat flow to the object from its surroundings via radiative heat transfer is proportional to the *first* power of the temperature difference ΔT.

Solution
Use Equation 13.47. The total heat flow to the object via radiative heat transfer is

$$\frac{dQ}{dt} = eA\sigma\left[\left(T_{\text{object}} + \Delta T\right)^4 - T_{\text{object}}^4\right]$$

Factor out T_{object} in the first term to obtain

$$\frac{dQ}{dt} = eA\sigma\left[T_{\text{object}}^4\left(1 + \frac{\Delta T}{T_{\text{object}}}\right)^4 - T_{\text{object}}^4\right] \quad (1)$$

Now expand the expression

$$\left(1 + \frac{\Delta T}{T_{\text{object}}}\right)^4$$

using the binomial theorem:

$$(1 + x)^4 = 1 + 4x + 6x^2 + 4x^3 + x^4$$

If x is small, you need keep only the first two terms of the expansion:

$$(1 + x)^4 \approx 1 + 4x$$

Thus you have

$$\left(1 + \frac{\Delta T}{T_{\text{object}}}\right)^4 \approx 1 + 4\frac{\Delta T}{T_{\text{object}}}$$

Substituting this into equation (1) for the total heat flow, you obtain

$$\frac{dQ}{dt} = eA\sigma\left[T_{\text{object}}^4\left(1 + 4\frac{\Delta T}{T_{\text{object}}}\right) - T_{\text{object}}^4\right]$$
$$= eA\sigma\left(4T_{\text{object}}^3 \Delta T\right) \quad (2)$$

If the surroundings are at a higher temperature, $\Delta T > 0$ K and the total heat flow is positive and represents heat flow to the object. Correspondingly, if the surroundings are at a cooler temperature, $\Delta T < 0$ K, the total heat flow is negative, indicating the object is an emitter radiating heat flow to the surroundings.

Equation (2) indicates that for small temperature differences between the object and its surroundings, the heat transfer per second (the heat flow via radiation processes) absorbed by the object from its surroundings is directly proportional to the temperature difference ΔT.

EXAMPLE 13.20 _____

The average human body temperature is 37 °C. If you assume that the human body is a blackbody radiator, estimate the total heat flow of the naked human body into a room at temperature 20 °C.

Solution
For a blackbody, the emissivity is $e = 1$. The temperature difference is

$$\Delta T = T_{\text{surroundings}} - T_{\text{object}}$$
$$= -17 \text{ K}$$

The absolute temperature of the human body is about

$$T_{\text{object}} = 273.15 \text{ K} + 37 \text{ K}$$
$$= 310 \text{ K}$$

We need to estimate the surface area of the human body. This might make an interesting experiment. Suppose we guesstimate the surface area as about 2.5 m².*

Using equation (2) from Example 13.19, you find the total heat flow is

$$\frac{dQ}{dt} = eA\sigma\left(4T_{\text{object}}^3 \Delta T\right)$$

_____

*Average values for the surface area of the human body as a function of height and weight can be found in *Geigy Scientific Tables*, edited by Cornelius Lentner (Medical Education Division, Ciba-Geigy Corp., West Caldwell, New Jersey, 1984), volume 3, page 329. The number used here corresponds to a body mass of 60 kg and a height of 1.80 m.

Making the appropriate substitutions,

$$\frac{dQ}{dt} \approx (2.5 \text{ m}^2) [(5.670 \times 10^{-8}(\text{W/m}^2 \cdot \text{K}^4)](4)(310 \text{ K})^3(-17 \text{ K})$$

$$\approx -2.9 \times 10^2 \text{ W}$$

The minus sign indicates that the total heat flow of the human is *from* the human (the system) to the cooler surrounding room.

13.12 THERMODYNAMIC PROCESSES

A thermodynamic process is any way that a system changes from one state of thermal equilibrium to another such state. Heat transfer and work are two processes that *change* the state of thermodynamic equilibrium of a system. For example, a gas may exist in thermal equilibrium at some specified temperature, pressure, and volume. Heat transfer and/or work can change these thermodynamic parameters to other values. To analyze the *way* the change occurs, we distinguish between two very different types of thermodynamic processes: quasi-static (reversible) processes and irreversible processes.

Quasi-static (Reversible) Processes

If a thermodynamic system undergoes a change from one state of thermal equilibrium to another slowly enough so at any instant the entire system essentially is in thermal equilibrium (though at different temperature, pressure, and/or volume), then we say the process is **quasi-static**. Strictly speaking, quasi-static processes do not exist: they are idealizations.

They are convenient fictions, much like the frictionless surfaces and ideal strings used so prolifically in classical mechanics. We use these artificial constructs to make calculations that *approximate* reality.

In order to *change* the temperature, pressure, and/or volume of the gas from some initial values to different final values, the system is going to have to depart from its initial state of thermal equilibrium. The critical factor here is the speed of the change. As long as the changes occur slowly enough so the whole system can adjust continuously to the change and pass through a succession of thermal equilibrium states, the process can be considered quasi-static. The meaning of slow enough depends on the particular thermodynamic system under consideration. For example, if a piston compressing a gas moves too quickly, causing turbulent eddies in the gas, then the process is not quasi-static. The point is to move the piston slowly enough during the compression so the gas can adjust to the change and, at any instant, be in a state of thermal equilibrium. Quasi-static processes are called **reversible processes** if it is assumed that at any instant the small differential changes in the system can be reversed.

With few exceptions, the thermodynamics we study in this text is **equilibrium thermodynamics**, which applies to *processes that are quasi-static and reversible* (we shall use the word quasi-static; it is understood that the processes also are reversible). We assume these conditions apply in our study of thermodynamics. Irreversible processes involve a more complicated thermodynamic formalism known as *nonequilibrium thermodynamics*.

There are four special types of quasi-static processes that are useful in equilibrium thermodynamics. The processes are characterized by the thermodynamic parameter that is kept constant during the process (while other parameters change):

a. **Isothermal process**: the *temperature* of the system is kept constant.
b. **Isobaric process**: the *pressure* of the system is kept constant.
c. **Isochoric process**: the *volume* of the system is kept constant.
d. **Adiabatic process** (also called an **isentropic process**): there is no heat transfer either to or from the system (the system is well insulated from the environment).

We study adiabatic processes in greater detail in Chapters 14 and 15.

Irreversible Processes

Any process that is not quasi-static and reversible is called **irreversible**.

In nature, all thermodynamic processes really are irreversible.

But to paraphrase the pigs in George Orwell's *Animal Farm*, some processes are more irreversible than others. For example, the bursting of a balloon (called a free expansion of a gas) is a classic example of an irreversible process (not quasi-static) and one to which we shall return later; the process does not spontaneously go the other way, nor does it occur slowly enough for the gas to be considered in a state of thermodynamic equilibrium at any instant during the expansion.

13.13 ENERGY CONSERVATION: THE FIRST LAW OF THERMODYNAMICS AND THE CWE THEOREM

As you know, energy is an important concept in physics. Here we formulate a unifying concept about **conservation of energy** and explore its relationship to something old, the CWE theorem from mechanics, and something new, the first law of thermodynamics.

The total energy of a system has two distinct contributions:

1. A macroscopic energy: the total *mechanical energy E* of the system, familiar to you from our study of classical mechanics (dynamics). The total mechanical energy is associated with the *macroscopic position and motion of the system as a whole*. The mechanical energy *E* is the sum of

   • the kinetic energy of translation of the center of mass point and the rotational kinetic energy about the center of mass point, as well as
   • potential energies associated with the position of the center of mass of the system in space, be they gravitational, elastic, or other potential energies yet to come (such as electrical potential energy).

Recall that the place where the zero of potential energy is located is arbitrary.* Once all the appropriate choices for the zeros of the potential energies are made, the system has a definite value for the total mechanical energy E when in a state of mechanical equilibrium. Because of this freedom of choice, it is only *changes* in the total mechanical energy that are physically significant when the system changes its state of mechanical equilibrium, say, through the action of forces on the system. In other words, the thing that is physically important is ΔE, not E itself, as reflected in the CWE theorem of Chapter 8.

2. The other contribution to the total energy is a vast collection of microscopic energies, known collectively as the **internal energy** U of the system. The internal energy is the sum of the individual kinetic and potential energies associated with the motions and interactions of all the individual particles (atoms and/or molecules) of the system itself. Many of these interactions involve rather complicated potential energy functions on a microscopic distance scale. In principle, after appropriate choices are made for the zeros of these microscopic potential energy functions, we can talk about a definite value for the internal energy of the system when it is in a state of thermodynamic equilibrium, but this is a rather complicated endeavor. When the system changes its state of thermodynamic equilibrium, it is only *changes* in the internal energy ΔU that are physically significant in thermodynamics. This is analogous to the situation in dynamics, where only changes in the total mechanical energy are physically significant. We will see that, with few exceptions (ideal gases, considered in Chapter 14), the *changes* in the internal energy are much easier to calculate than the actual value of the internal energy itself.

The total mechanical energy E of a system occasionally is referred to as *ordered* energy because of its macroscopic character. The internal energy U of a system is called *disordered* energy because of the microscopic nature of the motions of the system's many constituent particles.

Energy can be *transferred* to a system (from other systems) by two mechanisms or processes: (a) via heat transfer, instigated by temperature differences, and (b) via work, instigated by macroscopic forces. Let's look at each of these two processes in more detail.

Heat Transfer to a System

We have seen that heat transfer to a system can occur by three distinct mechanisms: conduction, convection, and radiation. For conduction and convection, the system gains energy by being brought into physical contact with another system at a higher temperature. For radiation, the transfer occurs via photons without the necessity of an intervening medium. Mechanisms for *heat transfer* are fundamentally microscopic (atomic and molecular) in character. The direction of the heat transfer always is from higher to lower temperatures; the reason for this unidirectional aspect of heat transfer will be explained when we investigate the consequences of the second law of thermodynamics (in Chapter 15).

*In some cases the zero is chosen at a particular location *by convention* (such as at $x = 0$ m for a mass on a spring, or at $r = \infty$ m for the general form of the gravitational potential energy).

Do not confuse the concept of heat transfer with the internal energy.

Heat transfer occurs because of a temperature difference. Heat transfer is accomplished by random molecular collisions and other essentially molecular interactions (by conduction, convection, or radiation).

As we have stated before, you *cannot* think of heat transfer as a fundamental property of the state of the system itself; a system in thermal equilibrium does *not* have an amount of heat, which is why we prefer not to use the word heat as a noun. On the other hand, a system in thermal equilibrium *can* be said to have a specific internal energy (at least in principle), just as the system can be said to have a definite temperature, pressure, and volume. The value of the internal energy thus is a characteristic of the state of the system itself. We distinguish between these concepts by saying that heat transfer Q is *not* a **state variable**; the internal energy U *is* a state variable, as are the temperature, pressure, and volume.

Physical Work Done *on* the System

Physical work W' done *on* the system can change the internal energy U as well as the total mechanical energy E of the system.

As an example of work resulting in a change of just the internal energy, consider a gas enclosed in an insulated container with a movable piston on one side, as shown in Figure 13.25. The insulation ensures that there is no heat transfer either to or from the gas (the system). An external agent (you!) compresses the gas by pushing on the piston, much like using a bicycle pump, thus doing work on the system.

The work done *on* the system (the gas) by the external agent (you) results in an increase in the temperature of the system, indicative of an increase in the internal energy of the gas. Although we might say, colloquially, that the ideal gas "heats up" (since its temperature increases), the increase in the internal energy and the rise in the temperature of the system do *not* come about because of heat transfer, but because of macroscopic work done on the system by the external agent.

It also is possible for the work done *on* the system to change the mechanical energy E of the system: this is just the CWE theorem of mechanics. We will return to explore this relationship again shortly. Thus the work done on the system is a *macroscopic* mechanism for transferring energy to a system; heat transfer to a system is fundamentally a *microscopic* mechanism for transferring energy to a system.

FIGURE 13.25 Compressing a gas by pushing on a piston.

Although heat transfer and work are quite distinct physical ideas, they share several common features: (1) both represent energy transfer to a system; and (2) neither is a state variable. Just as a system in thermal equilibrium cannot be said to have an amount of heat, the system in equilibrium cannot be said to have an amount of work. Once a system has gained the energy (either mechanical energy E or internal energy U) provided by either heat transfer or work processes, there is no way that the system can tell from what process or source the energy came. The energy transferred to the system by heat transfer Q and work W' goes into a common, dichotomous energy pool: the internal energy and the total mechanical energy of the system.

The situation is similar to your bank account. You can deposit money (like joules) into your checking or your saving account (like internal energy and mechanical energy) by two processes: via cash or checks (like heat transfer and work). Once the money is in each account, the account balance cannot distinguish between the means by which the money arrived; the dollars (and likely you!) care not how they got there. But the internal energy *can* be distinguished from the total mechanical energy (your checking account is distinct from your savings account), since the former is microscopic while the latter is macroscopic in character.

A system may gain energy via other mechanisms too—for example, by chemical or nuclear reactions or other, more exotic mechanisms.* Some of these alternative processes for the addition of energy are fundamentally microscopic and some are macroscopic; the former exotic forms of energy transfer to the system are considered as heat transfer, and the latter as work.

Fundamental Energy Conservation Law

We gather these ideas about energy into a fundamental *energy conservation law*. Energy is transferred *to* a system by (1) heat transfer Q and (2) work W' done *on* the system by external macroscopic, *nonconservative* forces. The total energy transfer to the system ($Q + W'$) appears as a *change* in the internal energy, ΔU, and/or as a *change* in the total mechanical energy of the system, ΔE (= $\Delta KE + \Delta PE$). Any work done by macroscopic *conservative* forces is already present in the total mechanical energy term ΔE, since this term includes changes in the potential energies associated with conservative forces. Thus the W' here is the work done by what we called (in Chapter 8) "other" macroscopic, nonconservative forces acting on the system.

Therefore our fundamental energy conservation law is

$$Q + W' = \Delta U + \Delta E \qquad (13.48)$$

This is the most general statement of energy conservation in classical physics.

Let us see how this statement about energy conservation relates to what we learned in "pure" classical mechanics and how it specializes to the "pure" thermodynamics of *simple* thermodynamic systems.

When we studied pure classical mechanics in Chapters 1–12, we *neglected* so-called thermal effects. That is, we restricted ourselves to situations where (1) there was no heat transfer to the system ($Q = 0$ J) *and* (2) no change in the internal energy ($\Delta U = 0$ J).

Recall that the CWE theorem states that total work done by *all* forces acting *on* the system is equal to the change in the macroscopic kinetic energy of the system:

Work done by all forces acting *on* the system = ΔKE

The forces were separated into those of conservative and non-conservative type, and the work done by each also segregated so that

| Work done by conservative forces *on* the system | + | Work done by other forces *on* the system (non-conservative forces) (this is W' above) | = ΔKE |

The work done by the conservative forces was defined to be the *negative* of the change in the potential energies associated with those conservative forces:

Work done by conservative forces *on* the system = $-\Delta PE$

and so

Work done by other forces acting on the system = $\Delta KE + \Delta PE$ (nonconservative forces, i.e., W')

$$W' = \Delta KE + \Delta PE$$

or

$$W' = \Delta E \qquad (13.49)$$

where E is the total mechanical energy of the system.

We now make the connection with the general energy conservation statement of Equation 13.48. With thermal effects neglected ($Q = 0$ J and $\Delta U = 0$ J), the most general statement of energy conservation, Equation 13.48, reduces to the CWE theorem of Equation 13.49. Section 13.14 explores the connection between the general statement (Equation 13.48) of energy conservation and the CWE theorem via a specific example.

In pure thermodynamics (the thermodynamics of *simple* systems), we consider only systems whose total mechanical energy does *not* change: $\Delta E = 0$ J. Therefore macroscopic conservative forces (with their associated potential energy terms) are irrelevant. The forces doing work on the system are then just "other" (nonconservative) forces acting on the system that do work W', such as the work done by the force we use to compress the gas in Figure 13.25. Thus the general statement of energy conservation (Equation 13.48) becomes (since $\Delta E = 0$ J)

$$Q + W' = \Delta U \qquad (13.50)$$

In pure thermodynamics, *it is useful to change the reference regarding the work term*. Instead of considering the work done *on* the system *by* surrounding external agents, as we did in mechanics, in pure thermodynamics we are more interested in the work done *by* the system *on* its surrounding environment.

---

*Recall that we restricted our treatment of thermodynamics in this text to *simple thermodynamic systems*, excluding by definition these more exotic mechanisms for transferring energy to the system.

Forces always appear in pairs (according to Newton's third law of motion). Each member of a force pair has the same magnitude, but the pair are in opposite directions and act on different systems. Hence the work W' done *on* the system by the surroundings is the *negative* of the work W done *by* the system *on* the surroundings:

$$\begin{array}{c} \text{Work done } \textit{on} \text{ the system} \\ \textit{by} \text{ the surroundings} \end{array} = \begin{array}{c} - \text{ Work done } \textit{by} \text{ the system} \\ \textit{on} \text{ the surroundings} \end{array}$$

$$W' = -W$$

Whoever thought prepositions could be so important in physics, of all places!*

Substitute $-W$ for W' in Equation 13.50.

Therefore, for pure thermodynamic systems, we have

$$Q - W = \Delta U$$

or

$$\boxed{Q = \Delta U + W} \qquad (13.51)$$

where W *is the work done by the system on its surroundings.* Equation 13.51 is the **first law of thermodynamics** for simple thermodynamic systems.

There are several things to note carefully about the statement of the first law of thermodynamics:

1. Stripped of everything else, the first law of thermodynamics is a statement of conservation of energy for pure (i.e., simple) thermodynamic systems.
2. The equation assumes there are no changes in the total mechanical energy E of the system. That is, there are no changes in the sum of the macroscopic kinetic and potential energies of the system. In this and the next two chapters on thermodynamics, we will only consider simple thermodynamic systems that have $\Delta E = 0$ J, so that we can use $Q = \Delta U + W$. The restriction $\Delta E = 0$ J is what we mean by pure thermodynamics. Essentially, the first law of thermodynamics says that heat transfer Q *to* the system can change the internal energy of the system and/or the system can do work on its surroundings.
3. The heat transfer Q is positive for heat transfer *to* the system; W is the work done *by* the system on its surrounding environment; and ΔU is the change in the internal energy of the system. The work term in the first law is the *negative* of the work done by "other" nonconservative forces, W', that we used in the CWE theorem of mechanics.
4. Since the internal energy U is a state variable and the heat transfer Q and work W are not state variables, it may seem remarkable that the change in the internal energy is equal to the difference between the heat transfer to the system and the work done by the system. As the system changes from one state of thermodynamic equilibrium (an initial state) to another such state (a final state), we shall see that both the heat transfer to the system and the work done by the system depend on the particular way the system interacts with its

environment in executing the change from the initial to the final state (such as via one of the specific types of processes mentioned in Section 13.12). But the change in the internal energy of the system ΔU is *independent of the way the change is made.* Once the change is accomplished, the system has no memory of how the energy was acquired or lost: via heat transfer or via work.

5. You should be aware that the sign convention for the work term in the first law of thermodynamics (Equation 13.51) is not universal; you may encounter other texts[†] that treat the work done *on* the system as positive, rather than the work done *by* the system. The convention we use is clearest for an introduction to thermodynamics. Regrettably, like politics, there is a certain amount of flip-flopping in any field.

QUESTION 9

Energy conservation is an important physical principle in physics. When environmentalists talk about energy conservation, is the term being used in the same sense as it is used in physics? Explain your answer.

13.14 THE CONNECTION BETWEEN THE CWE THEOREM AND THE GENERAL STATEMENT OF ENERGY CONSERVATION

For greater insight into the relationship between the general statement of energy conservation (Equation 13.48) and the specialized CWE theorem of Chapter 8 (Equation 13.49), we need to consider a system that is *not* a simple thermodynamic system, as we defined it in Section 13.1, but one that also has a change in its total mechanical energy. An inelastic collision meets this criterion.

In particular, consider a mass m, initially traveling at speed v on a frictionless surface, making a completely inelastic collision with an initially stationary mass M; see Figure 13.26. Both particles initially are at the same temperature and in thermal equilibrium with their environment. Take the system to be the two masses.

Total momentum is conserved in collisions. Hence the total momentum of the particles before the collision is equal to the total momentum after the collision:

$$\vec{P}_{\text{total bfr}} = \vec{P}_{\text{total aft}}$$
$$mv\hat{\imath} = (M + m)\vec{v}'$$

[†] Some formal, advanced texts in thermodynamics make the opposite choice for the sign convention of the work term in the first law of thermodynamics.

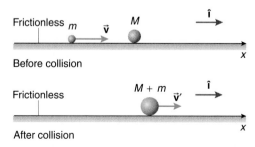

FIGURE 13.26 An inelastic collision of two particles.

---

*Such prepositions were critical in studying forces in mechanics, as you know!

where $\vec{v}'$ is the velocity of the composite particle after the collision. We solve for $\vec{v}'$:

$$\vec{v}' = \frac{m}{M + m} v\hat{i}$$

Kinetic energy is not conserved in such inelastic collisions, as we learned in Chapter 9. The *change* in the kinetic energy of the system is

$$\begin{aligned} \Delta KE &= KE_{aft} - KE_{bfr} \\ &= \frac{1}{2}(M + m)(v')^2 - \frac{1}{2}mv^2 \end{aligned}$$

Substituting for v' and simplifying, we find

$$\Delta KE = -\left(\frac{M}{m + M}\right)\frac{1}{2}mv^2 \qquad (13.52)$$

The change in the kinetic energy is negative: the kinetic energy of the system after the collision is less than it was before the collision.

We now can determine what happens to the missing kinetic energy. The CWE theorem of classical mechanics is of no help in solving this riddle of the missing kinetic energy. Indeed, if we apply the CWE theorem to this problem, we obtain a rather puzzling result. Recall that the CWE theorem states that the work done by *all* the nonconservative forces on the system is equal to the change in the total mechanical energy of the system:

$$W'_{all} = \Delta E$$

The forces acting on the system are (see Figure 13.27)

1. the weights of the two masses, and
2. the normal forces of the surface on the masses.

The force of m on M and the force of M on m during the brief time interval of the collision are *internal* to the system. They form a Newton's third law force pair, and together do zero total work.

The work done by the gravitational forces on the masses is zero since these forces are perpendicular to the change in the position vectors of the two masses as they collide; thus there is no change in the gravitational potential energy. The work done by the normal forces is zero since they are zero-work forces (normal forces, when they exist, always are perpendicular to any change in the position vector). Therefore the work done on the system by nonconservative forces is zero; but the change in the kinetic energy is negative. The CWE theorem (Equation 13.49) implies that zero is equal to a negative number! What is going on?

The resolution of the dilemma lies not with the CWE theorem, but with the general statement of energy conservation, Equation 13.48. The CWE theorem is a *special case* of this general statement of energy conservation that ignores thermal effects. Such thermal effects cannot be ignored here, and so the CWE theorem does not apply.

To resolve the dilemma of the missing kinetic energy, let's apply the general statement of energy conservation, Equation 13.48, to this problem:

$$Q + W' = \Delta U + \Delta E$$

Since both particles are initially at the same temperature and in thermal equilibrium with their environment, there is no heat transfer Q to the system: $Q = 0$ J. The quantity W' is the total work done on the system by all the nonconservative forces on the system, here just the normal force; but the normal force is a zero-work force, and so $W' = 0$ J. The work done by conservative forces is built into the change in total mechanical energy term via changes in the potential energy terms (here zero). The quantity ΔU is the change in the internal energy of the system (the change in the collective sum of all the microscopic kinetic and potential energies of the myriad particles making up the system). The change in the total mechanical energy of the system is ΔE and this represents

$$\Delta E = \Delta KE + \Delta PE$$

There was no change in the (macroscopic) potential energy of the system in the collision, indicating that the gravitational forces do no work here; the potential energy here is the gravitational potential energy, which is constant because the two masses do not change their vertical coordinates during the brief collision. Therefore the change in the mechanical energy is just the change in the kinetic energy:

$$\begin{aligned} \Delta E &= \Delta KE + PE \\ &= \Delta KE + 0 \text{ J} \end{aligned}$$

Making these substitutions into Equation 13.48, we have

$$0 \text{ J} + 0 \text{ J} = \Delta U + \Delta KE$$

Hence

$$\Delta U = -\Delta KE$$

Since the change in the kinetic energy is negative for our problem (see Equation 13.52), the change in the internal energy of the system is positive.

In other words, the missing kinetic energy appears as an increase in the internal energy of the system. The increase in the internal energy of the system typically manifests itself as an increase in the temperature of the system, but the increase in the internal energy also could be manifested by a phase change (e.g., melting) with no change in temperature. The increase in temperature (or phase change) occurs *as if* heat transfer to the system occurred, although *no heat transfer* actually does occur to the system of particles.

As the temperature of the system increases above the ambient environment because of the increase in internal energy, heat transfer *then* occurs from the system to the environment until they reach a common temperature.

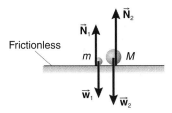

FIGURE 13.27 External forces on the system of colliding particles.

Thus we see that the CWE theorem *cannot* be applied willy-nilly to all systems; the theorem has limitations. It is a special case of the more fundamental statement of energy conservation, Equation 13.48.

13.15 WORK DONE BY A SYSTEM ON ITS SURROUNDINGS

It is always nice when something does work for us! The work done by a system such as a gas in a cylinder confined by a movable piston has enormous ramifications for our personal mobility, since it is the basis of the internal combustion engine. In this section we see how to calculate the work done *by* a system, one of the terms in the first law of thermodynamics, Equation 13.51.

Consider a gas at pressure P confined to a volume V by a movable piston, as shown in Figure 13.28. Let the temperature of the gas increase slightly, so that the gas expands, moving the piston a differential distance dx.

The differential work dW done by the gas *on* the piston is, from the definition of work,

$$dW = \vec{F} \cdot d\vec{r}$$

where $\vec{F}$ is the force of the gas on the piston and $d\vec{r}$ is the change in the position vector of the center of mass of the piston. The force $\vec{F}$ is provided by the pressure of the gas. With the coordinate system indicated in Figure 13.28, the force is

$$\vec{F} = PA\hat{\imath}$$

where A is the area of the piston. Recall that the change in the position vector $d\vec{r}$ always is written (in one dimension) as

$$d\vec{r} = dx\,\hat{\imath}$$

because the limits of integration account for the direction the piston actually moves along the *x*-axis. The differential work done by the gas is then

$$dW = PA\hat{\imath} \cdot dx\,\hat{\imath}$$
$$= PA\,dx$$

The product $A\,dx$ is the change in volume dV of the gas. Thus the differential work done by the gas is

$$dW = P\,dV \qquad (13.53)$$

The work done by the gas is *positive* for *increases* and *negative* for *decreases* in the volume of the gas. For finite changes in the volume, we integrate Equation 13.53.

The work done by the gas is

$$W = \int_{V_i}^{V_f} P\,dV \qquad (13.54)$$

Although we formulated this relationship for the work done by the gas using a one-dimensional change, Equation 13.54 is a general result. Since the arguments used to derive it could equally well apply to changes in the volume of liquids or solids, Equation 13.54 also can be used to calculate the work done by solids and liquids if they experience a change in volume caused by (1) thermal expansion or (2) changes of phase.

The pressure and volume of the gas are conveniently plotted on a *P–V* diagram. A gas in thermodynamic equilibrium is represented by a point on such a diagram. Let the gas, initially with volume V_i and pressure P_i, change quasi-statically to another volume V_f and pressure P_f. There are many ways (thermodynamic processes) that the gas can travel between its initial and final states through other states of thermodynamic equilibrium on the *P–V* diagram. We represent one such way schematically in Figure 13.29.

Whatever quasi-static path the gas follows from the initial state to the final state, the work W done by the gas has a convenient geometrical interpretation on a *P–V* diagram:

The work done by the gas is the area under the path on the *P–V* diagram. The area under the path on the *P–V* diagram has the dimensions of pressure times volume or, using SI units, joules:

$$\mathrm{Pa \cdot m^3 = (N/m^2) \cdot m^3 = N \cdot m = J}$$

PROBLEM-SOLVING TACTIC

13.6 Caution! Be sure the volume of the gas is in cubic meters and the pressure in pascals when calculating work (in joules) using the geometric approach on a *P–V* diagram. Many *P–V* graphs have the pressure in atmospheres and the volume in liters; if this is the case, then appropriate unit conversions to pascals and cubic meters must be made, so that the area under the curve emerges as joules.

Let's investigate several special paths connecting an initial and final state on the *P–V* diagram. These paths represent various special thermodynamic processes. We want to find the work done by the gas in each process.

FIGURE 13.28 A gas confined to a cylinder with a movable piston. The gas expands slightly, moving the piston.

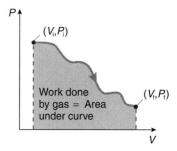

FIGURE 13.29 A gas changes from one state of thermodynamic equilibrium to another.

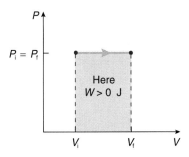

FIGURE 13.30 An isobaric process.

Work Done by the Gas in an Isobaric Process

Let the gas travel along a horizontal path on the P–V diagram, as shown in Figure 13.30. A horizontal path on a P–V diagram is an isobaric process (meaning constant pressure).

The work done *by* the gas is found from Equation 13.54. Since the pressure is constant (and equal to P_i) along an isobaric path, the integral becomes

$$W = P_i \int_{V_i}^{V_f} dV$$

Now we perform the integration.

The work done by the gas is
$$W = P_i(V_f - V_i)$$
$$= P_i \,\Delta V \qquad \text{(isobaric process)} \qquad (13.55)$$

Notice that if the gas expands, $V_f > V_i$, so that $\Delta V > 0$ m³ and the work done *by* the gas is *positive*. For a gas that is compressed, $V_f < V_i$, so that $\Delta V < 0$ m³ and the work done *by* the gas is *negative*. The work done *by* the gas is represented on the P–V diagram by the rectangular area under the isobaric path on the diagram. The area is positive or negative depending on whether the gas expands or is compressed.

Work Done by a Gas in an Isochoric Process

An isochoric process means constant volume. This is represented as a vertical line on a P–V diagram, as shown in Figure 13.31.

Since the volume does not change, Equation 13.54 indicates that the work done *by* the gas in such a process is *zero*. Notice that along an isochoric path segment there is zero area under the path on the P–V diagram. It makes no difference whether the

pressure change is positive or negative: the work done by the gas in an isochoric process is zero.

For an isochoric process, there is zero work done by the gas:
$$W = 0 \text{ J} \qquad \text{(isochoric process)} \qquad (13.56)$$

Work Done by a Gas in an Isothermal Process

In an isothermal process (constant temperature), *both* the pressure and volume change along the path. The initial and final pressures and volumes then are located along the same isotherm on the P–V diagram of the gas, indicated in Figure 13.32.

To calculate the work done by the gas, we begin again with Equation 13.54. The equation of state of the gas relates its pressure, volume, and temperature. For an *ideal* gas, the equation of state is the ideal gas law, Equation 13.20:

$$PV = nRT$$

We solve for the pressure:

$$P = \frac{nRT}{V}$$

Thus the work done by the gas is

$$W = \int_{V_i}^{V_f} P \, dV$$
$$= \int_{V_i}^{V_f} \frac{nRT}{V} \, dV$$

The number of moles n and the gas constant R are constants, and since the process is isothermal, the absolute temperature T also is constant. Hence we have

$$W = nRT \int_{V_i}^{V_f} \frac{dV}{V}$$
$$= nRT(\ln V) \Big|_{V_i}^{V_f}$$

Hence the work done by an ideal gas in an isothermal process is

$$W = nRT \ln \frac{V_f}{V_i} \qquad \text{(ideal gas isothermal process)} \qquad (13.57)$$

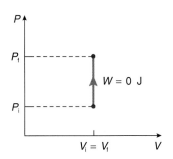

FIGURE 13.31 An isochoric process.

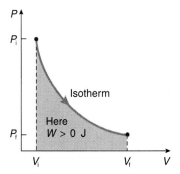

FIGURE 13.32 An isothermal process.

If $V_f > V_i$, the gas expanded and the work done by the gas is positive. Conversely, if the gas is compressed, the work done is negative, since the logarithm then is of a number less than 1.

As before, the work done by the gas is the area under the path on the P–V diagram and is positive or negative depending on whether the gas expands or is compressed.

EXAMPLE 13.21

With the exit valve of a tire pump blocked, you compress a gas isothermally at 300 K from an initial volume of 0.50 liter (at 1.00 atm pressure) to a final volume of 0.25 liter. What is the work done by the gas during its compression? Explain the sign of your answer.

Solution

The work done in an isothermal process is given by Equation 13.57:

$$W = nRT \ln \frac{V_f}{V_i}$$

Everything but the quantity of gas present is known. Use the ideal gas law, Equation 13.20, to determine the number of moles of the gas from the initial conditions:

$$PV = nRT$$

Solving for n, you have

$$
\begin{aligned}
n &= \frac{PV}{RT} \\
&= \frac{(1.00 \text{ atm})(1.013 \times 10^5 \text{ Pa/atm})\,(0.50 \times 10^{-3} \text{ m}^3)}{[8.315 \text{ J/(mol·K)}]\,(300 \text{ K})} \\
&= 2.0 \times 10^{-2} \text{ mol}
\end{aligned}
$$

Note that the volume ratio $V_f/V_i = 0.50$. The work done by the gas then is

$$
\begin{aligned}
W &= (2.0 \times 10^{-2} \text{ mol})[8.315 \text{ J/(mol·K)}](300 \text{ K})(\ln 0.50) \\
&= -35 \text{ J}
\end{aligned}
$$

The work done *by* the gas is negative; the negative result indicates that positive work was done *on* the gas by the agent (you) that compressed it.

13.16 WORK DONE BY A GAS TAKEN AROUND A CYCLE

Engines involve taking a gas around a closed path on a P–V diagram (see Figure 13.33), so that the process can be repeated many times with useful work done around each closed path. A closed curve on a P–V diagram is called a **cycle**, since the gas returns to its initial state (the initial point on the P–V diagram, regardless of where the process begins along the cycle).

Let the gas go around the cycle in the clockwise sense in Figure 13.33. As the gas expands from V_1 to V_2 in the figure, the work W_1 done by the gas is positive; as we have seen, this work is represented by the area under the curve on the P–V diagram. When the gas is compressed from volume V_2 back to V_1, along the lower part of the cycle, the work W_2 done by the gas is negative. This work is represented geometrically by the now negative area under the curve on the P–V diagram, as in Figure 13.34.

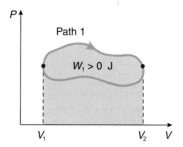

FIGURE 13.33 Positive work is done by the gas in expanding from volume V_1 to V_2.

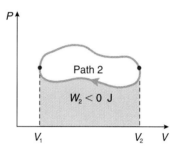

FIGURE 13.34 Negative work is done by the gas when compressed from volume V_2 back to V_1.

Hence the *total work* $W_{total} = W_1 + W_2$ done by the gas in executing the closed curve is represented by *the area enclosed by the cycle curve* on the P–V diagram.

If the cycle is completed in a clockwise manner, as in Figure 13.35, $W_{total} > 0$ J.

Since the gas has returned to the same place on the P–V diagram, the internal energy of the gas is *unchanged* by the whole procedure; that is, $\Delta U = 0$ J. The first law of thermodynamics, Equation 13.51, states that the total heat transfer Q_{total} to the system is equal to the sum of the change ΔU in the internal energy of the system and the work W_{total} done by the system:

$$Q_{total} = \Delta U + W_{total}$$

Because the change in the internal energy is zero for the complete cycle, the first law becomes

$$Q_{total} = W_{total} \qquad \text{(complete cycle)} \qquad (13.58)$$

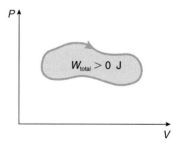

FIGURE 13.35 The total work done during the cycle is the area of the cycle on the P–V diagram and is positive if the path around the cycle is clockwise.

This means that the *total heat transfer* to the system is equal to the *total work* done by the system in the cycle. For the clockwise path indicated in Figure 13.35, the total work done *by* the system is positive; thus the total heat transfer around the cycle is positive (i.e., heat transfer was *to* the system).

QUESTION 10

What can be said about the work done *by* a gas (and the heat transfer to a gas) if the gas is taken counterclockwise around a cycle on a *P–V* diagram?

EXAMPLE 13.22

A gas is taken clockwise around a circular path on its *P–V* diagram between the volumes and pressures shown in Figure 13.36. What is the total work done by the gas around one cycle?

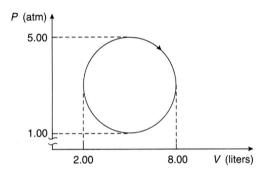

FIGURE 13.36

Solution

The total work done by the gas is the area of the cycle on the *P–V* diagram. Since the path is clockwise, the total work is positive. The area of a circle is π times the square of the radius. Use a pressure for one radius and a volume for the other, as shown in Figure 13.37.

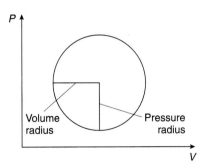

FIGURE 13.37

In order for the work to emerge in joules, the pressure must be expressed in pascals and the volume in cubic meters. Hence the total work done is

$$W_{\text{total}} = \pi(2.00 \text{ atm} \times 1.013 \times 10^5 \text{ Pa/atm})$$
$$\times (3.00 \text{ liters} \times 10^{-3} \text{ m}^3/\text{liter})$$
$$= 1.91 \times 10^3 \text{ J}$$

13.17 APPLYING THE FIRST LAW OF THERMODYNAMICS: CHANGES OF STATE

Here we explore the implications of the first law of thermodynamics for changes of state. When you boil water or melt ice, how much of the heat transfer to the water goes into work by the water (because of its expansion) and how much into changing its internal energy?

Vaporization (Boiling)

To change a system of mass m of liquid to vapor, the heat transfer Q to the system is m times the latent heat of vaporization L_v, as you learned earlier (Equation 13.29):

$$Q = mL_v$$

The process takes place at constant pressure (typically one atmosphere), and so the work done *by* the system is the work in an isobaric process (Equation 13.55):

$$W = P \Delta V$$

where ΔV is change in the volume—that is, the difference between the final volume (the volume of the vapor) V_{vapor} and the initial volume (the volume of the liquid) V_{liquid}:

$$W = P(V_{\text{vapor}} - V_{\text{liquid}})$$

Notice that this work is positive since $V_{\text{vapor}} > V_{\text{liquid}}$.

The first law of thermodynamics, Equation 13.51, states that

$$Q = \Delta U + W$$

Thus the change in the internal energy of the system is

$$\Delta U = Q - W$$

or

$$\Delta U = mL_v - P(V_{\text{vapor}} - V_{\text{liquid}})$$

To get a feeling for the relative numerical values of these quantities, consider the vaporization of $1.00 \text{ g} = 1.00 \times 10^{-3} \text{ kg}$ of water at 100 °C to steam at 100 °C. The heat transfer needed is $Q = mL_v$, where the latent heat of vaporization is found in Table 13.5:

$$Q = (1.00 \times 10^{-3} \text{ kg})(22.57 \times 10^5 \text{ J/kg})$$
$$= 2.26 \times 10^3 \text{ J}$$

To calculate the work done *by* the system in the vaporization, we need to find the volume of $1.00 \times 10^{-3} \text{ kg}$ of steam at 100 °C. We use the ideal gas law (Equation 13.20):

$$PV = nRT$$

$$V = \frac{nRT}{P}$$

A mole of water (H_2O) has a mass of 18 g. So $1.00 \times 10^{-3} \text{ kg} = 1.00 \text{ g}$ represents 1/18 mol = 0.0556 mol. The pressure is one atmosphere:

$$P = 1.000 \text{ atm}$$
$$= 1.013 \times 10^5 \text{ Pa}$$

and the absolute temperature at which the vaporization takes place is $T = 373$ K. The gas constant has the value $R = 8.315$ J/(mol·K). We substitute into the ideal gas law to find the volume of the vapor:

$$V_{vapor} = \frac{nRT}{P}$$

$$= \frac{(0.0556 \text{ mol})[8.315 \text{ J/(mol·K)}](373 \text{ K})}{1.013 \times 10^5 \text{ Pa}}$$

$$= 1.70 \times 10^{-3} \text{ m}^3$$

Since the density of water is 1.00×10^3 kg/m^3 = 1.00 g/cm^3, the volume of the liquid (1.00 g of water) is 1.00 cm^3, or

$$V_{liquid} = 1.00 \times 10^{-6} \text{ m}^3$$

Thus the work done *by* the water system in the vaporization is

$$W = P(V_{vapor} - V_{liquid})$$

where P is one atmosphere. That is, the work is

$$W = (1.013 \times 10^5 \text{ Pa})(1.70 \times 10^{-3} \text{ m}^3 - 1.00 \times 10^{-6} \text{ m}^3)$$

$$= 172 \text{ J}$$

The work done *by* the system is positive since the volume of the system increased.

According to the first law of thermodynamics,

$$Q = \Delta U + W$$

or

$$2.26 \times 10^3 \text{ J} = \Delta U + 172 \text{ J}$$

Thus the change in the internal energy of the system is

$$\Delta U = 2.09 \times 10^3 \text{ J}$$

Notice that most of the heat transfer went into the change in the internal energy of the system; indeed, 92% went to change the internal energy and only 8% went into work done by the system on its surroundings. In vaporization, much energy is needed to weaken the intermolecular bonds within the liquid.

Melting

This process is similar to the vaporization process. The heat transfer needed to melt a mass m is the product of the mass and the latent heat of fusion (Equation 13.29):

$$Q = mL_f$$

The work done *by* the system (since the process is accomplished at constant pressure) is

$$W = P(V_{liquid} - V_{solid})$$

The first law of thermodynamics states that

$$Q = \Delta U + W$$

and so the change in the internal energy of the system is

$$\Delta U = Q - W$$

$$= mL_f - P(V_{liquid} - V_{solid})$$

Once again, let us get a feeling for the sizes of the various terms by considering the melting of 1.00 g = 1.00×10^{-3} kg of ice at 0 °C to liquid water at 0 °C. The heat transfer needed to the ice is

$$Q = mL_f$$

where the latent heat of fusion of water is found in Table 13.5:

$$Q = (1.00 \times 10^{-3} \text{ kg})(3.335 \times 10^5 \text{ J/kg})$$

$$= 334 \text{ J}$$

To calculate the work, we need to know the difference in volume between 1.00 g of ice and 1.00 g of water. We can get this from the densities. The density of ice is 920 kg/m^3, and so 1.00×10^{-3} kg of ice has a volume of

$$V_{solid} = \frac{1.00 \times 10^{-3} \text{ kg}}{920 \text{ kg/m}^3}$$

$$= 1.09 \times 10^{-6} \text{ m}^3$$

The density of water at 0 °C is 1.00×10^3 kg/m^3, so that the volume of one gram of water is

$$V_{liquid} = 1.00 \times 10^{-6} \text{ m}^3$$

Thus the work done *by* the system in melting is

$$W = P(V_{liquid} - V_{solid})$$

$$= (1.013 \times 10^5 \text{ Pa})(1.00 \times 10^{-6} \text{ m}^3 - 1.09 \times 10^{-6} \text{ m}^3)$$

$$= -9 \times 10^{-3} \text{ J}$$

Notice that the work done *by* the system here is negative since the system *decreased* its volume in melting (the ice-to-water phase transition is one of just a few that have this characteristic). The work done by the system is quite small compared with the heat transfer to the system. Indeed the absolute value of the work done by the system is only about 0.003% of the heat transfer to melt the ice.

The first law of thermodynamics states that

$$Q = \Delta U + W$$

The change in the internal energy of the system is

$$\Delta U = Q - W$$

$$= 334 \text{ J} - (-9 \times 10^{-3} \text{ J})$$

The work term is very small compared with Q, and so

$$\Delta U \approx 334 \text{ J}$$

Essentially all of the heat transfer to the system in the melting goes into changing the internal energy of the system, rather than work done by it. This is a reasonable result since the change in the volume of the system in melting is much smaller than for vaporization.

CHAPTER SUMMARY

The fundamental SI unit for measuring temperature is the *kelvin* (K). The temperature of the *triple point* of water is defined to be 273.16 K, and so the size of a kelvin (degree) is 1/273.16 of the temperature of the triple point of water.

A temperature scale of convenience is the *celsius scale*. Temperatures $t_{celsius}$ on the celsius scale are related to those on the kelvin scale T by

$$T = t_{celsius} + 273.15 \text{ K} \tag{13.2}$$

The size of a celsius degree is equal to the size of a kelvin, and so temperature *changes* have the same numerical value using both temperature scales: $\Delta T = \Delta t_{celsius}$.

Two systems at the same temperature are said to be in *thermal equilibrium* with each other.

Heat is defined as that (energy) which is transferred between two systems simply because they are at *different temperatures*. To emphasize that heat really is a process between two systems rather than a characteristic property of either system, we prefer to use the term *heat transfer* instead of the single word heat as a noun. By convention, heat transfer Q *to* a system is considered positive; heat transfer *from* a system is negative.

The *zeroth law of thermodynamics* states that if system A and system B are each in thermal equilibrium with a third system C, then systems A and B are in thermal equilibrium with each other.

A temperature change causes a change in the size of materials. A change in length $\Delta\ell$ is found using

$$\Delta\ell = \alpha\ell\,\Delta T \tag{13.6}$$

where ℓ is the original length of the material and α is the temperature *coefficient of linear expansion*. The temperature *coefficient of area expansion* is twice the coefficient of linear expansion, so that changes in area are found using

$$\Delta A = (2\alpha)A\,\Delta T \tag{13.9}$$

The *coefficient of volume expansion* is three times the coefficient of linear expansion; thus, changes in volume are found using

$$\Delta V = (3\alpha)V\,\Delta T \tag{13.10}$$

Holes exhibit thermal expansion as if the hole were filled with the material that surrounds the hole.

An *ideal gas* has an *equation of state* that relates the pressure, volume, absolute temperature, and the quantity of gas:

$$PV = nRT \tag{13.20}$$

where n is the number of moles of gas and R is the *universal gas constant*:

$$R = 8.314\ 52 \text{ J/(mol·K)} \tag{13.19}$$

Real gases are approximately ideal gases at low pressures (up to several atmospheres).

Heat transfer Q to a system (expressed in joules) can change the temperature of the system and/or its phase. If no phase change occurs, the change in temperature is found using

$$Q = mc\,\Delta T \tag{13.22}$$

where c is the *specific heat* of the material:

$$c \equiv \frac{1}{m}\frac{dQ}{dT} \tag{13.21}$$

You also may use the molar specific heat:

$$c_{molar} = \frac{1}{n}\frac{dQ}{dT} \tag{13.24}$$

in which case the heat transfer to the system is

$$Q = nc_{molar}\,\Delta T \tag{13.23}$$

Heat transfer to a system at its freezing or boiling temperature changes the *phase* of the material without a change in its temperature. The heat transfer Q needed to change the phase of a mass m of a substance is

$$Q = mL \tag{13.29}$$

where L is the *latent heat* associated with the phase transition.

An ideal thermodynamic *reservoir* is a system that maintains a fixed temperature despite any heat transfer to or from it.

Heat flow is the time rate of heat transfer and is expressed in watts:

$$\frac{dQ}{dt}$$

Heat flow by *conduction* between two reservoirs at different temperatures is proportional to the temperature gradient

$$\frac{dT}{dx}$$

and is

$$\frac{dQ}{dt} = -kA\frac{dT}{dx} \tag{13.32}$$

where A is the area of thermal contact with each reservoir and of the conductor's cross section, and k is the *thermal conductivity* of the material connecting the two reservoirs. The steady-state heat flow is

$$\frac{dQ}{dt} = A\frac{T_H - T_C}{R} \tag{13.37}$$

where R is the *R-value* of the material connecting the warmer reservoir at temperature T_H with the cooler reservoir at temperature T_C. The total R-value of materials in series is the sum of the R-values. For materials in parallel, the heat flows add.

Heat transfer also may occur via radiation, in which case the heat flow radiated by a system at absolute temperature T is proportional to the fourth power of T, which is *Stefan's law*:

$$\frac{dQ}{dt} = -eA\sigma T^4 \tag{13.46}$$

where e is the *emissivity* of the object (a dimensionless number between 0 and 1), A is its surface area, and σ is the *Stefan–Boltzmann constant*. An object with an emissivity $e = 1$ is known as a *blackbody*.

The *internal energy* U of a system is the sum of the many microscopic kinetic energy and potential energy terms associated with all the individual particles (atoms and/or molecules) of the system. It is only *changes* in the internal energy that are physically significant on a macroscopic basis.

The most general statement of energy conservation in physics is

$$Q + W' = \Delta U + \Delta E \tag{13.48}$$

where Q is the heat transfer to a system, W' is the work done on the system by macroscopic nonconservative forces, ΔU is the change in the internal energy U of the system, and ΔE is the change in the total mechanical energy E of the system ($E = \text{KE} + \text{PE}$).

The CWE theorem of classical mechanics neglects thermal effects ($Q = 0$ J and $\Delta U = 0$ J in Equation 13.48):

$$W' = \Delta E \tag{13.49}$$

For "pure" thermodynamics, there is no change in the total mechanical energy of the system ($\Delta E = 0$ J). The *first law of thermodynamics*, a restatement of the general energy conservation principle for pure thermodynamic systems, then becomes

$$Q = \Delta U + W \qquad (13.51)$$

where W is the work done *by* the system ($= -W'$) on its surroundings.

The work done *by* a system is

$$W = \int_{V_i}^{V_f} P \, dV \qquad (13.54)$$

Geometrically, this is represented by the area under the path that the system follows on a *P–V* diagram between the initial and final states of thermodynamic equilibrium. The work is positive for expansions in volume and negative for compressions.

For an isobaric process, the work done by a system is

$$W = P_i \, \Delta V \qquad \text{(isobaric process)} \qquad (13.55)$$

For an isochoric process, the work done by the system is zero:

$$W = 0 \text{ J} \qquad \text{(isochoric process)} \qquad (13.56)$$

For an isothermal process, the work done by an ideal gas system is

$$W = nRT \ln \frac{V_f}{V_i} \qquad \text{(ideal gas isothermal process)} \qquad (13.57)$$

The work done by a system in executing a cycle on a *P–V* diagram is equal to the area of the cycle on the *P–V* diagram and is positive (negative) if the system executes the cycle in the clockwise (counterclockwise) fashion.

SUMMARY OF PROBLEM-SOLVING TACTICS

13.1 (page 594) If you recall that the sizes of the degrees on the celsius and fahrenheit scales differ by the factor 9 °F/5 °C or 5 °C/9 °F, depending on which way the comparison is made, you should be able to deduce either Equation 13.4 or 13.5 without memorizing them.

13.2 (page 600) Be careful with the units when using the ideal gas law. To use the value of $R = 8.314\ 52$ J/(mol·K), the pressure P must be in pascals, V in cubic meters, and n in moles.

13.3 (page 601) When using the ideal gas law, be sure to express the temperature in kelvin (*not* degrees celsius).

13.4 (page 602) Be careful in looking up the specific heat of a substance in tables in technical literature; other common energy and mass units also are used.

13.5 (page 604) There are no ready signals in Equations 13.29 and 13.30 to indicate whether Q should be positive or negative. The heat transfer Q is positive for heat transfer *to* a system and negative for heat transfer *from* a system.

13.6 (page 620) Caution! Be sure the volume of the gas is in cubic meters and the pressure in pascals when calculating work (in joules) using the geometric approach on a *P–V* diagram.

QUESTIONS

1. (page 588); 2. (page 590); 3. (page 590); 4. (page 594); 5. (page 598); 6. (page 598); 7. (page 603); 8. (page 613); 9. (page 618); 10. (page 623)

11. If a system is hot, can you say that it "has" a large amount of heat transfer? Explain your answer carefully.

12. The celsius and kelvin temperature scales yield quite different numerical values for the temperature of a system in thermodynamic equilibrium. Why then are temperature *changes* the same on both scales?

13. When two pieces of steel are being riveted together, it is common to use hot rivets. Explain why.

14. Why is it incorrect to use the celsius temperature scale in the ideal gas law rather than the kelvin scale, even though the size of a degree on both scales is the same?

15. Using a good dictionary such as the *Oxford English Dictionary*, trace the roots of the word *degree*.

16. For the common mercury laboratory thermometer, is the thermometric property the length of thin mercury column or the total volume of mercury in the mercury bulb at its base and in the column? Explain.

17. The tile floor in a bathroom is in thermal equilibrium with a fleecy bathroom rug on the floor. When you step on the tile floor, it *feels* much colder than the rug even though both are at the same temperature. Explain why.

18. Stainless steel pots occasionally are manufactured with a copper cladding on the bottom of the outside of the pot. In view of the relative thermal conductivities of these metals (see Table 13.6), why is this done? Is the reason purely aesthetic?

19. A common laboratory mercury thermometer initially at 20 °C suddenly is placed in boiling water to measure the temperature. The mercury column initially *descends* slightly before shooting upward. Explain why. See if you can observe this effect.

20. Explain how a blacksmith can fit an iron hoop as a rim over a wooden wheel so that the iron rim grips the wheel without additional fastenings.

21. If a pendulum clock is adjusted to keep accurate time at 20 °C, will the clock run slow, fast, or be accurate at 30 °C?

22. In northern climates, explain why it is good to tune pianos at least twice a year. At what times of the year should the tuning typically be done?

23. When home canning fruits and vegetables, jars are filled to within about one centimeter of the top rim. The lids are placed loosely on the jars and secured loosely with a screw cap. The jars then are covered completely with water and boiled vigorously for about 15 minutes. They then are removed and the cover is screwed on tight. Explain how this process creates a partial vacuum in the jar and how, when the jars cool, atmospheric pressure helps keep the jars sealed.

24. A tight-fitting, twist-off jar lid occasionally can be loosened by holding the lid under very hot water. Explain why.

25. Explain why a single block of ice melts more slowly than the same mass of ice in the form of ice cubes when both are placed in a warm room.

26. A heat-gun paint stripper can easily bubble old paint applied on wood surfaces but is unable to blister or bubble the same paint applied to metallic surfaces. Explain why.

27. Why is it more dangerous to be burned by steam at 100 °C than by boiling water at the same temperature?

28. Will coffee with cream cool faster than, slower than, or at the same rate as the same amount of black coffee in an identical cup?

29. Explain why the information on this highway sign is accurate during cold rainy or snowy weather:

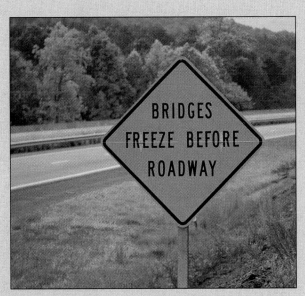

FIGURE Q.29

30. What physical property of a pot holder prevents you from getting burned when taking a scalding dish out of a hot oven?

31. Does the value of the coefficient of thermal expansion depend on the particular units used to measure the lengths?

32. If the temperature is measured using fahrenheit degrees, by what factor do the coefficients of linear expansion in Table 13.1 change?

33. The water at great depths in the Great Lakes has a temperature of about 4 °C year round. Explain why.

34. In northern climates, explain why it is advisable to drain all water from pipes in unheated summer cottages for the winter.

35. In the northern hemisphere, the Sun is above the local horizon for the greatest interval on the day of the summer solstice (~ 22 June) yet maximum average daily temperatures of lakes and the ocean and on land typically are in late July and early August. Explain what causes the time lag.

36. Explain why the measurement of the change in volume of a liquid when its temperature is raised is not as accurate using a calibrated, steel, graduated cylinder as when a graduated cylinder of pyrex glass is employed.

37. When fruit blossoms are in danger of freezing because of late spring frosts, farmers spray the trees and blossoms with water. Why might this procedure succeed in saving the blossoms from freezing?

38. A thermos bottle containing a cold beverage is shaken vigorously and the temperature of its contents rises slightly. Is there heat transfer to the system in this process? What causes the rise in temperature of the liquid?

39. When a mass m of ice melts at 0 °C to water at 0 °C, there is positive heat transfer to the system (equal to $Q = mL_{f\,water}$). Yet heat transfer was defined as energy transfer that arises solely because of a *difference* in temperature. Resolve this apparent dilemma.

40. Explain how it is possible to increase the temperature of a cup of coffee *without* heat transfer to it.

41. Would a retail gasoline dealer in petroleum be more inclined to want a delivery of petroleum from a wholesaler on a very hot afternoon or on a cool morning?

42. A thermometer placed in direct sunlight indicates a temperature higher than that of the surrounding air. Explain why.

43. Your index finger is a crude thermometer. Describe some circumstances under which your finger can tell approximately if two objects are at the same temperature and circumstances when the finger thermometer is quite inaccurate.

44. When a furnace increases the temperature inside a house, the volume of the gas inside the house does not change, nor does the pressure change. Does this violate the ideal gas law? Explain.

45. Imagine a thermodynamic system divided into smaller regions of various sizes. A thermodynamic parameter is called *intensive* if its value is independent of the mass of the sample in which it is measured. A thermodynamic parameter is *extensive* if its value depends on the mass of the region in which it is measured. Classify the following parameters as intensive, extensive, or *neither*: pressure; volume; internal energy; temperature; mass; work; heat transfer; molar mass.

46. Explain why it is easier to measure c_P rather than c_V for solids and liquids.

47. Write a paragraph to distinguish among temperature, heat transfer, and internal energy.

48. Your instructor asks you to sketch a graph of PV/T versus P for a sample of an ideal gas. Go to it. What is the value of the vertical intercept?

49. Does the length of the neck of a giraffe contract if the ambient temperature falls? What about the length of a snake?

50. A deep snowfall can protect many plants from freezing. Explain how.

51. Explain why a pressure cooker is able to cook beets in fifteen minutes while it might take two hours if they were simply boiled in a simple covered pot.

52. Why does freezing typically kill?

53. Can *you* be considered to be a thermodynamic reservoir?

54. Heat transfer causes so-called "thermal patterns" in snow. Cite several examples of common thermal patterns in snow.

55. What property of a system is measured by a thermometer?

56. Two systems in thermal contact have different values of the property measured by a thermometer. As a consequence, what is transferred between the two systems?

PROBLEMS

Sections 13.1 Simple Thermodynamic Systems
     13.2 Temperature
     13.3 Work, Heat Transfer, Temperature, and Thermal Equilibrium
     13.4 The Zeroth Law of Thermodynamics
     13.5 Thermometers and Temperature Scales
     13.6 Temperature Conversions Between the Fahrenheit and Celsius Scales*

1. The normal average human body temperature is 98.6 °F. What is the corresponding temperature on the celsius scale?

2. A student walks into the campus infirmary and Dr. Payne quickly assesses that the temperature of the poor soul is 103.5 °F. What is this temperature on the celsius scale?

3. Calculate the fahrenheit temperature corresponding to the absolute zero of temperature 0 K.

4. What kelvin temperature corresponds to 0 °F?

5. You awaken to the infernal clanging of your alarm clock and quickly get ready to trek off to another thrilling 8 A.M. physics class. A glance at the thermometer outside your dorm window indicates that the temperature is a bone-chilling −25 °C. Your roommate wants to know what the temperature is on the fahrenheit scale. Determine it.

•6. At what temperature is the numerical value the same on the celsius and fahrenheit scales?

•7. At what temperature is the numerical value the same on the kelvin and fahrenheit scales?

•8. A physicist devises a customized temperature scale that assigns a temperature reading of zero to the normal melting point of mercury (−39 °C) and a temperature reading of 100 to its boiling point (357 °C). (a) Formulate an equation to convert from degrees celsius to the degrees on this custom scale. (b) What temperature on the custom scale corresponds to 0 °C?

•9. At what temperature celsius is the sum of the celsius temperature, the kelvin temperature, and the fahrenheit temperature equal to 495.15?

•10. Wow are you hungry! You rush off to the Sweet Things ice cream shop and devour a 2000 Calorie ice cream sundae. Yum! (a) How many joules of energy does this food represent? (b) How many 10.0 kg cement blocks must you lift through a vertical distance of 2.00 m to do an equivalent amount of work on the blocks? This calculation indicates that it is much easier to lose weight simply by not eating highly Caloric foods rather than by exercising to "work it off."

Section 13.7 Thermal Effects in Solids and Liquids: Size

11. A steel beam of length 30 m is used to span a creek. Provisions must be made for changes in the length of the beam in the supporting structure. How much will the length of the beam change if the temperature variation between winter and summer ranges from −20 °C to 40 °C?

12. A copper electrical cable of length 0.500 km is strung over a river between two electrical pylons. If the temperature changes from −10 °C to +40 °C between winter and summer, by how much does the length of the cable change?

•13. A 100.000 m steel measuring tape was marked and calibrated at 20 °C. At a temperature of 30 °C, what is the percentage error in a 100.000 m distance when using this tape?

•14. An optical engineering firm needs to ensure that the separation of two mirrors is unaffected by temperature changes. Such a device is called a Fabry–Perot interferometer. The mirrors are attached to the ends of two bars of different materials that are welded together at one end as shown in Figure P.14. The surfaces of the bars in contact are lubricated. (a) Show that the distance ℓ does not change with temperature if $\alpha_1\ell_1 = \alpha_2\ell_2$, where α_1 and α_2 are the respective thermal coefficients of linear expansion. (b) If the materials are brass and steel, which material should be the shorter bar? (c) If the distance ℓ_1 is 2.66 cm at 20 °C, what is ℓ_2 and what is ℓ?

FIGURE P.14

•15. (a) Calculate the length of a pendulum that has a period equal to 1.000000 s. Assume *g* is exactly 9.81 m/s². (b) If the mass on the end of the pendulum is supported by means of a steel wire, what is the period of the pendulum if the temperature increases by 10 °C? (c) Does this result in the clock running slow or fast? (d) After what time interval will the clock differ from the correct time by one minute?

•16. A clock has a pendulum supported by a steel wire. The pendulum has a period of precisely 1.000 000 s at 0 °C. (a) As the temperature rises, will the clock run slow or fast? (b) Use the expression for linear expansion to derive an equation for the *change* in the period of the pendulum as a function of the temperature change from 0 °C. (c) The clock runs accurately at 0 °C. How many minutes per day will it gain or lose at 30 °C?

•17. A stretched wire precisely 1.000 m in length is tuned to play middle C (261.6 Hz). (a) If the tension in the wire is kept constant, by how much is it necessary to change the length of the wire so the note is flat by 0.20 Hz? (b) If the wire is made of steel, what temperature change is needed to accomplish this change in length? Temperature-induced changes in the tuning of stringed instruments are more complicated than this example indicates. An increase in the length of the string induced by a rise in temperature changes the tension in the string because there is more slack in the string.

•18. Imagine a hypothetical steel wire girdling the globe around the equator. Assume the Earth is a perfect sphere of radius 6370 km. The wire is at a uniform temperature of 25.0 °C. What temperature change is needed to increase the length of the wire by 1.0 km?

•19. The railroad and highway bridge over the dramatic gorge just downstream from Victoria Falls (also known as *mosi oa tunya*, "the smoke that thunders") in Zimbabwe was constructed at the turn of the 19th to 20th centuries. See Figure P.19. The steel bridge was fabricated in England and shipped in pieces for assembly on site. Working from each bank, the last section to be inserted was the center span. Unfortunately, it was about 10 cm too long to drop into place because of the sag of the halves of the bridge protruding from each bank. To pry apart the two ends of the bridge, long steel cables were attached to each and anchored inland. The cables were tightened during the hottest part of the day, when the temperature was about 25 °C. The following morning the temperature was 10 °C. The thermal contraction of the cables pried the two sections of the bridge apart just enough for the center span to be inserted. What length steel cable was used for each to contract 5 cm? Neglect any thermal contraction of the bridge structure itself.

Figure P.19

•20. A solid brass sphere has a radius of 5.00 cm at 20 °C. (a) Calculate the area of the sphere. (b) Calculate the volume of the sphere. (c) The temperature of the sphere is increased to 150 °C. Find the change in the radius, area, and volume of the sphere.

•21. Under a spreading chestnut tree, a farrier wants to put a steel rim on a wagon wheel. The wheel is 1.000 m in diameter but the rim is only 0.995 m in diameter. What temperature change of the rim is necessary for it to just fit over the wheel?

•22. (a) Calculate the surface area of a steel cubical box 0.500 m on a side. (b) If the temperature increases by 20 °C, by what percentage does the surface area increase? (c) By what percentage does the volume increase?

•23. The density of a solid is the amount of mass per cubic meter of the solid. Since a change in temperature changes the volume of a solid, the density also is temperature dependent. If we let ρ be the density of the solid at temperature T, show that the change in the density of the solid when its temperature changes by ΔT can be expressed as

$$\Delta\rho = -3\alpha\rho \, \Delta T$$

where α is the coefficient of linear expansion. Why is the minus sign necessary?

•24. On a chilly 0 °C day, you purchase 50.0 liters of gasoline at a pump that is assumed to measure volume accurately at all temperatures. You casually note that the thermal coefficient of volume expansion for gasoline is 0.0012 K^{-1}, duly posted on the pump to repel tort lawyers. (a) Will you get more or fewer *kilograms* of gasoline when you buy another 50.0 liters on a hot day at 30 °C? (b) What is the difference in the *volumes* of the two purchases in liters after both the purchases cool to 0 °C? Specify which volume is greater. Ignore evaporation effects.

•25. The area of a circle is $A = \pi r^2$. If the radius changes by a small amount Δr, show that the small change ΔA in the area of the circle is approximately

$$\Delta A = 2\pi r \, \Delta r$$

If the change in radius comes about because of a temperature change ΔT, show that the foregoing expression yields

$$\Delta A = (2\alpha)A \, \Delta T$$

where α is the (constant) coefficient of linear expansion. (Hint: Differentiate the expression for the area with respect to r and let the differentials represent the small changes in radius and area.) Thus Equation 13.9 is not restricted to a specific shape for the area.

•26. The volume of a sphere is $V = (4/3)\pi r^3$. Show that if the radius changes by a small amount Δr, the change in the volume ΔV is approximately

$$\Delta V = 4\pi r^2 \, \Delta r$$

If the change in radius occurs because of a change in temperature, show that the foregoing expression becomes

$$\Delta V = (3\alpha)V \, \Delta T$$

where α is the (constant) coefficient of linear expansion. (See the hint in Problem 25.)

•27. Use the graph in Figure 13.15 to determine the range for the average density of an object that will float in water at 10.0 °C yet sink in hot water at 90.0 °C. How might you construct a device of the appropriate average density that will demonstrate this effect?

•28. A thin rod of length ℓ and cross-sectional area A is at temperature T. The rod is clamped at both ends to *prevent* thermal expansion or contraction. If the temperature now is decreased, the tension in the rod must increase by an amount sufficient to produce the *same* fractional change in length as would occur if the rod could contract as a result of the temperature decrease. Show that the increase in the magnitude of the tension in the rod is

$$\alpha AE \,|\Delta T|$$

where E is Young's modulus (see Chapter 11).

•29. A 2.00 m steel rod is of radius 0.0050 m. The temperature of the rod is cooled from 100 °C to 20 °C. What force is needed to *prevent* the rod from changing its length?

Section 13.8 Thermal Effects in Ideal Gases

30. A 3.00 liter sample of an ideal gas has a pressure of 2.00 atm at 10 °C. If the volume remains constant, what is the pressure of the gas when the temperature is 40 °C?

31. A 2.50 mol sample of an ideal gas at 1.00 atm pressure has a temperature of 20 °C. What is the volume occupied by the gas?

32. A sample of an ideal gas occupies a volume of 4.00 liters at 20.0 °C with a pressure of 3.00 atm. How many moles are in the sample?

•33. For an ideal gas, show that the density of the gas at constant temperature is proportional to the pressure.

•34. An air bubble released from a scuba diver at a depth of 35 m rises to the surface. The water temperature is 5.0 °C in the vicinity of the diver and 20.0 °C at the surface. If you assume that the bubble is in thermal equilibrium with the water at each position, by what factor does the volume of the bubble change as it rises to the surface? Assume the air is an ideal gas.

•35. A spherical bubble of radius r_0 has a volume V_0 and pressure P_0 when just below the surface of a liquid of density ρ; see Figure P.35. Let F_0 be the magnitude of the buoyant force on the bubble when it is at this location. Assume P_0 is atmospheric pressure and that the temperature of the liquid and bubble do not change with depth. Let the y-axis be vertical, with $\hat{\jmath}$ directed up and the origin at the surface. Notice that positions

within the liquid correspond to negative values of y. (a) Show that the magnitude of the buoyant force on the bubble as a function of its vertical position $|y| \gg r_0$ in the liquid is

$$F_0 \,\frac{P_0}{P_0 - \rho g y}$$

(b) If the liquid is water, show that the buoyant force is half the surface value at a depth of 10.3 m.

•36. (a) For an ideal gas, show that the coefficient of volume expansion at constant pressure, defined by

$$\frac{1}{V}\left(\frac{dV}{dT}\right)_{\!P}$$

is the reciprocal of the absolute temperature. Evaluate the numerical value of the coefficient of volume expansion at constant pressure for an ideal gas at 0 °C. (b) If the gas is confined to a tube whose length is much greater than its diameter, what is the coefficient of linear expansion (at constant pressure) for the gas in the tube?

•37. The ambient temperature in outer space is about 3.0 K. Most of the matter in space is hydrogen gas at a density of about one atom per 3.0 m³. What is the pressure of this dilute gas if it is in thermal equilibrium at the ambient temperature?

•38. A florist has a 5.00×10^{-2} m³ tank of helium gas at a pressure of 100 atm. Spherical balloons are filled from the tank to a pressure of 1.20 atm and have a diameter of 30 cm. How many balloons can be filled from the tank?

•39. A sophisticated laboratory vacuum system creates a partial vacuum with a pressure of 2.00×10^{-6} atm in a 1.50×10^{-3} m³ chamber. The chamber has a temperature of 20.0 °C. How many gas particles are in the chamber?

•40. A cylinder with a diameter of 5.00 cm has a movable piston attached to a spring as shown in Figure P.40. The cylinder contains 1.50 liters of an ideal gas at 20.0 °C and 1.00 atm pressure. Under these conditions, the spring is unstretched. The temperature of the gas is increased to 100 °C and it is found that the spring compresses 3.50 cm. Determine the spring constant k.

FIGURE P.35

FIGURE P.40

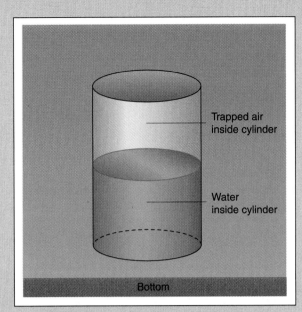

Trapped air
inside cylinder

Water
inside cylinder

Bottom

FIGURE P.41

•41. Early devices for going underwater were called diving bells. They were used for salvage purposes before the advent of modern diving equipment, but were quite limited in the depth they could attain. A cylindrical diving bell has a closed upper end and an open lower end. (See Figure P.41.) At the surface, air inside the cylinder has a temperature of 20.0 °C. The cylinder is lowered into the cold waters of Lake Superior. At what depth has the water risen halfway into the cylinder? Assume the gas in the submerged bell is in thermal equilibrium with the water at 4.0 °C.

Section 13.9 Calorimetry

42. What heat transfer to a 2.0 kg pot of water is necessary to raise its temperature from 20 °C to 100 °C for a coffee break? Neglect the mass of the pot.

43. What heat transfer to 2.0 kg of ice at 0 °C is needed to melt it to water at 0 °C?

•44. You have 0.50 kg of water at 20.0 °C. Assuming no heat transfer to the environment, how much ice at −10.0 °C should be added to chill the water to 0.0 °C with no ice remaining?

•45. A 0.500 kg block of ice at −10.0 °C is placed in a punch bowl with 4.00 kg of water (punch) at 20.0 °C. Assume no heat transfer to the surroundings and neglect the thermal effects of the punch bowl itself. (a) Does all the ice melt? If not, how much is left? (b) What is the final temperature of the punch plus ice system?

•46. A 2.0 kW electric heater is used in a well-insulated hot water tank of capacity 200 liters to raise the temperature of water from 5.0 °C to 60.0 °C. How long will it take to change the temperature of the water in the tank?

•47. An immersion heater supplies a constant 100 W to 0.200 kg of water initially at 10 °C. Assume there is no heat transfer to the surrounding environment. What time elapses until there is 0.190 kg of liquid water left?

•48. You are very particular about the temperature of beverages. You want to cool 2.00 liters of hot tea (consider it to be water) from a temperature of 98.0 °C to 1.0 °C by adding ice from a freezer at −20.0 °C. How much ice should be added to accomplish the task? Neglect any heat transfer to the environment.

•49. A typical household water heater has a capacity of about 0.50 m³ of water. The water is initially at a temperature of 15 °C and is warmed to 60 °C for the pleasure of long, hot showers. (a) What heat transfer is necessary to warm the water? (b) If the heating element is rated at 5.0 kW, how long will it take to warm the water?

‡50. Dense materials often have specific heats expressed in joules per kilogram-kelvin [J/(kg·K)]. Among the materials listed here, test the possibility that equal volumes of different materials might require equal heat transfer to raise the temperature by one kelvin. Calculate the specific heat of each in J/(m³·K). Present your results in an orderly table and state your conclusions. What conclusions can be drawn if a comparison is made among the specific heats of these substances in joules per mole-kelvin [J/(mol·K)]?

Substance	Specific heat [J/(kg·K)]	Density (kg/m³)	Molar mass (kg/mol)	Specific heat [J/(m³·K)]	Specific heat [J/(mol·K)]
Aluminum	896	2.70×10^3	27.0×10^{-3}		
Copper	386	8.9×10^3	63.5×10^{-3}		
Iron	481	7.8×10^3	55.8×10^{-3}		
Lead	128	11.0×10^3	207×10^{-3}		
Mercury	138	13.6×10^3	201×10^{-3}		

‡51. At very low temperatures, near 0 K, the specific heat [in J/(kg·K)] of solids is strongly temperature dependent and is given by the *Debye equation*:

$$c = \beta T^3$$

where β is a constant and T is the absolute temperature. What heat transfer is necessary to raise the temperature of a solid of mass m from 2.00 K to 5.00 K? Your result will be in terms of m and β.

Sections 13.10 Reservoirs
13.11 Mechanisms for Heat Transfer*

52. If the absolute temperature of a blackbody radiator is doubled, by what factor does the heat flow radiated by the body increase? Can the same be said if the celsius temperature of the blackbody radiator is doubled?

53. A mobile home is 3.0 m wide, 2.5 m high, and 25.0 m long. See Figure P.53. The average R-value for the walls, roof, and floor of the dwelling is 3.0 m²·K/W. If the temperature inside the home is kept at 20 °C, find the heat flow to the exterior on a cold winter night with an outside temperature of −20 °C.

FIGURE P.53

54. It is −20 °C outside but nice and toasty inside (20 °C) where you are diligently working on physics homework due tomorrow. The windowless outside wall in your dormitory cell has dimensions 2.40 m by 3.60 m and has an effective R-value of 6.0 m²·K/W. Calculate the heat flow through the wall.

•55. Two reservoirs, one at 0.0 °C and the other at 100.0 °C, are connected by means of two materials as indicated in Figure P.55. The relative R-values of the materials are specified. What is the temperature at the interface between the two materials?

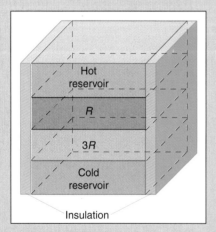

FIGURE P.55

•56. The metabolic rate of an average, clothed adult human being is about 100 W (this number corresponds to a diet of about 2000 kcal/day per human). This means that a human being is like a 100 W heater. Twenty-nine students and a professor are in a classroom 10.0 m long, 10.0 m wide, and 5.0 m high, with an air temperature initially at 20.0 °C. Assume the classroom is well insulated. Neglecting any heat transfer between the air and the walls, calculate how much the temperature of the air in the room rises over the course of a 55 min lecture. Assume the process takes place at constant volume in an airtight lecture room The specific heat of air at constant volume is about 20 J/(mol·K). A mole of air represents about 24.1 liters at room temperature and pressure. Neglect professorial histrionics.

•57. A wooden box of frozen fish is covered by equal thicknesses of cork and wood (with the cork on the inside of the box). The ratio of the thermal conductivities is

$$\frac{k_{wood}}{k_{cork}} = 3.0$$

The temperature of the inside of the box is −20 °C and the outside is at 25 °C. (a) When a steady state exists, what is the ratio of the heat flow through the wood to the heat flow through the cork? (b) What is the temperature of the surface at which the wood and cork are joined?

•58. A window pane has dimensions 1.00 m × 0.50 m × 2.0 mm. The inside of the window is at a comfortable 20 °C and the outside is a numbing −20 °C. (a) If the width and height are doubled, but not the thickness, by what factor is the heat flow changed? (b) If each dimension (height, width, and thickness) of the window is doubled, by what factor is the heat flow through the window changed? (c) If the height and thickness are doubled, leaving the width unchanged, by what factor is the heat flow changed?

•59. A solid rectangular block of lead has the dimensions 2.00 cm × 3.00 cm × 4.00 cm. A temperature difference of 100 °C can be applied to any pair of opposite faces that you choose. (a) Which choice of separation between the faces produces the largest heat flow? (b) Calculate the heat transfer in joules during 60.0 s between the opposite faces mentioned in part (a). (c) Which choice of separation between faces produces the least heat flow?

•60. Show that to convert the SI R-value to the English system R-value, the conversion factor is as indicated at the bottom of Table 13.7 on page 609.

•61. Warm-blooded animals maintain a constant body temperature, warmer than their surroundings. Ancient and medieval families occasionally built their dwellings above sheltered space for their animals. Everyone lived in the barn. In winter, the animals were a natural furnace, but not because of bovine flatulence. The result of Example 13.19 indicates that the heat flow from the animal to the surrounding environment is proportional to the surface area of the animal for a constant temperature difference. Making crude but reasonable assumptions, show that the heat flow likely is proportional to the two-thirds power of the mass m of the animal. That is,

$$\frac{dQ}{dt} \propto m^{2/3}$$

According to these assumptions, what is the ratio of the heat flow from a 1.20×10^3 kg steer compared with that from a 60 kg serf?

•**62.** Let T_n be the normal temperature in kelvin of the human body. When the body has a fever, let its temperature be $T_n + \Delta T$. Show that the ratio of the heat flow radiated by the body when it has a fever to that when the body temperature is normal is, approximately,

$$1 + \frac{4 \, \Delta T}{T_n}$$

Evaluate this ratio for a sick soul with a fever of 40 °C; normal body temperature is 37 °C.

•**63.** The *solar constant* is the amount of energy from the Sun we receive on the Earth during each second on a 1.000 m² area oriented perpendicular to the direction of the sunlight. The value of the solar constant is about 1.37 kW/m². Imagine sunlight illuminating an asphalt pavement as indicated in Figure P.63. The ambient temperature is 25 °C. What is the equilibrium temperature of the asphalt? Assume the asphalt is a blackbody.

FIGURE P.63

‡**64.** Copper rod A, 2.00 mm in diameter and 30.0 mm long, is joined to copper rod B, which is 3.00 mm in diameter and 20.0 mm long, making a conductor 50.0 mm long. The lateral area of the rods is fully insulated. The rods connect two reservoirs as indicated in Figure P.64. (a) Determine the temperature at the junction between the two rods. (b) Find the heat flow in each rod.

FIGURE P.64

Sections 13.12 Thermodynamic Processes
13.13 Energy Conservation: The First Law of Thermodynamics and the CWE Theorem
13.14 The Connection Between the CWE Theorem and the General Statement of Energy Conservation
13.15 Work Done by a System on Its Surroundings
13.16 Work Done by a Gas Taken Around a Cycle
13.17 Applying the First Law of Thermodynamics: Changes of State

•**65.** An ideal gas goes from an initial state of thermodynamic equilibrium (1) to a final state (2) via one of the three paths indicated on the *P–V* diagram in Figure P.65. What a gas! (a) Calculate the work done by the gas along path a. (b) Calculate the work done by the gas along path b. (c) Calculate the work done by the gas along path c. The gas now is taken from state (2) back to state (1) via one of the three paths. (d) Calculate the work done by the gas along path a. (e) Calculate the work done by the gas along path b. (f) Calculate the work done by the gas along path c.

FIGURE P.65

•**66.** Just for fun, you cause a sample of an ideal gas to move around a closed cycle on the *P–V* diagram beginning and ending at point A indicated in Figure P.66. (a) Find the total work done by the gas. (b) What is the change in the internal energy of the gas? (c) What is the total heat transfer to the gas in executing this cycle? If the gas is taken around the cycle in the opposite direction: (d) find the total work done by the gas; (e) find the change in the internal energy of the gas; and

FIGURE P.66

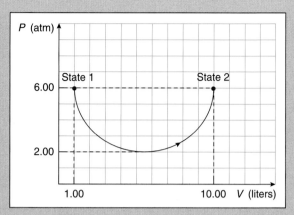

FIGURE P.68

(f) find the total heat transfer to the gas in executing the cycle in this direction.

•**67.** A gas on vacation takes a circle tour of a *P–V* diagram, beginning at point A as indicated in Figure P.67. (a) Find the total work done by the gas after one trip around the cycle. (b) What is the change in the internal energy of the gas in executing one cycle? (c) What is the total heat transfer to the system around the cycle? If the gas takes the circle tour in the opposite direction: (d) Find the total work done by the gas after one trip around the cycle. (e) What is the change in the internal energy of the gas in executing one cycle? (f) What is the total heat transfer to the gas around the cycle?

FIGURE P.69

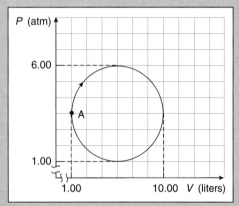

FIGURE P.67

•**68.** A gas is taken from thermal equilibrium state 1 to state 2 on a *P–V* diagram via the semicircular route indicated in Figure P.68. (a) Calculate the work done by the gas in going from state 1 to state 2. (b) What is the work done by the gas if it goes from state 2 back to state 1, retracing the semi-circular path?

•**69.** A gas is taken from thermal equilibrium state 1 to state 2 on a *P–V* diagram via a semicircular route as indicated in Figure P.69. (a) Calculate the work done by the gas in going from state 1 to

state 2. (b) What is the work done by the gas if it goes from state 2 back to state 1 retracing the semicircular path?

•**70.** A gas is taken around a closed cycle on a *P–V* diagram as indicated in Figure P.70. (a) Find the total work *W* done by the gas. (b) What is the change in the internal energy of the gas? (c) What is the total heat transfer to the gas in executing this cycle? If the gas is taken around the cycle in the opposite direction: (d) find the total work done by the gas; (e) find the change in the internal energy of the gas; and (f) find the total heat transfer to the gas in executing the cycle in this direction.

•**71.** Three (3.00) moles of an ideal gas is taken along the excursion on a *P–V* diagram indicated in Figure P.71. The first part of the path is an isobaric process and the second part of the path is isothermal. (a) What is the initial absolute temperature of the gas? (b) Is the work done by the gas positive or negative over the isobaric segment of the path? (c) Calculate the work done by the gas along the isobaric segment of the path. (d) What is the temperature of the gas along the isotherm? (e) Is the work done by the gas positive or negative along the isothermal segment of the path? (f) Determine the final volume V_f of the

FIGURE P.70

FIGURE P.72

FIGURE P.71

FIGURE P.76

gas so that the total work done by the gas is zero. Express your result in liters. (g) What is the final pressure P_f of the gas? Express your result in atmospheres.

•72. A gas undergoes processes that trace out the path indicated in the P–V diagram of Figure P.72, beginning at point A. (a) What is the total work done by the gas during one cycle? (b) What is the change in the internal energy of the gas during one cycle? (c) What is the total heat transfer to the gas during each cycle? (d) Answer the same questions if the cycle is executed in the opposite sense.

•73. A steel ball of mass 0.100 kg and specific heat 460 J/(kg·K) drops to a hard floor from a height of 10.00 m and rebounds with 0.7071 of its impact speed. (a) Show that the ball attains a height of 5.00 m after the impact. (b) Assume all kinetic energy lost by the ball in the impact is retained by the ball as an increase in its internal energy (as if heat transfer occurred to the ball, although none does). Calculate the rise in temperature of the ball caused by the single bounce. (c) If the bouncing is allowed to continue until the ball is at rest, calculate the total rise in the temperature of the ball. Make the same assumptions as in part (b).

•74. A snowball at 0 °C moving at speed v just barely misses you and smashes into a wall, coming to a complete halt. Assume the snowball retains the lost kinetic energy as a rise in its thermodynamic internal energy (as if heat transfer occurred to the snowball, although none does), with no heat transfer to the wall or the environment. Find the (large) speed v that makes the snowball melt completely into water at 0 °C.

•75. Victoria Falls on the border between Zimbabwe and Zambia in southern Africa has a height of about 130 m. The falls are visible in the background of Figure P.19. Assume that all the kinetic energy gained by the falling water appears (eventually) as an increase in the internal energy of the water (as if heat transfer occurred to the water, although none does). What is the temperature change of the water between the top and bottom of the falls?

•76. When your heart contracts to pump blood through your circulatory system, the blood pressure reaches a peak value about 140 mm Hg (called the systolic blood pressure value); when the heart muscle relaxes, the blood pressure falls to 80 mm Hg (called the diastolic blood pressure value). Your blood pressure is said to be 140/80, using medical parlance. The variation is indicated schematically in the *P–V* diagram of Figure P.76. The change in the volume of the heart is roughly 90 cm³ per contraction. (a) Convert these medical pressure and volume values to SI units. See Section 11.3. (b) From the *P–V* diagram, estimate the work done by the heart during each contraction cycle. (c) If a patient has a pulse rate of 80 heartbeats/min, what is the approximate time rate at which the heart is doing work (in watts)?

•**77.** A well-insulated waterbomb, with a mass of 10 kg and an average specific heat of 4186 J/(kg·K), is dropped from rest and undergoes free fall from the top of the Washington Monument (170 m). After the (intact) waterbomb comes to rest on impact, how much will its temperature change if you assume that its lost kinetic energy manifests itself completely as a change in the internal energy of the water (as if heat transfer occurred to the water, although none actually does)? Assume no heat transfer to the environment.

•**78.** You toss a lighted firecracker (of negligible mass) into an insulated steel box [of mass 1.8 kg and specific heat 460 J/(kg·K)] and then quickly seal the lid tight. Hang on. The firecracker explodes inside without rupturing the box and you note that the box warms from 20.0 °C to 22.0 °C, assuming no heat transfer to the environment. The air in the box (consisting mostly of N_2 and O_2) has an effective molar mass of 29 g/mol and a volume of 22.4 liters. The molar specific heat of air is about 20 J/(mol·K). (a) How many moles of air are in the

box? (b) Show that the box, rather than the air, has most of the energy released by the firecracker. (c) How much energy was released by the firecracker? (d) Name and describe the processes by which the energy of the exploding firecracker is transferred to the walls of the container.

•**79.** A steady flow of water vapor from a pot of boiling water passes into and out of an aluminum cup and eventually raises the temperature of the cup from 20 °C to 100 °C. (a) Explain why water vapor condenses in the aluminum cup. (b) When the cup reaches a temperature of 100 °C, explain why no more water vapor will condense in the cup. (c) If the metal cup has mass M and the condensed water in the cup has mass *m*, calculate the ratio *m/M*. Can measuring *m/M* give the latent heat of vaporization?

•**80.** A silver bullet [mass 0.010 kg with a specific heat of 233 J/(kg·K)] is fired by the Lone Ranger into a fixed block of lead [mass 1.00 kg with a specific heat of 128 J/(kg·K)] and comes to rest within it. Both objects initially are at a temperature of 30.0 °C. Tonto observes that the temperature of the total mass rises to 35.0 °C and also that the block (with the bullet in it) remains at rest. Assume there is no heat transfer to the external world. (a) How much energy is needed to change the temperature of the silver bullet? (b) How much energy is needed to change the temperature of the lead block? (c) What was the kinetic energy of the bullet before impact? (d) What was the speed of the bullet before impact?

INVESTIGATIVE PROJECTS

A. Expanded Horizons

1. Investigate the ways in which very low and very high absolute temperatures are measured experimentally.
   Robert Otani and Peter Siegal, "Determining absolute zero in the kitchen sink," *The Physics Teacher, 29,* #5, pages 316–319 (May 1991).
   Kurt Mendelssohn, *The Quest for Absolute Zero: The Meaning of Low Temperature Physics* (John Wiley, New York, 1977).
   Terry J. Quinn, *Temperature* (2nd edition, Academic Press, London, 1990).
   Thomas D. McGee, *Principles and Methods of Temperature Measurement* (Wiley, New York, 1988).
   James R. Leigh, *Temperature Measurement and Control* (Peregrinus, London, 1988).

2. After reading several of the following references, research and report on the various physical aspects of hyperthermia and hypothermia.
   William W. Forgey, *The Basic Essentials of Hypothermia* (ICS Books, Merrillville, Indiana, 1991).
   Albert A. Bartlett and Thomas J. Brown, "Death in a hot tub: the physics of heat stroke," *American Journal of Physics, 51,* #2, pages 127–132 (February 1983).
   G. M. Hahn, "Hyperthermia for the engineer: a short biological primer," *IEEE Transactions on Biomedical Engineering, BME-31,* #1, pages 3–8 (January 1984).
   Gilbert Nussbaum (editor), "Physical aspects of hyperthermia," *AAPM Monograph Number 8* (American Institute of Physics, New York, 1983).
   Best and Taylor's Physiological Basis of Medical Practice, edited by John B. West (12th edition, Williams and Wilkins, Baltimore, 1990).

3. Investigate aspects of heat transfer in medicine and biology.
   Avraham Shitzer and Robert C. Eberhart, *Heat Transfer in Medicine and Biology: Analysis and Applications* (Plenum, New York, 1985).
   T. H. Benzinger, "Thermodynamics of living matter: physical foundations of biology," *American Journal of Physiology, 244,* #6, pages R743–R750 (June 1983).

4. The average energy per day used by animals simply to act as "couch potatoes" (their basic metabolic rate) is expressed by *Kleiber's law.* The relationship indicates that the metabolic rate is proportional to the 0.75 power of the mass of the animal. Investigate this relationship and its relationship to the prediction of Problem 61.
   Knut Schmidt-Nielsen, *Animal Physiology: Adaptation and Environment* (5th edition, Cambridge University Press, New York, 1997).
   Malcolm S. Gordon, *Animal Physiology: Principles and Applications,* (Macmillan, New York, 1977), pages 70–75.

5. The significance of heat transfer by blackbody radiation was alluded to in the text. The story of the study of such radiation, leading to the discovery of Planck's constant *h*, is an interesting one and worth looking into in greater detail. (See Chapter 26 of this text.)
   Thomas S. Kuhn, *Black-Body Theory and the Quantum Discontinuity, 1894–1912* (University of Chicago Press, Chicago, 1987).
   Stephen T. Thornton and Andrew Rex, *Modern Physics for Scientists and Engineers* (Saunders, Fort Worth, Texas, 1993).

M. Russell Wehr, James A. Richards Jr., and Thomas W. Adair III, *Physics of the Atom* (4th edition, Addison-Wesley, Reading, Massachusetts, 1984), pages 71–85.

6. Early ideas about the nature of heat transfer included a hypothesis that heat was a substance (called caloric) transferred between hot and cold systems. Investigate the experiments of Benjamin Thompson (Count Rumford) (1753–1814) that disproved this hypothesis. (Count Rumford's legacy lives on with Count Rumford Baking Powder, still available in some supermarkets.)

    Arnold Arons, *Development of Concepts of Physics* (Addison-Wesley, Reading, Massachusetts, 1965), pages 411–415.

    Freeman J. Dyson, "What is heat?" *Scientific American*, 191, #3, pages 58–63, 188 (September 1954).

    Sanborn C. Brown, *Benjamin Thompson, Count Rumford: Count Rumford on the Nature of Heat* (Pergamon, Oxford, New York, 1967).

    Sanborn C. Brown, *Count Rumford: Physicist Extraordinary* (Greenwood Press, Westport, Connecticut, 1979).

    Sanborn C. Brown, *Benjamin Thompson, Count Rumford* (MIT Press, Cambridge, Massachusetts, 1979).

7. Investigate the reason that life is possible only over a fairly narrow range of temperatures.

    Harold Weinstock, "Thermodynamics of cooling a (live) human body," *American Journal of Physics*, 48, #5, pages 339–344 (May 1980).

    Frank H. Johnson, "Heat and life," *Scientific American*, 191, #3, pages 64–68, 188 (September 1954).

B. Lab and Field Work

8. With the assistance of your student health service, measure the human body temperature of a large sample of healthy students to a precision of at least 0.1 °C. You may need special permission from a committee at your university to conduct such an experiment since it involves human subjects. After deciding on an appropriate way to display the data, make three histogram graphs of the results, one for male students, one for female students, and one for all. Are there any distinctions between the histograms?

9. Devise an experiment to precisely measure the volume of 1.00 kg of water as a function of temperature between 0 °C and 100 °C. Note that the graph in Figure 13.15 is not what you would obtain using Equation 13.10. Hence the coefficient of volume expansion of water must be temperature dependent over this range of temperatures.

10. Butter melts at about 31 °C. Design an experiment to determine the melting point temperature more precisely and to determine whether there is a latent heat of fusion associated with the melting. Do your results depend appreciably on whether the raw material is sweet butter or lightly salted butter, or on the brand of butter?

11. Your physics department likely has a precision mercury laboratory thermometer as well as a platinum resistance thermometer (or other equivalent electronic thermometer). Use distilled water to make an ice–water slurry as well as a beaker of boiling water. Use these two fixed points to calibrate the two thermometers. Now measure the temperature of a beaker of warm water (at about 50 °C) with each and see if the two thermometers agree.

12. Using a narrow, long-necked bottle or flask and water, make a crude thermometer. Use an ice–water slurry and boiling water as the two calibration points. What is the precision of your thermometer?

C. Communicating Physics

13. Imagine you are asked by a local fourth-grade science teacher to explain the distinction between heat transfer and temperature, why masses in thermal equilibrium have a temperature but no heat, and the similarities and differences between heat transfer and work. Prepare a 15 minute talk for the class, remembering that the attention span of a fourth-grader is short, so that you will need to spice your talk with many simple demonstrations. As a community service, volunteer to give the talk at a local elementary school. You might be surprised (and aghast) at the number of elementary school science books that talk about heat transfer as if a mass has a "heat energy" when in thermal equilibrium.

"I CAN MAKE IT SIT, BEG, FETCH, AND PLAY DEAD JUST FINE, BUT I STILL NEED TO FINE-TUNE ITS COEFFICIENT OF VOLUME EXPANSION!"

Chapter 14
Kinetic Theory

Ludwig Boltzmann [1844–1906], who spent much of his life
studying statistical mechanics, died in 1906, by his own hand.
Paul Ehrenfest [1880–1933], carrying on the work, died similarly
in 1933. [So did another disciple, Percy Bridgman (1882–1961).]
Now it is our turn to study statistical mechanics.

Perhaps it will be wise to approach the subject cautiously.

*David L. Goodstein**

We have studied the mechanics of single particles and the interaction of several particles in collisions. When we approach the mechanics associated with the many particles in systems such as gases, liquids, and solids, we are faced with analyzing the dynamics of a *huge* number of particles. The dynamics of such many-particle systems is called *statistical mechanics* since, as we shall see, aspects of statistics come into play to facilitate the analysis.

The game involved in studying a system with a large number of particles is similar to what happens after every physics test. Of course, you are interested in your individual grade, but you also want to know the class average to see how you fit into the grand scheme.[†] The average characterizes the group as a whole. The average stems from the large number of individual grades via a suitable arithmetic, statistical procedure. In like manner, with huge numbers of particles, suitable averages of the physical characteristics and motions of individual particles provide information about the macroscopic behavior of the system as a whole. We forgo details about individual particles and concentrate on the collective behavior. So, borrowing a term from the founder of communism, Karl Marx (1818–1883), statistical mechanics may be viewed as a collectivization of physics.

The kinetic theory that we study in this chapter is a special aspect of the statistical mechanics of large numbers of particles. We begin with the simplest model for a monatomic ideal gas, and extend the results to molecular gases, even applying what we can of the theory to solids as well.

14.1 BACKGROUND FOR THE KINETIC THEORY OF GASES

One thing bears mentioning right away. It is *not* at all obvious that a gas consists of very tiny particles (atoms and molecules) moving around at high speeds. We certainly cannot directly see or sense that this is the case. Indeed, during the late 17th century, Newton himself imagined a very different model for a gas, retrospectively called a **static model**. In this model[‡] the corpuscular particles of the gas occupy fixed positions and fill the entire space between them, somewhat like foam squeegee-cubes. The corpuscular squeegees were imagined to expand and contract at their locations as the volume of the entire gas expands or contracts. With this model, Newton even was able to show that if the corpuscles repel each other with a force whose magnitude is inversely proportional to their separation, qualitatively similar to (but not quantitatively identical to) the way a spring resists compression, then the pressure is inversely proportional to the volume. The latter result neatly accounted for the isothermal pressure and volume experiments on gases made by Newton's contemporary, Robert Boyle (known as Boyle's law; see Section 13.8).

Other models of a gas emerged after Newton, but it was a **kinetic model**, first proposed by Daniel Bernoulli (1700–1782) in 1758, that was resurrected with great success during the 19th century by James Clerk Maxwell, Ludwig Boltzmann, and others. The kinetic model envisions a gas to be composed of many tiny particles, freely moving at high speed, whose pressure arises from the innumerable collisions of the particles with each other and with the walls of the container. Because the kinetic model gives a microscopic interpretation not only of pressure but also of absolute temperature and the specific heats of gases, it joined the pantheon of successful theories in physics and convinced many of the reality of the atomic view of matter. Nonetheless, like most physical theories, classical kinetic theory has its limitations and failures too, as we shall see.

Only in the 20th century did the rise of quantum mechanics account for the complex temperature dependence of the specific heats of gases and solids, as well as the structure of atoms and the mechanisms for molecular bonding. Fortunately, such ideas are not needed here for a kinematic understanding of key thermodynamic ideas. We shall find that the underlying microscopic motions of the particles in an ideal gas are mostly (but not entirely) within the reach of classical physics rather than quantum mechanics.

If we accept the atomic nature of matter, why do the particles in a solid resist both compression and tension? A brick is, after all, hard to compress and hard to pull apart. There must exist spring-like forces between the particles in solids; the particles must exist at equilibrium locations when not subjected to tension or compression. Nineteenth-century scientists thought such forces between the particles in solids (and liquids) likely were fundamentally electrical in nature, despite the fact that matter usually is electrically neutral. In view of our current view of the nuclear atom with its surrounding electrons, the ability of a solid to resist tension forces is all the more remarkable because electrons repel each other electrically. You might think that the atoms in solids should naturally fly apart of their own accord, which they clearly do not do because of molecular bonding and their overall electrical neutrality. On the other hand, the ability of solids and liquids to resist compression indicates that the particles likely are quite close together, packed cheek-to-cheek, if you will.

The trait that distinguishes gases from both liquids and solids is the ease with which gases can be compressed. The densities of gases are on the order of 1000 times smaller than those of liquids and solids. The low densities of gases (compared with solids and liquids) is a clue that the spacing of the particles in a gas likely is on the order of a factor of 10, the cube root of 1000, greater than in solids and liquids.[§]

Open a vial of perfume and soon the fragrance is detectable at some distance away from the vial. The same effect is apparent with obnoxious odors such as from cigarettes, cigars, skunks, or (embarrassingly) flatulence. This **diffusion** of smoke and odors through gases indicates that the particles in a gas likely are in motion, rather than occupying fixed positions, as in the static model of Newton. Gases also quickly occupy the entire volume allotted to them rather than settling down to just the bottom of a container in the manner of a liquid.

A consequence of the microscopic motion of the particles in a gas is extraordinary: the motion is *perpetual*, and for that we can be very glad. Yet on the macroscopic scale of our laboratories *no* motion is perpetual, thanks to frictional effects, as you can easily observe. The pressure of a gas in thermal equilibrium inside a container never runs down or decreases with time, with the gas particles gradually settling into the bottom of the container because of the gravitational force. If the pressure is due to the huge number of

---

*(Chapter Opener) *States of Matter* (Prentice-Hall, Englewood Cliffs, New Jersey, 1975), page 1.

[†] Grading certainly *is* a scheme!

[‡] Isaac Newton, *Principia*, Proposition XXIII, theorem XVIII.

[§] The cube root arises since volumes are the cube of a characteristic spatial dimension.

Daniel Bernoulli proposed a kinetic model of gases. He is also known for Bernoulli's principle of ideal fluids (Chapter 11).

collisions of the particles with each other and with the walls of the container, the collisions must be modeled as completely elastic.

Yet collisions between particles on a macroscopic scale are invariably inelastic. Put a bunch of BBs or tennis balls in motion inside a container in an orbiting spacecraft and the particles eventually decrease their speeds because of inelastic collisions with each other (and the walls of the container). So there is something fundamentally different about the dynamics of the small atomic and molecular particles that are the constituents of a gas, compared with the dynamics of collections of macroscopic masses such as the BBs or tennis balls. The forces involved in the collisions of the particles of the gas with each other and with the walls of the container must be conservative, briefly acting only during the short intervals of the collisions and essentially zero otherwise. Thus the total of all their microscopic mechanical energies—the internal energy—remains constant if the system is in a state of thermodynamic equilibrium.

The constancy of the pressure of a gas in thermal equilibrium also indicates that if the velocity of one of the particles in the gas changes because of a collision, there likely is another particle in the collection that, in another collision, makes a velocity change in the opposite direction, so that the development of winds or currents in the gas does not occur spontaneously. This principle of **detailed balancing** is a subtle concept that underlies much of kinetic theory.

The temperature of the gas must somehow be related to this molecular motion, because an increase in temperature, at constant volume, results in an increase in pressure, thus implying that the collisions of the particles with the walls of the container are more frequent and/or more violent. The heat transfer to the gas at constant volume, raising the temperature of the gas, thus must increase the translational kinetic energy of the particles, *provided* that the particles have no internal degrees of freedom* such as rotation or vibration to soak up the incoming energy in other ways. If such extra motions of the particles can be brought into play (we say these motions become active, excited, or populated), we expect the thermodynamic behavior of the gas, in particular its specific heat, to differ from that in the situation where the particles are treated merely as point particles. As we will see, this is indeed the case.

---
*We define this term more carefully later in the chapter.

One of the goals of kinetic theory is to discover the connection between temperature and pressure, and even a suitable average speed of the particles. For simplicity, we begin by assuming the gas particles are pointlike and see what results arise from studying the dynamics of such a collection. This is the kinetic theory of a monatomic ideal gas, a dilute gas whose particles are single atoms rather than molecules. The noble gases—helium, neon, argon, krypton, xenon, and radon—are examples of monatomic gases. Later, we will be able to extend the theory to include diatomic and polyatomic gases.

14.2 THE IDEAL GAS APPROXIMATION

In Chapter 13 we defined an ideal gas as one that obeyed the ideal gas equation of state:

$$PV = nRT \qquad (14.1)$$

where P is its pressure, V is its volume, n is the number of moles of the gas, R is the universal gas constant, and T is the absolute temperature of the gas. Real gases behave like an ideal gas if they are sufficiently dilute.

We describe an ideal gas as a suitable collection of pointlike particles and make the following assumptions:

1. The number of particles N in the gas is very large.

Of course any macroscopic mass of gas will satisfy this condition, because the number of particles in a mole of a gas is Avogadro's number ($N_A = 6.02 \times 10^{23}$ particles/mol). Even a trillionth (10^{-12}) of a mole has many particles:

$$10^{-12} \text{ mol} \times (6.02 \times 10^{23} \text{ particles/mol})$$
$$= 6.02 \times 10^{11} \text{ particles}$$

This is about 600 billion particles, three times the number of stars in an average-sized galaxy!

2. The volume V containing the gas is much larger than the total volume actually occupied by the gas particles themselves.

That is, if we let $V_{particle}$ be the volume of a single particle of the gas, then

$$NV_{particle} << V$$

This ensures that the separation of the particles in the gas is large compared with their size and that long-range forces between the particles are negligible. In other words, the particles are essentially pointlike and do not feel each other unless they collide.

3. The dynamics of the particles is governed by Newton's laws of motion.

You might wonder if the smallness of the particles requires quantum mechanics, but classical Newtonian physics is sufficient to describe most of their dynamics. The absence of long-range interactions (forces) means that the particles move freely (at constant velocity) between collisions with either the walls of the container or each other.

> 4. The particles are equally likely to be moving in any direction.

That is, the directions of the motions of the particles are random. There is no bulk motion, wind, or turbulence within the gas, which is consistent with our restriction to consider only simple thermodynamic systems.

> 5. The gas particles (be they atoms or molecules) interact with each other and with the walls of the container only via elastic collisions.

From our study of elastic collisions (Chapter 9), this means that both the total momentum and the total kinetic energy of the participants in each collision are conserved.

> 6. The gas is in thermal equilibrium with its surroundings.

This assumption is necessary because the thermodynamics we are considering is equilibrium thermodynamics. This assumption does not prohibit changes in the state of the gas; it simply means that any changes that occur are quasi-static, so that at any instant the gas is in a state of thermodynamic equilibrium.

> 7. The particles of the gas are identical and indistinguishable.

The gas is not a mixture of different constituents. This assumption is made merely to simplify the initial analysis before extending the theory to mixtures.

All pure gases, whether monatomic, diatomic, or polyatomic, that are sufficiently dilute exhibit behavior approaching this ideal gas approximation.

EXAMPLE 14.1 _____

One mole (1.00 mol) of helium gas is at a warm room temperature (300 K) and 1.000 atm pressure. How many helium atoms are in each cubic millimeter of the gas?

A cubic millimeter is not a very large volume!

Solution
Use the ideal gas law, Equation 14.1, to determine the volume of a mole of the gas. The pressure must be in pascals:

$$PV = nRT$$

$$V = \frac{nRT}{P}$$

$$= \frac{(1.00 \text{ mol})[8.315 \text{ J/(mol·K)}](300 \text{ K})}{1.013 \times 10^5 \text{ Pa}}$$

$$= 2.46 \times 10^{-2} \text{ m}^3$$

One cubic millimeter is a volume $V' = 1.00 \times 10^{-9}$ m^3. Since the particles are distributed uniformly throughout the container, the fraction of the particles that are in a cubic millimeter is the same as the volume ratio V'/V. One mole of gas has 6.02×10^{23} particles. Hence the number of helium atoms in one cubic millimeter is

$$N = \frac{V'}{V} N_A$$

$$= \frac{1.00 \times 10^{-9} \text{ m}^3}{2.46 \times 10^{-2} \text{ m}^3} (6.02 \times 10^{23} \text{ atoms})$$

$$= 2.45 \times 10^{16} \text{ atoms}$$

This is a huge number indeed. For comparison, the age of the universe (about 15 billion years) is only about 10^{17} s.

14.3 THE PRESSURE OF AN IDEAL GAS

From the ideal gas law, Equation 14.1, we already know that a sample of an ideal gas has the pressure

$$P = \frac{nRT}{V} \tag{14.2}$$

In the ideal gas approximation, the gas consists of a vast number of particles moving randomly and colliding occasionally with each other and with the walls of the container; we hope to account for the pressure, that is, derive an independent expression for the pressure, by examining the multitude of tiny impulses delivered to the walls of the container from the many collisions of the particles with the walls.

We consider the gas to be confined to a rectangular parallelopiped of volume V, as shown in Figure 14.1. Let A be the area of the right wall, perpendicular to the x-axis, and let ℓ be the length of the box along the x-axis.

We begin by examining the motion of one particle of the gas moving with a speed v. Let v_x be the x-component of the velocity of this particle. Assume that $v_x > 0$ m/s, as shown in Figure 14.2.

We need to assess the probability that the particle will collide with, say, the right-hand wall during a time interval Δt. If the particle is too far from the wall, the particle cannot reach

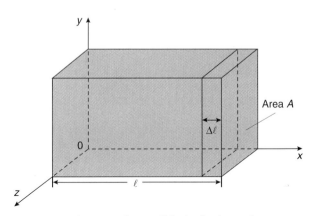

FIGURE 14.1 A rectangular parallelopiped volume of gas.

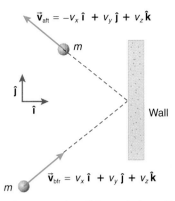

$$\vec{v}_{aft} = -v_x \hat{i} + v_y \hat{j} + v_z \hat{k}$$

$$\hat{j}$$
$$\hat{i}$$

Wall

$$\vec{v}_{bfr} = v_x \hat{i} + v_y \hat{j} + v_z \hat{k}$$

FIGURE 14.2 A gas particle collides with the wall of the container.

and collide with the wall during the interval Δt. The particle can travel a distance $\Delta \ell = v_x \Delta t$ along the x-axis during the interval Δt. Thus, if the particle is within a distance $\Delta \ell$ of the wall (see Figure 14.1) and moving toward increasing values of x (that is, $v_x > 0$ m/s), it will collide with the wall during the interval Δt. If the particle is further from the wall than the distance $\Delta \ell$, then it will not collide with the wall during the interval Δt even though the particle may be moving toward the wall.

Since the particle has an equal probability of being found anywhere within the container, the probability that it lies within $\Delta \ell$ of the right-hand wall is given by the ratio of the volume of a slab of thickness $\Delta \ell$ to the volume of the container:

$$\frac{\text{volume of slab of thickness } \Delta \ell}{\text{volume of the container}} = \frac{A \, \Delta \ell}{A \, \ell}$$

$$= \frac{\Delta \ell}{\ell}$$

Of all the particles in the slab, half are expected to have positive x-components for their velocity (in which case they will collide with the wall if they are within the slab) and half will have negative x-components for their velocity [in which case they will not collide with the wall even if they are within the slab, because they are moving away from the (right-hand) wall]. Thus the probability that a particle with velocity component v_x is within $\Delta \ell$ of the right-hand wall and moving toward it is

$$\frac{1}{2} \frac{\Delta \ell}{\ell}$$

or, after substituting $\Delta \ell = v_x \Delta t$,

$$\frac{v_x \Delta t}{2\ell} \qquad (14.3)$$

This also is the probability that a particle with an x-component of velocity v_x will collide with the right-hand wall of the container during an interval Δt.

Now we look at the dynamics of a collision of a particle with this wall. The collisions are elastic according to our assumptions (Section 14.2). Let m be the mass of each individual particle of the gas. The momentum of the particle before the collision with the wall is

$$\vec{p}_{bfr} = mv_x \hat{i} + mv_y \hat{j} + mv_z \hat{k}$$

Since the wall is fixed, its momentum is zero. After an elastic collision with the fixed wall, the x-component of the velocity of the particle is reversed (and becomes $-v_x$) while the y- and z-components of the velocity are unchanged. Thus the momentum of the particle after the collision is

$$\vec{p}_{aft} = -mv_x \hat{i} + mv_y \hat{j} + mv_z \hat{k}$$

The change in the momentum $\Delta \vec{p}$ of the particle is

$$\begin{aligned} \Delta \vec{p} &= \vec{p}_{aft} - \vec{p}_{bfr} \\ &= -mv_x \hat{i} + mv_y \hat{j} + mv_z \hat{k} - (mv_x \hat{i} + mv_y \hat{j} + mv_z \hat{k}) \\ &= -2mv_x \hat{i} \end{aligned}$$

The particle's momentum change comes from the impulse it received from the right-hand wall during its collision with this wall. However, because of Newton's third law of motion, the impulse given to the particle by the wall has the same magnitude but is directed opposite to the impulse given to the wall by the particle. Hence, *the impulse given to the wall by the particle* during the collision is

Impulse given to wall by particle in a single collision $= +2mv_x \hat{i}$

The probability that a particle with an x-component of velocity v_x actually makes such a collision during an interval Δt is given by Equation 14.3. Thus the impulse we can *expect* to be delivered to the wall during the interval Δt from a particle with an x-component of velocity v_x is

$$\begin{aligned} \text{Expected impulse given to wall} && \\ \text{in time interval } \Delta t \text{ from a particle} &= \frac{v_x \Delta t}{2\ell} 2mv_x \hat{i} \\ \text{with x-component of velocity } v_x && \\ &= \left(\frac{mv_x^2}{\ell} \Delta t \right) \hat{i} \end{aligned}$$

Now we have to sum over all N particles of the gas because they have various values for their x-components of velocity. Denote the particles with numerical subscripts: v_{1x} is the x-component of the velocity of particle 1, v_{2x} is the x-component of the velocity of particle 2, and so forth. Thus the *total impulse* delivered to the wall during the interval Δt by all the particles that collide with the wall during the time interval Δt is

$$\begin{aligned} \text{Total impulse} \\ \text{delivered to wall} \\ \text{during interval } \Delta t \end{aligned} = \left[\frac{m}{\ell} \left(v_{1x}^2 + v_{2x}^2 + \cdots + v_{Nx}^2 \right) \Delta t \right] \hat{i} \qquad (14.4)$$

There are N terms (a *great* number) in the sum of the squares of the x-components of the velocities. If we take the sum of the squares of the velocity components and divide by the total number of particles (N), we obtain an average value for the square of the x-component of the velocity:

$$\left\langle v_x^2 \right\rangle = \frac{v_{1x}^2 + v_{2x}^2 + \cdots + v_{Nx}^2}{N}$$

Thus

$$v_{1x}^2 + v_{2x}^2 + \cdots + v_{Nx}^2 = N\langle v_x^2 \rangle$$

Making this substitution into Equation 14.4, we find

$$\begin{matrix} \text{Total impulse} \\ \text{delivered to wall} \\ \text{during interval } \Delta t \end{matrix} = \left(\frac{m}{\ell} N \langle v_x^2 \rangle \Delta t \right) \hat{\mathbf{i}} \qquad (14.5)$$

The impulse given to the wall by the particles can be written as the *average total force* $\vec{\mathbf{F}}_{\text{ave total}}$ of all the particles on the wall, multiplied by the interval Δt during which the collisions took place:

$$\vec{\mathbf{F}}_{\text{ave total}} \Delta t = \text{Total impulse given to wall during interval } \Delta t$$

Substituting for the total impulse using Equation 14.5, we find

$$\vec{\mathbf{F}}_{\text{ave total}} \Delta t = \left(\frac{m}{\ell} N \langle v_x^2 \rangle \Delta t \right) \hat{\mathbf{i}}$$

Hence the magnitude of the average total force exerted on the wall by all the collisions is

$$F_{\text{ave total}} = \frac{m}{\ell} N \langle v_x^2 \rangle \qquad (14.6)$$

Since the particles are moving randomly throughout the volume without any bulk motion (wind, turbulence, or currents), there should be nothing sacred about the x-direction. Symmetry implies that the average value of the square of the x-component of the velocity must be equal to the average value of the square of the y- or z-components of the velocity of the particles; that is,

$$\langle v_x^2 \rangle = \langle v_y^2 \rangle = \langle v_z^2 \rangle$$

But the square of the velocity itself is $\vec{\mathbf{v}} \cdot \vec{\mathbf{v}} = v^2$:

$$v^2 = v_x^2 + v_y^2 + v_z^2$$

Also, the average value of a sum is the sum of the average values:

$$\langle v^2 \rangle = \langle v_x^2 + v_y^2 + v_z^2 \rangle$$
$$= \langle v_x^2 \rangle + \langle v_y^2 \rangle + \langle v_z^2 \rangle$$

The quantity $\langle v^2 \rangle$ is called the **mean square speed**. If we want a typical speed, we take the square root of the mean square speed and call it the **root-mean-square (rms) speed** v_{rms} of the particles of the gas:

$$v_{\text{rms}} \equiv \sqrt{\langle v^2 \rangle} \qquad (14.7)$$

Since the three averages of the velocity components are all equal because of symmetry, we have

$$\langle v^2 \rangle = 3 \langle v_x^2 \rangle$$

$$\langle v_x^2 \rangle = \frac{\langle v^2 \rangle}{3}$$

Hence Equation 14.6 for the magnitude of the average force exerted on the right-hand wall of the container can be written as

$$F_{\text{ave total}} = \frac{m}{\ell} N \frac{1}{3} \langle v^2 \rangle \qquad (14.8)$$

Since A is the area of this wall of the container, the *pressure* P exerted by the gas on this wall is

$$P = \frac{F_{\text{ave}}}{A}$$

We use Equation 14.8 for F_{ave}; the pressure becomes

$$P = \frac{m}{3A\ell} N \langle v^2 \rangle$$

But the product $A\ell$ is the volume V of the container in which the gas resides. Thus

$$P = \frac{m}{3V} N \langle v^2 \rangle \qquad (14.9)$$

Since $mN = mnN_A$, and $mN_A = M$, the molar mass, we can write Equation 14.9 in the form

$$P = \frac{nM}{3V} \langle v^2 \rangle \qquad (14.10)$$

Thus we see that the pressure of the gas arises from the impulses provided by the gas particles colliding with the walls of the container of the gas.

QUESTION 1 _____

In an elastic collision of a particle of an ideal gas with the wall of the container, the momentum of the particle *changes*. Yet we have seen that momentum is conserved in elastic (and inelastic) collisions. Why is the momentum of the particle not conserved?

EXAMPLE 14.2 _____

One mole (1.00 mol) of helium at temperature 300 K has a volume of 24.6 liters = 2.46×10^{-2} m^3 and a pressure of 1.00 atm. This volume can be calculated via the ideal gas law (see Example 14.1). Determine the mean square speed of the particles and the root-mean-square speed.

Solution
Use Equation 14.10, and solve for the mean square speed:

$$\langle v^2 \rangle = \frac{3PV}{nM}$$

One mole of helium has a molar mass of 4.00 g/mol = 4.00×10^{-3} kg/mol. One atmosphere pressure is 1.01×10^5 Pa. Making the substitutions, you find

$$\langle v^2 \rangle = \frac{3 (1.01 \times 10^5 \text{ Pa}) (2.46 \times 10^{-2} \text{ m}^3)}{(1 \text{ mol})(4.00 \times 10^{-3} \text{ kg/mol})}$$
$$= 1.86 \times 10^6 \text{ m}^2/\text{s}^2$$

The root-mean-square speed is the square root of the mean square speed:

$$v_{\text{rms}} = \sqrt{\langle v^2 \rangle}$$
$$= 1.36 \times 10^3 \text{ m/s}$$

The rms speed in m/s of this typical motion is quite large. For comparison, the speed of sound in helium under comparable conditions is about 980 m/s. When sound waves exist in the gas, the bulk motions of the particles caused by the sound waves are superimposed on the random thermal motions of the particles.

14.4 THE MEANING OF THE ABSOLUTE TEMPERATURE

We saw in the previous section that the pressure of a gas arises from the multitude of submicroscopic mechanical impulses resulting from the collisions of the gas particles with the walls of the container. In this section we discover a microscopic interpretation of the absolute temperature.

Rewriting Equation 14.9 slightly,

$$PV = \frac{Nm}{3} \langle v^2 \rangle \tag{14.11}$$

But wait. We also have an experimental relation for the product PV that came from laboratory measurements of gases. The kinetic theory expression for PV must be identical to the ideal gas law expression for PV, the equation of state, Equation 13.13:

$$PV = NkT$$

where k is Boltzmann's constant, 1.38×10^{-23} J/K. Thus the right-hand sides of Equations 14.11 and 13.13 must be identical, even though one came from kinetic theory and the other from the gas law:

$$NkT \equiv \frac{Nm}{3} \langle v^2 \rangle$$

Solving for the absolute temperature of the gas, we find

$$T = \frac{m}{3k} \langle v^2 \rangle \tag{14.12}$$

The *average* translational kinetic energy of a *single* particle of the gas is

$$KE_{ave} = \frac{1}{2} m \langle v^2 \rangle$$

Hence Equation 14.12 for the temperature can be written as

$$T = \frac{2}{3k} KE_{ave}$$

or

$$KE_{ave} = \frac{3}{2} kT \tag{14.13}$$

> The temperature of all the gas (a macroscopic quantity) is a manifestation of the average translational kinetic energy of each particle (a submicroscopic quantity).

> A measurement of temperature is a measurement of the average translational kinetic energy of any particle of the gas.

One of the triumphs of kinetic theory is this revelation of the meaning of the absolute temperature.

Equation 14.13 shows that the average translational kinetic energy of a particle in an ideal gas is the same for all particles and is equal to

$$\frac{3}{2} kT$$

> Two different gases at the same temperature have the same average translational kinetic energy per particle regardless of the masses of the particles of the two gases.

Thus a gas composed of more massive particles will have slower average particle speeds than a gas composed of smaller-mass particles at the same temperature, but the same average kinetic energy per particle.

From Equation 14.12, the mean square speed is

$$\langle v^2 \rangle = \frac{3kT}{m}$$

Hence the rms speed of the particles of the gas is

$$v_{rms} = \left(\frac{3kT}{m} \right)^{1/2} \tag{14.14}$$

Since $k = R/N_A$ from Equation 13.18, we can also write this as

$$v_{rms} = \left(\frac{3RT}{mN_A} \right)^{1/2} \tag{14.15}$$

Also, since m is the mass of an individual particle of the gas, the product mN_A is the molar mass M. Thus Equation 14.15 becomes

$$v_{rms} = \left(\frac{3RT}{M} \right)^{1/2} \tag{14.16}$$

The particles of a gas have a range of speeds. The average speed is found by taking the average of the speeds of all the particles at a given instant (remember that the speed is a positive scalar since it is the magnitude of the velocity):

$$\langle v \rangle = \frac{v_1 + v_2 + v_3 + \cdots + v_N}{N} \tag{14.17}$$

The rms speed is found from

$$v_{rms} \equiv \sqrt{\langle v^2 \rangle} = \left(\frac{v_1^2 + v_2^2 + v_3^2 + \cdots + v_N^2}{N} \right)^{1/2} \tag{14.18}$$

> The average speed is less than the rms speed.

The inequality arises because the squared numbers averaged in forming the rms speed typically are *larger* than the corresponding numbers used in calculating the average speed;* see Example 14.3.

---

*Even if most of the speeds are less than 1 m/s, the average speed is less than the rms speed, in this case because of the square root in v_{rms}. This can be verified by taking an example with all the particle speeds less than 1 m/s; see Problem 12.

QUESTION 2

The average speed of all the particles in a gas at a given instant is *not* zero, whereas the average velocity of all the particles *is* zero. Explain why.

QUESTION 3

A gas consists of a mixture of oxygen molecules (O_2) and nitrogen molecules (N_2). Which molecule has, on average, the greater rms speed?

EXAMPLE 14.3

Consider an 11 000-particle gas system with speeds distributed as follows:

1000 particles each with speed 100 m/s,
2000 particles each with speed 200 m/s,
4000 particles each with speed 300 m/s,
3000 particles each with speed 400 m/s, and
1000 particles each with speed 500 m/s.

Find the average speed and the rms speed of the particles.

Solution
The *average speed* is found using Equation 14.17:

$$\langle v \rangle = [1000\,(100\ \text{m/s}) + 2000\,(200\ \text{m/s})$$
$$+ 4000\,(300\ \text{m/s}) + 3000\,(400\ \text{m/s})$$
$$+ 1000\,(500\ \text{m/s})]/11\,000$$
$$= 309\ \text{m/s}$$

The *rms speed* is found from Equation 14.18:

$$v_{\text{rms}} = \{[1000\,(100\ \text{m/s})^2 + 2000\,(200\ \text{m/s})^2$$
$$+ 4000\,(300\ \text{m/s})^2 + 3000\,(400\ \text{m/s})^2$$
$$+ 1000\,(500\ \text{m/s})^2]/11\,000\}^{1/2}$$
$$= 328\ \text{m/s}$$

Note that the average speed is less than the rms speed.

EXAMPLE 14.4

A sample of helium gas is at a warm room temperature (300 K) and a pressure of 0.500 atm. What is the average kinetic energy of a particle of the gas?

Solution
The average kinetic energy of a particle in the gas is found from Equation 14.13,

$$\text{KE}_{\text{ave}} = \frac{3}{2}kT$$

where k is Boltzmann's constant. Notice that the average kinetic energy per particle depends only on the temperature, so that the pressure and identity of the particles are irrelevant. Substituting, you get

$$\text{KE}_{\text{ave}} = \frac{3}{2}(1.38 \times 10^{-23}\ \text{J/K})\,(300\ \text{K})$$
$$= 6.21 \times 10^{-21}\ \text{J}$$

EXAMPLE 14.5

A sample of helium gas at a warm room temperature (300 K) has a pressure of 1.00 atm. A sample of neon is under the same conditions. The molar mass of helium is 4.00 g/mol and that of neon is 20.2 g/mol.

a. Find the rms speed of the helium atoms and of the neon atoms.
b. What is the average kinetic energy per particle in each sample?

Solution
a. Following Problem-Solving Tactic 14.1, the molar mass of helium in SI units is 4.00 g/mol = 4.00×10^{-3} kg/mol. Use Equation 14.16 for the rms speed:

$$v_{\text{rms}} = \left(\frac{3RT}{M}\right)^{1/2}$$
$$= \left(\frac{3\,[8.315\ \text{J/(mol·K)}]\,(300\ \text{K})}{4.00 \times 10^{-3}\ \text{kg/mol}}\right)^{1/2}$$
$$= 1.37 \times 10^{3}\ \text{m/s}$$

Notice that the rms speed depends on the *absolute temperature* but not at all on the pressure.

The molar mass of neon in SI units is 20.2×10^{-3} kg/mol. Use Equation 14.16 for the rms speed once again:

$$v_{\text{rms}} = \left(\frac{3RT}{M}\right)^{1/2}$$
$$= \left(\frac{3\,[8.315\ \text{J/(mol·K)}]\,(300\ \text{K})}{20.2 \times 10^{-3}\ \text{kg/mol}}\right)^{1/2}$$
$$= 609\ \text{m/s}$$

Notice that the more massive neon atoms have significantly slower speeds than the less massive helium atoms at the same temperature.
b. According to Equation 14.13, the average kinetic energy per particle depends solely on the temperature. Since the two

samples are at the same temperature, they have the *same* kinetic energy per particle:

$$KE_{ave} = \frac{3}{2} kT$$

$$= \frac{3}{2} (1.38 \times 10^{-23} \text{ J/K}) (300 \text{ K})$$

$$= 6.21 \times 10^{-21} \text{ J}$$

The more massive neon atoms move more slowly than the helium atoms but have the same average kinetic energy.

EXAMPLE 14.6

At what temperature will the particles in a sample of helium gas have an rms speed of precisely 1.00 km/s?

Solution

Use Equation 14.16, and solve for the absolute temperature:

$$T = \frac{Mv_{rms}^2}{3R}$$

The molar mass of helium is 4.00 g/mol = 4.00×10^{-3} kg/mol. Thus the absolute temperature is

$$T = \frac{(4.00 \times 10^{-3} \text{ kg/mol}) (1.00 \times 10^3 \text{ m/s})^2}{3 [8.315 \text{ J/(mol·K)}]}$$

$$= 160 \text{ K}$$

Chilly.

14.5 THE INTERNAL ENERGY OF A MONATOMIC IDEAL GAS

The temperature of a monatomic ideal gas is a measure of the average kinetic energy per particle. This energy is associated with the translational motion of the particles, not with any internal motions such as vibrations and/or rotations of the particles, because it is the momentum of the whole particle that contributes to the total gas pressure. We can therefore imagine a monatomic ideal gas simply as a collection of pointlike particles. Thus, if N is the number of particles in the monatomic ideal gas, the internal energy U is N times the average kinetic energy per particle. Using Equation 14.13 for the average kinetic energy per particle, we find that the internal energy is

$$U = N \left(\frac{3}{2} kT \right) \quad \text{(monatomic ideal gas)} \quad (14.19)$$

The number of particles N is equal to the number of moles n times Avogadro's number N_A:

$$N = n N_A$$

Equation 14.19 thus becomes

$$U = n N_A \left(\frac{3}{2} kT \right) \quad \text{(monatomic ideal gas)}$$

But from Equation 13.18, the product $N_A k$ is the universal gas constant R. Thus the internal energy is

$$U = \frac{3}{2} nRT \quad \text{(monatomic ideal gas)} \quad (14.20)$$

The internal energy of a monatomic ideal gas is a function only of the absolute temperature T and the quantity n of gas.

14.6 THE MOLAR SPECIFIC HEATS OF AN IDEAL GAS

With a knowledge of the internal energy of a monatomic ideal gas based on kinetic theory, we can predict the molar specific heats of the gas.

Specific Heat at Constant Volume for a *Monatomic* Ideal Gas

If a differential amount of heat transfer dQ occurs to n moles of a gas at constant volume, there is a differential change dT in the temperature of the gas. The *molar* specific heat at constant volume is defined as (Equation 13.27)

$$c_V = \frac{1}{n} \left(\frac{dQ}{dT} \right)_V$$

We take differentials of the first law of thermodynamics, Equation 13.51:

$$dQ = dU + dW$$
$$= dU + P \, dV$$

However, the heat transfer here is at constant volume, and so $dV = 0 \text{ m}^3$ and there is no work done by the gas ($dW = P \, dV = 0$ J). Therefore

$$dQ = dU$$

and the molar specific heat at constant volume is

$$c_V = \frac{1}{n} \frac{dU}{dT} \quad (14.21)$$

We use Equation 14.20 for the internal energy of the monatomic ideal gas:

$$c_V = \frac{1}{n} \frac{d}{dT} \left(\frac{3}{2} nRT \right)$$

$$c_V = \frac{3}{2} R \quad \text{(monatomic ideal gas)} \quad (14.22)$$

Substituting the numerical value for R, we find

$$c_V = 12.47 \text{ J/mol}$$

Table 14.1 indicates experimental values of the molar specific heat at constant volume for various monatomic gases near room temperature; the values are satisfyingly close to the value predicted by using kinetic theory.

TABLE 14.1 Measured Molar Specific Heats of Various Gases at Approximately Room Temperature (300 K)

Gas	c_V (J/mol·K)	c_P (J/mol·K)	$c_P - c_V$ (J/mol·K)
Monatomic gases			
Helium (He)	12.52	20.79	8.27
Neon (Ne)	12.68	20.79	8.11
Argon (Ar)	12.45	20.79	8.34
Krypton (Kr)	12.45	20.79	8.34
Xenon (Xe)	12.52	20.79	8.27
Diatomic gases			
Hydrogen (H_2)	20.44	28.82	8.38
Nitrogen (N_2)	20.80	29.12	8.32
Oxygen (O_2)	20.98	29.37	8.39
Carbon monoxide (CO)	20.74	29.04	8.30
Polyatomic gases			
Carbon dioxide (CO_2)	28.17	36.62	8.45
Nitrous oxide (N_2O)	28.39	36.90	8.51
Hydrogen sulfide (H_2S)	27.36	36.12	8.76

Specific Heat at Constant Pressure for *Any* Ideal Gas

Now let's try predicting the specific heat at constant pressure for any ideal gas, not necessarily a monatomic gas. If differential heat transfer dQ occurs to n moles of an ideal gas at constant pressure, the temperature changes by dT. The molar specific heat at constant pressure is, from Equation 13.28,

$$c_P = \frac{1}{n}\left(\frac{dQ}{dT}\right)_P$$

The first law of thermodynamics indicates that

$$dQ = dU + dW$$
$$= dU + P\,dV$$

In this case work *is* done by the gas since its volume changes. Substituting for dQ in the expression for the molar specific heat at constant pressure, we have

$$c_P = \frac{1}{n}\left[\left(\frac{dU}{dT}\right)_P + P\left(\frac{dV}{dT}\right)_P\right]$$
$$= \frac{1}{n}\left(\frac{dU}{dT}\right)_P + \frac{1}{n}P\left(\frac{dV}{dT}\right)_P \qquad (14.23)$$

where the subscripts indicate that the derivatives are taken at constant pressure. For the first term, the subscript is immaterial, since for an ideal gas the internal energy depends only on the absolute temperature T, not the pressure or volume. From Equation 14.21, the first term on the right-hand side of Equation 14.23 is the molar specific heat at constant *volume*. Thus

$$c_P = c_V + \frac{1}{n}P\left(\frac{dV}{dT}\right)_P \qquad (14.24)$$

The ideal gas equation of state is $PV = nRT$, so that at constant pressure

$$\left(\frac{dV}{dT}\right)_P = \frac{nR}{P}$$

Substituting this into Equation 14.24, we find

$$c_P = c_V + R \qquad \text{(any ideal gas)} \qquad (14.25)$$

For any ideal gas, the molar specific heat at constant pressure is *greater* than the molar specific heat at constant volume by an amount equal to the gas constant.

There is an easy way to see (qualitatively) that the molar specific heat at constant pressure must be greater than the molar specific heat at constant volume. When heat transfer occurs to the system at constant volume, $dV = 0\text{ m}^3$ and no work is done by the gas: the heat transfer goes *solely* into the internal energy (a function of the absolute temperature T alone). However, when heat transfer to the gas occurs at constant pressure, some of the energy goes into the internal energy and some of it is manifested as work done by the gas. Since only a part of the heat transfer goes into internal energy, it is necessary to have more heat transfer at constant pressure than at constant volume in order to get a specified rise in temperature; hence $c_P > c_V$.

If we specialize to a *monatomic* ideal gas, for which we have found that $c_V = 3R/2$, the specific heat at constant pressure is

$$c_P = \frac{3}{2}R + R$$
$$= \frac{5}{2}R \qquad \text{(monatomic ideal gas)} \qquad (14.26)$$
$$= 20.79\text{ J/(mol·K)}$$

Experimental values of the specific heat at constant pressure for various monatomic gases are also tabulated in Table 14.1, and once again we see that the predictions of kinetic theory are very good.

The central assumption involved in deriving Equation 14.25 was that the internal energy U was a function of only the absolute temperature (and the quantity of the gas). Hence for *any* gas whose internal energy U is a function of T (and, of course, the quantity n of gas), the relationship between the molar specific heats at constant pressure and at constant volume is given by Equation 14.25:

$$c_P = c_V + R \qquad [\text{any gas for which } U = U(T)]$$

For such gases, the *difference* between the molar specific heats is equal to the gas constant R:

$$c_P - c_V = R \qquad [\text{any gas for which } U = U(T)] \qquad (14.27)$$

Note that the differences between the experimentally determined molar specific heats for each of many gases come close to this prediction of kinetic theory, as indicated in the last column of Table 14.1.

14.7 COMPLICATIONS ARISE FOR DIATOMIC AND POLYATOMIC GASES

In Table 14.1 we saw that kinetic theory predicts the molar specific heats of monatomic gases quite closely, at least for the temperatures at which the experiments were conducted (300 K, near room temperature). On the other hand, the table also indicates that diatomic and polyatomic gases have molar specific heats whose values are quite different from those of monatomic gases; yet the *differences* between the molar specific heat at constant pressure and that at constant volume are quite close to the value of the gas constant, R, as with monatomic gases. In view of the previous section, this indicates that the internal energy of diatomic and polyatomic gases also is a function of the absolute temperature (and the quantity n of gas present), but not the same function as Equation 14.20 for monatomic gases. How can we determine the temperature dependence of the internal energy for such more complex gases?

Table 14.1 indicates that the molar specific heats of diatomic gases and polyatomic gases are *greater* than the corresponding values for monatomic gases. Hence the heat transfer to diatomic and polyatomic gases must be going not only into kinetic energy associated with the motion of the center of mass of the particles (as measured by the temperature of the gas), but also into other types of motions *internal* to the particles of the gas themselves, such as rotations and vibrations about the center of mass of each particle. These types of motions we consider in the next section.

14.8 DEGREES OF FREEDOM AND THE EQUIPARTITION OF ENERGY THEOREM

The idea of **degrees of freedom**, as the name implies, suggests restrictions or limits. There are degrees of freedom associated with our behavior as children, teenagers, and adults. In classical physics, the term degrees of freedom refers to a pure number that quantifies the number of distinct quadratic **generalized coordinates** needed to specify the *microscopic* total mechanical energy of one particle of the system (be it an atom or molecule). The various contributions to the total mechanical energy of a particle are continually shared and exchanged among all the parts of the interacting particle system. The sum of the microscopic total mechanical energies of each of the particles is the total internal energy U of the many-particle system.

The phrase quadratic generalized coordinates might encompass traditional coordinates such as a squared position (as in the potential energy associated with Hooke's force law), and also squared velocity and angular velocity components (such as in the expressions for the translational and rotational kinetic energies). Consider a few macroscopic examples:

1. A pointlike bead confined to slide along a fixed straight wire in the horizontal plane is said to have 1 **active** degree of freedom; see Figure 14.3. The gravitational potential energy of the bead is constant and we can choose it to be zero. A single squared generalized coordinate, say that of the x-component of the velocity of the particle, is needed to specify the total mechanical energy of the particle, which in this case is simply the kinetic energy associated with the translation of the center of mass of the particle.

FIGURE 14.3 A pointlike bead on a wire has 1 degree of freedom.

FIGURE 14.4 A large bead that can spin about and slide along a wire has 2 degrees of freedom.

2. A large bead that can slide along the wire *and* also *rotate* about the axis of the wire has 2 active degrees of freedom; see Figure 14.4. One squared velocity component is needed to describe the kinetic energy associated with the translation of the center of mass, and a squared angular velocity component is needed to describe the kinetic energy associated with the spin of the bead about the axis of the wire.

3. A pointlike water droplet on a hot skillet has 2 active degrees of freedom, because it is confined to move only across the two-dimensional area of the skillet. Two squared velocity components are needed to specify the kinetic energy associated with the motion of its center of mass. If the droplet also spins about its center of mass, another degree of freedom comes into play (becomes active): one associated with the rotational kinetic energy of the drop about its center of mass.

In three dimensions, three squared velocity components are needed to specify the total mechanical energy (here purely translational kinetic energy) of a point particle; thus, a point particle in three dimensions has 3 active degrees of freedom.

The idea of a degree of freedom seems simple enough. We will see, however, that there are some subtleties associated with this concept that warrant further comment.

The importance of being able to specify the number of degrees of freedom associated with each particle in a thermodynamic system arises from what is called the **equipartition of energy theorem**. We will not prove the theorem here, since it is a consequence of more sophisticated ideas in statistical mechanics. Nonetheless, we state the theorem:

> For a system of particles in thermal equilibrium, the average energy associated with each active degree of freedom of a particle is the same. For a system with f active degrees of freedom, the average energy per particle is f times as great as for a system with a single degree of freedom.

We can use our model of an ideal gas to find out how much energy is associated with each active degree of freedom. We already have found that the internal energy of n moles of a monatomic ideal gas of point particles is

$$U = \frac{3}{2} nRT$$

Since these particles each have 3 active degrees of freedom, and there are $nN_A = N$ particles in n moles, the energy per particle may be written as

$$\frac{U}{N} = \frac{3nRT}{2N}$$

$$= 3\frac{RT}{2N_A}$$

$$= 3\frac{kT}{2}$$

where k, Boltzmann's constant, is conveniently related to R as

$$k = \frac{R}{N_A}$$

The average kinetic energy per particle of the monatomic ideal gas of point particles is thus

$$\frac{1}{2}kT$$

for each of its three active degrees of freedom.

Hence the average energy per particle associated with each active degree of freedom of a system is

$$\frac{1}{2}kT$$

We will see later that polyatomic gases have the 3 translational degrees of freedom *and* some for their possible internal motions like vibration and rotation. The expressions for their average internal energies per particle contain more than the foregoing three units of $kT/2$ and are of different form.

14.9 SPECIFIC HEAT OF A SOLID*

It is quite surprising that the equipartition of energy theorem can account for the molar specific heat of a *solid* with all its close neighbors and the restricted motion of its constituent particles.

A crude model of an ideal solid is to imagine each atom having an equilibrium position and connected to its neighbors by springs. The springs permit each atom to execute simple harmonic oscillations about its equilibrium position. The oscillations can be resolved into equivalent oscillations along each of the three Cartesian coordinate axes. These equivalent oscillations along the three directions likely have different spring constants and oscillation frequencies. In contrast to an ideal gas, here we have both potential and kinetic energies to account for. There are 2 degrees of freedom associated with *each* oscillator axis, arising from the squared generalized coordinates in the kinetic and potential energy terms of simple harmonic oscillation:

$$\text{KE} + \text{PE} = \frac{1}{2}mv_x^2 + \frac{1}{2}k_x x^2$$

Thus each atom in the solid has $3 \times 2 = 6$ active degrees of freedom. The equipartition of energy theorem states that the average energy per particle associated with each degree of freedom

is $kT/2$. The entire solid sample has a huge number N of these atoms. The total internal energy of a solid system of atoms is then

$$U = N\left[6\left(\frac{1}{2}kT\right)\right]$$

$$= 3NkT$$

Since $Nk = nN_A k = nR$, where n is the number of moles and N_A is Avogadro's number, the internal energy is

$$U = 3nRT \qquad \text{(ideal solid)} \qquad (14.28)$$

Now consider the implications of the first law of thermodynamics. The differential heat transfer dQ to the system is equal to the differential change dU in the internal energy plus the differential work dW done by the system on its surroundings:

$$dQ = dU + dW$$

Here $dW = P\,dV$, so that

$$dQ = dU + P\,dV$$

The work done by a solid when there is heat transfer to the system is quite small because the change in the volume of the solid is tiny, thanks to the small numerical value of the coefficient of thermal expansion (volume) for solids. If we ignore the work done by the solid, then

$$dQ \approx dU$$

This is equivalent to saying that there is no difference between the specific heat at constant volume and that at constant pressure for a solid. For differential heat transfer dQ to a system of n moles, the temperature rises by dT, and the *molar* specific heat is (from Equation 13.24):

$$c = \frac{1}{n}\frac{dQ}{dT}$$

But we found that if the work done by the solid is zero (constant volume) or negligible (at constant pressure), then $dQ \approx dU$, and so

$$c \approx \frac{1}{n}\frac{dU}{dT}$$

From Equation 14.28,

$$\frac{dU}{dT} = 3nR$$

and so the molar specific heat of the ideal solid is

$$c \approx \frac{1}{n}3nR$$

$$\approx 3R \qquad \text{(ideal solid)} \qquad (14.29)$$

$$= 24.95 \text{ J/mol·K}$$

This result implies that the molar specific heat of an ideal solid composed of point particles is constant and equal to $3R$. This remarkable result was first discovered by Pierre Dulong (1785–1838) and Alexis Petit (1791–1820) in their experiments on certain elemental metals in the early part of the 19th century, and thus is called the **law of Dulong and Petit**. Table 13.4

(p. 602) indicates that the molar specific heat at room temperature of many metallic elements is close to this prediction. Molecular solids cannot be regarded as point-particle solids, and so we do not expect the prediction of Dulong and Petit to apply to them [notice that the values for water and ice in Table 13.4 are not close to $3R = 24.95$ J/(mol·K)].

14.10 SOME FAILURES OF CLASSICAL KINETIC THEORY

Atoms are not truly point particles. The particles of a monatomic gas (single atoms) have finite size. So the question arises, what about the rotation of the particles? In classical physics it is possible to imagine painting a little dot or planting a flag on the atom to track its rotation. In classical physics it is possible, in principle at least, to follow the rotation of atoms (by watching the dot or flag) about three mutually perpendicular coordinate axes. Three squared angular velocity components are needed to describe the rotational kinetic energies about the three axes. Thus you might think that the monatomic gas particle has a total of 6 active degrees of freedom: 3 associated with the kinetic energy of the center of mass, and 3 with rotational kinetic energy about the center of mass. Maybe you also can imagine the atoms vibrating, twisting, and otherwise contorting.

If f is the number of degrees of freedom, classical kinetic theory and the equipartition of energy theorem predict that the average energy per particle is

$$f\left(\frac{1}{2}kT\right) \tag{14.30}$$

The internal energy is then

$$U = Nf\left(\frac{1}{2}kT\right)$$
$$= f\frac{nRT}{2} \tag{14.31}$$

The molar specific heat at constant volume is

$$c_V = \frac{1}{n}\frac{dU}{dT}$$
$$= \frac{1}{n}\frac{d}{dT}\left(f\frac{nRT}{2}\right)$$
$$= \frac{1}{n}f\frac{nR}{2}$$
$$= f\frac{R}{2} \tag{14.32}$$

Hence classical physics and the equipartition of energy theorem would attribute an average energy of at least

$$6\left(\frac{1}{2}kT\right) = 3kT$$

to each particle of an ideal monatomic gas. Using Equations 14.32 and 14.21, such an energy leads to a specific heat at constant volume of

$$6\frac{R}{2} = 3R$$

This is *twice* as large as experiments indicate it is, $3R/2$ (from Equation 14.22 and Table 14.1). This situation is a clear failure of classical theory. You then might try to argue that to the extent we regard the particles as points, the energies associated with such possible rotational motions scarcely compare with the other energies (such as the kinetic energy of translation). Pointlike atoms move only as points. With such an argument, you might think you could ignore the 3 extra (rotational) degrees of freedom and reduce the degrees of freedom to 3 (those associated with the translation of the pointlike particles), which yields the correct molar specific heat for monatomic gases, as we saw in Section 14.6.

The argument, however, is not legitimate. The equipartition of energy theorem associates an energy of $kT/2$ with each active degree of freedom, *regardless* of the contribution, however small, each makes to the total mechanical energy. The problem only can be resolved with quantum mechanics, which indicates that the little dots or flags we imagined to follow the rotation of such monatomic particles cannot, in fact, be so imagined.

> A monatomic gas atom has only 3 degrees of freedom associated with translational motion, each with its average energy $kT/2$; these atoms show no further degrees of freedom, rotation or otherwise, according to quantum mechanics.

What about common *diatomic gases* such as H_2, O_2, or N_2? Here the elementary particles of the gas are molecules. There are 3 translational degrees of freedom associated with the motion of the center of mass of each molecule. But things get more complicated because such molecules really can rotate and vibrate. Let us look at the effects of these separately. Such diatomic molecules can be imagined as dumbbells, as shown in Figure 14.5. The axis of the dumbbell is a symmetry axis of the molecule.

Rotation

In classical physics, three squared angular velocity components are needed to specify the rotational kinetic energies about the various axes of, say, a standard Cartesian coordinate system. If you choose the z-axis to represent the symmetry axis along which the two atoms lie, the angular velocity components are (a) the angular velocity component ω_z about the z-axis; (b) the angular velocity component ω_x associated with rotation of the molecule about the x-axis; and (c) the angular velocity component ω_y associated with rotation about the y-axis.

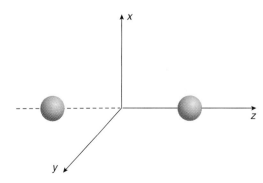

FIGURE 14.5 A diatomic molecule.

Quantum mechanics excludes the rotation about the axis of the molecule from consideration. An old (but incorrect) classical way to exclude this degree of freedom argues that the pointlike atoms have negligible moments of inertia about the symmetry axis because they lie on it; the rotational kinetic energy associated with the rotation about this axis thus is negligible. However, as with the rotation of monatomic gas particles, this argument still conflicts with the equipartition of energy theorem (and for the same reason): just because the moment of inertia and rotational energy are small, it does not dismiss the degree of freedom from the equipartition theorem.

> It is quantum mechanics that indicates that diatomic dumbbell molecules have only 2 (not 3) rotational degrees of freedom, associated with the rotational kinetic energy about the *x*- and *y*-axes.

Vibration

If you model the force between the two atoms in a diatomic molecule by means of a Hooke's force law between them, the oscillations of the spring mimic the vibration of the molecule along the symmetry axis, as shown in Figure 14.6. There are potential and kinetic energies associated with this oscillation, and so *2 vibrational* degrees of freedom exist.

Thus, for a diatomic molecule, there appear to be

3 degrees of freedom associated with the motion of the center of mass of each molecule,
2 rotational degrees of freedom associated with the rotational kinetic energies about the two significant rotational axes, and
2 vibrational degrees of freedom associated with the kinetic and potential energies of vibration along the symmetry axis.

This makes a total of 7 degrees of freedom for such molecules.

Now invoking the equipartition of energy theorem, we see that an energy *kT*/2 is associated with each degree of freedom, and the average energy per molecule is therefore

$$7\left(\frac{1}{2}kT\right)$$

The total internal energy of a system of *N* such molecules then would appear to be

$$U = N\left[7\left(\frac{1}{2}kT\right)\right]$$
$$= \frac{7}{2}nRT$$

Using Equation 14.21 for the molar specific heat at constant volume, we predict

$$c_V = \frac{1}{n}\frac{dU}{dT}$$
$$= \frac{7}{2}R$$
$$= 29.1 \text{ J/(mol·K)}$$

Consult the experimental values of c_V in Table 14.1. We see that the molar specific heat *at constant volume* for diatomic gases at room temperature is measured to be about 20.8 J/mol, which is very close to 5R/2, *not* 7R/2 as just predicted.

> Evidently, *only five* degrees of freedom are active for diatomic molecules at room temperature, since $c_V \approx 5R/2$.

Classical kinetic theory again is unable to account for this discrepancy in the number of degrees of freedom for a diatomic gas. Only quantum mechanics correctly explains why 5, not 7, degrees of freedom are available (see the next section): the vibrational degrees of freedom are not active or excited at room temperature. The permitted vibrational energies are quantized and there is insufficient energy available at room temperature to excite the vibrations. Effectively, at room temperatures, diatomic molecules translate and rotate but do not vibrate: they are like two masses connected by a very stiff spring or a massless rigid rod. We say the vibrational degrees of freedom are **frozen out** and are not active.

The temperature dependence of the molar specific heat of diatomic gases bears out this behavior; see Figure 14.7. At high temperatures the vibrational degrees of freedom become excited, resulting in the full 7 active degrees of freedom (translation, rotation, and vibration), and the molar specific heat at constant volume of diatomic gases approaches 7R/2. At very high temperatures, the molecule breaks apart. At room temperature, only 5 degrees of freedom are active (translation and rotation), and so the molar specific heat at constant volume is 5R/2.

The energies associated with the rotation also are quantized, but with a closer energy spacing than for the vibrational energies

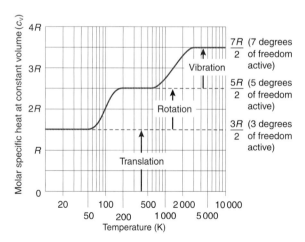

FIGURE 14.7 The molar specific heat at constant volume of a diatomic gas as a function of temperature.

FIGURE 14.6 Vibration along the symmetry axis of a diatomic molecule is modeled by a Hooke's law spring.

(see Section 14.11) so rotation is activated at a lower temperature than is vibration. At very low temperatures, even the rotational degrees of freedom are frozen out and the diatomic gas behaves like a monatomic gas with only 3 active degrees of freedom, those associated with the translational motion of the center of mass of the particles. The transitions between the various regions in Figure 14.7 are not abrupt, indicating that the average number of active degrees of freedom in these regions is not an integer, a phenomenon that cannot be explained classically.

In this text, we confine the temperatures of the gases to regions where the ratio γ of the molar specific heats of an ideal gas is the following: For a *monatomic* ideal gas, the ratio of the molar specific heats is

$$\gamma \equiv \frac{c_P}{c_V} = \frac{5R/2}{3R/2} = \frac{5}{3} = 1.67 \quad \text{(monatomic ideal gas)}$$
$$(14.33)$$

For *diatomic* ideal gases, the ratio of the molar specific heats is

$$\gamma \equiv \frac{c_P}{c_V} = \frac{7R/2}{5R/2} = \frac{7}{5} = 1.40 \quad \text{(diatomic ideal gas)}$$
$$(14.34)$$

The situation for triatomic and polyatomic molecular gases is still more complicated. Although we once again have 3 translational degrees of freedom, the situation becomes more involved for counting the various rotational and vibrational degrees of freedom; we will not consider such complex polyatomic molecular gases.

With solids, we saw that the molar specific heat of many metals at room temperature is close to $3R$, as predicted. But the molar specific heat of many molecular solids is not $3R$, indicating that something more complex is going on. Indeed, as with gases, the specific heat is a function of temperature; this is not what the classical arguments predict for an ideal solid. The specific heat varies from solid to solid and is a function of temperature, approaching zero as the absolute temperature approaches zero, as shown in Figure 14.8.

Once again, it is quantum mechanics that accounts for this behavior. As the temperature is lowered, various degrees of freedom evidently are frozen out because of quantum mechanical effects. Dulong and Petit found the specific heats of solids at high temperatures approach the value $3R$. At low temperatures, this is clearly not the case. As the temperature increases, many molecular solids *melt* before their specific heat approaches the value predicted by the law of Dulong and Petit.

We have seen how the temperature dependence of the specific heat of a substance is a macroscopic reflection of changes taking place at the microscopic level. We saw this in our study of gases, in the different specific heats of the same substance in its various phases (e.g., ice, water, steam), and in the latent heat associated with phase transitions. Dramatic temperature variations of the specific heat are clear indications that some new physics (quantum mechanics) is lurking. It is rather surprising that early hints of the need for a new mechanics at the atomic and molecular level (quantum mechanics) were signaled by the failures of classical kinetic theory; however, the problems were not readily understood by the wise minds that formulated classical thermodynamics.

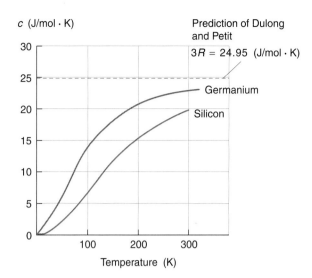

FIGURE 14.8 Molar specific heat of a solid as a function of temperature.

14.11 QUANTUM MECHANICAL EFFECTS*

In this section we present some of the features of quantum mechanics that serve to explain the peculiar behavior of the specific heat.

Think of a particle such as an atom confined to rattling around in a box of very small size (perhaps only a few atomic diameters across). According to quantum mechanics, particles confined to a box are not allowed to have just any amount of energy; in fact, most values of energy are forbidden. We say that the energy is *quantized* and that only certain values of the energy are permitted or allowed, as shown in Figure 14.9. The special energies that are allowed are given many equivalent names: **energy levels, energy states,** or **eigenstates of the energy.** There are forbidden values of energy (**energy gaps**) between these allowed values of the energy.

Let the difference between the permitted energies be called $E_{gap}(box)$;[†] energy added to the system must come in lumps of $E_{gap}(box)$. Thus $E_{gap}(box)$ represents a kind of bundle of energy: energy is added (or taken away) from a particle of the system only in lumps of $E_{gap}(box)$.

The larger the dimensions of the box, the more closely spaced the energy levels of a particle become. That is, $E_{gap}(box)$ becomes smaller and smaller as the dimensions of the box increase. For molecules in a box of macroscopic dimensions, the size of the energy gap $E_{gap}(box)$ is vanishingly small compared with the energy $kT/2$ associated with a degree of freedom in the equipartition of energy theorem; this is illustrated schematically in Figure 14.10.

[†] The differences between the permitted energies actually are not all the same for a particle confined to a box, but we make that assumption here for simplicity.

Energy

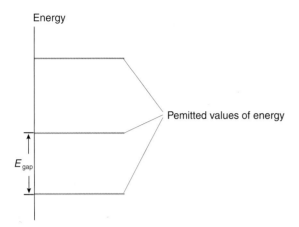

Pemitted values of energy

E_{gap}

FIGURE 14.9 Allowed energy values, with energy gaps between them.

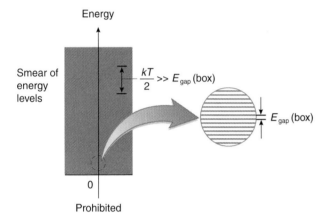

Energy

Smear of energy levels

$\frac{kT}{2} \gg E_{gap}$ (box)

E_{gap} (box)

0

Prohibited

FIGURE 14.10 For a particle in a box of macroscopic size, the size of the energy gap between allowed values of the energy is vanishingly small.

Thus, for particles inside a large box, essentially any translational kinetic energy greater than zero is permitted; negative values for the kinetic energy are not allowed since this implies a speed that is an imaginary number. Speeds must be real numbers because they are measurable quantities.

The rotational and vibrational energies of atoms in molecules also are quantized. For these cases, the magnitudes of the spacing of the permitted energy values, E_{gap}(vibration) and E_{gap}(rotation), are much larger than for the translational motion of the particles confined to a large box. Furthermore, the typical energy gap for vibrational energy levels, E_{gap}(vibration), is larger than for rotational energy levels, E_{gap}(rotation), as shown schematically in Figure 14.11.

The significance of the discrete nature of the energy levels becomes clear when we compare the spacing of the permitted energies with the energy $kT/2$ in the equipartition of energy theorem. If the energy $kT/2 \ll E_{gap}$, then it is not possible for the system to gain energy via this mechanism. We say that the degree(s) of freedom associated with this E_{gap} are frozen out and not active. In other words, the temperature is so low that those degrees of freedom are

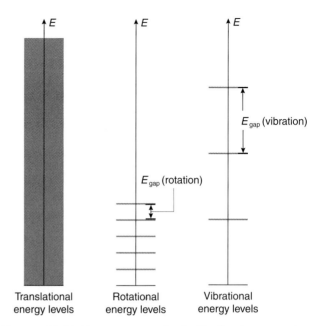

Translational energy levels

Rotational energy levels

E_{gap} (rotation)

Vibrational energy levels

E_{gap} (vibration)

FIGURE 14.11 Energy gaps associated with vibration are much larger than the gaps associated with rotation.

unreachable by the system. This means that the number of active degrees of freedom is a function of temperature.

As the temperature increases, $kT/2$ grows in value, and more degrees of freedom become accessible to the system. Thus at very low temperatures, a diatomic gas has only 3 active (translational) degrees of freedom. As the temperature increases and $kT/2$ approaches the size of E_{gap}(rotation), a diatomic gas then can populate 5 degrees of freedom (3 translational and 2 rotational). As the temperature increases further and $kT/2$ approaches the size of E_{gap}(vibrational), the system has 7 active degrees of freedom (3 translational, 2 rotational, and 2 vibrational). Of course, this assumes the gas does not condense at the low temperatures and does not disassociate at the high temperatures.

At very high temperatures (thousands of kelvin), further degrees of freedom come into play as the energy $kT/2$ becomes on the order of the sizes of the quantum mechanical energy gaps in typical atoms. At still higher temperatures (millions of kelvin), $kT/2$ is on the order of the energy gaps associated with particles in the nucleus, and (previously frozen) nuclear degrees of freedom join the others and become active.

It is not easy to describe the gradual way that the specific heat changes with temperature in Figure 14.7. The temperature regions where new degrees of freedom come into play actually are more complex than our treatment indicates.

For simplicity, we will avoid these transition regions and consider the number of degrees of freedom available to a system as an integer. In particular, we will say that the temperatures are such that

1. monatomic gases have 3 degrees of freedom, and

2. diatomic gases have 5 degrees of freedom.

14.12 AN ADIABATIC PROCESS FOR AN IDEAL GAS

> An **adiabatic process** involves zero heat transfer between a system and its surroundings.

We mentioned such a process briefly in Section 13.12, but deferred considering them until now. An adiabatic process is, of course, an idealization, since a perfect insulator does not exist. Nonetheless, the process is a useful idealization if, for example, (1) the system is well insulated from its surroundings to minimize heat transfer, or (2) the changes take place so rapidly that there is essentially no time for heat transfer to occur before the change is completed yet slow enough for the process to still be quasi-static. The compressions of gases that occur in engines and refrigerators are examples that may approximate the latter situation.

We investigate engines and refrigerators in the next chapter. Other examples of adiabatic processes include the rapid compressions and expansions that occur when sound waves propagate through a medium. Even clouds involve adiabatic processes: they form when moist rising air cools in an adiabatic expansion, enabling water vapor to condense into tiny fog droplets.

A quasi-static adiabatic process is one that takes place slowly enough for the system to be in thermodynamic equilibrium at any instant, but rapidly enough for heat transfer between the system and its surroundings to be negligible. This is commonly achievable.

The quasi-static adiabatic expansion of an ideal gas is represented on a P–V diagram as shown in Figure 14.12. Note that the pressure, volume, and temperature *all* change during an adiabatic process. We seek to discover how these thermodynamic parameters are related to each other at the beginning and end of such a quasi-static adiabatic process. Since the system is

Cloud formation involves an adiabatic expansion.

in thermodynamic equilibrium at any instant during the process, we assume the ideal gas law is valid at any instant.

We begin with the first law of thermodynamics in differential form:

$$dQ = dU + dW$$
$$= dU + P\,dV$$

In an adiabatic process the heat transfer to the system is zero, and so $dQ = 0$ J. Thus the first law of thermodynamics becomes

$$0 \text{ J} = dU + P\,dV \quad \text{(adiabatic process)} \quad (14.35)$$

> In other words, for an adiabatic process, the differential change in the internal energy of the gas is the negative of the differential work done by the gas:
>
> $$dU = -P\,dV \quad \text{(adiabatic process)} \quad (14.36)$$

From Equation 14.21, the differential change in internal energy can be written as

$$dU = nc_{\text{V}}\,dT$$

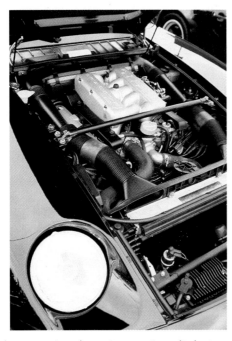

The quick compression of a gas in an engine cylinder is approximately adiabatic.

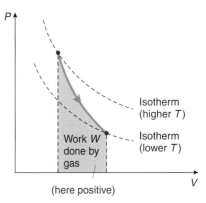

FIGURE 14.12 An adiabatic expansion of an ideal gas on a P–V diagram.

Making this substitution into Equation 14.35, we have

$$0 \text{ J} = nc_V \, dT + P \, dV$$

or

$$dT = -\frac{P \, dV}{nc_V} \quad (14.37)$$

The equation of state for an ideal gas is Equation 14.1:

$$PV = nRT$$

Since the pressure, volume, *and* temperature all change in an adiabatic process, if we take differentials of the equation of state, we get

$$P \, dV + V \, dP = nR \, dT \quad (14.38)$$

Using Equation 14.37 for dT, Equation 14.38 becomes

$$P \, dV + V \, dP = nR\left(-\frac{P \, dV}{nc_V}\right)$$

$$= -RP \frac{dV}{c_V} \quad (14.39)$$

Since the universal gas constant R is equal to the *difference* between the molar specific heats of an ideal gas (Equation 14.27), we can replace R in Equation 14.39 with

$$R = c_P - c_V$$

Thus Equation 14.39 becomes

$$P \, dV + V \, dP = -\left(\frac{c_P - c_V}{c_V}\right) P \, dV$$

$$= -(\gamma - 1) P \, dV \quad (14.40)$$

where γ is the ratio of the molar specific heats c_P/c_V.

Equation 14.40 simplifies to

$$V \, dP = -\gamma P \, dV$$

Rearranging this slightly, we find

$$\frac{dP}{P} = -\gamma \frac{dV}{V} \quad (14.41)$$

which is in a form suitable for integration. We integrate it from an initial pressure P_i and volume V_i to some final pressure P_f and volume V_f:

$$\int_{P_i}^{P_f} \frac{dP}{P} = -\gamma \int_{V_i}^{V_f} \frac{dV}{V}$$

$$\ln P_f - \ln P_i = -\gamma(\ln V_f - \ln V_i)$$

Collecting final and initial terms together,

$$\ln P_f + \gamma \ln V_f = \ln P_i + \gamma \ln V_i$$

Taking antilogarithms, we find

$$P_f V_f^\gamma = P_i V_i^\gamma \quad (14.42)$$

We see that in an adiabatic process, the product PV^γ remains constant:

$$PV^\gamma = \text{constant} \quad \text{(adiabatic process)} \quad (14.43)$$

If we couple this result with the equation of state for an ideal gas, $PV = nRT$, we can find similar relationships between the temperature and the volume and between the temperature and pressure:

1. We use Equation 14.43 and eliminate P with the ideal gas law; this yields

$$TV^{\gamma-1} = \text{constant} \quad \text{(adiabatic process)} \quad (14.44)$$

2. We use Equation 14.43 and eliminate V using the ideal gas law; this yields

$$P^{1-\gamma}T^\gamma = \text{constant} \quad \text{(adiabatic process)} \quad (14.45)$$

The work done by the gas during an adiabatic process can be found in the following way. In an adiabatic process $dQ = 0$ J, so that the first law of thermodynamics becomes

$$0 \text{ J} = dU + dW$$

or

$$W = -\Delta U \quad (14.46)$$

PROBLEM-SOLVING TACTIC

14.3 The work done by an ideal gas in a quasi-static adiabatic process can be found most directly by calculating the change in the internal energy of the system.

For an ideal gas, the internal energy is a function of temperature only. By finding the initial and final temperatures of the system, the change in the internal energy can be determined. The work done by the system then is just the negative of the change in the internal energy. Indeed, from Equation 14.21, we have $dU = nc_V \, dT$. Since the molar specific heat of an ideal gas is constant, we can integrate this to obtain

$$\Delta U = nc_V \, \Delta T \quad (14.47)$$

Employing Equation 14.46, we find that the work done by an ideal gas in an adiabatic process is

$$W_{\text{adiabatic}} = -nc_V \, \Delta T \quad (14.48)$$

QUESTION 4

An ideal gas is compressed adiabatically. What happens to the temperature of the gas? Since there is no heat transfer to the gas in an adiabatic process, what is the source of the energy that changes the temperature of the gas?

STRATEGIC EXAMPLE 14.7

A diatomic gas in an engine is initially at 300 K, 1.000 atm pressure, and occupies a volume 900 cm³. The gas is compressed adiabatically by a piston in a cylinder to 0.100 its initial volume.

a. Find the final pressure of the gas.
b. Find the final temperature of the gas.
c. What is the work done by the gas?

Solution

a. For an adiabatic process, use Equation 14.43 to find the final pressure:

$$PV^\gamma = \text{constant} \qquad \text{(adiabatic process)}$$

Hence

$$P_f V_f^\gamma = P_i V_i^\gamma$$

$$P_f = P_i \left(\frac{V_i}{V_f} \right)^\gamma$$

The volume ratio is 1/0.100 = 10.0. Since the gas is diatomic, the ratio γ of its molar specific heats is 1.40 (Equation 14.34). Hence you have

$$P_f = P_i (10.0)^{1.40}$$
$$= (1.000 \text{ atm}) \, 25.1$$
$$= 25.1 \text{ atm}$$

b. To find the final temperature, use Equation 14.44,

$$TV^{\gamma-1} = \text{constant} \qquad \text{(adiabatic process)}$$

Hence

$$T_f V_f^{\gamma-1} = T_i V_i^{\gamma-1}$$

$$T_f = T_i \left(\frac{V_i}{V_f} \right)^{\gamma-1}$$

The initial to final volume ratio is 10.0, and so

$$T_f = T_i (10.0)^{1.40-1}$$
$$= (300 \text{ K})(10.0)^{0.40}$$
$$= 7.5 \times 10^2 \text{ K}$$

In diesel engines, the increase in temperature accompanying an adiabatic compression is sufficient to make the fuel spontaneously explode without the necessity of the spark needed in ordinary gasoline engines.

c. To find the work done by the gas, use Equation 14.48, which stemmed from the change in the internal energy of the gas:

$$W_{\text{adiabatic}} = -nc_V \, \Delta T$$

You have to calculate the number of moles of gas present. Use the initial values of temperature, pressure, and volume in the ideal gas law to find n:

$$PV = nRT$$

Solve for n and remember to use pascals for the pressure:

$$n = \frac{PV}{RT}$$
$$= \frac{(1.013 \times 10^5 \text{ Pa}) (9.00 \times 10^{-4} \text{ m}^3)}{[8.315 \text{ J/(mol·K)}] (300 \text{ K})}$$
$$= 3.65 \times 10^{-2} \text{ mol}$$

The molar specific heat at constant volume for a diatomic gas is $5R/2 = 20.1$ J/(mol·K). Making these substitutions into the foregoing expression for the work, you find

$$W_{\text{adiabatic}} = -(3.65 \times 10^{-2} \text{ mol})[20.1 \text{ J/(mol·K)}]$$
$$\times (7.5 \times 10^2 \text{ K} - 300 \text{ K})$$
$$= -3.3 \times 10^2 \text{ J}$$

Negative work is done by the gas (since it is compressed), indicating that positive work was done *on* the gas by the piston.

CHAPTER SUMMARY

The *kinetic theory of an ideal gas* is based on the following assumptions:

1. The number of particles N in the gas is huge.
2. The volume V containing the gas is much larger than the total volume actually occupied by the gas particles themselves.
3. The dynamics of the particles is governed by Newton's laws of motion.
4. The particles are equally likely to be moving in any direction.
5. The gas particles (be they atoms or molecules) interact with each other and with the walls of the container only via elastic collisions.
6. The gas is in thermal equilibrium with its surroundings.
7. The particles of the gas are identical and indistinguishable.

The pressure of a gas arises from the myriad collisions of its constituent particles with the walls of the container. For an ideal gas of N particles each of mass m in a volume V, the pressure is

$$P = \frac{m}{3V} N \langle v^2 \rangle \qquad (14.9)$$

where $\langle v^2 \rangle$ is the average of the square of the speeds of the particles. The average kinetic energy per particle of the gas is

$$KE_{\text{ave}} = \frac{3}{2} kT \qquad (14.13)$$

where k is Boltzmann's constant and T is the absolute temperature of the gas. Hence the temperature is a measure of the average kinetic energy per particle.

The *root-mean-square (rms) speed* v_{rms} of the particles of mass m in the gas is

$$v_{rms} = \left(\frac{3kT}{m}\right)^{1/2} \qquad (14.14)$$

$$= \left(\frac{3RT}{M}\right)^{1/2} \qquad (14.16)$$

where R is the universal gas constant and M is the molar mass (the mass of one mole) of the gas.

The internal energy of an ideal gas is function of the absolute temperature T and the quantity n of gas. For a monatomic ideal gas, the internal energy is

$$U = \frac{3}{2}nRT \qquad \text{(monatomic ideal gas)} \qquad (14.20)$$

The number of *degrees of freedom* populated by (i.e., active in) a system indicates how many quadratic generalized coordinates are needed to specify the microscopic total mechanical energy of a particle.

The *equipartition of energy theorem* states that the average energy per particle is the same for each degree of freedom, and this energy is equal to

$$\frac{1}{2}kT$$

For an ideal gas with f active degrees of freedom, the average energy per particle is

$$f\frac{1}{2}kT \qquad (14.30)$$

The internal energy is

$$U = f\frac{nRT}{2} \qquad (14.31)$$

and the molar specific heat at constant volume is

$$c_V = f\frac{R}{2} \qquad (14.32)$$

An ideal monatomic gas has 3 active degrees of freedom per particle. Thus the molar specific heat at constant volume of a monatomic ideal gas is

$$c_V = \frac{3}{2}R \qquad \text{(monatomic ideal gas)} \qquad (14.22)$$

At room temperature, a diatomic gas has 5 active degrees of freedom; thus, its molar specific heat is

$$c_V = \frac{5}{2}R \qquad \text{(diatomic ideal gas)}$$

The difference between the molar specific heat at constant pressure and that at constant volume is equal to the universal gas constant:

$$c_P - c_V = R \qquad \text{[any gas for which } U = U(T)\text{]} \qquad (14.27)$$

This applies to *any ideal gas*, monatomic, diatomic, or polyatomic, for which the internal energy is a function of the absolute temperature, $U = U(T)$, independent of pressure and volume.

For an ideal gas undergoing a quasi-static adiabatic process,

$$PV^\gamma = \text{constant} \qquad (14.43)$$

where γ is the ratio of the molar specific heats:

$$\gamma \equiv \frac{c_P}{c_V}$$

We restrict the gas to a temperature range for which the value of γ is

$$\gamma = \frac{5}{3} = 1.67 \qquad \text{(monatomic ideal gas)} \qquad (14.33)$$

$$\gamma = \frac{7}{5} = 1.40 \qquad \text{(diatomic ideal gas)} \qquad (14.34)$$

The work done by an ideal gas in an adiabatic process is the negative of the change in the internal energy of the gas:

$$W = -\Delta U \qquad \text{(adiabatic process)} \qquad (14.46)$$

$$= -nc_V\Delta T \qquad \begin{array}{l}\text{(adiabatic process} \\ \text{for an ideal gas)}\end{array} \qquad (14.48)$$

SUMMARY OF PROBLEM-SOLVING TACTICS

14.1 **(page 646)** When using Equation 14.16, if R is expressed as 8.315 J/(mol·K), be sure the molar mass M is expressed in kilograms per mole rather than grams per mole.

14.2 **(page 646)** When using Equations 14.14–14.16, be sure you use the absolute temperature T in kelvin (not celsius).

14.3 **(page 656)** The work done by an ideal gas in a quasi-static adiabatic process can be found most directly by calculating the change in the internal energy of the system.

QUESTIONS

1. (page 644); 2. (page 646); 3. (page 646); 4. (page 656)

5. If the particles of an ideal gas made completely inelastic collisions with the walls of the container, what would happen to the pressure of the gas as a function of time?

6. Before Galileo (1564–1642), it was thought that a continual force was needed just to keep a system moving at constant velocity. Now you know from Newton's first law of motion that a system will continue at constant velocity indefinitely if there is zero total force on it. A similar question arises in kinetic

theory. Does a gas in thermal equilibrium need a continual positive heat transfer just to maintain its temperature? Explain your answer using ideas from the kinetic theory of gases.

7. One of the following statements is true, the other is false. Explain why each is which: (a) The temperature is a measure of the kinetic energy of a particle of a gas. (b) The temperature is a measure of the average kinetic energy of a particle of a gas.

8. At a given temperature, the particles of a gas have many different speeds rather than identical speeds. Give arguments to justify this statement.

9. (a) In what average direction do the gas molecules move in a closed room? (b) Explain why their average kinetic energy is not zero even though their average velocity is zero.

10. Explain why at low altitudes the temperature decreases with increasing altitude in the atmosphere of the Earth. How can this observation be reconciled with the fact that hot air rises by convection?

11. To triple the rms speed of the particles of an ideal gas, by what factor does the temperature need to be increased?

12. The pressure of an ideal gas is tripled while the volume is kept constant. By what factor does the average kinetic energy of a particle in the gas increase?

13. The volume of an ideal gas is tripled while the pressure is kept constant. By what factor does the average kinetic energy of a particle in the gas change?

14. The volume of a gas is decreased in an isothermal process. On the basis of kinetic theory, explain why the pressure of the gas increases.

15. The planet Mercury has a relatively small mass and is the planet closest to the Sun. Give some arguments that might explain why it has no atmosphere.

16. Some of the molecules in the upper atmosphere of the Earth have speeds that exceed the escape speed from the Earth. Does the atmosphere of the Earth gradually dissipate into space?

17. In order to evaporate from a liquid such as water, a molecule must have a certain minimum kinetic energy. (a) Explain why evaporation cools the remaining liquid. (b) How is it possible for a liquid to evaporate completely from an open dish?

18. Is it legitimate to talk about the temperature of a *single* molecule? If not, about how many molecules must be present before the concept of temperature becomes meaningful?

19. Explain why the molar specific heat at constant pressure for gases is greater than the molar specific heat at constant volume.

20. Is the molar specific heat at constant pressure for solids and liquids greater than the molar specific heat at constant volume? Explain.

21. Would you expect the *difference* between the specific heat at constant pressure and the specific heat at constant volume to be greater for liquids than for solids? Explain. (Hint: Think of the relative amounts of thermal expansion between solids and liquids.)

22. One thousand BBs, each of mass 10^{-4} kg, are in an evacuated cubical container of edge 10.0 cm. The container is in an orbiting spacecraft. The system is shaken vigorously and the average speed of a BB is 3.00 m/s. Is it legitimate to talk of a temperature of the BB system? If so, calculate the temperature. Is it likely that the particles will continue to move perpetually as they do in a real gas? Explain.

23. A gas-filled football is thrown by a quarterback to a wide receiver. Is the average velocity of the particles in the gas zero from the point of view of the quarterback? Explain.

24. On the basis of kinetic theory, explain why there is a minimum temperature but no maximum temperature. [This is true even though particles have an upper speed limit, the speed of light. However, this speed limit is not apparent at this stage in your study of physics; classical kinetic theory breaks down at relativistic speeds (speeds that approach the speed of light).]

25. If an ideal gas is simultaneously warmed (by heat transfer) and compressed, how is the molar specific heat for this process qualitatively related to the molar specific heats c_P and c_V? That is, is the molar specific heat for the process greater than both c_P and c_V, less than both, or greater than one and less than the other (which one?)? Explain your reasoning.

26. For an ideal gas, will the difference between the specific heat at constant pressure and that at constant volume be the same for all gases if the specific heats are *not* molar specific heats but the specific heat expressed as joules per kilogram-kelvin [J/(kg·K)]? Explain your answer.

27. A macroscopic mass m slides across a rough surface and comes to rest. On the basis of kinetic theory, where does the kinetic energy of the mass go? Why is it very unlikely that this energy ever will return of its own accord to become macroscopic kinetic energy of the mass m?

28. A thermometer is placed in a region that is a vacuum. Will the thermometer indicate 0 K? Explain why the thermometer will indicate the temperature of the material container enclosing the vacuum. Does this mean that a vacuum has a temperature? Is a vacuum *really* empty? Explain.

29. Explain why *only* the translational kinetic energy of the particles in a gas is proportional to the absolute temperature, and not other contributions to the internal energy such as rotations or vibrations of the gas particles.

30. The kinetic theory of an ideal gas assumes the total volume of the particles themselves is negligible. What effect might the finite volume of the particles have on the ideal gas law?

31. The kinetic theory of an ideal gas assumes that the particles of the gas exert no forces on each other except during the brief time intervals of collisions. What effect might you expect on the ideal gas law from the existence of long-range forces between the particles?

32. The difference between the molar specific heats of an ideal gas is equal to the universal gas constant:

$$c_P - c_V = R$$

Does this also apply for liquids and solids? If so, explain why. If not, explain why not.

33. In what ways are 100 third-graders in bumper cars, confined to a bumper car rink, similar to the particles in a gas? In what ways are they different? Is it possible to develop a kinetic theory of bumper cars?

34. Saturn's most massive moon, Titan, has an atmosphere. Explain why it is more likely for hydrogen gas (H_2) to be found in the Titan atmosphere than in the atmosphere of the much more massive Earth.

35. On the basis of kinetic theory, explain why the temperature of a gas decreases in an adiabatic expansion.

36. In an adiabatic expansion of a system, there is no heat transfer between the system and its environment, yet the temperature and pressure of the gas decrease. Explain how this occurs on the basis of kinetic theory.

PROBLEMS

Sections 14.1 Background for the Kinetic Theory of Gases
14.2 The Ideal Gas Approximation
14.3 The Pressure of an Ideal Gas
14.4 The Meaning of the Absolute Temperature

1. A mole of a gas contains Avogadro's number N_A of particles:

$$N_A = 6.022 \times 10^{23} \text{ particles/mol}$$

If you counted the particles in a mole at the rate of one particle per second, how many years would it take to complete the count? For comparison, the age of the universe is estimated to be about 15 billion years.

2. If 1.00 mol of particles is spread uniformly over the surface of the Earth, how many particles are on each square meter?

3. If the particles in 1.00 mol were dispersed to a density of one particle per cubic meter, what radius of sphere would contain these particles? Express your result in kilometers.

4. An ideal gas is at 1.00 atm pressure and 0 °C. Under these conditions, the number of particles per cubic *centimeter* is known as the *Loschmidt number*. Calculate the Loschmidt number.

5. Hydrogen fusion within the core of the Sun occurs at a temperature of about 15×10^6 K. What is the average kinetic energy of hydrogen atoms at this temperature? [The hydrogen atoms actually are completely ionized at this temperature, so this is the average kinetic energy of the nucleus (a proton).]

6. (a) In order to double the rms speed of the particles in a gas, by what factor must the temperature of the gas be increased? (b) Does this result depend on the temperature scale used? (c) Does this result depend on the units used to measure the speed of the particles?

7. A fast sprinter can attain a speed of 10 m/s. At what temperature do helium atoms have an rms speed of 10 m/s?

8. At what temperature do atoms of helium have an rms speed equal to 1.00% of the speed of light?

9. Calculate the rms speed of the particles in a gas of molecular hydrogen (H_2) at temperature 300 K. By what factor does this differ from the rms speed of oxygen molecules (O_2) at the same temperature?

•10. Show that equal volumes of gas at the same temperature and pressure have equal numbers of particles. This statement is called *Avogadro's law*.

•11. Show that the number of particles in a cubic millimeter of a gas at temperature 273 K and 1.00 atm pressure is 2.69×10^{16}. To get a feeling for the order of magnitude of this number, calculate the age of the universe in seconds assuming it is 15 billion years old.

•12. Six particles have speeds 0.10 m/s, 0.20 m/s, 0.30 m/s, 0.40 m/s, 0.50 m/s, and 0.60 m/s. Show that their average speed is less than their rms speed.

•13. A sample of oxygen gas (O_2) is at temperature 300 K and 1.00 atm pressure. One molecule, with a speed equal to the rms speed, makes a head-on elastic collision with your nose. Ouch! What is the magnitude of the impulse imparted to your schnozzle?

•14. For an ideal gas at temperature 300 K and 1.00 atm pressure, what are the dimensions of a cube that contains 1000 particles of the gas?

•15. To increase the rms speed of a gas by 1.0%, by what percentage must the temperature increase?

•16. Assume dust grains have masses about 10^9 times that of a helium atom. At temperature 300 K, what is the rms speed of a dust speck if you treat the dust as an ideal gas?

•17. One (1.00) mole of an ideal gas is at temperature 300 K and at 1.00 atm pressure. (a) What is the volume of the gas? (b) Estimate the average spacing between the particles of the gas.

•18. Five lonely gas molecules occupy a volume V. Their y- and z-velocity components are all zero and their x-velocity components are −250.0 m/s, −40.0 m/s, 40.0 m/s, 90.0 m/s, and 160.0 m/s. (a) Calculate their average *velocity*. (b) Calculate their average speed. (c) Calculate their rms speed.

•19. A census is made of the speeds of a random sample of 1010 oxygen molecules (O_2) inside a box. The results are presented in the following table:

Speed (m/s)	Number of particles with the given speed
100	40
200	125
300	180
400	215
500	180
600	125
700	75
800	40
900	20
1000	10

(a) Find the average speed of the particles. (b) Find the rms speed of the particles. (c) Make a graph of the number of particles versus the speed and indicate on the graph the locations of the average and rms speeds. (d) What is the temperature of the gas?

• **20.** An ideal gas of molar mass M is at temperature T. Show that if the temperature increases by a small amount ΔT, the increase in the rms speed of the gas particles is

$$\Delta v_{\text{rms}} = \frac{1}{2} \left(\frac{3R}{MT} \right)^{1/2} \Delta T$$

• **21.** Nitrogen (N_2) gas is at 300.0 K. If the temperature of the gas increases by 1.0 K, what is the increase in the rms speed?

• **22.** Raindrops of mass 1.00 mg fall vertically at a constant speed of 10.0 m/s, striking a horizontal skylight at the rate of 1000 drops/s and draining off. The window size is 15.0 cm × 25.0 cm. Assume the collisions of the drops with the window are completely inelastic. Calculate (a) the magnitude of the average force of the raindrops on the window, and (b) the resulting pressure developed by the raindrop collisions.

• **23.** One (1.00) mole of helium at 1.00 atm pressure and at temperature 0 °C is in a rigid container with a volume of 22.4 liters. (a) Calculate the rms speed of the particles in the gas. (b) The sample is warmed to 546 °C. By what factor does its pressure change? (c) By what factor does the rms speed of the particles change? (d) What heat transfer to the system produced the temperature change?

• **24.** The molar mass of nitrogen gas (N_2) is 28.0 g. (a) Find the mass of one nitrogen molecule. (b) Find the rms speed of a nitrogen molecule at a temperature of –23 °C. (c) Some hydrogen gas (H_2) (molar mass = 2.00 g/mol) also is present in the same container. What is the temperature of the hydrogen gas? (d) What is the rms speed of the hydrogen molecules? (e) What *new* temperature would cause the rms speed in part (b) to be greater by a factor of 2? (f) Does the *ratio* of the hydrogen to nitrogen rms speeds change with temperature?

• **25.** At what temperature do helium atoms have an rms speed equal to the escape speed from the surface of the Earth ($v_{\text{escape}} = 11.2$ km/s)?

• **26.** An energy unit commonly used in atomic, molecular, and nuclear physics is the *electron-volt*, abbreviated eV. One electron-volt is equivalent to 1.602×10^{-19} J. Show that the average translational kinetic energy of a particle in a gas at temperature 300 K is approximately 1/25 eV.

• **27.** At what temperature does $kT/2 = 1.00$ eV? (1 eV = 1.602×10^{-19} J) (See Problem 26.)

• **28.** Classically, at what temperature would the rms speed of helium atoms be equal to the speed of light? (Note: Particles with mass cannot attain speeds equal to the speed of light, as we will see when we study relativity in Chapter 25.)

• **29.** Calculate the rms speed of hydrogen gas (H_2) at temperature 300 K in the atmosphere and compare it with the escape speed from the Earth (11.2 km/s). Since hydrogen is the least massive gas, hydrogen molecules will have the highest rms speeds at a given temperature. How can this calculation explain why there is essentially no hydrogen gas in the atmosphere of the Earth?

• **30.** Consider helium gas at temperature 300 K near the surface of the Earth. (a) Calculate the average kinetic energy of one of the helium atoms. (b) Calculate the gravitational potential energy of a single helium atom near the surface of the Earth. Choose the zero of gravitational potential energy to be infinitely far away from the Earth. (c) What is the absolute value of the ratio of the gravitational potential energy of the helium atom to its average kinetic energy? Is it justifiable to neglect the gravitational potential energy in kinetic theory? Why or why not?

• **31.** A wind is howling at 30.0 m/s at a temperature of 25.0 °C. Consider a 1.00 kg mass of this moving air. Air has an effective molar mass of 29 g/mol; consider this value to be exact. (a) If the pressure is 0.97 atm, what is the density of the air? (b) How many joules of kinetic energy does the 1.00 kg mass of air have because of the wind? (c) How many joules of kinetic energy does the 1.00 kg mass of air have because of its internal energy? (d) Other than the difference in their magnitudes, what is the main distinction between the wind velocity and the thermal velocities?

• **32.** One (1.00) beautiful mole of helium gas at 300 K finds itself in a cubical box of sides 10.0 cm. (a) What is the rms speed of the particles of the gas? (b) If there were no collisions along the way, how long would it take the average helium atom in the container to travel from one side of the box to the other? (c) What is the pressure inside the container? (d) What is the magnitude of the average force that the particles exert on a side of the box?

• **33.** The temperature in outer space is about 2.7 K and the matter there consists mainly of isolated hydrogen atoms with a density of about 0.3 atoms per cubic meter. What is the pressure of this gas?

• **34.** The average kinetic energy of a particle in a gas at temperature T is given by Equation 14.13:

$$KE_{\text{ave}} = \frac{3}{2} kT$$

or

$$\frac{1}{2} m \langle v^2 \rangle = \frac{3}{2} kT$$

The special theory of relativity states that there is an upper limit on the speed of any particle: the speed of light $c = 3.00 \times 10^8$ m/s. For a gas of hydrogen *atoms*, the immediately preceding equation implies an upper limit on the temperature. Find the absolute temperature such that $\langle v^2 \rangle$ for a gas of hydrogen atoms is equal to the square of the speed of light. In fact, there is *no upper limit* on the temperature; so the classical expression for the kinetic energy cannot be valid for speeds approaching the speed of light.

• **35.** The isotope of uranium used in some nuclear fission reactors and bombs is uranium 235, with a molar mass of exactly 235 g/mol. The more common uranium isotope is uranium 238, with a molar mass of exactly 238 g/mol. One way of separating the isotopes is to form the gas uranium hexafluoride (UF_6) and exploit the rms speed difference of the two molecules made from the two uranium isotopes. This method of isotope separation is called gaseous diffusion and was first secretly employed in the Manhattan Project during World War II (at a site in Oak Ridge, Tennessee) to secure the material for the nuclear weapon used at Hiroshima near the end of the war with Japan;

the Nagasaki bomb was produced with plutonium made at the Hanford nuclear site in Washington State. Take the molar mass of fluorine (F) to be exactly 19 g/mol. Determine the *ratio* of the faster to slower rms speeds of the two molecules at room temperature (300 K). Does the ratio depend on the temperature?

36. A quantity of oxygen gas has a volume V, a temperature T, and pressure P_1. A quantity of nitrogen gas has the same volume V, identical temperature T, and pressure P_2. The two gases are mixed in a common volume of the same size V and have the same temperature T. Give arguments, based on kinetic theory, to show that the pressure of the mixture is $P_1 + P_2$. This result is known as *Dalton's law of partial pressures*: the total pressure of a mixture of gases is the sum of the partial pressures of the individual gases (occupying the same volume at the same temperature).

37. A helium atom has a mass approximately four times that of a hydrogen atom. Let a helium atom collide head-on with a hydrogen atom in a one-dimensional elastic collision. (a) If the two particles initially have equal kinetic energies, which atom (if either) gains kinetic energy in the collision? (b) If the two atoms initially have momenta of equal magnitudes, which (if either) gains energy in the collision?

Sections   14.5 The Internal Energy of a Monatomic Ideal Gas
    14.6 The Molar Specific Heats of an Ideal Gas
    14.7 Complications Arise for Diatomic and Polyatomic Gases
    14.8 Degrees of Freedom and the Equipartition of Energy Theorem
    14.9 Specific Heat of a Solid*
    14.10 Some Failures of Classical Kinetic Theory
    14.11 Quantum Mechanical Effects*
    14.12 An Adiabatic Process for an Ideal Gas

38. Evaluate the energy $kT/2$ per degree of freedom that appears in the equipartition of energy theorem for each of the following temperatures. Express your results in both joules and in another common energy unit known as the electron-volt (eV), defined as 1 eV $= 1.602 \times 10^{-19}$ J. (a) 3.0 K. This is approximately equal to the present ambient temperature of the radiation left over from the creation of the universe as a whole (the Big Bang). The temperature is just a bit above absolute zero. (b) 300 K. (c) The surface temperature of a cool star: 3000 K.

39. The ionization energy of the hydrogen atom is 13.6 eV $= 2.18 \times 10^{-18}$ J. This is the energy needed to remove the electron from the neutral atom. What temperature corresponds to this energy per degree of freedom according to the equipartition of energy theorem?

40. A nonspinning bead moves around a circular horizontal wire. How many degrees of freedom does the system have?

41. A nonspinning water droplet is free to slide around a frying pan. How many degrees of freedom does it have?

42. One (1.00) mole of a monatomic ideal gas at temperature 300 K occupies a volume of 5.00 liters. The gas now expands adiabatically till its volume is doubled. What is the final pressure of the gas?

43. One (1.00) mole of hydrogen gas (H_2) is warmed isobarically from an initial temperature of 273 K, doubling its volume. (a) What is the final temperature of the gas? (b) What is the heat transfer to the gas in the experiment?

44. A box of volume 1.00×10^{-3} m^3 has helium gas at a temperature of 300 K and a pressure of 1.00 atm. (a) What is the average kinetic energy per particle of the gas? (b) What is the total kinetic energy of the entire collection of atoms? This total is the internal energy of the gas.

45. How many moles of helium at temperature 300 K and 1.00 atm pressure are needed to make the internal energy of the gas 100 J?

46. Brrrr . . . it is freezing in the lab today. The temperature is 10 °C and the pressure is 1.00 atm. The lab is 5.0 m by 4.0 m by 3.0 m in size. A 1.0 kW heater is turned on to warm the room. (a) If no air escapes from the room, is the process at constant volume, constant pressure, or neither? (b) Assume the air is a diatomic gas. How long will it take the heater to raise the temperature of the room air from 10 °C to 20 °C?

47. The molar specific heat of a gas at constant pressure is measured to be 29.2 J/(mol·K). The molar mass of the gas is 28 g/mol. (a) What is the molar specific heat of the gas at constant volume? (b) What is the ratio γ of molar specific heats for the gas? (c) Is it possible to tell if the gas is monatomic or diatomic with the given information? If so, how? If not, what additional information is needed to make this determination?

48. An ideal gas has f degrees of freedom. (a) Show that the molar specific heat at constant volume is

$$c_V = \frac{f}{2} R$$

(b) Show that the molar specific heat at constant pressure is

$$c_P = \frac{f + 2}{2} R$$

(c) Show that the ratio of the molar specific heats is

$$\gamma = \frac{c_P}{c_V} = \frac{f + 2}{f}$$

49. Show that for a monatomic ideal gas

$$PV = \frac{2}{3} U$$

50. The *compressibility* κ of a substance is defined as

$$\kappa \equiv -\frac{1}{V} \frac{dV}{dP}$$

and represents the fractional change in volume for a given change in pressure. (a) Because of the minus sign in the definition, κ is positive. Explain why. (b) Show that if the gas is compressed *isothermally*, the compressibility is

$$\kappa_{isothermal} = \frac{1}{P}$$

(c) Show that if the gas is compressed *adiabatically*, the compressibility is

$$\kappa_{\text{adiabatic}} = \frac{1}{\gamma P}$$

Therefore

$$\gamma \kappa_{\text{adiabatic}} = \kappa_{\text{isothermal}}$$

●**51.** A truck is carrying a tank of helium gas of volume 0.250 m³ at a pressure of 100 atm. The temperature is 300 K. You observe the truck traveling at a speed of 100 km/h in the direction of increasing values of x. (a) What is the average value of the x-component of the velocity of the particles in the gas $\langle v_x \rangle$ according to you? (b) What are the average values of the y- and z-components of the velocity of the particles in the gas according to you? (c) Can you apply the equipartition of energy theorem to the particles of the gas? Why or why not? (d) *According to the driver of the truck*, what are the average values of the x-, y-, and z-components of the velocity of the particles of the gas? (e) Can the driver of the truck apply the equipartition of energy theorem to the particles of the gas? Why or why not?

●**52.** You compress an ideal gas isothermally in such a way that 500 J of work is done *by* you *on* the gas. (a) What is the change in the internal energy of the gas? (b) What was the heat transfer to the gas during this heroic feat?

●**53.** A well-insulated 4.00 liter box contains a partition dividing the box into two equal volumes as shown in Figure P.53. Initially, 2.00 g of molecular hydrogen gas (H_2) at 300 K is confined to the left-hand side of the partition, and the other half is a vacuum. (a) What is the rms speed of the particles in the gas? (b) What is the initial pressure of the gas? (c) The partition is removed or broken suddenly, so that the gas now is contained throughout the entire box. Assume that the gas is ideal. Does the temperature of the gas change? What is the change in the internal energy of the system? (d) When the gas reaches equilibrium, what is the final pressure?

Thin partition

Gas | Vacuum

Insulation

FIGURE P.53

●**54.** You expand n moles of an ideal gas adiabatically from temperature T_i to temperature T_f. Use the first law of thermodynamics to show that the work done by the gas is

$$-nc_V \, \Delta T$$

●**55.** A typical internal combustion engine compresses the gas by a factor of about 7. Assume that the initial temperature of the gas is 20 °C and it is initially at 1.00 atm pressure. Find the final temperature and pressure of the gas assuming an adiabatic compression. The gas is diatomic.

●**56.** A gas initially at temperature T_i, pressure P_i, and volume V_i expands to a final volume V_f via one of the following three types of processes: (1) isothermal; (2) isobaric; (3) adiabatic. (a) Sketch each process schematically on a P–V diagram. (b) In which process is the work done by the gas the greatest? Explain your reasoning. (c) In which process is the heat transfer to the gas the greatest? Explain your reasoning. (d) In which process is the change in the internal energy of the gas the greatest?

●**57.** A gas initially at temperature T_i, pressure P_i, and volume V_i has its pressure reduced to a final value P_f via one of the following types of processes: (1) isochoric; (2) isothermal; (3) adiabatic. (a) Sketch each process schematically on a P–V diagram. (b) In which process is the work done by the gas zero? (c) In which process is the work done by the gas the greatest? (d) Show that the ratio of the absolute magnitude of the heat transfer to the gas during the isothermal process to that during the isochoric process is

$$\left| \frac{Q_{\text{isothermal}}}{Q_{\text{isochoric}}} \right| = \frac{RP_i}{c_V \, |\Delta P|} \left| \ln \frac{P_i}{P_f} \right|$$

(e) In which process is the absolute magnitude of the change in the internal energy of the gas the greatest? Explain your reasoning.

●**58.** Begin with the definition of the work done by a gas:

$$W = \int_{V_i}^{V_f} P \, dV$$

Use Equation 14.43 for an adiabatic process and show that the work done by the gas in an adiabatic process can be expressed as

$$W_{\text{adiabatic}} = \frac{-1}{\gamma - 1} \left(P_f V_f - P_i V_i \right)$$

where γ is the ratio of the molar specific heats c_P/c_V.

●**59.** A 2.00 mol sample of a monatomic ideal gas undergoes an adiabatic expansion from a pressure of 2.00 atm at 300 K to a pressure of 1.00 atm. (a) Find the initial volume of the gas. (b) Find the final volume of the gas. (c) What is the final temperature of the gas? (d) What work is done by the gas?

●**60.** An ideal gas at 300 K is compressed adiabatically to half its initial volume. (a) What is the final temperature of the gas if it is monatomic? (b) What is the final temperature of the gas if it is diatomic?

◗**61.** At any point on a P–V diagram, show that the slope of an adiabatic curve through the point is γ times the slope of the isothermal curve through the same point, where γ is the ratio of the molar specific heats ($\gamma = c_P/c_V$). That is,

$$\left(\frac{dP}{dV} \right)_{\text{adiabatic}} = \gamma \left(\frac{dP}{dV} \right)_T$$

INVESTIGATIVE PROJECTS

A. Expanded Horizons

1. Investigate the distribution of speeds among the particles in a gas, known as the *Maxwell speed distribution*.

   Frederick Reif, *Statistical Physics*, Berkeley Physics Course, volume 5 (McGraw-Hill, New York, 1967), pages 231–246.

   Frederick Reif, *Fundamentals of Statistical and Thermal Physics* (McGraw-Hill, New York, 1965).

   R. P. Bonomo and F. Riggi, "The evolution of the speed distribution for a two-dimensional ideal gas: a computer simulation," *American Journal of Physics*, 52, #1, pages 54–55 (January 1984).

2. The *principle of detailed balancing* manifests itself in many aspects of physics. In this chapter we found that if the velocity of a particle in a gas changes in an collision, another particle of the gas likely experiences the opposite change, so that the average velocity of the particles in the gas is unchanged in thermal equilibrium (and is zero). Research other aspects of the principle of detailed balancing and formulate a general statement of the principle.

   Earle H. Kennard, *Kinetic Theory of Gases* (McGraw-Hill, New York, 1938), pages 55–58.

   Frank S. Crawford, "Using Einstein's method to derive both the Planck and Fermi–Dirac distributions," *American Journal of Physics*, 56, #10, pages 883–885 (October 1988).

3. The motion of a particle in a gas in thermal equilibrium can be modeled by means of what is called a *random walk*. Random walk problems can be formulated either in one, two, or three dimensions. A one-dimensional random walk problem is analogous to a drunk moving randomly from a lamppost along one dimension such as a narrow alley. The drunk takes steps of equal length d randomly either to the left or right. The problem is to determine the probability of the drunk being at various distances nd from the lamppost. Investigate random walk problems and their relevance to various problems in physics.

   Frederick Reif, *Fundamentals of Statistical and Thermal Physics* (McGraw-Hill, New York, 1965), pages 4–40.

4. *Globular clusters* are roughly spherical collections of hundreds of thousands of stars held together by their mutual gravitational attraction.* The clusters do not collapse because the stars are in motion (just as the solar system does not collapse because the planets are in motion). Pairs of stars in such clusters make "collisions" with each other that change the di-

   rections of their motion. They rarely actually collide; rather, the collisions are more like scattering events that send the stars off in different directions (much like a comet passing the Sun). These collisions can be modeled as elastic. Can one regard the collection of stars in such a cluster as an ideal gas? Develop a kinetic theory of a star cluster and thereby define a temperature and pressure of the star particle system. What would happen to the average kinetic energy of a star particle if the cluster as a whole shrank or expanded? Can a star evaporate from the cluster? How could you define the evaporation speed?

   Lyman Spitzer, *Dynamical Evolution of Globular Clusters* (Princeton University Press, Princeton, New Jersey, 1987).

   Martin Harwit, *Astrophysical Concepts* (3rd edition, Springer-Verlag, New York, 1997).

B. Lab and Field Work

5. In 1827 the English botanist Robert Brown (1773–1858) noticed through a microscope that tiny pollen grains suspended in water exhibit constant, randomly changing motions. Smoke particles in a gas exhibit similar behavior when examined in a small chamber. Such motion now is called *Brownian motion*. Investigate how Avogadro's number is determined experimentally from observations of Brownian motion. Design and perform such an experiment.

   John le P. Webb, "Einstein and Brownian motion—a student project," *Physics Education*, 15, #2, pages 116–121 (March 1980).

   Haym Kruglak, "Brownian motion—a laboratory experiment," *Physics Education*, 23, #5, pages 306–309 (September 1988).

   Robert Stoller, "Viewing Brownian motion with laser light," *American Journal of Physics*, 44, #2, page 188 (February 1976).

   Bill Reid, "Viewer for Brownian motion," *The Physics Teacher*, 29, #1, pages 52–53 (January 1991).

   F. Landis Markley and David Park, "Microscopic and macroscopic views of Brownian motion," *American Journal of Physics*, 40, #12, pages 1859–1860 (December 1972).

   Max Born, *Atomic Physics* (8th edition, Dover, Mineola, New York, 1989).

6. In 1929 Eduard Rüchhardt developed an ingenious way of measuring the ratio γ of molar specific heats by measuring the period of oscillation of a ball bearing sliding in a vertical glass tube inserted through a stopper into a large, gas-filled jar. The diameter of the glass tube is essentially the same as the ball bearing. Investigate this method and, if feasible, construct the apparatus and measure γ for a monatomic gas and for a diatomic gas.

   Mark W. Zemansky and Richard H. Dittman, *Heat and Thermodynamics* (7th edition, McGraw-Hill, New York, 1996).

7. Devise a kinetic theory for an ideal gas with particles confined to move in two dimensions. In particular, derive expressions for the pressure in terms of the rms speed of the particles.

---

*There are about sixty such globular clusters in our Milky Way Galaxy. The most easily visible such cluster in the northern celestial hemisphere is called M-13 in the constellation Hercules (a small telescope is needed to see it, since at most locations it is too dim to see with the naked eye). In the southern celestial hemisphere the globular cluster to see is called the ω Centauri cluster, visible to the naked eye as a small smudgy patch in the constellation Centaurus.

Design an experiment to test aspects of your model using small air pucks on a laboratory airtable. Summarize your theoretical results and experimental proposal. Perhaps your professor can secure the necessary equipment for you to perform your experiment.

Richard L. Liboff, *Kinetic Theory: Classical, Quantum and Relativistic Descriptions* (2nd edition, John Wiley, New York, 1998).

8. Write a proposal to investigate the dynamical behavior of a collection of BBs confined to a box in the orbiting space shuttle and how such a collection of particles differs from those of an ideal gas.

C. Communicating Physics

9. Your roommates play devil's advocates and say, "Atoms do not exist." How could you convince them otherwise? Write an essay describing the experimental evidence for the existence of atoms.

Edwin R. Jones Jr. and Richard L. Childers, "Observational evidence for atoms," *The Physics Teacher*, 22, #6, pages 354–360 (September 1984).

George Gamow, *Mr Tompkins Explores the Atom* (The University Press, Cambridge, England, 1961).

George Gamow, *Mr Tompkins in Paperback* (Cambridge University Press, Cambridge, England, 1993).

CHAPTER 15

THE SECOND LAW OF THERMODYNAMICS

A good many times I have been present at gatherings of people who, by the standards of traditional culture, are thought highly educated and who have with considerable gusto been expressing their incredulity at the illiteracy of scientists. Once or twice I have been provoked and have asked the company how many of them could describe the Second Law of Thermodynamics. The response was cold: it was also negative. Yet I was asking something which is about the scientific equivalent of: *Have you read a work of Shakespeare's*?

C. P. Snow (1905–1980) *

Air pollution from automotive exhaust, smoke emitted from coal-fired electrical power plants, and thermal emissions from nuclear power plants all have a significant impact on the environment. For decades scientists and politicians have struggled with finding solutions to these important problems. You might ask, why doesn't Congress just outlaw all such discharges and be done with them? Plug up the exhaust pipes, cap the chimneys, zero emissions, no more problem. It *is* a problem, and a big one at that. In this chapter we see why any such legislation, would be physically absurd; Congress might just as well try to repeal the law of universal gravitation or Newton's laws. In the two centuries of its existence, Congress has passed some ridiculous laws, but outlawing all emissions would take the cake. Should Congress or a politician even suggest such a solution, they would be demonstrating their complete ignorance of the second law of thermodynamics.

In this chapter we study this great integrating principle of physics. As C. P. Snow alludes to in the opening quotation, an understanding of the second law is a condition for scientific literacy, much as a knowledge of Shakespeare might be taken as a measure of literacy in the humanities. Hence we will look at the second law of thermodynamics carefully and closely to appreciate both the breadth of its applicability and the depth of its insight into the workings of nature.

15.1 WHY DO SOME THINGS HAPPEN, WHILE OTHERS DO NOT?

Why, sometimes I've believed as many as six impossible things before breakfast.

Lewis Carroll (1832–1898)[†]

Why is it that rocks do not jump up spontaneously and save us the trouble of lifting them? Wouldn't it be nice if pizzas warmed themselves: no need for ovens in the dorm. And of course we all want to win the lottery, but unfortunately it never seems to happen to us. Nuts.

One of the sacred truths of physics is the principle of energy conservation. Some things cannot happen in nature simply because they violate this fundamental principle. We first encountered energy conservation in our study of ("pure") mechanics with the CWE theorem (Chapter 8), which ignored thermal effects. In Chapter 13 we discovered that the CWE theorem in fact is a special case of a more general statement of energy conservation (Equation 13.48). The general statement then was restricted in other ways, so that we could consider simple ("pure") thermodynamic systems; the statement then was called the first law of thermodynamics (Equation 13.51).

Why a Rock Does Not Jump

A rock at the base of a cliff does not, of its own accord, jump spontaneously up to the top of the cliff; such an occurrence would violate the CWE theorem of mechanics. Recall that the CWE theorem states that

$$\text{Work done by nonconservative forces} = \Delta KE + \Delta PE$$

The rock initially resting at the bottom of the cliff has no kinetic energy; likewise, it has no kinetic energy when at rest at the top of the cliff. The change in the kinetic energy is zero: $\Delta KE = 0$ J. We choose the zero of gravitational potential energy to be at the initial location of the rock, as shown in Figure 15.1. With this choice of the origin, the rock has zero local gravitational potential energy at its initial position and its final potential energy is $+mgh$. The change in the local gravitational potential energy thus is $\Delta PE = PE_f - PE_i = mgh - 0$ J $= mgh$. If the rock spontaneously jumps to the top of the cliff without our effort (or force), the only force acting on the rock is the gravitational force; the work done by this force is accounted for by the change in the gravitational potential energy term in the CWE theorem. There is zero work done by other forces: there are no other forces. Therefore the CWE theorem here becomes

$$\text{Work done by nonconservative forces} = \Delta KE + \Delta PE$$

$$0 \text{ J} = 0 \text{ J} + mgh$$

Thus we can only have $h = 0$ m. In other words, the rock stays where it is. If we want to get the rock to the top of the cliff, we need to *transfer* energy to the system (the rock) via work: we have to lift the rock (doing work on the system), so that the work done by "other" forces (our force on the rock) is not zero. The work we do on the rock equals the change in its potential energy according to the CWE theorem (since here there is no change in the kinetic energy).

Why a Pizza Does Not Warm Itself

A cold pizza placed in an insulated container does not, of its own accord, spontaneously increase its temperature, thereby cooking itself. To do so would violate the principle of energy conservation in the form of the first law of thermodynamics. The first law, Equation 13.51, states that

$$Q = \Delta U + W$$

Since the system is isolated from its environment, there is no heat transfer to the pizza: $Q = 0$ J. Certainly the pizza does no work on the surroundings, since the pizza is isolated; thus, $W = 0$ J as well. Hence the first law becomes

$$0 \text{ J} = \Delta U + 0 \text{ J}$$

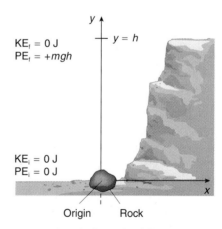

FIGURE 15.1 A rock at the base of a cliff does not jump up of its own accord; to do so would violate the CWE theorem.

---

*(Chapter Opener) *The Two Cultures and the Scientific Revolution* (Cambridge University Press, New York, 1961), pages 15–16.

[†]*Through the Looking Glass*, Chapter 5, Wool and Water (Boni and Liveright, New York, 1925).

and thus $\Delta U = 0$ J. The internal energy of the pizza does not change; its temperature must remain constant. If the pizza were to spontaneously warm itself, then $\Delta U \neq 0$ J and the first law of thermodynamics would be violated. To change the internal energy of the pizza (i.e., to cook it), we must effect heat transfer Q to the pizza system. We place it in a hot oven, so the energy exchange known as heat transfer is nonzero; the heat transfer to the pizza from the hot oven reservoir changes the internal energy of the pizza and thereby raises its temperature.

Why We Never Seem to Win the Lottery

Some things do not happen, not because they violate energy conservation, but because they are simply very unlikely. Such is the case with winning the lottery, at least for most individuals (including the author). *We* never seem to win the lottery; *somebody* wins the lottery but, unfortunately, it is not likely to be us. The reason is that winning the lottery is a very unlikely event for a given individual. It is not that it is impossible: it is just highly improbable. Everyone who purchases a ticket has an equal chance of winning. If we purchase many lottery tickets, the odds of winning improve: there are more ways for us to hit the jackpot because each ticket has an equal chance.

In thermodynamics, something similar occurs, with *vastly* longer odds than the typical lottery. Imagine something that is *permitted* according to energy conservation but still does not happen. Divide an isolated pizza into halves. Why can't one half of the pizza spontaneously warm the other half? Half of the pizza warms up at the (energy) expense of the other half. The energy given up by the cooling half warms the other half. For the pizza as a whole the total heat transfer is zero, since the heat transfer is from one half to the other half. If this were possible, we would end up with half a hot pizza and half a frozen pizza; half a hot pizza is better than none, after all! But, of course, we know from experience that this does not happen naturally. One half of the pizza will not spontaneously cool while warming the other half; it just does not happen.*

Notice that there is no violation of the principle of energy conservation; there is nothing in the first law of thermodynamics that prohibits this spontaneous process from taking place. It just does not happen. Why? A small child might say, "just because I said so," but we seek a more profound explanation! The process does not happen because it is so *exceedingly unlikely* that it effectively never occurs; it is comparable to one ticket in a vast lottery with a humongous number of tickets.

We know (from experiment!) that a hot pizza placed in thermal contact with a cold pizza results in two tepid pizzas: the hot one cools and the cold one warms. Energy is exchanged (heat transfer) until the two reach the same intermediate temperature. Heat transfer occurs from hot to cold objects, not from cold to hot objects. Why is this? We raised this question in Chapter 13 and finally answer it in this chapter.

The key to why these more subtle processes can or cannot take place lies *not* in the first law of thermodynamics (energy conservation), but in another fundamental law of thermodynamics: the **second law of thermodynamics**. The answer

- to our lack of lottery winnings,
- to our half-warmed, half-frozen pizza dilemma, and

- to the never observed spontaneous heat transfer from cold to hot objects

lies in the fact that such occurrences are exceedingly unlikely, so unlikely that these events effectively never happen to a given individual, to a given pizza, or in a given physical system. We shall see later in this chapter that "lotteries" exist in the natural world as well and are closely tied to heat transfer via a statistical interpretation of the second law of thermodynamics.

There are a few equivalent ways to state the second law of thermodynamics; we shall mention several and eventually show that they all are equivalent. We will then examine a statistical interpretation of the law. Several ways of stating the second law have to do with the workings of heat engines and refrigerator engines; it is difficult to see initially how this could relate to our half-hot, half-cold pizza problem or to the direction of heat transfer. But have faith while your appetite grows and Domino's responds (though no longer in thirty minutes, thanks to legal largesse); soon you will get a feast of insight into nature, if not pepperoni. But, as with Domino's, it may take longer than a thirty minute delivery time.

QUESTION 1

A rock cannot spontaneously jump up a flight of stairs, but you can decide to run up them anytime you want. What is the difference between you and the rock? Do you violate energy conservation when you spontaneously run up a flight of stairs?

15.2 HEAT ENGINES AND THE SECOND LAW OF THERMODYNAMICS

What does a heat engine do? Work, work, and more work. Just like students—and professors—but in a different sense! Here we try to examine the general features of all heat engines, not just particular ones like the one inside your mother's new Porsche or her John Deere tractor.

It is easy to change macroscopic physical work into an increase in the internal energy of the environment. For example, when a vehicle slows down, kinetic frictional forces do work on the vehicle, thereby changing (decreasing) its kinetic energy. The frictional work increases the internal energy of the brake lining and tires (they become warmer), resulting in heat transfer to the cooler environment and an eventual increase in the internal energy of the environment. The macroscopic ordered kinetic energy of the car is transformed into disordered internal energy of the environment.

The function of a **heat engine** is to turn this process around: to transform disordered internal energy of, say, a gas system into macroscopic work using the heat transfer between reservoirs at different temperatures.

We use the word **engine** to mean a system such as a gas that performs a closed cycle on a *P–V* diagram. If the cycle is performed clockwise, we call it a **heat engine**; if counterclockwise, we call it a **refrigerator engine** (or just a **refrigerator**).

---

*There also is the problem of how the pizza decides *which* half to warm and which to cool, since the halves can be formed in many ways.

The process involved with a heat engine is diagrammed schematically in Figure 15.2. Heat transfer Q_H occurs to a heat engine from a high-temperature reservoir; part of Q_H is transformed into macroscopic physical work W done *by* the heat engine, with the rest, Q_C, transferred to a cold-temperature reservoir (think of Q_C as exhaust).

If we consider the heat engine to be the thermodynamic system, the heat transfer Q_H to it is positive while Q_C is negative, since the convention is that heat transfer to a system is positive and heat transfer from a system is negative. The heat engine operates in a *cycle* such that it can repeat the process again and again. Whatever the heat engine actually does, since it is operated cyclically it returns to the same initial thermodynamic equilibrium state after completion of each cycle.

> Since the heat engine operates in a cycle, the change in the internal energy of the heat engine system over the complete cycle is zero.

If it were not zero, the engine would be blazing hot (or frozen) after many cycles.

The first law of thermodynamics implies that the total heat transfer to any engine during each cycle is equal to the total change in the internal energy of the engine plus the total work W it does:

$$Q_{total} = \Delta U + W$$

Since the change in the internal energy of the engine is zero around a cycle ($\Delta U = 0$ J),

$$Q_{total} = W \qquad (15.1)$$

The total heat transfer to a heat engine is the difference between the heat transfer *to* the heat engine from the hot reservoir ($|Q_H|$, where $Q_H > 0$ J for the heat engine) and the heat transfer *from* the heat engine to the cold reservoir ($|Q_C|$, $Q_C < 0$ J for the heat engine):

$$|Q_{total}| = |Q_H| - |Q_C| \qquad (15.2)$$

The absolute value signs are used for the following reason. The heat engine is the thermodynamic system under consideration. Heat

transfer Q_H to the heat engine is positive according to our convention that heat transfer to a system is positive. The exhaust heat transfer from the heat engine to the cold reservoir, Q_C, is *negative* for the heat engine system according to our sign convention. Thus, for the heat engine as a system, we could write Equation 15.2 without the absolute value signs as $Q_{total} = Q_H + Q_C$, since Q_C actually is negative for the heat engine system. But by writing $|Q_{total}|$ explicitly as $|Q_{total}| = |Q_H| - |Q_C|$, we perhaps avoid some confusion with signs.

> The fraction of the heat transfer to the heat engine from the hot reservoir that is transformed into work done by the heat engine during each cycle is a measure of the **efficiency** of the heat engine. The efficiency ε is defined as
>
> $$\varepsilon \equiv \frac{|W|}{|Q_H|} \qquad (15.3)$$

Since both W and Q_H are positive for the heat engine system, there really is no need for the absolute value signs, but we put them in anyway for reinforcement. The efficiency ε is a positive pure number between 0 and 1. Inspection of Equations 15.1 and 15.2 indicates that a **perfect heat engine** is one for which Q_C is zero, in which case $W = Q_H$. Thus a perfect heat engine has an efficiency of 1. *Nothing* in the *first* law of thermodynamics forbids a perfect heat engine, since the first law is a statement of energy conservation. A perfect heat engine simply converts *all* of the heat transfer $|Q_H|$ into work W with zero exhaust $|Q_C|$ in each cycle.

This is where the **second law of thermodynamics** comes into play.

> The second law of thermodynamics states that *perfect heat engines do not exist.*

This is another way of saying that heat engines with an efficiency of 1 do not exist. All heat engines must have finite, nonzero exhaust heat transfer Q_C from the heat engine to a cooler reservoir. This statement about the nonexistence of perfect heat engines is the first of several equivalent statements of the second law that we will examine.

You likely noted these things during this discussion:

1. The second law of thermodynamics says something *very different* from the first law.
2. The second law means that *the efficiency of real heat engines is always less than 1.* We will address the maximum efficiency that a heat engine can have in Section 15.3.
3. The second law means that no heat engine operating in a *cycle* can transform heat transfer $|Q_H|$ from a hot reservoir completely into work W. That is, there always is some unutilized exhaust, heat transfer $|Q_C|$ from the heat engine to a colder reservoir during the cycle.
4. The word *cycle* is essential in this regard. In a cycle, the heat engine or system returns to its initial state of pressure, volume, and temperature.

If the process is not a cycle, it is quite possible to make a complete conversion of heat transfer into work. For instance,

FIGURE 15.2 A schematic representation of the function of a heat engine.

consider the isothermal expansion of an ideal gas from some initial volume V_i to some final volume V_f, as shown in Figure 15.3. The work done by the gas is represented by the shaded area under the isotherm on the P–V diagram. Since the temperature of the ideal gas is unchanged in the isothermal process,* the change in the internal energy of the gas is zero. Thus the first law of thermodynamics, $Q = \Delta U + W$, becomes $Q = W$. The heat transfer to the gas from a reservoir at the temperature of the isotherm is equal to the work done by the gas. A complete (total) conversion of the heat transfer into work has been accomplished, with no exhaust heat transfer to a colder reservoir. This does *not* violate the second law of thermodynamics, because *the gas has not yet been returned to its initial thermodynamic state or condition.* The gas has not yet completed a cycle on the P–V diagram, and so there is no violation of the second law. To return the gas to its initial thermodynamic state, there *always* will be nonzero $|Q_C|$.

For the gas to return to its initial thermodynamic state, the gas must follow a closed path (a cycle) on a P–V diagram, which, according to our definition, makes the system an engine. The work done by the gas around such a closed path on a P–V diagram is represented by the area enclosed by the path on the diagram, as we saw in Chapter 13. Since we want the total work done by the gas to be positive, the gas must follow the closed path on the P–V diagram in the clockwise sense to be a heat engine, as shown in Figure 15.4. If the gas executes the closed path in the counterclockwise sense, we have a refrigerator engine and the total

---

*Recall that the internal energy of a given quantity of an ideal gas depends only on its absolute temperature.

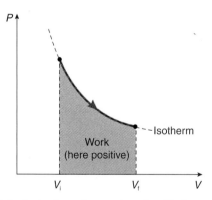

FIGURE 15.3 An isothermal expansion of an ideal gas.

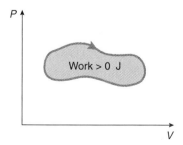

FIGURE 15.4 To do positive work, a heat engine must execute a cycle in the clockwise sense on a P–V diagram.

work done by the gas is negative, meaning that the environment did work on the gas system rather than the other way around.

Equation 15.3 indicates that the efficiency ε of any heat engine is the total work done by the system around one cycle divided by the heat transfer to the heat engine from the hot reservoir during the cycle:

$$\varepsilon = \frac{|W|}{|Q_H|}$$

Since the system (the gas) is taken around a complete cycle (making it an engine) and returns to its initial thermodynamic state, the change in its internal energy over one cycle is zero. The first law of thermodynamics, $Q_{total} = \Delta U + W_{total}$, becomes $|Q_H| - |Q_C| = 0$ J $+ W$. We use this expression for $|W|$ in Equation 15.3.

Hence the efficiency of *any* heat engine can be written as

$$\varepsilon = \frac{|Q_H| - |Q_C|}{|Q_H|}$$

$$= 1 - \frac{|Q_C|}{|Q_H|} \qquad \text{(any heat engine)} \quad (15.4)$$

EXAMPLE 15.1

During each complete cycle of the "Little Engine that Could," 2000 J of heat transfer accrues to the heat engine from a hot reservoir, with 1700 J exhausted by the heat engine to a colder reservoir. Determine the efficiency of the heat engine.

Solution
I think I can. . . I think I can. . . I think I can.

Method 1
Since the heat engine completes one cycle, the change in the internal energy of the heat engine is zero. Apply the first law of thermodynamics to the heat engine:

$$Q_{total} = \Delta U + W$$
$$= 0 \text{ J} + W$$

where Q_{total} is the total heat transfer to the heat engine around the cycle; this is

$$Q_{total} \equiv |Q_H| - |Q_C|$$
$$= 2000 \text{ J} - 1700 \text{ J}$$
$$= 300 \text{ J}$$

The total work done by the heat engine during one cycle is 300 J, since $Q_{total} = W$ as was just shown.

The efficiency of the heat engine is, from Equation 15.3,

$$\varepsilon = \frac{|W|}{|Q_H|}$$

$$= \frac{300 \text{ J}}{2000 \text{ J}}$$

$$= 0.150$$

Method 2

Use Equation 15.4 directly with the given information:

$$\varepsilon = 1 - \frac{\left| Q_C \right|}{\left| Q_H \right|} \quad \text{(any heat engine)}$$

$$= 1 - \frac{1700 \text{ J}}{2000 \text{ J}}$$

$$= 0.1500$$

The first method is preferable, since it displays greater understanding of the physics. Method 2 is tantamount to "plug-and-chug," which we try to avoid when possible.

15.3 THE CARNOT HEAT ENGINE AND ITS EFFICIENCY

In 1824 the French scientist Nicolas Léonard Sadi Carnot (1796–1832) brilliantly devised an idealized, theoretical thermodynamic cycle that now bears his name: the **Carnot cycle**. We determine the efficiency of this heat engine in this section, deferring the relationship between the Carnot heat engine and the second law of thermodynamics to Section 15.7.

> A Carnot cycle takes an ideal gas around a cycle consisting of alternating isothermal and adiabatic quasi-static processes, as shown on the *P–V* diagram in Figure 15.5.

According to Equation 15.4, to calculate the efficiency of the Carnot heat engine we need to find Q_C and Q_H. To do this we need to examine each of the four processes of the Carnot cycle individually with the gas as the system. We begin with the isothermal expansion.

Isothermal Expansion

This expansion is accomplished by placing the gas in thermal contact with a reservoir at temperature T_H. During the expansion there is (positive) heat transfer Q_H to the gas from the hot reservoir; the gas does positive work W_1, as indicated in Figure 15.6.

The work done by an ideal gas in an isothermal expansion was found in Chapter 13 to be (Equation 13.57)

$$W_1 = nRT_H \ln \frac{V_2}{V_1}$$

At the comparatively youthful age of 36, Carnot tragically contracted cholera and died within hours. Almost all of his personal papers and possessions were burned to prevent the spread of the disease, a tragedy for historians of science.

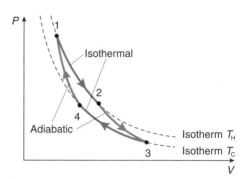

FIGURE 15.5 A Carnot cycle on a *P–V* diagram.

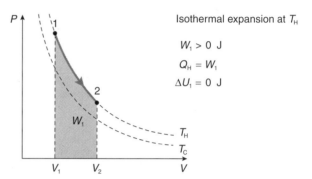

FIGURE 15.6 The work done by the gas in an isothermal expansion is positive.

This work is represented by the area under the isotherm between V_1 and V_2 on the *P–V* diagram of Figure 15.6. The internal energy of the ideal gas is purely a function of the temperature. Since the temperature is constant in an isothermal process, the internal energy is constant: $\Delta U_1 = 0$ J. The first law of thermodynamics, $Q = \Delta U + W$, here becomes

$$Q_H = 0 \text{ J} + W_1$$

$$= W_1$$

$$= nRT_H \ln \frac{V_2}{V_1} \qquad (15.5)$$

This indicates that *all* of the heat transfer Q_H to the gas is converted into the work W_1. This is *not* a violation of the second law of thermodynamics, because the gas has not yet completed a cycle on the *P–V* diagram.

Adiabatic Expansion

In an adiabatic process the system is totally insulated from its environment, so that there is zero heat transfer either to or from the system. Alternatively, the process takes place fast enough so that heat transfer has no time to occur, yet slow enough to be quasi-static and reversible. During an adiabatic process $Q = 0$ J. Positive work W_2 is done since the gas expands; as usual, the work W_2 is represented by the area under the adiabatic curve on the *P–V* diagram; see Figure 15.7.

Since $Q = 0$ J, the first law of thermodynamics, $Q = \Delta U + W$, becomes

$$0 \text{ J} = \Delta U_2 + W_2$$

$$\Delta U_2 = -W_2 \qquad (15.6)$$

FIGURE 15.7 The work done by the gas during an adiabatic expansion is positive.

Since W_2 is positive (the gas expands), the change in the internal energy of the gas is negative; the internal energy decreases. Since the internal energy of an ideal gas is directly proportional to the temperature, the temperature of the gas decreases to T_C, as indicated on the P–V diagram in Figure 15.7.

Isothermal Compression

Once again, since the process is isothermal the internal energy of the gas does not change: $\Delta U_3 = 0$ J. The gas does negative work W_3, as shown in Figure 15.8.

We use Equation 13.57 again for the work done by an ideal gas in an isothermal process:

$$W_3 = nRT_C \ln \frac{V_4}{V_3} \tag{15.7}$$

The work is negative because V_4 is less than V_3 (the logarithm of a number less than 1 is negative). The first law of thermodynamics, $Q = \Delta U + W$, here becomes

$$Q_C = 0 \text{ J} + W_3$$
$$= nRT_C \ln \frac{V_4}{V_3} \tag{15.8}$$

Since W_3 is negative, Q_C also is negative, indicating that the heat transfer is from the gas (the system) to the reservoir at temperature T_C during this process.

Adiabatic Compression

The final adiabatic compression takes the gas back to its initial state, as shown in Figure 15.9, so the cycle can be repeated. There is zero heat transfer to (or from) the system because this is an adiabatic process. The gas does negative work W_4 since the volume decreases, and this work is the negative area under this adiabatic curve on the P–V diagram.

The first law of thermodynamics, $Q = \Delta U + W$, here becomes

$$0 \text{ J} = \Delta U_4 + W_4$$
$$\Delta U_4 = -W_4 \tag{15.9}$$

Since W_4 is negative, ΔU_4 is positive, reflecting that the temperature of the gas increases in this process from T_C back to T_H.

Now bring the four parts of the cycle together and apply the first law to the *complete cycle*. Since the gas returns to its initial state, the total change in the internal energy of the gas is zero. That is,

$$\begin{aligned} \Delta U_{\text{total}} &= \Delta U_1 + \Delta U_2 + \Delta U_3 + \Delta U_4 \\ &= 0 \text{ J} + \Delta U_2 + 0 \text{ J} + \Delta U_4 \\ &= 0 \text{ J} \end{aligned}$$

This indicates that the two internal energy changes that occurred during the two adiabatic processes of the cycle are of equal magnitude but opposite sign. This has to be the case, since for an ideal gas the internal energy is a function of temperature alone: in one adiabatic process, the temperature of the gas decreased from T_H to T_C; and in the other adiabatic process, the temperature increased from T_C back to T_H. The first law applied to the whole cycle with the gas as the system therefore is

$$Q_{\text{total}} = W_{\text{total}} + \Delta U_{\text{total}}$$
$$Q_H + Q_C = W_1 + W_2 + W_3 + W_4 + 0 \text{ J}$$

Since Q_C, W_3, and W_4 are all negative for the gas system, we write this as

$$|Q_H| - |Q_C| = |W_1| + |W_2| - |W_3| - |W_4|$$

The left-hand side of the equation is the total heat transfer to the gas in the cycle (positive); the right-hand side is the total work done by the gas (also positive and represented by the area of the cycle on the P–V diagram in Figure 15.5).

FIGURE 15.8 The work done by the gas in an isothermal compression is negative.

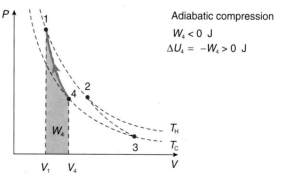

FIGURE 15.9 The work done by the gas during an adiabatic compression is negative.

We wrap things up by calculating the efficiency of the Carnot heat engine using Equation 15.4:

$$\varepsilon = 1 - \frac{|Q_C|}{|Q_H|}$$

We use the expressions for both Q_H and Q_C, Equations 15.5 and 15.8:

$$Q_H = nRT_H \ln \frac{V_2}{V_1}$$

Since Q_H is positive, $|Q_H|$ is the same thing.
For Q_C we found

$$Q_C = nRT_C \ln \frac{V_4}{V_3}$$

and this is *negative* since $V_4 < V_3$. We can write the absolute value of Q_C by flipping the numerator and denominator in the natural logarithm, since $\ln(a/b) = -\ln(b/a)$:

$$|Q_C| = nRT_C \ln \frac{V_3}{V_4}$$

To calculate the efficiency we need the ratio of $|Q_C|$ to $|Q_H|$:

$$\frac{|Q_C|}{|Q_H|} = \frac{nRT_C \ln(V_3/V_4)}{nRT_H \ln(V_2/V_1)}$$

$$= \frac{T_C \ln(V_3/V_4)}{T_H \ln(V_2/V_1)} \qquad (15.10)$$

This expression can be simplified further by making use of a result from Section 14.12. There we found that during an adiabatic process (Equation 14.44), the product $TV^{\gamma-1}$ remains constant. Hence, if we form the product $TV^{\gamma-1}$ at the beginning of the adiabatic process, the number obtained is the same as the corresponding product at the end of the process. In other words, for the adiabatic process from point 2 to point 3 (in the P–V diagram of Figure 15.7), write Equation 14.44 for point 2; this expression will have the same value at point 3:

$$\underset{\text{Point 2}}{T_H V_2^{\gamma-1}} = \underset{\text{Point 3}}{T_C V_3^{\gamma-1}} \qquad (15.11)$$

We do the same thing for the adiabatic process from point 4 back to point 1 on the P–V diagram of Figure 15.9:

$$\underset{\text{Point 1}}{T_H V_1^{\gamma-1}} = \underset{\text{Point 4}}{T_C V_4^{\gamma-1}} \qquad (15.12)$$

Dividing Equation 15.11 by Equation 15.12, we find the following remarkable result:

$$\frac{V_2}{V_1} = \frac{V_3}{V_4}$$

Thus the two natural logarithms in Equation 15.10 are the natural logarithms of the *same number*. Hence Equation 15.10 reduces to

$$\frac{|Q_C|}{|Q_H|} = \frac{T_C}{T_H} \qquad \text{(Carnot cycle only)} \qquad (15.13)$$

We use Equation 15.13 in Equation 15.4.

> The efficiency of the Carnot heat engine takes on the remarkably simple form
>
> $$\varepsilon_{\text{Carnot}} = 1 - \frac{T_C}{T_H} \qquad \begin{array}{l}\text{(Carnot heat engine only)}\\ \text{(use temperatures in kelvin)}\end{array} \qquad (15.14)$$

Notice that the efficiency of a Carnot heat engine increases and approaches 1 if $T_H \gg T_C$.

PROBLEM-SOLVING TACTIC

15.1 In Equation 15.14 for the efficiency of a Carnot heat engine, be sure to use absolute temperatures in kelvin, not celsius temperatures.

QUESTION 2

Equation 15.14 must use absolute temperatures. Is the equation still valid if one uses another temperature scale whose zero is at 0 K but whose degree size is different than a kelvin? For example, the *Rankine* temperature scale has its zero at 0 K but uses degrees equal in size to Fahrenheit degrees. Is Equation 15.14 also legitimate using temperatures on the Rankine scale?

EXAMPLE 15.2

A Carnot heat engine is operating between two reservoirs, one at 0 °C and the other at 100 °C. What is the efficiency of the Carnot heat engine?

Solution
The efficiency of a Carnot heat engine depends only on the *absolute* temperatures of the two reservoirs according to Equation 15.14. Hence

$$\varepsilon = 1 - \frac{T_C}{T_H}$$

$$= 1 - \frac{273 \text{ K}}{373 \text{ K}}$$

$$= 0.268$$

15.4 ABSOLUTE ZERO AND THE THIRD LAW OF THERMODYNAMICS

The equation for the efficiency of a Carnot heat engine (Equation 15.14) is used to define the zero point of the absolute temperature scale. The absolute zero of temperature is the temperature of a cold reservoir for which the efficiency of a Carnot heat engine is exactly 1. If $T_C = 0$ K, then according to Equation 15.13 the Carnot heat engine has zero exhaust heat transfer $|Q_C|$. Such a heat engine is a perfect heat engine; but such perfect heat engines do not exist according to the second law of thermodynamics.

Hence the absolute zero of temperature is an *unattainable* temperature; reservoirs with $T_C = 0$ K *cannot exist*. This conclusion is the **third law of thermodynamics**.

The absolute temperature scale (the kelvin scale we use in virtually all the equations of thermodynamics) thus is truly a thermodynamic temperature scale, because its zero of temperature is fundamentally connected with the second law; 0 K is a coldest temperature that exists only as a limit. The size of a kelvin degree does not have a fundamental thermodynamic significance; it simply was chosen to be equal to the size of the common and convenient celsius degree. Recall that the zero of temperature on the celsius scale was chosen for convenience and has no truly fundamental thermodynamic significance.

15.5 REFRIGERATOR ENGINES AND THE SECOND LAW OF THERMODYNAMICS

. . . hell must be isothermal, for otherwise the resident engineers and physical chemists (of which there must be some) could set up a . . . refrigerator to cool off a portion of their surroundings to any desired temperature.

Henry A. Bent*

There is another way to view the second law of thermodynamics that involves refrigerators.[†] The idea of a **refrigerator engine** is to effect heat transfer from a cold reservoir (the inside of what we commonly think of as a refrigerator) to a hot reservoir (the outside of the refrigerator, such as the room in which it is located). The heat transfer from the cold reservoir originally arises from heat transfer to this reservoir from warm objects that we place in the refrigerator compartment. The heat transfer from the cold interior to the warm exterior will not happen spontaneously (by itself), since the natural direction of heat transfer is from hot to cold objects. We have yet to explain this direction of heat transfer, but the explanation is coming soon.

A refrigerator engine (also called simply a refrigerator) accomplishes the unnatural heat transfer from a cold to a hot reservoir. We want the refrigerator engine to operate in a cycle so that the process can be done repetitively. Thus the change in the internal energy of the system (the refrigerator engine) around each complete cycle is zero. The process is represented schematically in Figure 15.10.

Work is done on the device or engine by external agents; in other words, the refrigerator engine system does *negative* work W according to our thermodynamic convention that work done *by* the system is positive. Since W is *negative*, the direction of energy transfer via work is *into* (or on) the engine. Heat transfer $|Q_C|$ occurs from the cold reservoir (the inside of the refrigerator) to the refrigerator engine and heat transfer $|Q_H|$ is exhausted from the refrigerator to the hot reservoir (the outside of the refrigerator). Since the refrigerator engine is the system, the heat transfer Q_C to the refrigerator engine (from the cold reservoir) is positive; the heat transfer Q_H from the refrigerator system (to the hot reservoir) is negative.

The first law of thermodynamics states that the total heat transfer to the refrigerator engine system during one cycle is equal to the change in the internal energy of the system plus the work done by the system. In other words,

$$Q_{\text{total}} = \Delta U + W$$

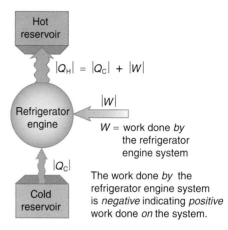

FIGURE 15.10 Schematic representation of a refrigerator engine operating between two reservoirs.

Since the change in the internal energy of the refrigerator engine system around a cycle is zero, $\Delta U = 0$ J, the first law becomes

$$Q_{\text{total}} = 0 \text{ J} + W$$
$$= W$$

For refrigerator engines, *both* Q_{total} and W are *negative* according to our sign conventions. To eliminate sign confusion, we use absolute values:

$$|Q_{\text{total}}| = |W| \tag{15.15}$$

The absolute value of the total heat transfer to the system is

$$|Q_{\text{total}}| = |Q_H| - |Q_C| \tag{15.16}$$

since $|Q_H|$ is larger than $|Q_C|$. Substituting $|W|$ for $|Q_{\text{total}}|$ from Equation 15.15 into Equation 15.16 and rearranging yields

$$|Q_H| = |Q_C| + |W| \tag{15.17}$$

in keeping with energy conservation.

The performance of a refrigerator engine is assessed by taking the ratio of the heat transfer to the refrigerator engine from the cold reservoir to the absolute value of the (negative) work that must be done by the system to accomplish the unnatural heat transfer from the colder to the hotter reservoir. The **coefficient of performance** K of a refrigerator engine is defined as

$$K = \frac{|Q_C|}{|W|} \tag{15.18}$$

where absolute values are used to avoid sign confusion.

For the refrigerator engine system Q_C is positive since the heat transfer Q_C is to refrigerator engine system (from the cold reservoir), so that the absolute value signs are superfluous for this quantity. But W itself is negative, because the work done *by* the refrigerator engine system is negative (meaning positive work must be done *on* the refrigerator engine system by external agents). The coefficient of performance K is a positive pure number. A good refrigerator engine is one that maximizes $|Q_C|$ while

---

The Second Law (Oxford University Press, Oxford, 1965), page 313.
[†]In the context of this section, freezers are equivalent to refrigerators.

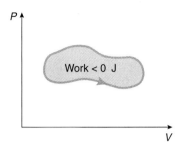

FIGURE 15.11 A refrigerator engine traverses a counterclockwise cycle on a *P–V* diagram and does negative work *W*.

minimizing $|W|$ and therefore has a large coefficient of performance. The examples we study in this chapter indicate that typical values for *K* are only in single digits. A **perfect refrigerator engine** is one that has $|W| = 0$ J and an *infinite* coefficient of performance.

Nothing in the first law of thermodynamics (energy conservation) prohibits a perfect refrigerator engine. Such a refrigerator engine simply has $|W| = 0$ J, so that $|Q_C| = |Q_H|$.

> *Perfect refrigerator engines do not exist; this is another way of stating the second law of thermodynamics.*

Among the implications of this statement of the second law are the following:

1. It is impossible to construct a system or device (a refrigerator engine) operating in a cycle to effect heat transfer from a cold reservoir to a hot reservoir without negative work being done *by* the device (i.e., positive work done *on* the system).
2. Since the refrigerator engine does *negative* work, its cyclical path on, say, a *P–V* diagram is traversed in the *counterclockwise* sense, as shown in Figure 15.11.

The coefficient of performance of a refrigerator engine is defined by Equation 15.18:

$$K = \frac{|Q_C|}{|W|}$$

But, as we found using the first law of thermodynamics (Equation 15.17), in one cycle $|Q_H| = |Q_C| + |W|$.

> Therefore the coefficient of performance of *any* refrigerator engine also can be written as
>
> $$K = \frac{|Q_C|}{|Q_H| - |Q_C|}$$
>
> $$= \frac{1}{\dfrac{|Q_H|}{|Q_C|} - 1} \qquad \text{(any refrigerator engine)} \quad (15.19)$$

The coefficient of performance of a refrigerator engine must be less than infinity; we address how large it can actually be in Section 15.7.

EXAMPLE 15.3

Two (2.00) kilograms of water at 20.0 °C are to be frozen into an ice cube at 0.0 °C using a refrigerator-freezer with a coefficient of performance equal to 4.00.

a. Determine the work done by the refrigerator-freezer.
b. Determine the heat transfer to the hot reservoir.

Solution

a. Take the water as the system. To cool the water from 20.0 °C to 0.0 °C, the heat transfer Q_1 to it is

$$Q_1 = mc_{\text{water}} \Delta T$$

Recall that to calculate ΔT you can use either the celsius or kelvin scales, since the sizes of the kelvin and celsius degree are the same. From Table 13.4, the specific heat of water is 4186 J/(kg·K). Thus

$$Q_1 = (2.00 \text{ kg})[4186 \text{ J/(kg·K)}](0.0 \text{ °C} - 20.0 \text{ °C})$$
$$= -1.67 \times 10^5 \text{ J}$$

The negative sign indicates that the heat transfer really is from the water, not to it.

To freeze the water involves an additional (negative) heat transfer Q_2 from it:

$$Q_2 = -mL_{\text{f water}}$$

Finding the latent heat of fusion of water from Table 13.5, you have

$$Q_2 = -(2.00 \text{ kg})(3.335 \times 10^5 \text{ J/kg})$$
$$= -6.67 \times 10^5 \text{ J}$$

So the total heat transfer from the water is

$$Q = Q_1 + Q_2$$
$$= -8.34 \times 10^5 \text{ J}$$

The heat transfer is from the water *to* the refrigerator engine. Thus positive heat transfer $Q_C = -Q$ accrues *to* the refrigerator engine:

$$Q_C = +8.34 \times 10^5 \text{ J}$$

This corresponds to Q_C in Figure 15.10. The coefficient of performance *K* is defined by Equation 15.18,

$$K = \frac{|Q_C|}{|W|}$$

$$4.00 = \frac{|8.34 \times 10^5 \text{ J}|}{|W|}$$

Hence

$$|W| = 2.09 \times 10^5 \text{ J}$$

The work W done *by* the refrigerator engine is *negative* because you (or another external agent) have to do work *on* the refrigerator engine for it to do its thing. So the work done by the refrigerator engine is

$$W_{\text{refrig}} = -2.09 \times 10^5 \text{ J}$$

b. The absolute value of the heat transfer from the refrigerator engine to the hot reservoir is found using Equation 15.17:

$$
\begin{aligned}
|Q_{\text{H}}| &= |Q_{\text{C}}| + |W| \\
&= 8.34 \times 10^5 \text{ J} + 2.09 \times 10^5 \text{ J} \\
&= 10.43 \times 10^5 \text{ J}
\end{aligned}
$$

The heat transfer to the hot reservoir system is positive, while that from the refrigerator engine system is negative. This energy is exhausted into the room (the hot reservoir), which is why a running refrigerator warms the kitchen.

15.6 THE CARNOT REFRIGERATOR ENGINE

A Carnot cycle executed in counterclockwise fashion on a *P–V* diagram is a **Carnot refrigerator engine**.

We previously showed that the absolute value of the heat transfer to a *Carnot* heat engine from the hot reservoir, $|Q_{\text{H}}|$, divided by that exhausted from the heat engine to the cold reservoir, $|Q_{\text{C}}|$, is equal to the ratio of the absolute temperatures of the reservoirs (Equation 15.13, slightly rearranged):

$$\frac{|Q_{\text{H}}|}{|Q_{\text{C}}|} = \frac{T_{\text{H}}}{T_{\text{C}}} \qquad \text{(Carnot cycle only)}$$

This is *not* true of other heat engine cycles. The result is true even if the Carnot cycle is performed in the counterclockwise sense on the *P–V* diagram. Thus, we substitute this ratio in Equation 15.19.

The coefficient of performance of a Carnot refrigerator engine is

$$K_{\text{Carnot}} = \frac{1}{\dfrac{T_{\text{H}}}{T_{\text{C}}} - 1} \qquad \begin{array}{l}\text{(Carnot refrigerator only)} \\ \text{(use temperatures in kelvin)}\end{array} \quad (15.20)$$

Notice that as the temperature difference between the two reservoirs *increases*, the coefficient of performance of a Carnot refrigerator engine *decreases*.

PROBLEM-SOLVING TACTIC

15.2 In Equation 15.20 for the coefficient of performance of a Carnot refrigerator engine, be sure to use absolute temperatures in kelvin, not celsius temperatures.

EXAMPLE 15.4 ⸻

A Carnot refrigerator engine is operating between temperatures −20 °C and 20 °C. What is its coefficient of performance?

Solution

The coefficient of performance of a Carnot refrigerator engine is found using Equation 15.20:

$$K_{\text{Carnot}} = \frac{1}{\dfrac{T_{\text{H}}}{T_{\text{C}}} - 1}$$

where absolute temperatures *must* be used. Making the temperature conversions and substitutions, you find

$$
\begin{aligned}
K_{\text{Carnot}} &= \frac{1}{\dfrac{293 \text{ K}}{253 \text{ K}} - 1} \\
&= 6.3
\end{aligned}
$$

15.7 THE EFFICIENCY OF REAL HEAT ENGINES AND REFRIGERATOR ENGINES

What is the most efficient heat engine or refrigerator that can operate between two given reservoirs? This is the problem that motivated Carnot when he began his research into the thermodynamics of engines at the beginning of the Industrial Revolution.

The Carnot heat engine, described in some detail in Section 15.3, is an idealized heat engine. The efficiency of a *real* heat engine operating between two reservoirs at temperatures T_{H} and T_{C} always is *less* than that of a Carnot heat engine operating between the same two reservoirs. In other words:

The efficiency of a Carnot heat engine is the maximum efficiency of a heat engine operating between the two given reservoirs.

We will prove this statement by contradiction. That is, we assume there is a prototype heat engine with an efficiency greater than the Carnot heat engine efficiency and then show that a violation of the second law of thermodynamics results.

Thus imagine we have a prototype heat engine (see Figure 15.12) with an efficiency ε greater than the Carnot heat engine

FIGURE 15.12 A prototype heat engine operating between two reservoirs.

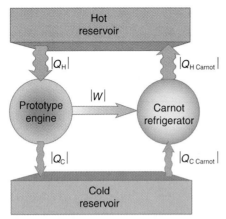

FIGURE 15.13 Use the prototype heat engine to run a Carnot refrigerator engine between the same two reservoirs.

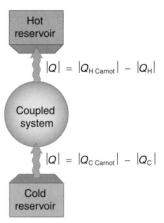

FIGURE 15.14 The total effect of the coupled prototype engine and Carnot refrigerator.

efficiency $\varepsilon_{\text{Carnot}}$ operating between the two given reservoirs at temperatures T_H and T_C. The prototype heat engine accrues heat transfer $|Q_H|$ from the hot reservoir, does work $|W|$, and exhausts heat transfer $|Q_C|$ to the cold reservoir. Now we use the work of the prototype heat engine to run a Carnot refrigerator engine operated between the same two reservoirs, as shown in Figure 15.13.

Heat transfer $|Q_{\text{C Carnot}}|$ occurs to the Carnot refrigerator engine from the cold reservoir.* Heat transfer $|Q_{\text{H Carnot}}|$ is from the Carnot refrigerator engine to the hot reservoir.† The work needed to run the Carnot refrigerator engine is provided by the prototype heat engine. By hypothesis, the prototype heat engine efficiency is greater than that of the Carnot heat engine operating between the same reservoirs; so from the definition of heat engine efficiency (Equation 15.3), we have

$$\frac{|W|}{|Q_H|} > \frac{|W|}{|Q_{\text{H Carnot}}|}$$

Thus

$$|Q_{\text{H Carnot}}| > |Q_H| \qquad (15.21)$$

Now we invoke the first law of thermodynamics. The work done by the prototype heat engine during one cycle is, from Equation 15.17,

$$|W| = |Q_H| - |Q_C|$$

(recall that in a cycle $\Delta U = 0$ J for any engine). The work done by the Carnot refrigerator engine is

$$|W| = |Q_{\text{H Carnot}}| - |Q_{\text{C Carnot}}|$$

The work done by the prototype heat engine is used to run the Carnot refrigerator engine, and so the two expressions for $|W|$ must be the same:

$$|Q_H| - |Q_C| = |Q_{\text{H Carnot}}| - |Q_{\text{C Carnot}}| \qquad (15.22)$$

Rearranging, we obtain

$$|Q_{\text{H Carnot}}| - |Q_H| = |Q_{\text{C Carnot}}| - |Q_C| \qquad (15.23)$$

The left-hand side of Equation 15.23 is positive according to the hypothesis (Equation 15.21); call it Q:

$$|Q_{\text{H Carnot}}| - |Q_H| = |Q_{\text{C Carnot}}| - |Q_C| \equiv |Q| \qquad (15.24)$$

The effect of the coupled system of the prototype heat engine and Carnot refrigerator engine is diagrammed schematically in Figure 15.14. The coupled system effects heat transfer $|Q|$ from a cold reservoir to a hot reservoir *with no net work done* by the system, since the work done by the prototype heat engine is used by the Carnot refrigerator. This combination is a perfect refrigerator, but *such perfect refrigerators do not exist* according to the second law of thermodynamics. Thus the quantity $|Q|$ in Equation 15.24 *must* be zero, and

$$|Q_H| = |Q_{\text{H Carnot}}|$$

which contradicts the hypothesis formulated in Equation 15.21.

Therefore the efficiency of the prototype heat engine is only at best identical to (but not greater than) the efficiency of the Carnot heat engine operating between the same two reservoirs.

A similar argument can be used to show the following:

> A Carnot refrigerator engine has the maximum coefficient of performance of any refrigerator engine operating between the same two reservoirs.

QUESTION 3

An appliance salesman receives a call from a customer to whom a refrigerator was just sold. The customer is complaining because, even though it is cold inside the refrigerator chamber, there is hot air coming into the kitchen from the vent strip at the bottom of the refrigerator. Is the refrigerator broken? Explain.

EXAMPLE 15.5

Two reservoirs have temperatures 20 °C and 300 °C.

a. What is the maximum possible efficiency for a heat engine operating between the two reservoirs?

---

*$Q_{\text{C Carnot}}$ is positive for the Carnot refrigerator engine system and negative for the cold reservoir system.
†$Q_{\text{H Carnot}}$ is negative for the Carnot refrigerator engine system and positive for the hot reservoir system.

b. What is the maximum value of the coefficient of performance of a refrigerator engine operating between the two reservoirs?

Solution

a. The maximum efficiency occurs with a Carnot heat engine operating between the two reservoirs. Equation 15.14 gives the efficiency of a Carnot heat engine:

$$\varepsilon_{\text{Carnot}} = 1 - \frac{T_C}{T_H} \qquad \text{(Carnot heat engine only)}$$

where the temperatures *must* be absolute temperatures. Therefore

$$\varepsilon_{\text{Carnot}} = 1 - \frac{293 \text{ K}}{573 \text{ K}}$$

$$= 0.489$$

b. The maximum value of the coefficient of performance is that of a Carnot refrigerator engine operating between the two reservoirs, given by Equation 15.20:

$$K_{\text{Carnot}} = \frac{1}{\dfrac{T_H}{T_C} - 1} \qquad \text{(Carnot refrigerator engine only)}$$

$$= \frac{1}{\dfrac{573 \text{ K}}{293 \text{ K}} - 1}$$

$$= 1.0$$

15.8 A NEW CONCEPT: ENTROPY

Thus far our study of thermodynamics has considered a number of thermodynamic state variables such as temperature, pressure, volume, and internal energy. A system in thermodynamic equilibrium has definite values for each of these physical quantities, which is why they are called state variables.

The internal energy of a system is a peculiar state variable. From a submicroscopic viewpoint it is possible, in principle, to talk about a specific value for the internal energy of the system in equilibrium. However, from a macroscopic, classical standpoint only *changes* in the internal energy are significant (and only they appear in the first law of thermodynamics).

Other thermodynamic concepts, such as the work W done by a system and the heat transfer Q to a system, are *not* thermodynamic state variables. As we have seen, a system in equilibrium *cannot* be said to have a certain amount of work or heat: there is no mathematical work function or heat function that is characteristic of the system when it is in thermal equilibrium. Heat transfer and work are meaningless concepts for a system in thermodynamic equilibrium. Work and heat transfer are energy transfers to or from a system, thus *changing* its state of thermodynamic equilibrium. We have to distinguish clearly between heat transfer and temperature, between work and internal energy, and between the submicroscopic internal energy and the macroscopic bulk motions of the system as a whole (mechanical energy).

We now come to a different classical physical concept known as the **entropy**.* The entropy is a state variable (as are the inter-

nal energy, volume, temperature, and pressure), which means that a system in equilibrium can be said to have a definite entropy.

> From a classical thermodynamic viewpoint, it is only *changes* in entropy that are significant when a system changes its state of thermodynamic equilibrium.

In this respect the entropy is similar to the concept of internal energy, since it is only changes in internal energy that are physically significant when a system changes its state of thermodynamic equilibrium. However, despite these similarities, the entropy is *not* the same thing as the internal energy, as we will see.

> The zeroth law of thermodynamics was concerned with thermal equilibrium. The first law of thermodynamics was concerned with energy conservation. The second law of thermodynamics is, as we will see, ultimately concerned with entropy changes.

Once we formulate just what is meant by an entropy change, we can restate the second law using the entropy concept and then show that the previous statements of the second law (the nonexistence of perfect heat engines and perfect refrigerator engines) are manifestations of this new formulation. We have our work[†] cut out for us!

Recall that a Carnot heat engine executes an alternating sequence of isothermal and adiabatic reversible processes that form a closed path on a P–V diagram, traversed in the clockwise sense. (A Carnot refrigerator engine also executes alternating isothermal and adiabatic processes, but the closed path is navigated in the counterclockwise sense on a P–V diagram.) We discovered that for a Carnot heat engine operating between two reservoirs at temperatures T_H and T_C, heat transfer $|Q_H|$ accrues to the engine along the isothermal path at T_H and heat transfer $|Q_C|$ is exhausted by the engine to the colder reservoir along the isothermal path at temperature T_C. No heat transfer occurs along the adiabatic paths. Specifically, we found (Equation 15.13)

$$\frac{|Q_H|}{|Q_C|} = \frac{T_H}{T_C} \qquad \text{(Carnot cycle only)}$$

or

$$\frac{|Q_H|}{T_H} = \frac{|Q_C|}{T_C} \qquad \text{(Carnot cycle only)} \qquad (15.25)$$

The *same* result applies to Carnot refrigerator engines, although in the interests of brevity, we did not show this explicitly. The heat transfers Q_H and Q_C have *opposite* signs for a heat engine system and for a refrigerator engine system. Using our convention that the heat transfer to a system is positive,

- for a Carnot heat engine, Q_H is positive and Q_C is negative;
- for a Carnot refrigerator engine, Q_C is positive and Q_H is negative.

Thus, for *either* a Carnot heat engine or a Carnot refrigerator engine, we can rewrite Equation 15.25 *without* the absolute value signs as

$$\frac{Q_H}{T_H} + \frac{Q_C}{T_C} = 0 \text{ J/K} \qquad \text{(Carnot cycle)} \qquad (15.26)$$

---

*The word stems from the Greek τροπή ("transformation") and was first proposed by Rudolph Clausius (1822–1888)

[†]We do *not* mean W in this sentence!

We derived this equation specifically for a Carnot cycle. To accomplish our objective of a general statement of the second law of thermodynamics, we need to generalize Equation 15.26 to *any* kind of (quasi-static) reversible cycle.

We represent an arbitrary quasi-static reversible cycle (closed path) on a *P–V* diagram as shown in Figure 15.15. We approximate the actual path on the *P–V* diagram by means of a succession of small, alternating isothermal and adiabatic paths.

Let Q_i be the heat transfer to the system along the isothermal at temperature T_i. Notice that the alternating isothermal and adiabatic paths along the upper part of the actual cycle on the *P–V* diagram can be paired with those on the lower part of the cycle to form abutting Carnot cycles on the diagram. The contributions to *Q/T* arising from the parts of the isotherms *common* to adjacent Carnot cycles sum to zero, since this part of the isotherm is traversed *twice*, once in each direction. So, along the common parts of each of the various isotherms, one Carnot cycle has a positive contribution *Q/T* and its abutter has a negative contribution of the same magnitude. In this way, the actual cycle can be approximated by the shaded series of Carnot cycles. Since Equation 15.26 can be written for each Carnot cycle, we write the following for the collection of Carnot cycles:

$$\sum_i \frac{Q_i}{T_i} = 0 \text{ J/K} \tag{15.27}$$

The approximation to the actual quasi-static reversible cycle improves by taking tiny *differential* isothermal and adiabatic paths, as indicated in Figure 15.16. Such differential isothermal and adiabatic paths and Carnot cycles then approach the actual cycle path on the *P–V* diagram as a limit. With this limiting procedure, Equation 15.27 becomes

$$\oint_{\substack{\text{reversible} \\ \text{cycle}}} \frac{dQ}{T} = 0 \text{ J/K} \tag{15.28}$$

where the elegant circle through the integral sign indicates that the integral is over a *reversible cycle* on the *P–V* diagram—that is, a closed path on the *P–V* diagram. The quantity

$$\frac{dQ}{T}$$

integrated around a reversible cycle on a *P–V* diagram is zero.

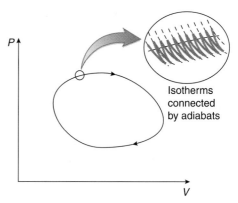

FIGURE 15.16 The actual cycle is approximated by a series of differential isothermal and adiabatic paths.

Now we consider a system undergoing a quasi-static reversible change from some initial equilibrium state to a final equilibrium state, as shown in Figure 15.17. There are many ways to get between these states; two arbitrary paths are indicated.

Suppose we go from the initial state to the final state via path (1) and then return to the initial state via path (2) in Figure 15.17. Since this is a reversible cycle, Equation 15.28 states that

$$\int_i^f \frac{dQ}{T} + \int_f^i \frac{dQ}{T} = 0 \text{ J/K}$$
$$\text{path (1)} \quad \text{path (2)}$$

Now we interchange the limits in the second integration, thus changing the sign of the integral:

$$\int_i^f \frac{dQ}{T} - \int_i^f \frac{dQ}{T} = 0 \text{ J/K}$$
$$\text{path (1)} \quad \text{path (2)}$$

or

$$\int_i^f \frac{dQ}{T} = \int_i^f \frac{dQ}{T} \tag{15.29}$$
$$\text{path (1)} \quad \text{path (2)}$$

Since paths (1) and (2) were *arbitrary* paths, Equation 15.29 means that the *change* in the quantity

$$\frac{dQ}{T}$$

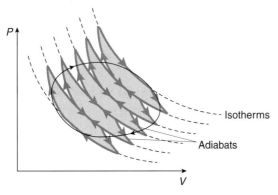

FIGURE 15.15 A cycle on a *P–V* diagram is approximated by abutting Carnot cycles.

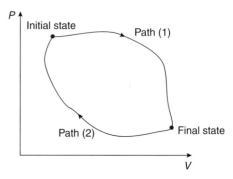

FIGURE 15.17 Proceed from an initial equilibrium state to a final equilibrium state via path (1) and return via path (2).

summed from an initial state to a final state, is *independent of the path* taken between the two states. This quantity

$$\frac{dQ}{T}$$

is defined to be the differential change dS in the entropy of the system when differential heat transfer dQ accrues to the system* at temperature T:

$$dS \equiv \frac{dQ}{T} \qquad (15.30)$$

For a reversible cycle, Equation 15.29 means that

$$\oint_{\substack{\text{reversible} \\ \text{cycle}}} dS = \oint_{\substack{\text{reversible} \\ \text{cycle}}} \frac{dQ}{T} = 0 \text{ J/K} \qquad (15.31)$$

The entropy change of a system between an initial equilibrium state and a final equilibrium state is found from

$$\int_i^f dS = \int_i^f \frac{dQ}{T}$$

$$\Delta S = \int_i^f \frac{dQ}{T} \qquad (15.32)$$
$$\text{reversible path}$$

The dimensions of entropy are those of heat transfer divided by temperature; in SI units, J/K.

Notice that the entropy change of a system as it changes from one thermodynamic equilibrium state to another is

- positive if heat transfer accrues *to* the system ($dQ > 0$ J);
- negative if the heat transfer is *from* the system ($dQ < 0$ J);
- zero for adiabatic processes ($dQ = 0$ J); such processes also are called *isentropic processes*, meaning constant entropy processes;
- zero if the system executes a cycle (since the initial and final states are the same and entropy is a state variable).

Equation 15.32 gives the entropy change of a system undergoing a quasi-static (reversible) process between initial and final equilibrium states. Such quasi-static (reversible) processes are idealizations; no such processes really exist in nature. So how

do we calculate the entropy change of a system undergoing a (realistic) *irreversible* change from an initial state to a final state?

Since the entropy is a state variable, the entropy change between two equilibrium states is the same regardless of how the change is executed.

Therefore, to *calculate* the entropy change between the two states for an irreversible process, we calculate the entropy change for *any* quasi-static (reversible) path connecting the two states. Table 15.1 summarizes the entropy changes that occur in various thermodynamic processes; see Examples 15.6–15.12.

TABLE 15.1 Summary of Entropy Changes in Various Thermodynamic Processes[†]

(a) Ideal gas in a quasi-static reversible process

$$\Delta S = nc_V \ln \frac{T_f}{T_i} + nR \ln \frac{V_f}{V_i}$$

(b) Melting or boiling a mass m:

$$\Delta S = \frac{mL}{T}$$

(where T is the absolute temperature of the phase transition and L is the latent heat associated with the phase transition; $L > 0$ J/kg)

(c) Freezing or condensation of a mass m:

$$\Delta S = -\frac{mL}{T}$$

(where T is the absolute temperature of the phase transition and is the latent heat associated with the phase transition; $L > 0$ J/kg)

(d) Warming or cooling solid or liquid of mass m:

$$\Delta S = mc \ln \frac{T_f}{T_i}$$

(where c is the specific heat of the substance)

(e) Heat transfer $|Q|$ *to* a reservoir at temperature T_C:

$$\Delta S = +\frac{|Q|}{T_C}$$

(f) Heat transfer $|Q|$ *from* a reservoir at temperature T_H:

$$\Delta S = -\frac{|Q|}{T_H}$$

(g) Heat engine or refrigerator engine in one complete cycle:

$$\Delta S = 0 \text{ J/K}$$

(h) Adiabatic process:

$$\Delta S = 0 \text{ J/K}$$

[†] Use absolute temperatures in all equations.

---

*As we have mentioned, since a system in thermal equilibrium *cannot* be said to have a certain amount of heat, there is no heat function associated with a system. The quantity dQ, therefore, represents an infinitesimal amount of heat transfer, and it is called an *inexact* differential dQ. It is quite amazing that the inexact differential dQ, when divided by the absolute temperature T, yields the change dS in the entropy of the system, which *is* a state function. The situation is analogous to transfers of energy via work. Think of the CWE theorem of pure mechanics. A differential amount of work done by all the forces on a system is equal to the differential change in the kinetic energy of the system; a system in equilibrium *cannot* be said to have a certain amount of work, but it *can* be said to have a certain amount of kinetic energy.

PROBLEM-SOLVING TACTIC

15.3 When using the equations in Table 15.1 to calculate entropy changes, be sure to use absolute temperatures in kelvin, not celsius temperatures.

QUESTION 4

Work and heat transfer are the energy transferred to a system by forces or a temperature difference, respectively. The differential entropy change dS is defined as

$$\frac{dQ}{T}$$

Is there value or a reason in dynamics for defining a corresponding concept

$$\frac{dW}{T}$$

or dW divided by some other corresponding parameter? On what basis do you make such decisions?

EXAMPLE 15.6

A 150 g ice cube at 0 °C melts into water at 0 °C. What is the entropy change of the mass?

Solution

The mass is the system. The process takes place at the constant temperature $T = 273.15$ K, and so the entropy change of the system is

$$\Delta S = \int_i^f \frac{dQ}{T}$$

$$= \frac{Q}{T}$$

where Q is the heat transfer to the ice to melt it. You must use temperatures in kelvin; notice how nonsensical it would be to use the 0 °C temperature in this equation. From our study of calorimetry,

$$Q = mL$$

where L is the latent heat of fusion of water: 3.335×10^5 J/kg. Hence

$$Q = (0.150 \text{ kg})(3.335 \times 10^5 \text{ J/kg})$$

$$= 5.00 \times 10^4 \text{ J}$$

The entropy change of the mass is

$$\Delta S = \frac{Q}{T}$$

$$= \frac{5.00 \times 10^4 \text{ J}}{273.15 \text{ K}}$$

$$= 183 \text{ J/K}$$

STRATEGIC EXAMPLE 15.7

a. Find the entropy change for an ideal gas when it changes its state of thermodynamic equilibrium from an initial state at temperature T_i and volume V_i to a final state at temperature T_f and volume V_f.

b. Specialize the result of part (a) for an isochoric process and for an isothermal process.

Solution

a. Imagine taking the ideal gas from its initial state to a final state via a quasi-static, reversible process. Since the entropy is a state variable, the entropy change will be the same regardless of how the gas proceeds between the two states. According to Equation 15.32, the entropy change of the gas is

$$\Delta S = \int_i^f \frac{dQ}{T}$$
reversible
path

The first law of thermodynamics, Equation 13.51, in differential form states that

$$dQ = dU + dW$$

According to Equation 14.21, the differential change in internal energy dU of an ideal gas is

$$dU = nc_V \, dT$$

and (from Equation 13.53)

$$dW = P \, dV$$

The first law of thermodynamics then can be written in differential form as

$$dQ = nc_V \, dT + P \, dV$$

The entropy change is

$$\Delta S = \int_i^f \frac{dQ}{T}$$

Substituting for dQ, you get

$$\Delta S = nc_V \int_i^f \frac{dT}{T} + \int_i^f \frac{P}{T} \, dV$$

But the ideal gas equation of state is (Equation 13.20)

$$PV = nRT$$

so that

$$\frac{P}{T} = \frac{nR}{V}$$

Substituting this into the second term of the entropy change, you obtain

$$\Delta S = nc_V \int_i^f \frac{dT}{T} + nR \int_i^f \frac{dV}{V}$$

Performing the integrals, you find

$$\Delta S = nc_V \ln \frac{T_f}{T_i} + nR \ln \frac{V_f}{V_i} \tag{1}$$

b. Notice that equation (1) simplifies to a single term if the two states of thermodynamic equilibrium are reached by various special processes:

1. *Isochoric process* (constant volume), as shown in Figure 15.18. Here $V_f = V_i$, and the second term in equation (1) is zero because $\ln 1 = 0$:

$$\Delta S = nc_V \ln \frac{T_f}{T_i} \tag{2}$$

(notice that $\Delta S < 0$ J/K if $T_f < T_i$—that is, if the system cools).

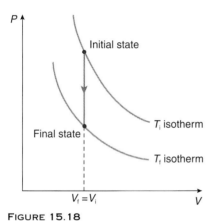

FIGURE 15.18

2. *Isothermal process* (constant temperature), as shown in Figure 15.19. Here $T_f = T_i$, and the first term in equation (1) is zero, again because $\ln 1 = 0$:

$$\Delta S = nR \ln \frac{V_f}{V_i} \tag{3}$$

(notice that $\Delta S < 0$ if $V_f < V_i$—that is, if the system is compressed).

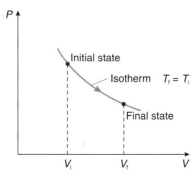

FIGURE 15.19

STRATEGIC EXAMPLE 15.8

An ideal gas is confined by a membrane to a volume V_i in the left-hand side of a well-insulated container of volume V_f, as shown in Figure 15.20a. The right-hand side of the container is initially empty (a vacuum). The entire system is thermally insulated from its environment. The membrane is punctured and the gas rushes to fill the entire container (see Figure 15.20b); this process is called a *free expansion* of a gas, much like the popping of a balloon, though the latter is less scientifically sanitized. Such a free expansion is an irreversible process; it is *not* a quasi-static (reversible) process. Find the entropy change associated with this free expansion of an ideal gas.

(a)   (b)

FIGURE 15.20

Solution
Begin by looking at the first law of thermodynamics; apply it to the gas as the system:

$$Q = \Delta U + W$$

That is, the total heat transfer to the system is the sum of the change in the internal energy of the system and the work done *by* the system on its surroundings. Since the system is thermally isolated by the insulation, there is no heat transfer to or from the system, and so $Q = 0$ J. No work is done by the gas on its surroundings either, because the gas is isolated; thus, $W = 0$ J. The first law then indicates that the change in the internal energy of the gas is zero. Since the internal energy of an ideal gas is purely a function of temperature (see Chapter 14), the temperature of the ideal gas is unchanged in the free expansion.

You want to calculate the entropy change of the system between the initial state (the gas confined to the volume V_i) and the final state (the gas occupying the entire volume V_f). Since the system (the gas) is thermally insulated from its environment, there is zero heat transfer to the system. That is, $dQ = 0$ J. You saw that the temperature T also does not change. It might appear, at first glance, that the entropy change of the system is zero since (from Equation 15.32)

$$\Delta S = \int_{\substack{i \\ \text{reversible} \\ \text{path}}}^{f} \frac{dQ}{T}$$

Since the temperature is constant, the integral becomes

$$\Delta S = \frac{1}{T} \int_{\substack{i \\ \text{reversible} \\ \text{path}}}^{f} dQ$$

With $dQ = 0$ J, this seems to imply that $\Delta S = 0$ J/K; *but this is not the case!* Why is that? *To calculate the entropy change using Equation 15.32, you must use a quasi-static reversible process* connecting the initial and final states of thermodynamic equilibrium. A free expansion is *not* such a process. Thus, to calculate the entropy change of the ideal gas in the free expansion, you use a quasi-static (reversible) process connecting the initial and final states of the gas and use *that* process to calculate the entropy change. That is, use equation (1) of Example 15.7, since it was derived for a quasi-static reversible process for an ideal gas:

$$\Delta S = nc_V \ln \frac{T_f}{T_i} + nR \ln \frac{V_f}{V_i}$$

Since the temperature of the ideal gas is unchanged in the free expansion, the first term is zero ($\ln 1 = 0$) and the entropy change of the system is

$$\Delta S = nR \ln \frac{V_f}{V_i}$$

So here is an example where $dQ = 0$ J, but $dS \neq 0$ J/K. Strange but true.

Note that the entropy change for the irreversible free expansion is positive since $V_f > V_i$ (the natural logarithm of a number greater than 1 is positive; the natural logarithm of numbers between 0 and 1 is negative).

EXAMPLE 15.9

Find the entropy change of a system of two reservoirs when heat transfer $|Q|$ occurs from the warmer reservoir at temperature T_H to the cooler reservoir at temperature T_C.

Solution
The process is irreversible. However, the entropy change is calculated by imagining slow quasi-static (reversible) heat transfer.

Recall that a reservoir is capable of losing or accruing any heat transfer without changing its temperature. Since the temperature of the reservoir is constant, the warmer reservoir has an entropy change

$$\Delta S = \int_i^f \frac{dQ}{T}$$
$$= -\frac{|Q|}{T_H}$$

The minus sign arises because the heat transfer is *from* the hotter reservoir. The entropy change of the hotter reservoir is *negative*.

The cooler reservoir also does not change its temperature when accruing heat transfer $|Q|$; the entropy change of the cold reservoir is

$$\Delta S = \int_i^f \frac{dQ}{T}$$
$$= \frac{|Q|}{T_C}$$

where the positive result arises since the heat transfer is *to* the cold reservoir. The entropy change of the cold reservoir is *positive*.

The *total* entropy change of both reservoirs is the sum of the two entropy changes:

$$\Delta S_{total} = \frac{|Q|}{T_C} - \frac{|Q|}{T_H} \tag{1}$$

Note that the total entropy change of the system of two reservoirs is positive, since the first term always is greater than the second term because $T_C < T_H$.

STRATEGIC EXAMPLE 15.10

A mass m_1 at temperature T_H is placed in thermal contact with another mass m_2 at temperature T_C. The entire system is isolated, so that heat transfer occurs only between the two masses. The final (common) equilibrium temperature of the system is T, where $T_C < T < T_H$. Find the total entropy change associated with the heat transfer between the hot and cold masses. The specific heats [in J/(kg·K)] of the masses are c_1 and c_2 respectively.

Solution
Once again, the process is irreversible. However, you can imagine each mass brought to the equilibrium temperature T by placing it in thermal contact with a succession of reservoirs with temperatures differing differentially between the initial temperature of the mass and the equilibrium temperature. In this way you construct a quasi-static (reversible) series of processes connecting the initial and final states of each mass.

The entropy change of the initially hot mass is found from

$$\Delta S_1 = \int_i^f \frac{dQ}{T}$$
$$= m_1 c_1 \int_{T_H}^{T} \frac{dT}{T}$$
$$= m_1 c_1 \ln \frac{T}{T_H}$$

where it is assumed that the specific heat of the substance is not a function of temperature. Since $T < T_H$, this entropy change is *negative* (the natural logarithm of a number between 0 and 1 is negative).

The entropy change of the initially cold mass is found in a similar way:

$$\Delta S_2 = \int_i^f \frac{dQ}{T}$$
$$= m_2 c_2 \int_{T_C}^{T} \frac{dT}{T}$$
$$= m_2 c_2 \ln \frac{T}{T_C}$$

where again it is assumed that the specific heat is not a function of temperature. This entropy change is *positive*, since $T > T_C$.

The *total* entropy change is the sum of the entropy changes:

$$\Delta S_{total} = \Delta S_1 + \Delta S_2$$
$$= m_1 c_1 \ln \frac{T}{T_H} + m_2 c_2 \ln \frac{T}{T_C} \tag{1}$$

The first term is negative and the second is positive. *In every case one finds that the total entropy change is positive*; in other words, the positive term is always larger than the negative term. This is yet another consequence of the second law of thermodynamics, as we shall see in the next section.

EXAMPLE 15.11

Find the entropy change associated with the mixing of identical liquids (or identical gases) at different temperatures.

Solution

Equation (1) of Example 15.10 also can be used to calculate the entropy change when two *identical liquids* (or identical gases) at different temperatures are mixed together and reach an equilibrium temperature. The word *identical* means mixing water with water, orange juice with orange juice, vodka with vodka, oxygen with oxygen, and so on (but not water with orange juice, orange juice with vodka, milk with coffee, or similar nonidentical liquids or gases). It is *not* obvious, but the mixing of *nonidentical* substances introduces an *additional term* into the total entropy change (a term known as the entropy of mixing); but to calculate this term is beyond the scope of an introductory course. You have earned the right to be grateful for that.

EXAMPLE 15.12

A 3.00 kg sample of water is warmed from 0 °C to 100 °C. What is the entropy change of the water?

Solution

Since $dQ = mc_{water} \, dT$, the entropy change is

$$\Delta S = mc_{water} \int_{T_i}^{T_f} \frac{dT}{T}$$
$$= mc_{water} \ln \frac{T_f}{T_i}$$

Make the appropriate substitutions and remember that absolute temperatures *must* be used:

$$\Delta S = (3.00 \text{ kg})(4186 \text{ J/kg·K}) \left(\ln \frac{373 \text{ K}}{273 \text{ K}} \right)$$
$$= 3.92 \times 10^3 \text{ J/K}$$

15.9 ENTROPY AND THE SECOND LAW OF THERMODYNAMICS

A theory is the more impressive the greater the simplicity of its premises is, the more difficult kinds of things it relates, and the more extended is its area of applicability. Therefore the deep impression which classical thermodynamics made upon me. It is the only physical theory of universal content concerning which I am convinced that, within the framework of the applicability of its basic concepts, it will never be overthrown (for the special attention of those who are skeptics on principle).

Albert Einstein (1879–1954)[*]

[*] *Albert Einstein: Philosopher-Scientist*, edited by Paul Arthur Schilpp (The Library of Living Philosophers, Evanston, Illinois, 1949), Volume VII, page 33.

We stated two versions of the second law of thermodynamics earlier in our study of heat engines and refrigerator engines:

1. *Perfect heat engines do not exist.*
2. *Perfect refrigerator engines do not exist.*

Now that we have a grasp on the peculiar and less intuitive thermodynamic state variable known as the entropy, we make a more general statement of the second law and show that it is equivalent to the two previous formulations.

> The most general statement of the second law of thermodynamics is as follows:
>
> 3. *The total entropy change of an **isolated** system is always greater than or equal to zero:*
>
> $$\boxed{\Delta S_{\text{total isolated}} \geq 0 \text{ J/K}} \qquad (15.33)$$
>
> The total entropy change is equal to zero only for reversible processes.

A few things need to be emphasized regarding this statement of the second law:

1. Notice that the statement concerns an *isolated* system; the entropy change of a system that is not isolated can very well be negative. If the system is not isolated, it can be incorporated into a new and larger isolated system by including appropriate surroundings. When the original nonisolated system is expanded to become an isolated system, the *total* entropy change of the now isolated system, as a whole, is *always* greater than or equal to zero.
2. The total entropy change is the algebraic sum of the individual entropy changes of the various components of the isolated system. The entropy changes of the individual components of the isolated system may be positive, negative, or zero. But the *total* entropy change (the algebraic sum of the individual entropy changes) of the *isolated* system is greater than or equal to zero.
3. The total entropy change of an isolated system is equal to zero only when the system undergoes a quasi-static reversible process.
4. The total entropy change of an isolated system in an *irreversible* process is always *greater than zero*.

Now let us show that our third version of the second law (Equation 15.33) incorporates our earlier versions.

Perfect Heat Engines Do Not Exist

A perfect heat engine is schematically represented in Figure 15.21. Such a perfect heat engine system would accrue positive heat transfer Q_H (> 0 J) from a reservoir at temperature T_H and do (positive) work W with no exhaust heat transfer Q_C (= 0 J) from the engine to a cooler reservoir at temperature T_C. The *isolated* system consists of the perfect heat engine *and* the two reservoirs.

We look at the entropy change of the isolated system for one cycle of the perfect engine. The entropy is a state variable, and so when the heat engine system returns to its initial thermodynamic

FIGURE 15.21 A perfect heat engine.

state, the entropy of the heat engine returns to its initial value. The entropy change of the heat engine itself is zero:

$$\Delta S_{\text{engine}} = 0 \text{ J/K}$$

The entropy change of the reservoir system at temperature T_H is

$$\Delta S_{\text{hot reservoir}} = -\frac{|Q_H|}{T_H}$$

which is negative because the heat transfer $|Q_H|$ is from this reservoir system.

For a perfect engine, no heat transfer occurs to the lower-temperature reservoir, and so the entropy change of the lower-temperature reservoir system is zero:

$$\Delta S_{\text{cold reservoir}} = 0 \text{ J/K}$$

The total entropy change of the *isolated* system is the sum of the individual entropy changes:

$$\Delta S_{\text{total isolated}} = \Delta S_{\text{engine}} + \Delta S_{\text{hot reservoir}} + \Delta S_{\text{cold reservoir}}$$

$$= 0 \text{ J/K} - \frac{|Q_H|}{T_H} + 0 \text{ J/K}$$

The total entropy change of the isolated system is *negative*. This violates the second law: the entropy change of an isolated system must be greater than or equal to zero. Thus perfect heat engines cannot exist.

Perfect Refrigerator Engines Do Not Exist

A perfect refrigerator engine is represented schematically in Figure 15.22. The perfect refrigerator engine effects heat transfer $|Q|$ from a reservoir at temperature T_C to a hotter reservoir at temperature T_H, doing no work in the process of each refrigerator cycle. The *isolated* system consists of the perfect refrigerator engine and the two reservoirs.

Let us calculate the entropy change for one cycle of the perfect refrigerator engine system. Since the entropy is a state variable and the refrigerator engine operates in a cycle, returning

to its initial state, the entropy change of the refrigerator engine in each cycle is zero:

$$\Delta S_{\text{refrig}} = 0 \text{ J/K}$$

The cold reservoir system suffers an entropy change

$$\Delta S_{\text{cold reservoir}} = -\frac{|Q|}{T_C}$$

which is negative because the heat transfer $|Q|$ is *from* this reservoir. The warmer reservoir system experiences an entropy change

$$\Delta S_{\text{hot reservoir}} = +\frac{|Q|}{T_H}$$

since the heat transfer $|Q|$ is *to* this reservoir. Thus the total entropy change of the *isolated* system is

$$\Delta S_{\text{total isolated}} = \Delta S_{\text{refrig}} + \Delta S_{\text{cold reservoir}} + \Delta S_{\text{hot reservoir}}$$

$$= 0 \text{ J/K} - \frac{|Q|}{T_C} + \frac{|Q|}{T_H}$$

This expression is negative, since $T_C < T_H$. The negative result violates the second law: the total entropy change of an isolated system must be greater than or equal to zero. Thus perfect refrigerator engines cannot exist.

FIGURE 15.22 A perfect refrigerator engine.

QUESTION 5

Explain why the freezing of water into ice does not violate the second law of thermodynamics. What is the isolated system?

EXAMPLE 15.13

A Carnot heat engine, in executing one cycle, uses 2000 J of heat transfer from a reservoir at 500 K, performs work, and effects heat transfer to a low-temperature reservoir at 300 K.

a. Find the efficiency of the Carnot heat engine.
b. Determine the work done by the heat engine in one cycle.
c. Calculate the entropy change of the heat engine and each reservoir during one cycle of the engine.
d. Find the total entropy change of the isolated system of the heat engine and two reservoirs and comment on the answer.

Solution

a. The efficiency of a Carnot heat engine is found using Equation 15.14:

$$\varepsilon_{\text{Carnot}} = 1 - \frac{T_C}{T_H} \qquad \text{(Carnot cycle only)}$$

$$= 1 - \frac{300 \text{ K}}{500 \text{ K}}$$

$$= 0.400$$

b. The work done by the heat engine is found from the definition of the efficiency, Equation 15.3:

$$\varepsilon = \frac{|W|}{|Q_H|}$$

where Q_H is the heat transfer to the engine from the hot reservoir. Since the efficiency is known from part (a), you have

$$0.400 = \frac{|W|}{2000 \text{ J}}$$

or

$$|W| = 800 \text{ J}$$

The work done by a heat engine is positive, so that

$$W = 800 \text{ J}$$

c. The entropy change of the heat engine system during one cycle is 0 J/K, because the heat engine returns to its initial thermodynamic state and entropy is a state variable.

The heat transfer from the hot reservoir system is −2000 J but, as a reservoir, no change in its temperature occurred. Hence the entropy change of the hot reservoir is negative:

$$\Delta S_H = -\frac{|Q|}{T_H}$$

$$= -\frac{2000 \text{ J}}{500 \text{ K}}$$

$$= -4.00 \text{ J/K}$$

Of the 2000 J of heat transfer to the engine from the hot reservoir, 800 J of work was done by the engine during each cycle. Hence the heat transfer to the cool reservoir is, according to the first law of thermodynamics (energy conservation),

$$2000 \text{ J} - 800 \text{ J} = 1200 \text{ J}$$

The cool reservoir—indeed, any reservoir—does not change its temperature. The entropy change of the cool reservoir system is positive, because the heat transfer to it is 1200 J. Hence the entropy change of the cool reservoir is

$$\Delta S_C = +\frac{|Q|}{T_C}$$

$$= \frac{1200 \text{ J}}{300 \text{ K}}$$

$$= 4.00 \text{ J/K}$$

d. The total entropy change of the isolated system consisting of the heat engine and two reservoirs is the sum of the individual entropy changes (with proper signs):

$$\Delta S_{\text{total isolated}} = 0 \text{ J/K} + 4.00 \text{ J/K} + (-4.00 \text{ J/K})$$

$$= 0 \text{ J/K}$$

The result is not surprising, since the cycle is a Carnot cycle (reversible) and the total entropy change of an isolated system is zero *only* for such reversible cycles.

STRATEGIC EXAMPLE 15.14

A 2.00 kg block of ice is taken out of a freezer at −20.0 °C, melts, and warms to the temperature of a large room at 25.0 °C. Calculate the entropy change of the ice and the entropy change of the room. Verify that the total entropy change of the (isolated) ice and room system is consistent with the second law of thermodynamics.

Solution

The entropy change of the ice must be calculated in three stages:

1. warming the ice from −20 °C to 0 °C;
2. melting of the ice at 0 °C to water at 0 °C; and
3. warming of the cold water at 0 °C to the temperature of the room.

Consider each stage:

1. Warming the Ice from −20 °C to 0 °C

The entropy change is

$$\Delta S = \int_i^f \frac{dQ}{T}$$

where the differential heat transfer dQ to warm the ice is

$$dQ = mc_{\text{ice}} \, dT$$

and c_{ice} is the specific heat of ice. Thus

$$\Delta S = mc_{\text{ice}} \int_{T_i}^{T_f} \frac{dT}{T}$$

where T_i is the initial absolute temperature of the ice and T_f is the final absolute temperature of the ice before melting. Integrating,

$$\Delta S = mc_{\text{ice}} \ln \frac{T_f}{T_i}$$

From Table 13.4, you have

$$c_{\text{ice}} = 2050 \text{ J/(kg·K)}$$

Making the appropriate substitutions, you find

$$\Delta S = (2.00 \text{ kg})(2050 \text{ J/kg·K}) \left(\ln \frac{273 \text{ K}}{253 \text{ K}} \right)$$

Thus to warm the ice to its melting point results in an entropy change of the ice of

$$\Delta S = 312 \text{ J/K}$$

2. Melting the Ice at 0 °C to Water at 0 °C

This process takes place at constant temperature. Therefore the entropy change is

$$\Delta S = \int_i^f \frac{dQ}{T}$$
$$= \frac{Q}{T}$$

where Q is the heat transfer to the ice to melt it. The latent heat of fusion of water is found from Table 13.5 to be

$$L_{f\,water} = 3.335 \times 10^5 \text{ J/kg}$$

Therefore

$$Q = mL_{f\,water}$$
$$= (2.00 \text{ kg})(3.335 \times 10^5 \text{ J/kg})$$
$$= 6.67 \times 10^5 \text{ J}$$

The entropy change in melting the ice then is

$$\Delta S = \frac{6.67 \times 10^5 \text{ J}}{273 \text{ K}}$$
$$= 2.44 \times 10^3 \text{ J/K}$$

3. Warming the Cold Water to Room Temperature

This process is similar to the warming of the ice, and so you can use the same equation as in part (1), replacing the specific heat of ice with the specific heat of water (see Table 13.4) and accounting for the different initial and final temperatures:

$$\Delta S = mc_{water} \ln \frac{T_f}{T_i}$$
$$= (2.00 \text{ kg}) (4186 \text{ J/kg·K}) \left(\ln \frac{298 \text{ K}}{273 \text{ K}} \right)$$
$$= 734 \text{ J/K}$$

The ice and room are an isolated system. The heat transfer to the ice is from the room. Let Q be the heat transfer to the room; Q is negative since the heat transfer is really *from* the room. The total heat transfer to the isolated ice–room system is zero:

$$Q + (mc_{ice} \Delta T_1 + mL_{f\,water} + mc_{water} \Delta T_2) = 0 \text{ J}$$

where ΔT_1 is the temperature change of the ice as it warms to its melting point, and ΔT_2 is the temperature change of the melted ice water as it warms to room temperature from the melting point. Thus

$$Q = -(mc_{ice} \Delta T_1 + mL_{f\,water} + mc_{water} \Delta T_2)$$

Inserting appropriate numerical values of the quantities, you get

$$Q = -\{(2.00 \text{ kg})[2050 \text{ J/(kg·K)}](20.0 \text{ K})$$
$$+ (2.00 \text{ kg})(3.335 \times 10^5 \text{ J/kg})$$
$$+ (2.00 \text{ kg})[4186 \text{ J/(kg·K)}](25.0 \text{ K})\}$$
$$= -9.58 \times 10^5 \text{ J}$$

The entropy change of the room, since its temperature is essentially unaffected and it acts as a reservoir, is

$$\Delta S = -\frac{|Q|}{T}$$
$$= -\frac{9.58 \times 10^5 \text{ J}}{298 \text{ K}}$$
$$= -3.21 \times 10^3 \text{ J/K}$$

The *total* entropy change of the ice-room system is the sum of all the entropy changes:

$$S_{total} = 312 \text{ J/K} + 2.44 \times 10^3 \text{ J/K} + 734 \text{ J/K}$$
$$- 3.21 \times 10^3 \text{ J/K}$$
$$= 2.7 \times 10^2 \text{ J/K}$$

The entropy change of the entire (isolated) ice–room system is greater than zero, consistent with the second law of thermodynamics.

15.10 THE DIRECTION OF HEAT TRANSFER: A CONSEQUENCE OF THE SECOND LAW

The entropy statement of the second law (Equation 15.33) finally enables us to explain why heat transfer is *from* hotter objects *to* cooler objects and not the other way around. If heat transfer $|Q|$ occurred spontaneously from a cold to a hot reservoir, represented schematically as shown in Figure 15.23, the total entropy change of the isolated system of reservoirs would be

$$\Delta S_{total\;isolated} = \Delta S_{cold\;reservoir} + \Delta S_{hot\;reservoir}$$
$$= -\frac{|Q|}{T_C} + \frac{|Q|}{T_H}$$

This expression is negative because $T_C < T_H$, and so the process *violates* the second law of thermodynamics; the process cannot occur.

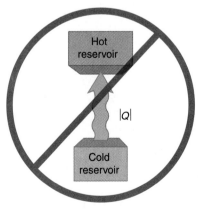

FIGURE 15.23 Spontaneous heat transfer from colder to warmer systems violates the second law of thermodynamics.

FIGURE 15.24 Spontaneous heat transfer from warmer to cooler systems conforms to the second law of thermodynamics.

Such unobserved heat transfer is just like a perfect refrigerator, which we know cannot exist.

For heat transfer $|Q|$ *from* a hot reservoir *to* a cold reservoir, as shown schematically in Figure 15.24, the total entropy change of the isolated system of two reservoirs is

$$\Delta S_{\text{total isolated}} = \Delta S_{\text{cold reservoir}} + \Delta S_{\text{hot reservoir}}$$
$$= +\frac{|Q|}{T_C} - \frac{|Q|}{T_H}$$

Since $T_H > T_C$, the total entropy change of the isolated system is indeed greater than zero, in conformity with the second law of thermodynamics.

Thus it is the second law of thermodynamics that finally explains why heat transfer occurs spontaneously from hotter to cooler systems and not the other way around.

15.11 A STATISTICAL INTERPRETATION OF THE ENTROPY*

In a real sense maximum disorder was our equilibrium.
T. E. Lawrence (1888–1935)[†]

In introducing the classical concept of entropy in Section 15.8, we mentioned that a physical system in equilibrium has a definite entropy, since it is a state variable (a property of the system). On a macroscopic basis, however, when a system *changes* its state of thermodynamic equilibrium, it is only changes in the entropy that are significant. The second law of thermodynamics is a statement about *changes* in the entropy of an *isolated* system.

The entropy a system has when in thermal equilibrium has a statistical interpretation. For simplicity, we use an analogy to explore such a statistical interpretation.[‡] We employ a macroscopic rather than a microscopic system, because the former is more familiar to you, and experiments using such a system are easier to visualize. Also, by using a macroscopic system we avoid quantum mechanical complications that might arise with microscopic

systems. The essential features of the macroscopic system we consider also have analogies at the microscopic level.

The system we consider is a collection of pennies in a box. Consider each penny as a particle with two sides: heads and tails. If the pennies are placed in a covered box and the container is shaken vigorously for a few moments, there are various possibilities for the number of heads and tails that will emerge (we exclude edge-on pennies). We begin with a system of just a few pennies; then we increase the number of pennies to see what happens for a system with many particles.

With a collection of four pennies, there are five possible states for the system after shaking the box:

1. 4 heads and 0 tails
2. 3 heads and 1 tail
3. 2 heads and 2 tails
4. 1 head and 3 tails
5. 0 heads and 4 tails

We call this set of possibilities a delineation of the various **macrostates** of the 4-penny system. There are five macrostates for the 4-penny system. Associated with *each* macrostate is a collection of **microstates**. A microstate specifies *which individual penny* has a head or a tail. Thus macrostate #1 (4 heads and 0 tails) has only one microstate: every penny must be heads. Macrostate #2 has four microstates: each of the four pennies could be the one that is tails, with the others heads. The various microstates corresponding to each 4-penny macrostate are delineated in Table 15.2. We also can make a histogram depicting the number of microstates associated with each macrostate of the 4-penny system; see Figure 15.25.

> Each of the microstates is equally likely to occur; but the macrostates are *not* equally likely to occur.

Indeed the number of *microstates* associated with a *given macrostate*, divided by the *total* number of *microstates*, is a measure of the *probability* of that macrostate occurring in a toss. Thus the macrostate with 2 heads and 2 tails is the most likely to occur, because it has the greatest number of associated microstates: 6 out of a total of 16 microstates; the probability of the 2-heads, 2-tails macrostate occurring is 6/16 = 0.375 (= 37.5%). The macrostate with all heads has just 1 microstate out of the 16 total possible microstates, so that the probability of all heads occurring is only 1/16 = 0.0625 (= 6.25%). Thus the probabilities of the various *macrostates* occurring are not all the same.

> After the box is shaken, the pennies are most likely to be found in the macrostate with the greatest number of associated microstates.

The microstates are like lottery tickets: each is equally likely to win. The number of lottery tickets you have is a macrostate; the more tickets you have, the greater the likelihood of your winning.

The statistical interpretation of entropy, developed by Ludwig Boltzmann during the late 19th century, defines the entropy of the system in a given macrostate to be Boltzmann's constant k times the natural logarithm (whose base is the pure number $e \equiv 2.718\,281\,828\ldots$) of the number of microstates Ω associated with a given macrostate.

[†] *Seven Pillars of Wisdom* (Doubleday, Doran, Garden City, New York, 1936), Chapter LIX, page 338.
[‡] Problem 65 explores a real example.

TABLE 15.2 Four-Penny Macrostates and Microstates

Macrostate (1): 4 heads (H) and 0 tails (T)

Penny 1	Penny 2	Penny 3	Penny 4
H	H	H	H
(1 microstate)			

Macrostate (2): 3 heads and 1 tail

Penny 1	Penny 2	Penny 3	Penny 4
H	H	H	T
H	H	T	H
H	T	H	H
T	H	H	H
(4 microstates)			

Macrostate (3): 2 heads and 2 tails

Penny 1	Penny 2	Penny 3	Penny 4
H	H	T	T
H	T	T	H
T	T	H	H
T	H	H	T
T	H	T	H
H	T	H	T
(6 microstates)			

Macrostate (4): 1 head and 3 tails

Penny 1	Penny 2	Penny 3	Penny 4
H	T	T	T
T	H	T	T
T	T	H	T
T	T	T	H
(4 microstates)			

Macrostate (5): 0 heads and 4 tails

Penny 1	Penny 2	Penny 3	Penny 4
T	T	T	T
(1 microstate)			

We define the entropy of a given macrostate to be

$$S \equiv k \ln \Omega \qquad (15.34)$$

where Ω is the number of microstates associated with the given macrostate.

The number of microstates for a system with many particles is staggeringly large. We shall see that for a 100-penny system, hardly a large number of particles, the total number of microstates is $2^{100} \approx 1.27 \times 10^{30}$, a fantastically huge number. This is one reason that the entropy is defined as the natural logarithm of the number of microstates.*†

Figure 15.25 is a histogram of the number of microstates associated with each macrostate of the 4-penny system. The entropy of each macrostate is Boltzmann's constant k times the natural logarithm of the number of microstates associated with that macrostate. The entropy of the system has a maximum at the macrostate that is most likely, the macrostate with the greatest number of associated microstates—namely, the macrostate with half heads and half tails.

Now we perform another experiment. We set the four pennies in a box, all with heads up initially, as shown in Figure 15.26.

---

*The *natural logarithm* (ln) is used, rather than logarithms to base 10 (log), because the natural logarithm emerges naturally from the detailed statistical mechanical analysis of a system with a large number of particles. This emergence is difficult to show here.

†The logarithm also ensures that the entropy of two systems, when combined into a single system, is the sum of the individual entropies. To see this, suppose we have two individual systems with entropies

$$S_1 = k \ln \Omega_1$$

and

$$S_2 = k \ln \Omega_2$$

where Ω_1 and Ω_2 are the respective numbers of microstates associated with the particular macrostates that the systems are in. If the two systems are brought together, the number of associated microstates is the product $\Omega_1 \Omega_2$. The entropy of the new system thus is

$$\begin{aligned} S &= k \ln(\Omega_1 \Omega_2) \\ &= k \ln \Omega_1 + k \ln \Omega_2 \\ &= S_1 + S_2 \end{aligned}$$

which is the sum of the individual entropies of the two systems.

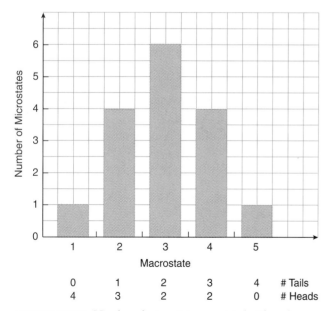

FIGURE 15.25 Number of microstates associated with each macrostate of a 4-penny system.

The epitaph for Ludwig Boltzmann in the Zentralfriedhof in Vienna, Austria, includes his famous entropy equation. The equation engraved on the stone, $S = k \log W$, is what we now write as $S = k \ln \Omega$.

FIGURE 15.26 One of the two 4-penny macrostates with smallest entropy; the other is the macrostate with four tails.

This macrostate has the smallest entropy (since there is only one microstate associated with this macrostate).

We shake the box and then examine the pennies to determine which macrostate the pennies are in. It is more likely that the system is in a macrostate with a higher entropy. In other words, in this physical process the system likely changed to a macrostate with greater entropy. Indeed, the most likely outcome is the macrostate of maximum entropy, because it corresponds to the macrostate with the greatest number of associated microstates.

Now we begin the experiment with the pennies arranged in the macrostate with maximum entropy (2 heads and 2 tails), as shown in Figure 15.27. In this case it is *unlikely* that the system will end up, after shaking, in a macrostate with significantly lower entropy; it is certainly *possible* for the system to end up in a state of lower entropy, but it is less likely to do so than to remain in the macrostate of maximum entropy.* So once a system

---

*Actually, for this system with just 4 pennies, the odds are 10/16 = 62.5% that the system will end up in a macrostate that is not the 2-heads, 2-tails macrostate, but these odds decrease dramatically as the number of pennies in the system increases (see the discussion that follows).

FIGURE 15.27 The macrostate with greatest entropy; half heads, half tails.

is in a state of maximum entropy, it most likely will remain there through natural processes (represented here by shaking the box).

The macrostates with the smallest entropy have only one associated microstate (4 heads and 0 tails, or 0 heads and 4 tails); thus, the exact behavior of each particle (penny) is known. These low-entropy macrostates are the most strictly organized; they have the greatest order. In contrast the macrostate with the greatest entropy (2 heads, 2 tails) has the largest number of associated microstates; there are six different ways the four particles (pennies) can be arranged consistent with this macrostate. Thus this macrostate is the least rigidly determined; it has the least order or, as it is generally phrased, the greatest disorder. This is why the entropy of the system is occasionally referred to as a measure of the disorder associated with the system. The greater the entropy, the greater the disorder.

These ideas can be extended to greater numbers of pennies. For example, for a 10-penny system there are 11 macrostates and $2^{10} = 1024$ microstates (see Table 15.3; Example 15.15 indicates how to determine the number of microstates for each macrostate). Each of the microstates is equally likely to occur, but the macrostates do not have equal probabilities of occurring. A histogram of the number of microstates associated with each macrostate for a 10-penny system is shown in Figure 15.28. The entropy of a macrostate is $S = k \ln \Omega$, where Ω is the number of microstates associated with the particular macrostate.

If the 10-penny system is prepared in a state of low entropy (say, the macrostate with all heads) and the box is shaken, the system is unlikely to remain in that macrostate. You would certainly never make a bet about it remaining in that state unless you really like long odds! In fact, the probability of remaining in that state is 1/1024, since each of the 1024 microstates is equally likely to occur. It is much more likely that the entropy of the system will increase and the system will be found in one of the macrostates of greater entropy, one of the macrostates near the half-heads, half-tails macrostate.

Correspondingly, if the system is in one of the macrostates of large entropy [for example, the macrostate with the greatest entropy (5 heads and 5 tails)], it is unlikely that the system will (after shaking the box) be found in a macrostate of low entropy (say all heads). It is *possible* to end up with 10 heads, but it is *quite unlikely*.

TABLE 15.3 Macrostates and Number of Microstates for a 10-Penny System

Macrostate	H	T	# microstates Ω
1	10	0	1
2	9	1	10
3	8	2	45
4	7	3	120
5	6	4	210
6	5	5	252
7	4	6	210
8	3	7	120
9	2	8	45
10	1	9	10
11	0	10	1
		Total	$1024 = 2^{10}$

FIGURE 15.28 **The number of microstates associated with each macrostate for a 10-penny system.**

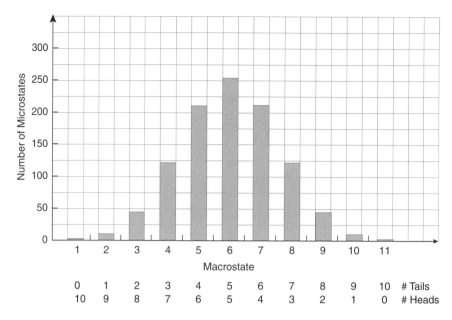

0	1	2	3	4	5	6	7	8	9	10	# Tails
10	9	8	7	6	5	4	3	2	1	0	# Heads

If we increase the number of pennies further (to say 50, 100, 1000, or many billions), several things happen to the shape of the histogram depicting the number of microstates associated with each macrostate; similar things happen to the shape of the corresponding entropy graph. Figure 15.29 presents the graph of the number of microstates associated with each macrostate for a 50-penny system, a 100-penny system, and a 1000-penny system. We note the following from these graphs:

1. The peak of the curve remains located at the macrostate with half the coins heads and half tails, since this is the macrostate with the greatest number of associated microstates and the greatest entropy.
2. The peak of the distribution becomes sharper and narrower as the number of pennies in the system increases. The distribution falls precipitously as we move to macrostates differing appreciably from the half-heads, half-tails macrostate.

> The implication of the sharp peak is that if a system with a large number of coins is prepared in a state of low entropy (say all heads) and then shaken, it becomes *exceedingly unlikely* that the system will remain in such a state.

Once the shaking occurs, the system of pennies emerges with a distribution that is essentially half heads and half tails. The system went from a macrostate of low entropy to a macrostate of high entropy: the entropy of the system increased.

> Likewise, once the system is in one of the macrostates near the peak, it is *exceedingly unlikely* that it will emerge from the shaking in a macrostate of low entropy (say all heads, or just a few heads or tails), for this entails a significant decrease in the entropy. Thus the system naturally maximizes its entropy.

The entropy form of the second law of thermodynamics is a statement about probabilities. An isolated system naturally will progress to a macrostate of higher entropy because such macrostates are much more probable. Thus, in the free expansion of an ideal gas, once the membrane is punctured it is *exceedingly unlikely* for the particles of the gas to remain confined to one side of the container. In principle, it is *possible* for the particles to remain on one side of the container, but it is so fantastically unlikely that we can say with absolute confidence that it *never* occurs.* Once the membrane is punctured, the macrostate corresponding to all the particles on one side of the box becomes an exceedingly unlikely macrostate. The macrostate with the particles distributed essentially uniformly throughout the container has the greatest number of associated microstates; it is the macrostate with the maximum entropy, and the system quickly and naturally evolves to that macrostate.

The game played in **statistical thermodynamics** is to find ingenious ways of counting the number of microstates associated with a given macrostate. A macrostate is, for example, a specified number of moles of a gas with a specific volume, pressure, and temperature. The entropy of the system then is found analogously to the way we found the entropy for our system of pennies. We leave such sophisticated counting techniques to later courses in statistical physics.

The relationship between the *statistical definition of the entropy* of a system in thermal equilibrium, Equation 15.34,

$$S = k \ln \Omega$$

and the *classical definition of differential entropy changes*, Equation 15.30,

$$dS = \frac{dQ}{T}$$

between different equilibrium states is difficult to fathom with the tools we have at our disposal. Alas, for that reason we must defer

---

*Using our penny system, it is similar to preparing a box with Avogadro's number of pennies initially all heads, shaking the box, and finding that all the pennies *still* are heads. It is possible but *exceedingly* unlikely.

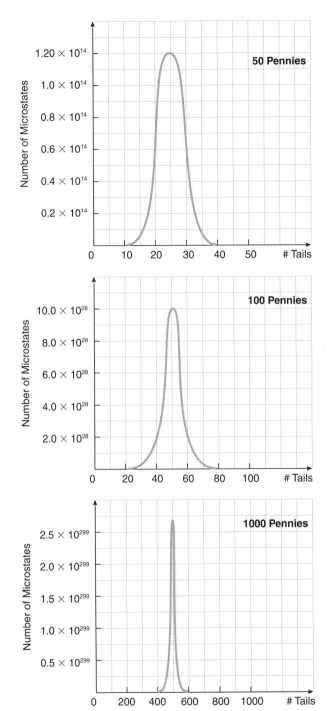

FIGURE 15.29 The number of microstates associated with each macrostate for a 50-penny, a 100-penny, and a 1000-penny system.

the connection between these apparently disparate ideas to your first course in statistical physics.

EXAMPLE 15.15

Consider a system of N pennies.

a. **Determine the total number of macrostates and microstates that exist for the system in terms of N.**

b. **Find an expression for the number of microstates associated with each macrostate.**

Solution

a. The macrostates are delineated by specifying the number of pennies that are heads and tails. Thus the macrostates are as follows:

Number of heads	Number of tails
N	0
$N - 1$	1
$N - 2$	2
$\vdots$	$\vdots$
2	$N - 2$
1	$N - 1$
0	N

There are $N + 1$ macrostates. Notice that for the 4-penny system, there are $4 + 1 = 5$ macrostates; for the 10-penny system, there are $10 + 1 = 11$ macrostates; and so on. To count the total number of microstates, we observe that each penny has two possible orientations: heads or tails. For N pennies, there are then a total of 2^N possible *microstates*. For the 4-penny system, there are $2^4 = 16$ microstates; for the 10-penny system, there are $2^{10} = 1024$ microstates. For a 100-penny system, the number of microstates is $2^{100} \approx 1.27 \times 10^{30}$. The number of microstates increases dramatically as the number of particles increases.

b. For a macrostate, it does not matter which pennies are heads or tails, only that there are a specified number of heads (say n) and tails ($N - n$). The microstates are penny-specific and delineate how many different ways we can have, say, n heads and $N - n$ tails, keeping track of the individual pennies (as we saw in our example with four pennies). The number of microstates associated with each macrostate involves calculating the number of *different* ways we can have, say, n heads (and $N - n$ tails). This is a problem in combinatorics. The number is the so-called binomial coefficient:

$$\binom{N}{n} \equiv \frac{N!}{n!\,(N - n)!}$$

where N is the total number of pennies, n is the number of heads, and $N - n$ is the number of tails. The binomial coefficient is what determines the number of microstates associated with each macrostate of the penny problem. Thus, for example, in the 10-penny collection, the number of microstates associated with the macrostate of 6 heads and 4 tails is

$$\binom{10}{6} \equiv \frac{10!}{6!\,4!}$$

$$= \frac{3\,628\,800}{(720)\,(24)}$$

$$= 210$$

In other words, there are 210 different microstates associated with the 6-heads, 4-tails macrostate.

15.12 ENTROPY MAXIMIZATION AND THE ARROW OF TIME*

It would be more impressive if it flowed the other way. [When commenting on the grandeur of Niagara Falls]
Oscar Wilde (1854–1900)[†]

It is easy to tell whether a movie is run backward or forward. Watching a video run backward is funny. We chuckle at seeing the process of water flowing uphill, watching the process of a bomb re-forming from its many fragments, noting the process by which an object appears magically out of flame and smoke while burning in reverse. A biographical movie run backward begins with a resurrection from the dead, with the individual gradually growing younger and smaller. Incredible. In the macroscopic world, the temporal direction of a process usually is easily recognized. This direction of time (forward or backward) is known as the **arrow of time**.

Yet, surprisingly, for some processes involving just a *few* particles, it is difficult if not impossible to determine the direction or arrow of time. A collision between two billiard balls can be viewed forward or backward in time with essentially no difference. Either view is quite plausible; there are no laughs either way (and both are, quite finally, pretty boring).

Likewise, a movie sequencing in time the results of a 2-penny or 3-penny shaking experiment is indistinguishable when run forward or backward; there is nothing funny about either direction. No arrow of time is apparent.

Correspondingly, if you could imagine a movie that shows the motion of the particles in a gas in *thermal equilibrium*, it would be difficult to tell whether the movie was running forward or backward. Likewise with a large number of pennies in a box: with the collection initially 50% heads and 50% tails, the direction of a movie sequencing the results of shaking the box is difficult to determine because the collection remains essentially

in a 50–50 distribution either way. No time arrow is apparent. So, for a system in *thermal equilibrium* (its entropy already maximized), the arrow of time likewise is not apparent.

Yet when a system with a large number of particles *changes* from one equilibrium state to another (from one permissible macrostate to another), the direction of time, or of a movie of the process, is easy to determine *if* the system is not already in the macrostate of maximum entropy (or one quite near it). The system naturally evolves *to* a macrostate of maximum entropy (the one with the greatest number of microstates), since that macrostate simply is more probable than any other. The system is exceedingly unlikely to do anything else. Hence in a movie of such a process or change, the direction or arrow of time *is* apparent.

If a many-particle system initially is in the state of maximum entropy (or one close to it), it is difficult to tell whether a movie is running forward or backward; the system is in thermal equilibrium and will not likely change to a macrostate of low entropy. However, if the many-particle system initially is in a macrostate of low entropy (say, all pennies heads) and is given the opportunity to *change* (such as by shaking the penny system), the probability is *overwhelming* that the system will change to a macrostate of maximum entropy. Thus it is easy to tell if a movie sequencing the change is running forward or backward.

A free expansion of gas can be analyzed in the same way. If the many-particle gas is confined initially to the left half of the container, this macrostate becomes *exceedingly unlikely* when the membrane is broken, and the system evolves or changes to the macrostate of highest entropy: the one with the particles equally likely to be found anywhere in the box. The arrow of time is very apparent, since the direction of a movie of the process is easily distinguished; running forward (see Figure 15.30a) the movie is believable but boring, whereas running backward (Figure 15.30b) it is unbelievable. Once in the state of maximum entropy, a movie of the system (in thermal equilibrium) has no discernable arrow of time.

Thus there is an intimate connection between the way isolated many-particle systems *change* macrostates and the arrow

---

[†]Attributed to Oscar Wilde by Daniel Patrick Moynihan, *Science, 197*, #4305, page 742 (19 August 1977); the quotation possibly is apocryphal, although quite in the spirit of Wilde's notorious visit to Niagara Falls in the late 19th century. Despite extensive searching, the author has not been able to trace this comment directly to Wilde's writings or to news reports of his visit.

It is quite easy to tell if a movie of a waterfall is being shown forward or backward.

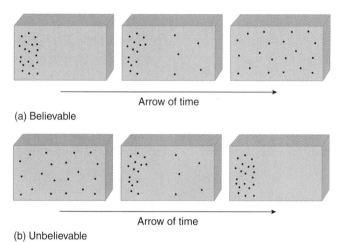

Arrow of time

(a) Believable

Arrow of time

(b) Unbelievable

FIGURE 15.30 Discerning the arrow of time.

of time. Changes to macrostates of *higher* entropy clearly indicate the direction of the arrow of time; the movie of the process is running forward. If the entropy change is *zero*, the arrow of time is not apparent; the system is in thermal equilibrium and is *exceedingly unlikely* to change its macrostate of maximum entropy. If the entropy change of the isolated system is *negative* (violating the second law of thermodynamics!), the direction of the arrow of time is reversed, and the movie becomes amusing and absolutely unbelievable. The second law of thermodynamics says a lot about why some things happen and why others do not.

15.13 EXTENSIVE AND INTENSIVE STATE VARIABLES*

Our study of thermodynamics has considered a number of thermodynamic state variables.

> The state variables whose values are independent of the size of the system are called **intensive**; those whose values are proportional to the size of the system are called **extensive**.

We summarize intensive and extensive state variables in Table 15.4.

TABLE 15.4 Intensive and Extensive State Variables

Intensive variables		Extensive variables	
Pressure	P	Volume	V
Temperature	T	Internal energy	U
Specific heat	c	Entropy	S
Latent heat	L	Mass	m
Work W		are *not* state variables	
Heat transfer Q			

You should not come away from thermodynamics thinking that these are the only thermodynamic state variables that exist. Many others also will be encountered if you pursue more advanced work in thermodynamics. Our study of thermodynamics necessarily has been introductory and concerned with *simple* thermodynamic systems (as defined in Chapter 13). The fundamental precepts of thermodynamics are applicable to *any physical system* with a large number of particles. In other words, thermodynamics is not just for gases! Thermodynamic ideas even can be applied to photons (particles of light) or *any* large collection of particles such as those in solids and liquids.

CHAPTER SUMMARY

We use the word *engine* to mean a system such as a gas that performs a closed cycle on a P–V diagram. If the cycle is performed clockwise, we call it a *heat engine*; if counterclockwise, we call it a *refrigerator engine*. The function of a *heat engine*, while executing a cycle, is to do useful work $|W|$ using heat transfer $|Q_H|$ from a high-temperature reservoir with heat transfer $|Q_C|$ exhausted to a colder reservoir. The *efficiency* ε of a heat engine is the absolute magnitude of the work done by the engine divided by the absolute magnitude of the heat transfer to the engine from the hot reservoir during each cycle:

$$\varepsilon \equiv \frac{|W|}{|Q_H|} \tag{15.3}$$

$$= 1 - \frac{|Q_C|}{|Q_H|} \quad \text{(any heat engine)} \tag{15.4}$$

The efficiency is a pure number between 0 and 1.

A *perfect heat engine* has zero heat transfer as exhaust ($|Q_C| = 0$) and an efficiency equal to 1. One statement of the *second law of thermodynamics* is as follows:

Perfect heat engines do not exist.

The function of a *refrigerator engine*, while executing a cycle, is to accomplish the unnatural heat transfer from a cold to a hot reservoir. The work done by a refrigerator engine during a cycle is negative. The *coefficient of performance* K of a refrigerator engine is the absolute magnitude of the heat transfer to the refrigerator engine from the cold

reservoir, $|Q_C|$, divided by the absolute magnitude of the work done by the refrigerator engine, $|W|$:

$$K \equiv \frac{|Q_C|}{|W|} \tag{15.18}$$

$$= \frac{1}{\dfrac{|Q_H|}{|Q_C|} - 1} \quad \text{(any refrigerator engine)} \tag{15.19}$$

A *perfect refrigerator engine* has an infinite coefficient of performance. Another statement of the second law of thermodynamics is the following:

Perfect refrigerator engines do not exist.

A *Carnot cycle* is an ideal cycle consisting of an alternating series of isothermal and adiabatic processes on a P–V diagram.

The efficiency of a Carnot heat engine operating between two reservoirs at absolute temperatures T_H and T_C is

$$\varepsilon_{\text{Carnot}} = 1 - \frac{T_C}{T_H} \quad \text{(Carnot heat engine only)} \tag{15.14}$$

Real heat engines have efficiencies *less* than a Carnot heat engine operating between the same two temperatures.

A Carnot refrigerator engine, operating between two reservoirs at absolute temperatures T_C and T_H, has a coefficient of performance

$$K_{\text{Carnot}} = \frac{1}{\dfrac{T_H}{T_C} - 1} \quad \text{(Carnot refrigerator engine only)} \tag{15.20}$$

Real refrigerator engines have coefficients of performance *less* than a Carnot refrigerator engine operating between the same two temperatures.

If a system changes from an initial state of thermodynamic equilibrium to a final equilibrium state, the entropy change ΔS of the system is defined to be

$$\Delta S = \int_i^f \frac{dQ}{T}$$

along a quasi-static reversible path between the two states. The entropy change is *independent of the path* connecting the two states on a P–V diagram. The entropy is a new state variable of a thermodynamic system, and (as for internal energy) it is only entropy *changes* that are significant in classical thermodynamics.

The entropy changes associated with various thermodynamic processes are summarized in Table 15.1 on page 681. Such entropy changes may be positive, negative, or zero.

The most general statement of the *second law of thermodynamics* (one that encompasses the two previous statements) is the following: *The total entropy change of an* **isolated** *system is always greater than or equal to zero:*

$$\Delta S_{\text{total isolated}} \geq 0 \text{ J/K} \qquad (15.33)$$

The total entropy change is equal to zero only for reversible processes.

The entropy of a system in thermal equilibrium has a statistical interpretation. The entropy is

$$S \equiv k \ln \Omega \qquad (15.34)$$

where k is Boltzmann's constant and Ω is the number of *microstates* associated with a given *macrostate* of the system. In any process, it is overwhelmingly probable that an isolated system with a large number of particles will change to a macrostate at or very near that with maximum entropy if the system is not already in that macrostate; any other change is exceedingly unlikely.

SUMMARY OF PROBLEM-SOLVING TACTICS

15.1 (page 674) In Equation 15.14 for the efficiency of a Carnot heat engine, be sure to use absolute temperatures in kelvin, not celsius temperatures.

15.2 (page 677) In Equation 15.20 for the coefficient of performance of a Carnot refrigerator engine, be sure to use absolute temperatures in kelvin, not celsius temperatures.

15.3 (page 682) When using the equations in Table 15.1 to calculate entropy changes, be sure to use absolute temperatures in kelvin, not celsius temperatures.

QUESTIONS

1. (page 669); 2. (page 674); 3. (page 678); 4. (page 682); 5. (page 686)

6. Explain why a rock cannot secure the energy to spontaneously jump to a higher elevation by simply decreasing its temperature.

7. Explain what this statement means: temperature is a state variable. Is time also a state variable?

8. Your un-air-conditioned dorm room is getting very hot on a sultry May afternoon. In desperation, you open the refrigerator door to cool off. Explain why leaving the refrigerator door open will not, in the long run, cool the room but actually will increase the temperature of the room.

9. Can a cool kitchen be heated by turning the oven on and leaving the oven door open? Why is this different from the previous question?

10. To increase the efficiency of power plants, it is desirable to increase the temperature of the hot reservoir. Explain how this increases the efficiency.

11. Two parents return from work on Friday only to discover that their 3-year-old prodigy has maximized chaos in the house, while the lazy baby-sitter stared blankly at a TV all day. Arrg! They subsequently spend the weekend restoring the house to a much-ordered macroscopic state. Explain why this does not violate the second law of thermodynamics.

12. Diesel heat engines compress air and fuel mixtures more than gasoline heat engines. Which heat engine compresses the gases to a higher temperature? What does this imply about the relative theoretical maximum efficiency of these heat engines?

13. Satellites in low Earth orbit experience kinetic frictional forces because of the atmosphere. Such satellites initially experience an *increase* in kinetic energy as they fall. Why? Does this violate energy conservation? Explain.

14. Explain what could be done to a mass of a material to increase its entropy. Explain what can be done to decrease its entropy. Does the latter violate the second law of thermodynamics? Explain.

15. How many isothermal curves and how many adiabatic curves pass through any given point on a P–V diagram? Explain your answer.

16. When you emerge from a swimming pool, your skin and your wet bathing suit spontaneously decrease in temperature. The effect is real, since you occasionally react to the temperature change by shivering. Does this violate the second law of thermodynamics? Why does the temperature decrease? Explain.

17. The presence of kinetic frictional forces increases the temperature of the surfaces in contact. Does the temperature increase occur because of heat transfer, work, and/or because of changes in internal energy that manifest themselves as a rise in temperature? Explain.

18. Can internal energy be converted into mechanical energy? Can mechanical energy be converted into internal energy? If so, give examples of each.

19. In its wisdom (?), Congress decrees that to prevent pollution, *no* exhaust is permissible from *any* engine. Explain why the honorable members desperately need to take a course in introductory physics.

20. In the internal combustion heat engine in most automobiles, what is the high-temperature reservoir and what is the low-temperature reservoir?

21. Winter is colder than summer. Yet automobile heat engines are *less* efficient in winter than in summer. Why is this so?

22. You are a remarkable thermodynamic system. You take in food, water, air, and sunlight, converting such disorganized materials into well-organized brains (well-ordered logic), brawn (well-ordered work), and occasionally braggadocio (well-ordered bragging). Does this violate the second law of thermodynamics? Explain.

23. In an unspecified thermodynamic process, the entropy of a system *decreases* by 1000 J/K. Does this violate the second law of thermodynamics? Explain.

24. A can containing nuts of various sizes, with the nuts initially dispersed randomly throughout the can, is shaken vigorously. The can is opened and the nuts are neatly arranged by size with largest on top and the smallest on the bottom. Explain how this occurs. Does this violate the second law of thermodynamics?

25. What factors make *real* heat engines and refrigerator engines less efficient than Carnot heat engines and refrigerator engines?

26. An ideal gas expands isothermally in a reversible process. Does the entropy of the gas increase, decrease, or remain the same? What about in an adiabatic expansion?

27. Are *we* heat engines? Explain your answer.

28. A sample of gas at volume V expands to volume 2V by either an isothermal or an adiabatic process. Which results in a gas at lower pressure? What happens to the entropy of the gas in each process?

29. Does the cooling that occurs with evaporation violate the second law of thermodynamics? Explain.

30. Use the results of Example 15.8, the concept of entropy changes, and the second law of thermodynamics to explain why the particles in a gas confined to a volume V do not spontaneously collect in, say, the left half of the container. Is this situation *absolutely impossible* in principle or just *exceedingly* unlikely?

31. In the free expansion of a gas there is no heat transfer to or from the gas, yet the entropy change of the isolated gas system is not zero. Explain why.

32. A scientist claims to have *once* observed a process that violated the second law of thermodynamics. Since science is predicated on the ability of scientists to *repeat* observations made by others, explain why the initial claim must be discounted.

33. A student states, "The second law of thermodynamics means that entropy always increases." Your professor marks the answer wrong. What must be added to the statement to make it correct?

34. To distinguish between the internal energy and the entropy, name a process involving an ideal gas for which (a) $\Delta U = 0$ J but $\Delta S \neq 0$ J/K; (b) $\Delta U \neq 0$ J but $\Delta S = 0$ J/K.

35. Qualitatively, which has the greater entropy: one kilogram of ice or one kilogram of liquid water? On what basis do you reach your conclusion?

36. Can it be said that the only long-term physical effect of your existence on the universe is to increase its entropy? Discuss.

37. The work done by a conservative force as a system moves between two locations in space is, by definition, independent of the path taken by the system. The entropy change of a system between two equilibrium states on a P–V diagram also is independent of the path taken. Yet work is *not* a state variable while entropy is. Discuss and explain, making reference to the concepts of potential energy and heat transfer.

38. In the free expansion of an ideal gas, which of the following thermodynamic quantities does *not* change?

entropy	pressure	internal energy
volume	density	

PROBLEMS

Sections 15.1 Why Do Some Things Happen, While Others Do Not?
    15.2 Heat Engines and the Second Law of Thermodynamics
    15.3 The Carnot Heat Engine and Its Efficiency
    15.4 Absolute Zero and the Third Law of Thermodynamics
    15.5 Refrigerator Engines and the Second Law of Thermodynamics
    15.6 The Carnot Refrigerator Engine
    15.7 The Efficiency of Real Heat Engines and Refrigerator Engines

1. A heat engine operates between the temperatures of -20 °C and 300 °C. What is the maximum theoretical efficiency for a heat engine operating between these temperatures?

2. Once upon a time there were plans to create a heat engine that would operate between temperature differences that exist between different depths of the ocean. Very deep ocean water has a temperature of about 4 °C while surface water (in tropical areas) has a temperature of about 20 °C. What is the maximum theoretical efficiency for a heat engine operating between these temperatures?

3. What is the maximum efficiency of a heat engine operating between the temperatures of 100 °C and 0 °C?

4. A refrigerator engine operates between the temperatures 1 °C (the inside of the refrigerator) and 25 °C (the outside of the refrigerator). What is the greatest possible value for the coefficient of performance of the refrigerator engine?

5. Consider your body to be a heat engine operating between a reservoir at 37 °C and room temperature at 20 °C. What is the

FIGURE P.6

greatest possible efficiency for a heat engine operating between these two temperatures?

6. Natural deposits of high-temperature steam exist within the Earth. See Figure P.6. Such geothermal sources have been suggested for running a heat engine. If the source of steam is at 200 °C and is exhausted to a reservoir at a temperature of 20 °C, what is the maximum theoretical efficiency of a heat engine operating between these two temperatures?

•7. A Carnot heat engine operates between a reservoir at 950 K and has heat transfer of 400 J exhausted to another reservoir at 300 K during each cycle. What is the amount of work done by the heat engine during each cycle?

•8. A patent application has been received from Professor Mediocratus, who claims to have developed a heat engine that uses 150 J of heat transfer from a reservoir at 150 °C with 50 J of heat transfer to a reservoir at −60 °C while doing 100 J of useful work. As a patent examiner, you have to decide whether to grant the patent or not. (a) Does the heat engine violate the first law of thermodynamics? (b) What is the claimed efficiency of the invented heat engine? (c) What is the efficiency of a Carnot heat engine operating between the same two temperatures? (d) Should the patent application be approved or denied? Explain your decision.

•9. A Carnot heat engine with an efficiency of 0.200 operates between two temperatures that differ from each other by 100 K. What are the temperatures between which the cycle operates?

•10. A refrigerator engine with a coefficient of performance of 3.0 accrues 1000 J of heat transfer from a warm pizza placed in the freezer compartment. What is the heat transfer to the room in which the refrigerator engine is located?

•11. Heat transfer of 200 J occurs to a refrigerator engine from a cold reservoir and heat transfer of 350 J occurs to a higher-temperature reservoir during each cycle. (a) What is the coefficient of performance of the refrigerator engine? (b) If the refrigerator engine performs five complete cycles each second, what power is required from external agents to operate the refrigerator engine?

•12. A Rube Goldberg heat engine uses 3000 J of heat transfer from a hot reservoir at 400 K and does 500 J of useful (?) work during

each cycle. (a) What heat transfer occurs to a cooler reservoir? (b) What is the efficiency of the Rube Goldberg heat engine?

•13. A 100 MW power planet (heat engine) operates with an efficiency of 0.350. The 100 MW refers to the rate at which energy is generated for useful purposes. (a) What is the heat flow to the environment? (b) The exhausted heat flow is to water drawn from a local river. To protect the aquatic environment in the river, the water returned to the river should be no more than 3.0 °C warmer than the water secured to do the cooling. What rate of water use is needed to accomplish this? Express your result in kg/s and m³/s.

•14. One (1.00) mole of an ideal diatomic gas (with $\gamma = 1.40$) initially at 20.0 °C and 1.00 atm pressure is taken around the following cycle (see Figure P.14):

> Path (1): an *isochoric* increase in pressure until the temperature of the gas is 150 °C and the pressure is P;
> Path (2): an *adiabatic* expansion until the pressure returns to 1.00 atm;
> Path (3): an *isobaric* compression until the gas reaches its original volume.

(a) What is the original volume of the gas at the beginning and end of the cycle? (b) What is the pressure of the gas at the completion of path (1)? (c) What is the volume of the gas at the completion of path (2)? (d) What is the temperature of the gas at the completion of path (2)? (e) Calculate the work done by the gas during each path of the cycle and the total work done by the gas. (f) Calculate the heat transfer to the gas during each path of the cycle and the total heat transfer to the gas over the cycle. (g) Find the efficiency of the cycle. (h) Calculate the maximum efficiency that a heat engine could have if it operated between the hottest and coldest temperatures encountered by this gas in this cycle.

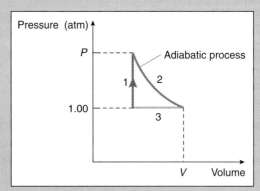

FIGURE P.14

•15. One (1.00) mole of an ideal diatomic gas (with $\gamma = 1.40$) initially at 20.0 °C and 1.00 atm pressure is taken around the following cycle (see Figure P.15):

> Path (1): an *isochoric* increase in pressure until the temperature of the gas is 150 °C and the pressure is P;
> Path (2): an *isothermal* expansion until the pressure returns to 1.00 atm;
> Path (3): an *isobaric* compression until the gas reaches its original volume.

(a) What is the original volume of the gas at the beginning and end of the cycle? (b) What is the pressure of the gas at the completion of path (1)? (c) What is the volume of the gas at the completion of path (2)? (d) What is the temperature of the gas at the completion of path (2)? (e) Calculate the work done by the gas during each path of the cycle and the total work done by the gas. (f) Calculate the heat transfer to the gas during each path of the cycle and the total heat transfer to the gas over the cycle. (g) Find the efficiency of the cycle. (h) Calculate the maximum efficiency that a heat engine could have if it operated between the hottest and coldest temperatures encountered by this gas in this cycle.

FIGURE P.15

• **16.** To increase the efficiency of a Carnot heat engine, does it make a difference whether the temperature of the hot reservoir is *increased* by 10 K or the temperature of the cold reservoir is *decreased* by 10 K? If there is a difference, which is better to do?

• **17.** To increase the coefficient of performance of a Carnot refrigerator engine, is it better to increase the temperature of the colder reservoir by, say, 10 K or decrease the temperature of the warmer reservoir by 10 K, or does it make any difference which you do?

• **18.** A Carnot heat engine has an efficiency $\varepsilon_{\text{Carnot}}$ and operates between two reservoirs at temperatures T_H and T_C. When the cycle is run in the reverse direction as a Carnot refrigerator engine, its coefficient of performance is K_{Carnot}. Show that

$$\varepsilon_{\text{Carnot}} K_{\text{Carnot}} = \frac{T_C}{T_H}$$

• **19.** A heat engine operates between two reservoirs at temperatures 700 K and 300 K. (a) What is the efficiency of a Carnot heat engine operating between the same two reservoirs? (b) If the actual efficiency of the heat engine is 1/3 the Carnot efficiency, how many joules of work does the heat engine perform if it accrues 1000 J of heat transfer from the high-temperature reservoir during each cycle?

• **20.** A small refrigerator engine produces 400 W of heat flow to the kitchen when 333 W is transferred to the refrigerator engine from the cooler reservoir. The inside of the refrigerator is at 0 °C and the kitchen in which it lies is at 25 °C. (a) What is the coefficient of performance of the refrigerator engine? (b) What is the power of the refrigerator engine?

• **21.** A peculiar "V-8" engine (see Figure P.21) has an efficiency of 0.250 when operating between reservoirs at temperatures 500 K and 300 K. (a) What is the theoretical maximum efficiency for

a heat engine operating between these two reservoirs? (b) If the "V-8" heat engine does 800 J of work per cycle, what heat transfer occurs to the engine from the high-temperature reservoir and what heat transfer is exhausted to the low-temperature reservoir during each cycle?

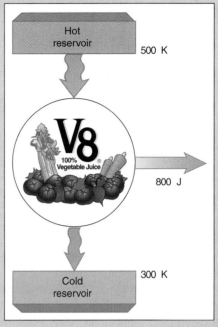

FIGURE P.21

• **22.** The coefficient of performance of a small refrigerator engine is 4.50 and its heat flow into the kitchen is 250 W. At what rate is work done by the refrigerator engine?

• **23.** A Carnot heat engine has an efficiency of 0.300 operating between a high-temperature reservoir at temperature T_H and a low-temperature reservoir at 20 °C. To increase the efficiency of the Carnot heat engine to 0.400, by how many kelvin should the temperature of the hot reservoir be increased?

• **24.** The heat flow to a Carnot heat engine from a reservoir at 100 °C is 5.00×10^3 W. The exhaust heat flow is to a reservoir at temperature 300 K. (a) At what rate is work being done by the heat engine? (b) What is the heat flow to the cooler reservoir?

• **25.** Nitrogen liquefies at temperature 77 K. To liquefy nitrogen already at 77 K, a refrigerator engine is operating in a room at temperature 300 K. (a) For each joule transferred from the gaseous nitrogen at 77 K, what is the *minimum* number of joules exhausted to the room? Do you need to know the latent heat of vaporization of nitrogen to answer this question? Explain why this is a *minimum* number of joules. (b) Repeat the calculation for a helium liquefier (helium liquefies at 4.2 K).

• **26.** A *heat pump* is essentially a refrigerator engine that uses the inside of a building as a hot-temperature reservoir in the winter, *or* the outside of the building as the hot-temperature reservoir in the summer. Neat! (a) For a Carnot refrigerator engine operating between reservoirs at 25 °C and 40 °C in the summer, what is its coefficient of performance? (b) For each joule of heat transfer from the cooler reservoir per cycle, how many joules of work are done by the Carnot refrigerator engine? Is this work positive or negative according to our conventions? (c) What additional information is needed to "size" or determine an appropriate heat pump for a given building?

•27. Air conditioners are essentially refrigerator engines that exhaust to the outside of a building rather than to the inside. Central air conditioners are rated by saying they have a certain tonnage capacity. To be specific, consider an air conditioning engine rated at 1-ton capacity. This tonnage terminology means that the heat transfer is equivalent to that required to freeze or melt one English ton (~ 909 kg) of ice at 0 °C during one day. (a) What heat transfer is required to melt 909 kg of ice? (b) If this is accomplished during one day, what is the heat flow to the ice in watts?

•28. The oceans contain a vast store of internal energy. (a) Estimate the heat transfer from the water if the temperature of one cubic kilometer of water is decreased by 1.00 K. Assume the oceans are pure water. (b) Explain why such heat transfer is not employed to do useful work via a heat engine.

•29. A thing-a-ma-bob-that-does-the-job has an efficiency $\varepsilon = 0.35$. Five hundred (500) joules of heat transfer occurs to a cool reservoir during each cycle. (a) What is the work done by the heat engine? (b) What is the heat transfer to the heat engine from the hot reservoir?

•30. A heat engine gizmo lifts a 5.00 kg package vertically at constant speed through a distance 1.50 m during each cycle of the heat engine. (a) What work is done by the heat engine during each cycle? (b) What is the change in the internal energy of the heat engine during one cycle? (c) What is the total heat transfer to the heat engine during each cycle? (d) If the engine exhausts 90.0 J of heat transfer per cycle, what is the heat transfer to the heat engine from the high-temperature reservoir? (e) What is the efficiency of the heat engine?

‡31. No two adiabatic curves can intersect on a *P–V* diagram. To show this, construct a hypothetical schematic *P–V* diagram for an ideal gas on which two different adiabatic curves intersect. Connect these adiabatic curves with an isothermal curve to form a cycle. Show that although a heat engine running around such a cycle satisfies the first law of thermodynamics, the engine violates the second law of thermodynamics.

Sections 15.8 A New Concept: Entropy
15.9 Entropy and the Second Law
of Thermodynamics
15.10 The Direction of Heat Transfer:
A Consequence of the Second Law

32. One (1.00) kilogram of ice melts at 0.00 °C to water at 0.00 °C. Find the entropy change of the mass.

33. One (1.00) kilogram of steam at 100 °C condenses into water at 100 °C. Calculate the entropy change of the mass. Does your result violate the second law of thermodynamics? Explain your answer.

34. Two (2.00) moles of a monatomic ideal gas is warmed slowly from 300 K to 400 K at constant volume. Determine the entropy change of the gas.

35. If 500 J of heat transfer occurs from a reservoir at 450 K to a reservoir at 300 K, what is the change in the entropy of the system of two reservoirs?

•36. A 0.100 kg ice cube at −10.0 °C is placed in a well-insulated container with 1.00 kg of water at 20.0 °C. (a) Find the final temperature of the ice and water system. (b) Calculate the entropy change of the warm water as it cools to the final temperature. (c) Calculate the entropy change of the ice as it warms to the final temperature. (d) What is the total entropy change of the isolated ice and water system?

•37. You throw an ice cube of mass 0.100 kg and temperature −5.00 °C into the ocean, which has a temperature of 10.00 °C. Assume the specific heat of salt water is the same as fresh water. (a) What is the entropy change of the ice? (b) What is the entropy change of the ocean? (c) What is the total entropy change of the ice–ocean system?

•38. Two cars each of mass 1500 kg traveling at 100 km/h in opposite directions collide head-on and come to a disastrous halt; see Figure P.38. Assume that the surrounding air and ground temperature remains fixed at 20 °C. Calculate the total entropy change of the cars and environment system that results from the collision.

FIGURE P.38

•39. A 2.00 kg ripe tomato falls 1.50 m to the ground. The tomato, air, and ground all are at a temperature of 300 K. What is the total entropy change of the tomato, air, and ground system?

•40. You are carefully carrying a 2.00 kg mass of copper at temperature 200 °C along a lakeside dock for no good reason. You stumble clumsily and the copper drops into the lake at temperature 5 °C. Sssssssst! (a) What is the entropy change of the copper? (b) What is the entropy change of the lake? (c) What is the total entropy change of the copper and lake system?

•41. Rather than graph the pressure versus the volume for a gas undergoing a cycle, it occasionally is useful to plot the absolute temperature *T* (on the vertical axis) versus the entropy *S* (on the horizontal axis), forming what is called a *T–S* diagram. (a) Make a schematic sketch of a Carnot cycle on such a *T–S* diagram. (b) What is the physical significance of the area enclosed by the cycle on the *T–S* diagram? (c) How can the efficiency of a Carnot heat engine be obtained *graphically* from a *T–S* diagram of the cycle?

•42. Show that the slope of an *isochoric* process on a *T–S* diagram (see Problem 41) is equal to

$$\frac{T}{nc_V}$$

where c_V is the molar specific heat at constant volume.

•**43.** Show that the slope of an *isobaric* process on a *T–S* diagram (see Problem 41) is equal to

$$\frac{T}{nc_p}$$

where c_p is the molar specific heat at constant pressure.

•**44.** An inventor, Charles LeTan, is attempting to convince a loan officer to make a several megadollar loan to Fly-by-Night Enterprises Ltd. to manufacture a heat engine that takes a gas around a cycle on a *P–V* diagram with the following consequences:

- total (net) heat transfer to the gas during *n* complete cycles $|Q_H| - |Q_C| = 9.21 \times 10^6$ J;
- total work done *by* the gas used in the machine during the same *n* cycles $= 10.0 \times 10^6$ J.

You have been hired as a technical consultant to the loan officer. Should you advise that the loan application be approved by the loan officer or not? Indicate the reasoning behind your decision.

•**45.** A heat engine with an efficiency of 0.250 operates between two reservoirs at 1000 K and 400 K. The power output of the heat engine is 1.00 kW. (a) What is the heat flow to the heat engine from the high-temperature reservoir? (b) What is the time rate of increase in entropy of the system consisting of the heat engine and the two reservoirs?

•**46.** One end of a large piece of copper is in thermal contact with a reservoir at temperature 100 °C and the other end is in thermal contact with another reservoir at 0 °C. The heat flow between the reservoirs through the copper is 500 W. (a) Determine the rate at which the entropy of each reservoir is changing with time. (b) Does the entropy of the copper rod change with time? Explain your answer.

•**47.** A 1250 kg car speeding at 150 km/h crashes completely in-elastically into a concrete bridge pylon when the temperature of the car, ambient air, and concrete is 20.0 °C. The car is totaled. (a) What happens to the kinetic energy of the car? (b) What is the entropy change of the car, bridge, and air system?

•**48.** One (1.00) mole of an ideal monatomic gas is initially at temperature 300 K and has a volume of 0.0100 m³. The volume of the gas increases to 0.0200 m³ by one of the following quasi-static reversible processes:

- isothermal expansion
- adiabatic expansion
- isobaric expansion

(a) Construct an *accurate* graph of the absolute temperature of the gas versus volume (a *T–V* graph) that represents *each* of these processes. (b) In which (if any) processes is the entropy change of the gas positive? Zero? Negative? Explain your reasoning. (c) For each process, sketch the path of another process that produces an identical entropy change in the gas.

•**49.** Since you are a warm-blooded mammal, your heat flow via radiation into the ambient environment (at temperature 293 K) is about 100 W when clothed. What is the increase in the entropy of the system consisting of you and the environment each day?

•**50.** Heat transfer from a reservoir at 150 °C is used to turn 1.00 kg of ice at 0 °C to steam at 100 °C, all at a pressure of 1.00 atm.

(a) What is the entropy change of the water? (b) What is the entropy change of the reservoir? (c) What is the entropy change of the system consisting of the water and reservoir? (d) Explain why the sign of the answer to part (c) is consistent with the second law of thermodynamics.

•**51.** One (1.00) mole of an ideal monatomic gas is taken around the cycle shown in Figure P.51. (a) What is the work done by the gas in going from point (1) to point (2)? (b) What is the work done by the gas in going from point (2) to point (3)? (c) What is the change in the internal energy of the gas in going from point (1) to point (3) along the path (1) → (2) → (3)? (d) What is the entropy change of the gas in going from point (1) to point (3)? (e) What is the change in the internal energy and entropy of the gas in one complete cycle?

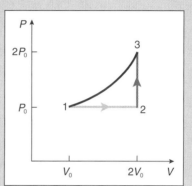

FIGURE P.51

•**52.** Heat transfer of 4500 J occurs from a reservoir at 100 °C to one at 20 °C. (a) What is the entropy change of the hot reservoir? Explain why your result does not violate the second law of thermodynamics. (b) What is the entropy change of the cool reservoir? (c) What is the entropy change of the system consisting of both reservoirs? (d) What is the maximum efficiency of a heat engine operating between the two reservoirs?

•**53.** Two (2.00) moles of a diatomic ideal gas expand adiabatically, doubling its volume while doing 2000 J of work. (a) What is the total heat transfer to the gas during the process? (b) Determine the change in the internal energy of the gas. (c) Determine the entropy change of the gas. (d) Did the temperature rise or fall? (e) Is there sufficient information given to determine the temperature change? If so, determine the temperature change. If not, what additional information is needed?

•**54.** A container with 1.00 liter of water at 20 °C is added to a well-insulated thermos jug holding 2.00 liters of water at 90 °C. Neglect heat transfer to the thermos itself. What is the increase in the entropy of the 3.00 liter water system?

•**55.** Heat transfer of 5000 J occurs to the (heat) engine of progress from a reservoir at 500 °C and 3500 J is exhausted to another reservoir at 20 °C during each engine cycle, while doing some useful work. (a) What work is done by the heat engine during each cycle? (b) What is the efficiency of the heat engine? (c) What is the efficiency of a Carnot heat engine operating between the same two reservoirs? (d) During one cycle of the heat engine, calculate the total entropy change of the system consisting of the two reservoirs and the heat engine.

•**56.** Two (2.00) moles of a diatomic ideal gas (with $\gamma = 1.40$) at 300 K and 3.00 atm pressure expands isothermally until the pressure drops to 2.00 atm. The gas then expands adiabatically until the pressure reaches 1.00 atm. (a) What is the initial volume of the gas? (b) What is the volume of the gas after the isothermal expansion? (c) Find the work done by the gas during the isothermal expansion. (d) Find the change in the internal energy of the gas during the isothermal expansion. (e) What is the final temperature of the gas? (f) Find the change in the internal energy of the gas during the adiabatic expansion. (g) What is the work done by the gas during the adiabatic expansion? (h) What was the total heat transfer to the gas during the entire expansion? (i) What is the total entropy change of the gas?

$\ddagger$57. Two containers labeled 1 and 2 each contain n moles of the same monatomic ideal gas and initially are isolated from each other. The two containers have the same pressure but different initial temperatures, T_1 and T_2. The two containers are brought into thermal contact with each other and the pressure is maintained at the same value P. What is the total change in entropy? Show that the total entropy change is not negative. Hint: Consider the following inequality:

$$(T_1 - T_2)^2 \geq 0 \text{ K}^2$$

$\ddagger$58. Two containers labeled 1 and 2 each contain n moles of the same monatomic ideal gas and initially are isolated from each other. Each container is at the same temperature T. The volume of container 1 is V_1 and that of container 2 is V_2. The two containers are brought into thermal contact with each other and the partition separating them is removed. What is the change in the entropy of the system of two gases when they are in thermal equilibrium? Show that the change in entropy is positive.

$\ddagger$59. A mass m initially at temperature T_1 is placed in thermal contact with an equal mass of the same material initially at temperature T_2. The combination is well insulated from the environment, so that heat transfer occurs only between the two masses. (a) Show that the entropy change of the isolated system of two masses is

$$\Delta S = mc \ln\left[\frac{(T_1 + T_2)^2}{4T_1 T_2}\right]$$

where c is the specific heat of the material in J/(kg·K). (b) Show that $\Delta S \geq 0$ J/K. Hint: Consider the following inequality:

$$(T_1 - T_2)^2 \geq 0 \text{ K}^2$$

Sections 15.11 A Statistical Interpretation of the Entropy *
15.12 Entropy Maximization and the Arrow of Time*
15.13 Extensive and Intensive State Variables*

•60. Consider a 6-penny system in a treasure box. (a) How many macrostates are there? (b) Determine the number of microstates associated with each macrostate of the 6-penny system. (c) Make a histogram of the number of microstates associated with each macrostate. (d) Calculate the quantity S/k for each macrostate. (e) What is the probability that six heads will result when the box is shaken?

•61. Suppose one penny in a 4-penny shaking experiment is a counterfeit two-headed penny (with no tails). How does this affect the number of macrostates and microstates? If such a penny is used unknowingly in a 4-penny experiment, how will the experimental results indicate that the system is not a true 4-penny system?

•62. Six pennies are in a box. The box is shaken repeatedly and the macrostate of the system (the number of heads and the number of tails) determined after each shake. Unbeknownst to you, one of the coins actually is glued to the bottom of the box. Estimate the number of times you need to shake the box before this becomes apparent from the survey of each macrostate after each shake, ignoring the locations of the pennies in the box.

•63. Take a pair of dice. When the dice are rolled, the possible total sums that emerge on the dice are 2, 3, 4, 5, 6, 7, 8, 9, 10, 11, and 12. These are the macrostates associated with a toss. (a) Construct a table indicating the various possible microstates associated with each macrostate. (b) What is the total number of microstates associated with a toss? (c) What is the probability of rolling a 7? (d) Construct a graph of S/k for each macrostate.

•64. Six eggs are to be placed in an egg carton with spaces for 12 eggs. (a) If the eggs are indistinguishable, is there any difference between a macrostate and microstate of the system? If so, indicate the number of microstates associated with each macrostate. (b) If the eggs are distinguishable, is there a difference between a macrostate and a microstate? If so, indicate the number of microstates associated with each macrostate.

•65. Six particles of a *very* dilute gas are confined to a box. Divide the box into a left half and a right half by an imaginary partition. (a) Specify the number of ways the six particles can be distributed between the two sides of the box. Let these be the macrostates of the system. Find the number of microstates associated with each macrostate. (b) What is the entropy of the macrostate in which all the particles are on the left side of the box? (c) What is the entropy of the most probable macrostate? Notice that the entropy of the most probable macrostate is greater than the entropy of the macrostate with all particles on one side of the box. The entropy change between these two macrostates is similar to the entropy change associated with a free expansion of gas initially confined to the left half of the box (see Example 15.8). The entropy change of a free expansion (of six particles) and that of this problem are the same order of magnitude but are not quantitatively identical, because this problem of simply distributing the particles between the two halves neglects energy considerations and the fundamental indistinguishability of the particles of a gas. Alas, the quantitative disagreement illustrates the difficulty in exploring the connection between the classical definition of an entropy change and that associated with statistical thermodynamics.

•66. Time to play cards with a standard 52 card deck with 13 cards of each suit (clubs, diamonds, hearts, and spades). (a) What is the probability that the first four cards dealt from a shuffled deck are the four aces? (b) What is the probability that the first four cards dealt are spades?

•67. Consider a row of ten lockers in a locker room. Define macrostates of the system corresponding to the following situations:
   1. two lockers immediately adjacent to each other
   2. two lockers with one other locker separating them

3. two lockers with two other lockers separating them
4. two lockers with three other lockers separating them
5. two lockers with four other lockers separating them
6. two lockers with five other lockers separating them
7. two lockers with six other lockers separating them
8. two lockers with seven other lockers separating them
9. two lockers with eight other lockers separating them

Let a microstate be a way each macrostate can be arranged within the row of ten lockers. (a) Delineate the number of microstates associated with each macrostate. (b) What is the total number of microstates? (c) Ten persons are assigned the lockers. Two of these individuals go to the locker room at the same time. What is the probability that they will have immediately adjacent lockers and, therefore, be in each other's way?

●**68.** A 4-penny system is shaken and examined 250 times; the head and tail distribution of the pennies are shown in following table:

Macrostate	Heads	Tails	Number of times the macrostate occurs
4	0	4	0
3	1	3	31
2	2	2	93
1	3	1	94
0	4	0	32

What can you conclude from the data?

●69. (a) Specify the macrostates for a 20-penny system, the number of microstates associated with each macrostate, and the value of S/k for each macrostate. (b) Make a graph of the number of microstates versus the macrostates. (c) Make a graph of S/k versus the macrostates.

INVESTIGATIVE PROJECTS

A. Expanded Horizons

1. Investigate the notion of perpetual motion machines and how they necessarily violate the laws of thermodynamics; thus they are impossible machines. Distinguish between so-called perpetual motion machines of the *first kind* and those of the *second kind*.
   Arthur W. J. G. Ord-Hume, *Perpetual Motion: The History of an Obsession* (St. Martin's Press, New York, 1977).
   Stanley W. Angrist, "Perpetual motion machines," *Scientific American*, 218, #1, pages 114–122 (January 1968).
   Henry Dircks, *Perpetuum Mobile* (Rogers and Hall, Chicago, 1916).
   Clifford B. Hicks, "Why won't they work?" *American Heritage*, XII, #3, pages 78–83 (April 1961).
   John Phin, *The Seven Follies of Science* (D. Van Nostrand, New York, 1906), pages 36–78.

2. Investigate and write a report about the famous Maxwell's demon, a hypothetical gremlin imagined by James Clerk Maxwell (1831–1879) as a way of violating the second law of thermodynamics.
   Harvey S. Leff and Andrew F. Rex, "Resource Letter MD-1: Maxwell's demon," *American Journal of Physics*, 58, #3, pages 201–209 (March 1990); this contains an extensive bibliography on the subject.
   Martin J. Klein, "Maxwell, his Demon, and the second law of thermodynamics," *American Scientist*, 58, #1, pages 84–97 (January–February 1970).
   Alfred Brian Pippard, *Elements of Classical Thermodynamics* (Cambridge, England, 1957), pages 99–100.
   George Gamow, *Mr Tompkins Explores the Atom* (Macmillan, New York, 1945), pages 1–19.
   Allen L. King, *Thermophysics* (W. H. Freeman, San Francisco, 1962), pages 250–251.
   Mark W. Zemansky, *Heat and Thermodynamics* (5th edition, McGraw-Hill, New York, 1968), pages 271–272.
   Richard P. Feynman, Robert B. Leighton, and Matthew Sands, *The Feynman Lectures on Physics* (Addison-Wesley, Reading, Massachusetts, 1963), Volume I, Chapter 46.

3. Investigate in further detail the relationship between the second law of thermodynamics and the arrow of time. Write a précis about the following discussion of the topic.
   Joel L. Lebowitz, "Boltzmann's entropy and time's arrow," *Physics Today*, 46, #9, pages 32–38 (September 1993).

Paul Davies, "The arrow of time," *Sky and Telescope*, 72, #3, pages 239–242 (September 1986).
David Layzer, "The arrow of time," *Scientific American*, 233, #12, pages 56–69, 148 (December 1975).
T. Gold, "The arrow of time," *American Journal of Physics*, 30, #6, pages 403–410 (June 1962).
Stephen Hawking, *A Brief History of Time* (Bantam, New York, 1988).
Richard Feynman, *The Character of Physical Law* (MIT Press, Cambridge, Massachusetts, 1965).

4. Write a short paper about the physics underlying the idea of *negative* absolute temperatures.
   Ralph Baierlein, "The meaning of temperature," *The Physics Teacher*, 28, #2, pages 94–96 (February 1990).

5. Investigate the energy and P–V cyclical action of automobile heat engines. Present an oral report on your findings to your class.
   Gene Waring, "Energy and the automobile," *The Physics Teacher*, 18, #7, pages 494–503 (October 1980).

B. Lab and Field Work

6. Design and perform an experiment to see if the temperature of an ideal gas remains the same in a free expansion.

7. Practical refrigerator engines make use of the *Joule–Kelvin effect* (also known as the porous-plug experiment): a gas seeping through a porous plug under the appropriate circumstances experiences a decrease in temperature. The effect also is used to liquefy various gases. Research the effect and, if appropriate materials are available, demonstrate the effect in the laboratory.
   Mark W. Zemansky, *Heat and Thermodynamics* (5th edition, McGraw-Hill, New York, 1968).

8. What measurements are needed to determine the coefficient of performance of a typical household refrigerator? Make the needed measurements and compare the refrigerator's coefficient of performance with that of a Carnot refrigerator engine operating between the same two temperatures.

9. Take a covered shoe box or some other such container and place 50 pennies in the box, all heads up. Perform an experiment to see how to shake the coins for them to reach their most probable distribution with the fewest shakes.

10. Roll a single die 100 times and keep a record of the number of times each face appears. Make a histogram of the results. How can you be sure the die is not loaded in some way? If the die is legitimate, what is the minimum numerical value of the sum of the values of the 100 throws? What is the maximum numerical value of the sum? Approximately, what is the most likely value for the sum?

C. Communicating Physics

11. Reread C. P. Snow's comment at the beginning of the chapter. Follow C. P. Snow's injunction and take up the challenge to "describe the Second Law of Thermodynamics" in language appropriate for a gathering of well-educated but scientifically illiterate peers.

12. Write a one-page essay describing what is meant by the term entropy in physics. Assume your audience is a class of humanists who know little of the world of physics.

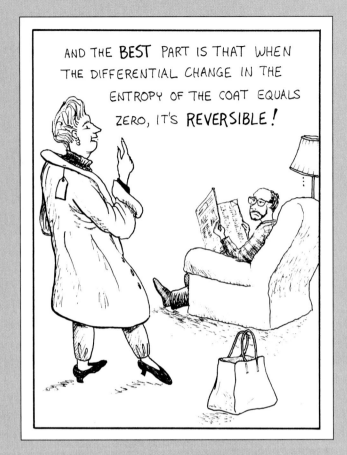

APPENDIX A
PROOFS OF THE GRAVITATIONAL SHELL THEOREMS

A.1 A MASS WITHIN A UNIFORM SPHERICAL SHELL

We seek to prove the following theorem:

> If a mass m is located anywhere within the volume enclosed by a thin uniform spherical shell of mass M, the total gravitational force of the shell on m is equal to zero.

Consider a pointlike mass m positioned anywhere *inside* a thin uniform spherical shell of radius R and mass M, as shown in Figure A.1. Draw two narrow, equal apex-angled, differential cones centered on the mass m, as indicated. These small conical constructions are like two narrow ice cream cones. The cones intercept masses dM and dM' on the spherical shell. The differential gravitational force $d\vec{F}$ of dM on m attracts m toward dM. The differential gravitational force $d\vec{F}'$ of dM' on m attracts m toward dM'.

We need to calculate the ratio of the magnitudes of these two forces on m. The vector addition of the two forces then will be quite easy, since they are in opposite directions because of how we constructed the differential cones.

If the apex angle of the cones is sufficiently small, the masses dM and dM' are essentially pointlike masses. Hence we use the point-mass gravitational force law deduced by Newton, Equation 6.3, to calculate the gravitational effects of dM and dM' on the mass m inside the shell. Thus the magnitude of the differential force of dM on m is

$$dF = \frac{Gm\, dM}{r^2}$$

where r is the distance of dM from the mass m inside the shell as shown in Figure A.1. Likewise, the magnitude of the differential force of dM' on m is

$$dF' = \frac{Gm\, dM'}{r'^2}$$

where r' is the distance between m and dM', also indicated in Figure A.1.

We take the ratio of the magnitudes of the two forces:

$$
\frac{dF}{dF'} = \frac{\dfrac{Gm\, dM}{r^2}}{\dfrac{Gm\, dM'}{r'^2}}
$$

$$
= \frac{dM}{dM'}\frac{r'^2}{r^2} \qquad (A.1)
$$

Since the shell is a *uniform* shell, the ratio of the masses intercepted by the differential cones is equal to the ratio of the areas intercepted:

$$\frac{dM}{dM'} = \frac{dS}{dS'}$$

We draw radii R from the center of the sphere to the two areas dS and dS', as shown in Figure A.2. The radii intersect the axis of the double cone, forming a large isosceles triangle with its apex angle at the center of the shell. The angle ϕ that the radii make with the common axis of the cones therefore is the same at area dS and dS'. Let dA be the *projected* area of dS that is perpendicular to the axis of the cone; let dA' be the corresponding projected area at dS'.

The angle between dS and dA is ϕ; likewise, the angle between dS' and dA' is ϕ. The projected areas are related to the areas on the shell itself by

$$dA = dS \cos \phi$$

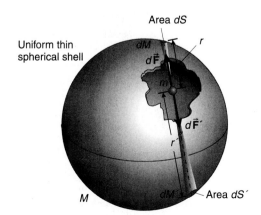

FIGURE A.1 A pointlike mass m inside a thin uniform spherical shell.

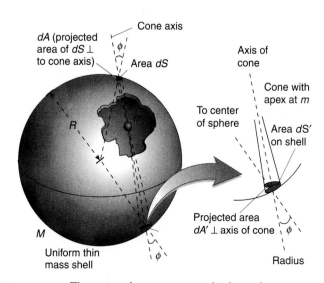

FIGURE A.2 The projected areas are perpendicular to the common axis of the differential cones.

and

$$dA' = dS' \cos \phi$$

So the ratio of the areas dS/dS' is the same as the ratio of the areas dA/dA'. Since the apex angles of the cones are equal, the sizes of the projected areas are proportional to the square of the distance they are from the apex of the cones; that is,

$$\frac{dA}{dA'} = \frac{r^2}{r'^2}$$

Hence the ratio of the masses intercepted by the cones is the same as

$$\frac{dM}{dM'} = \frac{dA}{dA'}$$
$$= \frac{r^2}{r'^2}$$

The ratio of the magnitudes of the two differential forces is then (from Equation A.1)

$$\frac{dF}{dF'} = \frac{dM}{dM'} \frac{r'^2}{r^2}$$
$$= \frac{r^2}{r'^2} \frac{r'^2}{r^2}$$
$$= 1 \qquad\qquad (A.2)$$

Thus the two differential forces on m have exactly the same magnitude, but point in opposite directions. The two differential forces are *not* a Newton's third law pair since they act on the *same* system, the mass m. Hence the vector sum of the two differential forces on m is *zero*. Although m is closer to dA than to dA', the more distant area dA' has more mass in just the right amount to compensate for its greater distance from m.

Now we sweep the differential cones around in all directions to take into account the effect of the entire spherical shell on m. *Every* pair of little differential masses at opposite ends of the differential cones produce zero total gravitational force on the mass m. Thus we reach the startling conclusion that the total gravitational force of the shell on mass m is zero, as indicated in Figure A.3.

Notice that this remarkable result depended *crucially* on the inverse square law nature of the gravitational force law; if the gravi-

tational force had any other power of the distance in the denominator, the distances in Equation A.2 would not cancel and there would be a nonzero total gravitational force on m unless it were at the very center of the shell. Nature was kind in giving us an inverse square law gravitational force.

A.2 A MASS OUTSIDE A UNIFORM SPHERICAL SHELL

We seek to prove the following theorem:

> If a pointlike mass m is located anywhere outside a thin uniform spherical shell of mass M, the total gravitational force of the shell on m is the same as that of a point mass, with a mass equal to the mass M of the shell, located at the center of the shell.

To prove this theorem, we have to calculate the gravitational force of a uniform spherical shell of mass M on a pointlike mass m lying a distance r from the center of the shell, where $r > R$, the radius of the shell (see Figure A.4).

It is necessary to do a vector integration here; we managed to avoid such an integration when the point mass was inside the sphere, but here it is unavoidable. Vector integrations can be difficult because we have to account both for the magnitudes *and* directions of the vectors. So we proceed with some caution; it will be beneficial for you to study this technique with some care, since vector integrations are encountered again later in our course of study. The typical procedure for performing a vector integration involves converting the vector integration into an ordinary scalar integration by subtly looking for ways to exploit or account for the directional effects of the vectors. This only comes with practice and patience.

The amount of mass on any square meter of the shell is known as a *surface mass density* σ. Since the shell is uniform, the surface mass density σ is constant and equal to the total mass of the shell divided by the surface area of the spherical shell:

$$\sigma = \frac{M}{4\pi R^2}$$

We begin by examining a thin differential ring of the shell, as shown in Figure A.5. Every point on the differential ring is the

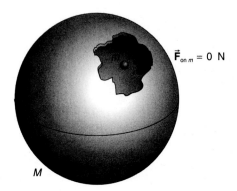

FIGURE A.3 The total gravitational force of the shell on mass m inside the shell is zero.

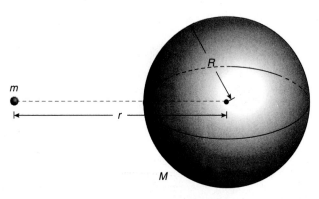

FIGURE A.4 A pointlike mass m outside a thin uniform spherical shell.

same distance s from the point mass m. The radius of the differential ring is $R \sin \theta$. The area of the differential ring is its circumference ($2\pi R \sin \theta$) times its width ($R\, d\theta$), or

$$dA = (2\pi R \sin \theta)(R\, d\theta)$$
$$= 2\pi R^2 \sin \theta\, d\theta$$

The mass dM on this ring of material is the product of the area dA and the surface mass density σ:

$$dM = \sigma\, dA$$
$$= \sigma\, (2\pi R^2 \sin \theta\, d\theta)$$

Each little segment of the ring produces a gravitational force on mass m that is directed toward that little piece of the ring, as shown in Figure A.6. Across the diameter of the ring is a twin segment of the ring. Each of the two twins produces a differential force on m. Note that the components of the two force vectors that are perpendicular to the line joining m and the center of the shell will vector sum to zero. The components of the force vectors of the twins directed *along* the line between

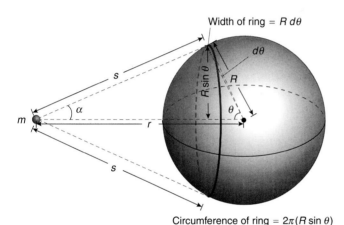

FIGURE A.5 A thin differential ring of the shell.

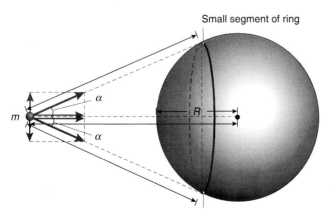

FIGURE A.6 Twin segments produce forces of equal magnitude on m. The force components perpendicular to the line joining m and the center of the shell vector sum to zero; the components along the line add.

m and the center of the shell will add, since they point in the same direction.

Thus, as we vector sum (integrate) around the ring, the components of the differential forces perpendicular to the line will pairwise vector sum to zero, while those along the line will simply add. This is certainly cause for celebration!

The *surviving* component of the differential force on m by the segment of material dM is

$$dF = \frac{Gm\, dM}{s^2} \cos \alpha$$

The $\cos \alpha$ term picks off the surviving part of the differential force as we vector integrate around the ring. Thus we have accounted for the directional effects of the vectors with the $\cos \alpha$ term. The resulting integration is now just an ordinary scalar integration for the *magnitude* of the total gravitational force of the shell on m.

If we substitute for dM, the expression for dF begins to look like a nightmare:

$$dF = Gm\, \frac{\sigma\, (2\pi R^2) \sin \theta\, d\theta}{s^2} \cos \alpha \qquad \text{(A.3)}$$

Now we integrate this expression over the shell. For all little rings anywhere on the spherical shell, the differential forces given by Equation A.3 simply add together because they all point in the same direction: toward the center of the shell.

It is convenient to express everything in terms of the distance s. It is not obvious that this is the thing to do, but using the variable s enables us to perform the integration easily and that is why we do it.

The law of cosines from trigonometry will help. In the triangle whose sides are r, s, and R in Figure A.5, the law of cosines implies that

$$s^2 = r^2 + R^2 - 2rR \cos \theta$$

As we integrate over the sphere, s and θ change but r and R are fixed constants. Taking differentials of the law of cosines equation, we find

$$2s\, ds = -2rR(-\sin \theta\, d\theta)$$

or

$$s\, ds = rR \sin \theta\, d\theta$$

Hence

$$\sin \theta\, d\theta = \frac{s\, ds}{rR}$$

We apply the law of cosines in a different way in the same triangle; this enables us to find an expression for $\cos \alpha$. Since

$$R^2 = s^2 + r^2 - 2sr \cos \alpha$$

we can solve for $\cos \alpha$:

$$\cos \alpha = \frac{s^2 + r^2 - R^2}{2sr}$$

Substituting the expressions for $\sin \theta \, d\theta$ and $\cos \alpha$ into the expression for dF in Equation A.3, we really get quite a mouthful:

$$dF = \frac{Gm\sigma(2\pi R^2)}{s^2} \frac{s \, ds}{rR} \frac{s^2 + r^2 - R^2}{2sr}$$

Simplifying this slightly, we obtain

$$dF = \frac{Gm\sigma\pi R}{r^2} \frac{s^2 + r^2 - R^2}{s^2} \, ds$$

Now we integrate to find the magnitude of the total force on m due to the entire shell. The limits on s are from $r - R$ to $r + R$ (see Figure A.6). The integral becomes

$$F = \frac{Gm\sigma\pi R}{r^2} \int_{r-R}^{r+R} \left[1 + \frac{(r^2 - R^2)}{s^2} \right] ds$$

The two integrals are not too bad to do, since both r and R are constants:

$$F = \frac{Gm\sigma\pi R}{r^2} \left[\int_{r-R}^{r+R} ds + (r^2 - R^2) \int_{r-R}^{r+R} \frac{ds}{s^2} \right]$$

Performing the integrations, we get

$$F = \frac{Gm\sigma\pi R}{r^2} \left[s - \frac{(r^2 - R^2)}{s} \right] \Bigg|_{r-R}^{r+R}$$

Inserting the limits of integration and simplifying, we find

$$F = \frac{Gm\sigma\pi R}{r^2} 4R$$

We rearrange this a bit, getting

$$F = \frac{Gm}{r^2} 4\pi R^2 \sigma$$

But the quantity $4\pi R^2 \sigma$ is the product of the surface mass density σ and the surface area of the shell. This is the total mass M of the shell. Hence the magnitude of the total force on m is

$$F = \frac{GmM}{r^2}$$

Voilà! Notice that this is the same expression as Equation 6.3 for the magnitude of the gravitational force between two pointlike masses separated by a distance r.

Thus the magnitude of the gravitational force of the shell on m is found by imagining the mass M of the shell to be located at its center, since the distance r is the distance between m and the center of the shell. Effectively, we can think of the mass of the shell as if it were at the center of the shell, as shown in Figure A.7. What a relief this result must have been to Newton as he wrestled with how to account for the gravitational effects of the shell on a mass m located outside the shell!

To satisfy Newton's third law of motion, the mass m exerts a gravitational force of the same magnitude but opposite direction *on the shell*, so that we have the third law force diagram shown in Figure A.8.

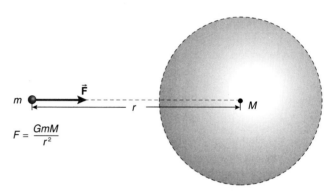

FIGURE A.7   The force of the shell on m can be found by imagining the entire mass of the shell to be at its center.

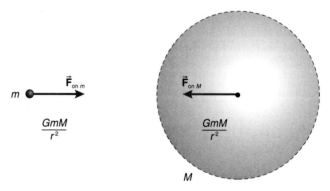

FIGURE A.8   Third law force diagram.

Answers to Problems

Chapter 1

1. 4.6403×10^{-5} m^3

5. peekaboos, nannygoats, military, millionaires, megaphones, microphones, megabucks, centipedes, microfilm, terrapin, megalopolis, deck of cards, microfiche, microwaves, decimate, microeconomics, to kill a mockingbird, terrible, terra firma, megalomaniacs, millinery, microscopes, megabytes

9. (a) 1.83×10^3 (b) 3.33×10^5 (c) 2×10^{11} (d) 3×10^{11} Parts (c) and (d) are the same order of magnitude.

13. (a) $\sim10^6$ m (b) $\sim10^5$ m

17. (a) A factor of 3 (b) A factor of 9

21. (a) A factor of $\sqrt{3} \approx 1.73$ (b) A factor of $3\sqrt{3} \approx 5.20$

25. 0.447 h·m/(mile·s)

29. 3.4×10^{22} molecules

33. 3.63 m^3

37. (a) 7.77×10^{-4} m^3 (b) 1.6×10^{-4} m^3

41. (a) 6.5×10^8 m^3 (b) 8.7×10^2 m on an edge (c) 3.6×10^8 m^2 (d) 19 km

45. 1.80×10^{12} furlongs/fortnight

49. (a) 9.74×10^{-4} kg (b) 8.0×10^{-5} m (c) 8.2×10^{-7} m^3 (d) 9.74×10^5 kg (e) 8.0×10^4 m (f) 1.0×10^7 m^2 (g) 9.4 m

53. (a) 5×10^{-28} kg/m^3 (b) 3×10^9 m

57. (a) ~12 minutes (b) ~3 years (c) ~10 minutes

61. 1.19×10^{57} proton masses

65. $\sim3 \times 10^7$ ties, 1×10^6 m^3

69. $\sim7 \times 10^8$ km^3 assuming an average depth of 2 km

73. $\sim$several percent

77.

Angle	Sine	Cosine	Tangent
78.0°	0.978	0.208	4.70
78.02°	0.9782	0.2076	4.713
78.024°	0.97823	0.20750	4.7143

Chapter 2

1.

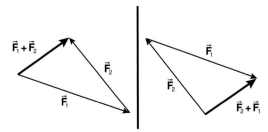

5. 100 m east, 141 m northwest

9. $(\vec{R} + \vec{r}) \cdot (\vec{R} - \vec{r}) = 0$ so $\vec{R} + \vec{r}$ is perpendicular to $\vec{R} - \vec{r}$. The vector $\vec{R} + \vec{r}$ is also perpendicular to $\vec{r} - \vec{R}$.

13. Answers are in Table 2.1.

17. $\dfrac{\hat{\imath} + \hat{\jmath} + \hat{k}}{\sqrt{3}}$

21.

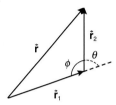

$r^2 = r_1^2 + r_2^2 - 2r_1 r_2 \cos \phi$

25. (a) 64.9° (b) 115.1°

29. Answers are given in the statement of the problem.

33. (a) 90° (b) 53° between the vector sum and the vector of magnitude 12. 37° between the vector sum and the vector of magnitude 16

37. $\alpha = -1.00$, $\beta = -10.00$, $\gamma = 9.00$

41. $\hat{k}$ or $\alpha\hat{k}$ where α is any scalar; the solution is not unique.

45. (a) 19.1 (b) $-8.2\,\hat{k}$

49. (a) 0 (b) $6\hat{\jmath} - 3\hat{k}$; The solution is not unique. (c) 0

53. A proof

57. (a) $-3.0\,\hat{\imath} + 3.0\,\hat{\jmath} - 4.0\,\hat{k}$ (b) 4.0 (c) 37°

61. Answers are given in the statement of the problem.

65. $\dfrac{5}{2\sqrt{t}}\,\hat{\imath} - 9\sqrt{t}\,\hat{\jmath}$

69. A proof

Chapter 3

1. (a) 0 m (b) 50 m (c) 4.2 m/s (d) 0 m/s

5. 4.0 weeks

9. 6.7 m/s, 2×10^{-2} s

13. Let $\hat{\imath}$ be in the direction the tanker is moving: (a) $\vec{a} = (-9.26 \times 10^{-3}$ m/s$^2)\,\hat{\imath}$ (b) 3.75 km

17. (a) 2.47 s (b) 24.2 m/s (c) 19.8 m/s (d) 25.1 m

21. (a) $v_x \approx -6.0$ m/s (b) Yes, when $t \approx 1.5$ s. (c) No. $x \approx 5.2$ m.

25. 3×10^{24} m/s^2

29. (a) 10 s (b) No. 10 m

33. (a) $\vec{a}_{\text{ave}} = (3.3 \times 10^{-2}$ m/s$^2)\hat{\imath}$, $\vec{a} = (3.3 \times 10^{-2}$ m/s$^2)\hat{\imath}$ (b) $\vec{a}_{\text{ave}} = (-6.7 \times 10^{-2}$ m/s$^2)\hat{\imath}$, $\vec{a} = (-6.7 \times 10^{-2}$ m/s$^2)\hat{\imath}$ (c) $\vec{a}_{\text{ave}} = (3.3 \times 10^{-2}$ m/s$^2)\hat{\imath}$, $\vec{a} = (3.3 \times 10^{-2}$ m/s$^2)\hat{\imath}$ (d) The barge begins at rest and accelerates to a speed 2.0 m/s when $t = 60$ s. The barge then begins to slow down, stops instantaneously when $t = 90$ s, reverses the direction it is moving and reaches a speed of 2.0 m/s in the opposite

direction when $t = 120$ s. Then the barge again begins to slow down and is brought to a stop when $t = 180$ s.

(e)

37. (a) When $t = 0$ s and 6.0 s (b) During the interval between $t = 1.0$ s and 4.0 s (c) $\vec{a} = (-2.0 \text{ m/s}^2)\,\hat{\imath}$ (d) 18 m

41. (a) 6.4×10^2 m/s^2 (b) 48 m, 6.2 s

45. (b) 12.0 s and 20.0 s (c) When $t = 12.0$ s, they first catch up to the back of the train. If they kept running they would pass the back of the train, but when $t = 20.0$ s, the back end of the accelerating train would pass them for the last time.

49. (a) 1.0 ms (b) 6.0×10^9 m/s^2

53. (a) Let $\hat{\imath}$ be upward with the origin at the point of release.
(b) 31.9 m (c) 5.10 s
(d)

(e)

(f)

57. 6.5 m

61. 1.7 m

65. 42 m

69. (a) Up (b) Up (c) ≈ 3.0 s (d) ≈ 25 m (e) ≈ 25 m

CHAPTER 4

1. (a) $\vec{r}_i = (-0.15 \text{ m})\hat{\imath}$, $\vec{r}_f = (0.26 \text{ m})\hat{\imath} + (0.15 \text{ m})\hat{\jmath}$
(b) average speed $= 7.5 \times 10^{-3}$ m/s, $\vec{v}_{ave} = (6.8 \times 10^{-3} \text{ m/s})\hat{\imath} + (2.5 \times 10^{-3} \text{ m/s})\hat{\jmath}$. $|\vec{v}_{ave}| = 7.2 \times 10^{-3}$ m/s.

5. Answers are given in the statement of the problem.

9. (a)

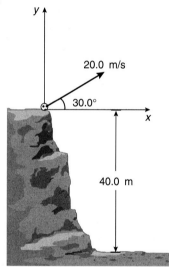

(b) $v_y(t) = 20.0 \text{ m/s} (\sin 30.0°) - gt$, $y(t) = 0 \text{ m} + 20.0 \text{ m/s} \times (\sin 30.0°)t - (1/2)gt^2$, $v_x(t) = 20.0 \text{ m/s} (\cos 30.0°)$, $x(t) = 20.0 \text{ m/s} (\cos 30.0°) t$ (c) 4.05 s (d) Using the coordinate system shown in part (a), the coordinates of the impact point are $x = 70.1$ m and $y = -40.0$ m.

13. 0.56 s and 2.09 s

17. (a)

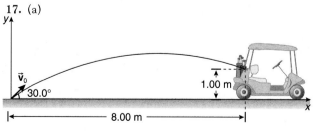

(b) 10.8 m/s (c) 0.856 s

21. (b) 35 m/s   (c) 23 m/s²

25. Answer is given in the statement of the problem.

29. (a) $(2gh)^{1/2}$ (b) $2h$ (c) $h/2$

33. A proof

37. (a) 36.9° to the current   (b) 120 s

41. (a) Choose $\hat{i}$ east and $\hat{j}$ north, $\vec{v} = (-1.50 \text{ m/s}) \hat{i} + (1.00 \text{ m/s}) \hat{j}$
(b) No

45. (a) Let $\hat{i}$ be east and $\hat{j}$ be north, $(-1095 \text{ km/h}) \hat{i} + (495 \text{ km/h}) \hat{j}$
(b) 1.20×10^3 km/h

49. 314.2 rad/s

53. 10 m

57. 4.4×10^8 m/s²

61. 18 km. If the radius decreases, the centripetal acceleration increases, so the curve must have a radius no smaller than 18 km.

65. (a) Second hand: 0.105 rad/s; minute hand: 1.75×10^{-3} rad/s; hour hand: 1.45×10^{-4} rad/s   (b) No   (c) Yes, perpendicular to the clock face directed into the clock   (d) 60   (e) 12

69. (a) 131 rad/s²   (b) 75.1 rev

73. (a) 25.0 rad/s²   (b) 3.52×10^5 rev   (c) 8.82×10^6 m/s², $a/g = 8.99 \times 10^5$ (d) 43.8 rad/s²

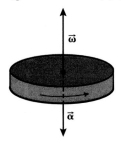

77. (a) A proof   (b) $\pi/4$ rad

81. (a) Let $\hat{k}$ be parallel to $\vec{\omega}$ (b) 6.28 s   (c) $\alpha_z = 0.159$ rad/s²
(d)

At capture, the areas under each curve are the same size, since both then have traveled through the same angle.

85. Answers are given in the statement of the problem.

CHAPTER 5 _____

1. 188 N

5. (a) 8.4 N, 73° clockwise from the +x-axis   (b) 2.1 m/s²

9. (a)

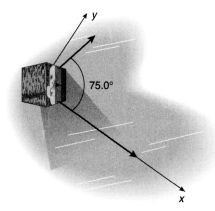

(b) $\vec{F}_{total} = (565 \text{ N}) \hat{i} + (241 \text{ N}) \hat{j}$, $F_{total} = 614$ N
(c) 7.7 m/s²   (d) $0.920 \hat{i} + 0.393 \hat{j}$, yes

13. (a)

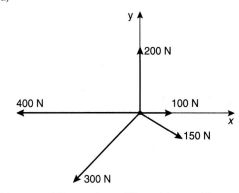

(b) $(-4.2 \text{ m/s}^2) \hat{i} - (1.0 \text{ m/s}^2) \hat{j}$ (c) 4.3 m/s²

17. 0.102 kg

21. (a)

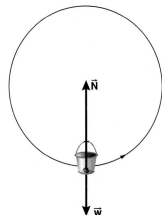

(b) 20.0 N toward the center of the circle   (c) 34.7 N vertically up   (d) It remains in contact with the bottom of the bucket since the normal force of the bucket bottom on the rock is not zero at this location. Its magnitude is 5.3 N, directed vertically down.

25. (a) 5.85°   (b) Yes   (c) 1.00 N

29. 10.2 kg

33. (a) 100 N vertically up

(b)

(c) Force of you on the rope, force of you on the surface, gravitational force of you on the Earth.   (d) 687 N   (e) 587 N   (f) the magnitude of the force you exert on the surface; 59.8 kg.

37. Answers are given in the statement of the problem.

41. (a)

(b) 1.48 m/s², minimum value   (c) 4.87 m/s

45. (a) 12.4 kN   (b) 1.13 kN. Since the oaf is accelerating upward, the normal force of the floor on the oaf is greater than the oaf's weight.

49. (a)

(b) 785 N   (c) 785 N

(d)

(e) 785 N   (f) 865 N
(g) 865 N. The scale measures the magnitude of the normal force, not the magnitude of the weight.

(h)

(i) 785 N   (j) 705 N
(k) 705 N

53. $\mu_s = 0.71$

57. $\mu_k = 0.09$

61. (a)

(b) 10 kg   (c) 49 N static friction

65. (a)

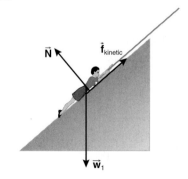

You will slip because the component of your weight down the plane is greater than f_{max}.

(b)

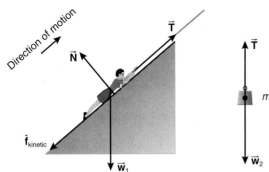

88 kg   (c) 7.3×10^2 N

69. 3.4 m/s^2

73. (a) The total force is directed toward the center of the circle since you have a centripetal acceleration in that direction. (b) 2.0 m/s   (c) 7.8 m   (d) Ground observer: the ball follows a parabolic trajectory. The horizontal component of the velocity of the ball is 2.0 m/s and the vertical component is equal to that given to the ball by the person on the merry-go-round; merry-go-round observer: the ball goes up and hooks to the right as the merry-go-round is spinning counterclockwise when seen from above.

77. (a) and (b)

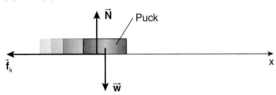

(c) $v_x(t) = v_0 - \mu_k gt$

(d) $t = \dfrac{v_0}{\mu_k g}$

(e) $x(t) = v_0 t - \dfrac{\mu_k g t^2}{2}$

(f) $\dfrac{v_0^2}{2\mu_k g}$

(g) 6.4×10^2 m

81. (a)

(b) The kinetic frictional force.   (c) Answer is given in the statement of the problem.

85. (a) and (b) Answer is given in the statement of the problem.

(c)

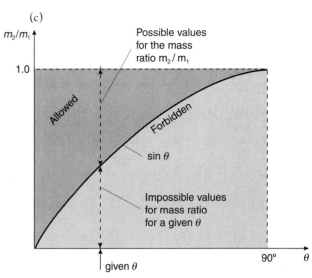

89. 0.0931 — The effects of pseudoforces are more apparent on Jupiter than on the Earth, since this ratio is almost 10%.

CHAPTER 6

1. 1.85×10^{-10} N

5. (a) $v = (GM_{Sun}/r)^{1/2} = 29.8$ km/s   (b) Nothing

9. (a) 1.05×10^{-8} N   (b) 1.05×10^{-8} N   (c) For the 0.400 kg sphere: 2.63×10^{-8} m/s^2; For the 4.00 kg sphere: 2.63×10^{-9} m/s^2

13. (a) Answer is given in the statement of the problem. (b) 0.16 m/s^2 (c) 1.2×10^2 m, 78 s

17. (a) $\rho = \left(\dfrac{9F}{4\pi^2 GR^4}\right)^{1/2}$

(b) No element is dense enough.   (c) 1.61 m, 3.93×10^5 kg

21. 6.43×10^{23} kg, $M_{Mars} = 0.108\, M_{Earth}$

25. (a) 5.00×10^6 km   (b) Outside the Sun (c) Perihelion: 1.471×10^8 km, aphelion: 1.521×10^8 km (d) 0.334 cm

29. 1.12

33. (a) Answer is given in the statement of the problem. (b) 0.140, 0.417

37. 1.034

41. 4.7 y

45. (a) 17.9 AU   (b) 0.968   (c) 35.2 AU   (d) ≈62

49. (a) 4.7×10^3 km from the center   (b) 1.7×10^3 km below the surface

53. (a) $\dfrac{GM^2}{64r^2}$

(b) $\dfrac{GM^2}{4r^2}$

(c) $\dfrac{17}{64}\dfrac{GM^2}{r^2}$

(d) $2\pi r/T$

(e) $T^2 = \dfrac{64\pi^2}{17GM} r^3$

57. 1.86×10^{32} kg, 93.5 solar masses

61. (a) 2.1×10^8 y   (b) About 24 times   (c) 1.9×10^{41} kg, 9.5×10^{10} solar masses

65. (a) 0.900 d, measured from the center of the Earth   (b) No

69. 34 km

73. -6.17×10^{13} m^3/s^2, -5.01×10^{15} m^3/s^2

CHAPTER 7

1. 50 N/m

5. 2.5×10^2 N/m

9. Answers are given in the statement of the problem.

13. (a) 40.0 rad/s   (b) 6.37 Hz

   (c) $x(t) = (0.050 \text{ m}) \cos\left[(40.0 \text{ rad/s}) t + \dfrac{\pi}{2} \text{ rad} \right]$

   (d) -0.0363 m

17. (a) $\omega = 11.5$ rad/s, $A = 0.201$ m, $\phi = -1.05$ rad   (b) 0.546 s
   (c) 2.31 m/s, 26.6 m/s^2

   (d)

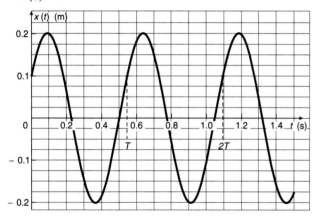

21. $v_x(t) = -0.500 \sin\left[(7.07 \text{ rad/s}) t - \dfrac{\pi}{2} \text{ rad} \right]$

     $a_x(t) = -3.54 \cos\left[(7.07 \text{ rad/s}) t - \dfrac{\pi}{2} \text{ rad} \right]$

graphs follow:

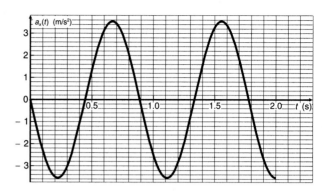

25. (a) 25.1 s, 18.0 s   (b) Approximately 10 s between the instants

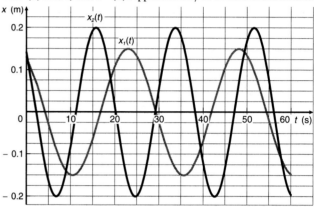

   (c) The times change as well as the intervals between them.

29. 0.496 m from the fixed end

33. (a) 1.30 Hz   (b) At the high point of the oscillation   (c) At the low point of the oscillation   (d) 0.147 m

37. (a) 1.5 m/s   (b) 23 m/s^2

41. $R \cos\left(\omega t - \dfrac{\pi}{6} \text{ rad} \right)$

45. (a) 10 s   (b) 0.10 Hz   (c) no

49. The equation of motion is not like (7.15).

53. 16.5 s

57. $\dfrac{T_1 T_2}{T_1 - T_2}$

61. 9.79 m/s^2

65. 526 km

69. An experiment

73. 14 Hz

CHAPTER 8

1. (a) 75.0 J   (b) 0 J   (c) 65.0 J   (d) −53.1 J

5. (a) The graph should be a downward-sloping straight line with a slope of −6.00 N/m, passing through the origin.   (b) 0 J

9. 2.4×10^2 J

13. (a) 360 J   (b) 0 J   (c) −62.5 J   (d) 62.5 J

17. Answer is given in the statement of the problem.

21. Answer is given in the statement of the problem.

25. Answer is given in the statement of the problem.

29. 7.91 km/s

33. (a) 7.91 km/s

    (b) $\dfrac{1}{\sqrt{2}} = 0.707$

37. 5.8 km

41. (a) $-\dfrac{GMm}{r}$

    (b) $\left(\dfrac{GM}{r}\right)^{1/2}$

    (c) $\dfrac{1}{2}\dfrac{GMm}{r}$

    (d) $-\dfrac{1}{2}$

45. (a) $(2gh)^{1/2}$ (b) $(gR)^{1/2}$ (c) 5/2   (d) $(2gh)^{1/2}$ (e) mgh
    (f) Only (e)

49. Answer is given in the statement of the problem.

53. (a) $-\dfrac{GMm}{r_1}$ $-\dfrac{GMm}{r_2}$

    (b) mv_1r_1, mv_2r_2 (c) Answer is given in the statement of the problem.

57. (a) Answer is given in the statement of the problem.
    (b) 3.33×10^{-3} m/s

61. Answer is given in the statement of the problem.

65. (a) 70.2 N/m   (b) 0.585 m

69. (a) 434 W   (b) 868 W

73. (a)

    (b) 3.1×10^3 J   (c) 8.9 m/s   (d) 20 m   (e) 0 J along both paths   (f) -3.1×10^3 J   (g) -1.4×10^3 W; No, since the velocity is changing.

77. 109 kW

81. (a) J/m   (b) $-C\,\hat{\imath}$ for $x > 0$ m, $C\,\hat{\imath}$ for $x < 0$ m
    (c)

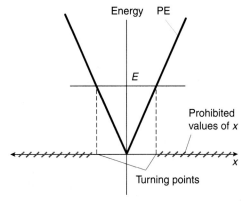

    (d) Two turning points   (e) The motion is oscillatory but is not simple harmonic oscillation. This force has a constant magnitude whereas the force in simple harmonic oscillation is a function of position.

CHAPTER 9

1. 3.3×10^2 km/h

5. 4.63 m/s = 16.7 km/h

9. 4.96 kg·m/s

13. (a)

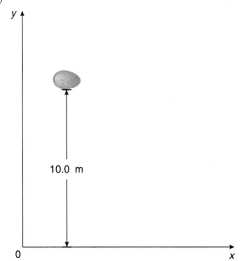

    (b) 98.1 J   (c) Yes, since the total force is constant.
    (d) $(-14.0\,\text{N}\cdot\text{s})\,\hat{\jmath}$

17. (a) $(15.0\,\text{N}\cdot\text{s})\,\hat{\imath}$ (b) $(7.50\,\text{N})\,\hat{\imath}$ (c) $(15.0\,\text{N}\cdot\text{s})\,\hat{\imath}$
    (d) $(3.75\,\text{N})\,\hat{\imath}$

21. 87 m/s²

25. 1.718

29. Answer is given in the statement of the problem.

33. (a) $(8.00\,\text{m/s})\,\hat{\imath}$ (b) It is an inelastic collision since the kinetic energy is not conserved. $\Delta KE = -2.40 \times 10^4$ J $\neq 0$ J   (c) No. The momentum of each particle in a collision is not conserved; only the total momentum of all the particles is conserved in a collision.

37. (a) A stupid one. Completely inelastic.
(b) $(1.67 \times 10^4 \text{ kg} \cdot \text{m/s}) \, \hat{\imath} + (1.11 \times 10^4 \text{ kg} \cdot \text{m/s}) \, \hat{\jmath}$
(c) The same as (b) (d) 10.0 m/s (e) $-1.01 \times 10^5 \text{ J}$

41. (a) Take $\hat{\imath}$ east and $\hat{\jmath}$ north, $(19 \text{ m/s}) \, \hat{\imath} - (4 \text{ m/s}) \, \hat{\jmath}$
(b) $-5 \times 10^4 \text{ J}$

45. (a) No. The speed of the car was 29.0 m/s. (b) 0.930 m/s

49. 17

53. A proof

57. $(-0.129 \text{ m}) \, \hat{\imath} + (0.257 \text{ m}) \, \hat{\jmath}$

61.

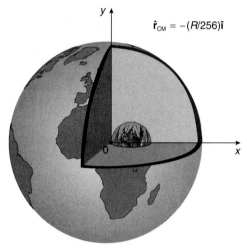

$\tilde{r}_{CM} = -(R/256)\hat{\imath}$

65. 0.71 m/s in a direction opposite to that of his daughter

69. (a) 1.8 m/s (b) $\Delta KE = -3 \times 10^4 \text{ J}$ (c) 1.8 m/s

CHAPTER 10 _____

1. 0 kg m²/s, orbital

5. Answers are given in the statement of the problem.

9. $4.78 \times 10^4 \text{ km}$

13. $7mR^2/4$

17. Answer is given in the statement of the problem.

21. (a) $1.36 \times 10^6 \text{ J}$ (b) 0.047

25. (a) Answer is given in the statement of the problem.
(b) $2.83 \times 10^{42} \text{ J}$ (c) Answer is given in the statement of the problem. (d) $-7.17 \times 10^{31} \text{ W}$, 1.87×10^5 times the luminosity of the Sun

29. (a) $7mR^2/5$ (b) $3mR^2/2$ (c) $5mR^2/3$

33. (a) $7.293 \times 10^{-5} \text{ rad/s}$ (b) $7.08 \times 10^{33} \text{ kg} \cdot \text{m}^2/\text{s}$ (c) The effect is to lengthen the day (the moment of inertia increases and decreases ω_{spin} because of conservation of angular momentum). (d) $1.99 \times 10^{-7} \text{ rad/s}$ (e) $2.64 \times 10^{40} \text{ kg} \cdot \text{m}^2/\text{s}$
(f) 3.73×10^6

37. $\dfrac{13}{10} \omega_0$

41. (a) Answer is given in the statement of the problem.

(b)

	KE$_{\text{rot}}$	KE$_{\text{CM}}$
highest	cylindrical shell	sphere
↓	spherical shell	disk
	disk	spherical shell
lowest	sphere	cylindrical shell

(c) First to last: sphere, disk, spherical shell, cylindrical shell.

45. (a) $1.96 \times 10^3 \text{ N} \cdot \text{m}$ (b) 7.34 rad/s^2 (c) No. The torque of the weight of the beam about the hinge varies with the orientation of the beam to the vertical. (d) 3.84 rad/s

49. $\left(\dfrac{1 - \varepsilon}{1 + \varepsilon} \right)^2$

53. (a) $4.67 \times 10^6 \text{ m}$ from the center of the Earth along the Earth–Moon line (b) $2.81 \times 10^{34} \text{ kg} \cdot \text{m}^2/\text{s}$
(c) $3.47 \times 10^{32} \text{ kg} \cdot \text{m}^2/\text{s}$ (d) $2.37 \times 10^{29} \text{ kg} \cdot \text{m}^2/\text{s}$
(e) $7.08 \times 10^{33} \text{ kg} \cdot \text{m}^2/\text{s}$

57. (a) The spherical shell (b) The solid sphere reaches the bottom of the incline before the spherical shell if both are released at rest from the same location.

61. $\dfrac{v_0/R}{1 + \dfrac{M}{2m}}$

The lazy-susan rotates in a direction opposite to that in which the mouse is going.

65. (a) $1.0 \times 10^{42} \text{ kg} \cdot \text{m}^2/\text{s}$ (b) $5 \times 10^{43} \text{ kg} \cdot \text{m}^2/\text{s}$ (c) 0.7 d

69. Answer is given in the statement of the problem.

73. 59°

77. 0.71 m

CHAPTER 11 _____

1. 8.0 kg

5. $1.2 \times 10^7 \text{ N/m}^2$

9. Answer is given in the statement of the problem.

13. 1.5 m

17. (a) $7.92 \times 10^3 \text{ m}$ (b) 0.93 km

21. (a) 1 (b) 10

25. 10.3 m

29. $0.833 \times 10^3 \text{ kg/m}^3$

33. $2 \times 10^{-3} \text{ kg}$

37. (a) It floats. (b) $3.00 \times 10^4 \text{ kg}$

41. (a)

(b) 0.120   (c) 4.2×10^{-2} kg   (d) 0.26 N

45. 2.7×10^3 kg/m^3

49. 5.3 cm

53. For 2 mm diameter straw and $h = 1.5$ cm, $\gamma = 0.074$ N/m.

57. 1.2×10^2 m/s

61. 9.0×10^{-3} m^3/s

65. (a) 19.8 m/s   (b) 156 liters/s   (c) 20.0 m

69. 26 m/s

Chapter 12 _____

1. 1.7×10^2 m

5. (a) 1.8×10^3 km   (b) Only the distance to the focus was found.
(c) Use three seismometers at different locations.

9. (a)

(b) 0.33 Hz

(c)

13. (a) and (b)

(c)

(d)

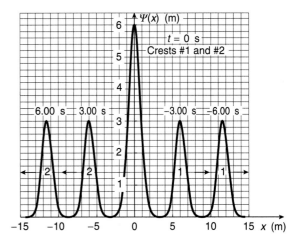

17. (a) 6.00×10^{-3} Pa   (b) 8.06 rad/m   (c) 8.06 wavelengths
    (d) 0.780 m   (e) 2.76×10^3 rad/s   (f) 439 Hz
    (g) 2.28×10^{-3} s   (h) 342 m/s
    (i)

(j)

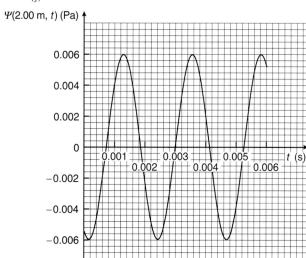

(k) -1.23×10^{-3} Pa. The negative result implies a rarefaction with an absolute pressure below the average value in the medium in which the sound is propagating.

21. (a) 3.00 Hz   (b) $\Psi(x, t) = (0.50 \times 10^{-2}$ m$) \cos[(31.4$ rad/m$) \times x - (18.8$ rad/s$) t]$)

25. 83 N

29. 3.95×10^{-2} s

33. 86 Hz

37. $I = \dfrac{P}{2\pi r\ell}$

41. (a) 3.1×10^2 W   (b) 1.6×10^2 m

45. 106 dB

49. (a) 2.180×10^3 Hz   (b) 3 m/s

53. (a) 1200 Hz   (b) 1.21×10^3 Hz   (c) 1.19×10^3 Hz

57. 30.2 kHz

61. (a) Mach number = 2.1   (b) 29°

65. Answer is given in the statement of the problem.

69. (a) The first (second) wavefunction represents a wave moving toward increasing (decreasing) values of x.   (b) The two will form a standing wave; $\lambda/2 = 0.349$ m   (c) 0.175 m, 0.524 m, 0.873 m

73. $\dfrac{v}{\alpha}$

77. (a)

(b)

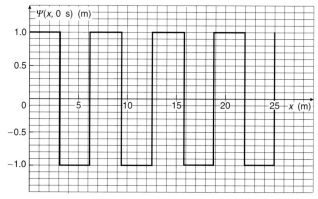

which is called a square wave.

(c) ≈ 6.3 m

CHAPTER 13

1. 37.0 °C

5. −13 °F

9. 50.00 °C

13. 0.012%

17. (a) 7.6×10^{-4} m   (b) $\Delta T = 63$ K

21. $\Delta T = 4 \times 10^2$ K

25. A proof

29. 1.5×10^4 N

33. A proof

37. 1.4×10^{-23} Pa

41. 9.3 m below the surface

45. (a) It all melts   (b) 8.40 °C

49. (a) 9.4×10^7 J   (b) 5.2 h

53. 3.9×10^3 W

57. (a) 1   (b) 14 °C

61. 7.4

65. (a) 3.04×10^3 J   (b) 1.52×10^3 J   (c) 2.28×10^3 J
    (d) -3.04×10^3 J   (e) -1.52×10^3 J   (f) -2.28×10^3 J

69. (a) -2.86×10^3 J   (b) 2.86×10^3 J

73. (a) A proof   (b) 0.11 K   (c) 0.213 K

77. 0.398 K

CHAPTER 14

1. 1.908×10^{16} y

5. 3.1×10^{-16} J

9. 1.93×10^3 m/s; The rms speed of the hydrogen molecules is four times faster than that of the oxygen molecules.

13. 5.14×10^{-23} N·s

17. (a) 2.46×10^{-2} m³   (b) ~3.4×10^{-9} m

21. 0.86 m/s

25. 2.01×10^4 K

29. 1.93 km/s; Although this rms speed is considerably less than the escape speed, some hydrogen molecules will have speeds exceeding the escape speed and can escape, if moving in the right direction, if no collisions occur. Heat transfer from the Earth continually replenishes the high-speed molecules, so they gradually escape. Such a process also explains the phenomenon of evaporation.

33. 1×10^{-23} Pa

37. (a) The helium atom loses KE while the hydrogen atom gains KE.   (b) The KE of each is unchanged.

41. 2

45. 2.67×10^{-2} mol

49. Answer is given in the statement of the problem

53. (a) 1.93×10^3 m/s   (b) 1.25×10^6 Pa   (c) No. $\Delta U = 0$ J.
    (d) 6.25×10^5 Pa

57. (a)

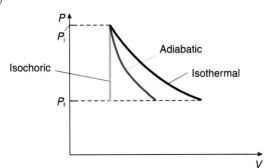

(b) The isochoric process   (c) The isothermal process
(d) Answer is given in the statement of the problem.
(e) The isochoric process

61. A proof

CHAPTER 15

1. 0.558

5. 0.055

9. 400 K and 500 K

13. (a) 186 MW   (b) 1.5×10^4 kg/s, 15 m³/s

17. The coefficient of performance increases more by warming the colder reservoir than by cooling the warmer reservoir.

21. (a) 0.400   (b) $|Q_H| = 3.20 \times 10^3$ J, $|Q_C| = 2.40 \times 10^3$ J

25. (a) 3.9 J; No   (b) 71 J

29. (a) 2.8×10^2 J   (b) 7.8×10^2 J

33. -6.05×10^3 J/K, No

37. (a) 141 J/K   (b) -136 J/K   (c) 5 J/K

41. (a)

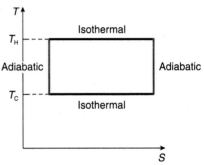

(b) The area enclosed by the cycle is equal to the total heat transfer to the gas system in one cycle.

(c) Take the ratio of $T_H - T_C$ to T_H:

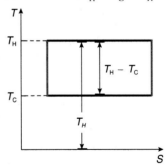

45. (a) 4.00 kW   (b) 3.50 W/K

49. 1.6×10^3 J/K

53. (a) $Q = 0$ J   (b) $\Delta U = -2000$ J   (c) $\Delta S = 0$ J/K   (d) The temperature fell.   (e) $\Delta T = -48.1$ K

57. $\Delta S = nc_P \ln\left[\dfrac{(T_1 + T_2)^2}{4T_1 T_2}\right]$

61. Shake the box many times and assess the experimental probability of securing each macrostate after a shake. If after many shakes, you *never* see a result with 0 heads, you can begin to become suspicious.

65. (a)

Number on left side	Number on right side	Number of microstates
6	0	1
5	1	6
4	2	15
3	3	20
2	4	15
1	5	6
0	6	1

(b) 0 J/K   (c) The macrostate with three particles in each half: 0 J/K, 4.137 × 10⁻²³ J/K

69. (a)

Macrostate Number of heads	Number of microstates Ω	$\dfrac{S}{k}$
0	1	0
1	20	3.00
2	190	5.25
3	1140	7.04
4	4845	8.49
5	15 504	9.65
6	38 760	10.57
7	77 520	11.26
8	125 970	11.74
9	167 960	12.03
10	184 756	12.13
11	167 960	12.03
12	125 970	11.74
13	77 520	11.26
14	38 760	10.57
15	15 504	9.65
16	4845	8.49
17	1140	7.04
18	190	5.25
19	20	3.00
20	1	0

(b) The graph of the number of microstates versus the macrostate is shown below:

(c) The graph of $\dfrac{S}{k}$ versus the macrostate is shown below:

Quotations Index

Reference Index

GENERAL INDEX

Note: f's, t's, and n's following page numbers refer to figures, tables, and footnotes, respectively.

CREDITS

C.2   Credits